Africa, with parts of Europe and Asia, from Apollo 11. (NASA.)

Physical geology

L. Don Leet HARVARD UNIVERSITY

Sheldon Judson PRINCETON UNIVERSITY

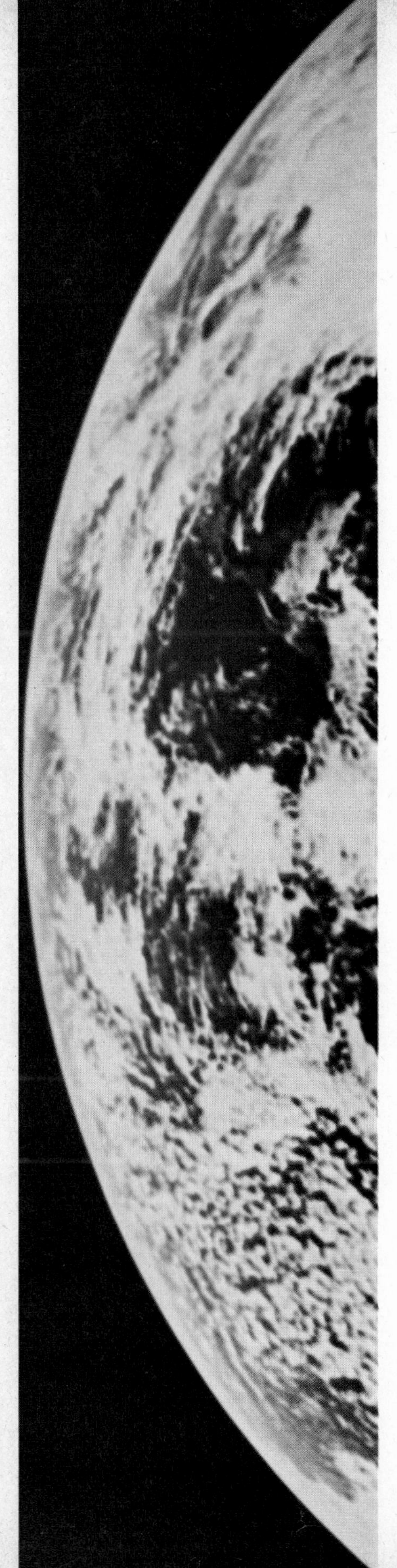

PHYSICAL GEOLOGY

4th edition

Prentice-Hall, Inc. *Englewood Cliffs, New Jersey*

L. DON LEET AND SHELDON JUDSON

Physical Geology, 4th edition

Printed in the United States of America.

0-13-669713-5

LIBRARY OF CONGRESS CATALOG CARD NUMBER *70-102280*

CURRENT PRINTING *10 9 8 7 6 5 4 3*

Prentice-Hall International, Inc. *London*
Prentice-Hall of Australia, Pty. Ltd. *Sydney*
Prentice-Hall of Canada, Ltd. *Toronto*
Prentice-Hall of India Private Ltd. *New Delhi*
Prentice-Hall of Japan, Inc. *Tokyo*

Preface

THE FOURTH EDITION OF THIS TEXTBOOK, LIKE ITS PREDECESSORS, HAS BEEN written for a first course in the subject, both for those who will go no further and for those who will take additional courses or even embark upon professional study in the earth sciences.

In the six years since the appearance of the third edition, the subject of physical geology has experienced a remarkable growth. Man has begun to explore the moon, and now we can examine at first hand the materials brought back from its surface; continental drift has been accepted by most workers, and now the corollaries of sea-floor spreading and plate tectonics bring new insights to earth history and processes; the interdependence of man and his environment is belatedly being recognized, and practitioners of an old science are faced with new questions. Furthermore, our knowledge of internal earth processes such as earthquakes has continued to expand, as has our understanding of surficial processes and features.

To bring the reader abreast of this new material we have added two new chapters: 22 (Man and His Environment) and 23 (The Moon, Earth Craters, Tektites). In addition, we have extensively revised Chapters 2 (Matter and Energy), 10 (Erosion on Hillslopes), 16 (Earthquakes), and 20 (Paleomagnetism, Continental Drift, and Plate Tectonics). We have also added Appendixes D (The Earth and Other Planets) and E (Tables of Measurement). Although we have made minor changes in the sequence of topics, we have attempted to maintain the general order of subjects used in previous editions and to preserve both their emphasis and approach.

All the maps and diagrams have been drawn anew to take advantage of the possibilities afforded by the introduction of color. In illustrating this edition we have missed the talented pen of Edward A. Schmitz, who worked with us on the preceding ones. Mr. William Taylor and his colleagues have proved able successors.

In the preparation of this edition we are pleased to acknowledge the patient and skillful assistance of David R. Esner and Cecil Yarbrough of the Project Planning Department of Prentice-Hall, Inc.

Leet has been assisted by his wife, Florence J. Leet, to an extent that could be fully acknowledged only by joint authorship. Judson acknowledges with pleasure the editorial assistance of Guenevar Pendray Knapp.

We are indebted to various colleagues who have given of their time and abilities in criticizing sections within their special fields. For the first three editions these colleagues included Sturgess Bailey, Marland Billings, Frances Birch, Bart Bok, Perry Byerley, Fred M. Bullard, Stanley N. Davis, Erling Dorf, William Drescher, Donald Eschman, Alfred G. Fischer, Clifford Frondel, C. S. Hurlbut, Jr., Robert Hargraves, Richard Jahns, George Kennedy, Marvin E. Kauffman, the late

Paul MacClintock, Richard B. Mattox, John R. Mosley, William Muehlberger, Willard H. Parsons, Ulrich Peterson, Eugene Robertson, Franklyn B. Van Houten, the late Henry Stetson, the late Frederik Thwaites, the late Stanley Tyler, Sidney E. White, and Keith P. Young. Many of their suggestions have carried over to this edition. In connection with this fourth edition we are further indebted to Julian D. Barksdale, Robert E. Boyer, Howard A. Coombs, Edward A. Hay, Earle F. McBride, Bates McKee, Stephen C. Porter, Harlow Shapley, and John T. Whetten. In addition, we have used the photographs and diagrams of many colleagues, as acknowledged in the text.

We acknowledge our indebtedness to all who have aided us, but of course reserve to ourselves alone responsibility for any errors in fact or judgment.

L. Don Leet
Sheldon Judson

Contents

6 Weathering and soils 104

7 Sedimentary rocks 130

8 Metamorphism and metamorphic rocks 156

9 Geologic time 172

10 Erosion on hillslopes 189

11 Running water 213

18 Deformation of the earth's crust 425

19 Mountains and mountain building 458

20 Paleomagnetism, continental drift, and plate tectonics 482

21 Natural resources 505

22 Man and his environment 535

23 The moon, earth craters, tektites 563

Physical geology

1

The nature and scope of physical geology

GEOLOGY IS THE SCIENCE OF THE EARTH, AN ORGANIZED BODY OF KNOWLEDGE about the globe on which we live—about the mountains, plains, and ocean deeps, about the history of life from slime-born amoeba to man, and about the succession of physical events that accompanied this orderly development of life.

Geology helps us to unlock the mysteries of our environment. Geologists explore the earth from the ocean floor to the mountain peaks to discover the origins of our continents and the encircling seas. They try to explain a land surface so varied that in than less than an hour a traveler can fly over the Grand Canyon of the Colorado River; over man-made Lake Mead; over Death Valley, the lowest point in the contiguous United States; and over Mt. Whitney, the highest point. They probe the action of glaciers that crawled over the land and then melted away over 0.5 billion years ago, and of some that even today cling to the high valleys and cover most of Greenland and Antarctica, remnants of a recent but receding ice age.

Geologists search for the record of life from the earliest one-celled organisms of ancient seas to the complex plants and animals of the present. This story, from the simple algae to the seed-bearing trees, from the primitive protozoa to the highly organized mammals, is told against the ever-changing physical environment of the earth.

For the earth has not always been as we see it today, and it is changing (though slowly) before our eyes. The highest mountains are built of material that once lay beneath the oceans. Fossil remains of animals that swarmed the seas 0.5 billion years ago are now dug from lofty crags. Every continent is partially covered with sediments that were once laid down on the ocean floor, evidence of an intermittent rising and settling of the earth's surface.

PHYSICAL AND HISTORICAL GEOLOGY We usually divide the study of geology into physical and historical geology. *Physical geology,* the subject of this book, covers the nature and properties of the materials composing the earth; the distribution of materials throughout the globe; the processes by which they are formed, altered, transported, and distorted; and the nature and development of the landscape.

Historical geology is a record of life on the earth from the earliest stirrings over 2 billion years ago to the flora and fauna of the present and to man himself. It is also a record of changes in the earth itself through 4 or 5 billion years—of advancing and retreating seas, of deposition and of erosion, of rocks fashioned into mountain ranges—the whole chronological story of how the processes of physical geology have operated.

We have all seen the forces of nature changing in a small way the face of our earth. Perhaps we have watched sea waves beating against a rocky headland or gentler waves sloshing the pebbles back and forth on the slope of a sandy beach. Perhaps we have felt the sting of swiftly moving sand grains against our ankles as we waded across a mountain stream or observed mud-charged rivulets flowing from a new-plowed field. And some of us have seen dust swirled from the parched plains. In these and countless other ways the earth's surface is being changed. Wind, water, and glacier ice erode the earth's surface and carry material to lower levels. Volcanoes erupt through unstable portions of

Figure 1.1 ***Running water does more than any other surface agent to fashion the face of the earth. The material in the banks of this small Wisconsin stream is being slowly eroded by the current. Eventually the products of this erosion will reach the Gulf of Mexico, there to be deposited in layers and perhaps to become rocks of some future time.*** (*Photograph by Wisconsin Conservation Department.*)

the earth's surface, and earthquakes shake the foundations of our continents. These processes are going on today.

THE PRESENT IS THE KEY TO THE PAST Modern geology was born in 1785 when James Hutton (1726–1797), a Scottish medical man, gentleman farmer, and geologist, formulated the principle that the same physical processes that are operating in the present also operated in the past. The principle is known to geologists as the *doctrine of uniformitarianism,* which simply means that the processes now operating to modify the earth's surface have also operated in the geologic past, that there is a uniformity of processes past and present.

Here is an example. We know from observations that modern glaciers deposit a distinctive type of debris made up of rock fragments that range in size from submicroscopic particles to boulders weighing several tons. This debris is jumbled, and many of the large fragments are scratched and broken. We know of no other agent but glacier ice that produces such a deposit. Now suppose we find in the New England hills, or across the plains of Ohio, or in the deep valleys of the Rocky Mountains, deposits that in every way resemble glacial debris but can find no glaciers in the area. We can still assume that the debris was deposited by now-vanished glaciers. On the basis of evidence like this, geologists have worked out the concept of the great Ice Age (see Chapter 13).

This example can be multiplied many times. Today, the great majority of

Figure 1.2 **The Barnard glacier in Alaska is streaked with rock debris, which it moves from higher to lower elevations.** (*Photograph by Bradford Washburn.*)

earth features and of the rocks exposed at the earth's surface are explained as the result of past processes similar to those of the present. We shall find that many of the conclusions of physical geology are based on the conviction that modern processes also operated in the past.

1.1 *Keys to geologic processes*

Time

Armed with Hutton's concept of uniformitarianism, geologists were able to explain earth features on a logical basis. But the very logic of the explanation raised a new concept for students of the earth. Past processes presumably operated at the same slow pace as those of today. Consequently, very long periods of time must have been available for the processes to accomplish their tasks. It was apparent that a great deal of time was needed for a river to cut its valley or for hundreds or thousands of feet of mud and sand to be deposited on an ocean bottom, hardened into rock, and raised above the level of the sea.

The concept of almost unlimited time in earth history is a necessary outgrowth of the application of the principle that "the present is the key to the past."

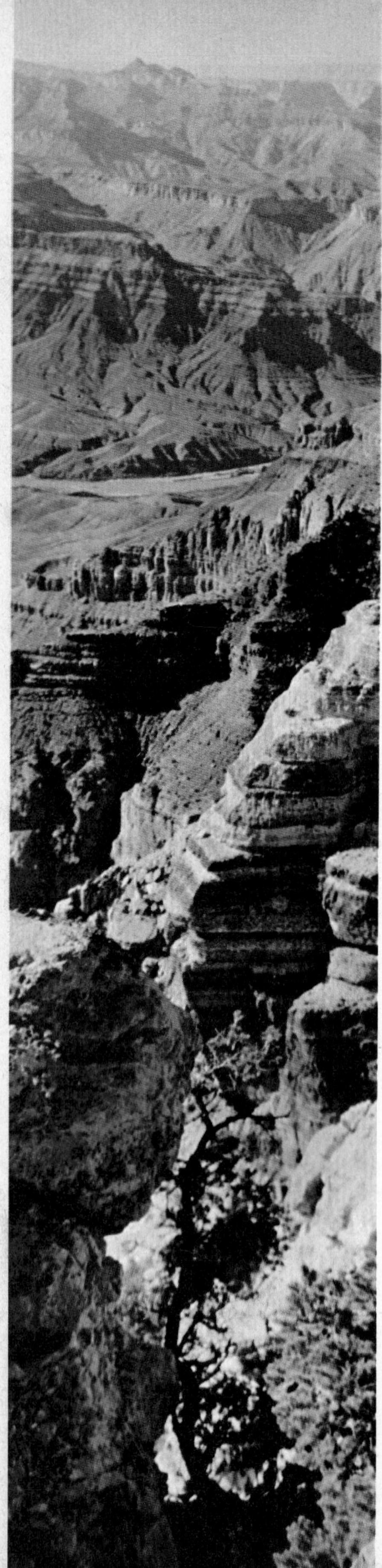

Figure 1.3 ***History in the Grand Canyon of the Colorado River, Arizona, reaches back some 2 billion years. The rocks in which this history is recorded have been exposed as the Colorado River and its tributaries have cut downward.*** *(Photograph from the Library of Congress.)*

For example, geologists know that mountains as high as the modern Rockies once towered over what are now the low uplands of northern Wisconsin, Michigan, and Minnesota. But only the roots of these mountains are left. The great peaks have long since disappeared. Geologists explain that the ancient mountains were destroyed by rain and running water, wind and creeping glaciers, landslides and slowly moving rubble, and that these processes acted essentially as they do now in our present-day world.

Now think of what this explanation means. We know from firsthand observation that streams, glaciers, and winds have some effect on the surface of the earth. But can such feeble forces level whole mountain ranges? Instinct and common sense tell us that they cannot. But this is where the factor of time comes into the picture. True, the small, almost immeasurable amount of erosion that takes place in a human lifetime has little effect. But when the erosion during one lifetime is multiplied by thousands and millions of lifetimes, it becomes clear that mountains can be destroyed. Time makes possible what seems impossible.

The distinguished American geologist Adolph Knopf has written:

If I were asked as a geologist what is the single greatest contribution of the science of geology to modern civilized thought, the answer would be the realization of the immense length of time. So vast is the span of time recorded in the history of the earth that it is generally distinguished from the more modest kinds of time by being called "geologic time."[1]

Change

The great extent of time included in the geologic calendar and the ceaseless operation of earth processes force us to admit that "as everlasting as the hills" refers to a permanency only in terms of human lifetimes. In the vastness of geologic time, everything changes, nothing is permanent.

We usually think of change through time in relation to living things, as in evolution. But inorganic things change and evolve as well as living things. Thus solid rock is changed by weathering at, and near, the earth's surface, and the original minerals of the rock are converted to new minerals within a soil that rests on still unaltered bedrock. Then water, wind, and ice conspire to move the soil minerals, to sort them, and to deposit them in a new environment as vast sheets of debris. These layers, in time, may be converted into rock quite different from the rocks from which they came.

This constant change comes about as an adjustment to new conditions, to changing environments. In a sense, it is an attempt to establish an equilibrium.

[1] Adolph Knopf, *Time and Its Mysteries*, ser. 3, New York University Press, New York, 1949.

Figure 1.4 ***Most of the time rivers stay within their banks, but it is their nature to flood from time to time. Here the Arno River has escaped its banks and flooded the Piazzo Duomo in Florence, Italy, on November 4, 1966.*** *(Photograph by Stefani.)*

Figure 1.5 ***Sometimes geologic processes cause human disasters. Search and rescue operations proceed in the rubble of a 12-floor apartment building tumbled by an earthquake July 29, 1967, in Caracas, Venezuela.*** *(Photograph by Venezuelan Presidential Commission.)*

Figure 1.6 ***This small volcanic cone in the crater of Vesuvius near Naples, Italy, belches out an ash-laden cloud of hot gases. This volcano is one of a chain of volcanoes, both active and extinct, that reaches from Mt. Etna in Sicily to north of Rome, far up the Italian boot. They all represent features built by molten material periodically extruded through the earth's crust.*** *(Photograph by Vincenzo Carcavallo.)*

Figure 1.7 ***Nearly a third of the land surface of the world is desert or near-desert. The processes of erosion are slower here than elsewhere but wind does move large amounts of material. Sometimes it is heaped up into dunes, as here in southern Saudi Arabia.*** *(Photograph by Arabian American Oil Company.)*

But nature is complex, and a shift toward equilibrium under one set of conditions sets up a new situation that demands still other changes.

Remember, then, that even the inanimate subjects of physical geology have been derived from some preexisting state and that they will change to a new form at some future time. The hard rock you hold in your hand may have flowed as lava from a now-extinct volcano millions of years ago and may form part of a fertile soil in some unknown future. The crag on a nearby hill may have formed an ancient shoreline, and, in some future existence, an as-yet-unborn stream may carry its fragments back to the ocean.

Energy

All change is an expression of energy, which is a common factor in all earth processes, both organic and inorganic. Thus different forms of matter result from differences in the energy that holds atoms together. Energy keeps materials molten so that they may break forth as a volcanic flow. Energy causes water to move and rock masses to slide in response to gravity. Solar energy supports life, produces changes in the weather, and helps keep the oceans in motion. All geology—and all existence—is a manifestation of energy.

1.2 *Gathering and using geologic data*

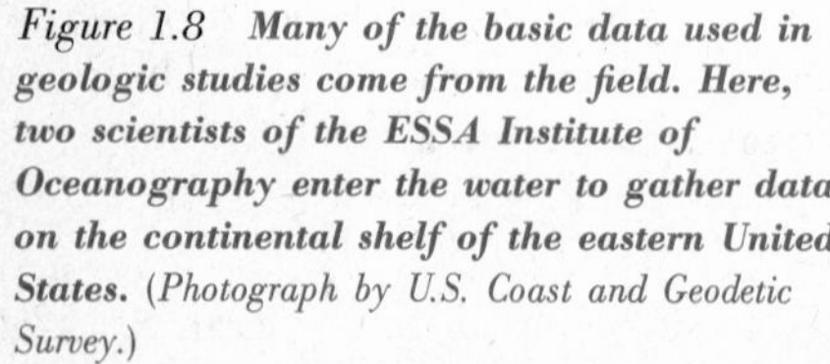

Figure 1.8 ***Many of the basic data used in geologic studies come from the field. Here, two scientists of the ESSA Institute of Oceanography enter the water to gather data on the continental shelf of the eastern United States.*** (*Photograph by U.S. Coast and Geodetic Survey.*)

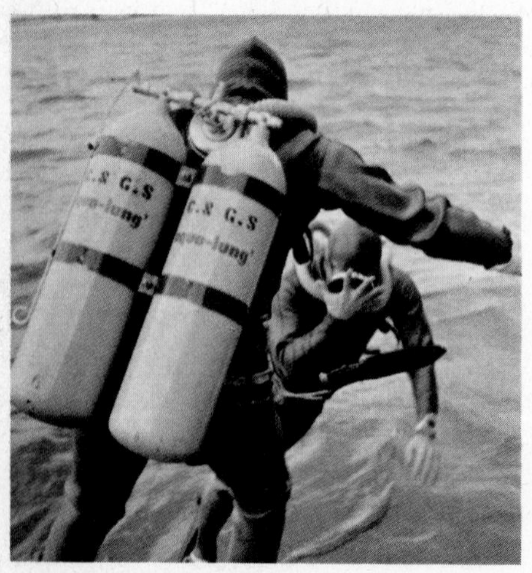

As the geologist traces the changing record of the earth, he gathers his facts in the out-of-doors—from the hills, the plains, and the glaciers, along the shore, and in the depths of the oceans. In short, the facts are not easy to come by, and gathering them often costs a great deal of time, manpower, and money. Moreover, there are many important facts that we cannot study in the laboratory. We cannot conjure up a volcanic eruption. We must wait for nature to provide one and hope that we can get to it when it occurs.

To make matters more difficult, the geologic record is often incomplete. This is true not only because it is difficult to gather the data but also because time may obscure the geologic record. And just as with the written record, so it is with the rock record: The farther back we go in time, the scantier our information becomes. Some of our geologic data may be obscured by burial, some may have been destroyed, and perhaps some were never recorded. In any event, much of the record is simply not available to us.

The very fact that the record is incomplete means that it is extremely important to use effectively all the data we do have. It means that our reasoning from the facts must be even more rigorous than if we had more complete data.

The geologist, then, must be prepared to produce a decision on the basis of very incomplete data. So he is anxious to make his information just as precise as he can, and the story of geology is in part the story of increasingly precise measurements and observations. Over the last 180 years, the science of geology has changed from an essentially qualitative science to a more

Figure 1.9 ***Petroleum, trapped in layers of sedimentary rock, is a product of past life. Offshore reserves are increasingly important and are tapped by wells drilled from platforms set in the sea, as is this one in the Arabian (Persian) Gulf.*** *(Photograph by Arabian American Oil Company.)*

quantitative one. In the past, the geologist was content with a superficial description of a rock layer, but now he uses refined techniques to determine the nature of the minerals that make up the rock. A century ago the only way a geologist could judge the violence of an earthquake, and even its approximate position, was to depend on subjective reports from widely scattered and untrained observers. Today, sensitive instruments manned by trained scientists enable us to measure an earthquake's size and to locate from afar its position on the earth's surface and the depth of the disturbance within the earth.

Although geologists have learned a great deal about this earth in a few generations, there is still much more to be learned. We have some fairly definite answers to a number of questions, but many problems exist for which we have but tentative answers. We need more facts and more precise data. Many questions remain completely unsolved, and there are undoubtedly many questions yet to be posed.

1.3 *The three rock families*

Geology is based on the study of rocks. We seek to know their composition, their distribution, how they are formed and how destroyed, why they are lifted up into continental masses and depressed into ocean basins.

Rock is the most common of all the materials on earth. It is familiar to us all. We may recognize it as the gravel in a driveway, the boulders in a stream,

Figure 1.10 **This aerial photograph shows the twisted pattern of deformed rocks. The beds, which show as bands 100 to 300 m wide, were once horizontal sedimentary rocks. They were deeply buried beneath the surface, tilted and folded by earth forces, and then exposed to view by subsequent erosion.** (*Photograph by Royal Canadian Air Force.*)

the cliffs along a ridge. And it is common knowledge that firm bedrock is exposed at the earth's surface or lies beneath a thin cover of soil or loose debris. Were we to examine rock material from locality to locality, we would begin to note the differences and the similarities of the various rock exposures. On the basis of our observations, we could begin to classify rocks into different groups.

Observations have led geologists to divide the earth's rocks into three main groups, based on mode of origin. These three types are *igneous, sedimentary,* and *metamorphic.* Later on, we shall discuss each type in detail, but here is a short explanation of all three.

Igneous rocks, the ancestors of all other rocks, take their name from the Latin *ignis,* "fire." These "fire-formed" rocks were once a hot, molten, liquidlike mass known as a *magma,* which subsequently cooled into firm, hard rock. Thus the lava flowing across the earth's surface from an erupting volcano soon cools and hardens into an igneous rock. But there are other igneous rocks now exposed at the surface that actually cooled some distance beneath the surface. We see such rocks today only because erosion has stripped away the rocks that covered them during their formation.

Most sedimentary rocks (from the Latin *sedimentum,* "settling") are made up of particles derived from the breakdown of preexisting rocks. Usually these particles are transported by water, wind, or ice to new locations where they are deposited in new arrangements. For example, waves beating against a rocky

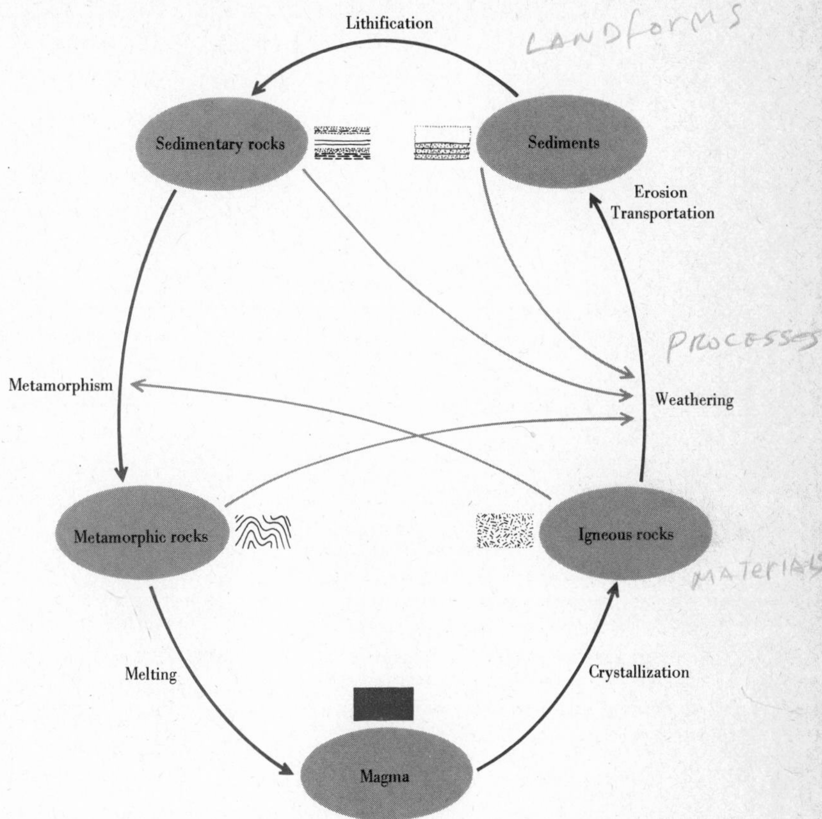

Figure 1.11 ***The rock cycle, shown diagrammatically. If uninterrupted, the cycle will continue completely around the outer margin of the diagram from magma through igneous rocks, sediments, sedimentary rocks, metamorphic rocks, and back again to magma. The cycle may be interrupted, however, at various points along its course and follow the path of one of the arrows crossing through the interior of the diagram.***

shore may provide the sand grains and pebbles for a nearby beach. If these beach deposits were to be hardened, we would have sedimentary rock. One of the most characteristic features of sedimentary rocks is the layering of the deposits that make them up.

Metamorphic rocks compose the third large family of rocks. Metamorphic, from the Greek words *meta,* "change," and *morphe,* "form," refers to the fact that the original rock has been changed from its primary form to a new form. Earth pressures, heat, and chemically active fluids beneath the surface may all be involved in changing an originally sedimentary or igneous rock into a new metamorphic rock.

The rock cycle

We have suggested that there are definite relationships among sedimentary, igneous, and metamorphic rocks. With time and changing conditions, any one of these rock types may be changed into some other form. These relationships form a cycle, as shown in Figure 1.11. This is simply a way of tracing out the various paths that earth materials follow. The outer circle represents the complete cycle; the arrows within the circle represent shortcuts in the system that can be, and often are, taken. Notice that the igneous rocks are shown as having formed from a magma and as providing one link in a continuous chain. From these parent rocks, through a variety of processes, all other rocks can be derived.

First, weathering attacks the solid rock, which either has been formed by the cooling of a lava flow at the surface or is an igneous rock that was formed deep beneath the earth's surface and then was exposed by erosion. The products of weathering are the materials that will eventually go into the creation of new rocks—sedimentary, metamorphic, and even igneous. Landslides, wind, running water, and glacier ice all help to move the materials from one place to another. In the ideal cycle, this material seeks the ocean floors, where layers of soft mud, sand, and gravel are consolidated into sedimentary rocks. If the cycle continues without interruption, these new rocks may in turn be deeply buried and subjected to heat, to pressures caused by overlying rocks, and to forces developed by earth movements. The sedimentary rocks may then change in response to these new conditions and become metamorphic rocks. If these metamorphic rocks undergo continued and increased heat and pressure, they may eventually lose their identity and melt into a magma. When this magma cools, we have an igneous rock again, and we have come full cycle.

But notice, too, that the complete rock cycle may be interrupted. An igneous rock, for example, may never be exposed at the surface and hence may never be converted to sediments by weathering. Instead, it may be subjected to pressure and heat and converted directly into a metamorphic rock without passing through the intermediate sedimentary stage. Other interruptions may take place if sediments, or sedimentary rock, or metamorphic rock are attacked by weathering before they continue to the next stage in the outer, complete cycle.

This concept of the rock cycle was probably first stated in the late eighteenth century by James Hutton, of whom we have already spoken:

> *We are thus led to see a circulation in the matter of this globe, and a system of beautiful economy in the works of nature. This earth, like the body of an animal, is wasted at the same time that it is repaired. It has a state of growth and augmentation; it has another state, which is that of diminution and decay. This world is thus destroyed in one part, but it is renewed in another; and the operations by which this world is thus constantly renewed are as evident to the scientific eye, as are those in which it is necessarily destroyed.*[2]

We can consider the rock cycle to be an outline of physical geology, as a comparison of Figure 1.11 with the table of contents of this book will show. Each step has its place; each is a part of the whole picture. But we must remember that in a more fundamental sense the rock cycle represents a response of earth materials to various forms of energy. We must realize that matter and energy are inseparable, that earth materials change and that features of the earth's face are altered in response to changes in energy.

[2] James Hutton, *Theory of the Earth*, vol. 2, p. 562, Edinburgh, 1795. Hutton's theory of the earth was first presented as a series of lectures before the Royal Society of Edinburgh in 1785. These lectures were published in book form in 1795. Seven years later, Hutton's concepts were given new impetus through a more readable treatment called *Illustrations of the Huttonian Theory* by John Playfair.

Figure 1.12 ***The earth is seen here from the*** **Apollo 8** ***spacecraft during man's first circumlunar voyage. The atmosphere which obscures the earth makes possible not only life but running water, glaciers, oceans, winds, and the familiar processes of weathering.*** *(Photograph by National Aeronautics and Space Administration.)*

1.4 *To worlds beyond*

We call the study of the rocks and history of this planet geology. But what about the study of other planets of the solar system, or their moons, or the planets of other stars?

Man has traveled to his earth's moon and back. In addition, his telescopes, cameras, and other instruments of remote sensing provide an ever-increasing store of data about the moon and about the sun's other planets and their moons. What do we call the study of these bodies? *Planetology* is the term sometimes used; *astrogeology* is also current, despite the ineptness of its derivation. But no matter what we call the study, we find that the techniques and principles of geology are applicable. Conditions, of course, are different on these extraterrestrial bodies from what they are on the earth. Nevertheless, as we shall find (particularly in Chapter 23), our earthbound experience and our knowledge

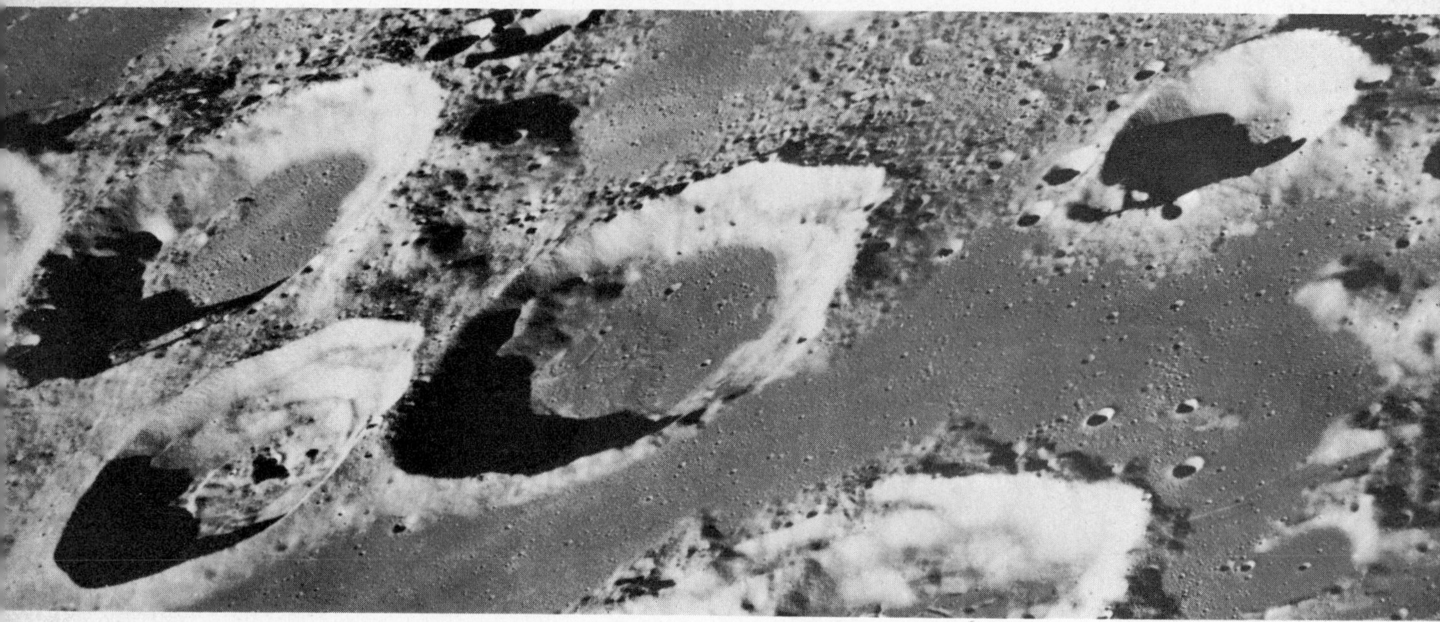

Figure 1.13 ***The principles and methods of geologic exploration are applicable to a study of the moon, some of whose history is already worked out. The three clustered craters are Magelhaens, Magelhaens A, and Colombo A.*** *(Photograph, from the* Apollo 8 *mission, by National Aeronautics and Space Administration.)*

of geology can be extended to help understand the materials, processes, and history of our distant neighbors. Indeed, the study of these bodies has become very much a human activity. Much of the study turns out to be what we know on earth as geology.

Supplementary readings

ALBRITTON, C. C., JR. (ed.): *The Fabric of Geology,* Addison-Wesley Publishing Company, Inc., Reading, Mass., 1963. A series of essays on the history and philosophy of geology.

HUTTON, JAMES: *Theory of the Earth, with Proofs and Illustrations* (2 vols.), Edinburgh, 1795 (facsimile reprint, Hafner Publishing Company, Inc., New York, 1959). The book that did more than any other to launch the discipline of geology.

MATHER, K. F.: *Source Book in Geology: 1900–1950,* Harvard University Press, Cambridge, Mass., 1967. This volume, taken along with the following one by Mather and Mason, contains extracts from some of the more important original writings in geology from the sixteenth century to 1950.

MATHER, K. F., and S. L. MASON: *A Source Book in Geology,* McGraw-Hill Book Company, Inc., New York, 1939.

PLAYFAIR, JOHN: *Illustrations of the Huttonian Theory of the Earth,* Cadell and Davies and William Creech, Edinburgh, 1802 (facsimile reprint, University of Illinois Press, Urbana, 1956). Playfair's exposition of Hutton's theory of the earth was influential in introducing it to the nineteenth-century naturalists.

2

Matter and energy

THE PHYSICAL UNIVERSE IS COMPOSED OF WHAT WE CALL MATTER, YET ONE OF the most elusive problems in science is to define matter precisely. It has long been customary to refer to states of matter as solid, liquid, or gas. And we say that matter has physical properties, such as color or hardness, as well as chemical properties, which govern its ability to change or to react with other bits of matter. But all this simply tells us *about* matter, not what matter is.

Centuries ago, Greek philosophers speculated on this problem, arguing at length over whether the smallest pieces into which anything could be divided were just miniatures of the original—microscopic drops of water, extremely small grains of sand, infinitesimal pieces of salt—or whether at some point down the line certain particles would be found that were joined together in different ways to form water, sand, salt, and all the other substances of the material world.

Today, we know that the second explanation is the true one. These particles are called *atoms.* So if we are to understand matter, we must first learn about the atoms and the ways in which they combine. This information will give us a basic understanding of what rocks actually are, why they differ, and how they can be changed.

2.1 Matter

Electric charge

All matter appears to be essentially electrical in nature. Some of the earliest ideas about what we now call electricity sprang from a very simple experiment—the rubbing together of a piece of amber and a piece of fur. After they were rubbed together, both the fur and the amber were found to be capable of picking up light pieces of other materials, such as feathers or wool. But the interesting thing was that materials *attracted* by the amber were *repelled* by the fur. So scientists decided that there must be two kinds of electricity. One was arbitrarily called positive and the other negative. This fact was first discovered by the Greeks about 600 B.C. Then, late in the sixteenth century, William Gilbert, personal physician to Queen Elizabeth I, proposed that the power responsible for this phenomenon be called electricity, from the Greek word *elektron,* meaning "amber."

We say that *like electric charges repel each other and unlike charges attract each other* (see Figure 2.1).

Atoms

Just what happens when amber is rubbed with fur? Particles pass from the fur to the amber, and the amber becomes negatively charged. We therefore reason that the particles that bring about this condition must be negatively charged. These negatively charged particles are called *electrons,* and they are fundamental particles of atoms.

Atoms cannot be assembled from electrons alone, however, for all electrons have negative electric charges and would not stick together by themselves. So scientists reasoned that there had to be some other particle with a *positive* charge. At last this particle was found and was called a *proton.* But in addition to having a positive charge, the proton differs from the electron in another respect: It acts like a much heavier unit of matter. Because any quantity of matter is arbitrarily described by a number called its *mass,* the proton is said to have greater mass than the electron. In fact, *we define the fundamental particles of atoms in terms of mass and electric charge* (see Table 2.1).

Scientists continued to chip away at atoms until finally they turned up a third particle, with a mass about equal to that of the proton but with no electric charge. This electrically neutral particle they called the *neutron.*

Protons, neutrons, and electrons are the fundamental particles that combine to form atoms. All matter is composed of atoms.

Nobody has yet come forward with an explanation of exactly what these fundamental particles are, but we do know that *energy* is intimately involved in their makeup.

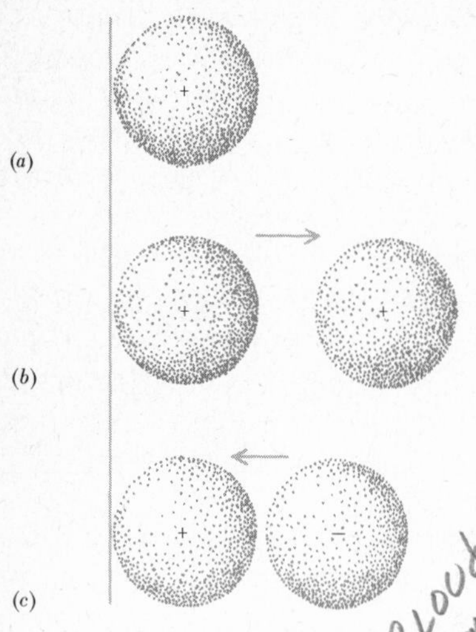

Figure 2.1 **Like charges of electricity repel each other; unlike charges attract each other. The positively charged sphere in (a) repels the approach of another positive charge (b) but attracts a negative charge (c).**

Table 2.1 **Fundamental particles**

	Electric charge	*Mass*
Electron	−1	0.00055
Proton	+1	1.00760
Neutron	0	1.00890

ATOMIC STRUCTURE All the information we have about the structure of atoms has been established by indirect observation. Physicists using doubly charged positive particles called *alpha particles*[1] to bombard atoms have made some interesting discoveries. For example, if these particles are shot at a target, such as a piece of metal made up of billions of atoms, no more than one alpha particle in 10,000 hits anything inside the target. It has been concluded, therefore, that the target must be largely open space.

Repeated tests have shown that *atoms contain a nucleus of protons and neutrons surrounded by a cloud of electrons.* They have also revealed that *a normal atom has as many electrons as it has protons.* The neutrons contribute to the mass of the atom, but they do not affect its electrical charge. The number of protons plus the number of neutrons in an atom constitute the atom's *mass number.* The negative electrons spin very rapidly around the nucleus, like planets around the sun (to keep from being pulled in by the attraction of the positive protons). In fact, they complete several thousand million million round trips per second, at a speed of hundreds of kilometers per second. Again like the planets revolving about the sun, electrons revolve around the nucleus at different distances.

ATOMIC SIZE The electrons form a protective shield around the nucleus and give *size* to the atom. Atomic dimensions are too small to have any real meaning as absolute numbers, but they yield some interesting comparisons in terms of relative size. In describing atomic dimensions,[2] we use a special unit of length, the *angstrom* (A), which is one hundred-millionth of a centimeter, 0.00000001 cm, usually written 1×10^{-8} cm.

[1] Two protons and two neutrons bound together.

[2] In dealing with very small or very large numbers it is convenient to express them in *powers of ten* as described in Appendix B.

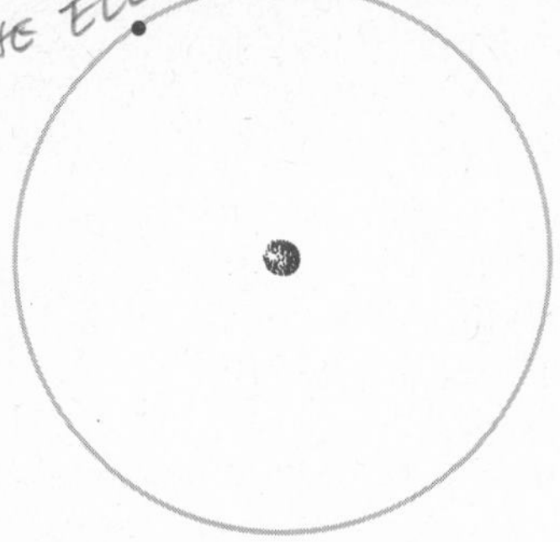

Atomic nuclei have diameters that range from a ten-thousandth to a hundred-thousandth of an angstrom—that is, from 10^{-4} to 10^{-5} A. Atoms of the most common elements have diameters of about 2 A, which is roughly 20,000 to 200,000 times the diameter of the nucleus. Again we see that matter consists mostly of open space. If the sun were truly the nucleus of our atomic model, the diameter of the atom would be greater than the diameter of the entire solar system.

ATOMIC MASS Although the nucleus occupies only about a thousandth of a billionth of the volume of an atom, it contains 99.95 percent of the atom's mass. In fact, if it were possible to pack a cubic centimeter with nothing but protons, this small cube would weigh more than 100 million tons at the earth's surface.

IONS *An ion is an electrically unbalanced form of an atom or group of atoms.* An atom is electrically neutral, but if it loses an electron from its outermost shell, the portion that remains behind has an extra unmatched positive charge. This unit is known as a positively charged ion. If the outermost shell gains an electron, the ion has an extra negative charge and is known as a negatively charged ion. As we shall see later, more than one electron may be lost or gained, leading to the formation of ions with two or more units of electrical charge.

Figure 2.2 **Schematic sketch of a hydrogen atom, which consists of one proton and one electron. This is the simplest atom.**

Figure 2.3 **Diagrammatic representation of an atom of helium. The nucleus consists of two protons and two neutrons and accordingly has a mass number of 4. There are two electrons (negative charges) to balance the positive charges on the two protons. Because there are two protons in the nucleus, this atom is in place 2 in the table of elements. The symbol ^4_2He indicates a mass number of 4, place 2 in the table of elements, and He for the name "helium." The nucleus of helium without any accompanying electrons is called an alpha particle.**

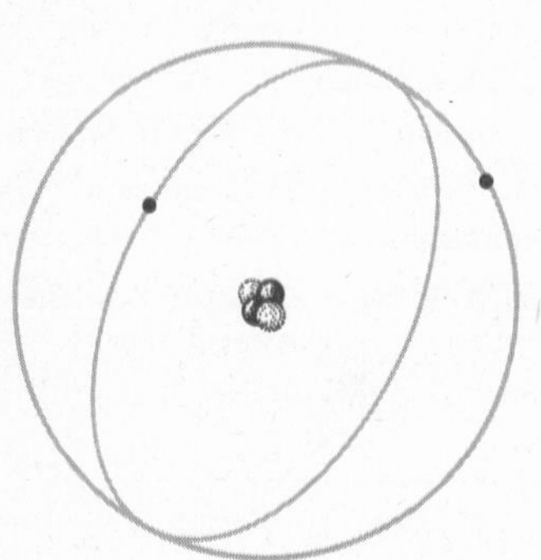

Elements

An atom is the smallest unit of an element. Eighty-eight elements have been found in nature, and 15 have been made in the laboratory. *Each element is a special combination of protons, neutrons, and electrons.* The distinguishing feature of each element is the number of protons in its nucleus. An element has an *atomic number* corresponding to the number of protons in its nucleus, and it is given a name and a symbol.

Element 1 is a combination of one proton and one electron (see Figure 2.2). Long before its atomic structure was known, this element was named hydrogen, or "water-former," from Greek roots meaning "water" and "to be born"—*hydro* and *gen*—because water is formed when hydrogen burns in air. The symbol of hydrogen is H. Because it has a nucleus with one proton, hydrogen assumes place 1 in the table of elements.

Element 2 consists of two protons (plus two neutrons in the most common form) and two electrons (see Figure 2.3). It was named helium, with the symbol He, from the Greek *helios,* "the sun," because it was identified in the solar spectrum before being isolated on the earth. Because of the two protons in its nucleus, helium takes place 2 in the table of elements.

Each addition of a proton, with a matching electron to maintain electrical balance, produces another element. Neutrons seem to be included more or less indiscriminately, though there are about as many neutrons as protons in the common form of many of the elements.

There are three classes of elements: metals, nonmetals, and metalloids.

Seventy-seven are classed as metals, 17 as nonmetals, and 9 as metalloids. These are keyed in the list of elements in Appendix A.

A metal typically shows a peculiar luster, called metallic luster; is a good conductor of electricity or heat; is opaque; and may be fused, drawn into wire, or hammered into sheets. A nonmetal lacks some or all of these properties. A metalloid has some metallic and some nonmetallic properties.

The elements appearing in nature begin with the lightest, hydrogen (place 1), and end with the heaviest, the metal uranium (place 92). Technetium (43), promethium (61), astatine (85), and francium (87) do not occur naturally. They have been produced only for an instant during radioactive decay.

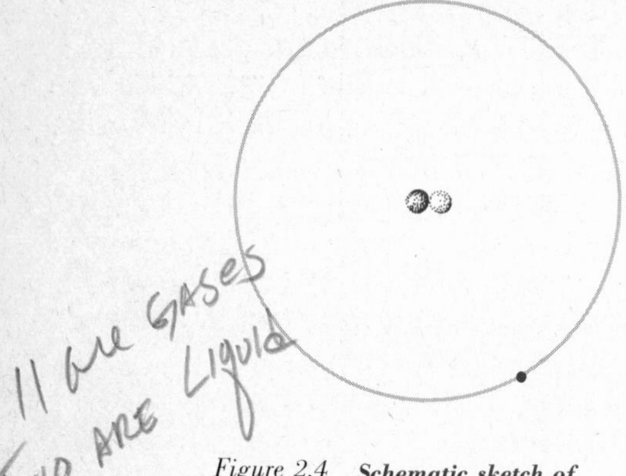

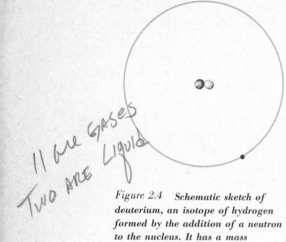

Figure 2.4 **Schematic sketch of deuterium, an isotope of hydrogen formed by the addition of a neutron to the nucleus. It has a mass number of 2.**

ISOTOPES Every element has forms that, though essentially identical in chemical and physical properties, have different masses. Such forms are called *isotopes*—from the Greek *iso,* "equal" or "the same," and *topos,* "place"—because each form has the same number of protons in the nucleus and occupies the same position in the table of elements. Isotopes show differences in mass as a result of differences in the number of neutrons in their nuclei. For example, hydrogen with one proton and no neutrons in its nucleus has a mass number of 1. When a neutron is present, however, the atom is an isotope of hydrogen with a mass number of 2, called deuterium. All elements have isotopes (see Figure 2.4).

Information about atomic structure is commonly expressed symbolically by

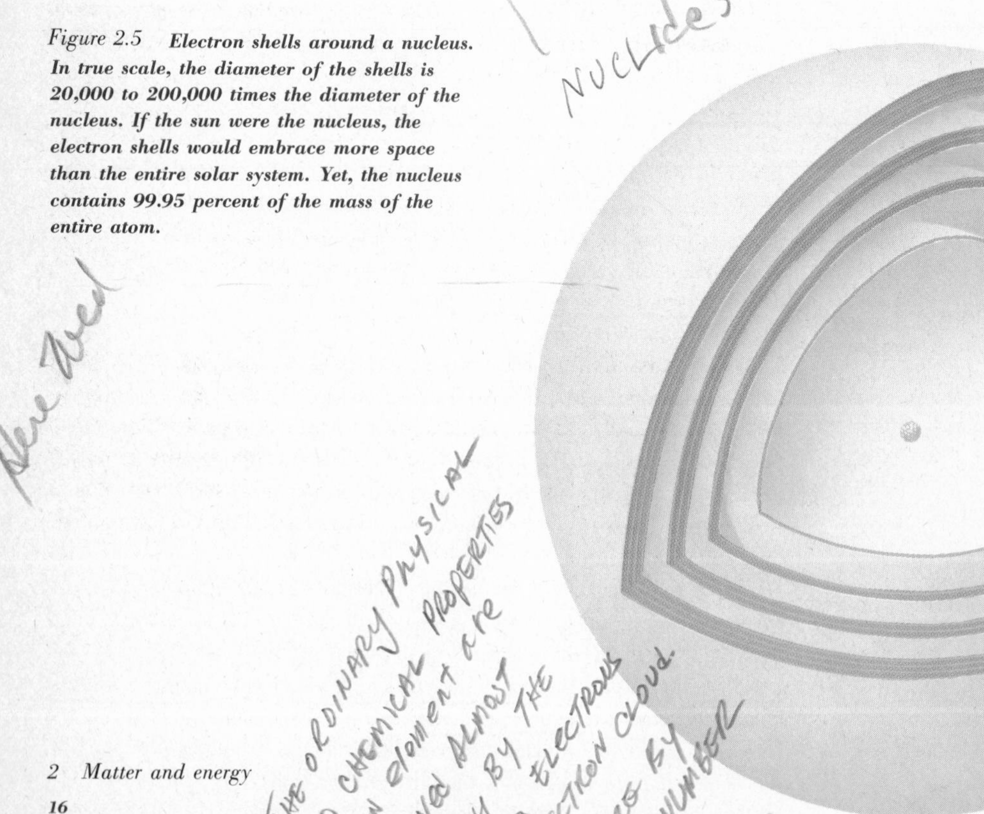

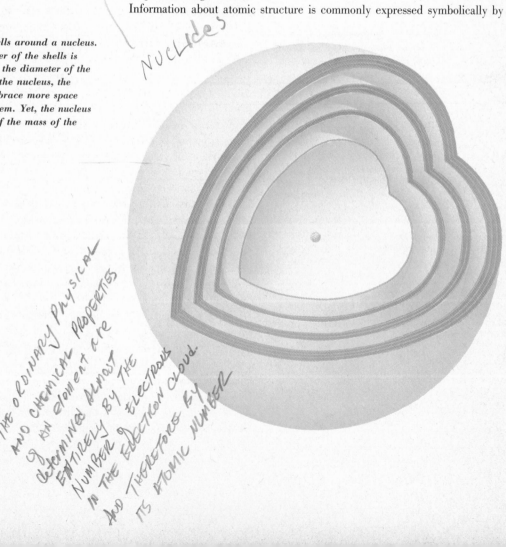

Figure 2.5 **Electron shells around a nucleus. In true scale, the diameter of the shells is 20,000 to 200,000 times the diameter of the nucleus. If the sun were the nucleus, the electron shells would embrace more space than the entire solar system. Yet, the nucleus contains 99.95 percent of the mass of the entire atom.**

a subscript number prefixed to the element's symbol to indicate the atomic number and by a superscript indicating the mass number. Thus hydrogen would be represented by ${}^{1}_{1}H$ and its isotope deuterium by ${}^{2}_{1}H$.

Many of the elements as they occur in nature are mixtures of isotopes. In general, regardless of the source of the element, the proportion of isotopes composing it is constant, unless special steps have been taken to separate them. Thus chlorine, whether occurring naturally or prepared in the laboratory, has an atomic weight of 35.457, which represents the average weight of a mixture of about three parts of ${}^{35}_{17}Cl$ to one part of ${}^{37}_{17}Cl$.

Energy-level shells

Each electron follows a definite radius of travel, called its energy-level orbit, around the nucleus. These orbits are arranged systematically at different distances from the nucleus and a specific amount of energy is required to maintain an electron at a given distance from its nucleus (see Figure 2.5). For convenient reference, these distances are sometimes referred to as *energy-level shells.*

In the building of successively more complex atoms, electrons are distributed in a systematic way among energy-level shells and corresponding subshells. The shells and subshells have certain energies and certain capacities for electrons. In the order of increasing radius, the shells are numbered 1, 2, 3, 4, 5, 6, and 7. The subshells are designated by *s, p, d,* and *f.* Shell 1 contains only one subshell, *s;* shell 2 contains only two subshells, *s* and *p;* shell 3 contains three subshells, *s, p,* and *d;* and shell 4 has four subshells, *s, p, d,* and *f* (see Figure 2.6). Theoretically, shell 5 should contain five subshells, shell 6 six, and shell 7 seven, but in known atoms in their normal or unexcited states no subshells but *s, p, d,* and *f* contain electrons. Each has a definite limit to the number of electrons it can hold: 2 for *s,* 6 for *p,* 10 for *d,* and 14 for *f.*

Figure 2.6 **Energy levels for the first four electron shells, to illustrate the principle that in the building of the elements shells of electrons are filled in the order of their level of energy: s of shell 1 first; then s of shell 2 and p of shell 2; then s of shell 3 and p of shell 3. Here a break in the sequence occurs, and s of shell 4 is filled before d of shell 3 is filled to finish out shell 3. The maximum number of electrons that can be contained in each s subshell is 2; in p, 6; in d, 10, in f, 14.**

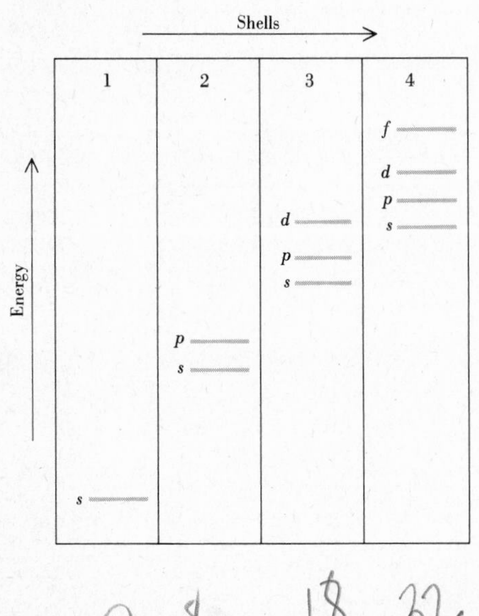

COMBINATIONS OF ATOMS When the outermost shell of an element has its full quota of electrons, its atoms do not combine with others. Helium atoms will not even combine with one another; they remain as separate atoms in the gas. This is because its outermost shell, shell 1, with a capacity for two electrons, is full. For neon, argon, krypton, xenon, and radon, the outermost shell, the $s + p$ subshells with a combined capacity for eight electrons, is also full. These elements are all inert gases.

Atoms of inert elements ordinarily do not combine with each other or with other elements. All other elements, because their outer shells are not filled to capacity, unite readily with each other or with other atoms.

Compounds

Compounds are combinations of atoms of different elements. Those formed by life processes are *organic compounds.* Others are *inorganic compounds.*

Figure 2.7 *In the formation of the compound halite, an atom of sodium, which has only one electron in its outermost shell, joins an atom of chlorine, which needs one electron to fill its outermost shell. The sodium's lone outermost electron slips into the vacant place to fill the chlorine's outermost shell.*

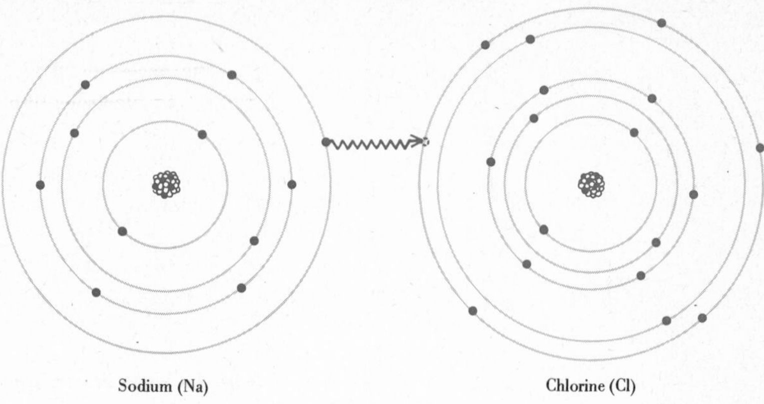

The number of electrons in the outer shell of an atom determines the manner and ease with which the atom can join with other atoms to form compounds. Compounds may be formed by different atoms losing or gaining electrons. For example, an atom of sodium has only one electron in its outermost shell but eight in the next shell. When a sodium atom and a chlorine atom join, the sodium's outermost electron slips into the vacant place in the chlorine's outermost shell. Chlorine gains the electron and becomes a negative ion, Cl^-, whereas sodium, by losing it, becomes a positive ion, Na^+ (see Figure 2.7). The result is called an *ionic compound.* The product is the compound halite, or common table salt, one of the world's most abundant substances. The chemical designation is simply a combination of the symbols for the elements in the proportions in which they are present, NaCl. (That is, one atom of sodium is joined with one atom of chlorine.) This unit is a *molecule* of salt. *A molecule is the smallest unit of a compound.*

When electrons are lost or added to form an ion, electrical forces between the nucleus and the electrons are thrown out of balance, and this affects the radius of the ion. For example, if sodium loses the electron from its shell 3, it has only two shells of electrons left. This is the same electronic supply that

Table 2.2 *Radii of atoms and ions with shell 2 filled*[a]

Atomic number	Element	Atom Symbol	Atom Radius, A	Ion Symbol	Ion Radius, A
8	Oxygen	O	0.60	O^{2-}	1.32
10	Neon	Ne	1.60		
11	Sodium	Na	1.86	Na^+	0.97
12	Magnesium	Mg	1.60	Mg^{2+}	0.66
13	Aluminum	Al	1.43	Al^{3+}	0.51
14	Silicon	Si	1.17	Si^{4+}	0.42
15	Phosphorus	P	1.08	P^{5+}	0.35
16	Sulfur	S	1.04	S^{6+}	0.30
17	Chlorine	Cl	1.07	Cl^{7+}	0.27

[a] Radii from Cornelius S. Hurlbut, Jr., *Dana's Manual of Mineralogy*, 17th ed., John Wiley & Sons, Inc., New York, 1961.

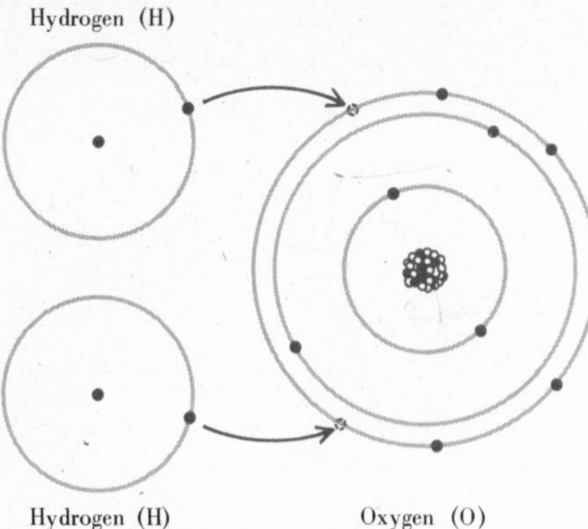

Figure 2.8 ***Two hydrogen atoms and one oxygen join to form water, H_2O, by a covalent bond. In this bond, the hydrogen electrons do double duty in a sense, filling the two empty places in the outer shell of oxygen, yet remaining at their normal distance from their hydrogen nuclei. The result is the formation of a molecule of water, the smallest unit that displays the properties of that compound.***

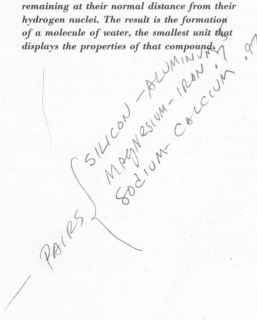

neon has, but the radius of the sodium ion, Na^+, is 0.97 A, whereas that of an atom of neon with the same number of electrons and shells is 1.60 A. The excess of unbalanced positive charge in the sodium ion pulls in its remaining electrons and shrinks the apparent radius of the ion. In fact, if two full shells of electrons always meant the same radius, the positive ions of the geologically important elements 11 through 17 would have about the same radius after all shell 3 electrons were lost. However, as more and more outer electrons are removed, the increasing relative power of the central positive charge pulls in the remaining electrons until the radii decrease progressively from the Na^+ ion's 0.97 A to the Cl^{7+} ion's 0.27 A (see Table 2.2).

The atoms of elements can combine in other ways to form compounds, as, for example, in the formation of a water molecule. The oxygen atom has six electrons in its outermost shell and therefore needs two more to achieve the stable number of eight. If two hydrogen atoms, each with a single electron, approach an oxygen atom, the hydrogen electrons in effect slip into the vacant slots in the outermost shell of the oxygen atom (see Figure 2.8). So the hydrogen and oxygen nuclei are sharing their electrons, in a way called a *covalent bond.* Again, the result is a compound that is different in every way from the elements themselves. This compound is water, whose symbol H_2O represents the elements that make it up and the proportions in which they are present. The combination is so perfect that water is one of nature's most stable compounds.

Because the oxygen atom has in effect gained two electrons, it takes on a negative charge. And each hydrogen atom, acting as though it had lost its electron, takes on a positive charge. As a result, a molecule of water acts like a small rod, with a positive charge on one end and a negative charge on the other (see Figure 2.9). These ends are referred to as a positive pole and a negative pole, because of the molecule's similarity to a bar magnet. The

Figure 2.9 ***The dipolar character of water. (a) The oxygen has, in effect, gained two electrons, hence a double negative charge, whereas the hydrogen atoms have each lost the effective service of an electron and represent positive charges. Accordingly, the water molecule acts like a small rod with a positive charge on one end and a negative charge on the other, as suggested in (b). Combinations of water molecules are suggested in (c).***

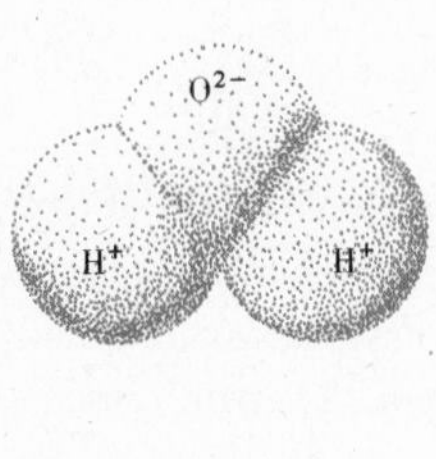

(a)

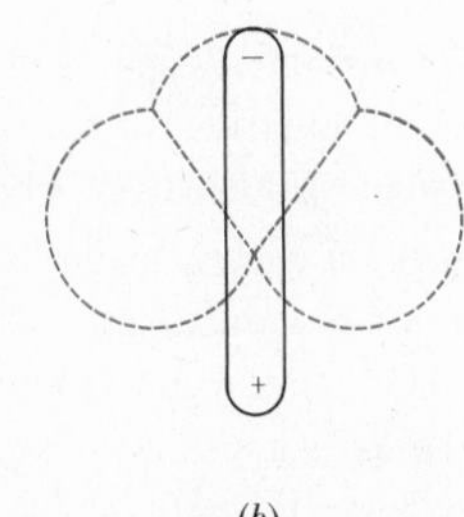

(b)

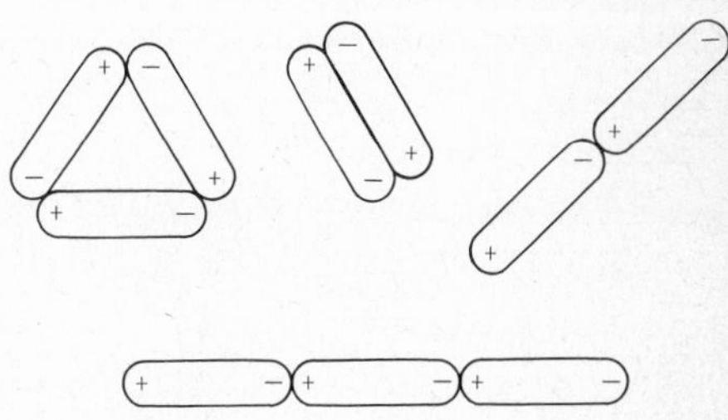

(c)

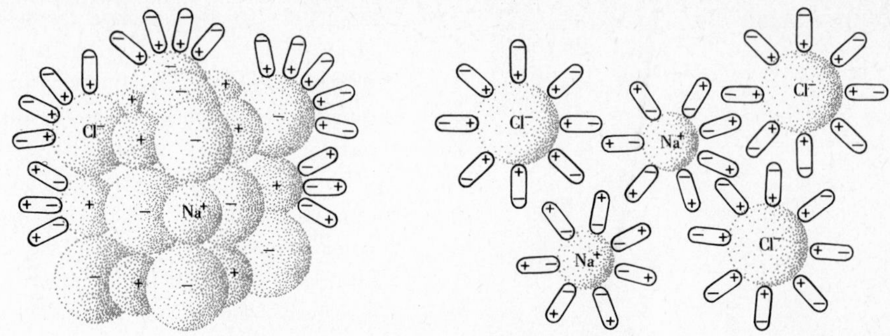

Figure 2.10 ***Mechanism by which water dissolves salt. Water dipoles attach themselves to the ions that compose the salt and overcome the ionic attractions that hold the salt together as a solid. Each Na^+ and Cl^- ion is then conveyed by a number of water dipoles into the body of the liquid.***

molecule, then, is a *dipole* ("two-pole"), and water is known as a *dipolar compound.* This fact gives water special properties that make it an extremely important agent in geological processes. The mechanism by which water dissolves salt (see Figure 2.10) is an illustration of the ease with which water dissolves various substances and participates in weathering and other geological activities.

Metallic bond

The atoms of metallic elements have a special kind of bonding which is responsible for metals being such good conductors of heat and electricity. In a piece of metal, the atoms of one element are closely packed together, but their outermost electrons are not shared or exchanged but are free to move around and connect to any atoms in the solid. These wandering electrons shared by any of the atoms in a piece of metal create the *metallic bond.* The relative freedom of movement of the electrons in this relationship accounts for the high level of conductivity of electricity and heat that is one of the characteristics of metals.

When there is no current, electrons jump randomly from atom to atom. When an electrical potential is applied to a good conductor, such as a copper wire, electrons flow through the wire without stopping to attach themselves to any particular atoms. They thus transfer electricity (or heat) throughout the whole wire.

Organization of matter

Matter is the substance of the physical universe. It is composed of atoms. Atoms are combinations of the fundamental particles protons, neutrons, and electrons (except for hydrogen, which is a combination of one proton and one electron). An atom is the smallest unit of an element. Eighty-eight elements have been found in nature. Four others have been identified as short-lived transients, and 11 have been made by man. Atoms of elements also have other forms, such as isotopes and ions.

Atoms are rarely found alone in nature. They are found in combination

Table 2.3 *The organization of matter*[a]

Element number	Atom[b]		
	Nucleus[c]		
	Protons	Neutrons[d]	Electrons[e]
1	1	—	1
2	2	—	2
3	3	—	3
.	.	.	.
.	.	.	.
.	.	.	.
103	103	—	103

[a] This table is continued in Table 3.5.
[b] Atoms of different elements combine to form *compounds* (but inert gases do not usually combine). The smallest unit of a compound is a *molecule.*
[c] The number of protons plus the number of neutrons equals the mass number.
[d] The number of neutrons varies. The gain or loss of neutrons in an atom results in *isotopes.*
[e] The gain or loss of electrons in an atom results in *ions.*

with each other and with other atoms. Atoms of different elements combine by exchanging or sharing electrons to form compounds. The smallest unit of a substance is a molecule (see Table 2.3). The 88 naturally occurring elements have been found combined in nature, forming numerous compounds. They have also been combined by man to form the millions of compounds we use today.

Water is the only natural substance to occur in the three states of solid, liquid, and gas.

2.2 Energy

Energy is a word that sounds familiar enough. We speak of energy-producing foods, of the energy needed to climb stairs, of energy radiated to us from the sun. Matter and energy are intimately related, but what is energy?

Energy is the capacity for producing motion. No form of matter is entirely devoid of motion. The motion may be in things we see, such as an automobile speeding down the highway, or in things we cannot see, such as atoms and molecules. All motion is produced by energy in one form or another.

In fact, we can think of the entire universe as a great bundle of energy. Heat and light from the sun represent energy in one form; the revolution of the earth around the sun represents energy in another form; and the chemical transformation of food into heat and body activity represents energy in yet another form.

In geological processes, through the motion of running water, energy sculptures the land; through the deformation of rocks, it builds mountains; and through the rupture of earth materials, it produces earthquakes. In fact, energy is involved in every geological process.

Energy manifests itself in different forms, which have descriptive names: potential, kinetic, heat, chemical, electrical, atomic, and radiant.

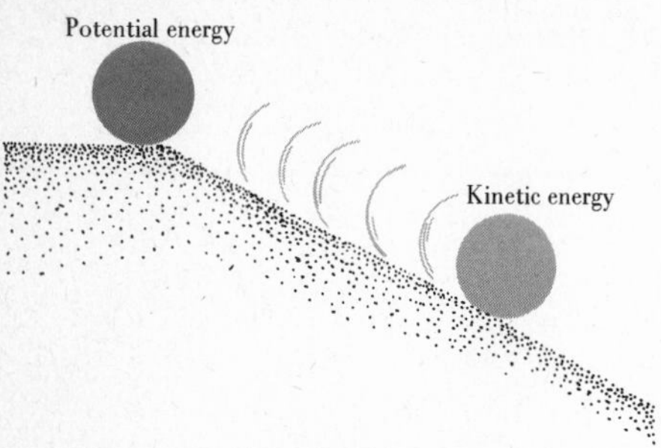

Figure 2.11 **The potential energy of a 10-ton boulder resting on a hillside is transformed into kinetic energy, the energy of movement, when the boulder is dislodged and rolls downhill.**

Potential energy

Potential energy is stored energy that is waiting to be used. Uranium, petroleum, coal, and natural gas are eagerly sought after because the potential energy they contain can be effectively released and put to work. Water in the clouds, or in lakes and reservoirs at high altitudes, has potential energy that is released when the water falls or runs downhill. A boulder poised on a hilltop has potential energy. Less obvious, but just as real, is the potential energy stored in the nucleus of every atom and in the molecules of every compound.

Kinetic energy

The potential energy of a 10-ton boulder resting on a hillside is transformed into the energy of movement when the boulder is dislodged and rolls downhill (see Figure 2.11). This energy of movement is called *kinetic energy. Every moving object possesses kinetic energy.*

The amount of kinetic energy possessed by an object depends on the mass of the object and on the speed with which it moves:

$$E_K = \frac{Mv^2}{2}$$

A 10-ton boulder rolling down a hill has more energy at the bottom of the slope than a pebble that has rolled down the same slope at the same speed. On the other hand, this same 10-ton boulder moving a few meters per hour would have less energy than if it were hurtling along at 30 m/sec (meters per second).

Running water owes much of its effectiveness as a geological agent to kinetic energy; so do waves beating on the coast. Massive glaciers creeping down a mountain slope do geological work by means of kinetic energy.

Heat energy

Heat energy is a special manifestation of kinetic energy in atoms. All atoms are moving constantly. They vibrate around fixed positions in solids and move about freely in liquids and more freely in gases, but all are in motion, and this movement produces the effect we call heat.

Heat applied to a solid can finally break its electron bonds so that the solid melts and becomes a liquid or, upon further heating, a gas. Extreme cold can change a gas into a liquid or a solid.

Everybody is familiar with ice, water, and steam. These are the compound H_2O in three different states: solid, liquid, and gas. What we call *temperature* is the degree of heat a substance has. It is actually an arbitrary number that represents the activity of atoms. Because the atoms are less active in ice, we say that ice has a lower temperature than water, in which they are moving more rapidly. If atoms were completely motionless, we would call the temperature *absolute zero.* Although man has been able to attain very low temperatures in matter, he has been unable to achieve absolute zero, which would be −273°C (degrees Centigrade).

Quantity of heat is measured in calories (cal). A calorie is the quantity of heat required to raise the temperature of 1 g (gram) of water by 1°C. (The food calorie is the quantity of heat required to raise 1,000 g of water 1°C.) The quantity of heat depends on the number of atoms that produce the temperature by their activity. We feel intuitively, for example, that a cup of boiling water and a bucket of boiling water represent different total quantities of heat even though they are at the same temperature. Evidence of this difference could be provided by the volume of ice melted by each. The filament of an incandescent light bulb is at a very high temperature, but the quantity of heat it produces would do a poor job of heating a home.

Heat is the avenue through which most forms of energy are applied to the needs of man. It is the agency through which he can change matter from one form to another. Research with supercold temperatures has opened new frontiers of science, such as space exploration. Supercold has enabled man to liquefy gases. It has given us the powerful space fuel liquid hydrogen, which becomes liquid at −240°C. Liquid hydrogen is a good rocket fuel because it burns easily, has the lowest atomic weight, and produces the fastest-moving exhaust. When oxidized by liquid oxygen, its thrust is 40 percent greater than that of the heavy kerosene–liquid oxygen fuel used by early space missiles.

Heat has supplied much of the energy involved in the formation of earth materials and the processes by which these materials have been altered throughout their history. An understanding of the nature of heat is fundamental to a full appreciation of the formation of igneous and metamorphic rocks, igneous activity, and the formation of mountains.

Chemical energy

When the atoms of elements combine to form compounds, or when compounds combine to form other compounds, the combining is called a chemical process, and the energy that is released or absorbed by the process is called *chemical energy.*

Chemical energy is the energy necessary to form compounds and involves only the outermost electrons of atoms. The manner and the ease with which atoms enter into compounds depend on the number of electrons each has in its outermost shell. Outermost electrons have been likened to arms that intertwine to link together different atoms to form compounds.

After a compound has formed, energy must be applied to break it up. When the elements are separated, they regain their separate identities and are free to seek new alliances with other elements to create new compounds. For instance, the molecular bonds of coal have to be broken before its carbon atoms can combine with the oxygen of the air. This chemical process is started by applying flame which agitates and finally loosens the bonds between the carbon atoms.

A molecule of water remains intact because its atoms are united by a powerful covalent bond. When frozen solid, or heated to temperatures at which many other compounds disintegrate, the molecule H_2O remains intact. Its atoms can be separated only by the application of electrical energy or certain chemicals, such as the element potassium. A small lump of potassium dropped into a container of water pulls the molecules apart so violently that they may actually explode.

Electrical energy

Scientists have found that electrons move readily through metals and have found methods of concentrating them to cause them to move under controlled conditions. A flow of electrons is called electric current and supplies us with the *electrical energy* that we use in countless ways.

Twenty-five percent of the world's electrical energy is being generated by hydroelectric plants. This does not use up our limited supply of fossil fuels as steam plants do. Hydroelectric plants use the energy of flowing water yet do not use up the water in the process.

Our first large hydroelectric power plant was built at Niagara Falls in 1895. It converted the potential energy of water at a high level into the kinetic energy of electricity as it fell. Today, one of the great hydroelectric developments of the world is being operated there by both Canada and the United States (see Chapter 21).

Hydroelectric power plants also generate electric power by using water stored behind great dams. This is directed into sluices which guide it to the rotary blades of turbines. The turbines' rotation spins electromagnets which generate current in stationary coils of wire. This current is put through transformers where the voltage is stepped up for transmission over power lines. Water that has done its work of spinning turbines and their generators to produce electricity is then returned to the river below the dam (see Figure 2.12).

Hydroelectric river dams store water for other purposes, too, such as irrigation, flood control, or recreation.

In 1961, construction of a new type of dam was begun for harnessing the flow of ocean tides near the mouth of the Rance estuary on the coast of France. It was designed so that a 9-m tide with a flow of water nearly equivalent to the flow of the Mississippi River will pour through 24 turbines generating 480,000 kw (kilowatts) of electricity with each rise and fall of the tide. A tidal-power dam is currently being talked about in Canada to harness the tides flowing in and out of the Minas Basin section of the Bay of Fundy in Nova Scotia.

Figure 2.12 ***In a hydroelectric plant, water's potential energy of position is used to generate electricity when it rushes down a sluice and is guided into the rotary blades of a turbine. The turbine's rotation spins electromagnets which generate current in stationary coils of wire. The current is put through adjoining transformers where the voltage is stepped up for transmission over power lines.*** (*After Wilson Mitchell,* Water, *p. 152, in* Life Science Library, *Time, Inc., New York, 1963.*)

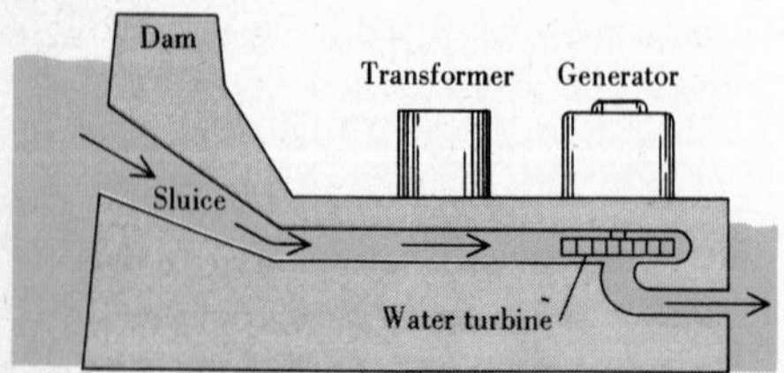

Atomic energy

Atomic energy is energy liberated by changes in the nuclei of atoms, as by fission of heavy nuclei or fusion of light nuclei into heavier ones with accompanying loss of mass.

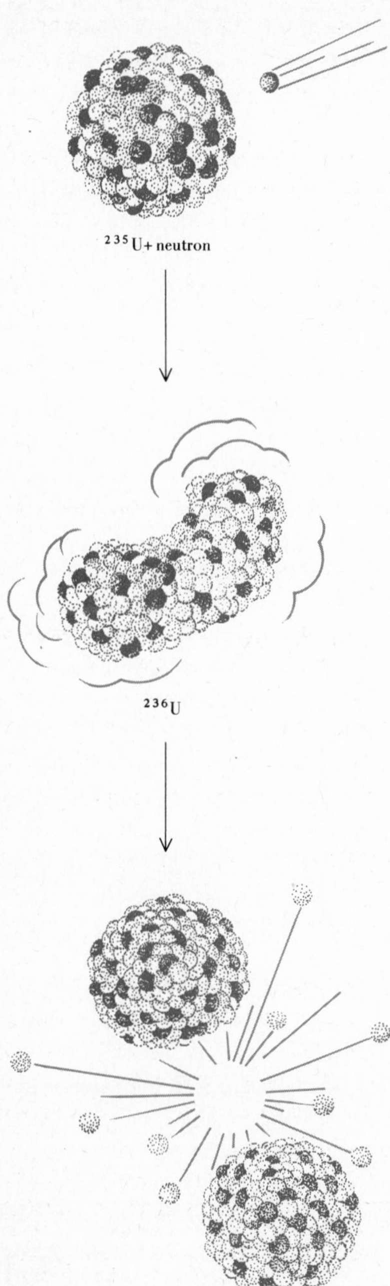

Figure 2.13 Energy released by the splitting of an unstable isotope of uranium. When ^{235}U captures a neutron, it becomes ^{236}U, which immediately splits into two lighter elements. These have no place in their nuclei for several of the neutrons. If such free neutrons are liberated in the presence of more ^{235}U, a chain reaction results and tremendous quantities of energy are released.

Man has been able to tap only a small fraction of the atom's potential energy. At present, we are using only that available from a few heavy elements, such as uranium. It is not the common isotope of uranium, with an atomic weight of 238, that fissions under neutron bombardment but rather the lighter isotope, uranium 235. This is found in nature intermixed with uranium 238 in the proportion of about 1 part in 140. Because the two isotopes are chemically identical, the ^{235}U could be separated only by a physical process capitalizing on the difference in mass. By making uranium into a gaseous compound and then pumping it against plates pierced with billions of tiny holes less than a millionth of a centimeter in diameter, the lighter ^{235}U moves through the holes producing a product with more than 90 percent concentration.

In nuclear fission, a slow-moving neutron smashes into the nucleus of the ^{235}U atom. It splits the nucleus into two new atomic nuclei of less mass, plus several neutrons (see Figure 2.13). Because the mass of the fission products is less than that of the original atom, some of the mass has been converted into energy. The fission of 1,000 g of $^{235}_{92}$U releases 8.23×10^{20} ergs of energy.

For nuclear fission to take place spontaneously, the piece of material must exceed a certain critical mass. That is, it must be so compressed that protons bombarding it cannot fail to hit a nucleus and split it. After the process is started, it proceeds by a chain reaction.

An atomic bomb consists of a few pounds of ^{235}U or ^{239}Pu surrounded by standard explosives. When detonated, the explosive charge squeezes the ^{235}U or ^{239}Pu until the critical density is reached and a self-sustaining chain reaction is started. In less than a millionth of a second, the solid mass is converted into a multimillion-degree blast of gas. Since the first atomic bomb was exploded in 1945, man has learned how to control the energy released by atomic fission and how to put it to use to serve his needs.

Nuclear fusion, a process the opposite of fission, results in even more high-powered atomic transformations. The two atomic lightweights—the hydrogen isotopes deuterium, $^{2}_{1}$H, and tritium, $^{3}_{1}$H—collide while in a superheated state to form one heavier nucleus of helium, $^{4}_{2}$He. The fused helium nucleus is not as heavy as the two lighter ones; some mass is converted into energy. Thus fusion, like fission, releases enormous energy. It occurs at such ultrahigh temperatures that man has been unable to make practical use of it. It awaits the development of some means of controlling it.

The sun's energy is believed to originate from the fusion of hydrogen into helium. These atomic reactions take place at temperatures of several million degrees and the process is self-controlling. If it operates too rapidly and is in danger of becoming an explosion, the gases expand. This expansion causes cooling, which lowers the rate of energy production and prevents explosion.

The sun is "burning" 4 million tons of hydrogen per second and producing an "ash" of helium. Even at this rate, it can keep on for 30 billion years.

Every second, the sun sends into space 1 million times as much energy as is stored in all our coal, petroleum, and natural gas fields. If we could duplicate and control the method by which this energy is released, we could revolutionize the world, and the possibility of doing just this is no more fantastic today than the concept of an atomic bomb was in 1900. If we could turn the hydrogen from a cubic mile of sea water into helium, it would provide enough energy to satisfy our needs at the present rate for 300 centuries.

Radiant energy

Radiant energy is electromagnetic waves radiating, or traveling, as wave motion. The travel of radiant energy can be described in language used for familiar water waves. The number of crests passing a given point in a second is called the *frequency;* the distance between adjacent crests is the *wavelength.* Frequency multiplied by wavelength equals the velocity with which the waves are traveling. Radiant energy of any wavelength travels through space at a speed of 3×10^9 m/sec.

A series of radiant energies arranged in order of wavelength is called a *spectrum.* These include gamma rays, X rays, ultraviolet rays, visible light, infrared rays, and radio waves.

Gamma rays have the shortest wavelength. They are also the most energetic, penetrating, and destructive. At maximum strength, a gamma ray packs several million times as much energy as visible light. If it hits a nucleus or knocks an electron loose, its energy is redistributed among other rays of lesser energy and longer wavelength.

Gamma rays from the heart of the sun are converted by collisions into *X rays* and *ultraviolet rays.* When X rays and ultraviolet rays hit other atoms, they excite them into giving off heat and light. When an atom is receiving energy, some of its electrons jump from inner to outer orbits, and this absorption of energy gives rise to a distinctive dark line in the spectrum. When excited electrons drop back toward their normal orbits, their atoms emit energy and form bright lines on the spectrum. Each combination of bright and dark lines is unique for each kind of atom and electron shift involved.

X rays are generated when electrons are knocked completely out of an orbit close to the nucleus. An electron that jumps between large and small orbits releases energy in the form of ultraviolet rays.

If electrons jump between nearby orbits, they generally give off visible light. Still smaller electron jumps can produce even longer waves of infrared, which we recognize as heat. Radio waves are the longest electromagnetic waves of all and are not generated by changes in electron orbits. They are set up by oscillations of electrons in conductors or by large-scale movements of ionized matter.

VISIBLE SPECTRUM Wavelengths of radiant energy between about 3,000 and 8,000 A produce color sensations in the human eye and constitute the *visible spectrum.* Wavelengths of about 7,100 A cause you to see red. If the wavelength is around 4,800 A, things look blue.

INVISIBLE SPECTRUM *Infrared waves* have lower frequencies and therefore greater wavelengths than do red waves. As a result, they are not scattered even by some of the hazes and fogs that block visible waves. Photographic film that is sensitive to infrared shows distant scenes with great clarity when many of the details are invisible to the eye, which responds only to wavelengths of the visible spectrum.

Ultraviolet waves are beyond violet on the spectrum in the direction of increasing frequency (shorter wavelength). If these waves were to pass unimpeded from the sun to the earth's surface, their great burning power would have a disastrous effect on life. In fact, the theory has been advanced that life did not develop on the earth until the last billion years because the primitive atmosphere transmitted ultraviolet radiation too efficiently. Our present atmosphere, however, has a sufficient supply of particles of gas and solid matter to absorb most of the short waves of the ultraviolet.

Radio waves are radiant energy exactly as light waves are, but they have greater wavelengths. The "short" waves of radio are about 10 million times as long as the waves of the visible spectrum. Broadcast-band radio waves are, on the average, about 100 million times as long.

RADIOACTIVITY When transformations occur in atomic nuclei, radiant energy is also emitted. Some isotopes are unstable and break down spontaneously, displaying what is called *radioactivity,* or the emission of radiant energy. This energy consists of alpha particles that are composed of two protons and two neutrons each, electrons, and occasional gamma rays that possess neither mass nor charge. The time required for half the nuclei of an isotope to disintegrate is called the *half-life* of that isotope. This may be a fraction of a second or millions of years. The best-known radioactive isotopes decay into other radioactive isotopes and finally into special isotopes of lead. Uranium 238 always decays to lead 206. Actinium 235 always decays to lead 207. Thorium 232 always decays to lead 208. Ordinary, or primary, lead has a mass number of 204.

Accordingly, the mass numbers of the different lead isotopes, 204, 206, 207, and 208, serve as labels indicating the past history of the lead: 204 was always lead, 206 was derived from uranium, 207 was derived from actinium, and 208 was derived from thorium. If uranium, actinium, or thorium is found in a rock associated with the lead produced by its decay, the age of the rock can be computed. Because the ratio of the amount of lead to the amount of the original radioactive mineral increases with the passage of time, this is sometimes called the *lead-ratio method* of measuring geologic time. Rocks approximately 3.3 billion years old have been dated by this method.

The physical nature of radiant energy is the same throughout the spectrum. In all sections it has the same velocity and the same electromagnetic nature. The descriptive names for sections are historical. They give a convenient classification according to the source of the radiation. The regions overlap, but the names indicate the common sources.

2.3 *Mass and energy*

It has been found that the mass of an atomic nucleus is less than the total mass of its components as separate particles. This is explained by the fact that when a nucleus forms, a small amount of mass disappears by changing into energy, which is radiated away (see Figures 2.14 through 2.17). The energy represented by this mass defect is called the *binding energy.* It is the amount of energy that must be supplied in order to break the nucleus into its component particles again.

The discovery of the equivalence of mass and energy is one of the most fundamental and significant events in the history of mankind.

Consider, for example, a helium nucleus of two protons and two neutrons:

	1 proton	1.00758	amu (atomic mass units)
	1 proton	1.00758	amu
	1 neutron	1.00893	amu
	1 neutron	1.00893	amu
Total	4 particles	4.03302	amu
	Helium nucleus	4.00280	amu
		0.03022	amu deficiency converted to energy when nucleus was formed

Figure 2.14 ***(a) Struck by an electric charge, an electron absorbs the energy and jumps from its normal orbit to a new but temporary one. When it returns, it gives off the absorbed energy as light. (b) The excited electron may jump more than one energy level and return in steps rather than all the way at one jump. Jumping from an outer level to an intermediate one emits longer wavelengths and the jump from an intermediate level to an inner one emits shorter wavelengths.*** (*After Wilson Mitchell,* Energy, *p. 77, in* Life Science Library, *Time, Inc., New York, 1963.*)

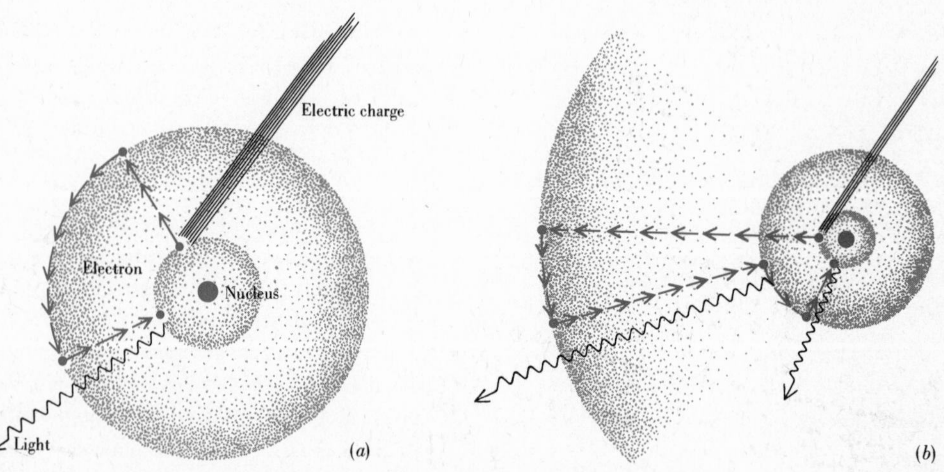

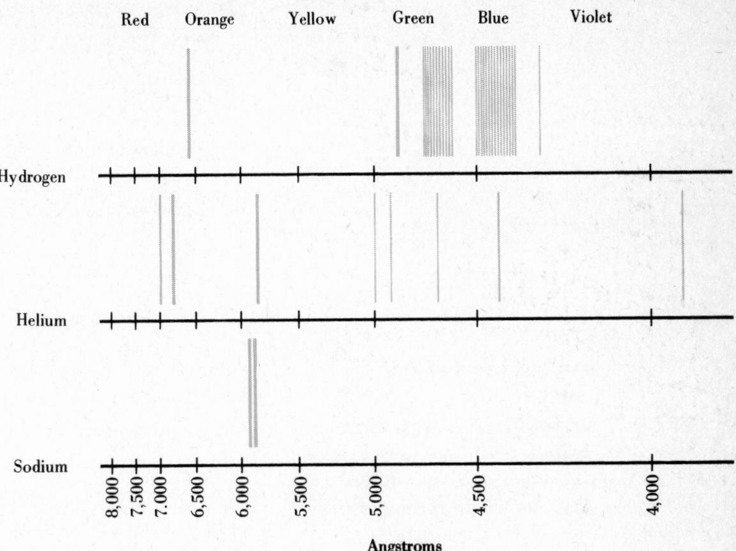

Figure 2.15 Spectra of hydrogen, helium, and sodium.

In 1905, Albert Einstein expressed the equivalence of mass and energy by the now-famous equation

$$E = mc^2$$

where E is the energy in ergs, m is the mass in grams, and c is the velocity of light in centimeters per second.

It was not until 1932 that experimental proof of this relationship was obtained. When ${}^7_3\text{Li}$ was bombarded with high-speed protons (${}^2_1\text{H}$), alpha particles (${}^4_2\text{He}$) were ejected from the lithium:

$$\underbrace{\overset{\text{Lithium}}{{}^7_3\text{Li}} + \overset{\text{hydrogen}}{{}^1_1\text{H}}}_{\textit{Mass: } 8.0241} \xrightarrow{\text{yielded}} \underbrace{\overset{\text{alpha particles}}{{}^4_2\text{He} + {}^4_2\text{He}}}_{8.0056}$$

The lost mass was 8.0241 − 8.0056 = 0.0185, which the two alpha particles were found to possess in the form of velocity of motion.

Figure 2.16 Spectrum of radiant energy.

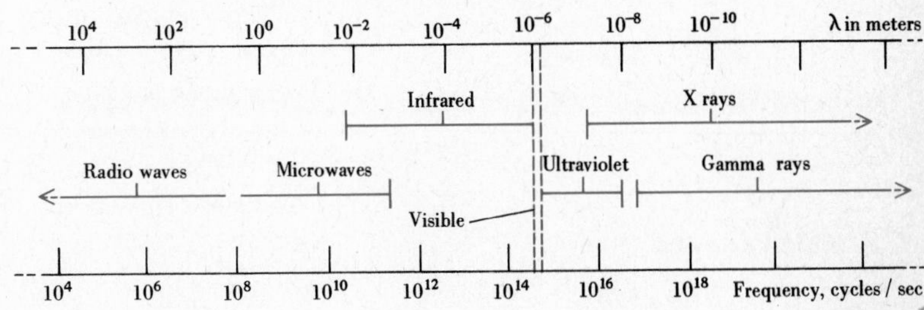

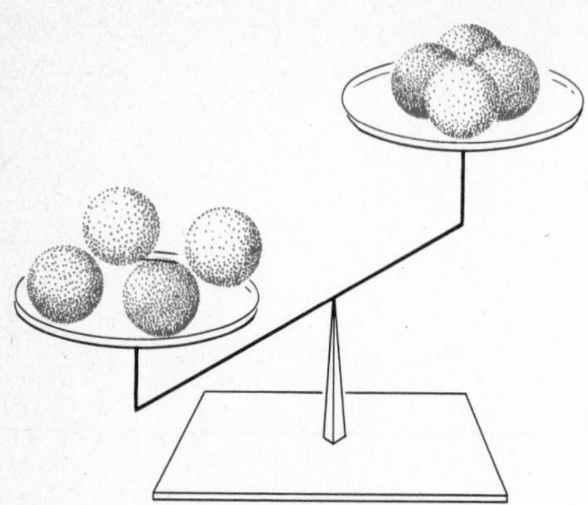

Figure 2.17 ***Equivalence of mass and energy. Two protons plus two neutrons as separate particles (left pan) total 4.03302 amu. When they combine to form an alpha particle, mass disappears in the form of energy, leaving the alpha particle with only 4.00280 amu (right pan). This lost energy (or mass), called the binding energy, must be applied to the alpha particle to break it into separate particles again.***

The quantities of energy released in reactions of this kind are almost inconceivably greater than those released by any other type of reaction involving similar quantities of matter. For example, 1 kg (kilogram) of matter, if converted entirely into energy, would give 25 billion kwh (kilowatt hours) of energy. In contrast, the burning of 1 kg of coal gives 8.5 kwh. If the conversion of mass into energy can be accomplished by nuclear technology, an almost limitless source of energy will become available.

Matter and energy transformations

Matter and energy can be transformed from one form to another. In fact, the entire universe owes its continued existence to matter and energy transformations. Without these, there would be nothingness. But no matter how they are transformed, the total amount *after* the transformation is the same as it was *before* the transformation. Matter and energy can be neither created nor destroyed by man.

In geology, matter and energy transformations occur when a liquid hardens into a solid rock and when one kind of rock is changed into another. Matter and energy transformations also occur during the formation of coal, petroleum, and gas. All geological processes consist of matter and energy transformations.

2.4 Gravity

Gravity is a universal force. On earth, it is the force that makes things fall to earth. In fact, it was long thought that this force did not apply to the heavenly bodies because they did not fall to earth. In 1666, when he was 24 years old, Isaac Newton proposed that gravity is a universal force: Every aggregation of matter attracts every other aggregation of matter by a force that is directly proportional to their respective masses and inversely proportional to the square of the distance between them. Newton reasoned that the moon does not fall to earth because its speed in orbit just balances the pull of the earth's and its own gravity. The universal law of gravity can be expressed as

$$F = \frac{mM}{r^2} G$$

where F is force, m and M are the masses, r is the distance between them, and G is the universal constant of gravitation. Newton's law provides an excellent summary of the observed motions in our solar system.

On the earth, masses can best be compared by a process called weighing. This actually compares the earth's attraction for one mass with its attraction for another. The *weight* of a body on earth is the force that gravity exerts upon it. This force is proportional to its mass, that is, to the quantity of matter it contains. This can be written

$$\frac{F}{m} = \frac{GM}{R^2}$$

where R is the earth's radius and M is its mass. G is the universal gravitational constant. The quantities R, G, and M do not change.

Acceleration of gravity

Gravity governs the fall of a body toward the earth and causes it to fall faster during each second. This is known as the acceleration due to gravity, represented by g. Falling freely in a vacuum, a body increases its speed by 980 cm/sec during each second of fall. This is written

$$g = 980 \text{ cm/sec}^2$$

which is read "g equals 980 centimeters per second per second."

A body falling in the earth's atmosphere does not increase in speed without limit, however. Friction of the atmosphere causes it to reach a terminal velocity and go no faster after that. For a man falling freely, this is about 200 km/hr.

The force we call weight is mass times the acceleration of gravity:

$$W = mg$$

On the moon, the acceleration of gravity is one-sixth of what it is on the earth, so a man would weigh one-sixth as much. But because the moon has no atmosphere, a body falling toward it would not have a terminal velocity as it does on earth.

Gravity anomalies

If the earth were a smooth, homogeneous, nonrotating sphere, gravity on its surface would be the same everywhere. It is not the same, however, because of (1) differences in elevation, which change the distance from the earth's center; (2) effects of rotation, which are different at different latitudes; and (3) differences in the density of materials in the earth's crust.

Because $W = mg$, and m for a given body remains the same, changes in g, the acceleration due to gravity, will be reflected in changes in the weight of a body. An instrument for making such measurements is called a *gravity meter.*

An acceleration of 1 cm/sec^2 has been defined as a unit called the *gal,* after Galileo. The acceleration of gravity at the earth's surface is around 980 gals, but changes as small as a ten-millionth of this may be significant. Consequently, the common or practical unit in which gravity differences are expressed is the *milligal* (mgal), or one-thousandth of 1 gal.

Measurements of the force of gravity have to be corrected for differences in altitude, latitude, and geological structure. Altitude of the point of measurement is taken into account by subtracting 0.3086 mgals for each meter of height above sea level or adding the same for each meter below sea level. Adjustment of an observed value to what it would be at sea level is known as the *free-air correction.*

The sea-level value of g varies from 978.049 gals at the equator to 985.221

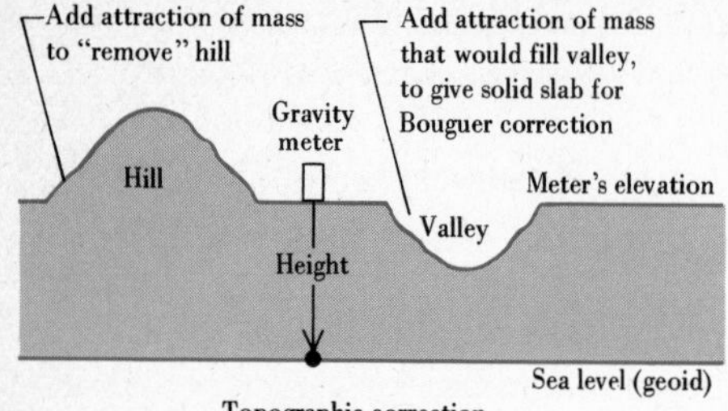

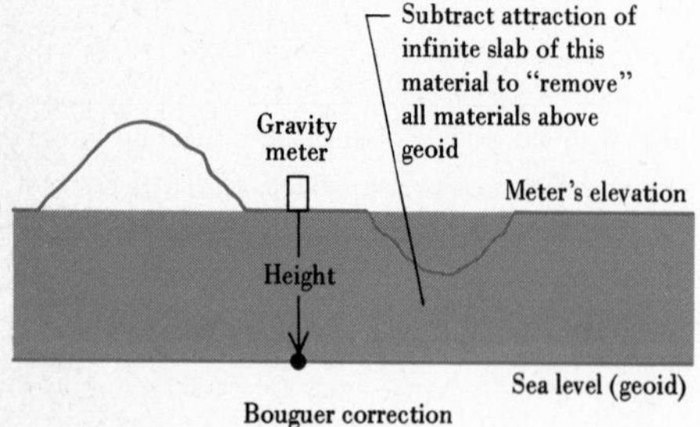

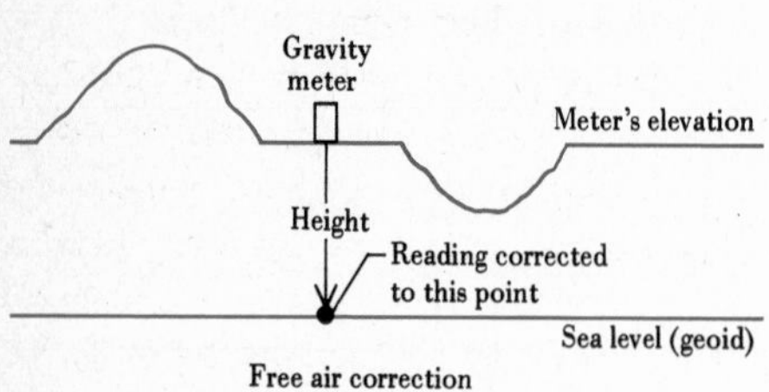

Figure 2.18 ***Correcting gravity-meter readings to a common level of reference at sea level. Because the hill rises above the meter, its attraction is upward and the meter would show a greater reading if the hill were not there. So the hill is "removed" by adding a figure to represent its attraction. In effect the topographic correction reduces the surface to a plane. The Bouguer correction leaves a reading that the meter should show if it were suspended with nothing but free air between it and sea level. The free-air correction leaves the reading at a value that it would have if the meter were at sea level vertically below the point at which it actually stands.***

gals at the poles because the centrifugal force of the earth's rotation tends to counteract the pull of gravity toward the center. Two kilometers away from the equator, the gravity is reduced by 1 mgal.

Corrections need to be made also for hills and valleys in the vicinity of the measuring point, as indicated in Figure 2.18.

After these factors have been evaluated, a number is obtained that represents the expected value of gravity at the point. If the measured value is different, this difference is called a *gravity anomaly.*

A sphere of salt 600 m in radius with its center 1,200 m below the surface and surrounded by sedimentary formations about 10 percent more dense causes gravity to be about 1 mgal less directly above it than at surface points beyond its range of influence. A buried granitic ridge under Kansas, with 1,500 m of relief, has a gravity anomaly of only 2 mgals. The effect of a volcanic pipe 3 km in diameter and of considerable depth may be 30 mgals.

There are significant negative gravity anomalies over ocean deeps, such as the Mindanao Deep off the Philippines, the Nero Deep off Guam, and the deep fronting Java and Sumatra in the East Indies. These have been cited as possibly caused by down-plunging currents involved in ocean floor spreading (see Chapter 15).

2.5 The earth's magnetism

All of us are familiar with the fact that the earth behaves as if it were a magnet and for that reason the compass needle seeks the north magnetic pole. It will pay us to review the earth's magnetism before examining its geologic implications.

We can picture the earth's magnetic field as a series of lines of force. A magnet, free to move in space, will align itself parallel to one of these lines. At a point north of the Prince of Wales Island at about 75°N and 100°W, the north-seeking end of the magnetic needle will dip vertically downward. This is the *north magnetic* or *dip pole.* Near the coast of Antarctica, at about 67°S and 143°E, the same end of our needle points directly skyward at the *south magnetic* or *dip pole.* Between these dip poles, the magnetic needle assumes positions of intermediate tilt. Halfway between the dip poles, the magnetic needle is horizontal and lies on the *magnetic equator.* Here the

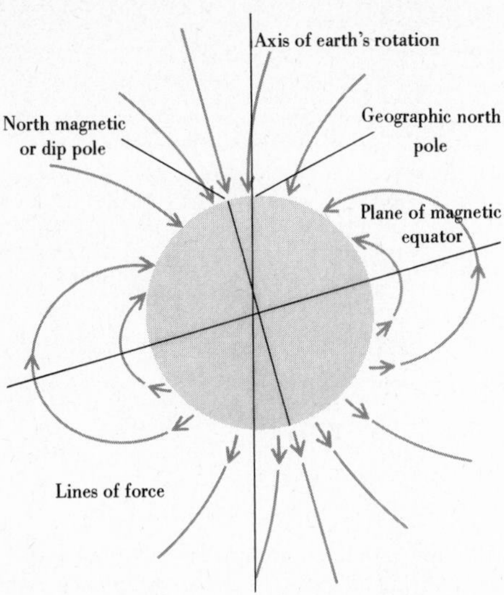

Figure 2.19 ***The earth's magnetic field can be pictured as a series of lines of force. The arrows indicate positions that would be taken by a magnetic needle free to move in space, located at various positions in the earth's field. The magnetic or dip poles do not coincide with the geographic poles, nor are the north and south magnetic poles directly opposite each other.***

intensity of the earth's field is least, and it increases toward the dip poles, where the field is approximately twice as strong as it is at the magnetic equator (see Figure 2.19).

The angle that the magnetic needle makes with the surface of the earth is called the *magnetic inclination* or *dip.* The north and south dip poles do not correspond to true north and south geographic poles, defined by the earth's rotation. Because of this the direction of the magnetized needle will in most instances diverge from the true geographic poles. The angle of this divergence between a geographic meridian and the magnetic meridian is called the *magnetic declination* and is measured in degrees east and west of geographic north (see Figure 2.20).

Secular variation of the magnetic field

As long ago as the mid seventeenth century it was known that the magnetic declination changed with time. Since then we have been able to demonstrate not only slow changes in declination but also changes in inclination and intensity. These changes in magnetism take place over periods measured in hundreds of years. Because they are detectable only with long historical records, these changes are called *secular changes,* from the Latin *saeculum,* meaning "age" or "generation," implying a long period of time.

Figure 2.21 shows changes in the declination and inclination records at Paris during Gallo-Roman times and from 1540 to 1950. Another long-range record, from 1573 to the present, is available from London and shows a pattern very similar to that of Paris from the sixteenth century on. Such records tempt us to suspect that these might represent worldwide, periodic variations in the earth's magnetic field. In truth, when records are compiled for the entire earth, it becomes apparent that the changes in the magnetic field are regional rather than global. The centers of greatest change wax and wane and also move generally westward at a rate averaging approximately one-fifth of a degree per year.

The geomagnetic pole

The magnetic field at the earth's surface can be considered as composed of three separate components. There is, first, a small component that seems to result from activity above the earth's surface and is sometimes referred to as the external field. Second, there is a quantitatively more important component, best described as if it were caused by a dipole—such as a simple bar magnet—passing through the center of the earth and inclined to the earth's axis

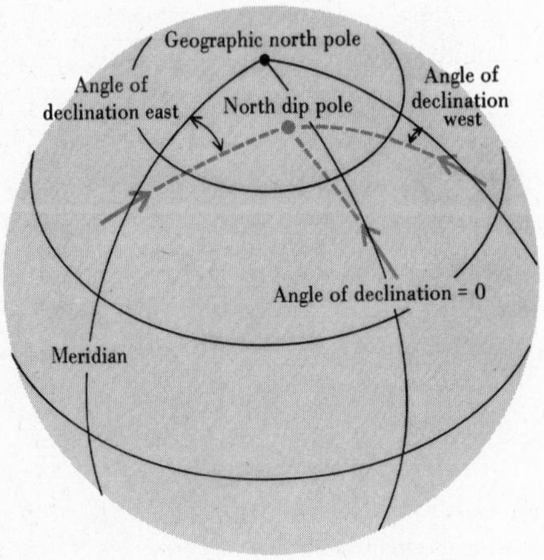

Geographic north pole
Angle of declination east
North dip pole
Angle of declination west
Angle of declination = 0
Meridian

Figure 2.20 ***Because the dip poles and geographic poles do not coincide, the compass needle does not point to true north. The angle of divergence of the compass from the geographic pole is the declination and is measured east and west of true north.***

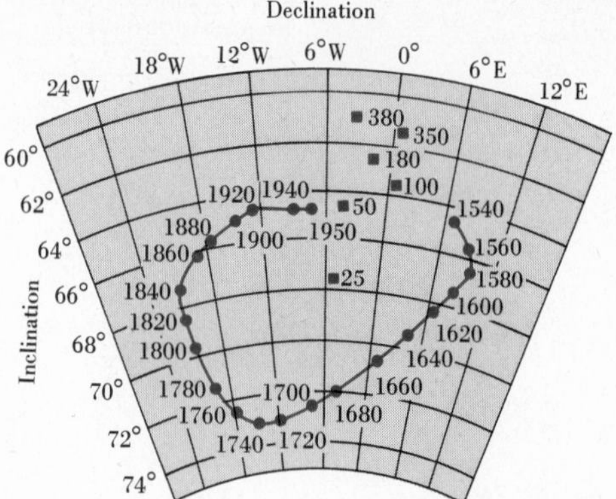

Figure 2.21 **Magnetic inclination and declination vary with time. At Paris we have a continuous record of these changes since 1540. The data for inclination and declination for Gallo-Roman times (colored squares) are based on magnetic measurements of archaeological materials.** (*Adapted from Emile Thellier.* "*Recherches sur le champ magnétique terrestre,*" L'Astronomie, *vol. 71, p. 182, May 1957, Fig. 64.*)

of rotation. Finally, there is what we refer to as the nondipole field, that portion of the earth's field remaining after the dipole field and the external field are removed.

The dipole best approximating the earth's observed field is one inclined 11.5° from the axis of rotation. The points at which the ends of this imaginary magnetic axis intersect the earth's surface are known as the *geomagnetic poles* and should not be confused with the magnetic or dip poles. The north geomagnetic pole is about 78.5°N, 69°W, and the south geomagnetic pole is exactly antipodal to it at 78.5°S and 111°E (see Figure 2.22).

Figure 2.22 **The geomagnetic poles are defined by an imaginary magnetic axis passing through the earth's center and inclined 11.5° from the axis of rotation. This magnetic axis is determined by a hypothetical, earth-centered bar magnet positioned to best approximate the earth's known magnetic field.**

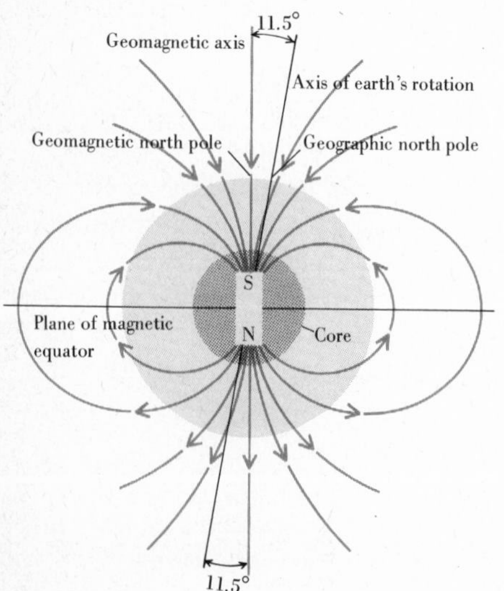

Magnetosphere

The earth's magnetic field performs a service for terrestrial life by warding off and trapping powerful radiation from space. This is in the form of electrons and very energetic ionized nuclei of atoms, mostly hydrogen. Some of the particles are believed to have been produced by cosmic-ray collisions in the atmosphere, but the major portion by violent sprays of particles from the sun in solar flares.

The particles are diverted into a great doughnut-shaped racetrack around the earth with the earth's axis under the doughnut's hole. They leak off and are turned back into space fairly rapidly. This region of trapped ions is called the *magnetosphere* (see Figure 2.23). It begins about 1,000 km above the earth and extends out to about 60,000 km. A similar band of charged particles has been reported encircling Jupiter.

Outline

Matter, the substance of the physical universe, is described in terms of atoms with mass and electric charge.

Electric charge is of two kinds, arbitrarily called positive and negative.

Atoms are composed of protons, neutrons, and electrons.

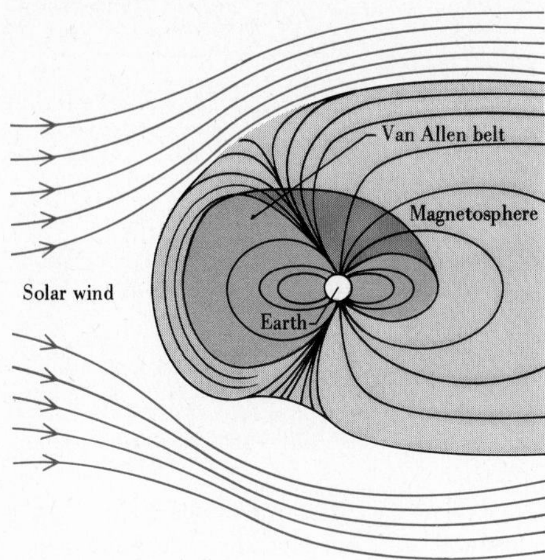

Figure 2.23 **The magnetosphere: a doughnut-shaped region where the earth's magnetic field traps and wards off powerful radiation from space. The inner portion was first detected by James A. Van Allen in interpreting measurements from early U.S. satellites in 1958. It is sometimes referred to as the Van Allen belt. Later satellite data defined the region from 1,000 km out to about 60,000 km now known as the magnetosphere.**

Atomic structure consists of a nucleus of protons and neutrons, surrounded by a cloud of electrons.

Atomic size, determined by the protective shield of electrons, for the most common elements is a diameter of about 2 A around a nucleus of 10^{-5} to 10^{-4} A.

Atomic mass is 99.95 percent in the nucleus.

Ions are electrically unbalanced forms of atoms or groups of atoms.

Elements are special combinations of protons, neutrons, and electrons.

Isotopes are forms of elements with the same number of protons in the nucleus but different mass because of their number of neutrons.

Energy-level shells are the radii of travel by electrons around the nucleus.

Combinations of atoms occur among those that do not have completely filled outermost shells of electrons.

Compounds are combinations of atoms of different elements.

Metallic bond is responsible for metals being such good conductors of heat and electricity.

Organization of matter combines protons, neutrons, and electrons into elements which combine to form numerous compounds.

Energy, the capacity for producing motion, is involved in every geological process.

Potential energy is stored energy waiting to be used.

Kinetic energy, possessed by every moving object, depends on the mass of the object and on the speed with which it is moving.

Heat energy is a special form of kinetic energy in atoms.

Chemical energy is that released or absorbed when atoms combine to form compounds or when compounds combine to form other compounds.

Electrical energy is produced by the flow of electrons.

Atomic energy is energy liberated by changes in the nuclei of atoms.

Radiant energy is electromagnetic waves radiating, or traveling, as wave motion.

Visible spectrum is wavelengths of radiant energy between about 3,000 and 8,000 A.

Invisible spectrum is all other wavelengths from gamma rays to radio waves.

Radioactivity is the spontaneous breakdown of unstable isotopes with the emission of radiant energy.

Mass and energy are equivalent.

Matter and energy transformations occur, but matter and energy can neither be created nor destroyed.

Gravity is a universal force by which every aggregation of matter attracts every other aggregation of matter.

Acceleration of gravity is the amount a freely falling body increases its speed of fall each second.

Gravity anomalies are differences between expected and measured values of gravity.

The earth's magnetism results from its behaving as if it were a magnet.

Secular variation of the magnetic field has been known through long-term changes in magnetic declination since the midseventeenth century.

The geomagnetic pole at the earth's surface does not coincide with the geographic pole.

The magnetosphere is a region above the earth which traps powerful radiation from space and wards it off.

Supplementary readings

LAPP, RALPH E.: *Matter,* in *Life* Science Library, Time, Inc., New York, 1963. Atomic structure, the elements, and the spectrum, extremely well illustrated, clearly written, and authoritative.

MASON, BRIAN: *Principles of Geochemistry,* 2nd ed., John Wiley & Sons, Inc., New York, 1960. A textbook which assumes that the reader is familiar with the principles of physics and chemistry as well as the fundamental concepts of geology.

PAULING, LINUS: *General Chemistry,* 2nd ed., W. H. Freeman and Company, San Francisco, 1953. An elementary textbook with exceptionally clear illustrations by Roger Hayward.

WILSON, MITCHELL: *Energy,* in *Life* Science Library, Time, Inc., New York, 1963. Top-quality treatment at elementary level.

THE SURFACE OF THE EARTH, THE PART ON WHICH WE LIVE, IS COMPOSED OF minerals—solid elements and compounds of relatively few elements (see Table 3.1). Minerals are everywhere about us. Almost any small plot of ground will offer numerous samples. They may occur in several forms, such as rocky outcrops, the soil of plowed fields, or the sands of a river bottom. Even some of the most common types are valuable enough to make mining them commercially worthwhile; the rarer minerals such as gold and silver have provided the basis of wealth and power since the dawn of civilization.

The word *mineral* has many different meanings in everyday usage. Some people use it to refer to anything that is neither animal nor vegetable, according to the old classification of all matter as animal, vegetable, or mineral. To prospectors and miners, mineral is an ore. And advertisers of pharmaceutical products associate the term with vitamins. But in our discussion, *mineral* will be used to refer to *a naturally occurring element or compound that has been formed by inorganic processes.* Later on, we shall expand this definition to make it more comprehensive.

3.1 Mineral composition

More than 2,000 minerals are known. Some are composed of only one element, such as diamond (C), copper (Cu), silver (Ag), gold (Au), mercury (Hg), and sulfur (S). Others are made up of relatively simple compounds of elements. For example, the mineral halite is composed of two elements, sodium and chlorine, in equal amounts. The chemical symbol for halite, NaCl, indicates that every sodium ion present is matched by one chlorine ion. The mineral pyrite, sometimes known as "fool's gold," is also composed of two elements, iron and sulfur, but in this mineral there are two ions of sulfur for each ion of iron, a relationship expressed by the chemical symbol for pyrite, FeS_2. Other minerals have a more complex composition. For example, sillimanite is Al_2SiO_5, and the white mica, muscovite, is $KAl_3Si_3O_{10}(OH)_2$.

A mineral's composition can vary slightly: an occasional substitution of other elements with atoms of similar size (see Section 2.1) does not create a new mineral. We can therefore say that *every mineral is composed of elements in definite or slightly varying proportions.*

3.2 Mineral structure

The orderly pattern that atoms of elements assume in a mineral is called its *crystalline structure*. In halite, positively charged ions of sodium alternate with negatively charged ions of chlorine. Each mineral has a unique crystalline structure that will distinguish it from another mineral even if the two are composed of the same element or elements. Consider the minerals diamond and graphite, for example. Each is composed of one element, carbon (see

3 Minerals

Figure 3.1 ***Different arrangements of atoms of carbon produce (a) the crystalline structure of diamond and (b) the crystalline structure of graphite.*** (*After Linus Pauling,* General Chemistry, *pp. 226–227, W. H. Freeman & Company, San Francisco, 1953.*)

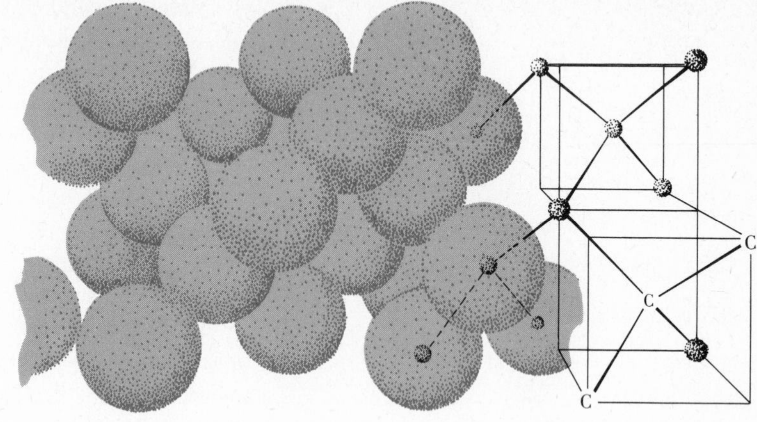

(*a*)

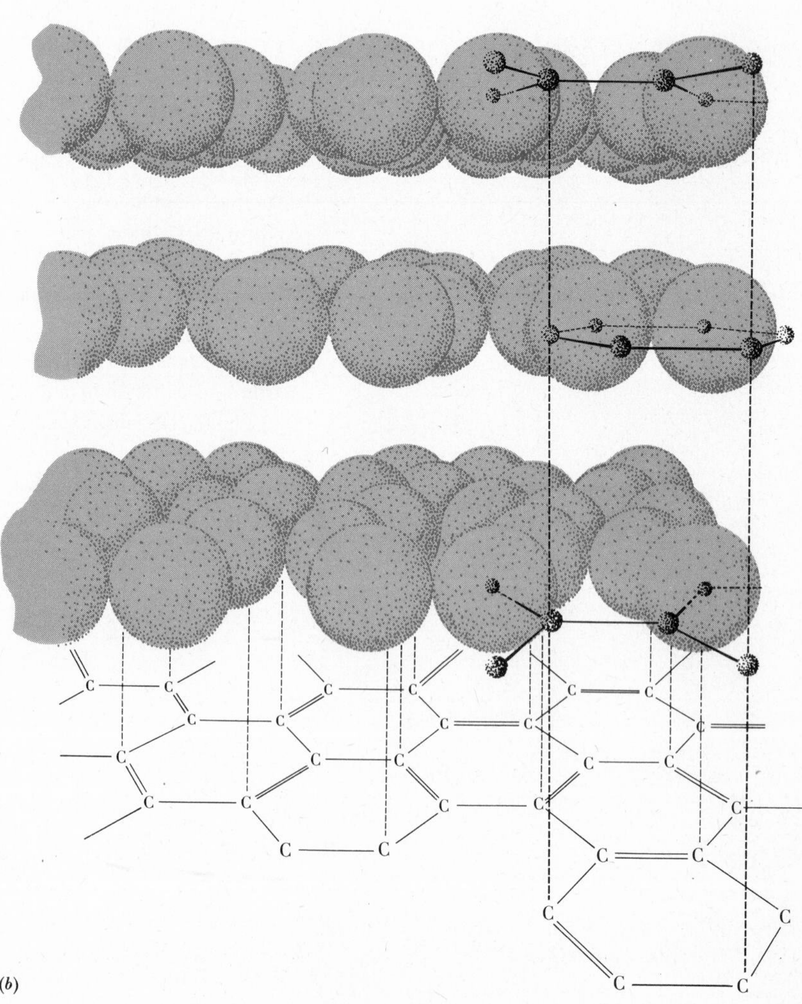

(*b*)

Table 3.1 ***Most abundant elements in the earth's crust***[a]

Element number	*Name and symbol*	*Volume %*
8	Oxygen (O)	93.77
19	Potassium (K)	1.83
11	Sodium (Na)	1.32
20	Calcium (Ca)	1.03
14	Silicon (Si)	0.86
13	Aluminum (Al)	0.47
26	Iron (Fe)	0.43
12	Magnesium (Mg)	0.29

[a] Based on Brian Mason, *Principles of Geochemistry*, 2nd ed., p. 46, John Wiley & Sons, Inc., New York, 1960.

Figure 3.1). In diamond, each atom of carbon is covalently bonded to four neighboring carbon atoms. This complete joining of all its atoms produces a very strong bond and is the reason diamond is so hard. In graphite, each atom of carbon is covalently bonded in a plane to three neighboring atoms. This bonding forms sheets or layers of carbon which are loosely joined and can be separated easily. Thus graphite is a soft substance.

Pyrite and marcasite are two other minerals with identical composition, FeS_2, but with different crystalline structures. In pyrite, ions of iron are equally spaced in all directions. In marcasite, they are not equally spaced. The difference in spacing accounts for their being two different minerals, although their composition is the same (see Figure 3.2).

Other minerals may have more complicated crystalline structure: They may contain more elements and have the elements joined in more complex patterns. The color, shape, and size of any given mineral may vary from one sample to another, but *the internal atomic arrangement of its component elements is identical in all specimens of a particular mineral.*

After taking all these factors into account, we find it necessary to include in our definition of a mineral not only the fact that it is a naturally occurring inorganic element or compound with a diagnostic chemical composition but also that it has a *unique orderly internal atomic arrangement of its elements.*

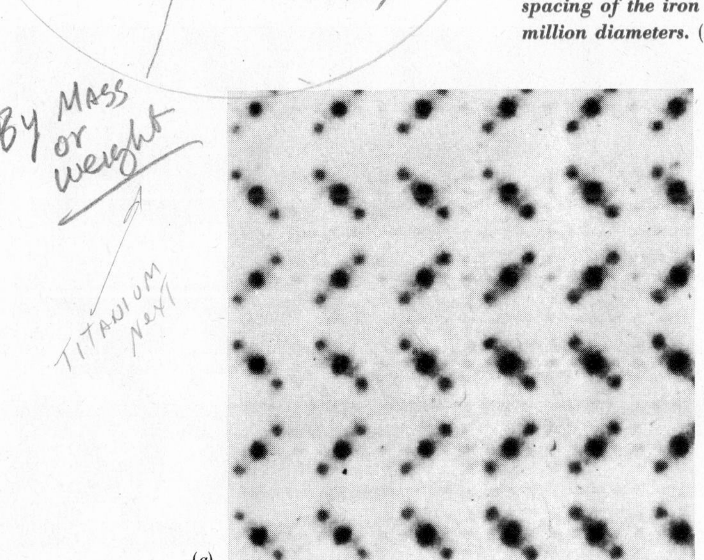

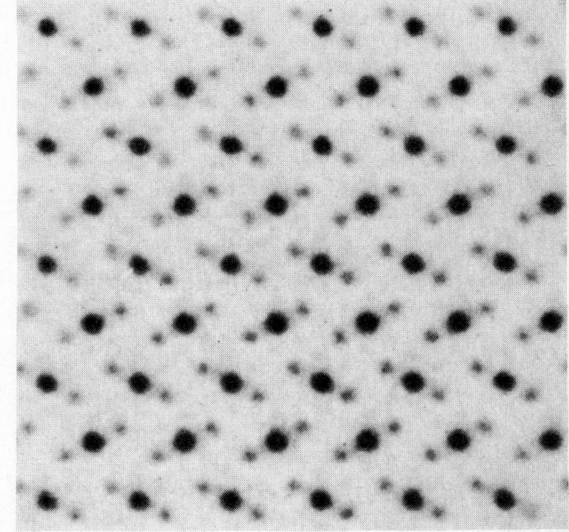

Figure 3.2 **X-ray photographs of atoms in pyrite and marcasite. (*a*) Pyrite shows the orderly arrangement characteristic of crystalline structure. The mineral is composed of iron and sulfur, FeS_2. The large spots are atoms of iron; the small ones are atoms of sulfur. Each atom of iron is bonded to two atoms of sulfur, and the spacing of iron atoms is the same in both directions of the plane of the photograph. Magnification is approximately 2.2 million diameters. (*b*) Marcasite, also FeS_2. Note the difference between the horizontal and vertical spacing of the iron atoms. Compare with pyrite. Magnification is about 2.8 million diameters.** (*Photographs by Martin J. Buerger.*)

3.3 *Identification of minerals*

All the properties of minerals are determined by the composition and internal atomic arrangement of their elements. So far, we have talked about chemical properties—the factors that account for the existence of so many different minerals from such a limited number of elements. We can identify minerals on the basis of their chemical properties, but their physical properties are the ones most often used. Physical properties include such things as crystal form, hardness, specific gravity, cleavage, color, streak, and striations.

Crystal Form

When a mineral grows without interference, it is bounded by plane surfaces symmetrically arranged, giving it a characteristic *crystal form.* Its crystal form is the external expression of its definite internal crystalline structure. The faces of crystals are defined by surface layers of atoms. These faces lie at angles to one another that have definite characteristic values, the same for all specimens of the same mineral. The size of the faces may vary from specimen to specimen, but the angles between them remain constant.

Every crystal consists of atoms arranged in a three-dimensional pattern that repeats itself regularly. In two dimensions, a similar technique is used in laying patterns of floor tile (see Figure 3.3). A combination of atoms whose repetition

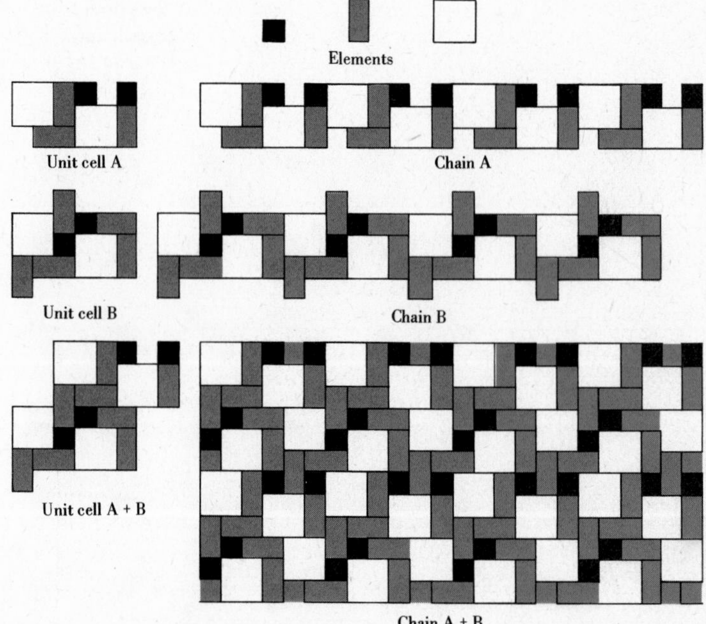

Figure 3.3 ***The pattern of a tile floor covering illustrates the repetition of basic units in a manner analogous to that in which atoms combine to form minerals. Three types of tile might be regarded as atoms or ions of three different elements. A simple combination of these elements is shown in unit cell A. A series of these unit cells, joined end to end, forms chain A. Another grouping, unit cell B, combined in a chain, forms chain B. One unit cell A combined with one unit cell B might be regarded as forming a more complex unit cell A + B. An entire pattern may then be formed which is simply a repetition of this unit cell. This is analogous to atomic patterns—even to the feature that the unit cell is arbitrarily defined. The cement that holds these "tiles" together in a mineral may be ionic bonding, covalent bonding, or a combination of the two.***

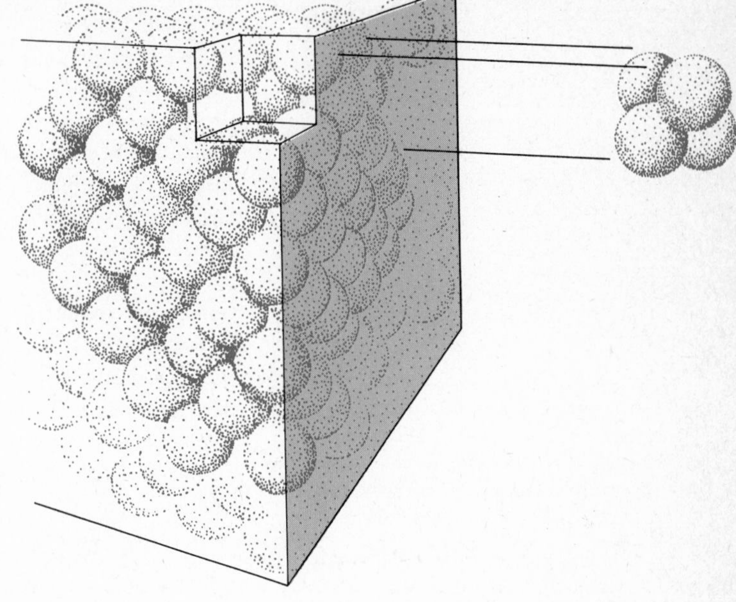

Figure 3.4 ***Arrangement of atoms in a crystal of copper. The small cube, with four copper atoms, is the unit cell. By its repetition the entire crystal is built up.*** (*After Linus Pauling,* General Chemistry, *p. 22, W. H. Freeman & Company, San Francisco, 1953.*)

Figure 3.5 ***Another atomic view of a copper crystal, showing small octahedral faces and large cubic faces.*** (*After Linus Pauling,* General Chemistry, *p. 23, W. H. Freeman & Company, San Francisco, 1953.*)

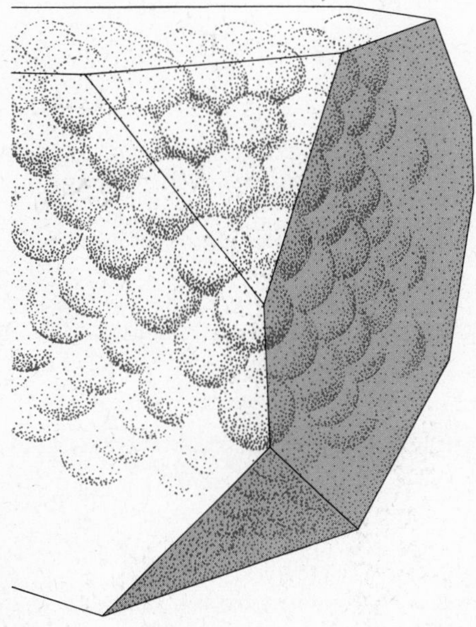

can produce a given mineral is called its *unit cell.* In a crystal of copper, all the atoms are alike, and its unit cell is four copper atoms arranged in a cube. This arrangement is limited by the size of the copper atoms. By repeating this unit cell in three dimensions, the entire crystal is built up (see Figure 3.4). In looking at the figure, remember that the atoms are greatly enlarged. If the crystal had sides only one-tenth of a millimeter long, there would be about 400,000 atoms in a row along each edge. The principal surface layers correspond to the faces of a cube; the smaller surface layer obtained by cutting off a corner of a cube is called an octahedral face. Native copper found in deposits of copper ore is often found in the form of crystals with cubic and octahedral faces (see Figure 3.5).

In crystals of halite, there are ions of two different kinds arranged in the regular pattern shown in Figure 3.6. The atomic arrangement is limited by the size of the ions. The smaller ones are those of sodium and the larger ones are those of chlorine. The unit cell of halite is a cube made up of six chlorine ions surrounding one sodium ion with all equidistant. The mineral is held together by ionic bonds. Repetition of the unit cell in three dimensions forms the cubic halite crystal.

The mineral quartz occurs in many rocks as irregular grains because its growth was constricted. Even in these irregular grains, however, the atoms are arranged according to their typical crystalline structure. And where conditions permitted the mineral to develop freely it has formed crystals which are

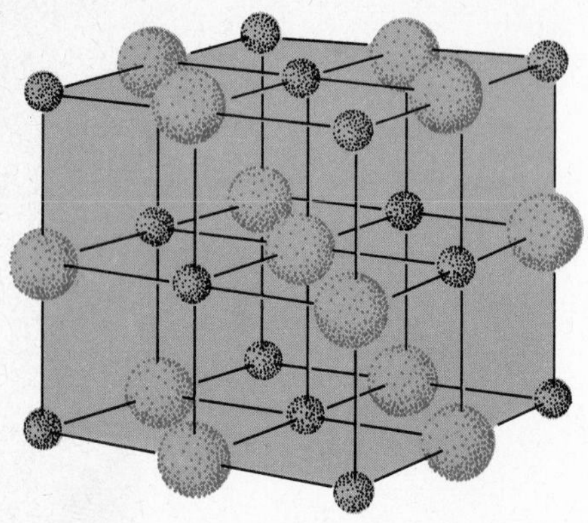

Figure 3.6 ***Arrangement of sodium ions with positive electrical charge,*** Na^+***, and chlorine ions with negative electrical charge,*** Cl^-***, to form the ionic compound*** NaCl***, common salt.*** Na^+ ***has a radius of 0.97 A, and*** Cl^- ***a radius of 1.81 A.***

always six-sided prisms. Whether an individual crystal of quartz is only a millimeter long or 25 cm, the faces of the prism always meet at the same angle (see Figure 3.7). This is an example of the constancy of interfacial angles. *The constancy of interfacial angles is due to the orderly internal arrangement of the atoms.* The pattern of packing determines angles between faces and is identical in all specimens of a particular mineral. Although faces of a crystal may be of different sizes and shapes because of malformation, the similarity is frequently evidenced by natural striations, etchings, or growths. On some crystals the similarity of faces of a form can be seen only after etching with acid.

There are six different systems of crystals, varying according to the angles between their faces. The systems are named in terms of their *axes* and are subdivided into 32 classes. An axis is an imaginary straight line that is drawn from the center of a face to the center of the opposite face (see Figure 3.8). The six crystal systems are described in Table 3.2.

Some elements or compounds may develop into several different cyrstal forms, producing different minerals. This assumption of two or more crystal forms by the same substance is called *polymorphism.* Each results from the conditions under which it was formed. Crystallized sulfur, for example, may exist as

Figure 3.7 ***Quartz crystals. Regardless of the shape or size of crystals, the angles between true crystal faces remain the same. Transverse striations on prism faces are most clearly seen on the faces of the two large crystals, which also carry blotches of foreign matter, but striations are present on the faces of the other crystals, too. The large, stubby crystal, from Dauphine, France, is about 35 cm tall; the other crystals are from Brazil.*** (*Harvard Mineralogical Collection. Photograph by Walter R. Fleischer.*)

Figure 3.8 ***The six crystal systems and their subclasses (opposite page).***

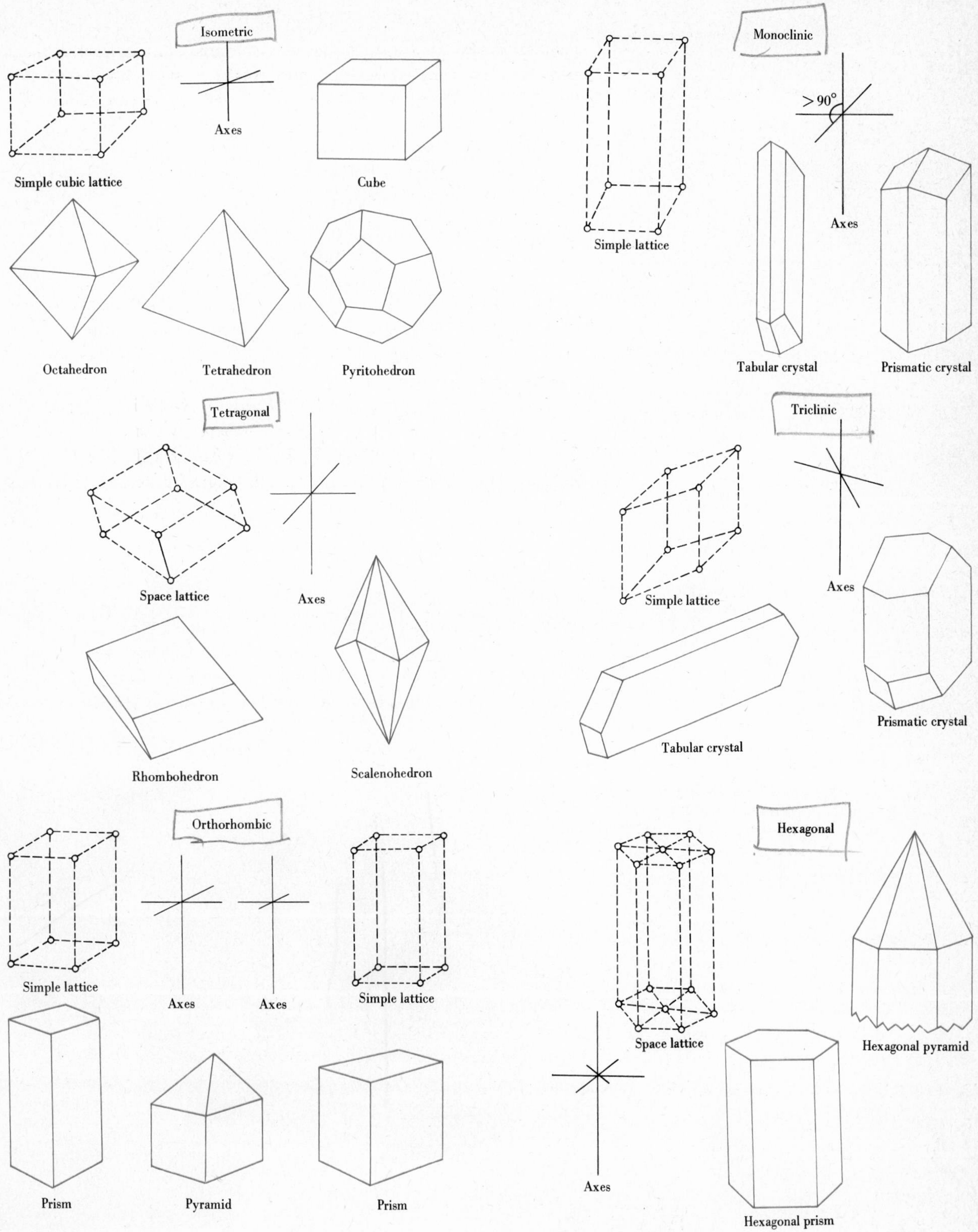

Isometric
Axes
Simple cubic lattice
Cube
Octahedron
Tetrahedron
Pyritohedron
Monoclinic
> 90°
Simple lattice
Axes
Tabular crystal
Prismatic crystal
Tetragonal
Space lattice
Axes
Rhombohedron
Scalenohedron
Triclinic
Simple lattice
Axes
Tabular crystal
Prismatic crystal
Orthorhombic
Simple lattice
Axes
Axes
Simple lattice
Prism
Pyramid
Prism
Hexagonal
Space lattice
Axes
Hexagonal pyramid
Hexagonal prism

Table 3.2 The six crystal systems

System	*Symmetry*	*Examples*
Isometric	Three equal-length axes at right angles to each other	Pyrite, FeS_2 Galena, PbS Halite, NaCl
Tetragonal	Two equal axes and a third either longer or shorter, all at right angles	Cassiterite, SnO_2 Zircon, $ZrSiO_4$
Orthorhombic	Three unequal axes, all at right angles	Corundum, Al_2O_3 Hematite, Fe_2O_3
Monoclinic	Three unequal axes, two at right angles and the third perpendicular to one but oblique to the other	Gypsum, $CaSO_4 \cdot 2H_2O$ Azurite, $Cu_3(CO_3)_2(OH)_2$ Jadeite, $NaAl(Si_2O_6)$
Triclinic	Three unequal axes meeting at oblique angles	Kyanite, Al_2SiO_5 Turquoise, $CuAl_6(PO_4)(OH)_8 \cdot 4H_2O$ Wollastonite, $Ca(SiO_3)$
Hexagonal	Three equal axes in the same plane intersecting at 60° and a fourth perpendicular to the plane of the other three	Quartz, SiO_2 Molybdenite, MoS_2 Ice, H_2O

rhombic octahedrons or as slender monoclinic needles. The internal arrangement of the atoms is different in the two types of crystals. Carbon's two common forms have been mentioned, graphite and diamond. The crystal form of diamond is an eight-sided figure called an octahedron (see Figure 3.9). Diamond belongs to the isometric system. The crystal form of graphite is a flat crystal with six sides. Graphite belongs to the hexagonal system. Although

Figure 3.9 ***A large, perfect crystal of diamond (uncut), which is the external expression of the orderly internal arrangement of atoms of carbon. Weight, 84 carats. From Kimberley, South Africa. Shortly after this picture was taken, the specimen was stolen from an alarm-guarded safe of the Harvard Mineralogical Collection. It has probably been cut up.*** (*Photograph by Harry Groom.*)

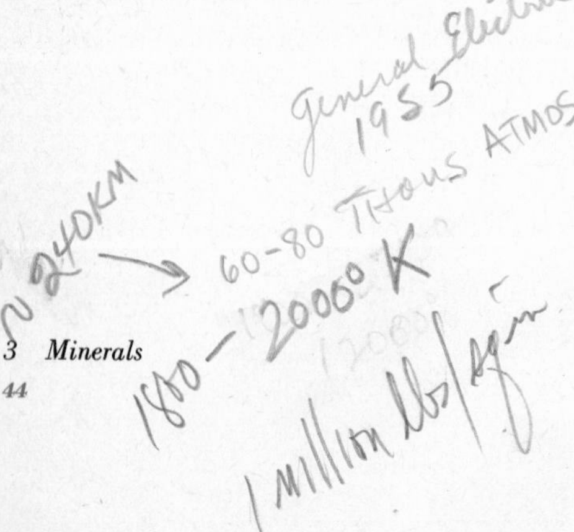

both minerals are composed of carbon, the difference in their crystal forms comes from the arrangement of their carbon atoms—one pattern in diamond, another in graphite. The crystal form of pyrite is a cube. It belongs to the isometric system. That of marcasite is a flattened tubular shape with orthorhombic symmetry. Both are composed of FeS_2. Here, again, the reason for the difference in crystal form lies in the internal arrangement of their atoms.

Every mineral has a characteristic crystal form, which is the external shape produced by its crystalline structure.

Hardness

Hardness is another physical property governed by the internal atomic arrangement of the elements of minerals. The stronger the binding forces between the atoms, the harder the mineral. The degree of hardness is determined by observing the relative ease or difficulty with which one mineral is scratched by another or by a fingernail or knife. It might be called the mineral's "scratchability." Hardness (*H*) ranges from 1 through 10:

Range	*Scratchability*
$H < 2\frac{1}{2}$	Will leave a mark on paper; can be scratched by fingernail
$2\frac{1}{2} < H < 3$	Cannot be scratched by fingernail; can be scratched by a penny
$3 < H < 5\frac{1}{2}$	Cannot be scratched by a penny; can be scratched by a knife
$5\frac{1}{2} < H < 7$	Cannot be scratched by a knife; can be scratched by quartz
$7 < H$	Cannot be scratched by quartz

To illustrate, if you pick up a piece of granite and try to scratch one of its light-colored grains with a steel knife blade, this light-colored mineral simply refuses to be scratched. But if you drag one of its light-colored grains across a piece of glass, a scratch is easily made. Clearly, then, these particular mineral grains in granite are harder than either steel or glass. But if you have a piece of topaz handy, you can reveal the vulnerability of these light-colored mineral grains, for although they are harder than either steel or glass, they are not as hard as topaz.

Minerals differ widely in hardness (see Appendix C). Some are so soft that they can be scratched with a fingernail. Some are so hard that a steel knife is required to scratch them. But diamond, which is the hardest mineral known, cannot be scratched by any other substance.

Specific gravity

Every mineral has an average weight per cubic centimeter. This characteristic weight is usually described by comparing it with the weight of the same volume of water. The number that represents this comparison is called the *specific gravity* of the mineral.

The specific gravity of a mineral increases with the mass of its constituent

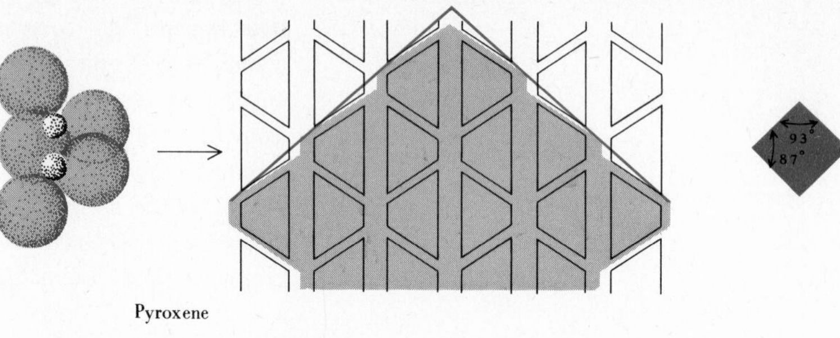

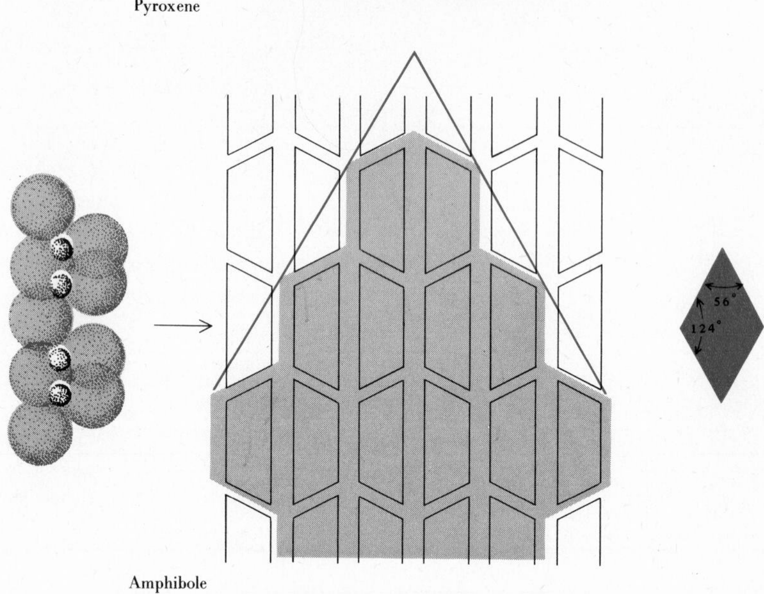

Figure 3.10 ***Atomic patterns explain the cleavage of pyroxenes and amphiboles. The difference in cleavage is one of the diagnostic features that help us to distinguish between the pyroxene augite and the amphibole hornblende (see Figure 3.16). These minerals supply a classic example of the influence of atomic structure on cleavage. On the left are end views of the unit cells of the pyroxene single chain and the amphibole double chain. On the right, these units (represented in outline only) are shown repeated in the patterns of the minerals. The minerals break between these unit cells into fragments represented by the shaded areas. The jogs of this drawing are of atomic dimensions in actual specimens and the surfaces of breakage appear perfectly smooth. The lengths of the unit cells control the cleavage angles.***

elements and with the closeness with which these elements are packed together in their crystalline structure.

Most rock-forming minerals have a specific gravity of around 2.7. The average specific gravity of metallic minerals is about 5. Gold has the highest specific gravities, ranging from 15 to 19.3.

It is not difficult to acquire a sense of relative weight by which to compare specific gravities. We learn to tell the difference between two bags of equal size, one filled with feathers and one filled with lead, and experience in hefting stones has given most of us a sense of the "normal" weight of rocks.

Cleavage

Cleavage is the tendency of a mineral to break in certain preferred directions along smooth plane surfaces. Cleavage planes are governed by the internal arrangement of the atoms (see Figure 3.10). They represent the directions in which the atomic bonds are relatively weak. Because cleavage is the breaking of a crystal between atomic planes, it is a directional property, and any plane throughout a crystal parallel to a cleavage surface is a potential cleavage plane.

Moreover, it is always parallel to crystal faces or possible crystal faces, because faces and cleavage both reflect the same crystalline structure.

Cleavage may be perfect, as in the micas, more or less obscure as in beryl and apatite, or entirely lacking as in quartz.

Cleavage is a *direction* of weakness, and a mineral sample tends to break along planes parallel to this direction. This weakness may be due to a weaker type of atomic bond, to greater atomic spacing, or frequently to a combination of the two. Graphite has a platy cleavage because of its relatively weak bond between the carbon layers. Diamond has but one bond joining all its carbon atoms together. Its cleavage takes place along planes having the largest atomic spacing.

Color

Although *color* is not a very reliable property in identifying most minerals, it is strikingly characteristic for a few. These include the intense azure blue of azurite, the bright green of malachite, and the pale yellow of sulfur. Magnetite is iron black, and galena is lead gray. The color of other minerals, such as quartz, can be quite variable because of slight impurities.

Minerals containing iron are usually "dark-colored." In geologic usage, *dark* includes dark gray, dark green, and black. Minerals that contain aluminum as a predominant element are usually "light-colored," a term that includes purples, deep red, and some browns.

Streak

The *streak* of a mineral is the color it displays in finely powdered form. The streak may be different from the color of the hand specimen. Although the color of a mineral may vary between wide limits, the streak is usually constant.

One of the simplest ways of determining the streak of a mineral is to rub a specimen across a piece of unglazed porcelain known as a *streak plate.* The color of the powder left behind on the streak plate helps to identify some minerals. Because the streak plate has a hardness of 7, it cannot be used to identify minerals with greater hardness.

Hematite, Fe_2O_3, may be reddish brown to black in color. Its streak is light to dark Indian red which becomes black on heating. Limonite, $FeO(OH) \cdot nH_2O$, sometimes known as brown hematite or bog-iron ore, has a color that is dark brown to black but a streak that is yellowish brown. Cassiterite, SnO_2, tin stone, is usually brown or black in color but has a white streak.

Striations

A few common minerals have parallel, threadlike lines or narrow bands called *striations* running across their crystal faces or cleavage surfaces. These can be seen clearly on crystal faces of quartz and pyrite, for example (see Figure

Figure 3.11 ***Cubic crystals of pyrite. Striations are clear on the large specimen. Note that those in adjacent faces are perpendicular to each other. The small specimen, about 5 cm wide, consists of three cubes intergrown.*** *(Harvard Mineralogical Collection. Photograph by Walter R. Fleischer.)*

3.11). Once again, this property is a reflection of the internal arrangement of the atoms and the conditions of growth of the crystals.

Other physical properties

Minerals have other physical properties that may be helpful in identifying them. These include magnetism, electrical properties, fluorescence, fusibility, solubility, fracture, tenacity, and luster (see Appendix C).

3.4 *Rock-forming minerals*

Although there are more than 2,000 minerals known, only a few of these are *rock-forming minerals,* the minerals that comprise most of the rocks of the earth's crust. Minerals are homogeneous crystalline materials, but they are not necessarily pure substances. Most rock-forming minerals have variable compositions caused by substitution of ions of some elements for other elements. These substitutions are distributed at random throughout the crystalline structure. Such replacement is called *solid solution.* The resultant mineral is a homogeneous material having the same physical properties throughout, but its composition is slightly variable.

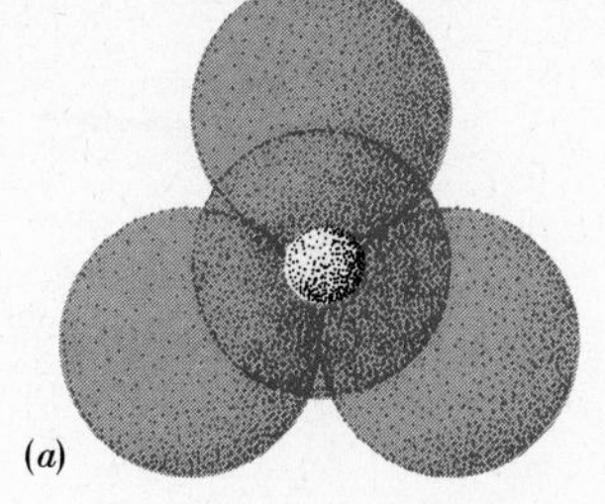

(a)

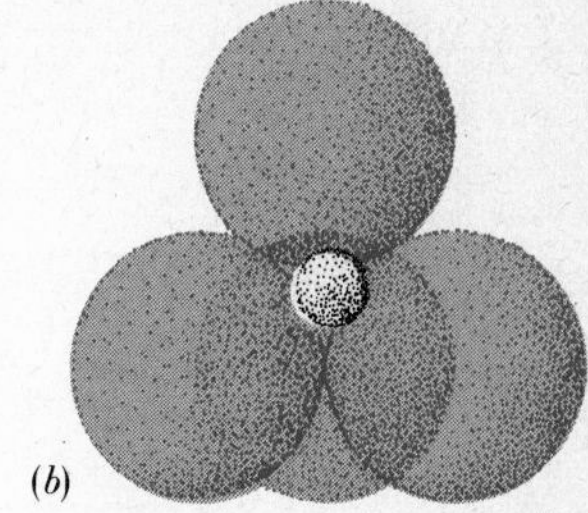

(b)

Figure 3.12 ***The silicon-oxygen tetrahedron*** $(SiO_4)^{4-}$***: (a) from above, (b) from the side. This is the most important complex ion in geology, because it is the central building unit of nearly 90 percent of the minerals of the earth's crust.***

Silicates

More than 90 percent of the rock-forming minerals are silicates, compounds containing silicon and oxygen and one or more metals. Each silicate mineral has as its fundamental unit, the *silicon–oxygen tetrahedron* (see Figure 3.12). This is a combination of one "small" silicon ion with a radius of 0.42 A surrounded as closely as geometrically possible by four "large" oxygen ions with a radius of 1.40 A forming a tetrahedron. The oxygen ions contribute an electric charge of −8 to the tetrahedron, and the silicon ion contributes +4. Therefore the silicon–oxygen tetrahedron is a complex ion with a net charge of −4. Its symbol is $(SiO_4)^{4-}$.

The most common of the rock-forming silicates are olivine, augite, hornblende, biotite, muscovite, feldspars, and quartz. They are listed and classified in Table 3.3.

Table 3.3 ***Silicate classification***[a]

Class[b]	*Arrangement of* SiO_4 *tetrahedra*	*Rock-forming minerals*
Nesosilicates	Isolated	Olivine
Sorosilicates	Double	Epidote
		The epidote family of hydrous calcium-aluminum-iron silicates common in crystalline metamorphic rocks
Inosilicates	Chains (single)	Augite
		The pyroxene family
	Chains (double)	Hornblende
		The amphibole family
Phyllosilicates	Sheets	Muscovite, white mica
		Biotite, black mica
Tectosilicates	Frameworks	Orthoclase
		Albite
		Anorthite
		Quartz

[a] From Cornelius S. Hurlbut, Jr., *Dana's Manual of Mineralogy*, 17th ed., John Wiley & Sons, Inc., New York, 1961.

[b] The prefixes are from the Greek *neso*, "island"; *soro*, "group"; *ino*, "chain" or "thread"; *phyllo*, "sheet"; *tecto*, "framework."

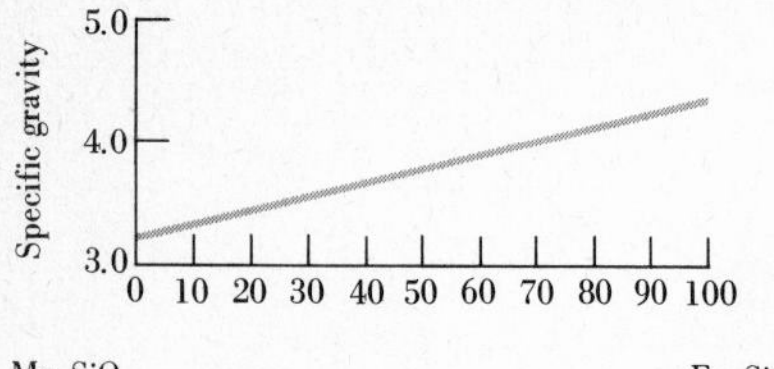

Figure 3.13 **Specific gravity of olivine.**

FERROMAGNESIANS In the first four of these rock-forming silicates—olivine, augite, hornblende, and biotite—the silicon–oxygen tetrahedra are joined by ions of iron and magnesium. Iron is interchangeable with magnesium in the crystalline structure of these silicates, because the ions of both elements are approximately the same size and have the same positive electric charge. These silicate minerals are known as *ferromagnesians,* from the joining of the Latin *ferrum,* "iron", with magnesium. All four ferromagnesians are very dark or black in color and have a higher specific gravity than the other rock-forming minerals.

Olivine is a nesosilicate because it is composed of isolated SiO_4 tetrahedra held together by positive ions of magnesium and iron. Its elements are so firmly held together by their ionic bonds that olivine exhibits no cleavage and is a relatively hard mineral, $6\frac{1}{2}$ to 7. Because there are no planes of weakness, olivine fractures when struck a blow.

Olivine is an example of a mineral that undergoes compositional changes by isomorphism. Its formula is $(Mg, Fe)_2SiO_4$. The proportions of magnesium and iron vary, so their symbols are in parentheses. The proportion of silicon and oxygen remains constant. Iron and magnesium substitute for one another quite freely in olivine's crystalline structure because they each have two electrons in their outer shell and their ionic radii are almost identical, 0.97 and 0.99 A. Because there may be varying amounts of magnesium and iron present, olivine is a solid solution mineral series. The end members of the series are forsterite and fayalite. When the positive ions are all magnesium, the mineral is forsterite, Mg_2SiO_4. When they are all iron, the mineral is fayalite, Fe_2SiO_4.

Olivine is geologically important as it comprises several percent of the crustal rocks at the surface and is believed to predominate in the heavier and deeper-seated rocks. Its specific gravity is 3.27 to 4.37, increasing with iron content (see Figure 3.13). Its color ranges from olive to grayish green, sometimes brown. This mineral usually occurs in grains or granular masses.

Augite has a crystalline structure based on single chains of tetrahedra, as shown in Figure 3.14, joined by ions of iron, magnesium, calcium, sodium, and aluminum. It is an inosilicate. It is dark green to black, with a colorless streak. Its hardness is 5 to 6 and its specific gravity ranges from 3.2 to 3.4. It has rather poor cleavage along two planes almost at *right* angles to each other. This cleavage angle is important in distinguishing augite from hornblende. A good way to recall it is to remember that *augite* rhymes with *right.*

Augite is the commonest variety of a group of minerals designated by the family name *pyroxenes* (see Appendix C) and characterized by a fundamental

(*a*)

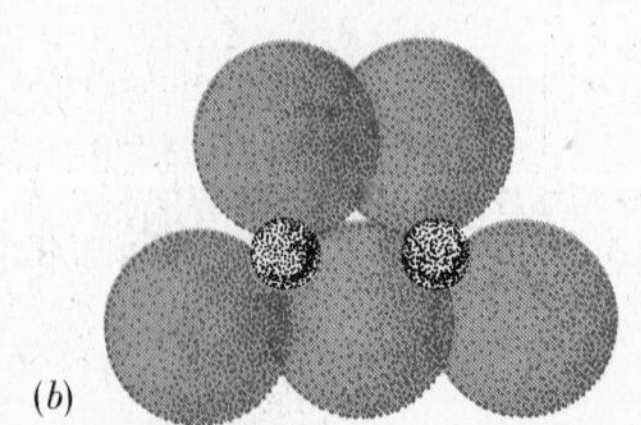

(*b*)

Figure 3.14 ***Single chain of tetrahedra viewed (a) from above, (b) from an end. Each silicon ion (small black sphere) has two of the four oxygen ions of its tetrahedron bonded exclusively to itself, and it shares the other two with neighboring tetrahedra fore and aft. The resulting individual chains are in turn bonded to one another by positive metallic ions. Because these bonds are weaker than the silicon–oxygen bonds that form each chain, cleavage develops parallel to the chains.***

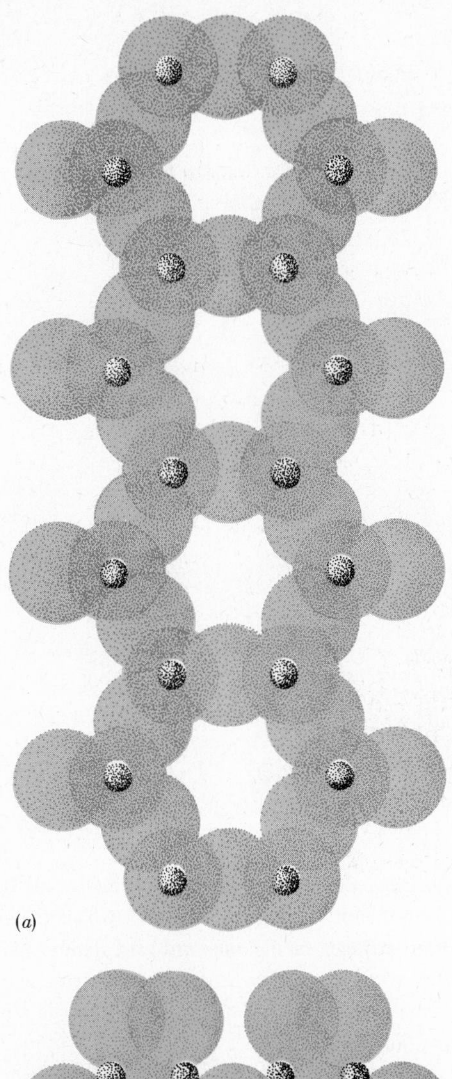

Figure 3.15 ***Double chain of tetrahedra viewed (a) from above, (b) from an end. The doubling of the chain of Figure 3.14 is accomplished by the sharing of oxygen atoms by adjacent chains.***

crystalline structure built on single chains of tetrahedra.[1] Pyroxenes crystallize in the orthorhombic and monoclinic systems and form a series in which members are closely analogous chemically to members of the amphibole group.

Hornblende has a crystalline structure based on double chains of tetrahedra, as shown in Figure 3.15, joined by the iron and magnesium ions common to all ferromagnesians and also by ions of calcium, sodium, and aluminum. It is an inosilicate. Hornblende's color is dark green to black, like that of augite; its streak is colorless; its hardness is 5 to 6 and its specific gravity is 3.2. Two directions of good cleavage meet at angles of approximately 56° and 124°, which helps distinguish hornblende from augite (see Figure 3.16).

Hornblende is an important and widely distributed rock-forming mineral. It is the commonest mineral of a group called the *amphiboles*, which closely parallel the pyroxenes in composition but contain hydroxyl (OH). Amphiboles are characterized by double chains of tetrahedra. The minerals classified as hornblendes are in reality a complex solid solution series from anthophyllite to hornblende (see Appendix C).

Biotite, black mica, is a potassium–magnesium–iron–aluminum silicate, essentially $K(Mg, Fe)_3(AlSi_3O_{10})(OH)_2$. It was named in honor of the French physicist J. B. Biot. Like other micas, it is a phyllosilicate constructed of tetrahedra in sheets, as shown in Figure 3.17. Each silicon ion shares three oxygen ions with adjacent silicon ions to form a pattern like wire netting.

[1]The term *tetrahedra* will be used throughout this chapter to refer to silicon–oxygen tetrahedra.

Figure 3.16 ***Cleavage of hornblende (a) compared with that of augite (b). The top "roof" of the hornblende specimen and the top and perpendicular left-hand faces of the augite are cleavage surfaces. Throughout each specimen, easiest breaking is parallel to these surfaces. On the front face of the augite are some "steps" outlined by cleavage planes. Such steps are the most common manifestation of cleavage, which seldom produces pieces as large as the 5-cm wide augite shown here.*** *(Harvard Mineralogical Collection. Photograph by Walter R. Fleischer.)*

(a) (b)

Figure 3.17 ***Tetrahedral sheets. Each tetrahedron is surrounded by three others, and each silicon ion has one of the four oxygen ions to itself, sharing the other three with its neighbors.***

The fourth, unshared oxygen ion of each tetrahedron stands above the plane of all the others. The basic structural unit of mica consists of two of these sheets of tetrahedra, with their flat surfaces facing outward and their inner surfaces held together by positive ions. In biotite, the ions are iron and magnesium. These basic double sheets of mica, in turn, are loosely joined together by positive ions of potassium.

Layers of biotite, or any of the other micas, can be peeled off easily (see Figure 3.18), because there is perfect cleavage along the surfaces of the weak potassium bonds. In thick blocks, biotite is usually dark green or brown to black. Its specific gravity is 2.8 to 3.2, and its hardness is 2.5 to 3.

NONFERROMAGNESIANS The other common rock-forming silicate minerals are known as *nonferromagnesians,* simply because they do *not* contain iron or magnesium. These minerals are muscovite, feldspars, and quartz. They are all marked by their light colors and relatively low specific gravities, ranging from 2.6 to 3.0.

Muscovite is white mica, so named because it was once used as a substitute for glass in old Russia (Muscovy). It has the same basic crystalline structure as biotite, but in muscovite each pair of tetrahedra sheets is tightly cemented together by ions of aluminum. As in biotite, however, the double sheets are held together loosely by potassium ions, along which cleavage readily takes

Figure 3.18 ***Mica cleavage. The large block (or "book") is bounded on the sides by crystal faces. Cleavage fragments lying in front of the block are of different thicknesses, as indicated by their degrees of transparency. The reference grid is made of 5-cm squares.*** *(Photograph by Walter R. Fleischer.)*

place. It is a phyllosilicate with a formula of, essentially, $KAl_2(AlSi_3O_{10})(OH)_2$. In thick blocks, the color of muscovite is light yellow, brown, green, or red. Its specific gravity ranges from 2.8 to 3.1, with hardness 2 to 2.5.

Feldspars are the most abundant rock-forming silicates. The name comes from the German *feld,* "field," and *spar,* a term used by miners for various nonmetallic minerals. Its name reflects the abundance of these minerals: "field minerals," or minerals found in any field. Feldspars make up nearly 54 percent of the minerals in the earth's crust.

The feldspars are silicates of aluminum with potassium, sodium, and calcium. They may belong to either the monoclinic or triclinic systems, but the crystals of different systems resemble each other closely in angles and crystal habit. They all show good cleavages in two directions, which make an angle of 90° or close to 90° with each other. Their hardness is about 6, and their specific gravity ranges from 2.55 to 2.76.

Feldspars are tectosilicates because all the oxygen ions in the tetrahedra are shared by adjoining oxygen ions in a three-dimensional network. However, in one-quarter to one-half of the tetrahedra, aluminum ions with a radius of 0.51 A and an electric charge of +3 have replaced silicon with its radius of 0.42 A and its electric charge of +4 in the centers of tetrahedra. The negative electric charge resulting from this substitution is corrected by the entry into the crystalline structures of K^{1+}, Na^{1+}, or Ca^{2+}.

The feldspars are *orthoclase* with potassium, and *plagioclase* with sodium or calcium (see Table 3.4). Orthoclase is named from the Greek *orthos,* "straight," and *klasis,* "a breaking," because the two dominate cleavages intersect at a right angle when a piece of orthoclase is broken (see Figure 3.19). Aluminum replaces silicon in every fourth tetrahedron, and positive ions of potassium correct the electrical unbalance. The introduction of aluminum in the tetrahedra cannot be regarded as solid solution. Aluminum is not a vicarious constituent of orthoclase whose percentage varies from sample to

Table 3.4 ***The feldspars***

Diagnostic positive ion	*Name*	*Symbol*	*Descriptive name*	*Formula*[a]
K^+	Orthoclase	Or	Potassium feldspar	$K(AlSi_3O_8)$
Na^+	Albite	Ab	Sodium feldspar	$Na(AlSi_3O_8)$
Ca^{2+}	Anorthite	An	Calcium feldspar	$Ca(Al_2Si_2O_8)$

[a] In these formulas, the symbols inside the parentheses indicate the tetrahedra. The symbols outside the parentheses indicate the diagnostic ions that are worked in among the tetrahedra.

sample or which may be wholly replaced by silicon. It is an essential constituent and is not replaceable by silicon without breakdown of the crystalline structure. Orthoclase is white, gray, or pinkish; its streak is white; and its specific gravity is 2.57.

Plagioclase (oblique-breaking) feldspars are so named because they have cleavage planes that intersect at about 86°. One of the cleavage planes is marked by striations. Plagioclase feldspars may be colorless, white, or gray, although some samples show a striking play of colors called opalescence.

The plagioclase feldspars albite and anorthite are end members of a solid solution series. The amount of aluminum varies in proportion to the relative amounts of calcium and sodium to maintain electrical neutrality; the more calcium, the greater the amount of aluminum. Here the variation in amount of aluminum may be properly regarded as ionic substitution. The number of ions of calcium that substitute for sodium is equaled by the number of aluminum ions substituting for silicon in the silicon–oxygen framework. Sodium can substitute quite easily for calcium in a mineral's crystalline structure because their ionic radii are almost identical, 0.97 and 0.99 A, even though there is a slight difference in electrons present in the outer shell—one in sodium and two in calcium.

Although the plagioclase feldspars have different compositions, they have the

Figure 3.19 ***Feldspar cleavage in specimens of orthoclase. The large block on the right and the small fragment on the black box (a 5-cm cube) show the cleavage planes at nearly 90°, a characteristic of feldspars.*** *(Photograph by Walter R. Fleischer.)*

same crystal form and for that reason are called *isomorphous.* At the same time, they can be thought of as solutions of albite in anorthite in varying proportions, or the reverse. For that reason, they have been called examples of solid solution. This term, however, does not imply that in the intermediate minerals there are molecules of either albite or anorthite as there would be of, say, sodium and chlorine in a solution of salt in water. Intermediate minerals simply have compositions that are conveniently described as mixtures of the pure end members of the series.

In albite, aluminum replaces silicon in every fourth tetrahedron, and positive ions of sodium correct the electrical unbalance. The specific gravity of albite is 2.62.

In anorthite, aluminum replaces silicon in every second tetrahedron, and positive ions of calcium correct the electrical unbalance. The specific gravity of anorthite is 2.76.

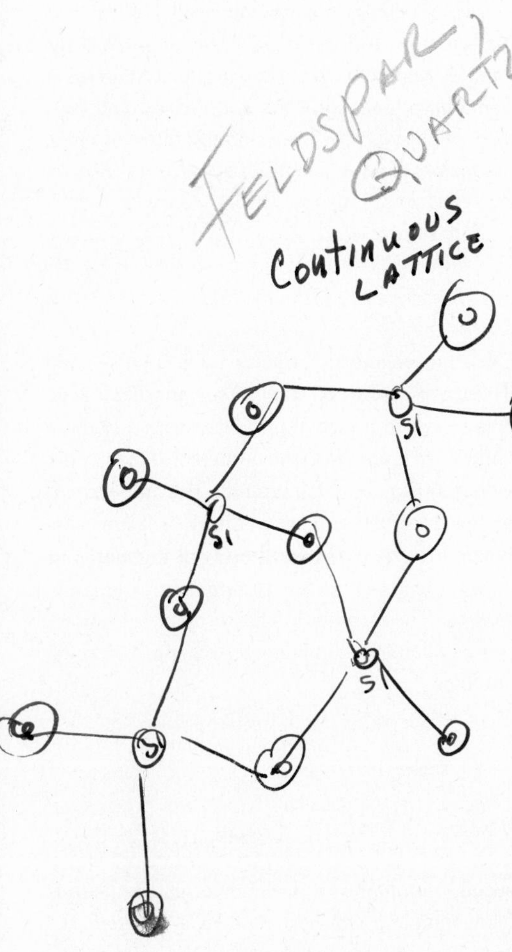

Quartz is the most common rock-forming silicate mineral that is composed "exclusively" of silicon–oxygen tetrahedra. It is a tectosilicate. Every oxygen ion is shared by adjacent silicon ions, which means that there are two ions of oxygen for every ion of silicon. This relationship is represented by the formula SiO_2. Because of its crystalline structure, quartz is a relatively hard mineral, 7. Because there are no planes of weakness in its crystalline structure, it does not exhibit cleavage but fractures along curved surfaces when struck a blow. Being composed of fairly light elements, its specific gravity is 2.65.

Of all the rock-forming minerals, quartz is most nearly a pure chemical compound. However, spectrographic analyses show that even in its most perfect crystals there are traces of other elements present. Quartz usually appears smoky to clear in color, but many less common varieties include purple or violet *amethyst,* massive rose-red or pink *rose quartz,* smoky yellow to brown *smoky quartz,* and *milky quartz.* These color differences are caused by traces of other elements present as minor impurities. They are not caused by, and do not affect, the crystalline structure of quartz.

Oxide minerals

Oxide minerals are formed by the direct union of an element with oxygen. These have relatively simple formulas compared to the complicated silicates. The oxide minerals are usually harder than any other class except the silicates, and they are heavier than others except the sulfides. Within the oxide class are the chief ores of iron, chromium, manganese, tin, and aluminum.

Some common oxide minerals are ice (H_2O), corundum (Al_2O_3), hematite (Fe_2O_3), magnetite (Fe_3O_4), and cassiterite (SnO_2).

Sulfide minerals

Sulfide minerals are formed by the direct union of an element with sulfur. The elements that occur most commonly in combination with sulfur are iron, silver, copper, lead, zinc, and mercury. Some of these sulfide minerals occur as commercially valuable ores, such as pyrite (FeS_2), chalcocite (Cu_2S), galena (PbS), and sphalerite (ZnS).

Carbonate and sulfate minerals

In discussing silicates, we found them to be built around the complex ion $(SiO_4)^{4-}$, that is, the silicon–oxygen tetrahedron. But two other complex ions are also of great importance in geology. One of these consists of a single carbon ion with three oxygen ions packed around it—the complex ion $(CO_3)^{2-}$. Compounds in which this ion appears are called carbonates. For example, the combination of a calcium ion with a carbon–oxygen ion produces calcium carbonate, $CaCO_3$, known in its mineral form as *calcite.* This mineral is the principle component of the common sedimentary rock, limestone. The other complex ion is $(SO_4)^{2-}$, a combination of one sulfur ion and four oxygen ions. This complex ion combines with other ions to form sulfates. For example, it joins with a calcium ion to form calcium sulfate, $CaCO_4$, the mineral *anhydrite.*

Mineraloids

Some substances do not yield definite chemical formulas upon analysis and show no sign of crystallinity. These are said to be *amorphous* and have been called *mineraloids.* A mineral may exist in a crystalline phase with a definite composition and crystalline structure, or, under certain conditions of formation, practically the same substance may occur as a mineraloid. The mineraloids are formed under conditions of low pressure and temperature and are commonly substances originating during the process of weathering of the materials of the earth's crust. They characteristically occur in mammillary, botryoidal, stalactitic, and similar-shaped masses. Their ability to absorb other substances accounts for their wide variations in chemical composition. Bauxite, limonite, and opal are examples of mineraloids.

3.5 *Organization of minerals*

We know that minerals are special combinations of elements or compounds (see Table 3.5), and now we can complete our definition of a mineral: (1) it is a naturally occurring element or inorganic compound in the solid state[2]; (2) it has a diagnostic composition; (3) it has a unique crystalline structure; and (4) it exhibits certain physical properties as a result of its composition and crystalline structure.

[2]Some mineralogists do not restrict the definition to the solid state but include such substances as water and mercury.

Table 3.5 ***The organization of common minerals***[a]

Elements	Compounds: Oxides, elements + O	Sulfides, elements + S	Carbonates, elements + CO_3 ion	Sulfates, elements + SO_4 ion	Silicates, elements + SiO_4 ion
Copper	Cassiterite	Chalcocite	Calcite	Anhydrite	Nonferromagnesian
Diamond	Corundum	Galena	Dolomite	Gypsum	Quartz
Gold	Hematite	Pyrite	Magnesite		Feldspars
Graphite	Ice	Sphalerite			Orthoclase
Iron	Magnetite				Plagioclase
Platinum					Albite
Silver					Anorthite
Sulfur					Muscovite
					Ferromagnesian
					Biotite
					Hornblende
					Augite
					Olivine

[a] Minerals may be either elements or compounds, though not all elements or compounds are minerals.

Outline

Minerals are naturally occurring elements or compounds that have been formed by inorganic processes.

Mineral composition includes elements in definite or slightly varying proportions.

Mineral structure is the internal orderly arrangement of its atoms, which is unique for each mineral.

Identification of minerals is determined by chemical properties and by physical properties including crystal form, hardness, specific gravity, cleavage, color, streak, and striations.

- *Crystal form* is the external shape produced by its crystalline structure.
- *Hardness* is governed by the internal atomic arrangement of the elements of a mineral.
- *Specific gravity* is a number which compares the weight of a given volume of a mineral with the weight of the same volume of water.
- *Cleavage* is the tendency of a mineral to break in certain preferred directions along smooth plane surfaces.
- *Color* is not a reliable property for identifying most minerals, but minerals containing iron are usually dark-colored and those containing aluminum are light-colored.
- *Streak* is the color of a mineral in finely powdered form.
- *Striations* are parallel threadlike lines or narrow bands running across crystal faces or cleavage surfaces.
- *Other physical properties* include magnetism, electrical properties, fluorescence, fusibility, solubility, fracture, tenacity, and luster.

Rock-forming minerals are 90 percent silicates, based on a complex ion called the silicon–oxygen tetrahedron.

Ferromagnesians olivine, augite, hornblende, and biotite contain iron and magnesium.

Olivine is a nesosilicate occurring in grains or granular masses, with characteristic olive color.

Augite is an inosilicate with single chains of tetrahedra and cleavage planes almost at right angles to each other.

Hornblende is an inosilicate with double chains of tetrahedra; it has directions of cleavage meeting at approximately 56° and 124°.

Biotite, black mica, is a phyllosilicate.

Nonferromagnesians are muscovite, the feldspars, and quartz.

Muscovite, white mica, is a phyllosilicate.

Feldspars, the most abundant rock-forming silicates, are tectosilicates.

Orthoclase is the potassium feldspar.

Albite, the sodium feldspar, and *anorthite,* the calcium feldspar, are plagioclase feldspars.

Quartz is the only rock-forming silicate mineral composed exclusively of silicon–oxygen tetrahedra.

Oxide minerals are formed by the direct union of an element with oxygen.

Sulfide minerals are formed by the direct union of an element with sulfur.

Carbonate minerals are built around the complex ion $(CO_3)^{2-}$.

Sulfate minerals are built around the complex ion $(SO_4)^{2-}$.

Mineraloids do not have definite compositions or crystalline structure.

Organization of minerals finds them naturally occurring combinations of elements or compounds in the solid state, each with a diagnostic composition and unique crystalline structure as well as certain physical properties.

Supplementary readings

HURLBUT, CORNELIUS S., JR.: *Minerals and Man,* Random House, Inc., New York, 1969. A lively, balanced presentation for the layman of expert information on the nature, origin, and properties of 150 of the world's principal minerals, together with stories of how they have been used throughout the ages, with 217 illustrations including 160 in superb color.

HURLBUT, CORNELIUS S., JR.: *Dana's Manual of Mineralogy,* 17th ed., John Wiley & Sons, Inc., New York, 1961. The latest edition of the textbook that has dominated this field for most of this century.

4

Igneous activity

FROM TIME TO TIME THROUGHOUT GEOLOGIC HISTORY, MOLTEN ROCK, WORKING ITS way to the surface through covering rock, has poured out or was blown onto the ground, there again to solidify into rock. Igneous activity consists of movements of molten rock inside and outside the earth. It also includes the variety of effects associated with these movements.

In some areas, molten rock was extruded through extensive fissures in the earth's surface. This activity, called *fissure eruptions,* built large plateaus. In other places, molten rock escaped to the surface through vents, and around these vents the ejected material accumulated to build up landforms that we know as *volcanoes.*

4.1 Volcanoes

A volcano may grow in size until it becomes a mountain (see Figure 4.1). Normally cone-shaped, it has a pit at the summit, which may be either a crater or a caldera. A *crater* is a steep-walled depression out of which volcanic materials are ejected. Its floor is seldom over 300 m in diameter; its depth may be as much as 100 m. A crater may be at the top of a volcano or on its flank. The much larger *caldera* is a basin-shaped depression, more or less circular, with a diameter many times greater than that of the included volcanic vent or vents. Most calderas, in fact, are more than 1,500 m in diameter; some are several kilometers across and several hundred meters deep.

Between eruptions, a volcano's vent may become choked with rock congealed from magma of a past eruption. Sometimes small jets of gas come out through cracks in this rock plug. A volcano is built by, and remains active because of, materials coming from a large deep-seated reservoir of molten rock. While in the ground, this molten rock is called magma. When extruded on the surface, it is called lava.

The world's largest volcano is Mauna Loa, on the island of Hawaii. It is 600 km around the base, and its summit towers nearly 10 km above the surrounding ocean bottom (see Figure 4.2). This and the rest of the island represent accumulations from eruptions that have gone on for more than 1 million years.

Volcanic eruptions

In its reservoir far below the earth's surface, magma is composed of elements in solution. Some are vaporized as magma approaches the surface. These volatile components play an extremely important role in igneous activity and are the primary agents in producing a *volcanic eruption.* As the magma nears the surface, volatiles tend to separate from the other components and migrate through them to the top of the moving mass. They accumulate if the volcanic vent is blocked. Pressure builds up until it can no longer be confined. Then it pushes out. If its temperature is 1000°C or higher, the gases expand several thousandfold as they escape, shattering rock that blocks the vent and throwing it and magma into the air. After the explosion, the magma still in the ground

Figure 4.1 ***Pavlof Volcano, Alaskan Peninsula.*** (*Official U.S. Navy photograph.*)

is left poorer in volatile components but may be fluid enough to pour out.

Volcanic eruptions are of different types, from relatively quiet outpourings of lava to violent explosions accompanied by showers of volcanic debris. The type of eruption depends to a great extent on the magma's composition and its volatile content. Lavas rich in ferromagnesians tend to produce relatively quiet flows, whereas those high in silica are more viscous and explosive.

Figure 4.2 ***Size of volcanoes.*** (*From Brian Bayly,* Introduction to Petrology, *Prentice-Hall, Inc., Englewood Cliffs, N.J., 1968.*)

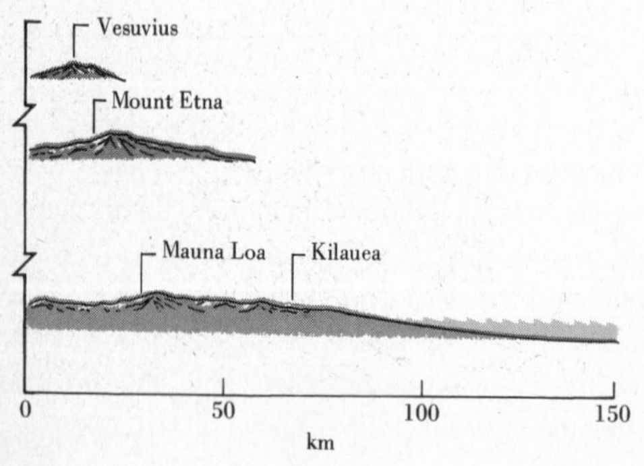

Composition of lavas

Lavas are of three principal types, classified primarily by their proportions of silica: acid, intermediate, and basic. The acid lavas have 70 percent or more of silica; intermediate lavas have 60 to 65 percent silica; and basic lavas have less than 50 percent.

All lavas have some similarities of composition, but no two volcanoes erupt lavas of exactly the same composition. In fact, the composition may vary from one eruption to another in the same volcano. Generally, however, over the course of a volcano's history it starts with basic lava, grades into intermediate, and ends its activity with dominantly acidic lava.

Volcanoes erupting primary basic lavas are confined to ocean basins. Four are on the Hawaiian Islands and three are on the Galapagos in the Pacific Ocean. In the Atlantic Ocean, they range from Iceland along the Midatlantic Ridge to Tristan de Cunha. Intermediate-type lavas are erupted by volcanoes ringing the Pacific and constituting the island arcs. In the Mediterranean Sea, a number of Italian volcanoes erupt lavas believed to have been contaminated by the ingestion of large quantities of limestone.

The surfaces of lava flows commonly have one of two contrasting shapes for which the Hawaiian names of *pahoehoe* and *aa* are used. In the pahoehoe type, sometimes known as "corded" or "ropy," the surface is smooth and billowy and frequently molded into forms which resemble huge coils of rope (see Figure 4.3). In the aa type, the surface of the lava is covered with a random mass of angular, jagged blocks. The surface that develops depends on the viscosity of the lava, with viscosity increasing as lavas become more acidic. Pahoehoe is the shape produced on the surface of a basic lava, and aa is the shape produced when the flow becomes sluggish because of an increase in silica content.

TEMPERATURE OF LAVAS Measurements of the temperature of lavas have been made at some volcanic vents. These show that basic lavas have an average temperature of around 1100°C. The hottest lavas known are Hawaiian lavas

Figure 4.3 **Pahoehoe lava.** (*Photograph by U.S. Geological Survey.*)

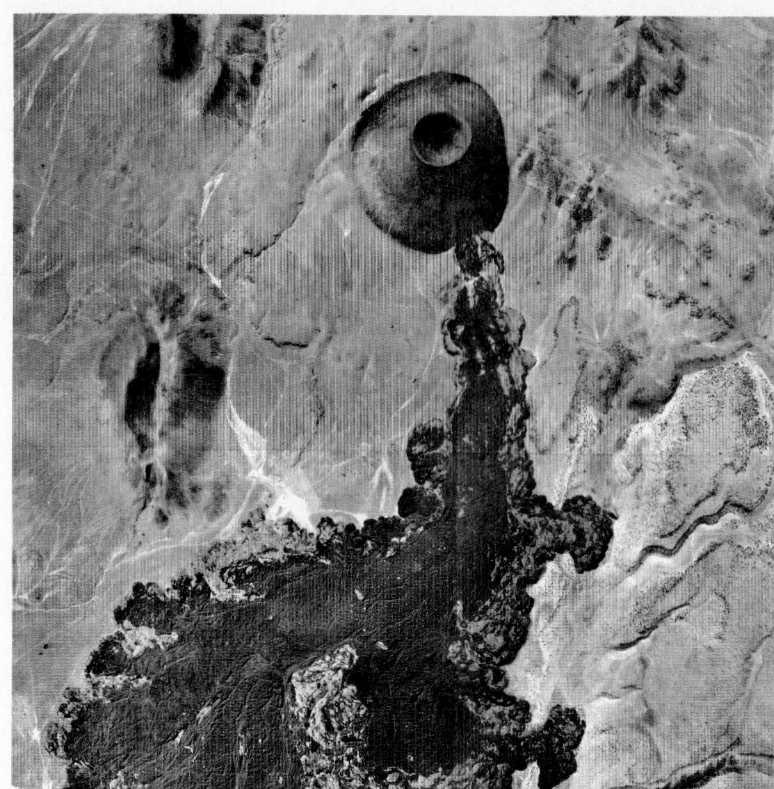

Figure 4.4 **A cinder cone with basaltic lava spreading from its base. Some older cones and flows can also be seen. San Francisco Mountains area, Coconino County, Arizona.** (*From Kirk Bryan Library of Geomorphology, Harvard University.*)

measured at Kilauea. Measurements were made in the lava lake Halemaumau, using Seger cones. These are of ceramic material constructed to melt at various temperatures. Seger cones were placed in pipes and then the pipes were lowered into the lake of lava. At a depth of 13 m, the lava had a temperature of 1175°C. This decreased to 860°C at a depth of 1 m and increased again to 1000°C at the lava surface. Temperatures as high as 1350°C were recorded above the lava surface where volcanic gases were reacting with atmospheric oxygen.

Frank A. Perret recorded temperatures in 1910 from the lava at Mt. Etna in Sicily as between 900 and 1000°C and on lavas in Vesuvius in 1916 to 1918 of from 1015 to 1040°C. Minimum temperatures observed on lavas are around 750°C.

VOLCANIC GASES As one might expect, taking an accurate sampling of volcanic gases is not an easy job. It is also difficult to decide whether the gases have come exclusively from the magma or partly from the surrounding rocks. We can, however, make a few generalizations from measurements at Kilauea in Hawaii.

Close to 70 percent of the volume of gases collected directly from a molten lake of lava was steam. Next in abundance were carbon dioxide, nitrogen, and sulfur gases, with smaller amounts of carbon monoxide, hydrogen, and chlorine. Even when gases other than steam make up only a small percentage of the total volume, their absolute quantities may be large. For example, in 1919 during the cooling of material erupted in 1912 from Katmai in Alaska, the total amount of hydrochloric acid released was estimated at 1,250,000 tons, and the total amount of hydrofluoric acid was approximately 200,000 tons.

When any igneous rock is heated, it yields some quantity of gases. Water vapor predominates, and measurements indicate that it constitutes about 1 percent of fresh—that is, unweathered—igneous rocks. Estimates of the average water content of actual magma range from about 1 to 8 percent, with the weight of opinion centering around 2 percent. A silicate melt will not hold more than about 11 percent of volatiles under any circumstances.

Pyroclastic debris

Fragments blown out by explosive eruptions and subsequently deposited on the ground are called *pyroclastic debris.* The finest of these constitute *dust,* which is made up of pieces of the order of 10^{-4} cm in diameter. When volcanic dust is blown into the upper atmosphere, it can remain there for months, traveling great distances. The following fragments settle around or near the volcanic crater:

Ash Fragments consisting of sharply angular glass particles. Smaller than cinders.

Cinders Small, slaglike, solidified pieces of magma $\frac{1}{2}$ to $2\frac{1}{2}$ cm across.

Lapilli Pieces about the size of walnuts.

Blocks Coarse, angular pieces of the cone or masses broken away from rock that blocks the vent.

Bombs Rounded masses that congeal from magma as it travels through the air.

Pumice Pieces of magma up to several centimeters across that have trapped bubbles of steam or other gases as they are thrown out (see Figure 4.5). After these solidify, they are honeycombed with gas-bubble holes that give them enough buoyancy to float on water.

FIERY CLOUDS During the Katmai eruption in 1912, a great avalanche of incandescent ash mixed with steam and other gases was extruded. Heavier than air, this highly heated mixture rolled down the mountain slope. Masses of such material are called *fiery clouds* (sometimes referred to by the French equivalent, *nuée ardente*). The volume of material extruded was so great that it covered a valley of 140 km^2, 30 m thick. For the next 10 years the steam and gases kept erupting from this extruded material through a great number of holes called *fumaroles,* and the area was given the name Valley of Ten Thousand Smokes.

Fiery clouds have characterized eruptions at Mt. Pelée on Martinique Island in the West Indies to such an extent that they have come to be known as Peléan types of eruption. Magma is expelled as pumice and ash. In the final stage of the eruption, a mass of viscous lava may accumulate in a domelike form sometimes

Figure 4.5 Pumice.

called a tholoid, when the gas content is reduced to where it no longer shatters the magma on reaching the surface.

On Mt. Pelée at a few minutes before 8 A.M. on May 8, 1902, a gigantic explosion occurred through one side of the volcano. A fiery cloud at temperatures around 800° C swept down the mountainside and engulfed the city of St. Pierre, wiping out its 25,000 inhabitants and many refugees from other parts of the island who had gathered there during the preceding days, when the eruption was building up with minor explosions and earthquakes (see Figure 4.6). Estimates of the death toll ran as high as 40,000.

By the middle of October, a dome of lava too stiff to flow had formed in the crater, and from it a spine was extruded like a great blunted needle with a diameter of 100 to 200 m and at its maximum a height of 300 m above the crater floor, "The Spine of Pelée."

Arenal[1] In 1968, Arenal, a volcano about 100 km northwest of San Jose, Costa Rica, was considered by its inhabitants to be extinct. It had not erupted since around 1530. At 5:30 P.M. on July 29, 1968, it began erupting with a series of explosions that lasted for 2 days. Three new craters formed along a line approximately east–west on the west side of the 1,600-m mountain (see Figure 4.7). The largest crater, from which the major explosions came, was 250 m wide at an elevation of about 1,100 m.

A Senor Arraya, his wife, and a companion were on their way to visit his daughter, son-in-law, and their six children in Tabacon. From a point about 3 km northwest of the volcano, he saw landslides on the side of the mountain where the lower crater was to form. These were followed immediately by a violent ground-shaking explosion that produced a great fiery cloud of ash which swept down upon them as a searing acrid-smelling gale. They were knocked down and severely injured. In all, 78 people died from burns and asphyxiation,

[1] Report by William G. Melson and Rodrigo Saenz R., "The 1968 Eruption of Volcano Arenal, Costa Rica: Preliminary Summary of Field and Laboratory Studies," Smithsonian Institution's Center for Short-Lived Phenomena, November 7, 1968.

Figure 4.6 **Ruins of St. Pierre, Martinique, shortly after its destruction by a fiery cloud on the morning of May 8, 1902.** (*Photograph by Underwood and Underwood.*)

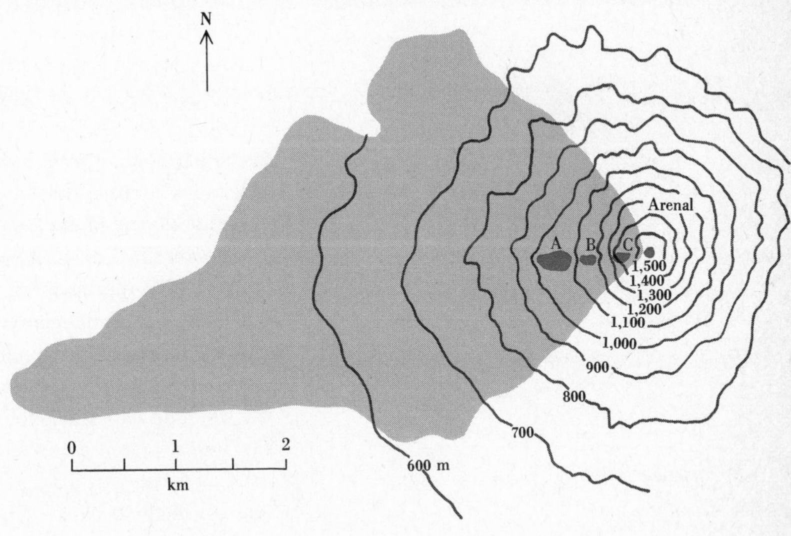

Figure 4.7 **Area devastated by fiery clouds from Mt. Arenal, Costa Rica, in an eruption which began on July 29, 1968. A, B, and C are the new craters that were formed.** (*Data from Smithsonian Institution's Center for Short-Lived Phenomena.*)

Figure 4.8 **Fiery clouds of eruption of Mt. Mayon, Philippine Islands, April 27, 1968.** (*Photograph by Smithsonian Institution.*)

including Senor Arraya's daughter and her family. The fiery cloud devastated vegetation and buildings over an area of 7 km^2, produced widespread craters from the impact of explosion-hurled blocks (see Figure 4.7), and was followed by flows of blocky ash. It has been estimated that up to 6 million m^3 of material were ejected during the 2 days of eruption.

A 4-day period of relative quiet was then followed by considerable ash and vapor emission for 1 week. Gas emanations from the newly formed craters dominated the interval from August 10 to September 14, when there was a renewal of explosions for 5 days. On September 19, an aa flow started from the lowest crater. It traveled down the valley of Quebrata Tabacon at rates estimated between 10 and 30 m/day and by October 17 was 1,000 m long and 25 m thick.

The Arenal eruptives appear to be similar to materials erupted April 20, 1968, at Mount Mayon in the Philippines (see Figure 4.8).

WORLDWIDE EFFECTS In 1783, Asama in Japan and Laki in Iceland had explosive volcanic eruptions. Large quantities of dust were blown into the upper atmosphere. Dry fogs in the stratosphere were recorded simultaneously at places as widely separated as northern Africa and Scandinavia. At one place, the density of the dry fog was so great that the sun was invisible until it had reached a position 17° above the horizon. The sun's effectiveness in heating the earth's surface was so reduced that the winter of 1783–1784 was one of the severest on record. Benjamin Franklin was the first person to connect the unusual weather with the volcanic eruptions. He published his ideas in May 1784.

During 1814 and 1815, the earth's temperature was reduced, following volcanic eruptions of Mayon on the Philippine Islands and Tambora on Sumbawa island, east of Java. The eruption of Tambora threw so much dust into the air that for 3 days there was absolute darkness for a distance of 500 km. With this dust and the dust erupted from Mayon in the atmosphere, the amount of the sun's heat reaching the earth's surface was significantly reduced. The year 1815 became known as the "year without a summer," marked throughout the world by long twilights and spectacular sunsets caused by the dust in the stratosphere.

The remarkable ability of volcanic dust to travel around the world was observed again in 1912. In June of that year, an observer from the Smithsonian Institution was in Algeria making measurements of the quantity of heat reaching the earth from the sun. During observations on June 19, he noticed streaks of dust lying along the horizon. These were joined by others, and in a few days the sky appeared "mackereled," although no clouds were present. Finally, the sky became so obscured that observations had to be discontinued. By June 29, the whole sky was filled with the dust, which persisted for months. At first, it was assumed that the condition was local, but reports gathered from many regions eventually indicated that the phenomenon was actually worldwide, caused by an eruption of Mt. Katmai on the Alaskan Peninsula. This eruption caused a 20 percent decrease in the amount of direct solar radiation reaching the earth's surface during the summer of 1912. The sun's rays, bouncing from dust in the upper atmosphere, caused an abnormal brightness of the sky. If the dust from Katmai had stayed in the atmosphere long enough, it would have been capable of so reducing the amount of heat received by the earth that the average world temperature would have dropped almost 7°C.

History of some volcanoes

A volcano is considered *active* if there is some record of its having erupted in historic time. If it has not done so, but shows notable lack of erosional alteration, pointing to eruption within quite recent geologic time, it is considered *dormant,* or merely "sleeping," and capable of renewed activity. If

Figure 4.9 **Vesuvius in eruption, 1906. The snow-clad slopes mark the modern volcano. To the left of this and in the foreground is the jagged remnant of Mt. Somma.** (*Photograph by A. & C. Caggiano.*)

a volcano not only has not erupted within historic time but also shows wearing away by erosion and no signs of activity (such as escaping steam or local earthquakes), it is considered *extinct.*

VESUVIUS Vesuvius, on the shore of the Bay of Naples, has supplied us with a classic example of the reawakening of a dormant volcano (see Figure 4.9). At the time of Christ, Vesuvius was a vine-clad mountain called Mt. Somma, a vacation spot in southwest Italy, favored by wealthy Romans. For centuries it had given no sign of its true nature. Then, in A.D. 63, a series of strong earthquakes shook the area, and around noon on August 24, 79, Vesuvius started to erupt. The catastrophe of that August day lay silent and impenetrable for nearly 17 centuries. When the remains of Herculaneum and Pompeii were discovered—cities that had been buried by the eruption—a story evolved to grip the world's imagination. Roman sentries had been buried at their posts. Family groups, in the supposed safety of subterranean vaults, had been cast in molds of volcanic mud cemented to a rocklike hardness, along with their jewels, candelabra, and the food that they had hoped would sustain them through the emergency.

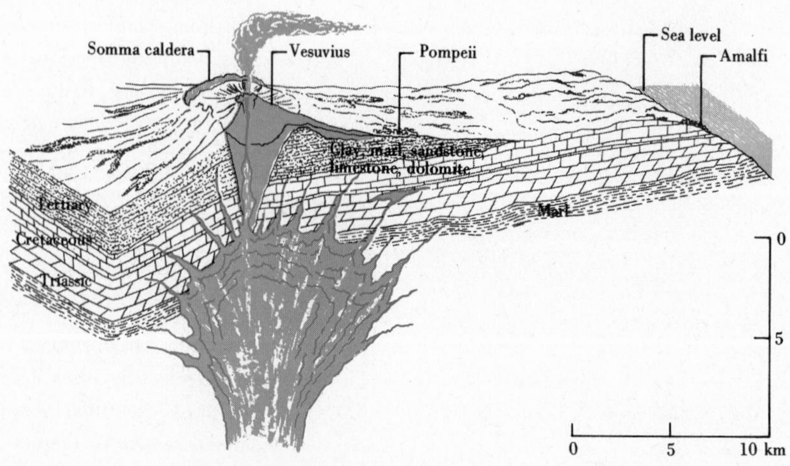

Figure 4.10 **Schematic reconstruction of supposed relationships in the Somma-Vesuvius volcano.** (*After Umbgrove,* The Pulse of the Earth, *p. 74, Martinus Nijhoff, The Hague, 1947.*)

Mt. Somma is believed to have erupted first about 10,000 years ago as a submarine volcano in the Bay of Naples. It then emerged as an island and finally filled in so much of the bay around it that it became a part of the mainland. It is the youngest volcano in that vicinity.

There must have been an exceedingly long interval of quiet before the eruption of the year 79, because no earlier historical records of volcanic activity exist. Then, in 79, part of the old Mt. Somma cone was destroyed and the new cone, Vesuvius, started. During this eruption, Pompeii was buried by pyroclastic debris; people died from asphyxiation due to gases from the ash and suffocation from the dust. Herculaneum was overwhelmed by mudflows of water-soaked ash.

Eruptions of pyroclastic debris occurred at intervals after 79. The longest period of quiescence lasted 494 years and was followed by an eruption in 1631 that poured out the first lava in historic time. This rejuvenation is believed to be attributable to chemical changes accompanying the ingestion of great quantities of limestone by magma while in a reservoir with a roof about 5 km below sea level (see Figure 4.10).

Since 1631, Vesuvius has erupted 14 times. Each eruption is the last of a series of events that repeat themselves in a cycle. Fred M. Bullard describes the cycle as follows:

The eruptive cycles vary in length, but the two latest cycles ran thirty-four and thirty-eight years respectively. The cycle begins with a repose period, averaging about seven years, in which only gases issue from the crater. The renewal of explosive activity begins with the building of small cinder cones on the crater floor. Outpourings of lava may also occur in the crater until the crater is gradually filled. Sometimes the lava flows spill over the top or issue from fissures in the crater rim, but such flows are of small volume and cause little damage. This type of moderate activity may continue for years (perhaps twenty to thirty years). When the cinder cones and the lava flows have filled the crater, the stage is set for the culminating eruption of the cycle. The column of lava now stands high in the throat of the volcano, and it is under tremendous pressure and saturated with gases. Finally, when the pressure becomes too great to be contained by the surrounding material, the eruption begins. Accompanied by sharp earthquakes and strong explosions, which give rise to great ash clouds, the cone splits. From the fractures, which frequently extend from the crater rim to the base, floods of lava pour out and flow rapidly down the side of the cone. These actions constitute the paroxysmal eruption which marks the end of a cycle. Such eruptions usually last for two or three weeks and are followed by a repose period which is the beginning of a new cycle.[2]

KRAKATOA One of the world's greatest explosive eruptions took place in 1883 at Krakatoa, in Sunda Strait between Java and Sumatra. Krakatoa had once been a single island consisting entirely of a volcanic mountain built up from the sea bottom. Then, at a remote period in the past, it split apart during an eruption. By 1883, after a long period of rebuilding, three cones had risen above sea level and had merged. These cones, named Rakata, Danan, and Perboewatan, and various unnamed shoals completed the outline of Krakatoa.

[2] Fred M. Bullard, *Volcanoes in History, in Theory, in Eruption*, p. 159, University of Texas Press, Austin, 1962.

Figure 4.11 ***A temporary island was formed at Metis Shoal of the Tonga Archipelago during a submarine eruption that began December 11, 1967. The island stayed above the surface for 58 days.*** *(From Smithsonian Institution's Center for Short-Lived Phenomena.)*

On August 26, 1883, a series of explosions began. The next day, at 10:20 A.M., a gigantic explosion blew the two cones Danan and Perboewatan to bits. A part of the island that had formerly stood 800 m high was left covered by 300 m of water. The noise of the eruption was heard on Rodrigues Island, 5,000 km across the Indian Ocean, and a wave of pressure in the air was recorded by barographs around the world. A great flood of water created by the activity drowned 36,500 persons in the low coastal villages of western Java and southern Sumatra.

Columns of ash and pumice soared kilometers into the air, and fine dust rose to such heights that it was distributed around the globe and took more than 2 years to fall. During that time, sunsets were abnormally colored all over the world. A reddish-brown circle known as Bishop's Ring, which was seen around the sun under favorable conditions, gave evidence not only of the continued presence of dust in the upper air but of the approximate size of the pieces—just under two thousandths of a millimeter. Since 1883, Krakatoa has revealed from time to time that it is actively rebuilding.

DISAPPEARING ISLANDS Submarine volcanoes like Krakatoa, which build themselves up above sea level, blow off their heads and then rebuild, produce the so-called disappearing islands of the Pacific. In 1913, for example, Falcon Island (20.4°S, 175.6°W), in the South Pacific, suddenly disappeared after an explosive eruption. On October 4, 1927, accompanied by a series of violent explosions, it just as suddenly reappeared. The island of Bogosloff (about 56°N, 168°W), in the Aleutians, was first reported in 1826 and has been playing hide-and-seek with map-makers ever since.

On December 11, 1967, a volcanic eruption started at Metis Shoal of the Tonga Archipelago. After 20 days, a new island about 700 m long and 100 m wide had poked up 20 m above the sea (see Figure 4.11). From the start of the activity, the island stayed above the surface for a total of 58 days. By February 1, it was a few "jagged rocks and water washing across." On February 19, there were "very high breakers on subsurface rocks" at the site, and by April 1 the shoal was completely under water and there were no breakers.[3]

PARÍCUTIN About 320 km west of Mexico City (19.50°N, 102.05°W) Parícutin sprang into being on February 20, 1943. Nine years later, it had become quite inactive, but during its life it was studied more closely than any other newborn vent in history.

Many stories of the volcano's first hours have been told. According to the version now generally regarded as the most reliable, Parícutin began about noon as a thin wisp of smoke rising from a cornfield that was being plowed by Dionisio Pulido. By 4 P.M., explosions were occurring every few seconds, dense clouds of ash were rising, and a cone had begun to build up. Within 5 days, the cone was 100 m high, and after 1 year it had risen to 425 m. Two days after the eruption began, the first lava flowed from a fissure in the field about 300 m north of the center of the cone. At the end of 7 weeks, this flow had advanced about 1.5 km. Fifteen weeks after the first explosion, lava had also begun to flow from the flanks of the cone itself.

After 9 years of activity, Parícutin abruptly stopped its eruptions and became just another of the many small "dead" cones in the neighborhood. The histories of these other cones, parasites of Tancítaro or of neighboring major volcanoes, undoubtedly parallel the story of Parícutin.

Other new volcanoes that have developed during historic time are Jorullo (18.85°N, 101.82°W) and Monte Nuovo (40.83°N, 14.10°E). Jurullo broke out in the middle of a plantation about 70 km southeast of Parícutin, in 1759. Monte Nuovo erupted in 1538, just west of Vesuvius.

HAWAII The Hawaiian Islands are peaks of volcanoes projecting above the ocean and strung out along a line running 2,400 km to the northwest. The Marquesas, Society, Tuamotu, Tubuai, Samoan, and other volcanic groups of the South Pacific Ocean form lines roughly parallel to the Hawaiian Islands.

At the northwestern end of the Hawaiian chain are the low Ocean and Midway islands. At the southeastern end is Hawaii, the largest of the group. Hawaii, 140 km long and 122 km wide, is the tallest deep-sea island in the world. The igneous activity that produced this group of islands apparently started at the northwestern end of the chain, where activity has now ceased, and worked southeastward to the present focus of most recent activity, on Hawaii itself.

Figure 4.12 ***Schematic drawing of the five volcanoes that have been built up from the sea floor to merge and form the island of Hawaii.***

The island of Hawaii is composed of five volcanoes—Kohala, Hualalai, Mauna Kea, Mauna Loa, and Kilauea (see Figure 4.12). Each volcano has developed independently, and each has its own geologic history. Lava from Mauna Kea has buried the southern slope of Kohala, and lava from Mauna Loa has buried parts of Mauna Kea, Haulalai, and Kilauea. The dimensions of the volcanoes

[3]Event Report, "The Submarine Volcanic Eruption and Formation of a Temporary Island at Metis Shoal, Tonga Islands," Smithsonian Institution's Center for Short-Lived Phenomena, June 15, 1968.

Table 4.1 **Volcanoes of the island of Hawaii**[a]

Name	Area, km^2	Percentage of island	Summit elevation, m
Mauna Loa	5,290	50.5	4,100
Mauna Kea	2,390	22.8	4,135
Kilauea	1,430	13.7	1,225
Hualalai	755	7.2	2,475
Kohala	610	5.8	1,650

[a] From H. T. Stearns and G. A. Macdonald, "Geology and Ground-Water Resources of the Island of Hawaii," p. 24, *Hawaii Division of Hydrography Bull. 9*, 1946.

are listed in Table 4.1. They represent portions of the volcanoes above sea level at the present time and do not take into account buried slopes.

Kohala has been extinct for many years, but Mauna Kea shows evidence of having been active in the recent geological past, though not within recorded history. Hualalai last erupted in 1801, and Mauna Loa was active about 6 percent of the time from 1832 to 1945. During the same interval, Kilauea was active about 66 percent of the time. Prior to that, extending back to 140 A.D., native legend tells of 40 to 50 eruptions of Kilauea.

Measurements made on tilt meters show that both Mauna Loa and Kilauea swell up during the period when magma is rising from below. Uplift reaches a maximum just before an eruption. After the eruption, the mountains shrink back again.

During the years 1912 to 1921, a point on the surface near the Hawaiian Volcano Observatory showed that Kilauea was slowly being elevated by amounts that totaled 0.6 m. Following the eruption of 1924, the same point had subsided 1 m by 1927.

Recent measurements on Kilauea[4] show that slow swelling over a period of months, about 16 on the average, produces a maximum uplift of 40 to 135 cm. Flank eruptions of the volcano allow sudden collapse of the summit over a period of days, amounting to 20 to 50 cm of subsidence per eruption.

The swelling extends over an area about 10 km in diameter centered on or just south of Kilauea caldera. From the area and the amount of swelling involved, it was computed that in the 1960 eruption each centimeter of uplift or subsidence involved deformation of 1 million m^3 of earth materials. Total volumes of 20 to 30 million m^3 have been computed for larger eruptions. Records of tilt can be used to forecast an impending eruption.

When Kilauea is active, magma rises up within the mountain and floods out as lava into a pit in the floor of the caldera. Occasionally, the lava flows out over the rim of the pit onto the caldera floor and gradually raises the level of the floor. Usually, however, the lava is confined to the pit, forming what is termed a *lava lake.* This lava lake may last for years and then disappear completely for equally long periods. The level of the lake falls when lava flows from the flanks of the volcano, both above and below sea level. From time to time, the system is drained and the caldera floor collapses. Then the magma rises again, lava floods into the caldera, and the process is repeated.

[4] R. W. Decker, D. P. Hill, and T. L. Wright, "Deformation Measurements on Kilauea Volcano, Hawaii," *Bull. Volcanol.*, vol. 29, pp. 721–732, 1966.

Figure 4.13 ***Profile of one of the world's most nearly perfect composite cones, Mayon on Luzon, Philippine Islands.*** *(Photograph from Gardner Collection, Harvard University.)*

Figure 4.14 ***Mt. Etna viewed from the sea near Catania, Sicily. The flat slopes to the left are those of a shield volcano. When Etna changed its eruptive habit late in its history, the explosive ejection of fragmental material built a 300-m pyroclastic cone on the summit of the broad shield of lava flows. These pyroclastics form the irregular and steepened slopes nearest the smoking vent.*** *(Photograph by Vittorio Sella.)*

Table 4.2 ***Active volcanoes of the world***[a]

Location	*Number*	
Circum-Pacific belt:		
Asia and the Southwestern Pacific		
Kamchatka	20	
Kurile Islands	33	
Japan	49	
Philippines	11	
Melanesia (New Guinea, New Britain, Admiralty Islands, Solomon Islands, Santa Cruz Islands, New Hebrides)	29	
New Zealand, Kermadec Islands, Tonga Islands, Samoa Islands	21	
	163	
North America		
Alaska and Aleutian Islands	36	
Western United States	1	
Mexico	11	
	48	
Central America		
Costa Rica	7	
Nicaragua	11	
El Salvador	7	
Guatemala	7	
	32	
South America		
Southern Andes	22	
Central Andes	8	
Northern Andes	11	
	41	284
Pacific Ocean:		
Hawaiian Islands	4	
Galapagos Islands	3	
	7	7
Antarctic:	2	2
Alpine-Himalayan belt:		
Canary Islands	3	
Mediterranean (Italy, Sicily, and Aegean Sea)	17	
Barren Island (Bay of Bengal)	1	
	21	21
Indonesia		
Sumatra	12	
Java	20	
Lesser Sunda Islands	20	
Banda Sea	8	
Celebes	6	
Sangihe Islands	5	
Moluccas	6	
	77	98
Atlantic Ocean:		
Iceland	26	
Azores	9	
Lesser Antilles	9	
St. Paul Rocks	1	
Cape Verde Islands	1	
	46	46
Indian Ocean:		
Reunion Island	1	
Heard Island	1	
	2	2
Africa:		
Ethiopia and Red Sea	6	
East Africa	7	
Central Africa	2	
West Africa	1	
	16	16
Total		455

[a] Fred M. Bullard, *Volcanoes in History, in Theory, in Eruption*, University of Texas Press, Austin, 1962.

Some observers have tried to show that this activity occurs in regular cycles, but the evidence so far has not been convincing. Apparently, the Hawaiian volcanoes behave in no regular, predictable manner.

Classification of volcanoes

Volcanoes are classified according to the materials that have accumulated around their vents. Thus we have shield volcanoes, composite volcanoes, and cinder cones.

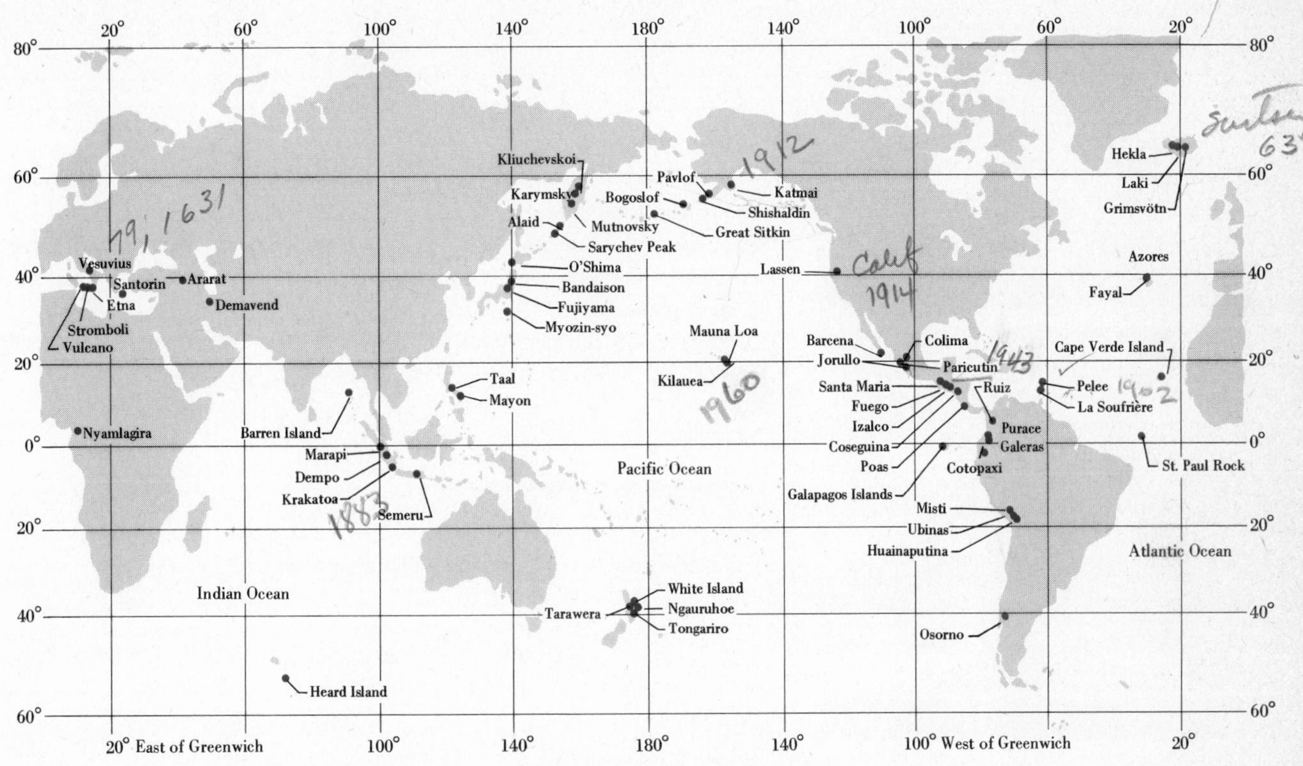

Figure 4.15 ***Location of some active volcanoes.*** (*After Fred M. Bullard,* Volcanoes in History, in Theory, in Eruption, *pp. 368–369, University of Texas Press, Austin, 1962.*)

When the extruded material consists almost exclusively of lava poured out in quiet eruptions from a central vent or from closely related fissures, a dome builds up that is much broader than it is high, with slopes seldom steeper than 10° at the summit and 2° at the base. Such a dome is called a *shield volcano.* The five volcanoes of the island of Hawaii are shield volcanoes.

Sometimes a cone is built up of a combination of pyroclastic material and lava flows around the vent. This form is called a *composite volcano* and is characterized by slopes of close to 30° at the summit, tapering off to 5° near the base. Mayon, on Luzon in the Philippines, is one of the finest examples of a composite cone (see Figure 4.13).

A single volcano may develop as a shield volcano during part of its history and as a composite volcano later. Mt. Etna is an example of such a volcano (see Figure 4.14).

Finally, small cones consisting mostly of pyroclastic debris, particularly cinders, are called *cinder cones.* They achieve slopes of 30° to 40° and seldom exceed 500 m in height. Many cinder cones have flows of basalt issuing from their base (see Figure 4.4). Parícutin, in Mexico, is an example of a cinder cone that has developed in modern times.

(a)

(b)

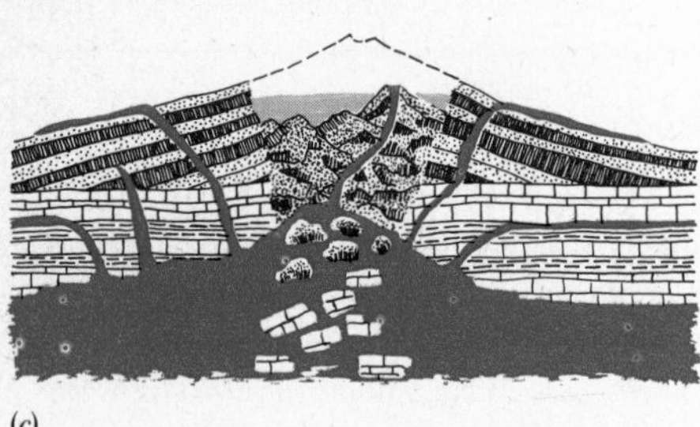

(c)

Figure 4.16 ***Sequence of events proposed by one hypothesis for the formation of a caldera. (a) An eruption begins with fiery clouds and dust clouds distributing materials on slopes and surrounding country. (b) Eruption continues. Part of cone is blown away, and lava flows join in draining the magma reservoir. (c) Most of cone collapses into the reservoir; later activity forms cinder cone in caldera.*** (*After H. Williams, "Calderas and Their Origin,"* Univ. Calif. Bull. Dept. Geol. Sci., *vol. 25, pp. 239–346, 1941.*)

Distribution of active volcanoes

We find evidence of volcanic eruptions in rocks of all ages. Apparently igneous activity has been going on throughout geologic time. Seemingly, no special geographic environment has particularly fostered such activity, which has occurred in the highest mountain ranges, on the bottom of the ocean, and on open plains.

There are 455 active volcanoes in the world today (see Figure 4.15). Their locations are listed in Table 4.2.

Formation of calderas

Calderas may be formed by explosion, by collapse, or by a combination of both (see Figure 4.16). It is often difficult to determine just which mechanism is responsible.

The caldera on Bandai (37.58°N, 140.05°E), on the island of Honshu, Japan, is one example of a caldera formed by explosion. After 1,000 years of dormancy, Bandai exploded on July 15, 1888, blowing off its summit and part of its northern slope. After the violent explosion had subsided, a caldera was discovered over $1\frac{1}{2}$ km in diameter, with walls 360 m high.

The caldera on Kilauea was probably formed by the collapse of the summit rather than by explosion. As great quantities of magma escaped from the reservoir beneath the volcano, support for the summit was withdrawn, and large blocks of it fell in, forming the caldera.

Crater Lake, in southern Oregon, lies in a basin that is an almost perfect example of the caldera shape. This is circular, with a diameter of a little more than 8.5 km, a maximum depth of 1,200 m, and surrounded by a cliff that rises 750 to 1,200 m. Crater Lake itself is about 600 m deep. The caldera was formed when the top of a symmetrical volcanic cone, Mt. Mazama, vanished during an eruption. Geologists have studied the deposits on the slopes and tried to piece together its history.

First, a composite cone was slowly built to a height of around 3,600 m. Then glaciers formed, moving down from the crest and grooving the slopes as they traveled. An explosive eruption occurred 6,600 years ago and the caldera was formed. Later activity built up a small cone inside the caldera, which protrudes above the surface of Crater Lake as Wizard Island (see Figure 4.17).

Not all observers agree on the origin of the Crater Lake caldera itself, however. The question is whether all or nearly all the missing material from the cone was actually blown out during an eruption or whether the caldera was created when the summit collapsed. The answer to the mystery should be provided by an analysis of the unconsolidated material found in the vicinity. Does this material consist of pyroclastics formed during an eruption, or does it consist of the broken remnants of Mt. Mazama's blown-off summit? The

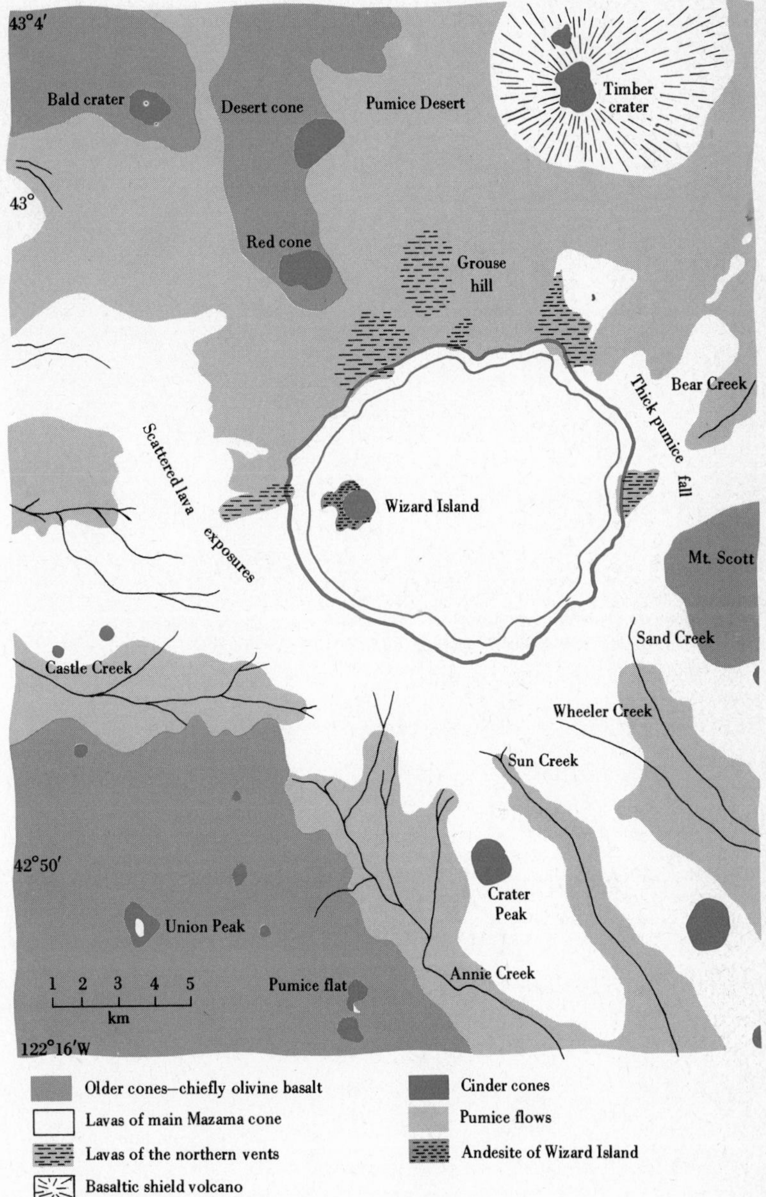

Figure 4.17 **Crater Lake, Oregon, with Wizard Island.**

problem is that the summit itself originally included pyroclastics and congealed lavas from earlier eruptions, and it is difficult to distinguish between the two. One investigator[5] has concluded that of the 70 km^3 of Mt. Mazama that disappeared, only 8 km^3 are represented in the materials now lying on the

[5]H. Williams, "Calderas and Their Origin," *Bull. Univ. Calif. Dept. Geol. Sci.*, vol. 25, pp. 239–346, 1941.

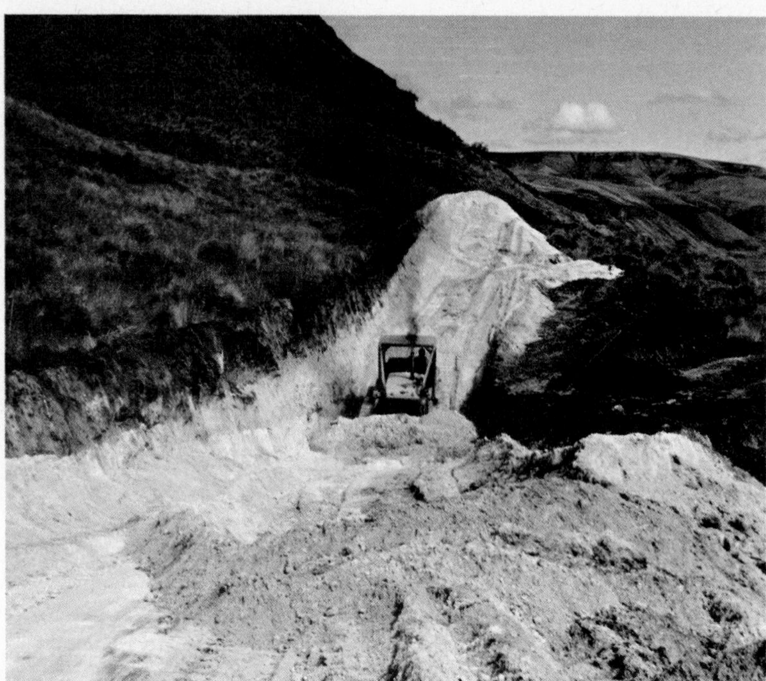

Figure 4.18 ***Pyroclastics from Mt. Mazama at a distance of 600 km.*** *(Photograph by U.S. Army Engineers.)*

immediate slopes, and that the rest of it was dropped into the volcano when the roof of an underlying chamber collapsed. This chamber may have been partially emptied by the ejection of large volumes of material during an eruption. H. Williams found evidence that ash spread over a radius of nearly 50 km. Others have reported Mt. Mazama ash as far east and northeast as Alberta and eastern Montana (see Figure 4.18), and some of the magma may have worked its way beneath the surface into adjoining areas. This explanation of the Crater Lake caldera has been challenged on the grounds that it is based on an invalid distinction between "old" and "new" pyroclastics in the debris that covers the area.

Evolution of volcanoes

Many volcanoes started as submarine volcanoes. Later in their history they became islands and eventually were surrounded by land masses. First in their eruptive cycles were basaltic lava flows, then andesitic lava flows, and then for thousands of years they produced only fragmental materials which built up their cones. In their final stages, a cone-disrupting explosion occurred with ejection of only fragmental materials. The cycle ends when acidic lava congeals in the volcanic vent.

First eruptions consist of basis low-viscosity lava which flows readily. With succeeding eruptions, lava becomes more acidic and more viscous. Eventually, lava becomes so viscous that only gases escape, causing eruptions of fragmental pyroclastic debris.

4.2 *Basalt plateaus*

On June 11, 1783, after a series of violent earthquakes on Iceland near Mt. Skapta, an immense outpouring of lava began along a 16-km line, the Laki Fissure. Lava poured into the Skapta River, drying up the water and overflowing the stream's channel, which was 150 to 200 m deep and 60 m wide in places. Soon the Skapta's tributaries were dammed up, and many villages in adjoining areas were flooded. The lava flow was followed by another 1 week later and a third on August 3. So great was their volume that they filled a former lake and an abyss at the foot of a waterfall. They spread out in great tongues 20 to 25 km wide and 30 m deep. As the lava flow diminished and the Laki Fissure began to choke up, 22 small cones formed along its length, relieving the waning pressures and serving as outlets for the final extrusion of debris.

This is the only authenticated instance within historical time of the mechanism known as *fissure eruption,* or *lava flood.* There is strong evidence, however, that floods of this sort occurred on a gigantic scale in the geological past. The rocks produced by lava floods are known as *flood basalts,* or *plateau basalts,* because of the tendency of the process to form great plateaus. The low viscosity that is required to make it possible for lava to flood freely over such great areas is characteristic only of lavas that have basaltic composition.

Iceland itself is actually a remnant of extensive lava floods that have been going on for over 50 million years and that have blanketed $\frac{1}{2}$ million km^2. The congealed lava is believed to be at least 2,700 m thick in this area. The Antrim Plateau of northeastern Ireland, the Inner Hebrides, the Faeroes, and southern Greenland are also remnants of this great North Atlantic, or Brito-arctic, plateau.

Of equal magnitude is the Columbia Plateau in Washington, Oregon, Idaho, and northeastern California (see Figure 4.19). In some sections, more than 1,500 m of rock have been built up by a series of fissure eruptions. Individual eruptions deposited layers ranging from 3 to 30 m thick, with an occasional exception of greater thickness. In the canyon of the Snake River, Idaho, granite hills from 600 to 750 m high are covered by 300 to 450 m of basalt from these flows. The Columbia Plateau has been built up during the past 30 million years. The principal activity took place a million years ago in northeastern California and Idaho, but some flows in the Crater of the Moon National Monument in southern Idaho, probably the most recent of U.S. fissure eruptions, are believed to have occurred within the last 250 to 1,000 years.

Other extensive areas built up by fissure eruptions include north-central Siberia, the Deccan Plateau of India, Ethiopia, around Victoria Falls on the Zambezi River in Africa, and parts of Australia.

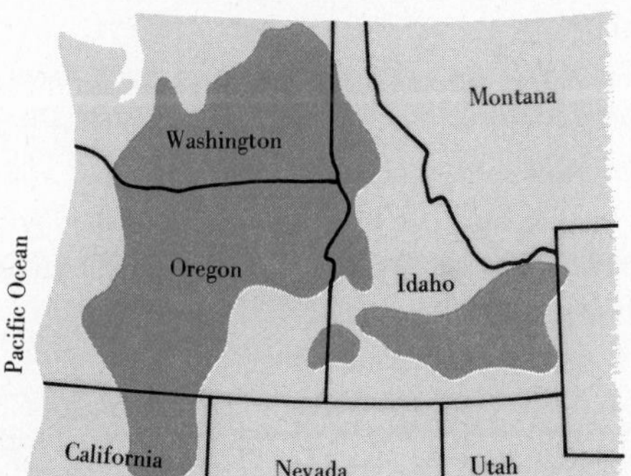

Figure 4.19 ***Map of Columbia Plateau, showing areas built by fissure eruptions.***

4.3 Igneous activity and earthquakes

Most volcanic eruptions are associated with earthquakes. In recent years, records of this seismic activity have given warnings of impending eruptions. The relationship between local earthquakes and eruptions is important enough to be described in detail in the following case.[6]

Raoul Island, a volcanic island, is 8 km across and is the largest of the Kermadec group. It is located approximately 1,000 km northeast of New Zealand. It rises from the Kermadec Ridge, just west of the Kermadec Trench, and lies in a very active zone of earthquake activity extending from Tonga Island to New Zealand. The only settlement on the island is a meteorological station maintained by the New Zealand Department of Civil Aviation and inhabited by nine men.

A seismograph for recording earthquakes was installed on the island in 1957 by the New Zealand Seismological Observatory and was maintained by the men of the meteorological station. The seismograph recorded earthquake activity but it could not be used by itself to determine the locations of earthquakes.

The first sign of unusual seismic activity began on November 10, 1964, when the first of a swarm of local earthquakes were recorded. In the 10 weeks before this, there had been only 1 earthquake. Then, within 4 hr, over 80 earthquakes/hr were being recorded. On November 11, earthquakes became less frequent, but the ground began to shake continuously, a phenomenon called *volcanic tremor.* Within 1 day, the volcanic tremors were large enough to mask records of small individual earthquakes. By November 13, tremor dropped off to about one-half its maximum intensity, and earthquakes were again being recorded at about 30 to 40/hr. At 21 hr 57 min Greenwich time, the largest earthquake of the series occurred. Raoul's seismograph was temporarily out of order, but this earthquake was large enough to be recorded on other seismographs, some as far away as North America. The U.S. Coast and Geodetic Survey determined it to be a moderate-sized earthquake and located its center at 20 km west of the island at a depth of 77 km. Volcanic tremor and frequency of earthquakes decreased then, until on November 20 only 10 to 15/hr were being recorded.

Suddenly, on November 20, just before 18 hr Greenwich time, the amplitude of volcanic tremor increased to 7 μm and there was a "sound of a big landslide" accompanied by a cloud of steam, followed by "another roaring noise" which heralded the appearance of a great column of black mud which shot up to 1,000 m in the center of the cloud, with rocks flying out of the column and falling back into the crater. During the eruption, a new crater 100 m in diameter was blown out within the main crater. Within 1 hr, the level of volcanic tremor settled down to about $\frac{1}{2}$ μm, with about 10 earthquakes/hr.

[6] R. D. Adams and R. R. Dibble, "Seismological Studies of the Raoul Island Eruption, 1964," *New Zealand J. Geol. Geophys.*, vol. 10, no. 6, Dec. 1967.

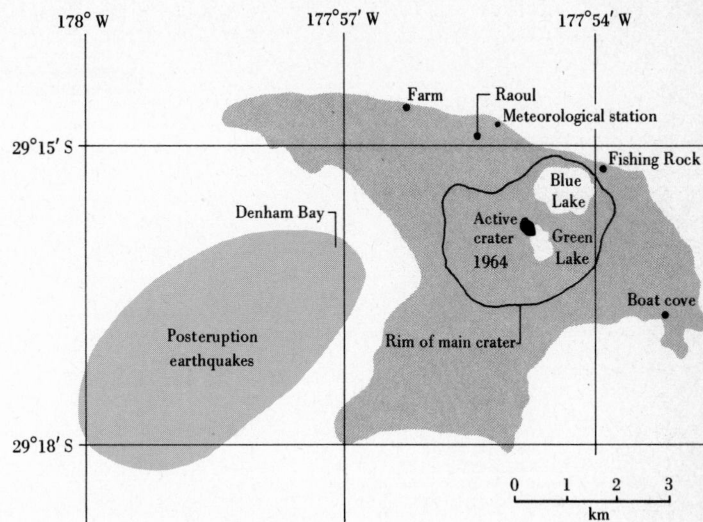

Figure 4.20 ***Map of Raoul Island, showing positions of seismographic stations and location of posteruption earthquakes.*** (*From R. D. Adams and R. R. Dibble, "Raoul Island Eruption, 1964,"* New Zealand J. Geol. Geophys., *vol. 10, no. 6., December 1967.*)

From November 23 to December 6, the island was evacuated. When the meteorological team returned, they were accompanied by 10 men from the New Zealand Department of Scientific and Industrial Research who brought along three portable seismographs. These, and Raoul's seismograph, were put in operation at once, and they showed that the frequency of earthquakes had dropped to between 2 and 4/hr. Thirty-two of the most clearly recorded posteruption earthquakes were to the west and southwest of the crater, beneath Denham Bay (see Figure 4.20). Some were at depths of less than 4 km and others at depths of 8 km. On December 12, the 10 men left the island with the portable seismographs. Since then, the meteorological staff has continued to operate the Raoul seismograph, but no renewal of unusual local activity has been reported.

The Raoul eruption followed the not unusual pattern of occurring when the associated earthquake activity was on the wane after an extremely sudden rise. Seismographs are being watched carefully in the hope they will forewarn of any future outbursts of volcanic activity.

The Arenal eruption in Costa Rica on July 29, 1968 (see Figure 4.7) was preceded by 10 hr of seismic activity. The warning was not recognized, however, because Arenal had not erupted since 1530. The preeruption earthquakes were estimated to represent about 10^{18} ergs of energy. The kinetic energy of the first great explosion was estimated at 4×10^{21} ergs.

Depth of magma source

Magma supplying eruptive materials of active volcanoes comes from deep within the earth. The depths of earthquakes and earth tremors associated with eruptions are giving us some clues to the depths of the magma.

Swarms of earthquakes and "harmonic tremor" with foci 60 km below the

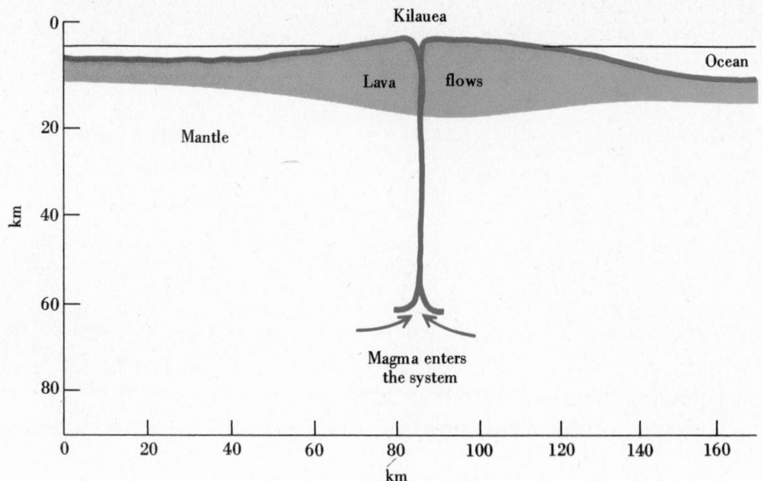

Figure 4.21 ***Schematic cross section of Kilauea.***

summit of Kilauea preceded the 1959 eruptions, and Eaton and Murata[7] suggest that 60 km may be the depth at which magma enters the "plumbing system" of Kilauea (see Figure 4.21).

The 1964 eruption on Raoul Island was preceded by an earthquake centered 20 km west of the island at a depth of 77 km. Later quakes were nearer and shallower, but there is a strong suggestion that the chain of events really started with magma at a depth of 77 km moving into the volcano's feeding conduit.

4.4 *The earth's heat*

The principal requirement for igneous activity is heat, and evidence has accumulated that there is a great deal of heat in the earth's interior. The earth's temperature increases with depth. This was first noted as early as the seventeenth century, when German miners were reporting a steady rise in temperature with increasing depth in their mine shafts. Further proof has now come from the temperature measurements made in other mines and from deep drill holes in the earth.

The increase in the earth's temperature with depth is called the *thermal gradient.* Reported rates of increase have varied from place to place even in areas far removed from igneous activity. The thermal gradient can range from less than 10°C/km to as much as 50°C/km. Excluding seasonal temperature changes that affect only the top few meters of the earth's surface, a thermal gradient of 30°C/km seems to be about average. These figures get

[7] J. P. Eaton and K. J. Murata, "How Volcanoes Grow," *Science*, vol. 132, pp. 925–938, 1960.

us "into hot water," so to speak, very quickly. Within 50 km of depth, a thermal gradient of 30°C/km would result in a temperature of 1500°C, believed to be high enough to melt any known rock, and at the earth's center, it would produce a temperature of 192,000°C, more than 40 times the temperature of the sun's atmosphere.

Heat travels extremely slowly through soil and rock. Measurements and calculations have shown that the earth's *heat flow,* which is the product of the temperature gradient and the thermal conductivity of earth materials, is reasonably constant. Its average value is 1.2 μcal (microcalories)/cm^2-sec, 1.2×10^{-6} cal/cm^2-sec. Present indications are that there is little difference in the heat flow from a number of different land and sea areas. The average heat flow represents about 40 cal/cm^2-year, which does not seem like much. But in its steady, quiet way it totals more energy in a year than all that released by seismic and igneous activity.

Speculations on the origin of the earth's heat

Most hypotheses of the earth's origin agree that it passed through a molten stage at some time in its early history before separating into its different layers. It was first supposed that the observed heat flow was due entirely to the original heat which remained after the molten earth solidified. Later it was suggested that no more than 20 percent of the observed heat flow comes from the original heat of the earth.

It has also been suggested that the rest of the heat must be caused by the radioactive decay of elements in the earth's crust. Calculations have indicated that the total amount of heat flowing out of the earth could be generated in a layer of granite 30 km thick, so the conclusion was drawn that the thermal gradient within the mantle must be quite low, as otherwise the flow of heat to the surface would be greater than the observed amount. However, this speculation also presents a problem in that radioactive elements are concentrated in the rocks of the continents but are absent in the rocks under the oceans. Because of this absence from the oceanic rocks, it might be expected that heat flow through the ocean floor would be considerably less than that observed through the continents. Measurements do not support this idea. Average heat flow is about the same whether through the continents or through the ocean floor.

Cause of igneous activity

Igneous activity requires magma, and magma requires that rocks be melted. Where does this melted rock come from? Earthquake records indicate that magma comes from the mantle. However, it is generally assumed that the mantle is composed of crystalline rock. This makes it necessary to explain the formation of local pockets of magma, because if it is solid the general internal temperatures within this zone must be below the melting point of rocks.

Geologists have had difficulty trying to explain the cause of igneous activity because they misinterpret what seismology tells them. They believe that the earth's mantle is solid—solid like the rocks on the surface—because it transmits the *S* waves of seismology. This is not necessarily a correct deduction. The material in the mantle "acts like a solid" in its transmission of *S* waves, but a material does not have to be crystalline to transmit these waves. A nonsolid but rigid material can also transmit them. All *S* waves require for transmission is rigidity, a resistance to a change in shape. This does not require that the material through which they are transmitted be crystalline. Elements in the mantle could be compressed to such a degree that their mobility is reduced and they cannot move about to unite with each other as in the solid state. These compressed elements could be rigid enough to transmit *S* waves.

Resorting to radioactivity as the source of the earth's heat seems forced and unnatural in the light of relatively unlimited supplies of primeval heat that could have been left over from the earth's formation. Radioactivity was principally a means for trying to explain local pockets of magma, yet radioactive elements have not been extruded at volcanic vents in the ocean basins. If we assume that mantle materials are not solid, they could be at an indefinitely high temperature.[8]

The ultimate causes of igneous activity are the same internal forces that elevate mountains, cause earthquakes, and cause metamorphism. Volcanic outpourings of lava and gases originate near the mantle's upper boundary. Hot magma squeezes upward and comes out through fissures in the earth's crust. It may remain for long periods in shallow chambers within the crust, or it may erupt at once. But science has not yet discovered just what triggers an eruption.

Outline

Igneous activity consists of movements of molten rock inside and outside the earth and the variety of effects associated with these movements.

- *Volcanoes* are surface piles of material that accumulated around vents during successive eruptions.
- A *volcanic eruption* is started and maintained by gases, of which steam is the most important.
- *Composition of lavas* has been designated by their proportions of silica as acid, intermediate, and basic.
- *Temperature of lavas* have been measured from 750 to 1175°C.
- *Volcanic gases* are two-thirds steam but include carbon monoxide and dioxide, nitrogen, sulfur, hydrogen, fluorine, and chlorine.
- *Pyroclastic debris* consists of fragments blown out by explosive eruptions and includes ash, cinders, lapilli, blocks, bombs, and pumice.
  - *Fiery clouds* are ash mixed with gases, are heavier than air, and roll down a volcano's side.
  - *Worldwide effects* include dry fogs in the stratosphere and reduction of solar heat reaching the earth's surface.

[8]Florence J. Leet and L. Don Leet, "The Earth's Mantle," *Bull. Seis. Soc. Am.*, vol. 55, pp. 619–625, 1965.

- *History of some volcanoes* shows they may be dormant for long periods and again become active.
  - *Vesuvius* was dormant for centuries and then in 79 A.D. started to erupt again.
  - *Krakatoa* erupted in 1883 with one of the world's greatest explosions.
  - *Disappearing islands* are submarine volcanoes that blow their heads off.
  - *Parícutin* sprang into being on February 20, 1943, and was active for 9 years.
  - *Hawaiian Islands* are peaks of volcanoes projecting above sea level and strung out along a line running 2,400 km to the northwest.
- *Classification of volcanoes* according to the materials that have accumulated around their vents designates them as shield, composite, and cinder cones.
- *Distribution of active volcanoes* locates 284 in the circum-Pacific belt, 98 in the Alpine–Himalayan belt, 7 in the Pacific Ocean, 46 in the Atlantic Ocean, 2 in the Indian Ocean, 16 in Africa, and 2 in the Antarctic, for a total of 455.
- *Formation of calderas* may be by explosion, by collapse, or by a combination of both.
- *Evolution of volcanoes* starts with basaltic lava flows, then andesitic lava flows, and then ejection of fragmental materials; the cycle ends when acidic lava congeals in the volcanic vent.

Basalt Plateaus are built by fissure eruptions.

Igneous activity and earthquakes seem to be closely related.

- *Depth of magma source* suggested by associated earthquakes may be 60 km at Kilauea and 77 km at Raoul Island.

The earth's heat is indicated by the thermal gradient to be more than adequate for igneous activity.

- *Speculations on the origin of the earth's heat* have included original heat and radioactivity.
- *Cause of igneous activity* seems to be the same internal force that elevates mountains, causes earthquakes, and causes metamorphism, but it is not known.

Supplementary readings

BULLARD, FRED M.: *Volcanoes in History, in Theory, in Eruption,* University of Texas Press, Austin, 1962. An authoritative treatment of all aspects of volcanic activity, extremely well organized and written.

COATS, ROBERT R., RICHARD L. HAY, and CHARLES A. ANDERSON (eds.): "Studies in Volcanology—A Memoir in Honor of Howell Williams," *Geol. Soc. Am. Mem.*, vol. 116, 1968. Seventeen monographs on various volcanological topics.

EPIS, RUDY C. (ed.): "Cenozoic Volcanism in the Southern Rocky Mountains," *Quart. Colorado School Mines,* vol. 63, July 1968. Fifteen papers on the subject, including data on eruptive centers not previously recognized and evidence for extensive Oligocene activity.

HERBERT, DON, and FULVIO BARDOSSI: *Kilauea: Case History of a Volcano,* Harper & Row, Publishers, New York, 1968. Layman's account of activity at Kilauea, with geological background and descriptions of measurements being made.

5

Igneous rocks

IGNEOUS ROCKS ARE FORMED FROM THE SOLIDIFICATION OF MAGMA. FROM TIME TO time, molten rock within the earth's crust works its way out to the surface. There it may pour or be blown forth from vents or fissures and solidify into rock, or it may remain trapped within the crust, where it slowly cools and solidifies. Ninety-five percent of the outermost 10 km of the globe is composed of rocks of igneous origin.

5.1 *Masses of igneous rocks*

During our discussion of igneous activity, we dealt primarily with the extrusion of lava and pyroclastic debris and with some of the landforms that result: basalt plateaus and volcanoes. These are surface masses of igneous rock.

When magma *within the crust* loses its mobility and ceases activity, it solidifies in place, forming igneous rock masses of varying shapes and sizes. Today, such rocks can be seen at the surface on continents where previously overlying rocks have been worn away by erosion.

The first internal part of a volcano to be exposed by erosion is the plug that formed when magma solidified in the vent. Revealed next are the channels through which the magma moved to the surface. Finally, in some regions of ancient activity, the crust has been so elevated and eroded that the reservoir that once stored the magma can now be seen as a solid rock mass at the surface.

As uplift (see Chapter 16) and erosion expose an extinct volcano's internal construction to view, igneous rock masses can be seen. Solidified offshoots of magma from the reservoir, having intruded themselves into other rocks within the crust, are included in these masses, which were not necessarily connected with the eruption of a volcano.

Plutons

All igneous rock masses that were formed when magma solidified within the earth's crust are called *plutons.* When rocks have a definite layering, we may speak of the magma that invades them as *concordant* if its boundaries are parallel to the layering or *discordant* if its boundaries cut across the layering.

Plutons are classified according to their size, shape, and relationship to surrounding rocks. They include sills, dikes, lopoliths, laccoliths, and batholiths (see Figure 5.1).

TABULAR PLUTONS A pluton with a thickness that is small relative to its other dimensions is called a *tabular* pluton.

Sills A tabular concordant pluton is called a *sill.* It may be horizontal, inclined, or vertical, depending on the attitude of the rock structure with which it is concordant.

Sills range in size from sheets less than 1 cm in thickness to tabular masses hundreds of meters thick. A sill must not be confused with an ordinary lava

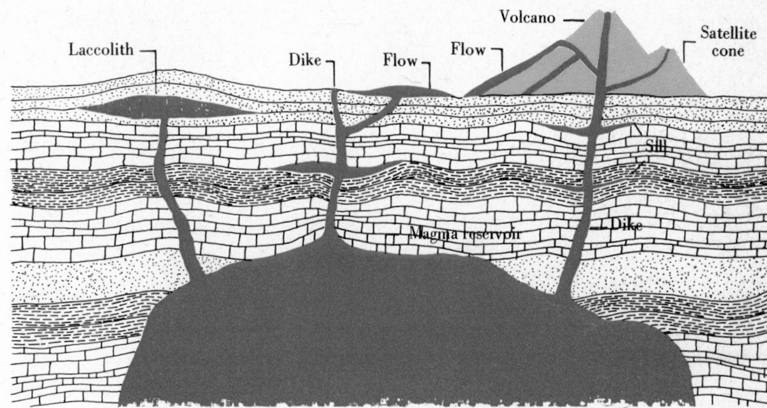

Figure 5.1 **Plutons and landforms associated with igneous activity.**

flow that has been buried by other rocks later on. Because a sill is an intrusive form—that is, it has forced its way into already existing rocks—it is always younger than the rocks that surround it. There are fairly reliable ways of distinguishing between the two types: A buried lava flow usually has a rolling or wavy-shaped top pocked by the scars of vanished gas bubbles and showing evidence of erosion, whereas a sill has a more even and unweathered surface. Also, a sill may contain fragments of rock that were broken off when the magma forced its way into the surrounding structures.

The Palisades along the west bank of the Hudson River near New York are the remnants of a sill that was several hundred meters thick. Here the magma was originally intruded into flat-lying sedimentary rocks. These are now inclined at a low angle toward the west.

Dikes A tabular discordant pluton is called a *dike* (see Figure 5.2). Dikes originated when magma forced its way through the fractures of adjacent rocks.

The width of individual dikes ranges from a few centimeters to many meters. The Medford dike near Boston, Massachusetts, is 150 m wide in places. Just how far we can trace the course of a dike across the countryside depends in part on how much of it has been exposed by erosion. In Iceland, dikes 15 km long are common and many can be traced for 50 km; at least one is known to be 100 km long.

As magma forces its way upward, it sometimes pushes out a cylindrical section of the crust. Today, as a result, we find exposed at the surface some roughly circular or elliptical masses of rock that outline the cylindrical sections of the crust. These solidified bodies of magma are called *ring dikes.* Large ring dikes may be many kilometers around and hundreds or thousands of meters deep. Ring dikes have been mapped with widths of 500 to 1,200 m and diameters ranging from 2 to 25 km.

Some dikes occur in concentric sets. These originated in fractures that outline an inverted cone, with the apex pointing down into the former magma source. These dikes are called *cone sheets.* In Scotland, the dip of certain cone sheets suggests an apex approximately 5 km below the present surface of the earth. Dikes are also found in approximately parallel groups called *dike swarms.*

Lopoliths Lopoliths are tabular concordant plutons shaped like a spoon, with both the roof and the floor sagging downward. A well-known example

Figure 5.2 **Basalt dike cutting through granite at Cohasset, Massachusetts.** (*Photograph by John A. Shimer.*)

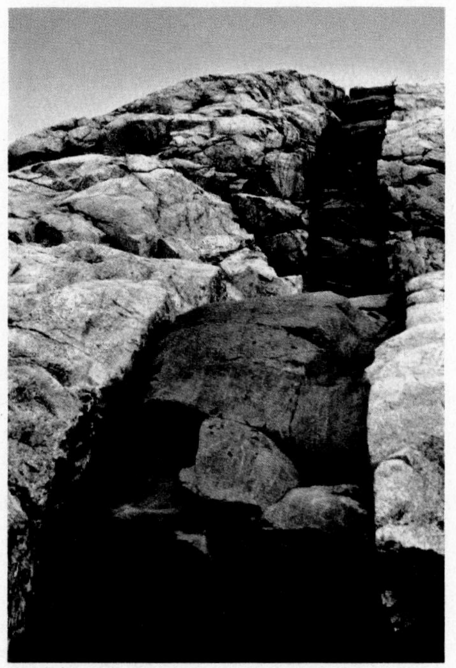

is the Duluth lopolith which outcrops on both sides of Lake Superior's western end and appears to continue beneath the lake. It has been computed to be 250 km across and 15 km deep, with a volume of 200,000 km^3.

Most lopoliths are composed of rock which has been differentiated into alternating layers of dark and light minerals, giving them the appearance of thinly bedded sedimentary rock.

MASSIVE PLUTONS Any pluton that is not tabular in shape is classified as a *massive pluton.*

Laccoliths A massive concordant pluton that was created when magma pushed up the overlying rock structures into a dome is called a *laccolith* (from the Greek *lakkos,* "a cistern," and *lithos,* "stone") (see Figure 5.3). If the ratio of the lateral extent of a pluton to its thickness is less than 10, the pluton is arbitrarily classed as a laccolith; if this ratio is more than 10, the pluton is classed as a sill. Obviously, because it is extremely difficult to establish the lateral limits of a pluton, in many cases it is best to use the term *concordant pluton,* supplemented by whatever dimensional details we can observe.

A classic development of laccoliths is found in the Henry, La Sal, and Abajo Mountains of southeastern Utah, where their features are exposed on the Colorado Plateau, a famous geological showplace.

Batholiths A large discordant pluton that increases in size as it extends downward is called a *batholith.* The term *large* in this connection is generally taken to mean a surface exposure of more than 100 km^2. A pluton that has a smaller surface exposure but that exhibits the other features of a batholith is called a *stock* (see Figure 5.4).

Batholiths are exposed thousands of meters above sea level, where they have been lifted by forces operating in the earth's crust. Thousands of meters of rock that covered the batholiths have been stripped away by the erosion of millions of years. We can observe these roots of mountain ranges in the White Mountains of New Hampshire and in the Sierra Nevada (see Figure 5.5).

Although batholiths provide us with some valuable data, they also raise a host of unsolved problems. All these problems bear directly on our understanding of igneous processes and the complex events that accompany the folding, rupture, and eventual elevation of sediments to form mountains. (We shall discuss these mountain-forming processes in Chapter 19).

We can summarize what we *do* know about batholiths as follows:

1 Batholiths are located in mountain ranges. Although in some mountain ranges no batholiths are exposed at the present time, we never find batholiths that are not associated with mountain ranges. In any given mountain range, the number and size of the batholiths seem to be directly related to the intensity of the folding and crumpling that have taken place. This does not mean, however, that the batholiths caused the folding and crumpling. Actually, there is convincing evidence to the contrary, as we shall see in some of the following features.

2 Batholiths usually run parallel to the axes of mountain ranges.

3 Batholiths have been intruded across the folds, indicating they were formed after the folding of the mountains, although the folding may have continued after the batholiths were formed.

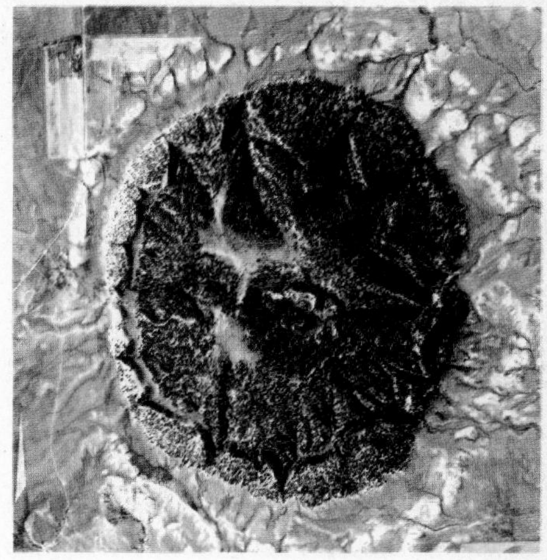

Figure 5.3 **Laccolith with igneous core still covered. Green Mountain Dome, Sundance Quadrangle, Crook County, Wyoming.** (*Photograph from Kirk Bryan Library of Geomorphology, Harvard University.*)

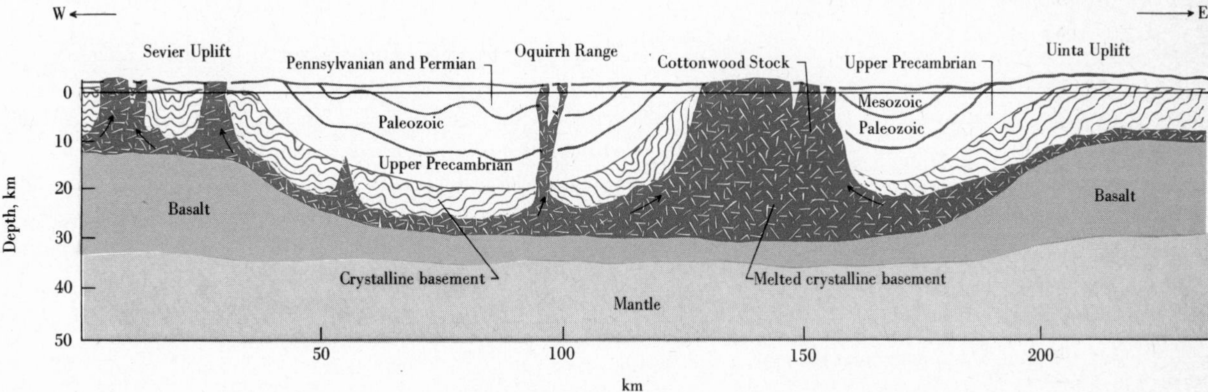

Figure 5.4 ***A. J. Eardley's schematic reconstruction of uplifts before block faulting in the Oquirrh-Uinta region of Utah showing the intrusion of a stock of silicic rock as the cause of the Cottonwood uplift.***

Figure 5.5 ***Weathering has exposed this portion of the Sierra Nevada batholith in Yosemite National Park, California.*** (*Photograph by W. C. Bradley.*)

Figure 5.6 ***Xenolith in Mt. Airy granite, North Carolina, which provides evidence in support of magmatic origin of this granite because the inclusion could not have retained its sharp edges and separate identity during granitization.*** *(Photograph by T. M. Gathright II.)*

4 Batholiths have irregular dome-shaped roofs. This characteristic shape is related to *stoping,* one of the mechanisms by which magma moves upward into the crust. As the magma moves upward, blocks of rock are broken off from the structures into which it is intruding. At low levels, when the magma is still very hot, the stoped blocks may be melted and assimilated by the magma reservoir. Higher in the crust, as the magma approaches stability and its heat has reduced, the stoped blocks are frozen in the intrusion as *xenoliths,* that is, "strange rocks" (see Figure 5.6).

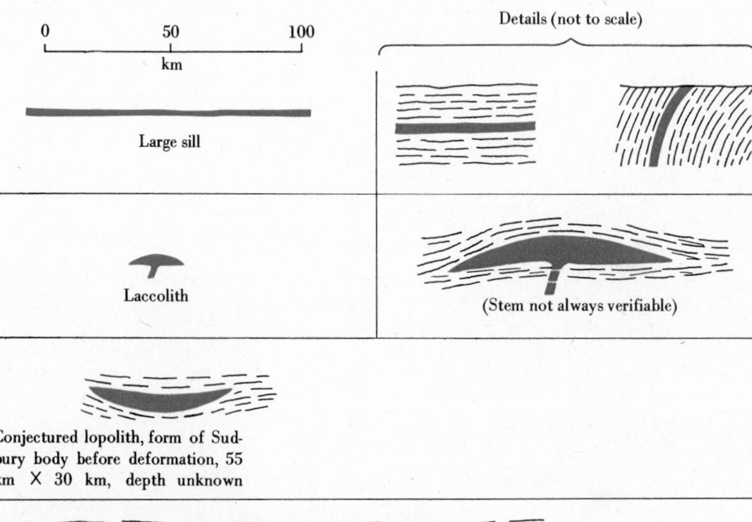

Figure 5.7 ***Concordant plutons.***

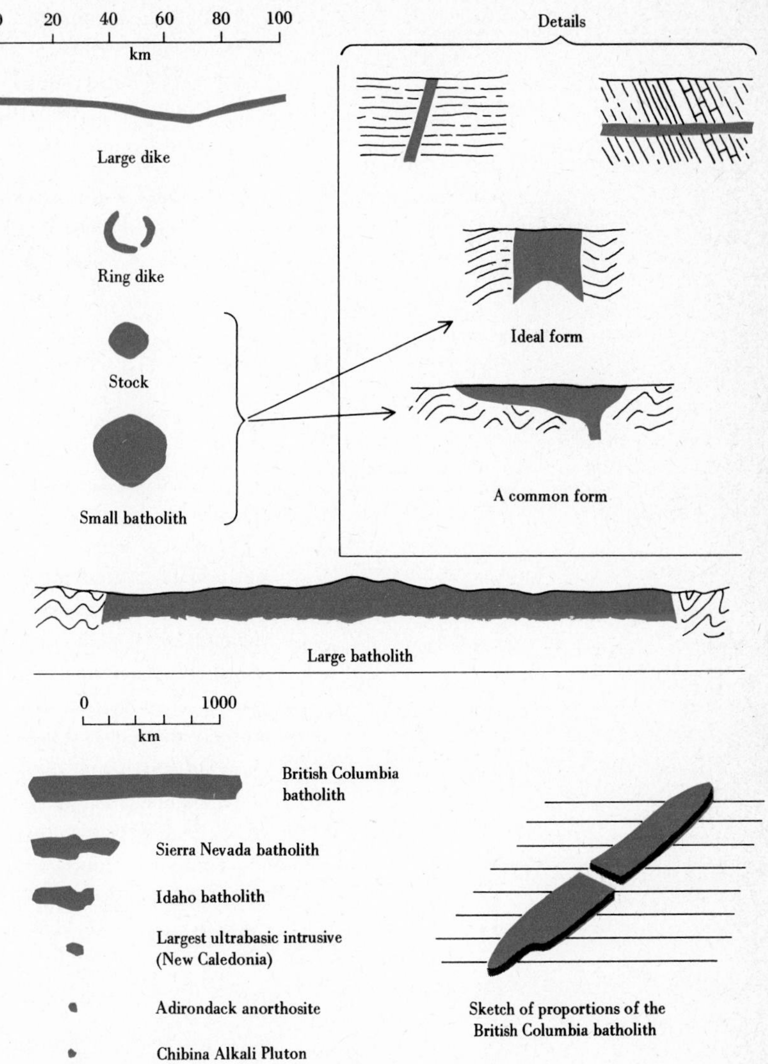

Figure 5.8 ***Discordant plutons.***

5 Batholiths are primarily composed of granite or granodiorite.

6 Batholiths give the impression of having replaced the rocks into which they have intruded, instead of having pushed them aside or upward. But if that is what really took place, what happened to the great volumes of rock that the batholiths appear to have replaced? Here we come up against the problem of the origins of batholiths—in fact, against the whole mystery of igneous activity. Some observers have been led to question even whether granitic batholiths were formed from true magmas at all. The suggestion has been made that the batholiths may have been formed through a process called *granitization,* in which solutions from magmas move into solid rocks, exchange ions with them, and convert them into rocks that have the characteristics of granite but have never actually existed as magma. We shall return to this highly controversial proposal in Chapter 8.

7 Batholiths contain a great volume of rock. The Sierra Nevada batholith of California is 650 by 60 to 100 km, and a partially exposed batholith in southern California and Baja California is probably 1,600 by 100 km.

8 Gravity measurements have been interpreted as indicating that the downward extent or "thickness" of many batholiths is some 10 to 15 km.[1]

Figures 5.7 and 5.8 summarize the conventional representation of the concordant and discordant plutons.

5.2 Formation of igneous rocks

Igneous rocks at the surface today were formed from magma. As we pointed out in Chapter 4, molten rock in the ground is called magma. When magma is extruded on the surface, it is called lava. And when solidified pieces of magma are blown out they are pyroclastic debris.

Pyroclastic debris eventually becomes hardened into rock through the percolation of ground water. In one sense, rocks formed in this way could be classified as sedimentary; but because they consist of solidified pieces of magma, we shall include them in our discussion of igneous rocks. Volcanic ash that has hardened into rock is called *tuff.* If many relatively large angular blocks of congealed lava are imbedded in a mass of ash and then hardened to rock, the rock is called *volcanic breccia.* If such included pieces are mainly rounded fragments, the rock is called *volcanic conglomerate.*

Magma, extruded as lava at the surface, cools and solidifies to form igneous rocks. The offshoots of magma that work their way into surrounding rock below the surface cool more slowly and solidify. Even the magma reservoir eventually cools and solidifies, but it takes much longer because it is a larger mass. *All igneous rocks were formed from the solidification of magma.*

Crystallization of magma

Magma solidifies through the process of *crystallization.* At first, magma is a *melt,* a liquid solution of elements at high temperature. After a decrease occurs in the heat that keeps the magma liquid, the melt starts to solidify. Bit by bit, mineral grains begin to grow. As this growth goes on, gases are released. Now we no longer have a complete liquid, but rather a liquid mixed with solid and gaseous materials.[2] As the temperature continues to fall, the mixture solidifies until igneous rock is formed.

Igneous rocks may consist of interlocking grains of a single mineral or a mixture of several. The most common rock-forming minerals are the silicates olivine, augite, hornblende, biotite, plagioclase, orthoclase, muscovite, and quartz.

[1] M. H. P. Bott and Scott B. Smithson, "Gravity Investigations of Subsurface Shape and Mass Distribution of Granite Batholiths," *Bull. Geol. Soc. Am.*, vol. 78, pp. 859–878, 1967.

[2] To reflect this changing picture, we define a magma as any naturally occurring silicate melt, whether or not it contains suspended crystals or dissolved gases.

Bowen's reaction principle

Magma is a solution of elements, but it does not crystallize the way ordinary solutions do. Most solutions of a given composition always crystallize into a solid of the same composition. This happens regardless of conditions during solidification. If crystallization of a magma were similar, it would always yield rock of the same composition. However, magma of a given composition may crystallize into a number of different kinds of rock.

In 1922, N. L. Bowen[3] of the Geophysical Laboratory of the Carnegie Institution of Washington, D.C., proposed that the differences in end products are due to the rate at which the magma cools and on whether early formed minerals remain in or settle out of the remaining liquid during its crystallization. He suggested that as a magma cools the first-formed minerals undergo continuous modification with the liquid remaining after they crystallized. He called this process *reaction.* Reaction is the key to magmatic crystallization.

Based on a number of laboratory experiments with silicate melts, Bowen was able to arrange the rock-forming minerals of igneous rocks into *reaction series.* He found that an important characteristic of a silicate melt is that as one mineral develops at a certain temperature it will be converted upon further cooling into a different mineral by reaction with the liquid around it. The reaction series are of two different types, continuous and discontinuous. In the *continuous reaction series,* some early-formed minerals are converted into new minerals by continuously changing their composition but not their crystalline structure. In the *discontinuous reaction series,* some early-formed minerals react with the melt to change to new minerals with different compositions and a different crystalline structure.

Taking a magma with olivine basaltic composition, Bowen arranged the rock-forming minerals into a discontinuous reaction series of the ferromagnesian minerals and a continuous reaction series of the feldspars. As this magma cools, olivine and the calcium feldspar, anorthite, are the first minerals to form. With further drop in temperature, these react with the melt. Olivine becomes augite and the calcium plagioclase grades into more sodic feldspar. If no minerals settle out, the melt may solidify to basalt or gabbro.

If some of the early-formed minerals settle out, however, in a process called *fractionation,* the reaction process will continue further and the remaining minerals react with the remaining melt. Augite becomes hornblende, and the calcium–sodium feldspar becomes the sodium feldspar, albite. The greater the degree of fractionation, the more extensive the reaction process. With a high degree of fractionation, the whole reaction series is gone through and later-formed minerals will be rich in silica (see Figure 5.9).

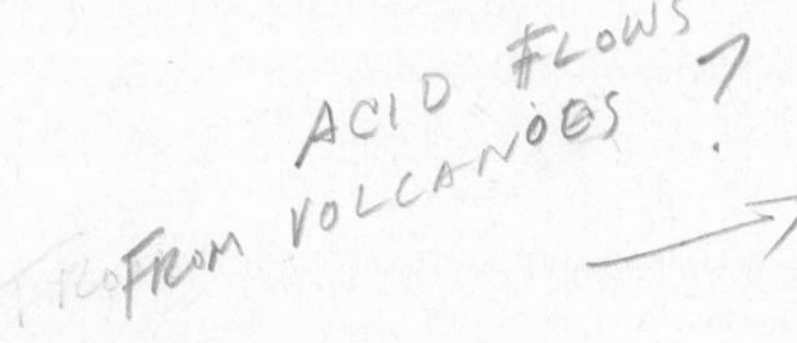

According to Bowen's reaction principle, the fractional crystallization of an olivine basaltic magma can lead to the formation of successively more siliceous rocks until finally a rock of granitic composition is reached.

Bowen's reaction principle explains how an olivine basaltic magma may

[3]N. L. Bowen, "The Reaction Principle in Petrogenesis," *J. Geol.,* vol. 30, pp. 177–198, 1922.

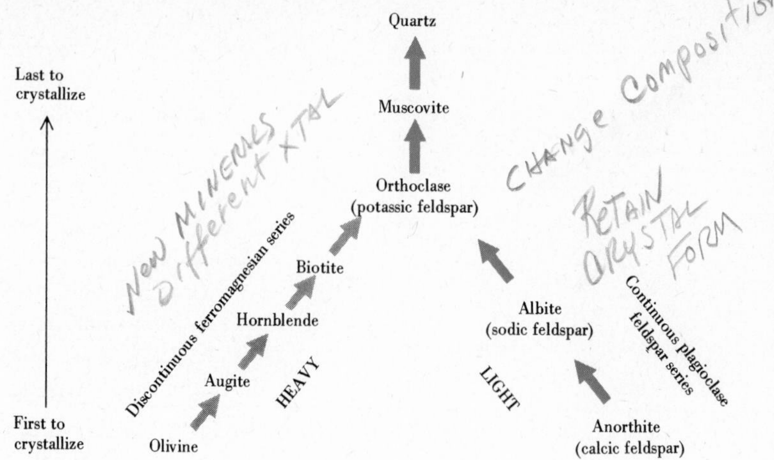

Figure 5.9 ***Bowen's reaction series.***

solidify as one rock type or may produce several rock types. It might crystallize to a gabbro which is composed largely of ferromagnesian minerals, or it might produce a granite which is composed predominantly of sialic minerals. The rock formed depends on the extent to which early formed minerals were removed from further reaction with the melt and on the rate of cooling.

Table 5.1 ***Magma crystallization and depth***[a]

Thickness, m	*Time required*
1	12 days
10	3 years
100	300 years
1,000	30,000 years
10,000	3,000,000 years

[a] R. A. Daly, *Igneous Rocks and the Depths of the Earth*, p. 63, McGraw-Hill Book Company, Inc., New York, 1933.

RATE OF CRYSTALLIZATION The rate at which a magma crystallizes influences the extent to which fractionation and reaction take place. When magma cools rapidly, there is no time for minerals to settle or to react with the remaining liquid. This occurs when a partially crystallized magma is extruded onto the surface or injected into thin dikes or sills. But when a large body of magma cools slowly, deep within the crust, a high degree of fractionation or chemical reaction may take place.

The rate of crystallization varies with depth. For example, a magma consisting largely of nonferromagnesians at 1100°C, exposed to the air on top, and ranging in thickness from 1 to 10,000 m, would solidify as shown in Table 5.1.

Support for Bowen's reaction principle

Support for Bowen's reaction series is found in certain masses of intrusive igneous rocks and in the composition of some volcanic eruptions. Some sills such as those of the Palisades in New York consist of igneous rock which shows fractional differentiation: Dense olivine-rich rocks are common near the floor of the sill and less dense quartz-bearing rocks at the top. Some sills in other areas show the same relationships. Fractional crystallization is believed to explain why more acidic eruptions occur late in a volcano's history. The eruptive history of the volcano Campo Bianco on Lipari Island just off the toe of Italy conforms to a pattern found in many other volcanoes. Its first eruptions consisted of basaltic lava. These were followed by eruptions of andesitic lava. Then, as the magma became more siliceous and also more viscous, explosive eruptions occurred with releases of gas shattering the magma into pumice and ash. No lava was emitted. Most recently, flows of obsidian appeared.

On Ascension and Easter Islands, the differentiation has been so effective that small amounts of siliceous rhyolite have been erupted.

Objections to Bowen's hypothesis

There are objections to Bowen's hypothesis. According to his reaction principle, only about 5 or 10 percent of an original basic magma could solidify into granite or granodiorite (a composition between that of granite and of diorite). This extremely small percentage cannot account for the large masses of these rocks found on the continents. Where are the earlier-formed more basic rocks? Also, it cannot explain why these rocks are found only on the continents and why the intermediate rock types such as diorite are relatively rare. It does not explain the great undifferentiated volumes of basalt or their prevalence in ocean basins.

5.3 *Texture of igneous rocks*

Texture, a term derived from Latin *texere,* "to weave," is a physical characteristic of all rocks. The term refers to the general appearance of rocks. In referring to the texture of igneous rocks, we mean specifically the size, shape, and arrangement of their interlocking mineral grains.

Granular texture

If magma has cooled at a relatively slow rate, it will have had time to develop grains that the unaided eye can see in hand specimens. Rocks composed of such large mineral grains are called *granular* (see Figure 5.10).

The rate of cooling, however, though important, is not the only factor that affects the texture of an igneous rock. For example, if a magma is of low viscosity—that is, if it is thin and watery and flows readily—large, coarse grains may form even though the cooling is relatively rapid, for in a magma of this sort, the ions can move easily and quickly into their rock-forming mineral combinations.

Aphanitic texture

The rate at which a magma cools depends on the size and shape of the magma body, as well as on its depth below the surface. For example, a small body of magma with a large surface area—that is, a body that is much longer and broader than it is thick—surrounded by cool, solid rock loses its heat more rapidly than would the same volume of magma in a spherical reservoir. And because rapid cooling usually prevents large grains from forming, the igneous rocks that result have *aphanitic textures.* Individual minerals are present, but they are so small that they cannot be identified without the aid of a microscope.

Glassy texture

If magma is suddenly ejected from a volcano or a fissure at the earth's surface, it may cool so rapidly that there is no time for minerals to form at all. The result is a *glass,* which by a rigid application of our definition is not really a rock but generally treated as one. Glass is a special type of solid in which the ions are not arranged in an orderly manner. Instead, they are disorganized, like the ions in a liquid. And yet they are frozen in place by the quick change of temperature.

Figure 5.10 ***Enlarged photograph of a piece of granular igneous rock, taken through a slice that has been ground to translucent thinness (known as a thin section). The photograph shows the rock to be composed of interlocking crystals of different minerals.***

Figure 5.11 **Enlarged photograph of a thin section of porphyritic igneous rock.**

Porphyritic texture

Occasionally, a magma cools at variable rates—slowly at first, then more rapidly. It may start to cool under conditions that permit large mineral grains to form in the early stages, and then it may move into a new environment where more rapid cooling freezes the large grains in a *groundmass* of finer-grained texture (see Figure 5.11). The large minerals are called *phenocrysts.* The resulting texture is said to be *porphyritic. Porphyry,* from the Greek word for "purple," was originally applied to rocks containing phenocrysts in a dark red or purple groundmass.

In rare cases, magma may suddenly be expelled at the surface after large mineral grains have already formed. Then the final cooling is so rapid that the phenocrysts become embedded in a glassy groundmass.

5.4 *Types of igneous rocks*

Several systems have been proposed for the classification of igneous rocks. All are artificial in one detail or another, and all rely on certain characteristics that cannot be determined in the field or from hand specimens. For our present

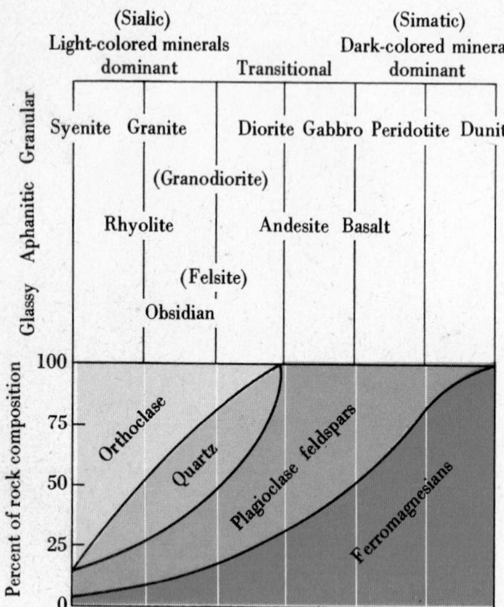

Figure 5.12 **General composition of igneous rocks is indicated by a line from the name to the composition chart: Granite and rhyolite consist of about 50 percent orthoclase, 25 percent quartz, and 25 percent plagioclase feldspars and ferromagnesian minerals. Granite is the most important granular rock and basalt the most important aphanitic rock.** (*Composition chart modified from L. Pirsson and A. Knopf,* Rocks and Rock Minerals, *p. 144, John Wiley & Sons, Inc., New York, 1926.*)

purposes, we shall emphasize texture and composition. Such a classification is entirely adequate for an introductory study of physical geology and even for many advanced phases of geology.

This classification appears in tabular form in Figure 5.12, together with a graph that shows the proportions of silicates in each type of igneous rock. The graph gives a better picture than the table of the *continuous progression from rock types in which light-colored minerals predominate to rock types in which dark-colored minerals predominate.* The names of rocks are arbitrarily assigned on the basis of average mineral composition and texture. Sometimes intermediate types are indicated by such names as granodiorite. Actually, there are many more igneous rocks than are shown in this figure.

Light-colored igneous rocks

The igneous rocks on the left side of the classification chart (Figure 5.12) are light both in color and specific gravity. They are sometimes referred to as *sialic rocks.*

It has been estimated that granites and granodiorites together comprise 95 percent of the igneous rocks of the continents that have solidified from magma. The origin and history of some granites are still under debate, but we use the term here to indicate composition and texture, not origin.

Granite is a granular rock (see Figure 5.13). Its mineral composition is as follows:

2 parts orthoclase feldspar + 1 part quartz + 1 part plagioclase feldspars + small amount of ferromagnesians = GRANITE

Rocks with the same mineral composition as granite but with an aphanitic rather than granular texture are called *rhyolite.*

The glassy equivalent of granite is called *obsidian* (see Figure 5.14). Although this rock is listed near the left side of the composition chart, it is usually pitch black in appearance. Actually, though, pieces of obsidian thin enough to be translucent turn out to be smoky white against a light background.

Dark-colored igneous rocks

Of the total volume of rock formed from magma that has poured out onto the earth's surface, it is estimated that 98 percent is basalts and andesites.

A popular synonym for basalt is *trap rock,* from a Swedish word meaning "step." This name refers to the tendency of certain basalts to form columns that look like stairways in some outcrops (see Figure 5.15). These are a product of the cooling process and become evident after weathering.

Basalt has aphanitic texture. Its mineral composition is as follows:

1 part plagioclase feldspars + 1 part ferromagnesians = BASALT

The granular equivalent of basalt is *gabbro. Peridotite* is a granular igneous rock that is composed largely of ferromagnesian minerals.

Figure 5.13 **Granite.** (*Photograph by Robert Navias.*)

Figure 5.14 ***Obsidian, the glassy equivalent of granite.*** (*Photograph by Robert Navias.*)

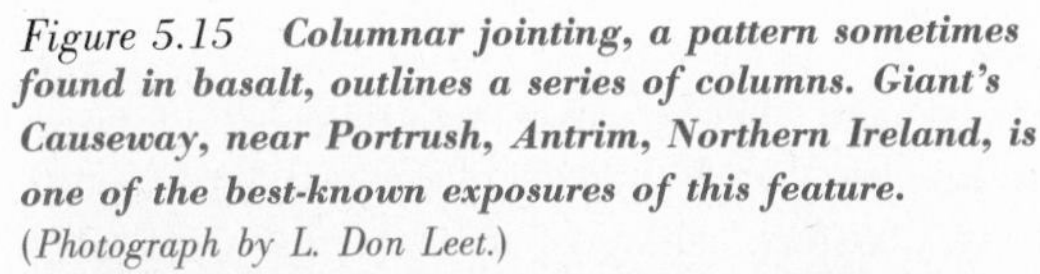

Figure 5.15 ***Columnar jointing, a pattern sometimes found in basalt, outlines a series of columns. Giant's Causeway, near Portrush, Antrim, Northern Ireland, is one of the best-known exposures of this feature.*** (*Photograph by L. Don Leet.*)

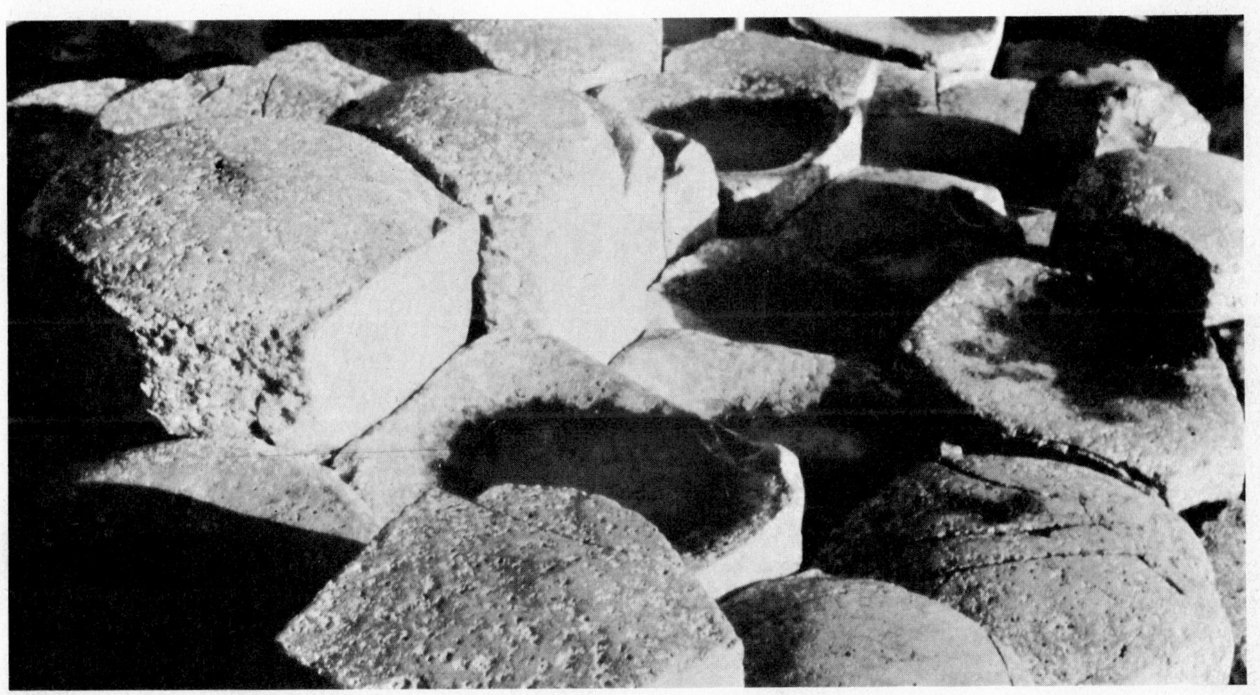

Intermediate types—composition

Igneous rock compositions blend continuously from one to another as we go from the light to the dark side of the classification chart. *Andesite* is the name given to aphanitic igneous rocks that are intermediate in composition between granite and basalt. These rocks were first identified in the Andes mountains of South America—hence the name andesite. Andesites are mostly found in areas around the Pacific Ocean. The granular equivalent of andesite is *diorite.*

Intermediate types—texture

Going from the top to the bottom of the chart in Figure 5.12, we find that the rock textures grade continuously from granular to aphanitic, whereas the composition remains the same. For example, if we read down along the first vertical rule, we find that granite, rhyolite, and obsidian become progressively finer-grained, although all three have essentially the same composition. The same is true of gabbro and basalt.

In addition to these textures, any of the rocks may have porphyritic texture. Essentially, this means that a given rock has grains of two distinctly different sizes: conspicuously large phenocrysts embedded in a finer-grained groundmass. When the phenocrysts constitute less than 25 percent of the total, the adjective *porphyritic* is used to modify the rock name, as in porphyritic granite or porphyritic andesite.

Figure 5.16 ***Orthoclase phenocrysts in a granite porphyry.*** *(Photograph by Robert Navias.)*

Table 5.2 ***Porphyritic rock and rock porphyry***

Less than 25% phenocrysts	*More than 25% phenocrysts*
Porphyritic granite	Granite porphyry
Porphyritic rhyolite	Rhyolite porphyry
Porphyritic diorite	Diorite porphyry
Porphyritic andesite	Andesite porphyry

When the phenocrysts constitute more than 25 percent, the rock is called a *porphyry* (see Figure 5.16). The composition of a porphyry and the texture of its groundmass are indicated by using as modifiers such rock names as granite porphyry or andesite porphyry. These relationships are summarized for the most common rocks in Table 5.2.

Pegmatite

The solutions that develop late in the cooling of a magma are called *hydrothermal solutions.* These crystallize into exceptionally granular igneous rock called *pegmatite* (from the Greek *pegma,* "fastened together"), which embodies the chief minerals to form from the hydrothermal solutions: potassium feldspar and quartz. So intimately intergrown are the grains of these minerals that they form what is essentially a single unit. The quartz is darker than the feldspar, and the overall pattern suggests the wedge-shaped figures of the writing of ancient Assyria, Babylonia, and Persia. As a result, this has become known as *graphic structure* (from the Greek *graphein,* "to write") (see Figure 5.17). The intimate association of the feldspar and quartz also led to the rock of which they are a part being named pegmatite.

Figure 5.17 ***Graphic granite.*** (*Photograph by Walter R. Fleischer.*)

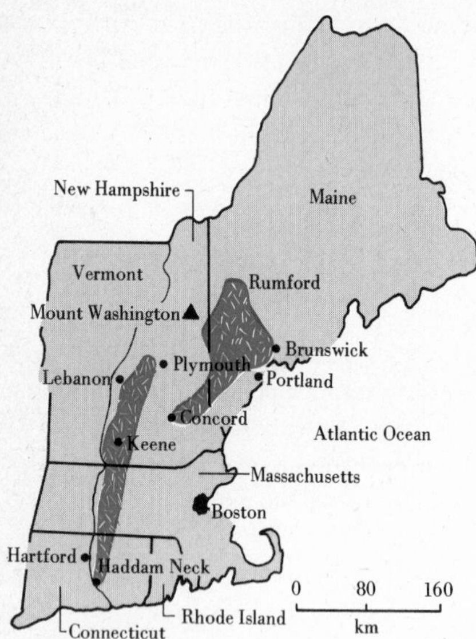

Figure 5.18 **Map of regions where pegmatites are extensively developed east of the Connecticut Valley in New England. Pegmatite areas are shown in color.**

Pegmatite is found in dikes at the margins of batholiths and stocks. The dikes range in length from a few centimeters to a few hundred meters and contain crystals of very large size. In fact, some of the largest crystals known have been found in pegmatite. Crystals of spodumene (a lithium mineral) that measure 12 m in length have been found in the Black Hills of South Dakota; crystals of beryl (a silicate of beryllium and aluminum) that measure 1 by 5 m have been discovered in Albany, Maine. Great masses of potassium feldspar weighing over 2,000 metric tons, yet showing the characteristics of a single crystal, have been mined from pegmatite in the Karelo-Finnish Soviet Socialist Republic.

Nearly 90 percent of all pegmatite is *simple pegmatite* of quartz, orthoclase, and unimportant percentages of micas. It is more generally called *granite pegmatite,* because the composition is that of granite and the texture that of pegmatite. The remaining 10 percent includes *complex pegmatites.* The major components of complex pegmatites are the same sialic minerals that we find in simple pegmatites, but in addition they contain a variety of rare minerals. These include lepidolite; tourmaline, best known as a semiprecious gem; topaz, also a gem; tantalite; and uraninite, sometimes called pitchblende.

Simple pegmatites are common in some regions and complex pegmatites in others. In southwestern New Zealand, for example, pegmatites are uniformly simple, but throughout the Appalachian regions of North America complex pegmatites are more abundant. Figure 5.18 shows the areas in New England where pegmatites are found.

Igneous rocks on the moon

Rocks brought back from the moon by the *Apollo 11* and *12* missions included two main igneous rock types, a vesicular basalt and a fine-grained gabbro. These and other lunar materials are discussed in Chapter 23.

Outline

Igneous rocks are formed from the solidification of molten matter.

Masses of igneous rocks are called plutons, which are classified according to size, shape, and relationships to surrounding rocks.

- *Sills* are concordant tabular plutons.
- *Dikes* are discordant tabular plutons.
- *Lopoliths* are tabular concordant plutons shaped like a spoon.
- *Laccoliths* are massive concordant plutons with domed tops.
- *Batholiths* are massive discordant plutons 10 to 40 km thick.

Igneous rocks at the surface today were formed from magma.

- *Magma* solidifies through the process of crystallization.
- *Bowen's reaction series* are incorporated in an hypothesis accounting for all igneous rocks coming from an olivine basaltic magma.
- *The rate of crystallization* is an important control over the rocks that form.
- *Limitations of Bowen's hypothesis* include failure to account for distribution of

granitic and basaltic rocks and large undifferentiated masses of granite and basalt.

Texture of igneous rocks is the size, shape, and arrangement of their interlocking mineral grains.

Granular texture includes large mineral grains from slow-cooling or low-viscosity magma.

Aphanitic texture from rapid cooling consists of individual minerals so small they cannot be identified without the aid of a microscope.

Glassy texture results from ions disorganized as in a liquid but frozen in place by quick cooling.

Porphyritic texture is a mixture of large mineral grains in an aphanitic or glassy groundmass.

Types of igneous rocks are arbitrarily defined in terms of texture and composition.

Light-colored igneous rocks, sometimes called sialic, are dominated by granites and granodiorites.

Dark-colored igneous rocks (basalts and andesites) constitute 98 percent of rock formed from magma that has poured out onto the earth's surface.

Intermediate types of composition are given arbitrary names, such as andesite and diorite, because igneous rock compositions blend continuously from one to another from the light to the dark side of the classification chart.

Intermediate types of texture are also given arbitrary names such as granite, rhyolite, and obsidian, because rocks of a given composition grade continuously from granular to aphanitic to glassy texture.

Pegmatite is an exceptionally granular rock formed by hydrothermal solutions late in the cooling of a magma.

Igneous rocks on the moon are similar in composition to basalt.

Supplementary readings

DALY, R. A.: *Igneous Rocks and the Depths of the Earth,* Hafner Publishing Company, New York, 1968. Reprint of the classic published by McGraw-Hill in 1933.

HESS, H. H., and ARIE POLDERVAART (eds.): *Basalts, the Poldervaart Treatise on Rocks of Basaltic Composition,* vols. I and II, Wiley-Interscience, Inc., 1967 and 1968. Twenty monographs on all aspects of basaltic rocks. Advanced in level, but the most comprehensive and authoritative compilation on this subject.

WAGER, L. R., and G. M. BROWN: *Layered Igneous Rocks,* W. H. Freeman and Company, San Francisco, 1967. Examples of crystal fractionation, with emphasis on the Skaergaard instrusion of East Greenland, but including many other localities with layered intrusions believed to be caused by bodies of homogeneous magma cooling and crystallizing slowly in large chambers with crystals accumulating on the floors of these chambers to form distinctive layers.

6

Weathering and soils

THE BLURRED INSCRIPTION ON A GRAVESTONE, THE CRUMBLING FOUNDATION OF AN ancient building, the broken rock exposed along a roadside—all tell us that rocks are subject to constant destruction. Marked changes of temperature, moisture soaking into the ground, the ceaseless activity of living things—all work to destroy rock material. This process of destruction we call *weathering*, and we define it as the changes that take place in minerals and rocks at or near the surface of the earth, in response to the atmosphere, to water, and to plant and animal life. Later on we shall extend this definition slightly, but for the time being it will serve our purpose.

Weathering leaves its mark everywhere about us. The process is so common, in fact, that we tend to overlook the way in which it functions and the significance of its results. It plays a vital role in the rock cycle, for by attacking the exposed material of the earth's crust—both solid rock and unconsolidated deposits—it produces the raw materials for new rocks (see Chapter 1 and Figure 1.12). The products of weathering are usually moved by water and by the influence of gravity, and less commonly by wind and glacier ice. They are then dropped, to settle down and accumulate in new places. The mud in a flooding river, for example, is really weathered material that is being moved from the land to some settling basin, usually the ocean. Sometimes, however, the products of weathering remain right where they are formed and are incorporated into the rock record. Certain ores, for example, such as those of aluminum, are actually old zones of weathering (see Chapter 21).

6.1 *Energy sources*

The energy that drives the weathering process comes both from within the earth and from outside. From time to time motions originating inside the earth elevate some portions of its surface above other portions. These are the same motions that express themselves in earthquakes (Chapter 16) and mountain building (Chapters 18 and 19). Whatever their cause they arrange the earth's surface materials so that gravity can be effective in the breakdown of rock material. Thus rock raised in mountain building has potential energy which may be transformed to kinetic energy if gravity is strong enough to pull it downward to a lower level. The shattering of this rock as it falls is a form of weathering.

We already have seen that there is a measurable amount of heat which flows from the earth's interior to its surface and is there dissipated (Section 4.4). Far more heat is received at the earth's surface from the sun. It is the distribution of this heat which causes the differential heating of the atmosphere and of the oceans. This differential heating brings about circulation of atmosphere and ocean, causes our weather and climate, and determines the pattern of organic activity. All these in turn help to modify the earth's surface materials—in short, to determine the process of weathering.

How much solar energy is available at the earth's surface? The sun radiates about 100,000 cal/cm^2-min for its surface. A plane at the edge of the earth's outer surface perpendicular to the sun's surface would receive approximately 2 cal/cm^2-min, which is 2×10^{-5}, the amount of heat generated by an equivalent area on the sun's surface. This figure is the *solar constant.*

If our hypothetical plane perpendicular to the sun's rays at the edge of the outer atmosphere has the same diameter as the earth, then geometry tells us that its area must be one-quarter that of the earth. If this be true, then we can express the solar constant as 0.5 cal/cm^2 of the earth's surface per minute. This is an average for the entire globe. But of this energy approximately one-third is unavailable to heat either the atmosphere or the earth. It is reflected back into space from clouds, or from the earth, or is scattered into space from particles within the atmosphere. Therefore, approximately 0.3 cal/cm^2-min is available to heat the atmosphere and the earth.

It is obvious that this energy is not evenly distributed over the earth's surface. Less is available toward either pole than is available toward the equator. We shall have occasion to return to this distribution later in this chapter when we consider the distribution of chemical weathering.

6.2 Types of weathering

There are two general types of weathering: *mechanical* and *chemical.* It is difficult to separate these two types in nature, for they often go hand in hand, though in some environments one or the other predominates. Still, for our purposes here it is more convenient to discuss them separately.

Mechanical weathering

Mechanical weathering, which is also referred to as *disintegration,* is the process by which rock is broken down into smaller and smaller fragments as the result of energy developed by physical forces. For example, when water freezes in a fractured rock, enough energy may develop from the pressure caused by expansion of the frozen water to split off pieces of the rock. Or a boulder moved by gravity down a rocky slope may be shattered into smaller fragments.

EXPANSION AND CONTRACTION RESULTING FROM GAIN AND LOSS OF HEAT Changes in temperature, if they are rapid enough and great enough, may bring about the mechanical weathering of rock. In areas where bare rock is exposed at the surface and is unprotected by a cloak of soil, forest or brush fires can generate enough heat to break up the rock. The rapid and violent heating of the exterior zone of the rock causes it to expand; and if the expansion is great enough, flakes and larger fragments of the rock are split off. Lightning often starts such forest fires and, in rare instances, may even shatter exposed rock by means of a direct strike.

The debate continues concerning whether variations in temperature from day to night, or from winter to summer, are great enough to cause mechanical

Figure 6.1 ***Differential weathering of rocks of varying resistance helps explain the topography of the Grand Canyon of the Colorado. The river flows in a steep-walled inner gorge cut in igneous and metamorphic rocks. The terraces and templelike forms have developed on alternating resistant and nonresistant sedimentary rocks.*** *(Photograph by John Shelton.)*

weathering. Theoretically, such changes in temperature cause disintegration. For instance, we know that the different minerals forming a granite expand and contract at different rates as they react to rising and falling temperatures. We expect that even minor expansion and contraction of adjacent minerals would, over long periods of time, weaken the bonds between mineral grains and that it would thus be possible for disintegration to occur along these boundaries.

In the desert we find fragments of a single stone lying close beside one another. Obviously the stone has split. But how? Many think the cause lies in expansion and contraction caused by heating and cooling.

But laboratory evidence to support these speculations is inconclusive. In one laboratory experiment, coarse-grained granite was subjected to temperatures ranging from 14.5 to 135.5°C every 15 min. This alternate heating and cooling was carried on long enough to simulate 244 years of daily heating and cooling. Yet the granite showed no signs of disintegration. Perhaps experiments extended over longer periods of time would produce observable effects. In any event, we are still uncertain of the mechanical effect of daily or seasonal temperature changes; if these fluctuations do bring about the disintegration of rock, they must do so very slowly.

FROST ACTION Frost is much more effective than heat in producing mechanical weathering. When water trickles down into the cracks, crevices, and pores of a rock mass and then freezes, its volume increases about 9 percent. This expansion of water as it passes from the liquid to the solid state sets up pressures that are directed outward from the inside of the rock. These pressures are great enough to dislodge fragments from the rock's surface. By the time the temperature has fallen to about $-22°C$, the resulting pressure may be as great as 2,100 kg/cm^2. This temperature is not unusually low and is experienced several times a year, even in temperate latitudes.

Under actual conditions, however, such great pressures are probably never produced by frost action, at least never close to the earth's surface. For an internal pressure of 2,100 kg/cm^2 to build up, a rock crevice would have to be completely filled with water and completely sealed off, and the containing rock would have to be strong enough to withstand pressures at least up to that point. But most crevices contain some air in addition to water and are open to either the surface or other crevices. Furthermore, no rock can withstand a pressure of 2,100 kg/cm^2 if the pressure is directed from within toward the outside.

And yet frost action is responsible for a great deal of mechanical weathering. Water that soaks into the crevices and pores of a rock usually starts to freeze at its upper surface, where it is in contact with the cooling air. The result is that, in time, the water below is confined by an ice plug. Then, as the freezing continues, the trapped water expands, and pressure is exerted outward. Rock may be subjected to this action several times a year. In high mountains, for example, the temperature may move back and forth across the freezing line almost daily.

The dislodged fragments of mechanically weathered rock are angular in shape, and their size depends largely on the nature of the bedrock from which they have been displaced. Usually the fragments are only a few centimeters in maximum dimension, but in some places—along the cliffs bordering Devil's Lake, Wisconsin, for instance—they reach sizes of up to 3 m.

A second type of mechanical weathering produced by freezing water is *frost heaving*. This action usually occurs in fine-grained, unconsolidated deposits rather than in solid rock. Much of the water that falls as rain or snow soaks into the ground, where it freezes during the winter months. If conditions are right, more and more ice accumulates in the zone of freezing as water is added from the atmosphere above and drawn upward from the unfrozen ground below, much as a blotter soaks up moisture. In time, lens-shaped masses of ice are built up, and the soil above them is heaved upward. Frost heaving of this sort is common on poorly constructed roads, and lawns and gardens are often soft and spongy in the springtime as a result of the soil's heaving up during the winter.

Certain conditions must exist before either type of frost action can take place: (1) there must be an adequate supply of moisture; (2) the moisture must be able to enter the rock or soil; and (3) temperatures must move back and forth across the freezing line. As we might expect, frost action is more pronounced in high mountains and moist regions where temperatures fluctuate across the freezing line, either daily or seasonally (see Figure 6.2).

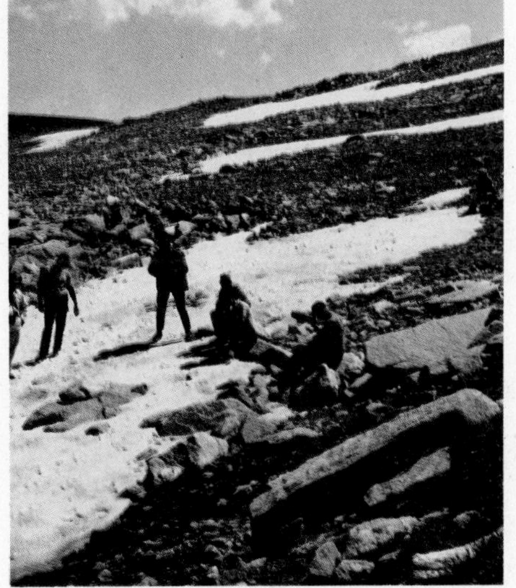

Figure 6.2 ***Vigorous frost action has pried off this granitic rubble from the underlying bedrock along the crest of the Beartooth Mountains near the Montana–Wyoming state line. In addition, frost action around the edges of snow banks has combined with the water of melting snow to create irregularities called*** **nivation hollows** ***on the slopes.*** *(Photograph by Sheldon Judson.)*

EXFOLIATION Exfoliation is a mechanical weathering process in which curved plates of rock are stripped from a larger rock mass by the action of physical forces. This process produces two features that are fairly common in the landscape: large, domelike hills, called *exfoliation domes*, and rounded boulders, usually referred to as *spheroidally weathered boulders*. It seems likely that the forces that produce these two forms originate in different ways.

Let us look first at the manner in which exfoliation domes develop. Fractures or parting planes, called *joints*, occur in many massive rocks. These joints are broadly curved and run more or less parallel to the rock surface. The distance between joints is only a few centimeters near the surface, but it increases to several meters as we move deeper in the rock (see Chapter 18). Under certain conditions, one after another of the curved slabs between the joints is spalled or sloughed off the rock mass. Finally, a broadly curved hill of bedrock develops, as shown in Figure 6.3.

Just how these slabs of rock come into being in the first place is still a matter of dispute. Most observers believe that as erosion strips away the surface cover the downward pressure on the underlying rock is reduced. Then, as the rock mass begins to expand upward, lines of fracture develop, marking off the slabs that later fall away. Precise measurements made on granite blocks in New England quarries provide some support for this theory. Selected blocks were accurately measured and then removed from the quarry face, away from the confining pressures of the enclosing rock mass. When the free-standing blocks

Figure 6.3 ***North Dome, Yosemite National Park, California, is an example of an exfoliation dome. The massive granite in this dome has developed a series of partings or joints more or less parallel to the surface. Rock slabs spall off the dome giving it its rounded aspect. The jointing probably originated as the granite expanded after the erosion of the overlying material.*** (*Photograph by William C. Bradley.*)

Figure 6.4 **Granite boulders are beginning to develop by spheroidal weathering in King's Canyon, California. Weathering is proceeding most rapidly along the joint system.** (*Photograph by U.S. Geological Survey.*)

were measured again, it was found that they had increased in size by a small but measurable amount. Massive rock does expand, then, as confining pressures are reduced, and this slight degree of expansion may be enough to start the exfoliation process.

Among the better-known examples of exfoliation domes are Stone Mountain, Georgia; the domes of Yosemite Park, California; and Sugar Loaf, in the harbor of Rio de Janeiro, Brazil.

Now let us look at a smaller-scale example of exfoliation: spheroidally weathered boulders. These boulders have been rounded by the spalling off of a series of concentric shells of rock (Figures 6.4, 6.5, and 6.6). But here the shells develop as a result of pressures set up within the rock by chemical weathering rather than by the lessening of pressure from above by erosion. We shall see later that when certain minerals are chemically weathered, the resulting products occupy a greater volume than the original material, and it is this increase in volume that creates the pressures responsible for spheroidal weathering. Because most chemical weathering takes place in the portions of the rock most exposed to air and moisture, it is there that we find the most expansion and hence the greatest number of shells.

Spheroidally weathered boulders are sometimes produced by the crumbling off of concentric shells. If the cohesive strength of the rock is low, individual grains are partially weathered and dissociated, and the rock simply crumbles away. The underlying process is the same in both cases, however.

Certain types of rocks are more vulnerable to spheroidal weathering than others. Igneous rocks such as granite, diorite, and gabbro are particularly susceptible, for they contain large amounts of the mineral feldspar, which, when weathered chemically, produces new minerals of greater volume.

Figure 6.5 **Spheroidally weathered granite boulders almost completely isolated from the bedrock.** (*Photograph by Sheldon Judson.*)

OTHER TYPES OF MECHANICAL WEATHERING Plants also play a role in mechanical weathering. The roots of trees and shrubs growing in rock crevices sometimes exert enough pressure to dislodge previously loosened fragments of rock, much as tree roots heave and crack sidewalk pavements (Figure 6.7).

More important, though, is the mechanical mixing of the soil by ants, worms, and rodents. Constant activity of this sort makes the soil particles more susceptible to chemical weathering (see below) and may even assist in the mechanical breakdown of the particles.

Finally, agents such as running water, glacier ice, wind, and ocean waves all help to reduce rock material to smaller and smaller fragments. The role of these agents in mechanical weathering will be discussed in later chapters.

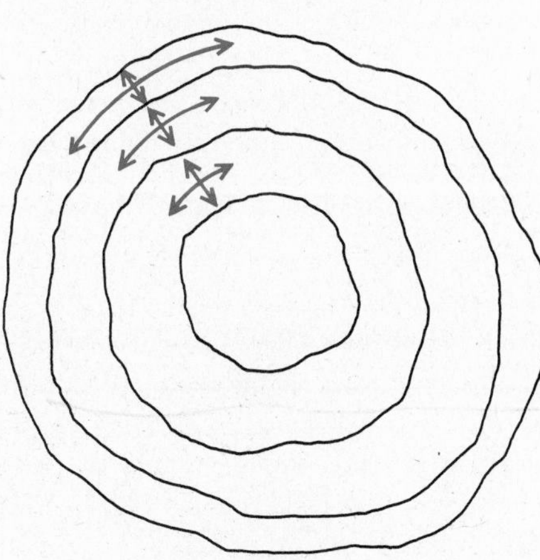

Figure 6.6 **This cross section through a spheroidally weathered boulder suggests the stresses set up within the rock. The stress is thought to develop as a result of the change in volume as feldspar is converted to clay. (See also Figures 6.4 and 6.5.)**

Figure 6.7 ***A white birch tree growing in a crevice pries a large block from a low rock cliff in Hermosa Park, Colorado.*** *(Photograph by U.S. Geological Survey.)*

Chemical weathering

Chemical weathering, sometimes called *decomposition*, is a more complex process than mechanical weathering. As we have seen, mechanical weathering merely breaks rock material down into smaller and smaller particles, without changing the composition of the rock. Chemical weathering, however, actually transforms the original material into something different. The chemical weathering of the mineral feldspar, for example, produces the clay minerals, which have a different composition and different physical characteristics from those of the original feldspar. Sometimes the products of chemical weathering have no mineral form at all, as the salty solution that results from the transformation of the mineral halite, common salt.

PARTICLE SIZE AND CHEMICAL WEATHERING The size of the individual particles of rock is an extremely important factor in chemical weathering, because substances can react chemically only where they come in contact with one another. The greater the surface area of a particle, the more vulnerable it is to chemical attack. If we were to take a pebble, for example, and grind it up into a fine powder, the total surface area exposed would be greatly increased. And as a result, the materials that make up the pebble would undergo more rapid chemical weathering.

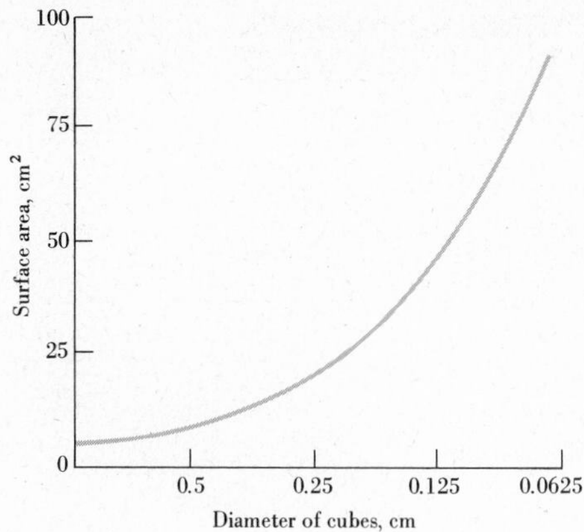

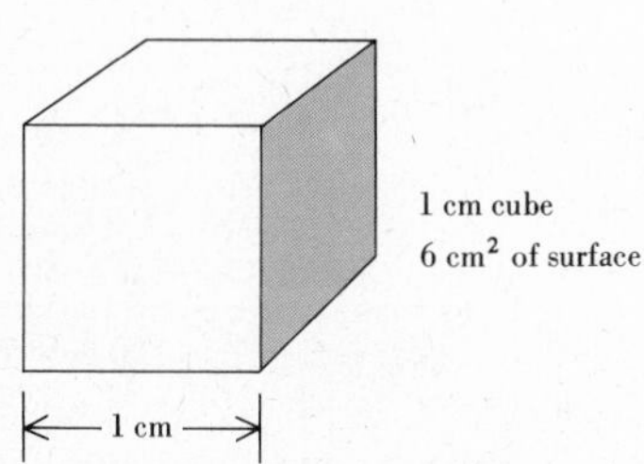

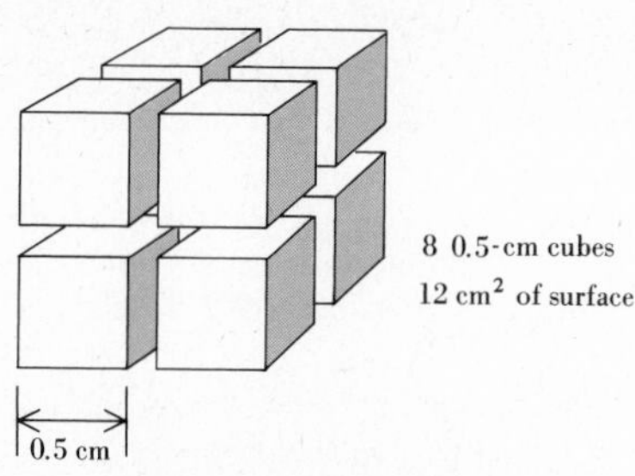

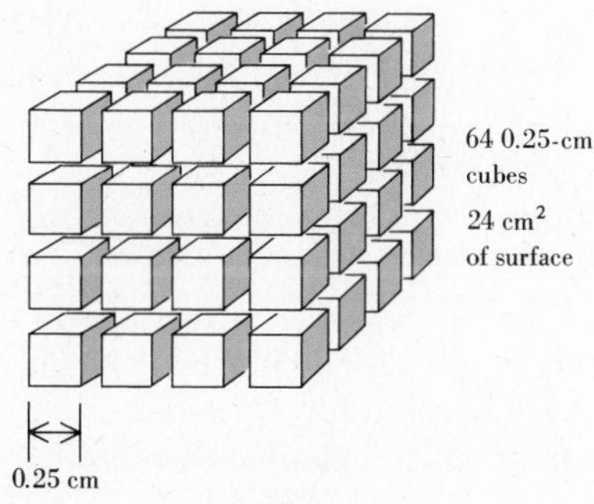

Figure 6.8 **Relation of volume, particle size, and surface area. In this illustration, a cube 1 cm square is divided into smaller and smaller units. The volume remains unchanged, but as the particle size decreases, the surface area increases. Because chemical weathering is confined to surfaces, the more finely a given volume of material is divided, the greater is the surface area exposed to chemical activity and the more rapid is the process of chemical weathering.**

Figure 6.8 shows how the surface area of a 1-cm cube increases as we cut it up into smaller and smaller cubes. The initial cube has a surface area of 6 cm^2 and a volume of 1 cm^3. If we divide the cube into smaller cubes, each .5 cm on a side, the surface area increases to 12 cm^2, though, of course, the total volume remains the same. Further subdivision into 0.25-cm cubes increases the surface to 24 cm^2. And if we divide the original cube into units 0.125 on a side, the surface area increases to 96 cm^2. As we have seen, this same process is performed by mechanical weathering: It reduces the size of the individual particles of rock, increases the surface area exposed, and thus promotes more rapid chemical weathering.

OTHER FACTORS IN CHEMICAL WEATHERING The rate of chemical weathering is affected by other factors as well—the composition of the original mineral, for example. As we shall see later, a mineral such as quartz (SiO_2) responds much more slowly to chemical weathering than does a mineral such as olivine [(Fe, Mg)$_2$$SiO_4$].

Climate also plays a key role in chemical weathering. Moisture, particularly when it is accompanied by warmth, speeds up the rate of chemical weathering; conversely, dryness slows it down. Finally, plants and animals contribute directly or indirectly to chemical weathering, because their life processes produce oxygen, carbon dioxide, and certain acids that enter into chemical reactions with earth materials.

The interrelation of some of these factors is shown in Figure 6.9. This is a generalized section from the pole to the equator and shows the fluctuation of precipitation, temperature, and amount of vegetation. At the same time the figure shows the relative depth of weathering as these three factors fluctuate from pole toward the tropical climates. Thus the weathering is most pronounced in the equatorial zone where the factors—precipitation, temperature, and vegetation—reach a maximum. Weathering is least in the desert and semidesert areas of the subtropics and in the far north. A secondary zone of maximum weathering exists in the zone of temperate climates. Here, both the precipitation and vegetation reach secondary maxima.

Chemical weathering of igneous rocks

In Chapter 5, we found that the most common minerals in igneous rocks are silicates and that the most important silicates are quartz, the feldspars, and certain ferromagnesian minerals. Let us see how chemical weathering acts on each of these three types of silicate.

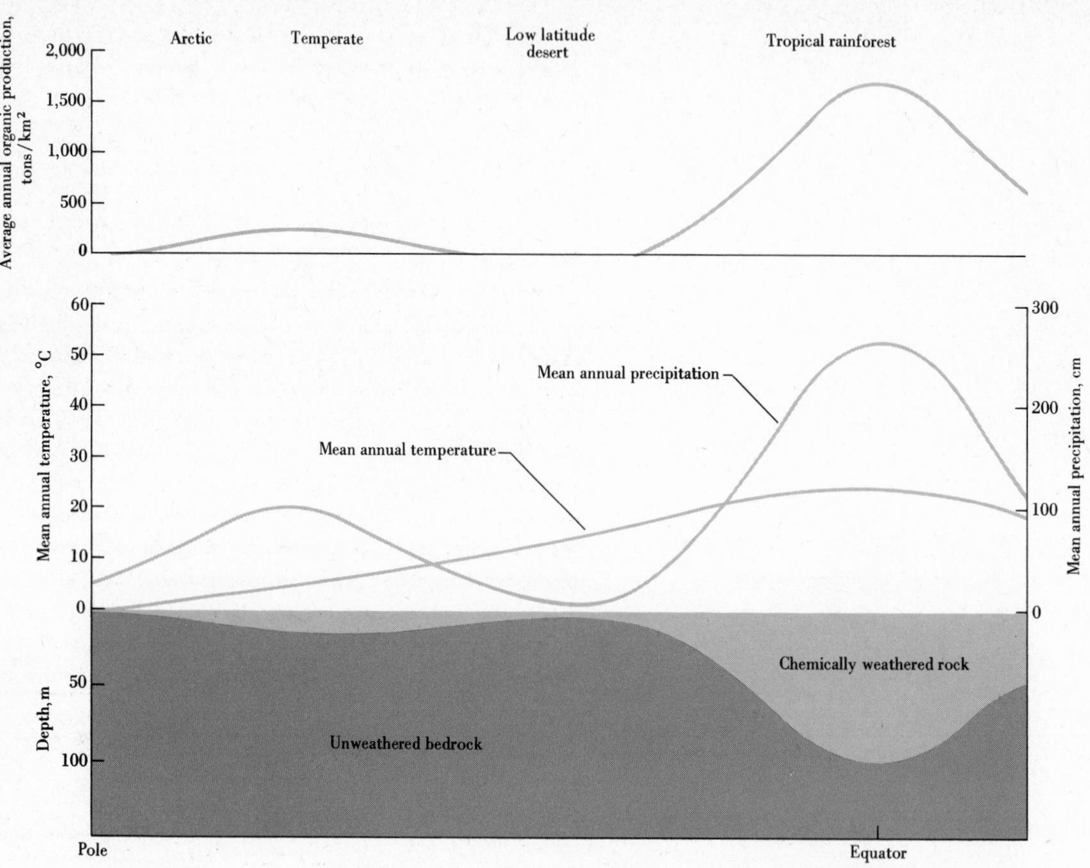

Figure 6.9 ***Variation in temperature, precipitation, and organic matter from the pole to the tropics is related to the depth of chemical weathering.***

WEATHERING OF QUARTZ Chemical weathering affects quartz very slowly, and for this reason we speak of quartz as a relatively stable mineral. When a rock such as granite, which contains a high percentage of quartz, decomposes, a great deal of unaltered quartz is left behind. The quartz grains (commonly called *sand grains*) found in the weathered debris of granite are the same as those that appeared in the unweathered granite.

When these quartz grains are first set free from the mother rock, they are sharp and angular, but because even quartz responds slowly to chemical weathering, the grains become more or less rounded as time passes. After many years of weathering, they look as though they had been abraded and worn by the action along a stream bed or a beach. And yet the change may have come about solely through chemical action. Indeed, the presence of silica in natural waters of lakes and rivers reminds us that the silicate minerals are soluble and that some of this may come from the chemical weathering of quartz.

WEATHERING OF FELDSPARS In the Bowen reaction series (see Section 5.2), we saw that when a magma cools to form an igneous rock such as granite the feldspars crystallize before the quartz. When granite is exposed to weathering at the earth's surface, the feldspars are also the first minerals to be broken down. Mineralogists and soil scientists still do not understand the precise process by which the feldspars weather, and some of the end products of this action—the clay minerals—offer many puzzles. But the general direction and results of the process seem fairly clear.

The clay minerals are made up chiefly of aluminum silicate derived from the chemical breakdown of the original feldspar. This combines with water to form hydrous aluminum silicate, which is the basis for another group of silicate minerals, the clays.

The decomposition of orthoclase is a good example of the chemical weathering of the feldspar group of silicates. In this instance a source of hydrogen ions is necessary to the weathering process and two substances play an important role in producing them. They are carbon dioxide and water. The atmosphere contains small amounts of carbon dioxide and the soil contains much greater amounts. Because carbon dioxide is extremely soluble in water, it unites with rain water and water in the soil to form the weak acid H_2CO_3, called *carbonic acid.* This ionizes to form hydrogen and bicarbonate ions as follows:

water *plus* carbon dioxide *yields* carbonic acid *yields* hydrogen ion *plus* bicarbonate ion

$$H_2O + CO_2 \rightarrow H_2CO_3 \rightarrow H^+ + (HCO_3)^-$$

When orthoclase comes in contact with hydrogen ions, the following reaction takes place:

orthoclase *plus* hydrogen ions *plus* water *yields* clay *plus* potassium ions *plus* silica

$$2K(AlSi_3O_8) + 2H^+ + H_2O \rightarrow Al_2Si_2O_5(OH)_4 + 2K^+ + 4SiO_2$$

In this reaction, the hydrogen ions from the water force the potassium out of the orthoclase, disrupting its crystal structure. The hydrogen ion combines with the aluminum silicate radical of the orthoclase to form the new clay minerals. (The process by which water combines chemically with other molecules is called *hydration.*) The disruption of the orthoclase crystal yields a second product, potassium ions. These may join with the bicarbonate ions formed by the ionization of the carbonic acid to form potassium bicarbonate. The third product, silica, is formed by the silicon and oxygen that are left after the potassium has combined with the aluminum silicate to form the clay mineral.

The action of living plants may also bring about the chemical breakdown of orthoclase. A plant root in the soil is negatively charged and is surrounded by a swarm of hydrogen ions (H^+). If there happens to be a fragment of orthoclase lying nearby, these positive ions may change places with the potassium of the orthoclase and disrupt its crystal structure (see Figure 6.10). Once again a clay mineral is formed, as in the equation above.

Now let us look more closely at each of the three products of the decomposi-

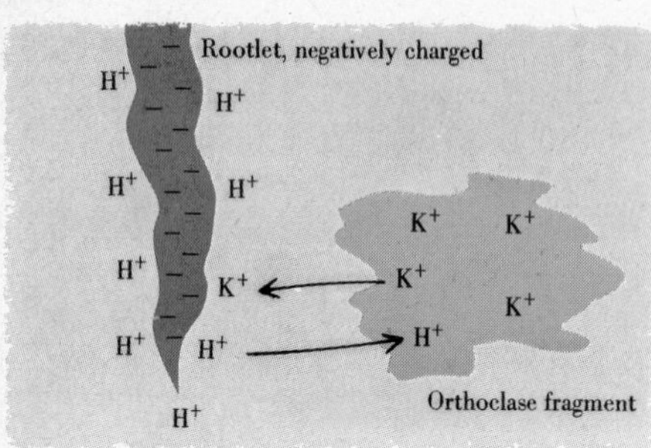

Figure 6.10 **The conversion of orthoclase to a clay mineral by plant roots. In this diagram, a swarm of hydrogen ions (positive) are shown surrounding a negatively charged plant rootlet. The suggestion has been made that a hydrogen ion from the rootlet may replace a potassium ion in a nearby orthoclase fragment and there bond with the oxygen within the original mineral, to begin the conversion of the orthoclase to clay. The potassium ion thus ejected replaces the hydrogen ion along the negatively charged rootlet and is eventually utilized in plant growth.** (*Redrawn from W. D. Keller and A. F. Frederickson, "Role of Plants and Colloidal Acids in the Mechanism of Weathering,"* Am. J. Sci., *vol. 250, p. 603, 1952.*)

tion of orthoclase: first, the clay minerals. At the start, these minerals are very finely divided. In fact, they are sometimes of colloidal size, a size variously estimated as between 0.2 and 1 μm.

Immediately after it is formed, the aluminum silicate may possibly be amorphous; that is, its atoms are not arranged in any orderly pattern. It seems more probable, however, that even at this stage the atoms are arranged according to the definite pattern of a true crystal. In any event, as time passes, the small individual particles join together to form larger crystals which, when analyzed by such means as X rays, exhibit the crystalline pattern of true minerals.

There are many different clay minerals and each has its own chemical behavior, physical structure, and evolution. Most of the clay minerals fall into three major groups: *kaolinite, montmorillonite,* and *illite.* The term *kaolinite* is derived from the Chinese *Kao-ling,* "High Hill," the name of the mountain from which the first kaolinite was shipped to Europe for ceramic uses. The mineral montmorillonite was first described from samples collected near Montmorillon, a town in west-central France. The name illite was selected by geologists of the Illinois Geological Survey in honor of their state.

Like the micas shown in Figure 3.17, the clay minerals are built up of silicon–oxygen tetrahedra linked together in sheets. These sheets combine in different ways with sheets composed of aluminum atoms and hydroxyl molecules. For this reason we refer to the clay minerals as hydrous aluminum silicates. In addition, montmorillonite may contain magnesium and some sodium and calcium, and illite contains potassium, occasionally with some magnesium and iron.

We still do not understand exactly what factors determine which clay minerals will form when a feldspar is weathered. Climate probably plays a key role, for we know that kaolinite tends to form as a result of the intense chemical weathering in warm, humid climates and that illite and montmorillonite seem to develop more commonly in cooler, drier climates. The history of the rock also seems to be influential. For example, when a soil forms from a sedimentary rock in which a clay has been incorporated, we often find that the soil contains the same type of clay as does the parent rock. The analysis of a number of sedimentary rocks and the soils developed on these rocks has shown that when illite is present in the original rock it is usually the dominant clay in the soil, regardless of climate. Clearly, then, both environment and inheritance seem to influence the type of clay that will develop from the chemical weathering of a feldspar.

Let us look back for a minute to the equation for the decomposition of orthoclase on page 113. Notice that the second product consists of potassium ions. We might expect that these would be carried off by water percolating through the ground, and that all of it would eventually find its way to the rivers and finally to the sea. Yet analyses show that not nearly as much potassium is present in river and ocean water as we would expect. What happens to the rest of the potassium? Some of it is used by growing plants before it can be carried away in solution, and some of it is absorbed by clay minerals or even taken into their crystal structure.

The third product resulting from the decomposition of orthoclase is silica, which appears either in solution (for even silica is slightly soluble in water) or as very finely divided quartz in the size range of the colloids. In the colloidal state, silica may exhibit some of the properties of silica in solution.

So far, we have been talking about the weathering of only orthoclase feldspar. But the products of the chemical weathering of plagioclase feldspars are very much the same. Instead of potassium carbonate, however, either sodium or calcium carbonate is produced, depending on whether the feldspar is the sodium albite or the calcium anorthite (see Table 6.1). As we found in Section 3.4 (see "Silicates"), the plagioclase feldspars almost invariably contain both sodium and calcium. The carbonates of sodium and calcium are soluble in water and may eventually reach the sea. We should note here, however, that calcium carbonate also forms the mineral calcite (see "Carbonate and sulfate materials," Section 3.4). Calcite, in turn, forms the greater part of limestone (a sedimentary rock) and marble (a metamorphic rock). Both limestone and marble are discussed in subsequent chapters.

WEATHERING OF FERROMAGNESIAN MINERALS Now let us turn to the chemical weathering of the third group of common minerals in igneous rocks: the ferromagnesian silicates. The chemical weathering of these minerals produces the same products as the weathering of the feldspars: clay, soluble salts, and finely divided silica. But the presence of iron and magnesium in the ferromagnesian minerals makes possible certain other products as well.

The iron may be incorporated into one of the clay minerals or into an iron carbonate mineral. Usually, however, it unites with oxygen to form hematite, Fe_2O_3, one of the most common of the iron oxides. Hematite commonly has a deep red color, and in powdered form it is always red; this characteristic

Table 6.1 ***Chemical weathering products of common rock-forming silicate minerals***

		Important decomposition products	
Mineral	*Composition*	*Minerals*	*Others*
Quartz	SiO_2	Quartz grains	Some silica in solution
Feldspars:			
Orthoclase	$K(AlSi_3O_8)$	Clay Silica	Some silica in solution Potassium carbonate (soluble)
Albite (sodium plagioclase)	$Na(AlSi_3O_8)$	Clay	Some silica in solution
Anorthite (calcium plagioclase)	$Ca(Al_2Si_2O_8)$	Silica Calcite (from Ca)	Sodium and calicum carbonates (soluble)
Ferromagnesians:			
Biotite Augite Hornblende	Fe, Mg, Ca silicates of Al	Clay Calcite Limonite Hematite Silica	Some silica in solution Carbonates of calcium and magnesium (soluble)
Olivine	$(Fe, Mg)_2SiO_4$	Limonite Hematite Silica	Some silica in solution Carbonates of iron and magnesium (soluble)

gives it its name, from the Greek *haimatitēs*, "blood-like." Sometimes the iron unites with oxygen and an hydroxyl ion to form *goethite*, FeO(OH), generally brownish in color. (Goethite was named after the German poet Goethe, in deference to his lively scientific interests.) Chemical weathering of the ferromagnesian minerals often produces a substance called *limonite*, yellowish to brownish in color and referred to in everyday language as just plain "rust." Limonite is not a true mineral because its composition is not fixed within narrow limits, but the term is universally applied to the iron oxides of uncertain composition that contain a variable amount of water. Limonite and some of the other iron oxides are responsible for the characteristic colors of most soils.

What happens to the magnesium produced by the weathering of the ferromagnesian minerals? Some of it may be removed in solution as a carbonate, but most of it tends to stay behind in newly formed minerals, particularly in the illite and montmorillonite clays.

SUMMARY OF WEATHERING PRODUCTS If we know the mineral composition of an igneous rock, we can determine in a general way the products that the chemical weathering of that rock will probably yield. The chemical weathering products of the common rock-forming minerals are listed in Table 6.1. These products include the minerals that make up most of our sedimentary rocks, and we shall discuss them again in Chapter 7.

6.3 Rates of weathering

Some rocks weather very rapidly, and others only slowly. Rate of weathering is governed by the type of rock and a variety of other factors, from minerals and moisture, temperature and topography, to plant and animal activity.

Rate of mineral weathering

On the basis of field observations and laboratory experiments, the minerals commonly found in igneous rocks can be arranged according to the order in which they are chemically decomposed at the surface. We are not sure of all the details, and different investigators report different conclusions, but we can make the following general observations:

1 Quartz is highly resistant to chemical weathering.

2 The plagioclase feldspars weather more rapidly than orthoclase feldspar.

3 Calcium plagioclase (anorthite) tends to weather more rapidly than sodium plagioclase (albite).

4 Olivine is less resistant than augite, and in many instances augite seems to weather more rapidly than hornblende.

5 Biotite mica weathers more slowly than the other dark minerals and muscovite mica is more resistant than biotite.

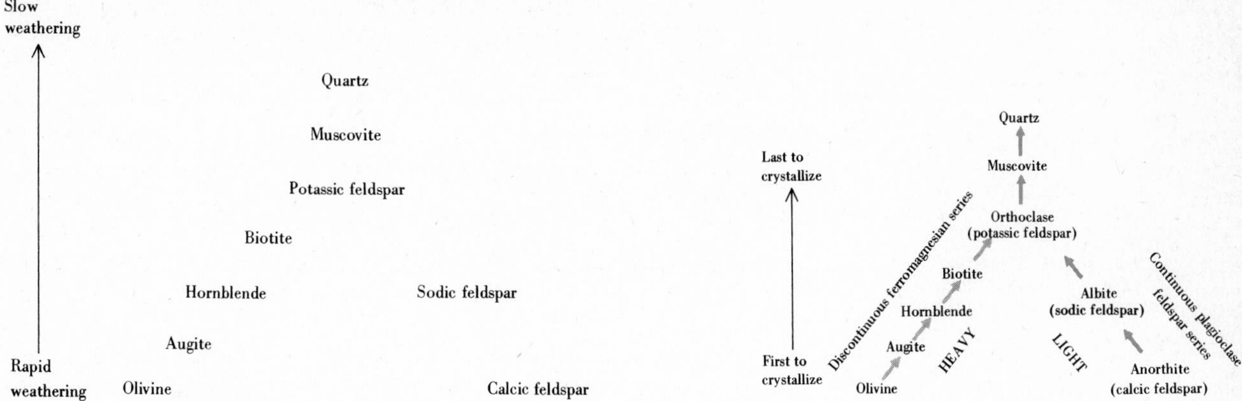

Figure 6.11 ***Relative rapidity of chemical weathering of the common igneous rock-forming minerals. The rate of weathering is most rapid at the bottom and decreases toward the top. Note that this table is in the same order as Bowen's reaction series (right). The discrepancy in the rate of chemical weathering between, for instance, olivine and quartz, is explained by the fact that in the zone of weathering olivine is farther from its environment of formation than is quartz. It therefore reacts more rapidly than quartz to its new environment and thus weathers more rapidly.***

Notice that these points suggest a pattern (Figure 6.11) similar to that of Bowen's reaction series for crystallization from magma, discussed in Section 5.2 (illustrated in Figure 5.9, shown as inset, above, for comparison). But there is one important difference: In weathering, the successive minerals formed do not react with one another as they do in a continuous reaction series.

The relative resistance of these minerals to decomposition may reflect the difference between the surface conditions under which they weather and the conditions that existed when they were formed. Olivine, for example, forms at high temperatures and pressures, early in the crystallization of a melt. Consequently, as we might expect, it is extremely unstable under the low temperatures and pressures that prevail at the surface, and it weathers quite rapidly. On the other hand, quartz forms late in the reaction series, under considerably lower temperatures and pressures. Because these conditions are more similar to those at the surface, quartz is relatively stable and is very resistant to weathering.

Now we can qualify slightly the definition of weathering given at the beginning of this chapter. We have found that weathering disrupts the equilibrium that existed while the minerals were still buried in the earth's crust and that this disruption converts them into new minerals. Following Parry Reiche,[1] we may revise our definition as follows: *Weathering is the response of materials that were once in equilibrium within the earth's crust to new conditions at or near contact with air, water, and living matter.*

Depth and rapidity of weathering

Most weathering takes place in the upper few meters or tens of meters of the earth's crust, where rock is in closest contact with air, moisture, and organic

[1]Parry Reiche, "A Survey of Weathering Processes and Products," *Univ. New Mexico Publ. Geol.*, no. 3, p. 5, 1950.

matter. But some factors operate well below the surface and permit weathering to penetrate to great depths. For instance, when erosion strips away great quantities of material from the surface, the underlying rocks are free to expand. As a result, parting planes or fractures—the joints that we spoke of earlier in the chapter—develop hundreds of meters below the surface.

Then, too, great quantities of water move through the soil and down into the underground, transforming some of the materials there long before they are ever exposed at the surface. Rock salt that is located deep below the surface in the form of sedimentary rock often undergoes exactly this transformation. If enough underground water is present, the salt is dissolved and carried off long before erosion can expose it.

Weathering is sometimes so rapid that it can actually be recorded. The Krakatoa eruption of August 1883, described in Chapter 4, threw great quantities of volcanic ash into the air and deposited it to a depth exceeding 30 m on the nearby island of Long-Eiland. By 1928, 45 years later, a soil nearly 35 cm deep had developed on top of this deposit, and laboratory analyses showed that a significant change had taken place in the original materials. Chemical weathering had removed a measurable amount of the original potassium and sodium. Furthermore, either mechanical weathering or chemical weathering, or both, had broken down the original particles so that they were generally smaller than the particles in the unweathered ash beneath.

Figure 6.12 ***Weathering of a marble headstone, burying ground of Christ Church, Cambridge, Massachusetts. The inscription, carved in 1818, is illegible. The monument illustrates the instability and rapid weathering of calcite (the predominant mineral in marble) in a humid climate.*** *(Photograph by Sheldon Judson, 1968.)*

In a more recent study, scientists from Ohio State University have demonstrated the nature and rate of soil development of unweathered material exposed to the atmosphere with the retreat of Muir glacier in Glacier Bay National Park in southeastern Alaska. Geologic and human records establish the successive positions of the retreating ice front since about 1700. Between this date and 1965, the front of the Muir glacier retreated approximately 65 km. The age of soils at various points along this line of retreat can thus be demonstrated. Over a period of about 250 years a soil about 35 cm thick has developed. Of this, the upper half is represented chiefly by the accumulation of organic material. The lower half, however, shows changes in the materials left by the retreating ice. After 250 years, virtually all the calcite and dolomite had been removed to a depth of at least 15 cm, and the soil acidity had increased markedly. The amount of iron oxide in the form of hematite had increased measurably, particularly in the lowest 10 cm.

Graveyards provide many fine examples of weathering that has occurred within historic time. Calcite in the headstone pictured in Figure 6.12 has weathered so rapidly that the inscription to the memory of Moses L. Gould, carved in 1818, is only partially legible after a century and a half. Examination of other marble slabs in the burying ground indicated that the earliest legible date was 1811.

Undoubtedly the rate of weathering has increased with time, for two reasons. First, continued weathering roughens the marble surface, exposing more and more of it to chemical attack and quickening the rate of decomposition. Second, as the number of factories and dwellings in Cambridge and neighboring towns increased, the amount of carbon dioxide in the atmosphere also increased. Consequently, rain water in the twentieth century carries more carbonic acid than it did in the nineteenth, and attacks calcite more rapidly.

Figure 6.13 ***Weathering of a slate headstone, burying ground of Christ Church, Cambridge, Massachusetts. The inscription date, 1699, testifies to the durability of the slate.*** *(Photograph by Sheldon Judson, 1968.)*

In contrast to the strongly weathered marble is a headstone of slate erected in 1699 (Figure 6.13). Two hundred and sixty-nine years later, the inscription is still plainly visible. Slate is usually a metamorphosed shale, which, in turn, is composed largely of clay minerals formed by the weathering of feldspars. The clay minerals in the headstone were originally formed in the zone of weathering. Slight metamorphism has since changed many of the clay minerals to muscovite, a white mica, and the shale to slate (Section 8.3). The muscovite, however, is relatively stable in the zone of weathering as indicated in Figure 6.11.

These examples show, then, that weathering often occurs rapidly enough to be measured during one's lifetime. Let us now turn to the process that moves these products of weathering: erosion.

6.4 *Rates of erosion*

We found that chemical and mechanical weathering produces certain materials from the rocks of the earth. It is this material that the agents of erosion move from one place to another until, finally, it reaches the settling basins of the world's oceans. Several agents are involved in this movement and they include water, ice, and wind. These agents are individually treated later (Chapters 10, 11, 12, 13, and 14), but here it would be instructive to pause and consider how much material is being removed from the continents and at what rate.

Sometimes ancient ruins provide an index to the rates of erosion. Thus Figure 6.14 is a photograph of the remains of a Roman cistern built 60 km north of Rome, Italy, in the second century A.D. The footings exposed at the base of the finished wall indicate the amount of erosion that has occurred here

Figure 6.14 **Ruins of a cistern built in the second century A.D. for a Roman villa about 60 km north of Rome. The exposed footings measure 1.3 m, as indicated by the tape. This is the amount of erosion which has taken place since the cistern was built.** (*Photograph by Sheldon Judson.*)

since the structure was built. The rate of erosion from then to the present averages about 30 cm/1,000 years.

This, however, is only a spot measurement at a specific place. What method can we use to measure the rates over large areas? One way is to measure the amount of material carried by a stream each year from its drainage basin. The amount of this material averaged over the area of the drainage basin gives an average figure of erosion for the river basin. Now, obviously, the rate of erosion is not the same at every place in the basin. In some places the material will be removed more rapidly than at others. Furthermore, there will be places at which deposition takes place and the material is temporarily halted on its way out of the basin. Nevertheless, this method gives us an average figure for a unit area within the drainage basin.

A stream carries material in solution as well as solid material in the form of sediments. Most of the solid material is buoyed up by the flow of water and is said to be carried in *suspension.* A relatively smaller amount is pushed and bounced along the stream bottom in what is known as *bed load* or *traction load.* Suspended and dissolved matter can be measured without too much difficulty. It is very difficult to measure traction load, but it is generally small and usually considered to be about 10 percent of the suspended load in the average stream.

With these facts in mind, we may ask how much material is carried annually by streams out of the drainage basin in which is located the ancient Roman cistern in Figure 6.14. The major stream here is the Tiber River. If we measure the load of the Tiber at Rome, we find that it is removing sediments on the average of about 7.5 million metric tons of solid material each year from the area upstream. This converts to a rate of 17 cm of erosion per 1,000 years over the entire basin. If we had data on bed load and dissolved load for the Tiber, the figures would be higher but probably not by very much.

Turning to the United States, we can examine Table 6.2, which describes the erosion going on in the major regions of the country. Not included here is the drainage to the Great Lakes and the St. Lawrence River. Also, the table does not include the area in the western states known as the Great Basin, where topography and rainfall are such that streams do not reach the sea. We see that the rates vary from region to region, but on the average it is approximately 6 cm/1,000 years.

Data from the Amazon River, the world's largest river, indicates that it is removing material from its basin at the rate of 4.7 cm/1,000 years. Another large tropical river, the Congo, is carrying enough material out of its basin each year to reduce its drainage basin by approximately 2 cm/1,000 years.

A single large drainage basin integrates many factors affecting the rapidity of erosion. One we should consider carefully is the effect of man on the rate of erosion. There is enough information from disciplines as diverse as archaeology and nuclear physics to indicate that when man occupies an area intensively and turns it to crop land, he increases the rate of erosion between 10 and 100 times that of a naturally forested or grassed area.

How much material is transported each year to the oceans? We have already cited some figures for portions of the earth. These figures, however, cover only 10 percent of the land's surface. Therefore, any estimate of the total amount

Table 6.2 *Rates of regional erosion in the United States*[a]

Drainage region	Drainage area[b], $km^2 \times 10^3$	Runoff, $m^3/sec \times 10^3$	Load tons, km^2/yr Dissolved	Solid	Total	Erosion, cm/1,000 yr	Area sampled, %	Av. yr of record
Colorado	629	0.6	23	417	440	17	56	32
Pacific Slopes, California	303	2.3	36	209	245	9	44	4
Western Gulf	829	1.6	41	101	142	5	9	9
Mississippi	3,238	17.5	39	94	133	5	99	12
South Atlantic and Eastern Gulf	736	9.2	61	48	109	4	19	7
North Atlantic	383	5.9	57	69	126	5	10	5
Columbia	679	9.8	57	44	101	4	39	<2
Total	6,797	46.9	43	119	162	6		

[a] After Sheldon Judson and D. F. Ritter, "Rates of Regional Denudation in the United States," *J. Geophys. Res.*, vol. 69, Table 3, p. 3399, 1964.
[b] Great Basin, St. Lawrence, and Hudson Bay drainage not considered.

of erosion of the earth's surface per unit time can be no better than that, a mere estimate. The information now at hand suggests, however, that something approaching 10 billion metric tons/year was being delivered each year to the ocean before man became an effective geological agent. This does not include wind-blown material, although it is thought to be negligible. Glacier ice is not considered but probably would not appreciably affect our figure. The rate of 10 billion metric tons/year coincides to a reduction of the continents of approximately 2.5 cm/1,000 years.

If we take into consideration man's effect on erosion, it is estimated that at present approximately 24 billion metric tons are moved annually by the rivers to the oceans. This is 2.5 times the amount of material moved before man's intervention.

Differential erosion

Differential erosion is the process by which different rock masses or different sections of the same rock mass erode at different rates. Almost all rock masses of any size weather in this manner. The results vary from the boldly sculptured forms of the Grand Canyon to the slightly uneven surface of a marble tombstone. Unequal rates of erosion are caused chiefly by variation in the composition of the rock. The more resistant zones stand out as ridges or ribs above the more rapidly weathered rock on either side.

A second cause of differential erosion is simply that the intensity of weathering varies from one section to another in the same rock. Figure 6.15 shows a sandstone memorial that has undergone mechanical weathering in certain spots. The rock is a coarse, red, homogeneous sandstone made up of quartz, feldspar, and mica, with some red iron oxides. The inner sides of the pillars have disintegrated so badly that the original fluting has been entirely destroyed and the underside of the horizontal slab has scaled noticeably. Notice that

Figure 6.15 **Differential weathering in a sandstone monument, burying ground of Christ Church, Cambridge, Massachusetts. Mechanical weathering has been most rapid on the shaded portions (detail at left). Here moisture remains longer and consequently frost action has been more effective than on the less shaded portions.** (*Photographs by Sheldon Judson, 1968.*)

Figure 6.16 **A generalized cross section in northeastern Pennsylvania and northwestern New Jersey in the vicinity of the Delaware Water Gap to show the relation of topography to the different rock types below the surface. See the text for discussion.**

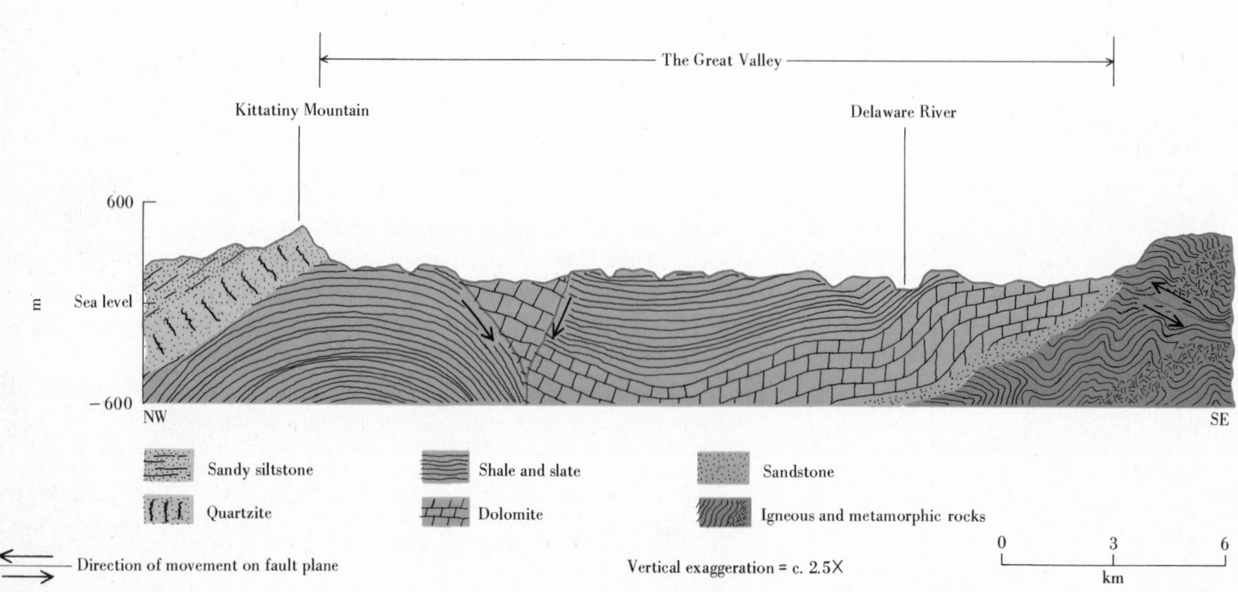

these are the areas least accessible to the drying action of the sun. Consequently, moisture has tended to persist here, and the frost action has pried off flakes and loosened individual grains.

On a larger scale, differential erosion is shown in the cross section in Figure 6.16. The section is drawn across a portion of northeastern Pennsylvania near the Delaware Water Gap. Note that the highest ridge is underlaid by a tough quartzite which weathers very slowly. The major portion of the Great Valley is underlaid by a shaley slate which weathers easily. In this same valley are still lower spots which are underlaid by a dolomitic limestone which is susceptible to rapid chemical weathering in this humid climate. To the southeast lie the New Jersey Highlands, held up by resistant units of crystalline rock, both igneous and metamorphic.

6.5 *Soils*

So far, we have been discussing the ways in which weathering acts to break down existing rocks and to provide the material for new rocks, but weathering also plays a crucial role in the creation of the soils that cover the earth's surface and sustain all life. In fairly recent years, the study of soils has developed into the science of *pedology* (from the Greek *pedon*, "soil," with *logos*, "reason," hence "knowledge of soil" or "soil science").[2]

Soil classification

In the early years of soil study in the United States, researchers thought that the parent material almost wholly determined the type of soil that would result from it. Thus, they reasoned, granite would weather to one type of soil and limestone to another.

It is true that a soil reflects to some degree the material from which it developed, and in some instances one can even map the distribution of rocks on the basis of the types of soil that lie above them. But as more and more information became available, it became apparent that the bedrock is not the only factor determining soil type. Russian soil scientists, following the pioneer work of V. V. Dokuchaev (1846–1903), demonstrated that different soils develop over identical bedrock material in different areas when the climate varies from one area to another. The idea that climate exerts a major control over soil formation was introduced in this country in the 1920s by C. F. Marbut (1863–1935), for many years chief of the U.S. Soil Survey in the Department of Agriculture. Since that time, soil scientists have discovered that still other factors exercise important influences on soil development. For instance, the

[2]In the United States, the term *pedology* is sometimes confused with words based on *ped* and *pedi*, combining forms meaning "foot," or with words based on *ped* and *pedo*, combining forms meaning "boy, child," as in pediatrics, the medical science that treats of the hygiene and diseases of children. Consequently, there is a tendency in this country to use *soil science* instead of *pedology*.

Figure 6.17 **The three major horizons of a soil. In many places it is possible to subdivide the zones themselves. Here the soil is shown as having developed from limestone.**

relief of the land surface plays a significant role. The soil on the crest of a hill is somewhat different from the soil on the slope, which, in turn, differs from the soil on the level ground at the foot of the hill; yet all three soils rest on identical bedrock. The passage of time is another factor. A soil that has only begun to form differs from one that has been developing for thousands of years, although the climate, bedrock, and topography are the same in each instance. Finally, the vegetation in an area influences the type of soil that develops there. One type of soil will form beneath a pine forest, another beneath a forest of deciduous trees, and yet another on a grass-covered prairie.

Exactly what is a soil? It is a natural, surficial material that supports plant life. Each soil exhibits certain properties that are determined by climate and living organisms operating over periods of time on earth materials and on landscape of varying relief. Because all these factors are combined in various ways all over the land areas of the globe, the number of possible soil types is almost unlimited.

And yet certain valid generalizations can be made about soils. We know, for example, that the composition of a soil varies with depth. A natural or artificial exposure of a soil reveals a series of zones, each recognizably different from the one above. Each of these zones is called a *soil horizon*, or, more simply, a *horizon*. The three major zones or horizons in a typical soil, shown in Figure 6.17, may be described, from the bottom upward, as follows.

C HORIZON The *C* horizon is a zone of partially disintegrated and decomposed rock material. Some of the original bedrock minerals are still present, but others have been transformed into new materials. The *C* horizon grades downward into the unweathered rock material.

B HORIZON The *B* horizon lies directly above the *C* horizon. Weathering here has proceeded still further than in the underlying zone, and only those minerals of the parent rock that are most resistant to decomposition (quartz, for example) are still recognizable. The others have been converted into new minerals or into soluble salts. In moist climates, the *B* horizon contains an accumulation of clayey material and iron oxides delivered by water percolating downward from the surface. In dry climates we generally find, in addition to the clay and iron oxides, deposits of more soluble minerals, such as calcite. This mineral, too, may have been brought down from above, but some is brought into the *B* horizon from below, as soil water is drawn upward by high evaporation rates. Because material is deposited in the *B* horizon, it is known as the "zone of accumulation" (see Figure 6.17).

A HORIZON The *A* horizon is the uppermost zone—the one into which we sink a spade when we dig a garden. This is the zone from which the iron oxides have been carried to the *B* horizon, and in dry climates it is the source of some soluble material that may be deposited in the *B* horizon. The process by which these materials have been moved downward by soil water is called *leaching*, and the *A* horizon is sometimes called the "zone of leaching" (Figure 6.17). Varying amounts of organic material tend to give the *A* horizon a gray to black color.

The three soil horizons have all developed from the underlying parent material. When this material is first exposed at the surface, the upper portion is subjected to intense weathering, and decomposition proceeds rapidly. As the decomposed material builds up, downward-percolating water begins to leach out some of the minerals and to deposit them lower down. Gradually, the *A* horizon and the *B* horizon build up, but weathering continues to go on, though at a slower rate now, on the underlying parent material, giving rise to the *C* horizon. With the passage of time, the *C* horizon reaches deeper and deeper into the unweathered material below, the *B* horizon keeps moving downward, and the *A* horizon, in turn, encroaches on the upper portion of the *B* horizon. Finally, a "mature" soil is built up.

The thickness of the soil that forms depends on many factors, but in the northern United States and southern Canada, the material that was first exposed to weathering after the retreat of the last ice sheet some 10,000 years ago is now topped by a soil 60 to 80 cm thick. Farther south, where the surface was uncovered by the ice at an earlier time, the soils are thicker, and in some places the processes of weathering have extended to 5 to 10 m below the present surface.

Some soil types

We can understand the farmer's interest in soil, but why is it important for the geologist to understand soils and the processes by which they are formed? There are several reasons.

First, soils provide clues to the environment in which they were originally formed. By analyzing an ancient soil buried in the rock record, we may be able to determine the climate and physical conditions that prevailed when it was formed.

Second, some soils are sources of valuable mineral deposits (see Chapter 21), and the weathering process often enriches otherwise low-grade mineral deposits, making them profitable to mine. An understanding of soils and soil-forming processes, therefore, can serve as a guide in the search for ores.

Third, because a soil reflects to some degree the nature of the rock material from which it has developed, we can sometimes determine the nature of the underlying rock by analyzing the soil.

But most important of all, soils provide the source of many of the sediments that are eventually converted into sedimentary rocks. And these, in turn, may be transformed into metamorphic rocks or, following another path in the rock cycle, may be converted into new soils. If we understand the processes and results of soil formation, we are in a better position to interpret the origin and evolution of many rock types.

The following pages will discuss three major types of soil. Two of them, the pedalfers and the pedocals, are typical of the middle latitudes. The third group, referred to as laterites, is found in tropical climates.

PEDALFERS A *pedalfer* is a soil in which iron oxides or clays, or both, have accumulated in the *B* horizon. The name is derived from *pedon,* Greek for

WARM HUMID

Podozol

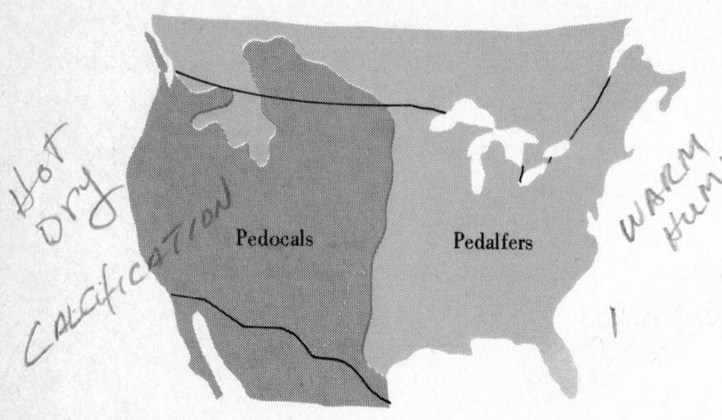

Figure 6.18 **Generalized distribution of pedalfer and pedocal soils in the United States and southern Canada. In the United States, the pedalfers have developed in the more humid climates to the east of the line that marks approximately 63 cm of precipitation per year. To the west of this line, where precipitation is generally less than 63 cm/year, pedocal soils predominate. In Canada and in the northern Rocky Mountains of the United States, temperature is more critical than precipitation in determining the distribution of the two soils. There pedocals occur in areas where the average annual temperature is 4.5° C or less.**

"soil," and the symbols Al and Fe for aluminum and iron. In general, soluble materials such as calcium carbonate or magnesium carbonate do not occur in the pedalfers. Pedalfers are commonly found in temperate, humid climates, usually beneath a forest vegetation. In the United States, most of the pedalfers lie east of a line that corresponds roughly to about 63 cm of rainfall per year. Northward in Canada, however, where the temperature is more important than the total rainfall in determining the distribution of pedalfers, the zone extends northwestward across Saskatchewan and Alberta coincident with a mean annual temperature of about 4.5°C or colder. Farther west, the pedalfers extend southward into the United States along the mountainous region of the Rockies, where the rainfall is somewhat higher and the temperatures lower than in the rest of western United States (see Figure 6.18).

In the formation of pedalfers, certain soluble compounds, particularly those that contain sodium, calcium, and magnesium, are rapidly removed from the *A* horizon by waters seeping into the soil from the surface. These soluble compounds proceed downward through the *B* horizon and are carried off by ground water. The less soluble iron oxides and clay are deposited in the *B* horizon, giving that zone a clayey character with a brownish to reddish color.

Using the information on the products of chemical weathering listed in Table 6.1, we can build up a picture of how a pedalfer develops from granite in temperate and humid areas. At some depth below the surface lies the unaltered granite. Directly above is the crumbly, partially disintegrated rock of the *C* horizon. Here we can still identify the minerals that made up the original granite, although the feldspars have started to decompose and have become cloudy, and the iron-bearing minerals have been partially oxidized.

Moving upward to the *B* horizon, the zone of accumulation, we find that here the feldspars have been converted to clay and that the material has a compact, clayey texture. Iron oxides or limonite stain the soil a reddish or brownish color. And because the grains of quartz released from the granite have undergone little change, we find some sand in this otherwise clayey zone.

The *A* horizon, a few centimeters thick, has a grayish to ashen color, for the iron compounds have been leached from this zone and now color the *B* horizon below. Furthermore, the texture of this zone is sandier, for most of the finer materials have also been moved downward to the *B* horizon, and the soluble salts have been largely dissolved and removed by water. The very top of the *A* horizon is a thin zone of dark, humic material. The final pedalfer soil is shown in Figure 6.19.

Notice in Figure 6.20 how the original minerals of the granite are transformed as weathering progresses. On the left we have quartz, plagioclase, and orthoclase, which were released directly from the granite. Then, as weathering progresses, we find that the amount of kaolinite increases at the expense of the original minerals. The initial rise in the amount of quartz and orthoclase

Forest

Forest litter
A horizon
Sandy gray loam
B horizon
Reddish-brown to yellowish sandy compact clay, grading into
C horizon
Rotted granite, still retaining original granitic minerals and texture, grading into
Unaltered granite
1.25 m

Figure 6.19 **A pedalfer soil that has developed on a granite. Note the transition from unaltered granite, upward through partially decomposed granite of the C horizon, into the B horizon, where no trace of the original granite structure remains, and finally into the A horizon, just below the surface.**

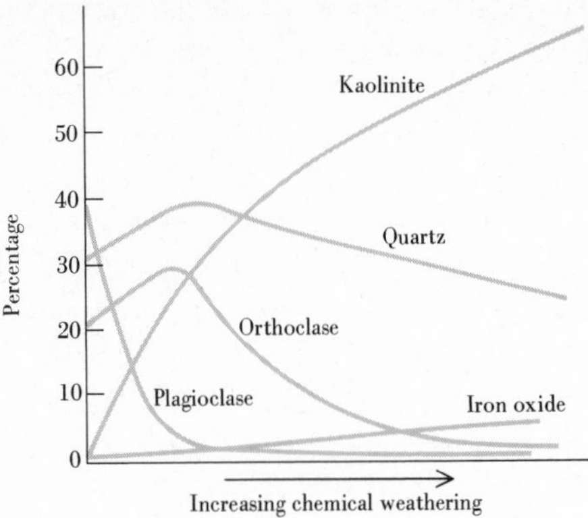

Figure 6.20 **Change in mineral percentages as a granite is subjected to increased chemical weathering. As weathering progresses, the orthoclase and plagioclase of the granite decrease in abundance and give rise to an increasing amount of clay (kaolinite). At the same time, the percentage of iron oxides increases at the expense of the original iron silicate minerals (not shown).** (*Redrawn from S. S. Goldich, "A Study of Rock Weathering,"* J. Geol., *vol. 46, p. 33, 1938.*)

indicates simply that these minerals tend to accumulate in the soil because of their greater resistance to decomposition. Iron oxides also increase with weathering as the iron-bearing silicates decompose.

There are several varieties of soils in the pedalfer group, including the red and yellow soils of the southeastern states, as well as *podsol* (from the Russian, "ashy gray soil"), and the gray-brown podsolic soils of the northeastern quarter of the United States and of southern and eastern Canada (Figure 6.21). Prairie soils are transitional varieties between the pedalfers of the East and the pedocals of the West.

PEDOCALS *Pedocals* are soils that contain an accumulation of calcium carbonate. Their name is derived from a combination of the Greek *pedon*, "soil," with an abbreviation for calcium. The soils of this major group are found in the temperate zones where the temperature is relatively high, the rainfall is low, and the vegetation is mostly grass or brush growth. In the United States, the pedocals are found generally to the west of the pedalfers; and in Canada, to the southwest.

In the formation of pedocals, calcium carbonate and, to a lesser extent, magnesium carbonate are deposited in the soil profile, particularly in the *B* horizon. This process occurs in areas where the temperature is high, the rainfall scant, and the upper level of the soil is hot and dry most of the time. Water evaporates before it can remove carbonates from the soil. Consequently, these compounds are precipitated as *caliche*, a whitish accumulation made up largely of calcium carbonate. *Caliche* is a Spanish word, a derivative of the Latin *calix*, "lime." The occasional rain may carry the soluble material down from the *A* horizon into the *B* horizon, where it is later precipitated as the water evaporates. Soluble material may also move up into the soil from below. In this case, water beneath the soil or in its lower portion rises toward the surface

Figure 6.21 **This podsol, a member of the pedalfer group of soils, occurs on Cape Cod, Massachusetts. The dark layer at the very top just beneath the plant cover is the humic zone in the upper part of the A horizon. The light gray, leached zone below makes up the bulk of the A horizon. The B horizon is thin and shows up dark in the photograph because of the iron oxides that have accumulated there. This zone grades down into the C horizon, the parent material, which is here an unconsolidated sandy deposit.** (*Photograph by Paul MacClintock.*)

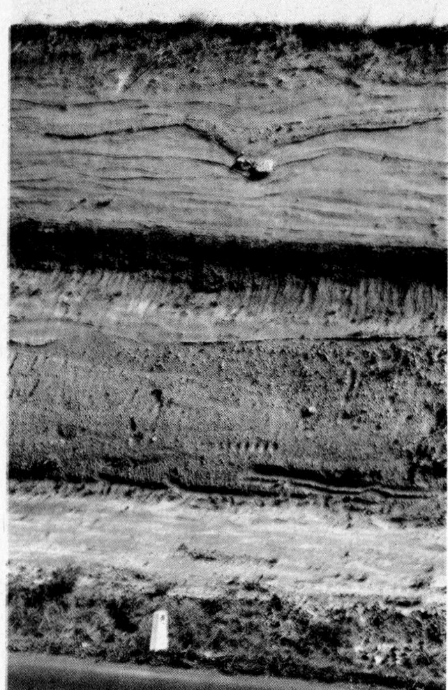

Figure 6.22 **Not all soils occur at the surface. Some are buried beneath younger material and are known as paleosols. The dark band in the center of this exposure is such a soil and measures over 1 m in thickness. It is developed on water-laid pyroclastic material and is buried by beds of similar volcanic material. At the top of the younger beds, the present surface soil has been almost completely destroyed by erosion processes accelerated by human activity. Crater of Baccano along the Via Cassia, north of Rome, Italy.** (*Photograph by Sheldon Judson.*)

through small, capillary openings. Then, as the water in the upper portions evaporates, the dissolved materials are precipitated.

Pedocals tend to develop under a growth of brush and grass, which also helps to concentrate the soluble carbonates by intercepting them before they can be moved downward in the soil. When the plants die, the carbonates are either added to the soil, where they are used by other plants, or are simply precipitated in the soil by high evaporation rates.

Because rainfall is light in the climates where pedocals form, chemical weathering proceeds only slowly, and clay is produced less rapidly than in more humid climates. For this reason, pedocals contain a lower percentage of clay minerals than do the pedalfer soils.

The pedocal group includes the black and chestnut-colored soils of southern Alberta and Saskatchewan in Canada and the northern plains of the United States, the reddish soils farther south, and the red and gray desert soils of the drier western states.

TROPICAL SOILS (LATERITES) The term *laterite* is applied to many tropical soils that are rich in hydrated aluminum and iron oxides. The name itself, from the Latin for "brick," suggests the characteristic color produced by the iron in these soils. The formation of laterites is not well understood. In fact, soil scientists are not even certain that the *A*, *B*, and *C* horizons characteristic of the pedalfers and the pedocals have their counterparts in the laterites, even though these soils do exhibit recognizable zones.

In the development of the laterites, iron and aluminum accumulate in what is presumed to be the *B* horizon. The aluminum is in the form of $Al_2O_3 \cdot nH_2O$, which is generally called *bauxite*, an ore of aluminum. This ore appears to be developed when intense and prolonged weathering removes the silica from the clay minerals and leaves a residuum of hydrous aluminum oxide, that is, bauxite. In some laterites, the concentration of iron oxides in the presumed *B* horizon is so great that it is profitable to mine them for iron, as is done in certain districts of Cuba.

The term *laterite* is most properly applied only to the zone in which iron and aluminum have accumulated. As we have seen, this is the zone that may be equivalent to the *B* horizon in more northerly soils. Overlying this zone, there is often a zone of crumbly loam, and below it is a light-colored, apparently leached zone that adjoins the parent material. Some soil scientists refer to these two zones as the *A* horizon and the *C* horizon, respectively.

Outline

Weathering is the response of surface or near-surface material to contact with water, air, and living matter.

Types of weathering are mechanical and chemical.

Mechanical weathering (*disintegration*) involves a reduction in the size of rock and mineral particles but no change in composition.

Chemical weathering (*decomposition*) involves a change in the composition of the material weathered. The rate of chemical weathering increases with the decrease in the size of particles and with increase in temperature and moisture. *Chemical weathering of igneous rocks,* which include quartz, feldspar, and ferromagnesian minerals, gives clay, iron oxides, quartz, and soluble salts.

Rates of weathering vary with material weathered and the environment. For example, olivine weathers chemically more rapidly than quartz, and limestone weathers very rapidly in a moist climate but very slowly in a dry climate.

Rates of erosion indicate that before man began to use the landscape intensively rivers transported about 10 billion metric tons of material to the seas annually. Today it is two or three times that, chiefly because of the presence of man.

Differential erosion is the process by which different rock masses or different sections of the same rock mass erode at different rates. For example, limestone is more resistant to erosion in a dry climate than is mudstone.

Soil is a naturally occurring surface material that supports life and generally is the product of weathering.

Soil zones are the *A, B,* and *C* horizons from the surface downward.

Soil types include *pedalfers* in moist temperate climates, *pedocals* from dry temperate climates, and *laterites* from moist tropical climates.

Supplementary readings

BUCKMAN, H. O., and N. C. BRADY: *The Nature and Properties of Soils,* 7th ed., The Macmillan Company, New York, 1969. A standard and successful introductory text.

DONAHUE, ROY L.: *Soils: An Introduction to Soils and Plant Growth,* Prentice-Hall, Inc., Englewood Cliffs, N.J., 1965. A basic introduction to the subject of soils and the interaction between soils and plants.

JUDSON, SHELDON: "Erosion of the Land—or What's Happening to our Continents?" *Am. Scientist,* vol. 56, pp. 356–379, 1968. An overview of worldwide erosion.

KELLER, WALTER D.: *Chemistry in Introductory Geology,* Lucas Brothers, Columbia, Mo., 1957. A compact statement of the role of chemistry in the geological processes, including weathering.

LEGGET, ROBERT F. (ed.): *Soils in Canada,* rev. ed., Special Publication, No. 3, Royal Society of Canada, University of Toronto Press, Toronto, 1965. A series of studies on geologic, pedologic, and engineering problems of Canadian soils.

REICHE, PARRY: *A Survey of Weathering Processes and Products,* University of New Mexico Press, Albuquerque, 1950. Despite its age, this remains an excellent summary of the subject.

Soil, U.S. Department of Agriculture Yearbook, 1957, Government Printing Office, Washington, D.C., 1957. Directed largely toward the management of soil.

7

Sedimentary rocks

WE HAVE ALL DUG OUR TOES INTO A SANDY BEACH, OR PICKED OUR WAY OVER THE gravels of a rushing stream, or perhaps slogged through the mud of a swamp. None of these—sand, gravel, or mud—immediately suggests hard, solid rock to us. Yet deposits of this sort, or materials very similar to them, are the "stuff" from which the great bulk of the rocks exposed at the earth's surface were formed. When we look down into the mile-deep Grand Canyon of the Colorado River in Arizona, we can see there layer upon layer of rocks that were once unconsolidated deposits of sand, gravel, and mud. Over the course of time, these loose sediments have been hardened into rocks that we call *sedimentary rocks.*

The story of sedimentary rocks begins with the weathering processes discussed in Chapter 6, for the products of chemical and mechanical weathering are the raw materials of sedimentary rocks. Streams, glaciers, wind, and ocean currents move the weathered materials to new localities and deposit them as sand, gravel, or mud. The transformation of these sediments into rock is the final step in the development of sedimentary rocks.

Some sediments, particularly sand and gravel, are consolidated into rock by a process that actually cements the individual grains. Subsurface water trickling through the open spaces leaves behind a mineral deposit that serves to cement the grains firmly together, giving the entire deposit the strength we associate with rock. Other sediments, such as fine deposits of mud, are transformed into rock by the weight of overlying deposits, which press or compact them into a smaller and smaller space.

The sedimentary rock that results from either of these processes may eventually be exposed at the earth's surface. If the rock was formed beneath the bottom of the ocean, it may be exposed either by the slow withdrawal of the seas or by the upward motion of the sea floor, forming new areas of dry land.

It is extremely difficult to work out a concise, comprehensive definition of sedimentary rocks. The adjective *sedimentary*, from the Latin *sedimentum,* means "settling." Therefore we might expect sedimentary rocks to be formed when individual particles settle out of a fluid, such as the water of a lake or an ocean. And many sedimentary rocks are formed in just that way. Fragments or minerals derived from the breakdown of rocks are swept into bodies of water where they settle out as unconsolidated sediments. Later, they are hardened into true rocks. But other rocks, such as rock salt, are made up of minerals left behind by the evaporation of large bodies of water, and these rocks are as truly sedimentary rocks as those formed from particles that have settled on an ocean floor. Still other sedimentary rocks are made up largely of the shells and hard parts of animals, particularly of invertebrate marine organisms.

Sedimentary rocks are often layered, or stratified. Unlike massive igneous rocks, such as granite, most sedimentary rocks are laid down in a series of individual beds, one on top of another. The surface of each bed is essentially parallel to the horizon at the time of deposition, and a cross section exposes a series of layers like those of a giant cake. True, some igneous rocks, such as those formed from lava flows, are also layered. By and large, however, *stratification is the single most characteristic feature of sedimentary rocks* (see Figure 7.1).

Figure 7.1 **Erosion has exposed and dissected these flat-lying beds of sedimentary rocks of mid-Tertiary age. The gentle, vegetated lower slopes are underlain by shales and the steep-walled cliffs are held up by more resistant sandstones. The center butte is the Glenn L. Jepsen Butte in the Castle Butte district of South Dakota.** (Photograph by C. C. O'Hara.)

7.1 Amounts of sediments and sedimentary rocks

About 75 percent of the rocks exposed at the earth's surface are sedimentary rocks or metamorphic rocks derived from them. Yet sedimentary rocks make up only about 5 percent by volume of the outer 15 km of the globe. The other 95 percent of the rocks in this zone are, or once were, igneous rocks (see Figure 7.2). The sedimentary cover is only as thick as a feather edge where it laps around the igneous rocks of the Adirondacks and the Rockies. In other places, it is thousands of meters thick. In the delta region of the Mississippi River, oil-drilling operations have cut into the crust nearly 7 km and have encountered nothing but sedimentary rocks. In the Ganges River

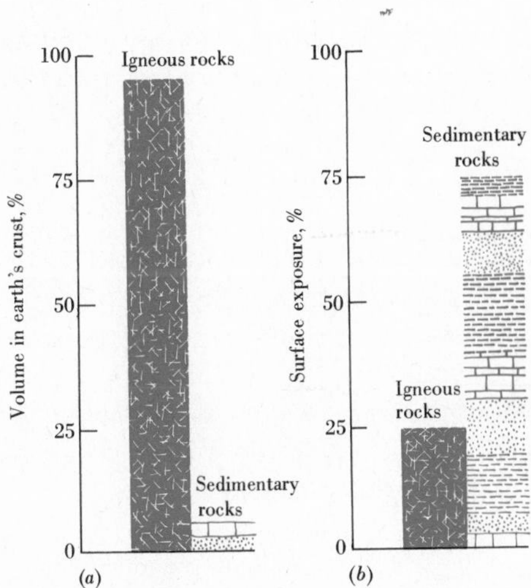

Figure 7.2 **Graphs showing relative abundance of sedimentary rocks and igneous rocks. (a) The great bulk (95 percent) of the outer 15 km of the earth is made up of igneous rocks; only a small percentage (5 percent) is sedimentary. In contrast (b), the areal extent of sedimentary rocks at the earth's surface is three times that of igneous rocks. Metamorphic rocks are considered with either igneous or sedimentary rocks, depending on their origin.**

Table 7.1 **Order of magnitude estimates by source of materials delivered annually to the oceans**[a]

Source	Metric tons/yr
Rivers	10 billion (10^{10})
Glaciers	100 million to 1 billion (10^8–10^9)
Wind	100 million (10^8)
Extraterrestrial	0.03 to 0.3 (3×10^{-2}–3×10^{-1})

[a] Estimated from various sources.

basin of India, the thickness of the sedimentary deposits has been estimated at between 13,500 and 18,000 m.

We have some estimates of the total mass of sedimentary rocks on the earth. Probably the best available estimate was made some time ago by Arie Poldervaart, who calculated their weight as $1{,}702 \times 10^{15}$ metric tons. Of this total, he estimated that 480×10^{15} metric tons are presently on the continents and that the rest are in the oceans.

We have seen that sedimentary rocks are formed from the materials weathered from preexisting rocks. Eventually this material reaches the deep ocean basin. What estimate can we make about the amount of material delivered each year to the oceans? The figures presented in Table 7.1 are order of magnitude estimates. Therefore, a statement of the approximate amount of material delivered annually to the oceans—where most sedimentary rocks form—is approximately 10^{10} metric tons. As Table 7.1 suggests, and as we have inferred in Section 6.4, the great bulk of this material is carried by the rivers. That contributed by wind, by ice, or by extraterrestrial sources does not appreciably change the total amount of material which is deposited in the oceans.

7.2 Formation of sedimentary rocks

We found in Chapter 5 that igneous rocks harden from molten material that originates some place beneath the surface, under the high temperatures and pressures that prevail there. In contrast, sedimentary rocks form at the much lower temperatures and pressures that prevail at or near the earth's surface.

Origin of material

The material from which sedimentary rocks are fashioned originates in two ways. First, the deposits may be accumulations of minerals and rocks derived either from the erosion of existing rock or from the weathered products of these rocks. Deposits of this type are called *detrital* (from the Latin for "worn down"), and sedimentary rocks formed from them are called *detrital sedimentary rocks.* Second, the deposits may be produced by chemical processes. We refer to these deposits as *chemical deposits* and to the rocks formed from them as *chemical sedimentary rocks.*

Gravel, sand, silt, and clay derived from the weathering and erosion of a land area are examples of detrital sediments. Let us take a specific example. The quartz grains freed by the weathering of a granite may be winnowed out by the running water of a stream and swept into the ocean. There they settle out as beds of sand, a detrital deposit. Later, when this deposit is cemented to form a hard rock, we have a sandstone, a detrital rock.

Chemically formed deposits are usually laid down by the precipitation of material dissolved in water. This process may take place either directly, through inorganic processes, or indirectly, through the intervention of plants or animals. The salt left behind after a salty body of water has evaporated is an example of a deposit laid down by inorganic chemical processes. On the other hand, certain organisms, such as the corals, extract calcium carbonate from sea water and use it to build up skeletons of calcite. When the animals die, their skeletons collect as a biochemical (from the Greek for "life," plus "chemical") deposit, and the rock that subsequently forms is called a *biochemical rock*—in this case, limestone.

Although we distinguish between the two general groups of sedimentary rocks—detrital and chemical—most sedimentary rocks are mixtures of the two. We commonly find that a chemically formed rock contains a certain amount of detrital material. In similar fashion, predominantly detrital rocks include some material that has been chemically deposited.

Geologists use various terms to describe the environment in which a sediment originally accumulated. For example, if a limestone contains fossils of an animal that is known to have lived only in the sea, the rock is known as a *marine* limestone. *Fluvial,* from the Latin for "river," is applied to rocks formed by deposits laid down by a river. *Aeolian,* derived from Aeolus, the Greek god of wind, describes rock made up of wind-deposited material. Rocks formed from lake deposits are termed *lacustrine,* from the Latin word for "lake."

Detrital and chemical, however, are the main divisions of sedimentary rocks based on the origin of material, and as we shall see later, they form the two major divisions in the classification of sedimentary rocks.

Sedimentation

The general process by which rock-forming material is laid down is called *sedimentation,* or deposition. The factors controlling sedimentation are easy to visualize. To have any deposition at all, there must obviously be something to deposit—another way of saying that a source of sediments must exist. We also need some process to transport this sediment. And, finally, there must be some place and some process for the deposition of the sedimentary material.

SOURCE OF MATERIAL In talking about the rock cycle (Chapter 1), we mentioned that igneous rocks are the ultimate source of the sediments in sedimentary rocks but that metamorphic rocks or other sedimentary rocks may serve as the immediate source.

In either case, after the rock material has been weathered, it is ready to be transported to some place of accumulation. Its movement is usually from a higher to a lower level. The energy for this movement is provided by gravity,

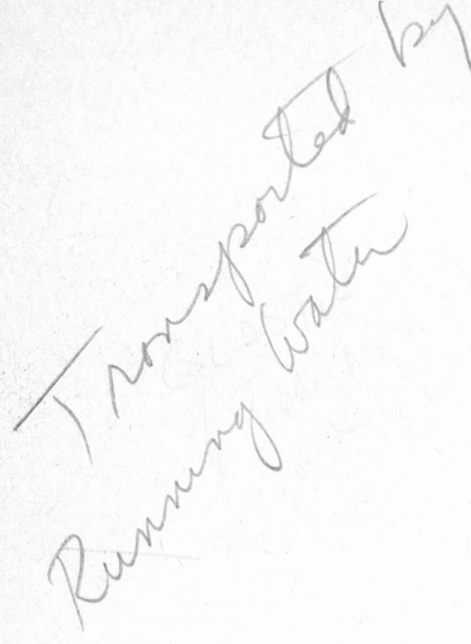

which makes possible not only the process of mass movement itself but also the activity of such agents of transportation as running water and glacier ice. If gravity were free to go about its work without opposition, it would long ago have reduced the continents to smooth, low-lying landmasses. But working against the leveling action of gravity are energies within the earth that elevate the continents and portions of the sea floor (see Chapters 18 and 19). By constantly exposing new areas of the earth's surface to weathering, movements of this sort ensure a continuing supply of material for the formation of sedimentary rocks.

The weathering of existing rock provides both detrital sediments and the soluble material that is eventually converted into chemical deposits. The nature of the rock that furnishes the sedimentary material has an effect on the nature of the sediments and therefore on the rocks that form from them. For instance, material derived from a granite highland differs from material eroded from a limestone plain. Furthermore, the type of weathering that takes place also influences the type of sediments that develop. Thus the *chemical* weathering of a granitic landmass produces clay, soluble salts, iron oxides, and silica, but the *mechanical* weathering of the same granitic landmass produces bits of broken granite and grains of the original quartz, feldspar, and ferromagnesian minerals that made up the granite.

METHODS OF TRANSPORTATION Water—in streams and glaciers, underground and in ocean currents—is the principal means of transporting material from one place to another. Landslides and other movements induced by gravity also play a role, as does the wind, but we shall look more closely at these processes in Chapters 10 through 15.

PROCESSES OF SEDIMENTATION Detrital material—material consisting of minerals and rock fragments—is deposited when its agent of transportation no longer has enough energy to move it farther. For example, a stream flowing along at a certain velocity possesses enough energy to move particles up to a certain maximum size. If the stream loses velocity, it also loses energy, and it is no longer able to transport all the material that it has been carrying along at the higher velocity. The solid particles, beginning with the heaviest, start to settle to the bottom. The effect is much the same when a wind that has been driving sand across a desert suddenly dies. A loss of energy accompanies the loss in velocity.

Material that has been carried along in solution is deposited in a different way: by precipitation, a chemical process by which dissolved material is converted into a solid and separated from the liquid solvent. As already noted, precipitation may be either biochemical or inorganic in nature.

Although at first glance the whole process of sedimentation seems quite simple, actually it is as complex as nature itself. Many factors are involved, and they can interact in a variety of ways. Consequently, the manner in which sedimentation takes place and the sediments that result from it differ greatly from one situation to another (see Figures 7.3 and 7.4).

Think of the different ways in which materials settle out of water, for instance. A swift, narrow mountain stream may deposit coarse sand and gravel along

Figure 7.3 **The environment of this quiet pond near Lexington, Massachusetts, favors the deposition of fine-grained sediments, largely mud. Compare with Figure 7.4.** (*Photograph by Sheldon Judson.*)

Figure 7.4 **Exposed to the direct attack of ocean surf, the environment of this cliffed California coastline favors the deposition of coarse sand and gravel. Compare with Figure 7.3.** (*Photograph by Sheldon Judson.*)

its bed, but farther downstream, as the valley widens, the same stream may overflow its banks and spread silt and mud over the surrounding country. A lake provides a different environment, varying from the delta of the inlet stream to the deep lake bottom and the shallow, sandy shore zones. In the oceans, too—those great basins of sedimentation—environment and sedimentation differ from the brackish tidal lagoon to the zone of plunging surf and out to the broad, submerged shelves of the continents and to the ocean depths beyond.

Mineral composition of sedimentary rocks

Sedimentary rocks, like igneous and metamorphic rocks, are accumulations of minerals. In sedimentary rocks, the three most common minerals are clay, quartz, and calcite, although, as we shall see, a few others are important in certain localities.

Rarely is a sedimentary rock made up of only a single mineral, although one mineral may predominate. Limestone, for example, is composed mostly of calcite, but even the purest limestone contains small amounts of other minerals, such as clay or quartz. The grains of many sandstones are predominantly quartz, but the cementing material that holds these grains together may

Figure 7.5 ***Quartz grains of varying sizes and shapes are cemented together by hematite. The large quartz fragment is about 1.2 cm long.*** *(Photograph by Willard Starks.)*

be calcite, dolomite, or iron oxide (see Figure 7.5). In general, we may say that most sedimentary rocks are mixtures of two or more minerals.

CLAY Chapter 6 described how clay minerals develop from the weathering of the silicates, particularly the feldspars. These clays subsequently may be incorporated into sedimentary rocks; they may, for example, form an important constituent of mudstone and shale. Examination of recent and ancient marine deposits shows that the kaolinite and illite clays are the most common clays in sedimentary rocks and that the montmorillonite clays are relatively rare.

QUARTZ Another important component of sedimentary rocks is silica, including the very common mineral quartz, as well as a number of much rarer forms such as chert, flint, opal, and chalcedony.

The mechanical and chemical weathering of an igneous rock such as granite sets free individual grains of quartz that eventually may be incorporated into sediments. These quartz grains produce the detrital forms of silica and account for most of the volume of the sedimentary rock sandstone. But silica in solution or in particles of colloidal size is also produced by the weathering of an igneous rock. This silica may be precipitated or deposited in the form of quartz, particularly as a cementing agent in certain coarse grained sedimentary rocks. Silica may also be precipitated in other forms, such as *opal,* generally regarded as a hydrous silica ($SiO_2 \cdot nH_2O$). Opal is slightly softer than true quartz and has no true crystal structure.

Silica also occurs in sedimentary rocks in a form called *cryptocrystalline.* This term (from the Greek *kryptos,* "hidden," and *crystalline*) indicates crystalline structure so fine that it cannot be seen under most ordinary microscopes. The microscope does reveal, however, that some cryptocrystalline silica has a granular pattern and that some has a fibrous pattern. To the naked eye, the surface of the granular form is somewhat duller than that of the fibrous form. Among the dull-surfaced or granular varieties is *flint,* usually dark in color. Flint is commonly found in certain limestone beds—the chalk beds of southern England, for example. *Chert* is similar to flint but tends to be lighter in color, and *jasper* is a red variety of granular cryptocrystalline form.

The general term *chalcedony* is often applied to the fibrous types of cryptocrystalline silica, which have a higher, more waxy luster than granular varieties. Sometimes the term is also used to describe a specific variety of brown, translucent cryptocrystalline silica. *Agate* is a variegated form of silica, its bands of chalcedony alternating with bands of either opal or some variety of granular cryptocrystalline silica, such as jasper.

CALCITE The chief constituent of the sedimentary rock limestone, calcite ($CaCO_3$), is also the most common cementing material in the coarse-grained sedimentary rocks. The calcium is derived from igneous rocks that contain calcium-bearing minerals, such as calcium plagioclase and some of the ferromagnesian minerals. Calcium is carried from the zone of weathering as calcium bicarbonate, $Ca(HCO_3)_2$, and is eventually precipitated as calcite, $CaCO_3$, through the intervention of plants, animals, or inorganic processes. The carbonate is derived from water and carbon dioxide.

OTHER MATERIALS IN SEDIMENTARY ROCKS Accumulations of clay, quartz, and calcite, either alone or in combination, account for all but a very small percentage of the sedimentary rocks, but certain other materials occur in quantities large enough to form distinct strata. The mineral *dolomite*, $CaMg(CO_3)_2$, for example, usually is intimately associated with calcite, though it is far less abundant. It is named after an eighteenth-century French geologist, Dolomieu. When the mineral is present in large amounts in a rock, the rock itself is also known as dolomite. The mineral dolomite is easily confused with calcite; and because they often occur together, distinguishing them is important. Calcite effervesces freely in dilute hydrochloric acid; dolomite effervesces very slowly or not at all, unless it is finely ground or powdered. The more rapid chemical activity results from the increase in surface area, an example of the general principle discussed in Section 6.2.

The feldspars and micas are abundant in some sedimentary rocks. In Chapter 6, we found that chemical weathering converts these minerals into new minerals at a relatively rapid rate. Therefore, when we find mica and feldspar in a sedimentary rock, chances are that it was mechanical, rather than chemical, weathering that originally made them available for incorporation in the rock.

Iron produced by chemical weathering of the ferromagnesian minerals in igneous rocks may be caught up again in new minerals and incorporated into sedimentary deposits. The iron-bearing minerals that occur most frequently in sedimentary rocks are hematite, goethite, and limonite. In some deposits, these minerals predominate, but more commonly they act simply as coloring matter or as a cementing material.

Halite (NaCl) and gypsum ($CaSO_4 \cdot 2H_2O$) are minerals precipitated from solution by evaporation of the water in which they were dissolved. The salinity of the water—that is, the proportion of the dissolved material to the water—determines the type of mineral that will precipitate out. The gypsum begins to separate from sea water when the salinity (at 30°C) reaches a little over 3 times its normal value. Then, when the salinity of the sea water has increased to about 10 times its normal value, halite begins to precipitate.

Pyroclastic rocks, mentioned in Chapter 4, are sedimentary rocks composed mostly of fragments blown from volcanoes. The fragments may be large pieces that have fallen close to the volcano or extremely fine ash that has been carried by the wind and deposited hundreds of kilometers from the volcanic eruption.

Finally, organic matter may be present in sedimentary rocks. In the sedimentary rock known as coal, plant materials are almost the only components. More commonly, however, organic matter is very sparsely disseminated through sedimentary deposits and the resulting rocks.

Texture

Texture refers to the general physical appearance of a rock—to the size, shape, and arrangements of the particles that make it up. There are two major types of texture in sedimentary rocks: clastic and nonclastic.

CLASTIC TEXTURE The term *clastic* is derived from the Greek for "broken" or "fragmental," and rocks that have been formed from deposits of mineral and rock fragments are said to have clastic texture (see Figure 7.5). The size and shape of the original particles have a direct influence on the nature of the resulting texture. A rock formed from a bed of gravel and sand has a coarse, rubblelike texture that is very different from the sugary texture of a rock developed from a deposit of rounded, uniform sand grains. Furthermore, the process by which a sediment is deposited also affects the texture of the sedimentary rock that develops from it. Thus the debris dumped by a glacier is composed of a jumbled assortment of rock material ranging from particles of colloidal size to large boulders. A rock that develops from such a deposit has a very different texture from one that develops from a deposit of wind-blown sand, for instance, in which all the particles are approximately 0.15 to 0.30 mm in diameter.

Chemical sedimentary rocks may also show a clastic texture. A rock made up predominantly of shell fragments from a biochemical deposit has a clastic texture that is just as recognizable as the texture of a rock formed from sand deposits (see Figure 7.6).

One of the most useful factors in classifying sedimentary rocks is the size of the individual particles. In practice, we usually express the size of a particle in terms of its diameter rather than in terms of volume, weight, or surface area. When we speak of "diameter," we imply that the particle is a sphere, but it is very unlikely that any fragment in a sedimentary rock is a true sphere. In geological measurements, the term simply means the diameter that an irregularly shaped particle *would* have if it were a sphere of equivalent volume. Obviously, it would be a time-consuming, if not impossible, task to determine the volume of each sand grain or pebble in a rock and then to convert these measurements into appropriate diameters. So the diameters we use for particles are only approximations of their actual sizes. They are accurate enough, however, for our needs.

Figure 7.6 ***Viewed under the microscope, some limestones are seen to have a clastic texture, as does this example from an ancient reef deposit in the Austrian Alps. The individual particles include the shells of one-celled marine animals called Foraminifera, as well as the fragments of other marine organisms and pellets of calcite. The particles are cemented together by clear calcite (dark). The black lines are also calcite that fills cracks in the rock. The area shown is 2.8 cm wide.*** (*Photograph by E. C. Bierwagen.*)

Table 7.2 ***Wentworth scale of particle sizes for clastic sediments***[a]

Wentworth scale Size, mm	Fragment	For next larger size, multiply by	Approximate equivalent, in.
	Boulder		
256			10
	Cobble	4	
64			2.5
	Pebble	16	
4			0.156
	Granule	2	
2			0.078
	Sand	32	
$\frac{1}{16}$			0.0025
	Silt[b]	16	
$\frac{1}{256}$			0.00015
	Clay[b]		

[a] Modified after C. K. Wentworth, "A Scale of Grade and Class Terms for Clastic Sediments," *J. Geol.*, vol. 30, p. 381, 1922.
[b] Dust.

Several scales have been proposed to describe particles ranging in size from large boulders to minerals of microscopic dimensions. The Wentworth scale, presented in Table 7.2, is used widely, though not universally, by American and Canadian geologists. Notice that although the term *clay* is used in the table to designate all particles below $\frac{1}{256}$ mm in diameter, the same term is used to describe certain minerals. To avoid confusion, we must always refer specifically to either "clay size" or "clay mineral," unless the context makes the meaning clear.

Because determining the size of particles calls for the use of special equipment, the procedure is normally carried out only in the laboratory. In examining specimens in the field, an educated guess based on careful examination usually suffices.

NONCLASTIC TEXTURE Some, but not all, sedimentary rocks formed by chemical processes have a nonclastic texture in which the grains are interlocked. These rocks have somewhat the same appearance as igneous rocks with crystalline texture. Actually, most of the sedimentary rocks with nonclastic texture do have a crystalline structure, although a few of them, such as opal, do not.

The mineral crystals that precipitate from an aqueous solution are usually very small in size. Because the fluid in which they form has a very low density, they usually settle out rapidly and accumulate on the bottom as mud. Eventually, under the weight of additional sediments, the mud is compacted more and more. Now the size of the individual crystals may begin to increase. Their growth may be induced by added pressure causing the favorably oriented grains

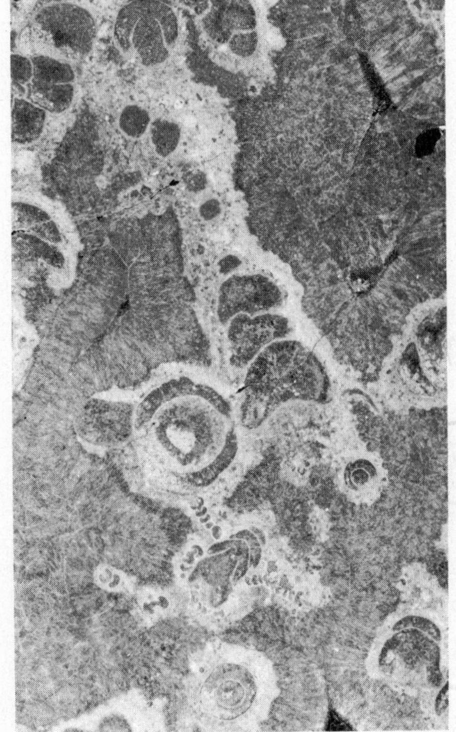

Figure 7.7 ***Microscopic examination reveals a sequence of events involved in the formation of this limestone. The globular forms are fossils of unknown modern affinities and are called Cheilosporites. Around them, white, halolike zones of calcite were deposited by algae. Most of the pores were then almost completely filled by calcite that grew with a fibrous habit. Finally, a few remaining pores were filled with a clear, coarse variety of calcite, the very dark patches at the top and bottom. This limestone, then, represents two periods in which calcite was deposited organically and two later periods of inorganic deposition. The area shown is .75 cm wide.*** (*Photograph by E. C. Bierwagen.*)

to grow at the expense of less favorably oriented neighboring grains. Or crystals may grow as more and more mineral matter is added to them from the saturated solutions trapped in the original mud. In any event, the resulting rock is made up of interlocking crystals and has a texture similar to that of crystalline igneous rocks. Depending on the size of the crystals, we refer to these non-clastic textures as fine-grained, medium-grained, or coarse-grained. A coarse-grained texture has grains larger than 5 mm in diameter, and a fine-grained texture has grains less than 1 mm in diameter.

Lithification

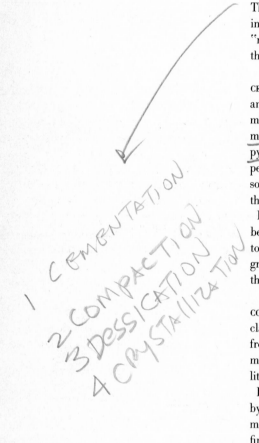

The process of *lithification* converts unconsolidated rock-forming materials into consolidated, coherent rock. The term is derived from the Greek for "rock" and the Latin "to make." In the following subsections, we shall discuss the various ways in which sedimentary deposits are lithified.

CEMENTATION In cementation, the spaces between the individual particles of an unconsolidated deposit are filled up by some binding agent. Of the many minerals that serve as cementing agents, the most common are calcite, dolomite, and quartz. Others include iron oxide, opal, chalcedony, anhydrite, and pyrite. Apparently, the cementing material is carried in solution by water that percolates through the open spaces between the particles of the deposit. Then some factor in the new environment causes the mineral to be deposited, and the former unconsolidated deposit is cemented into a sedimentary rock.

In coarse-grained deposits, there are relatively large interconnecting spaces between the particles. As we would expect, these deposits are very susceptible to cementation, because the percolating water can move through them with great ease. Deposits of sand and gravel are transformed by cementation into the sedimentary rocks sandstone and conglomerate, respectively.

COMPACTION AND DESICCATION In a fine-grained clastic deposit of silt-sized and clay-sized particles, the pore spaces are usually so small that water cannot freely circulate through them. Consequently, very little cementing material manages to find its way between the particles, but deposits of this sort are lithified by two other processes: compaction and desiccation.

In *compaction,* the pore space between individual grains is gradually reduced by the pressure of overlying sediments or by pressures resulting from earth movement. Coarse deposits of sand and gravel undergo some compaction, but fine-grained deposits of silt and clay respond much more readily. As the individual particles are pressed closer and closer together, the thickness of the deposit is reduced and its coherence is increased. It has been estimated that deposits of clay-sized particles, buried to depths of 1,000 m, have been compacted to about 60 percent of their original volume.

In *desiccation,* the water that originally filled the pore spaces of water-laid clay and silt deposits is forced out. Sometimes this is the direct result of compaction, but desiccation also takes place when a deposit is simply exposed to the air and the water evaporates.

CRYSTALLIZATION The crystallization of certain chemical deposits is in itself a form of lithification. Crystallization also serves to harden deposits that have been laid down by mechanical processes of sedimentation. For example, new minerals may crystallize within a deposit, or the crystals of existing minerals may increase in size. New minerals sometimes are produced by chemical reactions among amorphous, colloidal materials in fine-grained muds. Exactly how and when these reactions occur is not yet generally understood, but the fact that new crystals *have* formed after the deposit was initially laid down becomes increasingly apparent as we make more and more detailed studies of sedimentary rocks. Furthermore, it seems clear that this crystallization promotes the process of lithification, particularly in the finer sediments.

7.3 *Types of sedimentary rocks*

Classification

Having examined some of the factors involved in the formation of sedimentary rocks, we are in a better position to consider a classification for this rock family. The classification presented in Table 7.3 represents only one of many possible schemes, but it will serve our purposes very adequately. Notice that there are two main groups—detrital and chemical—based on the origin of the rocks and that the chemical category is further split into inorganic and biochemical. All the detrital rocks have clastic texture, whereas the chemical rocks have either clastic or nonclastic texture. We use particle size to subdivide the detrital rocks and composition to subdivide the chemical rocks.

Detrital sedimentary rocks

CONGLOMERATE A *conglomerate* is a detrital rock made up of more or less rounded fragments, an appreciable percentage of which are of granule size

Table 7.3 **Classification of sedimentary rocks**

Origin	*Texture*	*Particle size or composition*	*Rock name*
Detrital	Clastic	Granular or larger	Conglomerate
		Sand	Sandstone
		Silt and clay	Mudstone and shale
Chemical:			
Inorganic	Clastic and nonclastic	Calcite, $CaCO_3$	Limestone
		Dolomite, $CaMg(CO_3)_2$	Dolomite
		Halite, NaCl	Salt
		Gypsum, $CaSO_4 \cdot 2H_2O$	Gypsum
Biochemical	Clastic and nonclastic	Calcite, $CaCO_3$	Limestone
		Plant remains	Coal

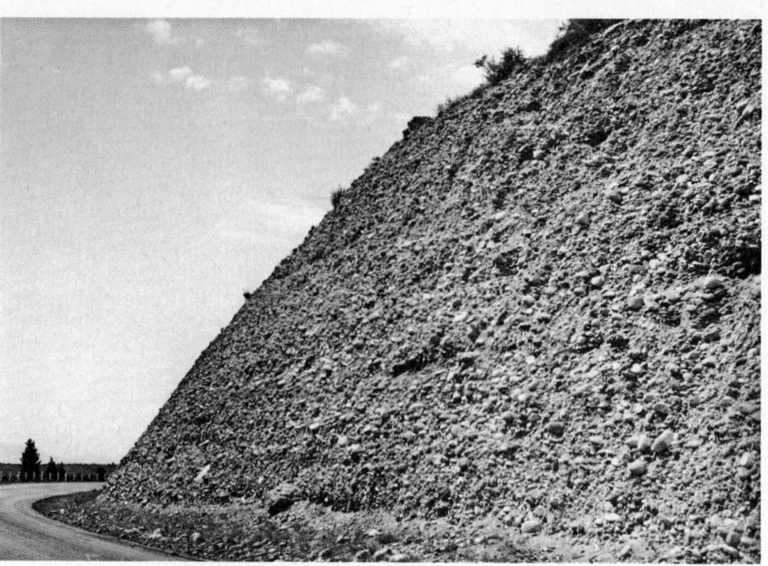

Figure 7.8 **A conglomerate is made up of rounded pebbles, as shown in this deposit of partially consolidated material in Estes Park, Colorado.** (*Photograph from Tozier Collection, Harvard University.*)

(2 to 4 mm in diameter) or larger. If the fragments are more angular than rounded, the rock is called a *breccia.* Another type of conglomerate is *tillite,* a rock formed from deposits laid down directly by glacier ice (see Chapter 13). The large particles in a conglomerate are usually rock fragments, and the finer particles are usually minerals derived from preexisting rocks (see Figures 7.8, 7.9, and 7.10).

SANDSTONE A *sandstone* is formed by the consolidation of individual grains of sand size (between $\frac{1}{16}$ and 2 mm in diameter). Sandstone is thus intermediate between coarse-grained conglomerate and fine-grained mudstone. Because the size of the grains varies from one sandstone to another, we speak of coarse-grained, medium-grained, and fine-grained sandstone.

Very often the grains of a sandstone are almost all quartz. When this is the case the rock is called an *orthoquartzite* (from the Greek meaning, here, "true," plus *quartzite*). The name *quartzose sandstone* is also used, as is *quartz arenite* (from *quartz* plus the Latin for "sand"). If the minerals are predominately quartz and feldspar, the sandstone is called an *arkose,* a French word for the rock formed by the consolidation of debris derived from a mechanically weathered granite. Another variety of sandstone, called *graywacke,* or *lithic sandstone,* is characterized by its hardness and dark color and by angular

Figure 7.9 **A breccia is a lithified deposit containing many angular fragments. This 12-cm-wide specimen is from eastern Nevada.** (*Princeton University Museum of Natural History. Photograph by Willard Starks.*)

Figure 7.10 ***Alternating beds of sandstone and conglomerate dip inland from the sea cliff at Lobos State Park, near Carmel, California.*** (*Photograph by Sheldon Judson.*)

grains of quartz, feldspar, and small fragments of rock set in a matrix of clay-sized particles. It is also called *lithic arenite.*

MUDSTONE AND SHALE Fine-grained detrital rocks composed of clay and silt-sized particles (less then $\frac{1}{16}$ mm in diameter) are termed either *mudstone* or *shale.* Mudstones are fine-grained rocks with a massive or blocky aspect, whereas shales are fine-grained rocks that split into platy slabs more or less parallel to the bedding. The particles in these rocks are so small that it is difficult to determine the precise mineral composition of mudstone and shale. We do know, however, that they contain not only clay minerals but also clay-sized and silt-sized particles of quartz, feldspar, calcite, and dolomite, to mention but a few.

Chemical rocks

LIMESTONE Limestone is a sedimentary rock that is made up chiefly of the mineral calcite, $CaCO_3$, which has been deposited by either inorganic or organic chemical processes. Most limestones have a clastic texture, but non-clastic, particularly crystalline, textures are common.

Biochemically formed limestones are created by the action of plants and animals that extract calcium carbonate from the water in which they live. The calcium carbonate may be either incorporated into the skeleton of the organism or precipitated directly. In any event, when the organism dies, it leaves behind a quantity of calcium carbonate, and over a long period of time thick deposits of this material may be built up. Reefs, ancient and modern, are well-known examples of such accumulations. The most important builders of modern reefs are algae, molluscs, corals, and one-celled animals—the same animals whose ancestors built up the reefs of ancient seas—the reefs, now old and deeply buried, that are often valuable reservoirs of petroleum.

Chalk (from the Latin *calx,* "lime") is made up in part of biochemically derived calcite in the form of the skeletons or skeletal fragments of microscopic oceanic plants and animals. These organic remains are found mixed with very fine-grained calcite deposits of either biochemical or inorganic chemical origin. A much coarser type of limestone composed of organic remains is known as *coquina,* from the Spanish for "shellfish" or "cockle," and is characterized by the accumulation of many large fragments of shells.

Inorganically formed limestone is made up of calcite that has been precipitated from solution by inorganic processes. Some calcite is precipitated from the fresh water of streams, springs, and caves, although the total amount of rock formed in this way is negligible. When calcium-bearing rocks undergo chemical weathering, calcium bicarbonate, $Ca(HCO_3)_2$, is produced in solution. If enough of the water evaporates, or if the temperature rises, or if the pressure falls, calcite is precipitated from this solution. For example, most *dripstone,* or *travertine,* is formed in caves by the evaporation of water that is carrying calcium carbonate in solution. And *tufa* (from the Italian for "soft rock") is a spongy, porous limestone formed by the precipitation of calcite from the water of streams and springs.

Although geologists understand the inorganic processes by which limestone is formed by precipitation from fresh water, they are not quite sure how important these processes are in precipitation from sea water. Some observers have questioned whether they operate at all. On the floors of modern oceans and in rocks formed in ancient oceans, however, we do find small spheroidal grains called *oölites*, the size of sand and often composed of calcite. These grains are thought to be formed by the inorganic precipitation of calcium carbonate from sea water. The term *oölite* comes from the Greek for "egg," because an accumulation of oölites resembles a cluster of fish roe. Cross sections show that many oölites, though not all, have grown up around a mineral grain, or around a small fragment of shell that acts as a nucleus. Some limestones are made up largely of oölites. One, widely used for building, is the so-called Indiana or Spergen limestone.

DOLOMITE In discussing the mineral dolomite, $CaMg(CO_3)_2$, we mentioned that when it occurs in large concentrations it forms a rock that is also called dolomite. The origins of extensive deposits of dolomite still are not well understood, but some uncertainties are beginning to disappear. There is no evidence that strata of dolomite develop by direct precipitation of the mineral in sea water. On the contrary, most dolomites appear to have been formed by replacement of preexisting deposits of calcite. Thus in some field situations, it can be shown that igneous intrusions accompanied by solutions—presumably rich in magnesium—have altered limestone, some of the calcium in the calcite having been replaced by magnesium to form dolomite.

But dolomites caused by metamorphism are unimportant when compared with the bulk of dolomites in the geologic column. There is now increasing agreement that most dolomite is in some way related to the local increase of the amount of magnesium in solution. Previously deposited calcite is modified by the movement through it of these magnesium-rich solutions. Thus field and laboratory observations show that in shallow-water intertidal zones evaporation of sea water may cause precipitation of calcium-bearing deposits, both calcite and sulfate, $CaSO_4$. Thus the waters may be increased by an order of magnitude in their content of magnesium relative to calcite. Such high-magnesium-content waters may then circulate through the underlying calcite deposits, replacing some of the calcium with magnesium and thus converting limestone to dolomite.

EVAPORITES An *evaporite* is a sedimentary rock composed of minerals that were precipitated from solution after the evaporation of the liquid in which they were dissolved. *Rock salt* (composed of the mineral halite, NaCl) and *gypsum* (composed of the mineral of the same name, $CaSO_4 \cdot 2H_2O$) are the most abundant evaporites. *Anhydrite* (from the Greek *anydros*, "waterless") is an evaporite composed of the mineral of the same name, which is simply gypsum without the water, $CaSO_4$. Most evaporite deposits seem to have been precipitated from sea water according to a definite sequence. The less highly soluble minerals are the first to drop out of solution. Thus gypsum and anhydrite, both less soluble than halite, are deposited first. Then, as evaporation progresses, the more soluble halite is precipitated.

In the United States, the most extensive deposits of evaporites are found in Texas and New Mexico. Here, gypsum, anhydrite, and rock salt make up over 90 percent of the Castile formation, which has a maximum thickness of nearly 1,200 meters. In central New York State there are thick deposits of rock salt, and in central Michigan there are layers of rock salt and gypsum. Some evaporite deposits are mined for their mineral content, and in certain areas, particularly in the Gulf Coast states, deposits of rock salt have pushed upward toward the surface to form salt domes containing commercially important reservoirs of petroleum (see Chapter 21).

COAL Coal is a rock composed of combustible matter derived from the partial decomposition of plants. We shall consider coal as a biochemically formed sedimentary rock, although some geologists prefer to think of it as a metamorphic rock because it passes through various stages.

The process of coal formation begins with an accumulation of plant remains in a swamp. This accumulation is known as *peat,* a soft, spongy, brownish deposit in which plant structures are easily recognizable. Time, coupled with the pressure produced by deep burial and sometimes by earth movement, gradually transforms the organic matter into coal. During this process, the percentage of carbon increases as the volatile hydrocarbons and water are forced out of the deposit. Coals are ranked according to the percentage of carbon they contain. Peat, with the least amount of carbon, is the lowest ranking; then come lignite or brown coal, bituminous or soft coal, and finally anthracite or hard coal, the highest of all the coals in its percentage of carbon.

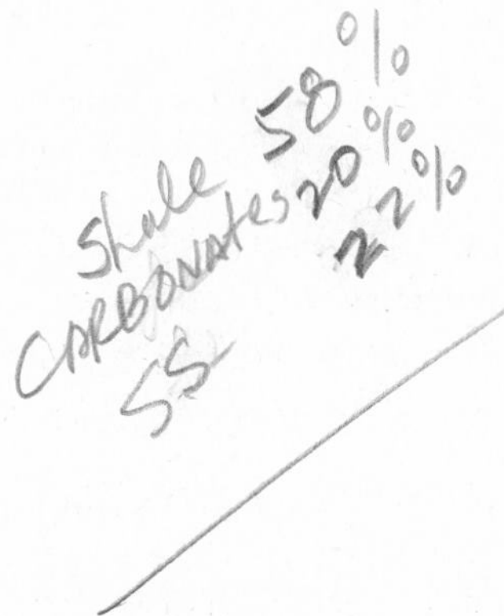

Relative abundance of sedimentary rocks

Sandstone, mudstone and shale, and limestone constitute about 99 percent of all sedimentary rocks. Of these, mudstone and shale are the most abundant. On the basis of extrapolation from measurements made in the field, the estimates of the percentages of mudstone and shale approximate 50 percent of all sedimentary rocks. Similar calculations for limestone and sandstone suggest that the limestone forms about 22 percent of these rock types and that sandstone accounts for the remaining 28 percent. These percentages, however, do not agree with theoretical determinations of relative abundances. These are based on the determination of the weathering products to be expected on the weathering of an average igneous rock. If these weathering products are assigned to the three major sedimentary rock types, then we find that shale should be considerably more important volumetrically than it appears to be on the basis of field measurements. On these theoretical grounds, mudstone and shale constitute approximately 75 percent of the three major sedimentary rock types. Sandstone and limestone are approximately of equivalent volume and together constitute the other 25 percent. The discrepancy between the two estimates has not yet been resolved.

Estimates have been made of the average chemical composition of the world's sediments. One of these estimates is presented in Table 7.4.

Table 7.4 ***Average composition of all sediments given as oxides in weight percent***[a]

Oxide	*Weight %*
SiO_2	44.5
TiO_2	0.6
Al_2O_3	10.9
Fe_2O_3	4.0
FeO	0.9
MnO	0.3
MgO	2.6
CaO	19.7
Na_2O	1.1
K_2O	1.9
P_2O_5	0.1
CO_2	13.4
Total	100.0

[a] From Arie Paldervaart, "Chemistry of the Earth's Crust," *Geol. Soc. Am. Spec. Paper,* no. 62, p. 132, 1955.

7.4 *Features of sedimentary rocks*

We have mentioned that the stratification, or bedding, of sedimentary rocks is their single most characteristic feature. Now we shall look more closely at this feature, along with certain other characteristics of sedimentary rocks, including mud cracks and ripple marks, nodules, concretions, geodes, fossils, and color.

Bedding

The beds or layers of sedimentary rocks are separated by *bedding planes,* along which the rocks tend to separate or break (see Figures 7.11, 7.12, and 7.13). The varying thickness of the layers in a given sedimentary rock reflects the changing conditions that prevailed when each deposit was laid down. In general, each bedding plane marks the termination of one deposit and the beginning of another. For example, let us imagine the bay of an ocean into which rivers normally carry fine silt from the nearby land. This silt settles out from the sea water to form a layer of mud. Now, heavy rains or melting snows may cause the rivers suddenly to flood and thereby pick up coarser material, such as sand, from the river bed. This material will be carried along and dumped into the bay. There it settles to the bottom and blankets the silt that was deposited earlier. The plane of contact between the mud and the sand represents a bedding plane. If, later on, the silt and the sand are lithified into shale and sandstone, respectively, the bedding plane persists in the sedimentary rock. In fact, it marks a plane of weakness along which the rock tends to break.

Bedding planes are usually horizontal, and when they are the bedding is called *parallel bedding.* Closely spaced horizontal bedding planes form *laminated bedding.* But some beds are laid down at an angle to the horizontal and such bedding is variously called *cross bedding* and *inclined bedding* (see Figure 7.14). Such nonhorizontal bedding can occur in several situations. Thus the bedding in sand dunes may have high angles on the leeward side of the dune (see Section 14.5). The deposits laid down at the growing edge of a delta may be inclined from 5 to 30° and such beds are usually given a special name, *foreset beds* (see Section 11.5). In flowing water, scouring of the stream floor by turbulent water may create small depressions in the channel deposits which later will be refilled by inclined beds. Alternatively, cross bedding can occur as flowing water passes to somewhat quieter conditions and drops its load in a way similar to that in which a stream builds a foreset bed in a delta.

When the particles in a sedimentary bed vary from coarse at the bottom to fine at the top, the bedding is said to be *graded bedding.* Such bedding is characteristic of rapid deposition from a water turbid with sedimentary particles of differing sizes. The largest particles settle most rapidly and the finest most slowly. Such deposits are called *turbidites,* a term from the Latin

To keep for Mon

Figure 7.11 **Layered limestone of Ordovician age at Trenton Falls, New York.** (*Photograph by Sheldon Judson.*)

Figure 7.12 **Dipping beds of the Clagett shale at Mud Creek Gap, Montana.** (*Photograph by U.S. Geological Survey.*)

Figure 7.13 **Exposure of essentially flat-lying Pierre shale of Cretaceous age (which ended about 65 million years ago) on the Fort Totten Indian Reservation in east-central North Dakota. The shale splits into small, thin slabs parallel to the bedding, a characteristic known as fissility. The fracture planes (approximately at right angles to the bedding) and the fissility give the exposure a blocky appearance.** (*Photograph by Saul Aronow, U.S. Geological Survey.*)

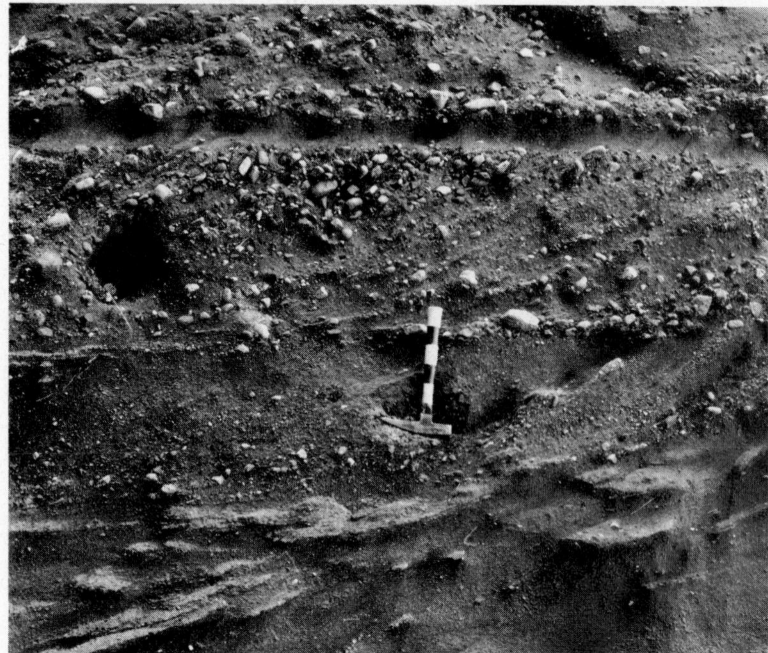

Figure 7.14 ***Inclined bedding in deposits of Pliocene age in the upper Magdalena Basin of Colombia. Note that the beds of sand and gravel dip to the left but that they are bounded by horizontal beds. The bands on the hammer handle are 5 cm wide.*** *(Photograph by Milton Howe.)*

turbidus for "disturbed," referring to the stirring up of sediments and water (see also Section 15.2).

Bedding may be distorted before the material becomes consolidated into firm rock. Thus unconsolidated sediments may sometimes slump or flow, as suggested in Figure 7.15. In addition, animals may burrow or tunnel through unconsolidated layers to produce *disturbed* or *mottled bedding.*

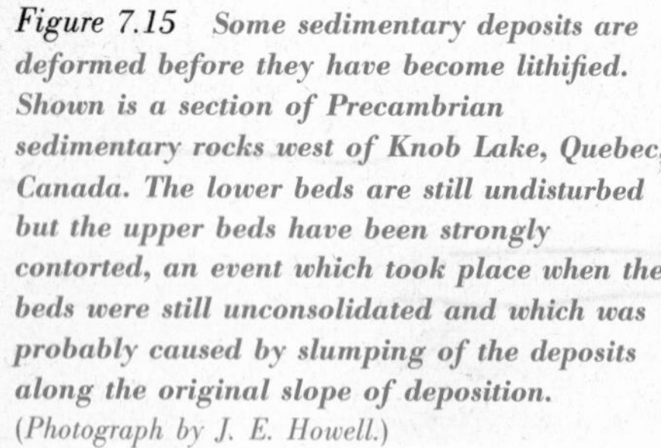

Figure 7.15 ***Some sedimentary deposits are deformed before they have become lithified. Shown is a section of Precambrian sedimentary rocks west of Knob Lake, Quebec, Canada. The lower beds are still undisturbed but the upper beds have been strongly contorted, an event which took place when the beds were still unconsolidated and which was probably caused by slumping of the deposits along the original slope of deposition.*** *(Photograph by J. E. Howell.)*

Mud cracks and ripple marks

Figure 7.16 ***Polygonal pattern of mud cracks resulting from dessication of modern fine-grained sediments in a playa lake in Nevada. Scale shown by pocket knife.*** (*Photograph by William C. Bradley.*)

Ripple marks are the little waves of sand that commonly develop on the surface of a sand dune, or along a beach, or on the bottom of a stream. *Mud cracks* are familiar on the dried surface of mud left exposed by the subsiding waters of a river. These features are often preserved in solid rock and provide us with clues to the history of the rock.

Mud cracks make their appearance when a deposit of silt or clay dries out and shrinks (see Figures 7.16 and 7.17). The cracks outline roughly polygonal areas, making the surface of the deposit look like a section cut through a large honeycomb. Eventually, another deposit may come along to bury the first. If the deposits are later lithified, the outlines of the cracks may be accurately preserved for millions of years. Then, when the rock is split along the bedding plane between the two deposits, the cracks will be found much as they appeared when they were first formed, providing evidence that the original deposit underwent alternate flooding and drying.

Ripple marks preserved in sedimentary rocks also furnish clues to the conditions that prevailed when a sediment was originally deposited. For instance, if the ripple marks are symmetric, with sharp or slightly rounded ridges separated by more gently rounded troughs, we are fairly safe in assuming that they were formed by the back-and-forth movement of water such as we find along a sea coast outside the surf zone. These marks are called *ripple marks of oscillation* (see Figure 7.18). If, on the other hand, the ripple marks are asymmetric, we can assume that they were formed by air or water moving more or less continuously in one direction. These marks are called *current ripple marks.*

Figure 7.17 ***Mud cracks preserved in shale at Mesa Verde Park, Colorado. The surface of the ledge shows the surface pattern and the small cliff exposes the structures in cross section.*** (*Photograph from Tozier Collection, Harvard University.*)

Figure 7.18 **Ripple marks of oscillation on a 35-cm-long slab of sandstone. The symmetrical nature of the ripples indicates that the current moved back and forth rather than continuously in one direction.** (*Photograph by Willard Starks.*)

Nodules, concretions, and geodes

Many sedimentary rocks contain structures that were formed only *after* the original sediment was deposited. Among these are nodules, concretions, and geodes.

A *nodule* is an irregular, knobby-surfaced body of mineral matter that differs in composition from the sedimentary rock in which it has formed. It usually lies parallel to the bedding planes of the enclosing rock, and sometimes adjoining nodules coalesce to form a continuous bed. Nodules average about 30 cm in maximum dimension. Silica, in the form of chert or flint, is the major component of these bodies. They are most commonly found in limestone or dolomite. Most nodules are thought to have formed when silica replaced some of the materials of the original deposit; some, however, may consist of silica that was deposited at the same time as the main beds were laid down.

A *concretion* is a local concentration of the cementing material that has lithified a deposit into a sedimentary rock. Concretions range in size from a fraction of a centimeter to a meter or more in maximum dimension. Most are shaped like simple spheres or disks, although some have fantastic and complex forms. For some reason, when the cementing material entered the unconsolidated sediment, it tended to concentrate around a common center point or along a common center line. The particles of the resulting concretion are cemented together more firmly than the particles of the host rock that surrounds it. The cementing material usually consists of calcite, dolomite, iron oxide, or silica—in other words, the same cementing materials that we find in the sedimentary rocks themselves.

Geodes, more eye-catching than either concretions or nodules, are roughly spherical, hollow structures up to 30 cm or more in diameter (see Figures 7.19 and 7.20). An outer layer of chalcedony is lined with crystals that project inward toward the hollow center. The crystals, often perfectly formed, are usually quartz, although crystals of calcite and dolomite have also been found and, more rarely, crystals of other minerals. Geodes are most commonly found in limestone, but they also occur in shale.

How does a geode form? First, a water-filled pocket develops in a sedimentary

Figure 7.19 ***Two geodes broken open to show their internal structure. The dark outer layer of chalcedony is lined with milky-to-clear quartz crystals that project inward toward a hollow center. These structures are most commonly found in limestone, where they apparently form by the modification and enlargement of an original void. The smaller geode was about 8 cm in diameter.*** *(Specimen from Geological Museum, Harvard University. Photograph by Walter R. Fleischer.)*

deposit, probably as a result of the decay of some plant or animal that was buried in the sediments. As the deposit begins to consolidate into a sedimentary rock, a wall of silica with a jellylike consistency forms around the water, isolating it from the surrounding material. As time passes, fresh water may enter the sediments. The water inside the pocket has a higher salt concentration than the water outside. To equalize the concentrations, there is a slow mixing of the two liquids through the silica wall or membrane that separates them. This process of mixing is called *osmosis.* So long as the osmotic action continues, pressure is exerted outward toward the surrounding rock. The original pocket expands bit by bit, until the salt concentrations of the liquids inside and outside are equalized. At this point, osmosis stops, the outward pressure ceases, and the pocket stops growing. Now the silica wall dries, crystallizes to form chalcedony, contracts, and cracks.

If, at some later time, mineral-bearing water finds its way into the deposit, it may seep in through the cracks in the wall of chalcedony. There the minerals are precipitated, and crystals begin to grow inward, toward the center, from the interior walls. Finally, we have a crystal-lined geode imbedded in the surrounding rock. Notice that the crystals in a geode grow inward, whereas in a concretion they grow outward.

Figure 7.20 ***Geodes in the Warsaw shale west of Keokuk, Iowa, show up as cavities, or, where they are unbroken, they give the rock surface a knobby appearance. The larger geode in the lower left corner measures about 10 cm in diameter.*** *(Photograph by L. M. Cline.)*

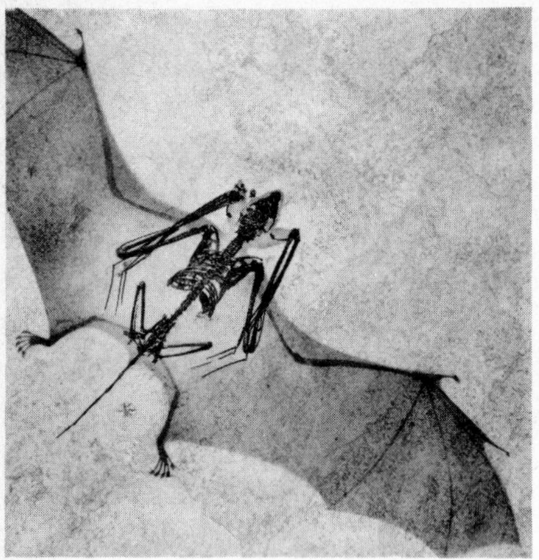

Figure 7.21 ***This fossil bat,*** **Icaronycteris index,** ***oldest known flying mammal, was found in the Green River formation of the early Eocene age in southwestern Wyoming. It measures 12 cm in length and its wing span (here restored) is 30 cm. Detailed study suggests that the bat was a young male that died on a summer evening while foraging for insects or small fish close to the surface of the lake in which the Green River formation was laid down. Just to the right of the tail, a small flower has been preserved.*** (*Princeton University Museum of Natural History. Photograph by Willard Starks.*)

Fossils

The word *fossil* (derived from the Latin *fodere,* "to dig up") originally referred to anything that was dug from the ground, particularly a mineral or some inexplicable form. It is still used in that sense occasionally, as in the term *fossil fuel* (see Chapter 21). But today the term *fossil* generally means any direct evidence of past life—for example, the bones of a dinosaur, the shell of an ancient clam, the footprints of a long-extinct animal, or the delicate impression of a leaf (see Figures 7.21, 7.22, and 7.23).

Fossils are usually found in sedimentary rocks, although they sometimes turn up in igneous and metamorphic rocks. They are most abundant in mudstone, shale, and limestone but are also found in sandstone, dolomite, and conglomerate. Fossils account for almost the entire volume of certain rocks, such as coquina and limestones formed from ancient reefs.

The remains of plants and animals are completely destroyed if they are left exposed on the earth's surface; but if they are somehow protected from destructive forces, they may become incorporated in a sedimentary deposit where they will be preserved for millions of years. In the quiet water of the ocean, for example, the remains of starfish, snails, and fish may be buried by sediments as they settle slowly to the bottom. If these sediments are subsequently lithified, the remains are preserved as fossils that tell us about the sort of life that existed when the sediments were laid down.

Fossils are also preserved in deposits that have settled out of fresh water. Countless remains of land animals, large and small, have been dug from the beds of extinct lakes, flood plains, and swamps.

The detailed story of the development of life as recorded by fossils is properly a part of historical geology, and although we do not have time to trace it here, in Chapter 9 we shall find that fossils are extremely useful in subdividing geologic time and constructing the geologic column.

Color of sedimentary rocks

Throughout the western and southwestern areas of the United States, bare cliffs and steep-walled canyons provide a brilliant display of the great variety of colors exhibited by sedimentary rocks. The Grand Canyon of the Colorado River in Arizona cuts through rocks that vary in color from gray, through purple and red, to brown, buff, and green. Bryce Canyon in southern Utah is fashioned of rocks tinted a delicate pink, and the Painted Desert farther south in Arizona exhibits a wide range of colors, including red, gray, purple, and pink.

Figure 7.22 ***Casts of burrows made by worms that lived in Devonian muds are shown on the underside of this slab from a locality near Ithaca, New York.*** (*Photograph by Sheldon Judson.*)

Figure 7.23 **Some sedimentary rocks are made up of extremely fine particles. This photograph by electron microscope shows the structure of a limestone of Paleocene age from Zumaya, Spain. The magnification is about 6,500×. The geometric markings represent coccoliths, interlocking plates of calcite that are developed by certain one-celled flagellate animals.** (*Photograph by S. Honjo.*)

The most important sources of color in sedimentary rocks are the iron oxides. Hematite (Fe_2O_3), for example, gives rocks a red or pink color, and limonite or goethite produces tones of yellow and brown. Some of the green, purple, and black colors may be caused by iron, but in exactly what form is not completely understood. Only a very small amount of iron oxide is needed to color a rock. In fact, few sedimentary rocks contain more than 6 percent of iron, and most contain very much less.

Organic matter, when present, may also contribute to the coloring of sedimentary rocks, usually making them gray to black. Generally, but not always, the higher the organic content, the darker the rock.

The size of the individual particles in a rock also influences the color, or at least the intensity of the color. For example, fine-grained clastic rocks are usually somewhat darker than coarse-grained rocks of the same mineral composition.

Sedimentary facies

If we examine the environments of deposition that exist at any one time over a wide area, we find that they differ from place to place. Thus the fresh-water environment of a river changes to a brackish-water environment as the river nears the ocean. In the ocean itself, marine conditions prevail. But even here the marine environment changes—from shallow water to deep water, for example. And as the environment changes, the nature of the sediments that are laid down also changes. The deposits in one environment show characteristics that are different from the characteristics of deposits laid down at the same time in another environment. This change in the "look" of the sediments is called a change in *sedimentary facies;* the latter word derives from the Latin for "aspect" or "form."

We may define sedimentary facies as an accumulation of deposits that exhibits specific characteristics and grades laterally into other sedimentary accumulations formed at the same time but exhibiting different characteristics. The concept of facies is widely used in studying sedimentary rocks and the conditions that gave rise to metamorphic rocks (see Chapter 8). The concept is generally not used in referring to igneous rocks, though there is no valid reason it should not be used.

Let us consider a specific example of facies. Figure 7.24 shows a coastline where rivers from the land empty into a lagoon. The lagoon is separated from the open ocean by a sandbar. The fine silts and clays dumped into the quiet waters of the lagoon settle to the bottom as a layer of mud. At the same time, waves are eroding coarse sand from a nearby headland outside the lagoon. This sand is transported by currents and waves and deposited as a sandy layer seaward of the sandbar. Different environments exist inside and outside the lagoon; therefore, different deposits are being laid down simultaneously. Notice that the mud and the sand grade into each other along the sandbar. Now, imagine that these deposits were consolidated into rock and then exposed to view at the earth's surface. We would find a shale layer grading into sandstone; that is, one sedimentary facies is grading into another.

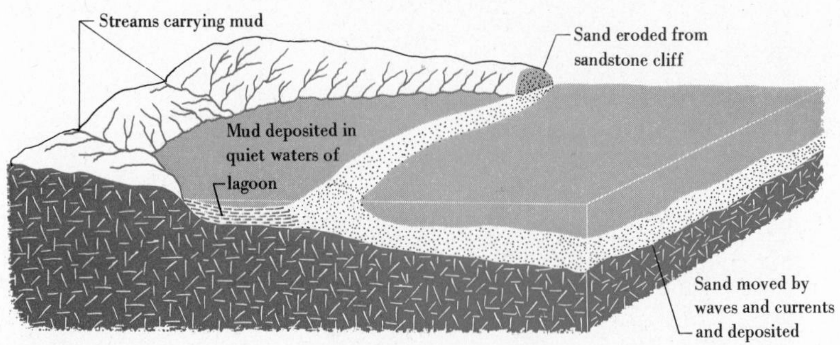

Figure 7.24 ***Diagram to illustrate a change in sedimentary facies. Here, the fine-grained muds are deposited in a lagoon close to shore. A sand bar separates them from sand deposits farther away from shore. The sand in this instance has been derived from a sea cliff and transported by waves and currents.***

But the picture is not always so simple. Figure 7.25 shows the distribution of sediments off the coast of southern California. Here, recent sediments range from sand, through a sandy mud, to mud, and in a few areas there are limy deposits. Where the sea floor is rocky, little or no recent sedimentation has taken place. Should the soft sediments become lithified, a sedimentary rock ranging from sandstone through sandy shale to shale and limestone would result.

Ancient sedimentary deposits show exactly this kind of variation in facies. Figure 7.26 pictures the actual pattern of rock types that have been identified in the middle portion of the Maquoketa formation in Illinois. Notice that the types include limestone, shale, limy shale, and a limy shale containing some sand.

Figure 7.25 ***Variation in facies of modern sediments off the coast of southern California.*** (*After Roger Revelle and F. P. Shepard, "Sediments of the California Coast," p. 246 in P. D. Trask, ed.,* Recent Marine Sediments: A Symposium, *Am. Assoc. Petrol. Geologists, Tulsa, 1939.*)

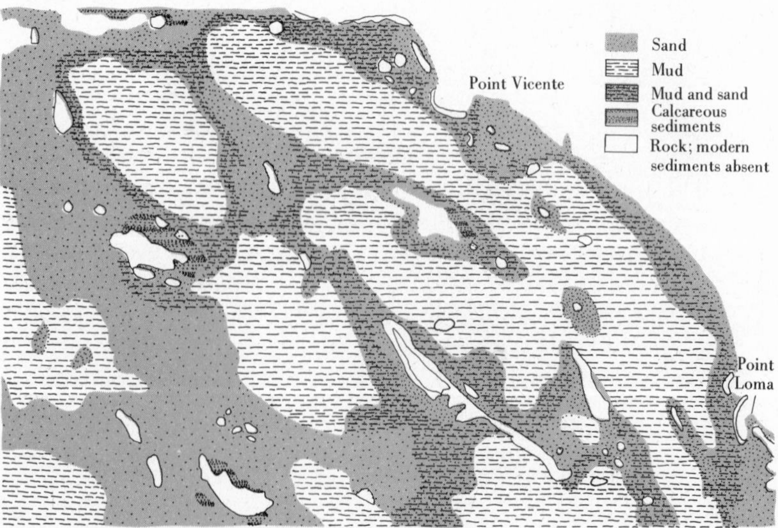

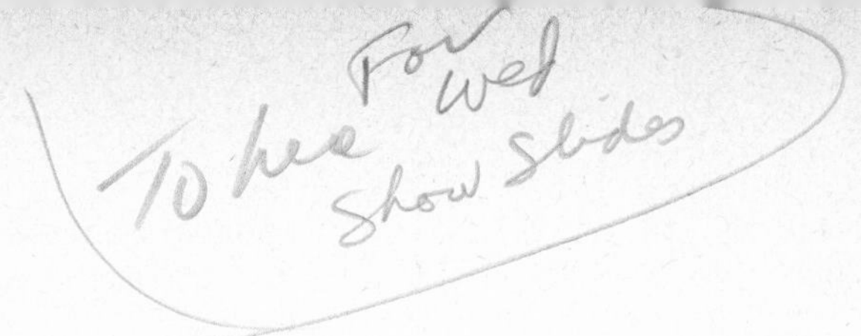

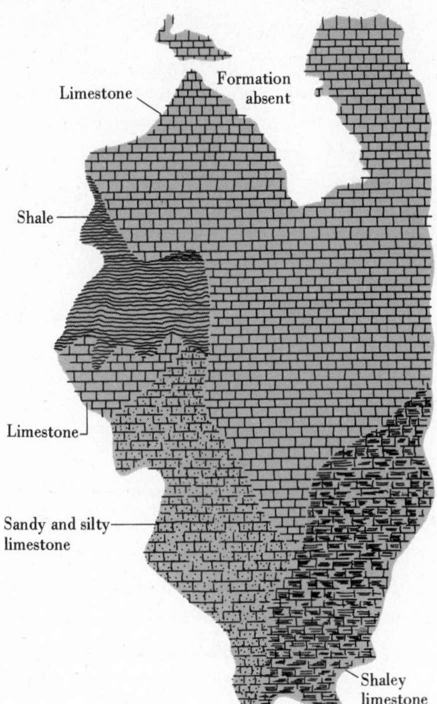

Figure 7.26 ***Variation in facies of ancient sediments comprising the Maquoketa formation in Illinois.*** (*After E. P. DuBois,* Illinois State Geol. Surv. Rep. Invest., *no. 105, p. 10, 1945.*)

Outline

Sedimentary rocks cover about 75 percent of the earth's surface and make up about 5 percent by volume of the outer 15 km of the solid earth.

Formation of sedimentary rocks takes place at or near the earth's surface.

Detrital material worn from the landmasses and *chemical* deposits precipitated from solution are the two chief types of sediments.

Sedimentation is the process by which rock-forming materials are laid down and the resulting deposits vary with the *source of material, the methods of transportation,* and the *processes of deposition.*

Clay, quartz, and *calcite* are the most common minerals in sedimentary rocks. Other minerals include *dolomite, goethite, hematite, limonite, feldspar, mica, halite,* and *gypsum.*

Texture depends on the size, shape, and arrangement of the particles. Texture may be *clastic* or *nonclastic.*

Lithification converts unconsolidated sediments to firm rock by *cementation, compaction, desiccation,* and *recrystallization.*

Types of sedimentary rocks include both *detrital* and *chemical* forms. Detrital rocks include *conglomerate sandstone, mudstone,* and *shale.* Chemical rocks include *limestone, dolomite, evaporites,* and *coal.* Most abundant are shale and mudstone, sandstone, and limestone, in that order. They form 99 percent of the sedimentary rock family.

Features of sedimentary rocks include *bedding, mud cracks, nodules, concretions, geodes,* and *fossils.*

Color of sedimentary rocks is due largely to small amounts of the iron oxide minerals and less importantly to organic matter.

Sedimentary facies refers to an accumulation of deposits that exhibits specific characteristics and grades laterally into other accumulations formed at the same time but showing different characteristics.

Supplementary readings

DEGENS, EGON T.: *Geochemistry of Sediments: A Brief Survey,* Prentice-Hall, Inc., Englewood Cliffs, N.J., 1965. A good place to begin if one is interested in the low-temperature geochemistry of sedimentary rocks.

KRUMBEIN, U. C., and L. L. SLOSS: *Stratigraphy and Sedimentation,* 2nd ed., W. H. Freeman and Company, San Francisco, 1963. Deals with the principles of correlation, nomenclature, and description of sedimentary rocks and of the processes of sedimentation.

LAPORTE, LEO F.: *Ancient Environments,* Prentice-Hall, Inc., Englewood Cliffs, N.J., 1968. An excellent treatment of the way sedimentary materials can be used to determine environments.

PETTIJOHN, F. J.: *Sedimentary Rocks,* 2nd ed., Harper & Row, Publishers, New York, 1957. A standard text on the subject.

8 Metamorphism and metamorphic rocks

MANY ROCKS EXPOSED AT THE EARTH'S SURFACE TODAY SHOW EVIDENCE OF CHANGE. At first glance, some of these rocks seem to resemble familiar igneous rocks, but then we discover that their mineral grains are arranged in a peculiar manner (see Figure 8.1). Other rocks have the same composition as limestone, but they seem to have developed larger mineral grains. Still others are strikingly different from both igneous rocks and sedimentary rocks. All these are metamorphic ("changed-form") rocks.

8.1 Metamorphism

Sedimentary and igneous rocks have been changed while in the solid state, in response to pronounced changes in their environment. These changes may bring about modifications *within* the rocks themselves through the process called *metamorphism.* Metamorphism occurs within the earth's crust, below the zone of weathering and cementation and outside the zone of remelting. In this environment, rocks undergo chemical and structural changes to adjust to conditions different from those under which they were originally formed.

We do not have much knowledge about the conditions of metamorphism because it takes place within the earth's crust. It cannot be seen or studied as can the processes of weathering, sedimentation, and igneous activity. However, a few generalizations can be made. During metamorphism, rocks are changed, but not all rocks respond to metamorphism in the same way. Also, the conditions of metamorphism vary.

Some rocks may begin to undergo metamorphism at shallow depths. These are usually rocks that are composed of minerals that were formed under nearly surface conditions. Other rocks, such as granite, that were formed at great depth will be little affected by shallow burial.

Agents of metamorphism

The term *metamorphism* is limited to changes that take place in the texture or composition of solid rocks. Metamorphism can occur only while a rock is solid, because after the rock's melting point has been reached, a magma is formed and we are in the realm of igneous activity. (Rock is considered solid when it is in the plastic state.)

Agents of metamorphism are heat, pressure, and chemically active fluids.

HEAT Heat, an essential agent of metamorphism, may even be *the* essential agent. Some geologists question whether pressure alone could produce changes in rocks without a simultaneous increase in temperature. In fact, they say that metamorphism appears invariably to be controlled by temperature.

PRESSURE Under the influence of pressure, changes take place to reduce the space occupied by the mineral components of a rock mass. Pressure may

Figure 8.1 **Contorted gneiss from Bedford, Westchester County, New York. Dark bands of ferromagnesian minerals trace out the pattern of distortion to which this rock was subjected by the processes of metamorphism.** (*Photograph by Walter R. Fleischer.*)

produce a closer atomic packing of the elements in a mineral, recrystallization of the mineral, or formation of new minerals.

When rocks are buried to depths of several kilometers, they gradually become plastic and responsive to the heat and deforming forces that are active in the earth's crust and mantle (see Figure 8.2). When plastic, they deform by intergranular motion, by the formation of minute shear planes within the rock, by changes in texture, by reorientation of grains, and by crystal growth. Turner[1] says that depth of burial combined with a thermal gradient of from 10 to 15°C/km explains "burial metamorphism." The type of the original rock, again, has an important effect on the results achieved by burial and deformation.

CHEMICALLY ACTIVE FLUIDS Hydrothermal solutions released in the solidification of magma often percolate beyond the margins of the magma reservoir and react on surrounding rocks. Sometimes they remove ions and substitute others. Or they may add ions to the rocks' minerals to produce new minerals. When chemical reactions within the rock or the introduction of ions from an external source cause one mineral to grow or change into another of different composition, the process is called *metasomatism.* The term describes all ionic transfers, not just those that involve gases or solutions from a magma. Fine-grained rocks are more readily changed than others because they expose greater areas of grain surface to chemically active fluids.

Some of the chemically active fluid of metamorphism is the liquid already present in the pores of a rock. It is believed that such pore liquid may often act as a catalyst; that is, it expedites changes without itself undergoing change. Sometimes the changes of metamorphism are essentially progressive dehydration.

[1] Francis J. Turner, *Metamorphic Petrology,* McGraw-Hill Book Company, Inc., New York, 1968.

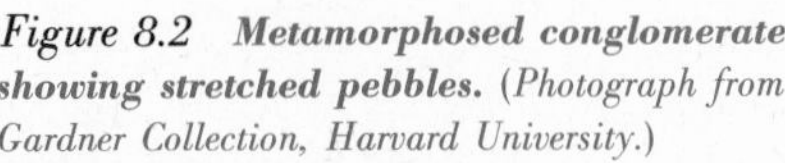

Figure 8.2 **Metamorphosed conglomerate showing stretched pebbles.** (*Photograph from Gardner Collection, Harvard University.*)

8.2 Types of metamorphism

Several types of metamorphism occur, but we shall concern ourselves here with the two basic ones: contact metamorphism and regional metamorphism.

Contact metamorphism

When magma is intruded into the earth's crust, it alters the surrounding rock. The alteration of rocks at or near their contact with a body of magma is called *contact metamorphism.* Minerals formed by this process are called *contact metamorphic minerals.* The type of reaction depends on the temperature, the composition of the intruding mass, and the properties of the intruded rock. At the actual surface of contact, all the elements of a rock may be changed or replaced by other elements introduced by the hydrothermal solutions escaping from the magma. Farther away, the replacement may be only partial.

Contact metamorphism occurs in zones called *aureoles* ("halos"), which seldom measure more than a few hundred meters in width and may be only a fraction of a centimeter wide. Aureoles are found bordering plutons (see Figure 8.3): sills, dikes, laccoliths, stocks, lopoliths, and batholiths. During contact metamorphism, temperatures may range from 300 to 800°C and load pressures from 100 to 3,000 atm (atmospheres).

Two kinds of contact metamorphic minerals are recognized: those produced by heating up the intruded rock and those produced by the hydrothermal solutions reacting with the intruded rock. Hydrothermal metamorphism develops at relatively shallow depths. It is only late in the cooling of a magma

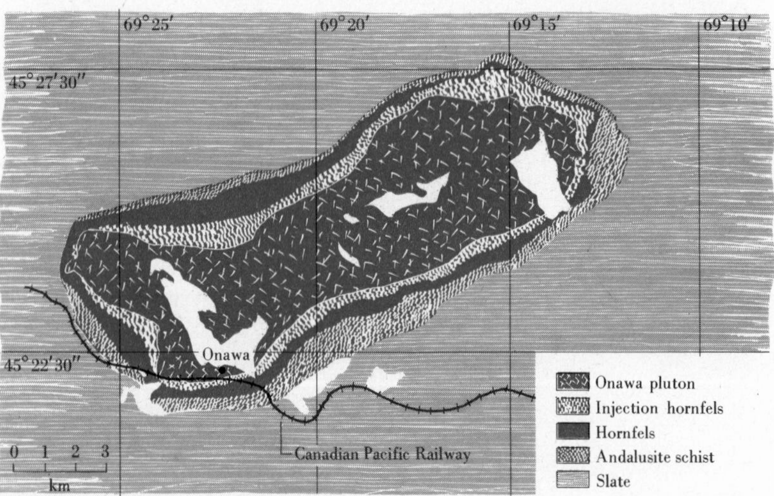

Figure 8.3 **Aureole of Onawa, Maine, granodiorite pluton.** (*After S. S. Philbrick,* Am. J. Sci., *vol. 31, pp. 1–40, 1936.*)

that large quantities of hydrothermal solutions are released, and these are released only as the body of magma nears the surface.

CONTACT METAMORPHIC MINERALS Many new minerals are formed during contact metamorphism. For example, when an impure limestone is subjected to thermal contact metamorphism, its dolomite, clay, or quartz may be changed to new minerals. Calcite and quartz may combine to form wollastonite. Dolomite may react with the quartz to form diopside. Aluminum, in the clay, will react to form corundum, spinel, or garnet. If carbonaceous materials are present, they may be converted to graphite. Many more minerals are produced by hydrothermal contact metamorphism. Solutions given off by the magma react to produce new minerals containing elements not present in the limestone. In this type of metamorphism, oxide and sulfide minerals are frequently formed to constitute ore deposits of economic importance (see Chapter 21).

Regional metamorphism

Regional metamorphism is developed over extensive areas, often involving thousands of square kilometers of rock thousands of meters thick. It is believed that regional metamorphism is related to huge masses of melted rock that form during the building of mountain ranges (see Chapter 19). This has not been proved. However, metamorphic rocks are found in the root regions of old mountains and in the Precambrian continental shields. Thousands of meters of rock have had to be eroded in order to expose these metamorphic rocks to view.

REGIONAL METAMORPHIC MINERALS During regional metamorphism, new minerals are developed as rocks respond to increases in temperature and pressure. These include some new silicate minerals not found in igneous and sedimentary rocks, such as sillimanite, kyanite, andalusite, staurolite, almandite, brown biotite, epidote, and chlorite. The first three of these new minerals are aluminosilicates and have the formula Al_2SiO_5. Their independent SiO_4 tetrahedra are bound together by positive ions of aluminum. *Sillimanite* develops in long, slender crystals that are brown, green, or white in color. *Kyanite* forms bladelike blue crystals. *Andalusite* forms coarse, nearly square prisms.

Staurolite is a nesosilicate composed of independent tetrahedra bound together by positive ions of iron and aluminum. It has a unique crystal habit that is striking and easy to recognize: it develops six-sided prisms that intersect either at 90°, forming a cross, or at 60°, forming an X (see Figure 8.4).

Garnets are a group of metamorphic nesosilicate minerals. All have the same atomic structure of independent SiO_4 tetrahedra, but a wide variety of chemical compositions is produced by the many positive ions that bind the tetrahedra together. These ions may be iron, magnesium, aluminum, calcium, manganese,

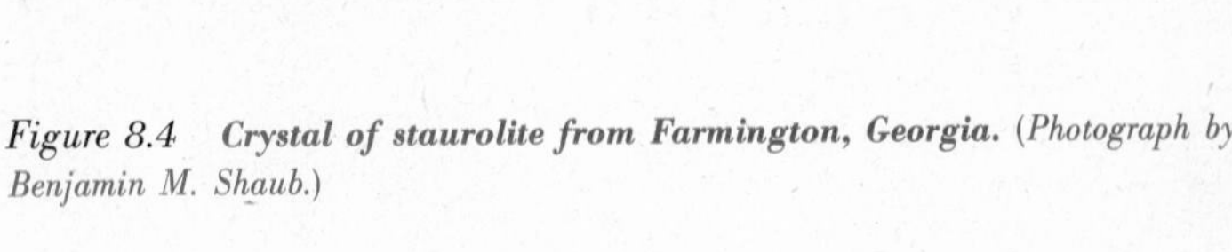

Figure 8.4 ***Crystal of staurolite from Farmington, Georgia.*** (*Photograph by Benjamin M. Shaub.*)

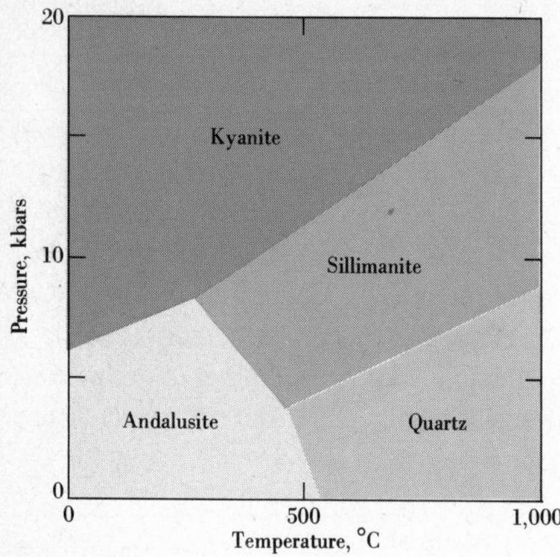

Figure 8.5 **Temperatures and pressures at which Al_2SiO_5 forms different minerals.** (*After H. Clark, E. Robertson, and F. Birch,* Am. J. Sci., *vol. 255, p. 255, 1957.*)

or chromium. But whatever the chemical composition, garnets appear as distinctive 12-sided or 24-sided fully developed crystals. Actually, it is difficult to distinguish one kind of garnet from another without resorting to chemical analysis. A common deep red garnet containing iron and aluminum is called *almandite.*

Epidote is a complex silicate of calcium, aluminum, and iron in which the tetrahedra are in pairs and these pairs are independent of each other. This mineral is pistachio green or yellowish to blackish green.

Chlorite is a phyllosilicate of calcium, magnesium, aluminum, and iron. The characteristic green color of chlorite was the basis for its name, from the Greek *chlōros,* "green" (as in chlorophyll). Chlorite exhibits a cleavage similar to that of mica, but the small scales produced by the cleavage are not elastic, like those of mica. Chlorite occurs either as aggregates of minute scales or as individual scales scattered throughout a rock.

REGIONAL METAMORPHIC ZONES Regional metamorphism may be divided into zones: high grade, middle grade, and low grade. Each grade is related to the temperature and pressure reached during metamorphism. High-grade metamorphism occurs in rocks nearest the magma reservoir, outside the zone of contact metamorphism. Low-grade metamorphism is found farthest away from the reservoir and blends into unchanged sedimentary rock.

Metamorphic zones are identified by using certain diagnostic metamorphic minerals called *index minerals.* Zones of regional metamorphism reflect the varied mineralogical response of chemically similar rocks to different physical conditions, and each index mineral gives an indication of the conditions at the time of its formation (see Figure 8.5).

The first appearance of chlorite, for example, tells us that we are at the beginning of a low-grade metamorphic zone. The first appearance of almandite is evidence of the beginning of a middle-grade metamorphic zone. And the first appearance of sillimanite marks a high-grade zone. Other minerals sometimes occur in association with each of these index minerals (see Table 8.1), but they are usually of little help in determining the degree of metamorphism of a given zone.

By noting the appearance of the minerals that are characteristic of each metamorphic zone, it is possible to draw a map of the regional metamorphism of an entire area. Of course, the rocks must have the proper chemical composition to allow these minerals to form.

Table 8.1 **Regional metamorphic minerals**

Zone	*Minerals*
Chlorite	Muscovite, chlorite, quartz
Biotite	Biotite, muscovite, chlorite, quartz
Garnet	Garnet (almandite), muscovite, biotite, quartz
Staurolite	Staurolite, garnet, biotite, muscovite, quartz
Kyanite	Kyanite, garnet, biotite, muscovite, quartz
Sillimanite	Sillimanite, quartz, garnet, muscovite, biotite, oligoclase, orthoclase

Low-grade regional metamorphism in progress

Studies on the cuttings and cores of wells in sedimentary deposits of the Imperial Valley of California have indicated that low-grade metamorphism is in progress on a regional scale and at relatively shallow depth.[2] Three wells were studied. Two of the wells are located 400 m apart in a geothermal field at the southeastern end of the Salton Sea. They are part of a group of 10 wells that have been drilled since 1960 to extract the elements contained in solution in the hot brines of the region and to tap the heat energy. The third was a dry oil test well drilled in 1963. It is located 35 km south-southeast of the geothermal field, in the Salton Trough (see Figure 8.6). Table 8.2 lists the wells and gives their depths and bottom-hole temperatures.

[2] L. J. Patrick Muffler and Donald E. White, "Active Metamorphism of Upper Cenozoic Sediments in the Salton Sea Geothermal Field and the Salton Trough, Southeastern California," *Bull. Geol. Soc. Am.*, vol. 80, pp. 157–182, 1969.

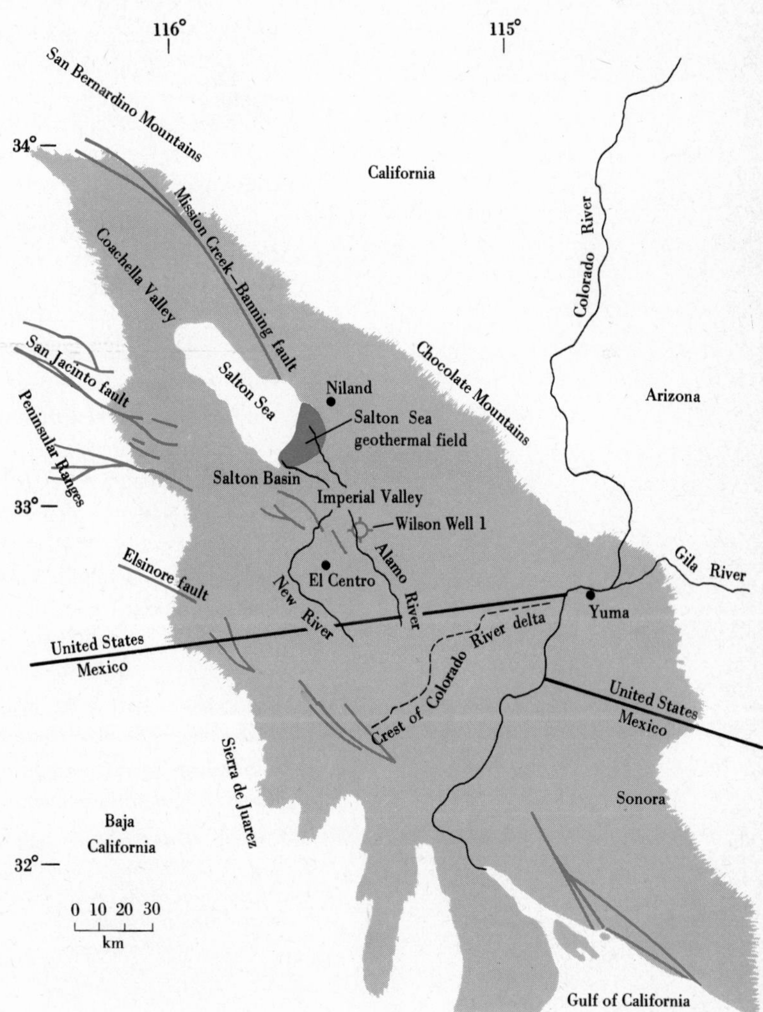

Figure 8.6 **Location of Salton Sea geothermal field.**

Table 8.2 ***Wells providing data on metamorphism***

Location	*Depth of well, m*	*Bottom-hole temperature, °C*	*Gradient, deg/15 m*
Sportsman No. 1 Geothermal Field	1,419	310	3
I.I.D. No. 1 Geothermal Field	1,570	328	3
Wilson No. 1 Salton Trough	4,033	260	1

In this region, the sedimentary deposits are geologically young. The oldest deposits in the trough were deposited in the early Pliocene age and those in the geothermal field during the Pleistocene age. They consist of poorly sorted sandstones and siltstones of the Colorado River delta. The original dominant minerals were quartz, calcite, subordinate dolomite, plagioclase, potassium feldspar, montmorillonite, illite, and kaolinite.

Samples from the three wells show increasing hardness and a regular sequence of mineral changes with depth as the temperature increases. Some detrital minerals drop out, and other minerals are formed. The rocks look like ordinary sedimentary rocks, but all have undergone mineralogical change at depth since deposition. This change is brought about by progressive metamorphism.

GEOTHERMAL FIELD The wells in the geothermal field produce a brine containing over 250,000 parts per million of dissolved solids, primarily Cl, Na, Ca, K, and Fe, plus a host of minor elements. Mineral transformations are due to hydrothermal metamorphism in the high-temperature environment. There is continuous transition from sediments through indurated sedimentary rocks to low-grade metamorphic rocks of the greenschist facies. The resultant rocks are more porous than most metamorphic rocks, probably because of the low pressures and short duration of metamorphism. No schistosity is developed. The rocks are analogous to products of burial metamorphism.

Montmorillonite is the dominant clay mineral. It is abundant in the surface sediments and to a depth of at least 40 m. Montmorillonite is converted to illite at essentially surface temperatures, that is, below 100°C. There is progressive change and montmorillonite is randomly interlayered with illite. Montmorillonite, both as a discrete phase and interlayered with illite, is converted to potassium mica at 210°C and is not found at greater temperatures.

Ankerite, a carbonate with dolomite crystalline structure, in which ferrous iron replaces part of the magnesium, is the first new mineral to form. It results from conversion of calcite or dolomite or both between 80 and 120°C. Major mineralogical transformations occur between 125 and 200°C. Through this interval, kaolinite, dolomite, and ankerite react to form chlorite, calcite, and

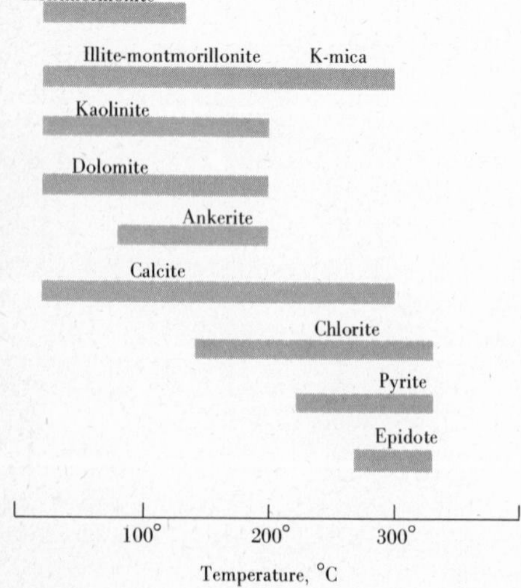

Figure 8.7 ***Mineral occurrence as a function of temperature in the Salton Sea geothermal field.*** (*Data from L. J. Patrick Muffler, and Donald E. White, "Active Metamorphism of Upper Cenozoic Sediments in the Salton Sea Geothermal Field and the Salton Trough, Southeastern California,"* Geol. Soc. Am. Bull., *vol. 80, Fig. 3, p. 165, 1969.*)

carbon dioxide, and decrease in abundance. At about 200° C, kaolinite, dolomite, and ankerite all disappear. Pyrite is first seen at 220° C and is conspicuous after 280° C. Epidote is first detected at 290° C, is sporadic to 300° C, and is then abundant. Calcite gradually decreases and is essentially absent at temperatures higher than 300° C. Chlorite, pyrite, and epidote, once formed, continue to the bottom of the well (see Figure 8.7). Potassium feldspar, plagioclase, and quartz are not changed and are found throughout the entire range.

OUTSIDE THE GEOTHERMAL FIELD Wilson No. 1 was a dry oil test well drilled in 1963. The temperature at the bottom of the well, a depth of 4,033 m, was 260° C, which represents an average gradient of 60° C/km. Mineralogical transformations occur over the same temperature ranges in this well as in the geothermal field, but because the thermal gradient is less, new minerals form at greater depths and extend over a much wider depth range. The temperatures encountered at the bottom of the well were not high enough for the development of epidote, pyrite, and hematite.

Montmorillonite interlayered with illite was not found at depths greater than 1,200 m. Kaolinite disappeared at approximately 2,400 m. Dolomite persisted to the bottom of the well but abruptly decreased in abundance below 3,750 m. It is suspected that dolomite is absent at a depth just a little greater than the bottom of this well.

Because the same mineralogical changes occurred at the same temperature in all three of the wells, it seems clear that temperature is the controlling factor.

8.3 Metamorphic rocks

Contact and regional metamorphic rocks are found in mountain ranges, at roots of mountain ranges, and on continental shields. The regional metamorphic rocks are by far the most widespread. They vary greatly in appearance, texture, and composition.

Even when different regional metamorphic rocks have all been formed by changes of a single uniform rock type such as a shale, they are sometimes so drastically changed that they are thought to be unrelated.

These and other characteristics of regionally metamorphosed shale are well illustrated in New Hampshire. Along the Connecticut River west of the White Mountains, rocks that were originally shales are found in a low-grade metamorphic zone as slate, with chlorite. Southeast of these rocks, the original shale is now found as phyllite, grading into schist. New metamorphic minerals appear one after the other toward the southeast: almandite, staurolite, and then sillimanite. These metamorphosed shales occur in a belt surrounding the White Mountain batholith (see Figure 8.8). The closer the area is to the batholith, the higher is its grade of regional metamorphism. This correlation is indicated by the presence of the index minerals found in the metamorphic rock.

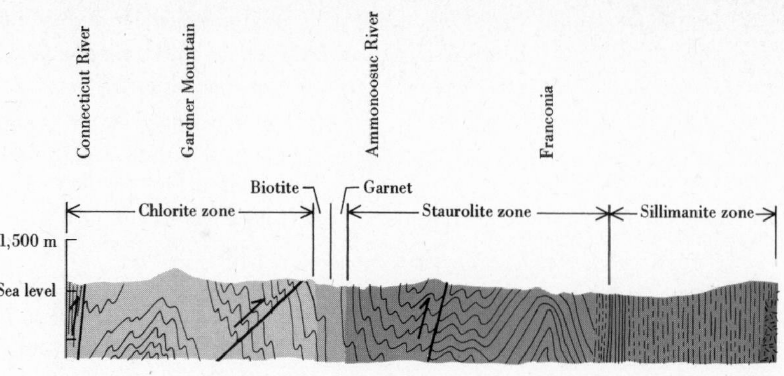

Figure 8.8 **Cross section of New Hampshire, showing metamorphic zones around the White Mountain batholith of the Older Appalachians. The length of the section is approximately 80 km.** (*After Marland P. Billings,* The Geology of New Hampshire. Part II: Bedrock Geology, *p. 139, Granite State Press, Inc., Manchester, 1956.*)

Texture of metamorphic rocks

In most rocks that have been subjected to heat and deforming pressures during regional metamorphism, the minerals tend to be arranged in parallel layers of flat or elongated grains. This arrangement gives the rocks a property called *foliation* (from the Latin *foliatus,* "leaved" or "leafy," hence consisting of leaves or thin sheets).

The textures most commonly used to classify metamorphic rocks are simply (1) *unfoliated* (either aphanitic or granular) and (2) *foliated.* Let us look first at the unfoliated textures. In rocks with aphanitic texture, the individual grains cannot be distinguished by the unaided eye, and these rocks do not exhibit rock cleavage. You will remember that we have used the term *cleavage* to describe the relative ease with which a mineral breaks along parallel planes. But notice here that we are using the modifier *rock* to distinguish rock cleavage from mineral cleavage. In rocks with granular texture, the individual grains are clearly visible, but again no rock cleavage is evident.

Rocks with foliated texture, however, invariably exhibit rock cleavage. There are four degrees of rock cleavage:

1 *Slaty* (from the French *esclat,* "fragment," "splinter"), in which the cleavage occurs along planes separated by distances of microscopic dimensions.

2 *Phyllitic* (from the Greek *phyllon,* "leaf"), in which the cleavage produces flakes barely visible to the unaided eye. Phyllitic cleavage produces fragments thicker than those of slaty cleavage.

3 *Schistose* (from the Greek *schistos,* "divided" or "divisible"), in which the cleavage produces flakes that are clearly visible. Here cleavage surfaces are rougher than in slaty or phyllitic cleavage.

4 *Gneissic,* in which the surfaces of breaking are from a few millimeters to a centimeter or so apart.

Types of metamorphic rocks

The many types of metamorphic rocks stem from the great variety of original rocks and the varying kinds and degrees of metamorphism. Metamorphic rocks

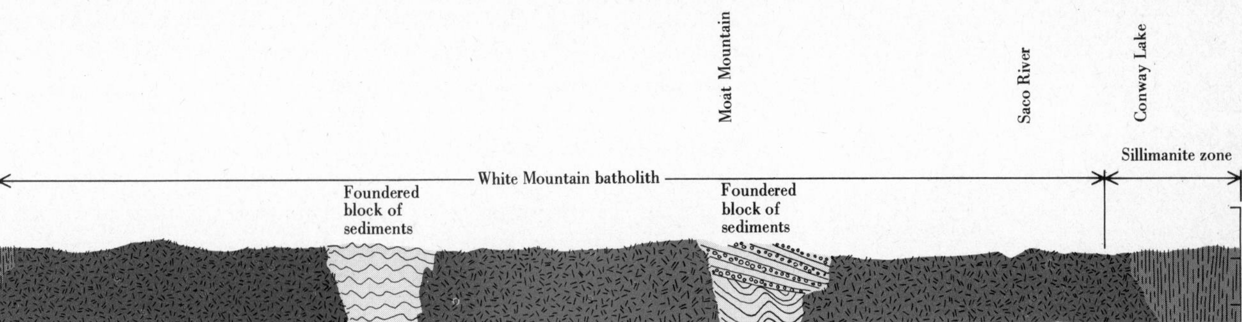

may be derived from any of the sedimentary or igneous rocks and other metamorphic rocks of lower grade.

Metamorphic rocks are usually named on the basis of texture. A few may be further classified by including the name of a mineral, such as chlorite schist, mica schist, and hornblende schist.

SLATE *Slate* is a metamorphic rock that has been produced from the low-grade metamorphism of shale or pyroclastic igneous rock (see Figure 8.9). It is aphanitic with a slaty cleavage caused by the alignment of platy minerals under the pressures of metamorphism. Some of the clay minerals in the original shale have been transformed by heat into chlorite and mica. In fact, slate is composed predominantly of small colorless mica flakes and some chlorite. It occurs in a wide variety of colors. Dark-colored slate owes its color to the presence of carbonaceous material or iron sulfides.

PHYLLITE *Phyllite* is a metamorphic rock with much the same composition as slate but whose minerals exist in larger units. Phyllite is actually slate that has undergone further metamorphism. When slate is subjected to heat greater than 250 to 300°C, the chlorite and mica minerals of which it is composed develop large flakes, giving the resulting rock its characteristic phyllitic cleav-

Figure 8.9 ***Vermont slate quarry, showing how rock cleavage controls breaking.*** *(Photograph from Gardner Collection, Harvard University.)*

Table 8.3 *Common schists*

Variety	*Rock from which derived*
Chlorite schist	Shale
Mica schist	
Hornblende schist	Basalt or gabbro
Biotite schist	
Quartz schist	Impure sandstone
Calc-schist	Impure limestone

age and a silky sheen on freshly broken surfaces. The predominant minerals in phyllite are chlorite and muscovite. This rock usually contains the same impurities as slate but sometimes a new metamorphic mineral such as tourmaline or magnesium garnet makes its appearance.

SCHIST Of the metamorphic rocks formed by regional metamorphism, *schist* is the most abundant. There are many varieties of schist, for it can be derived from many igneous, sedimentary, or lower-grade metamorphic rocks, but all schists are dominated by clearly visible flakes of some platy mineral, such as mica, talc, chlorite, or hematite. Fibrous minerals are commonly present as well. Schist tends to break between the platy or fibrous minerals, giving the rock its characteristic schistose cleavage.

Table 8.3 lists some of the more common varieties of schist, together with the names of the rocks from which they were derived.

Schists often contain large quantities of quartz and feldspar as well as lesser amounts of minerals such as augite, hornblende, garnet, epidote, and magnetite. A green schistose rock produced by low-grade metamorphism, sometimes called a *greenschist,* owes its color to the presence of the minerals chlorite and epidote.

AMPHIBOLITE *Amphibolite* is composed mainly of hornblende and plagioclase. There is some foliation or lineation due to alignment of hornblende grains, but it is less conspicuous than in schists. Amphibolites may be green, gray, or black and sometimes contain such minerals as epidote, green augite, biotite, and almandite. They are the products of medium-grade to high-grade regional metamorphism of ferromagnesian igneous rocks and of some impure calcareous sediments.

Figure 8.10 **Metamorphic rock showing alignment of previously unoriented minerals. The light-colored bands are mainly orthoclase and quartz; the dark streaks are biotite and other ferromagnesian minerals. The bulk composition is that of granite, but in contrast to the random mixing in granite, the minerals here are distributed in relatively systematic patterns.** (*Photograph by Robert Navias.*)

GNEISS A granular metamorphic rock, *gneiss* is most commonly formed during high-grade regional metamorphism (see Figure 8.10). A banded appearance makes it easy to recognize in the field. Although gneiss does exhibit rock cleavage, it is far less pronounced than in the schists.

In gneiss derived from igneous rocks such as granite, gabbro, or diorite, the component minerals are arranged in parallel layers: The quartz and the feldspars alternate with the ferromagnesians. In gneiss formed from the metamorphism of clayey sedimentary rocks such as graywackes, bands of quartz or feldspar usually alternate with layers of platy or fibrous minerals such as chlorite, mica, graphite, hornblende, kyanite, staurolite, sillimanite, and wollastonite.

MARBLE This familiar metamorphic rock, composed essentially of calcite or dolomite, is granular and was derived during the contact or regional metamorphism of limestone or dolomite. It does not exhibit rock cleavage. Marble differs from the original rock in having larger mineral grains. In most marble, the crystallographic direction of its calcite is nearly parallel; this is in response to the metamorphic pressures to which it was subjected. The rock shows no foliation, however, because the grains have the same color, and lineation does not show up.

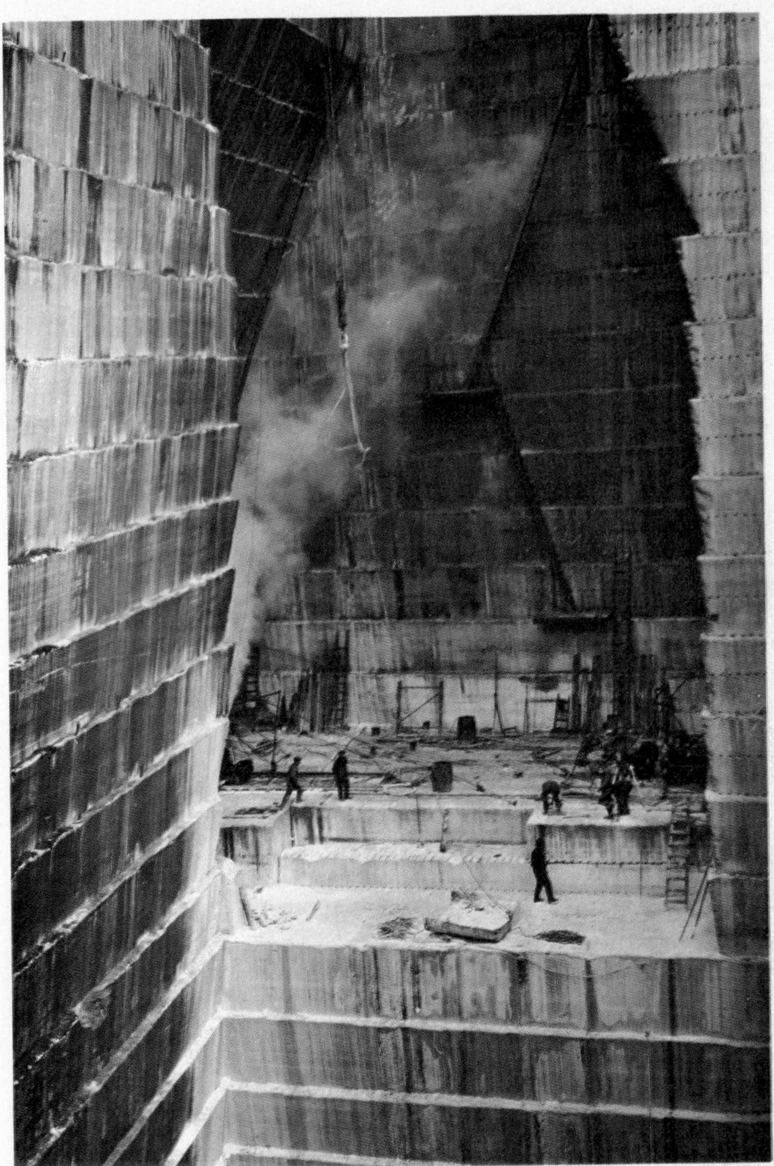

Figure 8.11 **Marble quarry showing how blocks are sawed out for dimension stone to be used in buildings.** (*Photograph by Benjamin M. Shaub.*)

Although the purest variety of marble is snow-white, many marbles contain small percentages of other minerals that were formed during metamorphism from impurities in the original sedimentary rock. These impurities account for the wide variety of color in marble. *Black marbles* are colored by bituminous matter; *green marbles* by diopside, hornblende, serpentine, or talc; *red marbles* by the iron oxide hematite; and *brown marbles* by the iron oxide limonite. Garnets have often been found in marble, as have rubies on rare occasions. The beautiful patterns of some marbles are often produced by the presence of fossilized corals in the original limestone.

Marble occurs most commonly in areas that have undergone regional metamorphism (see Figure 8.11), where it is often found in layers between mica schists or phyllites.

QUARTZITE The metamorphism of quartz-rich sandstone forms the rock quartzite. The quartz in the original sandstone has become firmly bonded by the entry of silica into the pore spaces.

Quartzite is unfoliated and is distinguishable from sandstone in two ways: There are no pore spaces in the quartzite, and the rock breaks right through the sand grains that make it up, rather than around them.

Quartzite may also be formed by percolating water under the temperatures and pressures of ordinary sedimentary processes working near the surface of the earth. Many quartzites, however, are true metamorphic rocks and may have been formed by metamorphism of any grade.

The structure of quartzite cannot be recognized without a microscope, but when we cut it into thin sections, we can identify both the original rounded sand grains and the silica that has filled the old pore spaces.

Pure quartzite is white, but iron or other impurities sometimes give the rock a reddish or dark color. Among the minor minerals that often occur in quartzite are feldspar, muscovite, chlorite, zircon, tourmaline, garnet, biotite, epidote, hornblende, and sillimanite.

Regional metamorphic facies

In Chapter 7, we defined the term *facies* as it is applied to sedimentary rocks. Actually, though, it has a broader application: In general, *a facies is an assemblage of mineral, rock, or fossil features reflecting the environment in which a rock was formed.*

Previously, we discovered that if at a given time there is an offshore sequence of sedimentation consisting of sands, shales, and limestones, we may speak of the sand facies, the shale facies, and the limestone facies of the rocks formed from those deposits. Similarly, metamorphic rocks that have been subjected to the same general range of temperature and pressure are said to belong to the same facies. A *metamorphic facies* is an assemblage of minerals that reached equilibrium during metamorphism under a specific set of conditions. Each facies is named after a common metamorphic rock that belongs to it, sometimes with a common mineral prefixed (such as epidote-amphibolite), and every metamorphic rock is assigned to a facies according to the conditions that attended its formation, not according to its composition.

It is not always possible to assign a rock to a particular metamorphic facies on the basis of a hand specimen, but it is always possible to do so after examining the other rocks in the region.

Three widely recognized regional metamorphic facies are shown in Table 8.4. They are:

1 Amphibolite facies, temperature 450 to 700°C; high-grade regional metamorphism.

2 Epidote-amphibolite facies, temperature 250 to 450°C; middle-grade regional metamorphism.

Table 8.4 **Regional metamorphic facies and related zones**[a]

Metamorphic temperature, °C	*Metamorphic facies*	*Metamorphic zones*
450–700	Amphibolite	Sillimanite Kyanite Staurolite
250–450	Epidote-amphibolite	Almandite (garnet)
150–250	Greenschist	Biotite Chlorite

[a] After Eskola and others. See Francis J. Turner, "Mineralogical and Structural Evolution of the Metamorphic Rocks," *Geol. Soc. Am. Mem.*, no. 30, 1948.

Table 8.5 ***Products of regional metamorphism***[a]

Original rock	Metamorphic rock: Chlorite zone	Biotite zone	Almandite zone	Staurolite zone	Sillimanite zone
Shale	Slate	Biotite Phyllite	Biotite-garnet Phyllite	Biotite-garnet-staurolite schist	Sillimanite schist or gneiss
Clayey sandstone	Micaceous sandstone	Quartz-mica schist	Quartz-mica-garnet schist or gneiss	Quartz-mica-garnet schist or gneiss	Quartz-mica-garnet schist or gneiss
Quartz sandstone	Quartzite	Quartzite	Quartzite	Quartzite	Quartzite
Limestone and dolomite	Limestone and dolomite	Marble	Marble	Marble	Marble
Basalt	Chlorite-epidote-albite schist (greenschists)	Chlorite-epidote-albite schist (greenschists)	Albite-epidote amphibolite	Amphibolite	Amphibolite
Granite	Granite	Granite gneiss	Granite gneiss	Granite gneiss	Granite gneiss
Rhyolite	Rhyolite	Biotite, schist, or gneiss	Biotite, schist, or gneiss	Biotite, schist, or gneiss	Biotite, schist, or gneiss

[a] After Marland P. Billings, *The Geology of New Hampshire. Part II: Bedrock Geology*, p. 139, Granite State Press, Inc., Manchester, 1956.

3 Greenschist facies, temperature 150 to 250°C; low-grade regional metamorphism.

To illustrate how a parent rock may change from one facies to another, let us assume that we have a mixture of sodium, calcium, iron, magnesium, and silicon in the proportions found in shale. If this mixture were subjected to a temperature of around 200°C and to low pressures, about 40 percent of the mixture by volume would form albite, 20 percent would form chlorite, and 23 percent would form epidote. This is the assemblage of minerals typical of the greenschist facies formed from the mixture of elements with which we started. If the same original mixture were heated to about 600°C, a different assemblage of minerals would be formed. This time, about 26 percent would form plagioclase feldspars, 72 percent would form hornblende, and 2 percent would form quartz. This assemblage, then, is typical of the amphibolite facies formed from the original mixture of elements. Table 8.5 shows the general relationships among zones, facies, and mineral assemblages in the progressive metamorphism of shale.

8.4 Origin of granite

The eighteenth-century geologist James Hutton once stated that granite was produced by the crystallization of minerals from a molten mass, and ever since, most geologists have accepted the magmatic origin of granite. However, several investigators have questioned this conclusion, suggesting instead that granite is a metamorphic rock produced from preexisting rocks by a process called *granitization*.

In discussing batholiths, we mentioned that one of the reasons for questioning the magmatic origin of granite was the mystery of what happened to the great mass of rock that must have been displaced by the intrusion of the granite batholiths. This so-called "space problem" has led some geologists to conclude that batholiths actually represent preexisting rocks transformed into granite by metasomatic processes.

Certain rock formations support this theory: These sedimentary rocks were originally formed in a continuous layer but now grade into schists and then into migmatites ("mixed rocks"), apparently formed when magma squeezed in between the layers of schist.

The concept of migmatite as a rock group was introduced by J. J. Sederholm in 1907.[3] It is a generic term which covers a large number of petrological combinations and is to be applied to rock formed from two principal ingredients: (1) a *paleosome,* or host of early-formed material, and (2) a *neosome,* or material introduced by permeation, metasomatism, or injection of liquids. The average composition of a migmatite is granitic. Its diagnostic characteristics are structure, texture, and broad regional relationships.

The migmatites, in turn, grade into rocks that contain the large, abundant feldspars characteristic of granite but that also seem to show shadowy remnants of schistose structure. Finally, these rocks grade into pure granite. The proponents of the granitization theory say that the granite is the result of extreme metasomatism and that the schists, migmatites, and granitelike rocks with schistose structure are intermediate steps in the transformation of sedimentary rocks into granite.

What mechanism could have brought about granitization? Perhaps ions migrated through the original solid rock, building up the elements characteristic of granite, such as sodium and potassium, and removing superfluous elements, such as calcium, iron, and magnesium. The limit to which the migrating ions are supposed to have carried the calcium, iron, and magnesium is called the *simatic front.* The limit to which the migrating ions are supposed to have deposited the sodium and potassium is called the *granitic front.*

At the middle of the twentieth century, geologists were carrying on an enthusiastic debate over the origin of granite, but they had reached agreement on one fundamental point: Various rocks with the composition and structure of granite may have different histories. In other words, some may be igneous and others metasomatic. So the debate between "magmatists" and "granitizationists" has been reduced to the question of what percentage of the world's granite is metasomatic and what percentage is magmatic. Those who favor magmatic origin admit that perhaps 15 percent of the granite exposed at the earth's surface is metasomatic. But the granitizationists reverse the percentages and insist that about 85 percent is metasomatic and only 15 percent of magmatic origin.

Field relationships in some cases suggest that great granite bodies have formed as second-generation igneous rocks in the cores of mountain ranges. This would involve remelting of deep portions of geosynclines and would leave no space problem or need for widespread metasomatism.

[3] J. J. Sederholm, "Om granit och gneiss," *Bull. Comm. Geol. Finland,* no. 98, Helsinki, 1907.

Outline

Metamorphism produces metamorphic rocks by changing igneous and sedimentary rocks while they are in the solid state.

Agents of metamorphism are heat, pressure, and chemically active fluids.

Heat may be the essential agent.

Pressure may be great enough to induce plastic deformation.

Chemically active fluids, particularly those released late in the solidification of magma, react on surrounding rocks.

Types of metamorphism are contact and regional.

Contact metamorphism occurs at or near an intrusive body of magma.

Contact metamorphic minerals include wollastonite, diopside, and some oxides and sulfides constituting ore minerals.

Regional metamorphism is developed over extensive areas and is related to the formation of some mountain ranges.

Regional metamorphic minerals include sillimanite, kyanite, andalusite, staurolite, almandite, garnet, brown biotite, epidote, and chlorite.

Regional metamorphic zones are identified by diagnostic index minerals.

Metamorphic rocks are found in mountain ranges, at mountain roots, and on continental shields.

Textures of metamorphic rocks are unfoliated and foliated.

Unfoliated rocks do not exhibit rock cleavage.

Foliated rocks exhibit rock cleavage classified as slaty, phyllitic, schistose, or gneissic.

Types of metamorphic rocks are many because of the variety of original rocks and the varying kinds of metamorphism.

Regional metamorphic facies is an assemblage of minerals that reached equilibrium during metamorphism under a specific set of conditions.

Origin of granite may be igneous, metamorphic, or by remelted sediments.

Supplementary readings

HARDER, ALFRED: *Metamorphism*, Methuen and Co., Ltd., London, 1932. A classic forerunner of modern treatments.

READ, H. H.: *The Granite Controversy*, Thomas Murby and Company, London, 1957. Eight addresses concerned with the origin of the granitic and associated rocks, delivered between 1939 and 1954.

TURNER, FRANCIS J.: *Metamorphic Petrology*, McGraw-Hill Book Company, Inc., New York, 1968. A general text for advanced students. Revises and replaces a large section of Turner and J. Verhoogen, embodying a survey of mineralogical aspects of metamorphism.

WINKLER, H. G. F.: *Petrogenesis of Metamorphic Rocks*, Springer-Verlag, Berlin, 1967. Stresses the mineralogical–chemical aspects of metamorphism, because "the question of the general relationship between the mineral associations, on the one hand, and temperature and pressure on the other, is the real core of the study of metamorphic rock" (Barth).

9

Geologic time

IN THE FIRST CHAPTER OF THIS BOOK WE STRESSED THE VASTNESS OF GEOLOGIC time and emphasized its importance to an understanding of the physical processes of geology. In this chapter, we shall look at some of the ways of measuring geologic time and at how the geologic time scale has been worked out.

As we begin to think about time, our first reaction is that it is definable without too much difficulty. But the more we think of what time really is, the more difficulty we have being precise about its definition. We begin to recognize, as Thomas Mann wrote in *The Magic Mountain*, that "Time has no division to mark its passage, there is never a thunder-storm or blare of trumpets to announce the beginning of a new month or year. Even when a new century begins it is only we mortals who ring bells and fire off guns."[1]

But we are mortals. And we may think of geologic time in two ways: relative and absolute. Relative time—that is, whether one event in earth history came before or after another event—disregards years. Whether a geologic event took place a few thousand years ago, a billion years ago, or at some date even farther back in earth history is reported in absolute time.

9.1 Absolute time

Relative and absolute time in earth history have their counterparts in human history. In tracing the history of the earth, we may want to know whether some event, such as a volcanic eruption, occurred before or after another event, such as a rise in sea level, and how these two events are related in time to a third event, perhaps a mountain-building episode. In human history, too, we try to establish the relative position of events in time. In studying U.S. history it is important to know that the Revolutionary War preceded the Civil War and that the Canadian–American boundary was fixed some time between these two events.

Sometimes, of course, events in both earth history and human history can be established only in relative terms. But our record becomes increasingly precise as we fit more and more events into an actual chronological calendar. If we did not know the date of the U.S.–Canadian boundary treaty—knew only that it was signed between the two wars—we could place it between 1783 and 1861. Recorded history, of course, provides us with the actual date, 1846.

Naturally, we would like to be able to date geologic events with precision. But so far this has been impossible, and the accuracy achieved in determining the dates of human history, at least written human history, will probably never be achieved in geologic dating. Still, we can determine approximate dates for many geologic events. Even though they may lack the precision of dates in recent human history, they are probably of the correct order of magnitude. We can say that the dinosaurs became extinct about 63 million years ago and that about 11,000 years ago the last continental glacier began to recede from New England and the area bordering the Great Lakes.

[1] Thomas Mann, *The Magic Mountain*, trans. H. T. Lowe-Porter, p. 225, Modern Library, Inc., New York.

Figure 9.1 ***The dark areas in this specimen from Grafton, New Hampshire, are made up of the uranium-bearing mineral gummite. Compare with Figure 9.2.*** *(Specimen in the Princeton University Museum of Natural History. Photograph by Willard Starks.)*

We earthlings use two basic units of time: (1) a day, the interval required for our globe (in the present epoch) to complete one revolution on its axis, and (2) a year, the interval required for the earth to complete one circuit of the sun. In geology, however, the problem is to determine how many of these units of time elapsed in the dim past when nobody was around to count and record them. Our most valuable clues in solving this problem are provided by the decay rates of radioactive elements.

Radioactivity

The nuclei of certain elements spontaneously emit particles and in so doing produce new elements. This process is known as *radioactivity.* Shortly after the turn of the twentieth century, researchers suggested that minerals containing radioactive elements could be used to determine the age of other minerals in terms of absolute time (see Figures 9.1 and 9.2).

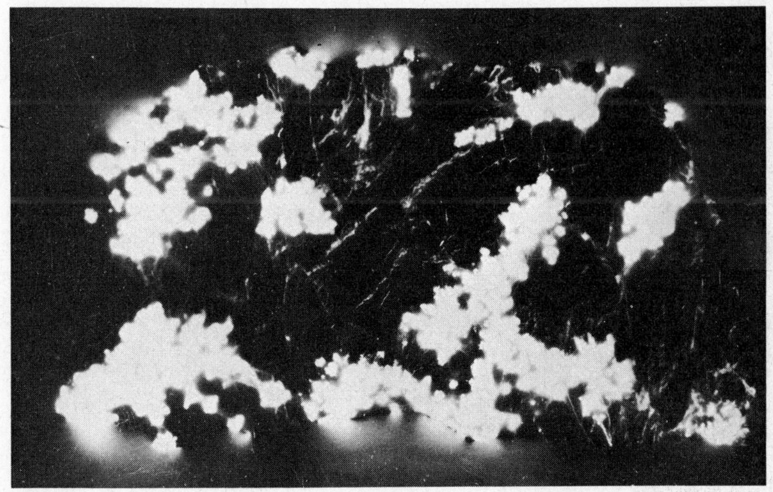

Figure 9.2 ***A photograph of the same specimen as that in Figure 9.1. In this photograph the specimen was placed on a photographic plate. As the uranium and some of its radioactive daughter products decayed, the plate was exposed to the emission of particles. The white areas mark the location of the uranium-bearing minerals.***

Table 9.1 ***History of 1 g of uranium 238***[a]

Age, yr × 10^6	*$^{206}_{82}Pb$ formed, g*	*$^{238}_{92}U$ remaining, g*
100	0.013	0.985
1,000	0.116	0.825
2,000	0.219	0.747
3,000	0.306	0.646

[a] Sir James Jeans, *The Universe Around Us*, p. 144, The Macmillan Company, New York, 1929.

Let us take a single radioactive element as an example of how this method works. Regardless of what element we use, we must know what products result from its radioactive decay. Let us choose uranium 238, $^{238}_{92}U$, which is known to yield helium and lead, $^{206}_{82}Pb$, as end products. We know, too, the rate at which uranium 238 decays. So far as we can determine, this rate is constant and unaffected by any known chemical or physical agency. The rate at which a radioactive element decays is expressed in terms of what we call its *half-life*—the time required for half of the nuclei in a sample of that element to decay. The half-life of uranium 238 is 4.51×10^9 years, which means that if we start with 1 g of uranium 238 there will be only 0.5 g left after 4,510 million years.

The history of 1 g of uranium 238 may be recorded as shown in Table 9.1. If you look carefully at the figures in the table, you will see that at any instant during this process there is a unique ratio of lead 206 to uranium 238. This ratio depends on the length of time decay has been going on. Theoretically, then, we may find the age of a uranium mineral by determining how much lead 206 is present and how much uranium 238 is present. The ratio of lead to uranium then serves as an index to the age of the mineral.

One of the basic assumptions we have made in applying radioactivity to age determination is that the laws governing the rate of decay have remained constant over incredibly long periods of time, but we are justified in wondering whether this assumption is valid. Geology supplies one piece of confirming evidence in the form of *pleochroic* ("many-colored") *halos.* These are minute, concentric, spherical zones of dark or colored material no more than 7.5×10^{-3} cm in diameter—the thickness of an ordinary sheet of paper—that form around inclusions of radioactive materials in biotite and in a few other minerals. If such a sphere is sliced through the center, the resulting sections show the pleochroic halos as rings. Each ring is a region in which alpha particles, shot out of the decaying radioactive mineral, came to rest and ionized the host material. The effect resembles that of light on a photographic film.

The energy possessed by an alpha particle depends on what element released it and on the stage of decay at which the particle was released. The pleochroic halos have radii that correspond to the energy of present-day alpha particles. Apparently, the energy of an alpha particle today is the same as it was hundreds of millions of years ago. This fact implies that the fundamental constants of nuclear physics that govern the travel of alpha particles have not changed.

This evidence, of course, does not prove explicitly that the rate of decay has always been the same. But if the laws governing the energy of the particles have not changed, it is reasonable to assume that related laws governing the rate of decay have also continued unchanged.

Uranium 238 is not the only element found useful in the age determination of rock material; there are others. But the same basic idea prevails, whatever element we choose. If we discover a mineral that contains one or more radioactive elements, we may be able (after proper chemical analysis) to determine how many years ago the mineral was formed. If the mineral was formed at the same time as the rock in which it is enclosed, the age of the mineral will also give us the age of the rock.

A radioactive element decays in one of several ways. The nucleus may lose an alpha particle. An alpha particle is the nucleus of the helium atom, which, as we saw in Chapter 2, consists of two protons and two neutrons. This process is called *alpha decay* and the mass of the element decreases by 4 (two protons plus two neutrons) and the atomic number decreases by 2 (two protons). An element undergoing *beta decay* loses an electron (a beta particle) from one of the neutrons of the nucleus. The neutron thereby becomes a proton and the atomic number is increased by 1. The mass—the sum of all protons and neutrons—remains the same. In a third type of decay, *electron capture decay*, the nucleus changes by picking up an electron from its orbital electrons. This electron is added to a proton within the nucleus, converting the proton to a neutron and decreasing the atomic number by 1.

Many elements, in decaying from a radioactive to a nonradioactive state, go through a series of transformations until one or more stable end products are reached. Thus uranium 238 begins to decay by alpha emission and continues with seven more alpha emissions and six beta emissions before lead 206 is produced.

Some elements may follow two or more different paths in their decay. Potassium 40 is an example. It may decay either by beta emission to form calcium 40 or by electron capture to form argon 40.

Methods of determining age by radioactivity have produced tens of thousands of dates for events in earth history, and new ones are constantly being reported. By 1970, rocks from a number of localities had been dated as approximately 3.3 billion years old. Field relations show that other, still older rocks exist. A granite about 3.2 billion years old from Pretoria, South Africa, contains inclusions of a quartzite, positive indication that older, sedimentary rocks existed before the intrusion of the granite. The exact age of the earth itself is still undetermined, but several lines of evidence converge to suggest an age of about 4.5 billion years.

Today's discoveries, then, have fully vindicated the assumptions made over a century and a half ago that geologic time *is* vast and that within earth history there *is* abundant time for slow processes to accomplish prodigious feats.

There are scores of radioactive isotopes, but only a few are useful in geologic dating (see Table 9.2). For older rocks, potassium 40, rubidium 87, uranium 235 and uranium 238 have proved most important. Potassium 40 has also

Table 9.2 **Some of the more useful elements for radioactive dating**

Parent element	*Half-life, yr*	*Daughter element*	*Types of decay*
Carbon 14	5,730	Nitrogen 14	Beta
Potassium 40	1,300 million	Argon 40	Electron capture
Rubidium 87	47,000 million	Strontium 87	Beta
Uranium 235	713 million	Lead 207	Seven alpha and four beta
Uranium 238	4,510 million	Lead 206	Eight alpha and six beta

Figure 9.3 ***This fragment of spruce log was part of a tree in a buried forest at Two Creeks, Wisconsin. Carbon-14 measurements of wood from the forest reveal that the trees died some 11,350 radiocarbon years ago, thereby establishing the date of the ice invasion. The fragment is about 20 cm long.*** *(Photograph by Willard Starks.)*

proved useful to date some of the more recent events of earth history, events younger than 2 million years. For very recent events, however, the radioactive isotope of carbon, carbon 14, ${}^{14}_{6}C$, has proved most versatile. It is usable only on organic material that is around 50,000 years old or less.

The carbon-14 method, first developed at the University of Chicago by Willard F. Libby, works as follows. When neutrons from outer space, sometimes called cosmic rays, bombard nitrogen in the outer atmosphere, they knock a proton out of the nitrogen nucleus, thereby forming carbon 14.

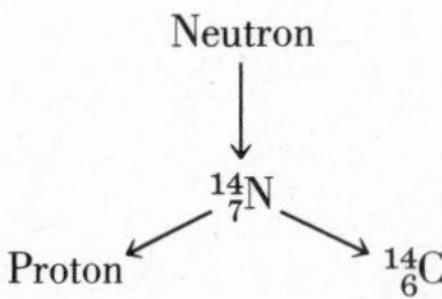

The carbon 14 combines with oxygen to form a special carbon dioxide, ${}^{14}CO_2$, which circulates in the atmosphere and eventually reaches the earth's surface, where it is absorbed by living matter. It has been found that the distribution of carbon 14 around the world is almost constant. Its abundance is independent of longitude, latitude, altitude, and the type of habitat of living matter.

The bulk of carbon in living material is the stable isotope carbon 12. Nevertheless, there is a certain small amount of carbon 14 in all living matter. And when the organism—whether it is a plant or an animal—dies, its supply of carbon 14 is, of course, no longer replenished by life processes. Instead, the carbon 14, with a half-life of about 5,730 years, begins spontaneously to change back to ${}^{14}_{7}N$ by beta decay. The longer the time that has elapsed since the death of the organism, the less the amount of carbon 14 that remains. So when we find carbon 14 in a buried piece of wood or in a charred bone, by comparing the amount present with the universal modern abundance, we

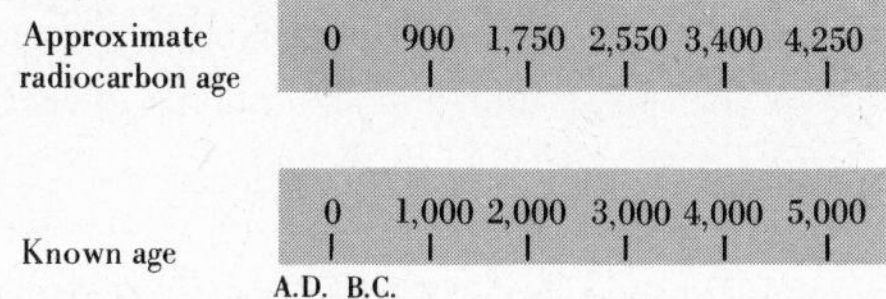

Figure 9.4 ***Radiocarbon analyses of tree rings of known age show that data postulate ages that are too young for the period from the birth of Christ back at least into the sixth millennium B.C. The discrepancy increases from zero to about 750 years in 5000 B.C.*** (*Data from E. K. Ralph and H. N. Michael,* Archaeometry, *vol. 10, pp. 3–11, 1967.*)

can calculate the amount of time that has elapsed since the material ceased to take in $^{14}CO_2$, that is, since the organism died.

We have seen that there are assumptions on which we base our evaluation of the validity of radioactive dating. These assumptions are reasonable and have some support from observation. Nevertheless, the more we use radioactive methods, the more sophisticated our understanding of them, which is another way of saying that we become aware of the problems involved. Let us take carbon 14 as an example of how increased knowledge reveals increased complexity.

When Libby did his initial work on carbon 14, its half-life was determined as 5,570 years. More refined measurements now indicate it to be about 5,730 years. By common agreement, however, 5,570 is still used as the "accepted value" and therefore all published dates should be adjusted upward 3 percent.

Beyond this, Hans Suess, while working with the U.S. Geological Survey, demonstrated that since about 1850 the amount of nonradioactive or "dead" carbon 12 poured into the atmosphere by the combustion of coal and petroleum has diluted the amount of carbon 14 to 98 percent of its original concentration before man began tampering with the atmosphere. Working in the opposite direction, man has created a great amount of new carbon 14 since 1945 when he began atmospheric nuclear explosions. In fact, this source of contamination has doubled the carbon-14 activity of the atmosphere.

Of more historical interest, perhaps, is the recent discovery that radiocarbon dates obtained on materials from about the time of Jesus Christ and going back for another 5,000 years, at least, are younger than we would expect when we compare them with carbon-14 dates of samples of known age dated independently of carbon 14. These samples have included historically dated material from the Mediterranean area, particularly from Egypt, and tree rings of the bristlecone pine from the U.S. West. This tree, which includes the world's oldest known living material, has provided us with a set of tree rings which goes back over 7,000 years and is securely tied to our modern solar calendar. When we subject tree rings older than 2,000 years to carbon-14 dating, the radiocarbon date is younger than the actual date. The farther back we go, the greater is the divergence. By the time, for example, the tree ring count reaches 3200 B.C., the radiocarbon age is only 2700 B.C. (see Figure 9.4). We do not yet know the cause of this divergence or how far back into time it may continue. But the divergence suggests that we are well-advised to use the term *radiocarbon years* to indicate that radiocarbon dates are not absolutely synonymous with dates in our solar calendar.

Despite these difficulties, the nearly 20,000 radiocarbon dates now available have revolutionized many aspects of archaeology as well as the study of the last 50,000 years of earth history.

Radioactivity and sedimentary rocks

Until recently, radioactive minerals suitable for dating geologic events were sought chiefly in the igneous rocks. These rocks were usually the uranium- and thorium-bearing minerals. Even today the great bulk of the radioactively dated rocks are igneous in origin. The development of new techniques, however, particularly the use of radioactive potassium, has extended radioactive dating to some of the sedimentary rocks.

Some sandstones and, more rarely, shales contain *glauconite,* a silicate mineral similar to biotite. Glauconite, formed in certain marine environments when the sedimentary layers are deposited, contains radioactive potassium, another geologic hourglass that reveals the age of the mineral and, hence, the age of the rock. The end products of the decay of radioactive potassium are argon and calcium. Age determinations are based on the ratio of potassium to argon.

Some pyroclastic rocks are composed mostly or completely of volcanic ash. Biotite in these rocks includes radioactive potassium and so offers a way of dating the biotite and sometimes the rock itself.

Sedimentation and absolute time

Another way of establishing absolute dates for sedimentary strata is to determine the rate of their deposit.

Certain sedimentary rocks show a succession of thinly laminated beds. Various lines of evidence suggest that, in some instances at least, each one of these beds represents a single year of deposition. Therefore, by counting the beds, we can determine the total time it took for the rock to be deposited.

Unfortunately, we have been able to link this kind of information to our modern calendar in only a very few places, such as the Scandinavian countries. Here the Swedish geologist Baron de Geer counted the annual deposits, or laminations, that formed in extinct glacial lakes. These laminations, called *varves* (see Figure 9.5 and Chapter 13), enable us to piece together some of the geologic events of the last 20,000 years or so in the countries ringing the Baltic Sea.

Much longer sequences of laminated sediments have been found in other places, but they tell us only the *total length of time* during which sedimentation took place, not *how long ago* it happened, in absolute time. One excellent example of absolute time sequence is recorded in the Green River shales of Wyoming (see Figures 9.6 and 9.7). Here, each bed, interpreted as an annual deposit, is less than 0.017 cm thick and the total thickness of the layers is about 980 m. These shales represent, then, approximately 6.5 million years of time.

9.2 Relative time

Before geologists knew how to determine absolute time, they discovered events in earth history that convinced them of the great length of geologic time. In putting these events in chronological order, they found themselves subdividing geologic time on a relative basis and using certain labels to indicate relative time. You have probably picked up a newspaper or magazine and have read

Figure 9.5 **Varves formed of alternating silt and clay near the mouth of Sherman Creek, Ferry County, Washington. Each pair of light layers is thought to represent a single year of sedimentation.** (*Photograph by U.S. Geological Survey.*)

Figure 9.6 **The Green River shale in Wyoming is composed of minute annual layers. Because each step in this block is 100 layers high, the entire block records 700 years of sedimentation. A portion of a fossil fish is seen on the large step on the left-hand side of the block. Counting of the layers indicates that the fish died 471 years after the formation of the lowest layer.** (*Specimen from Princeton University Museum of Natural History. Photograph by Willard Starks.*)

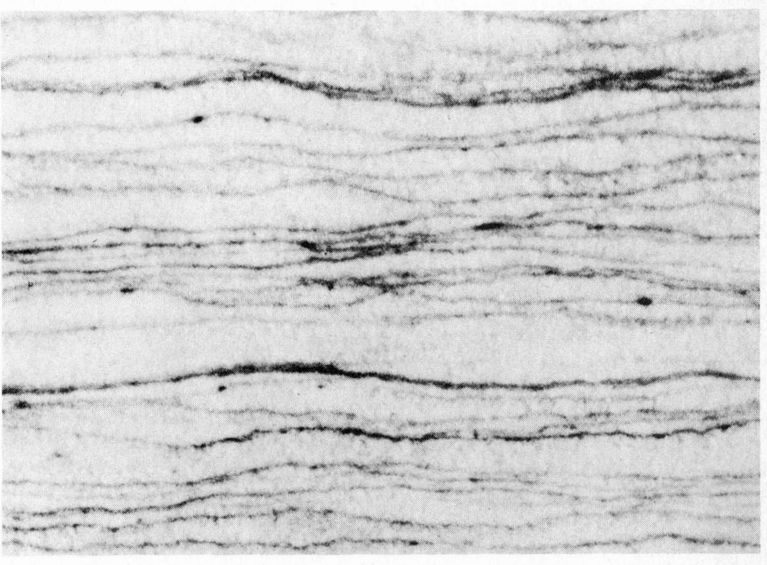

Figure 9.7 **An enlarged section of Green River shale showing, greatly magnified, the layered nature of the rock. The specimen represents about 50 years of time in a piece of rock about 0.5 cm thick. At this rate of accumulation, 1 m of Green River shale represents approximately 10,000 years.** (*Specimen from Princeton University Museum of Natural History. Photograph by Fred Anderegg.*)

of the discovery of a dinosaur that lived 100 million years ago during the Cretaceous period or of the development of a new oil field in strata formed 280 million years ago in the Pennsylvanian period. *Cretaceous* and *Pennsylvanian* are terms used by geologists to designate certain units of *relative geologic time.* In this section, we shall look at how such units have been set up and how absolute dates have been suggested for them.

Relative geologic time has been determined largely by the relative positions of sedimentary rocks. Remember that a given layer of sedimentary rock represents a certain amount of time—the time it took for the original deposit to accumulate. By arranging various sedimentary rocks in their proper chronological sequence, we are, in effect, arranging units of time in *their* proper order. Our first task in constructing a relative time scale, then, is to arrange the sedimentary rocks in their proper order.

The law of superposition

The basic principle used to determine whether one sedimentary rock is older than another is very simple, and it is known as the *law of superposition.* Here is an example. A deposit of mud laid down this year in, say, the Gulf of Mexico will rest on top of a layer that was deposited last year. Last year's deposit, in turn, rests on successively older deposits that extend backward into time for as long as deposition has been going on in the gulf. If we could slice through these deposits, we would expose a chronological record with the oldest deposit on the bottom and the youngest on top. This sequence would illustrate the law of superposition: *If a series of sedimentary rocks has not been overturned, the topmost layer is always the youngest and the lowermost layer is always the oldest.*

The law of superposition was first derived and used by Nicolaus Steno (1638–1687), a physician and naturalist who later turned theologian. His geologic studies were carried on in Tuscany in northern Italy and were published in 1669. On first glance the principle worked out by Steno is absurdly simple. For instance, in a cliff of sedimentary rocks, with one layer lying on top of another, it is perfectly obvious that the oldest is on the bottom and the youngest on top. We can quickly determine the relative age of any one layer in the cliff in relation to any other layer. The difficulty, however, lies in the fact that unknown hundreds of thousands of feet of sedimentary rock have been deposited during earth time, and there is no one cliff, no one area where *all* these rocks are exposed to our view. The rocks in one place may be older, or younger, or of the same age as those in some other place. The task is to find out how the rocks all around the world fit into some kind of relative time scale.

Correlation of sedimentary rocks

Because we cannot find sedimentary rocks representing all of earth time neatly arranged in one convenient area, we must piece together the rock sequence

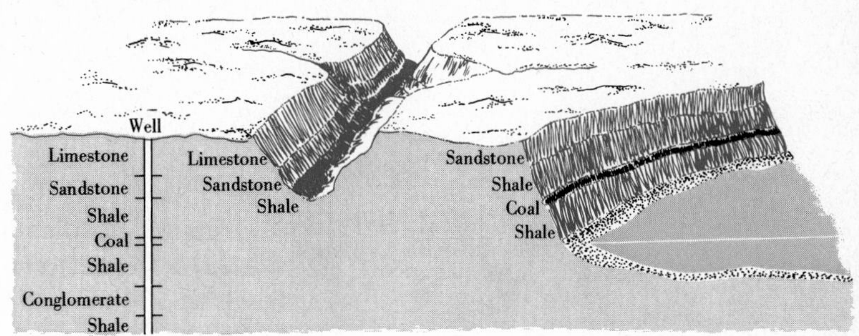

Figure 9.8 ***Diagram to illustrate the data that might be used to correlate sedimentary rocks (right) in a sea cliff with those (center) in a stream valley and with those (left) encountered in a well-drilling operation. (See Figure 9.9.)***

from locality to locality. This process of tying one rock sequence in one place to another in some other place is known as *correlation,* from the Latin for "together" plus "relate."

CORRELATION BY PHYSICAL FEATURES When sedimentary rocks show rather constant and distinctive features over a wide geographic area, we can sometimes connect sequences of rock layers from different localities. Figures 9.8 and 9.9 illustrate how this is done. Here is a series of sedimentary rocks exposed in a sea cliff. The topmost, and hence the youngest, is a sandstone. Beneath the sandstone we find, first, a shale, then a seam of coal, then more shale extending down to the level of the modern beach. We can trace these layers for some distance along the cliff face, but how are they related to other rocks farther inland?

Along the rim of a canyon that lies inland from the cliff we find that limestone rocks have been exposed. Are they older or younger than the sandstone in the cliff face? Scrambling down the canyon walls, we come to a ledge of sandstone that looks very much like the sandstone in the cliff. If it *is* the same, then the limestone must be younger, because it lies above it. The only trouble is that we cannot be certain that the two sandstone beds *are* the same, so we continue down to the bottom of the canyon and there find some shale beds very similar to the shale beds exposed in the sea cliff beneath the sandstone. We feel fairly confident that the sandstone and the shale in the canyon are the same beds as the sandstone and upper shale in the sea cliff, but we must admit the possibility that we are dealing with different rocks.

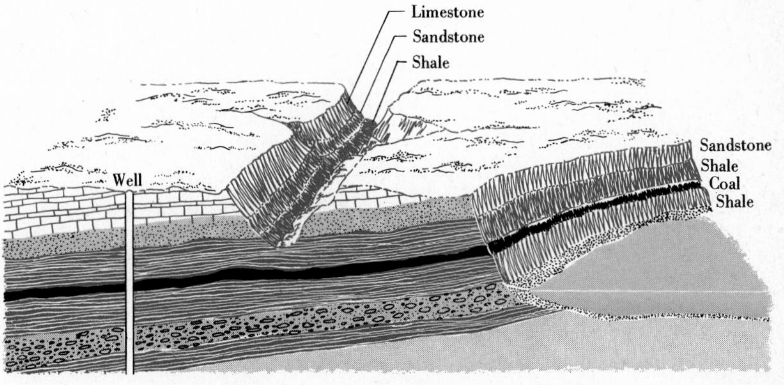

Figure 9.9 ***Similar lithologies and sequences of beds in the three different localities shown in Figure 9.8 suggest the correlation of rock layers shown in the diagram.***

In searching for further data, we find a well being drilled still farther inland, its rig cutting through the same limestone we saw in the canyon walls. As the bit cuts deeper and deeper, it encounters sandstone, then shale, then a coal seam, more shale, and then a bed of conglomerate, before the drilling finally stops in another shale bed. This sequence duplicates the one we observed in the sea cliff and also reveals a limestone and an underlying conglomerate and shale that we have not seen before. We may now feel justified in correlating the sandstone in the sea cliff with that in the canyon and that in the well hole: The limestone is the youngest rock in the area; the conglomerate and lowest shale are the oldest rocks. This correlation is shown in Figure 9.9.

Many sedimentary formations are correlated in just this way, especially when physical features are our only keys to rock correlation. But as we extend the range of our correlation over a wider and wider area, physical features become more and more difficult to use. Clearly, it is impossible to use physical characteristics to determine the relative age of two sequences of layered rocks in, say, England and the eastern United States. Fortunately, we have another method of correlation—a method that involves the use of fossils.

CORRELATION BY FOSSILS Around the turn of the nineteenth century, an English surveyor and civil engineer named William Smith (1769–1839) became impressed with the relationship of rock strata to the success of various engineering projects, particularly the building of canals. As he investigated rock strata from place to place, he noticed that many of them contained fossils. Furthermore, he observed that (no matter where found) some rock layers contained identical fossils, whereas the fossils in rock layers above or below were different. Eventually, Smith became so skillful that when confronted with a fossil he could name the rock from which it had come.

At about the same time, two French geologists, Georges Cuvier (1769–1832) and Alexandre Brongniart (1770–1847), were studying and mapping the fossil-bearing strata that surround Paris. They, too, found that certain fossils were restricted to certain rock layers, and they also had used the law of superposition to arrange the rocks of the Paris area in chronological order, just as Smith had done in England. Then Cuvier and Brongniart arranged their collection of fossils in the same order as the rocks from which the fossils had been dug. They discovered that the fossil assemblages varied in a systematic way with the chronological positions of the rocks. In comparing the fossil forms with modern forms of life, Cuvier and Brongniart discovered that the fossils from the higher rock layers bore a closer resemblance to modern forms than did the fossils from rocks lower down.

From all these observations, it became evident that the *relative age* of a layer of sedimentary rock could be determined by the nature of the fossils it contained. This fact has been verified time and again by other workers throughout the world. *It has become an axiom in geology that fossils are a key to correlating rocks and that rocks containing the same fossil assemblages are similar in age.*

But when we apply this axiom to actual situations involving the use of fossils for correlation, certain complications arise. For instance, it is obvious that in our modern world the distribution of living forms varies with the environment. This fact was presumably as true in the past as it is now. We found in discussing facies of sedimentary rocks that different sediments were laid down in different places at the same time. Plants and animals also reflect changes in environment, particularly if they happen to live on the sea bottom. Organisms living in an area where mud is being deposited are different from those living in an area where sand is being laid down. Thus we find somewhat different fossils in a bed of shale (formed from mud) than we do in a bed of sandstone (formed from sand).

If both the physical features and the fossils are different, how can we correlate the rocks in two different areas? There are two possible ways. First, we may actually be able to see that two different rock types, with their differing fossils, grade into each other laterally as we follow the beds along a cliff face. Second, we may find that a few fossils occur in both environments. In an oceanic environment, for instance, some forms float or swim over a wide geographic area that takes in more than one condition of deposition on the sea floor. The remains of these swimming or floating forms may settle to the bottom, to be incorporated in different kinds of sediment forming on the ocean floor.

9.3 *The geologic column*

Using the law of superposition and the concept that fossils are an index to time, geologists have made chronological arrangements of sedimentary rocks from all over the world. They have pictured the rocks as forming a great column, with the oldest at the bottom and the youngest at the top. The pioneer work in developing this pattern was carried out in the British Isles and in western Europe, where modern geology had its birth. There, geologists recognized that the change in the fossil record between one layer and the next was not gradual but sudden. It seemed as if whole segments recording the slow change of plants and animals had been left out of the rock sequence. These breaks or gaps in the fossil record served as boundaries between adjoining strata. The names assigned to the various groups of sedimentary rocks are given in the geologic column of Table 9.3.

Notice that the oldest rocks are called "Precambrian," a general term applied to all the rocks that lie beneath the Cambrian rocks. Although the Precambrian rocks represent the great bulk of geologic time, we have yet to work out satisfactory subdivisions for them because there is an almost complete absence of fossil remains, so plentiful in Cambrian and younger rocks. Without fossils to aid them, geologists have been forced to base their correlations on the physical features of the rocks and on dates obtained from radioactive minerals. Physical features have been useful for establishing local sequences of Precambrian rocks, but these sequences cannot be extended to worldwide subdivisions. On the other hand, radioactive dates are not yet numerous enough to permit subdivisions of the Precambrian rocks, although such dates will become increasingly important as more are assembled.

We have said that the geologic column was originally separated into different groups of rocks on the basis of apparent gaps in the fossil record. But as

Table 9.3 **The geologic column**

Era	System	Series	Some aspects of the life record: Some major events	Some aspects of the life record: Dominant life form[a]
Cenozoic	Cenozoic[b]	Pleistocene		Man
		Pliocene		
		Miocene	Grasses become abundant	Mammals
		Oligocene		
		Eocene	Horses first appear	Flowering plants
		Paleocene		
Mesozoic	Cretaceous	—[c]	Extinction of dinosaurs	Reptiles
	Jurassic		Birds first appear	Conifer and cycad plants
	Triassic		Dinosaurs first appear	Amphibia
Paleozoic	Permian	—[c]		
	Pennsylvanian		Coal-forming swamps	Spore-bearing land plants
	Mississippian			
	Devonian			Fish
	Silurian		First vertebrates appear (fish)	Marine invertebrates
	Ordovician			Marine plants
	Cambrian		First abundant fossil record (marine invertebrates)	
Precambrian	—[d]			Primitive marine plants and invertebrates One-celled organisms

[a] This column does not give the time range of the forms listed. For example, fish are known from pre-Silurian rocks and obviously exist today. But when the Silurian and Devonian rocks were being formed, fish represented the most advanced form of animal life.
[b] Some geologists prefer to divide the Cenozoic system into two systems, the Quaternary and the Tertiary. The Quaternary system of this division includes the Pleistocene epoch, and the Tertiary system includes the Paleocene through the Pliocene epochs.
[c] Many Mesozoic and Paleozoic series are distinguished but are not necessary here.
[d] Precambrian rocks are abundant, but worldwide subdivisions are not generally agreed upon.

geologic research progressed, and as the area of investigation spread from Europe to other continents, new discoveries narrowed the gaps in the fossil record. It is now apparent that the change in fossil forms has been continuous and that the original gaps can be filled in with data from other localities. This increase in information has made it more and more difficult to draw clear boundaries between groups of rocks, and yet, despite the increasing number of "boundary" problems, the broad framework of the geologic column is still valid.

9.4 *The geologic time scale*

The names in the geologic column refer to rock units that have been arranged in a chronological sequence from oldest to youngest. Because each of the units was formed during a definite interval of time, they provide us with a basis for setting up time divisions in geologic history.

Table 9.4 **The geologic timetable**[a]

Era	*Period*	*Epoch*	*Duration* (Yr × 10^6)	*Before present* (Yr × 10^6)
Cenozoic	Cenozoic	Pleistocene	1.5–2	
		Pliocene	5–5.5	
		Miocene	19	
		Oligocene	11–12	
		Eocene	15–17	
		Paleocene	11–12	
				65
Mesozoic	Cretaceous		71	
	Jurassic		54–59	
	Triassic		30–35	
				225
Paleozoic	Permian		55	
	Pennsylvanian		45	
	Mississippian		20	
	Devonian		50	
	Silurian		35–45	
	Ordovician		60–70	
	Cambrian		70	
				570
Precambrian	Precambrian time[b]			

[a] From W. B. Harland, A. Gilbert Smith, and B. Wilcock (eds.), "The Phanerozoic Time-Scale," *Quart J. Geol. Soc. London*, vol. 120S (suppl.), 1964.
[b] The oldest rocks are about 3.3 billion years old; the earth itself is 4 to 5 billion years old.

In effect, the terms we apply to time units are the terms that were originally used to distinguish rock units. Thus we speak either of Cambrian *time* or of Cambrian *rocks. When we speak of time units, we are referring to the geologic time scale. When we speak of rock units, we are referring to the geologic column.*

The geologic time scale is given in Table 9.4. Notice the terms *eras, periods,* and *epochs* across the top of the table. These are general time terms. Thus we can speak of the Paleozoic era, or the Permian period, or the Pleistocene epoch. In Table 9.3 the terms *system* and *series,* used as general terms for rock units, correspond to the time units period and epoch, respectively. There is no generally accepted rock term that is equivalent to the "era" of the geologic time scale; the term *group* is sometimes used.

Absolute dates in the geologic time scale

We have found that the geologic time scale is made up of units of relative time and that these units can be arranged in proper order without the use of any designations of absolute time. This relative time scale has been constructed on the basis of sedimentary rocks. As noted earlier, geologists have only recently been able to date some sedimentary rocks by radioactive methods, and most dates have come from the igneous rocks. How can the dates obtained from igneous rocks be fitted in with the relative time units from sedimentary rocks?

To insert the absolute age of igneous rocks into the geologic time scale, we must know the relative time relationships between the sedimentary and igneous rocks. The basic rule here is called the *law of crosscutting relationships,* which states that *a rock is younger than any rock that it cuts across.*

Let us consider the example given in Figure 9.10, a hypothetical section of the earth's crust with both igneous and sedimentary rocks exposed. The sedimentary rocks are arranged in three assemblages, numbered *1, 3,* and *5,* from oldest to youngest. The igneous rocks are numbered *2* and *4,* also from oldest to youngest.

The sedimentary rocks labeled *1* are the oldest rocks in the diagram. First, they were folded by earth forces; then, a dike of igneous rock was injected into them. Because the sedimentary rocks had to be present before the dike could cut across them, they must be older than the dike. After the first igneous intrusion, erosion beveled both the sedimentary rocks and the dike, and across this surface were deposited the sedimentary rocks labeled *3.* At some later time, the batholith, labeled *4,* cut across all the older rocks. In time, this

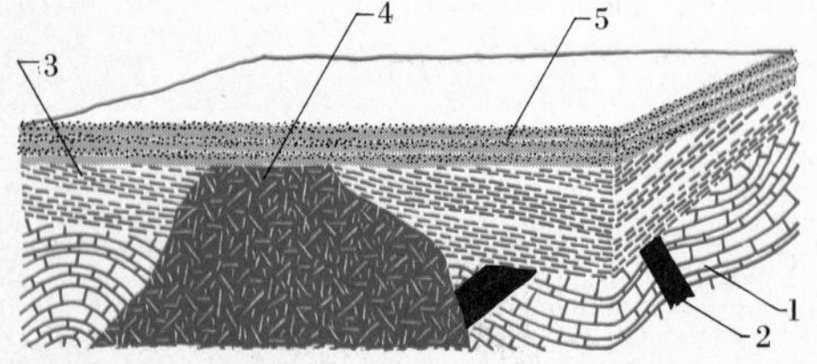

Figure 9.10 **This diagram illustrates the law of crosscutting relationships, which states that a rock is younger than any rock that it cuts. The rock units are arranged in order of decreasing age from 1 to 5. The manner in which radioactive ages of the igneous rock (2 and 4 of the diagram) are used to give approximate ages in terms of years for the sedimentary rocks (1, 3, and 5 of the diagram) is discussed in the text and in Table 9.5.**

Table 9.5 **Relative, absolute, and approximate ages of rocks**[a]

Event	Age Relative	Absolute, yr × 10^6	Approximate, yr × 10^6
Sedimentary rocks	*5*[b]		<230
Erosion			<230
Batholith	*4*	230	
Sedimentary rocks	*3*		>230, <310
Erosion			>230, <310
Dike	*2*	310	
Folding			>310
Sedimentary rocks	*1*		>310

[a] See Figure 9.10 and the text.
[b] Youngest.

batholith and the sedimentary rocks, *3*, were also beveled by erosion, and the sedimentary rocks labeled *5* were laid across this surface. We now have established the relative ages of the rocks, from oldest to youngest, as *1*, *2*, *3*, *4*, and *5*.

Now, if we can date the igneous rocks by means of radioactive minerals, we can fit these dates into the relative time sequence. If we establish that the batholith is 230 million years old and that the dike is 310 million years old, the ages of the sedimentary rocks may be expressed in relation to the known dates, and the final arrangement will be as shown in Table 9.5.

By this general method, approximate dates have been assigned to the relative time units of the geologic time scale, as shown in Table 9.4. These dates may be revised and refined as new techniques for dating develop. One of the most exciting developments in this field lies in the direct application of dating methods to the radioactive minerals of sedimentary rocks, as suggested earlier.

Outline

Geologic time can be expressed in either relative or absolute terms.

Absolute time is expressed in terms of years and is measured by the rate of decay of radioactive elements. Among the most useful elements are *carbon 14, potassium 40, rubidium 87, uranium 235,* and *uranium 238.*

Relative time has been determined largely by the relative position of sedimentary rocks to each other.

The rocks are arranged in proper chronological position according to the *law of superposition.*

Correlation of sedimentary rocks from one area to another allows the extension of a chronology of relative time from region to region and continent to continent. Correlation is based on *physical features* but more importantly on *fossils.*

The geologic column is a chronological sequence of rocks, from oldest to youngest. The units are rock units.

The geologic time scale is the chronological sequence of units of earth time represented by the rock units of the world geologic column. Absolute dates in the geologic time scale are based on the relation of radioactively dated igneous, sedimentary, and metamorphic rocks to the rocks of the geologic column. Position in the time scale of events dated from radioactive igneous and metamorphic rocks is determined by the relative position of the igneous and metamorphic rocks to the sedimentary rocks.

Supplementary readings

BERRY, WILLIAM B. N.: *Growth of a Prehistoric Time Scale*, W. H. Freeman and Company, San Francisco, 1968. A historical account of the development of the geologic time scale with emphasis on the role of organic revolution.

EICHER, DONALD L.: *Geologic Time*, Prentice-Hall, Inc., Englewood Cliffs, N.J., 1968. A short but comprehensive treatment of how geologic time is measured.

FAUL, HENRY: *Ages of Rocks, Planets and Stars*, McGraw-Hill Book Company, Inc., New York, 1966. A very good review of the use of radiometric methods of determining geologic time.

HARLAND, W. B., A. GILBERT SMITH, and B. WILCOCK (eds.): "The Phanerozoic Time-Scale," *Quart. J. Geol. Soc. London*, vol. 120S (suppl.), 1964. A symposium dedicated to Arthur Holmes, this volume summarizes expert opinion on the absolute ages assigned to geologic time divisions.

LIBBY, W. F.: "Radiocarbon Dating," *Am. Scientist*, vol. 44, no. 1, pp. 98–112, 1956. A summary of the method written by the man who discovered it.

STOKES, MARVIN A., and TERAH L. SMILEY: *An Introduction to Tree-Ring Dating*, University of Chicago Press, Chicago, 1968. A concise statement on a technique of restricted application.

TOULMIN, STEPHEN, and JUNE GOODFIELD: *The Discovery of Time*, Harper & Row, Publishers, New York, 1965. Two historians analyze the emergence of the concept of time from the pre-Christian Mediterranean world to the present.

ZEUNER, F. E.: *Dating the Past*, 4th ed., Methuen & Co., Ltd., London, 1958. A discussion of the diverse methods of geologic dating with an emphasis on their application to archaeological materials.

10

Erosion on hillslopes

THE EARTH'S SURFACE IS A COLLECTION OF SLOPES: SOME GENTLE, SOME STEEP; some long, some short; some cloaked with vegetation, some bare of plants; some veneered with soil, some with naked rock. They lead from hill crests downward to streams and rivers. Taken together they make up our landscape. Down these slopes move rain water and the water from melting snow to collect in rivulets, brooks, and rivers flowing to the ocean. Down them also travels soil and rock, which, via the rivers, finally gets to the sea, there to form sedimentary rocks. But because this material derives from the slopes themselves, it follows that they must be changing.

We have already learned that when rocks are exposed at the earth's surface, weathering immediately sets to work to establish an equilibrium between the rock and its new environment. Among the other factors joining forces with the processes of weathering is the direct pull of gravity. It acts to move the products of weathering, and even unweathered bedrock, to lower and lower levels. This movement of surface material caused by gravity is known as *mass movement.* Sometimes it takes place suddenly in the form of great landslides and rock falls from precipitous cliffs, but often it occurs almost imperceptibly as the slow creep of soil across gently sloping fields.

Water also plays a direct role in the movement of slope material. The impact of a raindrop, the flow of water in thin sheets or tiny rills, and the solution of solid material as water seeps into the ground all contribute to the erosion of hillslopes.

10.1 *Factors of mass movement*

Gravity provides the energy for the downslope movement of surface debris and bedrock, but several other factors, particularly water, augment gravity and ease its work.

Immediately after a heavy rainstorm, you may have witnessed a landslide on a steep hillside or on the bank of a river. In many unconsolidated deposits, the pore spaces between individual grains are filled partly with moisture and partly with air, and so long as this condition persists, the surface tension of the moisture gives a certain cohesion to the soil. When, however, a heavy rain forces all the air out of the pore spaces, this surface tension is completely destroyed. The cohesion of the soil is reduced, and the whole mass becomes more susceptible to downslope movement. The presence of water also adds weight to the soil on a slope, although this added weight is probably not a very important factor in promoting mass movement.

Water that soaks into the ground and completely fills the pore spaces in the slope material contributes to instability in another way. The water in the pores is under pressure, which tends to push apart individual grains or even whole rock units and to decrease the internal friction or resistance of the material to movement. Here again, water assists in mass movement.

Another factor, as we shall see, is air trapped beneath a rapidly moving mass of rock debris as it hurtles down steep slopes. This air, confined by the falling

Figure 10.1 **These badland slopes in Badlands National Monument, South Dakota, are almost devoid of vegetation. Gravity and rain wash are the major agents in slope maintenance.** (*Photograph by National Park Service.*)

material, acts as a cushion to reduce the friction of the debris with the ground and to make possible high-velocity movement of rock slides.

Gravity can move material only when it is able to overcome the material's internal resistance against being set into motion. Clearly, then, any factor that reduces this resistance to the point where gravity can take over contributes to mass movement. The erosive action of a stream, an ocean, or a glacier may so steepen a slope that the earth material can no longer resist the pull of gravity and is forced to give in to mass movement. In regions of cold climate, alternate freezing and thawing of earth materials may be enough to set them in motion. The impetus needed to initiate movement may also be furnished by earthquakes, excavations or blasting operations, the sonic booms of aircraft, or even the gentle activities of burrowing animals and growing plants.

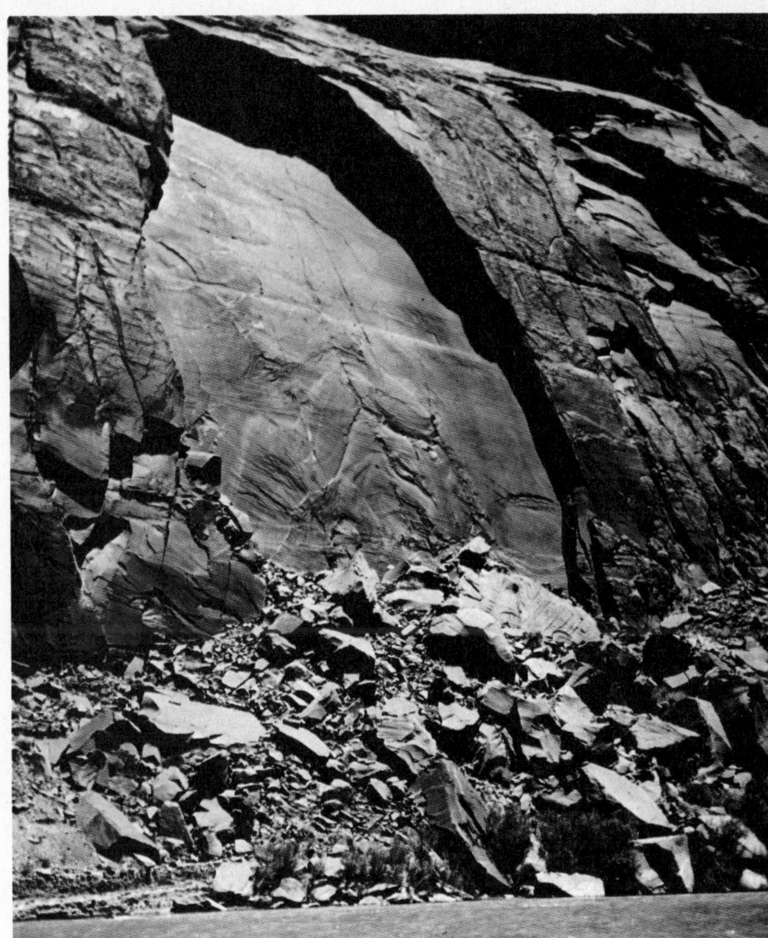

Figure 10.2 ***A large slab of massive Wingate sandstone has fallen and shattered at the foot of this precipitous cliff, where the San Juan River now begins to move away the fragments.*** (*Photograph by William C. Bradley.*)

10.2 *Behavior of material*

After material has been set in motion down a slope, it may behave as (1) an elastic solid or an aggregation of elastic solids, (2) a plastic substance, or (3) a fluid.

An elastic solid is a substance that undergoes a change in shape or volume when *stress* (defined as force per unit area) is applied to it, but the substance returns to its original condition when the stress is removed. An elastic solid breaks only when the stress applied to it is greater than its strength to resist.

A plastic substance undergoes continuous change of shape after the stress applied to it passes a critical point. If a slope is steep enough, masses of earth sometimes behave as a plastic substance, creeping slowly but continuously downhill under the influence of gravity.

Finally, a fluid offers little or no resistance to the stresses that tend to change its shape, and masses of earth sometimes behave in just this way, flowing down even the most gradual slope.

We could actually study any type of mass movement and classify it on the basis of these three types of movement, but we would have to assemble an excessive amount of technical data, and we would find the picture complicated by the fact that material often behaves in different ways during any one movement. Therefore we shall simply classify mass movement as either *rapid* or *slow.*

10.3 Rapid movements

Catastrophic and destructive movements of rock and soil, the most spectacular and easily recognized examples of mass movement, are popularly known as *landslides,* but the geologist subdivides this general term into slump, rock slides, debris slides, mudflows, and earthflows.

Landslides

Landslides include a wide range of movements, from the slipping of a stream bank to the sudden, devastating release of a whole mountainside. Some landslides involve only the unconsolidated debris lying on bedrock; others involve movement of the bedrock itself.

SLUMP Sometimes called *slope failure, slump* is the downward and outward movement of rock or unconsolidated material traveling as a unit or as a series of units. Slump usually occurs where the original slope has been sharply steepened, either artificially or naturally. The material reacts to the pull of gravity as if it were an elastic solid, and large blocks of the slope move downward and outward along curved planes. The upper surface of each block is tilted backward as it moves.

Figure 10.3 shows a slump beginning at Gay Head, Massachusetts. The action of the sea has cut away the unconsolidated material at the base of the slope, steepening it to a point where the earth mass can no longer support itself. Now the large block has begun to move along a single curving plane, as suggested in Figure 10.4.

After a slump has been started, it is often helped along by rain water collecting in basins between the tilted blocks and the original slope. The water drains down along the plane on which the block is sliding and promotes further movement.

ROCK SLIDES The most catastrophic of all mass movements are *rock slides*—sudden, rapid slides of bedrock along planes of weakness.

A great rock slide occurred in 1925 on the flanks of Sheep Mountain, along the Gros Ventre River in northwestern Wyoming, not far from Yellowstone

Figure 10.3 **The beginning of slump or slope failure along the sea cliffs at Gay Head, Massachusetts. Note that the slump block is tilted back slightly, away from the ocean. This slump block eventually moved downward and outward toward the shore along a curving plane, a portion of which is represented by the face of the low scarp in the foreground.** (*Photograph from Gardner Collection, Harvard University.*)

National Park (see Figure 10.5). An estimated 37 million m^3 of rock and debris plunged down the valley wall and swept across the valley floor. The nose of the slide rushed some 110 m up the opposite wall and then settled back, like liquid being sloshed in a great basin. The debris formed a dam between 68 and 75 m high across the valley, the dammed-up river creating a lake almost 8 km long. The spring floods of 1927 raised the water level to the lip of the dam, and in mid-May the water flooded over the top. So rapid was the downcutting of the dam that the lake level was lowered about 15 m in 5 hr. During the flood that followed, several lives were lost in the town of Kelly, in the valley below.

The Gros Ventre slide was a long time in the making, and there was probably nothing that could have been done to prevent it. Conditions immediately before the slide are shown in Figure 10.6(*a*). In this part of Wyoming, the Gros Ventre valley cuts through sedimentary beds inclined between 15° and 21° to the north. The slide occurred on the south side of the valley wall, where the beds dip into the valley. Notice that the sandstone bed is separated from the limestone strata by a thin layer of clay. Before the rock slide occurred, the sandstone bed near the bottom of the valley had been worn thin by erosion.

Figure 10.4 **This diagrammatic cross section shows the type of movement found in a slump similar to that pictured in Figure 10.3. A block of earth material along the steepened cliff has begun to move downward along a plane that curves toward the ocean.**

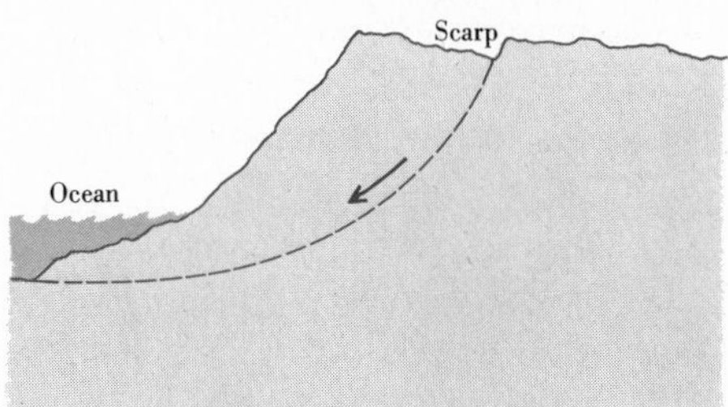

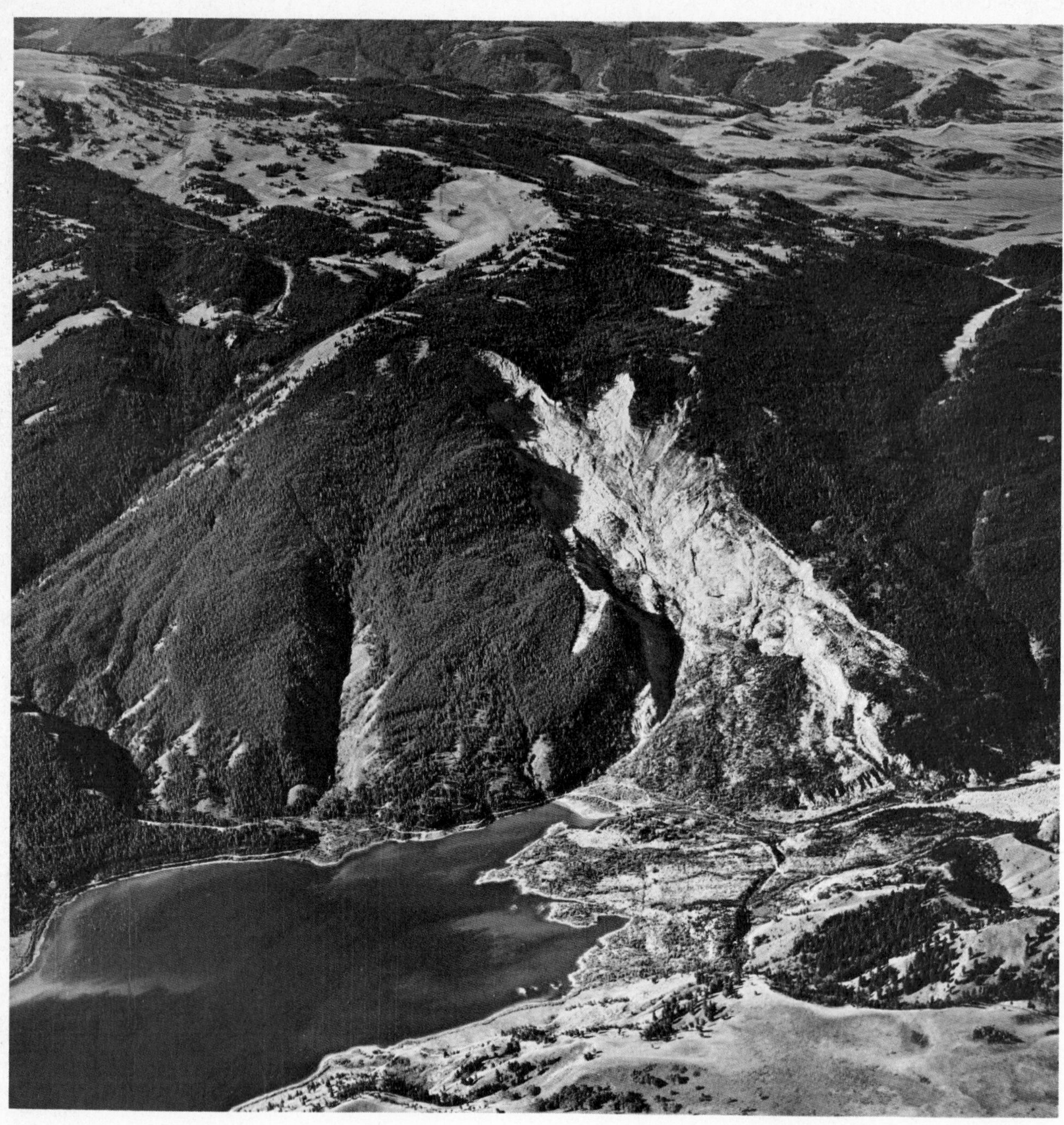

Figure 10.5 ***Aerial photograph of the Gros Ventre rock slide in northwestern Wyoming. The lake in the lower left of the picture has been dammed by a landslide that moved down into the valley of the Gros Ventre River. The area from which the material slid is about 2.5 km long and is well marked by the white scar down the center of the photograph. The vegetative cover of trees and bushes on the adjoining slopes has not yet reestablished itself in the slide area.*** *(Photograph by Austin Post, University of Washington.)*

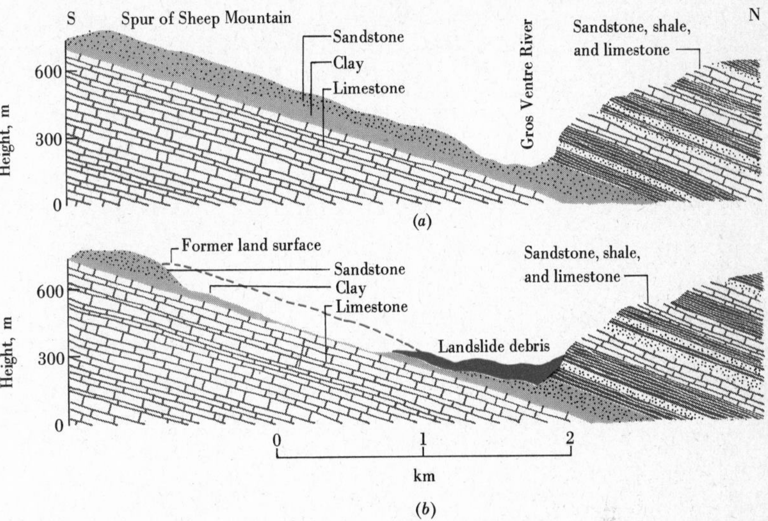

Figure 10.6 ***Diagrams to show the nature of the Gros Ventre slide. (a) Conditions existing before the slide took place; (b) the area of the slide and the location of the debris in the valley bottom. Note that the sedimentary beds dip into the valley from the south. The large section of sandstone slid downward along the clay bed.*** (*Redrawn from William C. Alden, "Landslide and Flood at Gros Ventre, Wyoming,"* Trans. AIME, *vol. 76, p. 348, 1928.*)

The melting of winter snows and the heavy rains that fell during the spring of 1925 furnished an abundant supply of water, which seeped down to the thin layer of clay, soaking it and reducing the adhesion between it and the overlying sandstone. When the sandstone was no longer able to hold its position on the clay bed, the rock slide roared down the slope. Figure 10.6(*b*) suggests the amount of material that was moved from the spur of Sheep Mountain to its resting place on the valley floor.

Another rock slide, in 1903, killed 70 people in the coal-mining town of Frank, Alberta, when some 36.5 million m^3 of rock crashed down from the crest of Turtle Mountain, which rises over 900 m above the valley. Mining activities may have triggered this movement, but natural causes were basically responsible for it.

As Figure 10.7 shows, Turtle Mountain has been sculptured from a series of limestone, sandstone, and shale beds that dip southwestward away from the valley. The diagram reveals four factors that contributed to the slide: (1) the steepness of the mountain, (2) the series of joints or fractures that dip down through the limestone strata, (3) the weak shale strata that underlie the limestone and form the base of the mountain, and (4) the coal-mining operations in the valley.

The steep valley wall enhanced the effectiveness of gravity, and the joints served as potential planes of movement. The weak shale beds at the base of the mountain undoubtedly underwent slow plastic deformation under the weight of the overlying limestone, and as the shale was deformed, the limestone settled lower and lower. The settling action may have been helped along by the coal-mining operations in the valley, as well as by frost action, rain, melting snows, and earthquake tremors that had shaken the area 2 years before. In any event, the stress finally reached the point where the limestone beds fractured and the great mass of rock hurtled down into the valley.

This time the rock material behaved in three different ways. The shales underwent plastic deformation, producing a condition of extreme instability on the mountain slope. When the strata that still held the limestone mass

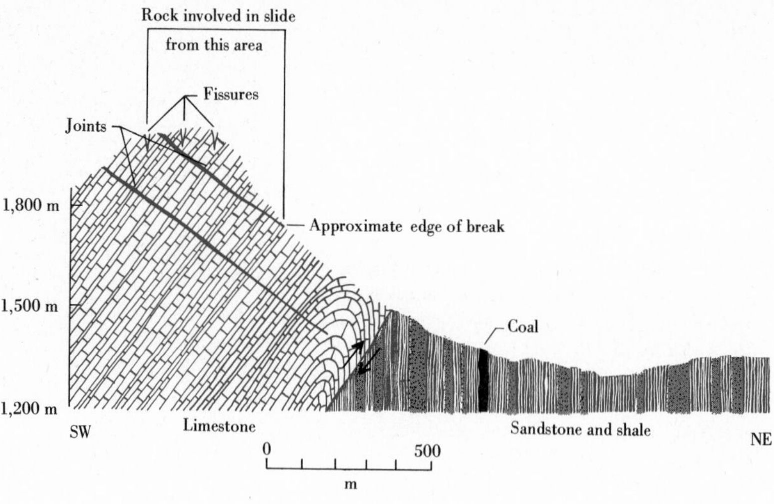

Figure 10.7 ***A cross section to show the conditions at Turtle Mountain that brought on the Frank, Alberta, landslide.*** *(Drawn on the basis of data furnished by Alan McGugan.)*

on the slope sheared in the manner of an elastic solid, the slide actually began. Once under way, the rock debris bounced, ricocheted, slid, and rolled down the mountain slope until it was literally "launched" into the air by a ledge of rock. Thereafter it arched outward and downward to the valley below. Once on the valley floor, it moved at high speed up onto the hills on the far side of the river. In this phase it moved like a viscous fluid with a series of waves spreading out along its front, and there is some evidence that air was caught and compressed beneath the debris, drastically reducing its friction with the ground and allowing speeds of at least 100 km/hr.

A more recent slide occurred in southwestern Montana, on August 17, 1959. An earthquake, whose focus was located just north of West Yellowstone, Montana, triggered a rock slide in the mouth of the Madison Canyon, about 32 km to the west. A mass of rock estimated to be over 32 million m^3 fell from the south wall of the canyon. It climbed over 100 m up the opposite valley wall and dammed a lake 8 km long and over 30 m deep. More than a score of people lost their lives in the Madison River campground area below the slide (see Figures 10.8 and 10.9). Survivors reported extremely powerful winds along the margins of the slide. In fact, some people were blown away, never to be seen, and a 2-ton automobile was observed to have been carried by the wind over 10 m before smashing against a row of trees. It seems clear that when the mass of rock fell to the valley bottom it trapped and compressed large quantities of air which served to provide a near frictionless base on which it spread across the valley. The high winds represented the extrusion of this mass of air from beneath the moving debris.

The Alaska earthquake of March 27, 1964 (Section 18.2) had many effects on the topography and shoreline of a large part of the southern portion of the state. Landslides were common and in inhabited places caused much damage. In the glaciated mountains of the Alaska range, rock slides were particularly spectacular. Austin Post of the University of Washington mapped some 78 earthquake-induced slides which dumped material down glacially

Figure 10.8 ***The Madison River Canyon before the landslide of August 1959. The slide began on the forested valley wall on the far side of the river. The debris clogged the river channel to form a lake. The spot where this fisherman once stood is now under 30 m of water. Compare with Figure 10.9.*** *(Photograph by Montana Power Co.)*

Figure 10.9 ***An aerial view of the Madison Canyon after the landslide. An outlet channel for the newly formed lake has been cut across the landslide dam.*** *(Photograph by William C. Bradley.)*

steepened mountain slopes onto the surfaces of the glaciers. The debris of the largest of these slides covered over 11 km^2 and the total debris on all glaciers amounted to more than 120 km^2. One large slide on the Sherman glacier near Cordova contributed 28 million m^3 of rock debris to the surface of the glacier. Assuming the density of the material to be about 2.3, the mass of rock material weighed 57.5 million metric tons.

In addition to the modern landslides described above, there are vestiges of prehistoric slides of great magnitude. One, at Saidmarreh, Iran, involved the movement of 4,245 million m^3 of rock for a distance of over 12 km. Data on this and some other landslides are given in Table 10.1.

Table 10.1 **Data on some large landslides**[a]

Landslide	*Volume of debris, $m^3 \times 10^6$*	*Distance moved, km*	*Minimum speed, km/h*	*Height climbed, m*
Elm, Switzerland, Sept. 11, 1881	11.3	1.4	160	100
Sherman, Cordova, Alaska, Mar. 27, 1964	28.3	4.3	185	145
Madison Canyon, Montana, Aug. 17, 1959	32.6	0.8	130	110
Gros Ventre River, Wyoming, June 23, 1925	36.5	1.3	150	105
Frank, Alberta, Canada, Apr. 29, 1903	36.8	1.7	175	120
Silver Reef, California, prehistoric	226	5.3	105	50
Blackhawk, California, prehistoric	283	8.0	120	60
Saidmarreh, Iran, prehistoric	4,245	12.8	340	460

[a] Compiled from various sources and arranged in order of volume of material moved.

DEBRIS SLIDES A *debris slide* is a small, rapid movement of largely unconsolidated material that slides or rolls downward and produces a surface of low hummocks with small, intervening depressions. Movements of this sort are common on grassy slopes, particularly after heavy rains, and in unconsolidated material along the steep slopes of stream banks and shorelines.

WARNINGS OF LANDSLIDES We usually think of a landslide as breaking loose without warning, but it is more accurate to say that people in the area simply fail to detect the warnings.

For example, a disastrous rock slide at Goldau, Switzerland, in 1806, wiped out a whole village, killing 457 people. The few who lived to tell the tale reported that they themselves had no warning of the coming slide but that animals and insects in the region may have been more observant or more sensitive. For several hours before the slide, horses and cattle seemed to be extremely nervous, and even the bees abandoned their hives. Some slight preliminary movement probably took place before the rock mass broke loose.

During the spring of 1935, slides took place in clay deposits along a German superhighway that was being built between Munich and Salzburg. The slides came as a complete surprise to the engineers, but for a full week the workmen had been murmuring. "Der Abhang wird lebendig" (the slope becomes alive).

Landslides like the one on Turtle Mountain are often preceded by slowly widening fissures in the rock near the upward limit of the future movement (see Figure 10.10).

Figure 10.10 **Warning of imminent landsliding is often recorded as cracks in the earth at the upper portions of the future slide area, as in this case near Smartville, California.** (*Photograph by U.S. Geological Survey.*)

There is some evidence that landslides may recur periodically in certain areas. In southeastern England, not far from Dover, extensive landslides have been occurring once every 19 to 20 years. Some observers feel that there may be some correlation between such periodical mass movement and periods of excessive rainfall. On steep slopes in very moist tropical or semitropical climates, for instance, landslides do seem to follow a cyclic pattern. First, a landslide strips the soil and vegetation from a hillslope. In time, new soil and vegetation develop, the old scar heals, and when the cover reaches a certain stage, the landsliding begins again. Although landslides may occur in cycles, our data are as yet far too scanty to support firm conclusions.

Mudflows

A *mudflow* is a well-mixed mass of rock, earth, and water that flows down valley slopes with the consistency of newly mixed concrete. In mountainous, desert, and semiarid areas, mudflows manage to transport great masses of material.

The typical mudflow originates in a small, steep-sided gulch or canyon where the slopes and floor are covered by unconsolidated or unstable material. A sudden flood of water, from cloudbursts in semiarid country or from spring thaws in mountainous regions, flushes the earth and rocks from the slopes and carries them to the stream channel. Here the debris blocks the channel until the growing pressure of the water becomes great enough to break through. Then the water and debris begin their course down-valley, mixing together with a rolling motion along the forward edge of the flow. The advance of the flow is intermittent, for sometimes it is slowed or halted by a narrowing of the stream channel; at other times it surges forward, pushing obstacles aside or carrying them along with it.

Eventually, the mudflow spills out of the canyon mouth and spreads across the gentle slopes below. No longer confined by the valley walls or the stream channel, it splays out in a great tongue, spreading a layer of mud and boulders that ranges from a few centimeters to a meter or more in thickness. Mudflows can move even large boulders weighing 80 metric tons or more for hundreds of meters across slopes as gentle as 5°.

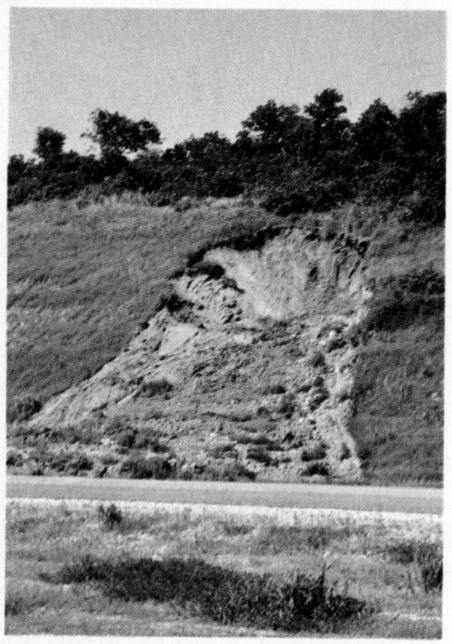

Figure 10.11 ***An earthflow in a roadcut near Dallas, Texas, shows the sharp scar high on the slope and, farther down, the area of soil movement by flow. Compare with Figure 10.12.*** (*Photograph by C. W. Brown.*)

Earthflows

Earthflows are a combination of slump and the plastic movement of unconsolidated material. They move slowly but perceptibly and may involve from a few to several million cubic meters of earth material. Some of the material behaves like an elastic solid, and some like a plastic substance, depending on its position in the moving mass.

The line at which a slump pulls away from the slope is marked by an abrupt scarp or cliff, as shown in Figure 10.11. Notice that the slump zone is made up of a series of blocks that move downward and outward, tilting the original surface back toward the slope. Farther down, the material tends to flow like a liquid, often beneath the vegetative cover. At the downslope limit of an earthflow, the sod often bulges out and fractures (see Figure 10.12). Earthflows occur in unconsolidated material lying on solid bedrock and are usually helped along by the presence of excessive moisture.

Figure 10.12 ***In this diagram of an earthflow, note the scarp and slump blocks that have pulled from the flow in the upper section. In the downslope section, the flowing has expressed itself at the surface by bulging and cracking the sod.***

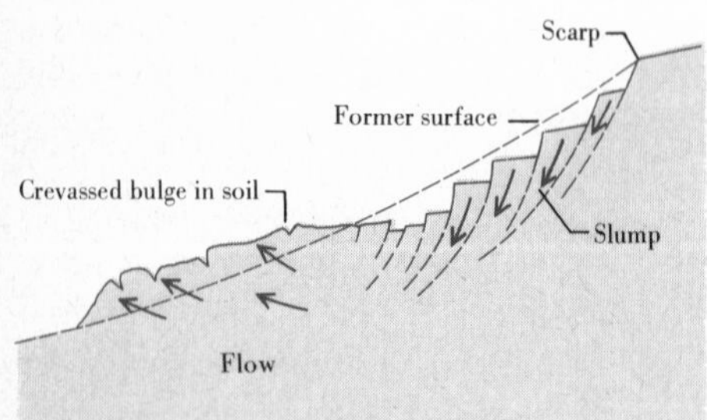

Talus

Strictly speaking, a *talus* is a slope built up by an accumulation of rock fragments at the foot of a cliff or a ridge. The rock fragments are sometimes referred to as *rock waste* or *slide-rock.* In practice, however, talus is widely used as a synonym for the rock debris itself.

In the development of a talus, rock fragments are loosened from the cliff and clatter downward in a series of free falls, bounces, and slides. As time passes, the rock waste finally builds up a heap or sheet of rock rubble. An individual talus resembles a half-cone with its apex resting against the cliff face in a small gulch. A series of these half-cones often forms a girdle around high mountains, completely obscuring their lower portions (see Figure 10.13). Eventually, if the rock waste accumulates more rapidly than it can be destroyed or removed, even the upper cliffs become buried, and the growth of the talus

Figure 10.13 ***A series of convergent talus cones have formed along the base of this cliff in the Sierra Nevada of California.*** (*Photograph by U.S. Geological Survey.*)

stops. The slope angle of the talus varies with the size and shape of the rock fragments. Although angular material can maintain slopes up to 50°, rarely does a talus ever exceed angles of 40°.

A talus is subject to the normal process of chemical weathering, particularly in a moist climate. The rock waste is decomposed, especially toward its lower limit, or toe, and the material there may grade imperceptibly into a soil.

10.4 *Slow movements*

Slow mass movements of unconsolidated material are more difficult to recognize and less fully understood than rapid movements, yet they are extremely important in the sculpturing of the land surface. Because they operate over long periods of time, they are probably responsible for the transportation of more material than are rapid and violent movements of rock and soil.

Before the end of the nineteenth century, William Morris Davis aptly described the nature of slow movements.

> The movement of land waste is generally so slow that it is not noticed. But when one has learned that many land forms result from the removal of more or less rock waste, the reality and the importance of the movement are better understood. It is then possible to picture in the imagination a slow washing and creeping of the waste down the land slopes; not bodily or hastily, but grain by grain, inch by inch; yet so patiently that in the course of ages even mountains may be laid low.[1]

Creep

In temperate and tropical climates, a slow downward movement of surface material known as *creep* operates even on gentle slopes with a protective cover of grass and trees. It is hard to realize that this movement is actually taking place. Because the observer sees no break in the vegetative mat, no large scars or hummocks, he has no reason to suspect that the soil is in motion beneath his feet.

Yet this movement can be demonstrated by exposures in soil profile (see Figure 10.14), by the behavior of tree roots, of large blocks of resistant rock, and of man-built objects such as fences and telephone poles (see Figure 10.15). Figure 10.16 shows a section through a hillside underlain by flat-lying beds of shale, limestone, clay, sandstone, and coal. The slope is covered with rock debris and soil. But notice that the beds near the base of the soil bend

[1] William Morris Davis, *Physical Geography*, p. 261, Ginn & Company, Boston, 1898.

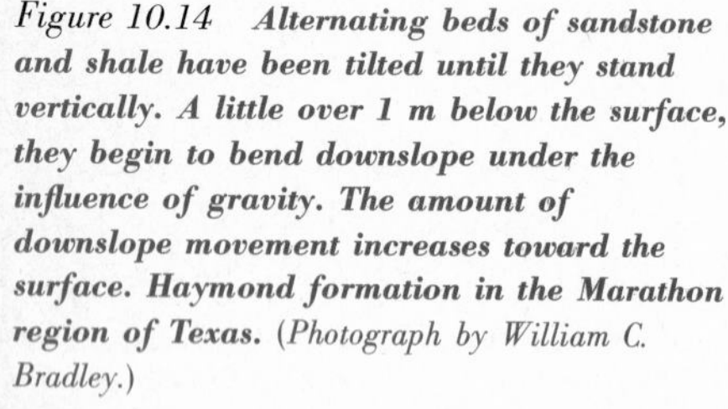

Figure 10.14 ***Alternating beds of sandstone and shale have been tilted until they stand vertically. A little over 1 m below the surface, they begin to bend downslope under the influence of gravity. The amount of downslope movement increases toward the surface. Haymond formation in the Marathon region of Texas.*** (*Photograph by William C. Bradley.*)

Figure 10.15 ***Telephone poles set in a talus slope south of Yellowstone National Park record movement of the slope material. Originally the poles stood vertically, but over a period of about 10 years the talus has moved and tilted the two nearest poles 5° and the most distant pole 8°.*** (*Photograph by Sheldon Judson, 1968.*)

downslope and thin out rapidly. These beds are being pulled downslope by gravity and are strung out in ever-thinning bands that may extend for hundreds of meters. Eventually, they approach the surface and lose their identity in the zone of active chemical weathering.

Figure 10.16 also shows other evidence that the soil is moving. Although when viewed from the surface the tree appears to be growing in a normal way, it is actually creeping slowly down the slope. Because the surface of the soil is moving more rapidly than the soil beneath it, the roots of the tree are unable to keep up with the trunk.

We can find other evidence of the slow movement of soil in displaced fences and tilted telephone poles and gravestones. On slopes where resistant rock layers crop up through the soil, fragments are sometimes broken off and distributed down the slope by the slowly moving soil.

Many other factors cooperate with gravity to produce creep. Probably the most important is moisture in the soil, which works to weaken the soil's resistance to movement. In fact, any process that causes a dislocation in the soil brings about an adjustment of the soil downslope under the pull of gravity. Thus the burrows of animals tend to fill downslope, and the same is true of cavities left by the decay of organic material, such as the root system of a dead tree. The prying action of swaying trees and the tread of animals and even of men may also aid in the motion. The end result of all these processes, aided by the influence of gravity, is to produce a slow and inevitable downslope creep of the surface cover of debris and soil.

Solifluction

The term *solifluction* (from the Latin *solum*, "soil," and *fluere*, "to flow") refers to the downslope movement of debris under saturated conditions. Soli-

Figure 10.16 ***The partially weathered edges of horizontal sedimentary rocks are dragged downslope by soil creep. The tree is also moving slowly downslope, as is evidenced by the root system spread out behind the more rapidly moving trunk.*** (*Redrawn from C. F. S. Sharpe and E. F. Dosch, "Relation of Soil-Creep to Earthflow in the Appalachian Plateaus,"* J. Geomorphol., *vol. 5, p. 316, December 1942, by permission of Columbia University Press.*)

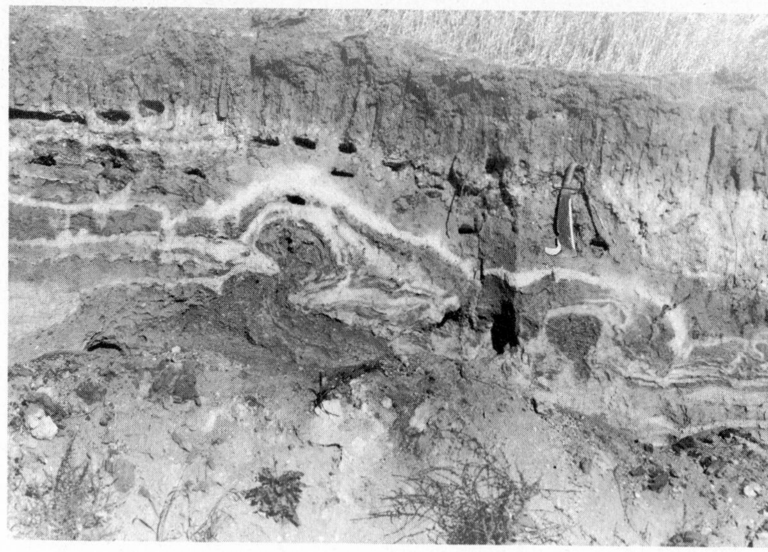

Figure 10.17 **Beds of silt, sand, and clayey gravel have been contorted by differential freezing and thawing during the more rigorous climates of glacial times. The gravel at the base of the exposure is clayey, gravelly till deposited directly by glacier ice. Above this are the highly contorted beds of water-laid sand (dark) and silt (light), which were originally flat-lying. The modern soil has developed across the contortions. The white of the silt bands is due to precipitation of calcium carbonate from soil water. The brush hook is about 0.5 m long. Exposure north of city of Devils Lake, east-central North Dakota.** (Photograph by Saul Aronow, U.S. Geological Survey.)

fluction is most pronounced in high latitudes where the soil is strongly affected by alternate freezing and thawing and where the ground freezes to great depths, but even moderately deep seasonal freezing promotes solifluction.

Solifluction takes place during periods of thaw. Because the ground thaws from the surface downward, the water that is released cannot percolate into the subsoil and adjacent bedrock, which are still frozen and therefore impermeable to water. As a result, the surface soil becomes sodden and water-laden and tends to flow down even the gentlest slopes. Solifluction is an important

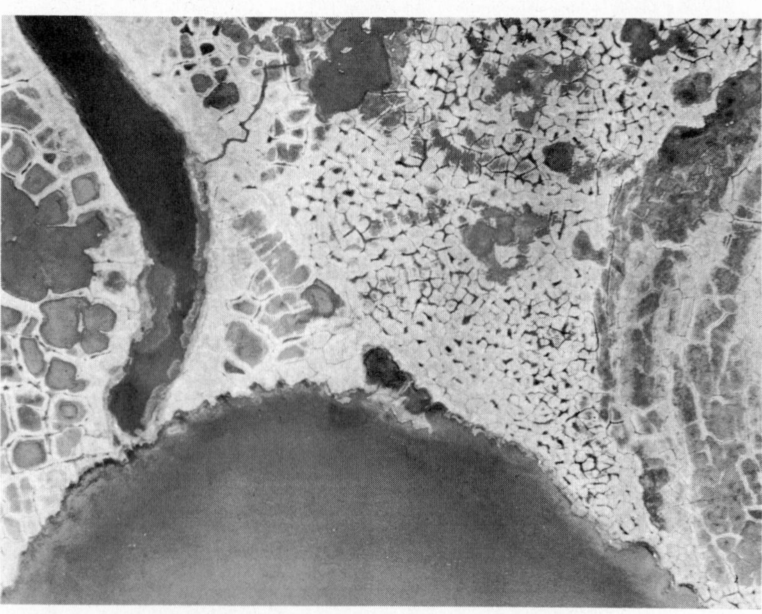

Figure 10.18 **Aerial photograph of polygonal patterns developed by ice wedges in the coastal plain of northern Alaska. Area shown is about 700 m wide.** (Photograph by U.S. Coast and Geodetic Survey.)

process in the reduction of land masses in arctic climates, where it transports great sheets of debris from higher to lower elevations.

During the glacier advances of the Pleistocene, a zone of intense frost action and solifluction bordered the southward-moving ice. In some places we can still find the evidence of these more rigorous climates preserved in distorted layers of earth material just below the modern soil (see Figure 10.17).

Frost action plays queer tricks in the soils of the higher elevations and latitudes. Strange polygonal patterns made up of rings of boulders surrounding finer material, stripes of stones strewn down the face of hillsides, great tabular masses of ice within the soil, and deep ice wedges that taper downward from the surface—all are found in areas where the ground is deeply frozen (see Figure 10.18). The behavior of frozen ground is one of the greatest barriers to the settlement of arctic regions. The importance of these regions has increased in recent years, and studies begun by Scandinavian and Russian investigators are now being intensively pursued by U.S. scientists.

Rock glaciers

Rock glaciers are tongues of rock waste that form in the valleys of certain mountainous regions. Although they consist almost entirely of rock, they bear a striking resemblance to ice glaciers. A typical rock glacier is marked by

Figure 10.19 ***A rock glacier in the Elk Mountains, Colorado.*** (*Photograph by U.S. Geological Survey.*)

a series of rounded ridges, suggesting that the material has behaved as a viscous mass (see Figure 10.19).

Observations on active rock glaciers in Alaska indicate that movement takes place through flow of interstitial ice within the mass. Favorable conditions for the development of rock glaciers include a climate cold enough to keep the ground continuously frozen, steep cliffs to supply debris, and coarse blocks that allow for large interstitial spaces.

10.5 *Direct action of water on hillslopes*

We have referred to the role of water in promoting mass movement. It also operates directly on slopes to move material downward to the channels of conventional streamways.

IMPACT OF RAINDROPS W. D. Ellison has made a study of the effectiveness of individual raindrops in erosion. A drop striking a wet, muddied field will splash water and suspended solid particles into the air. If there is any slope at all to the land, then the material splashed uphill will not travel as far as that splashed downhill. There will, then, be a net downslope motion of material by rain splash (see Figure 10.20).

Figure 10.20 ***A raindrop splashes onto a water-soaked surface, scattering fine particles of earth.*** (*Photograph by Wide World.*)

RAIN WASH Rain falling on a slope may soak into the ground or run off the slope. Vegetation will slow the runoff, promote infiltration, and generally reduce the erosive effectiveness of water flowing on the slopes, but if vegetation is lacking or sparse, rain water will rapidly flow down the slope. This flow, called rain wash, can become an effective agent of erosion.

A thin sheet of water may accumulate on a slope and become thicker as it proceeds downslope. This is sometimes referred to as *sheet wash* or *sheet flow*. Theoretically, a film of water would increase at a constant rate as it moved down a uniform slope, gathering more and more water. This increase does occur but not in a uniform sheet, for the simple reason that no slope is completely uniform. Initial irregularities, differing resistance to erosion, and varying lengths of flow may all contribute to the concentration of sheet wash into broad shallow troughs or channels which determine the location of miniature stream channels called *rills*. The rills are eroded along the axes of broad troughs, for here the water has its greatest velocity and hence its greatest erosive power. Most of us have seen the effect of such erosion on artificially created exposures along a roadway or on a slope newly prepared for planting.

SOLUTION It is clear that as water on the surface comes into contact with earth material it will begin to take mineral matter into solution. This will be true whether the water flows off the surface or soaks into the ground. The effect is to move material from the slopes, and we, therefore, must add solution as a third way in which hillslopes are reduced by the direct action of water.

10.6 *Rates of slope erosion*

The ways in which material moves down hillslopes have been generally known for a long time. Only in recent years have quantitative data begun to be available on the rates at which these movements take place and on the amounts of material which are involved. Here we look at some of these data.

Studies of the steep slopes in the rain forest of the mountains of New Guinea have been made by D. S. Simonett. He estimates that landslides alone remove material from the slopes into the valley bottoms at a rate for the area of 1 to 2 cm/1,000 years.

In the very different climate north of the arctic circle in extreme northern Sweden near the Norwegian border, we have exacting studies of slope erosion by the Swedish geographer Anders Rapp. There, in about 18 km^2 of the small valley known as Kärkevagge, Rapp made detailed observations of slope processes over a period of 9 years. About 1.5 million metric tons of material move annually. The results are summarized in Table 10.2.

Over a period of years, S. A. Schumm of Colorado State University has carried on studies of slope processes. One detailed study, in conjunction with R. J. Chorley of Cambridge University, reports on the fall of a large, 30,000-ton block of sandstone from a cliff in Chaco Canyon National Monument in New

Table 10.2 ***Movement of material per year in Karkevagge, Sweden, 1952–1960***[a]

Process	*Mass transfer*[b], *ton-m*		
Mass movement			
Rockfalls	19,565		
Avalanches	21,850		
Slides and mudflows	96,375		
Creep	8,000		
Subtotal		145,790	
Water			
Solution	90,000[c]		
Rain wash	—[d]		
Subtotal		90,000	
Total			235,790

[a] Adapted from Anders Rapp, *Geografis. Ann.*, vol. 42, p. 185, 1960.
[b] Mass transfer equals total tons moved times vertical distance moved. Thus a boulder that weighs 2 tons and descends a total *vertical* distance of 5 m would represent a mass transfer of 10 ton-m.
[c] The actual figure is 136,500, of which approximately 46,500 is estimated to have been derived from the atmosphere. The net loss to the slope is therefore 90,000.
[d] Negligible.

Mexico. This block had broken off from the main cliff along a vertical joint. It rested on shale. The rock, known as "Threatening Rock," was also called "Braced-up Cliff" by the local Indians in reference to the masonry at its base, laid, archaeologists tell us, in the eleventh century A.D. by the inhabitants of Pueblo Bonito in the hope that this would keep the rock from crashing into their village. In 1941 the rock did fall, although over six centuries after Pueblo Bonito had been abandoned. At the time of fall, the block stood 180 cm away from the cliff face.

Schumm and Chorley were able to show that, although the rock fell suddenly, the catastrophe was a long time in the making and that for centuries there had been slow movement of Threatening Rock away from the cliff face at a rate which increased with each passing year. On the basis of measurements kept by officials of the National Park Service and others, the rate of movement of the block was known from 1933 to the date of its fall, January 22, 1941. Extrapolation of these data back into time indicates that the rock became detached from the main cliff and began its slow movement toward final collapse about 550 B.C. This is graphed in Figure 10.21. From this the time required for each 30 cm of movement away from the cliff was estimated, as shown in Table 10.3. It could also be shown that in the final years before the fall the movement of the rock was greatest during the winter months, a fact attributed to the presumably moister conditions of the shale supporting the block during the winters.

In another study, Schumm has determined the rate of downslope movement of rock fragments on shaley hillslopes near Montrose, Colorado. Here Mancos shale of Cretaceous age contains thin beds of sandstone. These weather out of the shale to form platy fragments on the surface. A 7-year (September 1958 to September 1965) record of the movement of selected fragments on slopes

Figure 10.21 ***Threatening Rock in Chaco Canyon National Monument tumbled from a cliff face into the valley in 1941. The rock had been moving slowly away from the cliff for many years. Observations of the movement during the 9 years immediately prior to its fall are here extrapolated backward into time and indicate that movement began about 550 B.C.*** *(Adapted from S. A. Schumm and R. J. Chorley,* Am. J. Sci., *vol. 262, p. 1051, 1964.)*

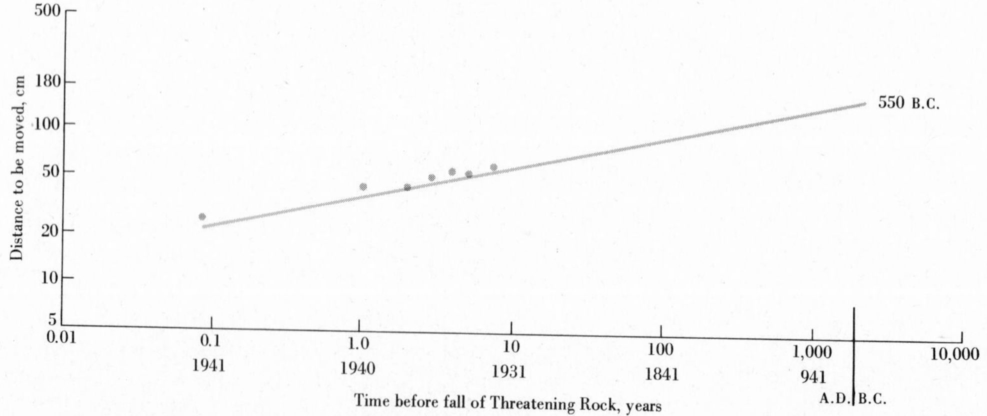

Table 10.3 **Rate of movement of Threatening Rock, Chaco Canyon, N.M.**

Movement, cm	*Years required*
First 30	1,600
Second 30	650
Third 30	200
Fourth 30	60
Fifth 30	8
Sixth 30	0.2

ranging from 3° to 40° has produced a quantitative statement of the rate of creep and has demonstrated the relation of the angle of slope to the velocity of movement. Figure 10.22 shows the relation of the mean rate of rock creep to the slope inclination for the 72 fragments that were followed through the 7-year study. Although the variation for individual specimens is wide, average rates of rock movements for 0.10 increments of the sine of the angle of surface slope (dark dots in Figure 10.22) fall very close to the fitted curve. The study demonstrates that in this case the average rate of creep is directly proportional to the sine of the slope angle, which is that component of gravity operating parallel to the hillslope. Inspection of Figure 10.22 shows that the average velocity of a sandstone fragment moving down the slope can be expressed as

Velocity (mm) = 100 × Sine of angle of slope

A different kind of study is reported by A. J. Eardley of the University of Utah. He has studied the relation of bristlecone pines to erosion rates in Cedar Breaks National Monument, Utah. The trees, which live for hundreds and thousands of years, betray the amount of erosion during their lifetimes by the amount of exposure of their roots. The depth of exposed roots on living trees is a measure of the amount the hillslope has been lowered since the trees

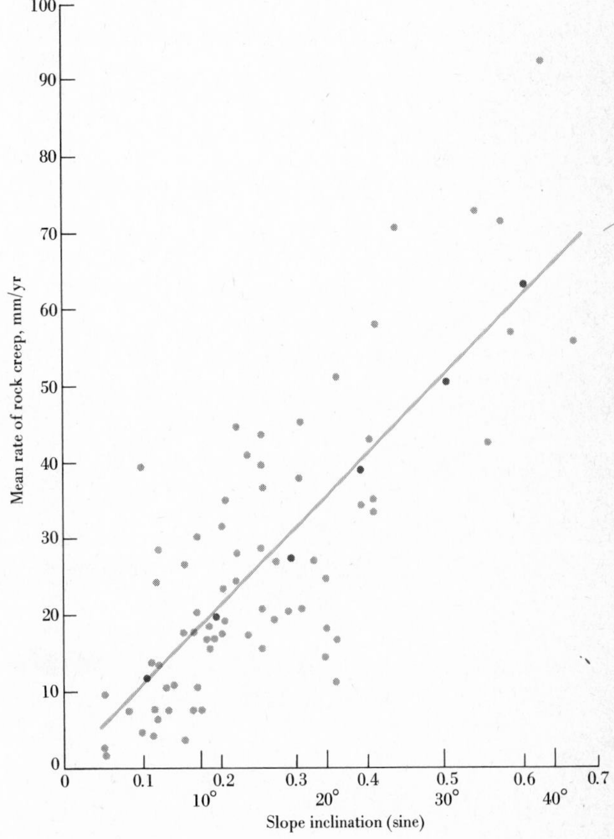

Figure 10.22 **Relation of angle of slope to rate of movement of fragments of thin sandstone beds weathered from the Mancos shale near Montrose, Colorado. Dots represent individual fragments. Darker-colored values indicate averages for fragments within 0.10 increments of the sine of the angle. The period of record is 7 years.** (*From S. A. Schumm,* Science, *vol. 155, p. 561, 1967. Copyright 1967 by the American Association for the Advancement of Science.*)

Figure 10.23 **This bristlecone pine in Cedar Breaks National Monument, Utah, is 2,840 years old. The man at the left is coring the tree through a strip of living bark. The man on the right is pointing to the level of the ground surface when the tree began to grow. The depth of exposure of the roots indicates the amount of lowering of land surface during the life of the tree.** (*Photograph by A. J. Eardley.*)

began to grow (see Figure 10.23). The study indicates that the slope processes move material in this arid land at rates which vary between 2 cm/1,000 years on slopes of 5° to about 10 cm/1,000 years on slopes of 30° that are underlaid by a crumbly, clayey limestone. Eardley found in Cedar Breaks, Utah, as Schumm had found in Colorado, that the rate of erosion was directly related to the angle of slope.

One final example will complete this brief survey of slope reduction. Careful excavation of the foundations of a Roman villa 60 km north of Rome by the Swedish archaeologist Carl-Eric Ostenberg shows that the 1,800-year-old structure has been measurably affected by the slow downward creep of shale and limestone bedrock. Ostenberg's excavation report provides data from which we can calculate that on a 7° slope material has been moving downslope at a rate capable of reducing the hillslope, at any one point, by 40 cm/1,000 years.

10.7 *Some implications stemming from erosion*

Throughout this chapter we have been considering the erosion of hillslopes where the direct influence of gravity, combined with some direct action of water on slopes, moves material downward to lower elevations. There are, of course, other ways that erosion takes place on the land, but the erosion we have described on hillslopes is quantitatively more important than any other type of erosion on the continents. This being so, this is a good place to pause and consider some of the implications of erosion.

Let us first recall the figures presented on erosion rates (Section 6.4) and the amount of material delivered annually to the oceans from the continents (Section 7.1). We found that the area of the coterminous United States is being eroded at approximately 6 cm/1,000 years. Assuming that the rate of erosion is constant and taking the average elevation of the United States as

690 m, the country would be reduced to a surface close to sea level in 12 million years. Or we could base an estimate of the time needed to reduce the continents of the world to near sea level on the 10 billion metric tons of material estimated to be received by the oceans each year. Again, assuming constant rates—this time, of sediment production—it would require 34 million years to wear down the continents.

These estimates, although interesting, probably lack reality. The assumption of a constant rate of erosion is questionable at best, and we have every reason to believe that as the land is reduced lower and lower toward sea level the rate of erosion slows. Furthermore, as we shall see in Chapters 17 and 19, the crust seems to "float" in a heavier plastic-behaving material at some depth below the surface. As lightweight material is moved from the continents to the ocean basins by erosion, the continents are buoyed upward, compensating, in part, for the loss in elevation. If this process—called *isostasy*—were to work perfectly, the time to reduce the continents would be some 10 times that which we have suggested above.

The important question for us, however, is not how long it takes to level continents by erosion, but rather "Does erosion ever really wear down the land of all the continents close to sea level?" The answer lies in the geologic record, which shows a continuous record of sedimentary rocks (and thus continuous sedimentation) far back into Precambrian time. Without emergent land masses to supply the sediments, there could be no sedimentation, no sedimentary rocks. We are led, then, to the conclusion that the continents for all practical purposes have always been with us. They may have changed in size, shape, and position, but they have not disappeared and they have not become so like pancakes that there are no sediments being produced. From this we must conclude that as the processes of weathering and erosion nibble away at the land, internal forces are pushing up the earth's crust to make material constantly available for erosion. Erosion can move volumes the size of continents, given enough time, but it does not flatten continents. The internal forces of the earth are treated in Chapters 4, 16, and 18. Here, we need only to recognize that these forces are constantly bending, breaking, and heaving the earth's crust, raising land above the oceans and exposing it to erosion.

In Chapters 11–15 we shall continue to consider in some detail the external agents that erode the land masses of the earth. We should remember that all these processes of erosion and transportation occupy a definite place in the rock cycle.

Outline

Erosion of hillslopes occurs as a direct result of gravity and by the action of water.

Mass movement of bedrock and unconsolidated material is in response to the pull of gravity. The moving material may behave as an elastic solid, as a plastic substance, or as a liquid. Movements are classified as either rapid or slow and may be influenced by saturation of material with water, by steepening of slopes by erosion, by earthquakes, and by the activity of animals, including man.

Rapid movements include landslides, mudflows, earthflows, and talus formation. *Slow movements* include creep, solifluction, and rock glaciers.

Water moves material on slopes by the direct impact of raindrops, by rain wash, and by solution.

Rates of slope erosion vary widely but can be measured in some places. Extrapolation of erosion rates leads to the implication that the continents are being raised to make earth materials continuously available for erosion.

Supplementary readings

ANDERSSON, J. G.: "Solifluction, a Component of Subaerial Denudation," *J. Geol.*, vol. 14, pp. 91–112, 1906. A classic paper relating to rapid erosion.

BLACK, ROBERT F.: "Permafrost," *Smithsonian Inst. Ann. Rept. 1950*, no. 4033. pp. 273–301, 1951. A readable summary of the nature and distribution of permanently frozen ground.

DALY, R. A., W. G. MILLER, and G. S. RICE: "Report of the Committee to Investigate Turtle Mt., Frank, Alberta," *Dept. Mines, Geol. Surv. Branch, Can., Mem.*, no. 27, 1912. An early report on a catastrophic landslide and one that still can be consulted with profit.

EARDLEY, ARMAND, and M. WILLIAM VIVAVANT: "Rates of Denudation as Measured by Bristlecone Pines, Cedar Breaks, Utah," *Utah Geol. Mineral. Surv. Spec. Studies*, no. 21, pp. 1–13, 1967. An effective demonstration of the application of botany to the determination of the rate of geological processes.

RAPP, ANDERS: "Recent Development of Mountain Slopes in Kärkevagge and Surroundings, Northern Scandinavia," *Geografis. Ann.*, vol. 42, pp. 1–200, 1960. An impressive quantitative study of erosion of slopes in a cold climate.

SCHUMM, S. A.: "Rates of Surficial Rock Creep on Hillslopes in Western Colorado," *Science*, vol. 155, pp. 560–561, 1967. Long-term measurements show the amount and nature of slow movement of slope material.

SHARPE, C. F. S.: *Landslides and Related Phenomena*, Columbia University Press, New York, 1938. A classification and nonquantitative description of mass movement.

SHREVE, RONALD L.: "The Blackhawk Landslide," *Geol. Soc. Am., Spec. Paper*, no. 108, pp. 1–47, 1968. A quantitative description of a prehistoric landslide and discussion of the "air cushion" concept of catastrophic mass movements.

WAHRHAFTIG, CLYDE, and ALLAN COX: "Rock Glaciers in the Alaska Range," *Geol. Soc. Am. Bull.*, vol. 70, pp. 383–436, 1959. A good description and discussion of a particular type of mass movement.

WASHBURN, A. L.: "Geomorphic and Vegetational Studies in the Mesters Vig District, Northeast Greenland," *Medd. Groenland*, vol. 166, pp. 1–60, 1965. A quantitative study of the movement of slope materials in an arctic environment.

WOOLEY, R. R.: "Cloudburst Floods in Utah, 1850–1938," *U.S. Geol. Surv. Water Supply Paper*, no. 994, 1946. Describes the mechanism of mudflows.

THROUGH MILLIONS OF YEARS OF EARTH HISTORY, AGENTS OF EROSION HAVE BEEN working constantly to reduce the land masses to the level of the seas. Of these agents, running water is the most important. Year after year, the streams of the earth move staggering amounts of debris and dissolved material through their valleys to the great settling basins, the oceans.

11.1 *World distribution of water*

The distribution of the earth's estimated water supply is given in Table 11.1. The great bulk of this water, over 97 percent, is in the oceans, and less than 3 percent is on the land. Atmospheric moisture is surprisingly low, about one-thousandth of 1 percent of the total. Even less is in the stream channels of the world at any one moment. We see, then, that the earth's water is in varying amounts at different places, yet it moves from one place to another. As it is written in Ecclesiastes 1:7, "All the rivers run into the sea, yet the sea is not full: unto the place from whence the rivers come, thither they return again." This is part of the circulation of water from land to ocean and back again to land that we call the *hydrologic cycle.*

11.2 *Hydrologic cycle*

The streams of the world flow to the sea at a rate that would fill the ocean basins in 40,000 years. These streams, flowing as they do, must have a continuing supply of water. Furthermore, it is clear that the oceans must themselves somehow lose water in order to make room for the continuous supply that comes to them.

The identification of the source of stream water is historically recent. Well into the eighteenth century, it was the general belief that streams were replenished by springs which drew their water by some complex system through the underground from the oceans. This belief was nourished by the assumption that rainfall was inadequate to account for the flow observed in rivers, for rivers ran continuously even though rain was intermittent. Furthermore, it was generally held that rain water could not soak into the ground and replenish springs, a conclusion that seems to owe its origin to Seneca (3 B.C.–A.D. 65), Roman statesman and philosopher, who based his conclusions on inconclusive observations while he tended his vineyards.

In the seventeenth century three different types of observation laid the base for the hydrologic cycle as we accept it and which is diagrammatically represented in Figure 11.2. In 1674 Pierre Perrault (1611–1680), a French lawyer and sometime hydrologist, presented the results of his measurements in the upper portion of the drainage basin of the Seine River. Over a 3-year span, he collected data on the amount of precipitation in this portion of the basin. At the same time he kept track of the amount of water discharged by the

11

Running water

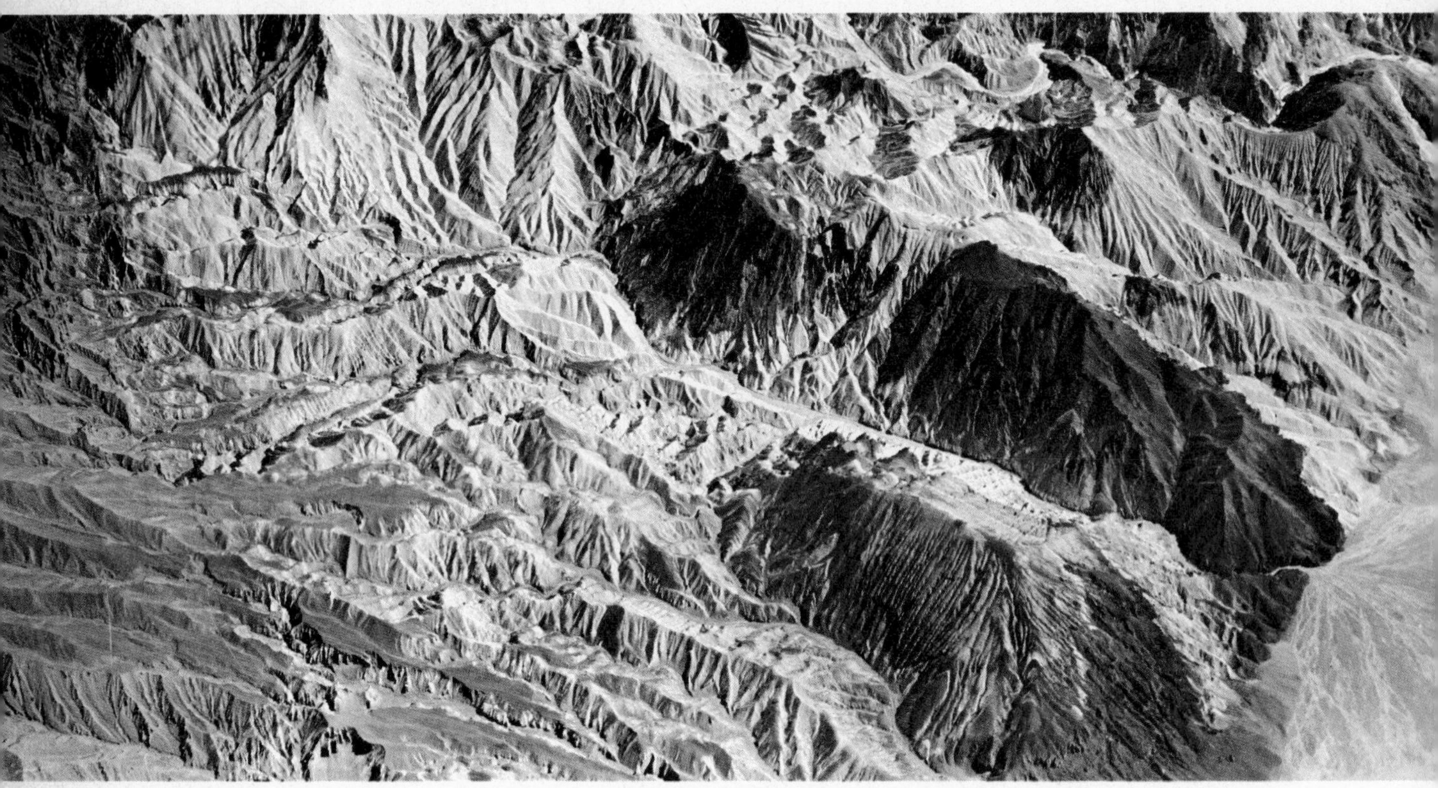

MARS HAS A STREAM (or CANYON) Valley 300 MI Long (MARINER 9)

Figure 11.1 **Stream valleys are the single most characteristic feature of the world's landscape. Even in the desert, as here on the slopes leading down into Death Valley, streams have fashioned an intricate system of valleys.** (Photograph by U.S. Army Air Corps.)

Table 11.1 **Distribution of the world's estimated supply of water**[a]

	Area		Volume		Percentage of total volume
	$km^2 \times 10^3$	$mi^2 \times 10^3$	$km^3 \times 10^3$	$mi^3 \times 10^3$	
World (total area)	510,000	197,000	—	—	—
Land area	149,000	57,500	—	—	—
Water in land areas					
Fresh water lakes	850	330	125	30	0.009
Saline lakes and inland seas	700	270	104	25	0.008
Rivers (average instantaneous volume)	—	—	1.25	0.3	0.0001
Soil moisture and vadose water	—	—	67	16	0.005
Ground water to depth of 4000 m	—	—	8,350	2,000	0.61
Ice caps and glaciers	19,400	7,500	29,200	7,000	2.14
Atmospheric moisture	—	—	13	3.1	0.001
World ocean	361,000	139,500	1,320,000	317,000	97.3
Total water volume (rounded)			1,360,000	326,000	100

[a] In part after R. L. Nace, *U.S. Geol. Surv. Circular,* no. 536, Table 1., 1967.

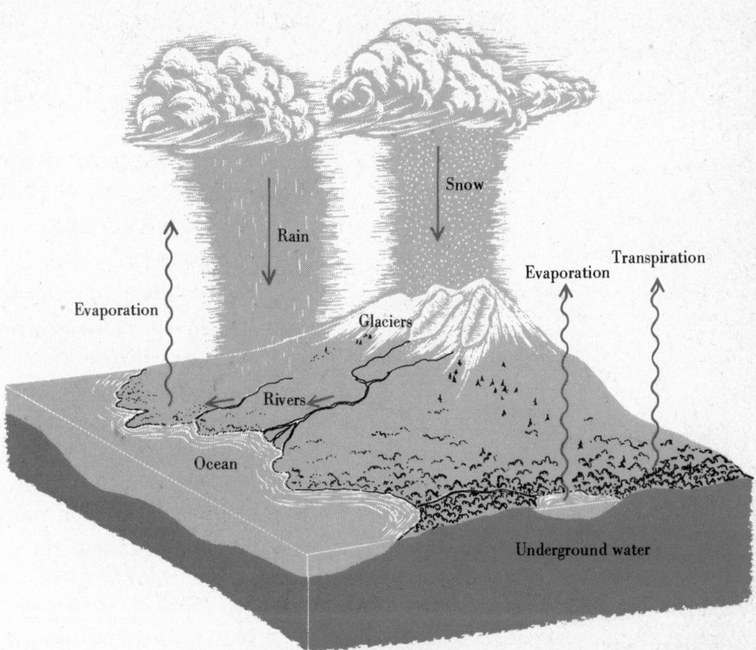

Figure 11.2 ***In the hydrologic cycle, water evaporated into the atmosphere reaches the land as rain or snow. Here it may be temporarily stored in glaciers, lakes, or the underground before returning by the rivers to the sea. Or some may be transpired or evaporated directly back into the atmosphere before reaching the sea.***

river below the portion of the river basin where he had data on precipitation. The results showed that the flow of the river was surprisingly low when compared with the total amount of water available from precipitation. In fact, the annual precipitation was six times the total volume of river flow. There was enough rainfall, then, to account for the flow of the Seine in this part of its basin.

At about the same time, the French physicist Edmé Mariotte (1620–1684) made more exact studies of discharge in the Seine River basin. His publications, appearing posthumously in 1684, verified Perrault's conclusions. Mariotte further demonstrated, by experimentation at the Paris Observatory, that seepage through the earth cover was less, but not much less, than the amount of rainfall. He also showed the increase in the flow of springs during rainy weather and the decrease during time of drought. It was then evident that the earth did permit penetration of moisture.

In 1693 Edmund Halley (1656–1742), of comet fame, provided data on evaporation in relation to rainfall. He roughly calculated the amount of water being discharged annually by rivers into the Mediterranean Sea and added this to the amount which fell directly on the sea's surface. He was then able to compute the approximate amount of water that was being evaporated back into the air each year from the surface of the Mediterranean Sea. He found that there was more than enough water being pumped into the atmosphere to feed all the rivers coming into the sea.

11.3 *Precipitation and stream flow*

After water has fallen on the land as precipitation, it follows one of the many paths that make up the hydrologic cycle. By far the greatest part is evaporated back to the air directly or is taken up by the plants and transpired (breathed back) by them to the atmosphere. A smaller amount follows the path of *runoff*, the water that flows off the land, and the smallest amount of precipitation soaks into the ground through *infiltration*.

Figure 11.3 shows how infiltration, runoff, and evaporation–transpiration vary in six widely separated localities in the United States. In the examples given, between 54 and 97 percent of the total precipitation travels back to the atmosphere through transpiration and evaporation. About 2 to 27 percent drains into streams and oceans as runoff, and between 1 and 20 percent finds its way into the ground through infiltration.

In Chapters 12, 13, and 15 we shall consider water at various stages of the hydrologic cycle. In Chapter 12 we look at the water that infiltrates into the ground; in Chapter 13 we study water that has been impounded on the land as glacier ice, a condition that represents a temporary halting of the water's progress through the cycle; Chapter 15 deals with water stored in oceans. In this chapter we shall concentrate on the nature and effects of runoff. Bearing in mind the ways in which water that falls as precipitation proceeds through the hydrologic cycle, we can express the amount of runoff by the following generalized formula:

$$\text{Runoff} = \text{Precipitation} - (\text{Infiltration} + \text{evaporation and transpiration})$$

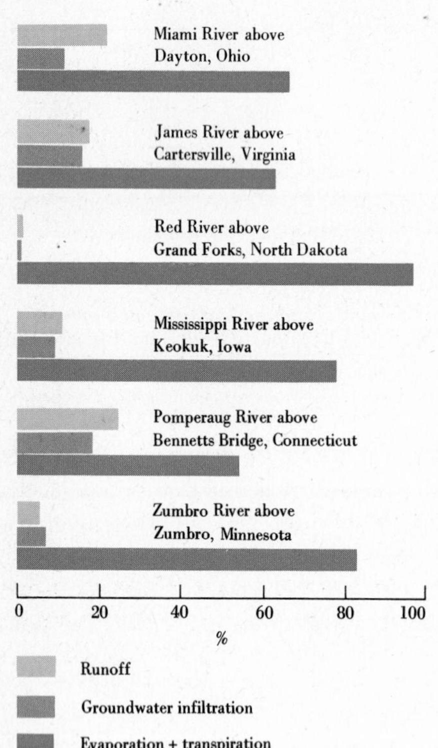

Figure 11.3 **Distribution of precipitation in selected drainage basins. Notice that in all cases 50 percent or more of all moisture that falls is returned to the atmosphere by evaporation and transpiration. Runoff from the surface is comparatively small, and infiltration of water in the underground is still less.** (*Data from W. G. Hoyt and others, "Studies of Relation of Rainfall and Run-off in the United States,"* U.S. Geol. Surv. Water Supply Paper, *no. 772, 1936.*)

L. B. Leopold and W. B. Langbein of the U.S. Geological Survey have reported some figures on the water budget of the hydrologic cycle in the United States (see Figure 11.4). There is an average of about 75 cm of precipitation over the country each year. Of this total, about 22.5 cm are carried as runoff by the streams to the ocean. Another 52.5 cm go back into the atmosphere by evaporation and transpiration. Of course, some water filters into the ground but all but a very small amount—about 0.25 cm—is returned to the surface where it disappears either as runoff or as evapotranspiration. These figures indicate that across the United States there is an apparent loss of 22.5 cm annually from the atmosphere. This is the runoff that goes to the ocean, but the debit is made up by the evaporation of moisture from the ocean and its transfer by the atmosphere to the continents.

We have some figures on the worldwide water budget. The average precipitation for the lands of the world is variously estimated and approximates about 80 cm/year. Estimates of runoff from the continents to the oceans vary. One estimate is given in Table 11.2. This does not include water in the form of ice that may get to the oceans. The figures indicate that 32.4×10^{15} liters of water flow annually to the sea each year. This is a runoff of 24.9 cm/unit area. The difference between runoff and precipitation is the evapotranspiration,

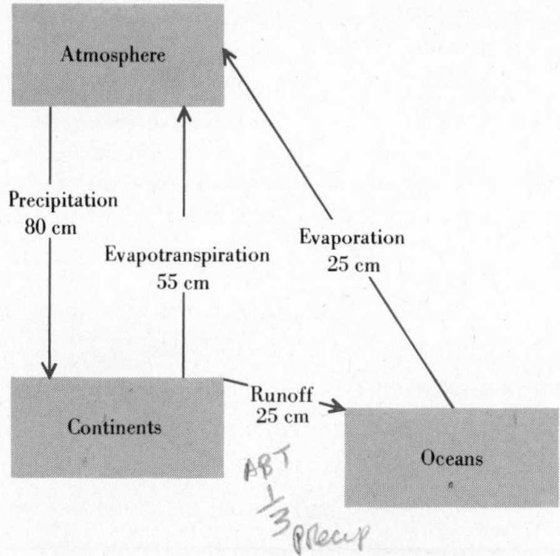

Figure 11.4 ***Water budget for the United States.*** *(Data from U.S. Geological Survey.)*

in this instance about 55 cm. It follows that the oceans contribute about 25 cm to the precipitation budget of the continents each year.

LAMINAR AND TURBULENT FLOW Consider now the manner in which water moves. When it moves slowly along a smooth channel or through a tube with smooth walls, it follows straight-line paths that are parallel to the channel or walls. This type of movement is called *laminar flow.*

If the rate of flow increases, however, or if the confining channel becomes rough and irregular, this smooth, streamlined movement is disrupted. The water in contact with the channel is slowed down by friction, whereas the rest of the water tends to move along as before. As a result (see Figures 11.5 and 11.6), the water is deflected from its straight paths into a series of eddies and swirls. This type of movement is known as *turbulent flow.* Water in streams usually flows along in this way, its turbulent flow being highly effective both in eroding a stream's channel and in transporting materials.

When a stream reaches an exceptionally high velocity along a sharply inclined stretch or over a waterfall, the water moves in plunging, jetlike surges. This type of flow, closely related to turbulent flow, is called *jet* or *shooting flow.*

VELOCITY, GRADIENT, AND DISCHARGE The velocity of a stream is measured in terms of the distance its water travels in a unit of time. A velocity of 15 cm/sec is relatively low, and a velocity of about 625 to 750 cm/sec is relatively high.

A stream's velocity is determined by many factors, including the amount of water passing a given point, the nature of the stream banks, and the *gradient* or slope of the stream bed. In general, a stream's gradient decreases from its headwaters toward its mouth; as a result, a stream's longitudinal profile is more or less concave toward the sky (Figure 11.7). We usually express the gradient of a stream as the vertical distance a stream descends during a fixed distance of horizontal flow. The Mississippi River from Cairo, Illinois, to the mouth of the Red River in Arkansas has a low gradient, for along this stretch the drop varies between 2 and 10 cm/km. On the other hand, the

Table 11.2 ***Runoff from the continents***[a]

	Area		*Annual runoff*			
			Total		*Depth per unit area*	
	km² × 10⁶	*mi² × 10⁶*	*liters × 10¹⁵*	*gal × 10¹⁵*	*cm*	*in.*
Asia	46.6	18.0	11.1	3.0	23.8	9.4
Africa	29.8	11.5	5.9	1.6	19.8	7.8
North America	21.2	8.2	4.5	1.2	21.1	8.3
South America	19.6	7.6	8.0	2.1	41.4	16.3
Europe	10.9	4.2	2.5	0.6	23.1	9.1
Australia	7.8	3.0	0.4	0.1	2.5	1.0
Total or (mean)	135.9	52.5	32.4	8.6	(24.9)	(9.8)

[a] Calculated from data from D. A. Livingstone, *U.S. Geol. Surv. Prof. Paper,* no. 440-G, 1963.

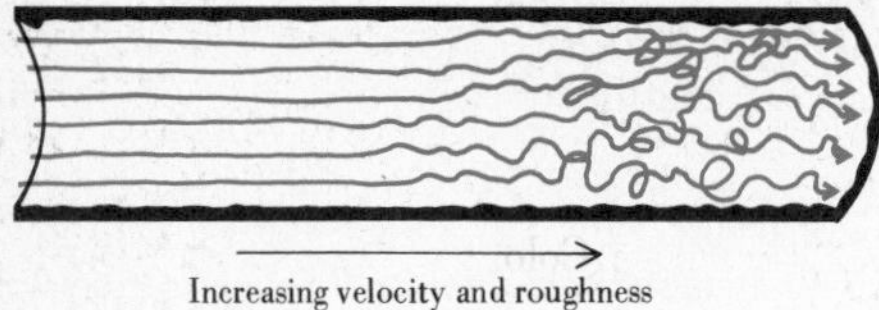

Figure 11.5 ***Diagram showing laminar and turbulent flow of water through a section of pipe. Individual water particles follow paths depicted by the colored lines. In laminar flow, the particles follow paths parallel to the containing walls. With increasing velocity or increasing roughness of the confining walls, laminar flow gives way to turbulent flow. The water particles no longer follow straight lines but are deflected into eddies and swirls. Most water flow in streams is turbulent.***

Figure 11.6 ***Most stream flow is turbulent. Here the Lewis River in Wyoming tumbles over falls and swirls and boils through a rough, irregular channel.*** (*Photograph by Sheldon Judson.*)

Figure 11.7 ***In longitudinal profile (from mouth to headwaters), a stream valley is concave to the sky. Irregularities along the profile indicate variations in rates of erosion.*** (*Redrawn from Henry Gannett, "Profiles of Rivers in the United States,"* U.S. Geol. Surv. Water Supply Paper, *no. 44, 1901.*)

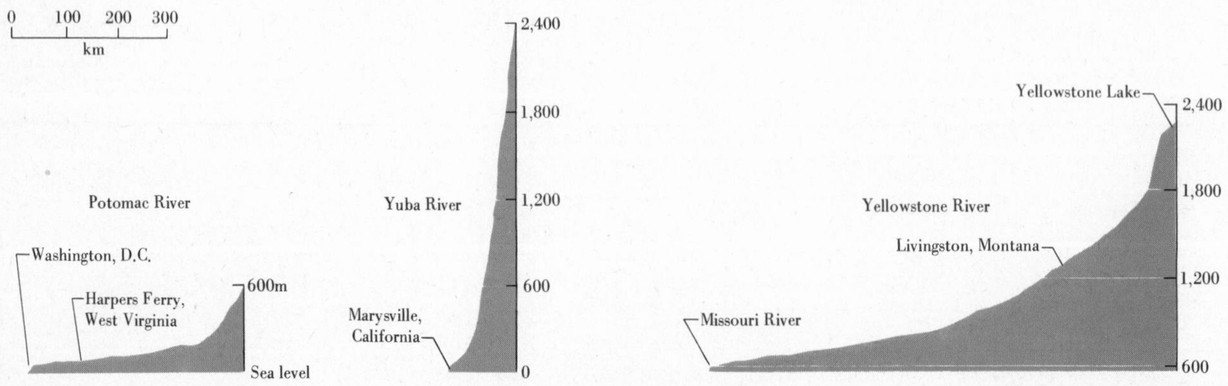

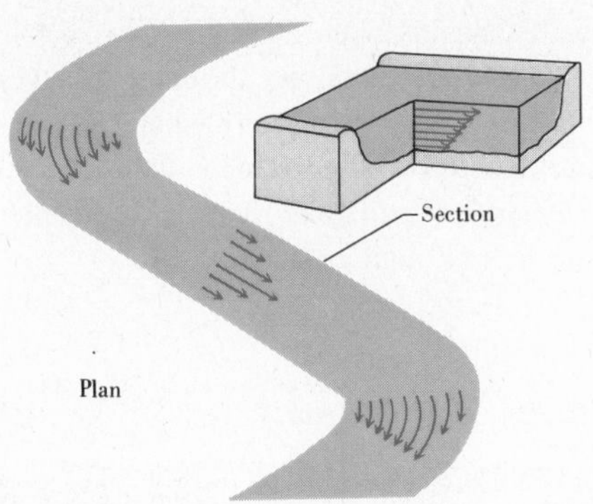

Figure 11.8 ***Velocity variations in a stream. Both in plan view and in cross section, the velocity is slowest along the sides and bottom of the stream channel, where the water is slowed by friction. On the surface, it is most rapid at the center in straight stretches and toward the outside of a bed where the river curves. Velocity increases upward from the river bottom.***

Arkansas River in its upper reaches through the Rocky Mountains in central Colorado has a high gradient, for there the drop averages 7.5 m/km. The gradients of other rivers are even higher. The upper 20 km of the Yuba River in California, for example, have an average gradient of 42 m/km; and in the upper 6.5 km of the Uncompahgre River in Colorado, the gradient averages 66 m/km.

The velocity of a stream is checked by the turbulence of its flow, friction along the banks and bed of its channel, and, to a much smaller extent, by friction with the air above. Therefore, if we were to study a cross section of a stream, we would find that the velocity would vary from point to point. Along a straight stretch of a channel, the greatest velocity is achieved toward the center of the stream at, or just below, the surface, as shown in Figure 11.8.

We have, then, two opposing forces: the *forward flow* of the water under the influence of gravity and the *friction* developed along the walls and bed of the stream. These are the two forces that create different velocities. Zones of maximum turbulence occur where the different velocities come into closest contact. These zones are very thin, because the velocity of the water increases very rapidly as we move into the stream away from its walls and bed; but within these thin zones of great turbulence, a stream shows its highest potential for erosive action (see Figure 11.9).

There is one more term that will be helpful in our discussion of running water: *discharge*, the quantity of water that passes a given point in a unit of time. In the United States it is usually measured in cubic feet per second, abbreviated cfs. Discharge varies not only from one stream to another but also within a single stream from time to time and from place to place along its course. Discharge usually increases downstream as more and more tributaries add their water to the main channel. Spring floods may so greatly increase a stream's discharge that its normally peaceful course becomes a raging torrent.

11.4 The economy of a stream

Elsewhere, we have seen that earth processes tend to seek a balance, to establish an equilibrium, and that there is, in the words of James Hutton, "a system of beautiful economy in the works of nature." We found, for example, that weathering is a response of earth materials to the new and changing conditions they meet as they are exposed at or near the earth's

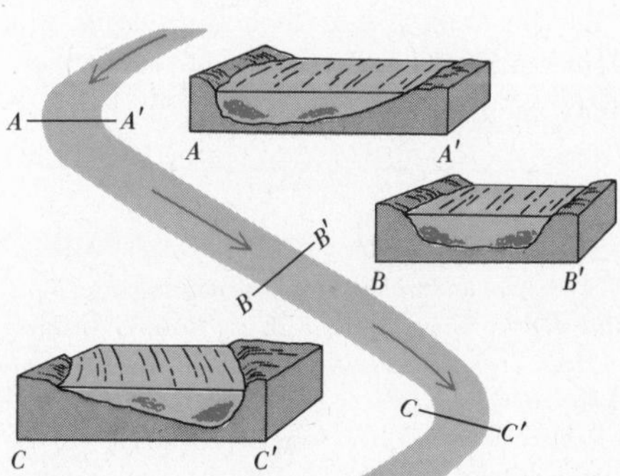

Figure 11.9 ***Zones of maximum turbulence in a stream are shown by the dark green areas in the sections through a river bed. They occur where the change between the two opposing forces—the forward flow and the friction of the stream channel—is most marked. Note that the maximum turbulence along straight stretches of the river is located where the stream banks join the stream floor. On bends, the two zones have unequal intensity; the greater turbulence is located on the outside of a curve.***

surface. On a larger scale, we found that the major rock groups—igneous, sedimentary, and metamorphic—reflect certain environments and that as these environments change, members of one group may be transformed into members of another group. These changes were traced in what we called the rock cycle (Figure 1.12). Water running off the land in streams and rivers is no exception to this universal tendency of nature to seek equilibrium.

Adjustments of discharge, velocity, and channel

Just a casual glance tells us that the behavior of a river during flood stage is very different from its behavior during the low-water stage. For one thing, a river carries more water and moves more swiftly in flood time. Furthermore, the river is generally wider during flood, its level is higher, and we would guess, even without measuring, that it is also deeper. We can relate the discharge of a river to its width, depth, and velocity as follows:

Discharge (cfs) = Channel width (ft) × channel depth (ft) × water velocity (ft/sec)

In other words, if the discharge at a given point along a river increases, then the width, depth, or velocity, or some combination of these factors must also increase. We now know that variations in width, depth, and velocity are neither random nor unpredictable. In most streams, if the discharge increases, then the width, depth, and velocity each increase at a definite rate. The stream maintains a balance between the amount of water it carries on the one hand, and its depth, width, and velocity on the other. Moreover, it does so in an orderly fashion, as shown in Figure 11.10.

Let us turn now from the behavior of a stream at a single locality to the changes that take place along its entire length. From our own observation, we know that the discharge of a stream increases downstream as more and more tributaries contribute water to its main channel. We also know that the width and depth increase as we travel downstream. But if we go beyond casual observation and gather accurate data on the width, depth, velocity, and discharge of a stream from its headwaters to its mouth for a particular stage of flow—for example, flood or low-water—we would find again that the changes follow a definite pattern and that depth and width increase downstream as the discharge increases (Figure 11.11). And, surprisingly enough, we would also find that the stream's *velocity* increases toward its mouth. This is contrary to our expectations, for we know that the gradients are higher upstream, which suggests that the velocities in the steeper headwater areas would also be higher. But the explanation for this seeming anomaly is simple: To handle the greater

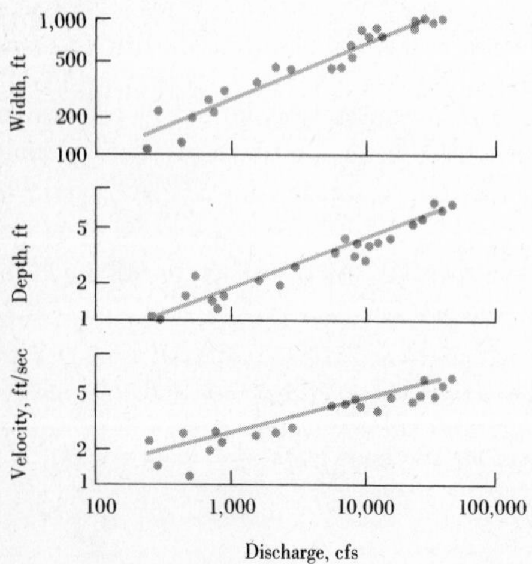

Figure 11.10 ***As the discharge of a stream increases at a given gauging station, so do its velocity, width, and depth. They increase in an orderly fashion, as shown by these graphs based on data from a gauging station in the Cheyenne River near Eagle Butte, South Dakota.*** (*Redrawn from Luna B. Leopold and Thomas Maddock, "The Hydraulic Geometry of Stream Channels and Some Physiographic Implications,"* U.S. Geol. Surv. Professional Paper, *no. 252, p. 5, 1953.*)

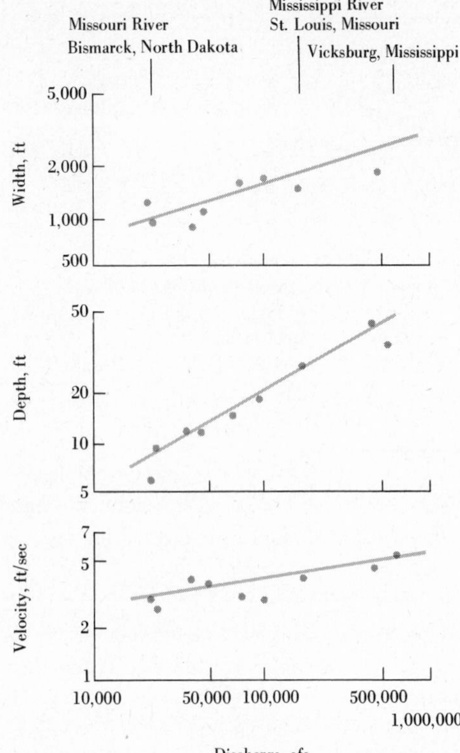

Figure 11.11 **Stream velocity and depth and width of a channel increase as the discharge of a stream increases downstream. Measurements in this example were made at mean annual discharge along a section of the Mississippi–Missouri river system.** (*Redrawn from Leopold; see Figure 11.10.*)

discharge downstream, a stream must not only deepen and widen its channel but must also increase its velocity.

Adjustments of gradient

The gradient of a stream decreases along its course from headwaters to mouth, producing an overall profile that is concave to the sky. If it were not for this gradual flattening of the profile, the increased discharge downstream would produce velocities fantastically higher than those observed in nature. A concave slope tends to decrease the rate at which stream velocity increases.

The actual profile of a stream is determined by the particular conditions the stream meets along its course. In its attempt to establish a balance between discharge on the one hand and channel characteristics, velocity, and gradient on the other, the stream reduces its gradient and increases its velocity, width, and depth as it flows downstream. The resulting profile or gradient is an expression of the equilibrium that is set up along each section of the stream.

Base level of a stream

Base level is a key concept in the study of stream activity. The *base level* of a stream is defined as the lowest point to which a stream can erode its channel. Anything that prohibits the stream from lowering its channel serves to create a base level. For example, the velocity of a stream is checked when it enters the standing, quiet waters of a lake. Consequently, the stream loses its ability to erode, and it cannot cut below the level of the lake. Actually, the lake's control over the stream is effective along the entire course upstream, for no part of the stream can erode beneath the level of the lake—at least until the lake has been destroyed. But in a geologic sense, every lake is temporary, and therefore, when the lake has been destroyed, perhaps by the downcutting of its outlet, it will no longer control the stream's base level, and the stream will be free to continue its downward erosion. Because of its impermanence, the base level formed by a lake is referred to as a *temporary base level.* But even after a stream has been freed from one temporary base level, it will be controlled by others farther downstream, and its erosive power is always influenced by the ocean, which is the *ultimate base level.* Yet, as we shall see in Chapter 15, the ocean itself is subject to changes in level, so even the ultimate base level is not fixed.

The base level of a stream may be controlled not only by lakes but also by layers of resistant rock and the level of the main stream into which a tributary drains (see Figure 11.12).

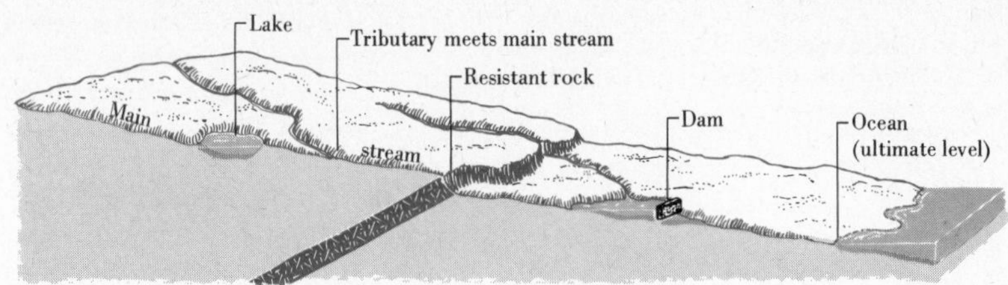

Figure 11.12 **Base level for a stream may be determined by natural and artificial lakes, by a resistant rock stratum, by the point at which a tributary stream enters a main stream, and by the ocean. Of these, the ocean is considered the ultimate base level; the others are temporary base levels.**

Figure 11.13 **A stream adjusts its channel to changing base level. See text for explanation.**

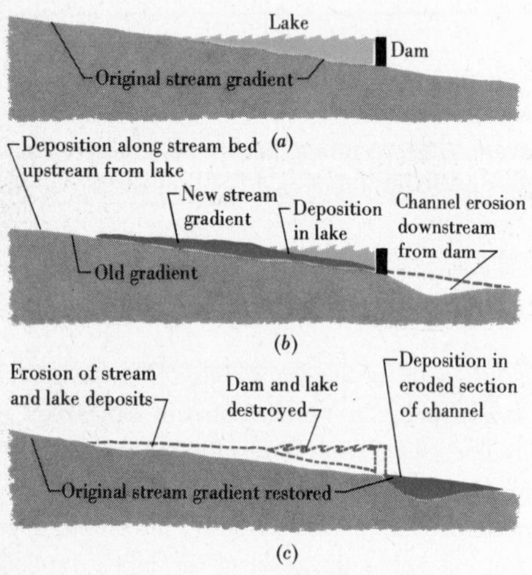

ADJUSTMENT TO CHANGING BASE LEVEL We have defined base level as the lowest level to which a stream can erode its channel. If for some reason the base level is either raised or lowered, the stream will adjust the level of its channel to adapt to the new situation.

Let us see what happens when we raise the base level of a stream by building a dam and creating a lake across its course. The level of the lake serves as a new base level, and the gradient of the stream above the dam is now less steep than it was originally. As a result, the stream's velocity is reduced. Because the stream can no longer carry all the material supplied to it, it begins to deposit sediments at the point where it enters the lake. As time passes, a new river channel is formed with approximately the same slope as the original channel but at a higher level (see Figure 11.13).

What happens when we *lower* the base level by removing the dam and hence the lake? The river will now cut down through the sediments it deposited when the lake still existed. In a short time, the profile of the channel will be essentially the same as it was before we began to tamper with the stream.

In general, then, we may say that a stream adjusts itself to a rise in base level by building up its channel through sedimentation and that it adjusts to a fall in base level by eroding its channel downward.

11.5 *Work of running water*

The water that flows along through river channels does several jobs: (1) It transports debris, (2) it erodes the river channel deeper into the land, and (3) it deposits sediments at various points along the valley or delivers them to lakes or oceans. Running water may help to create a chasm like that of the Grand Canyon of the Colorado, or in flood time it may spread mud and sand across vast expanses of valley flats, or it may build deltas like those at the mouths of the Nile and Mississippi Rivers.

The nature and extent of these activities depend on the kinetic energy of the stream, and this in turn depends on the amount of water in the stream and the gradient of the stream channel. A stream expends its energy in several ways. By far the greatest part is used up in the friction of the water with the stream channel and in the friction of water with water in the turbulent eddies we discussed above. Relatively little of the stream's energy remains to erode and transport material. Deposition takes place when energy decreases and the stream can no longer move the material it has been carrying.

Transportation

The material that a stream picks up directly from its own channel—or that is supplied to it by slope wash, tributaries, or mass movement—is moved downstream toward its eventual goal, the ocean. The amount of material that a stream carries at any one time, which is called its *load*, is usually less than its *capacity*—that is, the total amount it is capable of carrying under any given set of conditions (see Figure 11.14). The maximum size of particle that a stream can move measures the *competency* of a stream.

There are three ways in which a stream can transport material: by (1) solution, (2) suspension, and (3) bed load.

Figure 11.14 ***These converging rivers in British Columbia illustrate the different loads carried by two streams. The Frazer River enters from the upper right, milky with suspended sediment derived largely from the melting of mountain glaciers. Its load is high but probably somewhat less than capacity. The Thompson River, entering from the lower right, is relatively clear and carries a very small load, much less than its capacity.*** *(Photograph by Elliott A. Riggs.)*

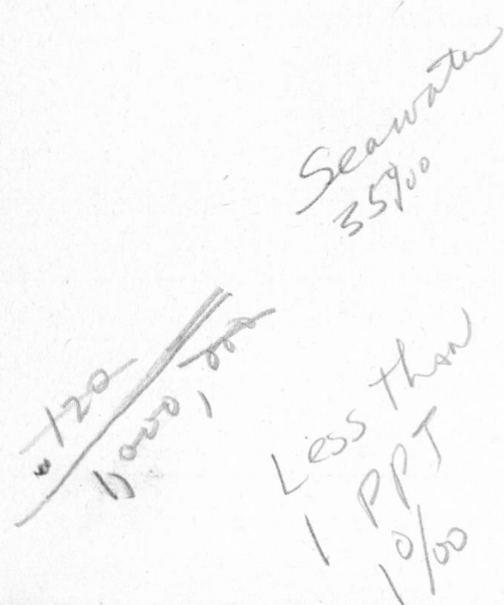

Figure 11.15 **The suspended load of a stream increases very rapidly during floods, as illustrated by this graph based on measurements in the Rio Puerco near Cabezon, New Mexico.** (*Redrawn from Luna B. Leopold and John P. Miller, "Ephemeral Streams—Hydraulic Factors and Their Relation to the Drainage Net,"* U.S. Geol. Surv. Professional Paper, *no. 282-A, p. 11, 1956.*)

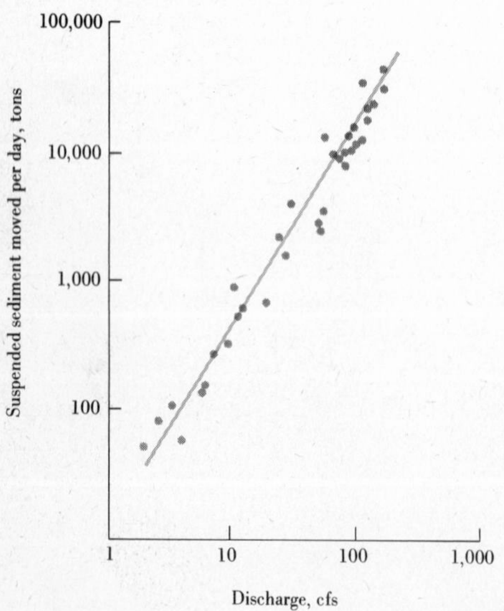

SOLUTION In nature, no water is completely pure. We have already seen that when water falls and filters down into the ground, it dissolves some of the soil's compounds. Then the water may seep down through openings, pores, and crevices in the bedrock and dissolve additional matter as it moves along. Much of this water eventually finds its way to streams at lower levels. The amount of dissolved matter contained in water varies with climate, season, and geologic setting and is measured in terms of parts of dissolved matter per million parts of water. Sometimes the amount of dissolved material exceeds 1,000 parts per million, but usually it is much less. By far the most common compounds found in solution in running water, particularly in arid regions, are calcium and magnesium carbonates. In addition, streams carry small amounts of chlorides, nitrates, sulfates, and silica, with perhaps a trace of potassium. It has been estimated that the total load of dissolved material delivered to the seas every year by the streams of the United States is nearly 300 million metric tons. The rivers of the world average an estimated 115 to 120 parts per million of dissolved matter, which means that annually they carry to the sea about 3.9 billion metric tons.

SUSPENSION Particles of solid matter that are swept along in the turbulent current of a stream are said to be in *suspension.* This process of transportation is controlled by two factors: (1) the turbulence of the water and (2) a characteristic known as *terminal velocity* of each individual grain. Terminal velocity is the constant rate of fall that a grain eventually attains when the acceleration caused by gravity is balanced by the resistance of the fluid through which the grain is falling. In this case, the fluid is water. If we drop a grain of sand into a quiet pond, it will settle toward the bottom at an ever-increasing rate until the friction of the water on the grain just balances this rate of increase. Thereafter, it will settle at a constant rate, its terminal velocity. If we can set up a force that will equal or exceed the terminal velocity of the grain, we can succeed in keeping it in suspension. Turbulent water supplies such a force. The eddies of turbulent water move in a series of orbits, and grains caught in these eddies will be buoyed up, or held in suspension, so long as the velocity of the turbulent water is equal to, or greater than, the terminal velocity of the grains.

Terminal velocity increases with the size of the particle, assuming that its general shape and density remain the same. The bigger a particle, the more turbulent the flow needed to keep it in suspension. And because turbulence increases when the velocity of stream flow increases, it follows that the greatest amount of material is moved during flood time when velocities and turbulence are highest. The graph in Figure 11.15 shows how the suspended load of a stream increases as the discharge increases. In just a few hours or a few days during flood time, a stream transports more material than it does during the much longer periods of low or normal flow. Observations of the area drained by Coon Creek, at Coon Valley, Wisconsin, over a period of 450 days, showed that 90 percent of the stream's total suspended load was carried during an interval of 10 days, slightly over 2 percent of total time.

Silt and clay-sized particles are distributed fairly evenly through the depth of a stream, but coarser particles in the sand-size range are carried in greater amounts lower down in the current, in the zone of greatest turbulence.

BED LOAD Materials in movement along a stream bottom constitute the stream's *bed load*, in contrast to its suspended load and solution load. Because it is difficult to observe and measure the movement of the bed load, we have little data on the subject. Measurements on the Niobrara River near Cody, Nebraska, however, have shown that at discharges between 200 and 1,000 cfs, the bed load averaged about 50 percent of the total load. Particles in the bed load move along in three ways: by saltation, rolling, or sliding.

The term *saltation* has nothing to do with salt. It is derived from the Latin *saltare*, "to jump." A particle moving by saltation jumps from one point on the stream bed to another. First, it is picked up by a current or turbulent water and flung upward; then, if it is too heavy to remain in suspension, it drops to the stream floor again at some spot downstream.

Some particles are too large and too heavy to be picked up, even momentarily, by water currents, but they may be pushed along the stream bed, and depending on their shape, they move forward either by *rolling* or by *sliding*.

Erosion

A stream does more than simply transport material that has been brought to it by other agents of erosion, for it is an effective agent of erosion in itself. In various ways, an actively eroding stream may remove material from its channel or banks.

DIRECT LIFTING In turbulent flow, as we have seen, water travels along paths that are not parallel to the bed. The water eddies and whirls, and if an eddy is powerful enough, it dislodges particles from the stream channel and lifts them into the stream. Whether or not this will happen in a given situation depends on a number of variables that are difficult to measure, but if we assume that the bed of a stream is composed of particles of uniform size, then the graph in Figure 11.16 gives us the approximate stream velocities that are needed to erode particles of various sizes, such as clay, silt, sand, granules, and pebbles. A stream bed composed of fine-sized sand grains, for example, can be eroded by a stream with a velocity of between 18 and 50 cm/sec, depending on how firmly packed they are. As the fragments become larger and larger, ranging from coarse sand to granules to pebbles, increasingly higher velocities are required to move them, as we would expect.

But what we might *not* expect is that as the particles become smaller than about 0.06 mm in diameter they do not become more easily picked up by the stream. In fact if the clay and silt are firmly consolidated, then increased velocities will be needed to erode the particles. The reason for this is that generally the smaller the particle, the more firmly packed the deposit and thus the more resistant to erosion. Moreover, the individual particles may be so small that they do not project high enough into the stream to be swept up by the turbulent water.

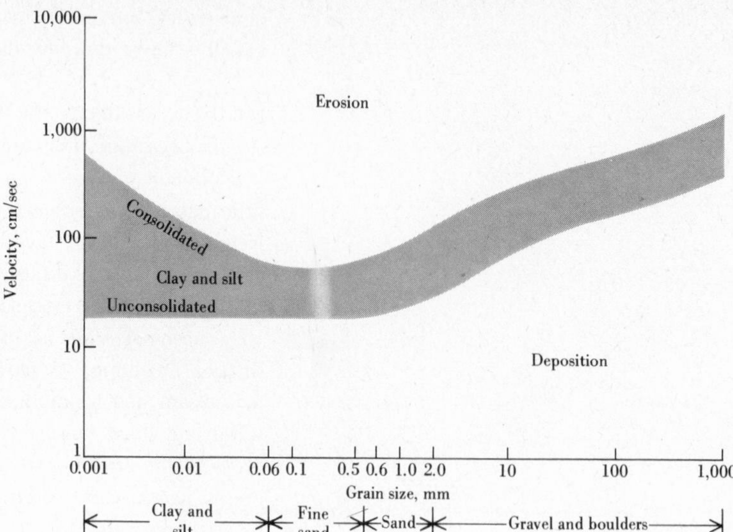

Figure 11.16 **The central band in this diagram shows the range in velocity at which turbulent water will lift particles of differing size off the stream bed. The width is affected by the shape, density, and consolidation of the particles.** (*From Ake Sundborg,* Geografis. Ann., *vol. 38, p. 197, Fig. 16, 1956.*)

ABRASION, IMPACT, AND SOLUTION The solid particles carried by a stream may themselves act as erosive agents, for they are capable of abrading (wearing down) the bedrock itself or larger fragments in the bed of the stream. When the bedrock is worn by abrasion, it usually develops a series of smooth, curving surfaces, either convex or concave. Individual cobbles or pebbles on a stream bottom are sometimes moved and rolled about by the force of the current, and as they rub together, they become rounder and smoother.

The impact of large particles against the bedrock or against other particles knocks off fragments, which are added to the load of the stream.

Some erosion also results from the solution of channel debris and bedrock in the water of the stream. Most of the dissolved matter carried by a stream, however, is probably contributed by the underground water that drains into it.

CAVITATION At very high stream velocities, above 8 or 10 m/sec, a highly effective erosive process known as *cavitation* sometimes comes into play. From the Latin meaning "hollow," *cavitation* refers to the sudden collapse of vapor bubbles in the water of a stream. If a bubble is in contact with the stream bed at the moment when the collapse occurs, an extremely strong impact is produced. Obvious difficulties frustrate measurement of the exact strength of the impact, but various experiments, as well as theoretical considerations, suggest that the minimum impact may be as much as 100 to 140 kg/cm^2—a very effective erosive force. In fact, in one experiment, cavitation eroded away nearly 0.5 m of a concrete spillway during a 23-hr period. Because very high velocities are needed to produce cavitation in streams, however, this process probably occurs only in the jet and shooting flow of waterfalls and rapids.

TIME OF MOST RAPID EROSION We have found that, other things being equal, the greater the stream's velocity, the greater its erosive power. Obviously, then,

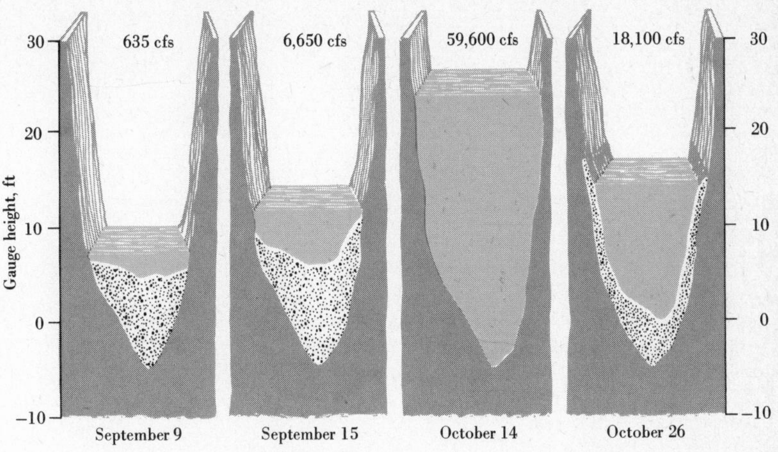

Figure 11.17 ***Changes in the channel of the San Juan River near Bluff, Utah, during a flood in 1941. On September 9 and 15, periods of moderate discharge, the river flowed on a bed of gravel within the larger bedrock channel. On October 14, flood waters swelled to 59,600 cfs and swept the gravel downstream, exposing the bedrock abrasion. With the subsidence of the flood, new gravel deposits brought from upstream began to build up the gravel bed of the river. The width of the river on September 9 was approximately 45 m.*** (*Redrawn from Leopold; see Figure 11.10.*)

the greatest erosive (and transporting) power of any stream is developed during flood time. When a stream is at flood stage, the water level rises and the channel is deepened. The fast-moving water picks up the layer of sand and gravel that usually lies on the bedrock of the stream channel during nonflood stages and sweeps it downstream. If the flood is great enough, the bedrock itself is exposed and eroded. A new layer of debris collects as the flood waters subside, but by that time great masses of material have been moved downstream toward the oceans, and the bedrock channel of the stream has been permanently lowered. Figure 11.17 illustrates this action.

In general, then, we may say that erosion is most effective during flood periods.

Deposition

As soon as the velocity of a stream falls below the point necessary to hold material in suspension, the stream begins to deposit its suspended load. Deposition is a selective process. First, the coarsest material is dropped; then, as the velocity (and hence the energy) continues to slacken, finer and finer material settles out. We shall consider stream deposits in more detail elsewhere in this chapter.

11.6 *Features of valleys*

Cross-valley profiles

Earlier in this chapter we mentioned the longitudinal profile of a stream (Figure 11.7). Now let us turn to a discussion of the cross-valley profile, that is, the profile of a cross section at right angles to the trend of the stream's valley. In Figure 11.18, notice that the channel of the river runs across a broad,

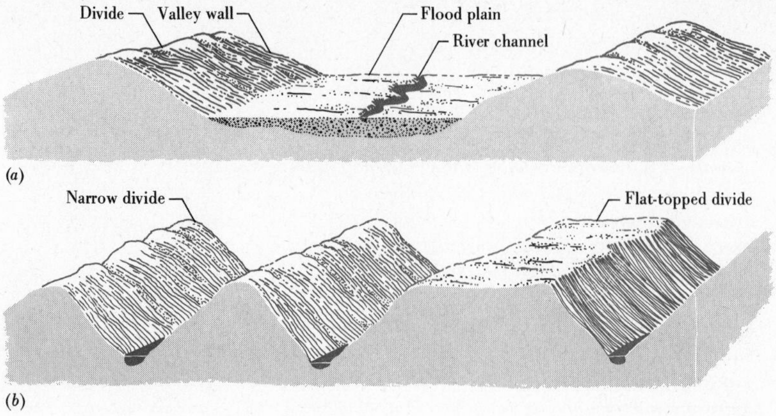

Figure 11.18 ***Cross-sectional sketches of typical stream valleys. The major features of valleys in cross section include divides, valley walls, river channel, and, in some instances, a flood plain. Divides may be either flat-topped or broadly rounded.***

relatively flat *flood plain.* During flood time, when the channel can no longer accommodate the increased discharge, the stream overflows its banks and floods this area. On either side of the flood plain, *valley walls* rise to crests called *divides,* separations between the central valley and the other valleys on either side. In B (Figure 11.18), no flood plain is present, for the valley walls descend directly to the banks of the river. This diagram also illustrates two different shapes of divide. One is broad and flat; the other is narrow, almost knife-edged. Both are in contrast to the broadly convex divides shown in A.

Drainage basins, networks, and patterns

A *drainage basin* is the entire area from which a stream and its tributaries receive their water. The Mississippi River and its tributaries drain a tremendous section of the central United States reaching from the Rocky to the Appalachian Mountains, and each tributary of the Mississippi has its own drainage area, which forms a part of the larger basin. Every stream, even the smallest brook, has its own drainage basin, shaped differently from stream to stream but characteristically pear-shaped, with the main stream emerging from the narrow end (see Figure 11.19).

Individual streams and their valleys are joined together into networks. In any single network the streams prove to have a definite geometric relationship in a way first detailed in 1945 by the U.S. hydraulic engineer Robert Horton. We can devise a demonstration of this relationship by first ranking the streams in a hierarchy. This order, shown in Figure 11.20, lists those small headwater streams without tributaries as belonging to the first order in the hierarchy. Two or more first-order streams join to form a second-order stream segment; two or more second-order segments join to form a third-order stream; and

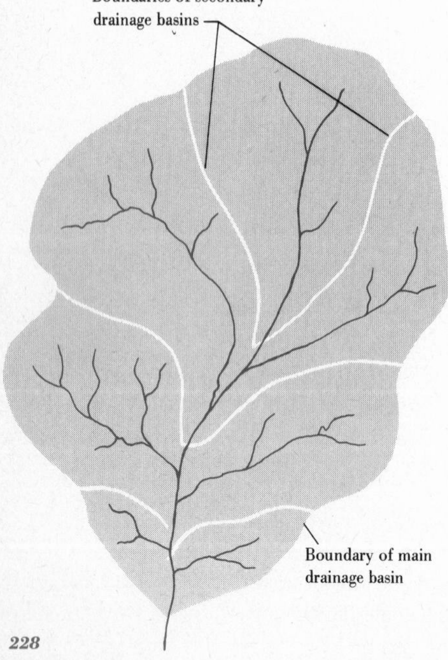

Figure 11.19 ***Each stream, no matter how small, has its own drainage basin, the area from which the stream and its tributaries receive water. This basin displays a pattern reminiscent of a tree leaf and its veins.***

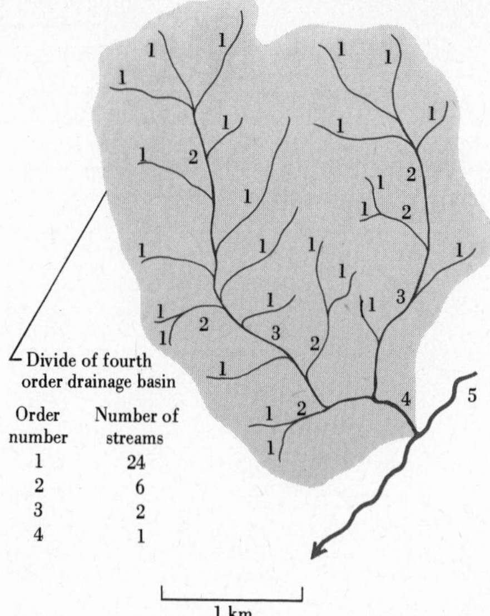

Order number	Number of streams
1	24
2	6
3	2
4	1

Figure 11.20 ***Method of designating stream orders.*** (*Based on the system devised by A. N. Strahler,* Proc. Intern. Geol. Congr., *section 13, part 3, p. 344, 1952.*)

so on. In other words a stream segment of any given order is formed by the junction of at least two stream segments of the next lower order. The main stream segment of the system always has the highest order number in the network, and the number of this stream is assigned to describe the stream basin. Thus the main stream in the basin shown in Figure 11.20 is a fourth-order stream and the basin is a fourth-order basin.

It is apparent from Figure 11.20 that the number of stream segments of different order decrease with increasing order. When we plot the number of streams of a given order against order, the points define a straight line on semilogarithmic paper (see Figure 11.21*a*). Likewise, if we plot order against the average length of streams of a given order or plot order against areas of

Figure 11.21 ***Relation between stream order and the number of streams, the mean length of streams, and the mean drainage area of streams of an area in central Pennsylvania. The system used to designate stream orders is that proposed by Robert Horton in 1945 and differs slightly from that illustrated in Figure 11.20.*** (*After Lucien Brush, Jr.,* U.S. Geol. Surv. Professional Paper, *no. 282-F., Fig. 97, 1961.*)

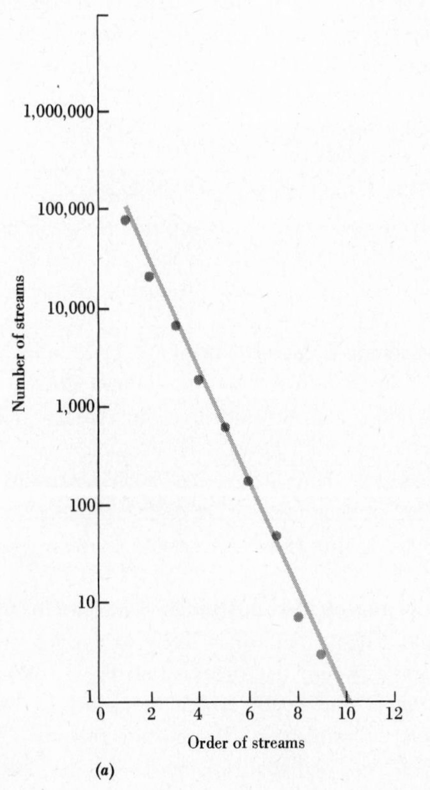

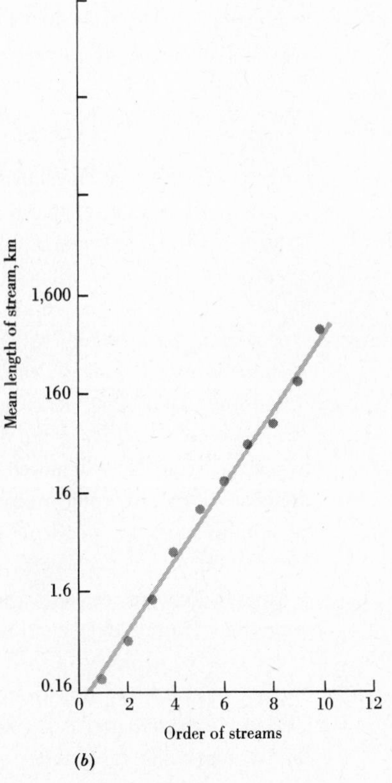

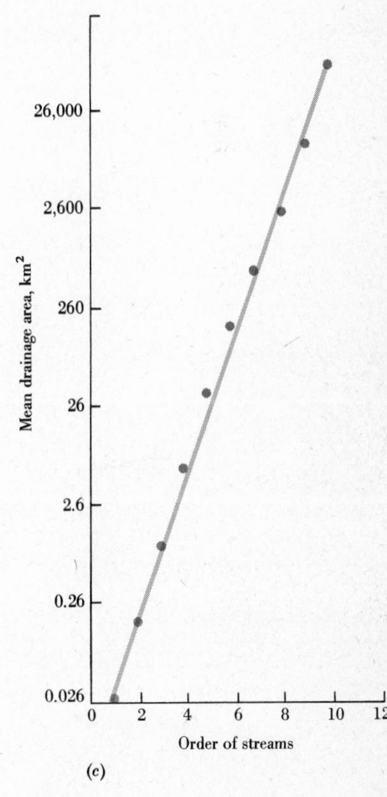

Table 11.3 ***Number and length of river channels of various sizes in the United States, excluding tributaries of smaller order***[a]

Order	*Number*	*Average length, km*	*Total length, km*	*Mean drainage area, including tributaries,* km^2	*River representative of size*
1[b]	1,570,000	1.6	2,512,000	2.6	
2	350,000	3.7	1,295,000	12	
3	80,000	8.5	680,000	59	
4	18,000	19	342,000	283	
5	4,200	45	189,000	1,348	
6	950	102	97,000	6,400	
7	200	219	44,000	30,300	Allegheny
8	41	541	22,000	144,000	Gila
9	8	1,243	10,000	683,000	Columbia
10	1	2,880	2,880	3,238,000	Mississippi

[a] From Luna B. Leopold, "Rivers," *Am. Scientist,* vol. 50, p. 512, 1962.
[b] The size of the order 1 channel depends on the scale of the maps used; these order numbers are based on the determination of the smallest order using maps of scale 1:62,500.

drainage basins of a particular order, we find a well-defined relationship (see Figure 11.21*b* and *c*). Using these relationships, the river channels and basins of the United States can be described as shown in Table 11.3.

We have seen that when streams are arranged in a hierarchical fashion the arrangement bears a definite relation to the number and lengths of the stream segments in each order and to the size of the basins of the stream segments of different orders. Beyond this there are overall patterns developed by stream systems which depend in part on the nature of the underlying rocks, including their arrangement, distribution, differential resistance to erosion, and even history.

The overall pattern developed by a system of streams and tributaries depends partly on the nature of the underlying rocks and partly on the history of the streams. Almost all streams follow a branching pattern in the sense that they receive tributaries; the tributaries, in turn, are joined by still smaller tributaries, but the manner of branching varies widely (see Figure 11.22).

A stream that resembles the branching habit of a maple, oak, or similar deciduous tree is called *dendritic,* "treelike." A dendritic pattern develops when the underlying bedrock is uniform in its resistance to erosion and exercises no control over the direction of valley growth. This situation occurs when the bedrock is composed either of flat-lying sedimentary rocks or massive igneous or metamorphic rocks. The streams can cut as easily in one place as another; thus the dendritic pattern is, in a sense, the result of the random orientation of the streams.

Another type of stream pattern is *radial:* Streams radiate outward in all directions from a high central zone. Such a pattern is likely to develop on the flanks of a newly formed volcano, where the streams and their valleys radiate outward and downward from various points around the cone.

A *rectangular* pattern occurs when the underlying bedrock is criss-crossed by fractures that form zones of weakness particularly vulnerable to erosion.

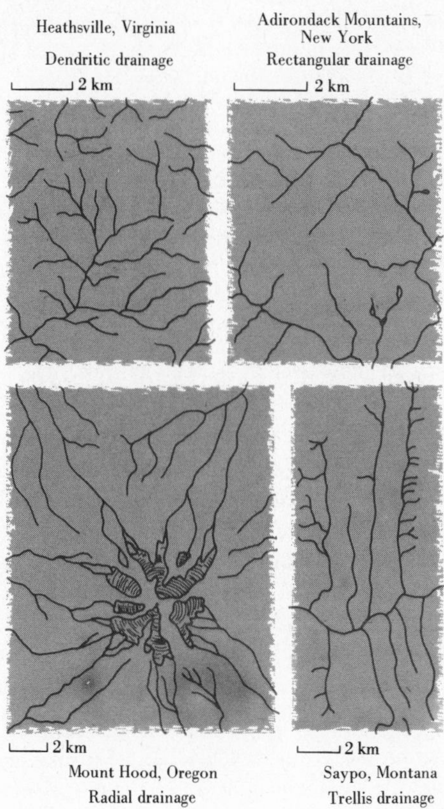

Figure 11.22 ***The overall pattern developed by a stream system depends in part on the nature of the bedrock and in part on the history of the area. See the text for discussion.***

The master stream and its tributaries then follow courses marked by nearly right-angle bends.

Some streams, particularly in a belt of the Appalachian Mountains running from New York to Alabama, follow what is known as a *trellis* pattern. This pattern, like the rectangular one, is caused by zones in the bedrock that differ in their resistance to erosion. The trellis pattern usually, though not always, indicates that the region is underlain by alternate bands of resistant and nonresistant rock.

Particularly intriguing features of some valleys are short, narrow segments walled by steep, rocky slopes or cliffs. These are called *water gaps.* The river, in effect, flows through a narrow notch in a ridge or mountain that lies across the course of the river, as shown in Figure 11.23.

Here is one way water gaps can form. Picture an area of hills and valleys carved by differential erosion from rocks of varying resistance. Then imagine sediments covering this landscape so that the valleys and hills are buried

Figure 11.23 ***An aerial photograph of a water gap in North Africa. The master stream cuts through the ridge of resistant rock.*** (*Photograph by U.S. Army Air Force.*)

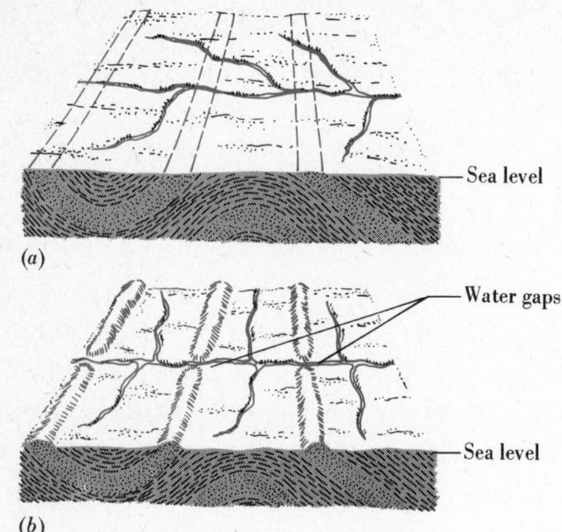

Figure 11.24 ***One way in which water gaps are thought to form is shown here. (a) A countryside underlain by folded rocks of different resistance to erosion has been beveled down to a surface of low relief, a peneplain. Across this country a stream, unaffected by underlying folds, is shown to flow. (b) The land has been uplifted, allowing the streams to cut downward. Erosion by the main stream has managed to keep pace with the uplift and the stream has cut through both resistant and nonresistant rock and passes through the resistant beds in water gaps. The tributary streams, unable to erode as rapidly as the main stream, have taken new courses along the less resistant beds and now enter the main stream nearly at right angles, thus producing a trellis pattern.***

beneath the debris. Any streams that flow over the surface of this cover will establish their own courses across the buried hills and valleys. When these streams erode downward, some of them may encounter an old ridge crest. If such a stream has enough erosional energy, it may cut down through the resistant rock of the hill. The course of the stream is thus superimposed across the old hill, and we call the stream a *superimposed stream.* Differential erosion may excavate the sedimentary fill from the old valleys, but the main, superimposed stream flows through the hill in a new, narrow gorge or water gap. An example of such a gap is that followed by the Big Horn River through the northern end of the Big Horn Mountains in Montana.

Water gaps may form in another way, and many of the rivers in the Appalachian Mountains provide examples. In this instance we visualize a countryside in which originally flat-lying rocks have been folded. Erosion may then bevel these deformed beds so that the surface is underlain by bands of rocks of differing resistance to erosion. If the erosion has proceeded far enough, the area may be reduced to a very flat surface, which we shall call a *peneplain* (see Section 11.8). At this stage the river may flow on its flood plain across the underlying rocks without regard for the hard and soft rocks beneath and probably will have a dendritic pattern. Later uplift of the area will cause the streams to cut downward. As the land rises, the main stream may be able to maintain its old course and to cut its valley into the bands of both resistant and nonresistant rock athwart its course. Tributary streams will etch out the more easily eroded materials and adjust their course to avoid the tougher rock of the emerging ridges. These ridges will be cut by the master stream and the notches thus formed will be water gaps (see Figure 11.24). Among the better known ones are the Delaware Water Gap, formed by the Delaware River along the boundary of northwestern New Jersey and northeastern Pennsylvania, and the water gaps of the Susquehanna River just west of Harrisburg, Pennsylvania.

A stream does not maintain a constant course through time, and one of the most interesting shifts in stream direction comes about as the result of *stream piracy,* or *stream capture.* In this process, one stream actually steals the headwater portions of a neighboring stream in the following manner. If one of two streams in adjacent valleys is able to deepen its valley more rapidly than the other, it may also extend its valley headward until it breaches the divide between the streams. When this happens, the more rapidly eroding stream captures the upper portion of the neighboring stream. The capturing

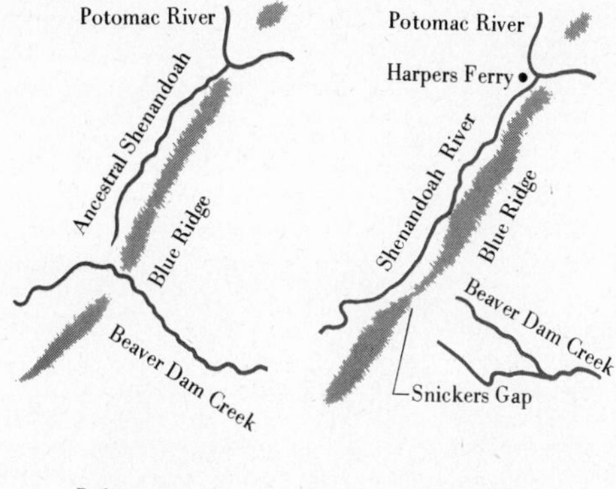

Figure 11.25 ***The Shenandoah River has captured the upper reaches of Beaver Dam Creek. The Shenandoah deepened its valley more rapidly than did the Beaver Dam and eventually breached the divide between them. The upper section of the Beaver Dam was thus diverted to the Shenandoah. Snickers is the abandoned water gap of the old Beaver Dam Creek through the Blue Ridge. The general name for such an abandoned water gap is "wind gap."***

stream is called the *pirate stream,* and the stream that has lost its upper section is called the *beheaded stream.*

Figure 11.25 diagrams this activity. At some time in the past, two streams, the Potomac River and Beaver Dam Creek, flowed through the Blue Ridge, each in its own water gap. The ancestral Shenandoah River joined the Potomac, as it does today, just before the Potomac entered its gap at Harpers Ferry. As time passed, the Shenandoah deepened its valley more rapidly than did Beaver Dam Creek and eventually extended itself headward through the divide, separating its valley from that of Beaver Dam Creek. Now the waters of the Upper Beaver Dam no longer flowed through the gap in the Blue Ridge but were diverted into the Shenandoah River. Here the Shenandoah is the pirate stream, and the shortened Beaver Dam Creek is the beheaded stream. At the time of capture, the old water gap of Beaver Dam Creek was abandoned and is now locally known as Snickers Gap. The general term, however, for an abandoned water gap such as this is *wind gap.*

Enlargement of valleys

We cannot say with assurance how running water first fashioned the great valleys and drainage basins of the continents, for the record has been lost in time. But we do know that certain processes are now at work in widening and deepening valleys, and it seems safe to assume that they also operated in the past.

If a stream were left to itself in its attempt to reach base level, it would erode its bed straight downward, forming a vertically walled chasm in the process. But because the stream is not the only agent at work in valley formation, the walls of most valleys slope upward and outward from the valley floor. In time, the cliffs of even the steepest gorge will be angled away from the axis of its valley.

As a stream cuts downward and lowers its channel into the land surface, weathering, slope wash, and mass movement come into play, constantly wearing away the valley walls, pushing them farther back (see Figure 11.26). Under the influence of gravity, material is carried down from the valley walls and dumped into the stream, to be moved onward toward the seas. The result is a valley whose walls flare outward and upward from the stream in a typical cross-valley profile (see Figure 11.27).

The rate at which valley walls are reduced and the angles that they assume depend on several factors. If the walls are made up of unconsolidated material that is vulnerable to erosion and mass movement, the rate will be rapid; but if the walls are composed of resistant rock, the rate of erosion will be very

Figure 11.26 **This gully in Wisconsin has been formed by the downward cutting of the stream combined with the slope wash and mass movement on the gully walls.** (*Photograph by U.S. Forest Service.*)

slow, and the walls may rise almost vertically from the valley floor (see Figure 11.28).

In addition to cutting downward into its channel, a stream also cuts from side to side, or laterally, into its banks. In the early stages of valley enlargement, when the stream is still far above its base level, downward erosion is dominant. Later, as the stream cuts its channel closer and closer to base level, downward erosion becomes progressively less important. Now a larger percentage of the stream's energy is directed toward eroding its banks. As the

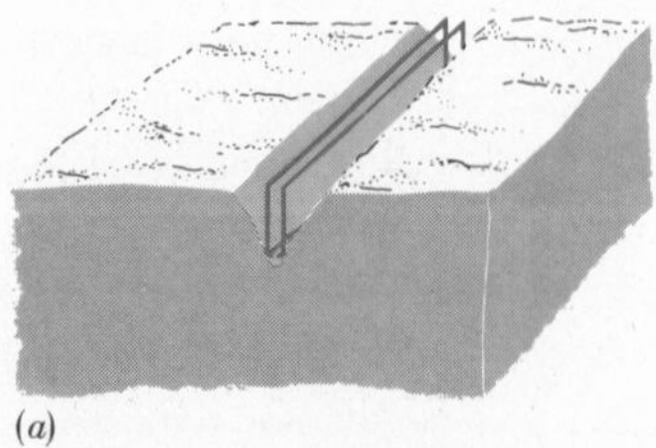

(*a*)

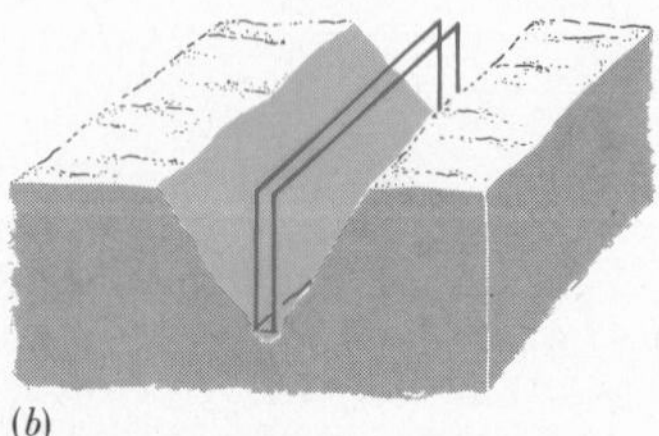

(*b*)

Figure 11.27 ***If a stream of water were the only agent in valley formation, we might expect a vertically walled valley no wider than the stream channel, as suggested by the colored lines in (a) and (b). Mass movement and slope wash, however, are constantly wearing away the valley walls, carving slopes that flare upward and away from the stream channel, as shown in the diagrams.***

Figure 11.28 ***Erosion by a small stream has fashioned vertically walled Labyrinth Canyon, Utah, in beds of massive sandstone. The stream is the major agent in formation of the valley. Mass movement and other agents of slope modification have yet to lower the angle of valley walls. Compare Figure 11.27.*** *(Photograph by William C. Bradley.)*

stream swings back and forth, it forms an ever-widening flood plain on the valley floor, and the valley itself becomes broader and broader.

How fast do these processes of valley formation proceed? In some instances we can measure the ages of gully formation in years and even months, but for large valleys we must be satisfied with rough approximations—and usually with plain guesses. For instance, one place where we can make an approximation of the rate of valley formation is the Grand Canyon of the Colorado in Arizona. We know that the Colorado River, downstream from the canyon, cuts sedimentary rocks of the late Miocene age and that therefore the river (and its canyon) must be younger than late Miocene. The Miocene epoch, now thought to have lasted for 19 million years, ended between 6.5 and 7.5 million years ago. Therefore, cutting of the canyon can be presumed to have begun sometime around this date. We also have information on when the canyon approached its present depth. This is deduced from the presence of a basaltic lava flow which partially blocked the canyon with a dam 165 m high. The age of the flow, based on potassium–argon dating, is about 1.2 million years. Since the eruption, the river has removed the lava dam and incised itself another 15 m into the older rocks beneath the lava.

From these facts we know that it has taken 1.2 million years for the river to cut down 180 m. This is a rate of 15 cm/1,000 years, which is the rate at which the Colorado River is now reducing its drainage basin (see Table 6.2). Before the lava flow occurred, the canyon had cut down about 1,600 m and had pushed back its valley walls some 6 to 20 km. If the cutting proceeded at the same rate as it did after the flow, then canyon formation would have begun some 10 million years earlier. It is difficult to say how close this figure is to reality, but it certainly seems reasonable to believe that it is the correct order of magnitude.

We shall return to the progressive development of a valley later, when we consider the cycle of erosion. First, let us examine some specific features of valleys.

Features of narrow valleys

WATERFALLS AND RAPIDS Waterfalls are among the most fascinating spectacles of the landscape. Thunderous and powerful as they are, however, they are actually short-lived features in the history of a stream. They owe their existence to some sudden drop in the river's longitudinal profile—a drop that may be eliminated with the passing of time.

Waterfalls are caused by many different conditions. Niagara Falls, for instance, is held up by a relatively resistant bed of dolomite underlain by beds of nonresistant shale (see Figure 11.29). This shale is easily undermined by

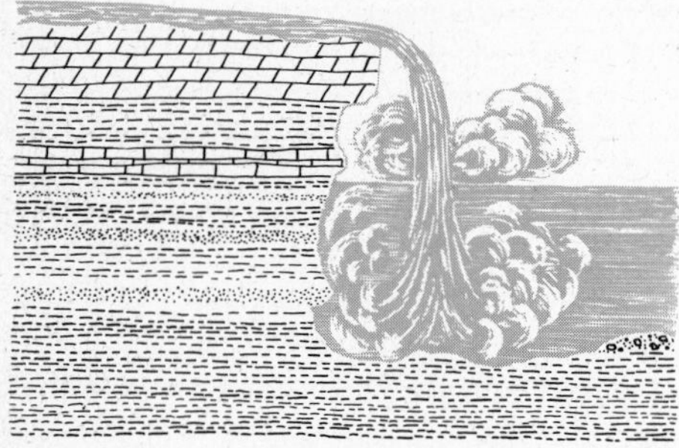

Figure 11.29 ***Niagara Falls tumbles over a bed of dolomite, underlain chiefly by shale. As the less resistant bed is eroded, the undermined ledge of dolomite breaks off, and the lip of the falls retreats.*** (*Redrawn from G. K. Gilbert,* Niagara Falls and Their History, *p. 213, American Book Company, New York, 1896.*)

the swirling waters of the Niagara River as they plunge over the lip of the falls. When the undermining has progressed far enough, the dolomite collapses and tumbles to the base of the falls. The same process is repeated over and over again as time passes, and the falls slowly retreat upstream. Historical records suggest that the Horseshoe or Canadian Falls (by far the larger of the two falls at Niagara) have been retreating at a rate of 1.2 to 1.5 m/yr, whereas the smaller American Falls have been eroded away 5 to 6 cm/yr. The 11 km of gorge between the foot of the falls and Lake Ontario are evidence of the headward retreat of the falls through time.

Yosemite Falls in Yosemite National Park, California, plunge 770 m over the Upper Falls, down an intermediate zone of cascades, and then over the Lower Falls. The falls leap from the mouth of a small valley high above the main valley of the Yosemite. The Upper Falls alone measure 430 m, nine times the height of Niagara. During the ice age, glaciers scoured the main valley much deeper than they did the unglaciated side valley. Then, when the glacier ice melted, the main valley was left far below its tributary, which now joins it after a drop of nearly half a mile (see Section 13.2).

The Victoria Falls of Africa's Zambezi River were caused by yet another set of conditions. Here the Zambezi drops over a resistant bed of lava 90 m to the gorge below. The gorge itself follows a zigzag course excavated by the retreating falls along zones of weakness in the rock. These zones are the result of intersecting fractures in the beds of lava.

Rapids, like waterfalls, occur at a sudden drop in the stream channel. Although rapids do not plunge straight down as waterfalls do, the underlying cause of formation is often the same. In fact, many rapids have developed directly from preexisting waterfalls (see Figure 11.30).

POTHOLES As the bedrock channel of a stream is eroded away, *potholes* sometimes develop (see Figure 11.31). These are deep holes, circular to elliptical in outline, and a few centimeters to meters in depth. They are most often observed in the stream channel during low water or along the bedrock walls, where they have been left stranded after the stream has cut its channel downward. Potholes are most common in narrow valleys, but they often occur in broad valleys as well.

A pothole begins as a shallow depression in the bedrock channel of a stream. Then, as the swirling, turbulent water drives sand, pebbles, and even cobbles round and round the depression, the continued abrasion wears the pothole ever deeper, as if the bedrock were being bored by a giant drill (see Figure

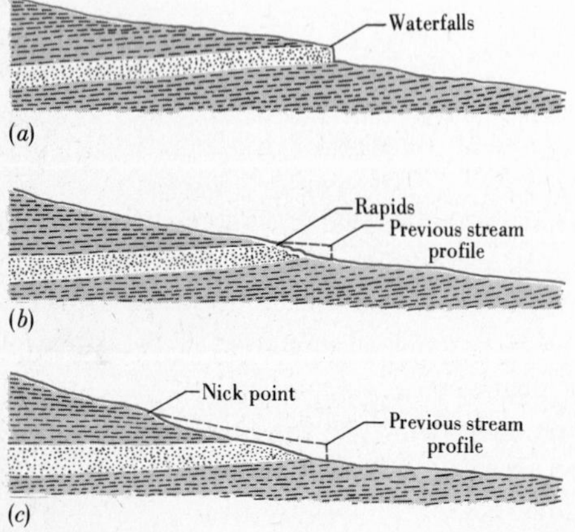

Figure 11.30 ***Rapids may represent a stage in the destruction of waterfalls, as suggested in this diagram.***

Figure 11.31 ***The rock walls of this stream are marked by a series of dissected potholes. Coarse sand and gravel caught in the eddies of turbulent water served as the cutting materials that carved the holes.*** *(Photograph by Sheldon Judson.)*

Figure 11.32 ***A cluster of pebbles in a pothole exposed at the low-water stage in the bed of the Blackstone River, Blackstone, Massachusetts. Pencil shows scale.*** *(Photograph by W. C. Alden, U.S. Geological Survey.)*

11.32). The initial depression may be caused by ordinary abrasion, by some irregularity in the bedrock, or by cavitation.

Features of broad valleys

If conditions permit, the various agents working toward valley enlargement ultimately produce a broad valley with a wide level floor. During periods of normal or low water, the river running through the valley is confined to its channel; but during high water, it overflows its banks and spreads out over the flood plain.

A flood plain that has been created by the lateral erosion and the gradual retreat of the valley walls is called an *erosional flood plain* and is characterized by a thin cover of gravel, sand, and silt a few meters or a few tens of meters in thickness. On the other hand, the floors of many broad valleys are underlain by deposits of gravel, sand, and silt scores of meters thick. These thick deposits are laid down as changing conditions force the river to drop its load across the valley floor. Such a flood plain, formed by the building up of the valley floor, or aggradation, is called a *flood plain of aggradation.* Flood plains of aggradation are much more common than erosional flood plains and are found in the lower reaches of the Mississippi, Nile, Rhône, and Yellow Rivers, to name but a few.

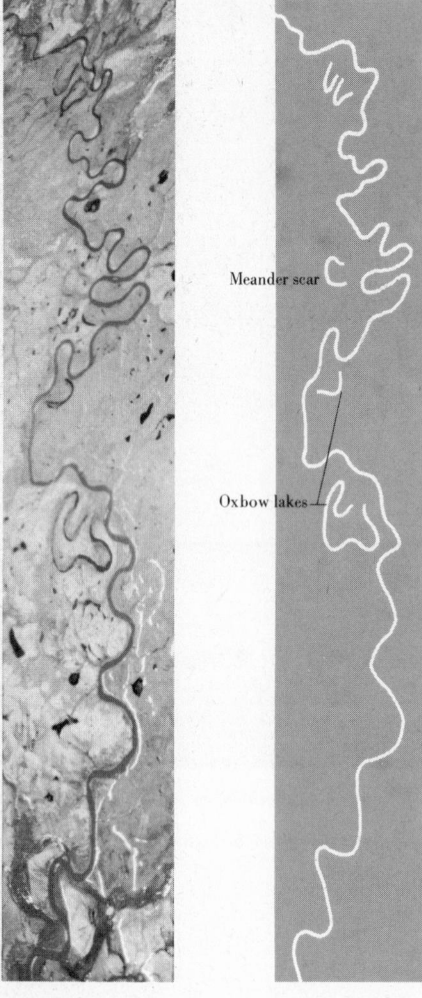

Figure 11.33 ***Aerial photograph and a drawing of meanders in an Alaskan stream. Note the oxbow lakes.*** *(Photograph by U.S. Army Air Force.)*

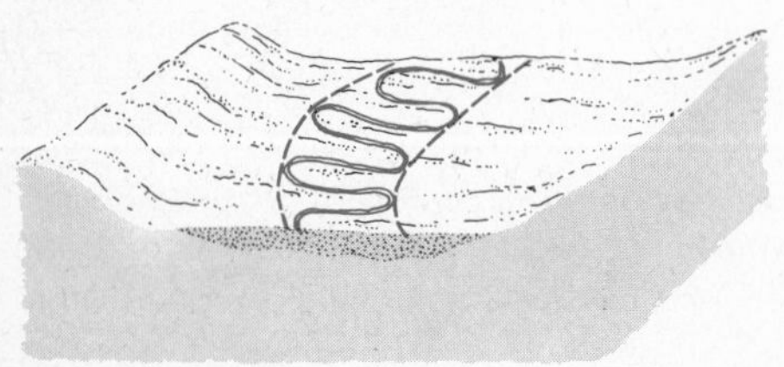

Figure 11.34 ***The meander belt is the portion of the flood plain that encloses a meandering stream.***

Both erosional flood plains and flood plains of aggradation exhibit the following characteristic features.

MEANDERS The channel of the Menderes River in Asia Minor curves back on itself in a series of broad hairpin bends. In fact, the very name of the river is derived from the Greek *maiandros*, "a bend." Today, all such bends are called *meanders*, and the zone along a valley floor that encloses a meandering river is called a *meander belt* (see Figures 11.33 and 11.34).

Both erosion and deposition are involved in the formation of a meander. First, some obstruction swings the current of a stream against one of the banks, and then the current is deflected over to the opposite bank. Erosion takes place on the outside of each bend, where the turbulence is greatest. The material detached from the banks is moved downstream, there to be deposited in zones of decreased turbulence—either along the center of the channel or on the inside of the next bend. As the river swings from side to side, the meander continues to grow by erosion on the outside of the bends and by deposition on the inside. Growth ceases when the meander reaches a critical size, a size that increases with an increase in the size of the stream.

Because a meander is eroded more on its downstream side than on its upstream side, the entire bend tends to move slowly down-valley. This movement is not uniform, however, and under certain conditions the downstream sweep of a series of meanders is distorted into cutoffs, meander scars, and oxbow lakes.

In its down-valley migration, a meander sometimes runs into a stretch of land that is relatively more resistant to erosion. But the next meander upstream continues to move right along, and gradually the neck between them is narrowed. Finally, the river cuts a new, shorter channel, called a *neck cutoff*, across the neck. The abandoned meander is called an *oxbow* because of its characteristic shape. Usually both ends of the oxbow are gradually silted in, and the old meander becomes completely isolated from the new channel. If the abandoned meander fills up with water, an *oxbow lake* results. Although a cutoff will eliminate a particular meander, the stream's tendency toward meandering still exists, and before long the entire process begins to repeat itself.

We found that a meander grows and migrates by erosion on the outside of the bend and by deposition on the inside. This deposition on the inside leaves behind a series of low ridges and troughs. Swamps often form in the troughs, and during flood time the river may develop an alternate channel through one of the troughs. Such a channel is called a *chute cutoff*, or simply a chute (see Figures 11.35 and 11.36).

The meandering river demonstrates a unity in ways other than the balance of erosion and deposition. The length of a meander, for example, is proportional to the width of the river, and this is true regardless of the size of the river. It holds for channels a few meters wide as well as those as large as that of

Figure 11.35 **This oxbow lake near Weslaco, Texas, was once a part of the meandering Rio Grande, which lies in the distance. An old bend in the river was cut through at the neck, became isolated from the river, and filled with water.** (*Photograph by Standard Oil Co. of New Jersey.*)

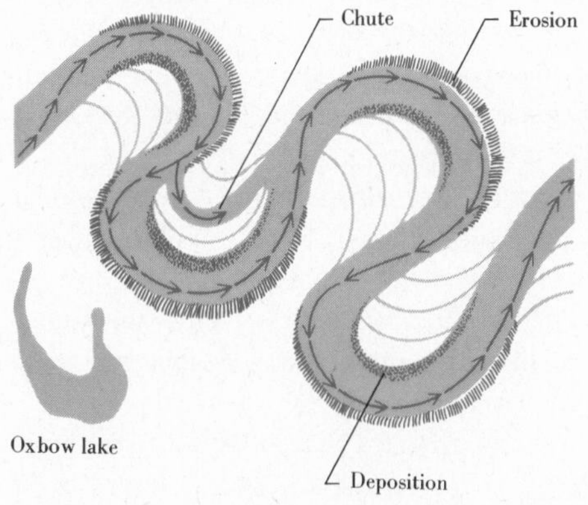

Figure 11.36 **Erosion takes place on the outside of a meander bend, whereas deposition is most marked on the inside. If the neck of a meander is eroded through, an oxbow forms. A chute originates along the inside of a meander where irregular deposition creates ridges and troughs as the meander migrates.**

the Mississippi River. As shown in Figure 11.37, this principle also is true of the Gulf Stream, even though this "river" is unconfined by solid banks. A similar relationship holds between the length of the meander and the radius of curvature of the meander (Figure 11.37). These relationships seem to be controlled largely by the discharge and the sediments carried. Increasing discharge causes an increase in the size of characteristics such as width, length, and wave length. For a given discharge, the wave length tends to be greater for streams carrying a high proportion of sand and gravel in its load than for those transporting mainly fine sands.

BRAIDED STREAMS On some flood plains, particularly where large amounts of debris are dropped rapidly, a stream may build up a complex tangle of converging and diverging channels separated by sand bars or islands. A stream of this sort is called a *braided stream.* The pattern most generally occurs when the discharge is highly variable and the banks easily eroded to supply a heavy load. It is characteristic of alluvial fans, glacial outwash deposits, and certain heavily laden rivers.

In general, the gradient of a braided stream is higher than that for a meandering stream of the same discharge, as indicated in Figure 11.38. This is apparently indicative of a tendency of a stream to increase its efficiency in order to transport proportionally larger loads. Thus, if a stream with a single channel divides into two channels, the new channels will have a higher gradient and a greater total width than the original channel carrying the same amount of water, and turbulence increases. The two new channels can now carry a greater load, particularly bed load, than could the single-channeled stream.

NATURAL LEVEES In many flood plains, the water surface of the stream is held above the level of the valley floor by banks of sand and silt known as *natural*

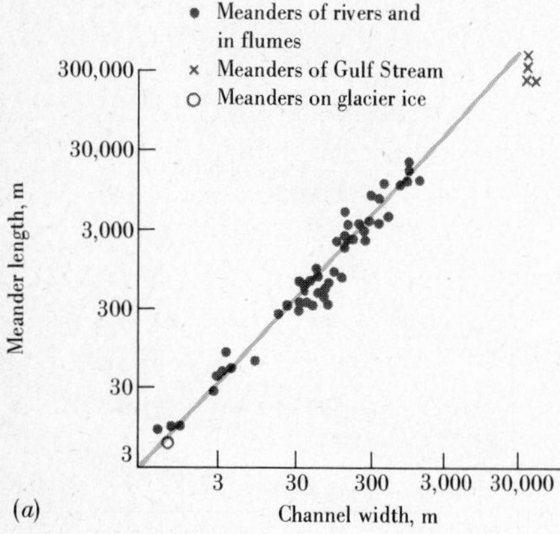

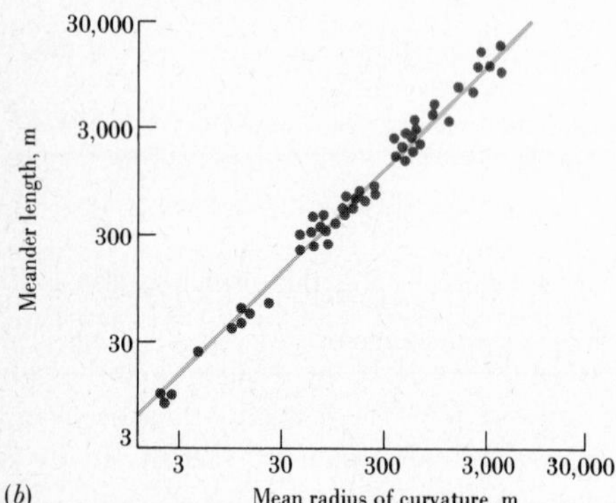

Figure 11.37 **(a) *Length of the meander increases with the widening meandering stream;* (b) *a similar orderly relationship exists between length of the meander and the mean radius of curvature of the meander.*** (*Redrawn from Luna B. Leopold and M. Gordon Wolman, "River Meanders,"* Geol. Soc. Am. Bull., *p. 773, 1960.*)

levees, a name derived from the French verb *lever,* "to raise." These banks slope gently, almost imperceptibly, away from their crest along the river toward the valley wall. The low-lying flood plain adjoining a natural levee may contain marshy areas known as *back swamps.* Levees are built up during flood time, when the water spills over the river banks onto the flood plain. Because the muddy water rising over the stream bank is no longer confined by the channel, its velocity and turbulence drop immediately, and much of the suspended load is deposited close to the river; but some is carried farther along, to be deposited across the flood plain. The deposit of one flood is a thin wedge tapering away from the river, but over many years, the cumulative effect produces a natural levee that is considerably higher beside the river bank than away from it (see Figure 11.39). On the Mississippi delta, for instance, the levees stand 5 to 6 m above the back swamps. Although natural levees tend to confine a stream within its channel, each time the levees are raised slightly, the bed of the river is also raised. In time, the level of the bed is raised above the level of the surrounding flood plain. If the river manages to escape from its confining walls during a flood, it will assume a new channel across the lowest parts of the flood plain toward the back swamps.

A tributary stream entering a river valley with high levees may be unable to find its way directly into the main channel, so it will flow down the back-swamp zone and may run parallel to the main stream for many miles before finding a place to enter. Because the Yazoo River typifies this situation by running 320 km parallel to the Mississippi, all rivers following similar courses are known as *yazoo-type* rivers.

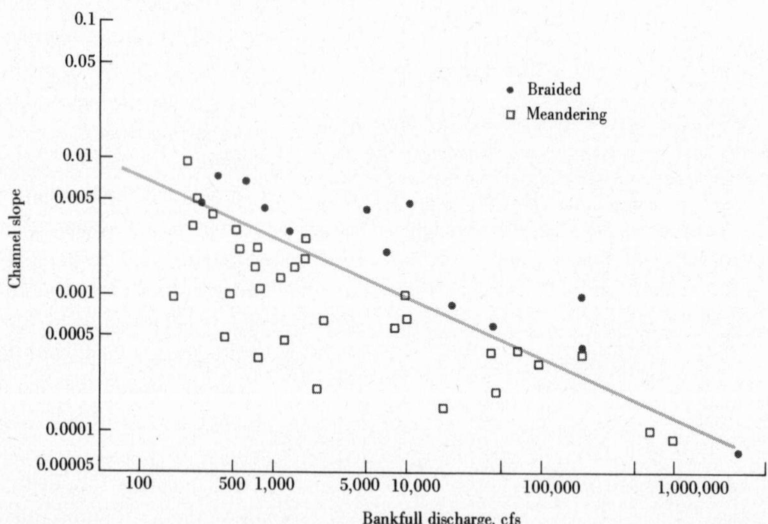

Figure 11.38 ***Meandering streams have lower gradients than do braided streams of the same discharge. In this diagram each symbol represents measurements at a single point on a stream.*** (*Adapted from Luna B. Leopold, M. Gordon Wolman, and John P. Miller,* Fluvial Processes in Geomorphology, *pp. 7–39, W. H. Freeman & Company, San Francisco, 1964.*)

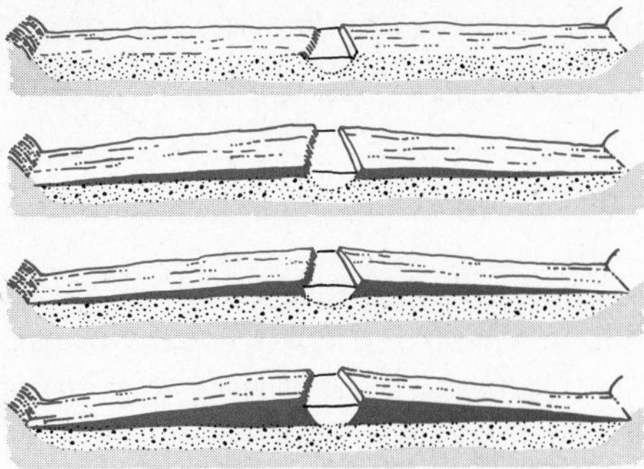

Figure 11.39 ***Natural levees, characteristic of many aggrading streams. These highly exaggerated diagrams show building up of wedge-shaped layers of silt that taper away from the stream banks toward valley walls. As the banks are built up, the floor of the stream channel also rises.***

FLOOD PLAINS AND THEIR DEPOSITS Virtually all streams are bordered by features called flood plains. These may range from very narrow (a few meters wide) to the flood plains whose widths are measured in kilometers, as is that of the lower Mississippi River.

The flood plain appears very flat, particularly when viewed in relation to the slopes of the valley walls that flank it. But the plain is not without its ups and downs. A large flood plain may have differences of relief of several meters and be marked by such features as natural levees, meander scars, and oxbow lakes.

The flood plain is made up of stream-carried sediments. Two general categories of deposits are common. One of these is made up of fine silts, clays, and sands that may be spread across the plain by a river which overflows its banks during flood. These are called *overbank deposits.* The other group of deposits is made of coarse material, gravel and sand, and is related directly to the channel of the stream. These deposits include chiefly the material that is deposited in the slack water on the inside of the bends of a winding or a meandering river. These bars of sand and gravel are called *point bars.*

We found, in discussing meanders, that the loops of a channel migrate laterally so that the river swings both across the valley and down-valley. As the river migrates laterally, it leaves behind coarse deposits in point bars and in the deeper portions of the abandoned channel. The rate of this channel migration varies widely, and studies by M. Gordon Wolman and Luna B. Leopold indicate that it can range from essentially no movement per year to as much as 750 m/yr along the Kosi River in India.

While erosion takes place on one bank of the river, deposition of sediments from upstream takes place on the point bars and, during flood, by overbank deposits—the river builds its flood plain in some places and simultaneously destroys it in others. Thus the flood plain becomes a temporary storage place for river sediments. How long a particular particle will, on the average, remain in the flood plain at any one spot varies widely with the flood plains of individual rivers. Again, studies by Wolman and Leopold on Seneca Creek, a small stream in Maryland, indicate that after a particle is entrapped in the flood plain it will stay there for over 1,000 years before it is put in motion again. Along the Little Missouri River in western North Dakota, studies by B. L. Everitt have shown that approximately $\frac{1}{2}$ million m^3 are eroded from the flood plain each year along each mile of river. A similar amount is added by deposition along point bars and in overbank flooding. In this instance, the average particle, once deposited in the flood plain, will stay there about 50 years before being eroded again.

We can view the flood plain, then, as a depositional feature which is more or less in equilibrium. This equilibrium may be disturbed in several ways. The stream may lose some of its ability to erode and carry sediment. A net gain of deposits will occur and the flood plain will build up. On the other hand, the flow of water or the gradient may increase, or the supply of sediments may decrease, and there will be a net loss of sediments as erosion begins to destroy the previously developed flood plain.

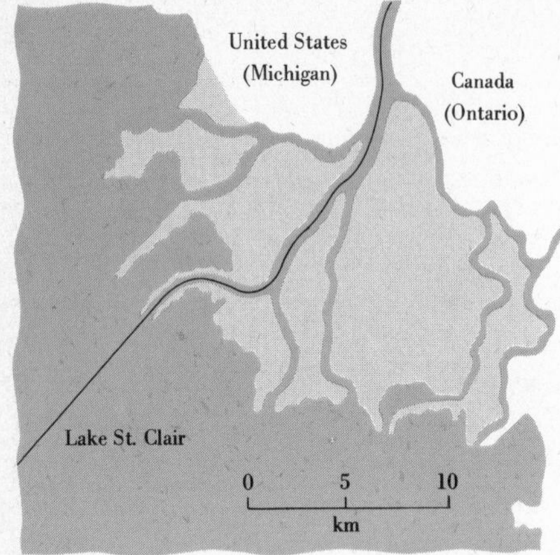

Figure 11.40 ***The delta of the St. Clair River in Lake St. Clair has the classic shape of a delta as well as its distributary channels.*** (*Redrawn from Leon J. Cole, "The Delta of the St. Clair River,"* Geol. Surv. Michigan, *vol. 9, part 1, Plate 1, 1903.*)

DELTAS AND ALLUVIAL FANS For centuries, the Nile River has been depositing sediments as it empties into the Mediterranean Sea, forming a great triangular plain with its apex upstream. This plain came to be called a *delta* because of the similarity of its shape to the Greek letter Δ (see Figures 11.40 and 11.41).

Whenever a stream flows into a body of standing water, such as a lake or an ocean, its velocity and transporting power are quickly stemmed. If it carries enough debris and if conditions in the body of standing water are favorable, a delta will gradually be built. An ideal delta is triangular in plan, with the apex pointed upstream and with the sediments arranged according to a definite pattern. The coarse material is dumped first, forming a series of dipping beds called *foreset beds.* But the finer material is swept farther along to settle across the sea or lake floor as *bottomset beds.* As the delta extends farther and farther out into the water body, the stream must extend its channel to the edge of the delta. As it does so, it covers the delta with *topset beds,* which lie across the top of the foreset beds (see Figure 11.42).

Figure 11.41 ***In this picture taken from*** **Gemini IV,** ***the delta of the Nile stands out because it is well vegetated and thus contrasts with the desert country that borders it.*** (*Photograph by NASA.*)

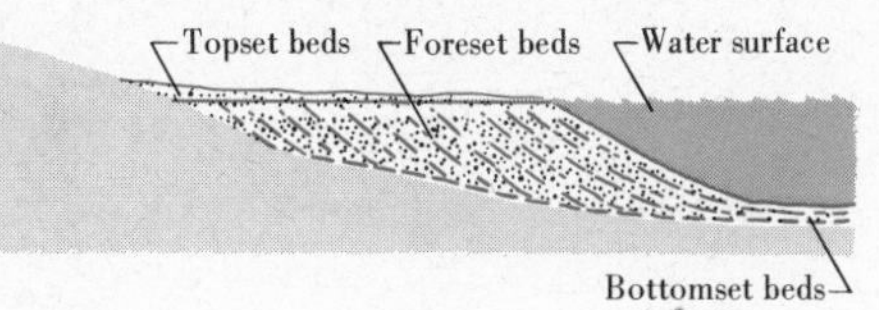

Figure 11.42 ***The ideal arrangement of sediments beneath a delta. Some of the material deposited in a water body is laid on the bottom of the lake or sea as bottomset beds. Other material is dumped in inclined foreset beds, built farther and farther into the body of water and partly covering the bottomset beds. Over the foreset beds the stream lays down topset beds.***

Very few deltas, however, show either the perfect delta shape or this regular sequence of sediments. Many factors, including lake and shore currents, varying rates of deposition, the settling of delta deposits as a result of their compaction, and the down-warping of the earth's crust, all conspire to modify the typical form and sequence.

Across the top of the delta deposits, the stream spreads seaward in a complex of channels radiating from the apex. These *distributary channels* shift their position from time to time as they seek more favorable gradients.

Deltas are characteristic of many of the larger rivers of the world, including the Nile, Mississippi, Ganges, Rhine, and Rhône. On the other hand, many rivers have no deltas, either because the deposited material is swept away as soon as it is dumped or because the streams do not carry enough detrital material to build up a delta.

An *alluvial fan* is the land counterpart of a delta. These fans are typical of arid and semiarid climates but they may form in almost any climate if conditions are right. A fan marks a sudden decrease in the carrying power of a stream as it descends from a steep gradient to a flatter one—for example,

Figure 11.43 ***An alluvial fan is the land counterpart of a delta. In this example in Death Valley, California, the streams flow only during the rare rains, carrying debris from the steep gulches along the cliff face. As the velocity of the streams is checked on the flat valley floor, material is deposited to form the alluvial fan.*** (*Photograph by Sheldon Judson.*)

Figure 11.44 **The Great Terrace of the Columbia River near Chelan, Washington.** (*Photograph by Austin Post, University of Washington.*)

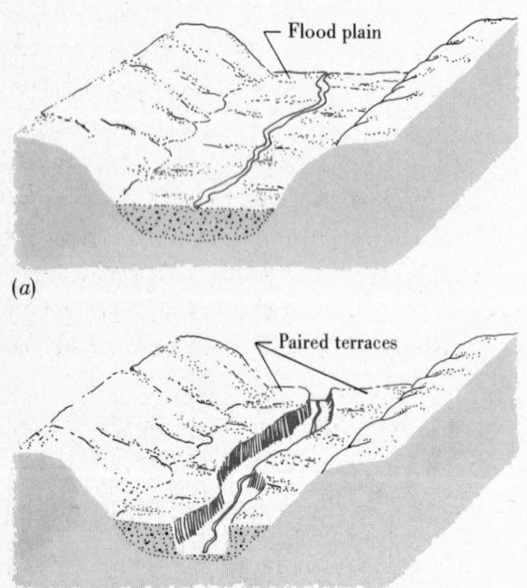

Figure 11.45 ***One example of the formation of paired terraces. In (a), the stream has partially filled its valley and has created a broad flood plain. In (b), some change in conditions has caused the stream to erode into its own deposits; the remnants of the old flood plain stand above the new river level as terraces of equal height. This particular example is referred to as a cut-and-fill terrace.***

when the stream flows down a steep mountain slope onto a plain. As the velocity is checked, the stream rapidly begins to dump its load. In the process, it builds up its channel, often with small natural levees along its banks. Eventually, as the levees continue to grow, the stream may flow above the general level. Then, during a time of flood, it seeks a lower level and shifts its channel to begin deposition elsewhere. As this process of shifting continues, an alluvial fan builds up (see Figure 11.43).

STREAM TERRACES A *stream terrace* is a relatively flat surface running along a valley, with a steep bank separating it either from the flood plain or from a lower terrace. It is a remnant of the former channel of a stream that now has cut its way down to a lower level (see Figure 11.44).

The so-called *cut-and-fill terrace* is created when a stream first clogs a valley with sediments and then cuts its way down to a lower level (see Figure 11.45). The initial aggradation may be caused by a change in climate that leads either to an increase in the stream's load or a decrease in its discharge. Or the base level of the stream may rise, reducing the gradient and causing deposition. In any event, the stream chokes the valley with sediment and the flood plain gradually rises. Now, if the equilibrium is upset and the stream begins to erode, it will cut a channel down through the deposits it has already laid down. The level of flow will be lower than the old flood plain, and at this lower level the stream will begin to carve out a new flood plain. As time passes, remnants of the old flood plain may be left standing on either side of the new one. Terraces that face each other across the stream at the same elevation are referred to as *paired terraces.*

Sometimes the downward erosion by streams creates *unpaired terraces* rather

Figure 11.46 ***Unpaired terraces do not match across the stream that separates them. Here is one way in which they may form. The stream has cut through unconsolidated deposits within the valley. As it eroded downward, it also swept laterally across the valley and created a sloping surface. Lateral migration was stopped locally when the stream encountered resistant bedrock (see arrows) beneath the softer valley fill. This bedrock not only deflected the stream back across the valley but also protected remnants of the valley fill from further stream erosion and allowed them to be preserved as terraces. Because they are portions of a surface sloping across the valley, however, no single remnant matches any other on the opposite side of the river.***

than paired ones. If the stream swings back and forth across the valley, slowly eroding as it moves, it may encounter resistant rock beneath the unconsolidated deposits. The exposed rock will then deflect the stream and prevent further erosion. A single terrace is left behind, with no corresponding terrace on the other side of the stream (see Figure 11.46).

Terraces, either paired or unpaired, may be cut into bedrock as well. A thin layer of sand and gravel usually rests on the beveled bedrock of these terraces.

11.7 *Evidence that a stream cuts its own valley*

So far, we have been assuming that every stream has created its own valley. True, we have seen that other processes, such as slope wash and mass movement, have helped in valley enlargement, but we have taken it for granted that they rely on the streams to transport the material they produce and to create and maintain slopes on which they can operate. This assumption has not always been accepted, however. In fact, during the early nineteenth century, geologists devoted a great deal of time to demonstrating that valleys were not great, original furrows in the earth's surface, along which the rivers flow merely for want of a better place to go. Today, we are confident that most valleys have been created by the streams that flow through them and by the processes that these streams have encouraged. Let us review some of the evidence for this belief.

Most of us have observed at first hand the direct results of running water during a heavy rain. For instance, we can watch the actual headward erosion of small gullies in miniature drainage basins. What we are observing here is really a small stream in the process of forming and extending its valley. Merely by dipping up a cupful of water we can see evidence of the load carried by the stream, and we know that this debris must have come from the drainage basin of the stream. Erosion is going on as we watch, and the valley is growing larger as the stream moves the eroded material down-valley.

We know from observing modern streams that abrasion wears and polishes the beds and banks of bedrock and sometimes drills potholes deep into the rock floor. When we find these marks of stream activity high above the channel of a modern stream, it is logical to assume that they were made by the stream when it was flowing at this higher level and that the stream has since cut downward and deepened its valley.

As we have seen, a terrace records the fact that a stream once flowed at a higher level. And when we find terraces covered with sand and gravel high above a modern stream, again it is reasonable to assume that the stream has cut its way down to a lower level.

Along the walls of many streams, we can observe beds of rocks formed from layers of sediments that were laid down in ancient seas, and we can match individual beds from one valley wall to the other. We know that these beds were laid down originally as continuous sheets and that some agent subsequently removed large sections of them. Barring evidence of crustal movement, we conclude that the stream flowing along the bottom of the valley has been responsible.

These and other arguments make it reasonably clear that a stream has created the valley through which it flows. No one argument or bit of evidence is final proof, of course. In fact, we know that some valleys were formed by earth movement (Chapter 18) and were deepened by glaciers (Chapter 13). Still, the weight of evidence points to the likelihood that most valleys have been formed by stream activity.

Figure 11.47 ***The canyon of the Yellowstone is typical of a youthful valley. The valley walls slope directly to the river banks to give it a V shape. The river rapids are also characteristic of a stream in a youthful valley.*** *(Photograph by National Park Service.)*

11.8 *Cycle of erosion*

From the beginning of the twentieth century, the concept that stream valleys, indeed entire regions, progress through a series of stages has influenced geologists in their attempts to explain the development of the landscape. This so-called *cycle of erosion* has been abused by over-enthusiastic advocates and bitterly attacked by critics. As in most violent controversies, the truth probably lies somewhere between the extremes.

The cycle of erosion does provide us with a qualitative description of river valleys and areas, but difficulties arise when we try to assign quantitative values to it. As a device that enables us to paint a word picture of a stream valley or a region, it has great value. And, just as important, the concept has led geologists into a quantitative investigation of the nature of streams and the development of landforms, an investigation that has only just begun.

The cycle of a stream valley

The cycle of a stream valley is divided into three main stages comparable to the ages of man: *youth, maturity,* and *old age.* Each of these stages has certain characteristics, but just as it is difficult to say at what moment a man passes from youth to maturity or from maturity to old age, so it is impossible to draw sharp lines between stages in the life of a stream valley.

Probably the most characteristic feature of a youthful stream valley is the stream's active and rapid erosion of its channel. At this stage, the valley walls come right down to (or almost to) the stream bank and form a V-shaped cross-valley profile. There is no flood plain, or only a very narrow one. The stream's gradient is steep, marked by falls and rapids, and tributaries tend to be few and small (see Figure 11.47).

In a mature valley, the rate of downward cutting slows. The gradient is smooth, and most of the falls and rapids have been eliminated. A flood plain begins to form as valley widening progresses more rapidly than valley deepening. Across this flood plain a stream may begin to meander. At this stage, the valley has reached its greatest depth (see Figure 11.48).

The distinction between maturity and old age is largely a matter of degree. In old age, valley widening, though slow, still dominates downward cutting, and the flood plain is wider than the meander belt. Oxbow lakes, meander scars, and natural levees are more common now than they were during the stream's maturity.

The factor that controls the progress of a stream valley from youth to old age is the base level. The closer a valley approaches base level, the further it has progressed through its aging cycle.

No one has ever observed a valley as it went through this complete cycle, of course, but there are valleys that represent different stages in the cycle.

Figure 11.48 ***A stream meanders across the flood plain of a small, mature valley near Steamboat Springs, Colorado.*** *(Photograph by Sheldon Judson.)*

And we sometimes find that a valley has reached different stages of development at different points along its course—maturity in its lower reaches, for example, and youth toward its headwaters.

INTERRUPTIONS IN THE CYCLE OF A STREAM VALLEY A stream valley moves smoothly through its life cycle only if the stream's base level remains constant. Any significant movement of the base level, either up or down, will interrupt the cycle. For example, if sea level falls, or if earth movements raise and warp the land, the stream will begin to cut its channel deeper and adjust its profile to the new base level. A youthful V-shaped valley, then, may develop within a broad, mature valley, setting off a new cycle of erosion. If the new base level remains constant for a long enough time, all evidence of the original cycle will be destroyed, and the valley will pass from youth to maturity to old age.

What happens when earth movements result in the uplifting of a mature stream valley? If the stream has progressed to the meandering stage before

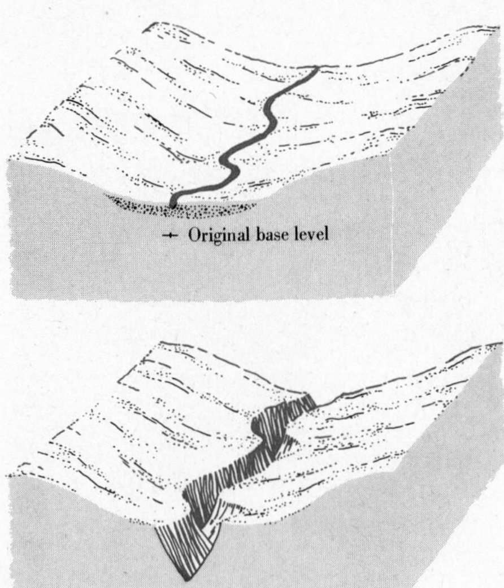

Figure 11.49 **The rejuvenation of a mature valley. In (a) a mature valley has formed in respect to the base level. In (b) the base level is pictured as having dropped. As a result, the stream has incised its channel, and a youthful valley has been formed within the older mature valley.**

uplift occurs, it may be able to incise its channel into the underlying bedrock in a pattern of *entrenched meanders.* In other words, the new valley will be youthful, but it will continue to follow the old pattern established by the stream before the change in base level occurred.

When a stream is forced by changing conditions to begin a new cycle of erosion, we say that it has been *rejuvenated* (see Figure 11.49).

Interruptions in the cycle of erosion may occur in still other ways. For example, if the base level rises, a river will deposit its load in an attempt to create a new profile. As a result, a youthful valley may become clogged with sediment and take on some of the features of old age. Such a valley, however, is not strictly within the old-age stage of the cycle of erosion, for its seeming age stems from deposition rather than erosion.

The cycle of a region

The concept of the cycle of erosion has been applied to the evolution of a whole region as well as to the evolution of a stream valley. And again the controlling factor is base level, although here we are dealing with a base level that includes a great many drainage basins. We can think of the base level of a region as a plane or a surface extending inland from the ocean and rising very gently beneath the major drainage basins. Erosion of the land works down toward this imaginary surface, though it is probably never reached.

As with a stream valley, so, too, the life cycle of a region progresses through youth, maturity, and old age. The regional cycle is thought to begin with the rapid uplift of a land mass, followed by a period of stability in which streams, mass movement, and slope wash combine to reduce the elevated mass ever lower toward the regional base level.

During the youthful stage of a region, its streams cut rapidly downward, and the difference in elevation between valley bottoms and hilltops (known as *relief*) increases rapidly. The streams have not yet extended into all portions of the land mass, and they are still separated by broad divides. Maturity begins as the streams and related agencies dissect the land more deeply, causing the broad divides of youth to disappear. With maturity, the region achieves its maximum relief, and the divides between the streams are narrowed until most of their surface is sloping (see Figure 11.50). The major streams have now adjusted their profiles to the conditions that exist along their courses, and they maintain this adjustment during the rest of the cycle. It is during maturity that the streams of a region etch out the greatest variety of land features from rocks of differing resistance and structure, molding valleys and lowlands in the areas underlain by the weakest rocks and hills, and leaving highlands in the areas underlain by the strongest rocks.

As a region passes slowly through maturity into old age, its streams wander freely across the wide valley, with little regard for differences in the resistance of the rocks that underlie the surface. The divides are now low and widely spaced. Theoretically the cycle progresses until the land has been reduced to a gently rolling plain very close to base level. Such a surface is termed a peneplain, from the Latin *paene,* "almost," with the English "plain." A

Figure 11.50 **Mature country near Death Valley, California. The land is intricately dissected and all in slope.** (*Photograph by Sheldon Judson.*)

few hills may rise above this surface, representing areas where erosion has not been able to reduce the original land mass completely. These residuals left unconsumed by erosion are called *monadnocks*, after Mt. Monadnock in New Hampshire, which has been interpreted as a residual standing above an ancient peneplain that has been uplifted and dissected. Monadnocks may survive either because they are composed of rock relatively resistant to erosion or because they stand in areas protected from the action of streams.

INTERRUPTIONS IN THE REGIONAL CYCLE A change in base level interrupts the cycle of erosion for a region just as it interrupts the cycle for a stream. If the land moves upward or the sea downward, a new cycle is set under way and the area is rejuvenated. Remnants of the earlier cycle will be preserved, however, until erosion in the new cycle succeeds in destroying them.

Objections to the cycle of erosion

This whole concept of the erosion cycle has been subjected to numerous criticisms—many of them well grounded. For instance, the cycle of erosion as originally conceived and as we have described it is the result of erosion in a humid climate. Yet humid climates cover only a fraction of the earth's surface. The erosion processes of other climates would presumably produce a somewhat different sequence of events. And we can ask whether the cycle of erosion as we have pictured it ever takes place in a desert climate (see Chapter 14), either hot or cold. Or we can ask what the cycle of erosion would be like in the humid tropics or in the steppes.

Also, the cycle of erosion is postulated as starting with rapid uplift, followed by a period of no movement during which the uplifted land mass progresses from youth through maturity to old age. But suppose the uplift were slow or intermittent and took place over many millions of years. Would the cycle of erosion then proceed in the manner we have described or in some other way?

If the cycle of erosion for a region carries through to completion—that is, to the formation of a peneplain—we should have examples of peneplains in the world today. Yet we have no examples of large areas that have been graded to the level of the modern seas. The nearest thing we have to a convincing example are surfaces of low relief that are now buried beneath younger rocks—as, for example, deep within the Grand Canyon or along the eastern seaboard. But even with these examples we are still in doubt, because we do not yet know enough about the nature of these ancient, buried surfaces.

These criticisms and others do throw into question certain details of the cycle of erosion, but the concept still serves the valuable purpose of indicating that streams and land masses go through a progressive development. Furthermore, it is very useful to have terms such as *youth, maturity,* and *old age* in describing the general nature of a valley or a region.

Outline

Running water has fashioned over 400 million km of stream channels, down which the bulk of the material eroded from the continents moves toward the oceans.

The *hydrologic cycle* is the path of water circulation from oceans to atmosphere to land to oceans.

Precipitation and *stream flow* (*runoff*) are related as follows:

Runoff = *precipitation* − (*infiltration* + *evaporation* and *transpiration*).

Flow is either *laminar* or *turbulent,* but turbulent is the general rule.

Water in a stream has *velocity,* flows down a *gradient,* and is measured in terms of *discharge.*

The *economy of a stream* is part of a dynamic equilibrium that characterizes earth processes.

It is clear that in a stream *discharge* = *width* × *depth* × *velocity*. The change of one characteristic will affect one or more of the others. In adjusting to an increased discharge downstream, the gradient decreases, and width, depth, and velocity increase.

Base level of a stream is the point below which it cannot erode. Lowering of the base level produces erosion and raising of the base level produces deposition.

Work of running water includes transportation, erosion, and deposition.

Transportation by a stream involves its *load* (the amount it carries at any one time, its *capacity* (the total amount it can carry under given conditions), and its *competence* (the maximum-size particle it can move under given conditions). Material is moved in solution, in suspension, and as bed load.

Erosion by a stream involves direct lifting of material, abrasion, impact, solution, and cavitation.

Deposition by a stream occurs with decreased velocity (hence decreased energy).

Features of valleys include those of the valley bottom, the drainage basin, and the river channel.

A cross-valley profile of a typical valley shows flood plain, valley walls, and divides.

A drainage basin is the area drained by a river and its tributaries. Individual streams and their valleys are joined in a definite geometric relationship.

The patterns developed by stream systems include *dendritic, radial, rectangular* and *trellis.* Some streams form *water gaps.* Some extend themselves by *piracy* of neighboring streams.

Enlargement of valleys is accomplished through downward and lateral erosion by the stream and by mass movement and water erosion on the valley walls.

Narrow valleys are marked by water falls, rapids, potholes, and steep gradients.

Broad valleys have meanders, braided streams, natural levees, flood plains, deltas, alluvial fans, and stream terraces.

Usually a *stream cuts the valley in which it flows,* as suggested by the load carried by the stream, the marks of stream erosion on bedrock, river terraces, and layers of rock which can be matched across the valley.

The *cycle of erosion* is the path through which stream valleys and regions are pictured as progressing. *Youth, maturity,* and *old age* are the three main stages. Interruptions in the cycle are caused by changes in base level.

Supplementary readings

DAVIS, WILLIAM MORRIS: "The Geographical Cycle," *Geograph. J.*, vol. 14, pp. 481–504, 1899. The classic statement of youth, maturity, and old age in the cycle of erosion.

DAVIS, WILLIAM MORRIS: "Base Level, Grade and Peneplain," *J. Geol.*, vol. 10, pp. 77–111, 1902. A basic paper on these three subjects.

HJULSTREM, FILIP: "Transportation of Detritus by Running Water," in *Recent Marine Sediments—A Symposium*, pp. 3–31, Parker D. Trask (ed.), American Association of Petroleum Geologists, Tulsa, 1939. A paper still worth referring to for one interested in the transport of sediments in a river.

HOYT, WILLIAM G., and WALTER B. LANGBEIN: *Floods*, Princeton University Press,

Princeton, N.J., 1955. A semipopular treatment of the subject.

KUENEN, P. H.: *Realms of Water,* John Wiley & Sons, Inc., New York, 1955. An inclusive discussion of the several forms of water and their role in geological processes.

LEOPOLD, LUNA B.: "Rivers," *Am. Scientist,* vol. 50, pp. 511–537, 1962. A brief summary of rivers as dynamic agents of the landscape.

LEOPOLD, LUNA B., M. GORDON WOLMAN, and JOHN P. MILLER: *Fluvial Processes in Geomorphology,* W. H. Freeman & Company, San Francisco, 1964. An advanced textbook on the subject.

LIVINGSTONE, DANIEL A.: "Data on Geochemistry," *U.S. Geol. Surv. Professional Papers,* no. 440-G, 1963. This paper documents the amount and nature of dissolved materials in the fresh waters of the world and is a basic reference for such information.

MORISAWA, MARIE: *Streams: Their Dynamics and Morphology,* McGraw-Hill Book Company, Inc., New York, 1968. A short book of intermediate level which is a readable summary of the hydrology and work of streams.

12

Underground water

TREMENDOUS QUANTITIES OF WATER LIE EXPOSED IN RIVERS, LAKES, OCEANS, and—in the solid state—in glaciers on the earth's surface. But beneath the surface, hidden from our sight, is another great store of water. In fact, in Table 11.1 we found that for the earth as a whole the water beneath the surface exceeded by more than 66 times the amount of water present in streams and in fresh water lakes. In the United States our streams each year discharge an estimated 30,000 km^3 of water into the oceans. In contrast, there lies beneath the surface of the country an estimated 7,575,000 km^3 of water. It is certainly true that underground reservoirs in the United States contain far more usable water than all our surface reservoirs and lakes combined, and we depend on this underground supply for about one-fifth of our total water needs.

12.1 *Basic distribution*

Underground water, subsurface water, and *subterranean water* are all general terms used to refer to water in the pore spaces, cracks, tubes, and crevices of the consolidated and unconsolidated material beneath our feet. The study of underground water is largely an investigation of these openings and of what happens to the water that finds its way into them. In Chapter 11, we found that most of the water beneath the surface comes from rain and snow that fall on the face of the earth. Part of this water soaks directly into the ground, and part of it drains away into lakes and streams and thence into the underground.

Zones of saturation and aeration

Some of the water that moves down from the surface is caught by rock and earth materials and is checked in its downward progress. The zone in which this water is held is known as the *zone of aeration,* and the water itself is called *suspended water.* The spaces between particles in this zone are filled partly with water and partly with air. Two forces operate to prevent suspended water from moving deeper into the earth: (1) the molecular attraction exerted on the water by the rock and earth materials and (2) the attraction exerted by the water particles on one another (see Figures 12.1 and 12.2).

The zone of aeration can be subdivided into three belts: (1) *belt of soil moisture,* (2) *intermediate belt,* and (3) *capillary fringe.* Some of the water that enters the belt of soil moisture from the surface is used by plants, and some is evaporated back into the atmosphere. But some water also passes down to the intermediate belt, where it may be held by molecular attraction (as suspended water). Little movement occurs in the intermediate belt, except when rain or melting snow sends a new wave of moisture down from above. In some areas, the intermediate belt is missing, and the belt of soil moisture lies directly above the third belt, the capillary fringe. Water rises into the capillary fringe from below, to a height ranging from a few cm to 2 or 3 m.

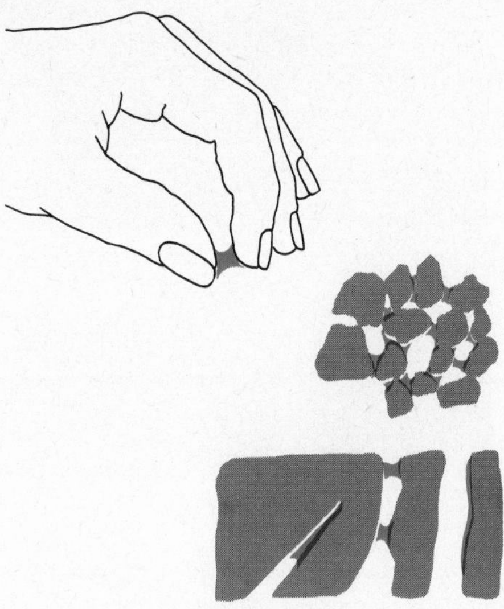

Figure 12.1 ***A drop of water held between two fingers illustrates the molecular attraction that holds suspended water in the zone of aeration. Surface tension of the water is great enough to prevent its downward movement to the zone of saturation.***

Beneath the zone of aeration lies the *zone of saturation.* Here the openings in the rock and earth materials are completely filled with *ground water,* and the surface between the zone of saturation and the zone of aeration is called the *ground-water table,* or simply the *water table.* The level of the water table fluctuates with variations in the supply of water coming down from the zone of aeration, with variations in the rate of discharge in the area and with variations in the amount of ground water drawn off by plants and human beings.

It is the water below the water table, within the zone of saturation, that we shall focus on in this chapter.

The water table

The water table is an irregular surface of contact between the zone of saturation and the zone of aeration. Below the water table lies the ground water; above it lies the suspended water. The thickness of the zone of aeration differs from one place to another, and the level of the water table fluctuates accordingly. In general, the water table tends to follow the irregularities of the ground surface, reaching its highest elevation beneath hills and its lowest elevation beneath valleys. Although the water table reflects variations in the ground surface, the irregularities in the water table are less pronounced.

In looking at the topography of the water table, let us consider an ideal situation. Figure 12.3 shows a hill underlain by completely homogeneous material. Assume that, initially, this material contains no water at all. Then a heavy rainfall comes along, and the water soaks slowly downward, filling the interstices at depth. In other words, a zone of saturation begins to develop. As more and more water seeps down, the upper limit of this zone continues to rise. The water table remains horizontal until it just reaches the level of the two valley bottoms on either side of the hill. Then, as additional water seeps down to the water table, some of it seeks an outlet into the valleys. But this added water is "supported" by the material through which it flows, and the water table is prevented from maintaining its flat surface. The water is slowed by the friction of its movement through the interstices and even, to some degree, by its own internal friction. Consequently, more and more water is piled up beneath the hill, and the water table begins to reflect the shape of the hill. The water flows away most rapidly along the steeper slope of the water table near the valleys and most slowly on its gentler slope beneath the hill crest.

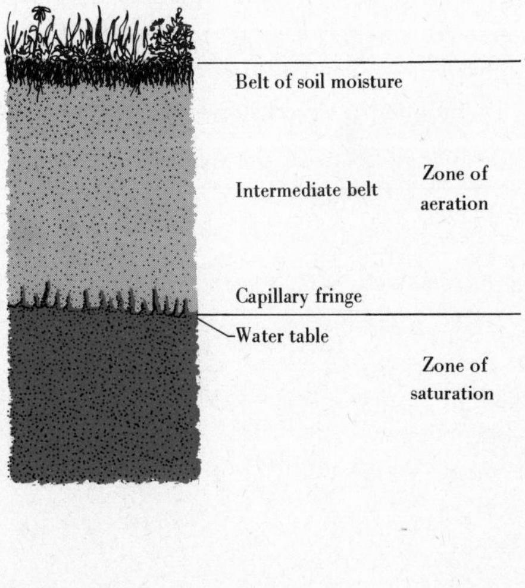

Figure 12.2 ***Underground water's two major zones: zone of aeration and zone of saturation. The water table marks the upper surface of the zone of saturation. Within the zone of aeration is a belt of soil moisture, the source of moisture for many plants. From here, also, some moisture is evaporated back to the atmosphere. In many instances, this belt lies above an intermediate belt where water is held by molecular attraction and little movement occurs except during periods of rain or melting snow. In the capillary fringe, just above the water table, water rises a few centimeters to better than a meter from the zone of saturation, depending on the size of the interstices.***

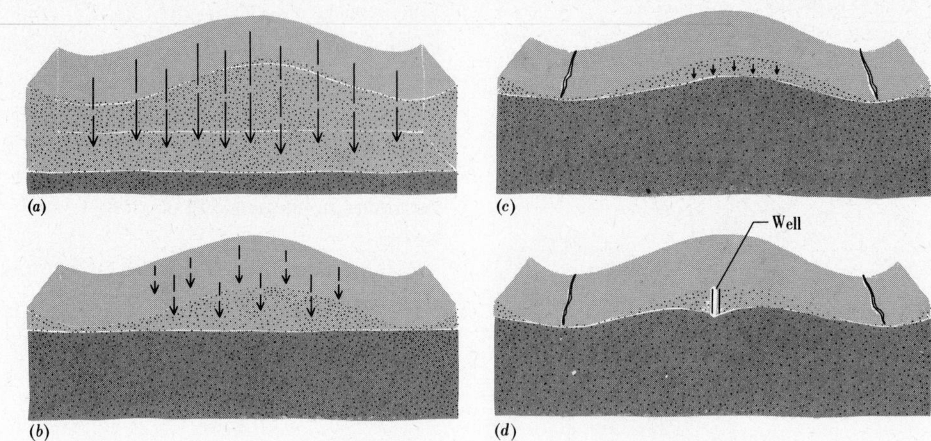

Figure 12.3 ***Ideally, the water table is a subdued reflection of the surface of the ground. In (a) and (b), the water table rises as a horizontal plane until it reaches the level of the valley bottoms on either side of the hill. Thereafter, as more water soaks into the ground, it seeks an outlet toward the valleys. Were the movement of the water not slowed down by the material making up the hill, it would remain essentially horizontal. The friction caused by the water's passing through the material (and even to some extent the internal friction of the water itself) results in a piling up of water beneath the hill; the bulge is highest beneath the crest and lowest toward the valleys (c). The shape of the water table may be altered by pumping water from a well (d). The water flows to this new outlet and forms a cone of depression.***

We can modify the shape of the ground-water surface by providing an artificial outlet for the water. For example, we can drill a well on the hill crest and extend it down into the saturated zone. Then, if we pumped out the ground water that flowed into the well, we would create a dimple in the water table. The more we pumped, the more pronounced the depression—called a *cone of depression*—would become.

Returning to our ideal situation, we find that if the supply of water from the surface were to be completely stopped, the water table under the hill would slowly flatten out as water discharged into the valleys. Eventually it would almost reach the level of the water table under the valley bottoms; then the flow would stop. This condition is common in desert areas where the rainfall is sparse.

12.2 *Movement of underground water*

The preceding chapter stated that the flow of water in surface streams could be measured in terms of so many feet per second. But in dealing with the flow of underground water, we must change our scale of measurement; for here, although the water does move, it usually moves very, very slowly. Therefore we find that centimeters or feet per day and, in some places, even centimeters or feet per year provide a better scale of measurement. The main reason for this slow rate of flow is that the water must travel through small, confined passages if it is to move at all. It will be worthwhile, then, for us to consider the porosity and permeability of earth materials.

Porosity

The *porosity* of a rock is measured by the percentage of its total volume that is occupied by voids or interstices. The more porous a rock is, the greater the amount of open space it contains. Through these open spaces, underground water must find its way.

Porosity differs from one material to another. What is the porosity of a rock made up of particles and grains derived from preexisting rocks? Here porosity is determined largely by the shape, size, and assortment of these rock-building units. A sand deposit composed of rounded quartz grains with fairly uniform size has a high porosity. But if mineral matter enters the deposit and cements the grains into a sandstone, the porosity is reduced by an amount equal to the volume of the cementing agent. A deposit of sand, poorly sorted with finer particles of silt and clay mixed in, has low porosity, because the smaller particles fill up much of the space between the larger particles.

Even a dense, massive rock such as granite may become porous as a result of fracturing. And a soluble massive rock such as limestone may have its original planes of weakness enlarged by solution.

Clearly then, the range of porosity in earth materials is extremely great. Recently deposited muds (called *slurries*) may hold up to 90 percent by volume of water, whereas unweathered igneous rocks such as granite, gabbro, or obsidian may hold only a fraction of 1 percent. Unconsolidated deposits of clay, silt, sand, and gravel have porosities ranging from about 20 to as much as 50 percent. But when these deposits have been consolidated into sedimentary rocks by cementation or compaction, their porosity is sharply reduced. Average porosity values for individual rock types have little meaning because of the extreme variations within each type. In general, however, a porosity of less than 5 percent is considered low; from 5 to 15 percent represents medium porosity; and over 15 percent is considered high.

Permeability

Whether or not we find a supply of fresh ground water in a given area depends on the ability of the earth materials to transmit water as well as on their ability to contain it. The ability to transmit underground water is termed *permeability*.

The rate at which a rock transmits water depends not only on its total porosity but also on the size of the interconnections between its openings. For example, although a clay may have a higher porosity than a sand, the particles that make up the clay are minute flakes and the interstices between them are very small. Therefore, water passes more readily through the sand than through the more porous clay simply because the molecular attraction on the water is much stronger in the tiny openings of the clay. The water moves more freely through the sand because the passageways between particles are relatively large and the molecular attraction on the water is relatively low. Of course, no matter how large the interstices of a material are, there must be connections between them if water is to pass through. If they are not interconnected, the material is impermeable.

A permeable material that actually carries underground water is called an *aquifer,* from the Latin for "water" and "to bear." Perhaps the most effective aquifers are unconsolidated sand and gravel, sandstone, and some limestones. The permeability of limestone is usually due to solution that has enlarged the fractures and bedding planes into open passageways. The fractured zones of some of the denser rocks such as granite, basalt, and gabbro also act as aquifers, although the permeability of such zones decreases rapidly with depth (see discussion of joints in Chapter 18). Clay, shale, and most metamorphic and crystalline igneous rocks are generally poor aquifers.

Because the flow of underground water is usually very slow, it is largely laminar; in contrast, the flow of surface water is largely turbulent. There is one exception, however: the turbulent flow of water in large underground passageways formed in such rocks as cavernous limestone.

In laminar flow, the water near the walls of interstices is presumably held motionless by the molecular attraction of the walls. Water particles farther away from the walls move more rapidly, in smooth, threadlike patterns, for the resistance to motion decreases toward the center of an opening. The most rapid flow is reached at the very center.

The energy that causes underground water to flow is derived from gravity. Gravity draws water downward to the water table; from there it flows through the ground to a point of discharge in a stream, lake, or spring. Just as surface water needs a slope to flow on, so must there be a slope for the flow of ground water. This is the slope of the water table, the *hydraulic gradient.* It is measured by dividing the length of flow (from the point of intake to the point of discharge) into the vertical distance between these two points, a distance called *head.* Therefore, hydraulic gradient is expressed as h/l, where h is head and l is length of flow from intake to discharge. Thus, if h is 10 m and l is 100 m, the hydraulic gradient is 0.1 or 10 percent.

An equation to express the rate of water movement through a rock was proposed by the French engineer Henri Darcy in 1856. What is now known as Darcy's law is essentially the same as his original equation. The law may be expressed as follows:

$$V = P\frac{h}{l}$$

where V is velocity, h head, l the length of flow, and P a coefficient of permeability that depends on the nature of the rock in question. But because h/l is simply a way of expressing the hydraulic gradient, we may say that in a rock of constant permeability the velocity of water will increase as the hydraulic gradient increases. Remembering that the hydraulic gradient and

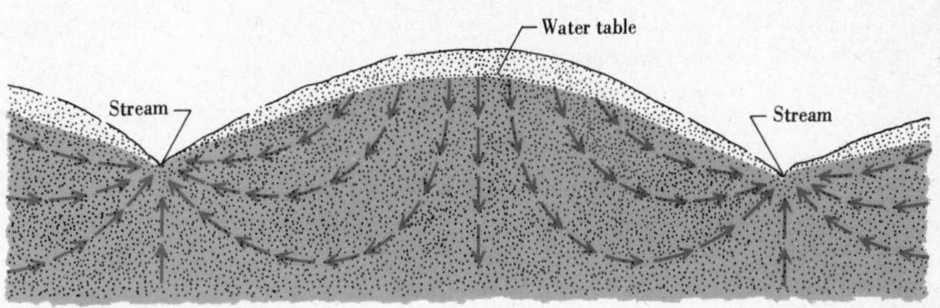

Figure 12.4 **The flow of ground water through uniformly permeable material is suggested here. Movement is not primarily along the groundwater table; rather, particles of water define broadly looping paths that converge toward the outlet and may approach it from below.** (*Redrawn from M. King Hubbert, "The Theory of Ground-Water Motion,"* J. Geol., *vol. 48, p. 930, 1940.*)

the slope of the groundwater table are the same thing, we may also say that the velocity of ground water varies with the slope of the water table. Other things being equal, the steeper the slope of the water table, the more rapid the flow. In ordinary aquifers, the rate of water flow has been estimated as not faster than 1.5 m/day and not slower than 1.5 m/year, though rates of over 120 m/day and as low as a few centimeters per year have been recorded.

The movement of underground water down the slope of the water table is only a part of the picture, for the water is also in motion at depth. Water moves downward from the water table in broad looping curves toward some effective discharge agency, such as a stream, as suggested in Figure 12.4. The water feeds into the stream from all possible directions, including straight up through the bottom of the channel. We can explain this curving path as a compromise between the force of gravity and the tendency of water to flow laterally in the direction of the slope of the water table. This tendency toward lateral flow is actually the result of the movement of water toward an area of lower pressure, the stream channel in Figure 12.4. The resulting movement is neither directly downward nor directly toward the channel but is, rather, along curving paths to the stream.

12.3 Ground water in nature

So far, we have assumed that ground water is free to move on indefinitely through a uniformly permeable material of unlimited extent. Actually, subsurface conditions fall far short of this ideal situation. Some layers of rock material are more permeable than others, and the water tends to move rapidly through these beds in a direction more or less parallel to bedding planes. Even in a rock that is essentially homogeneous, the ground water tends to move in some preferred direction.

Simple springs and wells

Underground water moves freely downward from the surface until it reaches an impermeable layer of rock or until it arrives at the water table. Then it begins to move laterally. Sooner or later it may flow out again at the surface of the ground in an opening called a *spring*.

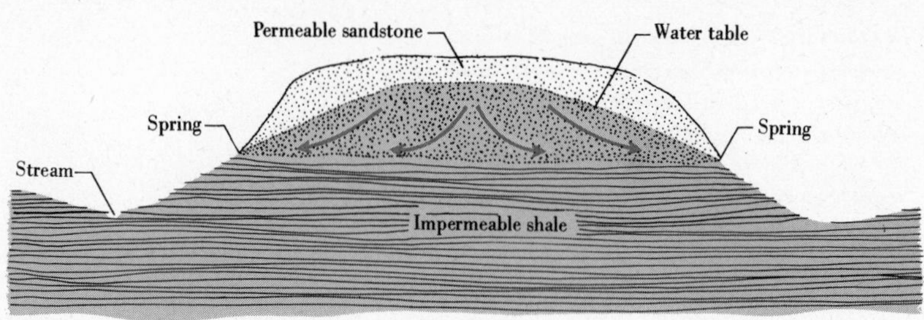

Figure 12.5 ***Nature seldom, if ever, provides uniformly permeable material. In this diagram, a hill is capped by permeable sandstone and overlies impermeable shale. Water soaking into the sandstone from the surface is diverted laterally by the impermeable beds. Springs result where the water table intersects the surface at the contact of the shale and sandstone.***

Springs have attracted the attention of men throughout history. In early days they were regarded with superstitious awe and were sometimes selected as sites for temples and oracles. To this day, many people feel that spring water possesses special medicinal and therapeutic values. Water from "mineral springs" contains salts in solution that were picked up by the water as it percolated through the ground. The same water pumped up out of a well would be regarded merely as hard and not very desirable for general purposes.

Springs range from intermittent flows that disappear when the water table recedes during a dry season, through pint-sized trickles, to an effluence of 3.8 billion liters daily, the abundant discharge of springs along a 16-km stretch of Fall River, California.

This wide variety of spring types is the result of underground conditions that vary greatly from one place to another. As a general rule, however, a spring results wherever the flow of ground water is diverted to a discharge zone at the surface (see Figure 12.5). For example, a hill made up largely of permeable rock may contain a zone of impermeable material, as shown in Figure 12.6. Some of the water percolating downward will be blocked by this impermeable rock, and a small saturated zone will be built up. Because the local water level here is actually above the main water table, it is called a *perched water table.* The water that flows laterally along this impermeable rock may emerge at the surface as a spring. Springs are not confined to points where a perched water table flows from the surface, and it is clear that if the main water table intersects the surface along a slope, then a spring will form.

Even in impermeable rocks, permeable zones may develop as a result of fractures or solution channels. If these openings fill with water and are intersected by the ground surface, the water will issue forth as a spring.

A spring is the result of a natural intersection of the ground surface and

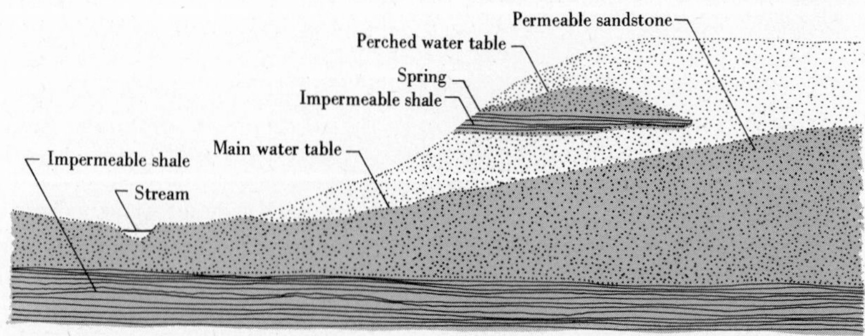

Figure 12.6 ***A perched water table results when ground water collects over an impermeable zone and is separated from the main water table.***

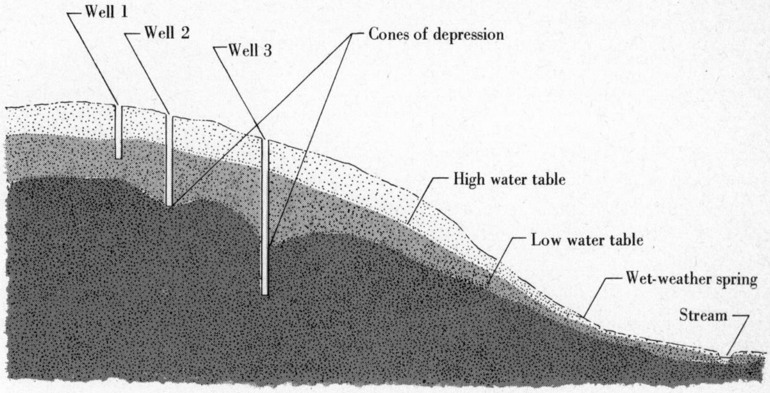

Figure 12.7 ***To provide a reliable source, a well must penetrate deep into the zone of saturation. In this diagram, well 1 reaches only deep enough to tap the ground water during periods of high water table; a seasonal drop of this surface will dry up the well. Well 2 reaches to the low water table, but continued pumping may produce a cone of depression that will reduce effective flow. Well 3 is deep enough to produce reliable amounts of water, even with continued pumping during low water-table stages.***

the water table, but a well is an artificial opening cut down from the surface into the zone of saturation. A well is productive only if it is drilled into permeable rock and penetrates below the water table. The greater the demands that are made on a well, the deeper it must be drilled below the water table. Continuous pumping creates the cone of depression previously described, which distorts the water table and may reduce the flow of ground water into the well (see Figure 12.7). Wells drilled into fractured crystalline rock, such as granite, may produce a good supply of water at relatively shallow depths, but we cannot increase the yield of such wells appreciably by deepening them, because the number and size of the fractures commonly decrease the farther down we go (see Figure 12.8).

Wells drilled into limestone that has been riddled by large solution passages may yield a heavy flow of water part of the time and no flow the rest of the time simply because the water runs out rapidly through the large openings. Furthermore, water soaking down from the surface flows rapidly through limestone of this sort and may make its way to a well in a very short time. Consequently, water drawn from the well may be contaminated because there has not been enough time for impurities to be filtered out as the water passes from the surface to the well. In sandstone, on the other hand, the rate of flow is slow enough to permit the elimination of impurities even within a very short distance of underground flow. Harmful bacteria are destroyed in part by entrapment, in part by lack of food and by temperature changes, and in part by hostile substances or organisms encountered along the way, particularly in the soil.

Figure 12.8 ***Wells may produce water from a fractured zone of impermeable rocks such as granite. The supply, however, is likely to be limited, not only because the size and number of fractures decrease with depth but also because the fractures do not interconnect.***

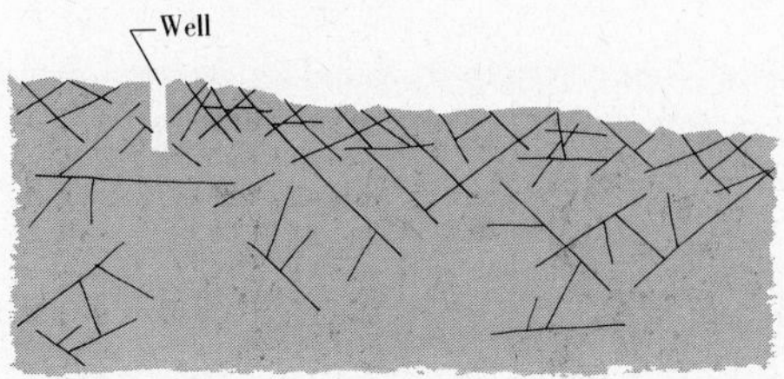

By applying Darcy's law, we can estimate the amount of water that a well will probably yield. The quantity of water passing through a given section in a unit of time is determined by the area of the section it passes through and by the velocity of the flow. Therefore,

$$V = Q/A$$

where V is velocity, Q the quantity of water per unit time, and A the area of the cross section through which the water flows. For h/l in the original equation for Darcy's law, we substitute the symbol i to represent hydraulic

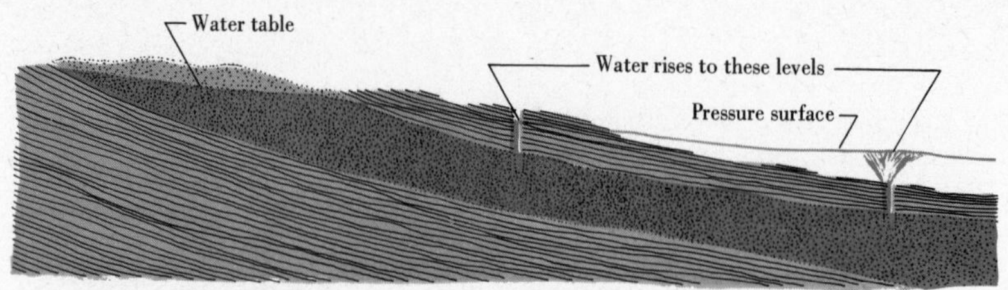

Figure 12.9 ***The wells in the diagram meet the conditions that characterize an artesian system: (1) an inclined aquifer, (2) capped by an impermeable layer, (3) with water prevented from escaping either downward or laterally, and (4) sufficient head to force the water above the aquifer wherever it is tapped. In the well at the right, the head is great enough to force water out at the surface.***

gradient. We then have a statement of Darcy's law that gives quantity of water in terms of permeability, hydraulic gradient, and area of cross section:

$$V = P\frac{h}{l} = Pi = Q/A \qquad \text{or} \qquad Q = PiA$$

Therefore, we can calculate the rate of discharge for a given well if we know the permeability of the bed, P, the hydraulic gradient, i, and the cross section through which the water passes, A; the last figure is taken as the area of the well wall receiving water from the surrounding rock. In actual practice, the most difficult problem in applying this formula is determining the hydraulic gradient.

Artesian water

Contrary to common opinion, artesian water does not necessarily come from great depths. But other definite conditions characterize an artesian water system: (1) the water is contained in a permeable layer, the aquifer, inclined so that one end is exposed to receive water at the surface; (2) the aquifer is capped by an impermeable layer; (3) the water in the aquifer is prevented from escaping either downward or off to the sides; and (4) there is enough head to force the water above the aquifer wherever it is tapped. If the head is great enough, the water will flow out to the surface either as a well or a spring (see Figure 12.9). The term *artesian* is derived from the name of a French town, Artois (originally called *Artesium* by the Romans), where this type of well was first studied.

A classic example of an artesian water system is found in western South Dakota, where the Black Hills have punched up through a series of sedimentary rocks, bending their edges up to the surface. One of these sedimentary rocks, a permeable sandstone of the Cretaceous age, carries water readily and is sandwiched between impermeable layers. Water entering the sandstone around the Black Hills moves underground eastward across the state, reaching greater and greater depths as it travels along. When we drive a well into this aquifer, we tap water that is under the pressure exerted by all the water piled up between the well and the Black Hills.

Increasing demands on this source of supply have been exceeding the replenishment rate for years. During a 35-year period the pressure fell off so sharply that the level of water in a well near Pierre, on the Missouri River, dropped nearly 100 m. From 1920 to 1935, the rate of decline was almost 2 m/year at the town of Chamberlain.

Thermal springs

Springs that bring warm or hot water to the surface are called *thermal springs, hot springs,* or *warm springs* (see Figure 12.10). A spring is usually regarded as a thermal spring if the temperature of its water is about 6.5°C higher than the mean temperature of the air. There are over 1,000 thermal springs in the western mountain regions of the United States, 46 in the Appalachian Highlands of the east, 6 in the Ouachita area in Arkansas, and 3 in the Black Hills of South Dakota.

Figure 12.10 ***Mammoth Hot Springs, Yellowstone National Park, are thermal springs which have built these terraces by depositing travertine.*** (*Photograph by U.S. Geological Survey.*)

Most of the western thermal springs derive their heat from masses of magma that have pushed their way into the crust almost to the surface and are now cooling. In the eastern group, however, the circulation of the ground water carries it to depths great enough for it to be warmed by the normal increase in earth heat (see "Thermal gradient" in the Glossary).

The well-known spring at Warm Springs, Georgia, is heated in just this way. Long before the Civil War, this spring was used as a health and bathing resort by the people of the region. Then, with the establishment of the Georgia Warm Springs Foundation for the treatment of victims of infantile paralysis, the facilities were greatly improved. Rain falling on Pine Mountain, about 2 miles south of Warm Springs, enters a rock formation known as the Hollis. At the start of its journey downward, the average temperature of the water is about 16.5°C. It percolates through the Hollis formation northward under Warm Springs at a depth of around 100 m and then follows the rock as it plunges into the earth to a depth of 1,140 m, 1 mile farther north. Normal rock temperatures in the region increase about 1.8°C/100 m of depth, and the water is warmed as it descends along the bottom of the Hollis bed. At 1,140 m, the bed has been broken and shoved against an impervious layer that turns the water back. This water is now hotter than the water coming down from above, and it moves upward along the top of the Hollis formation, cooling somewhat as it goes. Finally, it emerges in a spring at a temperature of 36.5°C.

Less than 1 mile away is Cold Spring, whose water comes from the same rainfall on Pine Mountain. A freak of circulation, however, causes the water at Cold Spring to emerge before it can be conducted to the depths and warmed. Its temperature is only about 16.5°C.

Figure 12.11 ***Old Faithful in Yellowstone National Park is America's most widely known geyser. The periodic eruption of a geyser is due to the particular pattern of its plumbing and its proximity to a liberal source of heat and ground water.*** *(Photograph by Barton W. Knapp.)*

Geysers

A *geyser* is a special type of thermal spring that ejects water intermittently with considerable force. The word *geyser* comes from the name of a spring of this type in Iceland, *geysir,* probably based on the verb *geysa,* "to rush furiously."

Although the details of geyser action are still not understood, we do know that, in general, a geyser's behavior is caused by the arrangement of its plumbing and the proximity of a good supply of heat. Here is probably what happens. Ground water moving downward from the surface fills a natural pipe, or conduit, that opens upward to the surface. Hot igneous rocks, or the gases given off by such rocks, gradually heat the column of water in the pipe and raise its temperature toward the boiling point. Now, the higher the pressure on water, the higher its boiling point, and because water toward the bottom of the pipe is under the greatest pressure, it must be heated to a higher temperature than the water above before it will come to a boil. Eventually the column of water becomes so hot that either a slight increase in temperature or a slight decrease in pressure will cause it to boil. At this critical point, the water near the base of the pipe is heated to the boiling point. The water then passes to steam and, as it does so, expands, pushing the water above

Figure 12.12 ***Thermal springs and small geysers characterize the West Thumb area of Yellowstone National Park.*** *(Photograph by Sheldon Judson.)*

it toward the surface. But this raising of the heated column of water reduces the pressure acting on it, and it, too, begins to boil. The energy thus developed throws the water and steam high into the air, producing the spectacular action characteristic of many geysers. After the eruption has spent itself, the pipe is again filled with water and the whole process begins anew.

We can compare this theoretical cycle with that of Old Faithful Geyser in Yellowstone National Park (see Figure 12.11). The first indication of a coming eruption at Old Faithful is the quiet flow of water in a fountain some 1 to 2 m high. This preliminary activity lasts for a few seconds and then subsides. It represents the first upward push of the column of water described in our theoretical case. This push reduces the pressure and thereby lowers the boiling point of the water in the pipe. Consequently, the water passes to steam, and less than 1 min after the preliminary fountain, the first of the violent eruptions takes place. Steam and boiling water are thrown 45 to 50 m into the air. The entire display lasts about 4 min. Emptied by the eruption, the tube then gradually refills with ground water, the water is heated, and in approximately 1 hr the same cycle is repeated. The actual time between eruptions at Old Faithful averages about 65 min but may be from 30 to 90 min.

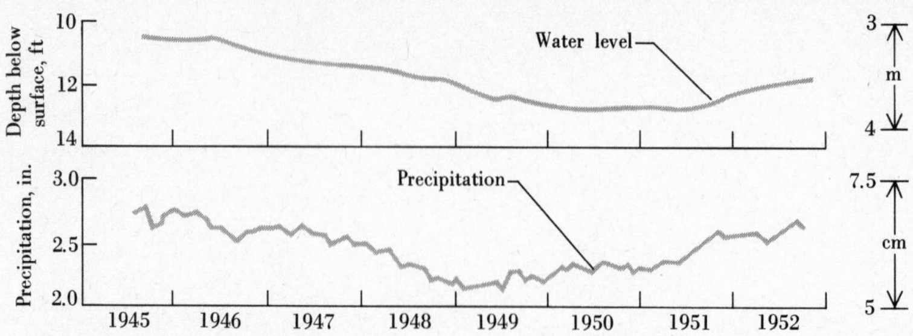

Figure 12.13 **Relationship between the water level in an observation well near Antigo, Wisconsin, and precipitation, as shown by records from 1945 to 1952. The water table reflects the changes in precipitation. The graphs represent 3-year running monthly averages. For example, 5.8 cm of precipitation for May means that precipitation averaged 5.8 cm/month from May 1947 to May 1950, inclusive.** (*From A. H. Harder and William J. Drescher, "Ground Water Conditions in Southwestern Langlade County, Wisconsin,"* U.S. Geol. Surv. Water Supply Paper, *no. 1294, 1954.*)

12.4 *Recharge of ground water*

As we have seen, the ultimate source of most underground water is precipitation that finds its way below the surface either through natural or artificial means.

Some of the water from precipitation seeps into the ground, reaches the zone of saturation, and raises the water table. Continuous measurements over long periods of time at many places throughout the United States show an intimate connection between water level and rainfall (see Figure 12.13). Because water moves relatively slowly in the zone of aeration and the zone of saturation, fluctuations in the water table usually lag a little behind fluctuations in rainfall.

Several factors control the amount of water that actually reaches the zone of saturation. For example, rain that falls during the growing season must first replenish moisture used up by plants or passed off through evaporation. If these demands are great, very little water will find its way down to recharge the zone of saturation. Then, too, during a very rapid, heavy rainfall, most of the water may run off directly into the streams instead of soaking down into the ground. A slow, steady rain is much more effective than a heavy, violent rain in replenishing the supply of ground water. High slopes, lack of vegetation, or the presence of impermeable rock near the surface may promote runoff and reduce the amount of water that reaches the zone of saturation. It is true, however, that some streams are themselves sources for the recharge of underground water. Water from the streams leaks into the zone of saturation, sometimes through a zone of aeration (see Figure 12.14).

Figure 12.14 **The ground water may be recharged by water from a surface stream leaking into the underground.**

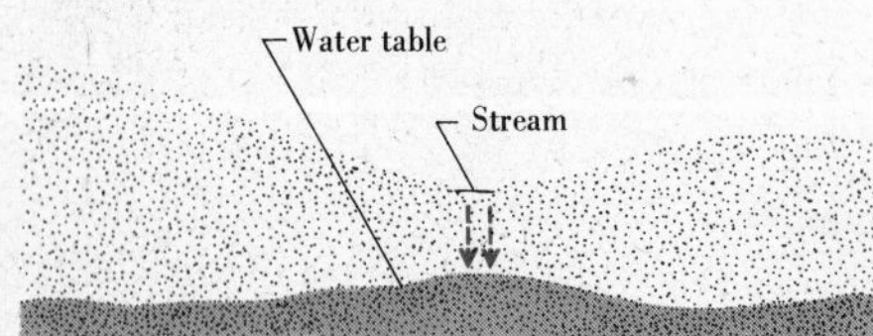

In many localities, the natural recharge of the underground supplies cannot keep pace with man's demands for ground water. Consequently, attempts are sometimes made to recharge these supplies artificially. On Long Island, New York, for example, water that has been pumped out for air-conditioning purposes is returned to the ground through special recharging wells or, in winter, through idle wells that are used in summer for air conditioning. In the San Fernando Valley, California, surplus water from the Owens Valley aqueduct is fed into the underground in an attempt to keep the local water table at a high level.

12.5 *Caves and related features*

Caves are probably the most spectacular examples of the handiwork of underground water. In dissolving great quantities of solid rock in its downward course, the water fashions large rooms, galleries, and underground stream systems as the years pass. In many caves, the water deposits calcium carbonate as it drips off the ceilings and walls, building up fantastic shapes of a material known as *dripstone* (see Figures 12.15 and 12.16).

Figure 12.15 **Stalactites grow downward, some to meet stalagmites growing upward from the cave floor. Carlsbad Caverns, New Mexico.** (*Photograph by National Park Service.*)

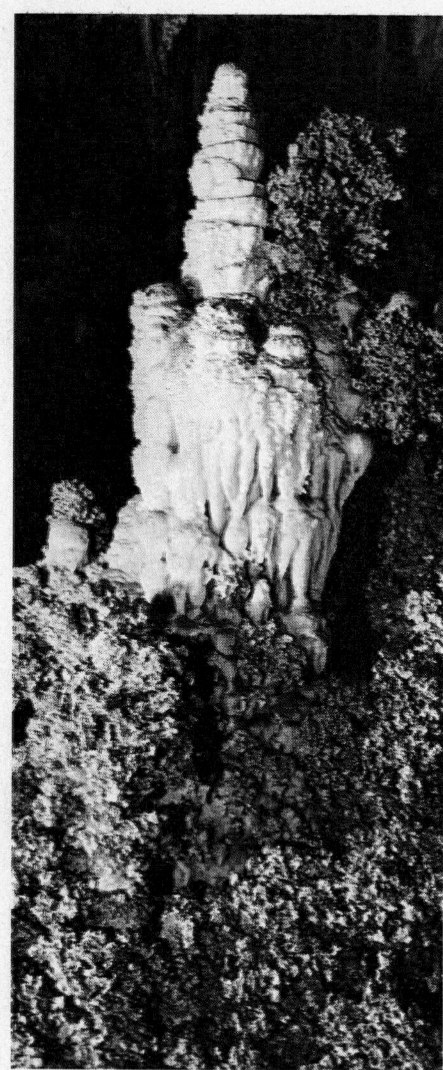

Figure 12.16 **Delicate corallike forms have been deposited on a massive stalagmite in the Queen's Chamber, Carlsbad Caverns, New Mexico.** (*Photograph by National Park Service.*)

Caves of all sizes tend to develop in highly soluble rocks such as limestone ($CaCO_3$), and small ones occur in the sedimentary rock dolomite [$CaMg(CO_3)_2$]. Rock salt (NaCl), gypsum ($CaSO_4 \cdot 2H_2O$), and similar rock types are the victims of such rapid solution that underground caverns usually collapse under the weight of overlying rocks before erosion can open them at the surface.

Calcite, the main component of limestone, is very insoluble in pure water. But when the mineral is attacked by water containing small amounts of carbonic acid, it undergoes rapid chemical weathering, and most natural water does contain carbonic acid (H_2CO_3), the combination of water with carbon dioxide. The carbonic acid reacts with the calcite to form calcium bicarbonate, $Ca(HCO_3)_2$, a soluble substance that is then removed in solution. If not redeposited, it eventually reaches the ocean.

Let us look more closely at underground water as it brings about the decay of calcite. Calcite contains the complex carbonate ion, $(CO_3)^{2-}$ (built by packing three oxygen atoms around a carbon atom) and the calcium ion, Ca^{2+}. These two ions are arranged in much the same way as sodium and chlorine in forming salt or halite. The weathering or solution of calcite takes place when a hydrogen ion, H^+, approaches a carbonate ion, $(CO_3)^{2-}$. Because the attraction of hydrogen for oxygen is stronger than the attraction of carbon for oxygen, the hydrogen ion pulls away one of the oxygen atoms of the carbonate ion and, with another hydrogen ion, forms water. The two other oxygen atoms remain with the carbon atom as carbon dioxide gas. The calcium ion, Ca^{2+}, now joins with two negative bicarbonate ions, $(HCO_3)^-$, to form calcium bicarbonate in solution. We can express these activities as follows:

Two parts water *plus* two parts carbon dioxide *yield* two parts carbonic acid

$$2H_2O + 2CO_2 \leftrightharpoons 2H_2CO_3$$

Two parts carbonic acid *yield* two hydrogen ions *plus* two bicarbonate ions

$$2H_2CO_3 \leftrightharpoons 2H^+ + 2(HCO_3)^-$$

Two hydrogen ions *plus* two bicarbonate ions *plus* calcite *yield* water *plus* carbon dioxide *plus* calcium bicarbonate in solution

$$2H^+ + 2(HCO_3)^- + CaCO_3 \leftrightharpoons H_2O + CO_2 + Ca^{2+} + 2(HCO_3)^-$$

The principal chemical reaction in the formation of caves is given above. Even though the reaction can be demonstrated in the laboratory and occurs in nature, its extension to the formation of caves is not as direct as it might first appear.

There is a strong difference of opinion as to whether caves form above the water table, at (or just below) the water table, or deep beneath the water table. Those who believe that caves form below the water table call upon some mechanism that would lower the water table and drain the cave, thus filling it with air. Drilling records, in fact, show that there are large openings in limestone that lie flooded with water beneath the water table. Presumably, if land keeps rising and erosion proceeds until valleys are deep enough to drain the caves in the surrounding countryside, then they will become filled with air and ready for discovery and exploration.

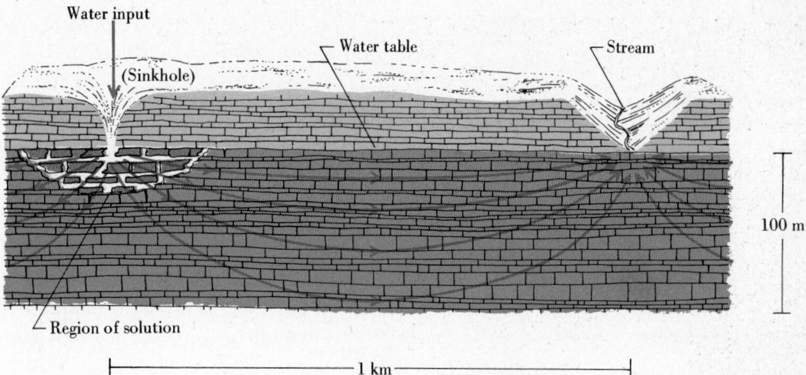

Figure 12.17 **Diagram through a small limestone aquifer showing approximate flow line and the location of region of solution. Vertical exaggeration, 4×. See the text for discussion.**
(*After John Thrailkill,* Bull. Geol. Soc. Am., *vol. 79, p. 39, Fig. 13, 1968.*)

We have seen that water and carbon dioxide will combine to provide a solution which will corrode limestone. Such a solution, seeping into the ground, will begin to dissolve limestone as soon as it meets it just below the soil. Here, the waters are rapidly neutralized and, seeping deeper, can no longer dissolve the rock. In this regard we can note that the waters seeping down from the surface into caves today are depositing rather than dissolving calcite. Yet it is also clear that the chambers we now call caves were formed at an appreciable depth below the surface. How, then, could this happen? The details of the answer are yet to be worked out, but studies by John Thrailkill of the University of Tennessee point a probable way to the answer. He suggests mechanisms by which the flow of ground water through limestone would allow concentrations of large amounts of fresh water undersaturated with respect to calcite. In such areas, probably close to the water table, solution of limestone would be possible.

One such mechanism is shown in Figure 12.17. In this diagram the water table is shown as being almost horizontal, a condition which is often encountered in limestone rock and is in contrast with the sloping water table, shown as an idealized case in Figure 12.4. The flow lines lead to the stream and the water table is one of these flow lines. Water reaches the underground chiefly via an enlarged opening at the surface. This opening could be formed by differential solution from the surface downward along particularly well-developed or favorably placed fractures in the limestone rock. Its presence allows the rapid injection of large amounts of surface water into the underground. Because it is injected rapidly and in large amounts, it may pass through the zone above the water table without coming into effective contact with limestone and remain undersaturated with respect to calcite as it joins the water. If this happens, there is a zone beneath the inlet and below the water table where ground water may still carry enough carbonic acid to effect solution in the underground.

Regardless of where and how caves are originally formed, we know that the weird stone formations so characteristic of most of them must have developed above the water table, when the caves were filled with air. These bizarre shapes are composed of calcite deposited by underground water that has seeped down through the zone of aeration. They develop either as *stalactites,* looking like stoney icicles hanging from the cave roof (Greek *stalactos,* "oozing out in drops") or as *stalagmites*, heavy posts growing up from the floor (Greek *stalagmos*, "a dropping" or "dripping"). When a stalactite and stalagmite meet, a *column* is formed.

A stalactite forms as water charged with calcium bicarbonate in solution seeps through the cave roof. Drop after drop forms on the ceiling and then falls to the floor. But during the few moments that each drop clings to the ceiling, a small amount of evaporation takes place; some carbon dioxide is lost, and a small amount of calcium carbonate is deposited. Over the centuries, a large stalactite may gradually develop. Part of the water that falls to the cave floor runs off, and part is evaporated. This evaporation again causes the deposition of calcite, and a stalagmite begins to grow upward to meet the stalactite hanging down from above.

On the ground surface above soluble rock material, depressions sometimes develop that reflect areas where the underlying rock has been carried away in solution. These depressions, called *sinkholes* or merely *sinks*, may form in one of two different ways. In one case the limestone immediately below the soil may be dissolved by the seepage of waters downward. The process of solution may be focused by local factors, such as a more abundant water supply or greater solubility of the limestone. Eventually a depression—a sink—evolves. Probably more commonly, sinks form when the surface collapses into a large cavity below. In either event, surface water may drain through the sinkholes to the underground, or if the sinks' subterranean outlets become clogged, they may fill with water to form lakes (Figure 12.18). An area with numerous sinkholes is said to display karst topography, from the

Figure 12.18 **St. Joseph's Well, a sinkhole filled with water, is in Clark County, Kansas.** (*Photograph by U.S. Geological Survey.*)

name of a plateau in Yugoslavia and northeastern Italy where this type of landscape is well developed.

12.6 Cementation and replacement

All underground water carries a certain amount of mineral matter in solution. Some of this matter, particularly the iron and silica compounds, is picked up as the water passes through the soil and down to the zone of saturation. The water also acquires calcium bicarbonate from the soil and from any limestone through which it passes. Later on, in new surroundings in the underground, this dissolved material may be deposited again. In fact, the common calcareous, siliceous, and iron oxide cements of the coarse-grained and medium-grained sedimentary rocks are derived in large part from minerals precipitated out of water that percolates beneath the surface.

Sometimes the dissolved material carried by underground water may replace other material. Thus silica in solution may replace the organic matter in buried logs and produce petrified wood, the original woody structure of the logs faithfully preserved in quartz, as in the Petrified Forest in Arizona.

Meteoric, juvenile, and connate water

Most underground water is derived from precipitation, and, fittingly, it is called *meteoric water,* from the Greek word meaning "high in the air." But some underground water originates within the earth itself, from sources related to igneous activity. It may appear at the surface during volcanic eruptions or simply accumulate below the surface after the crystallization of a magma. Appropriately known as *juvenile water,* this young, new water is entering into the hydrologic cycle for the first time.

The third type of underground water is *connate water,* water that was trapped in sedimentary deposits at the time of their formation. It takes its name from the Latin meaning "born at the same time." Actually, connate water was not *created* at the same time as the sedimentary rocks that hold it, but it did form a part of the original deposits. Thus sand laid down on an ancient sea floor may be converted to a sandstone, and if conditions permit, the sea water trapped during deposition may be sealed off in the rock. An example of connate water is the salt water that is often brought up from oil fields (Chapter 21).

Connate water in marine rocks is salty, reflecting the nature of its origin but differing in composition from modern sea water for several reasons. Part of the material carried in solution by the ancient sea water may have been precipitated as a cement, or it may have entered into chemical combination with other minerals in the sedimentary rock. Or the original connate water may have become somewhat diluted by the slow infiltration of fresh, unsalty ground water. Finally, it may be that the chemical composition of ancient seas differed from that of modern seas.

Not all salty water in the underground is connate, however. Fresh ground water may become salty as it flows down through beds of rock salt, and some salty water may seep from the ocean into the rocks along the shore.

12.7 Ground water as a natural resource

Use of ground water

By 1965 in the United States, man was withdrawing about 61 bgd (billion gallons per day) from the underground. This use represented nearly 20 percent of all water used from surface and subsurface sources in the country. The greatest amount of ground water is used for irrigation, as shown in Table 12.1, which also shows the amount of ground water used for various purposes in the United States and how these amounts have changed from 1950 to 1965.

Table 12.1 **Use of ground water in the United States**[a]

	Bgd	
Use	1950	1965
Rural (stock and domestic)	2.8	3.1
Municipal	3.6	8.0
Industrial	5.5	8.0
Irrigation	20.2	42.0
Total	32.1	61.1

[a] Data for 1950 from Jack B. Graham, *Water Well J.*, vol. 6, no. 3, p. 11, 1952. Data for 1965 from C. Richard Murray, *U.S. Geol. Surv. Circ.*, vol. 556, p. 12, fig. 12, 1968.

Because ground water is a replenishable natural resource, it stands in contrast to the nonreplenishable resources of the earth, such as coal, petroleum, copper, iron, and uranium, to name but a few (see Chapter 21). There is a constant recharging of the underground reservoir as waters seep into the ground from the surface. We may continue to use ground water without fear of serious depletion as long as withdrawal does not exceed the addition of new water over a period of years, but if we use more ground water than can be returned by nature, then we must try to balance our ground-water budget. For instance, we can return water to the zone of saturation after we have pumped and used it, instead of allowing it to run off as waste to surface streams. Or we can adopt practices that promote the infiltration of more surface water into the underground. As a final resort, we can either impose limitations on the amount of water pumped from the underground or restrict the uses to which ground water can be put.

Occurrence of ground water in the United States

There should be no need to emphasize that ground water does not occur in a vast, single underground reservoir across the country. Rather, ground water occurs in a great variety of individual aquifers. The amount and quality of ground water available at any one locality depend on the local geologic, climatic, and geographic conditions. Figure 12.19 is a map that shows the major ground-water areas of the United States and their general nature. It forms the basis for the discussion in the following section.

GROUND-WATER PROVINCES The great variations in climate, geography, and geology across the country produce a wide range of conditions under which ground water can be effectively developed. The country, however, can be subdivided into areas each of which is characterized by the similarity of the principal occurrences of ground water within it. Such as area is called a *ground-water province.* Local conditions, of course, cause variations within each province, but in general the boundaries of such a province enclose an

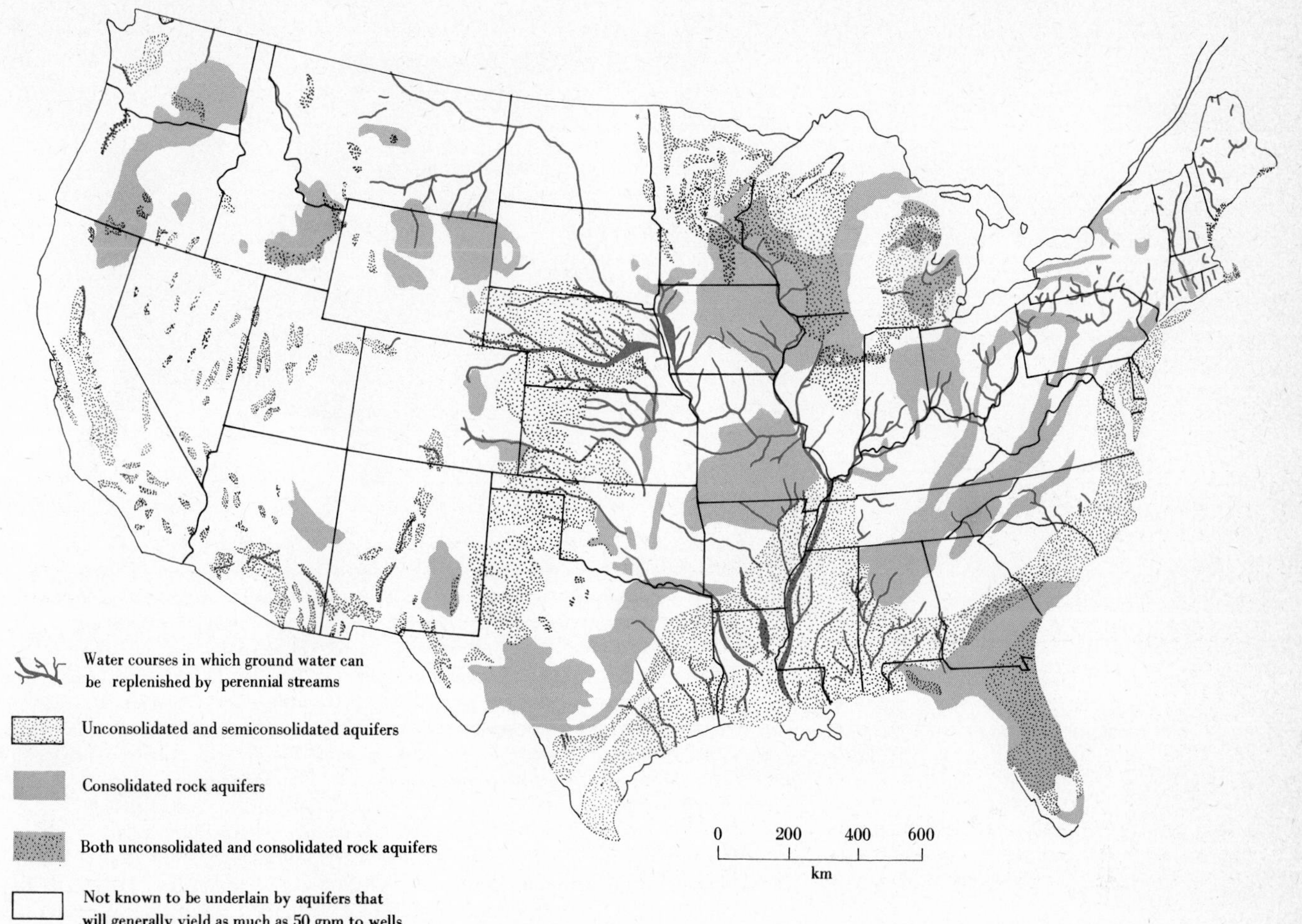

Figure 12.19 **The major aquifers of the United States.** (*Redrawn by permission from H. E. Thomas,* Conservation of Ground Water, *plate 1, McGraw-Hill Book Company, Inc., New York, 1951.*)

area of similar ground-water conditions. The ground-water provinces of the United States are grouped into four regions. The name for each province includes a geographic reference and usually a reference to the chief water-producing horizon or horizons. In the following paragraphs, provinces are identified by letter to key them to their fuller designation in Figure 12.20.

Coastal Plain region The coastal plain region comprises a belt of varying width along the Atlantic and Gulf coasts from Long Island, New York, to the Texas–Mexico boundary. Its landward margin marks the present known limit of deposits laid down by an invasion of this section of the United States by the ocean in Cretaceous time. Province A includes the entire region. Sedimentary strata of Cretaceous, Tertiary, and Quaternary age dip gently seaward. Water is found in the sandstone and porous limestone beds between more shaley layers. Down the dip of the beds from their intake area, artesian

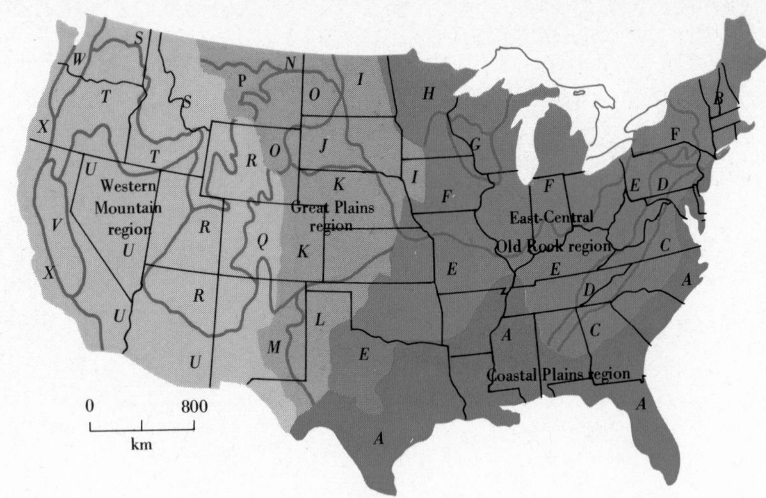

Figure 12.20 ***Ground-water provinces of the United States. Each area is distinguished by the similarity of the occurrence of ground water within it. The key to the provinces lettered on the map is as follows: A, Atlantic Coastal Plain province; B, Northeastern Drift province; C, Piedmont province; D, Blue Ridge–Appalachian Valley province; E, South-Central Paleozoic province; F, North-Central Drift–Paleozoic province; G, Wisconsin Paleozoic province; H, Superior Drift–Crystalline province; I, Dakota Drift–Cretaceous province; J, Black Hills Cretaceous province; K, Great Plains Pliocene-Cretaceous province; L, Great Plains Pliocene-Paleozoic province; M, Trans-Pecos Paleozoic province; N, Northwestern Drift–Cretaceous province; O, Montana Eocene-Cretaceous province; P, Montana Cretaceous province; Q, Southern Rocky Mountain province; R, Wyoming–Colorado Plateau province; S, Northern Rocky Mountain province; T, Columbia Plateau Lava province; U, Southwestern Basin province; V, Coastal Range province.*** (*Modified from O. E. Meinzer, "The Occurrence of Ground Water in the United States,"* U.S. Geol. Surv. Water Supply Paper, *no. 489, 1923.*)

pressure may be built up. Other ground water is found in stream and terrace deposits, particularly along the Mississippi valley.

East-Central Old Rock province Sedimentary rocks of the Paleozoic and Precambrian age underlie the greater part of this region. Across its northern portion are spread the much younger deposits left by the glaciers of the great Ice Age.

Province B includes New England and western and northern New York State. Here, bedrock, largely metamorphic and igneous, is overlaid by unconsolidated deposits. Small wells draw water of good quality from the fractured zone near the surface of the old rocks. Locally, sand and gravel of glacial origin yield large supplies.

In province C, folded and broken Paleozoic sedimentary rocks and igneous and metamorphic rocks of presumed Precambrian age supply good water to shallow wells and to springs. Locally, beds of Triassic sandstone produce moderately large amounts of ground water.

Provinces E, F, and G are underlain by relatively flat-lying sedimentary rocks of the Paleozoic age. The shales produce virtually no water, but shallow wells tap good and often abundant water in the sandstone and limestone beds. Water in deep wells tends to have a high mineral content and is generally unfit for use. Water occurs in the Paleozoic rocks under artesian conditions in many places. Glacial deposits in province F supply good water, sometimes in large amounts. Province G is the so-called "driftless area." Some portions never were glaciated and other parts have discontinuous deposits of ice advances long predating the last major ice advance in the area. Water is found in the Paleozoic rocks and in the sand and gravel washed from nearby glaciers of the last glaciation and deposited in the larger valleys.

Province H is underlain by igneous, metamorphic, and sedimentary rocks, largely of the Precambrian age. These are covered with glacially deposited material. Water supplies are meager from the bedrock but are important from the overlying unconsolidated material. Ground water tends to be highly mineralized in the western and northern portions of the province.

Great Plains region The large, wedge-shaped great plains region east of the Rocky Mountains extends southward from the Canadian boundary to southwestern Texas. Sedimentary strata of the Tertiary and Mesozoic (particularly Cretaceous) age underlie much of the region. In some areas, Paleozoic and Precambrian rocks are of importance.

In provinces I, J, K, N, O, and P, large amounts of water are found in permeable sandstone of the Cretaceous age. This water is generally high in mineral matter but it is still usable. Artesian wells may occur in each province, but are most frequent in I and to a lesser degree in J and K. The glacial material overlying the bedrock in province I supplies much water. Province J includes the Black Hills, where water comes from Precambrian crystalline rocks as well as from younger sedimentary strata. Elsewhere in the province, shallow wells generally produce only meager amounts of water, because the surface rock is in large part impermeable shale that must be pierced before a well reaches the deeper sandstone aquifers.

Provinces K and L are widely veneered with Tertiary sediments worn from the Rocky Mountains as they rose toward their present elevation. Large supplies of good water are obtained from wells in these deposits. In province K, some water is obtained from the sandstone beds of the Cretaceous age, whereas in province L, Triassic or Permian rocks bear water. In both provinces, Quaternary sand and gravel yield water in the larger valleys.

In most sections of province M, shallow wells yield only small supplies of highly saline water. However, flowing wells in Paleozoic limestone yield large amounts of water in the Roswell artesian basin of the Pecos Valley. Province N produces water from glacial deposits and at various places from the underlying late Mesozoic and early Tertiary sedimentary rocks. Province O is similar to N but has glacial material only in its northern and northeastern sections; this material does not produce water. Through large sections of province P, shales cropping out at the surface make shallow water difficult to find. Some production comes from deeper sandstone layers.

Western Mountain region In the western mountain region, ground water is found in all three major rock types, which vary widely in age. Province

Q embraces the southern Rocky Mountains, where ground water is obtained chiefly from springs issuing from the igneous and metamorphic rocks. Much ground water discharges directly to the streams, and water is found in shallow wells close to the stream banks. Province R includes essentially flat-lying sedimentary rocks of the Paleozoic, Mesozoic, and Tertiary age. Where these are deeply dissected by canyons, water supplies are meager. Elsewhere, however, wells in sandstone rocks may yield water in moderate amounts. In some instances, the wells are artesian.

The northern Rocky Mountains of province R present a wide variety of ground-water conditions. Most water, as in the southern Rockies, is supplied by springs and by direct discharge to the streams. Much water is also available, however, in the rocks, sand, and gravel of intermontane valleys. Province T is characterized by extensive lava flows of the Tertiary and Quaternary age. These are very permeable and make good aquifers. Locally, sand and gravel interbedded with the lava yield water, and valley fillings of sand and gravel are important.

Most water in province U comes from sand and gravel fill and alluvial fans of intermontane valleys, which are characteristic of the area. Some water is obtained from springs in the intervening mountains.

Provinces V and W are long, north-south valleys lying between highlands on either side. In V, most ground water comes from the complex alluvial cones that flank the valley sides, particularly on the east. In the northern section, sand and gravel along the valley axis are important. In province W, much water is available in the valley filling, but large supplies of surface water have not encouraged extensive development. Province X includes the young coastal ranges of the western margin of the United States. Demand for water in the southern portion of the section has brought about the development of ground water contained in alluvial cones and in the fill of the larger valleys.

Outline

Underground water is over 60 times as abundant as the water in streams and lakes.

Distribution is in the *zone of aeration* and an underlying *zone of saturation.* The irregular surface between them is the *ground-water table.*

Movement of underground water is usually by laminar flow.

- *Porosity* is the total percentage of void space to a given volume of earth material.
- *Permeability* is the ability of earth material to transmit water. Flow is driven by gravity according to *Darcy's law:* $V = P(h/l)$.

Ground water in nature moves in *aquifers.* It comes to the surface as springs and wells.

- *A spring* occurs when the ground surface intersects the water table.
- *Simple wells* draw water from the zone of saturation.
- *Artesian wells* are those in which gravity drives the water level above the height of the aquifer.
- *Thermal springs* derive their heat from the cooling of igneous rocks or by a normal increase of the earth's heat with depth.

Geysers are thermal springs marked by periodic, violent eruptions and controlled by a particular arrangement of underground passages.

Recharge of ground water is either natural or artificial.

Caves are usually created by the solution of limestone. The chemical reaction involves carbonic acid (formed by the combination of water and carbon dioxide) with limestone ($CaCO_3$).

Formation of caves is apparently in two stages: (1) the creation of chambers, galleries, and tunnels and (2) the decoration of these large openings by forms growing from the ceilings (*stalactites*) and forms growing from the floor (*stalagmites*). Solution of limestone develops an irregular "pitted" topography of depressions with no exterior surface drainage. This is called *karst topography* and is marked by numerous *sinkholes*.

Cementation and *replacement* by ground water affects earth materials. Unconsolidated sediments may be cemented by material precipitated from ground water. Material in solution may also be substituted for other material, replacing the original mineral matter with new mineral matter as in the silicification of petrified wood.

Meteoric, juvenile and *connate* waters are the three different types of underground waters.

Ground water is a replenishable natural resource which supplies about 20 percent of our water use. A ground-water province is an area in which ground-water conditions are similar.

Supplementary readings

BRETZ, J HARLAN: "Vadose and Phreatic Features of Limestone Caverns," *J. Geol.*, vol. 50, pp. 675–811, 1942. A paper which describes in considerable detail the features of caves formed above and below the water table.

DAVIS, STANLEY N., and ROGER J. M. DE WIEST: *Hydrogeology*, John Wiley & Sons, Inc., New York, 1966. A basic text on the geology of underground water.

HOLLAND, H. D., and others: "On Some Aspects of the Chemical Evolution of Cave Waters," *J. Geol.*, vol. 72, pp. 36–67, 1964. Some fundamental chemical properties of cave waters are considered.

HUBBERT, M. KING: "The Theory of Ground Water Motion," *J. Geol.*, vol. 48, pp. 785–944, 1940. A statement concerning the physics of the motion of ground water.

MEINZER, O. E.: "The Occurrence of Ground Water in the United States," *U.S. Geol. Surv. Water Supply Paper*, no. 489, 1923. This and the following report remain basic contributions to the subject of ground water.

MEINZER, O. E.: "Ground Water in the United States, A Summary," *U.S. Geol. Surv. Water Supply Paper*, no. 836-D, 1939.

THRAILKILL, JOHN: "Chemical and Hydrologic Factors in Excavation of Limestone Caves," *Geol. Soc. Am. Bull.*, vol. 79, pp. 19–46, 1968. A discussion of the chemical parameters involved in the development of caves in limestone.

TODD, D. K.: *Ground Water Hydrology*, John Wiley & Sons, Inc., New York, 1959. A basic text in the subject.

WHITE, DONALD E., JOHN D. HEM, and G. A. WARING: "Data of Geochemistry," *U.S. Geol. Surv. Professional Paper*, no. 440-F, 1963. A documentation of the amount and nature of dissolved constituents in ground waters of the world.

13

Glaciation

THE SEAS AND RIVERS OF MOVING ICE KNOWN AS GLACIERS HAVE ATTRACTED INQUISitive men deep into the Arctic, Antarctic, and mountainous regions of the world. There they have discovered that glaciers are active agents of erosion, transportation, and deposition and that these impressive masses of ice were far more widespread in the past than they are now. Geologists have learned, too, that the ice of the last great glacial period has modified and molded great stretches of landscape in what are now the temperate zones.

13.1 *Formation of glacier ice*

A *glacier* is a mass of ice that has been formed by the recrystalization of snow and that flows forward, or has flowed at some time in the past, under the influence of gravity. This definition eliminates the pack ice formed from sea water in polar latitudes and—by convention—icebergs, even though they are large fragments broken from the seaward end of glaciers.

Like surface streams and underground reservoirs, glaciers depend on the oceans for their nourishment. Some of the water drawn up from the oceans by evaporation falls on the land in the form of snow. If the climate is right, part of the snow may last through the summer without melting. Gradually, as the years pass, the accumulation may grow deeper and deeper, until at last a glacier is born.

In areas where the winter snowfall exceeds the amount of snow that melts away during the summer, stretches of perennial snow known as *snowfields* cover the landscape. At the lower limit of a snowfield lies the *snow line.* Above the snow line, glacier ice may collect in the more sheltered areas of the snowfields. The exact position of the snow line varies from one climatic region to another. In polar regions, for example, it reaches down to sea level, but near the equator it recedes to the mountain tops. In the high mountains of East Africa, it ranges from elevations of 4,500 to 5,400 m. The highest snow lines in the world are in the dry regions known as the "horse latitudes" between 20° and 30° north and south of the equator. Here the snow line reaches higher than 6,000 m.

Fresh snow falls as a feathery aggregate of complex and beautiful crystals with a great variety of patterns. All the crystals are basically hexagonal, however, and all reflect their internal arrangement of hydrogen and oxygen atoms (see Figure 13.1). Snow is not frozen rain; rather, it forms from the condensation of water vapor at temperatures below the freezing point.

After snow has been lying on the ground for some time, it changes from a light, fluffy mass to a heavier, granular material called *firn,* or *névé. Firn* derives from a Greek adjective meaning "of last year," and *névé* is a French word from the Latin for "snow." Solid remnants of large snow banks, those tiresome vestiges of winter, are largely firn.

Several processes are at work in the transformation of snow into firn. The first is sublimation, a general term for the process of a solid material passing into the gaseous state without first becoming a liquid. In sublimation, molecules

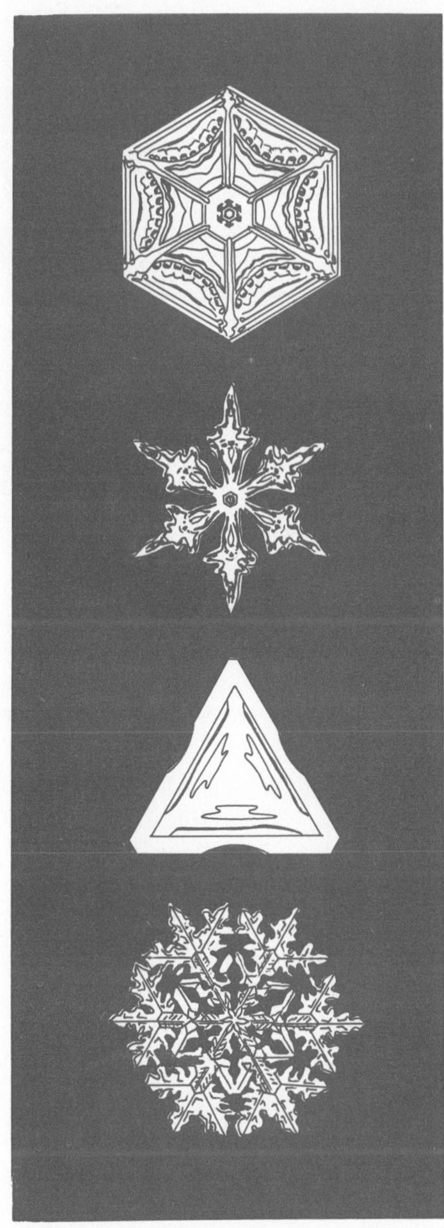

Figure 13.1 ***Snowflakes exhibit a wide variety of patterns, all hexagonal and all reflecting the internal arrangement of hydrogen and oxygen. It is from snowflakes that glacier ice eventually forms.***

of water vapor escape from the snow, particularly from the edges of the flakes. Some of the molecules attach themselves to the center of the flakes, where they adapt themselves to the structure of the snow crystals. Then, as time passes, one snowfall follows another, and the granules that have already begun to grow as a result of sublimation are packed tighter and tighter together under the pressure of the overlying snow.

Water has the unique property of increasing in volume when it freezes; conversely, it decreases in volume as the ice melts. But the cause and effect may be interchanged: If added pressure on the ice squeezes the molecules closer together and reduces its volume, the ice may melt. In fact, if the individual granules are in contact, they begin to melt with only a slight increase in pressure. The resulting meltwater trickles down and refreezes on still lower granules at points where they are not yet in contact. And all through this process the basic hexagonal structure of the original snow crystals is maintained.

Gradually, then, a layer of firn granules, ranging from a fraction of a millimeter to approximately 3 or 4 mm in diameter, is built up. The thickness of this layer varies, but 30 m seem to be average on many mountain glaciers.

The firn itself undergoes further change as continued pressure forces out most of the air between the granules, reduces the space between them, and finally transforms it into *glacier ice,* a true solid composed of interlocking crystals. In large blocks it is usually opaque and takes on a blue-gray color from the air and the fine dirt that it contains.

The ice crystals that make up glacier ice are minerals; the mass of glacier ice, made up of many interlocking crystals, is a metamorphic rock, for it has been transformed from snow into firn and eventually into glacier ice. Later, we shall see that glacier ice itself undergoes further metamorphism.

Classification of glaciers

The glaciers of the world fall into three principal classifications: (1) valley glaciers, (2) piedmont glaciers, and (3) ice sheets.

Valley glaciers are streams of ice that flow down the valleys of mountainous areas (see Figure 13.2). Like streams of running water, they vary in width, depth, and length. A branch of the Hubbard glacier in Alaska is 120 km long, whereas some of the valley glaciers that dot the higher reaches of our western mountains are only a few hundred meters in length. Valley glaciers that are nourished on the flanks of high mountains and that flow down the mountain sides are sometimes called *mountain glaciers* or *Alpine glaciers.* Very small mountain glaciers are referred to as *cliff glaciers, hanging glaciers,* or *glacierets.* A particular type of valley glacier sometimes grows up in areas where large masses of ice are dammed by a mountain barrier along the coast. Some of the ice escapes through valleys in the mountain barrier to form an *outlet glacier,* as it has done along the coasts of Greenland and Antarctica.

Piedmont glaciers form when two or more glaciers emerge from their valleys and coalesce to form an apron of moving ice on the plains below.

Ice sheets are broad, moundlike masses of glacier ice that tend to spread radially under their own weight. The Vatna glacier of Iceland is a small ice

Figure 13.2 **The Aletsch glacier chokes this Alpine valley leading away from the Jungfrau.** (*Photograph by the Swiss National Tourist Office.*)

sheet measuring about 120 by 160 km and 225 m in thickness. A localized sheet of this sort is sometimes called an *ice cap* (see Figure 13.3). The term *continental glacier* is usually reserved for great ice sheets that obscure the mountains and plains of large sections of a continent, such as those of Greenland and Antarctica. On Greenland, ice exceeds 3,000 m in thickness near the center of the ice caps. Ice in Antarctica averages about 2,300 m in thickness, and in Marie Byrd Land it is over 7,400 m.

Distribution of modern glaciers

Modern glaciers cover approximately 10 percent of the land area of the world. They are found in widely scattered locations in North and South America, Europe, Asia, Africa, Antarctica, Greenland, many of the north polar islands, and on the Pacific islands of New Guinea and New Zealand. A few valley glaciers are located almost on the equator. Mt. Kenya in East Africa, for instance, only $\frac{1}{2}°$ from the equator, rises over 5,100 m into the tropical skies and supports at least 10 valley glaciers.

The total land area covered by existing glaciers is estimated as 17.9 million km^2 of which Greenland and Antarctic ice sheets account for about 96 percent. The Antarctic ice sheet covers approximately 15.3 million km^2, and the Greenland sheet covers about 1,735,000 km^2. Small ice caps and numerous mountain glaciers scattered around the world account for the remaining 4 percent.

Figure 13.3 ***A small ice cap on Axel Heiberg Island, Northwest Territories, Canada.*** *(Photograph by Austin Post, University of Washington.)*

Nourishment and wastage of glaciers

When the weight of a mass of snow, firn, and ice above the snow line becomes great enough, movement begins and a glacier is created. The moving stream flows downward across the snow line until it reaches an area where the loss through evaporation and melting is so great that the forward edge of the glacier can push no farther. A glacier, then, can be divided into two zones: (1) *a zone of accumulation* and (2) *a zone of wastage.* Both are illustrated in Figure 13.4.

The position of the front of a glacier depends on the relationship between the glacier's rate of nourishment and its rate of wastage. When nourishment

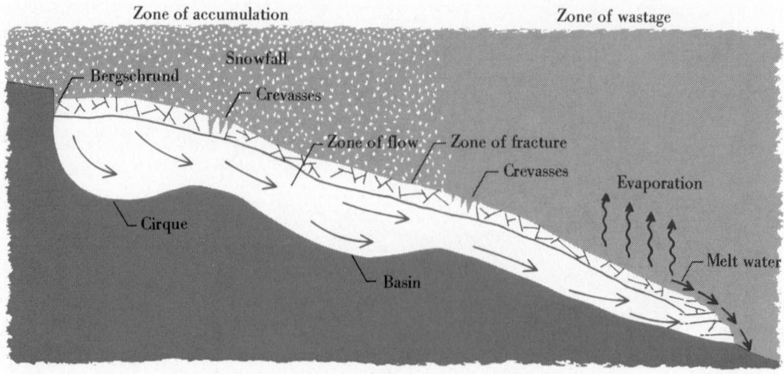

Figure 13.4 ***A glacier is marked by a zone of accumulation and a zone of wastage. Within a glacier, ice may lie either in the zone of fracture or deeper in the zone of flow. A valley glacier originates in a basin, the cirque, and is separated from the headwall of the cirque by a large crevasse, the bergschrund.***

just balances wastage, the front becomes stationary and the glacier is said to be in equilibrium. This balance seldom lasts for long, however, for a slight change in either nourishment or wastage will cause the front to advance or recede.

Today, most of the glaciers of the world are receding. With only a few exceptions, this process has been going on since the latter part of the nineteenth century, and since 1920 the rate has speeded up. A striking feature of modern glaciers is that they follow the same general pattern of growth and wastage the world over and serve as indicators of widespread climatic changes.

Valley glaciers are nourished not only in the zone of accumulation but also by great masses of snow that avalanche down the steep slopes along their course. In fact, according to one interpretation, avalanches caused by earthquakes have enabled a glacier to advance in a single month as far as it would have if it had been fed by the normal snowfall of several years.

Below the snow line, wastage takes place through a double process of evaporation and melting known as *ablation.* If a glacier terminates in a body of water, great blocks of ice break off and float away in a process called *calving.* This is the action that produces the icebergs of the polar seas.

Glacier movement

Except in rare cases, glaciers move only a few centimeters or at most a few meters per day. That they actually do move, however, can be demonstrated in several ways. The most conclusive test is to measure the movement directly, by emplacing a row of stakes across a valley glacier. As time passes, the stakes move down-valley with the advancing ice, the center stakes more rapidly than those near the valley walls. A second source of evidence is provided by the distribution of rock material on the surface of a glacier. When we examine the boulders and cobbles lying along a valley glacier, we find that many of them could not have come from the walls immediately above and that the only possible source lies up-valley. We can infer, then, that the boulders must have been carried to their present position on the back of the glacier. Another indication of glacier movement is that when a glacier melts it often exposes

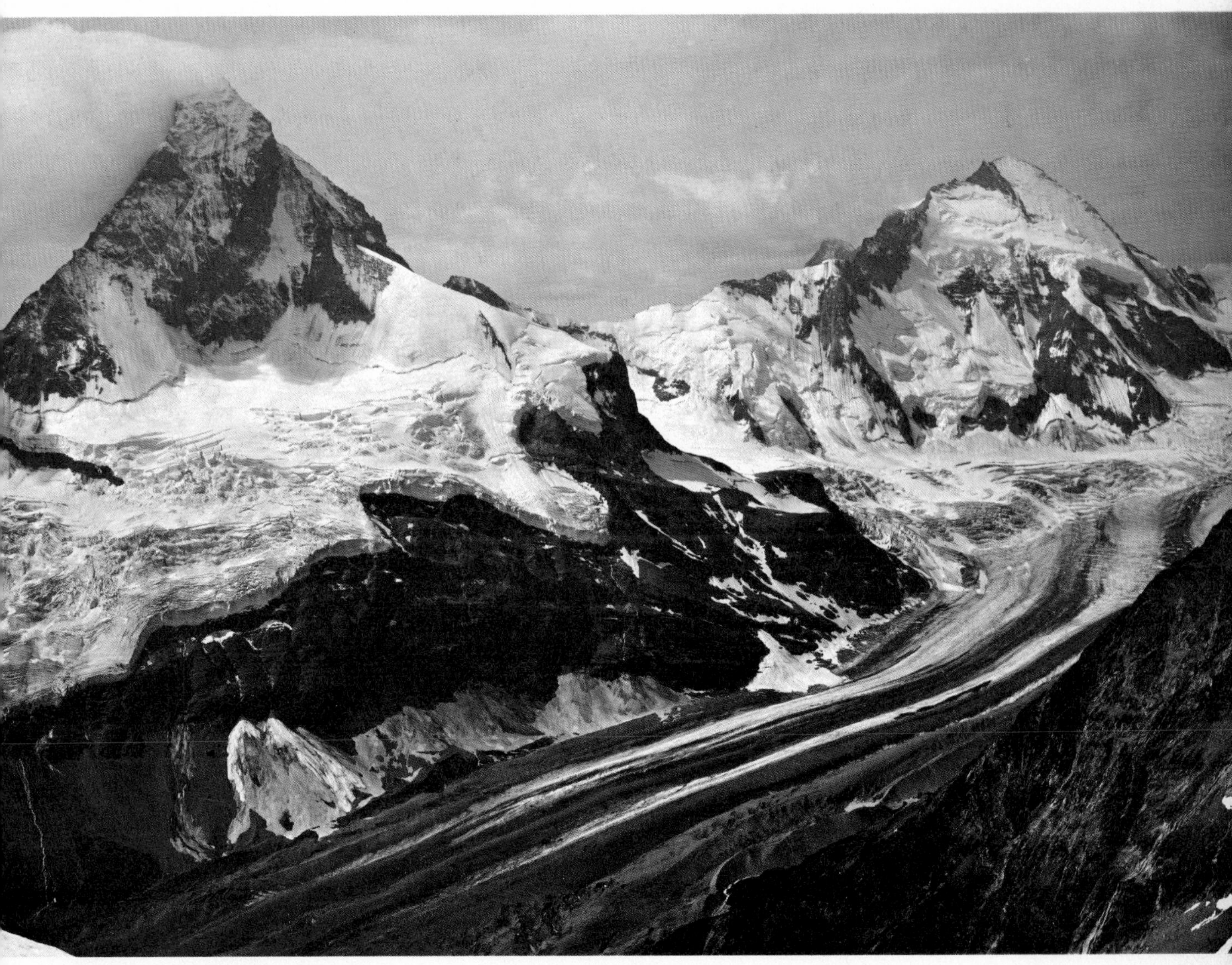

Figure 13.5 ***A glacier formed from snow gathered near Dent d'Herens, the 4,110-m peak on the right, flows past Matterhorn, the 4,440-m peak on the left, and carries rock debris down the valley.*** (*Photograph by Vittorio Sella.*)

a rock floor that has been polished, scratched, and grooved. It is simplest to explain this surface by assuming that the glacier actually moved across the rock floor, using embedded debris to polish, scratch, and groove it.

Clearly, then, a glacier does move (see Figure 13.5). In fact, different parts of it move at different rates. But though we know a good bit about how a glacier flows forward, certain phases are not yet clearly understood. In any event, we can distinguish two zones of movement: (1) an upper zone between 30 and 60 m thick, which reacts like a brittle substance—that is, it breaks sharply rather than undergoing gradual, permanent distortion, and (2) a lower zone, which, because of the pressure of the overlying ice, behaves like a plastic substance. The first is the *zone of fracture;* the second is the *zone of flow.*

As plastic deformation takes place in the zone of flow, the brittle ice above is carried along. But the zone of flow moves forward at different rates—faster in some parts, more slowly in others—and the rigid ice in the zone of fracture is unable to adjust itself to this irregular advance. Consequently, the upper part of the glacier cracks and shatters, giving rise to a series of deep, treacherous *crevasses* (see Figures 13.4 and 13.6).

A glacier attains its greatest velocity somewhere above the valley floor, in midstream, for the sides and bottom are retarded by friction against the valley walls and beds. In this respect the movement of an ice stream resembles that of a stream of water.

The mechanics of ice flow are still a matter of study—a study made difficult by the fact that we cannot actually observe the zone of flow, because it lies concealed within the glacier. Yet the ice from the zone of flow eventually

Figure 13.6 ***Crevassed surface of the Columbia glacier, Chugach Mountains, Alaska.*** (*Photograph by Austin Post, University of Washington.*)

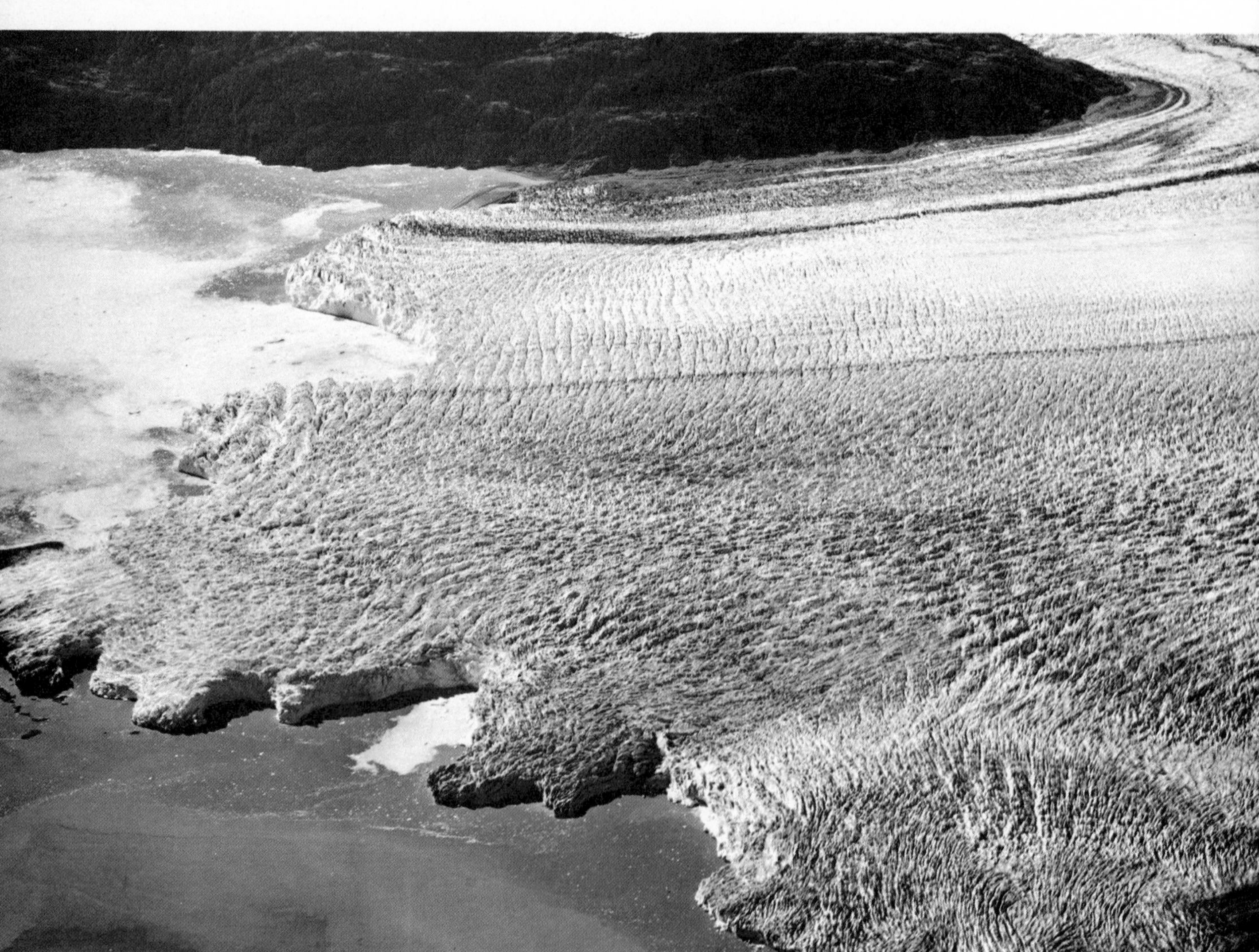

emerges at the snout of the glacier, and there it can be studied. We find that by the time it has emerged it is brittle, but it retains the imprint of movement by flow. The individual ice crystals are now several centimeters in size; in contrast, crystals in ice newly formed from firn measure but a fraction of a centimeter. We can conclude that the ice crystals have grown by recrystallization as they passed through the zone of flow. The ice at the snout is also marked by bands that represent shearing and differential movement within the glacier. Recrystallization has taken place along many of the old shear planes, and along others the debris carried forward by the ice has been concentrated. These observations suggest that some movement in the zone of flow has taken place as a result of shearing.

Measurements of the direction of flow of ice in modern glaciers show that in the zone of wastage the flow of ice is upward toward the ice surface at a low angle. In the zone of accumulation the direction of movement is downward from the ice surface at a low angle (see Figure 13.4).

However glacier ice may move in the interior of a glacier, there is also a movement which involves the entire ice tongue or sheet. The glacier literally slips along its base, moving across the underlying ground surface. This *basal slip* is added to the internal flow of the glacier to give the total movement of the ice body (see Figure 13.7).

Although most glacier motion is slow, some glaciers display a very rapid motion called a *surge*. Surges—or catastrophic advances—are now known from certain mountain and valley glaciers around the world (see Figure 13.8). The

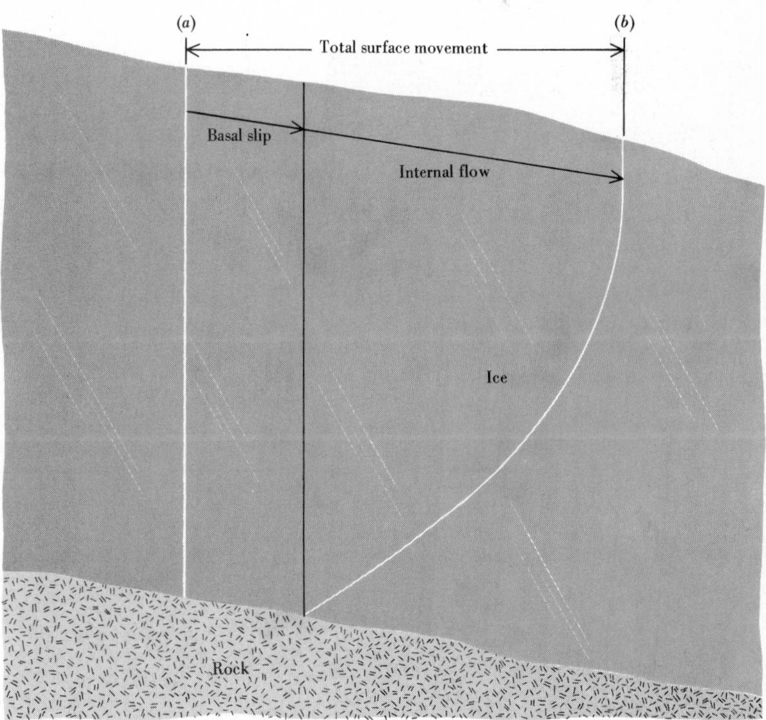

Figure 13.7 **Forward motion of a glacier consists in part of internal *flow* of the ice and in part of the slippage of the entire glacier over the base.** (*After Robert P. Sharp,* Glaciers, *Fig. 11, Oregon System of Higher Education, Eugene, 1960.*)

largest surge so far reported seems to be that of the Bråsellbreen, a portion of a roughly circular ice cap on Spitzbergen. Here the ice advanced up to 20 km along a 21-km front sometime between 1935 and 1938, as revealed by aerial photographs taken in those two separate years. The most rapid advance authenticated by actual observation is 110 m/day on the Kutiah glacier in northern India.

Glaciers surge because of an instability which allows gravity to tear them loose from the ground surface on which they lie. The cause of this instability is not yet completely understood, but it does involve the accumulation of an excess of water at the glacier's base. One way this water may accumulate is through a thickening of the ice and thus an increase of pressure in the glacier. This increase may be great enough at the base to allow some melting there.

Surges are as yet unrecorded from the Antarctic ice cap. Nevertheless, it has been estimated that should a surge occur there it might well involve the movement into the Antarctic Ocean of enough ice to raise the world's sea level between approximately 10 and 30 m during a period of a few years or few tens of years. The results would make some science fiction tame reading in comparison.

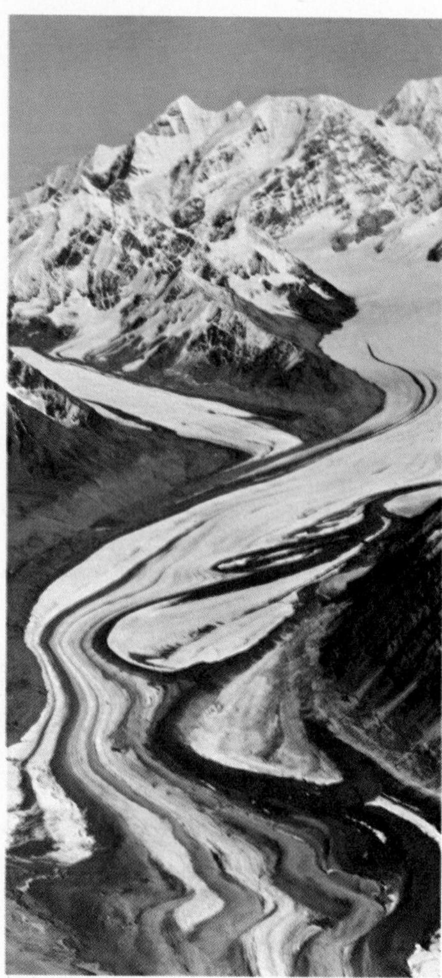

Figure 13.8 **Rapid surges along the Yanert glacier in the Alaska Range, Alaska, have displaced morainal loops down valley. The photograph was taken in 1964 and the latest surge at that time had been in 1942.** *(Photograph by Austin Post, University of Washington.)*

Temperatures of glaciers

All glaciers are cold, but some are colder than others. Therefore students of glaciers speak of "warm" and "cold" glaciers. A cold glacier is one in which no surface melting occurs during the summer months; its temperature is always below freezing. A warm glacier is one which reaches the melting temperature throughout its thickness during the summer season.

Obviously no glacier can exist above the melting point of ice, yet there are glaciers which hover at this temperature. All glaciers must form at a subfreezing temperature, and therefore we must ask how a glacier warms from its formation temperature to the melting point. The sun's heat cannot penetrate more than a few meters into the ice because it is such a poor conductor. Some other mechanism must operate to warm a glacier. This mechanism involves the downward movement of water from the surface of the glacier. There the heat of the sun melts surface ice and near-surface ice, and the melt water percolates downward into the glacier. When eventually it freezes it gives up heat at the rate of 80 cal/g of water. This amount of thermal energy can raise the temperature of 1 g of ice by 160°C or 160 g of ice by 1°C. Thus can a glacier be warmed.

13.2 *Results of glaciation*

Movement of material

Glaciers have special ways of eroding, transporting, and depositing earth materials. A valley glacier, for example, acquires debris by means of frost action, landsliding, and avalanching. Fragments pried loose by frost action clatter down from neighboring peaks and come to rest on the back of the glacier. And great snowbanks, unable to maintain themselves on the steep slopes of the mountainsides, avalanche downward to the glacier, carrying along quantities of rock debris and rubble. This material is either buried beneath fresh snow or avalanches or tumbles into gaping crevasses in the zone of fracture and is carried along by the glacier.

When a glacier flows across a fractured or jointed stretch of bedrock, it may lift up large blocks of stone and move them off. This process is known as *plucking*, or *quarrying*. The force of the ice flow itself may be strong enough to pick up the blocks, and the action may be helped along by the great pressures that operate at the bottom of a glacier. Suppose the moving ice encounters a projection of rock jutting up from the valley floor. As the glacier ice forces itself over and around the projection, the pressure on the ice is increased and some of the ice around the rock may melt. This meltwater trickles toward a place of lower pressure, perhaps into a crack in the rock itself. There it refreezes, forming a strong bond between the glacier and the rock. Continued movement by the glacier may then tear the block out of the valley floor.

At the heads of valley glaciers, plucking and frost action sometimes work together to pry rock material loose. Along the back walls of the collection basins of mountain glaciers, great hollows called *cirques* or *amphitheaters* develop in the mountainside (see Figure 13.9). As the glacier begins its

Figure 13.9 ***Cirques along the crest of the Wind River Mountains, Wyoming.*** *(Photograph by Austin Post, University of Washington.)*

movement downslope, it pulls slightly away from the back wall, forming a crevasse known as a *bergschrund.* One wall of the bergschrund is formed by the glacier ice; the other is formed by the nearly vertical cliff or bedrock. During the day, meltwater pours into the bergschrund and fills the openings in the rock. At night, the water freezes, producing pressures great enough to loosen blocks of rock from the cliff. Eventually, these blocks are incorporated into the glacier and are moved away from the headwall of the cirque.

The streams that drain from the front of a melting glacier are charged with *rock flour,* very fine particles of pulverized rock. So great is the volume of this material that it gives the water a characteristically grayish-blue color similar to that of skim milk. Here, then, is further evidence of the grinding power of the glacier mill.

Glaciers also pick up rock material by means of abrasion. As the moving ice drags rocks, boulders, pebbles, sand, and silt across the glacier floor, the bedrock is cut away as though by a great rasp or file. And the cutting tools themselves are abraded. It is this mutual abrasion that produces rock flour and gives a high *polish* to many of the rock surfaces across which a glacier has ridden. But abrasion sometimes produces scratches, or *striations,* on both the bedrock floor and on the grinding tools carried by the ice (see Figure 13.10). More extensive abrasion creates deep gouges, or *grooves,* in the bedrock. The striations and grooves along a bedrock surface show the direction of the glacier's movement. At Kelleys Island, in Lake Erie north of Sandusky, Ohio, the bedrock is marked by grooves 0.3 to 0.6 m deep and 0.6 to 1 m wide. In the Mackenzie Valley west of Great Bear Lake in Canada, grooves as wide as 45 m have been described with an average depth of 15 m and lengths ranging from several hundred meters to over 1 km.

Figure 13.10 **Striations on bedrock about 1.5 km west of Fair Haven, Vermont. The pen points about N, 10° W and marks the general direction of the younger of two sets of striations. The older set is oriented about N, 10° E.** *(Photograph by Paul MacClintock.)*

Erosional effects

The erosional effects of glaciers are not limited to the fine polish and striations mentioned above, however; for glaciers also operate on a much grander scale, producing spectacularly sculptured peaks and valleys in the mountainous areas of the world.

CIRQUES As we have seen, a cirque is the basin from which a mountain glacier flows, the focal point for the glacier's nourishment. After a glacier has disappeared and all its ice has melted away, the cirque is revealed as a great amphitheater or bowl, with one side partially cut away. The back wall rises a few scores of meters to over 900 m above the floor, often as an almost vertical cliff. The floor of a cirque lies below the level of the low ridge, separating it from the valley of the glacier's descent. The lake that forms in the bedrock basin of the cirque floor is called a *tarn.*

A cirque begins with an irregularity in the mountainside formed either by preglacial erosion or by a process called *nivation,* a term that refers to erosion beneath and around the edges of a snowbank. Nivation works in the following way. When seasonal thaws melt some of the snow, the meltwater seeps down to the bedrock and trickles along the margin of the snowbank. Some of the water works its way into cracks in the bedrock where it freezes again, producing pressures that loosen and pry out fragments of the rock. These fragments are moved off by solifluction, by rill wash, and perhaps by mass wasting, forming a shallow basin. As this basin gradually grows deeper, a cirque eventually develops. Continued accumulation of snow leads to the formation of firn; if the basin becomes deep enough, the firn is transformed into ice. Finally the ice begins to flow out of the cirque into the valley below, and a small glacier is born.

The actual mechanism by which a cirque is enlarged is still a matter of dispute. Some observers claim that frost action and plucking on the cirque wall within the bergschrund are enough to produce precipitous walls hundreds of meters in height. Others, however, point out that the bergschrund, like all glacier crevasses, remains open only in the zone of fracture, 60 m at most. Below that depth, pressures cause the ice to deform plastically, closing the bergschrund.

This debate has led to the development of the so-called *meltwater hypothesis* to explain erosion along the headwalls below the base of the bergschrund. The proponents of this theory explain that meltwater periodically descends the headwalls of cirques, melts its way down behind the ice and into crevices in the rock, and there freezes at night and during cold spells. The material thus broken loose is then removed by the glacier, and cirque erosion proceeds mainly by this form of headwall recession.

HORNS, ARÊTES, AND COLS A *horn* is a spire of rock formed by the headward erosion of a ring of cirques around a single high mountain. When the glaciers originating in these cirques finally disappear, they leave a steep, pyramidal

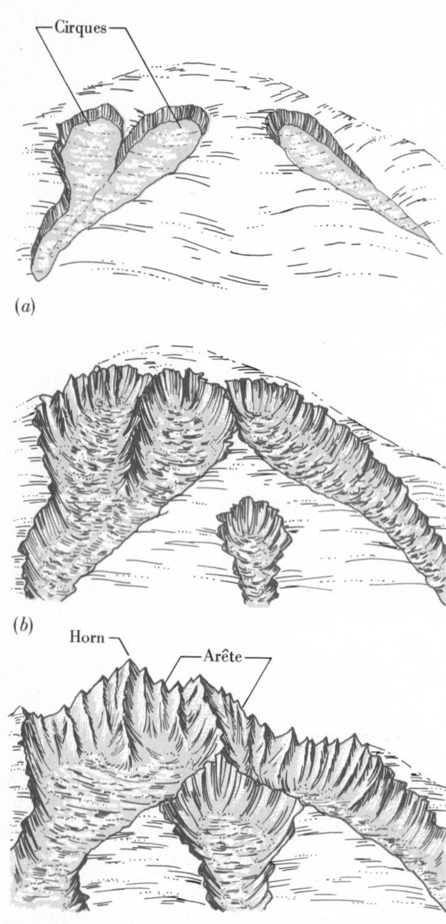

Figure 13.11 **The progressive development of cirques, horns, arêtes, and cols. In (a), valley glaciers have produced cirques, but because erosion has been moderate, much of the original mountain surface has been unaffected by the ice. The result of more extensive glacial erosion is shown in (b). In (c), glacial erosion has affected the entire mass and has produced not only cirques but also a matterhorn and jagged, knife-edged arêtes.** *(Redrawn from William Morris Davis, "The Colorado Front Range,"* Ann. Assoc. Am. Geog., *vol. 1, p. 57, 1911.)*

mountain outlined by the headwalls of the cirques. The classic example of a horn is the famous Matterhorn of Switzerland, (see Figures 13.5 and 13.11).

An *arête* (from the French for "fishbone," "ridge," or "sharp edge") is formed when a number of cirques gnaw into a ridge from opposite sides. The ridge becomes knife-edged, jagged, and serrated (see Figure 13.11).

A *col* (from the Latin *collum,* "neck"), or pass, is fashioned when two cirques erode headward into a ridge from opposite sides. When their headwalls meet, they cut a sharp-edged gap in the ridge.

GLACIATED VALLEYS Rather than fashioning their own valleys, glaciers probably follow the course of preexisting valleys, modifying them in a variety of ways; but usually the valleys have a broad *U-shaped* cross profile, whereas mountain valleys created exclusively by streams have narrow, V-shaped cross profiles. Because the tongue of an advancing glacier is relatively broad, it tends to broaden and deepen the V-shaped stream valleys, transforming them into broad, U-shaped troughs. And because the moving body of ice has difficulty manipulating the curves of a stream valley, it tends to straighten and simplify the course of the original valley. In this process of straightening, the ice snubs off any spurs of land that extend into it from either side. The cliffs thus formed are shaped like large triangles or flatirons with their apex upward and are called *truncated spurs* (see Figures 13.12 and 13.13).

Glaciers also give a mountain valley a characteristic longitudinal profile from the cirque downward. The course of a glaciated valley is marked by a series of *rock basins,* probably formed by plucking in areas where the bedrock was shattered or closely jointed. Between the basins are relatively flat stretches of rock that was more resistant to plucking. As time passes, the rock basins may fill up with water, producing a string of lakes that are sometimes referred to as *pater noster* lakes because they resemble a string of beads.

Hanging valleys are another characteristic of mountainous areas that have undergone glaciation. The mouth of a hanging valley is left stranded high above the main valley through which a glacier has passed. As a result, streams from hanging valleys plummet into the main valley in a series of falls and plunges. Hanging valleys may be formed by processes other than glaciation, but they are almost always present in mountainous areas that formerly supported glaciers and are thus very characteristic of past valley glaciation.

What has happened to leave these valleys stranded high above the main valley floor? During the time when glaciers still moved down the mountains, the greatest accumulation of ice would tend to travel along the central valley. Consequently, the erosive action there would be greater than in the tributary valleys, with their relatively small glaciers, and the main valley floor would be cut correspondingly deeper. This action would be even more pronounced where the main valley was underlain by rock that was more susceptible to erosion than the rock under the tributary valleys. Finally, some hanging valleys were probably created by the straightening and widening action of a glacier on the main valley. In any event, the difference in level between the tributary valleys and the main valley does not become apparent until the glacier has melted away.

Cutting deep into the coasts of Alaska, Norway, Greenland, Labrador, and

Figure 13.12 **A mountainous area before, during, and after glaciation.** (*Redrawn from William Morris Davis, "The Sculpture of Mountains by Glaciers,"* Scottish Geographical Magazine, *vol. 22, pp. 80–83, 1906.*)

Figure 13.13 **Little Cottonwood Canyon in the Wasatch Mountains of Utah shows the typical U-shaped profile of a glaciated valley. Compare with Figure 11.47.** (*Photograph by William C. Bradley.*)

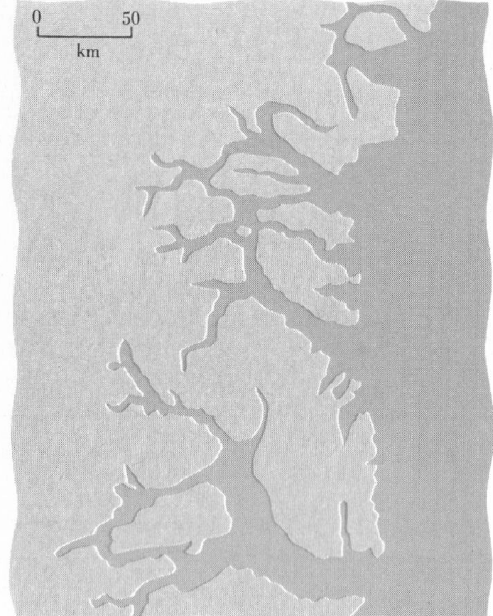

Figure 13.14 **Glaciated valleys have been flooded by the sea to produce these fiords along the coast of Greenland.** (*Redrawn from Louise A. Boyd and others, "The Fiord Region of East Greenland,"* Am. Geog. Soc. Spec. Publ., *no. 18, p. xii, 1935.*)

New Zealand are deep, narrow arms of the sea—*fiords* (see Figure 13.14). Actually, these inlets are stream valleys that were modified by glacier erosion and then partially filled by the sea. The deepest known fiord, Vanderford in Vincennes Bay, Antarctica, has a maximum depth of 2,287 m.

Some valleys have been modified by continental glaciers rather than by the valley glaciers that we have been discussing so far. The valleys occupied by the Finger Lakes of central New York State are good examples. These long, narrow lakes lie in basins that were carved out by the ice of a continental glacier. As the great sheet of ice moved down from the north, its progress seems to have been checked by the northern scarp of the Appalachian Plateau. But some of the ice moved on up the valleys that had previously drained the plateau. The energy concentrated in the valleys was so great that the ice was able to scoop out the basins that are now filled by the Finger Lakes.

ASYMMETRIC ROCK KNOBS AND HILLS Glacier erosion of bedrock in many places produces small, rounded, asymmetric hills with gentle, striated, and polished slopes on one side and steeper slopes lacking polish and striations on the opposite side. An assemblage of these undulating knobs is referred to as *rôches moutonnées,* from the French for "rocks" plus "curved." The now-gentle slope faced the advancing glacier and was eroded by abrasion. The opposite slope has been steepened by the plucking action of the ice as it rode over the knob (see Figure 13.15).

Large individual hills have the same asymmetric profiles as the smaller hills. Here, too, the gentle slope faced the moving ice.

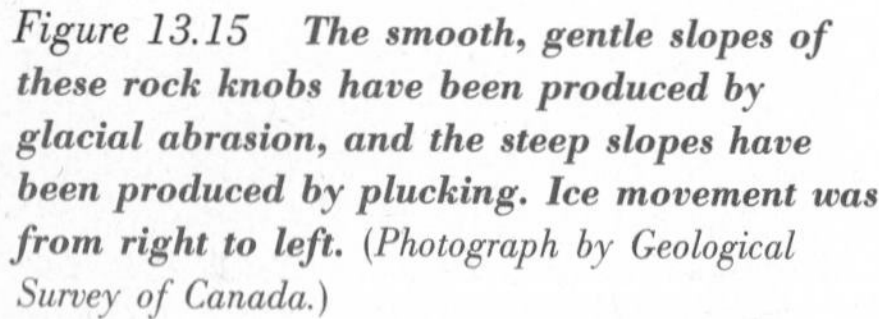

Figure 13.15 **The smooth, gentle slopes of these rock knobs have been produced by glacial abrasion, and the steep slopes have been produced by plucking. Ice movement was from right to left.** (*Photograph by Geological Survey of Canada.*)

Types of glacial deposits

The debris carried along by a glacier is eventually deposited, either because the ice that holds it melts or, less commonly, because the ice smears the debris across the land surface.

The general term *drift* is applied to all deposits that are laid down directly by glaciers or that, as a result of glacial activity, are laid down in lakes, oceans, or streams. The term dates from the days when geologists thought that the unconsolidated cover of sand and gravel blanketing much of Europe and America had been made to drift into its present position either by the sea or by icebergs. Drift can be divided into two general categories: *stratified* and *unstratified.*

DEPOSITS OF UNSTRATIFIED DRIFT Unstratified drift laid down directly by glacier ice is called *till.* It is composed of rock fragments of all sizes mixed together in random fashion, ranging all the way from boulders weighing several metric tons to tiny clay and colloid particles (see Figures 13.16 and 13.17). Many of the large pieces are striated, polished, and faceted as a result of the wear they underwent while being transported by the glaciers (see Figures 13.18 and 13.19). Some of the material picked up along the way was smeared across the landscape during the glacier's progress, but most of it was dumped when the rate of wastage began to exceed the rate of nourishment and the glacier gradually melted away.

The type of till varies from one glacier to another. Some tills, for instance, are known as *clay tills,* because clay-sized particles predominate, with only a scattering of larger units. Many of the most recent tills in northeastern and

Figure 13.16 **The range in the size of the particles composing till is very large. In this photograph of an exposure of till near Guilford, Connecticut, note the boulders and cobbles mixed with smaller particles ranging all the way down to colloid size. Spade gives scale.** (*Photograph by Sheldon Judson.*)

Figure 13.17 **This moraine composed of till lies along the East Rosebud River and records the last major advance of glaciers from the Beartooth Mountains, Montana. The largest boulders measure 3 m in maximum dimension.**

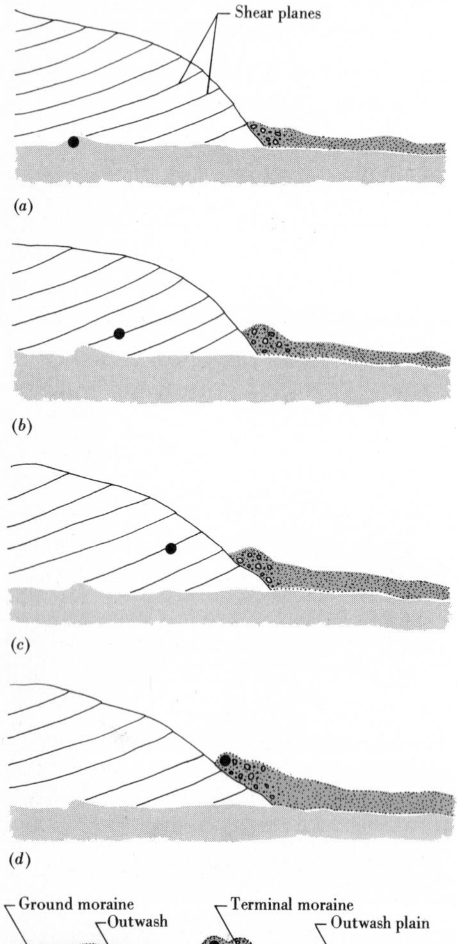

Figure 13.18 ***A sequence of diagrams to suggest the growth of a terminal moraine at the edge of a stable ice front. The progressive movement of a single particle is shown. In (a), it is moved by the ice from the bedrock floor. Forward motion of ice along a shear plane carries it ever closer to the stabilized ice margin, where finally it is deposited as a part of the moraine in (d). Diagram (e) represents the relation of the terminal moraine, ground moraine, and outwash after the final melting of the glacier.***

eastern Wisconsin are of this type, but in many parts of New England the tills are composed for the most part of large rock fragments and boulders. Deposits of this sort are known as *boulder tills,* or *stony tills.*

Some till deposits seem to have been worked over by meltwater. The materials have begun to be sorted out according to size, and some of the finer particles may even have been washed away. This is the sort of winnowing action we would expect to find near the nose of a melting glacier, where floods of meltwater wash down through the deposits.

Till is deposited by glaciers in a great variety of topographic forms, including moraines, drumlins, erratics, and boulder trains.

Moraines *Moraine* is a general term used to describe many of the landforms that are composed largely of till.

A *terminal moraine,* or *end moraine,* is a ridge of till that marks the utmost limit of a glacier's advance. These ridges vary in size from ramparts scores of meters high to very low, interrupted walls of debris. A terminal moraine forms when a glacier reaches the critical point of equilibrium—the point at which it wastes away at exactly the same rate as it is nourished. Although the front of the glacier is now stable, ice continues to push downward from above, delivering a continuous supply of rock debris. As the ice melts in the zone of wastage, the debris is dumped and the terminal moraine grows. At

Figure 13.19 **Stones from till are marked by facets and striations. The specimens are from northern Illinois. The smallest is about 5 cm across.** (*Photograph by Willard Starks.*)

the same time, water from the melting ice pours down over the till and sweeps part of it out in a broad flat fan that butts against the forward edge of the moraine like a giant ramp (Figure 13.18).

The terminal moraine of a mountain glacier is crescent-shaped, with the convex side extending down-valley. The terminal moraine of a continental ice sheet is a broad loop or series of loops traceable for many miles across the countryside.

Behind the terminal moraine, and at varying distances from it, a series of smaller ridges known as *recessional moraines* may build up. These ridges mark the position where the glacier front was stabilized temporarily during the retreat of the glacier.

Not all the rock debris carried by a glacier finds its way to the terminal and recessional moraines, however. A great deal of till is laid down as the main body of the glacier melts to form gently rolling plains across the valley floor. Till in this form, called a *ground moraine,* may be a thin veneer lying on the bedrock, or it may form a deposit scores of meters thick, partially or completely clogging preglacial valleys.

Finally, valley glaciers produce two special types of moraines. While a valley glacier is still active, large amounts of rubble keep tumbling down from the valley walls, collecting along the side of the glacier. When the ice melts, this debris is stranded as a ridge along each side of the valley, forming a *lateral moraine.* At its down-valley end, the lateral moraine may grade into a terminal moraine.

The other special type of deposit produced by valley glaciers is a *medial moraine,* created when two valley glaciers join to form a single ice stream; material formerly carried along on the edges of the separate glaciers is combined in a single moraine near the center of the enlarged glacier. A streak of this kind builds up whenever a tributary glacier joins a larger glacier in the main valley (see Figure 1.3). Although medial moraines are very characteristic of living glaciers, they are seldom preserved as topographic features after the disappearance of the ice.

Drumlins Drumlins are smooth, elongated hills composed largely of till. The ideal drumlin shape has an asymmetric profile with a blunt nose pointing in the direction from which the vanished glacier advanced, and with a gentler, longer slope pointing in the opposite direction. Drumlins range from about 8 to 60 m in height, the average being about 30 m. Most drumlins are between $\frac{1}{2}$ and 1 km in length and are usually several times longer than they are wide.

In most areas, drumlins occur in clusters, or *drumlin fields.* In the United States, these are most spectacularly developed in New England, particularly around Boston; in eastern Wisconsin; in west-central New York State, particularly around Syracuse; in Michigan (Figure 13.20); and in certain sections of Minnesota. In Canada, extensive drumlin fields are located in western Nova Scotia and in northern Manitoba and Saskatchewan; Figure 13.21 shows a drumlin field in British Columbia.

Just how drumlins were formed is still not clear. Because their shape is a nearly perfect example of streamlining, it seems probable that they were formed deep within active glaciers in the zone of plastic flow.

Erratics and boulder trains A stone or a boulder that has been carried from its place of origin by a glacier and left stranded on bedrock of different composition is called an *erratic.* The term is used whether the stone is embedded in a till deposit or rests directly on the bedrock. Some erratics weigh several tons, and a few are even larger. Near Conway, New Hampshire, there is a granite erratic 27 m in maximum dimension, weighing close to 9,000 metric tons. Although most erratics have traveled only a limited distance, many have been carried along by the glacier for hundreds of kilometers. Chunks of native copper torn from the Upper Peninsula of Michigan, for example, have been transported as far as southeastern Iowa, a distance of nearly 800 km, and to southern Illinois, a distance of over 960 km.

Boulder trains consist of a series of erratics that have come from the same source, usually with some characteristic that makes it easy to recognize their common origin. The trains appear either as a line of erratics stretching downvalley from their source or in a fan-shaped pattern with the apex near the place of origin. By mapping boulder trains that have been left behind by continental ice sheets, we can obtain an excellent indication of the direction of the ice flow (see Figure 13.22)

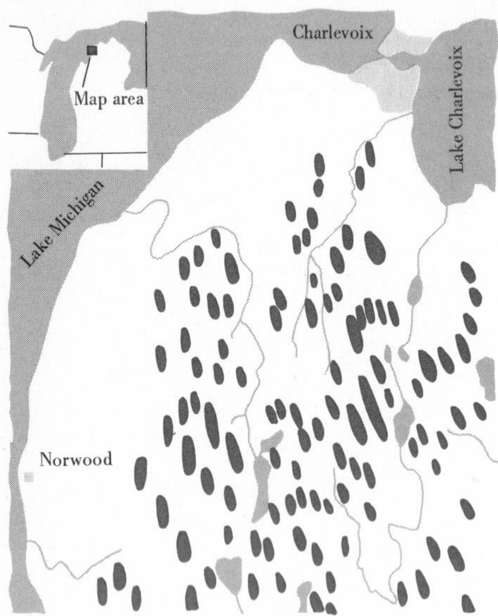

Figure 13.20 **Part of the drumlin field area near Charlevoix, Michigan. Ice moved toward the south-southeast.** (*Frank Leverett and F. B. Taylor, "The Pleistocene of Indiana and Michigan and the History of the Great Lakes,"* U.S. Geol. Surv. Monograph, *no. 53, p. 311, 1915.*)

DEPOSITS OF STRATIFIED DRIFT Stratified drift is ice-transported sediment that has been washed and sorted by glacial meltwaters according to particle size. Because water is a much more selective sorting agent than ice, deposits of stratified drift are laid down in recognizable layers, unlike the random arrangements of particles typical of till. Stratified drift occurs in outwash and kettle plains, eskers, kames, and varves—all discussed below.

Figure 13.21 **This aerial photograph shows the drumloidal patterns of hills and intervening grooves formed parallel to the flow of glacier ice on the Nechako Plateau, British Columbia. Weedon Lake is 9 km long. View is northeast in the direction of ice movement.** (*Photograph by U.S. Army Air Force.*)

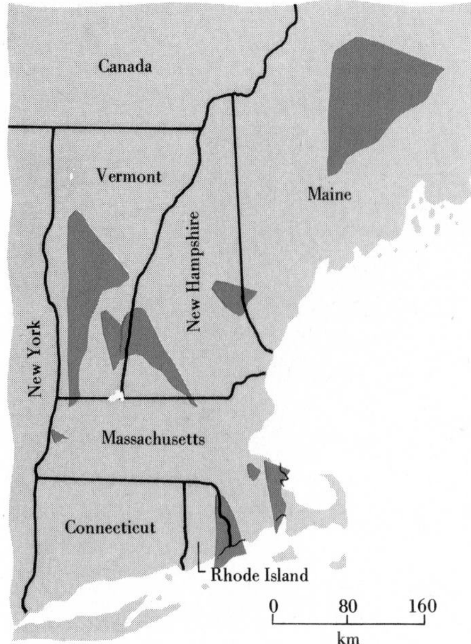

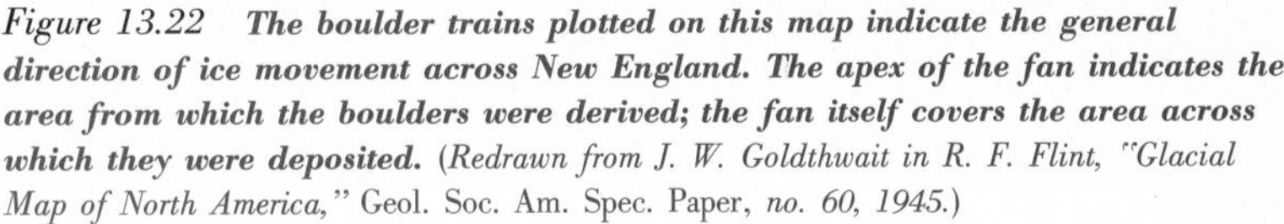

Figure 13.22 ***The boulder trains plotted on this map indicate the general direction of ice movement across New England. The apex of the fan indicates the area from which the boulders were derived; the fan itself covers the area across which they were deposited.*** (*Redrawn from J. W. Goldthwait in R. F. Flint, "Glacial Map of North America,"* Geol. Soc. Am. Spec. Paper, *no. 60, 1945.*)

Outwash sand and gravel The sand and gravel that are carried outward by meltwater from the front of a glacier are referred to as *outwash* (see Figure 13.23). As a glacier melts, streams of water heavily loaded with reworked till, or with material *washed* directly from the ice, weave a complex, braided pattern of channels across the land in front of the glacier. These streams, choked with clay, silt, sand, and gravel, rapidly lose their velocity and dump their load of debris as they flow away from the ice sheet. In time, a vast apron of bedded sand and gravel is built up that may extend for kilometers beyond the ice front. If the zone of wastage happens to be located in a valley, the outwash deposits are confined to the lower valley and compose a *valley train.* But along the front of a continental ice sheet, the outwash deposits stretch out for kilometers, forming what is called an *outwash plain* (see Figure 13.24).

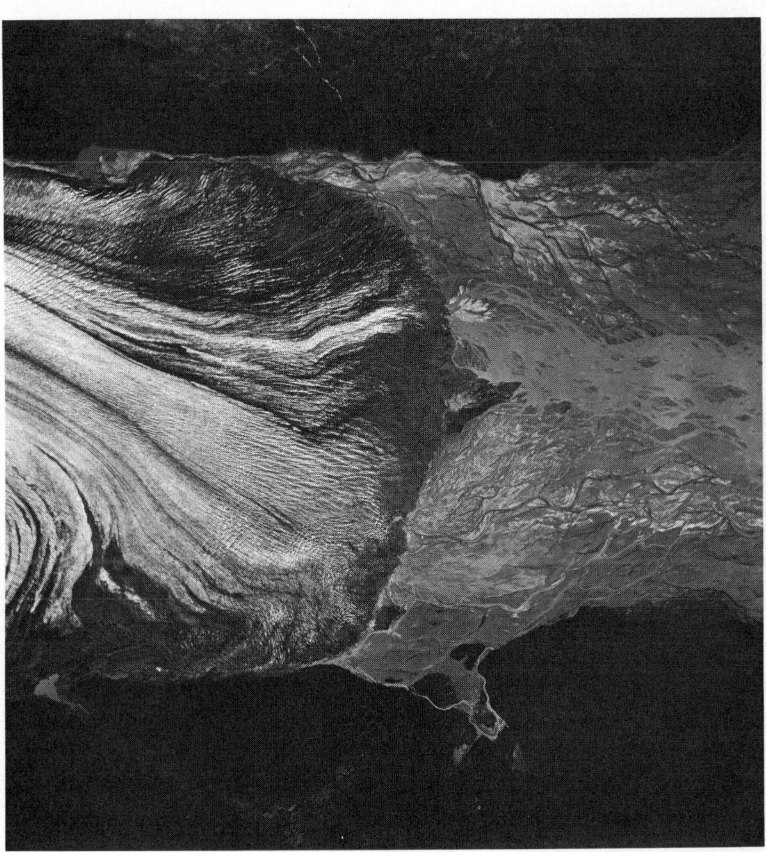

Figure 13.23 ***Outwash forms at the snout of Brady Range, Fairweather, Alaska.*** (*Photograph by Austin Post, University of Washington.*)

Figure 13.24 **These sand and gravel deposits represent the outwash from a now-vanished continental glacier. North of Otis Lake, Wisconsin.** (*Photograph by W. C. Alden, U.S. Geological Survey.*)

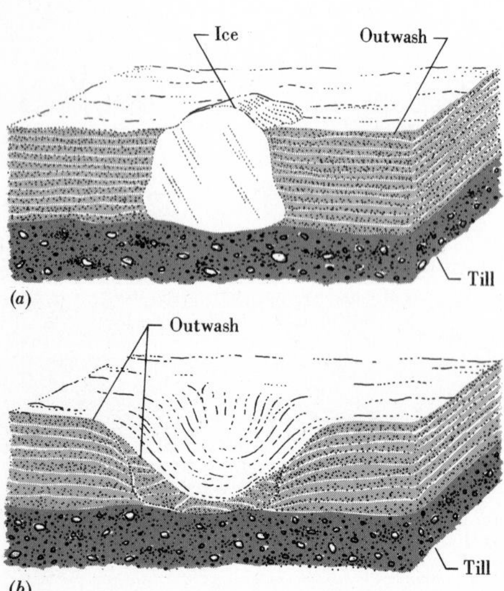

Figure 13.25 **Sequence in the formation of a kettle. A block of stagnant ice is almost buried by outwash in (a). The eventual melting of the ice produces a depression, as shown in (b).**

Kettles Sometimes a block of stagnant ice becomes isolated from the receding glacier during wastage and is partially or completely buried in till or outwash before it finally melts. When it disappears, it leaves a *kettle,* a pit or depression in the drift (see Figure 13.25). These depressions range from a few meters to several kilometers in diameter and from a few meters to over 30 m in depth. Many outwash plains are pockmarked with kettles and are referred to as *pitted outwash plains.* As time passes, water sometimes fills the kettles to form lakes or swamps, features found through much of Canada and the northern United States.

Eskers and crevasse fillings Winding, steep-walled ridges of stratified gravel and sand, sometimes branching and often discontinuous, are called *eskers* (see Figure 13.26). They usually vary in height from about 3 to 15 m, although a few are over 30 m high. Eskers range from a fraction of a kilometer to over 160 km in length, but they are only a few meters wide. Most investigators believe that eskers were formed by the deposits of streams running through tunnels beneath stagnant ice. Then, when the body of the glacier finally disappeared, the old stream deposits were left standing as a ridge.

Crevasse fillings are similar to eskers in height, width, and cross profile, but unlike the sinuous and branching pattern or eskers, they run in straight ridges. As their name suggests, they were probably formed by the filling of a crevasse in stagnant ice.

Kames and kame terraces In many areas, stratified drift has built up low,

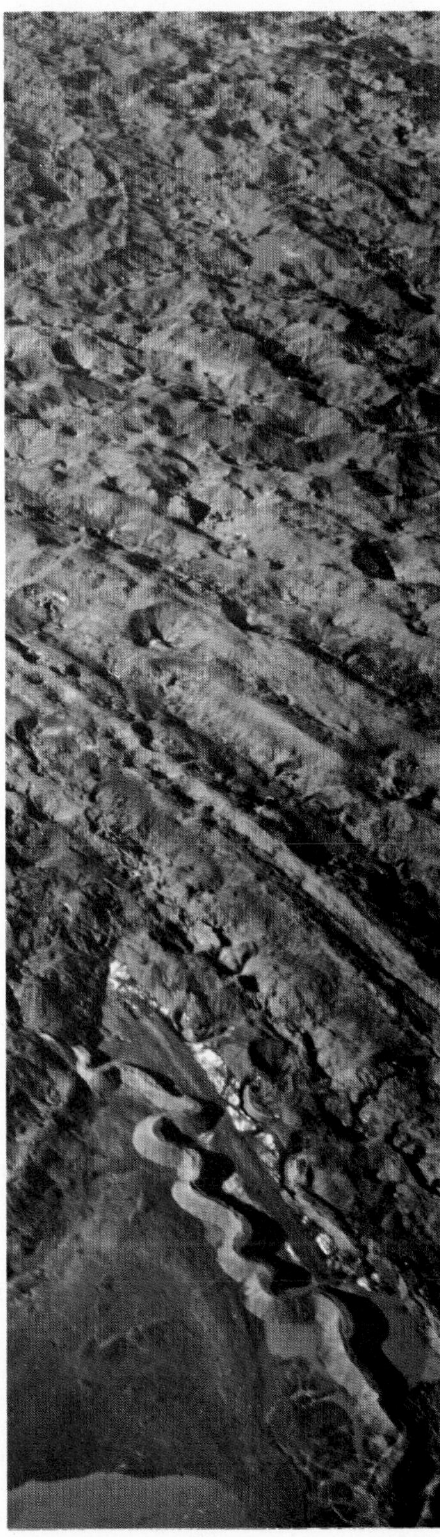

Figure 13.26 **An esker exposed by the retreat of the Malaspina glacier, Alaska. Note depressions in the areas of rough topography. They have formed by melting of ice blocks.** (*Photograph by Austin Post, University of Washington.*)

Figure 13.27 **This isolated kame in eastern Wisconsin was formed by the partial filling of an opening in stagnant glacier ice. The melting of the ice has left this steep-sided hill of stratified material.** (*Photograph by Raymond C. Murray.*)

Figure 13.28 **The sequence in the development of a kame terrace. (a) Ice wasting from an irregular topography lingers longest in the valleys. (b) While the ice still partially fills one of these valleys, outwash may be deposited between it and the valley walls. (c) The final disappearance of the ice leaves the outwash in the form of terraces along the sides of the valley.**

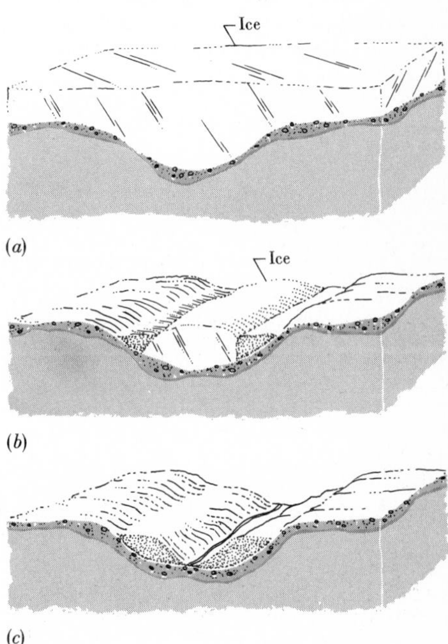

Table 13.1 ***Features of valley and continental glaciations***

Features	*Valley*	*Continental*
Striations, polish, etc.	Common	Common
Cirques	Common	Absent
Horns, arêtes, cols	Common	Absent
U-shaped valley, truncated spurs, hanging valleys	Common	Rare
Fiords	Common	Absent
Till and stratified drift	Common	Common
Terminal moraine	Common	Common
Recessional moraine	Common	Common
Ground moraine	Common	Common
Lateral moraine	Common	Absent
Medial moraine	Common, easily destroyed	Absent
Drumlins	Rare or absent	Locally common
Erratics	Common	Common
Kettles	Common	Common
Eskers, crevasse fillings	Rare	Common
Kames	Common	Common
Kame terraces	Common	Present in hilly country

relatively steep-sided hills called *kames*, either as isolated mounds or in clusters. Unlike drumlins, kames are of random shape and the deposits that compose them are stratified. They were formed by the material that collected in openings in stagnant ice. In this sense they are similar to crevasse fillings but without the linear pattern (see Figure 13.27).

A *kame terrace* is a deposit of stratified sand and gravel that has been laid down between a wasting glacier and an adjacent valley wall. When the glacier disappears, the deposit stands as a terrace along the side of the valley (see Figure 13.28).

Varves A *varve* is a pair of thin sedimentary beds, one coarse and one fine. This couplet of beds is usually interpreted as representing the deposits of a single year and is thought to form in the following way. During the period of summer thaw, waters from a melting glacier carry large amounts of clay, fine sand, and silt out into lakes along the ice margin. The coarser particles sink fairly rapidly and blanket the lake floor with a thin layer of silt and silty sand. But as long as the lake is unfrozen, the wind creates currents strong enough to keep the finer clay particles in suspension. When the lake freezes over in the winter, these wind-generated currents cease, and the fine particles sink through the quiet water to the bottom, covering the somewhat coarser summer layer. A varve is usually a few millimeters to 1 cm thick, though thicknesses of 5 to 8 cm are not uncommon. In rare instances, thicknesses of 30 cm or more are known.

COMPARISON OF VALLEY AND CONTINENTAL GLACIATION FEATURES Some of the glacial features that we have been discussing are more common in areas that have undergone valley glaciation; others usually occur only in regions that have been overridden by ice sheets; many other features, however, are found in both types of area. Table 13.1 lists and compares the features that are characteristic of the two types.

13.3 *Development of the glacial theory*

Geologists have made extensive studies of the behavior of modern glaciers and have carefully interpreted the traces left by glaciers that disappeared thousands of years ago. On the basis of their studies, they have developed *the glacial theory* that *in the past great ice sheets covered large sections of the earth where no ice now exists, and that many existing glaciers once extended far beyond their present limits.*

The beginnings

The glacial theory took many years to evolve, years of trying to explain the occurrence of erratics and the vast expanses of drift strewn across northern Europe, the British Isles, Switzerland, and adjoining areas. The exact time when inquisitive minds first began to seek an explanation of these deposits is shrouded in the past, but by the beginning of the eighteenth century, explanations of what we now know to be glacial deposits and features were finding their way into print. According to the most popular early hypothesis, a great inundation had swept these deposits across the face of the land with cataclysmic suddenness or else had drifted them in by means of floating icebergs. Then, when the flood receded, the material was left stranded in its present location.

By the turn of the nineteenth century, a new theory was in the air—the theory of ice transport. We do not know who first stated the idea or when it was first proposed, but it seems quite clear that it was not hailed immediately as a great truth. As the years passed, however, more and more observers became intrigued with the idea. The greatest impetus came from Switzerland, where the activity of living glaciers could be studied on every hand.

In 1821, J. Venetz, a Swiss engineer delivering a paper before the Helvetic Society, presented the argument that Swiss glaciers had once expanded on a great scale. It has since been established that from about 1600 to the middle of the eighteenth century there actually was a time of moderate but persistent glacier expansion in many localities. Abundant evidence in the Alps, Scandinavia, and Iceland indicates that the climate was milder during the Middle Ages than it is at the present, that communities existed and farming was carried on in places later invaded by advancing glaciers or devastated by glacier-fed streams. We know, for example, that a silver mine in the valley of Chamonix was being worked during the Middle Ages and that it was subsequently buried by an advancing glacier, where it lies to this day. And the village of St. Jean de Perthuis has been buried under the Brenva glacier since about 1600.

Although Venetz' idea did not take hold immediately, by 1834 Jean de Charpentier was arguing in its support before the same Helvetic Society. Yet the theory continued to have more opponents than defenders. It was one of the skeptics, Jean Louis Rodophe Agassiz, who did more than anyone else to develop the glacial theory and bring about its general acceptance.

Agassiz

Louis Agassiz (1807–1873), a young zoologist, had listened to Charpentier's explanation; afterward, he undertook to demonstrate to his friend and colleague the error of his ways. During the summer of 1836, the two men made a trip together into the upper Rhône valley to the Getrotz glacier. Before the summer was over, it was Agassiz who was convinced of his error. In 1837, he spoke before the Helvetic Society championing the glacial theory and suggesting that during a "great ice age" not only the Alps but much of northern Europe and the British Isles were overrun by a sea of ice.

Agassiz' statement of the glacial theory was not accepted immediately, but in 1840 he visited England and won the support of leading British geologists. In 1846 he arrived in the United States, where in the following year he became professor of zoology at Harvard College and later founded the Museum of Comparative Zoology. In this country, he convinced geologists of the validity of the glacial theory; by the third quarter of the nineteenth century the theory was firmly entrenched. The last opposition died with the turn of the century.

Proof of the glacial theory

What proof is there that the glacial theory is valid? The most important evidence is that certain features produced by glacier ice are produced by no other known process. Thus Agassiz and his colleagues found isolated stones and boulders quite alien to their present surroundings. They noticed, too, that boulders were actually being transported from their original location by modern ice. Some of the boulders they observed were so large that rivers could not possibly have moved them, and others were perched on high places that a river could have reached only by flowing uphill. They also noticed that when modern ice melted it revealed a polished and striated pavement unlike the surface fashioned by any other known process. To explain the occurrence of these features in areas where no modern glaciers exist, they postulated that the ice once extended far beyond its present limits.

Notice that the development of this theory sprang from a concept that we mentioned earlier: "The present is the key to the past." The proof of glaciation lies not in the authority of the textbook or the lecture. It lies in observing modern glacial activity directly and in comparing the results of this activity with features and deposits found beyond the present extent of the ice.

Theory of multiple glaciation

Even before universal acceptance of the glacial theory, which spoke of a single, great Ice Age, some investigators were coming to the conclusion that the ice had advanced and retreated not just once but several times in the recent geological past. By the early twentieth century, a broad fourfold division of the Ice Age, or Pleistocene, had been demonstrated in this country and in Europe. According to this theory, each major glacial advance was followed by a retreat and a return to climates that were sometimes even warmer than that of the present. In the United States, each glacial period is named for a midwestern state where deposits of that particular period were first studied or where they are well exposed: Nebraskan, Kansan, Illinoian, and Wisconsin (see Figure 13.29). Evidence from Alaska, the Sierra Nevadas, and western Europe indicates that one or two major ice advances preceded the Nebraskan.

This subdivision of the Ice Age was built on a great variety of evidence from widely scattered localities. All the evidence cited, however, points in the same direction: that in each period a glaciation took place, that it was followed by an interval during which the glacier wasted and disappeared, and that finally a younger glacier moved forward. Some of the lines of evidence used to support this concept are suggested in Figure 13.30. Needless to say, all this evidence is seldom present in any one exposure of glacial material, nor does our hypothetical example represent all the evidence that could be assembled to demonstrate multiple glaciation.

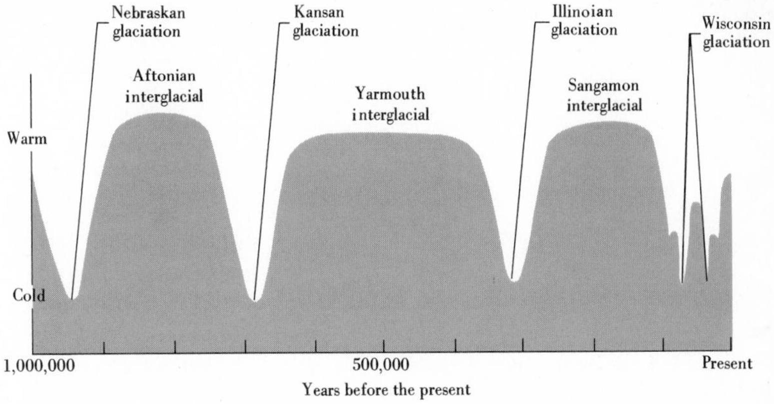

Figure 13.29 ***Investigation of the glacial and interglacial deposits shows that the great Ice Age or Pleistocene has been marked by four major advances of continental glaciers separated by periods of ice recession.***

On the basis of the evidence summarized in Figure 13.30, we can build up the following arguments in support of at least two major periods of glaciation:

1 Two different till types are present: a lower, gray till and an upper, red till. Because the composition of the two tills differs, they must have been deposited by ice from two different sources. The most logical explanation would be that the gray till was laid down first by glacier ice moving in from a given direction over rock material that was predominantly gray. Then a second glacier moved in from a different direction, over rock material that was predominantly red, and laid down the red till on top of the gray.

2 Cut into the surface of the gray till are old stream channels that must have been formed after the disappearance of the first glacier and before the appearance of the second.

3 The forest bed buried between the red and the gray till indicates that a forest must have grown up after the disappearance of the first glacier and before the appearance of the second.

4 Soils shown in Figure 13.30 indicate multiple glaciation in two different ways. In the first place, notice that the soil developed on the gray till is buried by red till. Again, the reasoning applied in numbers 2 and 3 also holds true here. The now-buried soil must have formed in an ice-free interval between two glaciations.

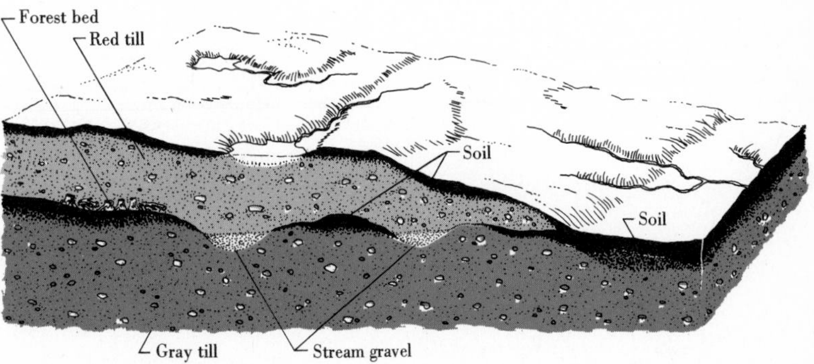

Figure 13.30 ***This diagram depicts some of the evidence that points to multiple glaciation. See the text for discussion.***

But there is also another convincing argument. Notice that the soil developed on the red till is much thinner than that developed on the *unburied* gray till. We can say that the thickness of a soil is, in a general way, an index to the length of weathering. Therefore, the unburied gray till must have undergone a longer period of weathering than did the red till. We can explain this difference in weathering periods if the red till is younger and was laid down long after the gray till was deposited. It then follows that there was an older glaciation represented by the deeply weathered till and a younger glaciation represented by the less deeply weathered till.

5 The topography of the two tills is different, indicating two distinct glaciations. Notice that there are water-filled kettles on the surface of the red till but not on the well-drained gray-till surface. Originally, the gray till was probably marked by kettles just as the red is now. But during the long nonglacial period, the kettles were filled, the low hills were worn down, and an efficient drainage system was built up. We can conclude that the gray till arrived long before the ice that deposited the red till and that a considerable period of time passed between the two glaciers.

Extent of Pleistocene glaciation

During the maximum advance of the glaciers of the Wisconsin age (the last of the four great ice advances during the Pleistocene), 39 million km^2 of the earth's surface—about 27 percent of the present land areas—were probably

Figure 13.31 **The extent of Pleistocene glaciation (white areas) in the northern hemisphere.** (*Redrawn from Ernst Anters, "Maps of the Pleistocene Glaciations,"* Geol. Soc. Am. Bull., *vol. 40, p. 636, 1929.*)

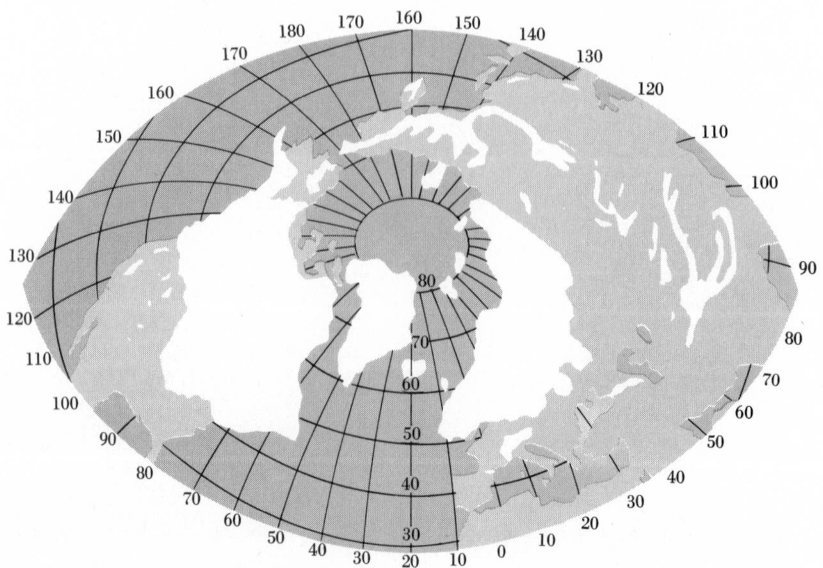

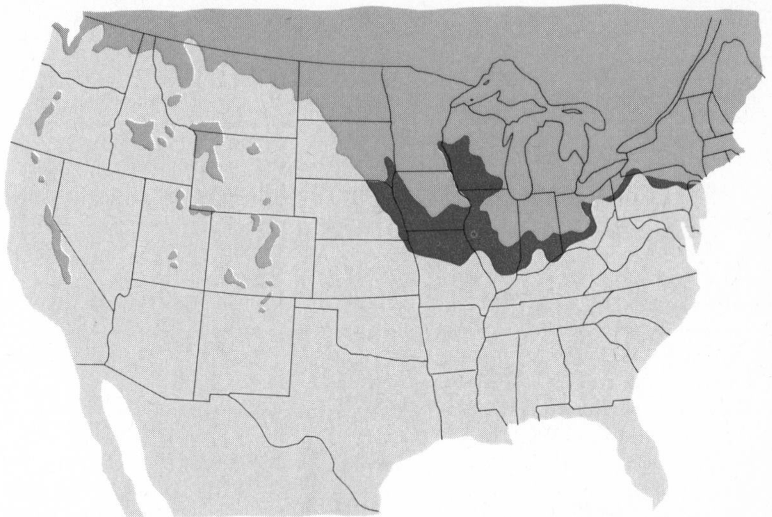

Figure 13.32 ***The extent of Pleistocene glaciation in the United States. Light green zones indicate area covered at various times during the Wisconsin glaciation. Dark green zone represents area glaciated during pre-Wisconsin stages but not covered by the later Wisconsin ice. The gray area was unglaciated.*** (*Modified from R. F. Flint, "Glacial Map of North America,"* Geol. Soc. Am. Spec. Paper, *no. 60, 1945.*)

buried by ice. Approximately 15 million km^2 of North America were covered. Greenland was also under a great mass of ice, as it is now. In Europe, an ice sheet spread southward from Scandinavia across the Baltic Sea and into Germany and Poland, and the Alps and the British Isles supported their own ice caps. Eastward in Asia, the northern plains of Russia were covered by glaciers, as were large sections of Siberia and of the Kamchatka Peninsula and the high plateaus of Central Asia (see Figure 13.31).

In eastern North America, ice moved southward out of Canada to New Jersey, and in the Midwest it reached as far south as St. Louis. The western mountains were heavily glaciated by small ice caps and valley glaciers. The southernmost glaciation in the United States was in the Sierra Blanca of south-central New Mexico. The maximum extent of the Wisconsin glaciation in the United States and the maximum limit of glaciation during the Pleistocene are shown in Figure 13.32.

13.4 *Indirect effects of glaciation*

The glaciers that diverted rivers, carved mountains, and covered half a continent with debris also gave rise to a variety of indirect effects that were felt far beyond the glaciers' immediate margins. Not all these effects are completely understood, even today.

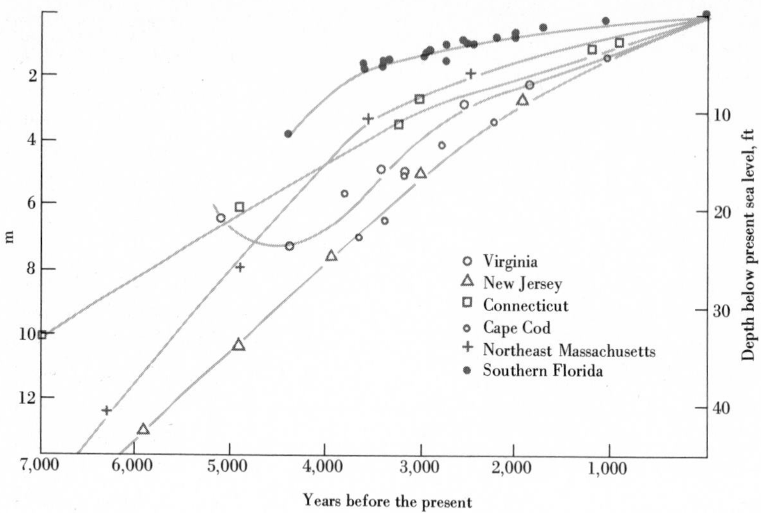

Figure 13.33 **Sea level has been rising for the last several thousand years. Here radiocarbon dates on samples of wood, shell, and peat originally deposited at or close to sea level have been combined to produce these curves of sea-level rise at various points along the eastern coast of the United States.** (*Redrawn from David Scholl and Minze Stuiver,* Geol. Soc. Am. Bull., *vol. 78, p. 488, 1967.*)

Changing sea level

The water that is now locked up in the ice of glaciers originally came from the oceans. It was transferred landward by evaporation and winds, precipitated as snow, and finally converted to firn and ice. If all this ice were suddenly to melt, it would find its way back to the ocean basins and would raise the sea level an estimated 60 m. A rise of this magnitude would transform the outline of the earth's land masses and would submerge towns and cities along the coasts. For the last several thousand years, melting glaciers have been raising the sea level (see Figure 13.33), and modern records of sea level indicate that the sea is still rising. Along the Atlantic coastline of the United States, the rate of rise prior to the early 1920s was about $8\frac{3}{4}$ cm/century. Thereafter, the rate of rise increased rapidly to about 60 cm/century. This sudden change in rate is coincident with a marked quickening of the retreat of present-day glaciers. It is tempting to see a cause and effect in this coincidence between the increased rate of the rise of sea level and the more rapid melting of glacier ice.

During most of the glacial periods of the past, water impounded on the land in the form of ice was more extensive than it is at present. Consequently, sea level must have been lower than it is now, and during the interglacial periods, when glaciers were even less widespread than they are today, the sea level must have been higher. It is a difficult job to estimate how far the sea level fell during a great glacial advance. To begin with, we have to know the total volume of ice at the height of the advance. To calculate that we must know what part of the earth's area was covered by ice and what the average thickness of the ice cover was. We can make a fairly good guess of the area that was covered by studying evidence that still exists, but to estimate the average thickness of the ice is extremely difficult. In any event, through a series of carefully controlled approximations, it has been estimated that during the maximum extent of Pleistocene glaciation the sea level was from 105 to

120 m lower than it is now. Most geologists accept this estimate, although some feel that it is too conservative. During the height of the Wisconsin glaciation, the last of the major ice advances, sea level is usually estimated as having been between 70 and 100 m lower than at present.

Pluvial periods

During the Ice Age, when glaciers lay across Canada and the northern United States and draped the flanks of most of the higher mountains of the West, the climate of the arid and semiarid areas of the western United States was quite different from what it is today. Glaciations produced *pluvial periods* (from the Latin *pluvia,* "rain") when the climate was undoubtedly not as moist as that of the eastern United States today but certainly more hospitable than we now find in this region. During any single pluvial period, rainfall was greater, evaporation less, and vegetation more extensive. At this time large sections of the southwestern states were dotted by lakes, known as *pluvial lakes* (Figure 13.34).

What was responsible for this pluvial climate? The presence of continental glaciers in the north is thought to have modified the general wind circulation of the globe. The belt of rain-bearing winds was moved to the south and the temperatures were lowered. Consequently, the rates of evaporation decreased and at the same time the amount of precipitation increased. When the ice receded, the climate again became very much what it is today.

Figure 13.34 ***During periods of glaciation, many basins of the western United States held permanent lakes where there are only intermittent water bodies today, Modern salt-water lakes are mere remnants of other basins formerly filled with lakes. The relative number and size of modern lakes (b) are compared with those existing during the glaciation of higher mountains and the northern United States (a).*** *(Redrawn from O. E. Meinzer, "Map of the Pleistocene Lakes of the Basin-and-Range Province and Its Significance,"* Geol. Soc. Am. Bull., *vol. 33, pp. 543, 545, 1922.)*

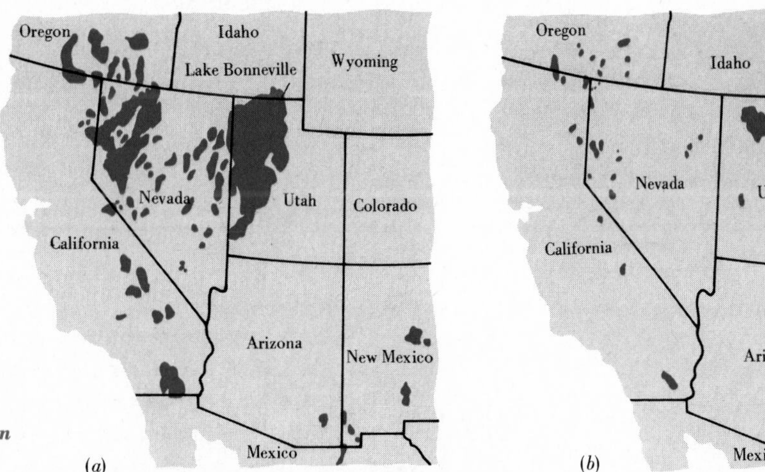

13.5 Pre-Pleistocene glaciations

So far, we have discussed only the glaciers that exist today and those that moved within the last million years or so, the Pleistocene. Geologists have found evidence that glaciers appeared and disappeared in other periods as well. The record is fragmentary, as we would expect, for time tends to conceal, jumble, and destroy the effects of glaciation.

Recent evidence indicates that glaciation and deglaciation so characteristic of the Pleistocene reaches back into late Miocene in Alaska. Glacial deposits in Antarctica have been dated by the potassium-argon method as 10 million years in age. Farther back in the geologic time scale it seems certain that there were other extensive glaciations. One of these was near the end of the Paleozoic era, probably in the Permian, 230 million years ago. Another was in the late Precambrian, over 600 million years ago. A third period of widespread glaciation may have occurred in the earliest Paleozoic, almost 600 million years ago. Later, in Chapter 20, we will refer more specifically to the late Paleozoic glaciation when we discuss the evidence bearing on continental drift.

This glaciation timetable is based on evidence found in rocks that were formed from sediments characteristic of glacier activity. A rock formed from the lithification of till is a tillite. Late Paleozoic tillities have been discovered in Africa, India, South America, and Australia. Lying beneath the tillite in many areas is the striated and polished pavement marking the path of the ancient glaciers.

13.6 Causes of glaciation

As Louis Agassiz did over 100 years ago, we can travel about the world today and observe modern glaciers at work, and we can reason convincingly that glaciers were more extensive in the past than they are at present. We can even make out a good case for the belief that glacier ice advanced and receded many times in the immediate and more remote geological past. But can we explain why glaciation takes place? The answer to that question is a simple "no." Still, we can examine the general problem and take stock of what we do know.

The geologic record has contributed some basic data that any theory of the causes of glaciation must take into account. Among these are the following:

1 *Periods of extensive glaciation have coincided with periods of high and extensive continents.* During most of geologic time the continents have been lower than they are today, and shallow seas have flooded across their margins. Such conditions were unfavorable for widespread glaciation. But for several million years the continents have been increasing in elevation, until now, in

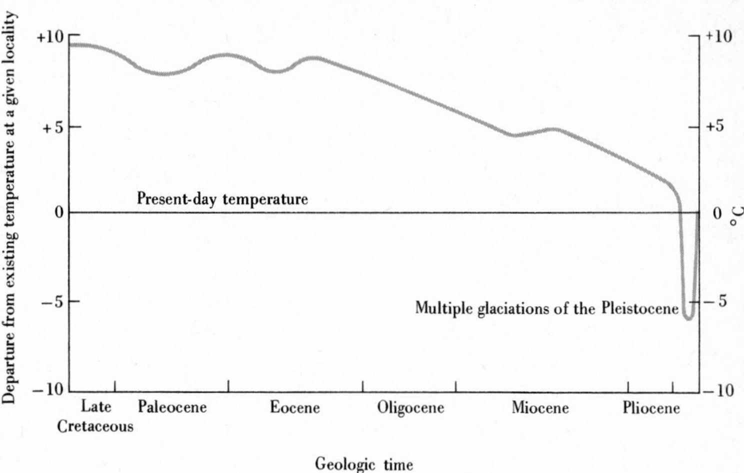

Figure 13.35 **Temperatures trended downward from the Late Cretaceous to the Pliocene. This general decline was followed by the several glacial and interglacial episodes, the details of which cannot be shown at the scale of this diagram. The values in degrees Centigrade represent approximate departures from today's temperatures of localities in the middle latitudes.** (*Data in part from Erling Dorf.*)

the Pleistocene, they stand on the average an estimated 450 m higher than they did in the mid-Cenozoic. We have already found that the last great Ice Age came with the Pleistocene. And we know, too, that the other great glaciations coincided with high continents.

2 *Glaciation is not due to a slow, long-term cooling off of the earth since its creation.* We have already found that extensive glaciation occurred several times during the geological past. But these glacial periods are unusual, for during most of geologic history the climate has been nonglacial.

3 *There has been a cooling of the earth's climate beginning during the Tertiary and climaxing with the glacial fluctuations of the Pleistocene.* Although the earth has been generally warm during most of its history, evidence now shows that its mean temperature had dropped an estimated 8 to 10°C from the Eocene to the end of the Pliocene and that the glacial and interglacial epochs of the Pleistocene are short-term fluctuations at the end of the long-term cooling (see Figure 13.35).

4 *The advance and retreat of glaciers have probably been broadly simultaneous throughout the world.* For instance, dating by means of radioactive carbon has demonstrated that geologically recent fluctuations of the continental glaciers in North America occurred at approximately the same time as similar fluctuations in Europe. Furthermore, observations indicate that the general retreat of mountain glaciers now recorded in North America and Europe is duplicated in South America.

One way to approach the problem is to consider that we are dealing with two great classes of causes. One is longterm and has caused the general decline of temperatures since the Eocene. The other is shorter range and accounts for the comings and going of the ice sheets during the Pleistocene. One operates on a scale of approximately 50 million years and the other on the scale of perhaps 100,000 years. In considering these two different types of causes, we can envisage the long-term cause as cooling the climate to the point where one or more other causes of shorter amplitude can trigger glaciation and deglaciation.

Long-term changes in climates

Geologic evidence shows that the earth's "normal" climate has been more equable than that of the present. Through most of the earth's history, climate has been warmer than the present and land masses presently in the poleward zones enjoyed average annual temperatures about 10°C higher than today. Sometime during the early Tertiary, probably in the Eocene, average temperatures began to drop. The evidence for this drop is of two types: paleontological and chemical.

Paleontological studies show that warmth-loving plants and animals lived closer to the poles in the past than they do now. For instance, Erling Dorf of Princeton University has compiled data based on fossil plants to show that the temperate forest belt of North America was 20° latitude farther north during the Eocene than it is today. At the same time the tropical forest belt had its northern boundary 10° to 12° farther north than it now is. Fossil plants of younger and younger geologic age show that the climate cooled steadily through time until the present distribution of plants was reached toward the end of the Pliocene.

The chemical laboratory confirms the paleontological evidence. In this instance our information comes from the changing ratio of oxygen 18 to oxygen 16. H. C. Urey of the California Institute of Technology first suggested the technique in 1947.

Oxygen 16, the most common of the six known isotopes of oxygen, is about 500 times as abundant as the next most abundant isotope, oxygen 18. The ratio $^{18}O/^{16}O$ does vary, however; and its variation in the calcite shells of certain marine animals allows the ratio to be used as a geologic thermometer. For instance, an organism depositing calcite at a temperature of 0°C will deposit slightly more ^{18}O in the $CaCO_3$ than will be deposited by an organism depositing $CaCO_3$ in water of 25°C. This is true even though the water in which the two live has the same ratio $^{18}O/^{16}O$. This means, then, that the ratio $^{18}O/^{16}O$ of organically produced $CaCO_3$ can be used as an index to the water temperature. Assuming that processes were the same in the geological past as they are today, the $^{18}O/^{16}O$ content of fossil shell material will form the basis for a determination of ancient water temperatures. Despite many difficulties inherent in the method, the data thus far collected indicate a general cooling of the oceans down to the beginning of the Pleistocene when the climatic fluctuations marked by widespread glaciation and deglaciation began.

CAUSES OF LONG-TERM CLIMATIC FLUCTUATION What could have caused the decline of temperatures during the last 50 or 60 million years? We can make a number of suggestions.

Increasing continentality Today the average height of continents is computed to be about 875 m. It is thought that this is about 450 m higher than it was in the beginning of the Eocene. Rising land masses would result in a general drop in temperature. It is estimated that about one-third of the 10°C decline in average temperatures might be accounted for by the increase in the height of continents since the Eocene.

Not only would the rise of continents lower global temperatures but this could also interfere with the patterns of worldwide atmospheric circulation. It is conceivable that the transfer of heat from equatorial latitudes toward the poles would be hindered and that the poleward latitudes would become colder. A corollary to this would be an interruption and change in the oceanic circulation and a lessened efficiency in the transfer of heat poleward.

Continental drift It has been suggested by some geologists that continental drift has positioned the landmasses in such a manner as to favor glaciation (see Chapter 20). Thus Antarctica centers on the southern pole and is in an ideal position for the growth of ice sheets. In the northern hemisphere, Europe, Asia, and North America are arranged around the Arctic Ocean which has restricted communication with the Pacific and the Atlantic. It is argued by some that this arrangement is conducive to the glaciation of the continents. We know, however, that the present pole–continent arrangements have not changed appreciably since the early Tertiary. Therefore, if continental drift is a factor in glaciation, its effect seems to lie in bringing the continents into their present relationship to the poles and to each other, thus facilitating the slow decline in temperature that eventually ended in the glaciations of the Ice Age.

Other Causes Some have suggested that an increase in the number of particles—cosmic dust—in space has decreased the amount of energy received from the sun by the earth. The suggestion is based on the fact that we know that there are great amounts of interstellar material forming vast clouds—dust clouds—in space. It is argued that perhaps the passage of the planetary system through such a cloud has caused glaciation. It is an argument that lacks proof or any forseeable way to test it.

Instead of dust in space it has been suggested that dust from volcanic eruptions has reduced the amount of solar energy that could penetrate the atmosphere and thus cause climatic cooling. The evidence indicates that there has been no appreciable variation in vulcanism through a large part of geologic time for the earth as a whole. We cannot, with any confidence, assign climatic changes of large scale to vulcanism.

There is, however, the possibility that the amount of heat received by the earth has varied appreciably through time. At the present, the amount of thermal energy received on an imaginary plane the diameter of the earth and normal to the sun's rays at the outer edge of the atmosphere is about 2 cal/cm^2-min. Measurements carried on since 1918 suggest that this figure has decreased approximately 3 percent in 50 years. Assuming that the solar radiation does change, some argue that a long-range decrease might account for a cooling of the earth's climate. All we can say is that there appears to have been a measurable change in the solar radiation during historic time. Larger changes *may* have happened in the past and thus affected climate.

Climatic changes of shorter duration

We have seen previously that the Ice Age was marked by the coming and going of glaciers that covered up to one-third of the earth's surface. The duration of these glacial and interglacial intervals has been much shorter than the long-term cooling of climate that we have discussed. The ratio of a glacial–interglacial cycle to the longer period of cooling which preceded the fluctuations of the Ice Age is roughly 1:500.

We are now confident that the climate had long been cooling before the first glaciation of the continents of the northern hemisphere took place. It is quite possible, however, that Antarctica supported an ice cap long before this, even as it does today in a period which we can call interglacial. Whatever the cause or causes of the several glaciations of the Pleistocene, it seems reasonable that they were superimposed on the long-term cooling which lasted some 50 million years. What caused these shorter but violent swings of climate?

CAUSES OF SHORT-TERM CLIMATIC FLUCTUATION Many hypotheses to explain the reasons for glaciations have been put forward since the fact of former glaciations was realized. Although all are interesting, and some ingenious, we shall consider only the theories that have recently gained the widest hearing in scientific circles.

The *heat-distribution hypothesis* is sometimes called the *Milankovitch* hypothesis after the Yugoslavian astronomer instrumental in its development. It holds that variations of the earth's position in relation to the sun produced climatic changes that brought on the great Ice Age. Three factors enter into the changing position of the earth in relation to the sun, each of which can be measured and its rate of change determined. One is the eccentricity of the earth's orbit around the sun. The earth is somewhat closer to the sun at some epochs than others. This motion has a period of about 92,000 years. A second factor is the obliquity of the ecliptic, which produces seasons and is determined by the angle that the earth's axis makes with the plane in which the earth circles the sun. This angle, or obliquity, changes through about 3° every 40,000 years. Finally, there is the precession of the equinoxes, which merely means that the axis of the earth wobbles because of the gravitational effect of the sun, the moon, and the planets. Like a giant top, the earth completes one wobble every 21,000 years. All these factors affect the relation of the earth to the sun—hence the amount of heat received at different places on the earth at any one time. The variation can be calculated backward in time to any desired date for any latitude. These calculations have been made, and a curve showing maxima and minima of heat received at the earth during the Pleistocene is the result. The periods of low heat receipt are said to coincide with and be the cause of the various ice advances.

The heat-distribution theory has long been ardently supported by many European workers and of recent years has gained a number of adherents among U.S. investigators. In some broad aspects it seems to fit the known glaciations of the Pleistocene. One of its weak points, however, is that the scheme does not allow for simultaneous glaciation and deglaciation in the northern and southern hemispheres. There is strong evidence that recent climatic changes in the two hemispheres have proceeded together and that in the late Pleistocene a similar coincidence also occurred. For the rest of the Ice Age, however, our data do not allow us to make a statement one way or the other concerning the simultaneity of climatic changes.

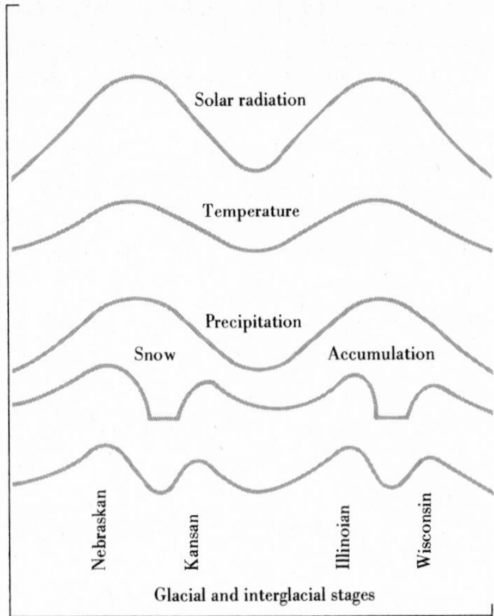

Figure 13.36 **The Simpson explanation of glaciation is based on variation in solar radiation through time and hence on variations of the amount of heat received at the earth's surface through time. The four major glaciations of the Pleistocene are accounted for by two peaks in solar radiation, as illustrated above and more fully explained in the text.** (*Redrawn from F. T. Thwaites after George Simpson, "World Climate During the Quaternary Period,"* Quart. J. Royal Meteorological Soc., *vol. 60, p. 432, 1934, by permission of the Royal Meteorological Society.*)

Solar radiation theory We have already noted in our discussion of the long-term cooling of climate during the Tertiary that variations in the sun's radiation may have occurred and caused climatic changes. The same argument can be applied to explain the shorter fluctuations of the Pleistocene—the glacial and interglacial epochs. Sir George Simpson did just this (see Figure 13.36).

Simpson based a theory on large-scale changes in the solar constant in the past. He assumed that increased solar radiation would raise temperatures, increasing planetary circulation of the atmosphere, precipitation, and cloudiness. The cloudiness and precipitation in low latitudes would produce a pluvial climate. In high latitudes, cloudiness would decrease the melting of increased snowfall. As a result, glaciers would begin to expand in both polar hemispheres and an ice age would begin. Continued rise of heat at the earth's surface, however, would eventually destroy these glaciers and an interglacial period would set in. As the radiation diminished, the process would reverse. Glaciers would form again and a second glaciation would begin, ending only when radiation had decreased to a point where precipitation was inadequate to support the glaciers. They would waste and the second glacial period would end.

This theory is unique in that it calls for glaciation with a rise of temperature and for two glaciations per cycle. If we assumed that two large fluctuations of temperature took place, the radiation curve would produce four glaciations. The theory has attractive points, including the possibility of accounting for four major ice advances in the Pleistocene and simultaneous worldwide fluctuations in climate.

Objections to Simpson's theory include the following:

1 Although we can observe fluctuations of solar radiation, we have no proof that the large fluctuations demanded by the theory have ever actually taken place. In fact, there is theoretical evidence to the contrary in connection with the self-controlled mechanism by which the sun is believed to generate its energy.

2 The theory suggests that two interglacial periods were warm and moist, whereas the middle one was cool and dry. There is no definite evidence in the geologic record indicating different types of interglacial periods. Furthermore, the theory suggests that the pluvial climates in southern latitudes did not coincide with glaciation, whereas the geologic evidence indicates that they were simultaneous for at least the late Pleistocene.

Ocean control hypothesis Oceanic circulation plays a vital role in climate and it may well have been critical in determining the fluctuations of the Pleistocene. Maurice Ewing and William Donn of Columbia University have emphasized that when the Arctic Ocean is permanently frozen, as it is today, there is no evaporation from the ocean, and therefore there is not enough precipitation in northern latitudes to produce snow in the quantities needed to create and nourish the great continental ice sheets of the immediate past.

William Lee Stokes of the University of Utah has suggested an ingenious method by which oceanic circulation might cause a coming and going of the glaciers. He reminds us that the Isthmus of Panama came into existence some time in the late Pliocene or earliest Pleistocene. Prior to this the Atlantic

Ocean connected directly to the Pacific across what is now dry land. Much of the heat carried by the Atlantic equatorial current moved westward into the Pacific instead of circling northward in the Gulf Stream. Now, if the continents of the northern hemisphere were arranged as they are today, then the Arctic Ocean should have been frozen and thus could not have provided moisture to the atmosphere to create and feed continental glaciers.

But as North America was joined with South America by the emergence of the Panamanian Isthmus, the equatorial current would be deflected northward and carry tropical heat with it. Eventually the Arctic Ocean would become ice-free and its water surface would provide through evaporation to the atmosphere precipitation in the form of snow needed to form and support continental glaciers. But once formed these glaciers, Stokes argues, would carry the mechanism of their own destruction. The glaciers would lower sea level, cool climate, and reduce the effectiveness of the Gulf Stream in keeping the Arctic free of perennial ice. Therefore the source of glacier nourishment would freeze over and as a result the glaciers would starve, dwindle, and disappear. With their disappearance and the accompanying climatic warming, a rising sea, and a more vigorous Gulf Stream, the ice cover would begin to disappear and eventually the ocean would become open water. At this time the stage would be set for the beginning of another glaciation. We can note here that measurements over the last 50 years show that the thickness of ice in the Arctic Ocean has decreased. Do we approach another time favorable to glacier formation?

Summary

We have touched on only a few of the theories advanced to explain glaciation, and there are many more we have not even mentioned. To date, we do not know which, if any, of these hold the right answer.

It does seem reasonable, however, to expect that the cause of Pleistocene glaciation is multiple. Clearly, the world climate has cooled slowly during the 50 million years or more from the early Tertiary to the Pleistocene. Whatever the cause or causes of this cooling we can theorize that the climate reached the point at which some other cause or causes took over and produced the glaciations and deglaciations of the Pleistocene, climatic fluctuations of a much shorter wavelength.

13.7 *Implications for the future*

If geology teaches us nothing else, it demonstrates that our globe is in constant change—that the face of the earth is mobile. Mountains rise, only to be laid low by erosion; seas lap over the continents; entire regions progress through various stages; and glaciers come and go. We still live in the Pleistocene Ice Age, a pinpoint in time that has been preceded by, and will be followed by, extensive climatic changes.

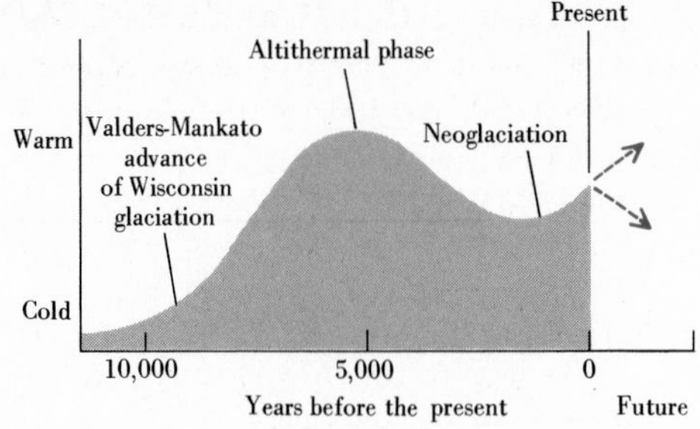

Figure 13.37 ***The climate during the last few thousand years has not been constant. Rather, it has been marked by the fluctuations generalized here. We can assume that the climate of the future will vary from that of the present. On this basis, we are either part way out of one glaciation or part way into another.***

Since the height of the last glacial invasion of the northern United States some 10,000 years ago, the climate has sometimes been warmer, sometimes colder than that of the present. Thus at one point there was a worldwide rise in temperature that produced a climate warmer than we are used to. This interval, known as the *Altithermal Phase* (that is, a period of great heat), reached its height about 6,000 years ago. Plants that are now found only in more southerly latitudes began to grow in northern areas. Many small glaciers in our western mountains disappeared completely, the glaciers in the Alps retreated, and the Greenland ice cap shrank (see Figure 13.37).

The Altithermal Phase gave way to a period in which the climate grew cooler again. Vegetation zones were pushed southward, and extensive areas in the higher latitudes and altitudes were stripped of their forests. Today we are apparently beginning to emerge from the latter phase, which some have called the "little Ice Age," or "neoglaciation."

What will happen in the future? We can be sure that the climate will grow either hotter or colder and that glaciers will either recede or advance in the ages ahead, just as they have in the past. But we cannot tell with certainty just what the changes will be and when they will occur. We can, though, predict what would happen at either of the two extremes: (1) complete deglaciation and (2) a resurgence of glacier ice on a continental scale.

If modern glaciers continue to waste away until they eventually disappear altogether, our decendants will be confronted with a flooding of the lowlands bordering the modern seas. As we have seen, if all the ice melted, sea level would rise an estimated 60 m. It is easy to foresee what would happen to our coastal cities. Furthermore, a climatic change that caused a complete deglaciation all over the world would alter the present pattern of our climatic belts. The northern United States would have a less rigorous climate, and there would be a general poleward expansion of the temperate and tropical zones. Such a development might affect the entire fabric of civilization.

If we dislike the thought of advancing seas and poleward shifts of warmer climate, we can contemplate the other alternative—a cooling of the climate, at least during the summer months. This shift would bring about the expansion of present glaciers, the birth of new ones, and a slow drainage of the harbors of the world.

We cannot say which way the world will go. At present, we are either part way out of one glaciation or part way into another. Whatever happens will have a profound effect on man's long-range future.

Outline

Glacier formation is a low-temperature metamorphic process that converts snow to firn to ice.

Glacier classification includes valley glaciers, piedmont glaciers, and ice sheets.

Nourishment of glaciers occurs in the *zone of accumulation.* The zone of wastage lies below the snowline where melting and evaporation exceed snowfall.

Glacier movement is usually a few centimeters per day but some glaciers move rapidly in *surges.* Below a 60-m *brittle zone,* glacier ice is in a *zone of flow.* The glacier moves by slipping along its base, by recrystallization of individual ice crystals, and by shearing.

Temperatures of glaciers vary. A warm glacier reaches the pressure melting point throughout during the summer. In a cold glacier no melting occurs even in the summer.

Results of glaciation are both erosional and depositional.

Erosion of the ground surface takes place by *plucking* or by *abrasion.* Features formed include *cirques, rôches moutonnées, horns, cols, arêtes, U-shaped valleys, hanging valleys, fiords, pater noster lakes, striations, grooves,* and *polish.*

Glacial deposits include unstratified material called *till,* which is found in *moraines* and *drumlins. Boulder trains* are special depositional features. *Stratified deposits* include *outwash* of sand and gravel which is usually found in *outwash plains, eskers, crevasse fillings, kames,* and *kame terraces. Varves* usually consist of clay and silt and form in glacial lakes.

The glacial theory was born in Switzerland and was first clearly stated by Louis Agassiz. *Proof of the glacial theory* rests on the principle of uniformitarianism.

Multiple glaciations mark the Pleistocene and four glaciations—Nebraskan, Kansan, Illinoian, and Wisconsin—are recognized in the central United States.

Indirect effects of glaciation include *changing sea level* and—in now-arid and semiarid regions—*pluvial periods.*

Pre-Pleistocene glaciations are known to have occurred in the late Paleozoic and the late Precambrian.

Causes of glaciation are not yet known, but they seem to involve a long, slow cooling of climate that began in the early Tertiary and a later set of short-term fluctuations which produced the glacial and interglacial epochs in the Pleistocene.

Implications for the future are several and the forecast is that change is certain.

Supplementary readings

AGASSIZ, LOUIS: *Studies on Glaciers,* Neuchâtel, 1840. Trans. and edited by A. V. Carozzi, Hafner Publishing Company, New York, 1967. No better exposition of the proof of glaciation can be found.

CHARLESWORTH, J. K.: *The Quaternary Era,* Edward Arnold & Co., London, 1957. A monumental and encyclopedic work.

EMBLETON, CLIFFORD, and CUCHLAINE A. M. KING: *Glacial and Periglacial Geomorphology,* Edward Arnold & Co., London, 1968. A very fine summary of the role of glaciers and ground ice in the fashioning of the landscape.

FLINT, RICHARD FOSTER: *Glacial and Pleistocene Geology,* John Wiley & Sons, Inc., New York, 1957. The standard U.S. text on the subject.

PATERSON, W. S. B.: *The Physics of Glaciers,* Pergamon Press Ltd., London, 1969. The only English summary of the state of our knowledge on the subject. Senior to first-year graduate level.

WEST, R. G.: *Pleistocene Geology and Biology,* John Wiley & Sons, Inc., New York, 1968. Focuses on Britain and northwestern Europe but is also a concise statement of general principles.

14 Deserts

ALL LIVING THINGS, INCLUDING HUMAN BEINGS, NEED MOISTURE TO SURVIVE, so it is not strange that the dry regions of the earth are sparsely populated and that our knowledge of these regions is scanty at best. Still, deserts and near-deserts cover nearly one-third of the land surface of the earth, and we cannot measure their importance solely in terms of population density.

14.1 Distribution and causes

Although there is no generally accepted definition of a desert, we can at least say that a desert is characterized by a lack of moisture—leading to, among other things, a restriction of the number of living things that can exist there. There may be too little initial moisture, or the moisture that does occur may be evaporated by extremely high temperatures or locked up in ice by extreme cold. Because we are not concerned here with polar deserts, we shall consider only those in the hotter climates. The distribution of middle- and low-latitude deserts is shown in Figure 14.1.

The deserts of the middle and low latitudes fall into two general groups. The first are the so-called *topographic deserts,* deficient in rainfall either because they are located toward the center of continents, far from the oceans, or more commonly because they are cut off from rain-bearing winds by high mountains. Takla Makan, north of Tibet and Kashmir in extreme western China, is an example of a desert located deep inside a continental landmass. The desert climate of large sections of Nevada, Utah, Arizona, and Colorado, on the other hand, is caused by the Sierra Nevada of California, which cut off the rain-bearing winds blowing in from the Pacific. A similar, though smaller, desert area in western Argentina has been created by the Andes mountains.

Much more extensive than the topographic deserts are the *tropical deserts,* lying in zones that range between 5° to 30° north and south of the equator. We can best understand their origin by looking at the general circulation of the earth's atmosphere.

Imagine our earth as completely covered by water. In such a situation we would find that the air moved in the manner depicted in Figure 14.2. The equator, where the greatest heating by the sun takes place, would be a belt of warm, rising air marked by low pressure. At the poles, lower temperatures would cause the air to settle and create a high-pressure zone. Other high-pressure zones of descending air, the subtropical high-pressure zones (the so-called "horse latitudes"), would lie in belts at about 30° north and south of the equator. And at about 60° north and south would be two more belts of lower atmospheric pressure. At the surface of the earth, air would move away from zones of high pressure toward zones of low pressure. Thus from the equatorward sides of each of the horse-latitude belts, air would move toward the equator. From the poleward sides, wind would blow generally toward low-pressure belts at about 60° north and south.

If it were not for the rotation of the earth, these surface winds would blow either directly south or directly north. But the earth's rotation introduces a

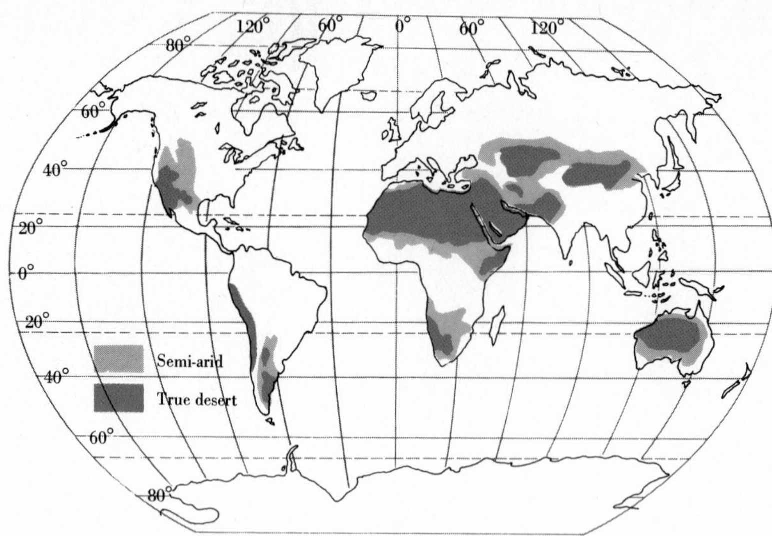

Figure 14.1 ***Deserts and near-deserts cover nearly one-third of the land surface of the earth. Middle- and low-latitude deserts, but not polar deserts, are shown here.***

factor known as the *Coriolis effect,* named after G. G. Coriolis, a nineteenth-century French mathematician who made the first extensive analysis of this phenomenon. The Coriolis effect influences everything that moves across the face of the earth—the atmosphere, ocean currents, birds in flight, aircraft, flowing streams, even an automobile speeding along a straight road.

We shall not analyze the reason for the effect, but the results can be simply stated.

Because of the Coriolis effect, anything moving in the northern hemisphere tends to veer to the right, and in the southern hemisphere any moving object tends to veer to the left.

Now apply this principle to the movement of air. If we stand at 30°N and face southward, in the direction toward which the air moves, the Coriolis effect will shift the air movement to our right—that is, to the west. These are the northeast trade winds. Standing at 30°S and facing northward, we find that the winds blowing toward the low pressure near the equator will veer to the left; they are known as the southeast trades. Remember that to apply the Coriolis effect you must face in the direction *toward* which the air moves. The winds, on the other hand, are named for the direction *from* which they move.

Now, to return to the origin of tropical deserts. At about 30° north and south of the equator, air descends in the subtropical high-pressure zones and spreads laterally across the surface toward both the equator and the poles. Under the influence of the Coriolis effect, the air currents moving toward the poles become the "prevailing westerly winds" of the middle latitudes, and those moving toward the equator become the trade winds of the low latitudes.

Now, the warmer the air becomes, the more moisture it can hold, and the

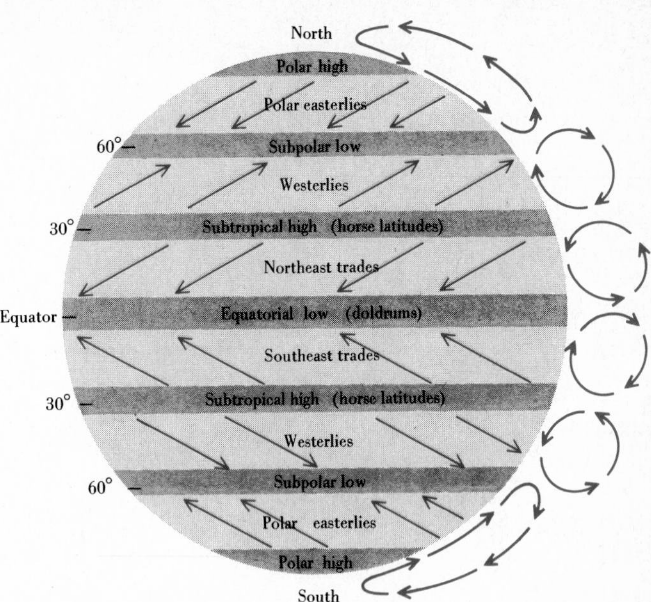

Figure 14.2 ***Idealized circulation of air on an earth presumed to have no landmasses. See the text for discussion.***

less likely it is to release moisture as precipitation. In the subtropical high-pressure belts, the air is heated as it descends over the warm land and tends to retain the moisture it contains. Consequently, the climate in these areas is relatively dry. The air that moves along in the trade winds continues to be heated as it enters warmer and warmer latitudes, and the dry belt is extended toward the equator. Finally, as the air approaches the equator itself, it becomes so heated that it rises to a higher altitude and is rapidly cooled. Then, unable to carry its great quantities of moisture, it releases the torrential rainfalls that are characteristic of the true tropics. This rising air eventually moves poleward, some of it descending again in the horse latitudes to begin its path all over again.

It is this continuing circulation of the air that creates the great deserts on either side of the equator. These include the Sahara Desert of North Africa; the Arabian Desert of the Middle East; the Victoria Desert of Australia; the Kalahari Desert of Bechuanaland in South Africa; the Sonora Desert of northwestern Mexico, southern Arizona, and California; the Atacama Desert of Peru and Chile; and the deserts of Afghanistan, Baluchistan, and northwestern India.

Some of the smaller deserts along the tropical coastlines have been created by the influence of oceanic currents bordering the continents. Winds blow in across the cool water of the ocean and suddenly strike the hot tropical landmass. There the air is heated, and its ability to retain moisture is increased. The resulting lack of precipitation gives rise to desert conditions, as along the coast of southern Peru and northern Chile, where the cool Humboldt current flows north toward the equator.

14.2 Climate

Rainfall

In the well-watered sections of the eastern United States and Canada and of western Europe, 75 to 150 cm of rain fall annually, but most deserts receive only 25 to 40 cm of rainfall per year, and in some there is less. Over much of the Sahara Desert, for example, as well as over the Kalahari, Sonora, and Atacama Deserts, the annual rainfall is usually less than 12 cm. At Lima, Peru, only 5 cm of rain falls during the average year, and Cairo, Egypt, does not receive even that much. Some parched areas often go for years without a single drop of rain. In fact, records at Calama, Chile, show a complete lack of rainfall for a 13-year period. Still there is probably no place in the world, even in the driest deserts, that does not receive some rain, even though years may pass between showers.

Rainfall in the deserts is both scanty and irregular; the scantier it is, the more erratic and unpredictable it is. Statistics on rainfall are therefore of little help in predicting when rain can be expected in a desert. Four years passed without rain in Iquique, Chile, but in the fifth year there was a single rainfall of 1.5 cm. We can say that the 5-year average rainfall was 0.3 cm, but the fact remains that the residents of this town enjoyed only one rain in 5 years.

Temperature

Temperatures in the desert swing from one extreme to the other in just a few hours. The air is heated rapidly during the day and cools rapidly at night, particularly in the tropical deserts. During a single 24-hr period in northern Tripoli, a daytime high of 37°C was followed by a nighttime low of 0.5°C, a variation of over 37°C. The highest temperature in the shade officially recorded in an American desert was 56.5°C, in Death Valley, California.

In the tropical deserts on either side of the equator, winter and summer are very much alike. Winter temperatures are not quite so high as in summer—and the nighttime temperature occasionally drops below freezing—but after the sun begins to shine, the temperature quickly rises to 15, 20, or even 25°C. In topographic deserts, farther away from the equator, however, the winters are often very severe. Thus at Urga, Mongolia, the mean temperature in July is 17°C, but in January it falls to −26°C. In fact, the mean temperature there is below freezing during 6 months of the year.

Winds

To add to the unpleasantness of the desert climate, violent winds often sweep across the parched earth. Unchecked by vegetation, they carry large clouds of dust to great heights above the surface and drive particles of sand along the ground. Locally, the winds are probably helped along by the rapid heating of the air during the day, which causes swift upward movements and reinforces the general movement of the surface winds that account for the deserts themselves.

14.3 *Weathering and soils*

Because of the lack of moisture in the desert, the rate of weathering, both chemical and mechanical, is extremely slow. Because most of the weathered material consists of unaltered rock and mineral fragments, mechanical weathering probably predominates (see Figure 14.3).

Some mechanical weathering is simply the result of gravity, such as the shattering of rock material when it falls from a cliff. Wind-driven sand also brings about some degree of mechanical weathering. Sudden flooding of a desert by a cloudburst moves material to lower elevations, reducing the size of the rock fragments and scouring the bedrock surface in the process. In almost every desert in the world, temperatures fall low enough to permit frost action. But here again the deficiency of moisture slows the process. Finally, the wide temperature variations characteristic of deserts cause rock materials to expand and contract and may produce some mechanical weathering.

This low rate of weathering is reflected in the soils of the desert. Seldom do we find extensive areas of residual soil, for the lack of protective vegetation permits the winds and occasional floods to strip away the soil-producing

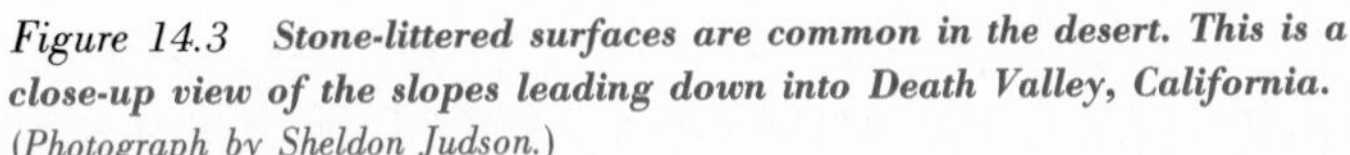

Figure 14.3 ***Stone-littered surfaces are common in the desert. This is a close-up view of the slopes leading down into Death Valley, California.*** *(Photograph by Sheldon Judson.)*

minerals before they can develop into true soils. Even so, soils do sometimes develop in local areas, but they lack the humus of the soils in moister climates, and they contain concentrations of such soluble substances as calcite, gypsum, and even halite because there is not enough water present to carry these minerals away in solution. In the deserts of Australia, rocklike concentrations of calcite, iron oxide, and even silica sometimes form a crust on the surface.

14.4 *Water*

Although rainfall is extremely sparse in desert areas, there is still enough water present to act as an important agent of erosion, transportation, and deposition. In fact, water is probably more effective than even the driving winds in molding desert landscape.

Running water

Very few streams flowing through desert regions ever find their way to the sea, and the few that do, such as the Colorado River in the United States and the Nile of Egypt, originate in well-watered areas where they receive enough water to sustain them through their long course across the desert. Most desert stream beds, however, are dry over long periods of time and flow only when an occasional flood comes along. Even then the flow is short-lived, for the water either evaporates rapidly or vanishes into the highly permeable rubble and debris of the desert. In some places, however, such as the western United States, broad desert plains slope down from the mountain ranges toward central basins, called *playas,* where surface runoff collects from time to time. But the *playa lakes* formed in these basins usually dry up in a short time or at best exist as shallow, salty lakes, of which Great Salt Lake is the best-known example.

Although the total rainfall in desert regions is scant and spottily distributed, the runoff from a single desert rain is often catastrophic. The very deficiency of rainfall is the main reason for the great effectiveness of water as a geological agent in the desert. Because there is not enough water to support a protective cover of vegetation, the runoff from the rare desert rain sweeps unimpeded across the surface. Rapid and intense, the flood waters pick up the loose rubble of a desert slope or channel and carry it along until they either reach a desert basin, evaporate, or sink into the ground. The runoff sometimes concentrates in vertically walled channels in dry regions—the *washes* or *arroyos* of the U.S. Southwest and the *wadis* of North Africa and the Near East (Figure 14.4). In other places, the water floods across the land in great sheets in a complex system of braided channels. In either case the water acts as an effective agent of erosion, transportation, and deposition.

Desert floods, then, are unlike the floods in humid areas. The typical desert flood, like the rain that produces it, is local in extent and of short duration. In moist regions, most floods arise from a general rain falling over a relatively

Figure 14.4 **In arid and semiarid regions many streams have fashioned vertically walled wadis or washes and carry water only intermittently. South of San Jon, New Mexico.** (*Photograph by Sheldon Judson.*)

long period of time; consequently, they affect large areas. Because of the widespread vegetative cover, these floods tend to rise and fall slowly. But on the bare ground of the desert, the runoff moves swiftly and floods rise and fall with great rapidity. These "flash" floods give little warning, and the experienced desert traveler has a healthy respect for them. He will never pitch camp on a dry stream floor, even though the stream banks offer protection from the wind. He knows that at any moment a surging wall of debris-laden water may sweep down the stream bed, destroying everything in its path.

Rainsplash

Even the individual raindrops falling on a barren surface are remarkably effective agents of erosion, for each drop tends to throw particles of unconsolidated material into the air. Careful measurements have shown that a heavy rainstorm may move as much as 100 metric tons of material per acre simply by means of rainsplash. On a level surface the particles are merely moved back and forth, but on an inclined surface they tend to move downslope (see Chapter 10).

Underground water

Because most underground water is derived from precipitation, it is not surprising that ground-water supplies are very meager and unreliable in desert areas. Even the water that does fall tends to evaporate before it can find its way down to underground reservoirs. If some of the surface water does manage to infiltrate into the zone of aeration, chances are that the capillary action of the rock particles will be strong enough to resist the influence of gravity. Consequently, the water is prevented from sinking down any deeper and is held in the zone of aeration as suspended water. Eventually, unless it is used by desert plants, this water is drawn back to the surface by evaporation and lost to the atmosphere.

And yet deserts are not completely devoid of ground water. Sometimes a relatively abundant supply is delivered by systems leading down from areas of higher rainfall. In the Sahara Desert, for example, the so-called Nubian sandstone brings artesian water to the oases of El Kharga and Dakhla, 1,050 km north of where the sandstone cuts beneath the surface in the Sudan. But even a supply of this sort is limited in amount, and dissolved salts sometimes make it undrinkable.

In general, then, the very arid regions of the earth have extremely limited supplies of ground water, and if these areas are to be developed for human habitation, some other source of water will have to be devised. In the semiarid deserts, on the other hand, where the rainfall may reach 25 cm or more per year, ground-water supplies are more dependable. But even here the rate of withdrawal must be carefully balanced against the rate of natural recharge if these supplies are to be used over any prolonged period of time.

14.5 *Work of the wind*

Although wind is less effective than water as an agent of erosion, it does play an important role in transporting earth materials in arid and semiarid regions—and in more humid areas as well. Moreover, even in parts of the world where the action of the wind is negligible today, we find evidence that it has been more effective at certain times in the past.

Movement of material

Wind velocities increase rapidly with height above the ground surface, just as the velocity of running water increases at levels above the channel floor. Furthermore, like running water, most air moves in turbulent flow. But wind velocities increase at a greater rate than water velocities, and the maximum velocities attained are much higher.

The general movement of wind is forward, across the surface of the land, but within this general movement the air is moving upward, downward, and from side to side. In the zone about 1 m above ground surface, the average velocity of upward motion in an air eddy is approximately one-fifth the average forward velocity of the wind. This upward movement greatly affects the wind's ability to transport small particles of earth material, as we shall see.

Right along the surface of the ground there is a thin but definite zone where the air moves very little or not at all. Field and laboratory studies have shown that the depth of this zone depends on the size of the particles that cover

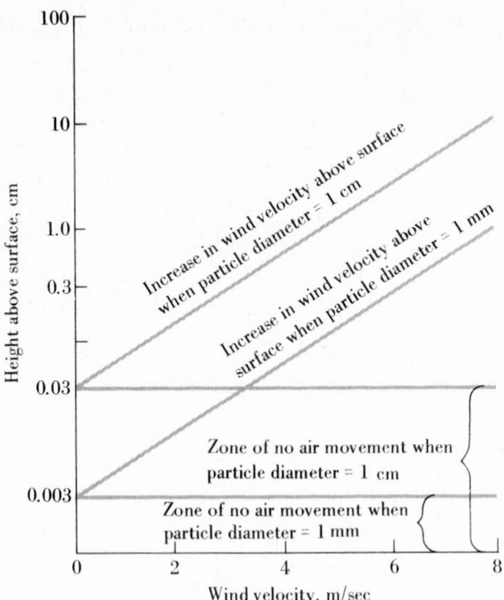

Figure 14.5 ***In a thin zone close to the ground there is little or no air movement, regardless of the wind velocity immediately above. This zone is approximately one-thirtieth the average diameter of surface particles. Two zones are shown in the graph: one for surfaces on which the particles average 1 mm in diameter, and one for surfaces with 1-cm particles. Diagonal lines represent the increase in velocity of a wind of given intensity blowing over surfaces covered with particles of 1 mm and 1 cm average diameter, respectively.*** (*Reproduced by permission from R. A. Bagnold,* The Physics of Blown Sand and Desert Dunes, *p. 54, Methuen & Co., Ltd., London, 1941.*)

the surface. On the average, the depth of this "zone of no movement" is about one-thirtieth the average diameter of the surface grains (see Figure 14.5). Thus over a surface of evenly distributed pebbles with an average diameter of 30 mm, the zone of no movement would be about 1 mm deep. This fact, too, has a bearing on the wind's ability to transport material.

DUSTSTORMS AND SANDSTORMS Material blown along by the wind usually falls into two size groups. The diameter of wind-driven sand grains averages between 0.15 and 0.30 mm, with a few grains as fine as 0.06 mm. All particles smaller than 0.06 mm are classified as dust.

In a true *duststorm* (see Figure 14.6), the wind picks up fine particles and sweeps them upward hundreds or even thousands of meters into the air, forming a great cloud that may blot out the sun and darken the sky. In contrast, a *sandstorm* is a low, moving blanket of wind-driven sand with an upper surface 1 m or less above the ground. Actually, the greatest concentration of moving sand is usually just a few centimeters above the ground surface, and individual grains seldom rise even as high as 2 m. Above the blanket of moving sand, the air is quite clear, and a man on the ground appears to be partially submerged, as though he were standing in a shallow pond.

Often, of course, the dust and sand are mixed together in a wind-driven storm, but the wind soon sweeps the finer particles off, and eventually the air above the blanket of moving sand becomes clear.

Apparently, then, the wind handles particles of different size in different ways. A dust-sized grain is swept high into the air, and a sand-sized grain is driven along closer to the ground. The difference arises from the strength of the wind and the terminal velocity of the grain.

We defined the terminal velocity of a grain as the constant rate of fall attained by the grain when the acceleration due to gravity is balanced by the resistance of the fluid—in this case, the air—through which the grain falls (see Section 11.5). Terminal velocity varies only with the size of a particle when shape and density are constant. As the particle size increases, both the pull of gravity and the air resistance increase too. But the pull of gravity increases at a faster rate than the air resistance: A particle with a diameter of 0.01 mm has a terminal velocity in air of about 0.01 m/sec; a particle with a 0.2-mm diameter has a terminal velocity of about 1 m/sec; and a particle with a diameter of 1 mm has a terminal velocity of about 8 m/sec.

To be carried upward by an eddy of turbulent air, a particle must have a terminal velocity that is less than the upward velocity of the eddy. Close to the ground surface, where the upward currents are particularly strong, dust particles are swept up into the air and carried in suspension. Sand grains, however, have terminal velocities greater than the velocity of the upward-moving air; they are lifted for a moment and then fall back to the ground. But how does a sand grain get lifted into the air at all if the eddies of turbulent air are unable to support it?

Movement of sand grains Careful observations, both in the laboratory and on open deserts, show sand grains moving forward in a series of jumps, in a process known as *saltation*. We used the same term to describe the motion of particles along a stream bed, but there is a difference: An eddy of water

Figure 14.6 **Silt-sized particles are swept high into the sky from the flood plain of the Knik River valley, Alaska. The Knik River is fed, in part, by active glaciers.** (*Photograph by William C. Bradley.*)

can actually lift individual particles into the main current, whereas wind by itself cannot pick up sand particles from the ground.

Sand particles are thrown into the air only under the impact of other particles. When the wind reaches a critical velocity, grains of sand begin to roll forward along the surface. Suddenly, one rolling grain collides with another; the impact may either lift the second particle into the air or cause the first to fly up.

Once in the air, the sand grain is subjected to two forces. First, gravity tends to pull it down to earth again, and eventually it succeeds. But even as the grain falls, the horizontal velocity of the wind drives it forward. The resulting course of the sand grain is parabolic from the point where it was first thrown into the air to the point where it finally hits the ground. The angle of impact varies between 10° and 16° (see Figure 14.7).

When the grain strikes the surface, it may either bounce off a large particle and be driven forward once again by the wind, or it may bury itself in the loose sand, perhaps throwing other grains into the air by its impact.

In any event, it is through the general process of saltation that a sand cloud is kept in motion. Countless grains are thrown into the air by impact and are driven along by the wind until they fall back to the ground. Then they either bounce back into the air again or else pop other grains upward by impact. The initial energy that lifts each grain into the air comes from the impact of another grain, and the wind contributes additional energy to keep it moving. When the wind dies, all the individual particles in the sand cloud settle to earth.

Some sand grains, particularly the large ones, never rise into the air at all, even under the impact of other grains. They simply roll forward along the ground, very much like the rolling and sliding of particles along the bed of a stream of water. It has been estimated that between one-fifth and one-quarter of the material carried along in a sandstorm travels by rolling and the rest by means of saltation.

Figure 14.7 **A sand grain is too heavy to be picked up by the wind but may be put into the air by saltation. Here a single grain is rolled forward by the wind until it bounces off a second grain. Once in the air, it is driven forward by the wind and is then pulled to the ground by gravity. It follows a parabolic path, hitting the ground at an angle between 10° and 16°.**

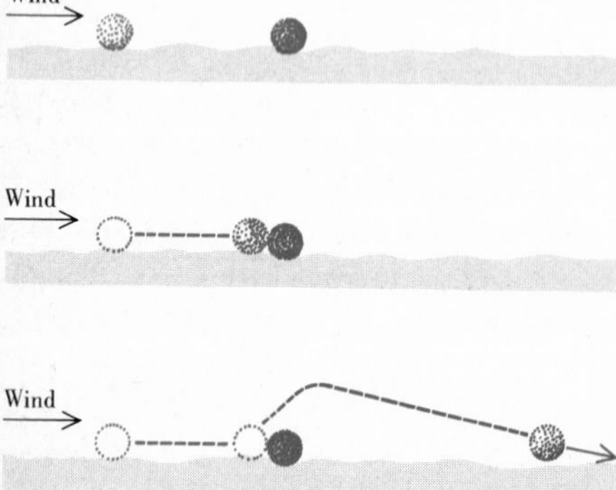

Notice that after the wind has started the sand grains moving along the surface, initiating saltation, the wind no longer acts to keep them rolling. The cloud of saltating grains obstructs the wind and shields the ground surface from its force; thus, as soon as saltation begins, the velocity of near-surface winds drops rapidly. Saltation continues only because the impact of the grains continues. The stronger the winds blow during saltation, the heavier will be the blanket of sand, and the less the possibility that surface grains will be rolled by the wind.

Movement of dust particles As we have seen, dust particles are small enough and have low enough terminal velocities to be lifted aloft by currents of turbulent air and to be carried along in suspension. But just how does the wind lift these tiny particles in the first place?

Laboratory experiments show that under ordinary conditions particles smaller than 0.03 mm in diameter cannot be swept up by the wind after they have settled to the ground. In dry country, for example, dust may lie undisturbed on the ground even though a brisk wind is blowing, but if a flock of sheep passes by and kicks loose some of the dust, a dust plume will rise into the air and move along with the wind.

The explanation for this seeming reluctance of dust particles to be disturbed lies in the nature of air movement. The small dust grains lie within the thin zone of negligible air movement at the surface. They are so small that they do not create local eddies and disturbances in the air, and the wind passes them by—or the dust particles may be shielded by larger particles against the action of the wind.

Some agent other than the wind must set dust particles in motion and lift them into a zone of turbulent air—perhaps the impact of larger particles or sudden downdrafts in the air movement. Irregularities in a plowed field or in a recently exposed stream bed may help the wind to begin its work by creating local turbulence at the surface. Also, vertical downdrafts of chilled air during a thunderstorm sometimes strike the ground with velocities of 40 to 80 km/hr and churn up great swaths of dust.

Erosion

Erosion by the wind is accomplished through two processes: *abrasion* and *deflation.*

ABRASION Like the particles carried by a stream of running water, saltating grains of sand driven by the wind are highly effective abrasive agents in eroding rock surfaces. As we have seen, wind-driven sand seldom rises more than 1 m above the surface of the earth, and measurements show that most of the grains are concentrated in the $\frac{1}{2}$ m closest to the ground. In this layer the abrasive power of the moving grains is concentrated.

Although evidence of abrasion by sand grains is rather meager, there is enough to indicate that this erosive process does take place. For example, we sometimes find fence posts and telephone poles abraded at ground level and bedrock cliffs with a small notch along their base. In desert areas the

Figure 14.8 **Ventifacts exhibit a variety of facets, pits, ridges, and grooves, as well as surface sheen. Scale is indicated by 5-cm squares.** (*Photograph by Walter R. Fleischer.*)

evidence is more impressive, for here the wind-driven sand has in some places cut troughs or furrows in the softer rocks. The knife-edge ridges between these troughs are called *yardangs,* a term used in the deserts of Chinese Turkestan, where they were first described; the furrows themselves are called *yardang troughs.* The cross profile of one of these troughs is not unlike that of a glaciated mountain valley in miniature, the troughs ranging from a few centimeters to perhaps 8 m in depth. They run in the usual direction of the wind, and their deepening by sand abrasion has actually been observed during sandstorms.

Figure 14.9 **A ventifact with three well-developed facets, on the floor of the Mojave Desert, California. The pocket knife to the right of the ventifact gives the scale.** (*Photograph by Sheldon Judson.*)

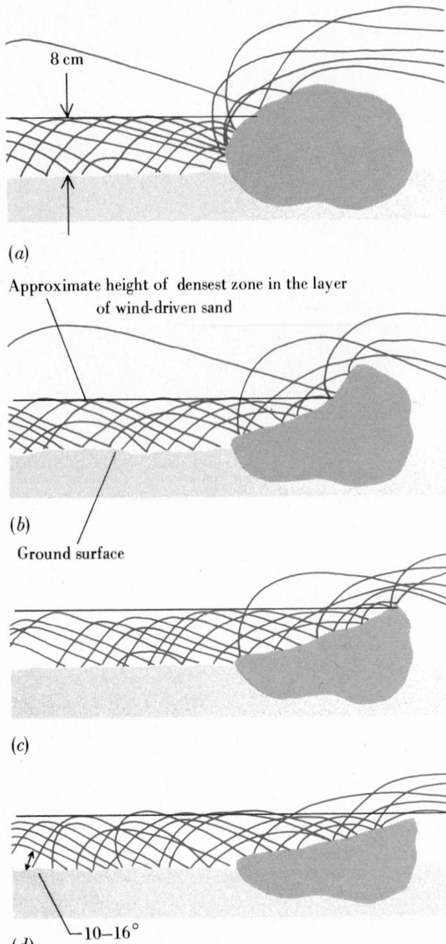

Figure 14.10 ***A facet on a ventifact is cut by the impact of grains of wind-driven sand.*** *(Redrawn from Robert P. Sharp, "Pleistocene Ventifacts East of the Big Horn Mountains, Wyoming,"* J. Geol., *vol. 57, p. 182, 1949.)*

The most common products of abrasion are certain pebbles, cobbles, and even boulders that have been eroded in a particular way. These pieces of rock are called *ventifacts*, from the Latin for "wind" and "made." They are found not only on deserts but also along modern beaches—in fact, wherever the wind blows sand grains against rock surfaces (see Figures 14.8, 14.9, and 14.10).

The surface of ventifacts is characterized by a relatively high gloss or sheen and by a variety of facets, pits, gouges, and ridges.

The face of an individual ventifact may display only 1 facet or 20 facets or more—sometimes flat, but more commonly curved. Where two facets meet, they often form a well-defined ridge, and the intersection of 3 or more facets gives the ventifact the appearance of a small pyramid. Apparently, the surface becomes pitted when it lies across the direction of wind movement at an angle of 55° or more; but it becomes grooved when it lies at angles of less than 55°.

But how is the wind able to cut more than one facet on a rock surface? If the original rock is somewhat angular and properly positioned in the wind's path, it may split the wind, so that two facets are cut simultaneously. Or if the wind changes its characteristic direction of movement, it may cut facets at various angles on the stone. Finally, underlying materials may be scoured away and the rock itself may shift position, or it may be moved by the force of very high winds, by the activities of animals, or by frost action. Whatever the cause, any shift in the position of the rock may expose a new surface to the abrasive action of wind-borne particles. Wind-splitting angles and variable winds probably create the multiple facets on large ventifacts, whereas movement of the rock itself is probably responsible for the variety of facets on ventifacts less than 25 cm in diameter.

DEFLATION Deflation (from the Latin for "to blow away") is the erosive process of the wind carrying off unconsolidated material. The process creates several recognizable features in the landscape. For example, it often scoops out basins in soft, unconsolidated deposits ranging from a few meters to several kilometers in diameter. These basins are known as *blowouts*, for obvious reasons (see Figure 14.11). Even in relatively consolidated material, the wind will excavate sizable basins if some other agency is at work loosening the material. We find such depressions in the almost featureless High Plains of eastern New Mexico and western Texas, where the bedrock is loosely cemented by calcium carbonate. Several times during the Pleistocene, the climate in this area shifted back and forth between moist and dry. During the moist periods, water dissolved some of the calcium carbonate and left the sandstone particles lying on the surface. Then, during the dry periods, the wind came along and removed the loosened material. Today, we find the larger particles piled up in sand hills on the leeward side of the basin excavated by the wind. The smaller dust particles were carried farther along and spread in a blanket across the plains to the east (see Figure 14.12).

In arid and semiarid country we sometimes see finely honey-combed rocks, and others called *pedestal rocks*, that have been fashioned into weird pillars resembling toadstools. Although wind has often been cited as the cause of

Figure 14.11 **Wind has excavated this blowout in unconsolidated sand deposits of Terry Andrae State Park, near Lake Michigan in eastern Wisconsin.** (*Photograph by Wisconsin Conservation Department.*)

these formations, differential weathering is primarily responsible. The wind has merely removed the loose products of weathering.

Deflation removes only the sand and dust particles from a deposit and leaves behind the larger particles of pebble or cobble size. As time passes, these stones form a surface cover, a *desert pavement* that cuts off further deflation.

Deposition

Whenever the wind loses its velocity, and hence its ability to transport the sand and dust particles it has picked up from the surface, it drops them back to the ground. The landscape features formed by wind-deposited materials are of various types, depending on the size of particles, the presence or absence of vegetation, the constancy of wind direction, and the amount of material available for movement by the wind. We still have a great deal to learn about this sort of deposit, but certain generalizations seem valid.

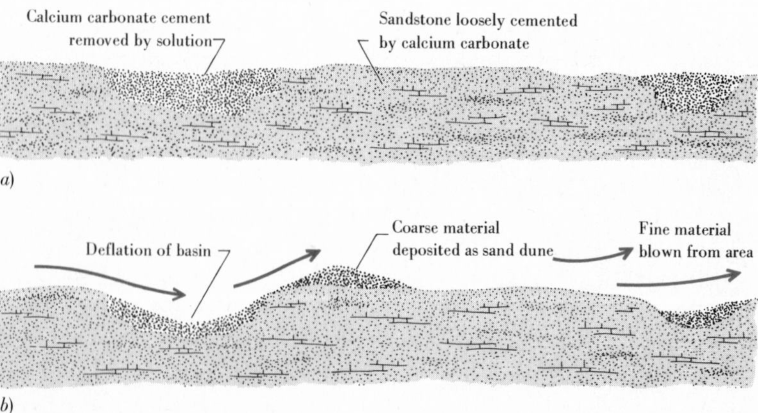

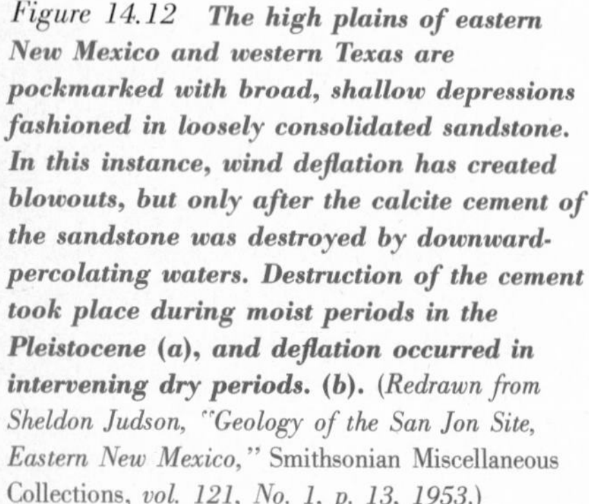

Figure 14.12 **The high plains of eastern New Mexico and western Texas are pockmarked with broad, shallow depressions fashioned in loosely consolidated sandstone. In this instance, wind deflation has created blowouts, but only after the calcite cement of the sandstone was destroyed by downward-percolating waters. Destruction of the cement took place during moist periods in the Pleistocene (a), and deflation occurred in intervening dry periods. (b).** (*Redrawn from Sheldon Judson, "Geology of the San Jon Site, Eastern New Mexico,"* Smithsonian Miscellaneous Collections, *vol. 121, No. 1, p. 13, 1953.*)

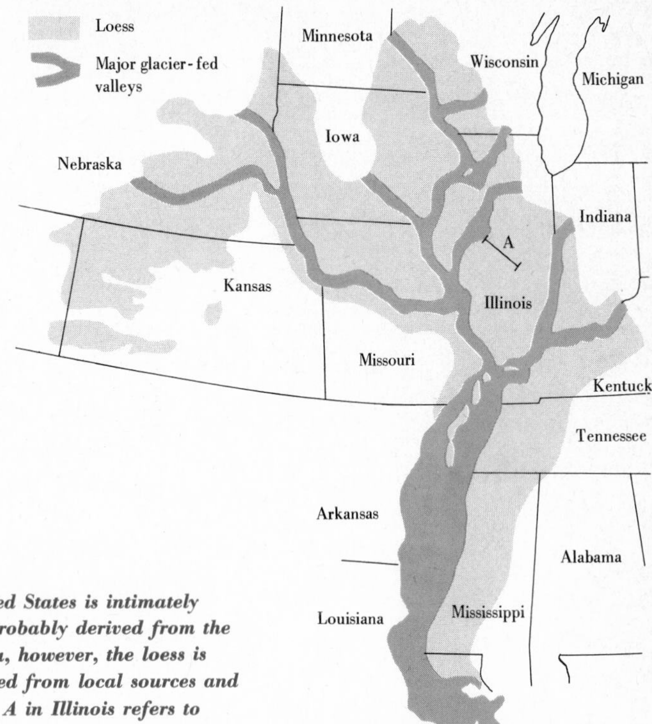

Figure 14.13 **The great bulk of the loess in the central United States is intimately related to the major glacier-fed valleys of the area and was probably derived from the flood plains of these valleys. In Kansas and parts of Nebraska, however, the loess is probably nonglacial in origin and has presumably been derived from local sources and the more arid regions to the west. The line of section marked A in Illinois refers to Figure 14.15.**

LOESS Loess is a buff-colored, unstratified deposit composed of small, angular mineral fragments. Loess deposits range in thickness from a few centimeters to 10 m or more in the central United States to over 100 m in parts of China. A large part of the surface deposits across some 0.5 million km^2 of the Mississippi River basin is made up of loess, and this material has produced the modern fertile soils of several midwestern states, particularly Iowa, Illinois, and Missouri (see Figure 14.13).

Most geologists, though not all, believe loess to be material originally deposited by the wind. They base their conclusion on several facts. The individual particles in a loess deposit are very small, strikingly like the particles of dust carried about by the wind today. Moreover, loess deposits stretch over hill-slopes, valleys, and plains alike, an indication that the material has settled from the air. And the shells of air-breathing snails present in loess strongly impugn the possibility that the deposits were laid down by water.

Many exposures in the north-central United States reveal that loess deposits there are intimately associated with till and outwash deposits built up during the great Ice Age. Because the loess lies directly on top of the glacial deposits in many areas, it seems likely that it was deposited by the wind during periods when glaciation was at its height rather than during interglacial intervals. Also, because there is no visible zone of weathering on the till and outwash deposits, the loess probably was laid down on the newly formed glacial deposits before any soil could develop on them (see Figure 14.14).

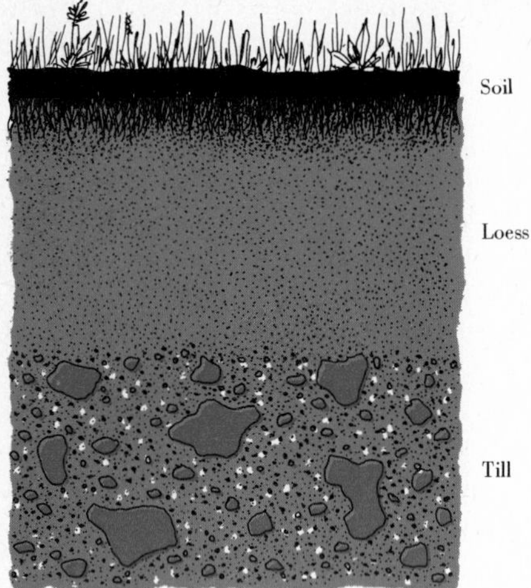

Figure 14.14 ***In many places, unweathered till is overlain by loess on which a soil zone has developed. The lack of a weathering zone on the till beneath loess often indicates rapid deposition of the loess immediately after the disappearance of the glacier ice and before weathering processes could affect the till. Not until loess deposition has slowed or halted is there time available to allow weathering and organic activity capable of producing a soil.***

Certain relationships between the loess deposits in the Midwest and the streams that drain the ancient glacial areas serve to strengthen the conclusion that there is a close connection between glaciation and the deposit of wind-borne materials. Figure 14.15, for example, shows that the major glacial streams cut across the loess belt and that the thickness of the loess decreases toward the east and away from the banks of the streams. Moreover, the mean size of the particles decreases away from the glacial streams. These facts can best be explained as follows. We know that loess is not forming in this area at the present, so we must look for more favorable conditions in the past. During the great Ice Age of the Pleistocene, the rivers of the Midwest carried large amounts of debris-laden meltwater from the glaciers. Consequently, the flood plains of these rivers built up at a rapid rate and were broader than they are today. During periods of low water, the flood plains were wide expanses of gravel, sand, silt, and clay exposed to strong westerly winds. These winds whipped the dust-sized material from the flood plains, moved it eastward, and laid down the thickest and coarsest of it closest to the rivers (see Figure 14.6).

Some geologists, however, have suggested that loess deposits, particularly those in the lower Mississippi valley, were not laid down by the wind but rather have been derived from fine-grained back-swamp deposits through a process called "loessification," the details of which we need not consider here. Suffice it to say that most investigators believe that the loess in the Mississippi valley was built up by the action of wind blowing across glacial outwash deposits and then scattering the fine particles over the landscape. In fact, much the same explanation is generally accepted for the great belt of loess that extends across France into Germany north of the Alps, on into the Balkan countries, and eastward across the plains of Poland and western Russia.

All loess, however, is not derived from glacial deposits. In one of the earliest studies of loess, it was shown that the Gobi Desert has provided the source material for the vast stretches of yellow loess that blanket much of northern China and that give the characteristic color to the Yellow River and the Yellow Sea. Much of the land used for cotton growing in the eastern Sudan of Africa is thought to be made up of particles blown from the Sahara Desert to the west. We have already seen that finely divided mineral fragments are swept

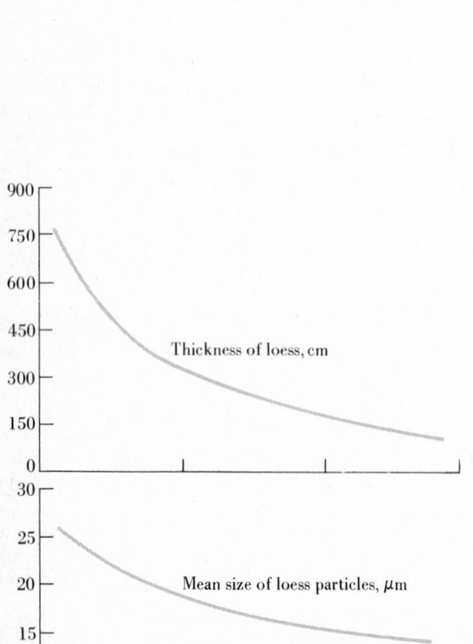

Figure 14.15 ***Loess related to the major glacier-fed rivers in the Midwest shows a decrease in thickness away from the rivers and a decrease in the size of individual particles. An example is shown in this diagram, based on data gathered along line A in Figure 14.13.*** (*Redrawn from G. D. Smith, "Illinois Loess—Variations in Its Properties and Distributions,"* Univ. Illinois Agricultural Experiment Station Bull., *no. 490, Figs. 5 and 6, 1942.*)

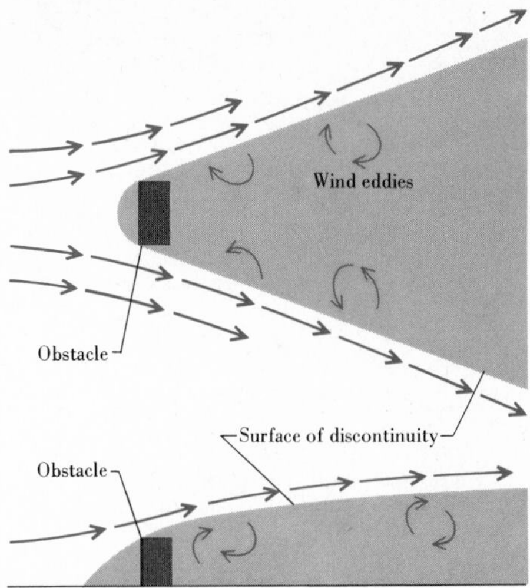

Figure 14.16 ***Wind is diverted around and over an obstacle to form a wind shadow (shaded area) within which wind velocity is low and air movement is marked by eddies. A surface of discontinuity separates the air within the shadow from the air outside.*** (*After R. A. Bagnold,* The Physics of Blown Sand and Desert Dunes, *p. 190, Methuen & Co. Ltd., London, 1941.*)

Figure 14.17 ***Because of its momentum, the sand in the more rapidly moving air outside the wind shadow either passes through the surface of discontinuity to settle in the wind shadow behind the obstacle or strikes the obstacle and falls in front of it.*** (*From Bagnold; see Figure 14.16.*)

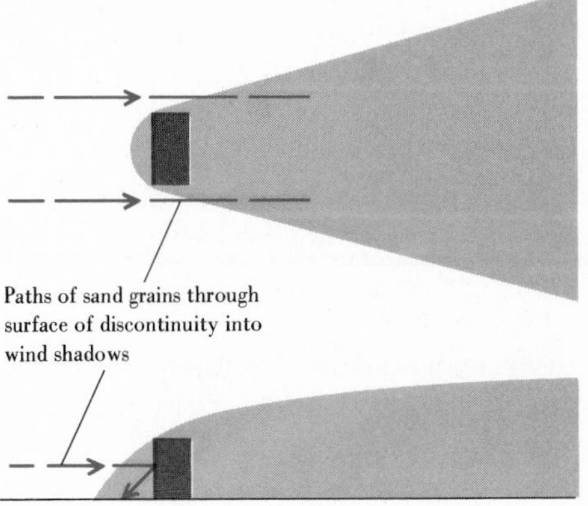

up in suspension during desert sandstorms and are carried along by the wind far beyond the confines of the desert. Clearly, then, the large amounts of very fine material present in most deserts would make an excellent source of loess.

SAND DEPOSITS Unlike deposits of loess, which blanket whole areas, sand deposits assume certain characteristic and recognizable shapes. Wind often heaps the sand particles into mounds and ridges called *dunes*, which sometimes move slowly along in the direction of the wind. Some dunes are only a few meters in height, but others reach tremendous sizes. In southern Iran, dunes have grown to 200 m with a base 1 km wide.

In Chapter 11 we found that as the velocity of a stream falls, so does the energy available for the transportation of material; consequently, deposition of material takes place. The same relation between decreasing energy and increasing deposition applies to the wind. But in dealing with wind-deposited sand we need to examine the relationship more closely and to explain why sand is deposited in the form of dunes rather than as a regular, continuous blanket.

The wind shadow Any obstacle—large or small—across the path of the wind will divert moving air and create a "wind shadow" to the leeward, as well as a smaller shadow to the windward immediately in front of the obstacle. Within each wind shadow the air moves in eddies, with an average motion less than that of the wind sweeping by outside. The boundary between the two zones of air moving at different velocities is called the *surface of discontinuity* (see Figure 14.16).

When sand particles driven along by the wind strike an obstacle, they settle in the wind shadow immediately in front of it. Because the wind velocity (hence energy) is low in this wind shadow, deposition takes place and gradually a small mound of sand builds up. Other particles move past the obstacle and cross through the surface of discontinuity into the leeward wind shadow behind the barrier. Here again the velocities are low, deposition takes place, and a mound of sand (a dune) builds up—a process aided by eddying air that tends to sweep the sand in toward the center of the wind shadow (see Figures 14.17 and 14.18).

Wind shadow of a dune Actually, a sand dune itself acts as a barrier to the wind; and by disrupting the flow of air, it may cause the continued deposition of sand. A profile through a dune in the direction toward which the wind blows shows a gentle slope facing the wind and a steep slope to the leeward. A wind shadow exists in front of the steep leeward slope, and it is here that deposition is active. The wind drives the sand grains up the gentle windward slope to the dune crest and then drops them into the wind shadow. The steep leeward slope is called the *slip face* of the dune because of the small sand slides that take place there.

The slip face is necessary for the existence of a true wind shadow. Here is how the slip face is formed. A mound of sand affects the flow of air across it, as shown in the topmost diagram of Figure 14.19. Notice that the wind flows over the mound in streamlined patterns. These lines of flow tend to converge toward the top of the mound and diverge to the leeward. In the zone of diverging air flow, velocities are less than in the zone of converging

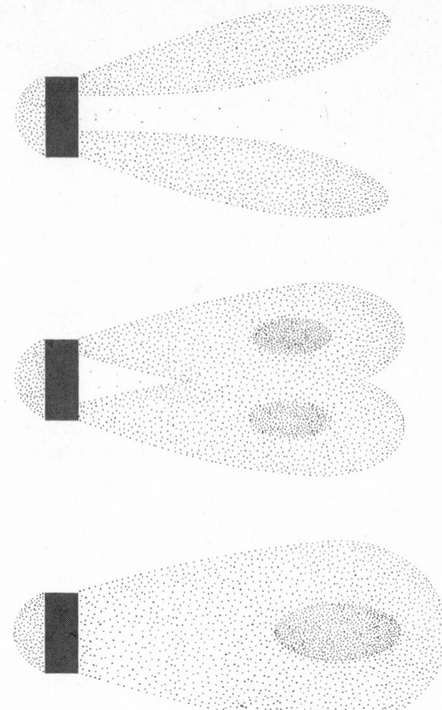

Figure 14.18 **Sand falling in the wind shadow tends to be gathered by wind eddies within the shadow to form a shadow dune, as shown in this sequence of diagrams.** (*From Bagnold; see Figure 14.16.*)

Figure 14.19 **The development of a slip face on a dune. Wind converges on the windward side of the dune and over its crest and diverges to the lee of the dune. The eventual result is the creation of a wind shadow in the lee of the dune. In this wind shadow, sand falls until a critical angle of slope (about 34°) is reached. Then a small landslide occurs, and the slip face is formed.** (*From Bagnold; see Figure 14.16.*)

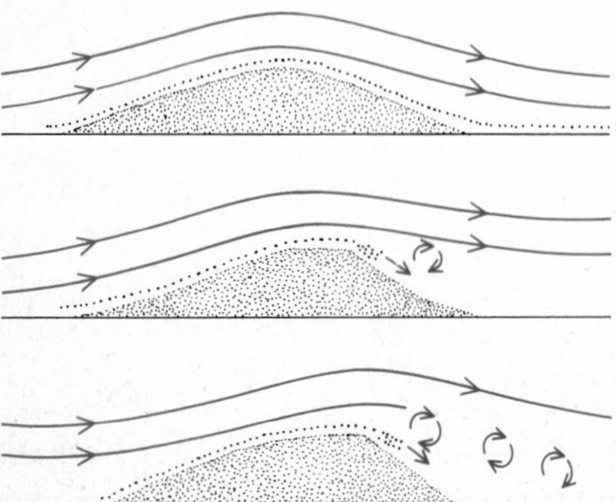

flow. Consequently, sand tends to be deposited on the leeward slope just over the top of the mound where the velocity begins to slacken. This slope steepens because of deposition, and eventually the sand slumps under the influence of gravity. The slump usually takes place at an angle of about 34° from the horizontal. A slip face is thus produced, steep enough to create a wind shadow in its lee (see Figure 14.20). Within this shadow, sand grains fall like snow through quiet air. Continued deposition and periodic slumping along the slip face account for the slow growth or movement of the dune in the direction toward which the wind blows.

Shoreline dunes Not all dunes are found in the deserts. Along the shores of the ocean and of large lakes, ridges of wind-blown sand called *fore dunes* are built up even in humid climates. They are well developed along the southern and eastern shores of Lake Michigan, along the Atlantic coast from Massachusetts southward, along the southern coast of California, and at various points along the coasts of Oregon and Washington (see Figure 14.21).

These fore dunes are fashioned by the influence of strong onshore winds acting on the sand particles of the beach. On most coasts, the vegetation is dense enough to check the inland movement of the dunes, and they are concentrated in a narrow belt that parallels the shoreline. These dunes usually have an irregular surface, sometimes pockmarked by blowouts (see the subsection on deflation, above).

Sometimes, however, in areas where vegetation is scanty, the sand moves inland in a series of ridges at right angles to the wind. These *transverse dunes*

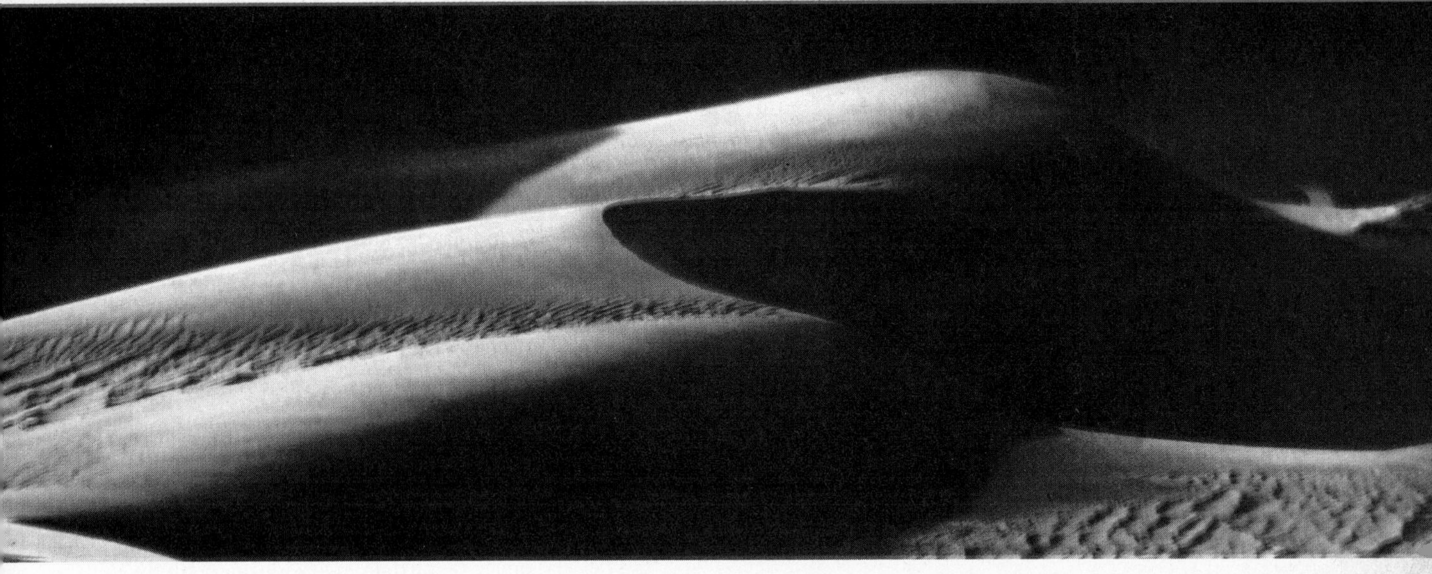

Figure 14.20 ***The slip face of a dune diagrammed in Figure 14.19 is seen here in nature. Slip faces on these dunes are in shadow. The gentler slopes upwind are well lighted and some show asymmetric sand ripples indicating movement of material up the slopes to the lips of the slip faces. Great Sand Dunes National Monument, near Alamosa, Colorado.*** *(Photograph by Macpherson Raymond, Jr.)*

Figure 14.21 ***These shoreline dunes form complex patterns behind the beach at Coos Bay, Oregon. The beach serves as a source of sand and this source is continuously renewed by longshore currents of ocean water. Onshore winds (from the left) drive the beach sand inland. The photograph shows about 450 m of shoreline.***

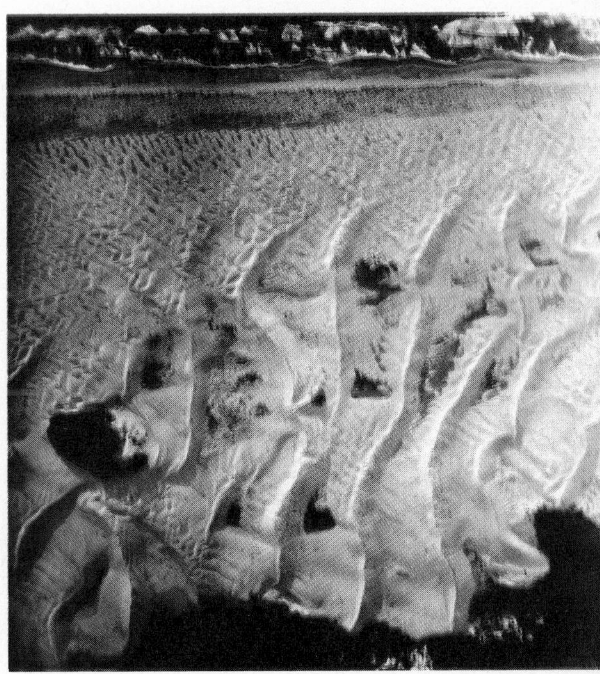

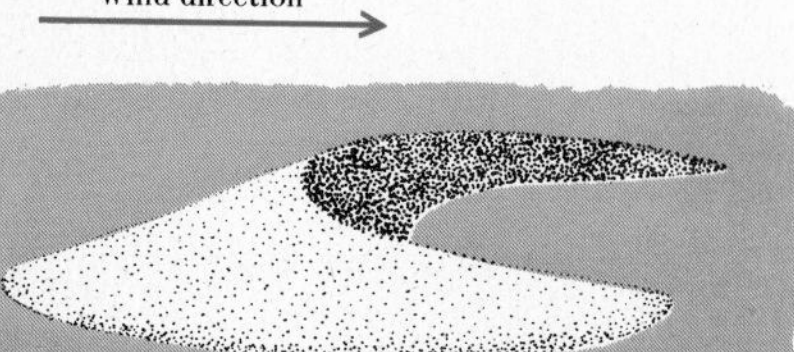

Figure 14.22 **A barchan is a crescent-shaped dune with its horns pointed downwind and its slip face on the inside of the crescent.**

exhibit the gentle windward slope and the steep leeward slope characteristic of other dunes. Transverse dunes are also common in arid and semiarid country where sand is abundant and vegetation sparse.

Barchans Barchans are sand dunes shaped like a crescent, with their horns pointing downwind. They move slowly with the wind, the smaller ones at a rate of about 15 m/year, the larger ones about 7.5 m/year. The maximum height obtained by barchans is about 30 m, and their maximum spread from horn to horn is about 300 m (see Figures 14.22 and 14.23).

Just what leads to the formation of a barchan is still a matter of dispute. Certain conditions do seem essential, however: a wind that blows from a fixed direction, a relatively flat surface of hard ground, a limited supply of sand, and a lack of vegetation.

Parabolic dunes Long, scoop-shaped, parabolic dunes look rather like barchans in reverse; that is, their horns point upwind rather than downwind. They are usually covered with sparse vegetation, which permits limited movement of the sand. Parabolic dunes are quite common in coastal areas and in various places throughout the southwestern states. Ancient parabolic dunes, no longer active, exist in the upper Mississippi valley and in central Europe.

Figure 14.23 **These barchans are moving across the Pampa de Islay, Peru, in the direction in which their horns point.** (*Photograph by Aerial Explorations, Inc.*)

Figure 14.24 ***These seif dunes are located in the Rub 'al Khali district of southern Saudi Arabia and are oriented parallel to the direction of the prevailing winds.*** *(Photograph by Arabian American Oil Co.)*

Longitudinal dunes Longitudinal dunes are long ridges of sand running in the general direction of wind movement. The smaller types are about 3 m high and about 60 m long. In the Libyan and Arabian Deserts, however, they commonly reach a height of 100 m and may extend for 100 km across the country. There they are known as *seif dunes,* from the Arabic word for "sword" (see Figure 14.24).

Stratification in dunes The layers of sand within a dune are usually inclined. The layers along the slip face have an angle of about 34°, whereas the layers along the windward slope are gentler.

Because the steeper beds along the slip face are analogous to the *foreset beds* in a delta (see Figure 11.42), we can use the same term in referring to them. These beds develop if there is a continuous deposition of sand on the leeward side of the dune, as in barchans and actively moving transverse dunes.

Backset beds develop on the gentler slope to the windward. These beds constitute a large part of the total volume of a dune, especially if there is enough vegetation to trap most of the sand before it can cross over to the slip face. *Topset beds* are nearly horizontal beds laid down on top of the inclined foreset or backset beds.

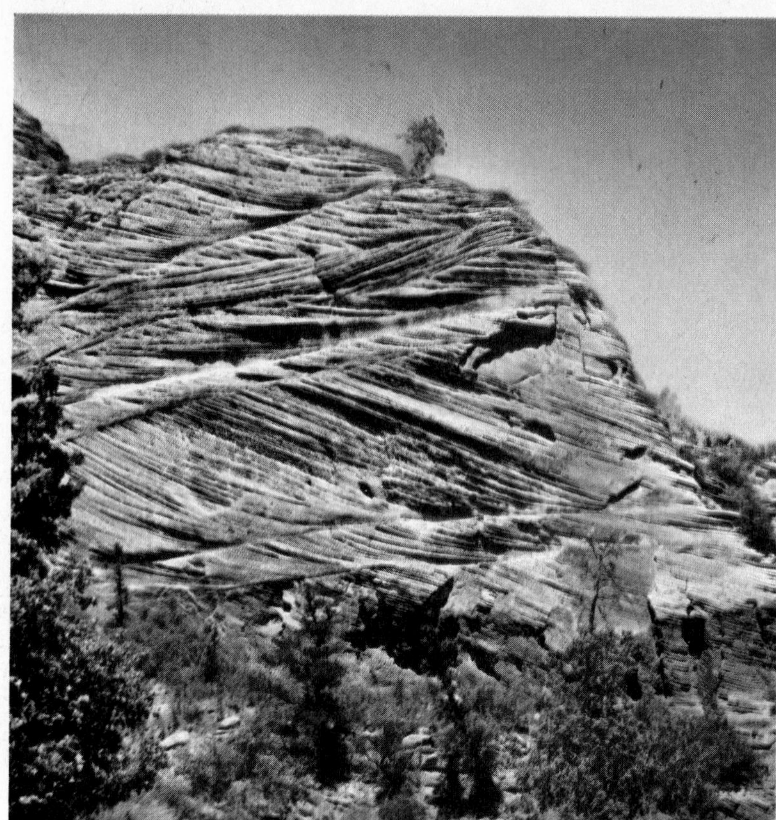

Figure 14.25 **Wind was an active agent in the geological past, and some sandstone has been formed by the lithification of dune deposits. Note the inclined bedding in this cliff of ancient wind-blown sand in Zion Canyon, Utah.** (*Photograph by Raymond C. Murray.*)

14.6 Desert landscapes

The sparse vegetative cover, the slow rate of weathering, and the skimpy soils of the desert produce a distinctive landscape. The vistas seem infinite, for no trees check the eye as it sweeps on to the distant horizon and up into the cloudless skies. There are occasional mountains and canyons, of course, but most of the country consists of open plains. Scattered over the plains are broad, wind-swept slopes of bare rock or rubble; flat-floored, salt-rich playas covered with fine-grained sediments; slowly moving sand dunes; and valleys floored with the deposits of occasional floods or mudflows. The desert plains are sometimes cut by deep, steep-walled canyons, such as the canyon of the Colorado and its tributaries. And here and there ranges of rock mountains rise abruptly from the plains, bordered by the slopes of alluvial fans leading down from the mountain gorges. Often a series of individual fans coalesce to form a ramplike apron of rubbly debris along the mountain front.

In the desert, the nature of the bedrock has a much more striking effect on the appearance of the landscape than it does in humid areas. Where there is

Figure 14.26 **In the desert, as here in the Grand Canyon of the Colorado in Arizona, the structure of the earth's crust is unobscured by soils and vegetation.** (*Photograph by U.S. Geological Survey.*)

plenty of moisture, the lush vegetation, well-developed soils, and unconsolidated debris cloak the slopes and obscure irregularities in the solid bedrock. Graceful curves lead from the hill crest down to valley floors. But in the desert the structure of the earth's crust is far more apparent, and the landscape gives a more faithful reflection of the unadorned bedrock (see Figure 14.26).

Figure 14.27 **Erosion carves a form similar in cross section to the pediment of a Greek temple, the triangular gable above the columns.**

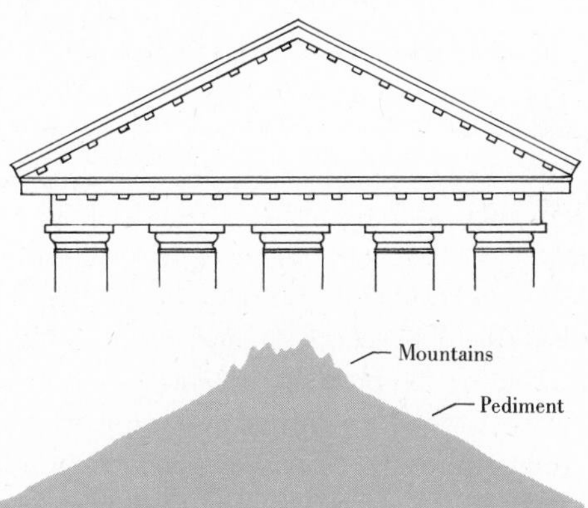

The arid cycle of erosion

In discussing the cycle of erosion in Chapter 11, we chose our examples largely from humid climates, but in a broad sense the concept is valid in arid regions as well, for the desert landscape goes through an erosional sequence.

PEDIMENTS We found that in a humid climate the end product of erosion is thought to be a gently rolling plain called a peneplain. In dealing with the arid and semiarid climates we are concerned with a somewhat different erosional form, the pediment. As soon as land is upraised, as in a mountain range or a high plateau, erosion begins to work. The steep faces of the upraised block are worn backward, and slopes are carved on the bedrock. These slopes are thinly veneered with gravel, and they grade downward toward desert streams or basins. These are the surfaces of erosion that we refer to as *pediments,* because in cross section they resemble the triangular unit at the front of Greek temples (see Figures 14.27 and 14.28).

Figure 14.28 **South of Phoenix, Arizona, these desert mountains are girdled by the sloping surfaces of stream-eroded pediments.** (*Photograph by William C. Bradley.*)

Figure 14.29 **Suggested comparison between the sequence of erosional stages in an arid climate and those in a humid climate. See the text for discussion.**

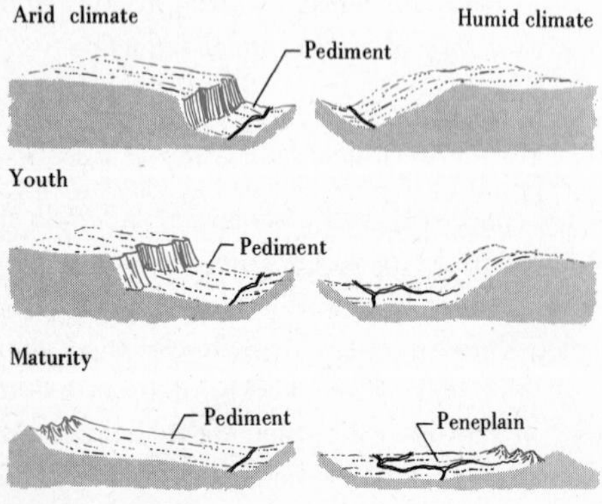

The precise manner in which pediments form is still not clear. We can say, however, that a steep cliff is eroded backward at a constant angle. Important in the process are mechanical weathering, slope wash, and rill wash. Intermittent streams may erode the front locally as they swing first to one side and then to the other after issuing forth from the highlands behind the cliff. In any event, the eroded material is carried down over the ever-widening pediment. With time, the mountain mass is destroyed or reduced to a low dome with perhaps a few projections of resistant rock rising from its surface.

During this process, there is a sequence of landforms, but the sequence differs somewhat from that in a humid climate. First, the pediment is present from the very beginning of the arid cycle, whereas the peneplain does not appear until old age in a humid climate. Desert areas progress through a series of stages as the cliff at the head of the pediment retreats at the expense of the landmass behind it. As the desert area becomes older, the pediments become more and more extensive in area and the plateau or mountain zones more and more restricted. The slope of the scarp at the head of the pediment and even the pediment itself do not become appreciably gentler from the beginning of the cycle to the end. This is in contrast to the slopes that develop in more humid climates, for these slopes appear to become less steep with the passage of time (see Figure 14.29).

Another difference between the cycle of erosion in an arid climate and that in a humid climate is brought about by the presence of broad desert basins that have no outlet to the sea. These basins were created by earth movements that dropped local land areas downward relative to adjacent mountain masses, as in large sections of the southwestern United States. The earth movements

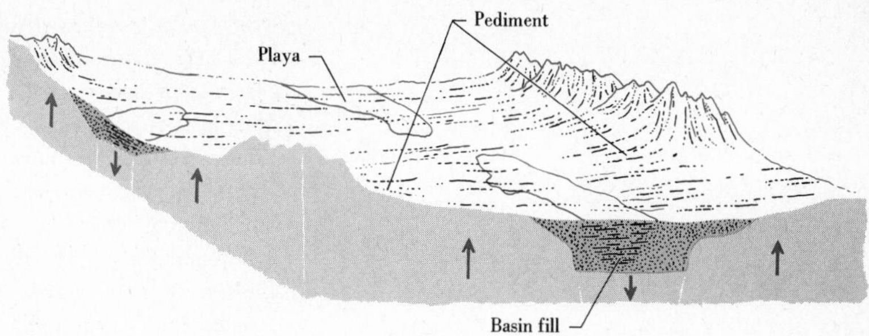

Figure 14.30 ***In large sections of the southwestern United States tectonic movements have produced great basins flanked by mountain ranges. Erosion has fashioned pediments around the flanks of the upthrown mountain blocks, and the products of this erosion have been dumped in the intervening basins. Intermittent lakes, or playas, often occupy the bottoms of these basins.***

are *not* caused by the climate, but the basins do persist because there is insufficient rainfall to establish stream valleys leading to the sea. Consequently, the landscape is leveled by a combination of erosion and deposition. The mountains are attacked by expanding pediments that grade into the basins, but the products of erosion cannot be removed from the basins except by wind. As fine sediments are deposited in the basins, therefore, they bury the lower slopes of the pediments (Figure 14.30). Eventually the basins are filled, and the upper portions of the pediments extend toward the divides, where a few small knobs or peaks mark the last vestiges of the mountains.

We must not infer that all arid lands are composed of mountains and intervening basins, however. In many deserts, drainage does reach the sea through continuous valley systems, and, here, pediments develop along the streams and move back into the landmass as the valley walls retreat.

Outline

Deserts cover one-third of the earth's land surface.

Distribution of deserts is divisible into *topographic* and *tropical* deserts. A topographic desert owes its aridity to the distance from a source of moisture or protection from rain-bearing winds by a mountain mass. A tropical desert lies between 5° and 30° north or south of the equator in one of two zones of subtropical high pressure developed by the planetary atmospheric circulation.

Rainfall is low (from near zero to 40 cm/year), unpredictable, and (when it comes) often torrential.

Temperature changes are large from day to night.

Wind, unchecked by vegetation, is an important geological agent.

Weathering, both mechanical and chemical, is slow because of lack of water.

Soils are rubbly and sometimes marked by crustlike accumulations of calcite, iron oxide, and silica.

Water, despite its scarcity, is abundant enough to create streamways in even the driest desert. Flow of streams is short-lived but often catastrophic. Ground water is also limited but present in some amounts in all deserts.

Work of wind is more effective in the desert than anywhere else on the lands. Even here, however, it is subordinate to the work of running water in shaping desert landscape.

Movement of material by wind depends in part on the size of particle. Dust-sized particles move differently than do sand-sized particles. Dust particles are carried high into the atmosphere in suspension. Sand moves along maintaining continuous contact with the ground or moves a few centimeters above the surface in saltation.

Erosion by wind consists of *abrasion,* of which *ventifacts* are the most common example, and *deflation,* of which *blowouts* are examples.

Deposition consists of dust deposits (*loess*) and sand deposits (usually *dunes*). Sand deposition in dunes begins in the *wind shadow* in the lea of an obstacle. Once established, the dune provides its own wind shadow. In the dune, material is moved up the windward slope and is deposited on the *steep slip face* in the dune's wind shadow. Dune types include *fore dunes, transverse dunes, barchans, parabolic dunes,* and *longitudinal dunes.*

Desert landscapes include *pediments,* gravel-covered surfaces formed by running water around the flanks of desert highlands.

Supplementary readings

BAGNOLD, R. A.: *The Physics of Blown Sand and Desert Dunes,* Methuen & Co., Ltd., London, 1941. The study of the movement of sand and dust by wind begins here.

BLACKWELDER, ELIOT: "Geomorphic Processes in the Desert," *Calif. Div. Mines Bull.,* no. 170, pp. 11–20, 1954. A general overview of the subject.

BRYAN, KIRK: "Erosion and Sedimentation in the Papago Country, Arizona," *U.S. Geol. Surv. Bull.,* vol. 730, 1922. An early paper on desert processes and forms, including one of the final statements on pediments.

COOPER, WILLIAM S.: "Coastal Dunes of California," *Geol. Soc. Am. Mem.,* no. 104, 1967. A clear description of dunes peculiar to coastlines.

DENNY, CHARLES S.: "Alluvial Fans in the Death Valley Region, California and Nevada," *U.S. Geol. Surv. Professional Paper,* no. 466, 1965. The nature and the processes of formation of alluvial fans are succinctly presented.

RITTER, DALE F.: "Terrace Development along the Front of the Beartooth Mountains," *Geol. Soc. Am. Bull.,* vol. 78, pp. 467–484, 1967. A clear statement of the origin and history of certain pedimentlike forms characteristic of large portions of the Rocky Mountains.

SCHULTZ, C. B., and J. C. FRYE, (eds.): *Loess and Related Eolian Deposits of the World,* University of Nebraska Press, Lincoln, 1968. Collection of papers by various authors.

SHARP, R. P.: "Pleistocene Ventifacts East of the Big Horn Mountains, Wyoming," *J. Geol.,* vol. 57, pp. 175–195, 1949. An excellent report on ventifacts.

15

Oceans and shorelines

MORE THAN 70 PERCENT OF THE EARTH LIES BENEATH THE OCEANS. THE UNDERstanding of this portion of the earth is of utmost importance to the geologist. It was in the oceans of the past that most of the sedimentary rocks were formed—rocks that today cover three-quarters of the continental landmasses—and the great ocean basins are clearly linked to the history and origin of the continents.

We have begun to piece together the picture of the ocean floors—their topography, composition and history, and the nature of the chemical and physical processes that operate across them. In this chapter we shall trace in briefest outline some of the facts that have been assembled and some of the problems that have arisen concerning the oceans and the ocean basins.

15.1 Ocean water

The northern hemisphere is sometimes referred to as the "land hemisphere," because north of the equator the oceans and seas cover only about 60 percent of the earth's surface, whereas in the southern hemisphere over 80 percent is flooded by marine waters. Between 45 and 70°N, the ocean occupies only 38 percent of the surface; in contrast, 98 percent of the earth's surface is covered by the ocean between 35 and 65°S (see Figure 15.1).

The greatest ocean depths so far recorded are from an area in the Pacific Ocean midway between the islands of Guam and Yap. Here the ocean bottom is more than 11,000 m below the surface of the water—considerably greater than the height of Mt. Everest, the world's highest mountain, about 8,840 m high. The average ocean depth is about 3,800 m and the mean elevation of the continents is only 840 m, as shown in Figure 15.3.

It has been estimated that the globe would be covered with a layer of water about $1\frac{1}{2}$ to 3 km thick if all the irregularities of the surface were eliminated. Such a situation has probably never existed in the past, nor need we worry about its occurring in the future. Modern oceans are confined to great basins and, presumably, so were the oceans of the past. We shall discuss the characteristics of these basins in some detail later in the chapter.

Origin of ocean water

We do not have a generally agreed upon explanation of the origin of the ocean's waters, but two different possibilities are generally referred to.

We first recognize that the vast amount of water in the hydrologic cycle is too great to have been derived from the weathering of the crustal rocks. It is true that there is water locked up as part of the crystal structure of minerals in the crust, but if we consider such rocks to be a source of the water, then the hydrogen in the water of the oceans really represents an "excess" if the equivalent of the outer few thousand meters of rock represents its source. The same can be said for chlorine, which combines with sodium to form the bulk

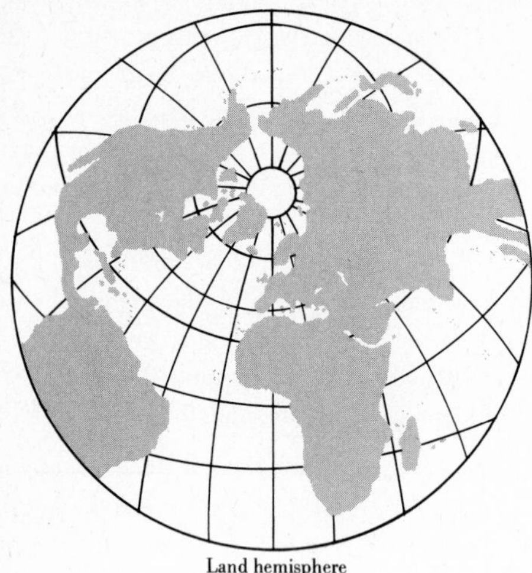

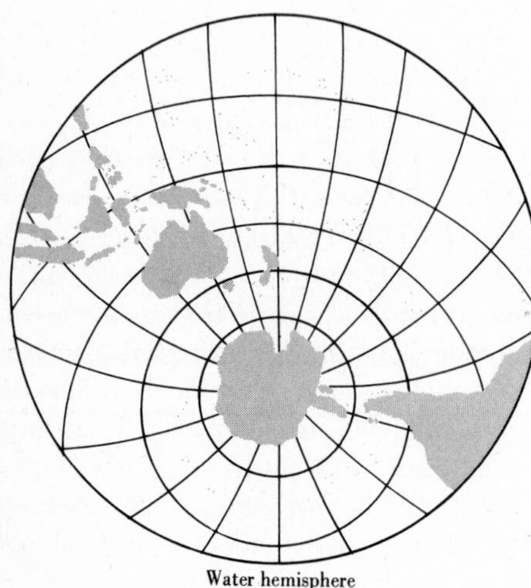

Figure 15.1 **On the land hemisphere map, centered on western Europe, land and sea are about evenly divided. But an indisputable predominance of the seas is revealed on the water hemisphere map, centered on New Zealand.**

Figure 15.2 **Exploration of the ocean floors is carried on from a variety of vessels. The Glomar Challenger shown here is a self-propelled deep-sea drilling ship.**

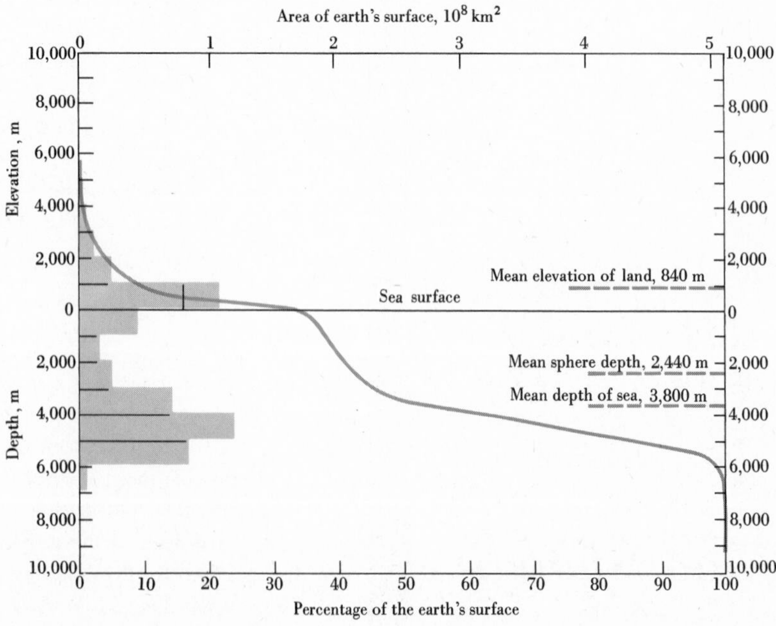

Figure 15.3 **The relative distribution of land and sea. Note that the mean ocean depth is 3,800 m, whereas the mean elevation of the land is only 840 m.** (*From H. U. Sverdrup, Martin W. Johnson, and Richard H. Fleming,* The Oceans, *p. 19, Prentice-Hall, Inc., Englewood Cliffs, N.J., 1942.*)

Table 15.1 ***The major constituents dissolved in sea water***[a]

Ion		Percentage of all dissolved material
Chlorine	Cl^-	55.04
Sodium	Na^+	30.61
Sulfate	SO_4^{2-}	7.68
Magnesium	Mg^{2+}	3.69
Calcium	Ca^{2+}	1.16
Potassium	K^+	1.10
Bicarbonate	HCO_3^-	0.41
Total		99.69

[a] From H. U. Sverdrup, Martin W. Johnson, and Richard H. Fleming, *The Oceans*, p. 166, Prentice-Hall, Inc., Englewood Cliffs, N.J., 1942.

of the salt of the ocean. Weathering of crustal rocks, even over hundreds of millions of years, could not have accounted for the chlorine and the hydrogen of the oceans. We must appeal to a larger source to explain their present volume in sea water.

One possibility is that volcanic eruptions have been continuously adding water and other elements from great depths below the surface, depths that exceed the zone of weathering. Such a mechanism could tap a much larger source of water and chlorine and over the long span of geologic time could account for the oceans of the world. This process, called *continuous degassing,* has the drawback that apparently most, if not all, of the material coming to the surface through volcanic eruptions is really recycled material and is not being added to the hydrologic cycle for the first time.

A second possibility assigns the accumulation of most of the oceanic volume to an early stage of earth history, a period for which we have little tangible evidence. In this instance, the excess hydrogen combined with oxygen and bubbled to the surface during the first half million years or less of earth time when temperatures were high and the present gross stratification of the earth was still incomplete. The excess chlorine came to the surface at the same time.

We do not as yet have sufficient evidence to choose between the two possibilities. In fact, both processes may have contributed to the creation of our oceans.

The nature of sea water

About half of the known elements have been identified in sea water, and many others undoubtedly await discovery. Included among the materials known to be dissolved in the sea are the chlorides that give sea water its familiar saltiness, all the gases found in the atmosphere, and a large number of less abundant materials, including such rare elements as uranium, gold, and silver (see Table 15.1 and Appendix D).

SALTS DISSOLVED IN SEA WATER Through millions of years of geologic time, the rivers of the world have been transporting tremendous quantities of dissolved material to the oceans. Some of this material, such as iron, silicon, and calcium, is used by plants and animals in the life processes and is removed from the water. As a result, the amount of these elements present in solution is less than we would expect to find, judging from the rate at which rivers are currently supplying them to the oceans. On the other hand, we find a relatively high percentage of the "salt" ions, notably Cl^-, even though rivers are presently supplying these materials at a relatively low rate. Salt ions have continued to collect in the sea water because plants and animals do not concentrate them and because they are extremely soluble.

Because the proportions of the various salt ions are relatively constant throughout the oceans, a measurement of any one of them enables us to compute the abundance of the others. The total concentration of salt ions—that is, the salinity of the sea water—varies from place to place, however. At the equator, heavy precipitation dilutes the sea water, reducing its salinity. But

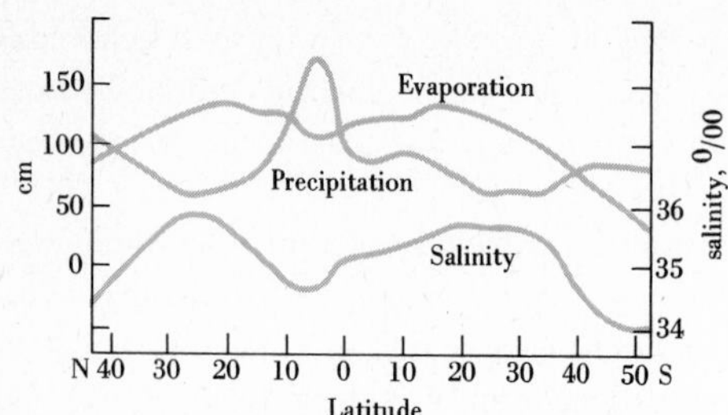

Figure 15.4 **Total salinity varies in the oceans from south to north across the equator. The low salinity of surface waters in the vicinity of the equator is attributed to the freshening effect of heavy tropical rains. North and south of this zone the rainfall decreases, and evaporation increases; as a result, the total salinity of the surface waters increases.** (*Redrawn from R. H. Fleming and Roger Revelle, "Physical Processes in the Oceans," in Parker D. Trask, ed.,* Recent Marine Sediments—A Symposium, *p. 88, American Association of Petroleum Geologists, Tulsa, 1939.*)

in the subtropical belts to the north and south, low rainfall and high evaporation tend to increase salinity, as indicated in Figure 15.4. In the Arctic and Antarctic areas, on the other hand, the melting of glacier ice reduces the saltiness of the seas.

GASES DISSOLVED IN SEA WATER Although all the gases found in the atmosphere are also present in water, probably the most important are oxygen and carbon dioxide. Near the surface of the oceans the water is saturated with both gases, but their concentration and relative proportions vary with depth. As the surface water circulates downward through the first few tens of meters, intense plant activity depletes the supply of carbon dioxide. At the same time, oxygen is given off by the plants. Consequently, this near-surface zone is deficient in carbon dioxide and tends to be oversaturated with oxygen.

Below the depth to which light can penetrate effectively, however, plant activity falls off and the amount of oxygen in solution decreases. Of the oxygen present, some is used by animals and some becomes involved in the oxidation of organic matter settling toward the bottom. At the same time, the relative amount of carbon dioxide increases, because there is no plant activity to deplete it. Thus with increasing depth, oxygen becomes relatively less abundant and carbon dioxide relatively more abundant.

Were it not for the slow circulation of sea water through the ocean basins, water at the greatest depths would be devoid of oxygen. Actually, at the bottom of some ocean basins, circulation of water is so slow that almost no oxygen is present, and the water has become stagnant. Here are high concentrations of hydrogen sulfide. This is true, for example, in the Black Sea below a depth of about 150 m and in many of the Norwegian fiords whose glacially deepened basins lie below the general level of the adjacent floor of the North Atlantic. Because of the almost complete absence of oxygen at the bottom of these basins, sediments lying there oxidize very slowly, if at all. Consequently, a high content of organic matter and the hydrogen sulfide compounds that are produced give them their characteristic black color. It is probable that petroleum was formed in some environment such as this in the ancient seas.

15.2 *Currents*

Sea water is in constant movement, in some places horizontally, in others downward, and in still others upward. Its rate of movement varies from spot to spot, but it has been estimated that there is a complete mixing of all the water of the oceans about once every 1,800 years. We must assume that movements similar to those of the present have been going on throughout the long history of the earth. By studying modern seas, we can gain an insight into the history recorded in sedimentary rocks that were once muds and sands on the floors of ancient seas.

The currents of ocean waters are caused by tides, by the differences in density of sea water, by wind, and by the rotation of the earth.

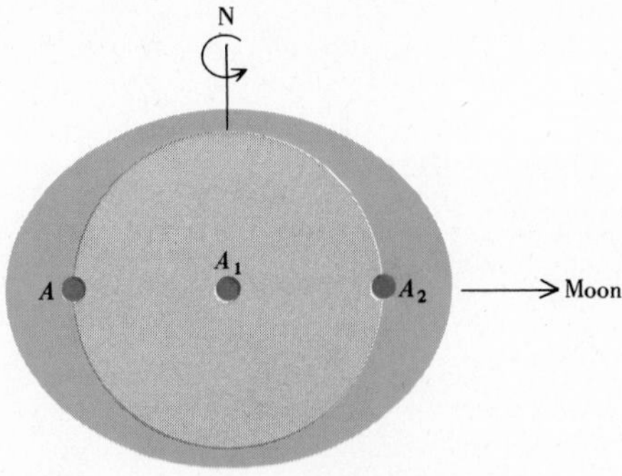

Figure 15.5 **Idealized diagram of tidal bulges. The earth is viewed as surrounded by a continuous ocean deformed into two bulges, one near the moon and the other on the opposite side of the earth. The gravitational pull of the moon is least at point *A* and greatest at point A_2.**

Tidal currents

The attractive forces that operate between the sun, the moon, and the earth set the waters of the ocean in horizontal motion to produce *tidal currents*. The speed of these currents may reach several kilometers per hour if local conditions are favorable. Velocities in excess of 20 km/hr develop during the spring tides in Seymour Narrows, between Vancouver Island and the mainland of British Columbia, and tidal currents of half this velocity are not uncommon. The swiftest currents usually build up where a body of sea water has access to the open ocean only through a narrow and restricted passage. Such currents are capable of moving particles up to, and including, those of sand size, and they may be strong enough to scour the sea floor.

In seeking an explanation of tides, we can first consider (1) the gravitational attraction of the moon on the earth, (2) the centrifugal force on the earth resulting from the motion of the earth–moon system around a common center of gravity, and (3) the period of earth rotation relative to the moon.

In the earth–moon system, the centrifugal force is balanced with gravitational pull and for the earth as a whole the pull of the moon is just equal to the centrifugal force. For different points on the earth, however, these forces are not necessarily in balance.

Consider a point on the earth's surface directly beneath the moon. Because it is closer to the moon, it is subject to a greater gravitational pull by the moon than is a point 12,800 km from the moon on the opposite side of the earth. Centrifugal force results from the motion of the earth–moon system around a center of mass 1,700 km beneath the earth's surface along a line connecting the centers of the earth and moon. This force is the same at all points on the earth's surface because all points describe the same motion around the center of gravity of the system. Remember, however, that the gravitational attraction of the moon on the earth differs with differing distances from the moon. At a point on the earth directly beneath the moon, the gravitational force is greater than the centrifugal force, and on the opposite side of the world the gravitational pull is less than the centrifugal force. There is, then, a pull toward the moon on one side of the earth and a pull away from the moon on the far side. Both solid earth and oceans respond to this stress—but the oceans more obviously than the solid earth (see Figure 15.5).

Figure 15.6 **The gravitational pull of the moon on the earth combines with the centrifugal force of the earth–moon system to create differential forces at various places on the earth. These tide-producing forces operate parallel to the earth's surface and are suggested in this diagram. See the text for discussion.**

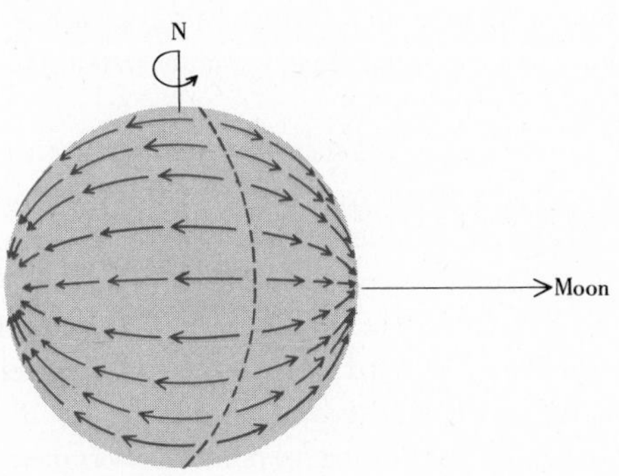

The earth rotates under these bulges. Because it takes about 24 hr and 50 min for a point on the earth to return to the same position in relation to the moon, it follows that there should be two high tides during that time and two low tides.

The amount of pull exerted away from the earth varies with geographic position on the earth, but in all places this is countered by the earth's gravitational field exerting a pull toward the center of the earth. Between these two pulls (one in toward the earth and the other away from it), a tractive force is set up which causes a lateral motion of the ocean waters. This is a tidal current. These tide-producing forces are shown in Figure 15.6.

A more extensive examination of the causes of tides will show that the sun's

Figure 15.7 ***Twice a day tides cause the withdrawal of ocean water and the exposure of this muddy bottom along a Maine inlet.*** (*Photograph by Sheldon Judson.*)

gravitational pull is also important, although because of the distance from the earth to the sun the effect is less than half that of the moon. When the sun, moon, and earth are in line, we experience the highest tides—the spring tides—and when the sun, earth, and moon are at right angles, the effects are least and neap tides occur.

In the United States we are most familiar with the semidiurnal, twice daily tides, or a modification of them (see Figure 15.7). But in some places, such as Manila in the Philippines, tides occur daily, one high and one low. Such variations occur because the plane of the moon's orbit around the earth does not coincide with that of the earth around the sun, and varying combinations of forces result from the differing relationships among the three bodies.

Density currents

The density of sea water varies from place to place with changes in temperature, salinity, and the amount of material held in suspension. Thus cold, heavy water sinks below warmer and lighter water; water of high salinity is heavier than water of low salinity and sinks beneath it; and heavy muddy water sinks beneath light, clear water.

In the Straits of Gibraltar, the water passage between the Atlantic and the Mediterranean, differences in density are partially responsible for a pair of currents flowing one above the other. The Mediterranean, lying in a warm, dry climatic belt, loses about 1.5 m of water every year through evaporation. Consequently, the saltier, heavier water of the Mediterranean moves outward along the bottom of the straits and sinks downward into the less salty, lighter water of the Atlantic. At the same time, the lighter surface water of the Atlantic moves into the Mediterranean basin. The water flowing from the Mediterranean

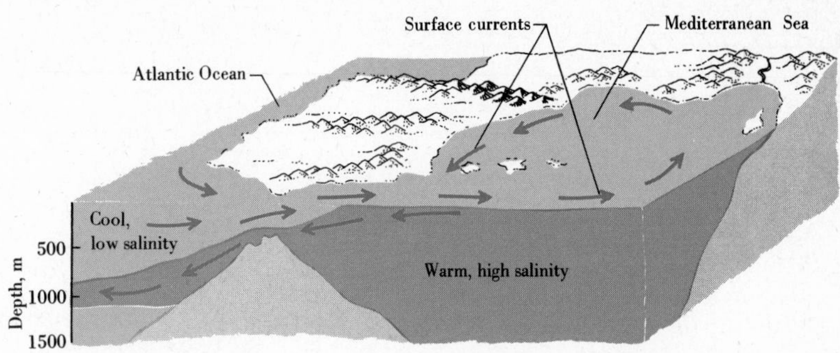

Figure 15.8 A density current flows from the Mediterranean Sea through the Straits of Gibraltar and spreads out into the Atlantic Ocean. High evaporation and low rainfall in the Mediterranean area produce a more saline and hence heavier water than the water of the neighboring Atlantic Ocean. As a result, Mediterranean water flows out through the lower portion of the straits, and lighter Atlantic water moves above it and in the opposite direction to replace it. The higher temperature of the Mediterranean water is more than counteracted by its greater salinity, and the water sinks to a level in the Atlantic somewhat lower than 1,000 m below the surface.

settles to a depth of about 1,000 m in the Atlantic and then spreads slowly outward beyond the equator on the south, the Azores on the west, and Ireland on the north. It has been estimated that as a result of this activity the water of the Mediterranean basin is changed once every 75 years (see Figure 15.8).

The density of water is also affected by variations in temperature. As a result of such variations, water from the cold Arctic and Antarctic regions creeps slowly toward the warmer environment near the equator. The cold, relatively dense water from the Arctic sinks near Greenland in the North Atlantic and can be traced to the equator and beyond as far as 60°S. This water is called North Atlantic Deep Water. Colder and denser water (Antarctic Bottom Water) moves downward to the sea floor off Antarctica and creeps northward, pushing beneath the North Atlantic Water. In fact the Antarctic water reaches well north of the equator before it loses its identity. Lighter than these two types of water is the Antarctic Intermediate Water, which sinks in the area of 50°S and flows northward above the North Atlantic Deep Water until it loses its identity about 30°N in the area where the Mediterranean water flows from east to west. These relationships are shown in Figure 15.9.

A third type of density current, known as a *turbidity current,* is caused by the fact that turbid or muddy water has a greater density than clear water and therefore sinks beneath it. We can demonstrate how these currents operate on a small scale in the laboratory and they have actually been observed in fresh-water lakes and reservoirs. Turbidity currents explain several different phenomena in the present-day oceans as well as in the oceans of the geological past.

Turbidity currents offer the most plausible explanation of certain deposits that have been studied from the deep ocean basins. Samples of sediments

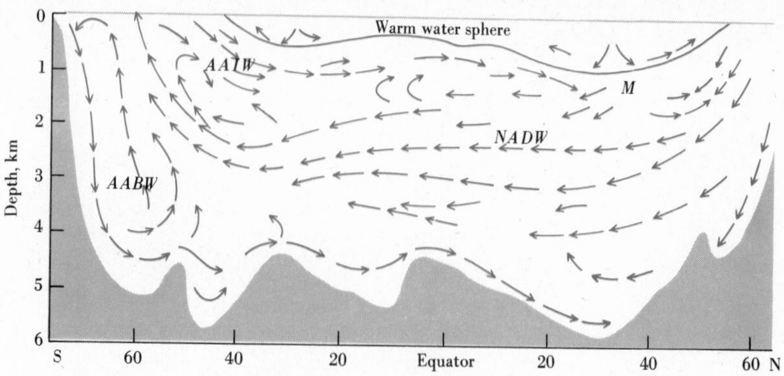

Figure 15.9 **Circulation of water in the Atlantic Ocean shown along a north–south profile as determined by temperature and salinity. NADW, North Atlantic Deep Water; AAIW, Antarctic Intermediate Water; and AABW, Antarctic Bottom Water.**

from these basins contain thin layers of sand that could not have been carried so far from shore by the slow drift of water along the ocean floor, but rapidly moving turbidity currents would be quite capable of moving the sand down the slopes of the ocean basins to the deep floors.

Analysis of other samples from the ocean floor reveals that particles become progressively finer from bottom to top of the deposit. The best explanation of this *graded bedding* is that the deposit was laid down by a turbidity current carrying particles of many different sizes and that the larger particles were the first to be dropped on the ocean floor. Such deposits are called *turbidites.*

Turbidity currents may be set in motion by the slumping and sliding of material along the slopes of the ocean basin under the influence of gravity, either by itself or aided by the jarring of an earthquake. Or the currents may be created by the churning up of bottom sediments under the influence of violent storms.

Major surface currents

The major movements of water near the ocean's surface occur in such currents as the Gulf Stream, the Japanese current, and the equatorial currents. These great currents are caused by a variety of factors, including the prevailing winds, the rotation of the earth, variations in the density of sea water, and the shape of ocean basins. Let us examine, by way of illustration, the surface currents of the Atlantic Ocean in both the northern and southern hemispheres, as shown in Figure 15.10.

The equatorial currents lie on each side of the equator, and they move almost due west. They derive their energy largely from the trade winds that blow constantly toward the equator, from the northeast in the northern hemisphere and from the southeast in the southern hemisphere. The westerly direction of the currents is explained by the Coriolis effect, discussed in Chapter 14. You will remember that this effect is produced by the rotation of the earth and that it causes moving objects to veer to the right in the northern hemisphere and to the left in the southern hemisphere. As the water driven by the trade winds moves toward the equator, it is deflected to the right (toward the west) in the northern hemisphere and to the left (also toward the west)

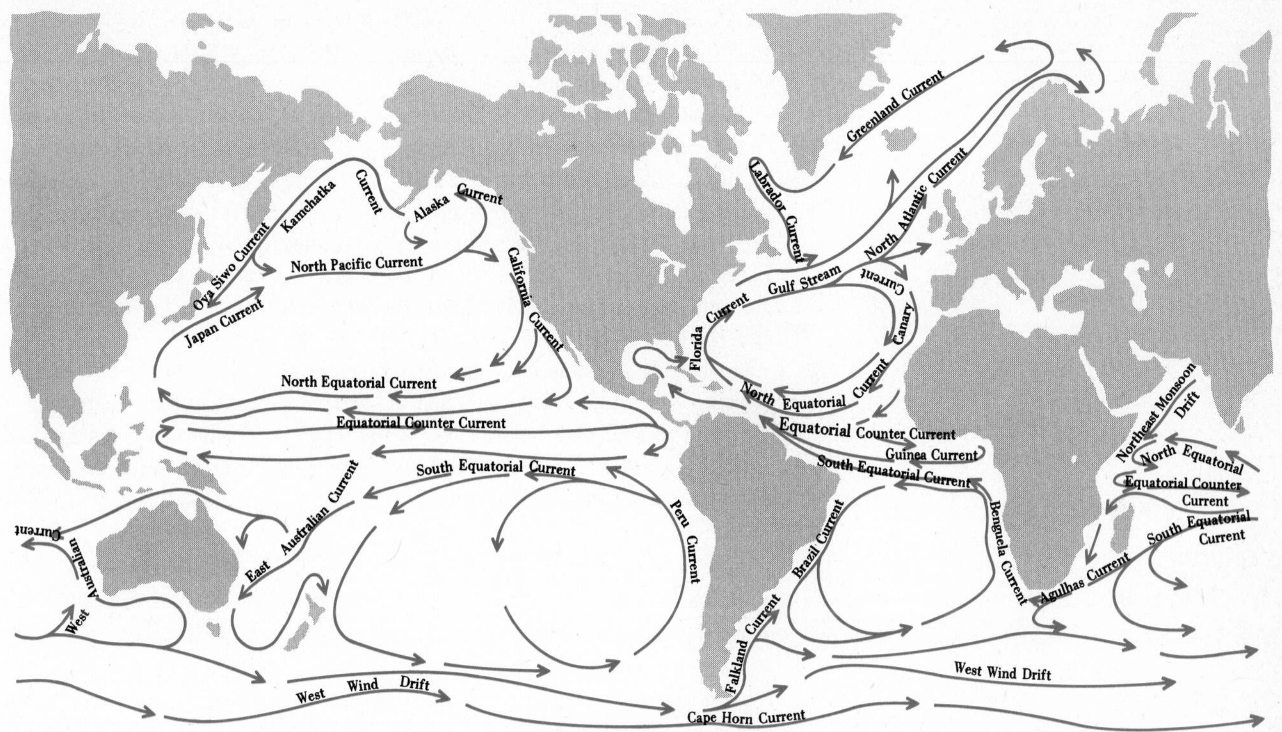

Figure 15.10 ***Major surface currents of the oceans of the world.***

in the southern. As a result, the north and south equatorial currents are formed. As these currents approach South America, one is deflected north and the other south. This deflection is caused largely by the shape of the ocean basins, but it is aided by the Coriolis effect and by the slightly higher level of the oceans along the equator where rainfall is heavier than elsewhere.

The north equatorial current moves into the Caribbean waters and then northeastward, first as the Florida current and then as the Gulf Stream. The Gulf Stream, in turn, is deflected to the east (to the right) by the Coriolis effect. This easterly movement is strengthened by prevailing westerly winds between 35 and 45°N, where it becomes the North Atlantic current. As it approaches Europe, the North Atlantic current splits. Part of it moves northward as a warm current past the British Isles and parallel to the Norwegian coast. The other part is deflected southward as the cool Canaries current and eventually is caught up against the northeast trade winds, which drive it into the north equatorial current.

In the South Atlantic, the picture is very much the same—a kind of mirror image of the currents in the North Atlantic. After the south equatorial current is deflected southward, it travels parallel to the eastern coast of South America as the Brazil current. Then it is bent back to the east (toward the left) by the Coriolis effect and is driven by prevailing westerly winds toward Africa. This easterly moving current veers more and more to the left until finally, off Africa, it is moving northward. There it is known as the Buenguela current, which in turn is caught up by the trade winds and is turned back into the south equatorial current.

The cold surface water from the antarctic regions moves along a fairly simple course, uncomplicated by large landmasses. It is driven in an easterly direction by the prevailing winds from the west. In the northern hemisphere, however, the picture is complicated by continental masses. Arctic water emerges from the polar seas through the straits on either side of Greenland to form the Labrador current on the west and the Greenland current on the east. Both currents subsequently join the North Atlantic current and are deflected easterly and northeasterly.

We need not examine in detail the surface currents of the Pacific and Indian Oceans. We can note, however, that the surface currents of the Pacific follow the same general patterns as those of the Atlantic. Furthermore, the surface currents of the Indian Ocean differ only in detail from those of the South Atlantic (see Figure 15.10).

15.3 The ocean basins

Most of the sea water surrounding the continents is held in one great basin that girdles the southern hemisphere and branches northward into the Atlantic, Pacific, and Indian Oceans. The Atlantic and Pacific Oceans, in turn, are connected with the Arctic Ocean through narrow straits. But the oceans still flood over the margins of the continents, for even this great, fingered basin cannot contain all the water of the seas.

Most geologists now agree that the continents of the world are composed largely of sialic rock overlying a layer of heavier, crystalline simatic rock. (For a discussion of the terms *sial* and *sima*, see Section 17.5.) The sialic layer is missing from the deep ocean basins, however, and the ocean floor is made up of simatic rock with a covering of sediments that ranges from zero to about 3 km, probably averaging about $\frac{1}{2}$ km.

The picture of the ocean basins that has emerged as the result of studies since World War II is more complicated than this. A discussion of the various layers in the ocean basin is found in Chapter 17. The nature and possible significance of magnetic variations are considered in Chapter 20.

15.4 Topography of the ocean floor

Continental shelves, slopes and rises, and ocean deeps

The margins of the continents lie flooded beneath the seas. The shallowest portions are called the *continental shelves.* The average width of these shelves is about 65 km, but there are many local variations. Along the western coast of South America, for example, the shelf is altogether missing, or at best a few kilometers in width. In contrast, the shelf reaches out 240 km off Florida, about 560 km off the coast of South America south of the Rio de la Plata, and from 1,200 to 1,300 km off the Arctic coasts of Europe and Russia.

The seaward edge of the continental shelves has an average depth of about 130 m, but it is commonly as deep as 180 m or as shallow as 90 m. Soundings indicate that the topography of the shelves changes somewhat from one place to another, but there seems to be a general lack of spectacular features. Some shelves, such as those off Labrador, Nova Scotia, and New England, show the marks of vanished continental glaciers, which deposited in some places and eroded in others. The surface of almost all the shelves is irregularly marked by hills, valleys, and depressions of low to moderate relief. Furthermore, soundings along the shelf bordering the eastern coast of North America show the presence of submarine terraces that record former lower levels of the ocean, just as higher levels are recorded by terraces stranded above sea level.

Generally, the shelves slope gently toward the ocean basins until they are abruptly terminated by the steeper *continental slopes* that lead down into the deep oceans. Many slopes, such as that off eastern North America, grade into somewhat gentler declivities, the *continental rises* which grade into the deep ocean floors (Figure 15.11). In other places, such as off the southern side of the Aleutian Islands or off western South America, the continental slopes lead directly down into the deep *trenches* in the ocean floor.

The greatest ocean depths occur in these trenches and not all of them are directly associated with continental margins. The trenches are great, arcuate or bow-shaped troughs on the sea floor. Several of these trenches dip to more than 9,000 m below the ocean surface, and they may reach 200 km in breadth and 24,000 km or more in length. Because arcuate chains of islands are located near these deep trenches, they are sometimes known as *island arc deeps.* Most of them occur in the Pacific Ocean (Figure 15.12).

SUBMARINE VALLEYS The surfaces of the continental shelves, slopes, and ocean deeps are all furrowed by submarine valleys of varying width, depth, and length, rather like the valleys of the continents. The origin of the deep, spectacular valleys, the *submarine canyons* that crease the continental slopes, is still in dispute, but the origin of the lesser valleys along the shallow continental shelves is fairly well understood.

Some of the valleys that cut across the continental shelves seem to be seaward extensions of large valleys on the adjoining land. One of the best-known

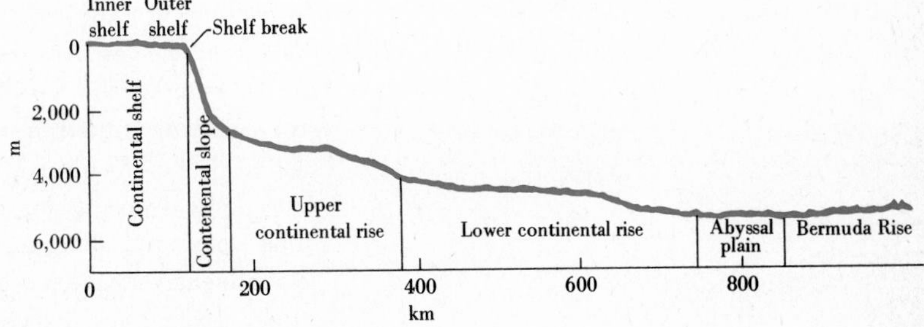

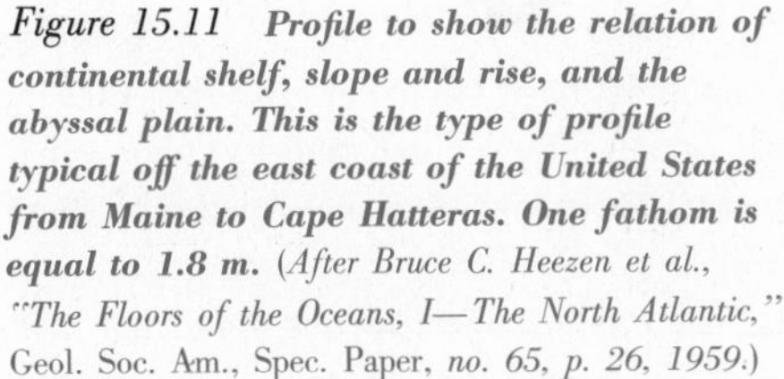
Figure 15.11 **Profile to show the relation of continental shelf, slope and rise, and the abyssal plain. This is the type of profile typical off the east coast of the United States from Maine to Cape Hatteras. One fathom is equal to 1.8 m.** (*After Bruce C. Heezen et al., "The Floors of the Oceans, I—The North Atlantic,"* Geol. Soc. Am., Spec. Paper, *no. 65, p. 26, 1959.*)

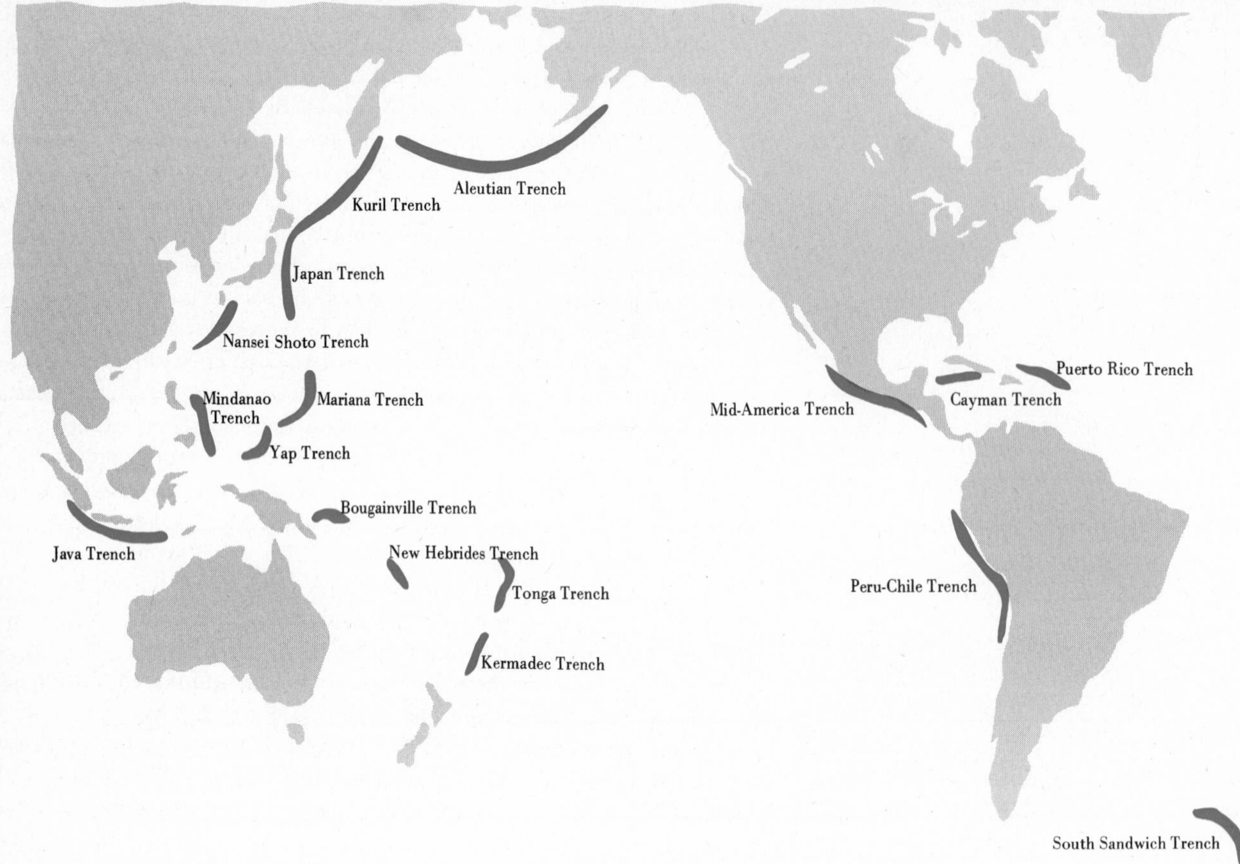

Figure 15.12 ***Deep-sea trenches outline the Pacific Ocean.***

submarine valleys is the submerged extension of the Hudson valley off the eastern coast of the United States. This valley extension is relatively straight, cuts down about 60 m into the continental shelf and widens from almost 5 km to approximately 24 km at its seaward end. It is speculated that this and similar valleys, such as those in the China Sea, were cut during periods of Pleistocene glaciation, when sea level was perhaps 100 m lower than it is at present and when large portions of the continental shelves were exposed to erosion by land streams.

In addition to these larger stream valleys, there are smaller troughs in the continental shelves that are presumed to have been cut by tidal scouring. Several occur off the northeastern coast of the United States. Other valleys were apparently cut on the continental shelves by glacier ice that may simply have deepened already-existing valleys.

More difficult to explain are the deep submarine canyons along the continental slopes and along the floor of the deep ocean basins. The canyons of the continental slopes have been known since the latter part of the nineteenth century. Today, even though soundings are scattered and incomplete, it seems certain that submarine canyons are characteristic of the continental slopes all around the world. Some canyons, such as those on the slopes off the Hudson and Congo Rivers, appear to be extensions of valleys on the land or on the continental shelves. Others seem to have no association with such valleys. In any event, these canyons have V-shaped cross profiles and gradients that decrease along the lower sections of the continental slopes. Their slightly winding courses may extend 235 km out to sea, as does the Congo canyon. Some canyons cut down more than 1 km beneath the surface of the continental slope.

Many theories have been advanced to explain the origin of these submarine canyons. Some investigators think that they were formed by earth movements, others conclude that they were carved out by turbidity currents, and still others believe erosion by tidal scour to be the cause. Then there are those who speculate that the canyons were cut during a vast lowering of the sea in the recent geological past or were fashioned by the emergence of submarine springs along the slopes or by submarine mudflows and landslides. Suffice it to say that no single explanation has yet proved satisfactory to a majority of investigators and that the origin of these canyons is still being debated.

Sonic depth measurements have revealed that other submarine valleys, some as deep as 3,600 m or more, run for great distances across the floors of the ocean out beyond the continental slopes. Valleys of this sort are known in both the Atlantic and the Pacific, but the data on them are too meager to permit exact descriptions, much less explanations of their origin.

Running generally from north to south along the center of the Atlantic Ocean is a broad zone, known as the Midatlantic Ridge, that rises above the general elevation of the deep ocean basin. A submarine valley up to 80 km wide and hundreds of meters deep follows along the crest of this ridge in the North Atlantic and also in the South Atlantic (Figure 15.13) as well as along its extension into the Indian and Pacific Oceans.

Oceanographers have suggested that this submarine valley is similar to the valley of the Dead Sea in the Near East and to the so-called Rift valley of Africa, which contains such lakes as Tanganyika and Nyassa. These land valleys have been caused by the down-dropping of sections of the earth's crust. The submarine valley on the crest of the Midatlantic Ridge (and its extensions) may have been formed in the same way.

SEAMOUNTS Dotting the ocean floors are drowned, isolated, steep-sloped peaks called *seamounts.* They stand at least 1,000 m above the surrounding ocean

Figure 15.13 ***A profile across the Atlantic Ocean from Cape Henry, Virginia, to the Spanish Sahara.*** (*From Bruch C. Heezen et al., "The Floors of the Oceans, I: The North Atlantic,"* Geol. Soc. Am. Spec. Paper, *no. 65, plate 22, 1959.*)

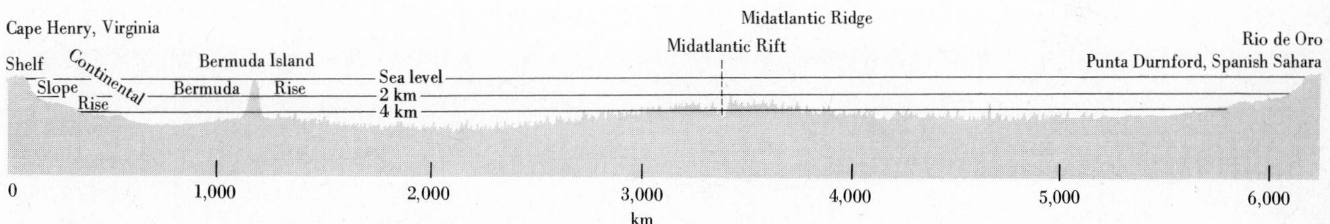

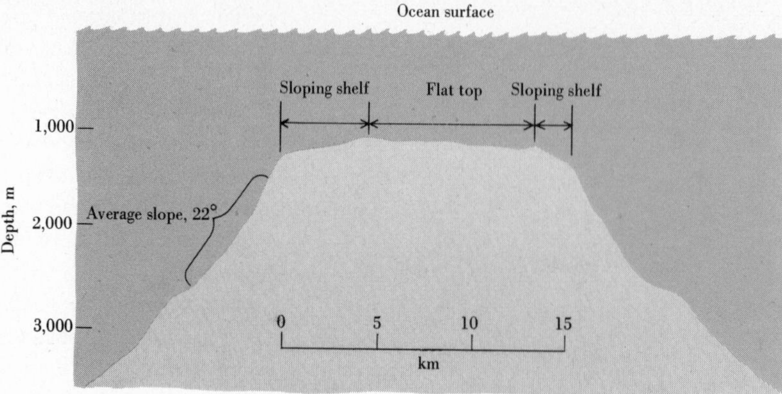

Figure 15.14 **Soundings in the Pacific Ocean during World War II revealed the presence of many flat-topped submarine mountains called guyots. The guyot in this profile is located at 8.8°N, 163.1°E.** *(Redrawn from H. H. Hess, "Drowned Ancient Islands of the Pacific Basin,"* Am. J. Sci., *vol. 244, p. 777, 1946.)*

floor, their crests covered by depths of water measured in hundreds, even thousands, of meters. Seamounts have been identified in all the oceans, but the greatest number by far are reported from the Pacific Ocean. There, 1,700 such peaks had been mapped by the late 1960s; it is estimated that 10 times this number remain to be discovered.

Most seamounts have sharp peaks, and insofar as we can tell, they are all of volcanic origin. Some of the peaks have flat tops called ***tablemounts*** or ***guyots*** in honor of Arnold Guyot, a Swiss–American geologist of the mid-nineteenth century. Guyots are thought to be volcanic cones whose tops have been cut off by the action of surface waves. Most evidence suggests that the guyots then sank beneath the sea, but it is also possible that they were drowned by a rise in sea level (see Figures 15.14 and 15.15).

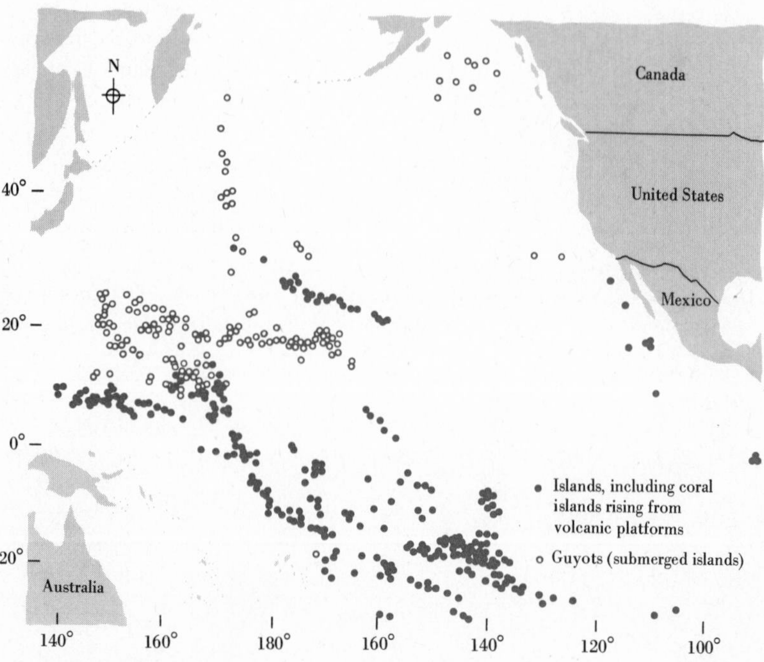

Figure 15.15 **Guyots of the Pacific Ocean basin occur in groups, as do the chains of present-day volcanic-based islands.** *(Redrawn from H. W. Menard, "Geology of the Pacific Floor,"* Experientia, *vol. 15, no. 6, pp. 210–212, 1959.)*

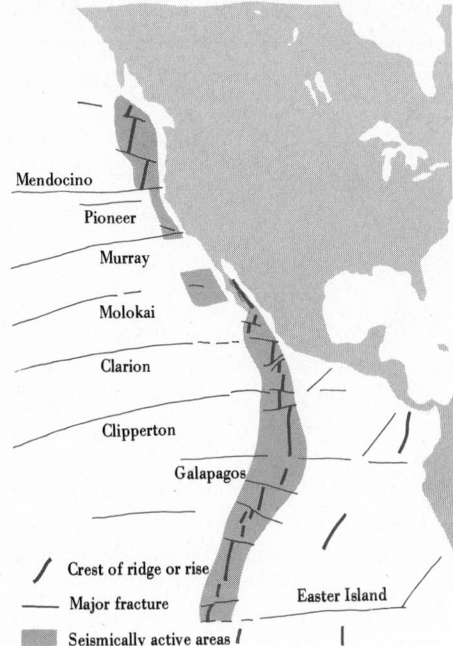

Figure 15.16 **Long fracture zones reach into the eastern Pacific Ocean. In addition, smaller fractures offset the oceanic rise of the eastern Pacific.** (*From H. W. Menard, "Some Remaining Problems in Sea Floor Spreading," in R. A. Phinney, ed.,* History of the Earth's Crust, *p. 112, Princeton University Press, Princeton, N.J., 1968. © 1968 by Princeton University Press. Redrawn by permission.*)

SUBMARINE RIDGES, RISES, AND FRACTURES Among the major features of the deep ocean basins are long, submerged ridges and rises. In general, ridges have steep sides and irregular topography. The rises differ in being broader and gentler in form. They rise thousands of meters above the deep ocean floor and in some places actually appear above the surface to form islands. These ridges and rises form a more or less integrated system of high topography that segments the deep oceans into smaller basins.

In addition to these two forms, long towering escarpments caused by earth movements scar some sections of the ocean floor. Thus the Mendocino, Murray, Clipperton, and Clarion fracture zones extend westward into the Pacific from the coast of the United States and Central America (see Figure 15.16). The vast Mendocino escarpment reaches heights of 2,400 m and extends 3,200 m into the central Pacific. Less well-known but similar fractures and related escarpments lie to the south of this system, and fracture systems of similar magnitude are reported along many of the oceanic ridges (Figure 15.17).

Figure 15.17 **The major rises and ridges of the oceans are cut by fractures.** (*After H. W. Menard, "The World Oceanic Rise–Ridge System," in* A Symposium on Continental Drift, *p. 110, Royal Society, London, Philosophical Trans. no. 1088, 1965.*)

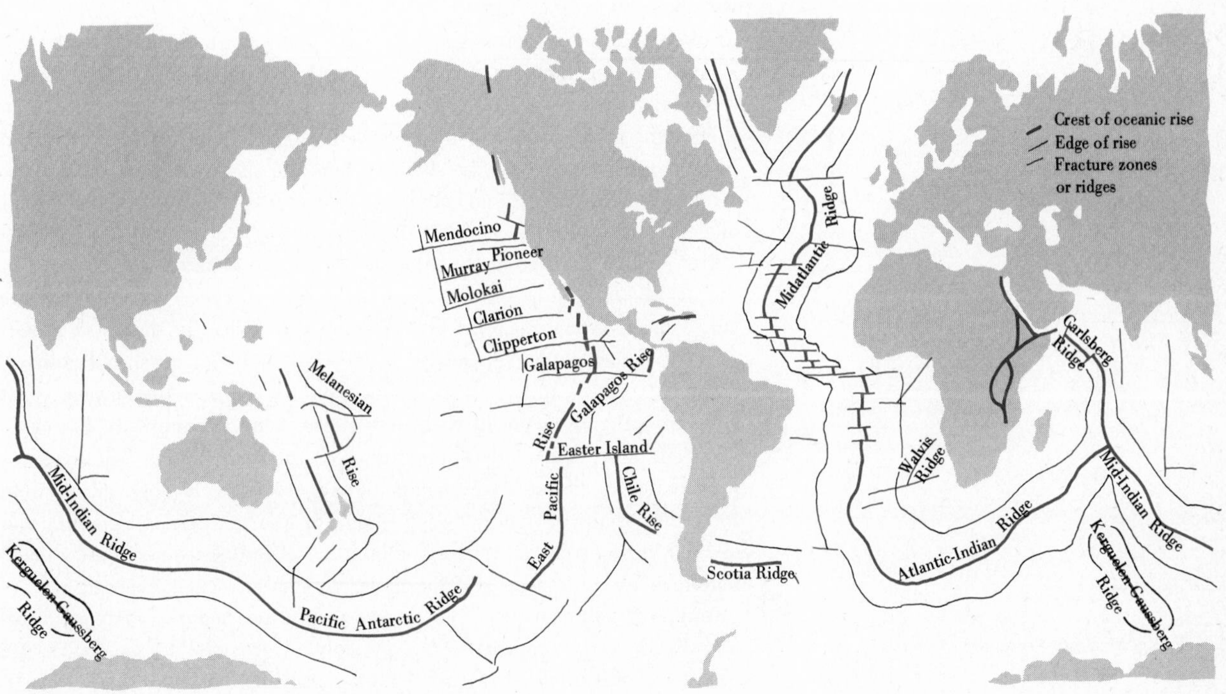

It now seems very probable that these major features, particularly the ridges, fractures, and trenches, are part of a vast, interconnected system. A discussion of their significance is deferred until Chapter 20, where we examine the question of continental drift.

15.5 *Sediments of the ocean*

In earlier chapters, we have spoken from time to time of the processes by which earth materials are weathered, eroded, transported, and finally deposited to be transformed into sedimentary rocks. The great ocean basins of the world constitute the ultimate collection area for the sediments and dissolved material that are carried from the land, and the great bulk of the sedimentary rocks found on our modern landmasses were once deposits on the ocean floors of the past. In this section we shall speak briefly of the sediments being laid down in the modern oceans—the sediments destined to become the sedimentary rocks of the future.

Deposits on the continental shelves

Theoretically, when particles of solid material are carried out and deposited in a body of water, the largest particles should fall out nearest the shore and the finest particles farthest away from the shore, in a neatly graduated pattern. But there are a great many exceptions to this generalization. Many deposits on the continental shelves show little tendency to grade from coarse to fine away from the shoreline. We would expect to find sand close to shore only —actually it shows up from place to place all along the typical continental shelf right up to the lip of the continental slope. It is particularly common in areas of low relief on the shelf. In fact, on glaciated continental shelves, sand mixed with gravel and cobbles comprises a large part of the total amount of deposited material.

Also common on the continental shelves are deposits of mud, especially off the mouths of large rivers and along the course of ocean currents that sweep across the river-laid deposits. Mud also tends to collect in shallow depressions across the surface of the shelves, in lagoons, sheltered bays, and gulfs.

Where neither sand nor mud collects on the shelves, the surface is often covered with fragments of rock and gravel. This is commonly the case on open stretches of a shelf, where strong ocean currents can winnow out the finer material, and off rocky points and exposed stretches of rocky shoreline. In narrow straits running between islands or giving access to bays, the energy of tidal and current movements is often so effectively concentrated that the bottom is scoured clean and the underlying solid rock is exposed.

Strangely, although the geologic record indicates many calcareous deposits laid down in the ancient seas, eventually to give rise to the rock we call limestone, only a few calcareous deposits are being built up on the continental shelves of modern seas. And most of these limy mud deposits are being built up in warmer waters, particularly near coral reefs. No satisfactory explanation accounts for the apparent deficiency.

Deposits on the continental slopes

Although we have less information about the deposits being laid down on the continental slopes than we have about the shelf deposits, evidence indicates that here, too, gravel, sand, mud, and bedrock are all found on the bottom. And deposition, it seems, is taking place even more rapidly on the slopes than on the shelves.

Deposits on the deep-sea floor

The deposits that spread across the floors of the deep sea are generally much finer than those on the slopes and shelves lying off the continents, although occasional beds of sand have been found even in the deeps (see Figure 15.18).

Deep-sea deposits of material derived from the continents are referred to as *terrigenous* ("produced on the earth") *deposits.* Those formed of material derived from the ocean itself are *pelagic* ("pertaining to the ocean") *deposits.*

Among the terrigenous deposits are beds of wind-borne volcanic ash that has fallen over the ocean and has settled through thousands of meters of water

Figure 15.18 **Map showing type and distribution of deep-sea sediments.** (*Redrawn from F. P. Shepard,* Submarine Geology, *2nd ed., Fig. 198, Harper & Row, Publishers, New York, 1963.*)

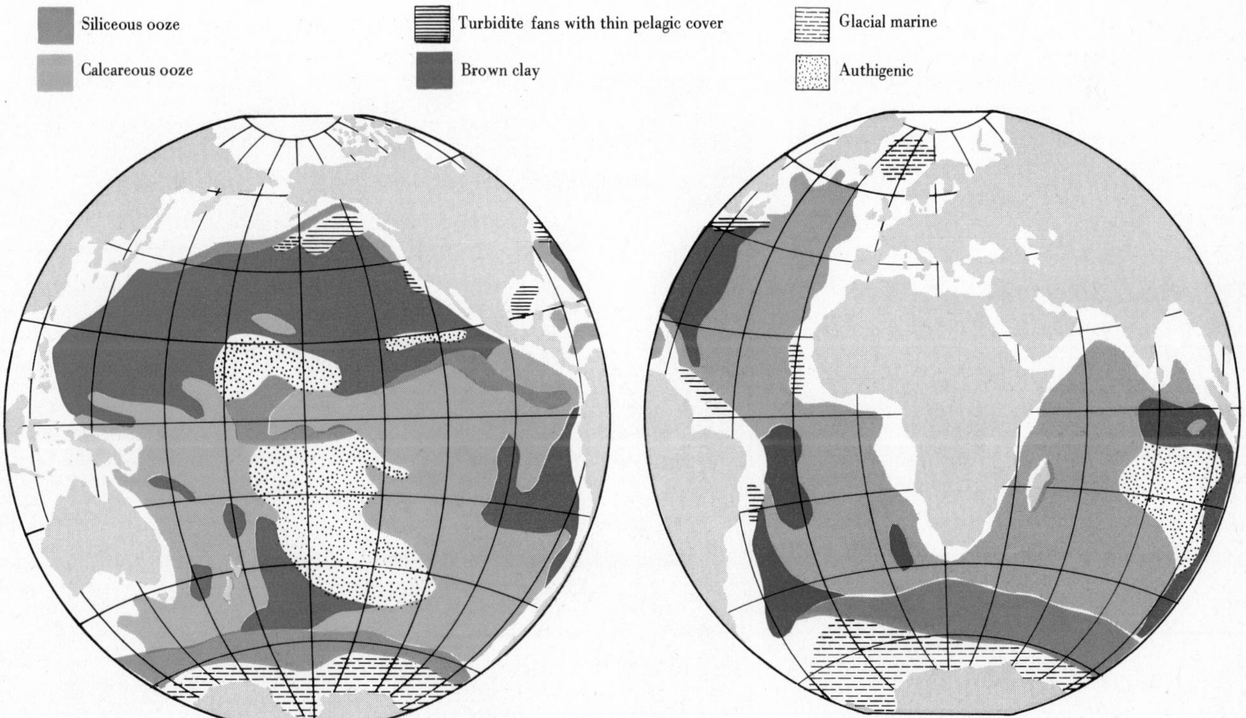

to the sea bottom. In the polar regions, silts and sands from glacier ice make up much of the bottom deposit, and around the margins of the continents we often find deep-sea mud deposits of silt and clay washed down from the landmasses. Much of the ocean bottom, especially in the Pacific, is covered by an extremely fine-grained deposit known as *brown clay*, which may have originated on the continents and then drifted out into the open ocean.

In the Antarctic and Arctic Oceans some of the deep ocean sediments are believed to have been ice-rafted from the land to the oceans by icebergs. These *glacial marine* deposits are chiefly silt containing coarse fragments. Near the landmasses are extensive layers of sand and coarse silt deposited by turbidity currents, the turbidites, discussed previously.

Most of the pelagic deep-sea deposits consist of *oozes*, sediments of which at least 30 percent by volume is made up of the hard parts of very small, sometimes microscopic, organisms. These hard parts are constructed of mineral matter extracted from the sea water by tiny plants and animals. With the death of the organism, the remains sink slowly to the sea floor. In composition, the oozes are either calcareous or siliceous.

Two calcareous oozes are common in the deep seas. In one—*globigerina ooze*—the limy shells of minute one-celled animals called *Globigerina* abound; in the other—*pteropod ooze*—the shells of tiny marine molluscs predominate. Globigerina oozes cover large portions of the floors of the Atlantic, Pacific, and Indian Oceans, and the greatest known concentration of pteropod ooze is in a long belt running from north to south in the Atlantic Ocean midway between South America and Africa.

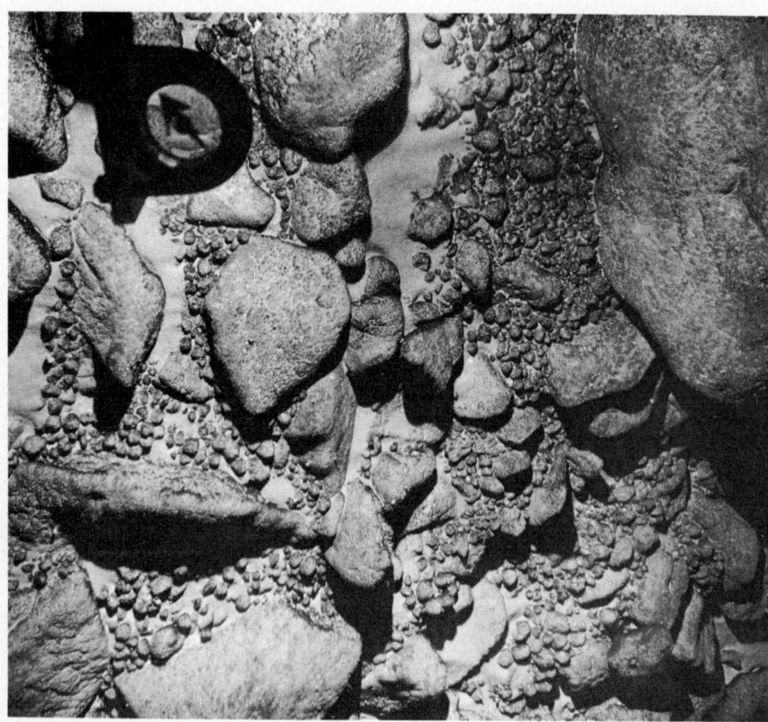

Figure 15.19 ***This photograph of the deep-sea floor was taken in the Atlantic Ocean at a depth of nearly 5,400 m. The objects are manganese nodules. The largest is about 15 cm in maximum dimension.*** *(Photograph by Lamont Geological Observatory.)*

Siliceous oozes include *radiolarian ooze,* made up largely of the delicate and complex hard parts of minute marine protozoa called *Radiolaria,* and by *diatomaceous ooze,* made up of the siliceous cell walls of one-celled marine algae known as *diatoms.* Radiolarian ooze predominates in a long east–west belt in the Pacific Ocean just north of the equator, and the greatest concentration of diatomaceous ooze occurs in the North Pacific and in the Antarctic Ocean.

Some of the pelagic deposits of the sea floor have apparently crystallized directly from sea water, and these we refer to as *authigenic* (from the Greek, "born on the spot"). In large areas of the southern Pacific Ocean, the silicate mineral *phillipsite* constitutes the great bulk of the sediments, but the mineral grains are too large to have been transported by the currents in the area of their present location; thus they are thought to have crystallized directly from sea water.

More spectacular authigenic deposits are the *manganese nodules* (see Figures 15.19 and 15.20). These average 25 cm in diameter and when cut open exhibit the patterns of concentric growth. Composed largely of oxides of manganese and iron, they contain in addition the oxides of many of the rarer elements.

Figure 15.20 ***This section through a manganese nodule shows that the concentric layers deposited as the nodule grew. The nodule, 6 cm across, was dredged from a depth of 5,400 m off the southeast coast of South Africa.*** *(Photograph by Raymond C. Gutschick.)*

It is estimated that they cover some 20 percent of the floor of the Pacific. Their composition and abundance have suggested a potential commercial use as an ore not only of manganese but also of such rarer elements as cobalt, titanium, and zirconium.

Rates of sedimentation

How fast do sediments accumulate in the oceans? We now have a number of determinations of rates of sedimentation. As yet these give us an answer to our question at only a few spots in the vast stretches of the world's oceans. Scanty as the data are, however, they give us a picture of slow sedimentation in the deep seas and very much more rapid accumulation in the waters of some continental shelves and upper slopes.

Our most reliable data are for the deep ocean floors where sedimentation is slow when compared to that on the slopes and shelves. Rates are determined from studies of cores of sediments obtained by sampling programs of the world's oceans (Figure 15.21). Radioactive carbon gives dates for the accumulation of sediments during the last 40,000 years or so. Rates of deposition for sediments older than this are obtained by extrapolation of the rates determined by carbon 14 or by the use of elements with longer half-lives. Another technique, paleomagnetism (see Chapter 20), can be applied to the longer cores and depends on the discovery of reversals of the earth's magnetic field imprinted on the sediments and on correlating these reversals with similar but dated reversals recorded in volcanic rocks on land.

In the deep Pacific and Indian Oceans sedimentation rates range between 1 mm and 1 cm/1,000 years. In the deep Atlantic the range is from about 1 to 10 cm with the average apparently around 2 or 3 cm/1,000 years. On the continental shelves and on the upper slopes, the rates are estimated to be, on the average, between 20 and 30 cm/1,000 years.

These figures of sedimentation are not directly comparable to equal volumes of rock on land. The oceanic accumulations have a very high water content and their dry specific gravity is low when compared with an equal volume of continental rock. A reasonable average value for the specific gravity of continental rock is 2.6 and something less than 1 for the unconsolidated sediments in the ocean. This means that 1 cm of accumulation on the sea floor is equal to about 0.3 cm of rock on the continents. With time, of course, oceanic sediments will probably become lithified and acquire the characteristics, including specific gravity, of solid rock.

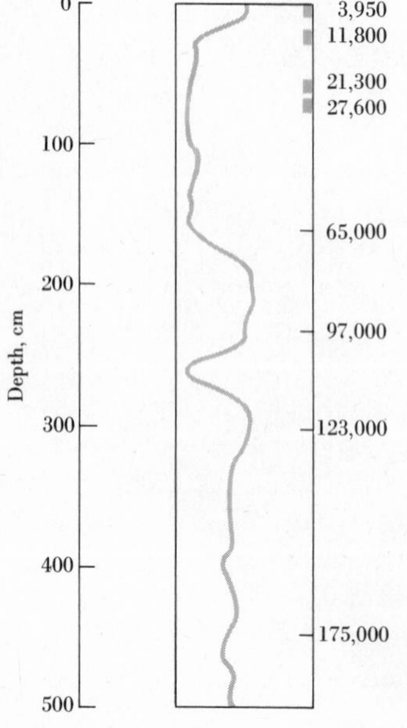

Figure 15.21 ***Rates of deposition in the deep sea are discovered from sediments in cores taken from the ocean floor. This is the record of a core from the Caribbean Sea halfway between Cuba and Colombia at a depth of 2,965 m. Numbers to the right of the boxes are ages in radiocarbon years. Ages to the right of the lines are determined by the protactinium–ionium method. Relative temperature of the water during deposition of sediments was determined by an analysis of micro fossils.*** (*From David B. Ericson et al., "The Pleistocene Epoch in Deep-Sea Sediments,"* Science, *vol. 146, p. 730, 1964. Copyright 1964 by the American Association for the Advancement of Science.*)

15.6 Shorelines

Not all of us have occasion to make a detailed study of the ocean currents or of the topography of the ocean floor, but most of us have many opportunities to observe the activity of water along the shorelines of oceans or lakes. The nature and results of wave action along such shorelines can be a drama of power and persistence.

The processes

The energy that works on and modifies a shoreline comes largely from the movement of water produced by tides, by wind-formed waves, and, to a lesser extent, by tsunami (see Chapter 16). Because we have already discussed tidal currents, we may now turn to the nature and behavior of wind-formed waves as they advance against a shoreline.

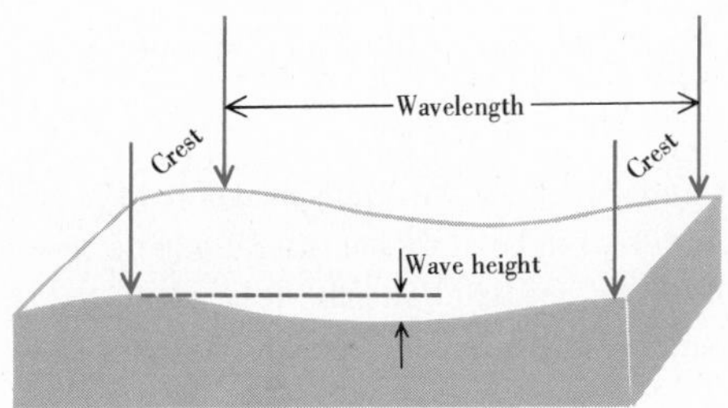

Figure 15.22 ***Diagrammatic explanation of terms used in describing waves of water.***

WIND-FORMED WAVES Most water waves are produced by the friction of air as it moves across a water surface. The harder the wind blows, the higher the water is piled up into long *wave crests* with intervening troughs, both crests and troughs at right angles to the wind. The distance between two successive wave crests is the *wave length,* and the vertical distance between the wave crest and the bottom of an adjacent trough is the *wave height* (Figure 15.22). When the wind is blowing, the waves it generates are called a *sea.* But wind-formed waves persist even after the wind that formed them dies. These waves, or *swells,* may travel for hundreds or even thousands of kilometers from their zone of origin.

We are concerned with both the movement of the wave form and the motion of water particles in the path of the wave. Obviously the wave form itself moves forward at a measurable rate. But in deep water, the water particles in the path of the wave describe a circular orbit: Any given particle moves forward on the crest of the wave, sinks as the following trough approaches, moves backward under the trough, and rises as the next crest advances. Such a motion can best be visualized by imagining a cork bobbing up and down on the water surface as successive wave crests and troughs pass by. The cork itself makes only very slight forward progress under the influence of the wind. Wave motion extends downward, until at a depth equal to about one-half the wave length it is virtually negligible. But between this level and the surface, water particles move forward under the crest and backward under the trough of each wave, in orbits that decrease in diameter with depth (Figure 15.23).

As the wave approaches a shoreline and the water becomes more shallow, definite changes take place in the motion of the particles and in the form of the wave itself. When the depth of water is about half the wave length, the bottom begins to interfere with the motion of water particles in the path

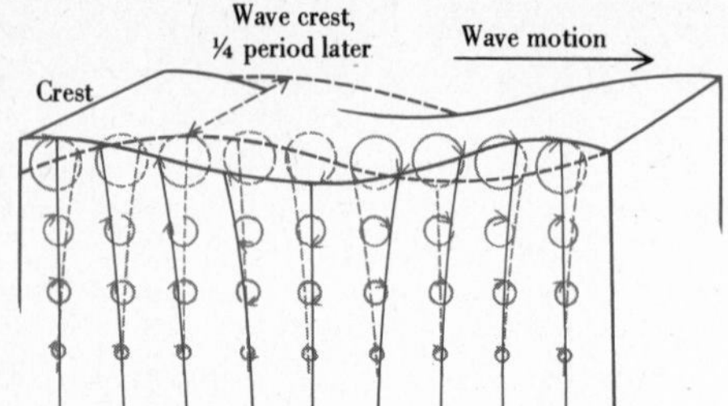

Figure 15.23 ***The motion of water particles relative to wave motion in deep water. Water particles move forward under the crest and backward under the trough, in orbits that decrease in diameter with depth. Such motion extends downward to a distance of about one-half the wave length.*** (*Redrawn from* U.S. Hydrographic Office Publ., *no. 604, Fig. 1.2, 1951.*)

of the wave and their orbits become increasingly elliptical. As a result, the length and velocity of the wave decrease and its front becomes steeper. When the water becomes shallow enough and the front of the wave steep enough, the wave crest falls forward as a breaker, producing what we call *surf*. At this moment, the water particles within the wave are thrown forward against the shoreline. The energy thus developed is then available to erode the shoreline or to set up currents along the shore that can transport the sediments produced by erosion.

WAVE REFRACTION AND COASTAL CURRENTS Although most waves advance obliquely toward a shoreline, the influence of the sea floor tends to bend or refract them until they approach the shore nearly head-on.

Let us assume that we have a relatively straight stretch of shoreline with waves approaching it obliquely over an even bottom that grows shallow at a constant rate. As a wave crest nears the shore, the section closest to land feels the effect of the shelving bottom first and is retarded, while the seaward part continues along at its original speed. The effect is to swing the wave around and to change the direction of its approach to the shore, as shown in Figure 15.24.

As a wave breaks, not all its energy is expended on the erosion of the shoreline. Some of the water thrown forward is deflected and moves laterally, parallel to the shore. The energy of this water movement is partly used up by friction along the bottom and partly by the transportation of material.

Refraction also helps explain why, on an irregular shoreline, the greatest energy is usually concentrated on the headland and the least along the bays. Figure 15.25 shows a bay separating two promontories and a series of wave

Figure 15.24 ***Wave crests that advance at an angle on a straight shoreline and across a bottom that shallows at a uniform rate are bent shoreward, as suggested in this diagram. Refraction is caused by the increasing interference of the bottom with the orbits of water-particle motion within the wave.***

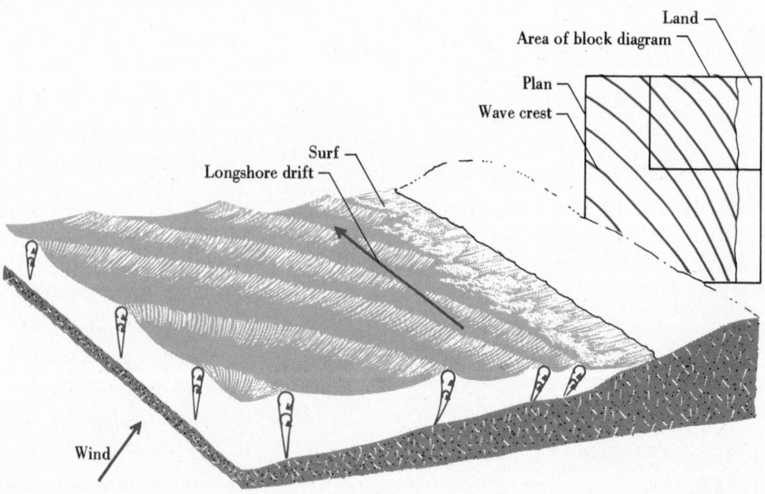

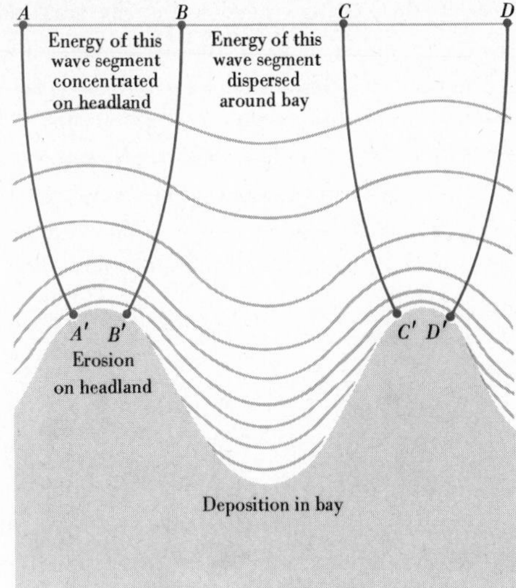

Figure 15.25 **Refraction of waves on an irregular shoreline. It is assumed that the water is deeper off the bay than off the headlands. Consider that the original wave is divided into three equal segments, A-B, B-C, and C-D. Each segment has the same potential energy. But observe that by the time the wave reaches the shore, the energy of A-B and C-D has been concentrated along the short shoreline of headlands A′-B′ and C′-D′, whereas the energy of B-C has been dispersed over a greater front (B′-C′) around the bay. Energy for erosion per unit of shoreline is therefore greater on the headlands than along the bay.**

crests sweeping in to the shore across a bottom that is shallow off the headland and deep off the mouth of the bay.

Where the depth of the water is greater than one-half the wave length, the crest of the advancing wave is relatively straight. Closer to shore, off the headlands, however, the depth becomes less than half the wave length, and the wave begins to slow down. In the deeper water of the bay it continues to move rapidly shoreward until there, too, the water grows shallow and the wave crest slows. This differential bending of the wave tends to make it conform in a general way to the shoreline. In so doing, the wave energy is concentrated on the headland and dispersed around the bay, as suggested in Figure 15.25.

A composite profile of a shoreline from a point above high tide seaward to some point below low tide reveals features that change constantly as they are influenced by the nature of waves and currents along the shore. All features are not present on all shorelines, but several are present in most shore profiles. The *offshore* section extends seaward from low tide. The *shore* or *beach* section reaches from low tide to the foot of the *sea cliff* and is divided into two segments. In front of the sea cliff is the *backshore,* characterized by one or more *berms,* resembling small terraces with low ridges on their seaward edges built up by storm waves. Seaward from the berms to low tide is the *foreshore.* Inland from the shore lies the *coast.* Deposits of the shore may veneer a surface cut by the waves on bedrock and known as the *wave-cut terrace.* In the offshore section, too, there may be an accumulation of unconsolidated deposits comprising a *wave-built terrace* (Figure 15.26).

Figure 15.26 **Some of the features along a shoreline and the nomenclature used in referring to them.** (*In part, after F. P. Shepard,* Submarine Geology, *2nd ed., p. 168, Harper & Row, Publishers, New York, 1963.*)

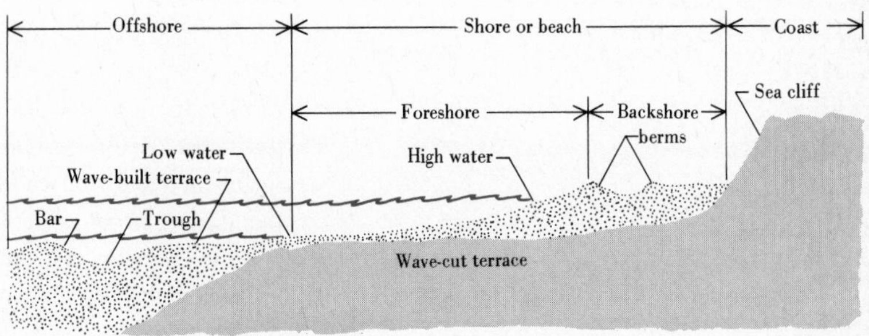

The shoreline profile is ever changing. During great storms the surf may pound in directly against the sea cliff, eroding it back and at the same time scouring down through the beach deposits to abrade the wave-cut terrace. As the storm (and hence the available energy) subsides, new beach deposits build up out in front of the sea cliff. The profile of a shoreline at any one time, then, is an expression of the available energy: It changes as the energy varies. This relation between profile and available energy is similar to the changing of a stream's gradient and channel as the discharge (and therefore the energy) of the stream varies (see Chapter 11).

Shoreline features

Not even shorelines have escaped man's constant desire for classification into neat pigeonholes, but to date no completely acceptable system of classification has been devised. For many years it was common practice to group shorelines as *emergent* or *submergent*, depending on whether the sea had gone down or had come up in relation to the landmass. Thus large sections of the California coast, having emerged from the sea during geologically recent times, would be termed emergent. Across the continent, the shoreline of New England indicates that it has been drowned by a slowly rising sea and would be referred to as submergent. The system has been criticized because some of the features that were thought to represent emergence of the land actually form where the land is being submerged. Conversely, along shorelines of submergence, features thought to characterize emergence may also develop.

Another attempt at a classification is based on the processes that form shorelines. Some have major features traceable to glacial erosion and others to glacial deposition. The system divided along these lines has much to recommend it. Nevertheless, we shall examine some of the individual shoreline features without attempting to fit them into an all-inclusive system.

Erosion and deposition work hand in hand to produce most of the features of the shoreline. An exception to this generalization is an offshore island that is merely the top of a hill or a ridge that was completely surrounded by water as the sea rose in relation to the land. But even islands formed in this way are modified by erosion and deposition.

FEATURES CAUSED BY EROSION *Wave-cut cliffs* are common erosional features along a shore, particularly where it slopes steeply down beneath the sea (see Figure 15.27). Here waves can break directly on the shoreline and thus can expend the greatest part of their energy in eroding the land. Wave erosion pushes the wave-cut cliff steadily back, producing a wave-cut terrace or platform at its foot. Because the surging water of the breaking waves must cross this terrace before reaching the cliff, it loses a certain amount of energy through turbulence and friction. Therefore, the farther the cliff retreats, and the wider the terrace becomes, the less effective are the waves in eroding the cliff. If sea level remains constant, the retreat of the cliffs becomes slower and slower.

Waves pounding against a wave-cut cliff produce various features, as a result of the differential erosion of the weaker sections of the rock. Wave action

Figure 15.27 **Waves have cut into volcanic rocks to produce these cliffs along the shore of Attu Island in the Aleutians.** (*Official U.S. Navy Photograph.*)

may hollow out cavities or *sea caves* in the cliff, and if this erosion should cut through a headland, a *sea arch* is formed. The collapse of the roof of a sea arch leaves a mass of rock, a *stack,* isolated in front of the cliff (see Figure 15.28).

FEATURES CAUSED BY DEPOSITION Features of deposition along a shore are built of material eroded by the waves from the headlands and of material brought down by the rivers that carry the products of weathering and erosion from the landmasses. For example, part of the material eroded from a headland may be drifted by currents into the protection of a neighboring bay, where it is deposited to form a sandy beach.

The coastline of northeastern New Jersey (Figure 15.29) illustrates some

Figure 15.28 **The sea has cut arches through this promontory. To the far right is a stack, a rock mass that erosion has cut off from the mainland. Arches State Park, California.** (*Photograph by Sheldon Judson.*)

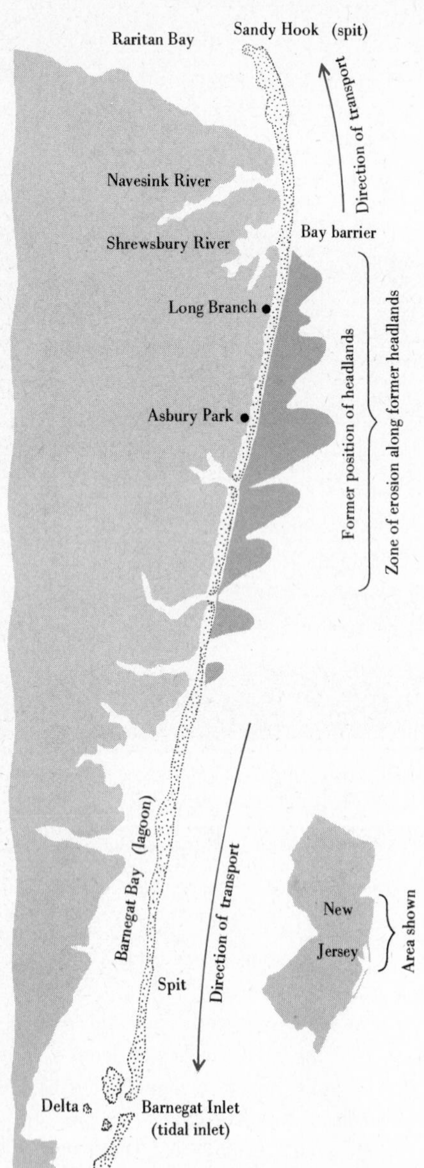

Figure 15.29 ***Erosion by the sea has pushed back the New Jersey coastline as indicated on this map. Some of the material eroded from the headlands has been moved northward along the coast to form Sandy Hook, a spit. To the south, a similar but longer feature encloses Barnegat Bay, a lagoon with access to the open ocean through a tidal inlet.*** (*After an unpublished map by Paul MacClintock.*)

of the features caused by deposition. Notice that the Asbury Park–Long Branch section of the coastline is a zone of erosion that has been formed by the destruction of a broad headland area. Erosion still goes on along this part of the coast, where the soft sedimentary rocks are easily cut by the waves of the Atlantic. The material eroded from this section is moved both north and south along the coastline. Sand swept northward is deposited in Raritan Bay and forms a long, sandy beach projecting northward, a *spit* known as Sandy Hook.

Just south of Sandy Hook, the flooded valleys of the Navesink River and of the Shrewsbury River are bays that have been almost completely cut off from the open ocean by sandy beaches built up across their mouths. These beaches are called *bay barriers.*

Sand moved southward from the zone of erosion has built up another sand spit. Behind it lies a shallow lagoon, Barnegat Bay, that receives water from the sea through a *tidal inlet,* Barnegat Inlet. This passage through the spit was probably first opened by a violent storm, presumably of hurricane force. Just inside the inlet a delta has been formed of material partly deposited by the original breakthrough of the bar and partly by continued tidal currents entering the lagoon.

Long stretches of the shoreline from Long Island to Florida, and from Florida westward around the Gulf Coast, are marked by shallow, often marshy lagoons separated from the open sea by narrow sandy beaches. Many of these beaches are similar to those that enclose Barnegat Bay, apparently elongated spits attached to broad headlands. Others, such as those that enclose Pamlico Sound at Cape Hatteras, North Carolina, have no connection with the mainland. These sandy beaches are best termed *barrier islands.* It has been suggested that these islands originated from spits detached from the mainland as large storms breached them at various points. Some geologists think that they may represent spits isolated from the mainland by a slowly rising sea level. Still a third possibility is that over a long period of time wave action has eroded sand from the shallow sea floor and has heaped it up in ridges that lie just above sea level.

Another depositional feature, a *tombolo,* is a beach of sand or gravel that connects two islands, or connects an island with the mainland. Numerous examples exist along the New England coastline and fewer off the west coast, although Morro Rock, a small, steep-sided island, is tied to the California mainland by a tombolo.

CORAL REEF SHORELINES In tropical and semitropical waters lying within a belt between about 30°N and 25°S, many shorelines are characterized by coral reefs of varying sizes and types. These reefs are built up by individual corals with calcareous skeletons, as well as by other lime-secreting animals and plants. The coral-reef shorelines are of three types: the *fringing reef*, the *barrier reef*, and the *atoll*. A fringing reef grows out directly from a landmass, whereas a barrier reef is separated from the main body of land by a lagoon of varying width and depth opening to the sea through passes in the reef. An atoll is a ring of low, coral islands arranged around a central lagoon.

The origin of atolls has been debated for well over a century, ever since Charles Darwin first advanced his explanation in 1842. Darwin postulated that an atoll begins as a fringing reef around a volcanic island. Because the island rests as a dead load on the supporting material, it begins to subside, but at a rate slow enough for the coral to maintain a reef. With continued subsidence, the island becomes smaller and smaller, and the actively growing section of the reef becomes a barrier reef. Then, with the final disappearance of the island below the sea, the upward-growing reef encloses only a lagoon and becomes a true atoll. In support of this theory are many volcanic islands in the Pacific now surrounded by barrier reefs. Furthermore, investigations on Bikini Island (an atoll) indicate that the volcanic rock core is surmounted by several hundred meters of coral rock. Finally, geophysical evidence suggests that there actually has been a subsidence of some of the volcanic islands of the Pacific.

The subsidence theory originally advanced by Darwin, however, does not explain the nearly constant depth of countless modern lagoons within atolls. In part to overcome this difficulty, the so-called "glacial control theory" has been advanced. Proponents of this explanation of atoll formation have postulated that volcanic islands were truncated at a lower sea level during one or more of the Pleistocene ice advances. Around the edges of such wave-planed platforms the coral reefs began to grow as the continental glaciers melted and sea level rose. The coral islands around the edge of the platforms would then ring lagoons of more or less constant depth. In summary, there is evidence in support of each theory. When the final answer is known, both may be applicable.

15.7 *Changes in sea level*

We can observe that the level of the sea in relation to the land changes daily as a result of tides and winds, but much slower fluctuations are no less real. In fact, in the geological past such changes have been very extensive and significant.

Eustatic and tectonic changes of sea level

It is easy enough to measure the daily changes in sea level caused by the rise and fall of tides along the coastline, but there are other movements so slow that they are revealed only by long-continued records of sea level or so local and rare that they are not generally recognized.

A change in sea level relative to land can be caused by the upward or downward movement of either the ocean or the land or by their combined movement. If the movement is confined to the ocean, change of level is *eustatic*, a term that refers to the static condition or continuing stability of the landmass. When the relative sea level changes because of land movement, the change is *tectonic* (from the Greek for "builder" or "architect"), reference to the movements that shape the earth's surface (see Chapter 18).

Often it is difficult, if not impossible, to distinguish between eustatic and tectonic movements of sea level by observing only a small section of the shore. Eustatic changes of level, however, are worldwide, and when recording stations over an extensive area report a long-continued movement of the sea, the movement can safely be termed eustatic. In contrast, tectonic movements tend to be local and spasmodic, controlled as they are by the forces that crumple and distort solid earth materials.

Eustatic movements may be caused in various ways. If the amount of water locked up in glaciers and lakes increases, the sea level falls; then, if the glaciers melt or if the lakes are drained, the sea level rises. Or sea level may rise or fall as a result of changes in the size of ocean basins, either because of continuing deposition of sediments on their floors or because of actual deforma-

Figure 15.30 ***In the recent geologic past, ocean waves beveled this platform in the middle distance across tilted rocks on the California coast south of San Francisco. Today the platform stands slightly above sea level as evidence of a change in level between land and sea. In this instance, geologists believe that the motion has been tectonic and that the land has moved upward relative to the sea. On the far headland waves beat against the cliffs and are cutting a lower bench related to the present-day level.*** (*Photograph by Sheldon Judson.*)

Figure 15.31 ***Solid pattern shows the approximate extent of encroachment of the sea over the continental borders during the Cenozoic. Stippled pattern shows the shorelines of emergence confined chiefly to the zone 35° north and south of the equator.*** (*From A. J. Eardley,* J. Geol. Educ., *vol. 12, no. 1, 1964.*)

tion by earth forces. Still another cause of eustatic change lies in the addition or removal of water from the earth's surface. Volcanoes are constantly adding new water to our atmosphere that eventually finds its way to the sea. And water is constantly being trapped in sedimentary deposits and incorporated in such minerals as clay, causing at least a temporary loss to the oceans.

Tectonic and eustatic changes are continuously taking place around the globe; mean sea level is endlessly rising and falling. Although recent changes in sea level are slight, we have ample evidence that they are actually occurring.

RECENT CHANGES IN SEA LEVEL On the basis of tidal measurements made at various points around the world, observers have concluded that a eustatic rise in sea level is taking place at the present time. The flooding of many river mouths from New England southward indicates that the rise has a long history, substantiated by the discovery of submerged stumps of ancient trees and primitive artifacts. The invasion of the sea is attributed to the melting of mountain glaciers and to the depletion of the Greenland and Antarctic ice caps. But there is still enough water stored in modern glaciers to raise the sea another 60 m (see Section 13.4).

Modern changes in sea level caused by tectonic movements have been observed in localities where the crust of the earth is known to be undergoing deformation at the present day. A striking example of movement of this sort in Japan is described in Chapter 18.

CHANGES OF SEA LEVEL IN THE GEOLOGICAL PAST Recent changes in sea level seem quite insignificant, however, compared to those of the geological past. Yet, the rate of sea-level change may never have been much more rapid than it is today, for when it is projected back over vast stretches of geological time, today's rate readily explains the presence of sea shells in lofty mountains and of shoreline terraces now submerged beneath the sea or raised above it (Figure 15.30).

As we have seen, sedimentary rocks constitute some 75 percent of all the rocks exposed at the earth's surface. Of these, the great majority are marine rocks formed from sediments laid down on the floors of ancient seas and now stranded above sea level. For over a century and a half, geologists the world over have studied the details recorded in these marine sedimentary rocks and have pieced together at least a part of the complex story of the advances and retreats of the ancient seas. The sequence of these events is a part of the story of historical geology. Suffice it to say here that the shape of the continents has been changing constantly through geologic time as the oceans spilled out of their basins, flooded slowly across the lands, and then withdrew—a process that has been repeated many times.

An example of a worldwide sea-level change recorded in the geologic record

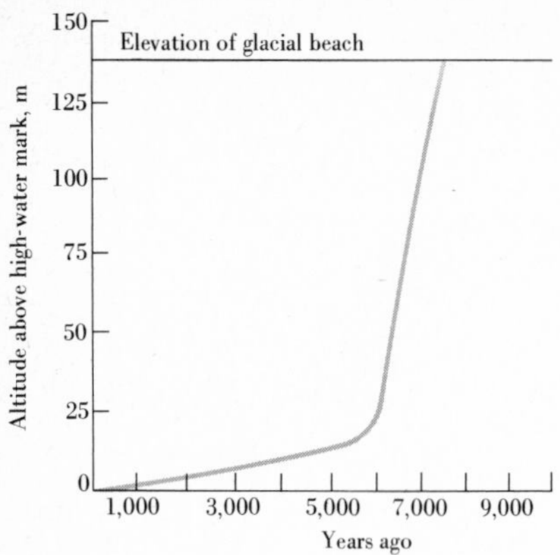

Figure 15.32 **Change between land and sea in northern Ungava, Quebec, Canada. Land has risen over 125 m since the formation of the last glacial beach in the area more than 7,000 years ago. This change was very rapid until about 6,000 years ago when it slowed abruptly and continued at a much lower rate to the present. The change in the relation between land and sea has been due to the isostatic recovery of the land after the disappearance of the ice sheet that had depressed it.** (*After B. Matthews,* Arctic, *vol. 20., no. 3, Fig. 7, p. 186, 1967.*)

has been presented by A. J. Eardley. He has shown that the sea level has risen in the Arctic and Antarctic and fallen in the equatorial regions since the Cretaceous (see Figure 15.31). The magnitude of this change seems to be in the range of 180 m.

Many of the modern coastal lands are marked by terraces arranged like great steps in a giant stairway. Each terrace indicates that the sea stood at this general level long enough to create, by erosion and deposition, a relatively smooth surface sloping gently seaward. Their position today above sea level indicates that the sea has retreated because of eustatic or tectonic movements, or both.

Continental glaciation has caused one type of earth movement leading to shoreline changes. The weight of a continental glacier is great enough to cause a downward warping of the land it covers. When the glacier melts and the load of ice is removed, the land slowly recovers and achieves its original balance once again (see Figure 15.32).

We have good evidence of the recoil of land following glacial retreat along the shores of the Gulf of Bothnia and the Baltic Sea. Accurate measurements show that from 1800 to 1918 the land rose at rates ranging from 0.0 cm/year at the southern end to 1.1 cm/year at the northern end. Studies indicate further that the land has been rising at a comparable rate for 5,000 years. Areas that were obviously sea beaches until recent times are now elevated from a few meters to 240 m above sea level. A comparison of precise measurements along railroads and highways in Finland made from 1892 to 1908 with measurements made from 1935 to 1950 shows that uplift there has proceeded at the rate of from 0.3 to 0.9 m/century. The greatest change has been on the Gulf of Bothnia, at about 64°N. Most of this movement has been caused by the recoil of the land following the retreat of the Scandinavian ice cap.

Similar histories of warping and recoil have been established in other regions, including the Great Lakes area in the United States. At one stage in the final retreat of the North American ice sheet, the present Lakes Superior, Michigan, and Huron had higher water levels than they do now and were joined together to form Lake Algonquin. Beaches around the borders of Lake Algonquin were horizontal at the time of their formation; today they are still horizontal south of Green Bay, Wisconsin, and Manistee, Michigan. But 290 km north, at Sault Sainte Marie, Michigan, the oldest Algonquin beach is 108 m higher than it is at Manistee.

Another example of land that has risen many meters following the retreat of a glacier is found in raised beaches along the Atlantic coast and in the effects left by an arm of the sea that penetrated beyond Montreal and Ottawa. Geologists predict that the ocean will be completely driven from 150-m-deep Hudson Bay by the time the landmass has recovered from the depression it was subjected to during the great Ice Age.

It is obviously a more difficult task to establish that sea levels were once lower than at present, because the evidence lies deep beneath the modern seas, but recent investigations of the sea bottom have revealed hints that the sea level has not always been as high as it is at present. These features include flat-topped submarine mountains, great submarine valleys, and submerged terraces.

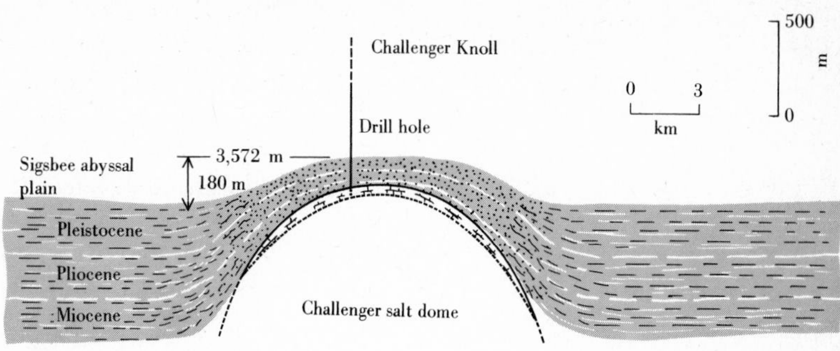

Figure 15.33 ***The crest of the Challenger Knoll lies 3,572 m below the surface of the sea. A hole drilled into the knoll encountered the sequence shown here. The materials of the salt dome were formed in fairly shallow water and thus indicate a change of level in excess of 3,500 m. The salt in the dome is thought to be Jurassic in age. Oil was found beginning with the Pliocene sediments.*** (*Redrawn from National Science Foundation,* Initial Reports of the Deep Sea Drilling Project, *vol. 1, Fig. 26, p. 110, U.S. Government Printing Office, 1969.*)

A spectacular change in sea level was demonstrated in August 1968, when the oceanographic vessel *Glomar Challenger* drilled into the bottom of the Gulf of Mexico at a depth of about 3,510 m. The bit penetrated some 140 m into the sea floor and encountered marine formation of rock salt and gypsum, deposits which generally form close to sea level. This indicates an upward change in relative sea level of the order of 3,000 m or more since the deposits were formed, probably during the Jurassic (Figure 15.33).

Outline

Oceans cover more than 70 percent of the earth's surface.

Ocean water has probably originated by "degassing" of the earth's interior. This may have been fairly continuous through volcanic eruptions or confined to an early high thermal stage of the earth.

Dissolved material gives the oceans their saltiness. The material is moved through the rock cycle from lands to oceans and back to lands.

Currents in the oceans include *tidal currents, wind-driven currents,* and *density currents.* Density currents occur because of differences in salinity, temperature, and turbidity of the water.

One worldwide *ocean basin* contains the world's oceans and branches northward as the Pacific, Atlantic, and Arctic Oceans.

Topography of the ocean floor is divisible into *continental shelves, slopes* and *rises,* and *ocean deeps.* These features are marked by one or more of the following: submarine valleys, seamounts, ridges, rises, fractures, and deep-sea trenches.

Sediments in the oceans are deepest on the shelves and slopes and thinnest on the deep ocean floors. They are divided into pelagic and terrestrial, the latter dominating the shelves and slopes.

Shorelines acquire characteristics generated by the energy of wind-driven waves.

Erosional features include cliffs, stacks, caves, arches, and tidal inlets.

Depositional features include pits, beaches, bay barriers, barrier islands, and tombolos.

Coral-reef shorelines occur in tropical and subtropical waters. The shorelines of atolls are related to a changing sea level.

Changes in *sea level* may have either *eustatic* or *tectonic* causes.

Supplementary readings

CARSON, RACHEL: *The Sea Around Us,* rev. ed., Oxford University Press, New York, 1961. A popular, accurate, and moving account.

FAIRBRIDGE, RHODES W. (ed.): *The Encyclopedia of Oceanography,* Reinhold Publishing Corporation, New York, 1966. You will find most pertinent items here, although the level of difficulty among items varies widely.

HILL, M. N. (ed.): *The Sea,* Wiley-Interscience, New York, 1963 and 1964. An extensive survey, in three volumes, containing papers by many authors; fairly advanced.

NEUMANN, GERHARD, and WILLARD J. PIERSON, JR.: *Principles of Physical Oceanography,* Prentice-Hall Inc., Englewood Cliffs, N.J., 1966. A standard text requiring in parts some mathematical ability.

SHEPARD, F. P.: *Submarine Geology,* 2nd ed., Harper & Row, Publishers, New York, 1963. A good introductory statement.

SVERDRUP, H. U., MARTIN W. JOHNSON, and RICHARD H. FLEMING: *The Oceans,* Prentice-Hall, Inc., Englewood Cliffs, N.J., 1942. Despite its age, this remains a remarkable repository of oceanographic information.

TUREKIAN, KARL K.: *Oceans,* Prentice-Hall, Inc., Englewood Cliffs, N.J., 1968. A brief, concise introduction to the oceans with some emphasis on their chemical aspects.

VON ARX, W. S.: *An Introduction to Physical Oceanography,* Addison-Wesley Publishing Company, Reading, Mass., 1962. A basic text demanding some mathematical ability.

ZENKOVICH, V. P.: *Processes of Coastal Development* (Trans. by D. G. Fry from the Russian), Oliver & Boyd, London, 1967. A very fine survey of coastal features and their origin.

16 Earthquakes

EARTHQUAKES HAVE PROFOUND EFFECTS ON MANY PEOPLES AND ON WORKS OF MAN. They also cause changes of the earth's surface, and give us controlling data on the structure and nature of the earth's interior.

One of the most significant measurements in physical geology is that some earthquakes have sources as deep as 700 km, *but no earthquakes are deeper.* Another is their geographical distribution, which is illustrated in Figures 16.1 and 16.2.

16.1 Seismology

The scientific study of earthquakes is called seismology, from the Greek *seismos,* "earthquake," and *logos,* "reason" or "speech." At the turn of the twentieth century, there were approximately a dozen men in the world who would have been classified as professional seismologists. The subject matured, however, until by midcentury there were close to 400 seismograph stations in the world that had seismologists recording and studying earthquakes and other ground vibrations. Data from seismology have become an integral part of physical geology in its growth from a descriptive "natural history" to a science that includes geophysics, a category that cuts across former groupings of knowledge labeled physics and geology.

16.2 Effects of earthquakes

Earthquakes are interesting to most people because of their effects on human beings and on manmade structures. Their geological effects may also be profound. Of all the earthquakes that occur every year, only one or two are likely to produce spectacular geological effects such as landslides or the elevation or depression of large landmasses. A hundred or so may be strong enough near their sources to destroy human life and property. The rest are too small to have serious effects.

Fire

When an earthquake occurs near a modern city, fire can be a greater hazard than the shaking of the ground. In fact, fire has caused an estimated 95 percent of the total loss caused by some earthquakes. This was dramatically demonstrated in Tokyo and Yokohama in 1923.

A minute and a half before noon on September 1, 1923, an earthquake occurred whose center was under Sagami Bay, 80 km from Yokohama and 110 km from Tokyo. The vibrations spread outward with such energy that they caused serious damage along the Japanese coast over an area 150 km long and 100 km wide.

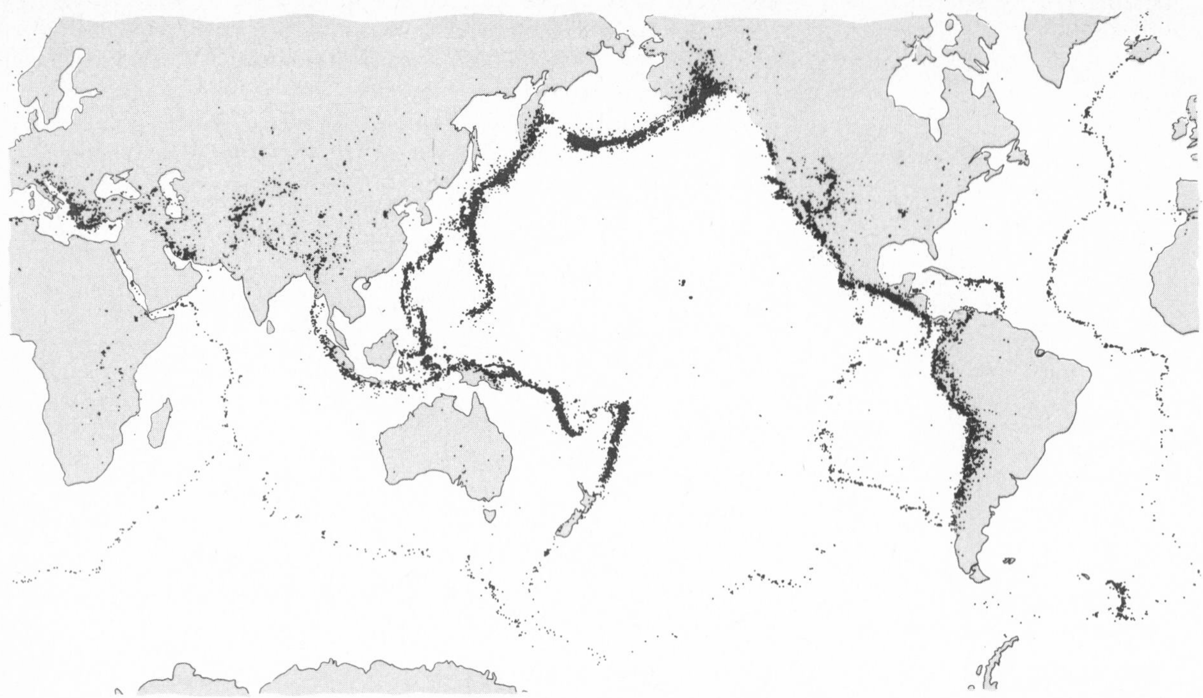

Figure 16.1 ***Locations of earthquakes, 1961 through 1967, with focal depths of 0 to 100 km.*** (*Data from U.S. Coast and Geodetic Survey, ESSA.*)

At the Imperial University in Tokyo, Professor Akitsune Imamura was sitting in his office at the Seismological Institute when the earthquake occurred. His account is one of the few accurately documented eyewitness reports of an earthquake from within a zone of heavy damage carefully recorded by an observer who understood what was happening.

At first, the movement was rather slow and feeble, so I did not take it to be the forerunner of so big a shock. As usual, I began to estimate the duration of the preliminary tremors, and determined, if possible, to ascertain the direction of the principal movements. Soon the vibration became large, and after 3 or 4 seconds from the commencement, I felt the shock to be very strong indeed. Seven or 8 seconds passed and the building was shaking to an extraordinary extent, but I considered these movements not yet to be the principal portion. At the 12th second from the start, according to my calculation, came a very big vibration, which I took at once to be the beginning of the principal portion. Now the motion, instead of becoming less and less as usual, went on increasing in intensity very quickly, and after 4 or 5 seconds I felt it to have reached its strongest. During this epoch the tiles were showering down from the roof making a loud noise, and I wondered whether the building could stand or not. I was able accurately to ascertain the directions of the principal movements and found them to have been about NW or SE. During the following 10 seconds the motion, though still violent, became somewhat less severe, and its character

Figure 16.2 **Locations of earthquakes, 1961 through 1967, with focal depths of 100 to 700 km.** (*Data from U.S. Coast and Geodetic Survey, ESSA.*)

gradually changed, the vibrations becoming slower but bigger. For the next few minutes we felt an undulatory movement like that which we experience on a boat in windy weather, and we were now and then threatened by severe aftershocks. After 5 minutes from the beginning, I stood up and went over to see the instruments. . . . Soon after the first shock, fire broke out at two places in the University, and within one and a half hours our Institute was enveloped in raging smoke and heat; the shingles now exposed to the open air as the tiles had fallen down due to the shock, began to smoke and eventually took fire three times. . . . It was 10 o'clock at night before I found our Institute and Observatory quite safe. . . . We all, 10 in number, did our best, partly in continuing earthquake observations and partly in extinguishing the fire, taking no food or drink till midnight, while four of us who were residing in the lower part of the town lost our houses and property by fire.

Within 30 min after the beginning of the earthquake, fire had broken out in 136 places in Tokyo. In all, 252 fires were started and only 40 were extinguished. Authorities estimated that at least 44 were started by chemicals. A 20-km/hr wind from the south spread the flames rapidly. The wind shifted to the west in the evening and increased to 40 km/hr, and then shifted to the north. These changes in wind direction added greatly to the extent of the area burned. Within 18 hr, 64 percent of the houses in Tokyo had burned. The fires died away after 56 hr, with 71 percent of the houses consumed—a total of 366,262. The spread of fire in Yokohama, a city of 500,000 population, was even more rapid. Within 12 hr, 65 percent of the structures in the city had burned, and eventually the city was completely destroyed (see Figure 16.3).

Besides destroying buildings, fire also killed many people. A spectacular example occurred in Tokyo in an area of 1 million m^2 of open ground on the eastern bank of the Sumida River. People gathered there with their belongings, seeking refuge from the circling fires. By 4 P.M. of September 1, men, women, and children were so closely packed that movement was almost impossible. Meanwhile, fire had closed in from three sides, pinning them against the river and blanketing them with sparks and suffocating fumes. Suddenly, a fire whirlwind approached, the result of the rapid heating of the air by the fires, superposed on unstable meteorological conditions. It had the characteristics of a true tornado. The central tube, with winds of incredible violence whirling around and upward, drawing smoke, flames, and debris from the surrounding fires, swept over the area. When it had passed, 38,000 were left dead. Only about 2,000 persons, who had been close to the river on the southern part of the ground, survived. One report states that the majority of the dead were terribly burned but that many showed no effects of the heat on their skin or clothing. They were apparently suffocated as the tornado sucked up the breathable air and replaced it with smoke and fumes.

Final government statistics for the entire section of Japan devastated by this earthquake reported 99,333 killed, 43,476 missing, and 103,733 injured, with a total of 576,262 houses completely destroyed.

One reason for rapid spread of fire after an earthquake is that the vibrations often disrupt the water system in the area. In San Francisco, for example, some 23,000 service pipes were broken by the great earthquake of 1906. Water pressure throughout the city fell so sharply that when the hoses were

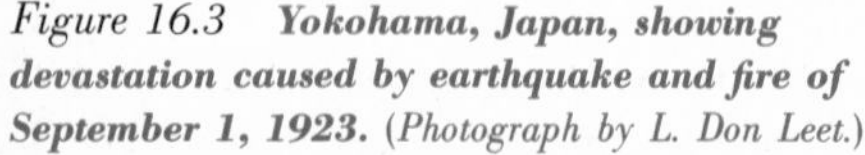

Figure 16.3 **Yokohama, Japan, showing devastation caused by earthquake and fire of September 1, 1923.** (*Photograph by L. Don Leet.*)

attached to fire hydrants only a small stream of water trickled out. Since that time, a system of valves has been installed to isolate any affected area and keep water pressure high in unbroken pipes in the rest of the city.

Damage to structures

Modern, well-designed buildings of steel-frame construction have withstood the shaking of even some of the most severe earthquakes. In the Japanese earthquake of 1923, for example, the Mitsubishi Bank building in Tokyo escaped with no structural failure even though it was surrounded by many badly damaged structures. The 43-story Latino-Americano tower in Mexico City rode the waves of the earthquake of July 28, 1957, undamaged, whereas surrounding buildings suffered greatly.

Chimneys have often been particularly sensitive to earthquake vibrations because they tend to shake in one direction while the building shakes in another. Consequently, chimneys often break off at the roof line. Two small earthquakes in New Hampshire in 1940 severed dozens of chimneys but caused no other damage. In contrast, tunnels and other underground structures are little affected by the vibrations of even the largest earthquakes, because they move as a unit with the surrounding ground.

In some earthquakes the extent to which buildings are affected by vibrations depends in part on the type of ground on which they stand. For example, in the San Francisco earthquake of 1906, buildings set on water-soaked sand, gravel, or clay suffered up to 12 times as much damage as similar structures built on solid rock nearby.

It has been found recently that the duration as well as the intensity of ground motion is a factor in causing damage to buildings.[1] Continued motion may cause damage even if it does not do so at the start. Reinforced concrete buildings start to show hairline shear cracks at the beginning of damaging motion. With continuing motion, the cracks become enlarged, and eventually disintegration results if shaking continues long enough. Repeated strong movements will bring destruction even to steel buildings.

In the Alaskan earthquake of March 27, 1964, the duration of damaging motion was approximately 3 min, or three times as long as the damaging shaking in San Francisco in 1906. Many buildings withstood the early vibrations only to collapse in the later stages. Had the buildings in Alaska been subjected to vibrations for only 1 min, many probably would not have collapsed.

Some of the buildings in Alaska that suffered severe damage had been built to conform to the earthquake provisions of the Uniform Building Code recommended by the Structural Engineers Association of California (see Figure 16.4). These provisions were intended to safeguard against major structural failures during earthquakes. However, they were formulated on building damage resulting from California earthquakes. No California earthquake ever had damaging shaking exceeding 1 min in duration. It will therefore be

[1]Fergus J. Wood (ed.), *The Prince William Sound, Alaska, Earthquake of 1964 and Aftershocks,* vol. II, U.S. Coast and Geodetic Survey, Washington, D.C., 1967.

Figure 16.4 **Contrast in building response to the Prince William Sound earthquake, Alaska, 1964.** (*Photograph by Coast and Geodetic Survey.*)

necessary to establish new building codes that will take into account the stresses involved in long-lasting damaging shaking.

The duration of strong shaking is also blamed for causing extensive landslides that destroyed many homes in Anchorage.

Seismic sea waves

If ever you are fortunate enough to be basking in the sun on Waikiki Beach and the water suddenly pulls away from the shore and disappears over the horizon, do not start picking up sea shells or digging clams—a seismic sea wave is coming, and the withdrawal of the water is the first warning of its approach.

Some submarine earthquakes abruptly elevate or lower portions of the sea bottom, setting up great sea waves in the water. The same effect may also be produced by submarine landslides at the time of a quake. These giant waves are called *seismic sea waves*, or *tsunami*, the Japanese term that has the same form for both singular and plural. Seismic sea waves may also be generated by volcanic eruptions, as in 1883, when Krakatoa erupted. These waves killed 36,500 in the East Indies.

Seismic sea waves have devastated oceanic islands and continental coastlines from time to time throughout history. The Hawaiian Islands have been hit 37 times since their discovery by Captain Cook in 1778.

On April 1, 1946, a severe earthquake occurred at 53.5°N, 163°W, 130 km southeast of Unimak Island, Alaska, in the Aleutian Trench. Here the ocean is 4,000 m deep. Minutes after the earthquake occurred, waves more than 33 m high smashed the lighthouse at Scotch Cap, Unimak Island, killing five persons. Four and one-half hours later, the first seismic sea wave from this quake reached Oahu, Hawaii, after traveling 3,600 km at 800 km/hr. At the time, marine geologist Francis P. Shepard and his wife were living in a seashore cottage on northern Oahu and were wakened by a loud hissing noise. They dashed to the window just in time to see waters of the ocean boiling up over a high ridge and heading toward their house. Shepard grabbed his camera, but when he got to the door, much to his disappointment, he saw the water retreating rapidly oceanward. The sea's level quickly dropped 10 m. It was then that he realized he had seen a seismic sea wave. Several more times after this, the water surged up over high-tide levels and then was sucked back into its basin.

Starting in the Alaskan waters where the earthquake occurred, these seismic sea waves had spread out much as waves do when a rock is thrown into a pond. But these waves were tremendous in length. Their crests were about 160 km apart. They swept out into the Pacific Ocean, moving at terrific speeds of over 800 km/hr. As they passed ships, however, these waves went unnoticed. They were about 1 m high in water 3,000 m deep, with 160 km between crests. Their effect was similar to the ground level rising 1 m as you walk 160 km; you just would not notice it. But when they reached Oahu and other Pacific shores, the effect was dramatic. The energy that moved thousands of meters of water in the open ocean became concentrated on moving a few meters of water at a shallow shore. There the water curled into giant crests that increased in height until they washed up over shores meters above high tide: 12 m on Oahu and 18 m on Hawaii.

These same seismic sea waves swept on down the Pacific. Eighteen hours after their launching, they reached Valparaiso, Chile, where they still had enough energy to cause 2-m rises of the water after traveling 13,000 km. Some even returned to hit the other side of Hawaii 18 hr later. In fact, tide gauges showed that seismic sea waves sloshed around the Pacific basin for days after the earthquake was over.

This 1946 seismic sea wave was one of the most destructive to hit Hawaii. It killed 159 people and injured 163. It demolished 488 homes and damaged 936, with a loss estimated at $25 million. This one came without warning and was rated the worst natural disaster in Hawaii's history, but it was the last destructive seismic sea wave to surprise the Hawaiian Islands.

SEISMIC SEA-WAVE WARNING SYSTEM In 1948, the U.S. Coast and Geodetic Survey established a seismic sea-wave warning system for Hawaii (see Figure 16.5). It operates continuously recording visible seismographs at its seismological observatory in Honolulu. These are equipped with an automatic alarm system that sounds whenever an unusually strong earthquake is recorded. Tide stations have also been set up to detect waves with characteristics of seismic

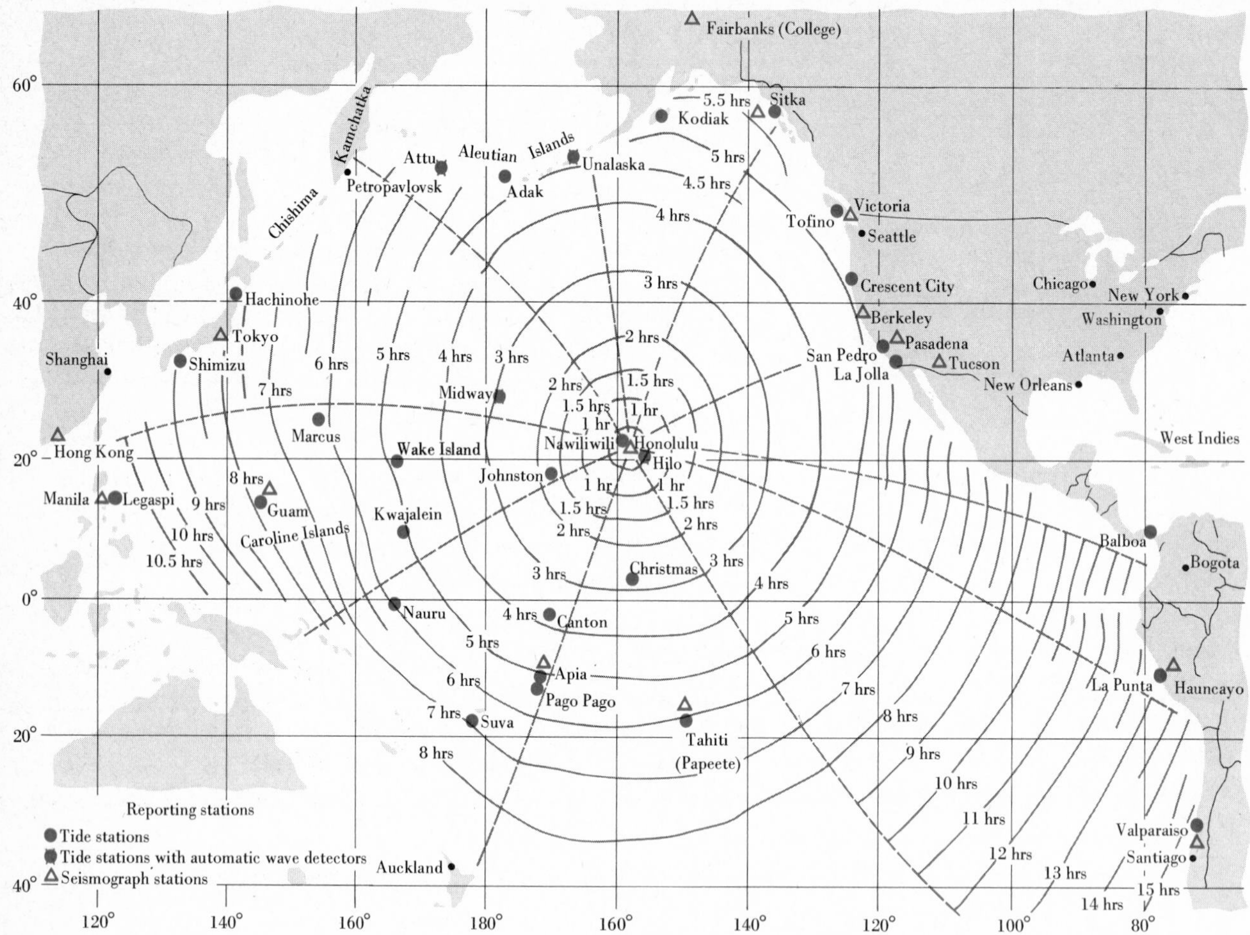

Figure 16.5 ***Seismic sea-wave travel times to Honolulu.***

sea waves. These gauges filter out the normal tides and the wind waves.

Whenever a strong earthquake is recorded at Honolulu, the alarm rings. Requests are then made for immediate readings from other seismograph stations around the Pacific. Within approximately 1 hr, the earthquake's location is determined. If it is found that the earthquake was in the Pacific Ocean or near its perimeter, tide gauges are then checked to see whether they show the existence of a seismic sea wave. If they do, an estimate of its arrival time is made, an alert is sounded, and people are warned to evacuate coastal areas.

The seismic sea-wave warning system proved very effective on November 4, 1962, when there was an earthquake under the sea off Kamchatka Peninsula at 17 hr 7 min Greenwich time. Within about 1 hr, with the help of reports from seismograph stations in Alaska, Arizona, and California, Honolulu located the earthquake at 51°N, 158°E. Reports from tide stations at Attu and Dutch Harbor (Aleutian Islands) indicated that seismic sea waves had been started by the quake. Honolulu thereupon computed the time it would take for the first wave to reach Oahu and Hawaii, as well as other Hawaiian Islands. It was due at Honolulu at 23 hr 30 min Greenwich time, 6 hr and 23 min after the quake occurred off Kamchatka. In a little over 3 hr, Midway reported that it was covered by 3 m of water as the first wave raced over it. At Honolulu and Hilo, the waves were not as large as those of 1946, though damage was about $800,000. But not a single life was lost, thanks in great measure to the warning system.

The Chilean seismic sea wave of May 22, 1960, was the most destructive in recent history, causing deaths and extensive damage in Chile, Hawaii, the Philippines, Okinawa, and Japan. As usual, Honolulu had computed the location and time of the earthquake, so when word was flashed that seismic sea waves were racing out over the Pacific, they issued warnings of the dangers, urging evacuation of coastal areas and correctly predicting the arrival time of the first waves. These arrived on schedule 6 hr after the warnings were broadcast, 15 hr after they started off South America. Through failure of many people to heed the warnings, however, 61 lives were lost in Hilo, mostly by crushing or drowning; 282 were injured enough to require hospitalization and medical care. Two hundred and twenty-nine dwellings were destroyed or severely damaged (see Figure 16.6), and total damage was estimated at $22 million. In Japan, no general seismic sea wave warning was issued, for it was not known that a seismic sea wave of such distant origin could be so destructive. About 8 hr after hitting Hawaii, more than 22 hr after the earthquake that started them, the seismic sea waves roared up onto the coasts of Honshu and Hokkaido. There, more than 17,000 km from where they started, they brought death to 180 people, significantly affected the homes and livelihood of 150,000, and caused damage estimated close to $500 million. There were

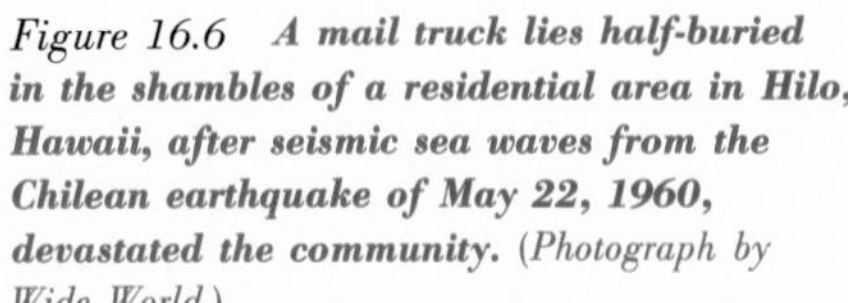

Figure 16.6 **A mail truck lies half-buried in the shambles of a residential area in Hilo, Hawaii, after seismic sea waves from the Chilean earthquake of May 22, 1960, devastated the community.** *(Photograph by Wide World.)*

20 deaths in the Philippines. Damage along the western coast of the United States was $500,000. The waves also did considerable damage in New Zealand. All Chilean coastal towns between the thirty-sixth and forty-fourth parallels were destroyed or severely damaged.

Fortunately, only an extremely small fraction of all submarine or coastal earthquakes cause seismic sea waves.

Landslides

In regions where there are many hills with steep slopes or areas with special soil conditions sensitive to vibration, earthquakes are often accompanied by landslides (see Figure 16.7). These slides occur within a zone seldom exceeding 40 to 50 km in radius, though the very largest earthquakes have affected areas as far away as 150 km. The Alaskan quake of 1964 did this.

One of the worst earthquake-caused landslides on record occurred on June 7, 1692. More than 20,000 lives were lost and much property destroyed in a large section of the then-bustling town of Port Royal, Jamaica. The whole waterfront was launched into the sea, together with several streets of two- and four-story brick homes. The brick homes and other buildings had been built on loose sands, gravel, and filled land. Shaken loose by the quake, the underlying gravel and sand gave way and slid into the sea; two-thirds of the town, consisting of the government buildings, wharf, streets, homes, and people went with it.

In the province of Kansu, China, in deposits of loess, an earthquake on

Figure 16.7 **Turnagain Heights, Anchorage, Alaska, showing slide damage caused by the 1964 earthquake.** (*Photograph by U.S. Coast and Geodetic Survey, ESSA.*)

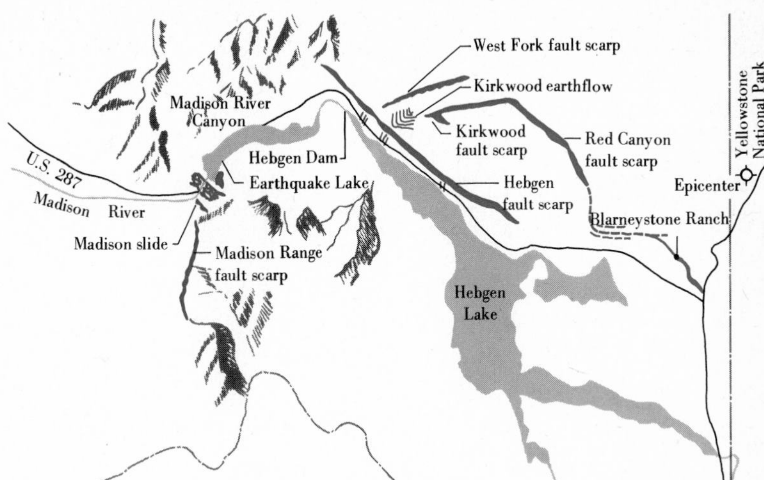

Figure 16.8 **Region of Hebgen Lake, Montana, earthquake of August 17, 1959.** (*From John H. Hodgson,* Earthquakes and Earth Structure, *Prentice-Hall, Inc., Englewood Cliffs, N.J., 1964.*)

December 16, 1920, caused some of the most spectacular landslides on record. The death toll was 100,000. Great masses of surface material moved nearly 2 km, and some of the blocks carried along undamaged roads, trees, and houses.

In the vicinity of the Japanese earthquake of 1923, one large slide moved down a valley as if it were a wall of water at a speed of 1.5 km/min, destroying a village and a railroad bridge at the mouth of the valley.

Several landslides occurred in the vicinity of Hebgen Reservoir, Montana, when the area was shaken by an earthquake just before midnight on August 17, 1959. The largest slide dammed the Madison River to form Earthquake Lake as shown in Figure 16.8. (See also Section 10.3 and Table 10.1.)

Cracks in the ground

One of man's most persistent fears about earthquakes is that the earth is likely to open up and swallow everyone and everything in the vicinity. Such fears have been nourished by a good many tall tales, as well as by pictures like Figure 16.9, which shows shallow cracks in a pavement left unsupported by the slumping of a canal bank. One account of the Lisbon earthquake of November 1, 1755, claimed that about 40 km from Lisbon the earth opened up and swallowed a village's 10,000 inhabitants with all their cattle and belongings and then closed again. The story probably got its start when a landslide buried some village in the area.

In Japan widespread ideas about the earth's ability to swallow people may have sprung from an allusion in one of the Buddhist scriptures. There it is stated that when Devadatta, one of Sakya Muni's disciples, turned against his Master and even made attempts on his life, Heaven punished him by consigning him to Hades, whereupon the ground opened and immediately swallowed him up.

It is true that landslides do bury people and buildings, and under special conditions they may even open small, shallow cracks. In California, in 1906,

Figure 16.9 **Cracks in pavement caused by an earthquake in Tokyo, Japan, 1923. Pictures of this kind are sometimes used to support the popular superstition that the earth yawns open during an earthquake. That is not what happened here. A hard-surface road lost its support when the bank of an adjacent canal collapsed.** (*Photograph by L. Don Leet.*)

a cow did fall into such a crack and was buried with only her tail protruding. But there is no authenticated case in which solid rock has yawned open and swallowed anything.

Changes in land level

Some earthquakes have been accompanied by changes in the level of the land over broad areas. The land surface has sunk in some places, risen in others, and often has tilted.

Vertical displacements in excess of 10 m have been recorded in some earthquakes and areas of thousands of km^2 may be affected. Details of some examples are discussed in Section 18.1.

Sound

When an earthquake occurs, the vibrations in the ground often disturb the air and produce sound waves that are within the range of the human ear. These are known as *earthquake sounds.* They have been variously described, usually as low, booming noises. Very near the source of an earthquake, sharp snaps are sometimes audible, suggesting the tearing apart of great blocks of rock. Farther away, the sounds have been likened to heavy vehicles passing rapidly over hard ground; the dragging of heavy boxes or furniture over the floor; a loud, distinct clap of thunder; an explosion; the boom of a distant cannon; or the fall of heavy bodies or great loads of stone. The true earthquake sound, of course, is quite distinct from the rumble and roar of shaking buildings, but in some cases the sounds are probably confused.

16.3 *Worldwide standard seismograph network*

In 1961, the U. S. Coast and Geodetic Survey began establishing a worldwide network of 125 standard seismograph stations (SSWWS—see Figure 16.10). Since then, seismological knowledge has been expanding rapidly due to the quantity and quality of earthquake measurements being made. Also, computers have helped to speed the processing of data for determining earthquake locations, depths of foci, and size. The U.S. Coast and Geodetic Survey now locates approximately 6,000 quakes/year, whereas 10 years before the number had been one-tenth that. This increase does not mean that the seismicity of

Figure 16.10 ***Stations of the worldwide standard seismograph network.***

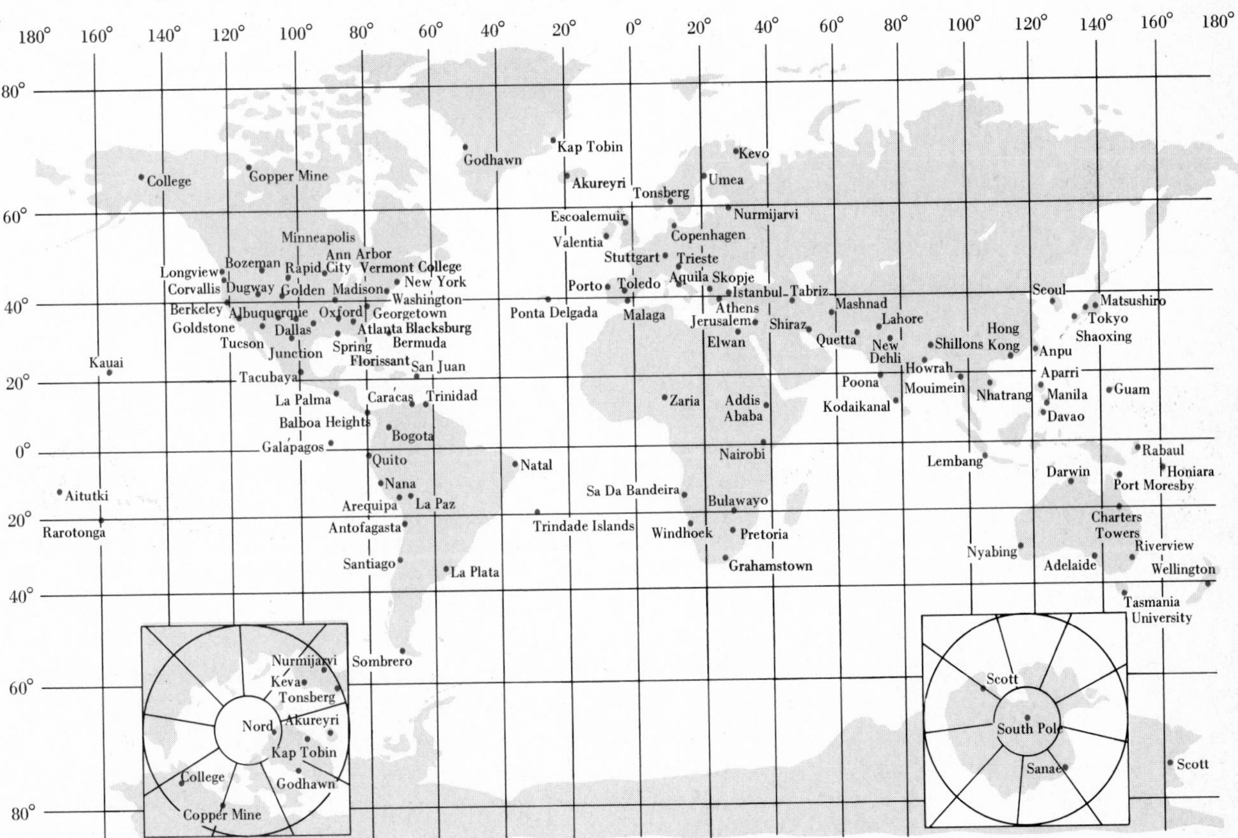

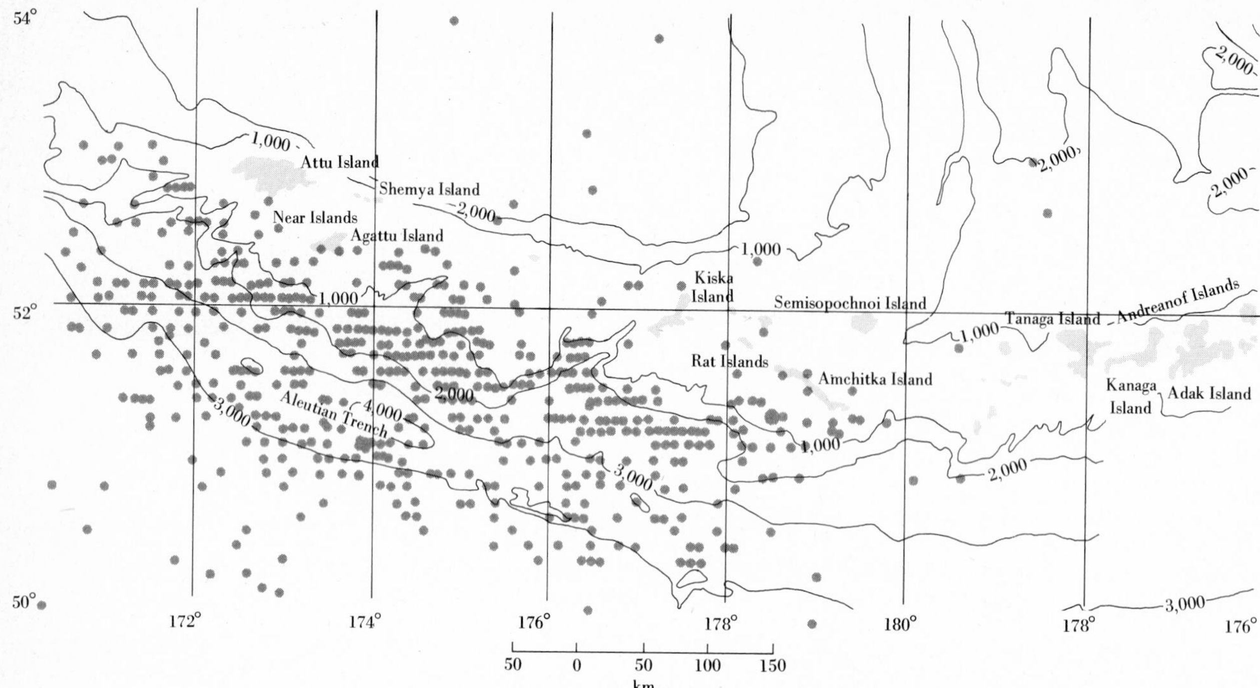

Figure 16.11 **Aftershocks of the earthquake of February 4, 1965, in the Rat Islands of the Aleutian Islands.**

the earth has increased. It only appears to have done so because the improved equipment in the increased number of recording stations is gathering more and better data.

From this worldwide seismic network, and from other seismograph stations, are coming measurements that give us new perspective on sequences of earthquake activity associated with large shocks and also information on the size of the area involved. For example, after the Rat Island earthquake of February 4, 1965, 1,300 aftershocks were recorded during the first 45 days. These occurred over an area roughly 600 km long by 300 km wide (see Figure 16.11). It has been found that the largest earthquakes and their aftershocks involve areas up to 200,000 km^2.

Figure 16.12 **Diagram showing the positions of the focus and epicenter of an earthquake.**

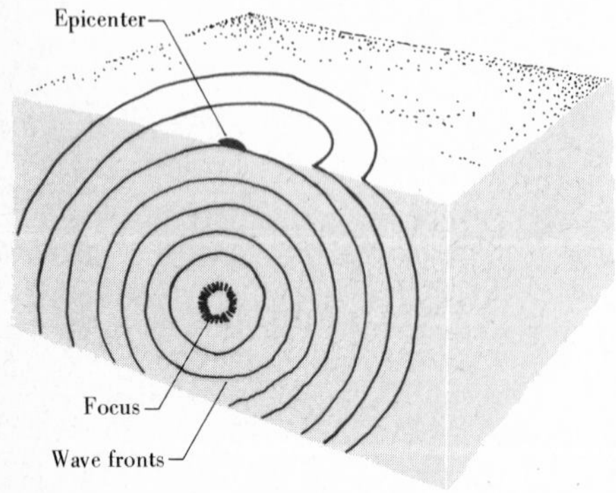

Earthquake focus

In seismology, the term *focus* is used to designate the source of a given set of earthquake waves. Just what is this source? As we know, the waves that constitute an earthquake are generated by the rupture of earth materials. When these waves are recorded by instruments at distant points, their pattern indicates that they originated within a limited region. Most sources have dimensions

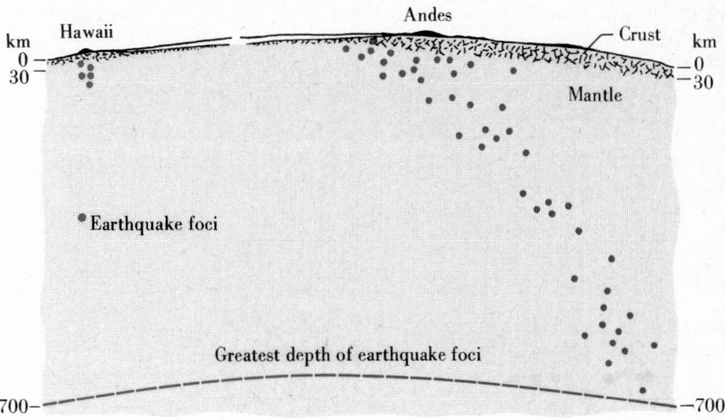

Figure 16.13 ***Earthquake foci under Hawaii and South America.***

probably closer to 50 km in length and breadth than to 5 or 500 km. A few of the largest earthquakes may involve up to 1,000 km. But trying to fix these dimensions more accurately offers a real problem that has not yet been solved.

In any event, the focus of an earthquake is usually at some depth below the surface of the earth. An area on the surface vertically above the focus is called the *epicentral area*, or *epicenter*, from the Greek *epi*, "above," and *center* (see Figure 16.12).

Foci have been located at all depths down to 700 km, a little more than one-tenth of the earth's radius (see Table 16.1). On some continental margins,

Figure 16.14 ***Earthquake foci between 600 and 700 km in depth.***

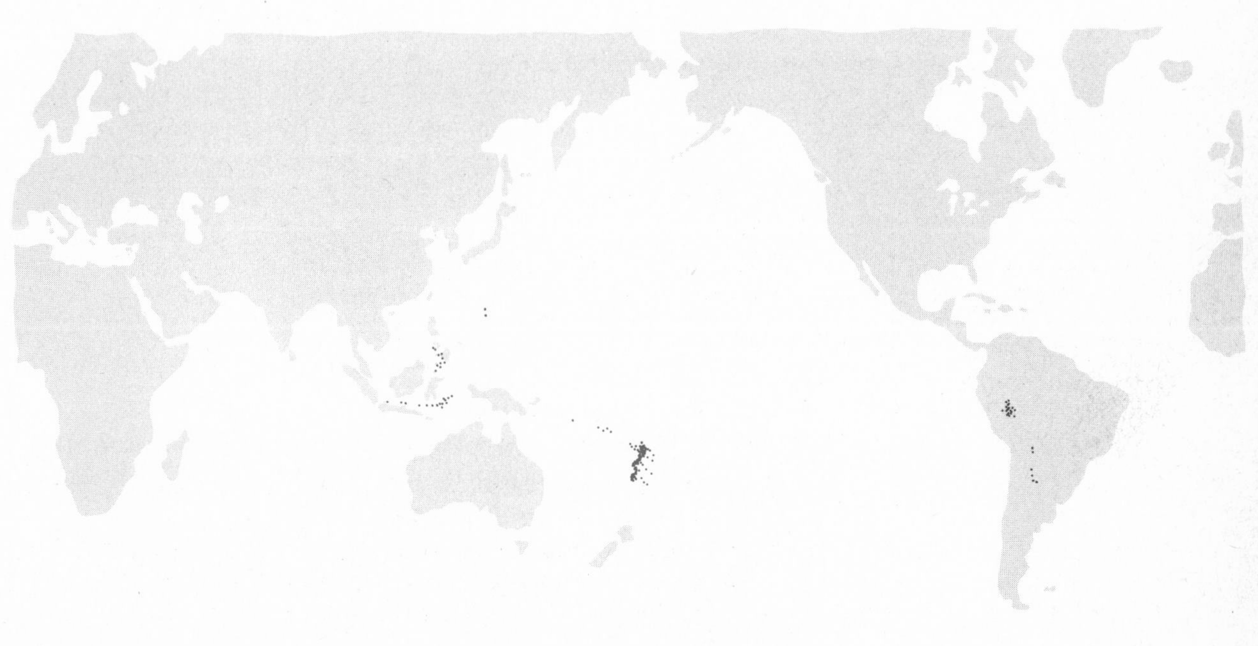

Table 16.1 ***Focal depths, 1961 through 1968***[a]

Depth, km	*Number of foci*
Less than 75	27,650
100 ± 25	3,533
150 ± 25	1,942
200 ± 25	974
250 ± 25	447
300 ± 25	223
350 ± 25	196
400 ± 25	200
450 ± 25	205
500 ± 25	292
550 ± 25	488
600 ± 25	448
650 ± 25	154
More than 650 ± 25	8
Total	36,760

[a] Summarized from U.S. Coast and Geodetic Survey's preliminary determination of epicenter data. Personal communication from Leonard M. Murphy, Chief, Seismology Division.

they have clustered along a plane dipping toward the interior of the continent, as in Figure 16.13. The deepest have been limited to the Tonga–Fiji area and the Andes, as shown in Figure 16.14.

Earthquake intensity

How to specify the size of an earthquake has always posed a problem. Before the development of instrumental seismology, some of the early investigations of earthquakes led to various attempts to describe the intensity of the shaking at a specific place. A missionary in some remote region would keep a diary of earthquakes rated as weak, strong, or very strong. This was at best a personal scale.

In 1883, an intensity scale was developed which combined the efforts of M. S. de Rossi of Rome and F. A. Forel of Lausanne, Switzerland. For half a century, the Rossi–Forel intensity scale was widely used throughout the world. According to their scale, earthquake effects were classified in terms of 10 degrees of intensity. It had definite limitations, however, from a scientific standpoint. For example, the definition of the sixth degree of intensity included "general awakening of those asleep; general ringing of bells; oscillation of chandeliers; stopping of clocks; some startled persons leave their dwellings." But an earthquake that produced those effects in Italy or Switzerland might not even wake the baby in Japan, or it might cause a stampede in Boston.

Objections to the Rossi–Forel intensity scale led L. Mercalli of Italy to set up a new scale in 1902. This was modified in 1931 by H. O. Wood and Frank Neumann and is the scale currently in use by the U.S. Coast and Geodetic Survey to evaluate earthquake intensity. It has 12 degrees of intensity and takes into account varying types of construction:

I Not felt except by a very few under specially favorable circumstances. (I, Rossi–Forel scale)

II Felt only by a few persons at rest, especially on upper floors of buildings. Delicately suspended objects may swing. (I to II, Rossi–Forel scale)

III Felt quite noticeably indoors, especially on upper floors of buildings, but many people do not recognize it as an earthquake. Standing motorcars may rock slightly. Vibration like passing of truck. Duration estimated. (III, Rossi–Forel scale)

IV During the day, felt indoors by many, outdoors by few. At night, some awakened. Dishes, windows, doors disturbed; walls make creaking sound. Sensation like heavy truck striking building. Standing motorcars rocked noticeably. (IV to V, Rossi–Forel scale)

V Felt by nearly everyone, many awakened. Some dishes, windows, etc., broken; a few instances of cracked plaster; unstable objects overturned. Disturbances of trees, poles, and other tall objects sometimes noticed. Pendulum clocks may stop. (V to VI, Rossi–Forel scale)

VI Felt by all, many frightened and run outdoors. Some heavy furniture moved; a few instances of fallen plaster or damaged chimneys. Damage slight. (VI to VII, Rossi–Forel scale)

VII Everybody runs outdoors. Damage negligible in buildings of good design and construction; slight to moderate in well-built ordinary structures; considerable in poorly built or badly designed structures; some chimneys broken. Noticed by persons driving motorcars. (VIII, Rossi–Forel scale)

VIII Damage slight in specially designed structures; considerable in ordinary, substantial buildings, with partial collapse; great in poorly built structures. Panel walls thrown out of frame structures. Fall of chimneys, factory stacks, columns, monuments, walls. Heavy furniture overturned. Sand and mud ejected in small amounts. Changes in well water. Persons driving motorcars disturbed. (VIII+ to IX−, Rossi–Forel scale)

IX Damage considerable in specially designed structures, well-designed frame structures thrown out of plumb; great in substantial buildings, with partial collapse. Buildings shifted off foundations. Ground cracked conspicuously. Underground pipes broken. (IX+, Rossi–Forel scale)

X Some well-built wooden structures destroyed; most masonry and frame structures destroyed with their foundations; ground badly cracked. Rails bent. Landslides considerable from river banks and steep slopes. Shifted sand and mud. Water splashed (slopped) over banks. (X, Rossi–Forel scale)

XI Few, if any, (masonry) structures remain standing. Bridges destroyed. Broad fissures in ground. Underground pipelines completely out of service. Earth slumps and land slips in soft ground. Rails bent greatly.

XII Damage total. Waves seen on ground surfaces. Line of sight and level distorted. Objects thrown upward into air.

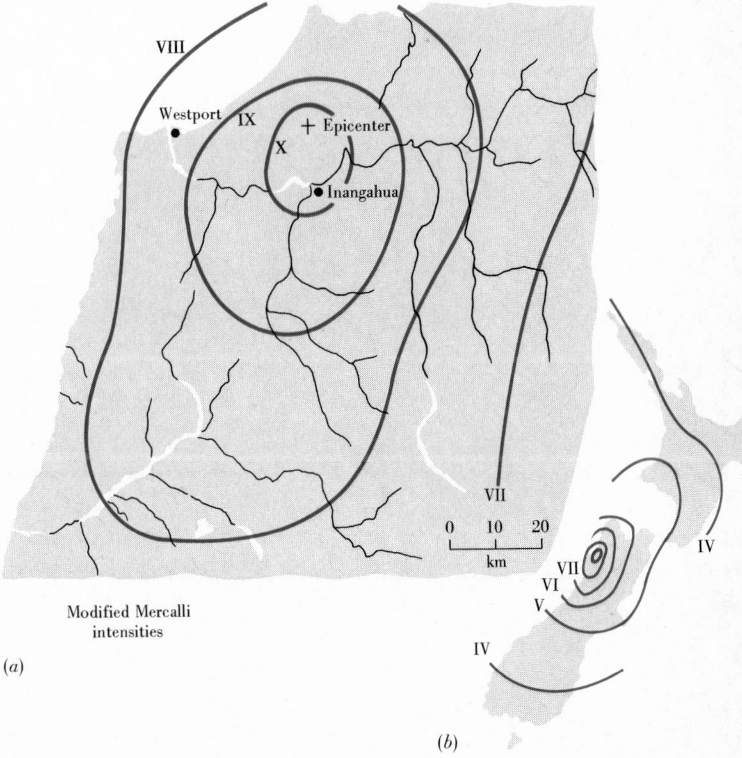

Figure 16.15 **(a) Isoseismic map of epicentral region of Inangahua earthquake, New Zealand, May 24, 1968; (b) isoseismic map of New Zealand for the same earthquake.**

By means of postcards, letters, and interviews, investigators make surveys of the effects caused by each earthquake. Then they determine what places had been shaken by about equal amounts. These are plotted on a map and the points connected by *isoseismic lines,* or lines of equal shaking. Figure 16.15 shows an isoseismic map for the New Zealand earthquake which occurred in May 1968.

Earthquake magnitude and energy

Having to rely on the impressions of people is a very unsatisfactory way of compiling accurate information on the actual size of an earthquake. Therefore a scale was devised in 1935, based on instrumental records.[2] By computation based on the amount of motion in certain waves, it ascribed to each earthquake a number called the earthquake's *magnitude,* an index of the quake's energy at its source. This was refined in 1956 by B. Gutenberg and C. F. Richter and in 1967 by an International Committee on Magnitude.

According to this scale, shallow earthquakes near inhabited areas can have the following effects: One of magnitude 2.5 would be just large enough to be felt nearby; one of magnitude 4.5 or over is capable of causing some very local damage; one of 6 or over is potentially destructive. A magnitude 7 or over represents a major earthquake.

Some earthquakes which are major catastrophes in a human and property sense are not of the greatest magnitude. Near the coast of northern Peru, at 9.2°S, 78.8°W, on May 31, 1970, a great quake killed an estimated 50,000 persons and left 800,000 homeless while causing $230 million damage. Hualas Canyon towns were flooded by burst dams and buried under landslides and mudslides. The quake was felt along 1,000 km of Peru, yet its magnitude was 7.8, less than that of any of the quakes listed in Table 16.2.

The largest magnitude assigned to an earthquake to date has been 8.6. Four of that size have occurred: (1) in Alaska, September 10, 1899; (2) in Colombia, January 31, 1906; (3) in Asia, August 15, 1950; and (4) in Alaska, March 27, 1964. The only earthquake in history that might have been larger, judging from the reported effects, was near Lisbon, Portugal, on November 1, 1755. Possibly the magnitude of that earthquake was between 8.7 and 9.0. An earthquake with a magnitude of over 10 should theoretically be perceptible

[2]C. F. Richter, "An Instrumental Earthquake Magnitude Scale," *Bull. Seis. Soc. Am.*, vol. 25, pp. 1–32, 1935.

Table 16.2 ***Great earthquakes, 1899 through 1970***

Date	*Location*	*Magnitude*
September 10, 1899	Alaska, 60°N, 140°E	8.6
January 31, 1906	Colombia, 1°N, 82°W	8.6
August 17, 1906	Chile, 33°S, 72°W	8.4
January 3, 1911	Tien Shan, 44°N, 78°E	8.4
December 16, 1920	Kansu, 36°N, 105°E	8.5
March 2, 1933	Japan, 39°N, 145°E	8.5
August 15, 1950	Asia, 29°N, 97°E	8.6
May 22, 1960	Chile, 38°S, 73.5°W	8.4
March 28, 1964	Alaska, 61.1°N, 147.7°W	8.6

in scattered areas over the entire earth, but such an occurrence has never been reported.

An earthquake of magnitude 5 releases approximately the same amount of energy as did the first atomic bomb when it was tested on the New Mexico desert, July 16, 1945. A megaton nuclear device releases the same amount of energy as an earthquake of magnitude 6. Of course, the energy is applied in quite different ways—highly concentrated in the device, widely dispersed in the earthquakes—and the results are correspondingly different. The energy released by an earthquake of magnitude 8.6 is 3 million times as great as that of an earthquake of magnitude 5, or a million million (10^{12}) times the energy of the smallest earthquake.

16.4 *Distribution of earthquakes*

The first extensive statistical study of earthquakes revealed that from year to year there were wide variations in the total energy released, as well as in the number of individual shocks. For example, 1906 showed 6 times the average energy released between 1904 and 1952 and 40 times the minimum; 1950 was another very large year, second only to 1906.

Most of the energy released is concentrated in a relatively small number of very large earthquakes. A single earthquake of magnitude 8.4 releases just about as much energy as was released, on the average, each year during the first half of the twentieth century. It is not unusual for the energy of one great earthquake to exceed that of all the others in a given year or of several years put together. Nine great earthquakes listed in Table 16.2 represented nearly a quarter of the total energy released from 1899 through 1968.

The maximum energy released by earthquakes becomes progressively less as depth of focus increases. The nine largest earthquakes from 1899 through 1968 were relatively shallow and had magnitudes of 8.6, 8.6, 8.4, 8.4, 8.5, 8.5, 8.6, 8.4, and 8.6. The five largest intermediate-depth shocks over the same interval had magnitudes of 8.1, 8.2, 8.1, 7.9, and 8.0, and the three largest deep shocks had magnitudes of only 8.0, 7.75, and 7.75. This trend suggests that the force required to rupture earth materials decreases as depth becomes greater.

Table 16.3 **Average annual number and magnitude of earthquakes, 1904–1946**[a]

	Magnitude	*Average number*
Actually observed		
Great	7.7–8.6	2
Major	7.0–7.7	12
Potentially destructive	6.0–7.0	108
Estimates based on sampling special regions	5.0–6.0	800
	4.0–5.0	6,200
	3.0–4.0	49,000
	2.5–3.0	100,000

[a] B. Gutenberg and C. F. Richter, *Seismicity of the Earth*, Princeton University Press, Princeton, N.J., 1949.

The average annual number of earthquakes from 1904 through 1946 was estimated by Gutenberg and Richter as shown in Table 16.3. This places the number of earthquakes large enough to be felt by someone nearby at more than 150,000/year. Gutenberg and Richter estimated that the total number of earthquakes "may well be of the order of a million each year." Seismograph stations throughout the world are now supplying data leading to the location of close to 6,000 earthquakes/year.

Locations for earthquakes have been noted from seismographic records since 1899. Figures 16.1 and 16.2 show their distribution 1961 through 1967, the first years of standardized networks and computer-programming techniques. These active zones were not significantly different in the past.

Earthquakes tend to occur in belts or zones marked also by active volcanoes. *The earthquakes that occur in the most active zone, around the borders of the Pacific Ocean, account for a little over 80 percent of the total energy released throughout the world.* The greatest activity is near Japan, western Mexico, Melanesia, and the Philippines. The loop of islands bordering the Pacific has a high proportion of great shocks at all focal depths.

Fifteen percent of the total energy released by all earthquakes is in a zone that extends from Burma through the Himalaya Range, into Baluchistan, across Iran, and westerly through the Alpine structures of Mediterranean Europe. This is sometimes called the Mediterranean and trans-Asiatic zone. Earthquakes in this zone have foci aligned along mountain chains. That leaves 5 percent of the energy being released throughout the rest of the world. Narrow belts of activity follow the oceanic ridge systems (see Figure 16.1).

From time to time, observers have tried to correlate the occurrence of earthquakes with sun spots, the tides, the position of heavenly bodies, and other phenomena. But they have all ignored the facts in one way or another. Occasionally, someone steps forward to predict earthquakes on the basis of

Figure 16.16 **Seismic risk map for the United States, issued January 1969, shows earthquake damage zones of reasonable expectancy in the next century.** *(Developed by U.S. Coast and Geodetic Survey, ESSA.)*

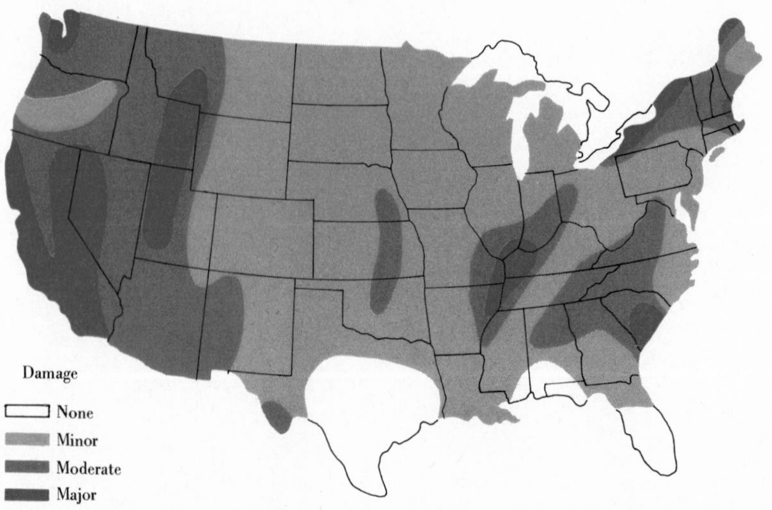

some correlation that he fancies he has made. Usually, however, he keeps his supporting data secret. One such prophet stated that on each of certain days during 3 months of one year there would be "an earthquake in the southwest Pacific." Because he did not specify magnitude, time, or place in this highly active zone, statistically he could not miss.

A map showing zones of earthquake expectancy in the United States has been issued by the U.S. Dept. of Commerce.[3] It is shown in Figure 16.16. The zones were outlined after a 2-year study of 28,000 earthquakes and show where earthquakes may be expected in the next century.

Seismicity of Alaska and Aleutian Islands

Before instruments were developed, reports of earthquakes were dependent on population distribution. Alaska was first settled in 1783 by Russians on Kodiak Island, who reported the first earthquake as occurring south of the Alaskan peninsula on July 27, 1788. From then through 1964, there were 1,410 earthquakes felt by persons living in the area. Information on early earthquakes was obtained from historical records, newspapers, and ships' logs.

[3]U.S. Department of Commerce news, ESSA release ES-1, Jan. 14, 1969.

Figure 16.17 ***Earthquakes in Alaska with a magnitude of 6 or greater, 1899 to 1964.***

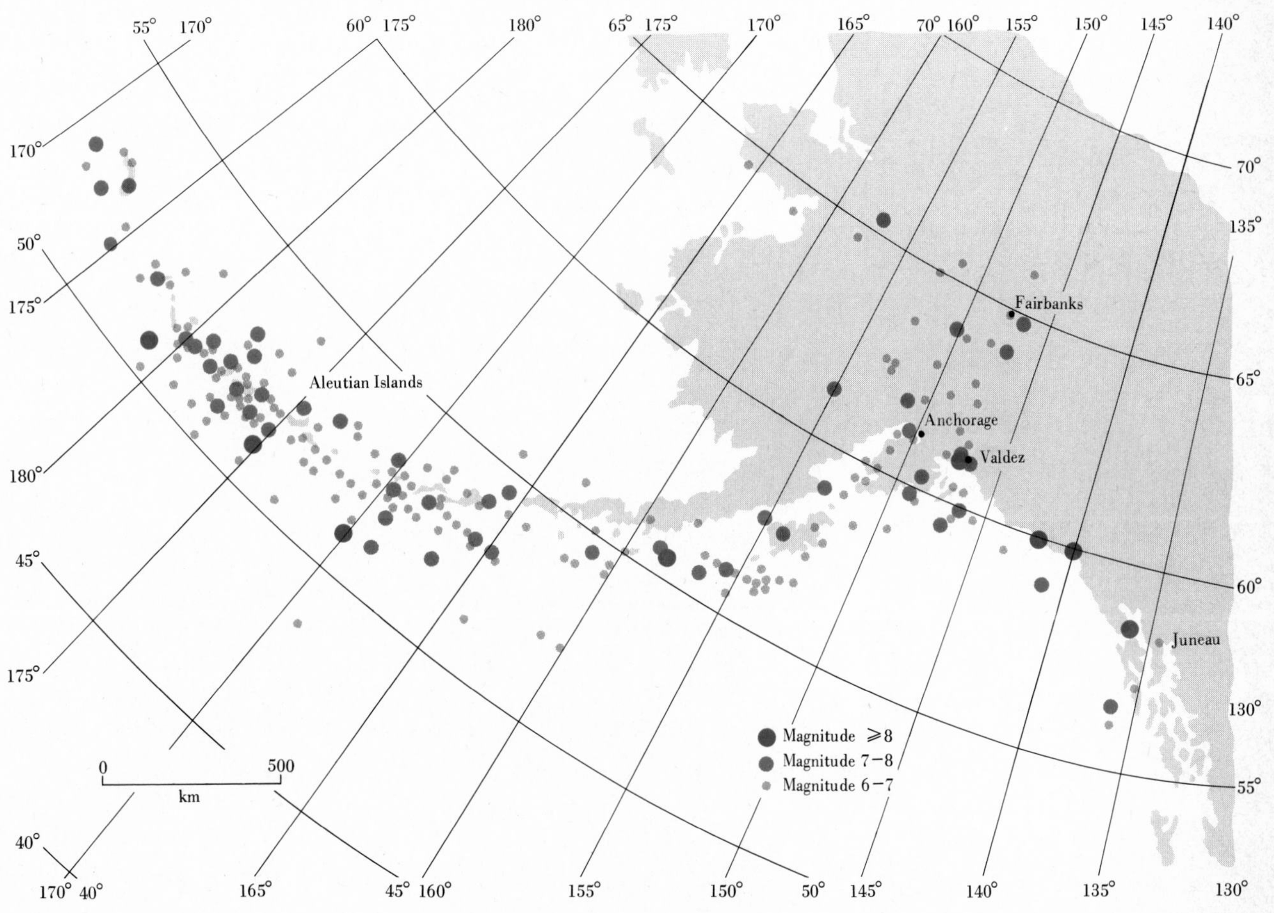

From 1899 through 1964, there were 298 instrumentally recorded earthquakes of magnitude 6 or greater (see Figure 16.17).

THE PRINCE WILLIAM SOUND EARTHQUAKE OF 1964 On March 27, 1964, at 5:36 P.M. Alaskan standard time (or 3 hr 36 min 13 sec Greenwich mean time, March 28, 1964), an earthquake of magnitude 8.6 rocked the Prince William Sound area in south-central Alaska. It released twice as much energy as the great San Francisco earthquake of 1906 and was the most destructive earthquake ever recorded on the North American continent—one of our greatest natural disasters.

Its center was located at 61.1°N, 147.7°W, at a depth of 33 km (see Figure 16.18). The earthquake was felt over 1.3 million km^2 and major damage to buildings, ports, and transportation occurred over an area of 130,000 km^2. Property damage was estimated at $311,192,000. Damage at Anchorage, the hardest hit, was estimated at $86 million. In spite of all the damage, there were only nine lives lost in Anchorage and six elsewhere from building collapse and landslides. If it had been earlier in the day, the casualties could have been much higher.

A seismic sea wave generated by the earthquake accounted for 98 deaths in Alaska; 11 in Crescent City, California; and 1 in Seaside, Oregon.

Although 10 percent of the land area of Alaska was involved in the effects

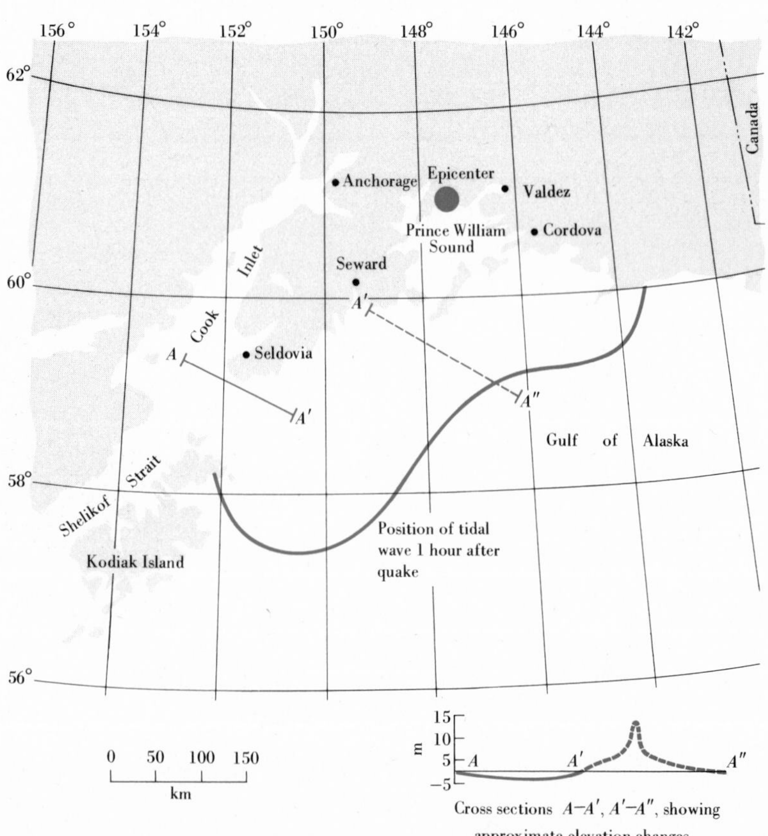

Figure 16.18 ***Epicenter of Alaskan earthquake of March 27, 1964. Elevation changes and the position of a tidal wave 1 hr after the quake are also shown.***

of the earthquake, this was occupied by 50 percent of Alaska's total population and included all its major seaports, highways, and railroads.

As soon as news of the disaster reached the U.S. Coast and Geodetic Survey in Washington, they took immediate action. They organized the largest project ever undertaken to obtain data about an earthquake. Investigations were conducted simultaneously from the land, sea, and air. Portable seismographs were sent into the area to record aftershocks. Survey teams determined the amounts and location of surface movements and alteration of the sea bottom.

Figure 16.19 ***Uplift and subsidence associated with quake of March 27, 1964.***

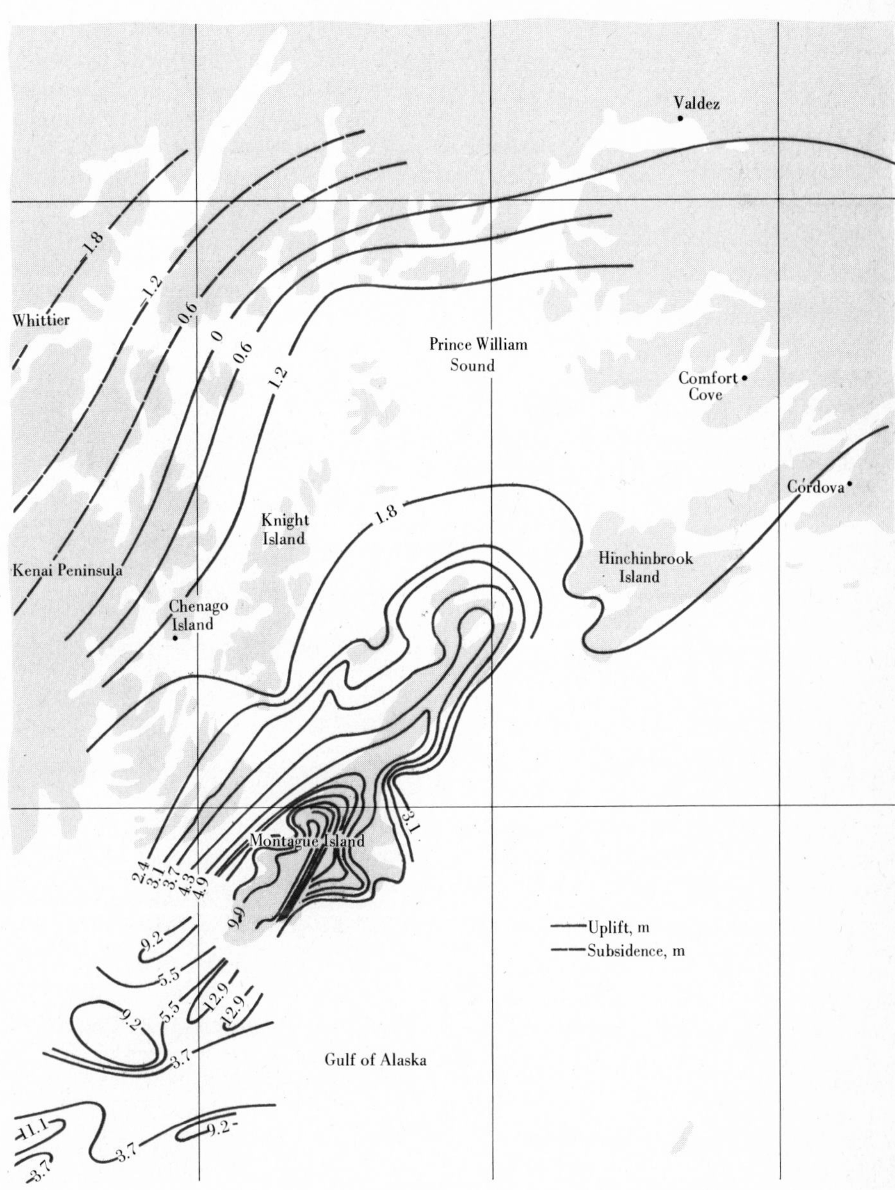

Engineers studied building damage. Much was learned from these investigations about the effects resulting from the unusually prolonged and violent shaking.

Many of the towns in the affected region had been built on unconsolidated deposits. As a result, landslides occurred, causing extensive damage to buildings and other structures. Displacements of the land and sea bottom modified shorelines and changed sea-floor topography. Because Alaska is dependent to a large extent on marine commerce for its survival, the U.S. Coast and Geodetic Survey acted swiftly to determine which sea lanes were navigable and to chart new lanes where necessary.

Surveys showed that permanent vertical and horizontal displacements occurred across considerable distances (see Figure 16.19). At the entrance to Prince William Sound, 150 km from the epicenter, uplifts of the sea bottom exceeding 13 m were discovered (see Figure 16.20). The entire land area from Anchorage to Seward, a distance of 150 km, subsided 1 to 2 m. Horizontal displacements resulted in changes of distances between points. These generally ranged from 1 to 2 m. The absolute amount and direction of movement could not be accurately determined, as reference points were also disturbed. Surveys across Montague Strait indicated that Montague and Latouche Islands were

Figure 16.20 ***Hanning Bay Fault Cove, northeast of Macleod Harbor, Montague Island, Alaska. Top photograph was taken by the U.S. Forest Service on June 8, 1959. The sharp break in the vegetation in the extreme lower right marks an old fault line. Bottom photograph was taken on July 7, 1964, by the U.S. Coast and Geodetic Survey. Note the fault line extending across the harbor entrance and the amount of uplift. Comparison of the photographs shows the uplift and its effect in damming the outlet to the lake on the lower right. Note also the channels being cut through deltaic deposits of the two major streams. This embayment lies about midway between Hanning Bay and Macleod Harbor, approximately 20 km northeast of Cape Cleare.***

5 to 7 m closer together after the quake than they had been in 1933. The general pattern of horizontal and vertical movements suggests turbulence of the earth's crust rather than simple displacement.

The earthquake was of sufficient magnitude to trigger the SSWWS at College, Alaska, and 8 min later the alarm sounded in the Honolulu Observatory, Hawaii. Within 1 hr, Honolulu had determined a preliminary epicenter and coastal cities were warned of an impending inundation.

The seismic sea wave was the first major one associated with an earthquake whose epicenter was on land. However, it was started when submarine landslides and vertical displacements disturbed a large area of the sea bottom. It was highly destructive around the Gulf of Alaska, causing extensive damage to many of Alaska's principal harbors: Anchorage, Cordova, Kodiak, Seward, Valdez, and Whittier. It also caused damage to the west coast of Canada and Crescent City, California. It accounted for the greatest loss of life.

In Anchorage, 120 km from the center of the quake, a substantial portion of earthquake damage was attributable to landslides. The largest and most damaging slide occurred in the Turnagain Heights section (see Figure 16.21). This is a residential area built on a bluff overlooking an arm of Cook Inlet. The vibrations triggered multiple slumps in the bluff. Downward and outward

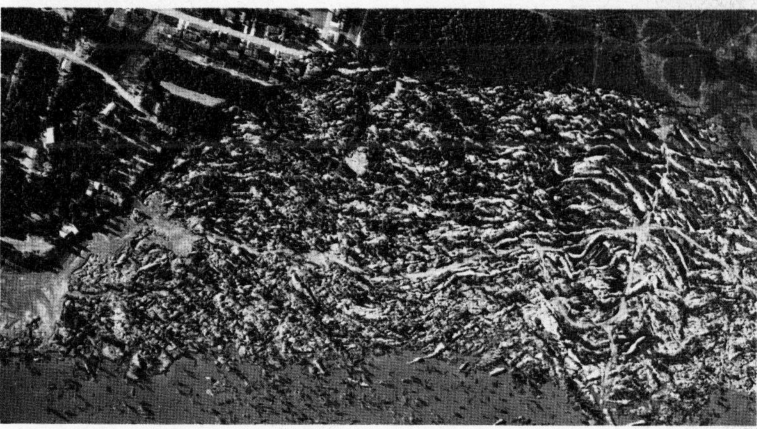

Figure 16.21 **Turnagain Heights, Anchorage, Alaska. Top photograph was taken on August 12, 1961; bottom photograph on July 26, 1964. Bootlegger Cove clay, which has little bearing strength when wet, underlies most of Anchorage. Earthquake-produced ground vibrations triggered multiple slides in the bluff area adjacent to Knik Arm. The slide covered an area about 2,400 m long and up to 400 m wide. The slide stopped when enough material collected at the toe to prevent further sliding. Clearing of debris and leveling of the ground has been started (lower left in bottom photograph). Note also the roadways through the slide area.** (*Both photographs courtesy of U.S. Coast and Geodetic Survey, ESSA.*)

Figure 16.22 **Seward, Alaska. The tsunami of March 27, 1964, hit the Seward port section with enough force to hurl this fishing vessel across the road. Note also the displaced buoy and other debris.** (*Photograph by U.S. Coast and Geodetic Survey, ESSA.*)

Figure 16.23 **Seward, Alaska: at left in September 1948; below in August 1964. Extensive shoreline changes and railroad damage resulted from the 1964 earthquake, with 1,500 m of waterfront slumping into Resurrection Bay, tsunami, and fire. The steep and unstable newly formed shoreline was backed by many tension cracks.** (*Both photographs by U.S. Coast and Geodetic Survey, ESSA.*)

movement occurred over a 2,400-m section of coastline extending east–west. The slide extended inland approximately 200 m at the east end and 400 m at the west. The edge of the bluff slid more than 200 m into the inlet. Some blocks moved intact as much as 170 m. Eyewitness accounts indicated that the sliding retrogressed from the edge of the bluff, with movements beginning approximately 2 min after the start of the earthquake. A total of 75 homes were destroyed in the Turnagain Heights section.

Valdez, 80 km from the epicenter, suffered extensive damage. The town was built on an alluvial fan that had grown out into the head of a deep, steep-sided fiord. The earthquake caused a major landslide which brought total destruction to its entire waterfront. There were extensive shoreline changes, and even late in 1964 the land was found to be still shifting. Plans were made to abandon the town and relocate it approximately 6.5 km northwest on more stable ground.

Seward, 150 km from the epicenter, suffered extensive damage. Waterfront for a length of 1,500 m slid into Resurrection Bay. This slide, coupled with the destructive effect of an 8-m seismic sea wave, left no usable docks or piers. About 90 percent of Seward's industry was obliterated and the rest was damaged. Investigations revealed that the northwest section of the bay was 27 to 30 m deeper than previously charted (see Figures 16.22 and 16.23).

Kodiak, on Kodiak Island, 450 km from the epicenter, had most of its business section and port facilities destroyed by the seismic sea wave.

AFTERSHOCKS A total of 7,500 shocks were instrumentally recorded in the months that followed. These occurred in a belt 300 km wide, stretching 800 km from the main shock southwestward to a region off the southwest coast of Kodiak Island. Their focal depths ranged between 20 and 60 km.

Seismic and volcanic activity

Volcanoes' locations relative to major earthquake belts have been coming into focus as one of the results of the increased data on number and quality of earthquake locations. At the scene of Chile's 1960 earthquake series, there is a chain of 16 active volcanoes paralleling the coast. One of these, Puyehue, erupted the day after the largest earthquake of the series, and 10 were variously active during the following year. Figure 16.24 shows the affected region and the volcanoes.

It has been estimated that approximately 6 percent of the world's large earthquakes occur in the Aleutian Islands and continental Alaska. The Aleutian Islands arc is more than 3,200 km in length and contains 76 volcanoes, of which 36 have been active since 1760. Two large earthquakes occurred off the adjacent coast after the 1912 eruption of Katmai.

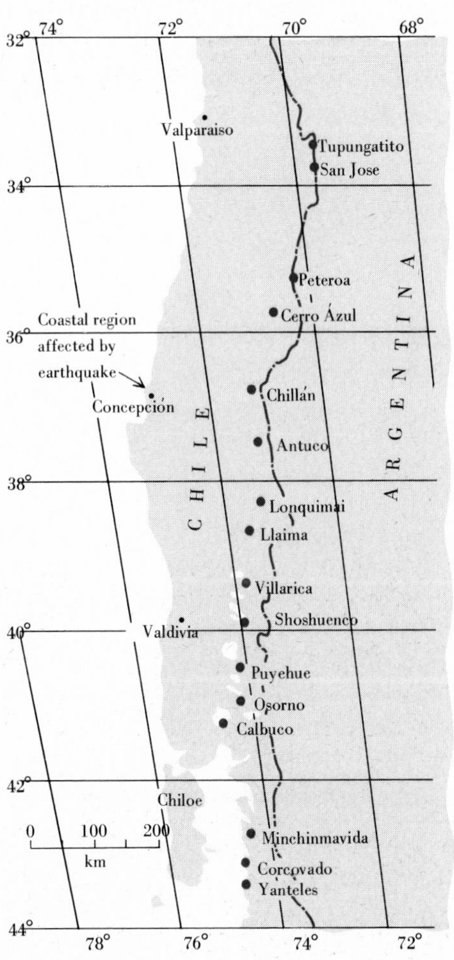

Figure 16.24 **Location of volcanoes of Chile in the region affected by the earthquakes of 1960.**

It has been fashionable for decades to think of "volcanic earthquakes" as a special feature of explosive eruptions, small in energy and seldom damaging. But there is now emerging a realization that volcanoes and earthquakes may have a common ultimate cause in deep movements of mantle materials, and the coincidence of belts of major earthquake activity with belts which also include active volcanoes supports the idea.

16.5 Cause of earthquakes

Centuries ago, people believed that mysterious shakings of the earth were caused by the restlessness of a monster that was supposed to be supporting the globe. In Japan, it was first a great spider and then a giant catfish; in some parts of South America, it was a whale; and some of the North American Indians decided that the earth rested on the back of a giant tortoise.

The lamas of Mongolia had another idea. They assured their devout followers that after God made the earth, He had placed it on the back of an immense frog, and every time the frog shook his head or stretched one of his feet, an earthquake occurred immediately above the moving part (see Figure 16.25). This was a major advance in earthquake theory, for at least it tried to explain the local character of earthquakes.

Figure 16.25 **Cause of earthquakes, according to the lamas of Mongolia. It was believed that when the frog lifted a foot, there was an earthquake immediately above the part that moved.**

The Greek philosopher Aristotle (384–322 B.C.) held that all earthquakes were caused by air or gases struggling to escape from subterranean cavities. Because the wind must first have been forced into the cavities, he explained that just before an earthquake the atmosphere became close and stifling. As time passed, people began to refer to "earthquake weather," and to this day some people insist that the air turns humid and stuffy when an earthquake is about to occur. A Japanese scientist, Omori, checked on this belief by investigating the conditions that preceded 18 catastrophic earthquakes in Japan between 1361 and 1891. He found that the weather was fair or clear on 12 occasions, cloudy on 2, rainy or snowy on 3, and rainy and windy once, but that humid and sultry "earthquake weather" had never just preceded an earthquake.

Aristotle's idea that earthquakes were caused by imprisoned gases has long been abandoned as we learned more about the behavior of gases, the structure of the earth's crust, and finally the depths at which earthquakes occur. What, then, *does* cause earthquakes? Most of them are caused by deforming forces in the earth. *The immediate cause of an earthquake is the sudden rupture of earth materials distorted beyond the limit of their strength.*

Elastic rebound hypothesis

After the California earthquake of 1906, it was generally accepted that the immediate cause of earthquakes was faulting due to rupture of rocks of the earth's crust. The displacements along the San Andreas fault provided an

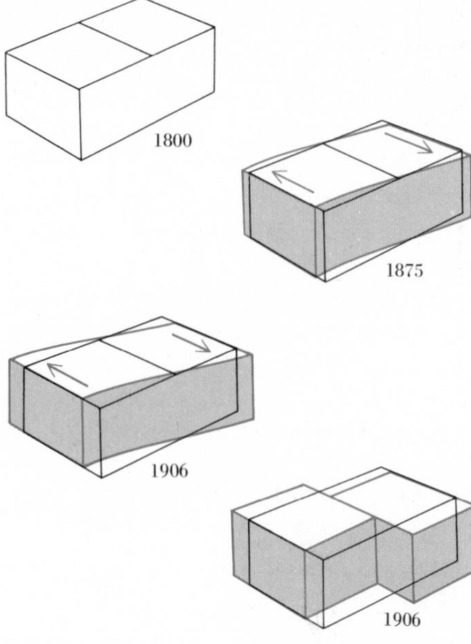

Figure 16.26 ***Reid's proposed representation of movements on the San Andreas fault associated with the 1906 earthquake, illustrating the elastic rebound hypothesis.***

unusual opportunity to study the mechanics of faulting and led to the formulation of Reid's *elastic rebound hypothesis.* Surveys of part of the fault that broke in 1906 had been made during the years preceding the earthquake. H. F. Reid of Johns Hopkins University analyzed these measurements in three groups: 1851 to 1865, 1874 to 1892, and 1906 to 1907. The first two groups showed that the ground was twisting in the area of the fault. The third showed displacements that occurred at the time of the earthquake. From these, Reid reconstructed a history of the movement.

Although there was no direct evidence, Reid assumed that the rocks in the earth's crust in the vicinity of the fault had been storing elastic energy at a uniform rate over the entire interval and that the region had started from an unstrained condition approximately a century before the earthquake. As the years passed, a line which in 1800 had cut straight across the fault at right angles was assumed to have become progressively more and more warped. When the relative movement on either side of the fault became as great as 6 m in places, the strength of the rock was exceeded and rupture occurred. The blocks snapped back toward an unstrained position, driven by the stored elastic energy (see Figure 16.26).

The San Andreas fault runs roughly from northwest to southeast. Land on the western or Pacific Ocean side of the fault has moved northwest relative to land on the eastern side. In 1906, a section of the fault 450 km long broke. The strains and adjustments were greatest within a zone extending 10 km on each side of the fault. Imagine a straight line 20 km long crossing this zone at right angles to the fault, which was in the center of the zone. After the earthquake, this line was broken at the fault and was shifted into two curves. The broken ends were separated by 6 m at the break, but the other ends were unmoved. Actually, fences, roads, and rows of vegetation provided short sections of lines by which the displacement at the fault could be gauged (see Figure 16.27).

The idea that earthquakes were caused by faulting as a result of elastic rebound was based on observed surface movements. It was also based on our knowledge of how rocks behave when they have been subjected to deforming forces. In the laboratory, rocks have been subjected to pressures that are equivalent to the pressures in the earth's crust. Under these pressures, the rocks gradually change shape, but they resist more and more as the pressure builds up, until they finally reach the breaking point. Then they tear apart and snap back into unstrained positions. This snapping back is called elastic rebound and was considered by Reid as the mechanism behind the generation of earthquakes.

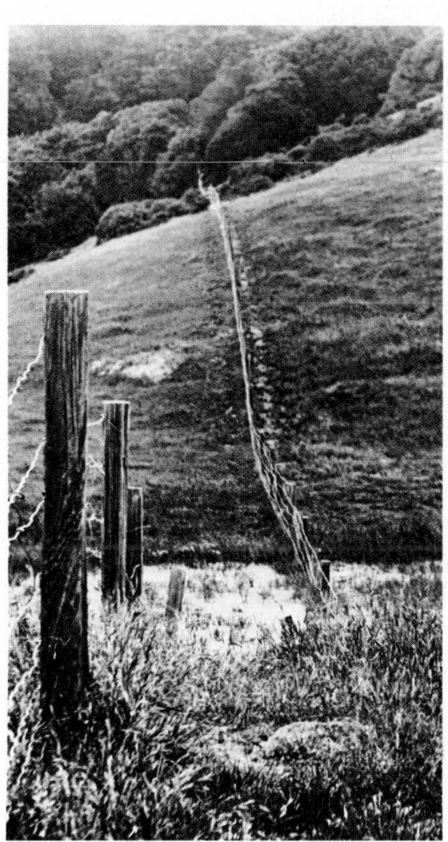

Figure 16.27 ***Looking along a fence that crossed the San Andreas fault at approximately right angles. Before the earthquake of April 18, 1906, this fence was a continuous straight line. It is shown here as it appeared after the earthquake.*** (*Photograph by J. C. Branner.*)

Evidence against elastic rebound hypothesis

Until recently, it was believed that all earthquakes were caused by faulting due to elastic rebound. However, elastic rebound assumes that earthquakes originate from distortion and rupture of elastic rocks capable of storing strain energy until it is released abruptly at the time of rupture. From laboratory measurements, it has been found that this is a reasonable assumption for rocks similar to those at the earth's surface. But with the increased pressures of burial under 20 or 30 km of overburden, all rocks deform plastically, and plastic rocks do not store strain elastically, nor can they rebound. Therefore, the elastic rebound hypothesis in its original form would apply only to earthquakes with surface or extremely shallow foci.

In the 1920s, it became apparent that at least some foci were at depths where plastic deformation takes place. The first thoughts were that "deep focus" earthquakes might represent a different mechanism of rupture than "shallow focus" earthquakes. However, there were so many features of earthquake records that indicated no really fundamental difference in the wave-generating mechanisms that an alternative suggestion was made involving the property of matter called *viscosity*, the ratio of deforming force to *rate* at which a substance changes shape in response to the force. A material may have such high viscosity that although it deforms plastically, it does this slowly enough to have at the same time a buildup of force enough to allow rupture before complete adjustment has been accomplished through flow.

In the 1950s, Percy Bridgman at Harvard University documented by laboratory measurements the mechanism of rupture of earth materials under pressures equivalent to depths of hundreds of kilometers. He described the deformation as taking place steadily by plastic flow, but interrupted at times when the material ruptures, or in his words "lets go to get a fresh hold." Thus in an environment of continuous plastic deformation, there would be sudden "jerks," or rupture, capable of causing earthquakes.

Since 1906, improvements in seismological recording and interpretation have accumulated an impressive number of measurements showing that an overwhelming number of earthquakes have their foci in the mantle and regions of the lower crust where earth materials are plastic. Therefore, the elastic rebound hypothesis would no longer seem adequate to explain most earthquakes.

Although a section of the San Andreas fault supplied data from which the elastic rebound hypothesis was formulated, it has since become the scene where measurements are being made that show surface faulting is occurring *after* earthquakes rather than at the instant of an earthquake, which also goes to disprove Reid's hypothesis.

An earthquake occurred in the Parkfield–Cholame region on June 27, 1966.[4] Figure 16.28 shows the region. During the following year, a special program

[4]C. R. Allen and S. W. Smith, "Parkfield Earthquake of June 27–29, 1966, Pre-earthquake and Post-earthquake Surficial Displacement," *Bull. Seis. Soc. Am.*, vol. 56, pp. 955–967, 1966.

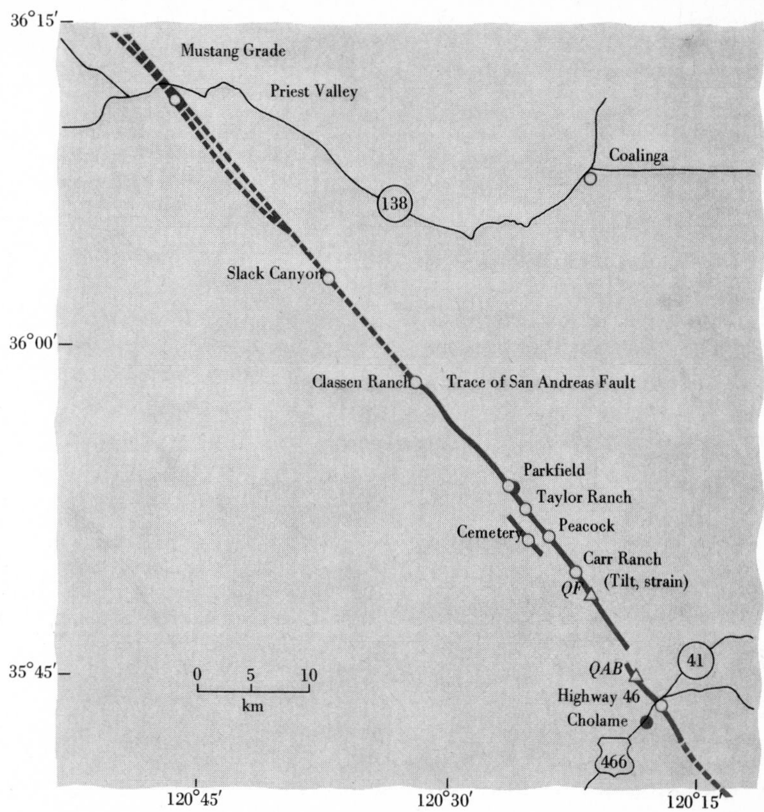

Figure 16.28 ***The Parkfield-Cholame region.***

of measurements was undertaken that produced some extremely significant data.

The earthquake occurred at night (21 hr 26 min Pacific Daylight Time) and nothing is known of surface faulting at that time. *There may have been none.* When the first inspection was made 10 hr later, a white line at the Highway 46 locality was offset 4.5 cm. By evening, the displacement had increased to 6.4 cm and by the following noon to 7.5 cm. During the following months, the average rate of creep decreased from 10 mm/day 2 days after the earthquake to 0.17 mm/day 1 year later. The total average displacement during the year was about 20 cm over a 30-km section of the fault. Most of this occurred in 4- to 6-day intervals of rapid creep *following* aftershocks (see Figure 16.29).

Surface displacements are a *result* of earthquake-causing mechanisms and are not themselves the vibration-generating movements. In other words, faulting does not cause earthquakes.

Plastic surge

The mantle-boiling hypothesis outlined in Chapter 19 explains the focal mechanism of an earthquake as a surge of plastic material. The same mechanism

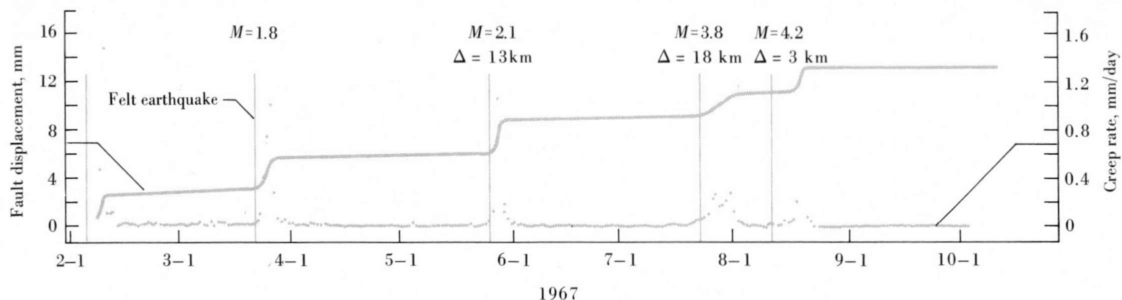

Figure 16.29 ***Cumulative displacement across the San Andreas fault measured by a taut-wire device is shown by the solid line. Daily creep rates are shown by dotted lines. All earthquakes reported as felt in the vicinity of Carr Ranch, the observing station, are shown by vertical lines.*** (*After Stewart W. Smith and Max Wyss,* Bull. Seis. Soc. Am., *vol. 58, pp. 1955–1973, 1968.*)

could be associated with convection currents in the mantle. The magnitude of the earthquake and the distribution of the aftershocks are determined by the volume of material involved in the movement. There is no limitation on intervals between successive shocks at one place. Aftershocks could be located in clusters rather than along fault lines.

The mantle-boiling hypothesis explains the lower limit of focal depths as marking the zone below which, during the earth's history, there has been no movement of material.

Outline

Earthquakes are vibrations caused when earth materials have been distorted until they rupture.

Seismology is the scientific study of earthquakes.

Effects of earthquakes include fire, damage to structures, seismic sea waves, landslides, cracks in the ground, changes in land level, and sound.

- *Fire* has caused an estimated 95 percent of the total loss caused by some earthquakes.
- *Damage to structures* depends on intensity and duration of the shaking.
- *Seismic sea waves* are now being detected in time to warn threatened communities.
  - *Seismic sea-wave warning system* is operated by the U.S. Coast and Geodetic Survey.
- *Landslides* often accompany the largest earthquakes.
- *Cracks in the ground* in the sense of yawning chasms in rock have not happened.
- *Changes in land level* have been widespread and considerable in some of the largest quakes.
- *Sound* from the earth has been observed.

Worldwide standard seismograph network is supervised by the U.S. Coast and Geodetic Survey.

Earthquake focus is a term used to designate the source of a given set of earthquake waves.

Earthquake intensity scales are used to estimate the amount of shaking at different places.

Earthquake magnitude evaluates the amount of energy at the earthquake's source.

Distribution of earthquakes is described in terms of energy released, of which about 80 percent occurs around the borders of the Pacific Ocean.

Seismicity of Alaska and Aleutian Islands has been estimated at 6 percent of the world's total.

The Prince William Sound earthquake of 1964 was the most destructive earthquake ever recorded on the North American continent.

Aftershocks occurred in a belt 300 km wide, stretching 800 km southwestward from the main shock.

Seismic and volcanic activity may have a common ultimate cause in deep movements of mantle materials.

Cause of earthquakes is the sudden rupture of earth materials distorted beyond the limit of their strength.

Elastic rebound hypothesis proposed snapping back of distorted and ruptured rocks as the cause of earthquakes.

Evidence against elastic rebound hypothesis includes observations that fault movements take place *after* earthquakes in some places.

Plastic surge explains many of the features of earthquakes.

Supplementary readings

GUTENBERG, BENO, and CHARLES F. RICHTER: *Seismicity of the Earth and Associated Phenomena,* Hafner Publishing Company, New York, 1965. A facsimile of the 1954 edition of this first and still significant treatment of the subject.

PAGE, ROBERT (ed.): *Proceedings of the Second United States–Japan Conference on Research Related to Earthquake Prediction Problems,* Lamont Geological Observatory, Palisades, N.Y., 1966. Forty-eight brief summaries of studies related to earthquake prediction.

SMITH, STEWART W., and MAX WYSS: "Displacements on the San Andreas Fault Subsequent to the 1966 Parkfield Earthquake," *Bull. Seis. Soc. Am.*, vol. 58, pp. 1955–1973, 1968. Describes pioneering measurements showing that surface displacements along the fault are a *result* of earthquake-causing mechanisms and are not themselves the vibration-causing movements.

WOOD, FERGUS J. (ed.): *The Prince William Sound, Alaska, Earthquake of 1964 and Aftershocks,* vol. I, 1966; vol. II, 1967; vol. III, 1969; U.S. Department of Commerce, Environmental Science Services Administration, Coast and Geodetic Survey, Washington, D.C., Publication 10-3. The definitive reports on this greatest earthquake ever recorded on the North American continent, making it also the most thoroughly studied.

17

Interior of the earth

INFORMATION ABOUT THE EARTH'S INTERIOR COMES FROM STUDYING THE RECORDS of earth waves. In Chapter 16 we learned that some large earthquakes involve violent displacements of the earth's crust. These movements use up some of the energy released at the time of the earthquake. The rest is carried to great distances by earth waves. These may travel through the whole earth and around its entire surface, if the earthquake is large enough. At great distances, sensitive instruments are needed to detect their passage.

By studying the manner in which earth waves travel out from earthquakes, supplemented by the study of waves generated by dynamite and nuclear explosions, geologists have assembled a wealth of information about the structure of the globe from surface to center. But before we look at the results of these studies, let us review some of the methods used to obtain information.

17.1 *Early instrumental observations*[1]

The earliest instrument used to detect an earthquake was built around A.D. 136. Its design is credited to a Chinese philosopher named Chang Hêng. His instrument was said to resemble a wine jar about 2 m in diameter. Evenly spaced around this were eight dragon heads under each of which was a toad with head back and mouth open. In each dragon's mouth was a ball (see Figure 17.1). When there was an earthquake, one of the balls was supposed to fall into a toad's mouth. There is no known record of what was inside the jar. Speculations over the centuries have assumed a pendulum of some sort which would swing as the ground moved, knocking the ball out of a dragon's mouth in line with the swing. Chang Hêng had the notion that if a frog on the south side caught a ball, the earthquake had happened to the north of the instrument. His instrument was intended to record the occurrence of an earthquake and show the direction of its origin from the observer. It was a seismoscope, as it had no provision for a written record of the motion.

In 1703, J. de la Haute Feuille had the same general idea. He proposed a seismoscope design that he felt would respond to the tilting of the earth's surface at the time of an earthquake. His instrument consisted of a bowl of mercury so placed that when the earth's surface tilted mercury would spill over the rim and fall into one of eight channels. Each channel represented a principal direction of the compass and the spilled mercury was to show the direction of the earthquake, the amount of spillage the earthquake's intensity. However, there is no record that this seismoscope was ever built.

The first European to use a mechanical device for studying earthquakes was Nicholas Cirillo. He used his device to observe the motion of the ground from a series of earthquakes in Naples in 1731. His instrument was a simple pendulum whose swing indicated the amplitude of ground motion. He made observations at different distances from earthquakes to see how the motion died out with distance. He found that the amplitude decreased with the inverse square of the distance.

[1] James Dewey and Perry Byerly, "The Early History of Seismometry (to 1900)," *Bull. Seis. Soc. Am.*, vol. 59, pp. 183–227, 1969.

Figure 17.1 Chang Hêng's seismoscope as visualized by Wang Chen-To.

In Italy, in 1783, there occurred a series of earthquakes in Calabria which resulted in 50,000 deaths. These stimulated the development of other mechanical devices for studying such events. It also led to the appointment of the first "earthquake commission." Shortly after the first large quake, a clockmaker and mechanic named D. Domenico Salsano had an instrument operating in Naples approximately 320 km from Calabria. It consisted of a long common pendulum with a brush attached which was supposed to trace out the motion with slow-drying ink on an ivory slab, the first attempt to preserve a written record. His instrument was also equipped with a bell which was to ring when motions were large enough. It has been reported that the bell did ring on several occasions. A similar instrument was reported in use in Cincinnati around the time of the 1811–1812 New Madrid, Missouri, earthquakes.

A series of small earthquakes near Comrie, Scotland, led to the development of an inverted-pendulum seismometer by James Forbes in 1844. His instrument consisted of a vertical metal rod with a movable mass on it. This was supported by a steel wire. The stiffness of the wire and the position of the mass could be adjusted to alter the free period of the pendulum. Forbes was trying to give his pendulum a long period so that it would remain stationary when the ground moved during an earthquake. At one end of the metal rod, he attached a pencil which came into contact with a stationary, paper-lined dome above it, to preserve a written record of the movement (see Figure 17.2). By moving the pencil some distance from the mass, he was able to magnify the motion of the pendulum two or three times. This instrument, a crude seismograph, was not successful because friction between the writing pencil and the recording surface greatly reduced the sensitivity of the instrument and prevented it from recording even local earthquakes that were felt.

Figure 17.2 Forbes' seismometer. The screws, E, acting on the support, D, were to help set the pendulum in an upright position.

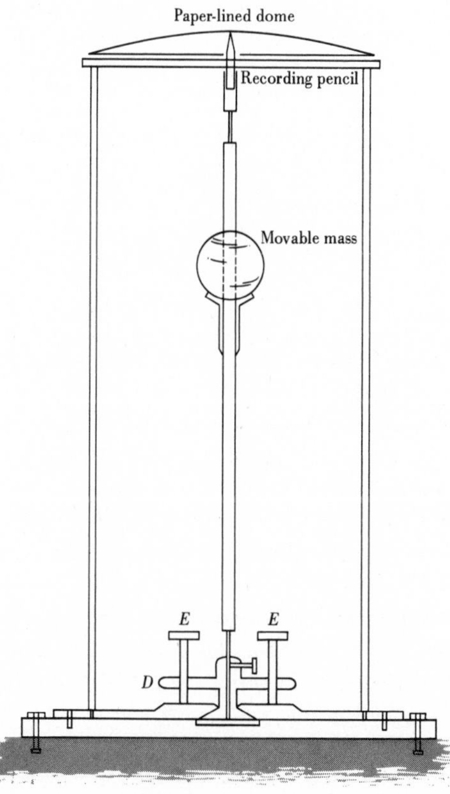

The first record of a distant earthquake was obtained accidentally by Ernst von Rebeur-Paschwitz at Potsdam, Germany, on April 17, 1889.[2] The earthquake was felt in Tokyo about 1 hr before it was recorded in Germany. The motion was too small and was on a time scale that was too compressed to give much information. However, it showed that the energy from an earthquake could travel halfway around the world. And because the record gave the time of day, it indicated how long the waves took to travel that far, how long the motion lasted, and also that the motion of the ground had a horizontal component.

In 1939, at the Harvard Seismograph Station we repeated the von Rebeur-Paschwitz measurement of tidal tilts with a horizontal pendulum. Figure 17.3 shows one of the records made at that time. It covered the interval from December 4 to 16, 1939, and shows how tides in the Gulf of Maine affect the ground 50 km inland at Harvard. It also shows the traces of two earthquakes similar to the record obtained by von Rebeur-Paschwitz. The first earthquake's epicenter was in Mexico at a distance of 3,470 km, and the second was in northern Japan at a distance of 9,760 km.

[2]E. von Rebeur-Paschwitz, "The Earthquake of Tokyo, April 18, 1889," *Nature*, vol. 40, pp. 294–295, 1889.

17.2 *Modern seismology*

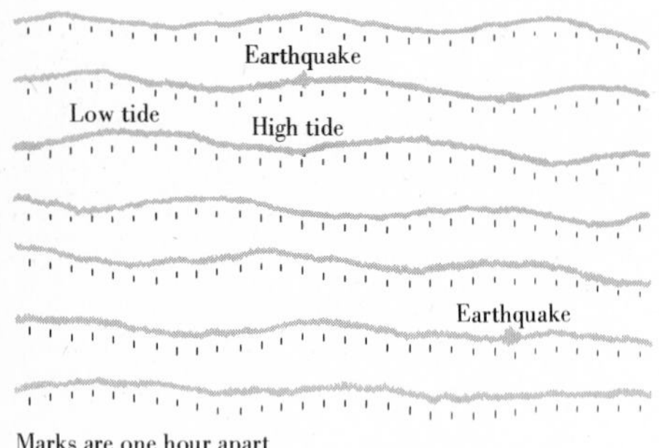

Figure 17.3 **A repetition of von Rebeur-Paschwitz' measurements that led to recording distant earthquakes. A horizontal pendulum record showing tilting of the ground 50 km inland from tides in the Gulf of Maine.** (*Taken at the Harvard Seismograph Station.*)

With the discovery that earth waves could travel to such great distances from an earthquake, modern seismology was born. The first records of earthquake motion registered on the tilt instrument were too fuzzy to be analyzed accurately. They were too small in amplitude and the events of 1 hr were compressed into the space of 6 mm. To obtain meaningful data about earthquakes, it was necessary to develop seismographs with increased recording speed and sensitivity so that the fuzzy lines would be spread out and enlarged.

A group of British professors teaching in Japan in the late nineteenth century initiated the activities that developed into all aspects of seismology as we find it today. Chief among these in his influence on the entire field was John Milne. He was the one mainly responsible for developing a seismograph capable of picking up and recording intelligible signals from distant earthquakes. Also, Milne was responsible for promoting the first worldwide network of seismograph stations for systematic, continuous registration and interpretation of their records. By 1900, Milne's seismographs were operating on all the inhabited continents (see Figure 17.4). Sixteen stations were regularly recording and sending their data to Milne.

Figure 17.4 **The Milne horizontal seismograph. Light from L was reflected by M through the intersection of two crossed slits onto photographic paper. The lower illustration is a top view of the instrument with its outer case removed. T is a flexible wire holding up the boom. The weight, W, was pivoted on the boom.**

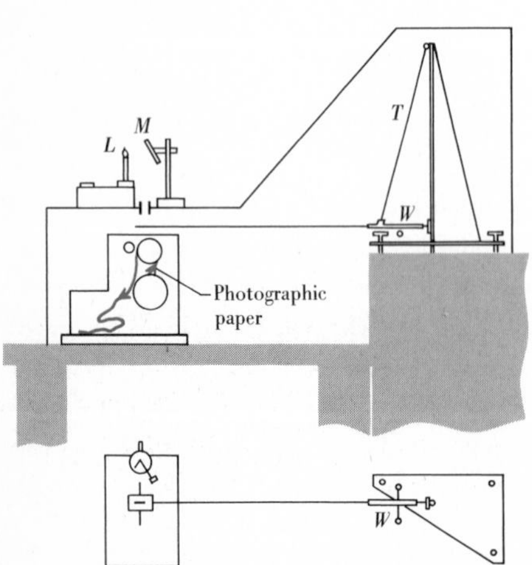

Modern instruments

A modern seismograph consists of three basic parts: an inertia member, a transducer, and a recorder. The inertia member is a weight suspended by a wire, or spring, so that it acts like a pendulum, but so constructed that it can move in only one direction. The inertia member tends to remain at rest as earth waves pass by. It has to be damped so that the mass will not swing freely.

The transducer is a device which picks up the relative motion between the mass and the ground. It converts this into a form which can be recorded. It may be a mechanical lever or an electrodynamic system. In one electrodynamic system, a coil of wire moves back and forth in a magnetic field. This movement creates an electric current which passes through a galvanometer to be recorded on a sheet of paper.

To record motion in all directions, it is necessary to have three seismographs. One records vertical motion and two record horizontal motion in directions at right angles to each other. A well-equipped seismograph station will have a set of three components: vertical, north–south, and east–west.

A mass on a spring will oscillate if displaced and then released. The time required to complete one oscillation is called the *period.* If the ground under this system oscillates with a shorter period, the mass hangs still in space, or nearly so. It then serves as a point of reference from which to measure the earth's motion. The inertia members are built to stand still during the passage

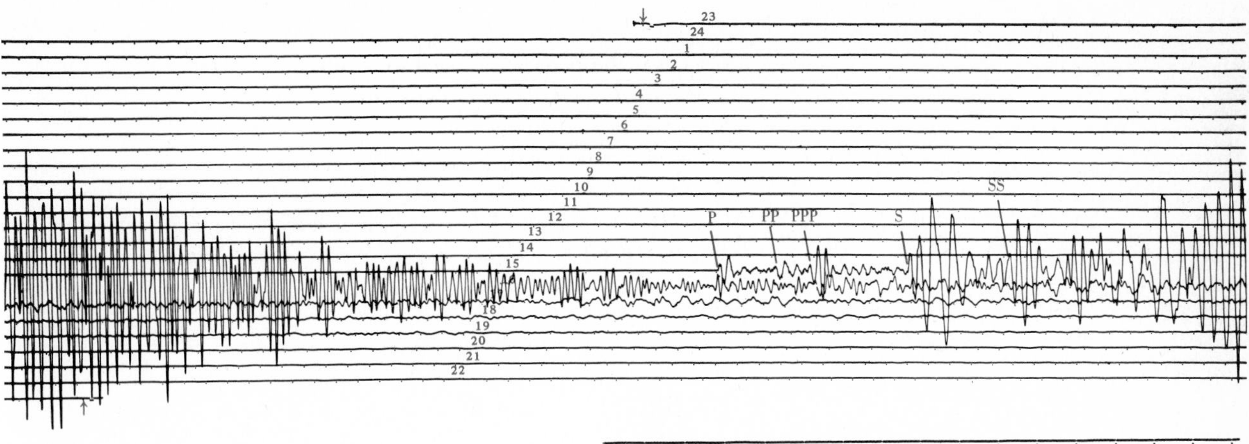

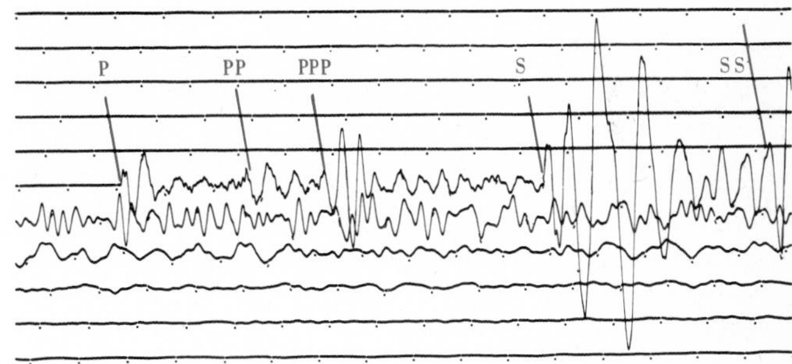

Figure 17.5 **Long-period record from Alert in northern Canada, for May 18 and 19, 1962, with an enlargement of part of it at right. Numbers down the middle of the record indicate hours. The phases PP, PPP, and SS are explained in Figure 17.9.** (*After John H. Hodgson,* Earthquakes and Earth Structure, *Prentice-Hall, Inc., Englewood Cliffs, N.J., 1964.*)

of waves of a certain selected period range. Generally, two sets are used, one for short-period waves of 5 sec or less and one for long-period waves of 5 to 60 sec. Because the motion of the ground at distant stations is microscopic, the motion must be magnified to make a visible record. Therefore, the transducer includes a system to magnify the ground motion.

A standard recorder is a cylindrical drum on which is wrapped a sheet of recording paper. It rotates at a constant speed and by a helical drive moves sidewise as it turns. This produces a continuous record which is a series of parallel lines. These lines appear straight when there is no ground motion, but if there is motion the lines move in response. A clock-controlled device jogs the recording line briefly once per minute and for a longer time each hour. These jogs on the record identify the time of day. Each sheet of paper is good for 24 hr of continuous recording (see Figure 17.5).

17.3 Earth waves

When earth materials rupture and cause an earthquake, some of the energy released travels away by means of earth waves. The manner in which earth waves transmit energy can be illustrated by the behavior of waves on the surface of water.

A pebble dropped into a quiet pool creates ripples that travel outward over the water's surface in concentric circles. These ripples carry away part of the energy that the pebble possessed as it struck the water. A listening device at some distant point beneath the surface can detect the noise of impact. The noise is transmitted through the body of the water by sound waves, far different from surface waves and not visible by ordinary means.

Just as with water-borne waves, there are two general classes of earth waves: (1) *body waves,* which travel through the interior of the mass in which they are generated, and (2) *surface waves,* which travel only through the surface.

Body waves

Body waves are of two general types: push–pull and shake. Each is defined by its manner of moving particles as it travels along.

Push–pull waves, more commonly known as sound waves, *can travel through any material—solid, liquid, or gas.* They move the particles forward and backward; consequently, the materials in the path of these waves are alternately compressed and rarefied. For example, when we strike a tuning fork sharply, the prongs vibrate back and forth, first pushing and then pulling the molecules of air with which they come in contact. Each molecule bumps the next one, and a wave of pressure is set in motion through the air. If the molecules next to your eardrum are compressed at the rate of 400 times/sec, you hear a tone that is called middle A.

Shake waves can travel only through materials that resist a change in shape. These waves shake the particles in their path at right angles to the direction of their advance. Imagine that you are holding one end of a rope fastened to a wall. If you more your hand up and down regularly, a series of waves will travel along the rope to the wall. As each wave moves along, the particles in the rope move up and down, just as the particles in your hand did. In other words, the particles move at *right angles* to the direction of the wave's advance. The same is true when you move your hand from side to side instead of up and down.

Surface waves

Surface waves can travel along the surface of any material. Let us look again at the manner in which waves transmit energy along the surface of water. If you stand on the shore and throw a pebble into a quiet pool, setting up surface waves, some of the water seems to be moving toward you. Actually, though, what is coming toward you is *energy* in the form of waves. The particles of water move in a definite pattern as each wave advances: up, forward, down, and back, in a small circle. We can observe this pattern by dropping a small cork into the path of the waves (see Figure 17.6).

When surface waves are generated in rock, one common type of particle motion is just the reverse of the water-particle motion—that is, forward, up, back, and down.

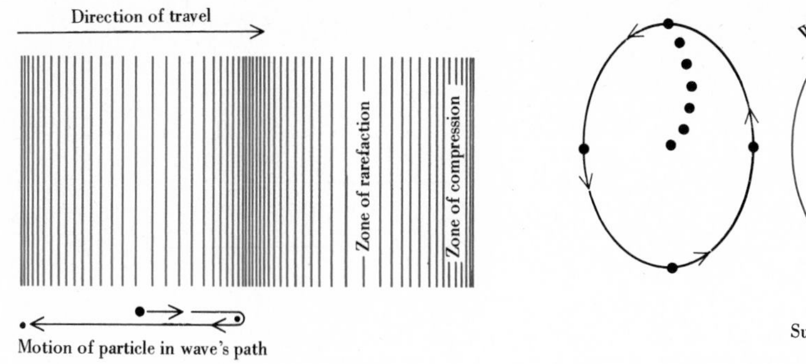

Push-pull (compressional) wave

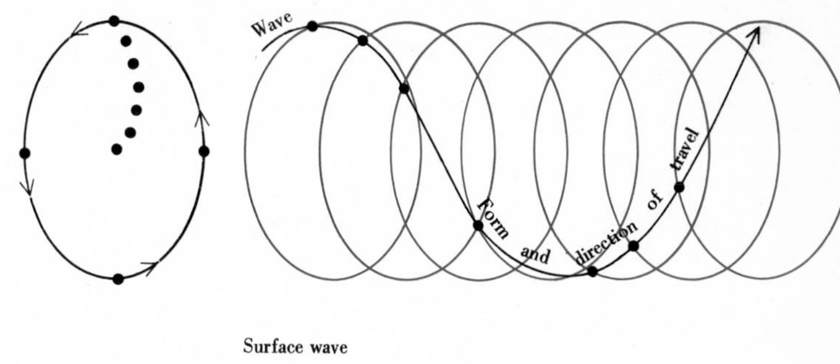

Surface wave

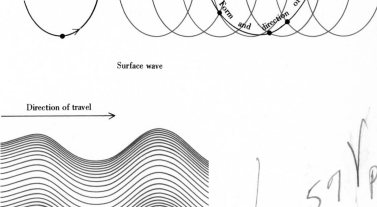

Direction of travel

Shake (shear) wave

Figure 17.6 ***Motion produced by three earthquake wave types.***

17.4 *Records of earthquake waves*

The first records of earthquake motion registered on a tilt instrument were too fuzzy to be analyzed accurately because the events of 1 hr were compressed into the space of 6 mm. But when the drum speed was increased to spread the events of 1 hr over a space of 1 m, a sharp pattern emerged.

This pattern consisted of three sets of earth waves. The first waves to arrive at the recording station were named primary; the second to arrive were named secondary; and the last to arrive were named large waves. The abbreviations *P*, *S*, and *L* are commonly used for these three types. Closer study revealed that the *P* waves are push–pull waves and travel with a speed determined by the bulk modulus and rigidity of the material, as defined in Section 18.2, as well as the density of the material. It also revealed that the *S* waves are shake waves and travel with a speed determined by the rigidity and the density of the material.

When V is the velocity of P waves, B the bulk modulus, G the rigidity, and d the density, then

$$V^2 = \frac{B + \frac{4}{3}G}{d}$$

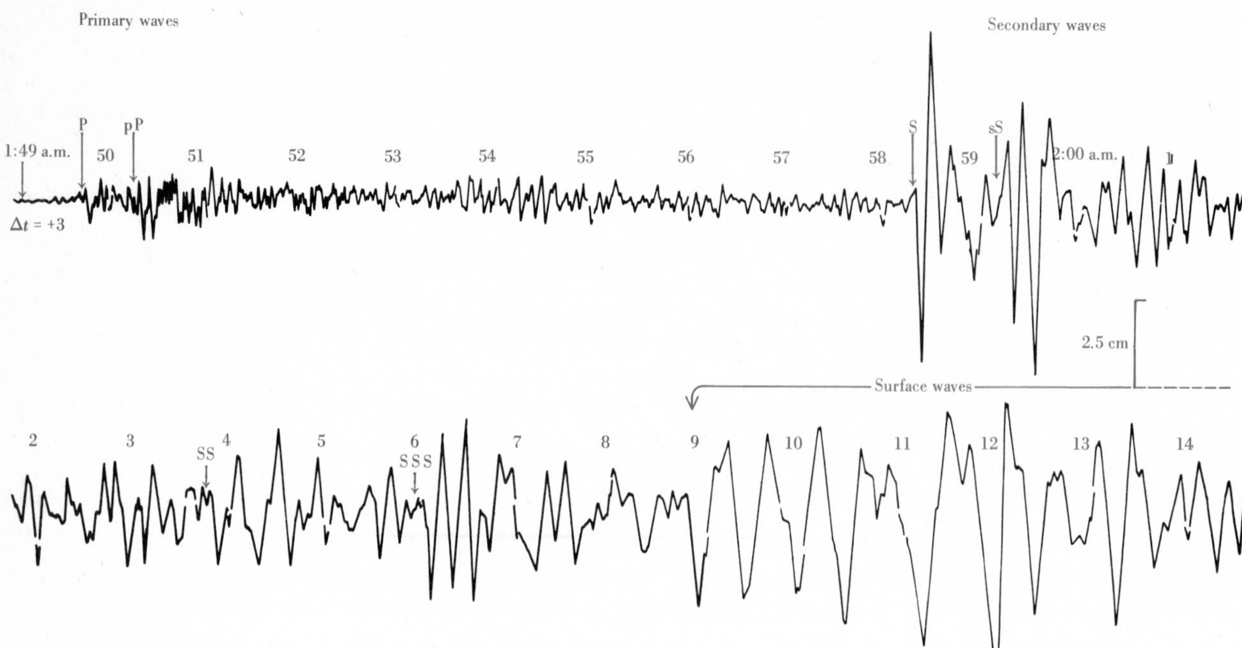

Figure 17.7 ***A record of an earthquake on a seismogram at Harvard, Mass., on which 1 min is spread over more space than was given to 3 hr in Figure 17.3. All waves started in Rumania at the same instant. They arrived as indicated above because of different speeds and paths. The distance was 7,445 km.***

And when v is the velocity of S waves, G the rigidity, and d the density, then

$$v^2 = \frac{G}{d}$$

The P and S waves travel from the focus of an earthquake through the interior of the earth to the recording station. The L waves are surface waves that travel from the area directly above the focus, along the surface to the recording station.

Table 17.1 ***Sample timetable for P and S***

Distance from source, km	*Travel time*				*Interval between P and S (S − P)*	
	For P		*For S*			
	Min	*Sec*	*Min*	*Sec*	*Min*	*Sec*
2,000	4	06	7	25	3	19
4,000	6	58	12	36	5	38
6,000	9	21	16	56	7	35
8,000	11	23	20	45	9	22
10,000	12	57	23	56	10	59
11,000	13	39	25	18	11	39

The P waves arrive at a station before the S waves because, although they follow the same general paths of travel, they go at different speeds. The push–pull mechanism by which P waves travel generates more rapid speed than does the shake mechanism of the S waves. The S waves travel at about two-fifths the speed of P waves in any given earth material. The L waves are last to arrive, because they travel at slower speeds and over longer routes (see Figure 17.7).

Time–distance graphs

As data accumulated, Milne began plotting the arrival time of the first P, S, and L waves from felt earthquakes. He set up time–distance graphs for each wave. He found that all three waves took longer and longer to travel to greater and greater distances. He also found that the time between the arrival of each wave type increased as the distance from the source increased.

TRAVEL TIMES If earth waves traveled through a material of uniform elasticity, their paths would be straight lines as they went at constant speed to greater and greater distances. However, if there is a progressive increase in elasticity with depth, they will gradually increase their speed and their paths will become smooth curves. When there is a sharp increase in elasticity, the wave directions are changed and this is reflected by breaks in the line on a time–distance graph.

Distance is sometimes expressed as the number of kilometers for the length of a great circle arc between two surface points, or as the number of degrees for the angle at the earth's center subtended by that arc. For example, one-quarter of the way around the earth is 10,000 km, or 90°.

LESS THAN *11,000* KM From thousands of measurements the world over, it has been learned that P, S, and L waves have regular travel schedules for distances up to 11,000 km. From an earthquake in San Francisco, for example, we can predict that P will reach El Paso, 1,600 km away, in 3 min 22 sec and S in 6 min 3 sec; P will reach Indianapolis, 3,220 km away, in 5 min 56 sec and S in 10 min 48 sec; P will reach Costa Rica, 4,800 km away, in 8 min 1 sec and S in 14 min 28 sec.

The travel schedules move along systematically out to a distance of 11,000 km, as shown in Table 17.1.

BEYOND *11,000* KM Beyond 11,000 km, however, something happens to the schedule, and the P waves are delayed. By 16,000 km, they are 3 min late. When we consider that up to 11,000 km we could predict their arrival within seconds, a 3-min delay becomes significant.

The fate of the S waves is even more spectacular: They disappear altogether, never to be heard from again.

When the strange case of the late P waves and the missing S waves was first recognized, seismologists became excited, for now they realized that they were not just recording earthquakes but were developing a picture of the interior of the earth.

Locating earthquakes

We now have timetables for earth waves for all possible distances from an earthquake. These are represented in the graphs of Figures 17.8 and 17.9. Data of this sort are the essential tools of the seismologist.

When the records of a station give clear evidence of the *P*, *S*, and *L* waves from an earthquake, the observer first determines the intervals between them, as shown in Table 17.2. Next, he plots these times on the edge of a strip of paper, using the graph's time scale: (1) at zero to represent *P*, (2) at 2 min 38 sec, (3) at 4 min 16 sec, (4) at 8 min 54 sec, and (5) at 13 min 34 sec. He then tries these five marks on different parts of the graph, always

Figure 17.8 **Travel-time curves for earth waves.** (*After Harold Jeffreys and K. E. Bullen.*)

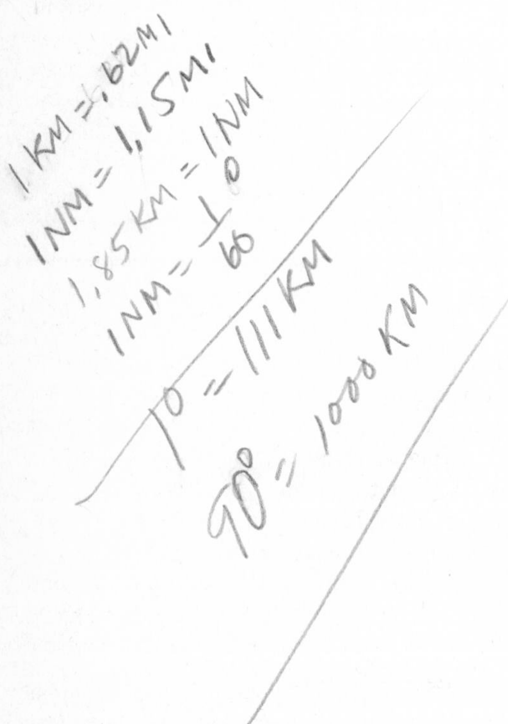

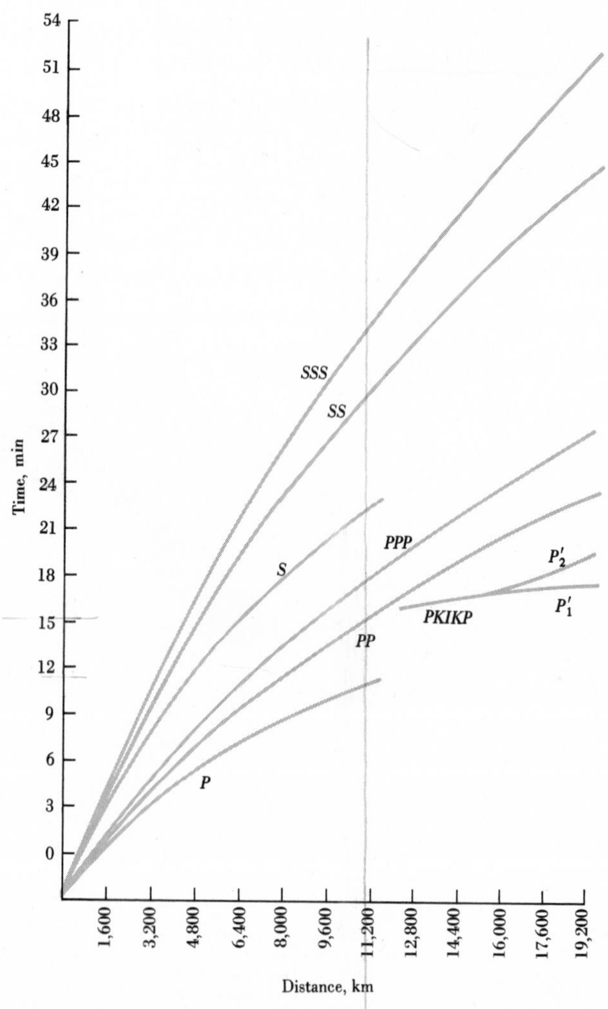

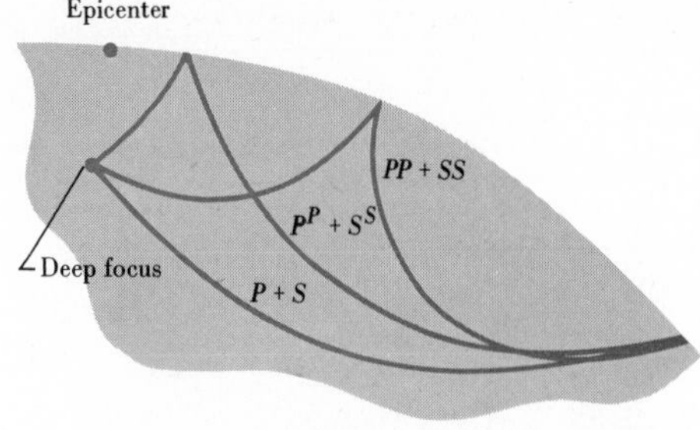

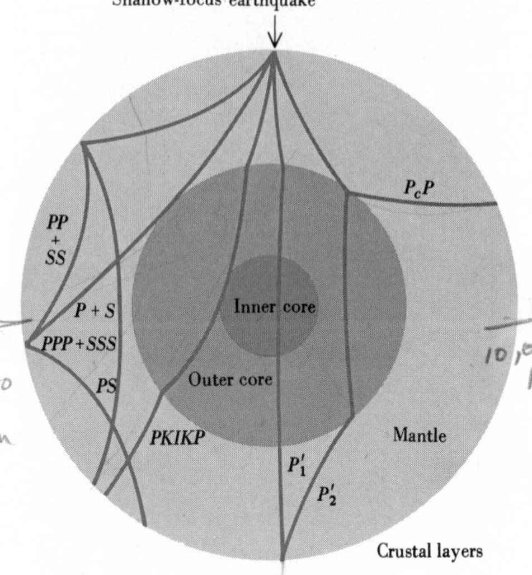

Figure 17.9 The paths of some of the more common earthquake waves.

Table 17.2

	Time			*Interval, time after P*	
	Hr	*Min*	*Sec*	*Min*	*Sec*
P	15	09	06		
	15	11	44	2	38
	15	13	22	4	16
	15	18	00	8	54
	15	22	40	13	34

keeping the first mark on the *P* line. At one place, and only one, will the marks each fall on a line of the graph. This will happen at the place which corresponds to the quake's distance, in this case 7,440 km, or 67°. The graph also shows that *P* travels 7,440 km in 10 min 52 sec. He subtracts this travel time from the arrival time of *P* and has the time of the earthquake at its source:

P arrived	15 hr 09 min 06 sec
P had traveled	10 min 52 sec
P started at	14 hr 58 min 14 sec

This process is carried out for several seismograph stations that have recorded the quake. Then an arc is drawn on a globe to represent the computed distance from each station. The point where the arcs all intersect indicates the center of the disturbance (see Figure 17.10).

This example was based on a record taken at the station in Alert, Canada,[3] Palisades, New York, recorded it at 3,580 km and Pasadena, California, at 2,675 km. The U.S. Coast and Geodetic Survey located it in south-central Mexico, where they reported extensive property damage, 3 deaths, and 16 injured.

Although this whole procedure is essentially very simple, some have found its accuracy hard to believe. On December 16, 1920, seismologists all over the world found the record of an exceptionally severe earthquake on their seismographs. Each of them computed the distance of the quake and sent the information along to the central bureaus where the various reports were assembled. The next day, the location of the earthquake was announced to the press, unlike many lesser shocks that fail to make the news. The announcement stated that a very severe earthquake had occurred at 5 min 43 sec after 12:00, Greenwich time, December 16, 1920, in the vicinity of 35.6°N, 105.7°E. That placed it in the province of Kansu, China, about 1,600 km inland from Shanghai, on the border of Tibet. This area is densely populated but quite isolated. No reports of damage came in, however, and the matter was soon forgotten by the general public. But it was not forgotten by the members of the press, who were sure they had been misinformed. Then, 3 months later, a survivor staggered into the range of modern communications with a story of a catastrophe in Kansu on the day and at the time announced, a catastrophe that had killed an estimated 100,000 persons and had created untold havoc by causing the great landslides described in Chapter 16.

In contrast, however, on another occasion the news of an earthquake traveled faster than the waves themselves. On August 20, 1937, at 6:59 P.M., an earthquake occurred at Manila, in the Philippines, and the news story was transmitted with unusual promptness. It was flashed to North America and found its way to the Boston office of a news agency. An operator there picked up a phone and called the Harvard Seismograph Station at Harvard, Massachusetts, 13,440 km from Manila, 1 hr after the earthquake happened, to inquire whether the disturbance had been recorded. The conversation took place 10 min before the earthquake's surface waves reached the station.

[3]John Hodgson, *Earthquakes and Earth Structure*, Prentice-Hall, Inc., Englewood Cliffs, N.J., 1964.

Figure 17.10 ***Seismologists at the U.S. Coast and Geodetic Survey's National Earthquake Information Center locating an earthquake by means of a large globe.***

17.5 *Structure of the earth's interior*

Studies of the travel habits of waves through the earth and of surface waves around the earth have given us information about the structure of the globe from its surface to its center. These studies have been made possible by our knowledge of the speed of these earth waves and of their behavior in different materials. For example, waves travel at greater speeds through simatic materials than through sialic materials.

When earth waves move from one kind of material to another, they are refracted and reflected (see Figure 17.11). These waves have revealed several places within the earth where there is a change in material's physical properties. These could be due to changes in composition, atomic structure, or atomic state. The boundary where such a change takes place is called a *discontinuity.*

For body waves to reach greater and greater distances on the surface, they must penetrate deeper and deeper into the earth's interior. Thus in traveling from an earthquake in San Francisco to a station at Dallas, a surface distance of 2,400 km, the body waves penetrate to 480 km below the surface. This holds true for any other 2,400-km surface distance. To reach a station 11,200 km away, the body waves dip into the interior to a maximum depth of 2,900 km and bring out information from that depth.

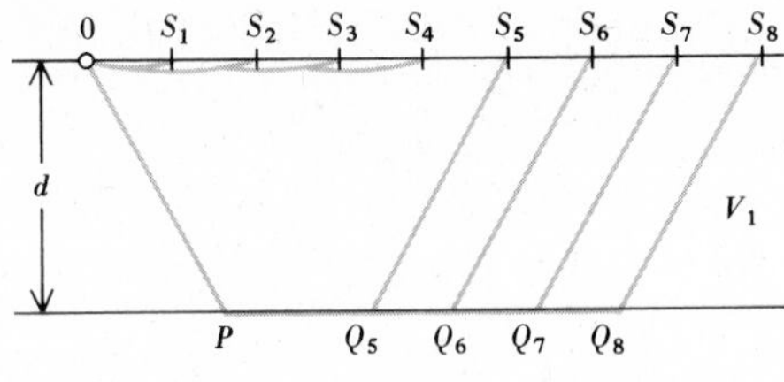

Figure 17.11 **Paths of refracted waves used to determine the thickness of a single layer. If velocity in V_2 material at depth d is greater than V_1, waves from O arrive before any others at stations S_1, S_2, S_3, S_4 by way of surface paths. Waves from O through P to Q_5S_5, Q_6S_6, Q_7S_7, and Q_8S_8 take the lead after S_5, and are first to arrive. By plotting time-distance relationships here, seismologists measure V_1, V_2, and d.**

On the basis of data assembled from studies of the travel habits of earth waves, the earth has been divided into three major zones: *crust, mantle,* and *core.*

The earth's crust

Information on the earth's crust comes primarily from observations of the velocities of *P* and *S* waves from local earthquakes (within 1,000 km), dynamite, and nuclear blasts. One of the first things revealed is that the earth's crust is solid rock. Early in the history of crustal studies, a seismologist in Yugoslavia, A. Mohorovičić, made a study of records of the earth waves from an earthquake on October 8, 1909, in the Kulpa Valley, Croatia. He observed two *P* waves and two *S* waves and concluded that the first *P* and *S* had encountered something that caused some of their energy to be reflected back to the surface. He also concluded that velocities of *P* and *S* waves increased abruptly below a depth of about 50 km. This abrupt change in the speed of *P* and *S* waves indicated a change in material and became known as the *Mohorovičić discontinuity*. For convenience, it is now referred to as the *Moho*. The Moho marks the bottom of the earth's crust and separates it from the mantle.

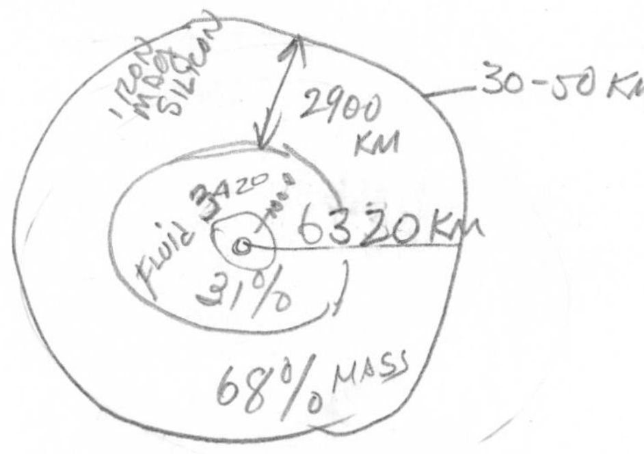

THE CRUST OF THE CONTINENTS The depth to the Moho varies in different parts of the continents. However, it is generally greater under mountain regions than under lowlands. It is also not uniform in composition. These relationships suggest that isostatic compensation may be accomplished by a little of both the Pratt and the Airy hypotheses (see Section 19.2).

In the United States, the continental crust has an average thickness of 33 km but varies between 20 and 60 km (see Figure 17.12). Under the intermountain plateaus and the Great Basin Ranges, it averages less than 30 km; the Great Valley of California, 20 km; under the Sierra Nevada, more than

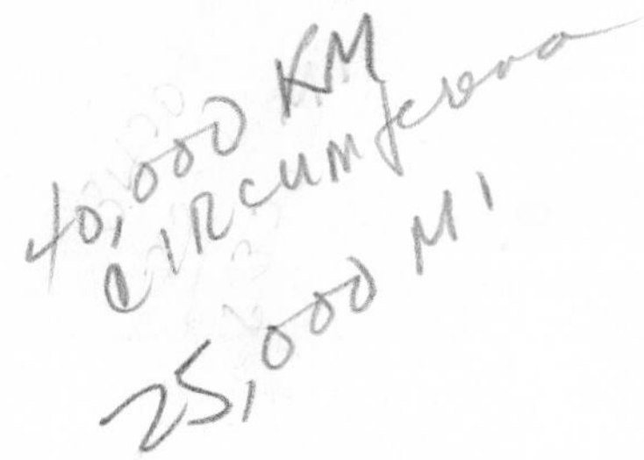

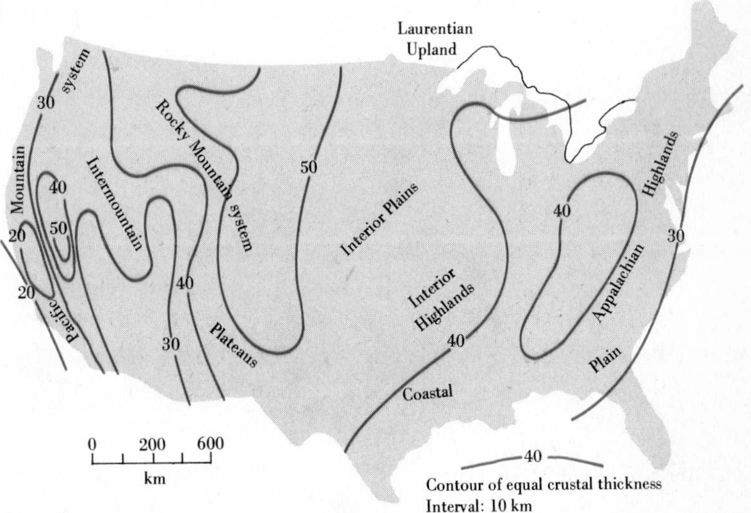

Figure 17.12 **Crustal thickness determined from seismic data.** (*After L. C. Pakiser and I. Zietz.*)

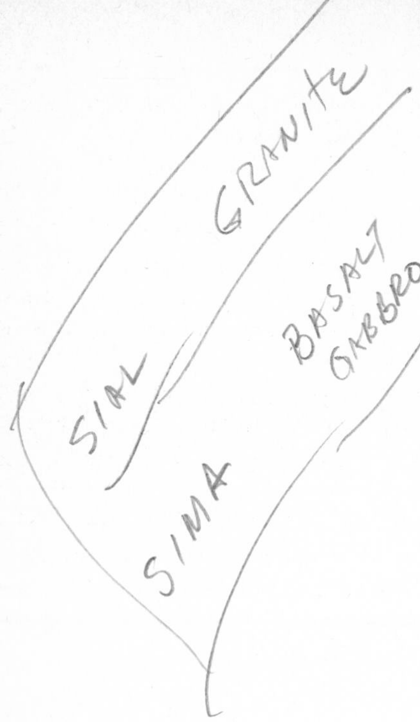

50 km. It is thickest under the eastern front of the Rocky Mountain ranges. The crust under New England is about 36 km thick.

It has been difficult to get precise data on the earth's crust from the waves of earthquakes. Waves from dynamite blasts, however, with precisely known locations on the surface, and times of detonation, have filled in some of the details. In 1941, the Harvard Seismograph Station determined the structure of the continental crust in New England. Analysis of many blast records revealed that in New England the continental crust has three layers, each one with different elastic properties, indicating different rock types. The layered structure of the crust is fairly representative of other sections of the continents. Table 17.3 summarizes the data.

The average velocity of waves traveling through the upper crustal layer is similar to the velocity expected for granite, granodiorite, or gneiss. Because these rocks are rich in silica and aluminum, we say the material of the upper crust is sialic in composition; the term *sial, si* for silicon and *al* for aluminum, is generally used in speaking of this layer which is found only on the continental areas of the earth. The second and third layers are more and more basic in composition. The third is believed to be basalt. These darker, heavier rocks are sometimes designated collectively as *sima,* a name coined from *si* for silicon and *ma* for magnesium. A simatic layer encircles the earth, underlying the sial of the continents and is believed to be the outermost rock layer under deep permanent ocean basins. The velocities of waves traveling through the crust indicate that layering and composition are similar for all the continents.

THE CRUST UNDER OCEANS Our knowledge of the structure of the crust beneath the oceans is based on observations of rocks exposed on volcanic islands and on studies of the velocities of *L* waves from earthquakes, supplemented by dynamite-wave profiles.

The types of rock found on islands help to determine the edge of the Pacific basin. The *andesite line* (see Figure 17.13) has on its ocean side rocks composed primarily of basalt, whereas on the other side they are principally andesite. This has been viewed as the dividing line between oceanic and continental crusts.

On the basis of seismic wave velocities, it appears that the crust under the

Table 17.3 Earth's crust under New England[a]

Thickness, km		Velocity, km/sec P	S	Rock type
16	Layer 1	6.1	3.5	Sialic
13	Layer 2	6.8	3.9	Intermediate
7	Layer 3	7.2	4.3	Simatic
	Moho			
	Top of mantle	8.4	4.6	

[a] L. D. Leet, "Trial Travel Times for Northeastern America," *Bull. Seis. Soc. Am.*, vol. 31, pp. 325–334, 1941.

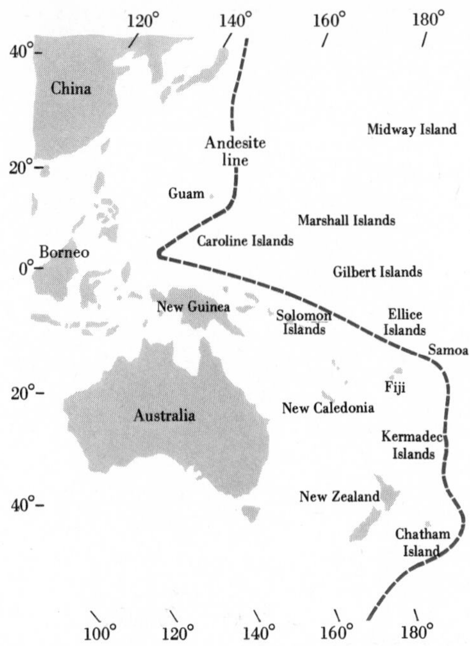

Figure 17.13 **Position of the andesite line in the southwestern Pacific Ocean. This line marks the border of the Pacific basin in a geologic sense. On the Pacific side of the line, young eruptive rocks are basaltic; on the other side, they are principally andesitic. Islands east of the line are isolated or grouped volcanic peaks; west of the line they have the characteristic structure of folded continental mountain ranges.** *(After R. A. Daly.)*

Pacific basin is not layered and is appreciably thinner than the crust of the continents. Its thickness also varies but averages about 5 km. It is thinner under the Gulf of California and in the northeast Pacific.

The crust under the Pacific Ocean is composed of simatic rocks. The composition of the crust under the Atlantic and Indian Oceans is still a subject of debate, mainly because we lack reliable information on the speeds of waves traveling through it. Simatic rocks underlie at least parts of the Indian and Atlantic Oceans and some investigators are convinced that the sialic layer is missing.

The crust under Hawaii The crust under the island of Hawaii was surveyed by seismic refraction methods in 1964.[4] It was concluded that it is 10 km thick under the summit of Kilauea and 12 to 15 km under other parts of the island.

The crust under Hawaii appears to be divided into two layers. The upper layer, 4 to 8 km thick, is probably an accumulation of lava flows that form the bulk of the island. The lower layer, also 4 to 8 km thick, is probably the original basaltic crust modified by a complex intrusive system associated with the central vents and rift zones of the island.

The mantle

Below the earth's crust is a second major zone, the *mantle,* which extends to a depth of approximately 2,900 km into the interior of the earth. Our knowledge of the mantle is based in part on evidence supplied by the behavior of *P* and *S* waves recorded between 1,100 and 11,000 km.

At the Moho, the speeds of *P* and *S* waves increase sharply, an indication that the composition of the material suddenly changes. We have no direct evidence of the new material's nature, but the change in speed suggests that it may contain more ferromagnesian minerals than the crust. For more than half a century, the majority of scientists have accepted the idea that the mantle is solid, because it is capable of transmitting *S* waves. In an attempt to explain mountain-building processes and the tendency of the earth's crust toward isostasy, some observers have emphasized that the mantle material undergoes slow flow as it adapts to changing conditions on the surface. Some have suggested that at least the upper portion of the mantle may consist of elements arranged in a random pattern, in other words, not crystalline. They feel that a disorderly atomic arrangement might permit the material of the mantle to flow more readily than it would if it were crystalline.

There are also variations in the composition of the upper mantle. We know this from variations in the velocity of P_n, the wave that travels through the upper portion of the mantle. Speeds of P_n vary from 7.5 to 8.5 km/sec. This difference may be due also in part to variation in elastic properties of the mantle material.

There is a worldwide discontinuity in the mantle at a depth of 550 to 600 km, where the velocities of *P* and *S* waves increase sharply to create the *20°*

[4]David P. Hill, "Crustal Structure of the Island of Hawaii from Seismic-Refraction Measurements," *Bull. Seis. Soc. Am.*, vol. 59, pp. 101–130, 1969.

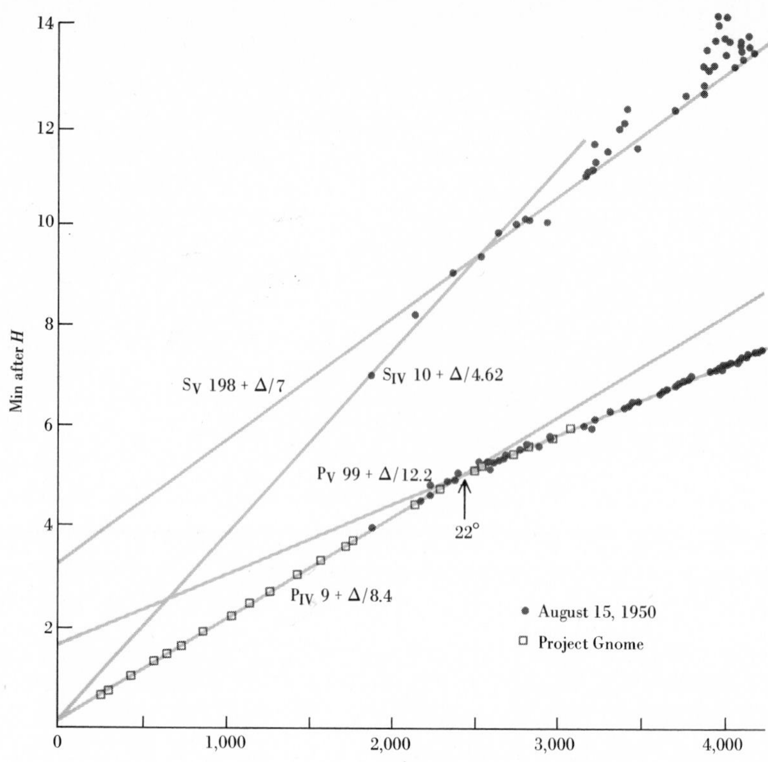

Figure 17.14 ***Time–distance graph from measurements of a nuclear test, Project Gnome in New Mexico, and a large earthquake, August 15, 1950, in Asia. This indicates that below a depth of 550 km the velocity of P waves jumps from 8.4 to 12.2 km/sec.*** (*Interpretation by L. Don Leet.*)

discontinuity (see Figure 17.14), so-called because it becomes apparent on earth-wave records at stations 20° (2,200 km) from an earthquake focus.

What produces this change in the mantle? It may be a rearrangement of atoms under pressure or a change in the kinds of atoms present. We do not know exactly, yet. Whatever the new material or state of matter that produces the change, it seems to be substantially uniform from 600 to 2,900 km, the inner limit of the mantle.

The core

We come now to the core, a zone that extends from the 2,900-km inner limit of the mantle to the center of the earth at a depth of 6,320 km. An analysis of seismographic records from earthquakes 11,200 km or more distant reveals that the core has two parts, an outer zone 2,200 km thick and an inner zone with a radius of 1,260 km.

In traveling between two points on the surface 11,200 km apart, we know that *P* and *S* waves penetrate 2,900 km into the interior. But after they go deeper than that, they enter a material that delays *P* and eliminates *S* altogether. There have been various suggestions made to explain these observations. The most widely favored for half a century postulated that the outer core is liquid. By definition, *S* waves are capable of traveling only through substances that possess rigidity. Because rigidity is a property of the solid state, it was assumed for years that the core, because it did not transmit *S* waves, was in a liquid state. Actually, the measurements indicate only that it is not solid in the usual sense of the term. The temperature and pressure at that depth are impossible for us to imagine, so we should not try to postulate its composition or state from what we are familiar with here at the surface. Radical atomic changes may account for this discontinuity.

C. G. Knott, a noted British scientist,[5] pointed out that under the tremendous pressures of the interior it is unrealistic to suggest that matter is in a liquid state. He said, "it is not easy to find an explanation in terms of any ordinary theory of the constitution of matter," and he "purposely refrained from speaking of this central core as being liquid, since that word connotes properties which may not be possessed by the material at the earth's core."

Measurements along with current knowledge are presently inadequate to produce a solution. Just what the state or material of the core may be, we do not know.

As the earth swings around the sun, it behaves like a sphere with a specific gravity of 5.5, but geologists have found that the average specific gravity of rocks exposed at the surface is less than 3.0. And even if rocks with this same specific gravity were squeezed under 2,900 km of similar rocks, their specific gravity would increase to only 5.7. Geophysicists have computed that the specific gravity of the core must be about 15.0 in order to give the whole globe an average of 5.5. To meet this requirement, it has been suggested that the core may be composed primarily of iron, possibly mixed with about 8 percent of nickel and some cobalt, in the same proportions that have been found in metallic meteorites.

Outline

Interior of the earth is known to us from studying the records of earth waves.

Early instrumental observations were mostly by seismoscopes, but the first record of a distant earthquake was made in 1889.

Modern seismology had its beginnings with a group of British professors teaching in Japan in the late nineteenth century.

- *Modern instruments* consist of an inertia member, a transducer, and a recorder.

Earth waves carry away some of the energy released when earth materials rupture.

- *Body waves* travel through the interior of the earth.
- *Surface waves* travel along the surface.

Records of earthquake waves are characterized by three sets of earth waves, *P*, *S*, and *L*.

- *Time–distance graphs* are the basis for analyzing the history of earth waves.
  - *Travel times* are governed by the materials through which waves have traveled.
  - *Travel schedules less than 11,000 km* move along fairly systematically.

[5]C. G. Knott, "The Propagation of Earthquake Waves Through the Earth, and Connected Problems," *Proc. Roy. Soc. Edinburgh*, vol. 39, part II, no. 14, pp. 157–197, 1919.

Beyond 11,000 km, P is delayed and *S* disappears.
Locating earthquakes requires data from several stations.
Structure of the earth's interior is divided into crust, mantle, and core.
The earth's crust is separated from the mantle at a discontinuity called the Moho.
The crust of the continents in the United States varies between 20 and 60 km in thickness.
The crust under oceans averages about 5 km.
The crust under Hawaii is 10 km thick under Kilauea and 12 to 15 km under other parts of the island.
The mantle extends to 2,900 km in depth.
The core was once thought to have a liquid outer zone and a solid center.

Supplementary readings

CLARK, S. P.: *Structure of the Earth,* Prentice-Hall, Inc., Englewood Cliffs, N.J., 1970. An excellent and up-to-date treatment of the subject.

DEWEY, JAMES, and PERRY BYERLY: "The Early History of Seismometry," *Bull. Seis. Soc. Am.,* vol. 59, pp. 183–227, 1969. Describes the first instruments tried for recording ground motion from earthquakes, the first successful record of a distant earthquake, and early research that was the forerunner of all aspects of modern seismology.

HART, PEMBROKE J. (ed.): "The Earth's Crust and Upper Mantle," Geophysical Monograph 13, American Geophysical Union, Washington, D.C., 1969. A collection of papers for the more advanced student.

HODGSON, JOHN H.: *Earthquakes and Earth Structure,* Prentice-Hall, Inc., Englewood Cliffs, N.J., 1964. A good account written for the nonspecialist.

LEET, L. DON, and FLORENCE J. LEET: *Earthquake—Discoveries in Seismology,* Dell Publishing Company, New York, 1964. A layman's account of a number of topics in seismology.

18

Deformation of the earth's crust

THE CRUST OF THE EARTH IS CONTINUOUSLY CHANGING. WE HAVE ALREADY discussed some of the changes that are brought about by erosion and deposition, but the crust is being changed in other, more fundamental ways. It is being deformed by forces acting within it (see Figure 18.1). These help to maintain the surface of the land above sea level and work to offset the destructive effects of erosion.

18.1 *Evidence of deformation of the earth's crust*

Evidence of deformation can be seen everywhere. Sediments that were deposited on the bottom of the sea are now found hardened into rocks in mountainous areas high above sea level. They contain fossils that testify to their sedimentary origin. Far below sea level, miners often find pieces of trees or other plants embedded in layers of coal. This shows that these beds, now deeply buried, were at one time at the earth's surface.

Deforming forces elevate and depress large landmasses. They also work to distort the earth's crust. Rocks which are rigid at the surface become plastic at depth and respond to deforming forces by folding (see Figure 18.2). These shapes could not have been produced in a rigid rock.

Abrupt movements of the earth's crust

Crustal deformation has occurred at the time of some large earthquakes. This has resulted in measurable displacements during historic time. Surveys reveal that during some earthquakes large portions of the earth's crust have been warped, tilted, or moved horizontally. Earthquakes have also altered the earth's surface by triggering landslides.

Three major earthquakes in 1811 and 1812 were centered near New Madrid, Missouri, on December 16, January 23, and February 7. These were accompanied by changes in topography over an area of 8,000 to 13,000 km.^{2} The crust sank from 1.5 to 6 m in places, forming swamps and lakes and drowning forests. The largest lake to form was Reelfoot Lake in Tennessee. In 1891, in the provinces of Mino and Owari, Japan, a great chunk of the earth's crust moved upward a maximum of 3 m and sideward 6 m. After a series of large earthquakes in 1899 at Yakutat Bay, Alaska, a geological survey party found beaches raised by as much as 16 m. Subsidence was also associated with these same earthquakes in such a way as to indicate a tilting of a large section of the earth's crust.

During the summer of 1954, a topographic survey was made of Fairview and Dixie Valleys in Nevada. Then on December 16 of the same year there were two earthquakes centered in that area. A survey run again in 1955 showed the valleys twisted from 2 to 3 m out of their former positions, and the crust was warped for kilometers around.

The Prince William Sound earthquake of March 27, 1964, was accompanied in Alaska by considerable vertical and horizontal displacements. These oc-

Figure 18.1 **Rocks tilted almost vertically: quartzite in the Great Smoky Mountains.** (*Photograph by L. B. Gillett.*)

Figure 18.2 **Photograph of a thin section of deformed rock 50 cm across from the Auburn Mine near Virginia, Minnesota. The miniature forms seen here mimic many of the structural features produced in rock layers during the process of mountain building.** (*Photograph by R. C. Gutschick.*)

curred over more than 120,000 km.^{2} The U.S. Coast and Geodetic Survey found that there was a general south to north rotation of a portion of the earth's crust centered at the northern coast of Prince William Sound. A maximum uplift of 13 m occurred off the southwestern tip of Montague Island. The largest subsidence of 2 m occurred 43 km north of Glennallen. Maximum horizontal movements were found between Montague and Latouche Islands. These were 4.5 to 6 m closer together after the earthquake than before. Surveys showed that the southwest side of Montague Island moved northwest with respect to Latouche Island. Figure 18.3 shows horizontal movements that occurred and the areas involved. One of the largest changes took place at Valdez when the earthquake triggered a landslide which deepened the harbor as much as 100 m in one place.

VERTICAL DISPLACEMENTS Vertical displacements during one earthquake cannot account for the amount of elevation of the earth's crust associated with mountains or for the large depressions of sedimentary layers found in deep coal mines. However, if displacements occur often enough throughout geologic time,

Figure 18.3 ***Relative horizontal movement of earth's crust near Prince William Sound as determined by 1964 surveys of U.S. Coast and Geodetic Survey.***

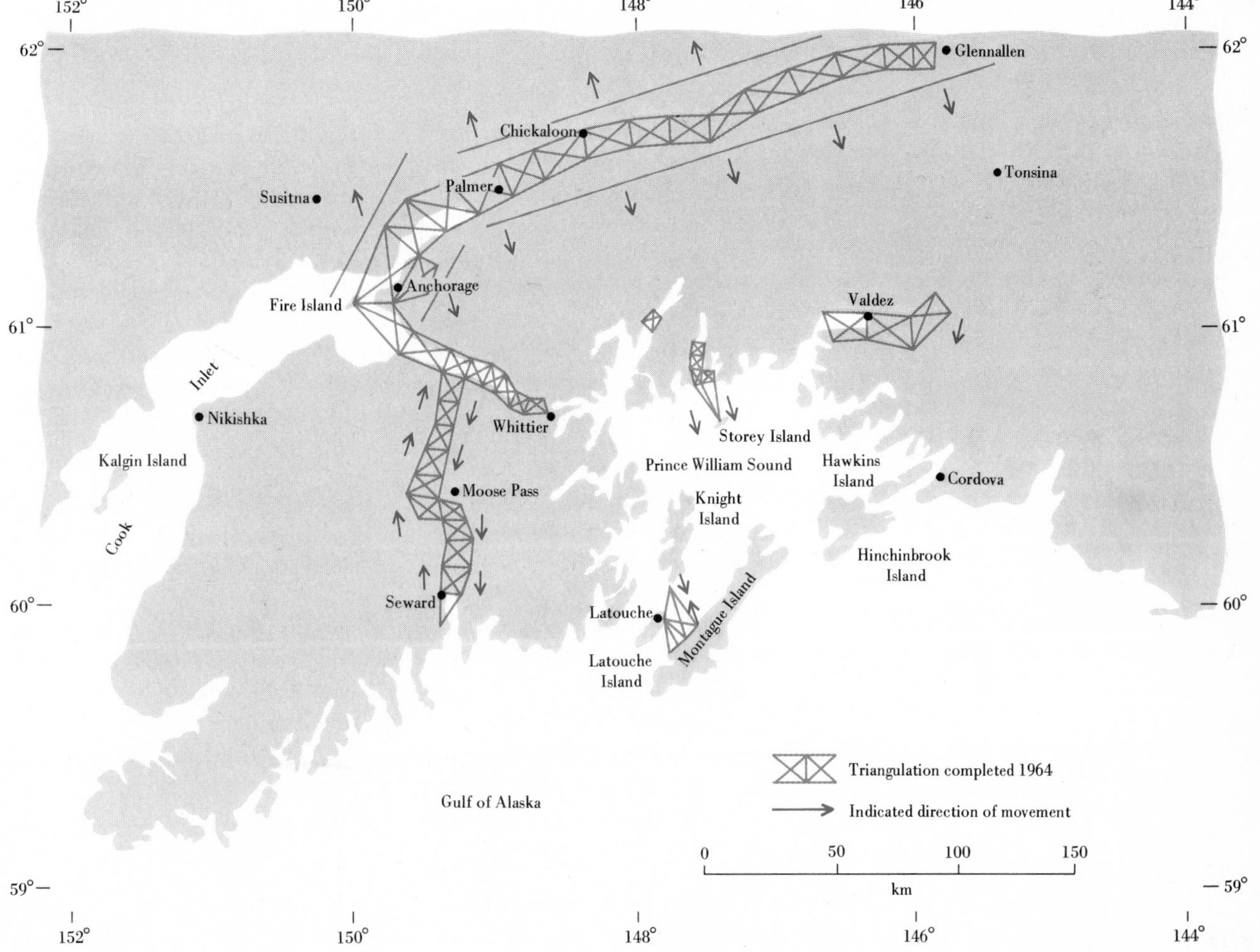

Figure 18.4 ***Drawing showing elevated cliff on Sagami Bay, Japan. Inset, Lithophaga shells.***

the total can become significant. For instance, on the shore of Sagami Bay, Japan, not far from Yokohama, a cliff reared up during an earthquake on September 1, 1923. The amount of movement was measured by using some marine bivalves called *Lithophaga* ("rock eaters") as a reference. These little animals scoop out small caves for their 5-cm shells at mean sea level and spend their lives waiting for the sea to bring food at each rise of the tide. After the 1923 earthquake, rows of *Lithophaga* were found starved to death 5 m above the waters that used to feed them. Other rows of *Lithophaga* holes in this same cliff were found and correlated with quakes in A.D. 1703, 818, and 33 (see Figure 18.4). The total elevation over that 2,000-year interval was 15 m. At this rate, the elevation over a geologic time interval of 2 million years would be 15,000 m.

HORIZONTAL DISPLACEMENTS The San Andreas fault is a large scar in the earth's crust approximately 1,000 km long extending in a nearly straight line from Cape Mendocino southeastward to the Gulf of California (see Figure 18.5). Displacements have occurred along portions of the fault at the times of some earthquakes but not over its entire length during any one quake (see Figure 18.6). At the time of the earthquake of April 18, 1906, there were horizontal displacements along 400 km of the northerly end from Point Arena to San Juan Bautista. The maximum slippage was 7 m on the shore near Tomales Bay. This displacement died out in both directions.

Rocks on both sides of the San Andreas fault have been studied, and some geologists believe formations can be correlated which were joined across the fault as a single unit 150 million years ago. Estimates of the total displacement run in the scores of kilometers up to as many as 200.

AT LEAST 15KM

Figure 18.5 ***A section of the San Andreas fault in the Indio Hills, California; view looks northwest. An abrupt movement of 6 m along part of this fault occurred at the time of the San Francisco earthquake of 1906.*** *(Photograph by Spence Air Photos.)*

Slow movements

Not all crustal movements are accompanied by earthquakes. Slow movements of the crust are going on today. These are being measured along faults in California by the Earthquake Mechanism Laboratory of the Environmental Science Service Administration. They use cross-fault strain meters which show that deformation occurs in zones up to 10 m across. Creep is in progress along the Hayward fault from Richmond to south of Fremont, along the Calaveras fault in the Hollister area, and along the San Andreas fault from San Juan Bautista to Cholame (see Figure 18.7). The time and amount of movement are shown in Figure 18.8. It is interesting to note that at San Juan Bautista and at the Cienega Winery the creep occurs as distinct events, whereas the pattern of movement is different at Stone Canyon Observatory. Creep does not occur simultaneously at San Juan Bautista and Cienega Winery but may be several days apart. The difference in time at which creep starts suggests that it propagates from San Juan Bautista to the Cienega Winery at a rate of approximately 10 km/day.[1] The average movement along a 100-km section of the San Andreas in the vicinity of Parkfield is 5 cm/year, with the Pacific side of the fault moving northwestward.[2]

Mt. Suribachi on Iwo Jima, Japan, was the scene of a famous flag raising

[1]Robert D. Nason, "Preliminary Instrumental Measurements of Fault Creep Slippage on the San Andreas Fault, California," *Earthquake Notes*, vol. 40, pp. 6–10, March 1969.

[2]Stewart W. Smith and Max Wyss, "Displacement on the San Andreas Fault Subsequent to the 1966 Parkfield Earthquake," *Bull. Seis. Soc. Am.*, vol. 58, pp. 1955–1973, 1968.

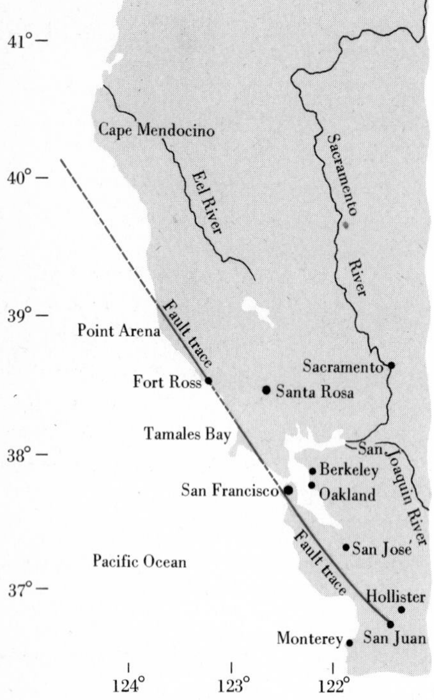

Figure 18.6 ***Section of the San Andreas fault which slipped at the time of the earthquake in 1906.*** *(After John H. Hodgson,* Earthquakes and Earth Structure, *p. 15, Prentice-Hall, Inc., Englewood Cliffs, N.J., 1964.)*

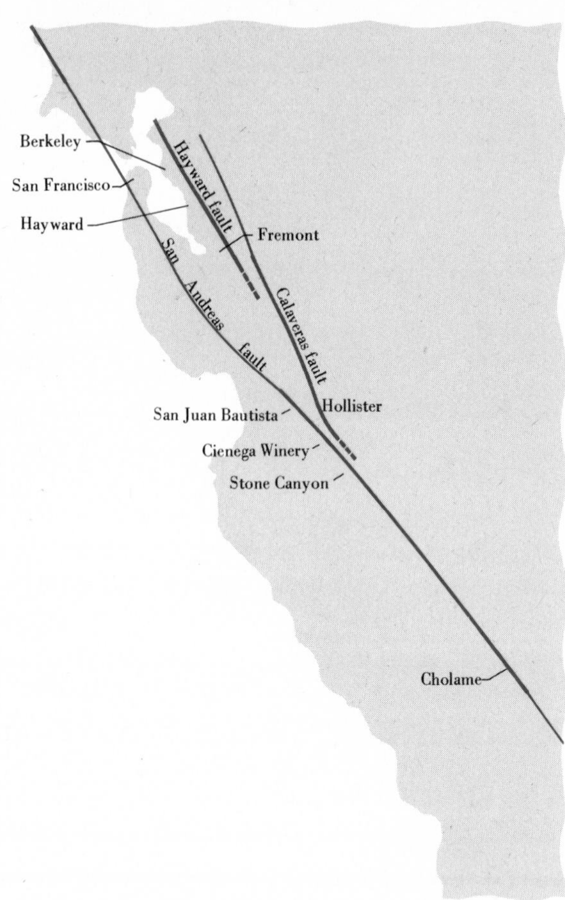

Figure 18.7 ***Sections of the San Andreas, Hayward, and Calaveras faults, showing places of creep measurement. Heavier lines indicate sections of faults undergoing active movement.***

Figure 18.8 ***Fault movement at Cienega Winery, San Juan Bautista, and Stone Canyon (18 km southeast of Cienega Winery).***

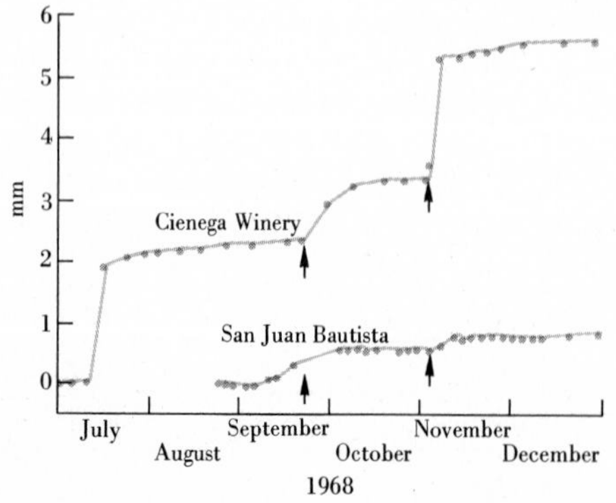

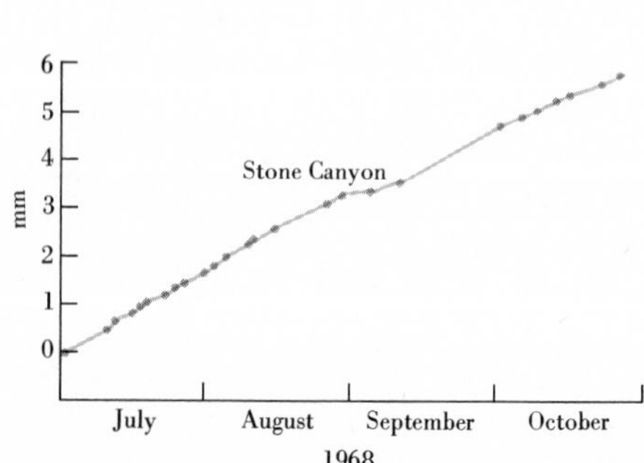

by the U.S. Marines in 1946. It was classed as a dormant volcano, but during the next 23 years, according to geologist Takeyo Kosaka, pressure from underlying magma caused it to rise 7 m.

18.2 *Kinds of deformation*

Deformation of the earth's crust can be described in terms of change of volume, change of shape, or a combination of both. The kinds of deformation are illustrated in Figure 18.9. If a material changes shape without a change of volume, the deformation is called *shear.* An analogy for pure shear is supplied by staggering a deck of cards. If the bottom card is held fixed and the others are slid forward by amounts proportional to their distances from the bottom, the shape of the space occupied by the deck is altered, but its volume (the total amount of cardboard) remains the same.

Stress–strain

The effects of some deforming forces on earth materials have been studied in the laboratory by measuring the strain produced by application of stress. For example, if a cylindrical specimen of rock is stretched, the amount of its extension can be plotted against the force producing it. If its original length is l_0, and the force used to stretch it is P, the increase in length Δl leaves it at a new length l. The strain, sometimes called unit strain, is the change in length per unit length, or $\Delta l/l_0$. The stress, or unit stress, is the total force P divided by the area of the cross section of the specimen, or P/A (see Figure

Figure 18.9 ***Deformation may produce change in volume without change in shape, or change in shape without change in volume, or a combination of the two.***

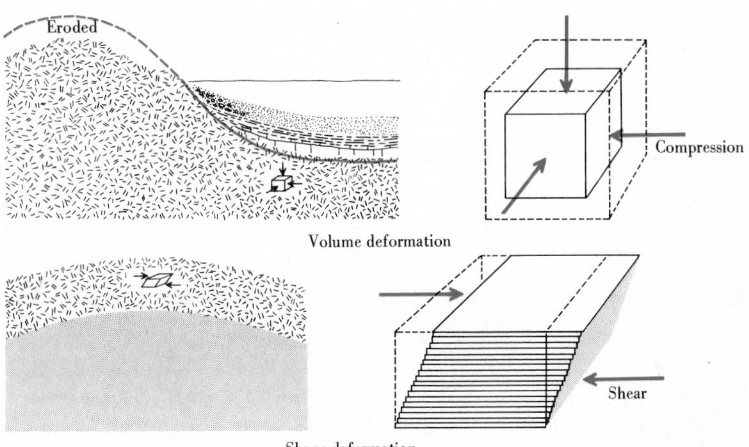

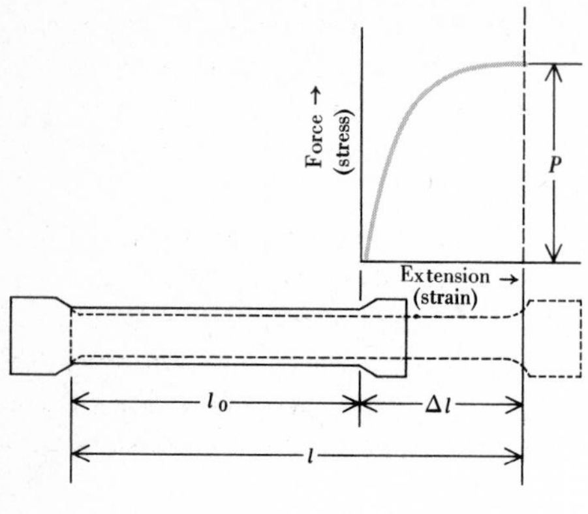

Figure 18.10 **Deformation of a cylindrical specimen of rock by extension. See the text for discussion.**

18.10). Graphs of stress vs. strain are usually set up with unit stress and unit strain as the axes.

In most materials, as stress increases, for a time strain is proportional to stress. If stress is removed, strain goes back to zero. This is the range of elastic deformation. Elasticity equals stress/strain. Then, when the stress reaches the yield point, the strength of the material has been overcome. Deformation becomes plastic; that is, with no additional stress, strain continues to increase. Plasticity equals stress/rate of strain. When stress is removed, the plastic part of the deformation is not recovered (see Figure 18.11).

ELASTIC DEFORMATION Elastic deformation is recoverable. It involves a property of rock called elasticity. A solid is deformed elastically if it can recover its size and shape after the deforming force has been removed. An elastic solid resists deformation up to a certain point. Its resistance to a change in volume without a change of shape is called its *bulk modulus.* This is defined as the ratio of stress to strain and is given the symbol *B.*

$$B = \frac{\text{Increase in force per unit area}}{\text{Change in volume per unit volume}}$$

A material's resistance to elastic shear is called *rigidity* and is measured by the *shear modulus, G.*

$$G = \frac{\text{Force per unit area}}{\text{Relative displacement of planes-unit distance apart}}$$

When deformation is purely elastic, the strain is a linear function of the stress and the stress–strain graph is a straight line. The slope of this line, stress/strain, defines the material's elasticity and is called the *elasticity modulus.* The stretching or compressing of a cylindrical specimen results in an associated narrowing or widening as well as change in length. The ratio of change in diameter per unit diameter, $\Delta d/d_0$, to change in length per unit length, $\Delta l/l_0$, is called *Poisson's ratio,* $l_0 \Delta d/d_0 \Delta l$.

Experiments have shown that rocks that deform elastically become plastic when their strength is reached but do not lose their elasticity during plastic deformation. If the deforming stress is removed, the elastic portion of the deformation is recovered. Only the plastic deformation is permanent.

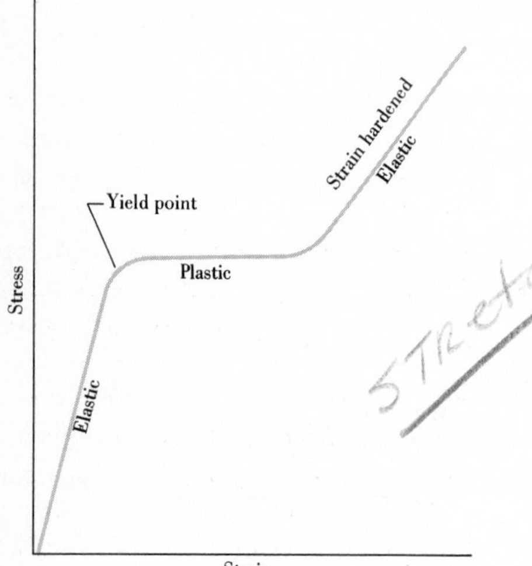

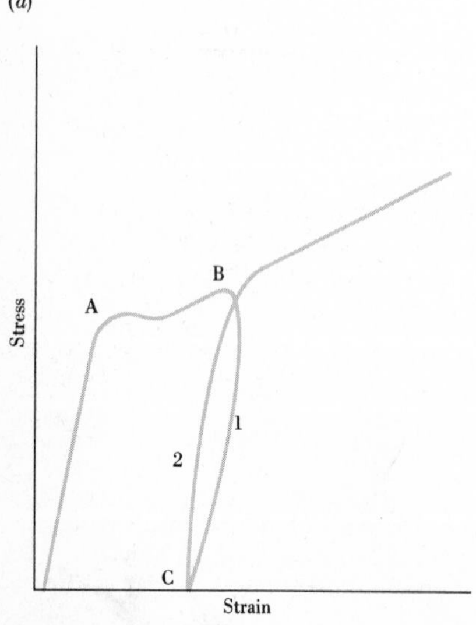

Figure 18.11 **Behavior of material under stress. (a) As stress increases, for a while strain is proportional to stress. This is the range of elastic deformation. Beyond the yield point, deformation becomes plastic for a while until the material again resists and stress must be increased to produce more strain. (b) After deformation to point B, if the stress is removed, the elastic part of the deformation is recovered, but the rest is not. A retardation of the recovery is called hysteresis. If stress is again applied, deformation picks up where it left off at B and continues as though there had been no interruption.**

Table 18.1 ***Upwarping in regions of complete or partial deglaciation***[a]

Area	*Maximum observed uplift, m*
Scotland	30
Iceland	120
Greenland	146+
Novaya Zemlya	100
New Zealand, South Island	100±
Antarctica	100
Norway and Sweden	275+
Eastern Canada–Labrador	270+
Newfoundland	137

[a] R. A. Daly, *Strength and Structure of the Earth*, Prentice-Hall, Inc., Englewood Cliffs, N.J., 1940.

Retardation of recovery Rocks are said to be deformed elastically if the deformation disappears when the stress is removed. However, they do not ordinarily regain their former shape the instant the stress is removed. There is a time lag over which recovery is made. This retardation of recovery is called *hysteresis.*

It has been suggested that a demonstration of hysteresis is provided on a grand scale by basining of the crust under loads of glacial ice and subsequent recovery upon melting of the ice. The meltings and upwarpings from the last ice advance began over 20,000 years ago, but recovery is still progressing in areas where the ice had completely disappeared as much as 10,000 years ago. Table 18.1 lists some areas where upwarping has occurred and the amounts recovered so far.

PLASTIC DEFORMATION Plastic deformation is permanent. It involves a property of rock called viscosity. A material that is deforming plastically does so by flow along an indefinitely large number of shear planes. If the rate of the flow is proportional to the stress causing it, the material is said to be viscous, and if we plot the rate of flow against the stress, the slope of the graph is the viscosity (Figure 18.12). Viscosity equals stress/rate of flow. Viscosity is an important property in some geological processes. It is the property that governs the ability of magma to flow during igneous activity and is the property that enables the mantle to adjust to crustal loads. It is measured experimentally by observing rate of flow, the force needed to turn a paddle wheel at constant speed, or the rate of sinking of standard spheres (see Figure 18.13).

We do not know the absolute viscosity of materials in the earth's interior. This is difficult to determine experimentally and its effects are difficult to visualize. However, the effect of viscosity may be illustrated by considering a granite magma with a viscosity of hard pitch, 10^{13} times that of water. A sphere of gneiss 2 m in diameter would sink in this magma about 10 cm/day. A sphere of 4 m in diameter would sink four times as fast.

TIME FACTOR Within the earth, materials show a duality of response. They transmit earthquake waves elastically but move plastically during the formation of mountains and other surface features. Here, an important factor that cannot be reduced to a graph is time. Material may be elastic for short-time stresses in transmitting earthquake waves but plastic over a longer time.

On a short time scale, a tuning fork can be formed from pitch to ring as though it were purest steel. Yet, given time, it will flow into a formless blob from its own weight. A steel bar supported at its ends and loaded in the middle will appear to have strength entirely adequate to support the load but will in time slowly bend. And some metamorphic rocks testify to the deformation

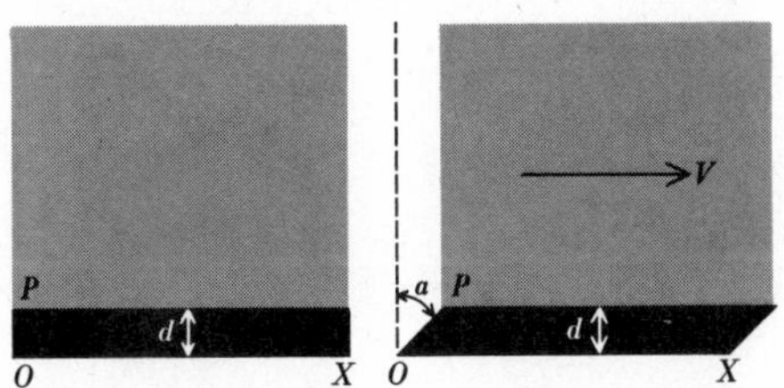

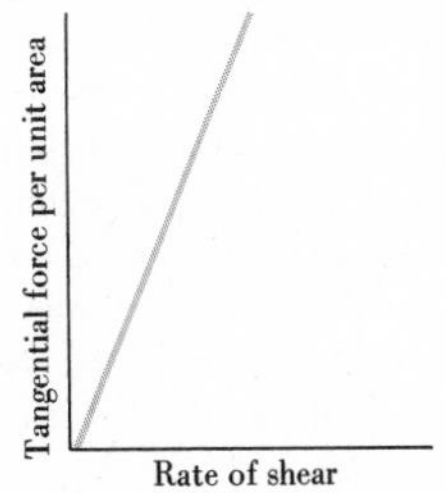

Figure 18.12 ***Demonstration of viscosity. If an oil film of thickness* d *rests on a surface* OX *under a block* P, *and* P *is displaced sidewise with a velocity* V, *the change in the angle* a *which* OP *makes with the vertical can be used to measure the shear (change in shape) of the oil. If the rate of shear is proportional to the force causing it, the deformation is said to be viscous.***

of rocks under prolonged periods of stress and increased temperature. On geologic time scales, the rocks of the crust have shown themselves to be weak.

18.3 Types of stress

There are different types of stress: tension, compression, and shear. Tension is a stretching stress and can increase the volume of a material. Compression tends to decrease the volume. A shear stress produces changes in shape. Any stress carried beyond a material's strength can cause rupture.

Strength of rocks

Rocks possess strength. *Strength* is the stress at which a material begins to be permanently deformed. A special kind of permanent deformation is rupture.

Rocks possess several types of strength, as they respond differently to different stresses. Laboratory test results for rocks generally give compression strength, shear strength, and tension strength (Figure 18.14). These have shown that a rock's strength in tension is smaller than its strength in compression. When the difference between these two is large, we say the material is brittle. When it is small, we say the material is ductile. As the confining pressure increases, a brittle rock tends to become ductile. It does not rupture in the ordinary sense of disintegration but flows along an indefinite number of shear planes. Rupture in tension may be important near the earth's surface, but it will not occur at great depth. The strength in compression for brittle rocks is about 30 times larger than the strength in tension.

How a rock mass responds to stress depends on temperature and depth of burial. At depth, rocks will respond to stress by flowing before they rupture, whereas at the earth's surface they will break. The stresses at the earth's surface are quite different from those at depth, where temperatures and pressures increase considerably. Whether a rock breaks or flows under a given stress depends on the prevailing temperatures. Rocks are generally weaker at high temperatures than they are at low ones. If a rock is surrounded by equal pressure on all sides, it tends to increase in strength. Such pressure strengthens a rock and increased temperature weakens it.

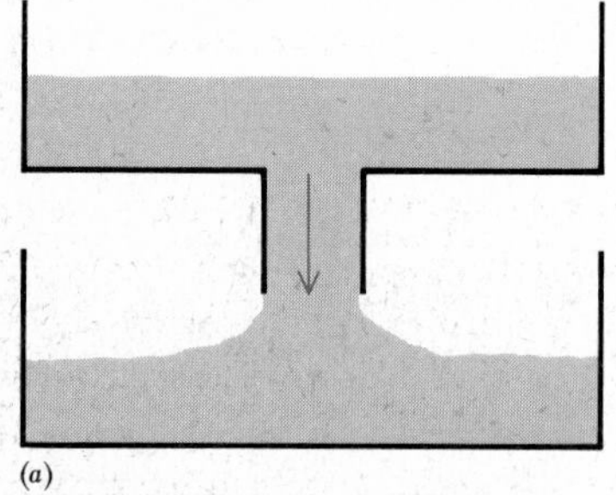

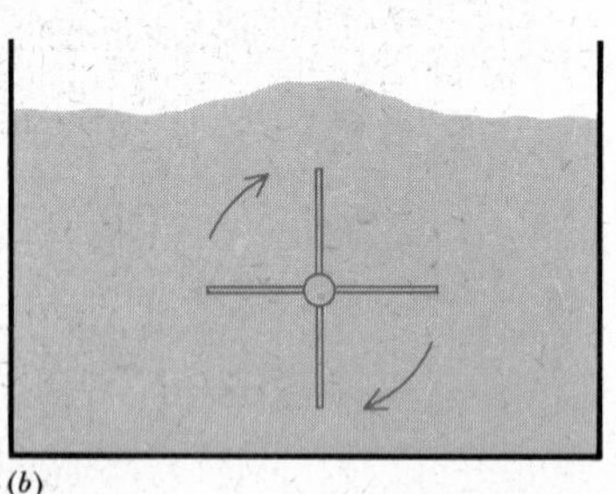

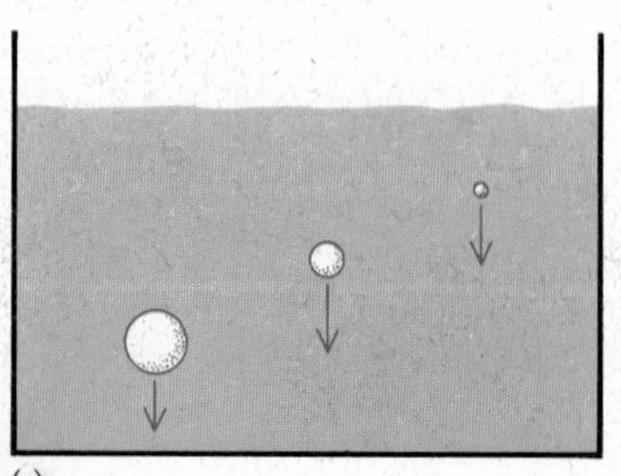

Figure 18.13 **Methods of measuring viscosity: (a) rate of flow, (b) force necessary to turn paddle wheel at known speed, and (c) rate at which spheres of the same material but different radii fall through the fluid. For method c, $x = 0.22gr^2(d - d')/v$, where x is the velocity of the sphere when the motion is steady (terminal velocity), g is the acceleration of gravity, d is the density of the sphere, d' is the density of the fluid, r is the radius of the sphere, and v is the viscosity of the fluid. In a granite magma with the viscosity of hard pitch (10^{13} times that of water, or 10^{10} times that of glycerine), a sphere of gneiss with a diameter of 2 m would sink about 10 cm/day. A similar sphere of twice the diameter would sink four times as fast.**

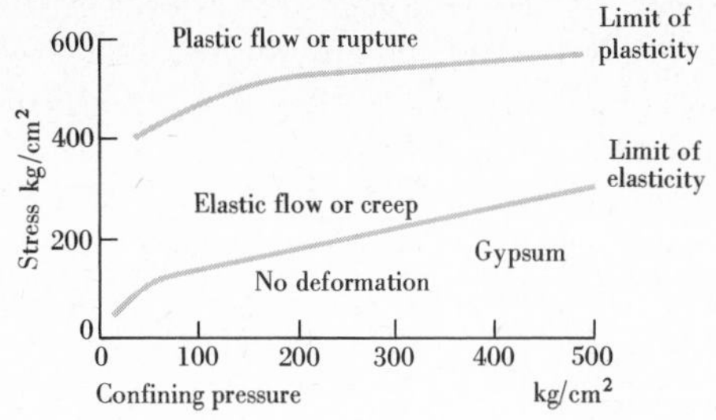

Figure 18.14 **Stress vs. confining pressure diagram for gypsum.** (*After J. Goguel,* Mém. carte géol. France, *p. 530, 1943.*)

Different rocks have different compressive strengths. Hard, brittle, competent rocks such as sandstone and limestone have a different strength than clay or shale.

18.4 *Structural features*

Structural geology deals primarily with deformed masses of rock, their shapes, and stresses that caused the deformation. In describing structural features, the concept of *relief* is sometimes helpful. This is familiar in connection with the ground surface, where it denotes the difference in elevation between the highest and lowest points in a specified area. For example, the topographical relief of California is 4,400 m. This is the difference between the elevation of Mt. Whitney and Death Valley.

Figure 18.15 **Measuring dip with a Brunton compass.**

(a)

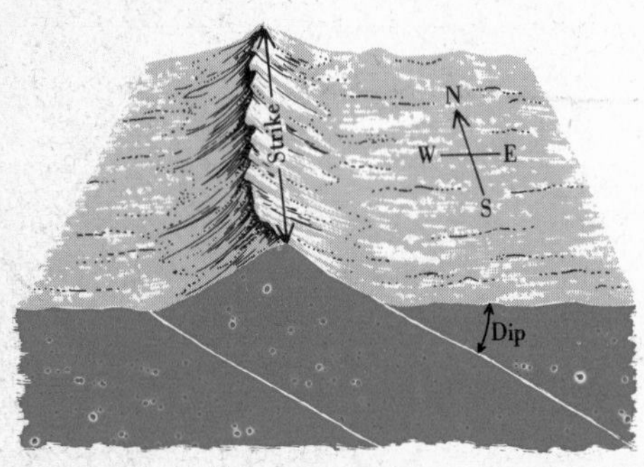

(b)

Figure 18.16 **Dip and strike.** (a) Photograph showing outcropping edges of tilted beds in southwestern Colorado, a few kilometers east of Durango. (b) Sketch illustrating terms used to describe the attitude of these beds. The beds strike north and dip 30° east. (*Photograph by Soil Conservation Service, U.S. Department of Agriculture.*)

Structural relief is the difference in elevation of parts of a deformed stratigraphic horizon and is a measure of the extent of deformation. Sometimes the shape of a rock mass is determined by drilling and sometimes by hypothetical reconstruction of eroded portions.

Sedimentary rocks and some igneous rocks form in approximately horizontal layers. When these are found tilted, folded, or broken, they provide evidence of deformation of the earth's crust. In some areas, erosion has stripped away as much as thousands of meters of uplifted portions of the crust, revealing structures which were once buried deeply. The greatest depths from which we have information by drilling are a few thousand meters. Deformation features include folds, faults, joints, and unconformities.

In describing the attitudes of structural features, geologists have found it convenient to use two special terms: dip and strike. These are more easily described with reference to layered rocks. If a rock layer is not horizontal, the amount of its slope is called its *dip* (see Figure 18.15), measured by specifying the acute angle that the layer makes with the horizontal. The dip is measured in the direction of the greatest amount of inclination. Its *strike* is the course or bearing of the outcrop of an inclined bed or structure on a level surface. If the rock layer is tilted so that it disappears below the surface but protrudes somewhat because of resistance to weathering, as in Figure 18.16, strike is the direction in which the resulting ridge runs. A bed that dips either east or west has a north–south strike, usually designated simply as north. A bed that dips either to the north or to the south has an east–west strike.

Folds

Folds are a common feature of rock deformation. They are produced by compressive stress (see Figure 18.17). They range in size from microscopic wrinkles in a piece of metamorphic rock to huge folds involving thousands of meters of thickness for distances of hundreds of kilometers. Folds are seldom isolated structures but generally occur in closely related groups. Sometimes they are very broad and sometimes tight and narrow. Sometimes they are tilted to one side and sometimes to one end.

In some areas, sedimentary layers are only slightly bent, whereas in others, usually associated with mountain structures, they are intensely deformed. The difference in the kind of folding is mainly dependent on the amount of deforming stress in relation to the strength of the rocks.

MECHANISM OF FOLDING Folding can be classified as concentric or flow. *Concentric folding* is basically an elastic bending of an originally horizontal sheet with all internal movements parallel to a *basal plane*, which is the lower boundary of the fold. It occurs in surface layers of the crust. The size of a concentric fold is determined by the thickness of the beds involved and by their elastic properties. Beds are shortened during concentric folding, but the thickness of the beds and the volume remain the same. Large-amplitude folds form from thick strata and small-amplitude folds from thin strata. The distance between crests is also controlled by the thickness of the beds.

Figure 18.17 Syncline and adjacent anticline in sandstone beds of the Moccasin formation along the river road to Goodwin's Ferry, Giles County, Virginia. (*Photograph by T. M. Gathright II.*)

Flow is a type of deformation that occurs in rocks when they are in the plastic state. It involves internal movement that may take any direction. It occurs in weak rocks near the surface and in strong rocks at depth. Flow is due to conditions of confining pressures and high temperatures. It is the only true plastic deformation of rocks.

MONOCLINES Monoclines are relatively simple examples of deformation involving an elastic bending of sedimentary layers. A *monocline* is a double flexure connecting strata at one level with the same strata at another level. Extensive horizontal layers are bent down and pass beneath younger horizontal

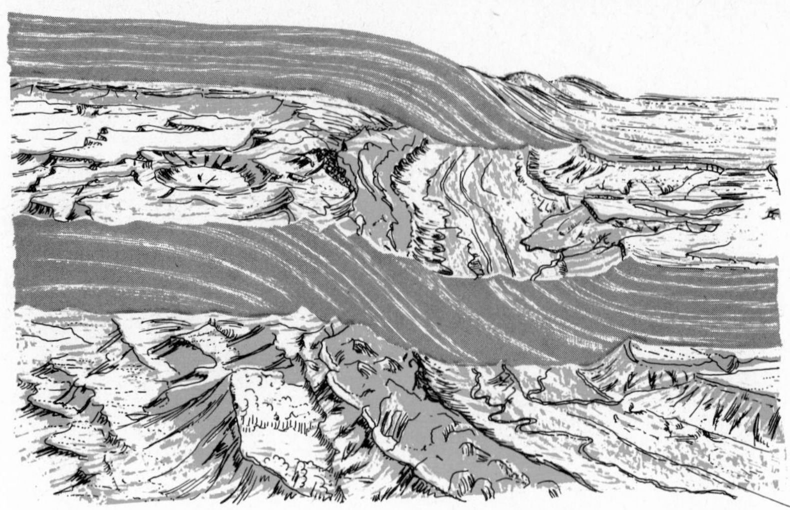

Figure 18.18 **The Waterpocket monocline in southern Utah. In the drawing, an imaginary cut has been introduced to show the subsurface structure, and the eroded beds have been restored in the background.** (*After John S. Shelton,* Geology Illustrated, *Fig. 88, W. H. Freeman and Company, San Francisco, 1966.*)

Figure 18.19 **An anticline in the Appenines east of Assisi, Italy.**

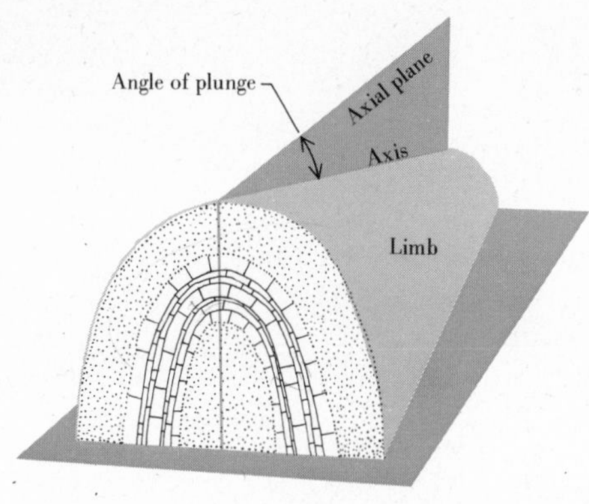

Figure 18.20 ***The parts of a fold, shown on a plunging anticline, but applicable to a syncline also. The axis is the line joining the places of sharpest folding. The axial plane includes the axis and divides the fold as symmetrically as possible. If the axis is not horizontal, the fold is said to be plunging, and the angle between the axis and the horizontal is the angle of plunge. The sides of a fold are called the limbs.***

strata. The flexed layers did not accumulate with the bend in them. As bending progressed, erosion worked to strip away the higher portion.

There are many monoclines in the Colorado Plateau. Some involve displacements as great as 4,000 m and lengths of 250 km. The "Waterpocket Fold" in southern Utah is one of the better known. It involves displacements of close to 2,000 m. Estimates of the rate of its deformation range from a few centimeters to possibly a few meters per century. Figure 18.18 shows a section of the Waterpocket monocline with its eroded beds restored. Its strike is approximately northwest, and there is a maximum dip of 45° toward the northeast.

ANTICLINES AND SYNCLINES Local arching of layered rocks is the most common form of a structure called an *anticline* (see Figure 18.19). The two sides of the fold, called its *limbs,* dip away from each other. The *axis* of an anticline is the direction of an imaginary line drawn on the surface of a single layer parallel to the length of the fold. The axes of most geological folds are inclined. The angle of dip of its axis is the *plunge* of the fold. Parts of a fold are illustrated in Figure 18.20.

Anticlines can be spectacular when their anatomy has been exposed by erosion (see Figure 18.21). The oldest rocks are at the core and the youngest on the outer flanks.

A *syncline* is a downward fold, the opposite of an anticline. The limbs dip toward the central axis. Synclines are best seen in mountainous areas where there has been uplift along with the folding. The youngest rocks are in the center of the fold, with successively older rocks farther out.

Anticlines and synclines commonly occur together in elongate groups like belts of wrinkles in the earth's outer skin. Anticlines and synclines are said to be symmetrical if their opposite sides have approximately equal dips and asymmetrical if one limb is steeper than the other. Types of folds are listed in Table 18.2 and illustrated in Figure 18.22.

Figure 18.23 shows synclines, anticlines, and faults underlying a section of the Berkeley Hills and adjacent ranges of the middle coast ranges of California near San Francisco. The Wildcat fault marks the edge of the hills, and Wildcat Creek appears to follow the axis of a syncline. Elsewhere, there is less marked connection between surface forms and subsurface structure, except for a general trend of the principal streams along the strike of the structures.

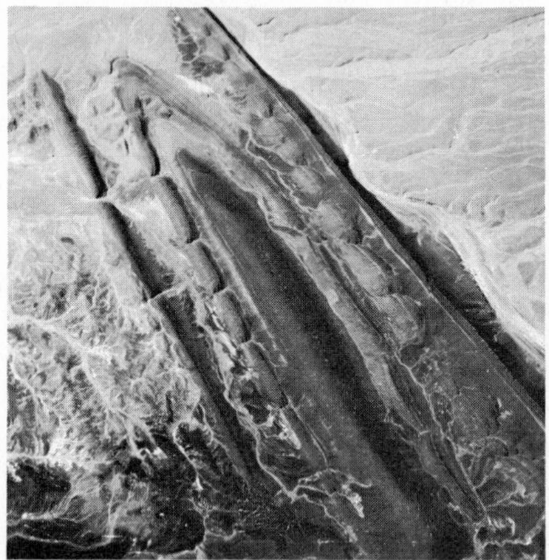

(*a*)

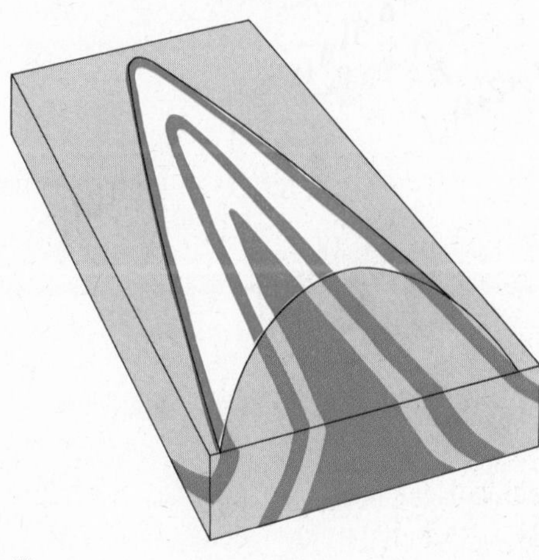

(*b*)

Figure 18.21 ***(a) Aerial photograph showing resistant beds of an eroded plunging anticline. (b) Sketch illustrating the relationship of outcrops to fold. Spanish Sahara, Africa.*** *(Photograph by U.S. Army Air Force.)*

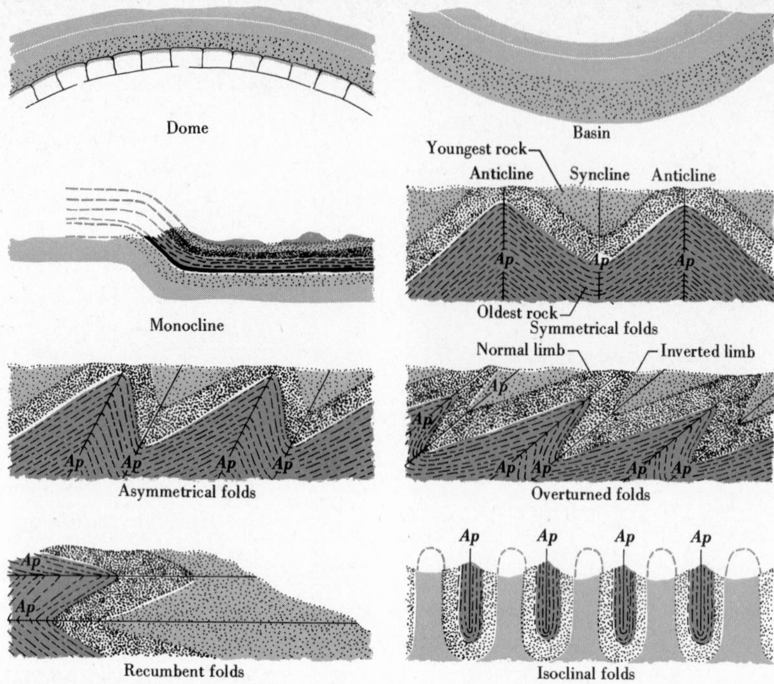

Figure 18.22 ***Types of folds.***

STRUCTURAL DOMES Structural domes come in many sizes. They may be circular or elongated. Circular domes are formed when the active force is vertical, with horizontal stress equal and constant. Elongated domes are formed when the horizontal stress is not uniform.

Structural domes are the result of pressure acting upward from below to

Table 18.2 ***Types of folds***

Name	*Description*
Anticline	Upfold or arch
Syncline	Downfold or trough
Monocline	Local steepening of an otherwise uniform dip
Dome	An anticline roughly as wide as it is long, with dips in all directions from a high point
Basin	A doubly plunging syncline
Asymmetrical	The strata of one limb of the fold dip more steeply than those of the other
Overturned	Limbs tilted beyond the vertical; both dip in the same direction, though perhaps not the same amount
Recumbent anticline	The beds on the lower limb are upside down
Recumbent syncline	The upper limb is inverted
Isoclinal	Beds on both limbs are nearly parallel, whether fold is upright, overturned, or recumbent

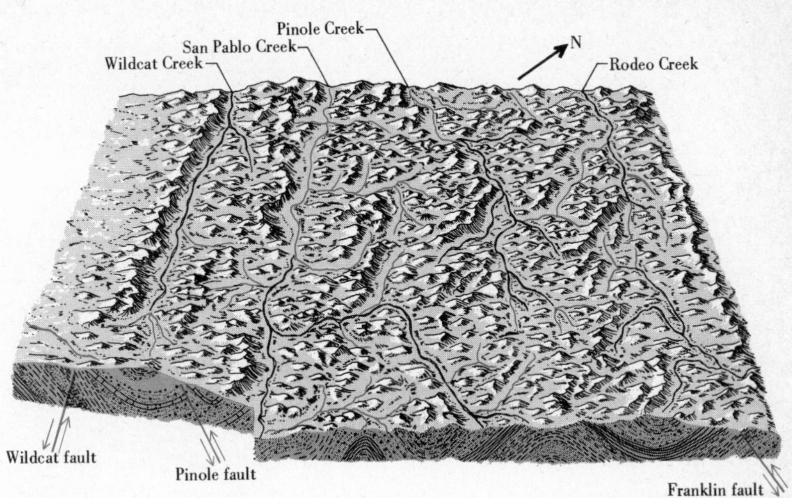

Figure 18.23 **Synclines, anticlines, and faults underlying a section of the Berkeley hills and adjacent ranges, near San Francisco, of the middle coast ranges of California. The Wildcat fault marks the edge of the hills, and Wildcat Creek appears to follow the axis of a syncline. But elsewhere there is less marked connection between surface forms and subsurface structures, except for a general trend of the principal streams along the strike of the structures. The location of the corner of block near Franklin Fault is 122°07′ W, 37°57′N.**

produce an uplifted portion of the crust with beds dipping outward on all sides. Where erosion has removed the top of a dome, concentric ridges are formed by one or more relatively resistant beds. The troughs are formed by erosion of weaker rocks. It is possible to project the dip and strike of the beds upward over the dome and reconstruct its whole shape before erosion removed parts.

Some structural domes are formed by the upwelling of plastic material such as salt or magma. It is impossible to know whether slight folding started the upwelling or whether hydrostatic adjustment of loads on the crust was the primary cause. When there is a local thickening or thinning of the crust, the plastic material starts to rise and continues as a result of the unequal static load. Most domes show marked upward bending of the surrounding beds against the stock, often accompanied by faulting. The beds above are domed and stretched by the push from below and often show an intricate pattern of normal tension faults. A structural dome produced by an igneous intrusion is illustrated in Figure 18.24.

Deformation of rock by fracture

Masses of rock may be broken, with or without movement along the cracks. If there has been no slippage along the fracture surface, the crack is called a *joint.* If there has been displacement, the structural feature is called a *fault.*

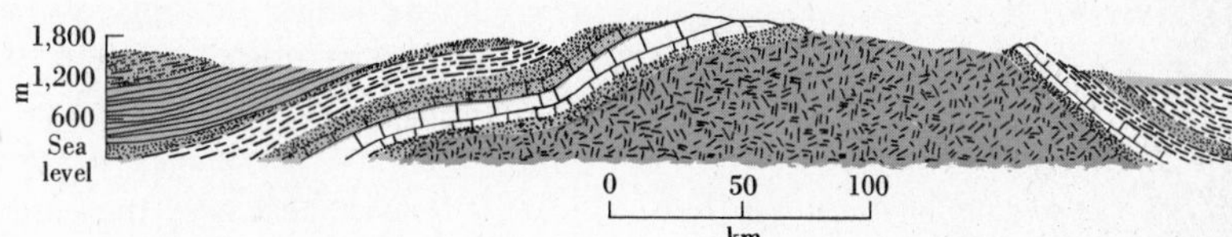

Figure 18.24 **Section across the Black Hills in South Dakota and Wyoming, illustrating a structural dome produced by the intrusion of an igneous mass after deposition of sediments.** (*After N. H. Darton.*)

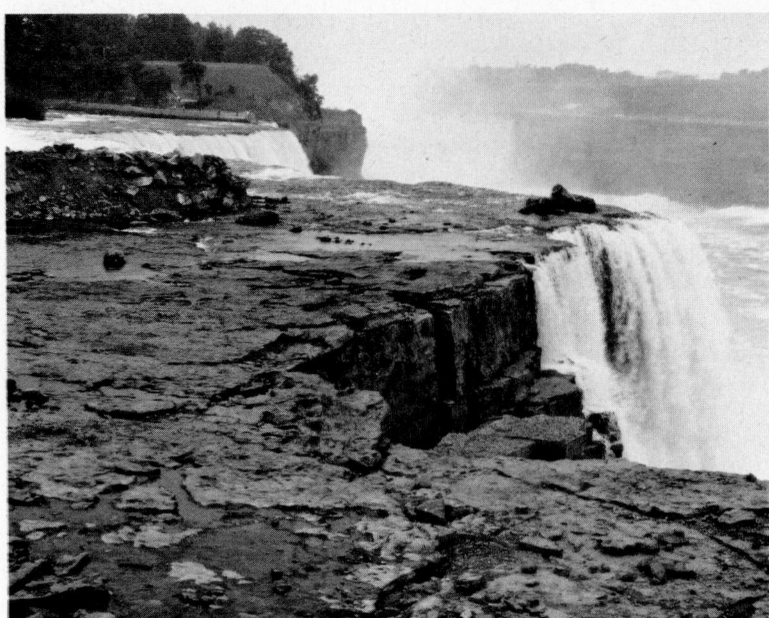

Figure 18.25 **The sawtooth shape of the American Falls of Niagara Falls results from the prevailing joint pattern.** (*Photograph by State of New York Power Authority during unwatering of June 1960 for Niagara power project.*)

JOINTS Joints are found in all kinds of rock and are the most common feature of rocks exposed at the surface. They usually occur in sets, the spacing between them ranging from just a few centimeters to a few meters. As a rule, the joints in any given set are almost parallel to one another, but the whole set may run in any direction—vertically, horizontally, or at some angle (see Figure 18.25). Most rock masses are traversed by more than one set of joints, often with two sets intersecting at approximately right angles. Such a combination of intersecting joint sets is called a *joint system.* A regional pattern of joint systems may occur over areas of hundreds of square kilometers in a given type of rock exposed at the surface. As a rule, the kinds of rock have a marked influence on what joint systems develop. A massive sandstone, a graywacke, and a sandy shale each show characteristic jointing directions and spacing.

It is rarely possible to determine a joint set's age or origin. It is generally accepted, however, that different sets in a joint system were probably made at different times and under different conditions. In sedimentary rocks, one set could develop while the rock was being compressed by the weight of overlying rocks during burial and consolidation. Another set could be produced when pressure is released during unloading by erosion. In other cases, sets may be developed during deformation by tension. Some joint sets are known to be formed during the cooling of a mass of igneous rock and others during movements of magma.

Columnar jointing Columnar jointing is a special pattern of jointing found in some masses of basalt. It consists of sets of cracks produced by the mechanism of cooling. In certain lava flows and shallow intrusive bodies, joints are formed as the rock shrinks in cooling. Dikes or sills cool by fairly uniform loss of heat over the entire surface that is in contact with cold wall rock. This allows shrinkage about many equally spaced centers, not unlike that which

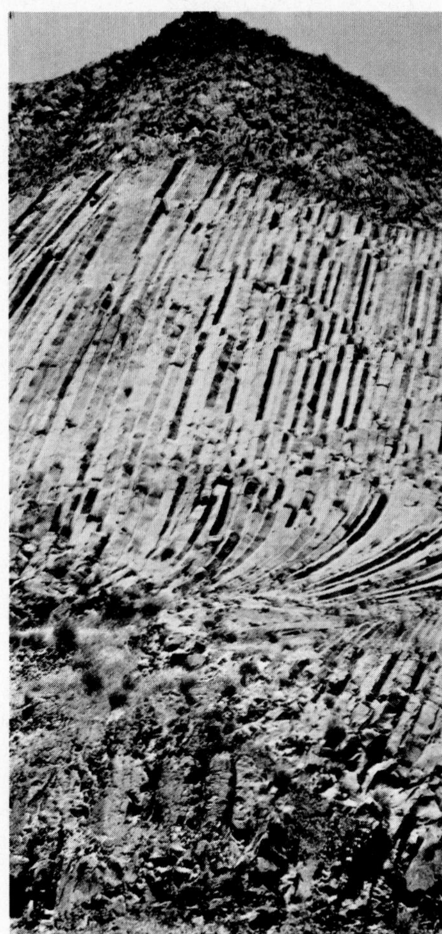

Figure 18.26 **Basalt sill and columnar jointing on the north shore of Snake River, Washington State. Note the abrupt changes in column orientation.** (*Photograph by Corps of Engineers, Walla Walla District. Courtesy of Harlan E. Moore.*)

produces mud cracks. Such contraction means tension between the centers, which results in a system of short straight cracks separating each center of contraction from its neighbors. These form a polygonal pattern, ideally hexagonal, but sometimes four- or five-sided because of irregularities in cooling. As cooling progresses into the body, contraction advances with it and each column may grow until it reaches the opposite side.

Columnar jointing is not clearly developed in all masses of basalt. It seems to be most characteristic of tabular masses, where columns form across the narrow dimension of the mass. The columns are often broken into sections of varying lengths by transverse joints (see Figure 18.26).

One of the best-known areas in the world where columnar jointing can be seen is the Giant's Causeway, near Portrush, Antrim, Northern Ireland. It is also well developed in Devil's Post Pile National Monument, Sierra Nevada, California, and along the Snake River in the state of Washington.

Sheeting A pattern of essentially horizontal joints is called *sheeting* (see Figure 18.27). Here, the joints occur fairly close together near the surface, but less and less frequently the deeper we follow them down, until they seem to disappear altogether a few tens of meters below the surface. But even at considerable depth the rock shows a tendency to break along surfaces parallel to the surfaces of sheeting above. This type of jointing is especially common in masses of granite, and engineers often put it to good use in planning blasting operations.

Joints formed during movements of magma Some joint systems are formed during movements of magma within the crust. Moving magma exerts pressure on adjacent rocks, developing cracks. Magma is then injected into the cracks as they open, enlarging them. This produces such structural features as dikes and sills.

Dikes and sills are thought to form before the magma reaches the surface and is extruded. After extrusion begins, pressure is reduced. Because extrusion takes less pressure than intrusion, the magma wells out instead of intruding into adjacent rocks.

Dikes are exposed at the earth's surface after softer rocks have been eroded away. They vary in width from a few centimeters to kilometers across and may be many kilometers in length. The largest dike known is in Rhodesia. It is 480 km long and up to 8 km wide.

Dikes occur in different patterns: ring dikes, cone sheets, and swarms. *Ring dikes* and *cone sheets* consist of a concentric arrangement of dikes formed by a definite center of stress. They occupy fractures around a center where magma has pushed up a section of the earth's crust. Dikes are filled by upwelling magma (see Figure 18.28). Ring dikes have diameters ranging from 2 to 25 km, with individual dikes sometimes as much as 500 to 1,000 m wide. Ring dikes dip away from the magmatic body that cracked the crust to provide avenues of weakness along which they moved into place. Cone sheets outline inverted cones with sides dipping toward the magmatic body. From their angle of dip, it is possible to estimate the depth below the surface to the body of magma which supplied them.

In some places, dikes occur in roughly parallel groups called *dike swarms.* Some dikes in a swarm are up to 400 km in length. Dike swarms differ

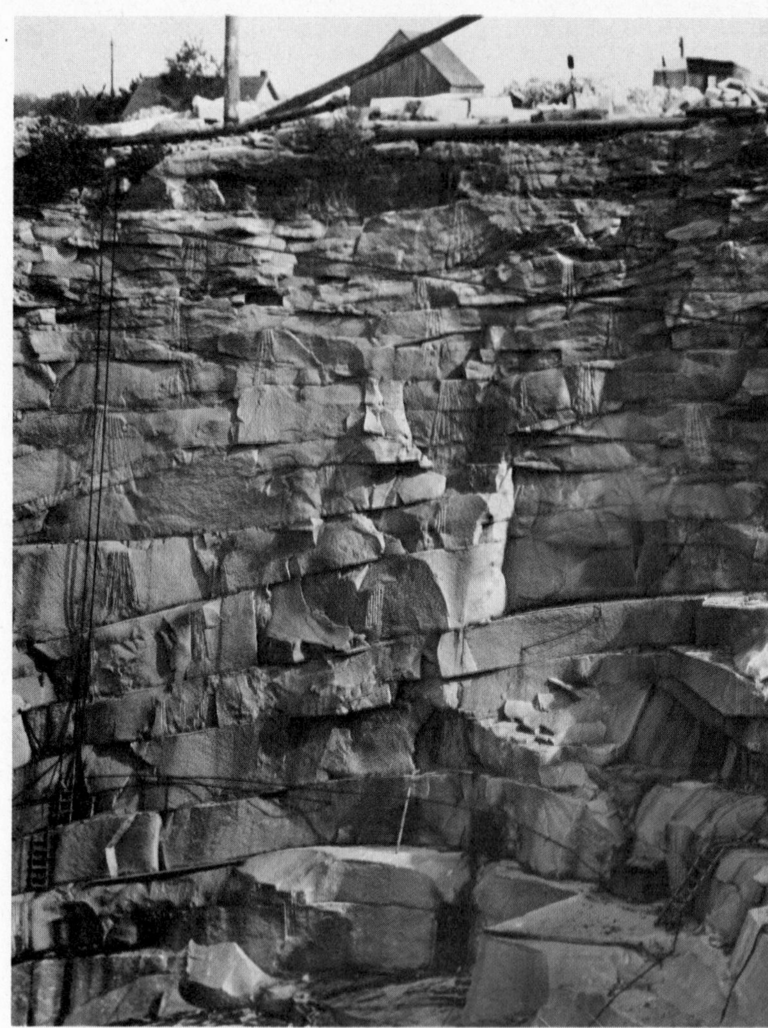

Figure 18.27 **Sheeting in granite.**
(*Photograph by Dale.*)

Figure 18.28 **Ring dikes and cone sheets.**
(*After E. M. Anderson and H. Jeffreys,* Proc. Roy. Soc. Edinburgh, *vol. 56, 1936.*)

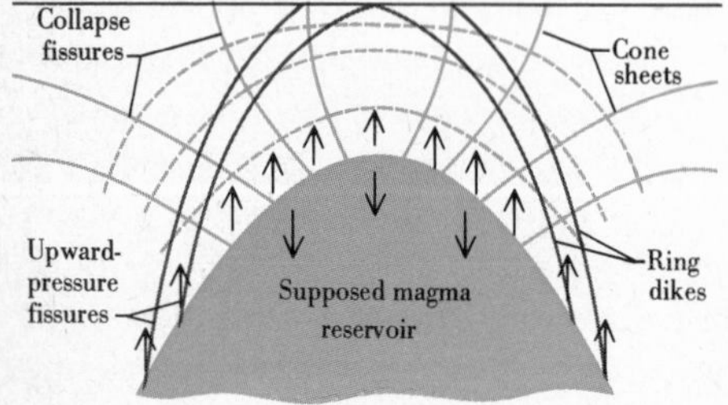

from ring dikes in their pattern, suggesting that they originate in a different kind of stress field.

Some dike swarms are caused by an elastic release of pressure. This has been documented in several places in Scotland. The Caledonian orogenic belt contains dike swarms running parallel to the main folding axis, which trends northeast–southwest (see Figure 18.29). The trend of the dikes suggests that they were formed by an elastic release after compression. Dike swarms are also associated with strike–slip faults in Scotland. These include the Great Glen fault, the Highland Boundary fault, and the Southern Upland fault, which trend northeast–southwest. Their direction suggests that they were formed by a north–south compression. The dike swarms trend east–west, perpendicular to the compressive stress. This again suggests that the cracks in the earth's crust resulted from an elastic release of compression.

Some dike swarms are associated with the folding process but do not involve an elastic release of compression. They are the result of tension. Such a case

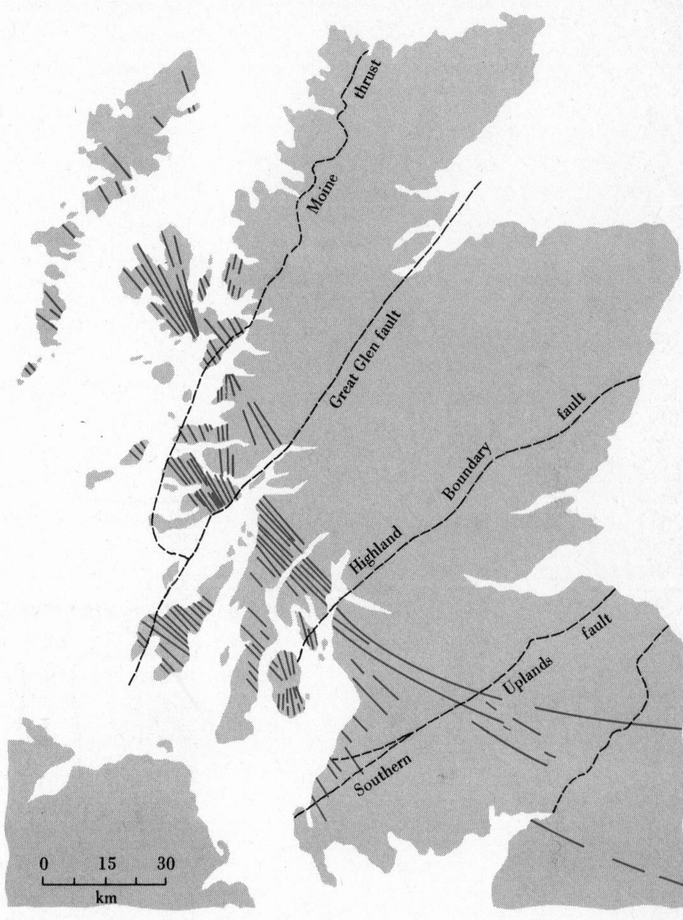

Figure 18.29 ***Tertiary dike swarms in Scotland.*** (*After J. E. Richey,* Trans. Geol. Soc. Edinburgh, *vol. 13, p. 393, 1939.*)

Figure 18.30 ***Dike swarms on the eastern Greenland coast.*** (*After L. R. Wager,* Med. om Grønland, *vol. 134, 1947.*)

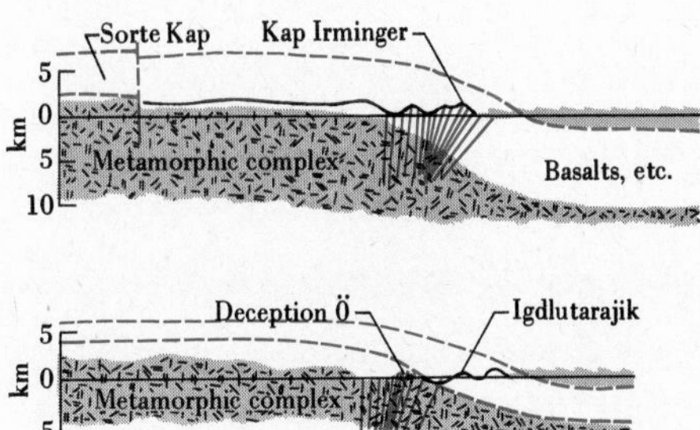

has been reported in the bend of a monocline on the east coast of Greenland.[3] The monocline consists of a sheet of basalt blanketing metamorphic rocks stretching many hundreds of kilometers and varying in dip. The dike swarm is located on the convex side of the flexure and is absent in the concave part (see Figure 18.30). Where the flexure dips 55°, there are more than 70 dikes/km across the structure. When the dip is 12°, there are only about 15 dikes/km, and where the dip is 7°, there are only a few dikes. The dikes dip at about right angles to the basalt sheet and their position indicates that they fill fissures developed by tension as the crust bent. The monocline has a maximum relief of at least 8 km.

Faults

Faults are deformation by rupture in which the sections on each side of the break move relative to each other. They may be of any size, theoretically,

[3] L. R. Wager and W. A. Deer, "A Dike Swarm and Crustal Flexure in East Greenland," *Geol. Mag.*, vol. 75, pp. 39–46, 1938.

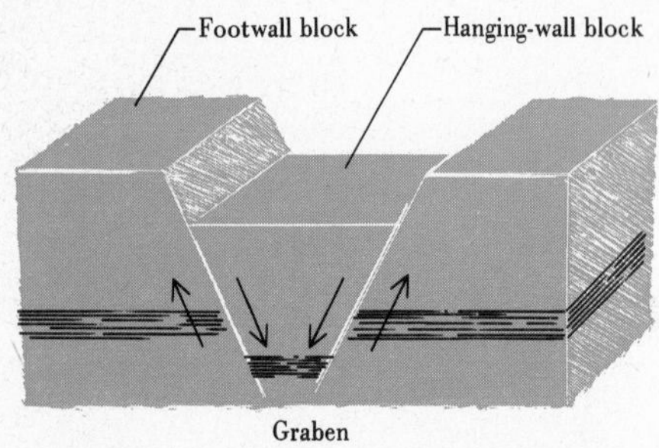

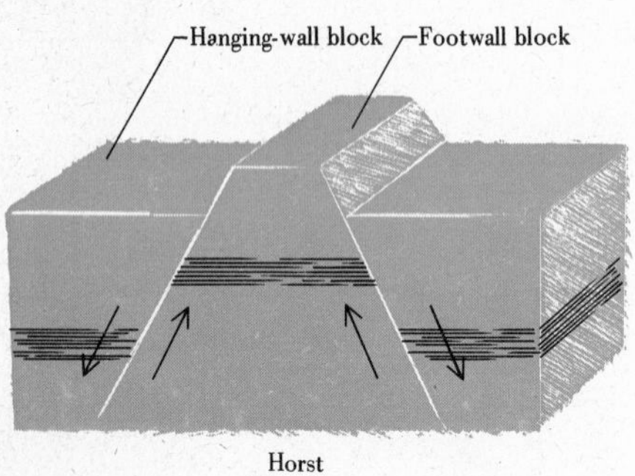

Figure 18.31 *A graben and a horst.*

but here we shall discuss only the types that are large shearing surfaces in the earth's crust.

A useful basis for the classification of faults is the nature of the relative movement of the rock masses on opposite sides of the fault. If displacement is in the direction of dip, it is called a *dip–slip fault.* The sections separated by the fault are named as they were by miners who encountered faults underground. The block that is overhead is called the *hanging wall,* and the one beneath is the *foot wall.* Dip–slip faults are classified according to the relative movement of these blocks. If the hanging wall seems to have moved downward in relation to the foot wall, it is called a *normal fault.*

Normal faults are a result of tensional stresses due to push from below that causes the surface to stretch and rupture. A *graben* is an elongated, trenchlike structural form bounded by parallel normal faults created when the block that forms the trench floor moved downward relative to the blocks that form the sides. The term comes from the German for "trough" or "ditch" and is the same in both singular and plural. A *horst* is an elongated block bounded by parallel normal faults in such a way that it stands above the blocks on both sides. The term comes from the German and is used figuratively for a crag or height. The type of movement involved in formation of a graben and a horst is illustrated in Figure 18.31.

A dip–slip fault in which the hanging wall appears to have moved upward in relation to the foot wall is a *thrust fault.* It is the result of largely horizontal compressive stresses.

A fault along which the movement has been predominantly horizontal is called a *strike–slip fault,* because the slipping has been parallel to the strike of the fault. Such faults have also been called *transcurrent faults.* Strike–slip faults are further designated as *right-lateral* or *left-lateral,* depending on how the ground opposite you appears to have moved when you stand facing the fault. Types of faults are listed in Table 18.3 and shown in Figure 18.32.

The directions of movement involved in faulting are entirely relative. It is convenient to indicate one block as having moved upward or downward or as moving left or right. However, the absolute direction of movement usually cannot be determined, and the best that can be done is to indicate relative movements. During the upward warping of a large region, for example, all blocks could have moved upward, but some may have moved less than others. These lagging blocks could be said to have dropped relative to their neighbors

Table 18.3 *Types of faults*

Type	*Predominant relative movement*
Strike–slip	Parallel to the strike
Right lateral	Offset to the right
Left lateral	Offset to the left
Dip–slip	Parallel to the dip
Normal	Hanging wall down
Thrust	Hanging wall up
Oblique–slip	Components along both strike and dip
Hinge	Displacement dies out perceptibly along strike and ends at a definite point

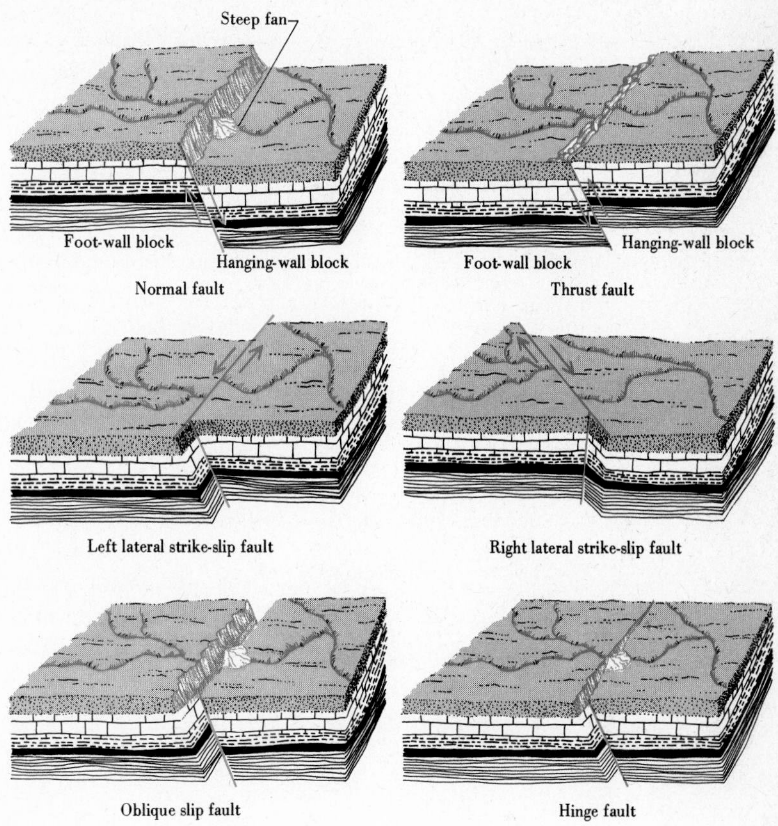

Figure 18.32 *Types of faults.*

even though all parts at the finish were at higher elevations than when they started.

NORMAL FAULTS In the earth's crust there are large zones which have repeatedly been disturbed by large-scale normal faulting. The faults form horsts and rift valleys. The most famous, the African rift zone, extends over 6,000 km in a north–south direction. In some areas, volcanoes are found along the zone where magma has worked its way up along fault surfaces. In Europe, there are several zones of normal faults, including the Upper Rhine valley rift zone and the Roer valley rift. The maximum total displacement of the Roer valley rift is 2,000 m. The dip of its faults ranges from 40° to 70°. The greater the displacement, the smaller the dip. In the western part of the United States, we find an extensive zone of normal faulting in the Basin and Range province. Innumerable smaller normal fault zones can be found all over the world.

The Basin and Range province consists of tilted fault blocks. Its eastern limit is the Wasatch fault. The main faults in the vicinity of Wasatch Mountain dip 50° to 55°. Net slip on major fault planes in the Wasatch Range is from

2,200 to 2,600 m. Farther west are the faulted Oquirrh Range, the Stansbury Range, and the Cedar Range. Each is a tilted fault block some 30 km in breadth. The western limit is the great normal fault which limits the tilted Sierra Nevada block on the west.

THRUST FAULTS Thrust faults are usually closely associated with the folding process. Big, steep thrust faults are characteristic of the marginal stress of mountain chains and in their uplifted central blocks of crystalline rocks. Rocks above the fault plane seem to have been pushed up and over those below. Along the fault plane, rocks are considerably crushed and sheared. Thrust faulting is more clearly seen in eroded regions where metamorphic rocks that were once deeply buried now overlie sedimentary rocks which were formed near the surface. Old rocks generally are thrust over younger ones.

The total movement along a thrust fault can only be estimated in most cases, but some involve total vertical displacements as much as 1,000 m, combined with horizontal movements of tens of kilometers. The dips of thrust faults vary. They may be less than 10° or up to 60°.

Some well-known regions of thrust faulting are in the Alps, the southern Appalachians, the central and northern Rocky Mountains, and southern Nevada. Besides thrust faults of considerable displacement, every deformed region contains numerous small thrust faults (see Figure 18.33). These may have displacements as small as 2 m, which do not usually appear on geologic maps unless they occur in a mining area.

Thrust faults are often associated with asymmetric anticlines where rupture

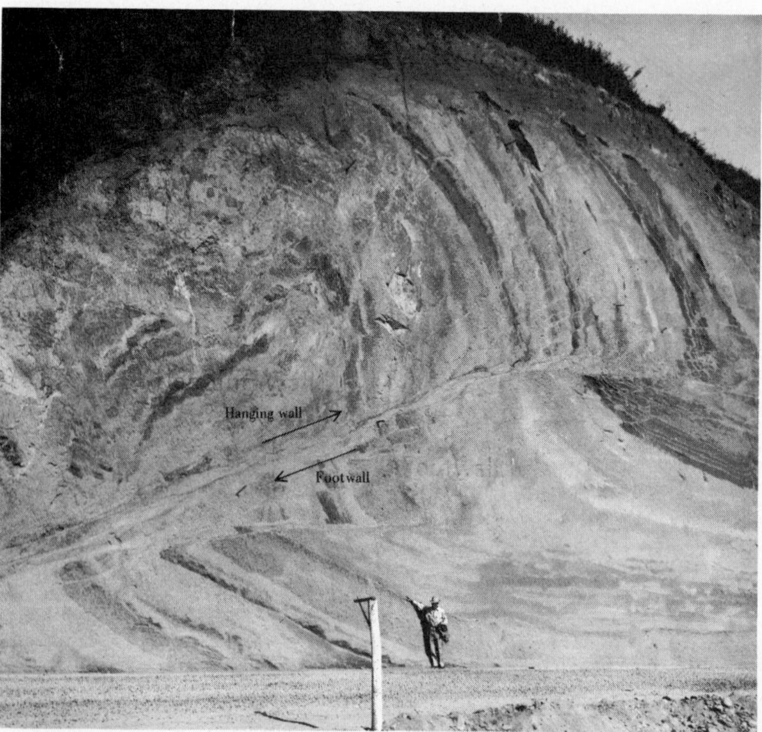

Figure 18.33 ***Thrust fault on the Pan American Highway 40 km northwest of San Salvador.*** *(Photograph by Thomas F. Thompson.)*

occurs in the highest part of the fold. A thrust fault of this type is a superficial structure because it extends down only to the basal plane and occurs in brittle rocks near the earth's surface. Some are found in the Valley and Ridge province of the Appalachians, with the best known being the Cumberland thrust in Virginia (see Figure 18.34).

STRIKE–SLIP FAULTS (TRANSCURRENT) Strike–slip faults are found all over the world. A well-known example of left-lateral strike–slip faulting is the Great Glen fault, which intersects Scotland from coast to coast. There is a string of lakes, including Loch Ness, along its eroded outcrop, which is marked by a belt of crushed, sheared rock up to 1.5 km in width. Horizontal displacements along the fault of as much as 100 km have been suggested on the basis of correlating geological structures on both sides. It dates from the Upper Paleozoic.

Several left-lateral strike–slip faults rip across folds of the Jura Mountains not far from Lake Geneva and Lake Neuchatel. They clearly originated during the folding process.

The San Andreas fault in California is a right-lateral strike–slip fault that has been traced for nearly 1,000 km. Terrace deposits cut by the fault show offsets of as much as 10 km, and stream channels show shifts of 25 km within relatively short geological intervals. The total strike–slip has been debated. Evidence has been interpreted as showing a displacement of 16 km since the Pleistocene, of 370 km since late Eocene time, and 580 km since Precretaceous time. The Garlock fault separates two clusters of dike swarms which seem pretty clearly to have been intruded as a single event. They are now separated by 80 km in a sense that makes the Garlock a left-lateral strike–slip fault.

TRANSFORM FAULTS Another basis for the classification of faults is their relationship to other structural features.[4] It has been suggested that deformation of the earth's crust produces major structural forms: mountains, deep sea trenches, midocean ridges, and strike–slip faults with large horizontal displacements. J. Tuzo Wilson proposed that these features are interrelated, that they do not come to dead ends but link in continuous networks girdling the

[4] J. Tuzo Wilson, "A New Class of Faults and Their Bearing On Continental Drift," *Nature*, vol. 207, pp. 343–347, 1965.

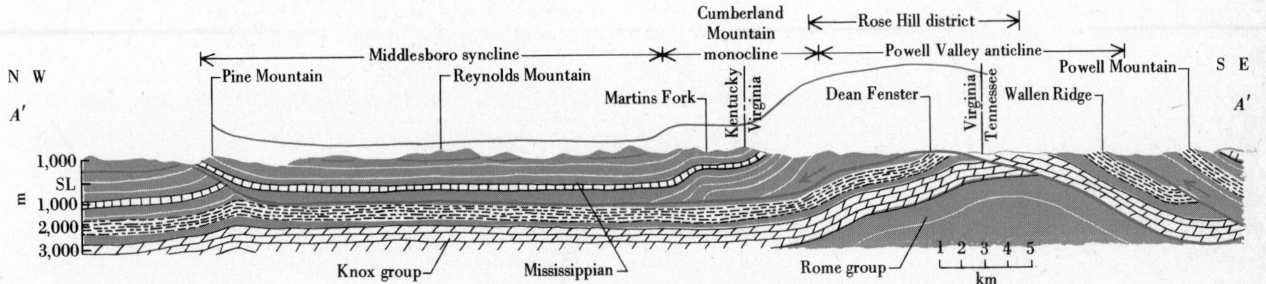

Figure 18.34 ***Cumberland overthrust, Valley and Ridge Province of the Appalachians.*** (*After R. L. Miller and J. O. Fuller,* Bull. Virginia Geol. Survey, *vol. 71, 1955.*)

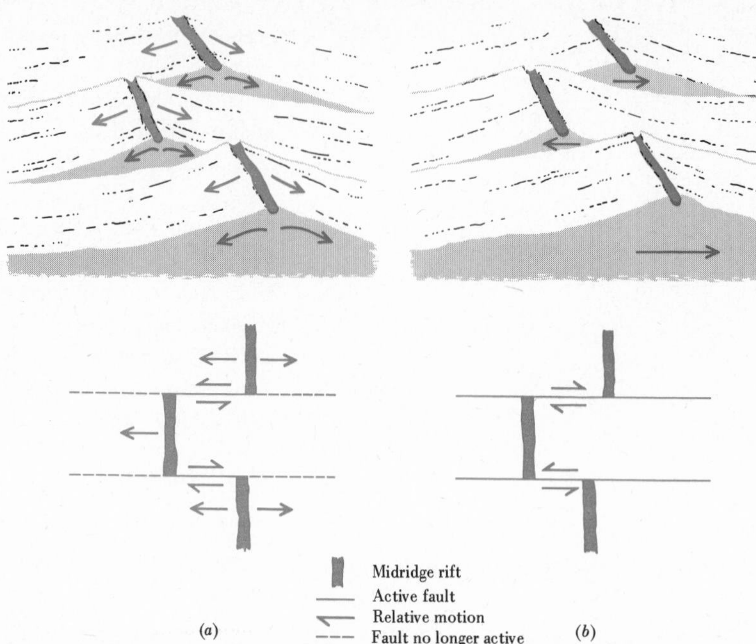

Figure 18.35 **Comparison between transform and transcurrent faults. The perspective sketch and plan in (a) show a ridge-to-ridge transform fault. New crustal material is postulated as being continuously added at the crest and transferred laterally away from the crest. There is no net displacement of the forms of the ridge segments with time. Transcurrent faults are shown in perspective and plan in (b). The ridge fragments are increasingly offset with time.**

globe, and that any major feature at its apparent termination may change into one of the other types. A junction where this takes place is called a *transform.* It has been proposed that the name *transform fault* be applied to a class of faults connecting other major structural features. Figure 18.35 suggests the relationships that exist along a transform fault connecting two segments of a fragmented oceanic ridge, and compares them with the relationships along a transversely faulted ridge. Analysis of earthquakes along the faults connecting segments of ridges indicates the motion to be as shown in Figure 18.35(*a*). This is exactly opposite to that which would result from a simple offsetting of the ridge by transcurrent faults, as shown in Figure 18.35 (*b*). The original offset of the ridges along transform faults may well have been due to transcurrent faulting. But with the establishment of transform motion the ridge crests do not change position with time. In the case of continued transcurrent faulting the offset increases with time. Many of the faults across the ocean ridges have been interpreted as transform faults, as for example those of the equatorial Midatlantic Ridge shown in Figure 18.36. Later, in Chapter 20, we will reconsider the question of transform faults. We should point out now, however, that the transform fault as discussed here demands the creation of new oceanic crust along the crest of the ridge and its lateral

Figure 18.36 **Equatorial section of the Midatlantic Ridge showing ridge crest and fracture zones.** (*After L. R. Sykes, "Seismological Evidence for Transform Faults, Sea Floor Spreading and Continental Drift," in R. A. Phinney, ed.,* The History of the Earth's Crust, *Fig. 10, Princeton University Press, Princeton, N.J. 1968. © 1968 by Princeton University Press, Reprinted by permission.*)

movement away from it. As we shall also see in Chapter 20, the counterpart of this would be the destruction of crust at other localities.

Unconformities

There is no known place on earth where sedimentation has been continuous throughout geologic time. During the formation of a continent, large sections of the crust have been raised out of the shallow seas in which its rocks were formed, subjected to erosion, and then lowered again to levels where deposition of sediments is renewed. Activity of this sort produces a buried erosion surface with younger rocks overlying older rocks. Some surfaces of this kind can be seen today because of another cycle of uplift. A buried erosion surface separating two rock masses, the older of which was exposed to erosion for a long interval of time before deposition of the younger, is called an *unconformity*.

The time represented by an unconformity is important geologic evidence of the history of a region, marking an interval when the surface was above the sea and sediments were not being deposited. Some unconformities represent gaps of a few thousand years and others as many as 400 million years.

There are three principal types of unconformities: angular unconformity, disconformity, and nonconformity.

ANGULAR UNCONFORMITY An unconformity in which the older strata dip at an angle different from that of the younger strata is called an *angular unconformity*. On a wall of Box Canyon near Ouray, Colorado, can be seen some layers of Precambrian sedimentary rocks dipping at nearly 90°. Above them are some nearly horizontal Devonian sedimentary rocks. The older rocks were deposited under the waters of the ocean and were then folded and uplifted

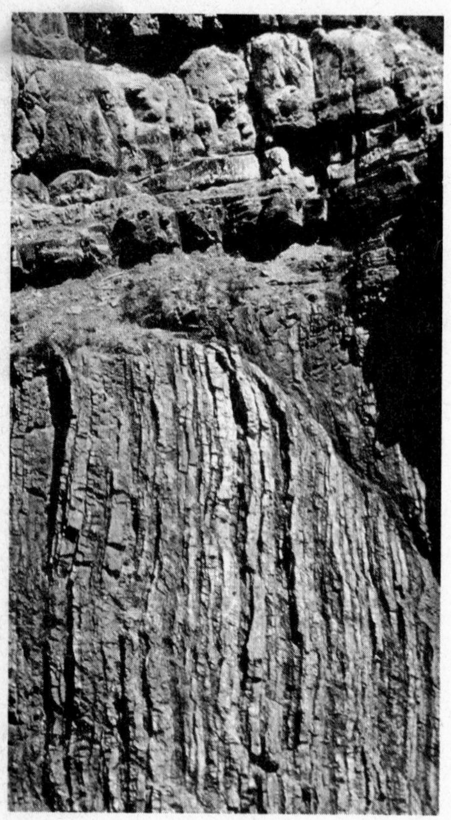

Figure 18.37 ***A striking unconformity between Precambrian sedimentary rocks that were twisted from their original horizontal position before deposition of overlying Devonian beds. In Box Canyon, near Ouray, Colorado.*** (*Photograph by Kirtley F. Mather.*)

above the water. While exposed at the surface, the tilted beds were beveled by erosion. Then, as time passed, these tilted and eroded rocks sank again beneath the ocean, where they were covered by new layers of sediments. Both were later elevated and exposed to view (see Figure 18.37). On the basis of fossil evidence, we know that the second sedimentation began during the Devonian, indicating that more than 180 million years elapsed between the two periods of sedimentation. The Precambrian rocks meet the younger Devonian rocks at an angle, so this unconformity is an angular unconformity.

DISCONFORMITY An unconformity with parallel strata on opposite sides is called a *disconformity.* It is formed when layered rocks are elevated and exposed to erosion and then lowered to undergo further deposition without being folded. Careful study and long experience are required to recognize a disconformity, because the younger beds are parallel to the older ones. Geologists rely heavily on fossils to correlate the times of deposition of beds above and below a disconformity. Fossils are a useful tool because certain ones lived during a definite geologic time range.

Figure 18.38 ***A nonconformity between light-colored granite and dark overlying Table Mountain sandstone on the Cape of Good Hope, South Africa. The cave, about 6 m above sea level, was cut by waves when sea level was higher than it is now.*** (*Photograph by R. A. Daly.*)

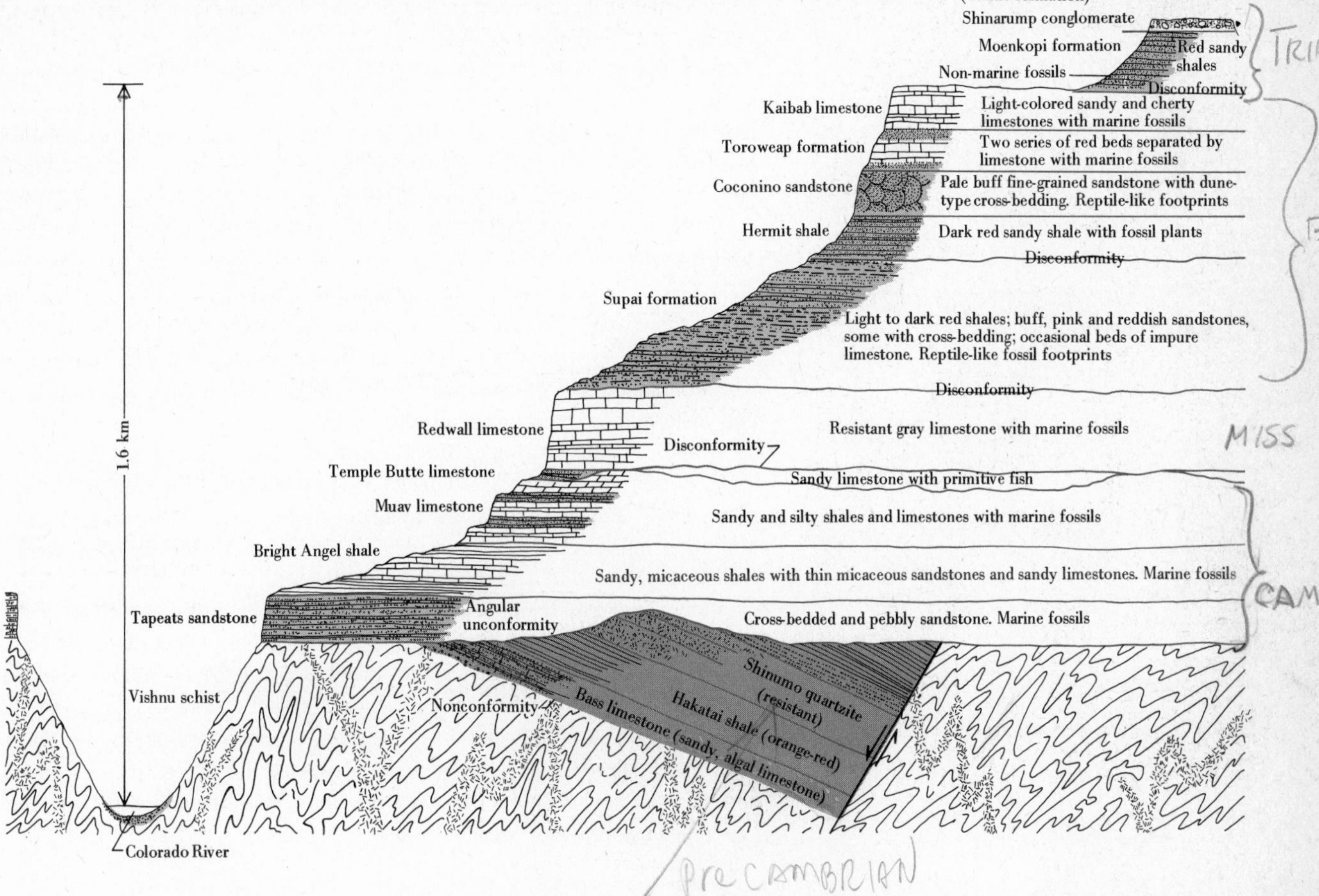

Figure 18.39 ***Diagrammatic representation of one wall of the Grand Canyon, showing unconformities.*** (*After John Shelton,* Geology Illustrated, *W. H. Freeman and Company, San Francisco, 1966.*)

NONCONFORMITY An unconformity between profoundly different rocks is called a *nonconformity*. Nonconformities are formed where intrusive igneous rocks or metamorphic rocks are exposed to erosion and then downwarped to be covered by sedimentary rocks. A structure of this sort is illustrated in Figure 18.38, which shows a sandstone deposit lying on top of an eroded surface of granite. Field studies have revealed that pieces of weathered granite occur in the bottom layers of the sandstone and that cracks in the granite are filled with sandstone. This evidence supports the view that the granite did not come into its present position as an intrusion after the sandstone was formed.

UNCONFORMITIES IN THE GRAND CANYON The Grand Canyon of the Colorado exposes several unconformities (see Figure 18.39). This region has undergone three major sequences of uplift and erosion, subsidence and deposition, crustal deformation and crustal stability. These have produced three major rock units: (1) Vishnu schist, (2) Grand Canyon series, and (3) Paleozoic series.

The Vishnu schist is the base rock of the canyon. It is composed of highly metamorphosed sedimentary rock and some volcanics. It contains no fossils but may be 2 billion years old.

Resting unconformably on the Vishnu schist in some places are tilted beds of the Precambrian Grand Canyon series. The beds consist of 75 m of Bass limestone, 180 to 240 m of sandy shale, and 300 m of relatively resistant beds of Shinumo sandstone, topped off by 600 m of shaly sandstone. It has been estimated that an additional 3,000 m of Grand Canyon sediments were deposited on the beds before the region was elevated, deformed, and then eroded to the surface on which Paleozoic sediments were deposited. An uplift of at least 3,600 m had to take place to bring the lowest Grand Canyon layers above sea level. Erosion then removed all but the lower sections of some tilted blocks. It is not possible at the present time to say how long it took to complete the downwarping of the Vishnu schist, deposition, sedimentation, deformation, elevation, and erosion.

When the Grand Canyon series was eroded to a surface of low relief and then submerged, Paleozoic sediments were deposited. These sediments are found today in horizontal layers along the upper part of the canyon walls, indicating uplifting of the region without deformation. The Paleozoic series contain at least four discontinuities. The longest gap is between the Muav limestones and the Temple Butte formation. The Muav limestones were deposited in Cambrian time. The Temple Butte formation immediately overlying these and essentially parallel to them contains fossils of a primitive armored fish which occur only in Devonian rocks. Accordingly, there was no deposition during Ordovician or Silurian time, a lapse of approximately 80 million years.

In areas where all the Grand Canyon series had been eroded away before deposition of Paleozoic sediments, we find the "great unconformity." Its time gap is estimated as several hundred million years.

18.5 *Broad surface features*

Every continent has a nucleus of Precambrian rocks. These are called *shields*, and the continents have grown around them. The oldest rocks were formed about 3 billion years ago along chains of volcanoes. By about 2 billion years ago, platforms of the earliest rocks had developed on which sediments accumulated and geosynclines developed. The full range of mountain-building cycles has been operating since then. Deformation has produced the continental land features we find today—the plains, the plateaus, and the mountains.

North America's shield, the Canadian shield (Figure 18.40), is exposed over an appreciable part of the northeastern portion of the continent. It covers more than 5 million km^2 and includes all of Labrador; much of Quebec, Ontario, and the Northwest Territories; the northeastern parts of Manitoba and Saskatchewan; and part of the Arctic Islands. Exposure at the surface could not have taken place without elevation—in this case brought about by a broad warping of the basement complex.

Figure 18.40 **Broad surface features of North America.**

Bordering the shield are the interior lowlands, extending to eastern Ohio on the east, the Ouachita Mountains on the south, and the Arctic Ocean on the north. They are composed of sedimentary rocks, the oldest having been deposited in seas lapping up onto the shield during Cambrian time. On 14 different occasions, portions of this continental lowland sank beneath the ocean and later rose above the water. So it was a restless area during this building of the continent, and the rocks bear the traces of unsettled times. Gently tilted in places, folded into small domes and basins in others, their thickness varies to more than 3,000 m in areas of exceptional subsidence and deposition; still, they lie relatively low and flat and form plains.

The sedimentary rocks surrounding the interior lowland plains have been broadly upwarped into plateaus, although the rocks still lie nearly flat. Still farther from the shield the plateaus merge into mountain ranges, which represent deformation at its greatest. As much as 15,000 m of sediments accumulated here through the formation of basining in the basement complex; they then consolidated into sedimentary rock. Intense folding, rupture, and elevation of the entire region resulted in mountain ranges such as the Appalachians and the Rockies.

Plains are regions of low relief composed of nearly horizontal sedimentary rocks. Plateaus are higher regions of nearly horizontal strata with a relief generally over 150 m. Plateaus may originate by uplift or by deposition of lava flows. Mountains encompass a wide range of geological structures and relief. These include fault block mountains, thrust mountains, and mountain ranges. Youthful mountains stand today thousands of meters high, whereas the older mountain chains that have been attacked by many more years of erosion have low relief.

Broad warps

Some large sections of the continents are composed of horizontal strata which have been gently bent upward or downward. In North America, these strata are found in the lowland between the Rocky Mountains and the Appalachians. They are distinguished only by mapping the elevation of certain rock layers. Data for the mapping were obtained from many field studies or from the borings obtained from thousands of drill holes. Regional downwarping forms basins; upwarping forms plateaus.

The Great Basin, 470,000 km^2 between the mountains, is centered in Nevada and includes adjacent parts of California, Oregon, and Utah. Its relief is close to 3,900 m.

Outline

Evidence of deformation of the earth's crust is best preserved in sedimentary and metamorphic rocks.

- *Abrupt movements of the earth's crust* have occurred at the times of some large earthquakes.
  - *Vertical displacements* have been such as to amount to 15,000 m if continued at the same rate for 2 million years.
  - *Horizontal displacements* have been estimated as great as 200 km in 150 million years.
- *Slow movements* are being measured instrumentally in several places today and on the San Andreas fault near Parkfield they average 5 cm/year.

Kinds of deformation are described in terms of change of volume, change of shape, or a combination of both.

- *Stress–strain* relationships are used to define a material's response to stress.
  - *Elastic deformation* is recoverable, plastic deformation is not.
    - *Retardation of recovery* is recalled hysteresis.
  - *Plastic deformation* is permanent and involves a property of rock called viscosity.
  - *Time factor* sometimes results in a material being elastic for a short time but plastic over a longer time.

Types of stress are tension, compression, and shear.

- *Strength of rocks* is the stress at which they begin to be permanently deformed.

Structural features are the shapes of deformed masses of rock, including folds, faults, joints, and unconformities.

- *Folds* are a common feature of rock deformation and are produced by compressive stress.
  - *Mechanism of folding* is elastic bending or plastic flow.
  - *Monoclines* are double flexures.

Anticlines and *synclines* occur together in elongate groups like belts of wrinkles in the earth's outer skin.

Structural domes result from pressures acting upward.

Deformation of rock by fracture produces joints and faults.

Joints usually occur in sets.

Columnar jointing is sets of cracks produced by the mechanism of cooling certain tabular plutons to outline columns.

Sheeting is a pattern of essentially horizontal joints.

Joints formed during movements of magma include dikes and sills.

Faults are deformation by rupture in which the sections on each side of the break move relative to each other.

Normal faults have the hanging wall apparently dropping relative to the foot wall.

Thrust faults have the hanging wall apparently moving higher than the foot wall.

Strike–slip faults have movement predominantly horizontal parallel to the strike of the fault.

Transform faults are strike–slip faults that terminate in a mountain range, midocean ridge, or deep sea trench.

Unconformities are buried erosion surfaces separating rock masses where the older was exposed to erosion for a long interval of time before deposition of the younger.

Angular unconformity is one in which the older strata dip at an angle different from that of the younger.

Disconformity is an unconformity with parallel strata on opposite sides.

Nonconformity is an unconformity between profoundly different rocks.

Unconformities in the Grand Canyon show three major sequences of uplift and erosion, subsidence and deposition, crustal deformation and crustal stability.

Broad surface features are basins, plains, plateaus, mountains, and Precambrian shields with continents built around them.

Broad warps have produced basins and plateaus.

Supplementary readings

BILLINGS, MARLAND P.: *Structural Geology*, 2nd ed., Prentice-Hall, Inc., Englewood Cliffs, N.J., 1954. A textbook designed to follow an introductory course in physical geology.

DE SITTER, L. U.: *Structural Geology*, McGraw-Hill Book Company, Inc., New York, 1956. A textbook presupposing a certain familiarity with the elements of structural geology and with its terminology.

SPENCER, EDGAR W.: *Introduction to the Structure of the Earth*, McGraw-Hill Book Company, Inc., New York, 1969. An up-to-date survey of the subject.

UMBGROVE, J. H. F.: *The Pulse of the Earth*, 2nd ed., Martinus Nijhoff, The Hague, 1947. A treatise that must be rated a classic for its impact on the broad philosophical aspects of structural geology.

19

Mountains and mountain building

MOUNTAINS ARE SPECTACULAR PRODUCTS OF DEFORMING FORCES ACTING ON THE earth's crust (see Figure 19.1). They are called *mountains* because their peaks stand from a few hundred to a few thousand meters above the surrounding terrane. Therefore, one thing they have in common is elevation above their immediate surroundings. The highest mountain in the world, Mount Everest, stands 8,700 m above sea level.

Mountains are born, grow, achieve old age, are worn down, and die. This all takes time. The oldest known rocks on the earth's surface today are believed to be the roots of some ancient mountains. These form the shield areas of the continents. In North America, the oldest mountains are found in southwestern Minnesota, where granite gneisses have been dated as 3.55 billion years old.[1] Throughout geologic history, mountains have been formed at various times and places. Every continent has mountains, and there may be no place on the land where there have not been mountains at one time or another.

Looking around us today, we see many high mountains still withstanding weathering, mass movements, and the work of running water, wind, and glaciers. These *external* processes have been continually at work leveling the land for millions of years. We might expect that they would have reduced all the land surfaces of the earth to great plains long ago, yet we see many high mountains still standing. We must conclude, therefore, that there are *internal* forces at work that counteract the effects of the leveling processes by continually renewing the height of the land.

Mountains differ in history and in age. They may occur as single isolated peaks or in lines or groups. The name *mountain range* has been given to an elongated series of mountain peaks considered to be a part of one connected unit (see Figure 19.2), such as the Appalachian Range, the Sierra Nevada, and the Front Range. A broad belt of mountain ranges is called a *cordillera.* The American Cordillera extends continuously from Alaska to Cape Horn and is subdivided into the North American Cordillera and the South American Cordillera. The North American Cordillera includes the Coast Ranges, the Sierra Nevada, the Cascades, the Great Basin Ranges, and the Rocky Mountains. Mountains, ranges, and cordilleras are all descriptive terms and we can get into complications if we use them to imply anything about their structural features or processes of formation.

The development of mountains is called *orogeny,* from the Greek *oros,* "mountain", and *genesis,* "to come into being." Orogenic forces determine where and when surface irregularities appear in the earth's crust. The geological processes that build mountains take millions of years—years during which the internal forces of the earth are waging a constant battle against erosion. To achieve the elevation of a Mt. Everest, for example, internal forces must push the land up faster than erosion can wear it away. But even if these forces succeed in elevating a mountain range, it is only a passing triumph, for external forces continually attack the elevated mass and in time will manage to reduce it nearly to sea level once again.

Mountains are the dominant surface features of the continents of the world, North America, South America, Eurasia, Africa, Australia, and Antarctica (see

[1]E. J. Catanzano, "Zircon Ages in Southwestern Minnesota," *J. Geophys. Res.*, vol. 68, pp. 2045–2048, 1963.

Figure 19.1 **Distorted rocks 4,000 m high in the Bernese Alps, Switzerland.** (*Photograph by Bradford Washburn.*)

Figure 19.2 **A mountain range near Sion, Switzerland; view southwest down the Rhône valley.** (*Photograph by Bradford Washburn.*)

Figure 19.3). Each continent has a core of old land that has been relatively stable for the past 500 million years, but the margins of the continents have been scenes of active mountain building.

Geologically "young" mountains include the Andes, Alps, Himalayas, and North American Cordillera. Young mountains are characterized by high elevations and active deformation going on today. Some geologically "old" mountains include the Green Mountains of Vermont and the Catskills. Age is indicated by their low, round shapes and their Precambrian rocks.

In a general way, the major landmasses of the earth, except for about one-twentieth of the total land surface, are situated on the side of the globe opposite water-covered areas. This pattern becomes less striking when we realize that some mountains are actually submerged beneath the waters.

19.1 Kinds of mountains

It is very difficult to set up a clear-cut classification of mountains because most have a variety of structural features, but it is helpful to do the best we can by grouping certain types that are dominated by one characteristic. On this basis, there are four main kinds of mountains: volcanic, fold, fault-block, and broadly upwarped (see Figure 19.4). Mountains built up by the extrusion of lava and pyroclastic debris are called *volcanic mountains* (see Chapter 4). Mountains consisting primarily of elevated folded sedimentary rocks are called *fold mountains.* Where faulting is dominant and has produced large elevated crustal blocks bounded by faults, these are called *fault-block mountains.* The rocks composing the fault blocks may be either flat-lying or folded. When broad upwarping of the crust has elevated rocks to mountainous relief, the resulting landforms are *broadly upwarped mountains.* Upwarped mountains may also have been folded during some previous cycle in their history.

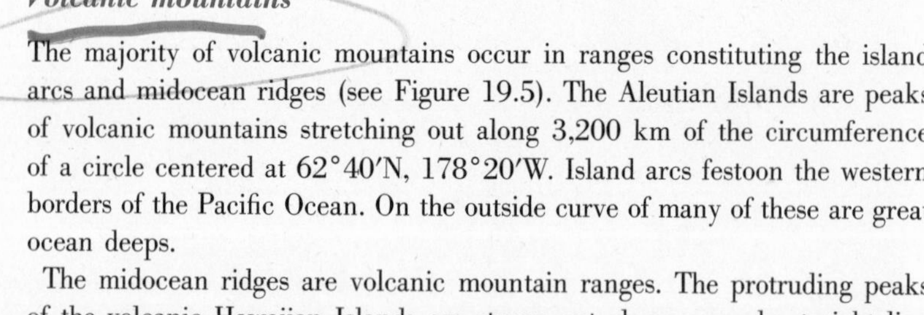

Volcanic mountains

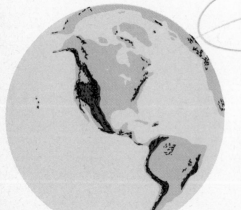

The majority of volcanic mountains occur in ranges constituting the island arcs and midocean ridges (see Figure 19.5). The Aleutian Islands are peaks of volcanic mountains stretching out along 3,200 km of the circumference of a circle centered at 62°40′N, 178°20′W. Island arcs festoon the western borders of the Pacific Ocean. On the outside curve of many of these are great ocean deeps.

The midocean ridges are volcanic mountain ranges. The protruding peaks of the volcanic Hawaiian Islands are strung out along a nearly straight line 2,400 km long, running to the northwest. Parallel to this line are other island chains of the Pacific: the Marquesas, Society, Tuamotu, Tubuai, and Samoa.

Under the Atlantic Ocean, from Iceland to the Antarctic, is a belt of moun-

Figure 19.3 ***Areas more than 1,000 m above sea level.***

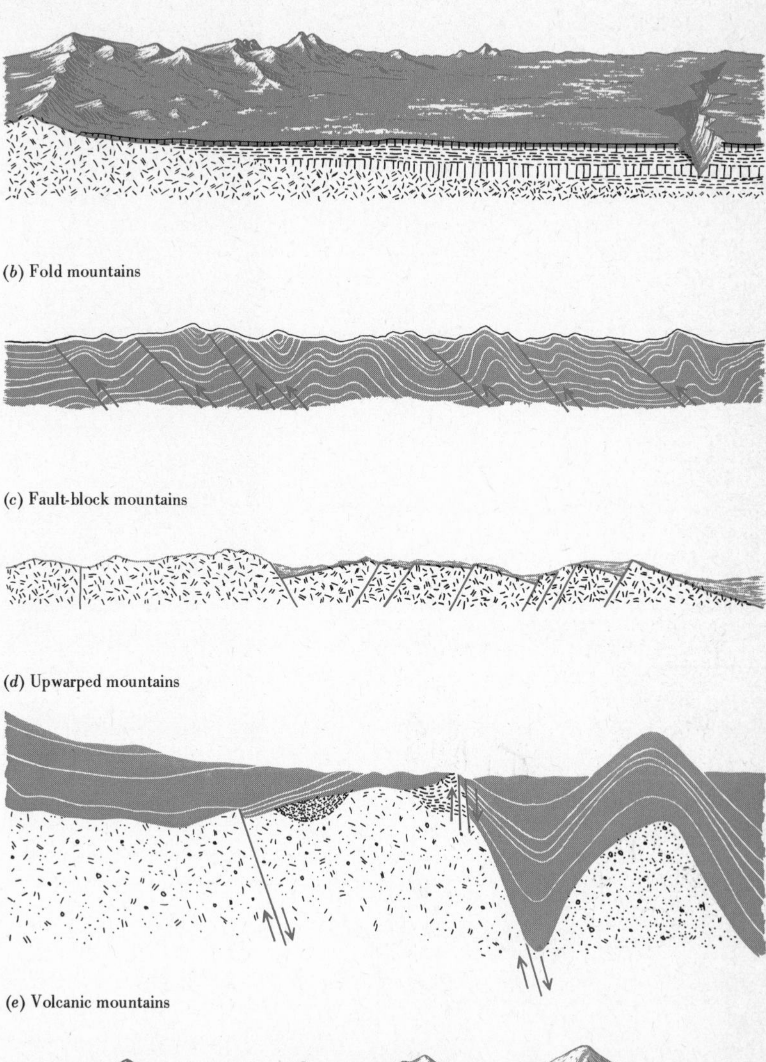

Figure 19.4 ***Plateau and kinds of mountains.***

tains called the *Atlantic Ridge* or *Midatlantic Ridge*, which roughly parallels the outlines of the continents and is nearly midway between them. It stands as much as 1,800 m above the ocean bottom and is covered in places by 2,700 m of water. In a few places, its peaks protrude above the water to form islands such as the Azores, St. Helena, and Tristan da Cunha. There are similar ridges under the Indian Ocean. The *Kerguelen–Gaussberg Ridge* runs between India and Antarctica, and the *Carlsberg Ridge* runs from the Gulf of Aden south

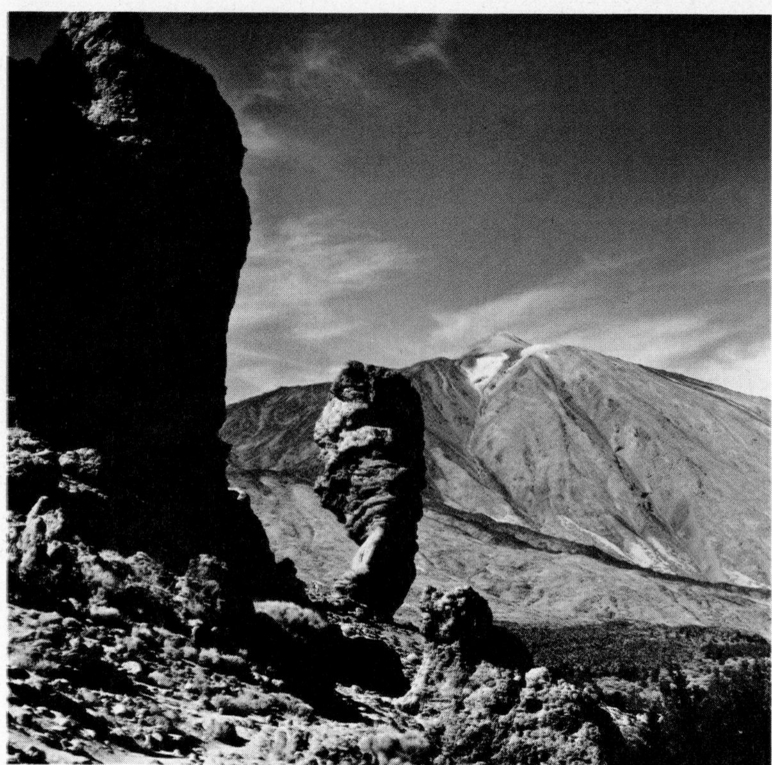

Figure 19.5 ***Teide volcano, the highest mountain in the Atlantic basin, rises 3,730 m above sea level on the island of Tenerife, Canary Islands. It last erupted in 1909. The jagged erosional spires of breccia and dike rock rising in the foreground are remnants of an ancient caldera wall.*** *(Photograph by Richard S. Fiske, U.S. Geological Survey.)*

of Arabia to join Kerguelen–Gaussberg near the Maldive Islands southwest of India. Over 64,000 km of midoceanic ridges have been mapped.

Few volcanic mountains are found on the continents. Isolated peaks such as Kenya, 5,100 m high, and Kilimanjaro, 5,800 m high, are found rising on the plain in Africa. In South America, volcanoes constitute many peaks of the Andes. The Cascades are a range of volcanic mountains in Washington and Oregon. The highest peaks include Mounts Hood, Adams, Scott, Rainier, and Shasta.

Volcanic mountains seem to have their origin connected to deep faults, which extend below the earth's crust to the mantle to supply their building materials.

Fold mountains

Folded mountain ranges are complex structures involving all the mountain-building forces of folding, faulting, and igneous activity. The Rockies, Alps, Himalayas, Juras, Urals, and Carpathians are some of the world's folded mountain ranges. They provide us with spectacular examples of deformed rocks

Figure 19.6 **Folded and elevated sediments form these mountains in the Rockies.** (*Photograph by D. B. Dowling.*)

(see Figure 19.6). These can be seen where streams have cut through them as the land moved upward during elevation. Anticlines and synclines are numerous. Along the borders of these mountains, faults are numerous, and there are few unbroken folds. Some of the faults are normal faults on the flanks of anticlines. Many are low-angle thrust faults that have pushed masses of rock up and over other portions of the crust for many kilometers. Some thrust faults can be traced for hundreds of kilometers along their strike, until they eventually die out. The land bordering some of these mountain ranges now stands as plateaus—broad high-standing regions deeply cut by streams and composed of sedimentary rocks relatively little deformed.

GEOSYNCLINES The folded mountain ranges of the world have developed from thick deposits of sediments that accumulated in geosynclines. A *geosyncline* is a large sediment-filled basin that formed from deformation of the earth's crust. Geosyncline means literally "earth syncline." Usually scores of kilometers wide and hundreds of kilometers long, some geosynclines have accumulated sediments 15,000 m or more in thickness, but their fossils, texture, ripple marks, and other features indicate that the water in which they accumulated was not deep—around 300 m at most, it is estimated. So there must have been a slow sinking of the crust, as deposition of sediments more or less kept pace.

Formation of a geosyncline We know that the development of fold mountain ranges is always preceded by the formation of geosynclines. But this is not a simple process that takes place all at once—a huge basin on the earth's surface does not suddenly develop into which the sediments pour until it is filled up. Rather, the formation of a geosyncline involves a slow, continuous downwarping of the earth's crust, with the deposition of sediments going on at the same time. The growing weight of the sediments tends to deepen the trough, but not enough to make room for the thicknesses known to have accumulated.

As a corollary to the problem of how the geosyncline sinks, there is the necessity for elevation of adjacent land to supply the staggering quantities of sediments to fill the geosyncline. Then, the geosyncline must also be elevated if it is to become a mountain range. There seems to have been continuous movement of the crust, sometimes down, sometimes up, in the making of mountains.

The sediments in a geosyncline eventually sink to levels where they become surrounded by denser rock, and their own buoyancy sets a limit to the depth to which they can sink under their own weight. For example, assume that we begin with a trough 300 m deep and that we pour sediments into it. The sediments will push the bottom of the trough down into the denser substratum until it is 750 m below sea level. But by that time, the sediments will have filled the trough to sea level, and there is no room to add any more. The whole system has become isostatic, or in balance, and the thickness of the sediments cannot be increased just by load.

Geosynclines have certain features that are common to all. Sediments were delivered to them from bordering lands. The rocks that formed in the geosynclines were composed of types of sediment that accumulate only in shallow water—probably not deeper than 300 m. Sometimes at the edge of a geosyncline sediments actually piled up above sea level. For example, strata in the mountains of south-central Pennsylvania include 2,700 m of layers that were originally deposited over a great delta plain. These grade northwestward into limestone that was deposited in a shallow sea. These rocks constitute only a part of a deposit that reached 12,000 m in thickness. Such a thick deposit of sediments could have accumulated only if deforming forces slowly deepened the geosyncline while sedimentation kept pace with the deepening.

Deformation of geosynclines For many years, geologists thought that the folding and faulting of the sedimentary layers in geosynclines occurred only after the total thickness of thousands of meters accumulated. However, recent data have indicated that folding and faulting occur continuously while sediments are accumulating.

Rocks at the surface are brittle, and they break before they flow. But under heat and deep burial, they become plastic and change their shape and volume by folding and slow flow. When sediments are buried deep enough, they melt. Expansion of these melted rocks causes the whole overlying mass to rise (see Figure 19.7). Near the edges of the geosyncline the rocks are squeezed upward and outward along great thrust faults; in the central area, they are pushed upward to form an intermountain plateau.

Crustal shortening For many years it has been generally thought that folded and thrust-faulted mountains have involved shrinking of the earth's crust. Some believe the shrinking occurred in the entire crust of the earth and others that it was limited to the upper part of the crust. There are some geologists who see the folding and thrusting as a result of sedimentary rocks sliding along an unyielding basement complex. But Marland P. Billings concluded that "field geology indicates that the basement rock often participates in the folding."[2]

Zones of geosynclines Mountain ranges were not formed by the deformation

[2]Marland P. Billings, "Diastrophism and Mountain Building," *Geol. Soc. Am. Bull.*, vol. 71, p. 369, 1960.

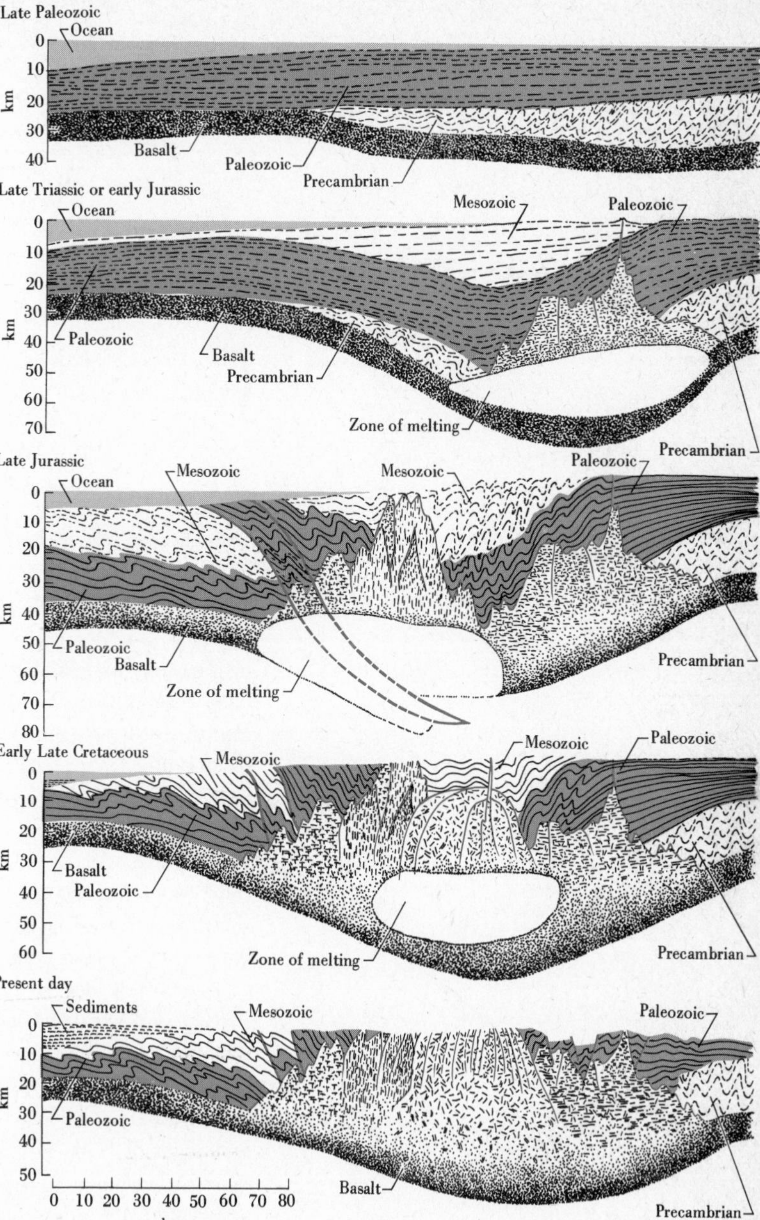

Figure 19.7 ***Cross sections illustrating diagrammatically the sequence of events explaining mountain building from a geosyncline.*** (*Modified from Paul C. Bateman, "Geologic Structure and History of the Sierra Nevada,"* UMR Jour., *no. 1, Fig. 4, p. 128, 1968.*)

of *one* geosyncline but rather by the deformation of many. Two great elongated belts of roughly parallel geosynclines once occupied the region where the Appalachians now stand. Strung out along the Atlantic side of the continent from Newfoundland to Alabama, they extended 2,400 km in length and 560 to 640 km in width. During the Paleozoic, they were supplied with sediments from elevated landmasses in what is now the central part of the United States and from volcanic landmasses in what is now the Atlantic Ocean.

One zone of geosynclines bordered the continent itself and contained deposits

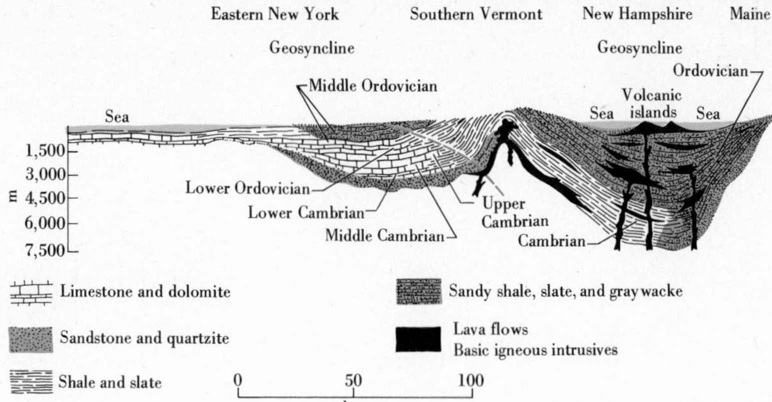

Figure 19.8 **Possible conditions in Appalachian geosynclines across New York, Vermont, and New Hampshire at the close of Ordovician time.** (*After G. M. Kay, "Development of the Northern Allegheny Synclinorium and Adjoining Regions,"* Bull. Geol. Soc. Am., *vol. 53, pp. 1601–1658, 1942.*)

of limestone and sandstone. This zone was nonvolcanic. The other zone was offshore and was supplied by clay sediments from volcanic island arcs (see Figure 19.8). This zone was marked by igneous activity. Moreover, the detailed history of each individual geosyncline within these zones was to a degree independent of the histories of its neighbors.

THE APPALACHIANS The Appalachian Mountains are a belt of mountainous relief extending along the eastern coast of the United States. They do not show the relief or signs of current deformation that are found in the North American Cordillera, and so they are thought of as older mountains. It was once believed that they were all deformed and elevated at the close of the Paleozoic in what was called the Appalachian Revolution. As data have accumulated, however, it has been decided that the Appalachian history consists of more than one episode, involving different sections of the range, the northern or New England Appalachians and the southern Appalachians, essentially north and south of New York City, respectively.

In the northern Appalachians, the mountains had been elevated to their greatest heights before the end of the Paleozoic. By 80 million years ago, they had been eroded to an area of low relief. Then, 20 million years ago, the region began to rise slowly once more, and now it has attained the form of a broad arch, with ridges and peaks standing out as the present Appalachians.

The southern Appalachians consist of a belt of closely folded and faulted Paleozoic rocks extending from the west slope of the Blue Ridge across the Valley and Ridge province to a well-defined structural front west of which the folds are gentle and open, with few if any major faults. The present ridges and valleys in Virginia and western Pennsylvania resulted from erosion of parallel anticlines and synclines of great extent, involving strata ranging in age from Cambrian to early Permian.

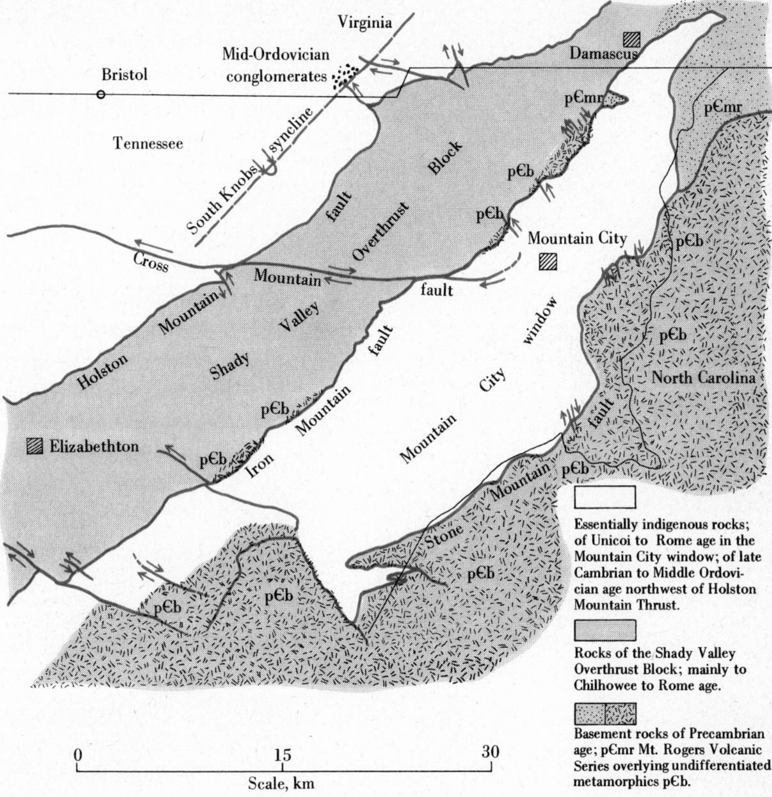

Figure 19.9 ***The Holston Mountain–Iron Mountain thrust forms a major portion of what remains of the extensive fault that occurs along the northwest base of the Blue Ridge for 160 km. In southwestern Virginia and northeastern Tennessee it cuts beds no younger than Middle Ordovician; hence it considerably predates the Appalachian Revolution, traditionally assigned to the late Paleozoic.*** (*After Byron N. Cooper, "Profile of the Folded Appalachians of Western Virginia,"* UMR Jour., *no. 1, Fig. 5, p. 39, 1968.*)

The region has been studied more or less continuously for 125 years but is still rather poorly known. However, from a careful new analysis of their structures, Byron N. Cooper has found some relationships that expand our ideas about the processes involved in the making of mountains.[3]

Cooper concluded that sedimentation and deformation were going on at the same time within the Appalachian geosynclines (see Figure 19.9). The structural features of the Appalachians were produced while the sediments were being deposited and not late in the geosyncline's history. In other words, deformation was not confined to a grand climax at the close of the Paleozoic after 10,000 m of sediments had been deposited, as was long supposed.

Cooper has presented evidence that some of the major Appalachian structures

[3]Byron N. Cooper, "Profile of the Folded Appalachians of West Virginia," *Univ. Missouri Rolla J.*, April 1968.

were formed during middle Ordovician and others at different times, the oldest in the Cambrian and the youngest in Mississippian and Pennsylvanian time. He found evidence of preconsolidation folding that supports his contention that mountain-building forces were active during sedimentation.

Cooper believes that the dominant forces involved in creating the folds and faults of the Appalachians were vertical. He points out that differential downwarping could produce the synclines while deposition was in progress. Thrust faulting along the margins of the geosynclines could have been initiated by a bordering zone of differential subsidence. Tangential compressive stresses became active late in the geosyncline's history to elevate the already deformed strata to mountainous heights.

The more closely we look at the folded Appalachians, the more it becomes clear that mountain building is not a series of separate events: deposition of thousands of meters of sediments, then folding, then elevation—but the deformation is continuous. The earth's surface today is compounded of features in varying stages of the universal sequence of crustal writhing.

NORTH AMERICAN CORDILLERA Along the western border of North America other geosynclinal zones similar to those in the east were also being filled with sediments during the Paleozoic. They extended more than 4,800 km from Alaska to Mexico, and their combined width reached a maximum of 1,600 km. Deformation and elevation began in these geosynclines at least as early as it did in the Appalachian belts, but it was more extensive and is continuing at the present time. Their deformation produced a variety of structures and these constitute the North American Cordillera.

Modern counterparts of these ancient geosynclinal zones are thought to exist today between the Pacific border of Asia and arcs of volcanic islands off the continental coast.

THE COAST RANGES Western California has been the scene of nearly continuous deformation and sedimentation from the middle of the Mesozoic. Until near the end of the Jurassic, this edge of the continent was the site of downwarping with massive sedimentation and outpourings of lava. Strata 20,000 m thick accumulated before elevation of the region stopped deposition. Then, during latest Jurassic and early Cretaceous, thick sediments were again deposited and extensive deformation also took place. This continued into the Cenozoic. In late Miocene, a series of basins and highs developed, which fragmented the topography in the central Coast Ranges. Miocene strata, both sedimentary and volcanic, are marked by facies and thickness changes from place to place. Lower and middle Pliocene sedimentary beds are markedly deformed. At some places they stand vertically or were even overturned. Midpliocene was the time of most active deformation of the Coast Ranges, but there is evidence that some of the strike–slip faults in the area, such as the San Andreas and San Gabriel, had their beginnings in Midmiocene some 6 million years ago. This region is still undergoing deformation today.

The Coast Ranges are a series of northwest-striking mountain ranges with intervening valleys. They consist of broken and deformed sedimentary and volcanic rocks ranging in age from Precambrian through Paleozoic. The oldest

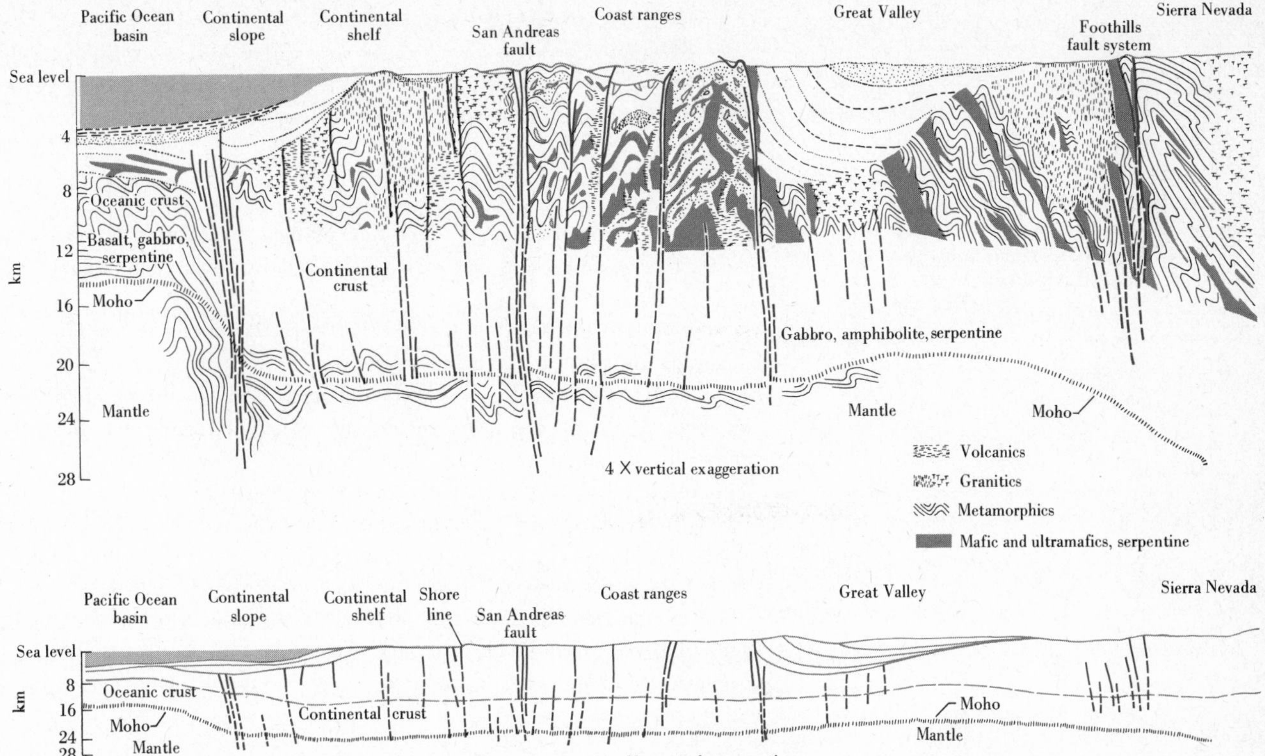

Figure 19.10 ***Schematic tectonic cross section from the Pacific Ocean to the Sierra Nevada foothills.*** *(After John C. Crowell, "The California Coast Ranges,"* UMR Jour., *no. 1, Fig. 1, p. 135, 1968.)*

rocks show evidence of having undergone several cycles of deformation. Folds and faults are numerous and structure is so complex that relatively few have been mapped. Deformed rocks are largely of the same age as those of the Great Valley and in part as those of the Sierra Nevada (see Figure 19.10).

Some of the ranges still retain patches of Pleistocene and early Recent alluvial terraces that testify to the recency of their uplift. Their strata reveal that valleys and ridges have moved vertically as much as 1,000 m during the Pliocene and Pleistocene. Movement along strike–slip faults has been independent, though aligned with the trends of the ranges.

Fault-block mountains

Fault-block mountains occur in many parts of the world, frequently adjacent to fold mountain ranges. Although produced at approximately the same time as the fold mountains, they are quite a different type.

Fault-block mountains are bounded by faults and are large uplifted sections of the earth's crust. Their rocks, originally deposited in geosynclines, have sometimes been folded during an earlier cycle of deformation before they were broken into blocks and uplifted. The elevation of a block of the earth's crust to mountainous heights requires successive movements along the faults over millions of years. In the North American Cordillera, the fault-block mountains

Figure 19.11 **Schematic section across the Basin and Range country of Nevada, Colorado Plateau, and Colorado Piedmont.** (*After A. K. Lobeck,* Physiographic Map of the United States, *The Geographical Press, Columbia University, New York, 1922.*)

began to be elevated about the same time as their neighboring plateaus and fold mountains, indicating that regional deformative forces were acting.

THE GREAT BASIN AND RANGE PROVINCE The Great Basin and Range province of Nevada and western Utah is bordered on the west by the Sierra Nevada Range and on the east by the Colorado Plateau and the Rocky Mountains (see Figure 19.11). These are tectonically different regions but they are genetically related.

The Great Basin and Range topography developed in a region formerly involved in mountain building. During late Mesozoic and early Tertiary time, the region was uplifted and able to supply enormous volumes of sediments to the Rocky Mountain and Coast Range geosynclines on either side. This unloading and loading of adjacent areas affected crustal and mantle equilibrium. In late Tertiary and Quaternary, distension and collapse of the crust developed (see Figure 19.12). Igneous activity occurred. This seems to have

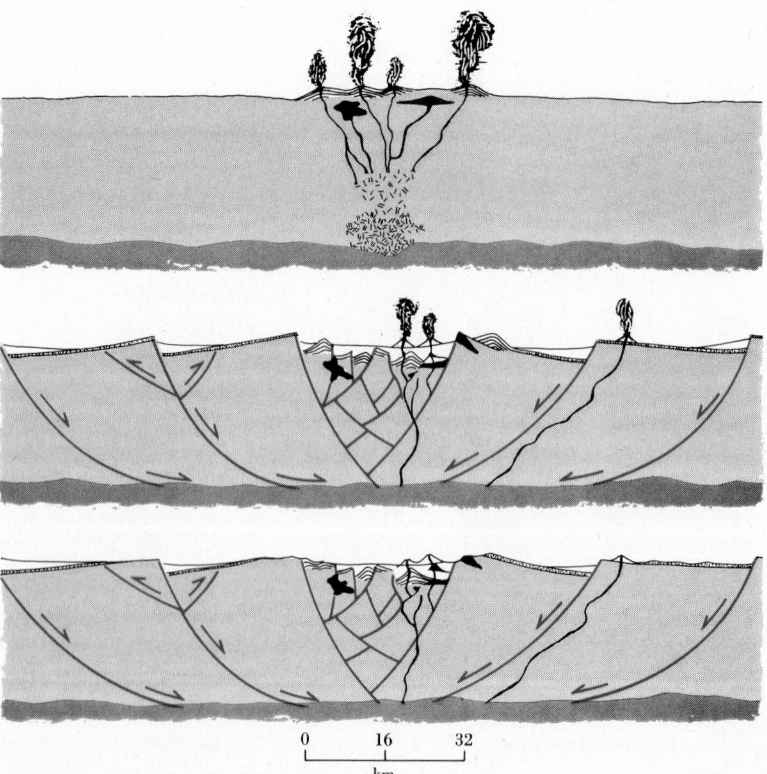

Figure 19.12 **Model of Great Basin and Range structure.** (*After J. H. Mackin in R. J. Roberts, "Tectonic Framework of the Great Basin,"* UMR Jour., *no. 1, Fig. 3, p. 103, 1968.*)

been a result of rather than the cause of the Great Basin and Range structure. Strike–slip faults are also a significant feature of the province.

The pattern of breakup of the earth's crust differed from place to place. In northwestern Nevada, the valleys are broad and are mostly covered with volcanic rocks. In north-central and northwestern Nevada, the ranges are closely spaced and are composed of Paleozoic and Mesozoic rocks. In western Utah, the ranges are narrow and the valleys extremely broad. Volcanic rocks are sparse and the ranges are mostly composed of late Paleozoic rocks. The crust broke up into thousands of blocks, forming basins and ranges. Some classic examples of fault-block mountains are in this area. The fault-block mountains are bounded on at least two opposite sides by high-angle faults dipping from 45° to 70°. Tilting is pronounced, causing differences in elevation of adjacent sides. Bordering the Colorado Plateau is the Wasatch Range, which has a net slip of 2,200 to 2,600 m and a dip of 50°W. The Stansbury and Cedar Ranges are tilted fault blocks some 32 km in breadth. Other fault-block ranges in the area include the Oquirrh, Stillwater, and Ruby–East Humboldt Ranges. The ranges reach elevations of up to 4,000 m.

THE SIERRA NEVADA The Sierra Nevada is a tilted fault-block mountain range. It is located in eastern California and is approximately 100 to 125 km wide and 600 km long. This huge block of the earth's crust has broken loose and pushed up on the east, giving it a tilt westward. Its long, gentle western slope is overlapped by Upper Cretaceous and Cenozoic deposits in the Great Valley. The eastern side is marked by steep escarpments. On its eastern ridge stand many peaks from 3,900 to 4,200 m high. Its highest peak is Mount Whitney, which stands 4,350 m above sea level.

During late Paleozoic and Mesozoic, geosynclinal sediments were accumulating and being deformed. Granitic plutons invaded the region in three episodes: (1) 183 to 210 million years ago, (2) 124 to 135 million years ago, and (3) 80 to 90 million years ago. These plutonic rocks constitute the Sierra Nevada batholith, which is in tightly folded but weakly metamorphosed sedimentary and volcanic strata of Paleozoic and early Mesozoic age (see Figure 19.13). The batholith is in the central Sierra Nevada and is located in the axial region of a large, complexly faulted synclinorium. Geosynclinal sedimentation stopped after the Mesozoic, and the region was uplifted and eroded to great depths. The region was at a virtual standstill during most of the Eocene and Oligocene, when it may have been no more than 1,000 to 1,500 m high. During the Oligocene, volcanic eruptions occurred in the north. In middle or late Pliocene, the range was uplifted and tilted to the west. Faulting along the eastern border began during the Pliocene, about 10 million years ago. In the Pleistocene, glaciers shaped the ridges and peaks and deepened the valleys.

The Sierra Nevada has a root of granitic-type rocks that is 60 km deep. This was probably formed during the Mesozoic when plutonic rocks were being emplaced.

The Sierra Nevada borders the Great Valley of California, which includes the Sacramento and San Joaquin valleys. This region has been the scene of modern subsidence and sedimentation. Westward extensions of rocks of the Sierra Nevada form the "basement" on which sedimentation began in the Great

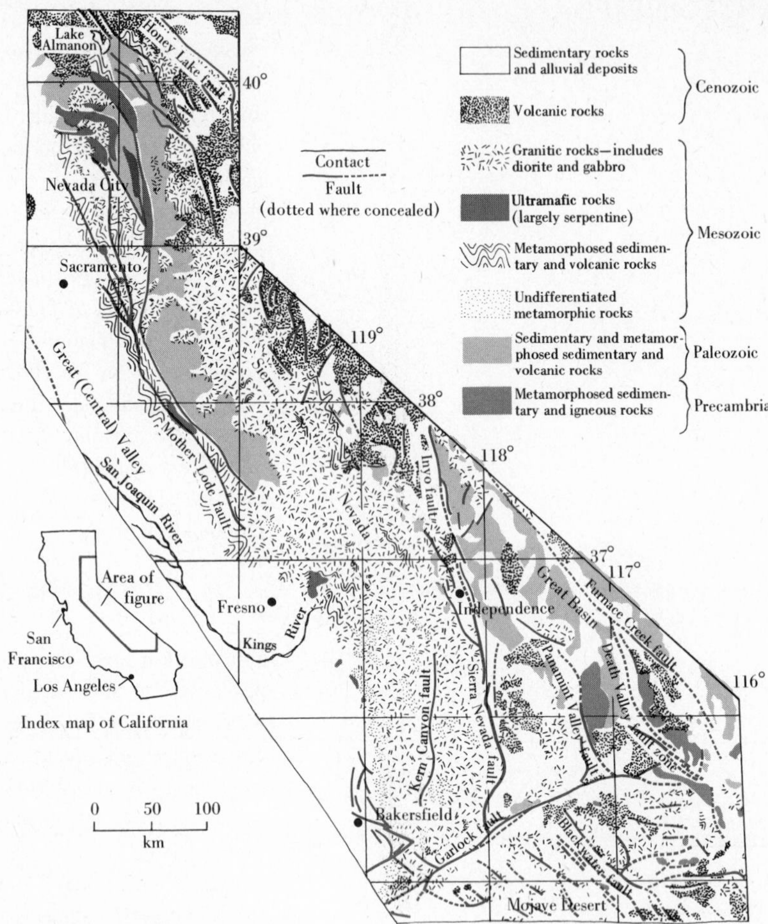

Figure 19.13 ***Geologic map of the Sierra Nevada.*** *(After Paul C. Bateman, "Geologic Structure and History of the Sierra Nevada,"* UMR Jour., *no. 1, Fig. 1, p. 122, 1968.)*

Valley during the late Jurassic. An immense pile accumulated without a significant stratigraphic break through most of the Cretaceous. In the Sacramento valley, deposits have reached thicknesses as great as 12,000 m, with lesser thickness of from 1,000 to 4,000 m in the San Joaquin valley to the south. Strata are warped into a broad syncline.

Upwarped mountains

Many mountains owe their present height and appearance to broad upwarping of the earth's crust. Mountains formed in this manner include the Ozarks, the Adirondacks, the present Appalachians, the Black Hills, the highlands of Labrador, and many other mountains in other parts of the world.

Some upwarped mountains have had an earlier history similar to present fold mountains. Then they were worn down and later reelevated. The Adirondack Mountains, for example, were completely eroded by the Cambrian. They were then covered by a relatively thin veneer of sediments. Eventually, broad

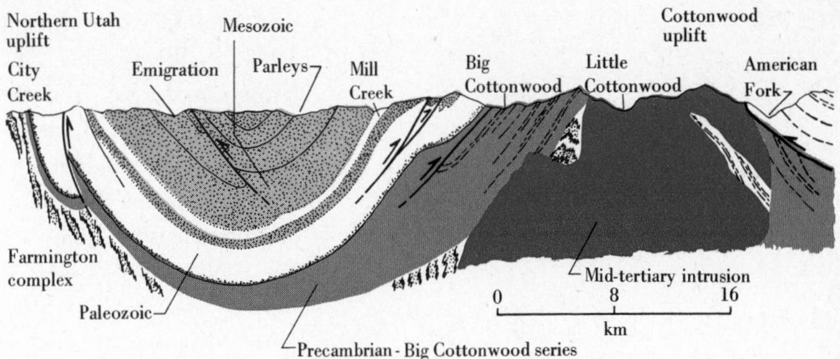

Figure 19.14 ***Cross section in the Wasatch Mountains from the Cottonwood uplift to the northern Utah uplift.*** *(After A. J. Eardley, "Major Structures of the Rocky Mountains of Colorado and Utah,"* UMR Jour., *no. 1, Fig. 6, p. 87, 1968.)*

upwarping reelevated the Adirondack region and brought it into the influence of erosional processes once more. The present Adirondacks are carved from ancient Precambrian rocks after a thin covering of Paleozoic sediments had been removed.

ROCKY MOUNTAINS The major structures of the Rocky Mountains in Colorado and Utah are broadly upwarped mountains (see Figure 19.14). This region of the crust was uplifted during the Cenozoic. Rocks were elevated so high that Precambrian basement rocks still tower above their surroundings. A thin veneer of sedimentary rock of the Paleozoic, Triassic, and Jurassic ages is on the flanks of the mountains. These strata have been folded and thrust-faulted but are believed to be secondary features of the uplifted region. The mountains vary in size from 80 to 160 km long to 16 to 40 km wide. They are oval and irregular in shape. The mountains are separated by intermontane valleys in which sediments have accumulated since the late Cretaceous.

19.2 *Gravitational attraction of mountains*

A mass standing high above its surroundings should exert a gravitational attraction that could be computed and measured. One device for measuring this attraction makes use of a plumb bob suspended on a plumb line. Like every other object on the globe, the suspended bob is pulled by gravity. On the surface of a perfect sphere with uniform density, the bob would be pulled straight down, and the plumb line would point directly toward the center of the sphere. But if there is any variation from these ideal conditions—that is, if there are surface irregularities on the sphere—the plumb line will be deflected as the bob is attracted by their concentrations of mass.

In 1749 Pierre Bouger found that plumb bobs were deflected by the Chimborazo Mountain in the Andes, but by amounts much less than calculated.[4] In 1849, in the Pyrenees, F. Petit actually found the plumb bob appeared to be deflected away from the mountains. Similar discrepancies between calculated and measured values of gravitational attraction of mountains were also observed in the middle of the nineteenth century by British surveyors in India. They were using the plumb line to sight stars, in an attempt to fix the latitude of Kaliana, near Delhi in northern India, and of Kalianpur, about 600 km due south. They observed that the difference in latitude between these two stations was 5°23′37.06″. Then they checked the difference directly by standard surveying methods. The difference computed from these measurements was 5°32′42.29″. There was a discrepancy of 5.23″, or about 150 m, between measurements. That may not seem very much over a distance of 600 km, but it was too large to be explained by errors of observation. Scientists then concluded that the plumb line at Kaliana had been deflected more by the attraction of the Himalaya Range than it had been at Kalianpur farther south. Actually, however, the discrepancy should have been three times as large, assuming that the mountains were of the same average density as the surrounding terrane and that they were resting as a dead load on the earth's crust.

Two quite different explanations were proposed to explain the discrepancies between computed and observed values of gravity. The first was made by G. B. Airy in 1855. He regarded the earth's crust as having the same density everywhere and suggested that differences in elevation resulted from differences in the thickness of the outer layer. Hydrostatic equilibrium was achieved by lighter material floating in a denser substratum. The depth of compensation is variable. Continents, mountains, and other topographic features are in equilibrium because they have roots like icebergs and are floating in a denser material.

J. H. Pratt, on the other hand, proposed that all portions of the earth's crust have the same total mass above a certain uniform level, called the level of compensation. Any section with an elevation higher than its surroundings would have a proportionately lower density. These ideas of Airy and Pratt are illustrated in Figure 19.15.

Isostasy

In 1889, C. E. Dutton[5] suggested that different portions of the earth's crust should balance out, depending on differences in their volume and specific gravity (see Figure 19.16). He called this *isostasy*. In his words,

> *If the earth were composed of homogeneous matter its normal figure of equilibrium without strain would be a true spheroid of revolution; but if heterogeneous, if some parts were denser or lighter than others, its normal figure would no longer be spheroidal. Where the lighter matter was accumulated there would be a*

[4]Pierre Bouger, *La Figure de la Terre*, p. 391, Paris, 1749.

[5]C. E. Dutton, *Bull. Phil. Soc. Washington*, vol. 11, p. 51, 1889. Reprinted in *J. Washington Acad. Sci.*, vol. 15, p. 359, 1925; also in *Bull. Natl. Res. Council (U.S.)*, vol. 78, p. 203, 1931.

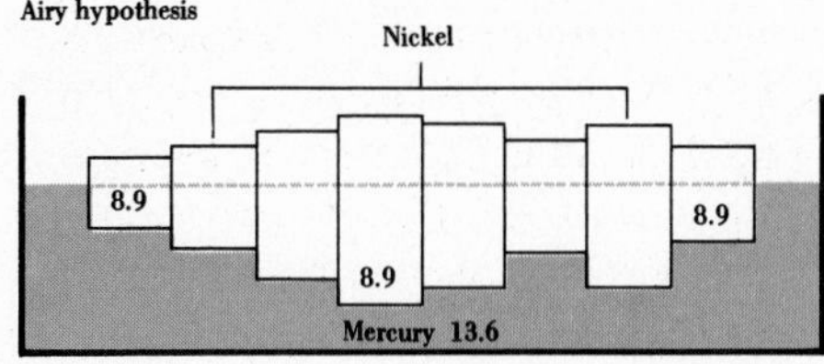

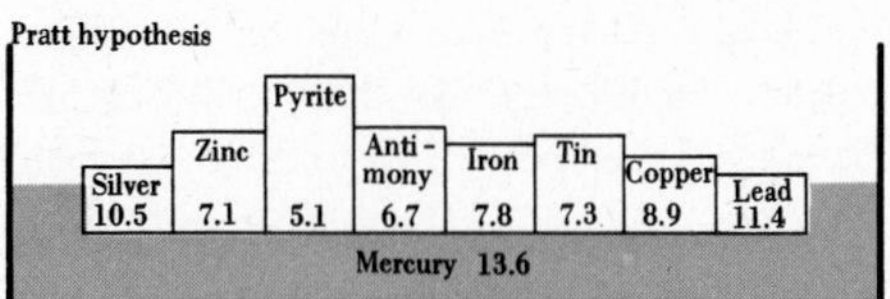

Figure 19.15 ***The Airy and Pratt explanations for why mountains stand high.***

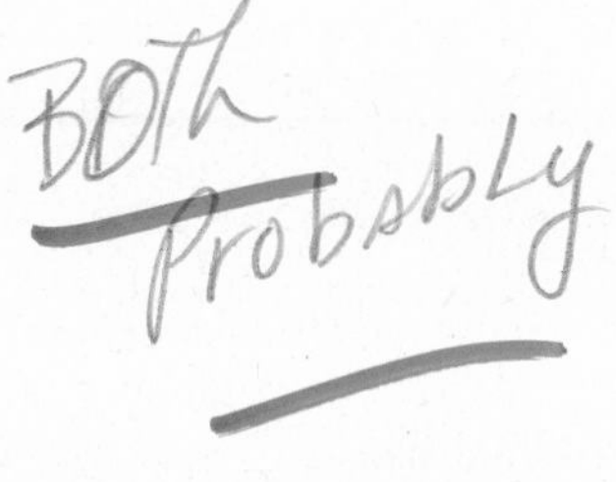

tendency to bulge, and where the denser matter existed there would be a tendency to flatten or depress the surface. For this condition of equilibrium of figure, to which gravitation tends to reduce a planetary body, irrespective of whether it be homogeneous or not, I propose the name isostasy *(from the Greek* isostasios, *meaning "in equipoise with"; compare* isos, *equal, and* statikos, *stable). I would have preferred the word isobary, but it is preoccupied. We may also use the corresponding adjective* isostatic. *An isostatic earth, composed of homogeneous matter and without rotation, would be truly spherical. . . .*

Mountain elevation

If gravity were the only force acting on the earth's surface, all masses of surface rocks would be standing today at heights governed by their thickness and the ratio of their specific gravity to that of the rocks supporting them. Mountains are great masses of sedimentary rocks once deeply buried but now standing high above adjacent rocks. One of the problems of geology is to explain why such masses stand high.

Three possible explanations have been suggested. One is that the crust underlying mountains is strong enough to support them as dead loads. Laboratory experiments, however, have shown that no rocks are strong enough to support the weight of even comparatively low hills. Consequently, we must conclude that the crust beneath the mountains is not by itself capable of supporting the dead weight of mountains.

Another possible explanation is that mountains retain their elevation because the forces that originally formed them are still active. Certainly in areas where mountain-building forces still make their presence known through earthquakes and changes of level it is quite reasonable to assume that they are contributing to the continued elevation of mountains. If fact, it would be difficult to point with assurance to any region on the globe that lacks at least some kind of active internal force.

A third explanation for the continued elevation of mountains is that they

Figure 19.16 **Hypothesis explaining isostasy.** (*After R. C. Daly,* Strength and Structure of the Earth, *p. 61, Prentice-Hall, Inc., Englewood Cliffs, N.J., 1940.*)

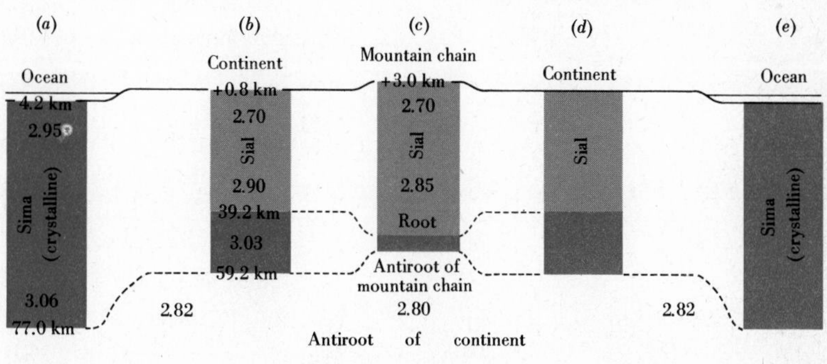

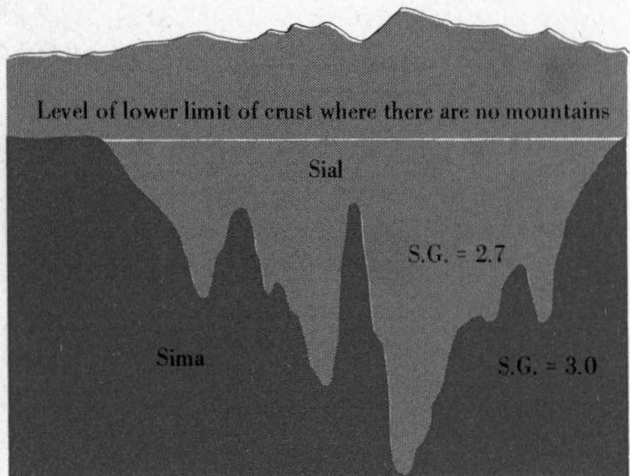

Figure 19.17 **Schematic representation of mountain roots.** (*After R. C. Daly,* Strength and Structure of the Earth, *p. 60, Prentice-Hall, Inc., Englewood Cliffs, N.J., 1940.*)

are isostatic in relationship to surrounding portions of the crust. That is, mountains are merely the tops of great masses of rock floating in a substratum, as icebergs float in water (see Figure 19.17). Such a situation requires a substratum of rock, at not too great a depth, that will flow to adjust itself to an excess load. This rock, however, need not be a liquid in the ordinary sense of the word. It could be rock in a state not unlike that of silicone putty, which can be shaped into a ball that bounces but which under a very slight load—even under its own weight—gradually loses its shape entirely. Hence this rock would be rigid enough for stresses of short duration but would have very little real strength over a long period of time. A small piece of art gum nine-tenths as dense as silicone putty would, in time, sink into the putty until it was floating with only one-tenth of its volume above the surface. Likewise, a mountain range with an average specific gravity of 2.7 (that of granite) can sink into a layer of plastic simatic rock with a specific gravity of 3.0 until the range is floating with a "root" of about nine-tenths and a mountain of one-tenth its total volume.

19.3 Cause of mountain building

Colossal amounts of energy are required to deform and elevate large portions of the earth's crust into mountain ranges. To explain mountain building, we must account for the source of the energy. Also, we have to take into account other surface features as well as the continents, the oceans, and the midocean ridges.

Continents and oceans differ in elevation, composition, and structure. Continents are made up of complex structures of varying age; however, they seem to have been built up on a basaltic layer which is the crust which underlies all the oceans and extends all around the earth.

It is difficult to comprehend the enormous deforming forces that could produce these varied surface features. For one thing, we are limited by our ignorance about the behavior of surface materials under the heat and pressures of the interior. We are also unable to grasp the meaning of the role of geologic time.

The search for the cause of mountain building is far from complete. It seems clear, however, that the sources of energy available include the heat left over from the creation of the earth, radioactive heat from the decay of elements, and gravity. Three hypotheses centering on the search for a more precise source of energy are reviewed below. They are (1) thermal contraction, (2) convection currents, and (3) mantle boiling. One or more of these hypotheses may account for the events recorded in the earth's mobile crust. We will wish to remember them when, in Chapter 20, we consider some of the more recent suggestions about mountain building, including *continental drift, sea floor spreading,* and *plate tectonics.*

Thermal contraction hypothesis

Evidence that mountain ranges are composed of sedimentary layers that have been compressed led to the formulation of the thermal contraction hypothesis. It is based on the idea that heat is lost during the cooling of the earth. Loss of heat causes a decrease in the volume of the earth, and the crust is compressed as it adjusts to the earth's shrinking interior.

Regardless of the details of the earth's origin, there is evidence that at one stage it was a molten sphere. As it cooled, its radius decreased and its surface area shrank. The thermal contraction hypothesis states that the original heat energy present in the earth at its formation has provided the source of energy for mountain building.

Harold Jeffreys, a world-renowned British geophysicist, supported this hypothesis with computations based on seismological evidence and on the behavior of rocks subjected to heat and pressure in the laboratory. He came to certain conclusions on the history of the earth's cooling: It is solid to a depth of 2,900 km, but the temperature of this solid material varies. Near the surface, the rocks are undergoing no further cooling, but from a depth of a few kilometers to about 650 km the rocks are still cooling. Below 650 km, the rocks are still as hot as they were when first solidified.

On the basis of these temperature conditions, we can picture three zones. The inner and outer zones are fixed in volume because they are not losing heat, but the middle zone is cooling and shrinking as it cools. The outer zone attempts to adjust itself to the shrinking zone below and in this process becomes squeezed. The squeezing buckles it into low basins where sediments may accumulate and high places to provide the sedimentary materials. When the buckling of the crust pushes the sediments in the geosyncline to a depth at which the temperatures melt them, the melted sediments expand and produce the elevation of the geosyncline to form mountains.[6]

The thermal contraction hypothesis has been criticized on the grounds that it fails to provide enough shrinkage of the crust. Some geologists believe that

[6]Harold Jeffreys, *The Earth,* 4th ed., pp. 303–310, Cambridge University Press, New York, 1959.

Figure 19.18 ***Convection currents and cells.***

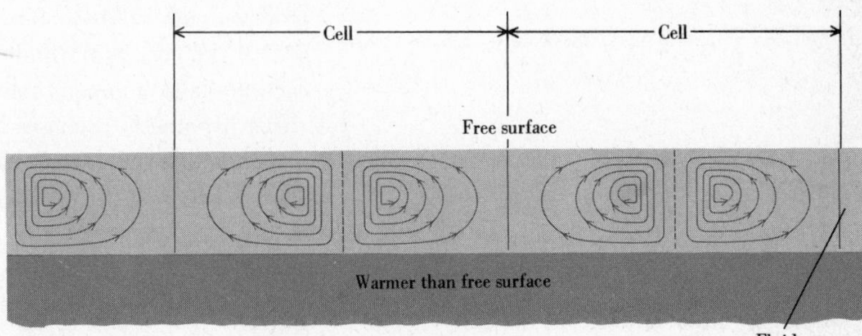

the mantle is solid and from this conclude that the earth cannot have lost significant amounts of heat in the last 2 billion years, at least not enough heat to counterbalance the great volume of material we find in mountains.

Another criticism involves the distribution of mountain ranges in time, in direction, and in place. Some feel that a contracting crust would shorten every great circle the same amount and that mountains should therefore be more evenly distributed than they are. Underlying this objection is the assumption that heat is lost uniformly through the earth's crust; not taken into account is the localized heat and material loss due to volcanic eruptions.

Convection-current hypothesis

Convection currents within the earth have also been suggested as an explanation of mountain building. Convection is a mechanism by which heat is transferred from one place to another by the movement of particles. Its operation is illustrated in heating a pan of water on a stove: The water at the bottom heats fastest, rises (because it is lighter), and is replaced by cooler water sinking to the bottom. Under proper conditions, a pattern of circulation called a *convection current* is established. Convection currents normally occur in pairs, each called a *convection cell.*

It has been proposed that convection currents originating at the core may be circulating in the mantle of the earth (see Figure 19.18). The materials of the mantle do not behave like water, but they do flow plastically if subjected to enough force. They have some strength, even though it may not be great. So to establish convection currents, the heat and expansion at the core must build up until the push to get out is greater than the strength of the overlying mantle. It has been computed that if we assume the mantle to be in equilibrium at a given time, then heating at the core would expand the nearby mantle material until it starts to move upward. For about 25 million years it would gradually "speed up" until it was moving upward 12 cm/year. After currents reached the base of the crust, they would be deflected horizontally, move along the base of the crust, and eventually plunge back toward the core again. They are supposed to drag the crust downward, forming geosynclines. After the geosynclines have deepened to 15,000 m, it is proposed that the currents crumpled and folded the sediments that were by now softened by immersion in hot mantle rocks. Then the convection currents lost their drive and slowed to a stop. Meanwhile, melting in the geosynclines caused local expansion and produced the uplift to form mountains. A quiescent period followed, while heat at the core again built up where the strength of the overlying rocks was exceeded and another cycle could start.

This hypothesis proposes heat as the source of the energy. In its support are offered calculations that the temperature of the earth's core must be enormously high to keep the material liquid (see Chapter 17) under pressures of 1 to 3 million atm. Therefore, heat in ample supply is presumably available for driving the convection mechanism. Also, folded and deformed rocks show that they yield plastically under the right conditions of heat and pressure, so it requires no great extrapolation from laboratory and field data to picture the materials of the mantle as capable of plastic movement assumed by the convection theory.

On the other hand, seismological evidence shows that the entire mantle has not been stirred by convection currents, regardless of the plausibility of arguments that it could be. Increasingly precise data on the travel times of waves in the mantle are becoming available through studies of earth waves generated by earthquakes and nuclear explosions. These data now clearly establish that there are zones in the mantle structure. There could be no such layering in the mantle if convection currents had stirred it, mixing the materials between the core and the top.

There are geologists who believe that the convection hypothesis also fails to account for the geographical distribution of mountains, or for their dates of formation, in even a rough approximation.

Mantle-boiling hypothesis

In 1965, L. Don Leet and Florence J. Leet proposed a hypothesis to explain the cause of earthquakes, which could also account for the formation of mountains, volcanoes, and other surface features. It is called the mantle-boiling hypothesis.[7]

The mantle-boiling hypothesis assumes that the earth's mantle is not solid but rather a combination of atoms and molecules randomly mixed at indefinitely high temperature. These are prevented by pressure-maintained viscosity from combining to form solids, liquids, or gases, the three familiar states of matter at the earth's surface. The name *soliqueous* was coined to designate this state of matter.

Very early in the earth-forming process, the outside material at high temperature, but under little or no pressure, immediately cooled and formed compounds and minerals. This initial scum was the primitive crust. Steam and other gases assembled below this and escaped through it, cooling and removing material from the outer mantle. This process is continuing today and in 5 billion years has brought to the surface the gases of our atmosphere, the water of the oceans, and the rocks that form the continents.

The removal of material has occurred within the upper 600 km of the earth's interior. Escape of material at the surface lowers pressures below and permits formation of steam, other gases, and magma. These move up and out. The process is like a very, very slow boiling of extremely viscous stuff. Below about 600 km, there has been no movement of material since the earth was formed.

Figure 19.19 **Mantle-boiling hypothesis.**

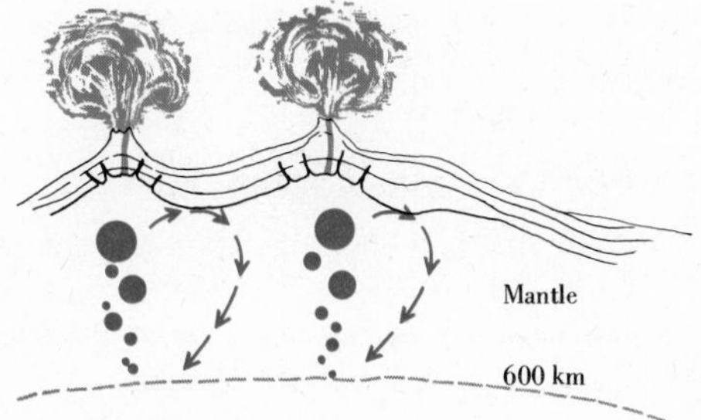

The boiling of mantle material is reflected in igneous activity and both the heaving and settling of the crust, which are continuous (Figure 19.19). It forms and renews the world's mountains, continents, and ocean deeps, in constant conflict with erosion and sedimentation. The ultimate source of energy is primeval heat. The mechanism is one of shallow, numerous, and randomly distributed convection currents.

Support for the boiling-mantle hypothesis includes the complex and continuous

[7]L. Don Leet and Florence J. Leet, "The Earth's Mantle," *Bull. Seis. Soc. Am.*, vol. 55, pp. 619–625, 1965.

up-and-down history of mountain making and rock deformation throughout time, as well as the patterns of earthquake occurrence described in Chapter 16, their association with volcanic activity, and their exclusive location within the outer tenth of the earth's radius.

Outline

Mountains stand from a few hundred to a few thousand meters above the surrounding terrane and are the dominant surface features of the continents of the world.

Kinds of mountains are volcanic, fold, fault-block, and broadly upwarped.

- *Volcanic mountains* form the island arcs and midocean ridges.
- *Fold mountains* are complex structures involving all the mountain-building forces of folding, faulting, and igneous activity.
  - Fold mountains have developed from thick deposits of sediments that accumulated in *geosynclines.*
    - *Formation of a geosyncline* involves a slow, continuous downwarping of the earth's crust, with the deposition of sediments going on at the same time.
    - *Deformation of geosynclines* occurs continuously while sediments are accumulating; deeply buried rocks melt and expand, causing the whole overlying mass to rise.
    - *Zones of geosynclines* developed into mountain ranges.
  - *The Appalachians* had different histories in the northern and southern sections.
  - *North American Cordillera* formed from geosynclinal zones extending 4,800 km from Alaska to Mexico.
  - *The Coast Ranges* have been the scene of nearly continuous deformation and sedimentation from the middle of the Mesozoic.

Fault-block mountains are bounded by faults and are large uplifted sections of the earth's crust.

- *The Great Basin and Range province* of Nevada and western Utah has some classic examples of fault-block mountains.
- *The Sierra Nevada* is a tilted fault-block mountain range.
- *Upwarped mountains* owe their present height and appearance to broad upwarping of the earth's crust.
  - *The Rocky Mountains* in Colorado and Utah are broadly upwarped mountains.

Gravitational attraction of mountains can be computed and measured.

- *Isostasy* is the principle that different portions of the earth's crust should be in balance.
- *Mountain elevation* may be maintained because the forces that originally formed them are still active or because they are isostatic.

Cause of mountain building must account for the source of great amounts of energy as well as the complex sequences of events in time.

- *Thermal contraction hypothesis* postulates compression in the crust as it adjusts to a shrinking earth's interior.
- *Convection-current hypothesis* has convection cells maintained by heat of the core deflected by the crust and dragging it along and down to form geosynclines.

Mantle-boiling hypothesis states that the mechanism is one of shallow, numerous, and randomly distributed convection currents in the outer 600 km of the earth's mantle.

Supplementary readings

CLARK, THOMAS H., and COLIN W. STEARN: *Geological Evolution of North America,* 2nd ed., The Ronald Press Company, New York, 1968. A textbook of historical geology nicely balanced with topics sometimes isolated as physical geology but with direct long-range significance: geochronology, the geosynclinal concept, pole wandering, and continental drift, with orogeny illustrated by specific areas.

A Coast to Coast Tectonic Study of the United States, V. H. McNutt Colloquium Series 1, *Univ. Missouri Rolla J.,* April 1968. Seven monographs by specialists in each region.

KING, PHILIP B.: *The Evolution of North America,* Princeton University Press, Princeton, N.J., 1958. A treatment of selected regions, mostly from the middle part of the continent, chosen to illustrate principles of continental evolution.

VAN BEMMELEN, R. W., with introduction by RAYMOND C. MOORE: *Mountain Building,* Martinus Nijhoff, The Hague, 1954. An examination of some mountain-building hypotheses in the light of measurements in the Indonesian area; one of the best involving interrelated volcanic activity, seismicity, gravity anomalies, and chemically distinct suites, correlated with geosynclines. Local in detail but worldwide in application, by a world leader in the field.

ZEN, E-AN, WALTER S. WHITE, JARVIS B. HADLEY, and JAMES B. THOMPSON, JR. (eds.): *Studies of Appalachian Geology—Northern and Maritime,* Wiley-Interscience, New York, 1968. Thirty-three monographs, representative samples of data, and interpretations that make this one of the better studied and understood orogenic belts of the world.

20

Paleomagnetism, continental drift, and plate tectonics

EROSION OF CONTINENTS, CREATION OF MOUNTAINS, AND SHIFTING SHORELINES of the seas are changes of the earth that can be documented and visualized without overtaxing our credulity. But now we must consider change on a different scale, challenging not only our sense of direction but our concept of geographic permanency as well. We wish to investigate whether the north and south poles have shifted position through time—and if so, why? We shall examine the question of *continental drift:* whether or not our landmasses have been rent from their moorings and have wandered over the globe's surface. This will lead us in turn to a consideration of *plate tectonics,* which pictures the earth's crust as a jigsaw puzzle of rigid plates jostling one another, growing in some places and decaying in others.

In Chapter 2 we referred briefly to the earth's magnetic field and defined some of the terms useful in describing it. We begin this chapter with a more extensive discussion of magnetism, and particularly of magnetism of the geological past, *paleomagnetism.* The data derived from paleomagnetic studies not only gave the subject of continental drift a new degree of scientific respectability just after World War II but also provided proponents of continental drift with some of their strongest arguments. Furthermore, they have helped demonstrate the youthfulness of the ocean basins as compared with the continents, and have been instrumental in the development of the idea of plate tectonics.

20.1 *Cause of the earth's magnetism*

The cause of the earth's magnetism has remained one of the most vexing problems of earth study. A completely satisfactory answer to the question is still forthcoming.

We have already indicated that the earth's magnetic field is composed both of internal and external components (Chapter 2). The external portion of the field is due largely to the activity of the sun. This activity affects the ionosphere and appears to explain magnetic storms and the northern lights. The changes and effects of the external field may be rapid and dramatic, but they have little effect on the internal field of the earth, which is of greatest concern to us.

William Gilbert, who first showed that the earth behaves as if it were a magnet, suggested that the earth's magnetic field results from a large mass of permanently magnetized material beneath the surface. The idea is attractive not only because large quantities of magnetic minerals have been found in the earth's crust but also because geologists think that the earth's core is made largely of iron (Chapter 17).

Close examination reveals that the average intensity of the earth's magnetization is greater than that of the observable crustal rocks. We must therefore look deeper for the source of magnetism.

In looking deeper, the first difficulty we face is that materials normally magnetic at the earth's surface lose their magnetism above a certain temperature. This temperature is called the *Curie temperature* and varies with each material. The Curie temperature for iron is about 760°C; for hematite, 680°C; for magnetite, 580°C; and for nickel, 350°C. The temperature gradient for the earth's crust is estimated to average about 30°C/km. At this rate of increase, the temperature should approximate the Curie temperature of iron at about 25 km below the surface and exceed the Curie temperatures for most normally magnetic materials. Therefore, below 25 km we would not expect earth materials to be magnetic, and permanent magnetism can exist only above this level.

On the other hand, if all of the earth's magnetism were concentrated in the crustal rocks, then the intensity of magnetism of these rocks would have to be some 80 times that of the earth as a whole. And yet we know that the magnetic intensity of the surface rocks is less than that of the earth's average intensity.

From this observation we must conclude that the earth's magnetic field is not due to permanently magnetized masses either at depth or near the surface.

Some physicists have suggested that the rotation of the earth accounts for the earth's magnetic field. This explanation, like that of permanent magnetization, has met with insuperable difficulties.

The theory of earth magnetism most widely entertained at present is that the earth's core acts as a self-exciting dynamo. One model of the earth's core pictures its outer portion as a fluid composed largely of iron. This core therefore not only is an excellent conductor of electrical currents but also exists in a physical state in which motions can easily occur. Electromagnetic currents are pictured as generated and then amplified by motions within the current-conducting liquid. The energy to drive the fluid is thought to come from convection, which in turn results from temperature differences.

The dynamo theory further requires that the random convective motions and their accompanying electromagnetic fields be ordered to produce a single united magnetic field. It is thought that the rotation of the earth can impose such an order. The dynamo theory of earth magnetism still remains a theory, but so far it has proved the most satisfactory explanation of the earth's magnetism.

20.2 *Paleomagnetism*

Some rocks, such as iron ores of hematite or magnetite, are strongly magnetic. Most rocks, however, are only weakly so. Actually, the magnetism of a rock is located in its individual minerals, and we would be more correct to speak of the magnetism of minerals rather than of the rock. By convention, however, we refer to rock magnetism. This magnetism is referred to as the rock's *natural remanent magnetism* (NRM). This remanent magnetism may or may not agree with the present orientation of the earth's field and may have been acquired by the rock in many different ways. Identifying, measuring, and interpreting the different components of a rock's NRM forms the basis of paleomagnetism, the study of the earth's magnetic field in the geological past.

Let us examine some of the ways an igneous rock acquires its magnetism.

As a melt cools, minerals begin to crystallize. Those which are magnetically susceptible acquire a permanent magnetism as they cool below their Curie temperatures. This magnetism has the orientation of the earth's field at the time of crystallization. It is called *thermo remanent magnetism* (TRM). This remanent magnetism remains with the minerals—and hence with the rock—unless the rock is reheated past the Curie temperatures of the minerals involved. This new heating destroys the original magnetism; and when the temperature again drops below the Curie temperatures of the magnetic minerals, the rock acquires a new TRM.

The NRM of our igneous rock may include other components. One of these is an *induced magnetism* arising from the present magnetic field of the earth. This induced magnetism is parallel to the earth's present field, but it is weak when compared with the TRM of the rock.

Sedimentary rocks acquire remanent magnetism in a different way than do igneous rocks. Magnetic particles such as magnetite tend to orient themselves in the earth's magnetic field as they are deposited. This orientation is retained as the soft sediments are lithified. This magnetism, known as *depositional remanent magnetism* (DRM), records the earth's field at the time the rock particles were deposited. Of course, the sedimentary rock, like the igneous rock, may also acquire an induced magnetism reflecting the current magnetic field.

VIRTUAL GEOMAGNETIC POLES Paleomagnetic studies define the earth's magnetic field at various localities at different moments in geologic time. Instead of expressing the data on declination and inclination for a given locality, we usually express them in terms of equivalent pole positions. We refer to these poles as *virtual geomagnetic poles.* These are different from the dip poles and the geomagnetic poles we have already discussed. The pole consistent with the magnetic field as measured at any one locality is the virtual geomagnetic pole of that locality. It differs from the geomagnetic pole because it refers to the field direction of a single observational station, whereas the geomagnetic pole is the best fit of a geocentric dipole for the entire earth's field. Inasmuch as it is impossible to describe the entire earth's field at various times in the past, the virtual geomagnetic pole is commonly used in expressing paleogeomagnetic data.

GEOMAGNETIC INTENSITY, VARIATIONS, AND REVERSALS In Chapter 2 we found that there has been a secular variation in the position of the geomagnetic pole over the last several hundred years. In addition to this change in pole position, we also know that there has been a variation in intensity of the earth's magnetic field. Thus by 1965 the intensity had decreased nearly 6 percent from the time it was first successfully analyzed in 1835 by K. F. Gauss, a German mathematician.[1]

We have been able to extend our knowledge of the variation in the field's intensity by techniques first successfully used by the French physicist Emile Thellier. Samples that can be precisely dated by historical or radiocarbon

[1]The measured decline was from 8.5×10^{25} to 8.0×10^{25} gauss (G).

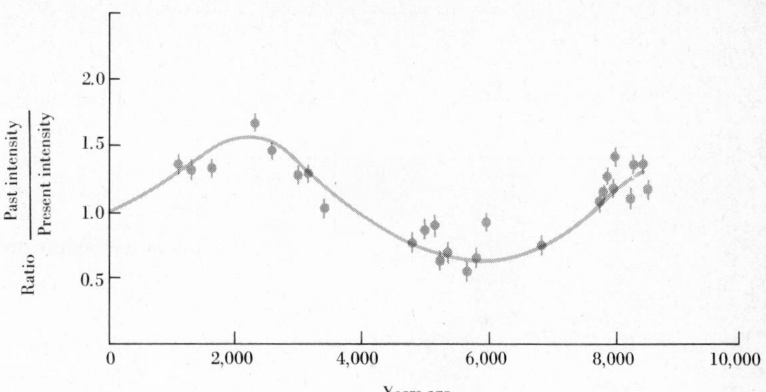

Figure 20.1 **Variation in the intensity of the earth's magnetic field as determined from archaeological samples in Czechoslovakia. Each circle represents one sample, or a group of samples of the same age. The vertical bars represent standard deviation from the mean ratio.** (*After V. Bucha,* Archaeometry, *vol. 10, p. 20, 1967.*)

techniques show that the intensity of the field has been decreasing since about the time of Christ, when it was about 1.6 times its present intensity. Previously it had risen from a low value of about half the present intensity around 3500 B.C., this having been preceded by a steady decline from an intensity of about 1.5 times that of the present around 6500 B.C. (Figure 20.1).

If the decay of magnetic intensity were to continue at its present rate, it has been estimated by Allan Cox, Stanford University geophysicist, that the intensity would reach zero point about 2,000 years from now. Thereafter the direction of the magnetic field would reverse itself. There is no assurance that this will happen, and it may just as well fluctuate as it has for the last several thousand years. But if we turn farther back into earth history, we find that the poles have been reversed, not once, but many times. So we assume that such a change is possible in the future.

We now have a great deal of information about past reversals of the magnetic field. That is to say, if we were able to take our ordinary magnetic compass back into time, we would find some periods in which the north-seeking end of the needle would point toward the north pole as it does today. But there would be many other times during which the needle would point to the south pole instead. The present polarity is called "normal." A polarity at 180° to it is called "reversed." A succession of normal and reverse fields can be pieced together for the last 4 million years or more. We call the longer intervals of dominance by a particular polarity a "polarity epoch" and we have named them after distinguished students of earth magnetism. During each polarity epoch are shorter periods in which the polarity is in the opposite direction and these time intervals are referred to as "polarity events" (see Table 20.1 and Figure 20.2). We will come back shortly to the use of this information in the study of continental drift.

Table 20.1 **Intervals for which the earth's magnetic field was of normal polarity during the last 4.5 million years**[a]

Years ago $\times 10^6$	
0.00–0.007	2.11–2.13
0.01–0.110	2.43–2.80
0.12–0.69	2.90–2.94
0.89–0.95	3.06–3.32
1.61–1.63	3.70–3.92
1.64–1.79	4.05–4.25
1.95–1.98	4.38–4.50

[a] Modified from Allan Cox, "Geomagnetic Reversals," *Science*, vol. 168, pp. 237–245, Fig. 4, 1969.

Results of paleomagnetic studies

Clearly, if we can measure the TRM or the DRM of a rock and relate it to the earth's present field, we can determine to what extent the orientation of the earth's magnetic field has varied at that spot through time.

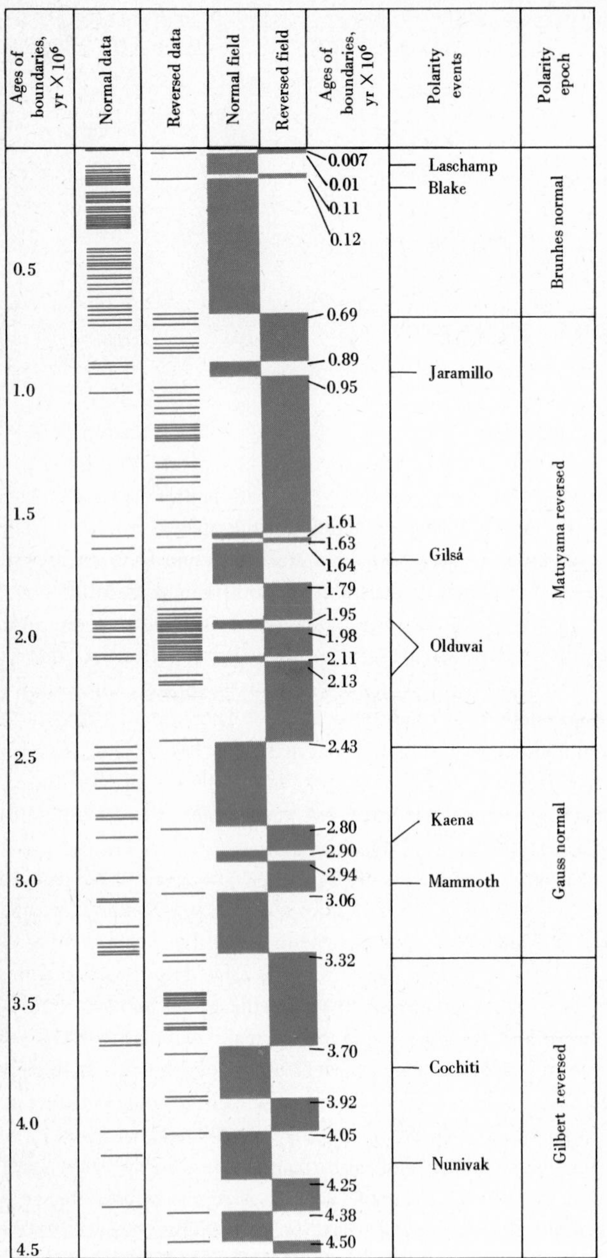

Figure 20.2 ***A time scale for geomagnetic reversal during the last 4.5 million years has been based on extrusive igneous rocks. Each horizontal line in the two left-hand columns represents a rock sample whether its polarity is normal or reversed. In the "field normal" column, normal-polarity intervals are shown in color; in the "field reversed" column, reversed-polarity intervals are shown in color.*** (*Modified from Allan Cox, "Geomagnetic Reversals,"* Science, *vol. 168, Fig. 4, 1969.*)

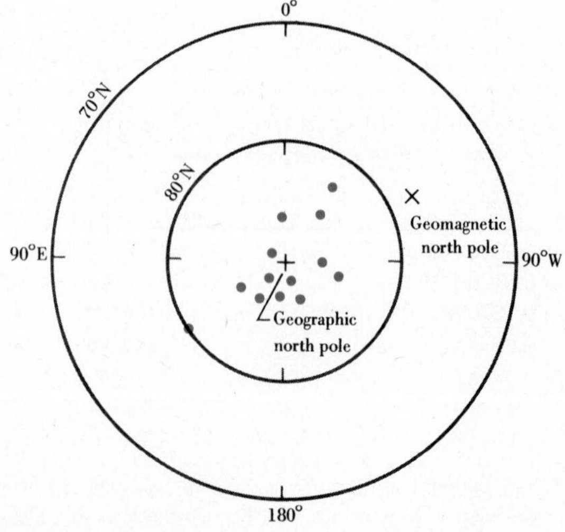

Figure 20.3 ***The virtual geomagnetic poles determined by magnetic measurements of earth materials of Pleistocene and Recent age cluster around the modern geographic pole rather than around the present geomagnetic pole.*** (*Redrawn from Allan Cox and R. R. Doell, "Review of Paleomagnetism,"* Geol. Soc. Am. Bull., *vol. 71, p. 734, Fig. 17, 1960.*)

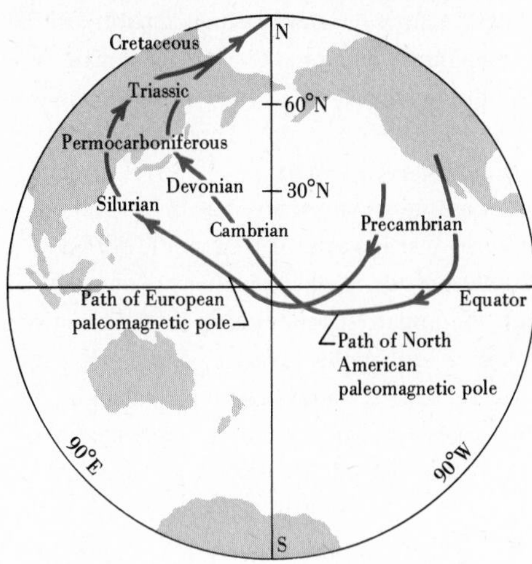

Figure 20.4 ***Paleomagnetic measurements of rocks from North America and Europe show the paths followed by the magnetic poles of these two continents from Precambrian times to the present.*** (*Redrawn from Allan Cox and R. R. Doell, "Review of Paleomagnetism,"* Geol. Soc. Am. Bull., *vol. 71, p. 758, Fig. 33, 1960.*)

Studies of ancient pole positions assume that the earth's field has been dipolar and, further, that this dipole has approximated the earth's axis of rotation. A consequence of these assumptions is that the earth's geographic poles must have coincided in the past with the earth's geomagnetic poles. Clearly this is not so at present. Why should we think that it was true in the past?

A partial answer to the question lies in the dynamo theory of earth magnetism. If the dynamo theory is correct, then theoretical considerations suggest that the rotation of the earth should orient the axis of the magnetic field parallel to the axis of rotation.

Observational data support the theoretical considerations. When we plot the virtual geomagnetic poles of changing fields recorded over long periods at magnetic observatories around the world, we find a tendency for these pole positions to group around the geographic poles. More convincing are the paleomagnetic poles calculated on the basis of magnetic measurements of rocks of Pleistocene and Recent age. These materials, including lava flows and varves, reveal pole positions clustered around the present geographic pole rather than the present geomagnetic pole (see Figure 20.3).

As a result of theoretical and observational data, therefore, most authorities feel that the apparent, present-day discrepancy between magnetic and rotational poles would disappear if measurements were averaged out over a span of approximately 2,000 years. The same principle would apply for any 2,000-year period throughout geologic time. Thus when we speak of a paleomagnetic pole, we have some confidence that it had essentially the same location as the true geographic pole of the time.

Paleomagnetic data derived from rocks of Tertiary age indicate, fairly conclusively, no significant shift of the geomagnetic poles from the Oligocene to the present. The farther backward we go in time beyond the Oligocene, however, the more convincing becomes the case for a changing magnetic pole and, on the basis of the above discussion, for a changing geographic pole. The magnitude of this change is suggested in Figure 20.4.

Figure 20.4 shows that the paleomagnetic poles for Europe and North America were in the eastern Pacific during Precambrian time. Thereafter they moved southwestward and crossed the equator into the southern hemisphere before moving northwestward toward Asia and eventually to the position of the present globe.

The extensive migration of the magnetic poles (and by extension the geographic poles), combined with the observation that the paths of polar migration of different continents fail to coincide, raises tantalizing questions. We find ourselves, in fact, faced with the entire concept of continental drift.

20.3 *Continental drift*

In considering the theory of continental drift we should bear in mind that two general types of movements may be involved: movement of individual continents in relation to one another and movement of the earth's poles. Actually, if the latter movement did take place, most students feel that a slippage of the earth's crust and upper mantle has produced an apparent motion of the poles. In other words, the magnetic and rotational poles remain fixed within the earth, but different points on the earth's crust would at different times be located at the polar positions as the crust moved. Such motion might involve the crust as a single unit or fragments of it.

The first coherent theory that our continents have moved as individual blocks was presented a few years before World War I by Alfred Wegener (1880–1930), a German meteorologist and a student of the earth. Wegener, as many before and after him, was intrigued with the apparent relationship in form between the opposing coasts of South America and Africa. Was it possible that these two landmasses were once part of the same general continent but have since drifted apart? Wegener's emphatic "yes" and his extensive documentation of the lateral motion of continents started a spirited discussion that has continued to the present.

The case for continental drift

WEGENER'S THEORY In 1912, Wegener published in detail his theory of continental drift. He pictured the dry land of the earth as included in a single, vast continent that he named *Pangaea* (from the Greek for "all" and "earth"). This primeval continent, he argued, began to split asunder toward the end of the Mesozoic. These fragments began a slow drift across the earth's face, and by the Pleistocene they had taken up their positions as our modern continents.

The idea of continental drift has been argued now for more than 50 years. The early evidence for the theory was drawn almost entirely from the geologic record. In the midtwentieth century, the development of geophysical techniques brought new data into the discussion of continental drift. Let us look at some of the evidence.

THE SHAPE OF CONTINENTS Anyone observing the map of Africa and South America is struck by the jigsaw puzzle match of the two continents. If we fit the eastern nose of South America into the large western bight of Africa, the two continents have a near perfect match. One distinguished geologist, while admitting to skepticism of continental drift, found this fit so credible that he opined, "If the fit between South America and Africa is not genetic, surely it is a device of Satan for our frustration."[2]

If one examines the outlines of the other continents, one can perhaps persuade himself that these landmasses, too, can be reassembled into a single large landmass as suggested by Wegener. If one fits the continents at a level of about 900 m below sea level, one obtains a very impressive match indeed, as shown in Figure 20.5.

EVIDENCE FROM PALEOMAGNETISM We pointedly began this chapter with a consideration of the earth's magnetism, for it has been the increasing body

[2]Chester Longwell, "My Estimate of the Continental Drift Concept," in *Continental Drift, a Symposium*, S. W. Carey (ed.), p. 10, Tasmania University Press, Hobart, Australia, 1958.

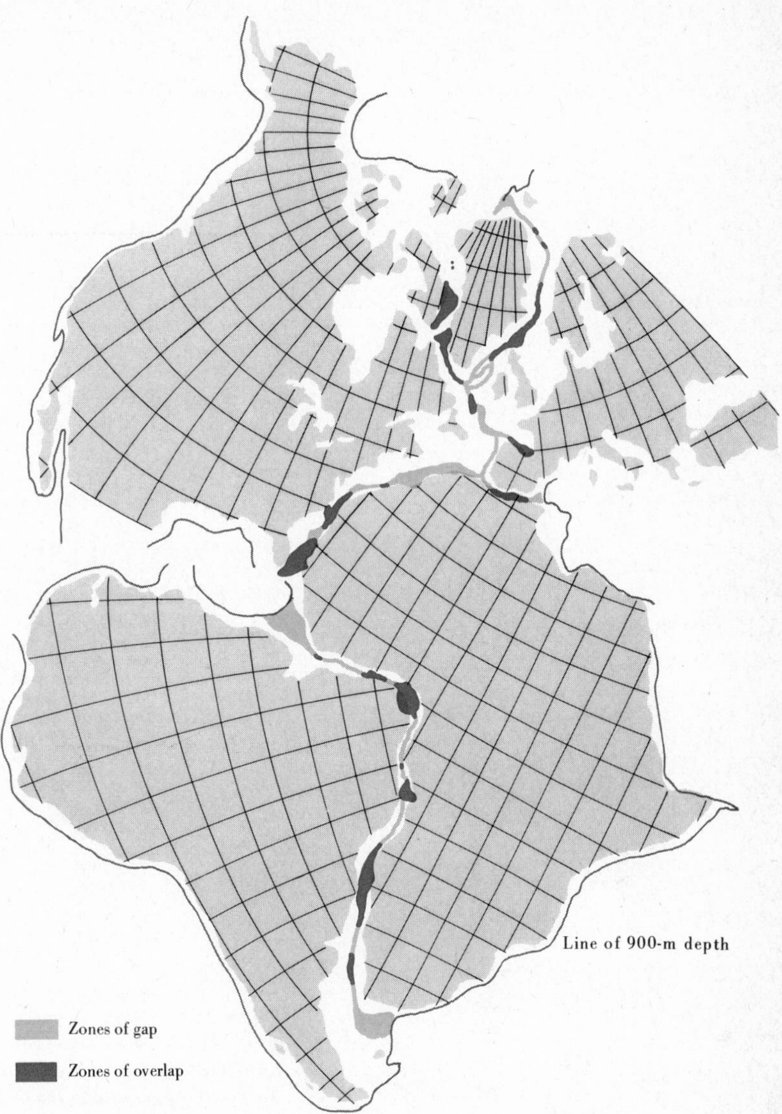

Figure 20.5 ***A statistically determined fit of North America, South America, Africa, and Europe at the 900-m depth in the ocean. Zones of overlap are shown in the dark tone; zones of gap are shown in the lighter tone.*** (*After Edward Bullard et al., "The Fit of the Continents Around the Atlantic,"* Phil. Trans. Roy. Soc. London, *no. 1088, Fig. 8, 1965.*)

of information on rock magnetism and paleomagnetism that has kindled new interest in the old theory of continental drift. Figure 20.4 shows how the paleomagnetic poles have shifted during the last 500 million years. For the two continents represented, North America and Europe, there is a divergence between the ancient poles—that is, the virtual paleomagnetic poles—of these two landmasses as we go back in time. The case of North America and Europe

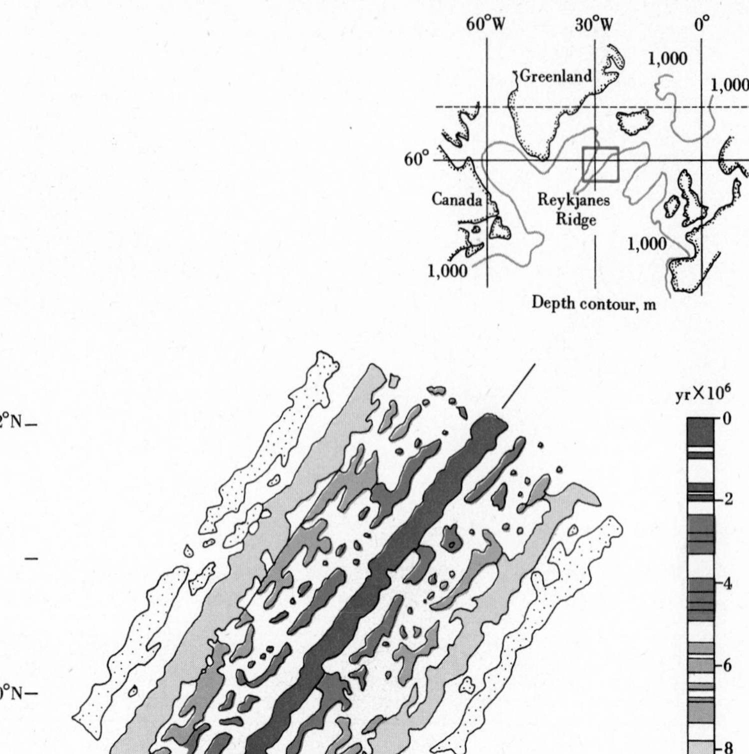

Figure 20.6 **Magnetic anomalies over the Reykjanes Ridge southwest of Iceland. The patterned areas display normal polarity and are separated by areas of reversed polarity. Note that the belts of normal and reversed polarity are generally symmetric to the axis of the Midatlantic Ridge. Ages of the various polarity belts are based on a time scale derived from a study of magnetic anomalies on land and on the sea floor.** (*After F. J. Vine, "Magnetic Anomalies associated with Ocean Ridges," in R. A. Phinney, ed.,* The History of the Earth's Crust, *Fig. 6. Princeton University Press, Princeton, N.J., 1968. © 1968 by Princeton University Press. Reprinted by permission.*)

is not unique. The pattern of polar migration, as sketched from paleomagnetic information, is different for each continent.

The divergence of the paths of polar migration among continents can be explained by the shifting of the landmasses in relation to each other. The course of polar migration suggests that the continental movement consists of a general drift of these continents away from each other, and in some instances there is an additional rotation of continents.

Magnetic measurements at sea have revealed a characteristic pattern of

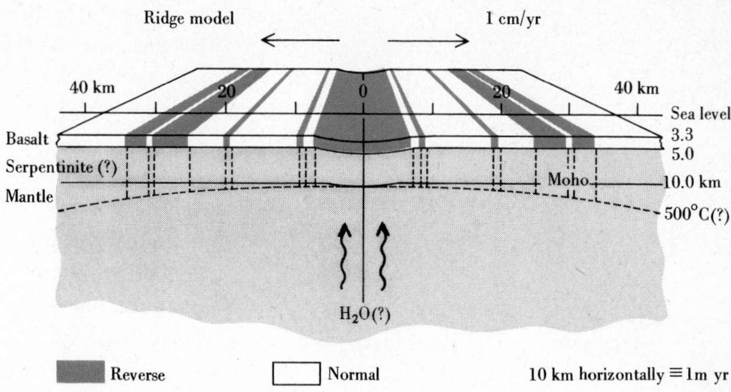

Figure 20.7 ***A diagram to suggest the development of magnetic anomalies related to ocean ridges. New oceanic crust, cooling below the Curie point, forms along the axis of the ridge and assumes the magnetic characteristics prevailing at the time of its formation. As the crust moves laterally away from the ridge crest, changes in the magnetic field of the earth are retained in strips of the oceanic crust. Serpentinite, a rock made up chiefly of the mineral serpentine, is thought by many to form the lower portion of the oceanic crust.*** (*After F. J. Vine, "Sea-Floor Spreading—New Evidence,"* J. Geol. Educ., *vol. 17, Fig. 4, 1969.*)

anomalous magnetic intensities. These magnetic anomalies are most often associated with ocean ridges and are arranged in stripes of alternating intensity parallel to the ridge. An example is provided by the magnetics of the Reykjanes Ridge, a portion of the Midatlantic Ridge southwest of Iceland. Figure 20.6 shows the stripes parallel with the ridge and shows also a bilateral symmetry, the axis for which is the rift valley along the crest of the ridge.

It was two geophysicists, Fred Vine and D. H. Matthews, who suggested that the alternating stripes of high- and low-intensity magnetism in reality were zones of normal and reverse polarity in the rock of the sea floor. From this suggestion, now widely accepted, have flowed some intriguing concepts about the sea floor, the drift of continents, and the building of mountains.

SEA-FLOOR SPREADING The Vine-Matthews hypothesis leads us to a consideration of *sea-floor spreading.* This is a mechanism which is thought to involve the active spreading of the sea floor outward, away from the crests of the main ocean ridges. As material moves from the ridge, new material is thought to replace it along the ridge crest by welling upward from the mantle. As this mantle material cools below the Curie temperature, it would take on the magnetization of the earth's field at that time. Room for the new material would be made along the ridge by the continued pulling apart of the crust and its movement laterally away from the ridge. At times of change in the earth's polarity, then, newly added material along the crest would record this change. Spreading laterally, it would carry this record with it, to be followed at a later time by the record of the next polarity change. Figuratively, we can visualize this as a gigantic magnetic tape preserving a history of the earth's changing magnetic polarity. A diagram to suggest the mechanism is shown in Figure 20.7.

The rates at which the sea floor appears to be spreading vary. In the North Atlantic it is about 1 cm/year and in the South Atlantic about 2.3 cm/year. Along the Juan de Fuca Ridge off California, the rate is 3 cm/year, and on the east Pacific rise, it is calculated to be 4.6 cm/year, and may reach a rate of 9 cm/year in some sections.

Sea-floor spreading indicates that the ocean floor is a dynamic system and in terms of continental drift it suggests that expansion of ocean basins may also affect the location and position of continents. Furthermore it raises many questions, such as what drives the ocean floor and where do the moving ocean floors go. The proposed answers to these questions bear on continental drift but we will defer their consideration until we examine the subject of plate tectonics (Section 20.4). In the meantime let us take up additional geologic data bearing on continental drift.

EVIDENCE FROM ANCIENT CLIMATES Much geologic evidence cited in support of continental drift and polar wandering is based on the reconstruction of climates of the past. Their use is justified by the fact that modern climatic belts are arranged in roughly parallel zones whose boundaries are east–west and which range from the tropical equatorial climates to the polar ice climates, as suggested in Figure 20.8. Although climates at various times in earth history have been both colder and warmer than the present, we assume that basic climatic controls have remained the same and, therefore, that the climatic belts of the past have always paralleled the equator and been concentric outward from the poles.

Evidence from the geologic record allows us to map in a very general way the distribution of ancient climates. The pattern of some ancient climatic zones suggests that they were related to poles and an equator with locations different from those of today and that, therefore, the landmasses and the poles have varied relative to each other since those climates existed.

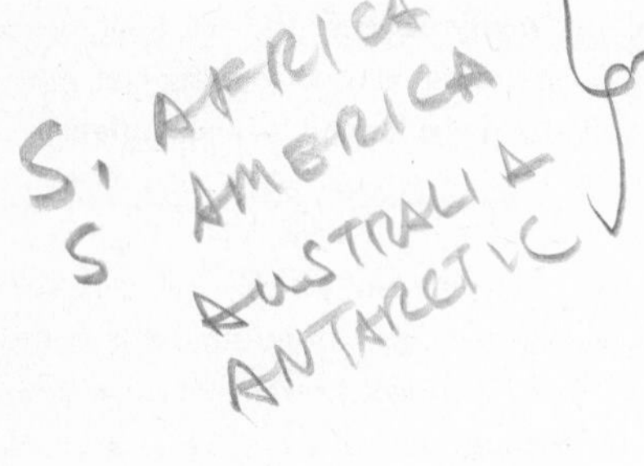

Late Paleozoic glaciation On the Indian peninsula lies a sequence of rocks known as the Gondwana system, reaching in age from the late Paleozoic to the early Cretaceous. Beds of similar nature and age are recorded in South Africa, Malagasy, South America, the Falkland Islands, Australia, and Antarctica. In these other localities they are known by other names, but we can still refer to them as belonging to the Gondwana system.

Geologists who have worked on the Gondwana formations have discovered many similarities among the rocks of the various continents, despite their wide geographic separation. Some of these similarities are so striking that many accept only one interpretation: The various southern lands must once have been part of a single landmass, a great southern continent, early called *Gondwanaland* by these geologists.

The distribution of ancient glacial deposits is one of the most convincing

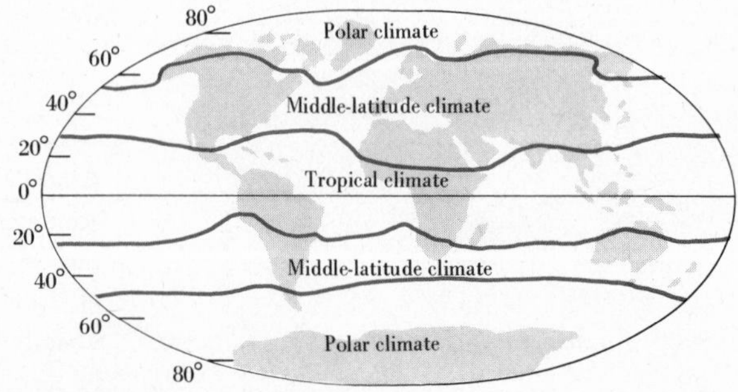

Figure 20.8 ***Present-day climatic boundaries are arranged concentrically around the poles and thus are approximately parallel to lines of latitude.***

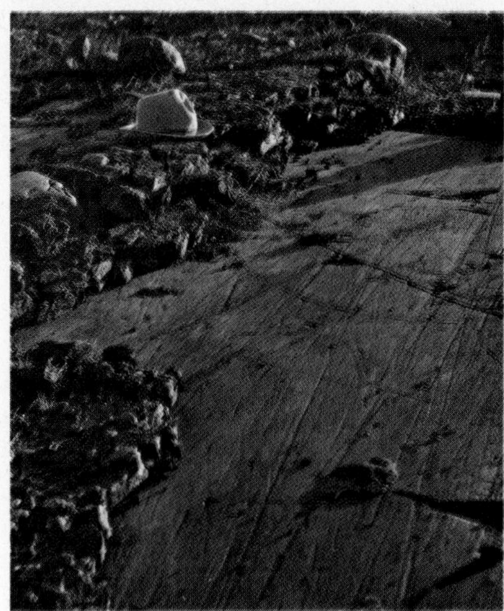

Figure 20.9 **The Dwyka tillite of South Africa is the deposit of a continental ice sheet of late Paleozoic age. Here the bedrock floor, striated by the Paleozoic glacier, passes beneath the tillite.** (*Photograph by R. B. Young.*)

lines of evidence for continental drift in these southern lands. During late Paleozoic time, continental ice sheets covered sections of what are now South America, Africa, the Falkland Islands, India, and Australia. In southwestern Africa, deposits related to these ancient glaciers are as much as 600 m thick. In many places, the now-lithified deposits (tillites) rest on older rocks striated and polished by these vanished glaciers (see Figure 20.9).

Plot the distribution of these deposits and the direction of ice flow on a map (see Figure 20.10) and one can make two immediate observations. First, these traces of Paleozoic ice sheets occur in areas where no ice sheets exist now or have existed during the glacial epochs of the Pleistocene ice age. Only an occasional towering peak may bear the scars of modern or Pleistocene ice. Continental ice sheets cannot exist in these latitudes today. Second, the direction of glacier flow is such that we can imagine the ice of Africa and South America to have been part of the same ice sheet when the two continents were one.

These observations have led some students of the earth to two conclusions. First, to account for glaciers in present tropical and subtropical areas, they locate the South Pole at this time somewhere in or near southern Africa. Second, to account for, among other things, the apparent continuity of glacier flow, these same students have postulated a single southern continent. This continent later split into several sections that drifted apart to form the modern landmasses.

Ancient evaporite deposits and coral reefs Evaporites, sedimentary rocks composed of minerals that have been precipitated from solutions concentrated by the evaporation of the solvents, are generally accepted as evidence of an arid climate (see Chapter 7). The ancient evaporite deposits represent the great arid belts of the past. The present "hot arid" belts are located in the zones of subtropical high pressure centered at about 30° north and south of the equator. In the northern hemisphere, an "evaporite belt" has shifted through time from a near polar location in the Ordovician and Silurian to its present position in the modern desert belts. This again suggests a relative motion of pole and landmasses of the past as compared with present-day conditions.

Turning to another line of evidence, we find that corals depict a climate

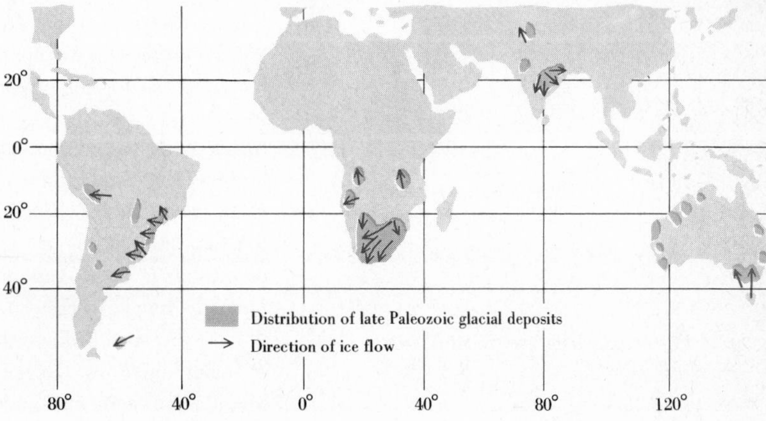

Figure 20.10 **Direction of movement of late Paleozoic ice sheet and distribution of known late Paleozoic tillites.**

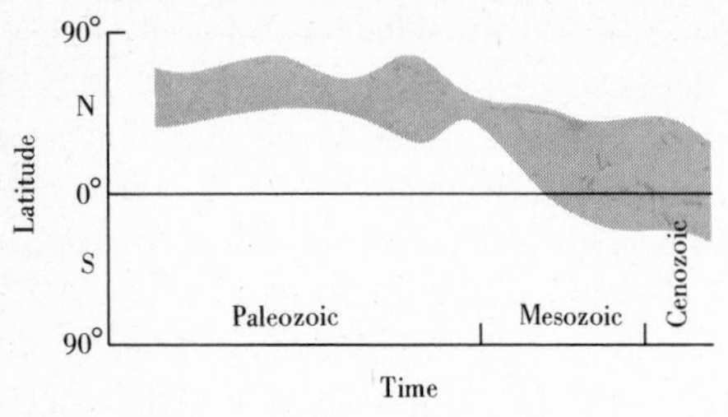

Figure 20.11 **_The growth of coral reefs today is restricted to warm, equatorial waters between 30° N and 30° S. During the Paleozoic and early Mesozoic, the reef belt was displaced far northward, as suggested by this diagram._** (*Adapted from Martin Schwarzbach,* Climates of the Past, *p. 214, Fig. 123, D. Van Nostrand Company, Inc., Princeton, N.J., 1963.*)

shifting geographically through time. Today, true coral reefs are restricted to warm, clear marine waters between 30° north and south of the equator. If we assume ancient reef-forming corals had similar restrictions, then plotting their distribution in the past will show the distribution of tropical waters of the past and the location of the past equator. Doing this, we find that reef-forming corals did not approximate their present distribution until halfway through the Mesozoic. Prior to that time they lay well north of the present equator, as shown in Figure 20.11.

PLANTS, REPTILES, AND CONTINENTAL DRIFT Shortly after the disappearance of the southern hemisphere's late Paleozoic ice sheets, an assemblage of primitive land plants became widespread. This group of plants, known as *Glossopteris* flora, named for the tonguelike leaves of the seed fern *Glossopteris* (see Figure 20.12), has been found in South America, South Africa, Australia, India, and within 480 km of the South Pole in Antarctica. The *Glossopteris* flora is very uniform in its composition and differs markedly from the more varied contemporary flora of the northern hemisphere. Some geologists have argued that the uniformity of the *Glossopteris* flora could not have been achieved across the wide expanses of water now separating the different collecting localities. In other words, in one way or another, there must have been continuous or near-continuous land connections between now separate continents. To some, this suggests that a single continent with a single uniform flora has been split apart into smaller continents that have since migrated to their present position. As we shall see, this conclusion has not gone unchallenged.

Among the vertebrate fossils of the late Paleozoic and earliest Mesozoic we find a great number of different reptilian types. Two of them provide us with arguments for continental drift. *Mesosaurus,* a toothed early reptile of the late Permian, lived in the water and thus far is known only from Brazil and South America. Although it was aquatic, most paleontologists do not believe that it could have made the trip across the South Atlantic. If this is so, then *Mesosaurus* may offer evidence for a closer proximity of South America and Africa and thus for continental drift. A second reptile, *Lystrosaurus,* dates from the Triassic period. *Lystrosaurus,* which is about a meter in length, was adapted to an aquatic but nonmarine environment, and its remains are reported from South Africa and from Antarctica's Alesandra Range. It is argued that *Lystrosaurus* could not have made the journey across the Antarctic Ocean separating South Africa from Antarctica, and that this points to the former physical connection of the two continents.

OTHER EVIDENCE Some anciently formed mountain chains now terminate abruptly at the continental margins. Join the continents together, and some of these geological structures match up between the two landmasses. Thus the Cape Mountains of South Africa are thought to be the broken extension of the Sierra de la Ventana of Argentina in one direction and of the Great Dividing Range in eastern Australia in the other direction. The entire stretch is cited by some adherents of continental drift as a once-continuous chain of mountains now segmented and separated. Similarly, the Appalachian Mountain system ends in the sea on the northern shore of Newfoundland.

Is its extension to be found in the orogenic belts of the British Isles and western Europe? Many geologists think so.

P. M. Hurley and his colleagues at the Massachusetts Institute of Technology have carried on extensive geochronologic studies of Precambrian igneous and metamorphic rocks in West Africa and the eastern bulge of South America. Radiometric age determinations of many samples show distinct belts of roughly similar age on the two continents. If the continents are shifted back together, the several provinces of Precambrian rocks of differing ages on the rejoined landmasses match fairly well (Figure 20.13). We are tempted to believe that rocks of similar age in Africa were once continuous with rocks of the same age in South America. Since they have formed they have been separated so that we see them on two widely separated continents.

Planetary winds (Figure 14.2) probably always existed. It would be interesting to see if ancient aeolian deposits might give some indication of such major wind belts as the northeast and southeast trades of the past. Sand beds accumulated on the face of a sand dune dip in the direction that the wind blows

Figure 20.12 ***These leaves of the fossil plant* Glossopteris *come from strata of Permian age in Australia. The* Glossopteris *flora is found also in South America, Africa, India, and Antarctica. The widespread occurrence of this very uniform flora has been used as evidence both by opponents and proponents of continental drift. Diameter of detail is 2.5 cm.*** *(Specimen from Paleobotanical Collections, Princeton University. Photograph by Willard Starks.)*

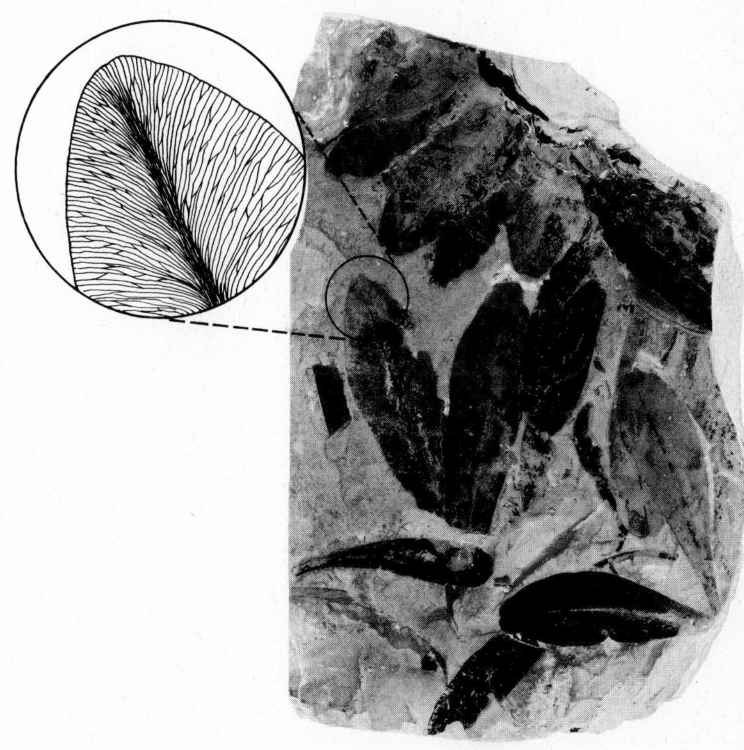

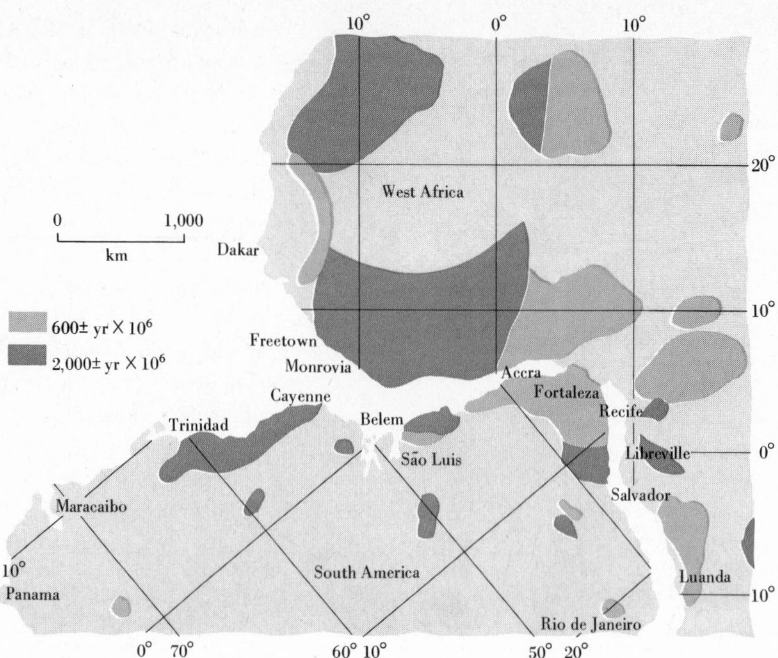

Figure 20.13 ***Rocks of two different ages mark the opposing shield areas of South America and Africa. If the two continents are fitted together, the zones of the same age match across the line of fit.*** (*After P. M. Hurley and J. R. Rand, "Review of Age Data in West Africa and South America," in R. A. Phinney, ed.,* The History of the Earth's Crust, *Fig. 1, Princeton University Press, Princeton, N.J., 1968.*

(Figure 14.19). We can measure such dip in ancient wind-deposited sediments. If enough of these directions are available, the direction of the dune-forming wind can be determined statistically. Preliminary studies on presumed aeolian deposits of late Paleozoic age in Wyoming, Utah, Arizona, and England indicate that these areas fell within the northeast trade wind belt when the earth's pole (as suggested by the paleomagnetic evidence) was located on the East China coast.

WINDS LATITude

The case against continental drift

So far, we have considered only the case in support of the theory of continental drift and wandering poles and nothing of the case against this hypothesis. Let us now look at some of the evidence on the other side.

GONDWANALAND We have already referred to Gondwanaland as a great southern continent. Supporters of the concept of drifting continents believe that this continent broke into pieces and that these pieces moved to the present positions of the southern continents. But those who first proposed the concept

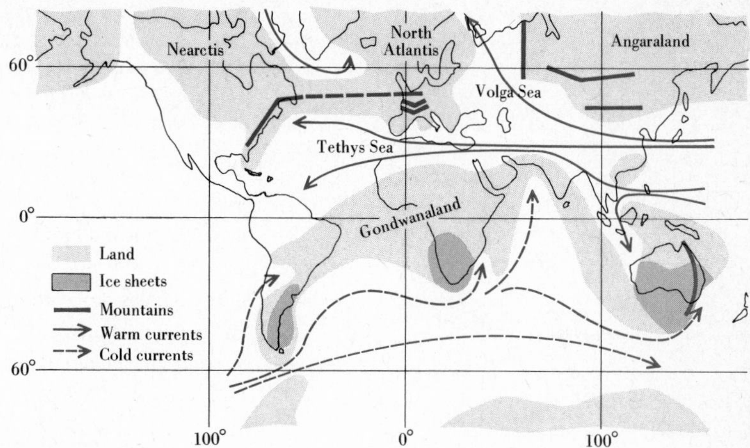

Figure 20.14 ***Some scientists reject the idea of drifting continents. They have pictured Gondwanaland as a once-continuous southern continent, portions of which foundered into the deep oceans, leaving fragments to stand as South America, Africa, Australia, and India. This point of view is reflected in a reconstruction of late Paleozoic geography.*** (*Redrawn from C. E. P. Brooks,* Climate Through the Ages, *p. 248, Fig. 29, McGraw-Hill Book Company, New York, 1949.*)

of Gondwanaland had no idea of suggesting such a radical explanation for the present geography as that demanded by continental drift.

The initial concept of Gondwanaland to explain the similarity of geologic events in now widely separated areas pictured the southern continent as a continuous landmass stretching from South America through Africa and India to Australia. Sometime after the late Paleozoic, large portions of this vast east–west landmass foundered, leaving only fragments of Gondwanaland to form the present continents and the subcontinent of India. One of the best known of these reconstructions is reproduced in Figure 20.14.

The difficulty of explaining how great masses of the earth's crust could sink to create ocean basins has been as difficult for some geologists to accept as has been the idea of shifting continents. To avoid problems of both explanations, some have proposed that the southern lands were connected by narrow land bridges and that Gondwanaland was thus made up of what is now South America, Africa, India, Australia, and Antarctica, hooked together with narrow bridges of land. Such a theory allows for the permanency of the continents and requires only that narrow (although hypothetical) land bridges need to be postulated and then drowned. Supporters of this theory are faced with the dilemma that thus far there is no geophysical evidence from the southern oceans suggesting the presence of drowned "lightweight" land bridges.

THE FOSSIL EVIDENCE Some paleontologists have questioned whether there is any evidence of climatic belts having changed their position in time and hence of poles having moved or continents having shifted. They cite the evidence of fossil plants and animals.

One reconstruction of past climate based on the distribution of fossil plants

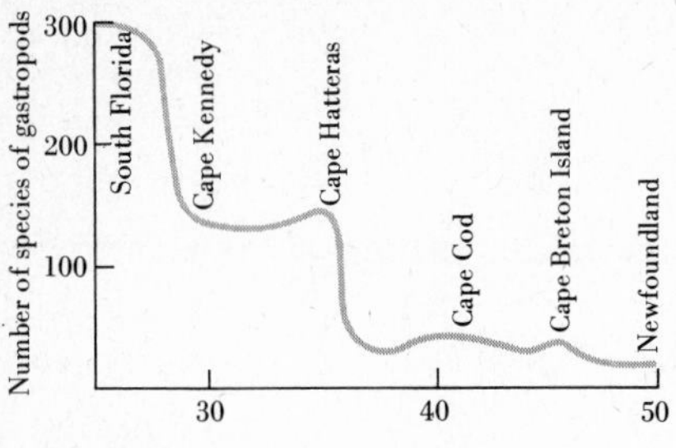

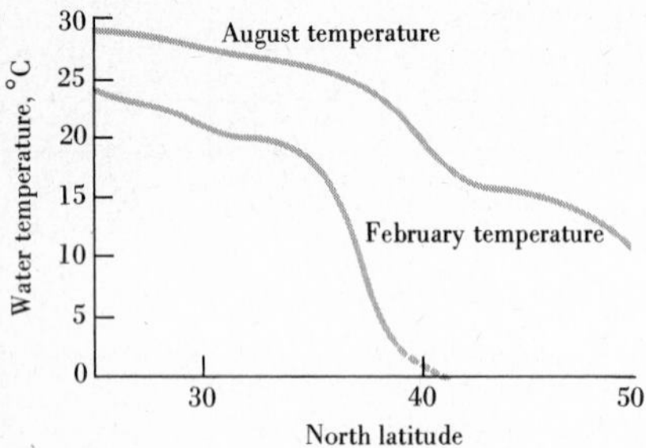

Figure 20.15 ***The number of different species of living things decreases with decreasing temperature. This change is called a diversity gradient. It is illustrated here in the relation between number of species of gastropods, temperature of water, and latitude along the eastern coast of North America.*** (*Adapted from A. G. Fischer, "Latitudinal Variations in Organic Diversity,"* Evolution, *vol. 14, no. 1, p. 69, Fig. 10, 1960.*)

suggests that the climatic zones of the Eocene were arranged with boundaries parallel to the modern equator. If so, then there is no need to postulate a shift in poles since the early Cenozoic. But proponents of a polar shift find this not unreasonable. Other evidence indicates to them that the greatest part of polar movement had already taken place by the Eocene.

In a recent attempt to refine the analysis of fossils as climatic indicators, the existence of diversity within fossil groups has been used. This technique is based on the fact that not only is the distribution of an individual species controlled by climate and hence by latitude but so are the numbers of different forms within any taxonomic group. For instance, the numbers of species in a given genera decreases with decreasing temperature, and the latter, in turn, is related to latitude (see Figure 20.15). Statistical analysis demonstrated that the diversity gradient slopes downward toward the lower temperatures of the poles.

The technique of diversity gradients has been applied to some bivalved marine animals (brachiopods) of Permian age, which are known to have a diversity gradient. Statistical analysis of the distribution of the brachiopods suggests to some workers that the north pole was in essentially the same position during the Permian period as it is today (see Figure 20.16). Reexamination of the distribution data and of the statistical techniques, however, leads other investigators[3] to the conclusion that the biologic diversity gradient of the brachiopods confirms that the north pole was located differently during the Permian than it is today—a conclusion earlier arrived at on the basis of paleomagnetic evidence.

RELIABILITY OF EVIDENCE The reliability of evidence cited in support of continental drift has been questioned by many workers. The paleowind directions discussed above, for example, are thought by some to be not only inconclusive but actually based on deposits not aeolian.

Some of the evidence from the fossil record is attacked as too fragmentary and open to too many interpretations to be diagnostic. The use of *Mesosaurus* as representing evidence for continental drift is one such example.

Paleomagnetic evidence is currently regarded as one of the most conclusive arguments in favor of continental drift. But some argue that the evidence may not be as strong as it first appears, and they point out that this evidence may be overextrapolated. For instance, we said earlier that discussions of paleomagnetism assume a dipolar magnetic field for the earth. Yet, the more ancient

[3]The first serious challenge to the use of diversity gradients as support for the case against continental drift came from Robert Duncan, who, as a junior at Princeton University, presented a report at the annual meetings of the American Geophysical Union in Washington D.C. in April, 1970, an event that reminds us that one does not have to be an established scholar to contribute to the fund of knowledge.

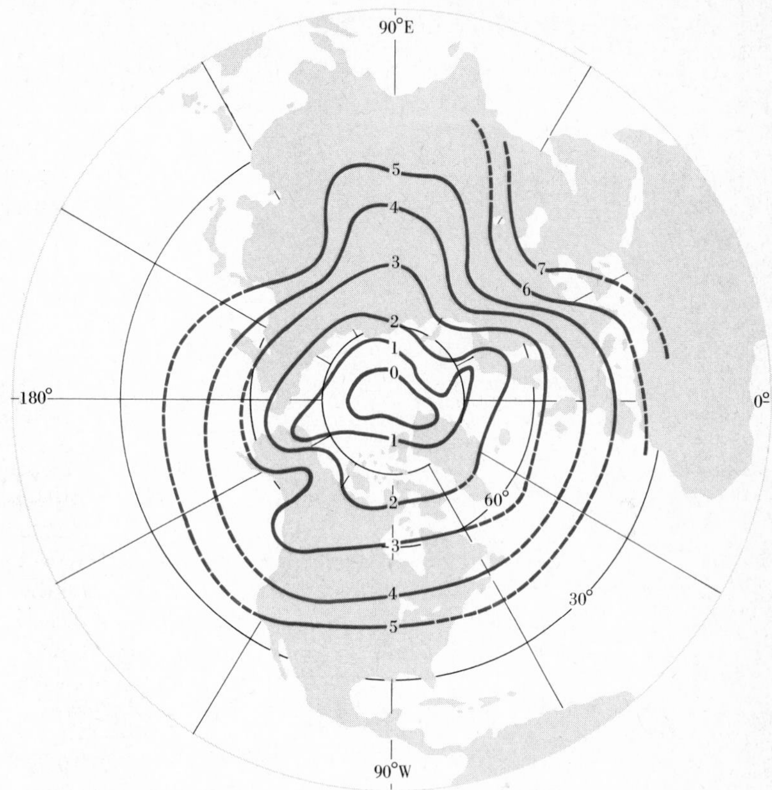

Figure 20.16 **According to one analysis the number of genera of Permian brachiopods decreases toward the position of the modern geographic pole in the northern hemisphere. This pattern suggests to some that there has not been a motion of the pole since the late Paleozoic.** (*Redrawn from F. G. Stehli and C. E. Helsley, "Paleontologic Technique for Defining Ancient Pole Positions,"* Science, *vol. 142, p. 1058, Fig. 3, 1963.*)

fields could have had another form—four poles instead of two, perhaps. If so, the entire paleomagnetic argument in support of continental drift would have to be reevaluated.

Status of the problem

Prior to World War II, most of the supporters of the theory of continental drift lived in the southern hemisphere, particularly in South Africa. Geologists and geophysicists of the northern hemisphere, particularly in the United States and western Europe, felt that there was little or no support for the motion of continents or the wandering of the poles. In fact, it was hardly respectable to entertain the possibility of such movement.

After World War II, the techniques of geophysics, and particularly those of paleomagnetic measurement, began to bring new data into the discussion. This new evidence reopened the entire question of continental drift for American and European geologists. By 1970, a vast amount of data bearing on continental drift had accumulated. Most workers in the field now feel that the available evidence favors the reality of continental drift. Many, having accepted that continental drift has occurred, have turned to how it might have happened—and to what implications drift might have for our understanding of earth history and earth mechanics. In the following section we take up this thread of thought and consider the concept of plate tectonics.

20.4 Plate tectonics

Recent workers have pointed out that the earth's surface can be divided into large units or plates. It is suggested that the origin, movement, and interrelation of these plates can be coordinated with sea-floor spreading, earthquake belts, crustal movement, and continental drift. Let us examine this hypothesis.

The world system of plates

At least six major plates are recognized (the Indian, Pacific, Antarctic, American, African, and Eurasian plates), as shown in Figure 20.17. The boundaries of these plates are loci of present-day earthquakes, volcanic activity, and crustal movement. Consider for example the Pacific plate. On its southern and southeastern side it is bounded by an oceanic rise which undergoes active spreading. Elsewhere it is marked by island arcs, deep-sea trenches, and tectonic and

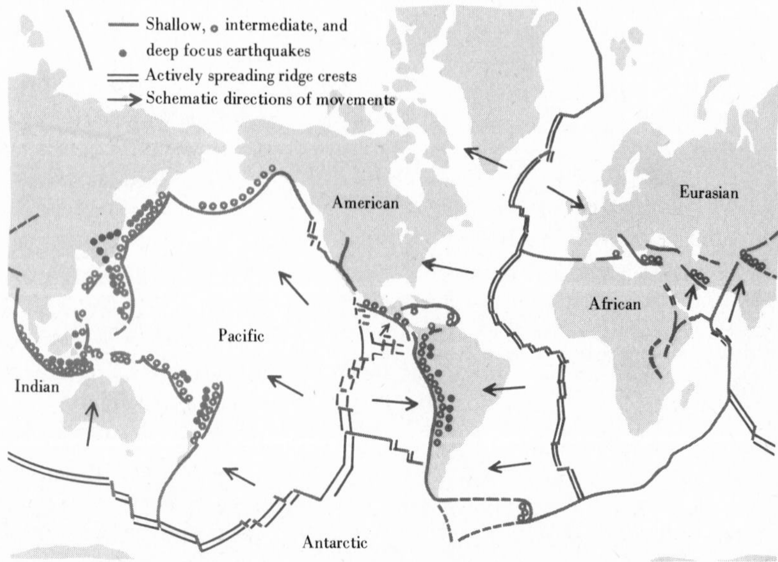

Figure 20.17 **Six major crustal plates are outlined by actively spreading ridge crests, trench systems of the Pacific, faults, and the mountain-building zone of the Mediterranean-Himalayan section. See the text for discussion.** (*Compiled by F. J. Vine from various sources, "Sea-Floor Spreading—New Evidence,"* J. Geol. Educ., *vol. 17, Fig. 1, 1969.*)

volcanic activity. Its northeastern side along western United States is defined by the San Andreas fault, where the Pacific plate moves generally northwestward and jostles the American plate in the process.

The relation of the plates to the distribution of earthquakes is striking when one compares Figure 20.17 with Figures 16.1 and 16.2. In general, the shallow quakes mark the ocean ridges; deep earthquakes are missing in these zones. Both shallow and deep quakes are found along the arcs and trenches and along young mountain systems such as the Himalayas.

The plates may include both continental and oceanic areas, as for example the American plate which embraces North and South America and Greenland as well as the western Atlantic Ocean. On the other hand, the Pacific plate is restricted to oceanic area.

The movement of plates

We see in Figure 20.17 that some of the sutures between adjacent plates are zones of active sea-floor spreading, a process described in Section 20.3. Along such zones, therefore, crustal material is thought to be continuously forming and, to make room for it, the sea floor is pictured as moving laterally away from the ridge crest. The details of the movement in the ridge zones have been in part worked out. Thus the fractures that cut across the ridge and offset are thought to be the transform faults described in Chapter 18 and illustrated in Figure 18.35. An analysis of the first movement of the crust during earthquakes on the ridge shows two types of movement. Along the rifts that mark the axes of the oceanic ridges motion during earthquakes indicates tensional stress, normal faulting, and down-dropping of the blocks associated with the grabens along the ridges. In contrast is the movement along the fractures offsetting the ridges. Along these structures, analysis of earthquake data shows that the movement is of a strike-slip nature on a plane at right angles to the ridge. Movement is away from the ridge crests, as shown in Figure 18.36 of the equatorial section of the Midatlantic Ridge.

If plates are actually moving away from each other through a process of sea-floor spreading then a number of questions arise. Here are two: What drives the ocean floor? Where does the moving ocean floor go?

Most students of the problem appeal to convection currents (described in Chapter 19) to provide a mechanism for moving of the ocean floor. They visualize the ocean ridges as lying over a rising convection current. The facts that seismic and volcanic activity are generally high on the ridges and that heat flow from the interior is higher at the ridge crest than at most other places add appeal to the suggestion. The current is thought to lie in the upper mantle well below the Moho and to represent ascending limbs of adjacent convection cells. At some point below the Moho the rising current is pictured as splitting into two currents flowing laterally away from the midoceanic ridge.

What happens to the old sea-floor crust as it is displaced from the ridge crest by younger and younger material? Those who explain the motion by convection see the crust and a part of the upper mantle as carried laterally by convection currents toward the margins of the ocean basins. This is supported not only

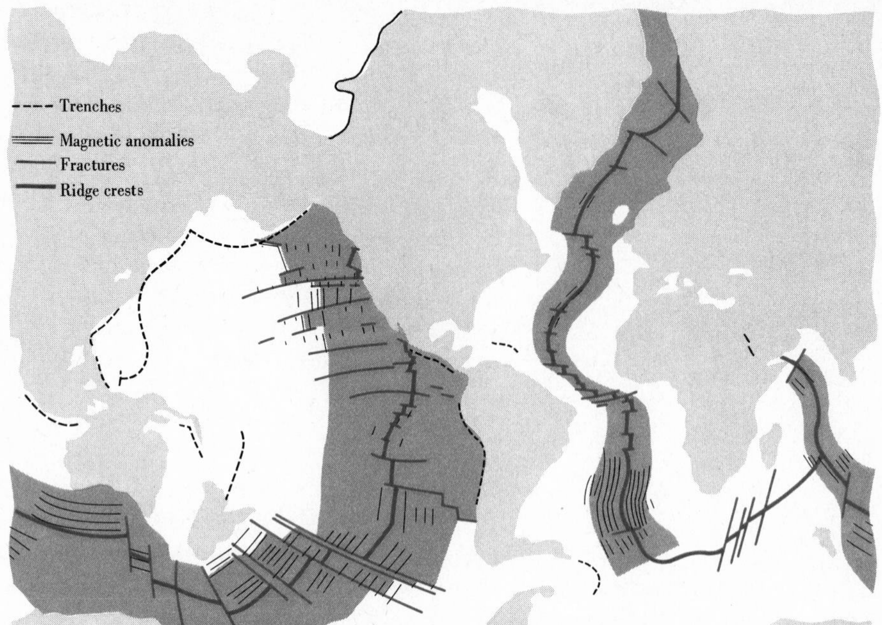

Figure 20.18 **The shaded portion of the basins is thought to represent the area of oceanic crust formed within the last 65 million years.** (*From F. J. Vine, "Sea-Floor Spreading—New Evidence,"* J. Geol. Educ., *vol. 17, Fig. 10, 1969.*)

by the paleomagnetic measurements already discussed but also by studies of oceanic sediments. These indicate that the sedimentary deposits are not only thinner in the zones of the midoceanic ridges but also that as they become thicker away from the ridge the basal layers of the sedimentary accumulations become older. This all suggests that the older sea floor lies farthest from the oceanic ridges.

If we follow this line of reasoning we would expect to find that belts of varying width along the active ridges mark the most recently formed oceanic crust. If we accept the reconstruction shown in Figure 20.18, then an area approaching 50 percent of the ocean floor has been formed during the Cenozoic—in approximately the last 65 million years. In this regard it is interesting to note that by the beginning of the 1970s no rock older than 150 million years had been taken from the deep ocean basins. Furthermore, present-day sedimentation rates in the oceans can account for the sedimentary accumulations there within the last 100 to 200 million years. All this leads to the suggestion that the ocean basins, as we know them, are young. It would then follow that although ocean basins have existed through most of geologic time they have been continuously reconstituted. Furthermore, their shapes and their geographic locations have been shifting constantly.

The idea of plate tectonics can be used to integrate some of the major features of the ocean basins. Thus the midoceanic ridges and rises, with their central rift valleys, and a sedimentary cover which is young and thin (or lacking) reflect the upwelling of mantle material along the ascending currents of adjacent convection cells. The large fracture systems which offset the midoceanic ridges are transform faults. These faults come into being as the rising convection currents finally diverge laterally, carrying with them the overlying

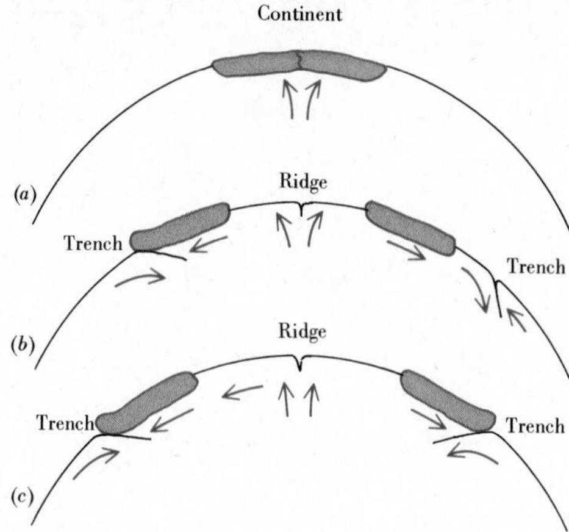

Figure 20.19 **This sequence illustrates what might be expected if mantle material upwells beneath a continent. In (a), a continent is shown as rifting begins. In (b), the sea floor has opened between the two fragments of the original landmass. The leading edge of one is shown overriding a trench and that of the other is still separated from an oceanic trench. Diagram (c) shows a still further stage of sea-floor spreading and continental drift. Both landmasses are shown overriding trenches.** (*After F. J. Vine, "Sea-Floor Spreading—New Evidence,"* J. Geol. Educ., *vol. 17, Fig. 2, 1969.*)

crustal plates and upper mantle. As the currents cool and turn downward, they carry with them the oceanic crust, with its overlying sediments, as well as the underlying upper mantle. To many students of the problem, the deep, arcuate trenches of the oceans represent the downturning of the convection currents. Great crustal plates, then, are seen as moving within a worldwide framework. The leading edges of the plates are consumed in the trench-island arc systems of the Pacific basin and along the Mediterranean–Himalayan belt of Eurasia. Along the trailing edges, new crust is generated in the active ridges. The drifting of continents, then, becomes a consequence of the movement of crustal plates and the appearance and disappearance of oceanic crust.

ADDITIONAL SPECULATIONS If the mantle wells up beneath a continent we can expect that it will be rifted and that the two continental fragments will drift apart as a new ocean basin forms. Such a possibility is suggested in Figure 20.19. As the ocean basins enlarge and the continents, "tied" to the underlying but moving upper mantle, drift away from each other their leading edges may be fronted by trenches. The continents, being lighter, override the trenches.

A modern example may be present in the western margin of South America. Here Pacific oceanic crust appears to spread eastward toward the continent. South America's margin is marked by a deep ocean trench, a narrow continental shelf, and the towering Andes. Earthquakes occur, becoming deeper in origin inland, the deepest being about 700 km below the surface. The earthquake foci lie along the Benioff zone that dips beneath the South American continent.[4]

The elevation of the Andean chain is interpreted in part as the result of underflow of oceanic crust. Seismic and igneous activity are seen as related to the motion between the Pacific and American plates along the Benioff zone. Farther north, the Mesozoic history of western United States has been interpreted in terms of the underflow of 2,000 km of Pacific material beneath the American plate. The early Paleozoic mountain-building episodes in Newfoundland and Scotland have been ascribed by some workers to the interplay between two plates, one underthrusting another.

Outline

Earth magnetism and its changes as recorded in rocks is closely tied to the case in favor of continental drift.

Cause of magnetism is both internal and external in relation to the earth. A small part is external, caused by the sun. The largest part of the earth's field originates within the earth and is best explained as caused by circulation within the earth's liquid core, which acts as if it were a giant dynamo.

The *direction* and *intensity* of the magnetic field varies with time. If the intensity decreases far enough it may eventually pass a zero point and the

[4]A seismic zone dipping beneath a continent or continental margin and having a deep-sea trench as its surface expression is called a *Benioff zone* after Hugo Benioff, an American seismologist who first described the feature.

earth's field will become polarized in a sense the reverse of that of the present.

Paleomagnetism is the study of the earth's field in past time.

Thermal remanent magnetism (TRM) occurs in igneous rock-forming minerals as they cool below the Curie temperature.

Depositional remanent magnetism (DRM) forms as magnetic particles align parallel to the earth's field at time of deposition. DRM and TRM reflect the magnetic field existing at time of rock formation.

The *virtual geomagnetic pole* is the equivalent pole position representing the magnetic field as recorded at one observation point. The discrepancy between geomagnetic poles and geographic poles is thought to average out in time. The ancient poles are thought to have been, like the modern poles, dipolar. The ancient poles have migrated thousands of miles since the Precambrian, and the path of polar migration is different for each continent. Direction of the earth's magnetic field reversed itself many times in the geological past.

Continental drift is supported by the data provided by paleomagnetism, the shape of continents, sea-floor spreading, ancient climatic patterns, and disrupted geological structures. Alternative explanations to drift involve land bridges and differing interpretations of the climatic evidence.

Plate Tectonics is a hypothesis which postulates that the earth's crust is broken into large blocks that are being renewed in some zones and destroyed in others. Proponents of the hypothesis see the plates as driven by convection currents in the upper mantle and feel that the hypothesis integrates continental drift, sea-floor spreading, seismic zones, and major tectonic activity.

Supplementary readings

COX, ALLAN: "Geomagnetic Reversals," *Science*, vol. 168, no. 3864, pp. 237–245, 1969. The chronology of magnetic reversals.

COX, ALLAN, and R. R. DOELL: "Review of Paleomagnetism," *Geol. Soc. Am. Bull.*, vol. 71, pp. 645–768, 1960. A good review of the principles and assumptions involved.

JACOBS, J. A.: *The Earth's Core and Geomagnetism*, Pergamon Press Inc., New York, 1963. A 137-page book, over half of which is devoted to magnetism. Mathematical discussions are largely confined to appendices.

MORGAN, W. J.: "Rises, Trenches, Great Faults and Crustal Blocks," *J. Geophys. Res.*, vol. 73, pp. 1959–1982, 1968. A classic paper on plate tectonics.

PHINNEY, ROBERT A. (ed.): *The History of the Earth's Crust: A Symposium*, Princeton University Press, Princeton, N.J., 1968. Contains papers on, among other subjects, continental drift and sea-floor spreading.

SCHWARZBACH, M.: *Climates of the Past*, D. Van Nostrand Company, Inc., Princeton, N.J., 1963. A scholarly treatment of climates of the geological past, the evidence for which is often applied to the discussion of continental drift.

TAKUCHI, H., S. VYEDA, and H. KANAMORI: *Debate about the Earth*, Freeman, Cooper & Company, San Francisco, 1967. A stimulating discussion of continental drift.

VINE, F. J.: "Sea-Floor Spreading—New Evidence," *J. Geol. Educ.*, vol. 17, pp. 6–16, 1969. A concise, easily understood statement of the state of knowledge about sea-floor spreading.

21 Natural resources

ALL THROUGH THIS BOOK, WE HAVE BEEN TALKING ABOUT ENERGY, ELEMENTS, minerals, and rocks, about geological processes and the landforms they produce. By this time you have a wealth of detailed information about the materials and processes on which the science of geology rests. But what practical use can you make of this information? How is it related to your everyday life?

Our modern civilization demands a great range of materials and tremendous amounts of energy. We use more of these than others have at any other time in the history of the human race. And our demands are increasing. We depend on materials for our buildings, transportation, communication, work, and security. We depend on energy to make these materials available. Where do these materials and energy come from? Most come from deposits in the earth's crust. These deposits have been formed by nature working through the very geological processes we have been discussing. They constitute our earth's natural resources.

21.1 *Distribution of natural resources*

The world's natural resources are concentrated in relatively small areas and are not evenly distributed throughout the crust. They have been concentrated by the geological processes we have been studying, such as weathering, running water, igneous activity, and metamorphism. But only a few geologic environments have favored their formation, and they are sporadically distributed. Also, after they are used, they are gone forever. Unlike plants, we cannot grow another crop. A deposit that has been located and mined may soon be exhausted unless it has huge reserves.

Our need for natural resources

Natural resources are the foundation of our industrial civilization. Our living standards and power depend to a great extent on them. However, no nation has within its boundaries or under its control all the natural resources it needs to maintain its way of life. The United States, though richly endowed, is still dependent on others for some of its needs. Its supply of energy from the fossil fuels, oil, gas, and coal, as well as water power, is excellent, but it is still dependent on trade with other nations to obtain certain minerals. Of the most critical minerals, we lead the world in the production of 8, have a sufficient supply of 12 more, and lack 4 (see Table 21.1). Our increasingly complex machines are often completely dependent on small quantities of rare minerals, such as exceptionally heat-resistant ones. Therefore, a lack of a few grams of one mineral may bring to a halt the use of tons of another. We are in a situation similar to that in which the loss of a nail caused the loss of a shoe, a horse, and a kingdom.

Table 21.1 **Production of natural resources, 1967**[a]

Resource	Total world production[b]	Percent of world total: Produced by U.S.	Percent of world total: Leading producer, if not U.S.
Metals			
Molybdenum	64,750 MT	61.4	
Uranium oxide	16,900 MT	49.1	
Magnesium	183,000 MT	48.3	
Tellurium	127,000 MT	48.2	
Vanadium	9,644 MT	46.7	
Aluminum	7,451,000 MT	39.8	
Cadmium	12,409,000 kg	31.4	
Titanium, Ilmenite	2,710,000 MT	31.3	
Iron ore	625,370,000 MT	13.7	U.S.S.R., 26.9
Silver	260,915,000 oz t	12.3	Mexico, 14.5
Zinc	4,916,000 MT	10.1	Canada, 23.0
Lead	2,914,000 MT	9.9	U.S.S.R., 13.7
Mercury	242,000 units[c]	9.8	Spain, 20.8
Gold	45,614,000 oz t	3.5	South Africa, 66.9
Nickel	413,000 MT	3.0	Canada, 51.1
Manganese	17,073,000 MT	0.1	U.S.S.R., 42.2
Nonmetals			
Mica	140,000 MT	76.6	
Vermiculite	366,000 MT	68.8	
Native sulfur	11,150,000 MT	63.9	
Phosphate rock	78,703,000 MT	45.8	
Diatomite	1,564,000 MT	39.8	
Feldspar	1,987,000 MT	31.5	
Salt	118,262,000 MT	29.9	
Barite	3,508,000 MT	24.4	
Potash	15,400,000 MT	19.5	
Coke (metallurgical)	304,978,000 MT	19.2	U.S.S.R., 22.6
Coal	2,725,521,000 MT	18.8	
Gypsum	46,626,000 MT	18.3	
Fuels			
Gas, natural	28,611,206,000,000 ft^3	—	North and Central America, 70
Petroleum, crude	12,889,705,000 brl	—	Western hemisphere, 41.4; free world, 82.0
No significant U.S. production			
Diamonds			
Industrial	35,295,000 c	—	Eastern hemisphere, 99.2
Gem	9,093,000 c	—	Eastern hemisphere, 97.4
Columbium–Tantalum	9,458,000 kg	—	Brazil, 52.2
Chromite	5,094,000 MT	—	Eastern hemisphere, 99.3
Strontium minerals	14,293 MT	—	United Kingdom, 70.0
Figures withheld			
Cobalt	—	—	—

[a] Mineral Industry Surveys, U.S. Department of the Interior, Bureau of Mines, November 15, 1968. Of 51 commodities for which fairly complete world data were available, the United States ranked first in the production of 24; U.S.S.R. ranked first for 11.

[b] Abbreviations: MT, metric tons; kg, kilograms; oz t, troy ounces; brl, barrels; c, metric carats.

[c] Mercury is measured in units of 76-lb flasks.

21.2 Mineral deposits

Mineral deposits are generally classed as metallic or nonmetallic. Metallic deposits are mined for such elements as gold, silver, copper, iron, lead, zinc, and aluminum. Nonmetallic or earthy deposits include salt, phosphate, coal, limestone, gems, and even sand, gravel, and rock. The quality of mineral deposits varies from mine to mine and even within a given mine. Mineral deposits very rarely come out of the ground ready to be used, so besides being mined the materials must be processed.

Figure 21.1 **Panning for gold on the Anderson River, a tributary to the Fraser River, British Columbia's Sierra Cascade Mountains. The prospector partially fills the pan with water and throws in a shovelful of dirt. He picks out the pebbles and stirs the mass until clay-sized particles are dislodged and can be sloughed away in the muddied water. He partially fills the pan with water again and gives it a slightly eccentric circular motion to build up a wave that slops over the edge each time, carrying with it a little sand. He continues this process until only the specks of gold, which have greater specific gravity, remain in the pan.** (*Photograph by Elliott A. Riggs.*)

Mining

Mining involves the removal of materials from the earth. It is done in several ways, depending on the size and type of deposit. Quarrying is generally limited to surface removal of rock. This can be cut into dimension stone or blasted and crushed. Placer mining is associated with the separation of valuable metals from sand and gravel deposits (see Figure 21.1). The sluice box is used in rich deposits. Alternatively, large quantities of sand and gravel may be worked by machines to obtain small quantities of a valuable metal. Open pit or strip mining is confined to huge deposits fairly close to the surface. The operation first requires the removal of worthless overburden. Twenty-three percent of U.S. bituminous coal is strip mined. Underground mining involves thick tabular bodies of mineral deposits. Mines that were once shallow are now being worked at greater and greater depths. The deeper the mine, the more expensive the operation. However, with the development of better equipment, new techniques, better transportation, or higher prices, some abandoned mines have been reopened.

Only a few metals and precious gems were being mined for many centuries. The mining depended mainly on men's muscles, and laborers toiled under extremely dangerous and harsh conditions, using only primitive tools. Despite this, mining was widespread. With a growing population and demand for a wider variety of basic minerals and metals, mining was stimulated. The greatest surge came with the advent of the Industrial Age. New materials were required in great quantities and large markets were created. Miners were supplied with better tools and power.

Ore deposits

When metallic minerals occur in concentrations that can be worked at a profit, they are called *ore deposits.*

SEPARATION OF VALUABLE METAL Although a metal is more concentrated by nature in an ore deposit than in another place, it still has to be separated

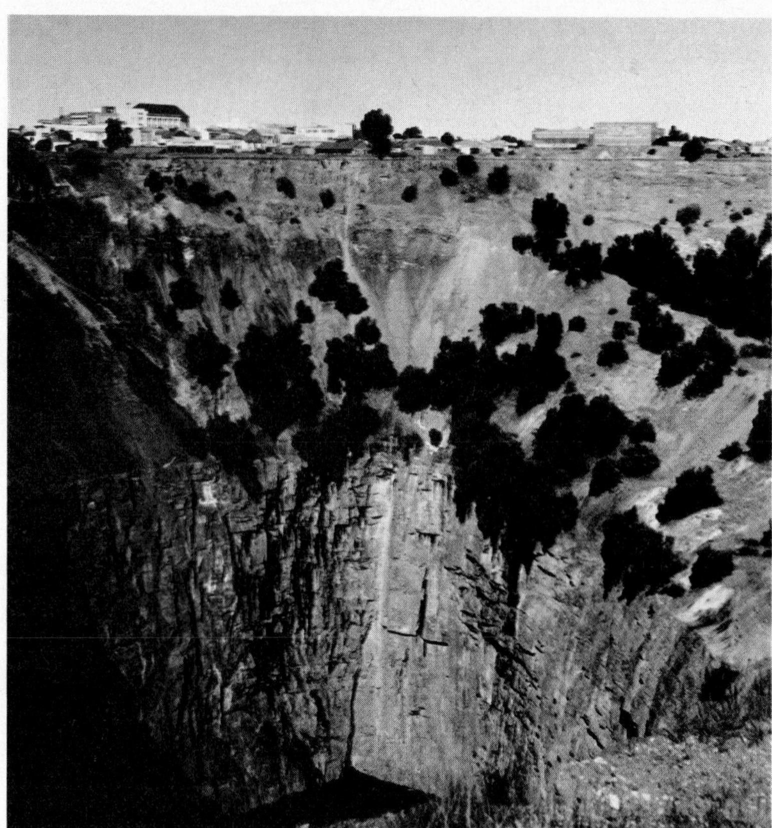

Figure 21.2 ***Abandoned pit of Kimberley diamond mine in the rock of a volcanic neck near Kimberley, South Africa. Water now stands in the pit to within 200 m of ground level. Before it was abandoned, the mine was developed to a depth of 1,200 m.*** *(Photograph by Cornelius S. Hurlbut, Jr.)*

from the ore. This may be done in different ways. The simplest are used in extracting metals which are found in an uncombined state.

Gold-bearing quartz is pulverized and the pulverized material is then treated with mercury, in which gold dissolves. This is then easily separated from the quartz because of its higher specific gravity. Heating and distilling of the mercury leaves a residue of gold. The ore after this first process may be treated again with a cyanide solution and gold recovered from the solution by electrolysis or treatment with zinc.

Some metals are combined with other elements, such as sulfur, oxygen, or carbon, in the crystalline structure of a mineral. This mineral may be mixed with unwanted minerals, rock, or earthy materials called *gangue,* from which it must be separated.

The most common ore minerals of copper are sulfides. These are found disseminated through rocks. Concentration from this gangue begins by a process called *flotation.* The rock is finely ground and treated with a mixture of water and special oil. The water wets the silicate minerals and the oil wets the sulfide minerals. When air is blown through the mixture, it forms a froth of the oil and sulfide minerals, which is skimmed off. The concentrated sulfide is then roasted in a furnace with air passing through. The air removes some of the sulfur as sulfur dioxide. However, some unwanted minerals such as

Figure 21.3 ***Underground gold mine in gold-bearing quartz vein a little over 1 m wide, Bralorne, British Columbia.***

FeO and SiO_2 are still left. Limestone is then mixed and heated with the roasted ore to serve as a *flux.* The iron oxide and silica combine with the limestone and form a slag, leaving the melted metal, which is drawn off.

Gold Gold is a rare element used principally in coinage and jewelry. It usually occurs in the uncombined state widely distributed through other materials in small amounts. It is found most commonly in hydrothermal deposits in sialic igneous rocks, particularly those that are rich in quartz (see Figure 21.3). It is also found concentrated in placer deposits where it had been weathered out of its place of original deposition and transported along stream channels, accumulating in low places because of its specific gravity.

Figure 21.4 ***Native gold. A unique hornlike mass from California.*** *(Harvard Mineralogical Collection. Photograph by Harry Groom.)*

The gold discovered on the western slopes of the Sierra Nevada Mountains in California in 1848 was concentrated in placers so rich that fortunes were made simply by panning it by hand. Deposits of lower grade are uncovered by modern hydraulic giants which wash away the barren material that overlies pay dirt and sluice the gold-bearing gravels into mercury-lined boxes where the gold is trapped. Deposits may also be worked by dredges where the gravel is below the water level of ponds. Some deposits contain only a few cents' worth in each cubic meter, but even these have been dredged at a profit.

Since 1879, the Homestake mine in the Black Hills at Lead, South Dakota, has been the leading U.S. gold producer. Its shafts now extend 1,600 m below the surface. Great reserves of medium-grade ore remain to be mined. Gold mines have operated at a profit in a deposit with less than 10 g/metric ton.

Silver Silver is widely distributed in small amounts in ore deposits of other metals. Native silver is principally found in the oxidized zone of ore deposits. The principal mineral containing silver is argentite, Ag_2S. Argentite is found in veins where it has been deposited by hydrothermal solutions. It may also be found in deposits concentrated through weathering processes. Significant silver production comes from ores mined primarily for lead, copper, and zinc.

Since 1885, over $1 billion of silver, lead, and zinc have been mined in the area around Coeur d'Alene, Idaho. This district is first in silver, second in lead, and fourth in zinc production in the United States. The Sunshine mine is the second largest producer of silver in the world.

Iron Iron manufacture in the United States began on the eastern seaboard. Bogs and swamps surrounding the coastal settlements of the New World contained iron deposits which had been leached from the rocks by streams. The water carried the iron down into lowlands where it became embedded in the marshes. Through natural chemical change, this formed an iron ore of commercial value for that era. It could be extracted by baking out the moisture. Within 1 year after the founding of the colony at Jamestown, Virginia, in 1607, the settlers had sent back to England enough of this crude bog ore to produce 17 tons of iron. In 1644, at Saugus, Massachusetts, the first usable iron product was manufactured. It was a $1\frac{1}{2}$-kg cast-iron cooking pot.

The iron and steel industry has grown tremendously since then. It is the basis of all our other great industries. It depends on many natural resources, of which iron is the most important. This is supplemented by other elements which are used as alloys to harden or soften steel. Fuels such as coal and oil are required in great quantities to fire the furnaces that purify the iron and also cook the steel. Limestone is quarried as a flux for purifying iron in the refining process. When heated, it removes the impurities sulfur and phosphorus by absorbing them.

With industrial growth, the bog deposits were soon exhausted, and 70 years ago the increased need of the iron and steel industry for new sources of iron

Figure 21.5 **Mining taconite in an open pit at Babbitt, Minnesota. The ore, containing 25 percent iron, is one of the hardest rocks in the world. The operation is like that of a rock quarry producing crushed stone.** (*Photograph by Hercules Powder Company, courtesy Reserve Mining Company.*)

ore focused interest on the Lake Superior region, which contains a number of iron ranges with high-grade iron ore. These deposits were concentrated by weathering, which enriched exposed iron sediments. The Mesabi Range has been the world's leading producer of iron ore for over half a century. Slightly tilted beds are worked by surface methods.

About nine-tenths of the iron ore in the United States occurs as hematite (Fe_2O_3). In the Lake Superior district, the zone that has still not been leached by weathering or circulating ground water consists of a mineral called *taconite,* containing chert with *hematite, magnetite* (Fe_3O_4) (see Figure 21.5), *siderite* ($FeCO_3$), and *hydrous iron silicates.* The iron content of taconite averages about only 25 percent, but in the zone that has been leached, most of the iron has been oxidized to hematite, which produces ores of from 50 to 60 percent iron. This high-grade ore is nearly exhausted in some ranges, but more and more low-grade ore is being used. Commercial methods have now been developed for recovering iron from the taconite of this district (see Figure 21.6). Great stretches of original unleached rock have been added to our iron reserve.

The newest reserves of iron ore on the North American continent, near 55°N, 67°W, on the Labrador–Quebec border, also consist of deposits enriched by weathering. Other deposits are being worked in sedimentary formations of Silurian age, known as the Clinton beds, which outcrop across Wisconsin and New York, and along the southern Appalachians. These beds are being mined extensively in the Birmingham district of Alabama (see Figure 21.7). The primary unleached ores from the Clinton beds are often high in $CaCO_3$ and contain 35 to 40 percent iron. But after the $CaCO_3$ has been leached out by weathering, they may contain as much as 50 percent iron.

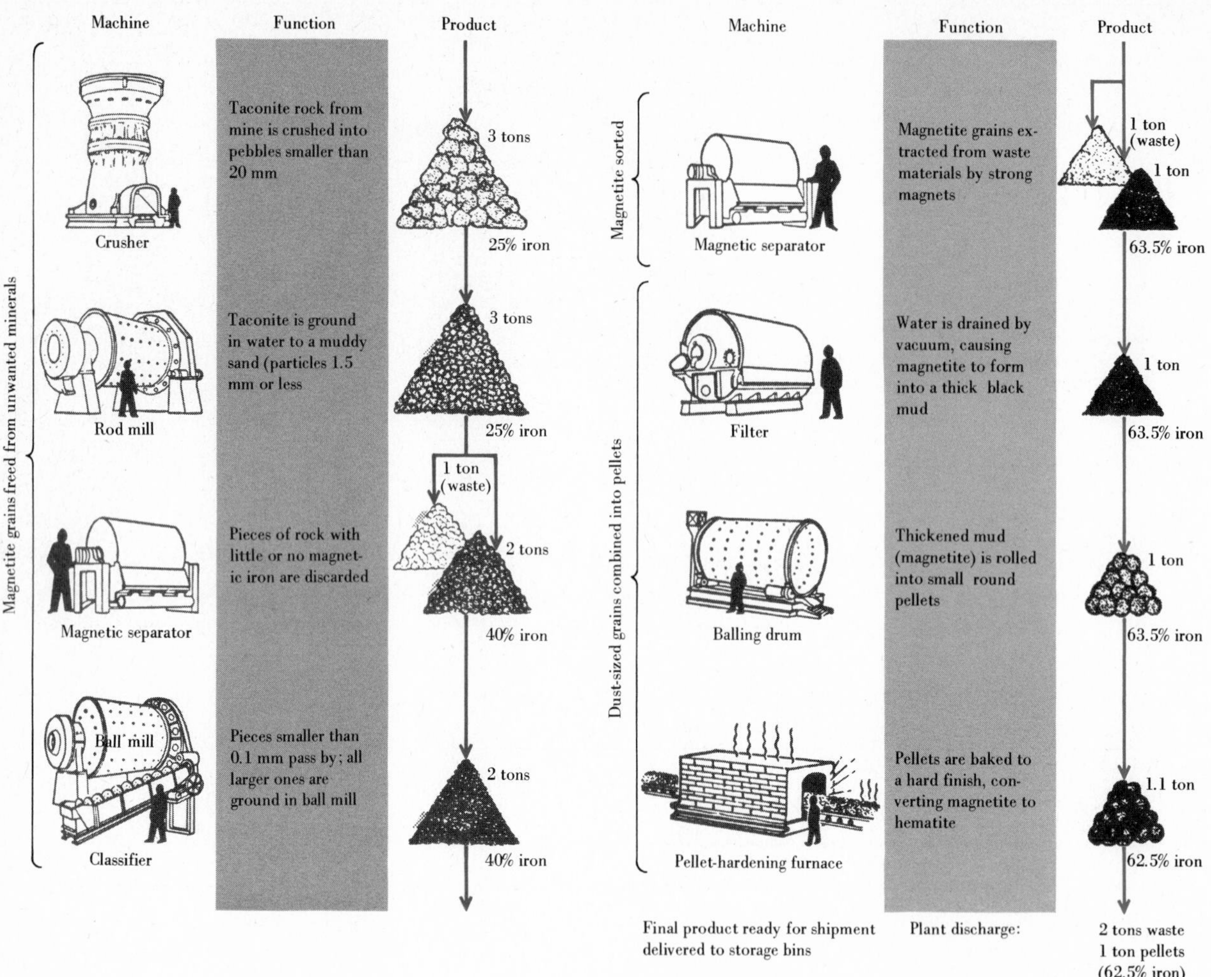
Machine
Function
Product
Magnetite grains freed from unwanted minerals
Crusher
Taconite rock from mine is crushed into pebbles smaller than 20 mm
3 tons
25% iron
Rod mill
Taconite is ground in water to a muddy sand (particles 1.5 mm or less
3 tons
25% iron
1 ton (waste)
Magnetic separator
Pieces of rock with little or no magnetic iron are discarded
2 tons
40% iron
Ball mill
Classifier
Pieces smaller than 0.1 mm pass by; all larger ones are ground in ball mill
2 tons
40% iron
Machine
Function
Product
Magnetite sorted
Magnetic separator
Magnetite grains extracted from waste materials by strong magnets
1 ton (waste)
1 ton
63.5% iron
Dust-sized grains combined into pellets
Filter
Water is drained by vacuum, causing magnetite to form into a thick black mud
1 ton
63.5% iron
Balling drum
Thickened mud (magnetite) is rolled into small round pellets
1 ton
63.5% iron
Pellet-hardening furnace
Pellets are baked to a hard finish, converting magnetite to hematite
1.1 ton
62.5% iron
Final product ready for shipment delivered to storage bins
Plant discharge:
2 tons waste
1 ton pellets
(62.5% iron)

Figure 21.6 ***Graphic presentation of Reserve Mining Company's methods of processing crushed taconite ore into pellets containing 62.5 percent iron. Babbit, Minnesota.*** (*Courtesy Reserve Mining Company.*)

Copper Copper occurs both uncombined and in combination with other elements in minerals such as *chalcopyrite* ($CuFeS_2$), *bornite* (Cu_5FeS_4), *chalcocite* (Cu_2S), and *enargite* (Cu_3AsS_4). Most copper deposits consist of concentrations created by hydrothermal solutions. Chalcocite deposits, however, are usually the result of secondary enrichment. In some regions, igneous activity has built up deposits of copper but not in great enough concentrations to be worked. Ground water has worked on some of these, dissolved the copper, and carried it down to be deposited in an enriched zone.

At Bingham Canyon, Utah (see Figure 21.8), is a spectacular open-pit mining operation that recovers at a profit an ore containing as little as four-tenths of 1 percent of copper and averaging about 1 percent. The ore is a sialic porphyry that contains finely disseminated sulfides concentrated by secondary

Figure 21.7 ***Underground mine of Clinton iron ore near Birmingham, Alabama.*** *(Courtesy Tennessee Coal and Iron Division, U.S. Steel Corporation.)*

enrichment. The benches of the mine are from 15 to 20 m high and not less than 20 m wide. The operation covers $3\frac{1}{2}$ km^2 and contains about 250 km of standard-gauge railroad track, most of which is moved continually to meet operating needs. In 1952, this mine was producing more than 250 million kg of copper per year, about 30 percent of all the copper mined in the United States.

Copper occurs in the pure metallic state in one of the United States' more remarkable ore deposits on the Keweenaw Peninsula of northern Michigan. The peninsula extends 160 km into Lake Superior. It consists of beds of ancient lava flows, conglomerates, and sandstones that have been folded and glaciated. The whole series dips toward the north. The copper is found in veins intersecting the igneous rock and as a cementing material in the conglomerates. This region was the world's greatest producer of copper before Butte, Montana, was developed. At Butte, the total length of deep underground workings has been reported as 1,600 km. Low-grade ore has already produced over 9 billion kg of copper, silver, lead, zinc, and gold.

At Morenci, Arizona, an impoverished zone is as deep as 65 m. Beneath this, an enriched zone extends about 300 m farther down. Underlying the enriched zone is unaltered bedrock, often too low-grade to mine.

Other metals The leading producer of *lead* and *zinc* is the tristate district of Missouri, Kansas, and Oklahoma. The ore deposits are low-grade. Galena (PbS) is the ore of lead and sphalerite (ZnS) of zinc. These are often found together, in irregular veins and pockets in limestone and chert. Lead and zinc

Figure 21.8 **Copper mine at Bingham Canyon, Utah. The benches are 15 to 20 m high and not less than 20 m wide.** (*Photograph by Rotkin, P.F.I.*)

valued at over $1 billion have been mined from a 5,200 km^2 area. In southeastern Missouri, galena disseminated through flat-lying dolomitic limestone beds has made the world's greatest deposits of lead.

The world's largest deposit of *molybdenum* is in the Rocky Mountains at Climax, Colorado. Due largely to this deposit, the United States produces about 90 percent of the world's output. Molybdenum is derived from the mineral molybdenite (MoS_2), which forms as an accessory mineral in certain granites and pegmatites. At Climax, it occurs in quartz veinlets in silicified granite.

With the arrival of the atomic age, search began for deposits of uranium ore. The highest grade is *uraninite* (UO_2). The principal deposits found to date are in the Belgian Congo, in Great Bear Lake in Canada, and in Czechoslovakia. In the United States, the greatest supply of uraninite so far discovered is at the Mi Vida mine, which is located 65 km southeast of Moab, Utah. This is in the center of Colorado Plateau. Workable deposits have also been found on the Colorado Plateau in Arizona, Colorado, New Mexico, and other sections of Utah.

Uraninite occurs as a constituent of granitic rocks and pegmatites and also as a placer mineral.

Another mineral that is a source for uranium is *carnotite,* a soft yellow weathering product. It comes principally disseminated in cross-bedded sandstones found in southwest Colorado and adjoining districts of Utah.

Uranium prospectors use a Geiger counter, an instrument that makes an audible click every time it is hit by a particle released by the spontaneous decay of uranium. Because this is a simple, inexpensive device, much of the prospecting for this mineral is being done by laymen with a little spare cash and time. In fact, Geiger counters can actually be purchased in some California supermarkets.

Measured in terms of metric tons, the uranium reserves now known are not great, but measured in terms of their potential energy, they probably equal the reserves of coal, oil, and gas combined.

Aluminum, although it is one of the most common elements in the earth's crust, almost always occurs in feldspars and other silicates from which it cannot be extracted economically by any process now known. Fortunately, however, under tropical conditions, weathering breaks the feldspars down into clay minerals; they in turn become hydrous oxides of aluminum and iron. The soils produced by this activity are sometimes known as "laterites" (see Chapter 6), and the aluminum ore is called *bauxite.* The principal deposits of bauxite in the United States occur in Arkansas.

PROSPECTING FOR ORE DEPOSITS A search for ore deposits is always guided by a hypothesis of some kind. This may be based on dreams or phases of the moon. It may be an assumption that because a certain species of spoofberry bush once grew on the surface above a copper mine, anybody who finds a spoofberry bush can expect to find a copper deposit below it. Or it may be based on the latest ideas in geophysics. But the hypothesis is there.

Figure 21.9 **Magnetite from Magnet Cove, Arkansas. This magnetic mineral is the component of taconite that is removed by the processes outlined in Figure 21.6, then converted to hematite, from which iron is recovered.** (*Photograph by Benjamin M. Shaub.*)

On the other hand, luck has often played a large role in even the most carefully planned prospecting. It did so in Nevada, where a painstaking program laid out to pan for gold was rendered ineffective by a soft, sticky, bluish mineral. When somebody recovered from the disappointment of not finding gold and had the bluish "clay" assayed, it turned out to be an important silver ore, argentite, a silver sulfide. This discovery led to the development of the fabulous Comstock Lode. Many mining camps have a story of a prospector who threw a hammer at a fox and had it bounce off an outcrop of rock. When he went to retrieve the hammer, he found ore in the outcrop. Another popular account of accidental finding tells of a prospector tracking down his stray mule and finding the mule grazing near an outcrop of ore. And rich deposits of gold and copper were discovered in rock cuts made for westward-driving railroad lines in the early days.

Known occurrences are the best guide to conditions under which to look for a particular ore. Even the most highly instrumented methods of modern geophysics or geology are best adapted to extending old ore deposit outlines or looking for new deposits under geological conditions similar to those where old ones were formed.

Geophysical methods of prospecting for ore deposits are indirect, for the most part: They determine geological structures favorable for the presence of desired minerals. By making magnetic measurements, however, the mineral magnetite can be located directly because of its magnetism (see Figure 21.9). But even this method is more often used indirectly, to locate minerals that commonly have magnetite associated with them. Iron ores, such as hematite and limonite, though themselves nonmagnetic, have enough magnetite associated with them to permit detection and mapping. Magnetic measurements aided in finding gold-bearing beds in the Witwatersrand, South Africa. The gold is scattered through steeply dipping beds of Precambrian quartz conglomerate that extend for 160 km from east to west. Parallel to these conglomerates are some shales rich in magnetite. When the Witwatersrand ran into barren ground some years ago, a geologist and geophysicist worked together to see if the gold-bearing zone could be found again. They measured the vertical intensity of the magnetic field over extensive areas, tracing what they believed to be the magnetite-bearing shales associated with the gold-bearing conglomerates. From their measurements, they were able to map the area containing the magnetite. Drilling confirmed their location of the magnetite and then of the gold-bearing formation. These were both under a cover of 600 m.

Seismic methods have been used successfully in mapping bedrock under unconsolidated deposits. This method has paid off in some areas where low places in the bedrock were sought as possible locations for placer accumulations of heavy minerals. But seismic methods have not worked well in most ore-finding problems because they cannot distinguish one mineral from another.

Nonmetallic deposits—diamonds, asbestos, and salt

Diamonds are the most valuable of the gemstones. In general, the most valuable of these are flawless stones that are colorless or have a "blue-white" color. A faint straw-yellow color detracts much from the value. In recent years, diamonds have been put to industrial uses. Fragments of diamond crystals are used to cut glass. Fine diamond powder is used in grinding and polishing diamonds and other gemstones. Wheels are impregnated with diamond powder for cutting rocks and other hard materials. Steel bits set with the cryptocrystalline variety, carbonado, make diamond drills used in geological exploratory work. The diamond is also used in wire drawing and in tools for truing grinding wheels.

Diamonds are most commonly found in placer deposits in the sands and gravels of stream beds where they have been transported after they were freed from their original environment by weathering. They have been preserved by their chemical inertness.

Ninety-five percent of the world's output of diamonds comes from the African continent. The first African diamonds were found in the gravels of the Vaal River, South Africa, in 1867. Four years later, diamonds were found in soils weathered from peridotite near the present town of Kimberley, south of the Vaal River. Most mines in the Kimberley region were originally open pits, working surface deposits of the altered peridotite in ancient volcanic necks. Later, underground methods were used in the same mine. The Kimberley mine was developed to a depth of over 1,000 m before it was abandoned. The Belgian Congo furnishes over 50 percent of the world's supply of diamonds from placer deposits. Diamonds also come from India and Brazil, where they are produced from gravel deposits.

Materials made from *asbestos* are extremely versatile. They can withstand fire, insulate against heat and sound, are light in weight, and can be made

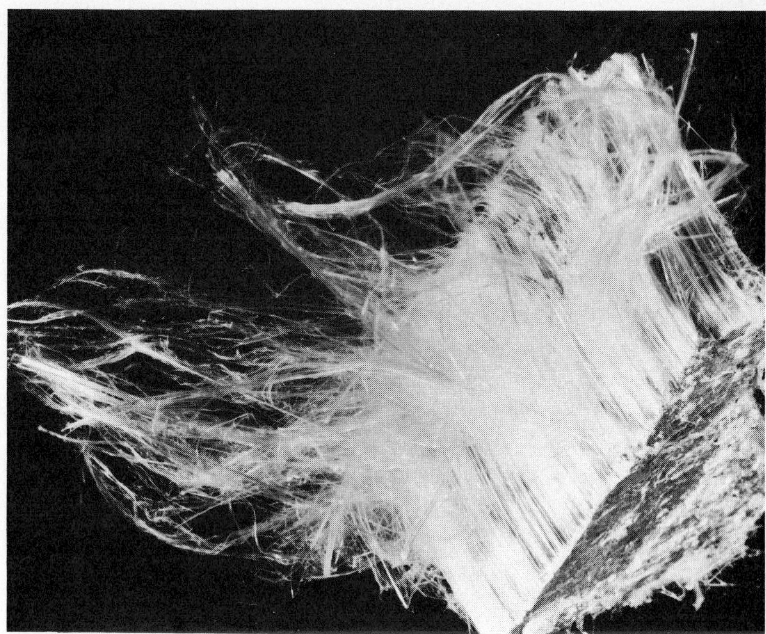

Figure 21.10 **Chrysotile asbestos, a fibrous variety of serpentine from Thetford, Quebec.** (*Photograph by Benjamin M. Shaub.*)

into pliable fabrics which resist soil, corrosion, and vermin. The most common asbestos mineral is *chrysotile*, which is an alteration product of some magnesium silicates, especially olivine, pyroxene, and amphibole. Chrysotile deposits are found in both igneous and metamorphic rocks, sometimes in such concentration as to make up the entire rock mass. Chrysotile forms soft, silky, flexible fibers (see Figure 21.10). The long fibers are woven into yarn for use in brake linings and heat-resistant tapes and cloth. The United States, which is the largest user of asbestos, imports 90 percent of its needs from Quebec, Canada. In the United States, chrysotile can be found in parts of Vermont, New York, New Jersey, and Arizona.

Figure 21.11 **Schematic diagram of a salt dome.**

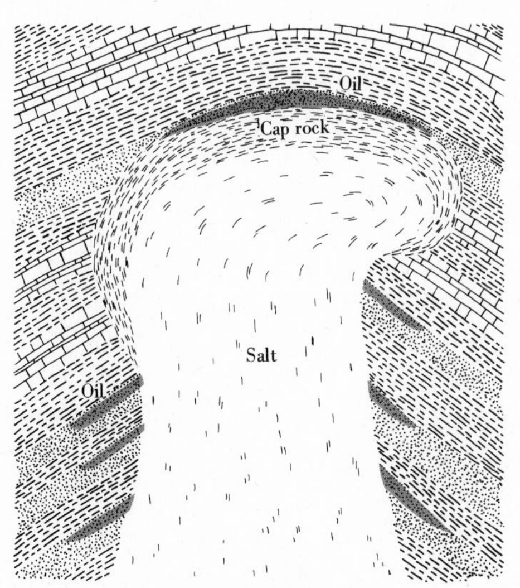

Salt, NaCl, is essential to life and fortunately it is one of the most abundant substances in the world. It is produced commercially from brine, salt beds, and salt domes. Salt beds were formed by the natural evaporation of water from enclosed salt-water bodies. Subsequently they were covered by sediments and sometimes buried to great depths. Salt beds vary in thickness from 1 m to over 30 m and are mined at various depths.

In Louisiana and Texas salt is produced from salt domes. These are great masses of salt that appear to have pushed their way up through overlying sedimentary strata while in a plastic state. More than 100 salt domes have been located. Details of shape vary somewhat from one dome to another, but they all tend toward a cylindrical shape with a top diameter of 2 km. Some have pushed up to within a few hundred meters of the surface, but others stopped several thousand meters down. Reservoirs of oil and gas are often associated with salt domes (see Figure 21.11).

Salt, limestone, coal, and oil are four of the five basic raw materials of the chemical industry. The fifth is sulfur. The largest known reserves are in sedimentary deposits. Among its most useful and familiar compounds are hydrogen sulfide (H_2S), sulfur dioxide (SO_2), sulfuric acid (H_2SO_4), and gypsum plaster (calcium sulfate). About 87 percent of the sulfur produced goes into the making of sulfuric acid, which is the most used intermediate compound in chemical processing.[1]

Rock deposits

Many rocks are valuable in their natural condition, and are usable without having to undergo changes. Stone has been used for several thousand years as a building material and is still being so used today. Dimension-stone quarries cut blocks of granite, slabs of marble, and various sizes and shapes of limestone.

Rock has grown in importance during the past half century as its uses have expanded. In our modern world, every mode of transportation depends in some degree on crushed rock. It provides the base for thousands of kilometers of modern highways, ballast for railways, bases for landing fields, and jetty stone for harbor facilities. With increasing demands, techniques have been developed for removing it from the ground by blasting and crushing it into usable sizes (see Figure 21.12).

Some rocks have commercial value because of their chemical properties. *Limestone,* for example, is used as a flux in purifying metals. It is also used to neutralize acids in the processing of sugar, to correct the acidity of soil, and to supply calcium to plants. Limestone that contains limited amounts of impurities serves as the raw material in the manufacture of cement; the impurities give cement its characteristic hardness. The type known as Portland cement consists of 75 percent calcium carbonate (limestone), 13 percent silica, and 5 percent aluminum oxide, along with the silica and alumina that are normally present in clays or shales. Some manufacturers add the right percentages of impurities to the limestone; others use deposits called *cement rock,* in which the impurities occur naturally.

Phosphate rock is a popular term used for sedimentary rocks that contain high percentages of phosphate, usually in the form of the mineral *apatite,* calcium fluophosphate. Phosphate rock is derived from accumulations of animal remains and chemical precipitation from sea water. It is extremely important as a source of agricultural fertilizer. The Rocky Mountain states have high-grade phosphate deposits with reserves estimated at 6 billion tons, enough to last for many centuries. Reserves in Idaho run close to 5 billion tons. Other deposits are being mined in Florida and Tennessee.

Water

Water use in the United States in 1965 averaged about 1.25 billion m^3/day, up 15 percent from 1960, according to the U.S. Geological Survey. Average use per person per day varies from state to state by a factor as great as 70 times, depending largely on the amount of irrigated land, but it comes out to 6.4 m^3. Approximately one-fourth of the total was consumed

[1] Christopher J. Pratt, "Sulfur," *Scientific American,* pp. 63–72, May 1970.

Figure 21.12 **Modern methods break over 200,000 tons of rock from a quarry face with 25,000 kg of explosive.** (*Courtesy New York Trap Rock Corp.*)

or made unavailable for further use because of evaporation or use in manufactured products. About 0.25 billion m^3/day came from ground-water sources and 1 billion m^3/day from surface sources.

Water is a natural resource that many people take for granted, but in areas where there have been large increases in population, maintaining a sufficient water supply often requires great ingenuity and dramatic measures. Southern California is such a region. It is arid, with a 90-year average annual rainfall at Los Angeles of only 37.5 cm. Yet after World War II, an 11,700-km^2 area grew in population to nearly 10 million by 1969, with 17 million expected by 1995.

The Metropolitan Water District of Southern California met the skyrocketing needs first by building a 387-km aqueduct from Parker Dam and Lake Havasu on the Colorado River (see Figure 21.13). This carried 4 million m^3/day across the Rocky Mountains. The system included 147 km of 5-m-diameter tunnels. To this is being added "the biggest aqueduct project of all time," a California Aqueduct system starting at Oroville Dam on the Feather River north of Sacramento and distributing water to areas in northern and central as well as southern California. It consists of a complex system of great dams, reservoirs, pumps, canals, tunnels, and pipelines, and crosses the San Andreas fault three times. The system is scheduled to start delivery over 720 km to southern California in the early 1970s. And by 1973, it is expected that this water supply will be supplemented by the world's largest desalination plant on a man-made island off the Orange County coast.

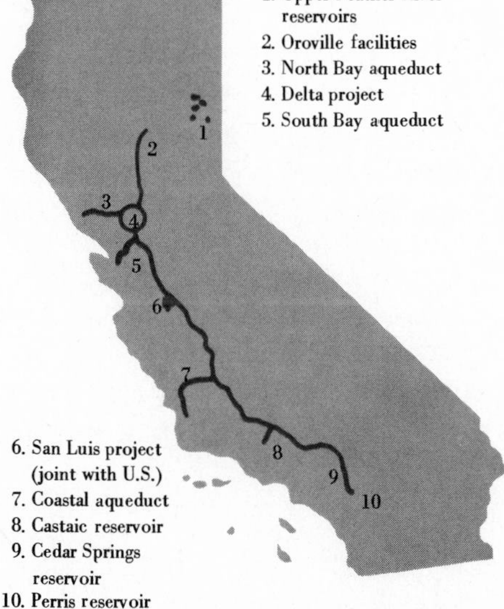

Figure 21.13 **California's state water project.**

21.3 *Sources of energy available today*

In our present state of knowledge, we can draw on only a few sources of energy—principally coal, oil, gas—which are the very sources that have been known to man for centuries. And even these can be employed only if they have been concentrated by geological processes.

Chemical energy

The discovery of fire and its uses was one of man's first steps along the road to civilization, and even today fire still serves as the basis of civilization. Fire is a type of combustion in which oxygen combines chemically with the carbon and other elements of organic substances to produce heat and light. Of the various substances that can be used as fuels, wood is the least efficient. Coal, oil, and gas are far more efficient, for they represent energy that has been concentrated by the decay of organic material. These three fuels are often referred to as *fossil* fuels, for obvious reasons. During the decay process, the less combustible components are driven off, leaving behind the highly combustible elements carbon, hydrogen, and oxygen.

When organic fuels are burned, great quantities of stored chemical energy are released in the form of heat energy. This heat may be either used directly or converted into other forms, such as electrical energy. Steam turbines are the principal means of converting heat energy released by fossil and nuclear fuels into the kinetic energy needed to drive power generators. They account for 75 percent of the world's electric power. Most of the rest is hydroelectric.

COAL *Coal* is the end product of vegetable matter that accumulated in the swamplands of the earth millions of years ago. The size of a coal bed depends on the extent of the original swamp and the amount of vegetable matter that collected in it. Unlike most beds, these were not formed by the usual sedimentary processes. Matter accumulated right in the area where it is found today. Plants grew in swamps, died, then were buried and chemically altered after burial.

We distinguish among grades of coal on the basis of their carbon content, which increases the longer the material has undergone decay and burial. The plant matter from which coal has developed contains about 50 percent carbon. *Peat* (which is not coal) is the first stage in the decay process and contains about 60 percent carbon. The lowest grade of coal, *lignite,* has about 70 percent carbon and 20 percent oxygen. Lignite is gradually converted to *subbituminous* and then *bituminous* coal, which is soft coal containing about 80 percent carbon and 10 percent oxygen. Lignite and the bituminous coals are considered to be sedimentary rocks. At a later stage, when bituminous coal is metamorphosed, it becomes *anthracite,* 95 to 98 percent carbon.

Coal occurs in beds interlayered with shale and sandstone (see Figure 21.14).

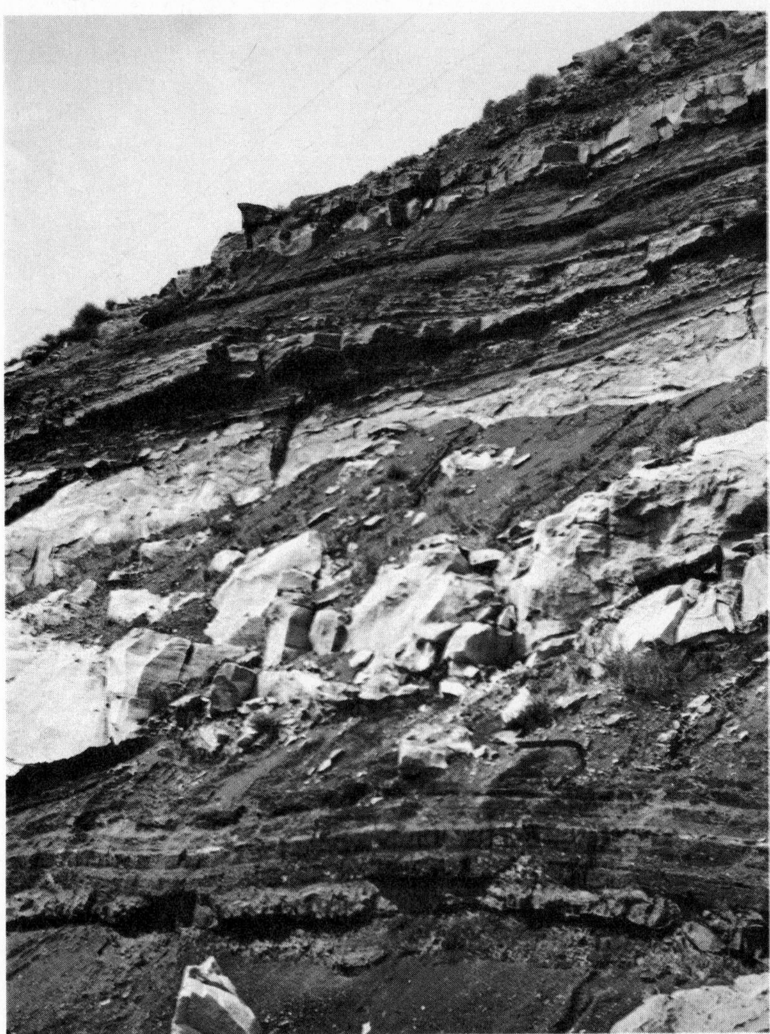

Figure 21.14 **Coal seams form part of a series of sedimentary rocks. The darkest layers are coal and shale; the lightest are limestone. Near Cokedale, Colorado.** (*Photograph by Tozier.*)

The Appalachian Plateau has some of the greatest known coal seams. One of high-grade bituminous coal averages over 2 m in thickness and underlies nearly 40,000 km^2 of southwestern Pennsylvania and adjacent West Virginia. The anthracite deposits around Scranton and Wilkes-Barre in eastern Pennsylvania have been affected by earth movements so they tend to be folded and often deeply buried, making mining and extraction difficult and costly. But they have one favorable feature: They are relatively free of dangerous gases.

Although carbon is by all odds the most important element in coal, as many as 72 elements have been found in some deposits. The ash formed by the bituminous coals of West Virginia consists of about 1 percent each of sodium, potassium, calcium, aluminum, silicon, iron, and titanium. And there are 26 metals present in concentrations ranging from 0.01 percent, including lithium, rubidium, chromium, cobalt, copper, gallium, germanium, lanthanum, nickel, tungsten, and zirconium.

Coal is also important as the source of coke used in the steel industry. In fact, one-fourth of the coal produced every year is used for this purpose. The coke is burned in blast furnaces, where it supplies carbon, which combines with the oxygen of iron ores to free the metallic iron. In the future, coal may become even more valuable as a source of coke than as a direct source of heat. At present, over half the coal mined goes to generate electricity.

OIL AND GAS Coal is rapidly being replaced as a fuel by more efficient, easier-to-handle oil and gas. Fortunately, there are also large supplies of these fuels in the United States, and it is to them that we owe much of our industrial progress and high standard of living. Great Britain, on the other hand, although her coal and iron ore enabled her to pioneer the industrial revolution, now has to buy large quantities of oil and gas reserves from other countries, since she lacks the large sedimentary basins where oil and gas accumulate.

What are oil and gas? Oil and gas are the remains of organic matter that has been reduced by decay to a state in which carbon and hydrogen are the principal elements. These elements are combined in a great variety of ways to form molecules of substances called *hydrocarbons.* The distinguishing feature of the molecule of each hydrocarbon is the number of carbon atoms it contains. One carbon atom combined with four hydrogen atoms, for example, forms a molecule of a gas called methane, CH_4. Two carbon atoms combined with six hydrogen atoms form a molecule of a gas called ethane, C_2H_6. Various hydrocarbons are listed in Table 21.2.

Natural deposits of oil contain many kinds of hydrocarbons mixed together. They are separated by an industrial process called *fractional distillation,* based on the principle that light molecules are volatilized more readily than heavy molecules. As early as 600 B.C., Nebuchadnezzar, king of Babylon, was building roads that consisted of stones set in asphalt. The asphalt was nothing more than the hydrocarbons left behind where natural oil had seeped to the surface and lost its lighter components by evaporation.

Source beds Most petroleum (from the Latin *petra,* "rock," and *oleum,* "oil"; hence, "rock oil") and natural gas have developed from organic remains originally deposited in a marine sedimentary environment. A modern example

Table 21.2 **Petroleum products**

Product	*Boiling range, °C*	*No. C atoms in molecule*	*Uses*
Gas	Below 32	1–4	Fuel gas, carbon black, rubber
Gasoline	38–205	4–12	Motor fuel
Naphthas	52–205	7–12	Solvents
Kerosine[a]	205–315	12–15	Tractor fuel, diesels, heating, printing ink
Fuel oil	205–370	15–18	Furnace fuel, diesels
Lubricating oil	Above 345	16–20	Lubrication
Petrolatum	Above 345	18–22	Lubrication, salves
Wax	Melts 52–55	20–34	Waterproofing, candles
Asphalt	Residue	—	Road making, roofing

[a] Spelling adopted by American Society for Testing Materials.

Figure 21.15 **Core samples lined up for examination.** *(Photograph by Ohio Oil Co., Inc.)*

of such an environment is the Black Sea. Here the water circulates very slowly, and the bottom sediments contain as much as 35 percent organic matter, in contrast to the 2.5 percent that is normal for marine sediments. When the putrefaction of organic remains takes place in an environment of this sort, the product is a slimy black mud known as *sapropel* (from Greek *sapros,* "rotten," and *pelagos,* "sea"). Petroleum and natural gas are believed to develop from the sapropel through a series of transformations not unlike the stages in coal's development from peat.

Three conditions are required for the development of a deposit of petroleum or natural gas: (1) source beds where the hydrocarbons can form, (2) a relatively porous and permeable reservoir bed into which they can migrate, and (3) a trap at some point in the reservoir bed where they can become imprisoned.

The most important source beds are generally believed to be marine shales, although certain limestones, particularly if they form a reef, may also serve the purpose (see Figure 21.16). There are also extensive beds of shales formed from fresh-water deposits, such as the Eocene lake deposits in Utah, Colorado, and Wyoming. These *oil shales,* when heated, have yielded from 5 to 10 gal of oil per ton and constitute important fuel reserves.

Figure 21.16 **Oil and gas trapped in an ancient coral reef, surrounded by impermeable shales. Some reefs are believed to have contributed animal remains as a source material for petroleum as well as reservoir rocks for storing the naturally distilled hydrocarbons.**

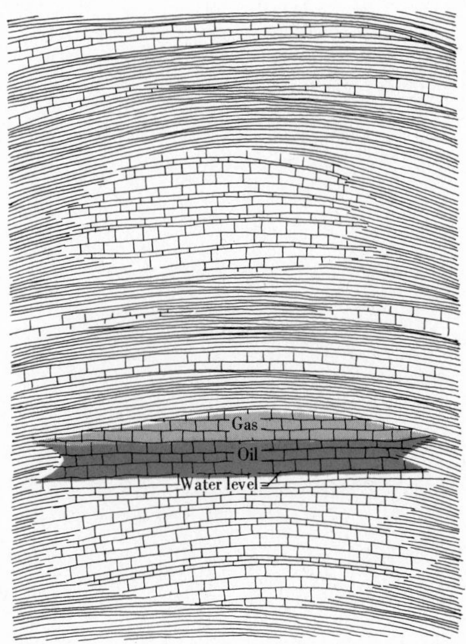

LOCATION OF TRAPS Just where we shall find a reservoir of petroleum or natural gas depends on the laws that govern the migration of these substances to reservoir rocks. Unfortunately, we do not yet understand just what these laws are, although several empirical relationships have been established.

Simple gravity seems to explain the location of many occurrences. According to the *gravitational theory,* if oil, gas, and water are present in a reservoir bed, the oil and gas, being lighter than water, will rise to the top, with the gas uppermost (see Figure 21.16). If the reservoir is trapped in a dome or an anticline capped by an impermeable formation, the oil and gas will accumulate along the crest of the anticline or dome. This *anticlinal theory* of accumulation, one aspect of the gravitational theory, has proved to be a valuable guide to prospectors and has led to a substantial volume of production.

A corollary of the gravitational theory is that if no water is present, the oil will gather in the trough of a syncline with the gas above it. Some reservoirs of this type have led careless critics of geological methods to point with scorn to the demonstrably successful anticlinal theory.

Another structure that is important in the gravitational theory is the *stratigraphic trap,* formed when oil and gas in the presence of water are impeded by a zone of reduced permeability as they migrate upward. This situation may develop, for example, along old shorelines or in ancient sandbars, where facies change horizontally from sand to clay. Or the upward progress of the oil and gas through a permeable reservoir bed may be blocked by an impermeable bed at an unconformity or at a fault.

Significant oil deposits throughout the world, excluding those under Russian control, are indicated on Figure 21.17.

METHODS OF FINDING OIL AND GAS Because concentrations of oil and gas seem to develop only in thick masses of marine sediments, prospectors limit their

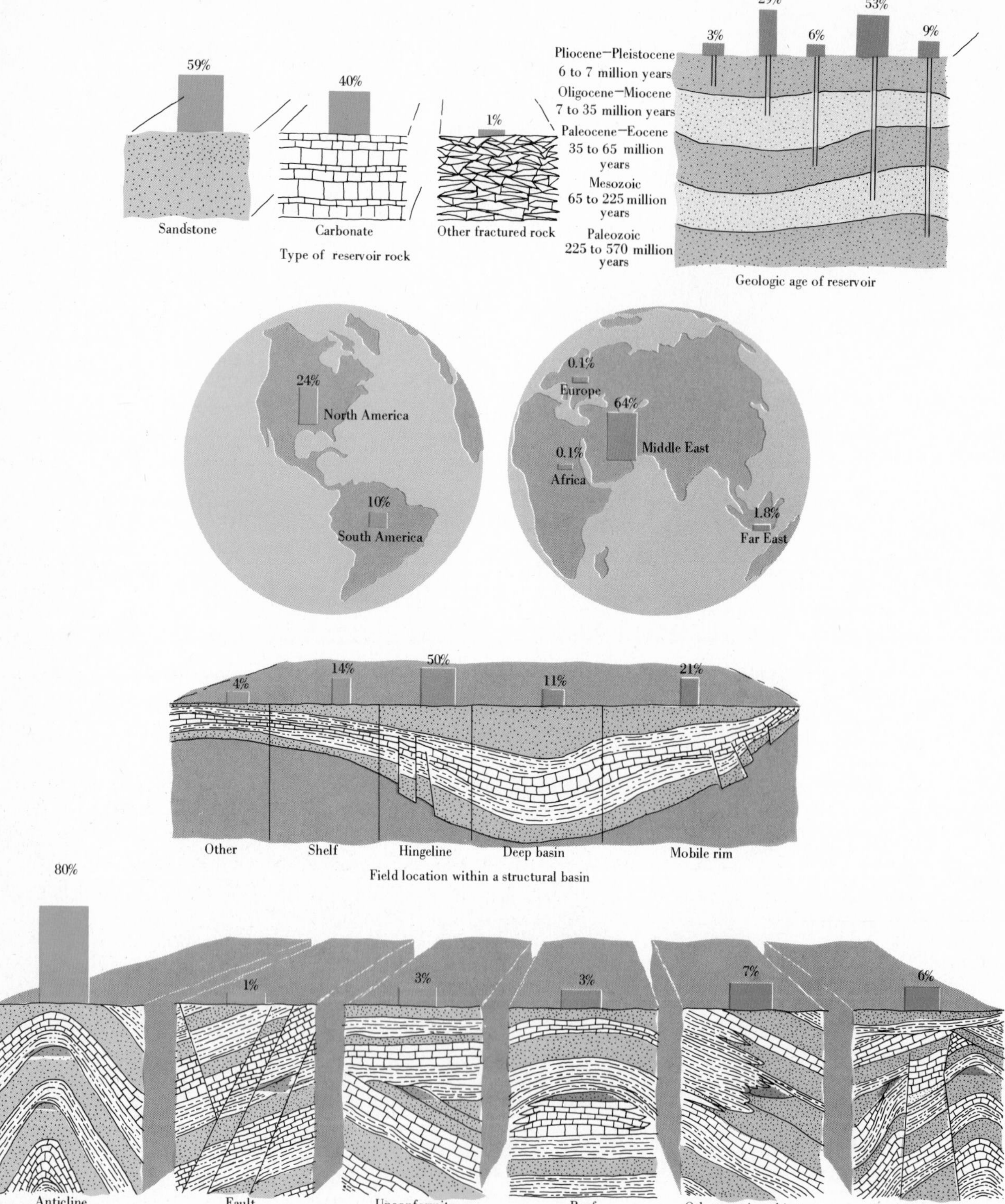
59%
40%
1%
Sandstone
Carbonate
Other fractured rock
Type of reservoir rock
3%
29%
6%
53%
9%
Pliocene–Pleistocene
6 to 7 million years
Oligocene–Miocene
7 to 35 million years
Paleocene–Eocene
35 to 65 million
years
Mesozoic
65 to 225 million
years
Paleozoic
225 to 570 million
years
Geologic age of reservoir
24%
North America
10%
South America
0.1%
Europe
64%
Middle East
0.1%
Africa
1.8%
Far East
4%
14%
50%
11%
21%
Other
Shelf
Hingeline
Deep basin
Mobile rim
Field location within a structural basin
80%
1%
3%
3%
7%
6%
Anticline
Fault
Unconformity
Reef
Other stratigraphic traps
Combination of types
Type of oil trap

Figure 21.17 **Where world's oil reserves (excluding Russian-controlled areas) are estimated to be.** (*From data prepared by G. M. Knebel, Standard Oil Company of New Jersey, and Guillermo Rodriguez-Eraso, Creole Petroleum Corporation. After drawing by W. Holloway.*)

search to sedimentary rock formations. They have mapped and tested most of the anticlines that crop out at the surface in the United States and Canada and are now concentrating on rock structures beneath the surface. They use several methods in their search, including (1) core drilling, (2) seismic prospecting, and (3) gravity prospecting. These methods reveal whether or not there is a structure that is likely to trap oil and gas, but they do not give direct evidence of the presence of an actual reservoir.

In core drilling, several closely spaced holes are drilled into the surface to reveal the structure of the underlying sedimentary beds. On the basis of the core samples, the beds are matched from hole to hole, and the height of each above sea level is determined. Then each of the beds is carefully plotted, and a map of the entire structure is built up.

Seismic prospecting is based on our knowledge of earthquake waves. Small dynamite blasts are set off in shallow holes about 15/m deep, and the waves generated travel into the interior and are reflected back to the surface, where they are recorded (see Figures 21.18 and 21.19). If they originate in a zone of shale, for example, and encounter a bed of sandstone, some of them bounce back to the surface where instruments pick them up and register the time of their arrival. The depth to which they have traveled can then be computed, and through a series of such measurements an entire structure can be plotted.

Gravity prospecting makes use of variations in the specific gravity of rocks underlying the surface. If a bed lies in a horizontal position beneath the surface, sensitive gravity meters will give a constant reading for the force of gravity all along the surface above the bed. But if the bed dips or rises, the gravity-meter readings reflect the changing structure. When the readings suggest the presence of anticlines, faults, or other structures in which oil might accumulate, test wells are drilled to determine whether reservoirs actually exist beneath the surface.

Special techniques have been developed for using seismic methods to locate

Figure 21.18 **Seismic prospecting for oil-bearing structures. Dynamite is shot in a shallow hole from one of the trucks on the left. After the vibrations are reflected from buried rocks, they are recorded on seismographs operated from the other truck. The rig on the right drills the shot holes. Bar-S Ranch, West Texas.** (*Photograph by Humble Oil and Refining Company.*)

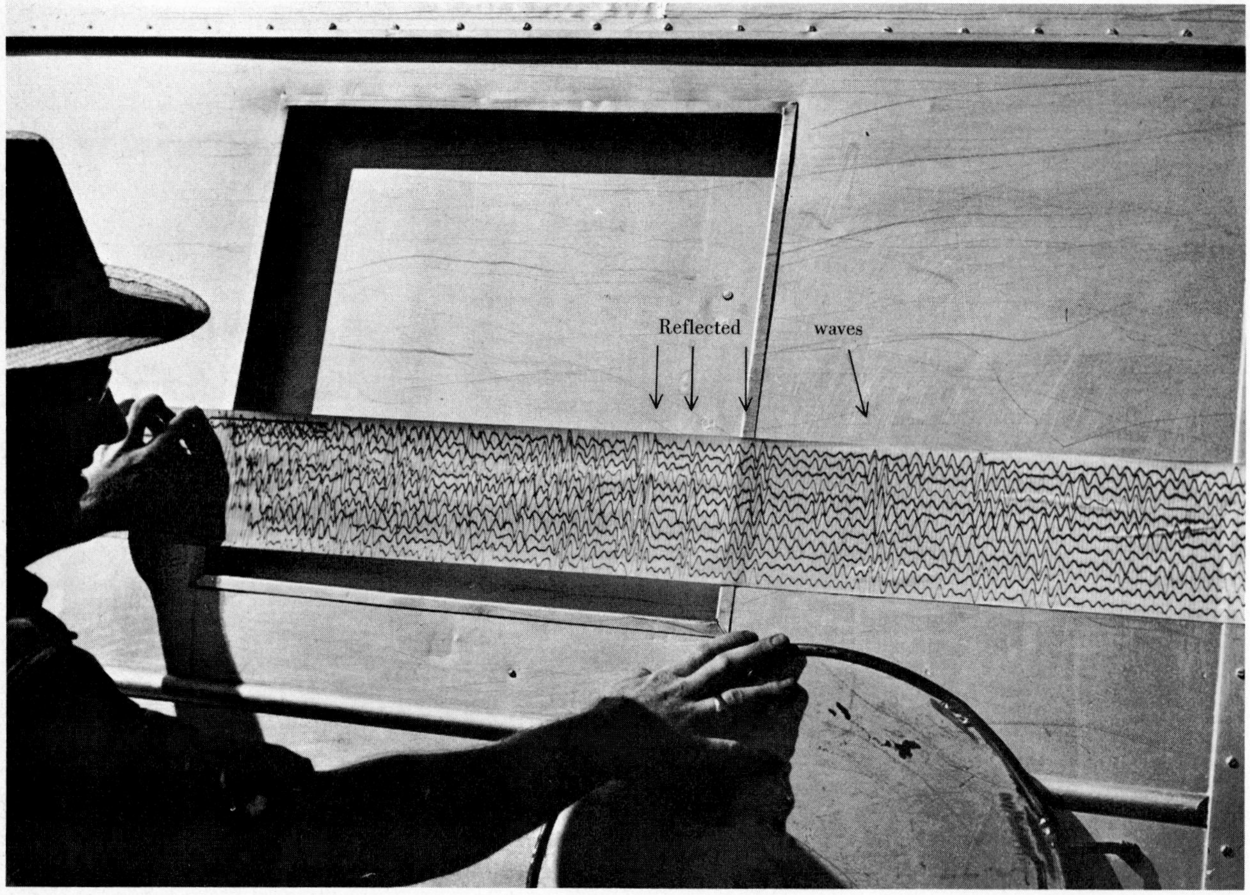

Figure 21.19 ***A reflection seismograph record being inspected by operator after shot, in Oklahoma. Final interpretation of this record is made in the computing offices of the Carter Oil Company.*** *(Photograph by Standard Oil Company of New Jersey.)*

potential traps under water-covered areas. Figure 21.20 illustrates one of these in which energy sources are towed in water and signals are picked up on equipment towed in a streamer cable.

Offshore drilling and production have advanced dramatically in recent years.[2] From a beginning in 1948, offshore operations have increased and by 1968 over $7 billion of oil was produced (see Figure 21.21). This represented about 1 million barrels/day, or more than 10 percent of the total U.S. daily production. Production is expected to increase at a rate of about 15 percent/year until 1975.

[2]T. D. Barrow, "Oil and Gas Extraction Technology," in *Mineral Resources of the World Ocean Symposium at Newport, Rhode Island, July 12, 1968,* Humble Oil and Refining Co., Houston, Texas, 1968.

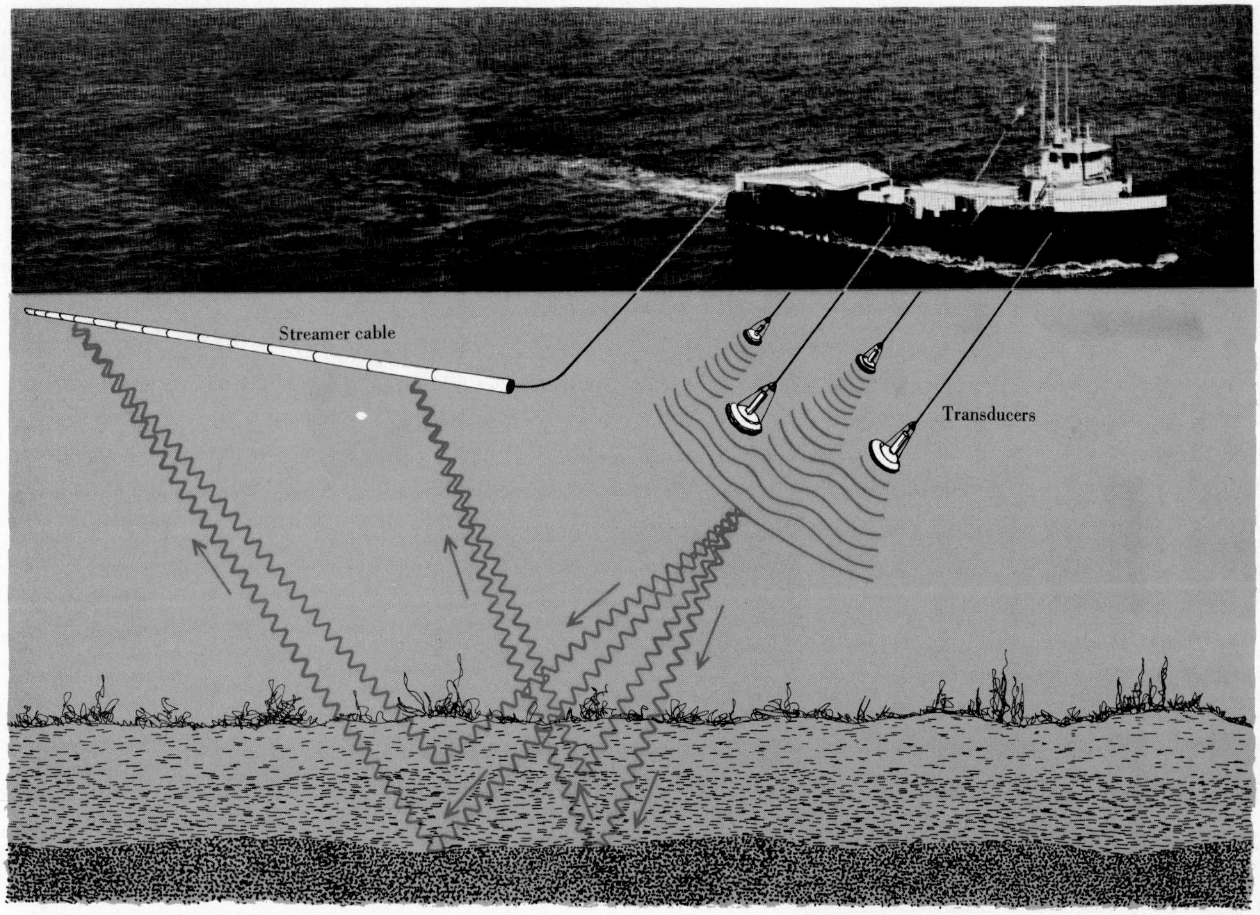

Figure 21.20 **Search for oil-bearing formations beneath the ocean floor, using marine vibrations.** (*Courtesy Continental Oil Company and Westinghouse Air Brake Company Drilling Equipment Division's publication* Core Driller.)

Figure 21.21 **These huge platforms are constructed onshore and moved to the offshore site by barges. There they are launched and set in place. They can accommodate on the deck space all the drilling equipment and housing facilities for some 50 men, together with a heliport. Production equipment replaces the rig following initial development. More than 20 wellbores can be drilled from such a platform.** (*Courtesy T. D. Barrow, Humble Oil and Refining Company.*)

Figure 21.22 **Niagara power development area photographed from a height of 6,600 m above the Canadian terrain, looking northeast toward Lake Ontario. The city of Niagara Falls, N.Y., is in the center. The falls are at the left bottom of the photograph. The Robert Moses Power Plant and the pumped-storage reservoir opposite the Sir Adam Beck Generating Station and a portion of its storage reservoir are near the upper left.** *(Power Authority of the State of New York Photograph No. PA7-26040; reproduced by permission.)*

Hydroelectric energy

Hydroelectric power accounts for about one-fourth of the world's electric energy and some of the developments are among the great energy sources of the world. A good example is at Niagara Falls, where the Niagara River drops 53 m. It also drops 18 m through cataracts above the falls and 23 m through various rapids below the falls. Energy from these 94 m of total fall is converted into electricity.

A unique combination of geological factors made this possible. The area is underlain by nearly horizontal dolomite, shale, limestone, and sandstone. By processes described in Chapter 11, the falls developed and have eroded headward since the ice age. Today, they constitute spectacular scenic features and power sources (see Figures 21.22 and 21.23). An international treaty preserves them by assuring that diversions of water for power do not reduce flow during

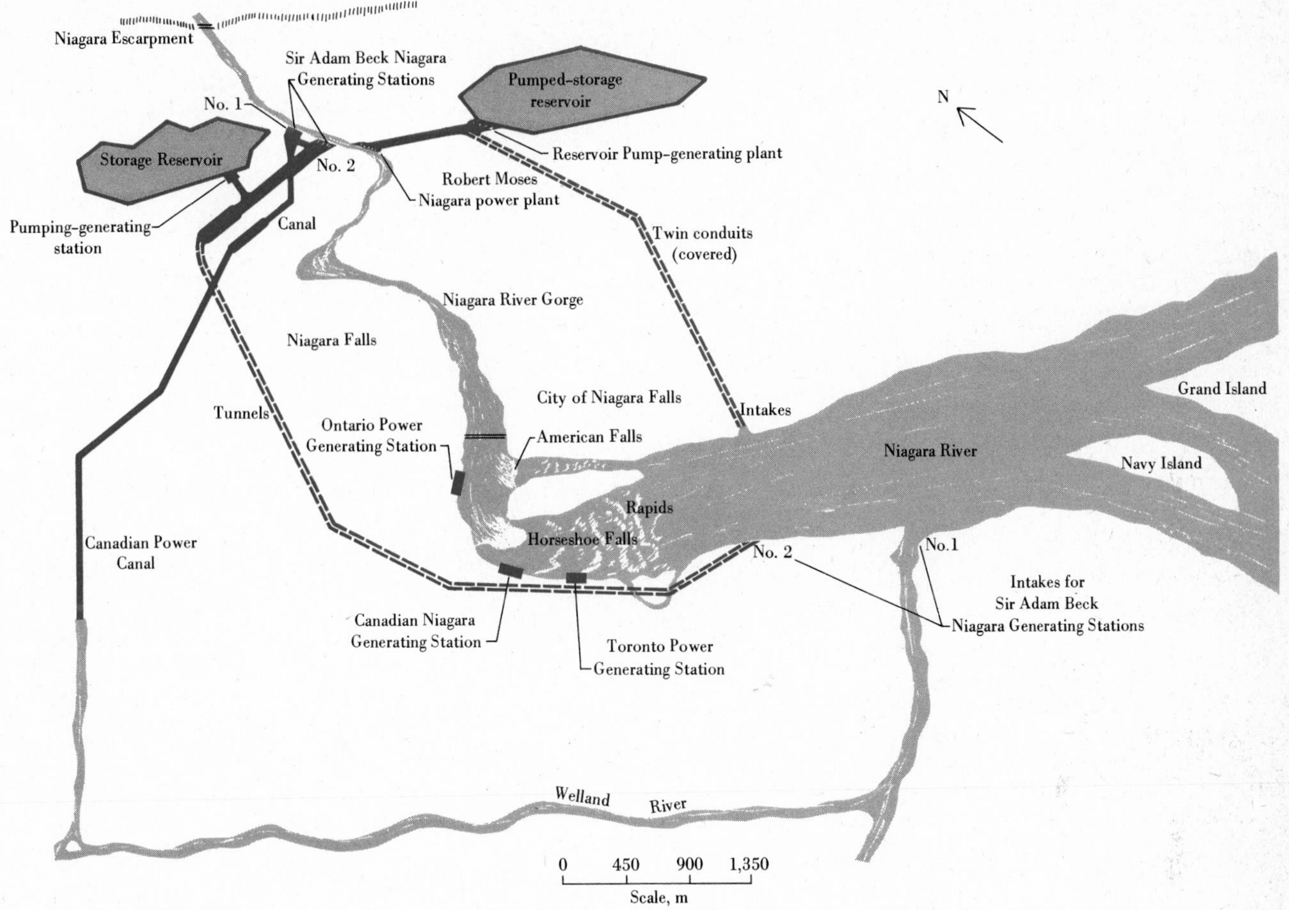

Figure 21.23 ***Niagara power development area.***

daylight hours below 28,000 m^3/sec in the tourist season from April 1 to October 31. Increased water available at night is pumped into reservoirs and used during the daytime to maintain power generation.

Atomic energy

Atomic energy will undoubtedly be our greatest source of energy in the future. Coal, oil, and gas supply only the chemical energy stored in the electrons of atoms; atomic fuels release the much greater energy that is locked in atomic nuclei. Because mass and energy are interchangeable, the nucleus of an atom, which contains 99.95 percent of its mass, contains almost all of its total energy. In fact, if atomic nuclei could interact with one another in a way that would release their inner sources of energy, the reaction would produce 1 million times more energy than is released by ordinary energy-producing chemical reactions.

Table 21.3 ***The world's known reserves of energy (exclusive of water power)***

Countries	*Oil,*[a] *brl* $\times 10^6$	*Natural gas,*[a] *cubic m* $\times 10^9$	*Coal,*[b] *MT* $\times 10^9$	*Uranium oxides,*[c] *MT* $\times 10^3$
North America				
Canada	8,620	1,632	1,234	—
United States	44,000	7,793	3,839	145
Latin America				
Argentina	4,000	249	N	—
Brazil	1,000	142	1	—
Chile	130	71	2	180
Colombia	1,750	85	27	—
Mexico	8,000	340	—	—
Venezuela	14,000	765	N	—
Europe				
Austria	180	12	77	—
France	150	204	17	32
Italy	250	178	1	—
Netherlands	275	2,331	—	—
Norway	8,500	85	—	—
Poland	20	8	180	—
United Kingdom	7	991	190	—
West Germany	584	337	422[d]	—
Yugoslavia	240	99	N	—
Africa				
Algeria	30,000	2,833	N	—
Egypt	5,000	142	0	—
Libya	29,200	850	—	—
Nigeria	5,000	142	N	—
South Africa	0	0	56	130
Middle East				
Iran	55,000	6,062	N	—
Iraq	27,500	539	0	—
Kuwait	68,000	1,105	0	—
Neutral Zone	15,000	99	—	—
Saudi Arabia	138,000	1,416	0	—
Trucial Coast	30,500	622	—	—
Asia				
Australia	2,500	357	167	8.5
China	19,600	101	998	S
India	715	57	79	9.5
Indonesia	18,000	65	N	—
Pakistan	45	567	N	—
U.S.S.R.	55,000	14,983	60	S
All other countries	30,705	1,324	—	—
Total	621,471	46,676	7,294	—

[a] *International Petroleum Encyclopedia*, Petroleum Publishing Company, Tulsa, Oklahoma, 1971. Oil shales and tar sands not included.
[b] Modified from *McGraw-Hill Encyclopedia of Science*, McGraw-Hill Book Company, Inc., New York, 1965, vol. 8, p. 454.
[c] U.S. Bureau of Mines, 1965, quoted in Brian Skinner, *Earth Resources*, Prentice-Hall, Inc., Englewood Cliffs, N.J., 1969, p. 125.
[d] Includes both East and West Germany.
Key: N = negligible; S = significant; — = negligible or information lacking.

So far, however, this great store of nuclear energy has been released from the atoms of only a few elements. One of these is uranium, a naturally unstable element. In 1938, it was discovered that when the isotope uranium 235 captures a neutron to form uranium 236, its nucleus becomes unstable and splits apart. Two other elements are produced whose protons total the 92 of uranium but whose mass numbers do not total 236 (see Figure 2.13). Clearly, some of the neutrons have escaped, carrying with them tremendous amounts of energy. When this process is initiated in a pile of uranium, the neutrons that have escaped from one atom hit the nuclei of other atoms and set up a chain reaction. It was a large-scale reaction of this sort that produced the first atomic bomb explosion on July 16, 1945. Since then, scientists have been developing methods of controlling the reaction so that the energy released can be used for constructive purposes. In fact, nuclear energy is now driving submarines and producing electric power, and in the near future it will be used to propel airplanes.

A score of years after the first nuclear chain reaction was produced, there were nearly 400 nuclear reactors in the world. Most of these were for research, but others for power generation are now being installed.

To generate power, three basic units are needed: an atomic reactor, a heat exchanger, and an electric turbine generator. The atomic reactor consists of a radioactive core which heats water to around 290°C. This is under pressure to keep it from turning to steam. In the heat exchanger this hot water transfers its heat to a second water system, where steam forms to operate a turbine. This drives a generator, which develops electricity that is put into power lines. The steam is then condensed to water and pumped back to the heat exchanger for another round.

Energy reserves

It is generally thought to be safe to assume that the fossil fuel deposits will last at least another century and a half. The fuels of the future, it seems assured, will be nuclear fuels, with their enormous efficiency.

Table 21.3 lists estimates of the world's energy reserves, exclusive of water power.

Resources of the ocean

For centuries the seas have provided man with food, but only recently have we turned from the land to the ocean in search of sources of energy and minerals.

OIL AND GAS Earlier in this chapter we noted the extension of the exploration for oil and gas to the oceans. By 1970, 25 countries were producing oil and gas from subsea fields. In addition, the coastal waters of 75 countries were being actively explored, and drilling was being conducted by 42 countries.

Offshore fields accounted for about 15 percent of the world's 1970 production and the petroleum industry expects this to reach about 30 percent of the world's production by 1980.

In 1970, water depths of about 120 m marked the maximum practical depth for petroleum production. One prediction suggests that it will be technically feasible to produce petroleum from water depths of 1800 m by 1980. The areas of greatest potential for the occurrence of petroleum offshore are included within the continental shelves, slopes, and rises (see Figure 21.24). This represents an area of 75 million km^2, about half the area of dry land. Over one third of this oceanic area is estimated to be within the area of the shelf, a zone whose seaward margin is covered by approximately 200 m of water. Estimates of offshore petroleum reserves lack precision, but it is generally agreed that the reserves are large and may well exceed those on dry land.

MINERAL RESOURCES In 1970, the petroleum industry accounted for 90 percent of the income based on oceanic mineral resources. This percentage will increase in the years to come. Despite the economic dominance of petroleum, however, production has begun or is imminent for several other commodities.

Mining beneath the ocean's surface is either by entry through a shaft from land (natural or artificial) or by dredging. By 1970, the value of the minerals so produced was about 1 percent of the onshore production of these minerals.

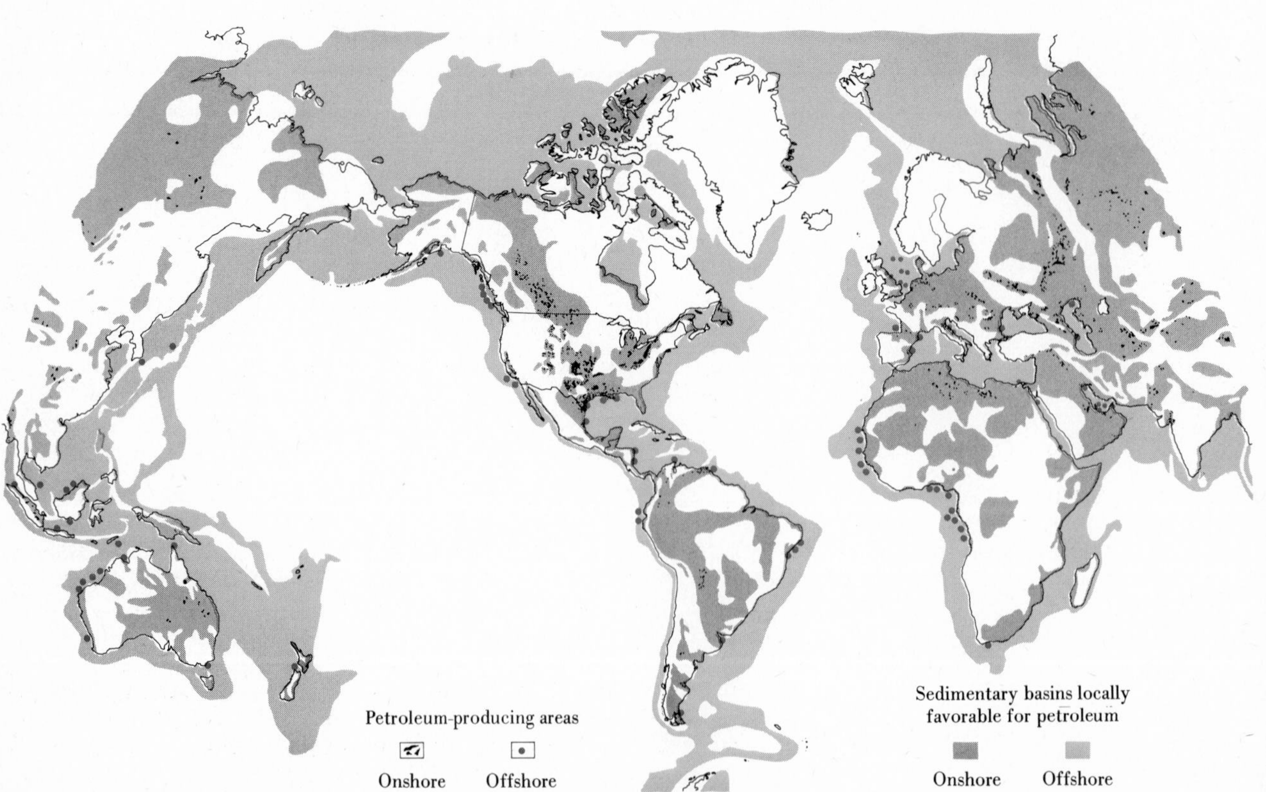

Figure 21.24 **Areas favorable to the accumulation of petroleum both on the continents and subsea. Only small portions of the areas shown as favorable actually contain producible accumulations.** (*After V. E. McKelvey et al., "World Subsea Mineral Resources,"* U.S. Geol. Surv. Misc. Geol. Invest., *Map I-632, sheet 3, 1969.*)

Although the known subsea reserves still remain small it is quite probable that, as with petroleum, the potential is very high.

Manganese oxide nodules have received considerable attention as an economic resource. These deposits most commonly occur on the deep sea floors but some occur near land and in shallower waters, as in the Baltic Sea and on the Blake Plateau off southeastern United States. The composition of the nodules is variable. Thus manganese may range from 13 to 14 percent, iron from 2 to 7 percent, nickel from 0.1 to 1 percent, cobalt from less than 0.1 to 1 percent and copper from 0.1 to 0.5 percent. Estimates of total resources on the ocean floor vary between 90×10^9 and 1.7×10^{12} tons. Whatever the actual amount, however, recoverable amounts can be expected to be much smaller.

Phosphate is widely distributed in the oceans, particularly along the shelves and slopes. Concentrations are not high enough, however, to make ocean phosphates competitive as yet with onshore deposits.

In some places muds with high metallic content have been reported from the marine environment. Most of the occurrences are in areas of high heat flow associated with volcanic and structural activity. An example occurs in the Red Sea in a basin called the Atlantis II deep, just west of Mecca. It is estimated that the upper 10 m of mud in the basin contains 2.9 million tons of zinc and 1.1 million tons of copper, as well as smaller amounts of other metals.

Dredging for *sand* and *gravel* as well as for *shell* is feasible and profitable, particularly from shallow waters close to the markets.

Tin in the form of cassiterite occurs in drowned stream placer deposits but operations off west Africa have not proved economically feasible.

Outline

Natural resources supply civilization with a great range of materials and tremendous amounts of energy.

Distribution of natural resources is governed by geological processes.

Our need for natural resources is great, for they are the foundation of our industrial civilization.

Mineral deposits are generally classed as metallic and nonmetallic.

Mining involves the removal of materials from the earth.

Ore deposits are concentrations of metals that can be worked at a profit.

Separation of valuable mineral removes unwanted material called gangue.

Prospecting for ore deposits is guided by known occurrences.

Nonmetallic deposits include diamonds, asbestos, and salt.

Rock deposits are valuable in their natural condition.

Water is a natural resource that many people take for granted.

Sources of energy available today are still principally ones that have been known to man for centuries.

Chemical energy from fossil fuels accounts for 75 percent of the electric power generated in the world.

Coal is decayed vegetable matter that accumulated in swamps.
Oil and gas are the decayed remains of organic matter.
Location of traps depends on laws that govern migration of petroleum and natural gas.
Methods of finding oil and gas are indirect.
Hydroelectric energy generates about 25 percent of our electric power.
Atomic energy is expected to be the great source of the future.
Energy reserves of the fossil fuels are expected to last another century and a half.
Resources of the oceans are becoming increasingly important.
Oil and gas are the most important offshore deposits.
Sand, gravel, and *tin* are examples of currently exploited rock and mineral deposits in the ocean.
Manganese nodules, metalliferous muds, and *phosphates* are examples of subsea deposits that may be exploited in the near future.

Supplementary readings

CRUMP, LULIE H., and PHILLIP N. YASNOWSKY: "Supply and Demand for Energy in the United States by States and Regions, 1960 and 1965. Part 4: Petroleum and Natural Gas Liquids," *U.S. Dept. Interior Inf. Circ.*, no. 8411, Washington, D.C., 1969. An authoritative statement on the subject.

GUILLION, EDMUND A. (ed.): *Uses of the Seas,* Prentice-Hall, Inc., Englewood Cliffs, N. J., 1968.

JOHNSTONE, SYDNEY J., and MARGERY G. JOHNSTONE: *Minerals for the Chemical and Allied Industries,* 2nd ed., Barnes & Noble, Inc., New York, 1961. Correlated information on mineral producers regarding the grades of product acceptable and the uses for 77 minerals and related substances. A comprehensive, well-organized reference source.

MC KELVEY, V. E., and FRANK F. H. WANG: "World Subsea Mineral Resources: Preliminary Maps," *U.S. Geol. Surv. Misc. Geol. Invest.,* Map I-632, 1969.

MC KELVEY, V. E., and others: "Subsea Mineral Resources and Problems Related to the Development," *U.S. Geol. Surv. Circ.,* no. 619, 1969. This and the maps referenced immediately above form the best summary of oceanic mineral resources currently available.

PARK, CHARLES F., JR., in collaboration with MARGARET C. FREEMAN: *Affluence in Jeopardy,* Freeman, Cooper & Company, San Francisco, 1968. A book for the layman on the place of minerals and energy in a modern industrial economy. Points out where critical shortages of several minerals may well develop by 1980.

RIDGE, JOHN D. (ed.): "Ore Deposits of the United States," *Am. Inst. Mining Met. Petrol. Engrs,* Inst. Metals Div. Spec. Rept., Ser., 2 vols., 1968. Eighty chapters on individual mines and districts and a summary of 162 papers published between 1933 and 1967 on various aspects of ore genesis, with emphasis on additions to or modifications of ore-formation hypotheses regardless of the type of deposit.

SKINNER, BRIAN J.: *Earth Resources,* Prentice-Hall, Inc., Englewood Cliffs, N.J., 1969. An up-to-date survey of the origin and supply of mineral and energy resources.

22

Man and his environment

THROUGHOUT THIS BOOK WE HAVE SEEN EXAMPLES OF THE RELATION OF MAN TO his environment, whether it be the supplies of water underground, the effect of agriculture on rates of erosion, or the catastrophes caused by earthquakes or volcanic eruptions. In the last chapter we looked specifically at the origin and distribution of economically useful minerals and of energy sources. In this chapter we focus more directly on the intimate relations between man and his physical environment.

In thinking about the geological processes we are impressed with the fact that although their results are large their rates are generally slow. We have thus adopted the nineteenth-century view that small changes over long periods of time have produced grand results. In thinking thus it is natural that we should look at man as an ineffectual agent in changing the earth. In one way this is correct; in another way it is not.

We are disposed to consider man as part of nature, and thus to view his actions as restricted, at least in part, by the physical environment. This is obviously true and yet as long as man has been a numerous species on the earth he has also influenced the other elements in his environment. For example, as he has moved from a hunting to an agricultural society man has changed the vegetative cover of over a third of the earth's land surface, and in so doing has modified the nature of the soil beneath. A century ago the marriage of technology with science triggered a new series of changes. These changes include, among other things, the flow and quality of both underground and surface water, and the quality of the atmosphere; they also affect large portions of the oceans. Their magnitude creates new problems, forcing man to adapt to an environment which he has in part created and yet which he cannot seem to control.

What follows is a discussion of some of the things that man has been able to do to his environment. We understand the causes and some of the effects of these changes, although we have yet to find effective social answers to their control. In addition, we will look at some catastrophic natural phenomena such as earthquakes and volcanoes, which we not only cannot control but which we cannot yet predict.

22.1 *Running water*

In Chapter 11 we considered how water is carried in streams and rivers acting to fashion the landscape. Some of the principles discussed in that chapter apply to our examination of the ways in which man can affect streams and their channels.

Changes in stream flow

THE HYDROGRAPH The hydrograph of a river shows the variation of stream flow with time as indicated in Figure 22.1. In the example given we see a period of high flow, peaking in a flood, and a generally low-water stage which

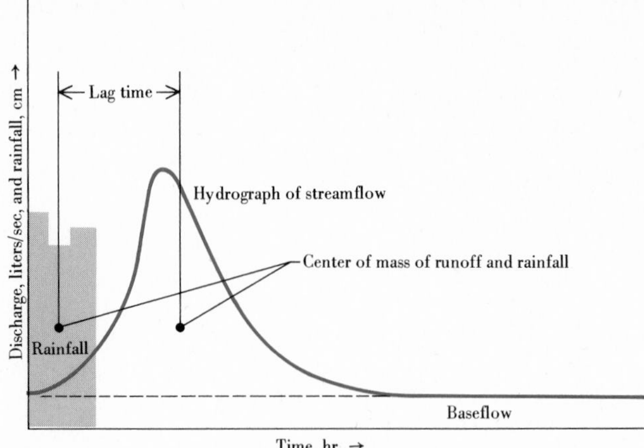

Figure 22.1 **Hypothetical hydrograph showing significant characteristics.** (*After Luna B. Leopold, "Hydrology for Urban Land Planning,"* U.S. Geol. Surv., *Circ. 554, Fig. 1, p. 3, 1968.*)

represents that portion of stream flow attributable to ground-water recharge, a flow referred to as *base flow.* In addition, the figure indicates a period of rainfall responsible for the high-water or flood stage which follows by an interval of time called the *lag time.* The shape of the hydrograph for different streams (and even for different places on the same stream) varies with a number of natural factors, including the infiltration rate, the relief, the geology, and the vegetative cover. We are interested here in the way man affects this flow in ways both planned and unplanned.

FLOOD CONTROL Flood control depends upon modifying the hydrograph by lowering the flood peak, increasing the lag time between the precipitation and flood crest, and spreading the flood flow over a longer period of time. A flood control dam, then, is designed to store water during periods of high runoff and let the excess water out slowly to downstream areas. The result is to reduce the flood crest downstream and spread the discharge of flood waters over a longer time interval, as suggested in Figure 22.2.

URBANIZATION AND SUBURBANIZATION Building of cities and their suburbs affects stream flow and changes the hydrograph, although in the opposite direction than do flood control dams. Agglomerations of buildings with their associated roads, streets, sidewalks and paved parking areas achieve the following effects: (1) the amount of water sinking into the underground is curtailed in proportion to the amount of area sealed off by surface veneers of buildings and pavements. Even the areas of ground left open tend to be less permeable because of its compaction by extensive human activity; (2) the rate and amount of surface runoff increases; (3) the ground-water level falls because of the decreased infiltration. The effect on stream flow is shown in Figure 22.3 and can be characterized as follows: (1) the time lag between precipitation and flood peak is shortened because water is not slowed on its way to the stream channel by a vegetative cover nor does any appreciable amount sink into the ground; (2) the peak is higher because the stream must carry more water in a shorter period of time; and (3) the base flow is generally lower because of a decreased supply of ground water on which the base flow depends. The net result of

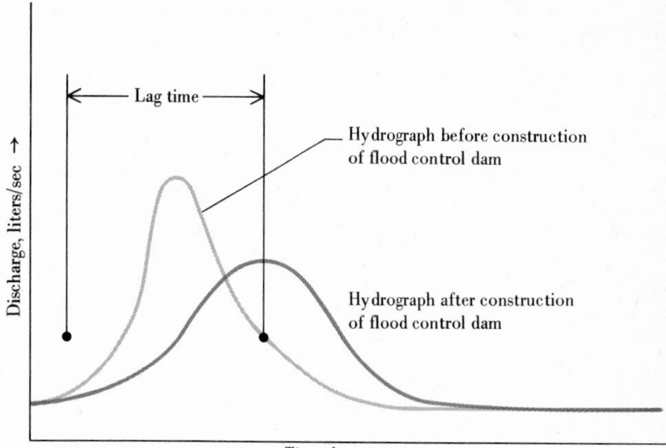

Figure 22.2 **Hypothetical hydrograph to show the effect on stream flow of a flood control dam upstream from the hydrograph station.**

these changes is a *flashy* stream, one which has a low base flow and a high, short flood peak.

CHANGES IN VEGETATIVE COVER Changes in vegetative cover can effect changes in the hydrograph. In a general way, the decrease in the vegetative cover acts in the same direction as does an increase in urbanization.

Studies of the effect of selective forest cutting on stream flow have been carried on by the U.S. Forest Service in the Fernow Experimental Forest near Parsons, West Virginia. The results, not entirely unexpected, are several. First, the removal of the trees increases the flow of streams during the growing season, chiefly because the water ordinarily transpired back into the atmosphere by trees makes its way to the streamways instead. The increase in stream flow is approximately proportional to the percentage of tree cover removed, the greatest increase—over 100 percent—occurring with complete cutting of the forest cover. Concurrently with the increase of stream flow, the flashiness of streams increases. As the tree cover replaces itself the flow decreases, the flashiness of streams declines, and a return to stream-flow conditions similar to those before cutting is projected to be approximately 35 years.

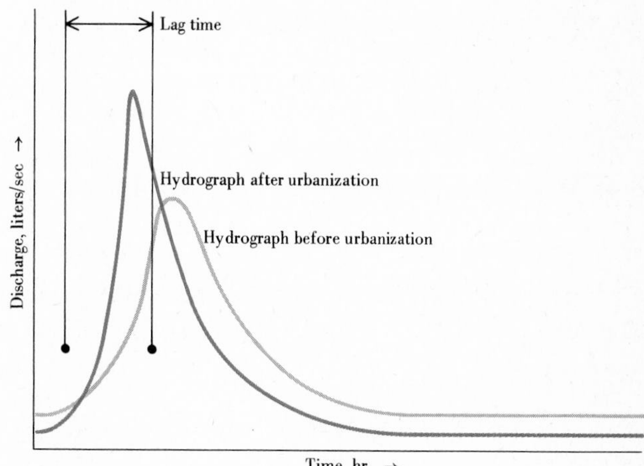

Figure 22.3 **Hypothetical hydrograph showing the effect of urbanization on stream flow.** (*After Luna B. Leopold, "Hydrology for Urban Land Planning,"* U.S. Geol. Surv., *Circ. 554, Fig. 1, p. 3, 1968.*)

Changes in stream channels

The changes in stream channels associated with man-induced changes in stream flow are not too well documented or understood as yet, but some suggestions can be made.

In Chapter 11 we found that the stream channel is adjusted to the stream discharge and this was expressed as discharge = width × depth × velocity. Observation shows that a stream rises to the limit of its bank on the average of once every 1.5 to 2 years. This stage of flow is called the *bankfull stage.* The stream channel is adjusted to handle the size of flow which occurs every 1.5 to 2 years. When flow exceeds this amount, flooding begins. Studies reported by Luna Leopold of the U.S. Geological Survey indicate that with an increase of urbanization the bankfull stage of flow increases. If an area is 50 percent urbanized, experience indicates that the bankfull stage of discharge increases by a factor of 2.7 over that of the same stream in an unurbanized condition. An example is shown in Table 22.1.

In the table, the stream has increased the depth and width of its channel as the discharge for bankfull stage increased. The bankfull stage increased because of increase in runoff rates caused by urbanizing one-half of the drainage area. This erosion of the channel produces sediments that are moved downstream, where they may be deposited if conditions permit.

Erosion of a stream bed can be artificially induced in other ways. Thus it has been found that a stream below a dam will enlarge its channel over a considerable distance and redeposit the products of channel erosion farther downstream. This erosion is in part due to the fact that the dam provides a settling basin in which stream deposits are stored. The stream section below the dam is deprived of its normal sedimentary load. When erosion takes place immediately below the dam there is a lack of sediment to replace this normal loss through point bar, building, or overbank deposition.

Changes in water quality

The most obvious change that man has brought to the rivers of the world is their pollution, a change directly caused by the planned (or unplanned) use of the rivers as a sewerage system. Increasing technology, industrialization, and population have produced an increasing amount of refuse of all kinds.

Table 22.1 ***Stream flow and urbanization***

	Before urbanization	*After 50% urbanization*
Discharge at bankfull stage	1.5 m^3/sec	4.0 m^3/sec
Velocity	0.75 m/sec	0.75 m/sec
Depth of channel	0.5 m	0.9 m
Width of channel	4.0 m	6.0 m
Area of drainage basin	2.5 km^2	2.5 km^2

Table 22.2 **Pollution-caused fish kills in the United States, 1961–1968**[a]

	1961	1962	1963	1964	1965	1966	1967	1968
Number of states reporting	45	37	38	40	44	46	40	42
Total estimated number of fish killed $\times 10^3$	15,910	7,118	7,860	18,387	11,784	9,115	11,591	15,236
Average size of kill $\times 10^3$	6.5	5.7	7.8	5.5	4.3	5.6	6.5	6.0
Largest kill reported $\times 10^3$	5,387	3,180	2,000	7,887	3,000	1,000	6,549	4,029

[a] Data from Federal Water Pollution Control Administration.

Where rivers are available they become a handy and, at first, inexpensive way to remove our cultural refuse. This load has proved more than many can handle and still maintain their original quality. That pollution of water has its effect on the life process is shown by Tables 22.2 and 22.3.

SOLID LOAD Agriculture, more than any other human activity, increases the rate at which sediments are delivered to the streams. It is true that construction activities produce a higher rate of sediment production but these are isolated events and over the long run the continuous use of land for farming is volumetrically more important in sediment production.

There is a great deal of information showing the effect of the clearing of land and of farming practices on the production of sediments. For instance, one of the effects of forest clearance in the Fernow Experimental Forest referred to earlier in this chapter was to increase the particles delivered to the stream. With complete commercial cutting of the watershed, the streams carried as much as 3.7×10^3 more solid materials than they did in areas in which no lumbering occurred. In Mississippi studies on sediment production from land under differing uses produce the startling conclusions shown in Figure 22.4. Land under cultivation produced 10^3 times more sediment than did similar land under mature stands of trees. M. Gordon Wolman has shown the effect of changing land use on the production of sediments in the Piedmont area of Maryland from about 1800 to the present (Figure 22.5). Interestingly, urbanization has reduced sediment production to the preagricultural levels.

The increased volume of sediments move on to the streams, which then must handle them as best they can. The results vary with the stream. In general, however, increased solid loads increase the turbidity or muddiness of a stream,

Table 22.3 **Fish kill by source of pollution in the United States, 1968**[a]

Source of pollution	*Fish killed*	*Average kill*	*Game fish*	*Nongame, including commercial fish*
Agriculture	422,000	4,240	152,000	270,000
Industry	6,398,000	5,675	415,000	5,983,000
Municipalities	6,952,000	7,585	320,000	6,632,000
Transportation	880,000	9,155	430,000	459,000
Other	584,000	1,995	19,000	565,000
Total	15,236,000		1,336,000	13,900,000

[a] Data from Federal Water Pollution Control Administration.

Figure 22.4 *Variation in sediment yields from individual watersheds in northern Mississippi under different types of land use and changing amounts of precipitation.* (*After S. J. Ursic and F. E. Dendy, "Sediment Yields from Small Watersheds under Various Land Uses and Forest Covers,"* Proc. Federal Inter-Agency Sedimentation Conference, 1963, *U.S. Department of Agriculture, Misc. Publ. 970, Fig. 2, p. 51, 1965.*)

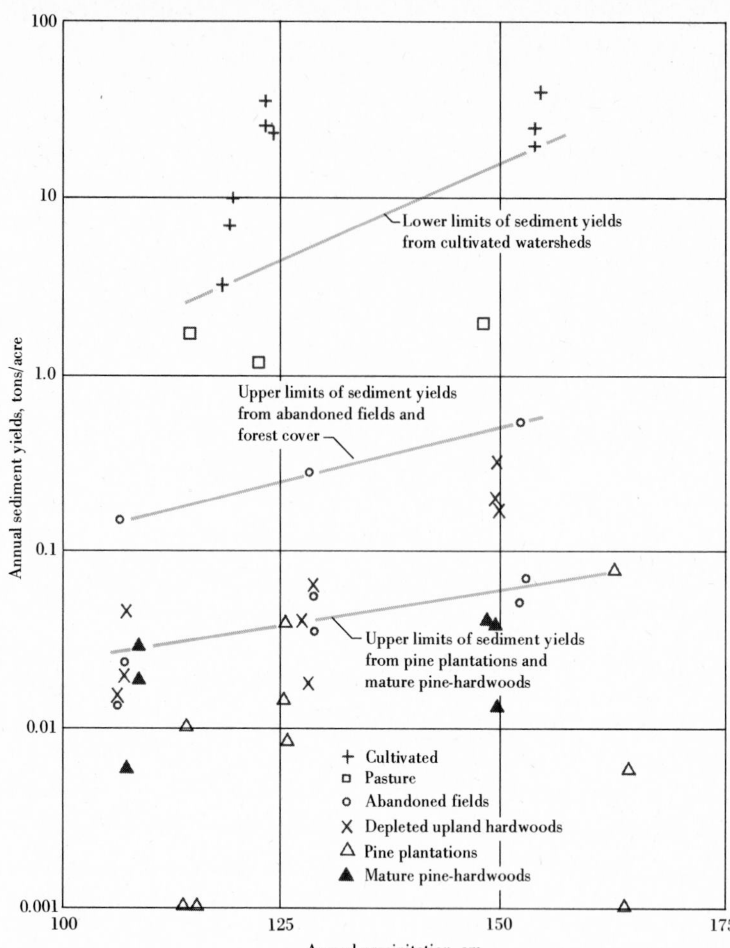

Figure 22.5 *A sequence of land use changes and sediment yield beginning prior to the advent of extensive farming and continuing through a period of construction and subsequent urban landscape. Based on experience in the Middle Atlantic region of the United States.* (*After M. G. Wolman, "A Cycle of Sedimentation and Erosion in Urban River Channels,"* Geograf. Ann., *vol. 49-A, Fig. 1, p. 386, 1967.*)

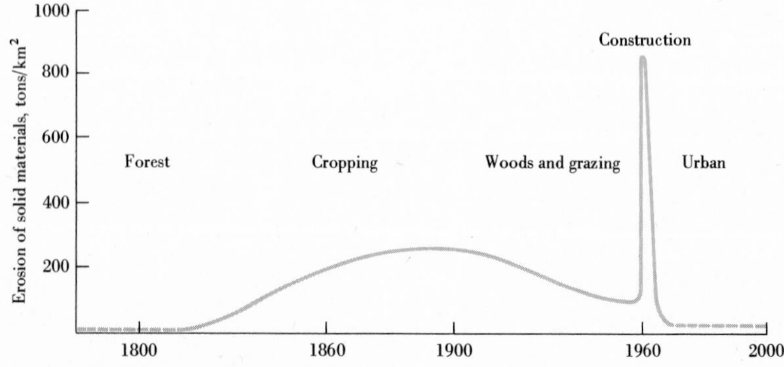

decrease the depth of light penetration, and can cause silting up of valley bottoms.

To agriculture, deforestation, urban construction, and grazing we must add mining and industry as producers of excess sediments. These activities may bring other undesirable features with them in addition to increased sediment. For instance, the tailings of many coal mines contain large amounts of sulfur which in turn can be converted to undesirable compounds, including sulfuric acid.

DISSOLVED LOAD We found in Chapter 11 that the average content of dissolved material carried by streams is 120 ppm on a world-wide average. Most of this comes from the chemical decomposition of earth materials during the normal processes of weathering. Some is recycled from the oceans through the atmosphere to the streams, and some is contributed by man.

That part of the dissolved load contributed by man covers a wide range of materials. Some are toxic to river life, some serve as nutrients promoting explosive organic proliferation, and a few are merely an affront to human sensibilities.

Figure 22.6 illustrates the increase of dissolved solids in the Passaic River at Little Falls, N.J., from 1948 through 1963, a period of increasing industrialization and urbanization. It is not, however, just the increase of the dissolved load itself but the various factors involved in the increase that cause problems.

Among the most bothersome of the stream pollutants are the phosphorus and nitrogen that come from fertilizers, from sewage (both treated and raw), and, in the case of phosphorus, from detergents. These additions to streams spark a complicated set of reactions in the chemistry, biologic processes, and ecology of the stream, all usually undesirable (from the human point of view) and all leading to a stream of a different character than it was in its natural state.

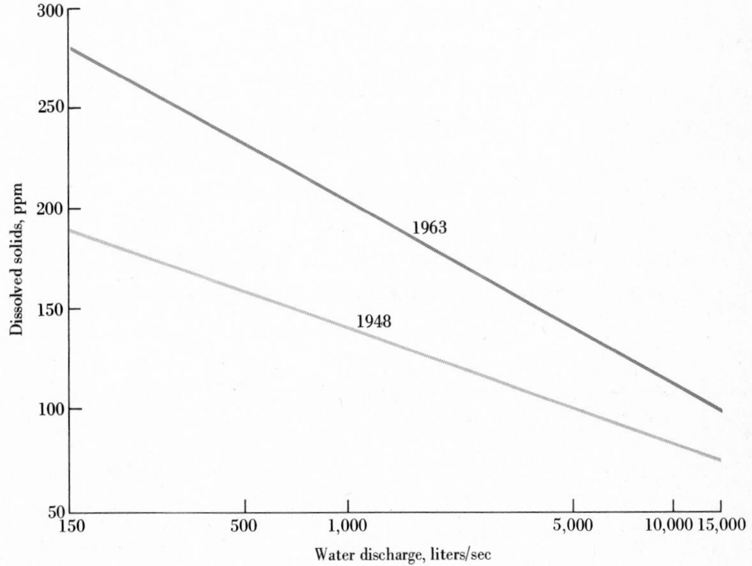

Figure 22.6 **Increase in dissolved solids in relation to discharge in the Passaic River at Little Falls, N.J., from 1948 to 1963.** (*After Peter W. Anderson and S. D. Faust, "Quality of Water in the Passaic River at Little Falls, N.J., as Shown by Long Range Data,"* U.S. Geol. Surv., *Professional Paper 525-D, Fig. 2, p. 215, 1965.*)

We will see later that these same elements are critical in the quality of lake water as well.

Pesticides, particularly those such as DDT which are based on chlorinated hydrocarbons, have received a great deal of attention in recent years. We deal here with the process of biological concentration of the DDT. Fairly dilute concentrations of the chemical in water may build up to toxic concentrations through the food chain from the lowest forms to the carnivorous forms. The species toward the top of the food chain may have concentrated the pesticide (usually in their fatty tissues), to a level 10^3 to 10^4 times that of the concentration in the stream. It was this situation which, in 1970, led to the virtual banning of DDT use by the Department of Health, Education, and Welfare.

The list of chemical pollutants is as varied as the products of our industrial society. They are present in the streams because we continue to use the channels of water as a sewer system. Until we adopt other ways of cleaning our nests the streams will continue to be polluted.

THERMAL POLLUTION The use of river water for industrial cooling purposes, particularly by electric power plants, has led us to recognize that the discharge of waste heat into our streams can lead to unwanted effects on the aquatic life. We dignify this human activity as *thermal pollution,* a problem faced not only by rivers but by lakes, underground water, and shallow sea waters as well (see Figure 22.7).

Thus far it has been the electric power industry that has done most to introduce us to the dangers of thermal pollution. Presently it uses about three-fourths of all waters used in the United States for industrial cooling. Thus far, however, thermal pollution has become important only in local areas. This is not to say that electric power may not pose greater problems in the future. One estimate suggests that by the year 2,000 the industry will be producing so much waste heat that it will require, if river water is used as a coolant, approximately one-third of the country's runoff to get rid of the excess energy. This appears totally unacceptable, and some other solution must be found.

The primary effect of thermal pollution is to raise water temperatures to a point at which the native aquatic life can no longer survive. There are other effects as well. For instance, increased temperatures of river waters can produce

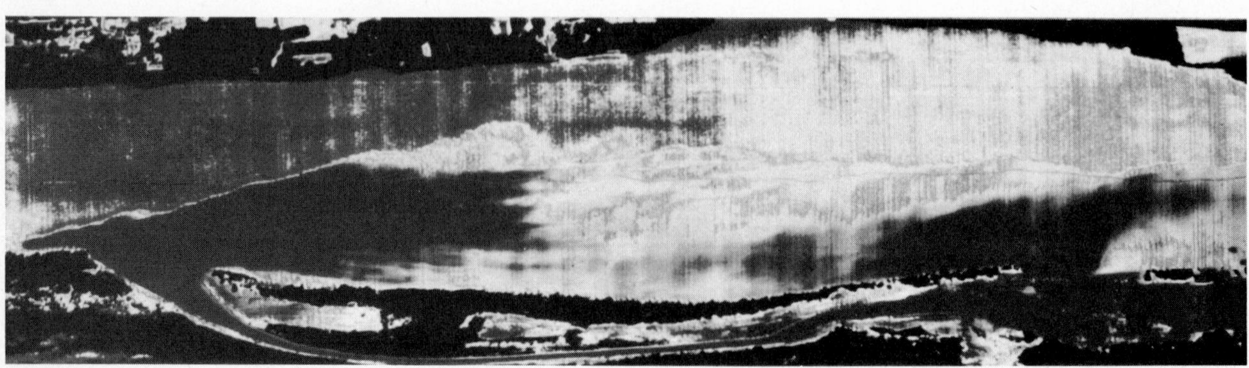

Figure 22.7 ***Thermal pollution of the Connecticut River by a nuclear power plant near Haddam, Connecticut, as evidenced by infrared photography. In this color-enhanced version of a photograph made with an infrared camera, higher temperatures of the water appear as darker tones. Thus the power-plant discharge through the canal at the bottom of the picture appears dark gray: it has a temperature of about 34°C. The plume of heated water trails downstream from the mouth of the effluent canal, mixing with the river water, which has a normal temperature of 25°C and shows a lighter gray.*** *(Photograph by HRB-Singer, Inc., for the U.S. Geological Survey.)*

local fog conditions as cool air sweeps across the warmed water. Such conditions could lead one to such exercises as "How many power plants will be needed on New York's East River to keep LaGuardia airfield perpetually fogbound?"

Industrial cooling is a dramatic way to change the temperature of river water. But man can bring about these changes in another way. Studies of streams in urbanized areas on Long Island have been carried out by E. J. Pluhowski of the Johns Hopkins University. Comparing those streams in urbanized zones with a control stream still in an undeveloped area, Pluhowski found that streams most affected by man's activities had summer temperatures from 5.5°C to 8°C above those of the control stream. Conversely, stretches of streams affected by man were found to be 2.2°C to 5.5°C colder in winter than those unaffected by man.

Some of these variations are due to the nature of runoff between unurbanized and urbanized zones. During the summer months runoff from heated pavements can raise temperature of stream water appreciably, as much as 6°C as the result of a single storm, whereas an unurbanized stream shows little variation during the storm and subsequent runoff. Streams in urbanized zones are fed more by surface runoff than are streams in unurbanized zones, which depend more heavily on ground water for their flow. In winter the surface runoff is colder than the groundwater, which tends to reflect the mean annual temperature. The stream in an urbanized area, therefore, is colder during the winter than is the stream in an unurbanized area.

22.2 Lakes

Viewed against the large span of geologic time lakes are only temporary features. They are disruptions in the general drainage system of the earth's surface. Some of these disruptions may last for hundreds of thousands of years, others for a few tens or few hundreds of years. Despite their fleeting existence in the geological record they are important to man. Here we will consider briefly the life cycle of a lake and then look at the ways in which man may hasten the demise of a lake.

Life cycle of a lake

The basin which holds a lake may form in one of several ways. Glaciers may scour bedrock, or pile up debris across a streamway; earth movements may downdrop a portion of the earth's crust; volcanic craters and calderas may hold the waters of a lake; landslides or volcanic flows may dam valleys; the surface may subside because of solution of rock in the underground; streams may cut off meanders to form crescent-shaped, oxbow lakes; lagoons may be cut off from the sea by coastal processes; wind may scour a basin later flooded by water, and man may build dams.

Whatever its origin, a lake begins to die as soon as it is born. Death is due

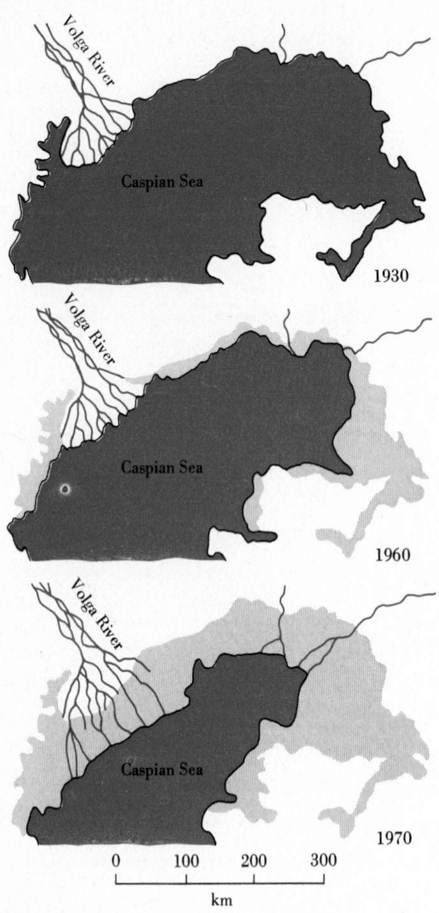

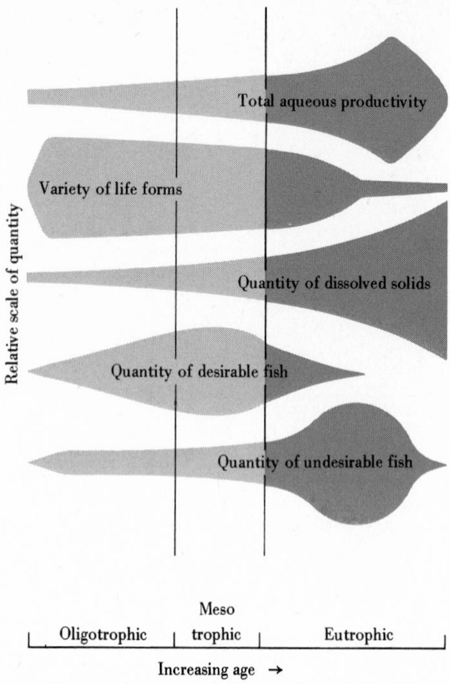

Figure 22.9 **Aging indicators as a lake progresses from oligotrophic to eutrophic.** (*After U.S. Department of the Interior,* Lake Erie Report, *Federal Water Pollution Control Administration, Great Lakes Region, Fig. 3-2, p. 32, 1968.*)

Figure 22.8 **The Caspian Sea is really a large saltwater lake. These three maps of its northern section show how it has decreased in size from 1930 to 1970. This decrease is due in part to a natural decline in rainfall, also in part to man, who has used large amounts of water from the Volga River for irrigation, thus decreasing the river's discharge into the Caspian.** (*After A. E. J. Engel, "Time and the Earth,"* Am. Scientist, *vol. 57, Fig. 12, p. 499, 1969.*)

Figure 22.10 **In two generations man-induced pollution has caused Lake Erie to advance to the eutrophic stage.** (*After U.S. Department of the Interior,* Lake Erie Report, *Federal Water Pollution Control Administration, Great Lakes Region, Fig. 3-1, p. 32, 1968.*)

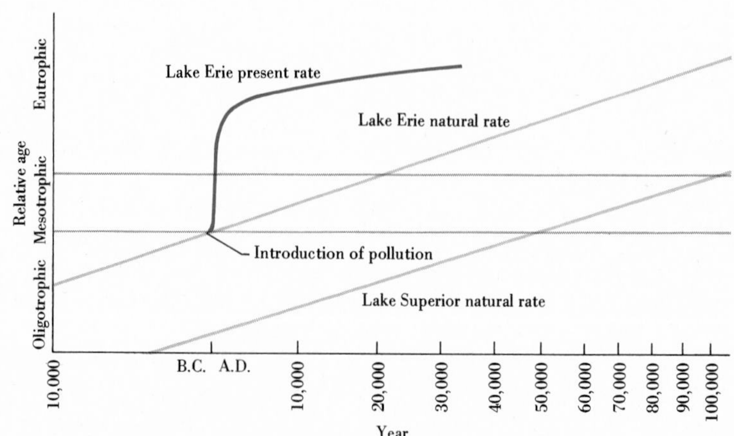

primarily to two different processes, which may act concurrently. First the outlet of the lake becomes lowered by erosion, cutting the dam downward and lowering the level of the lake behind it. Second, a lake fills up with sediment. Some of the filling comes from the solid load delivered to the lake by the streams that feed it; some is due to precipitation of material from solution by either organic or inorganic processes, and some is the accumulation of the remains of plant and animal life. A third fate of a lake is to be dried up because of insufficient water, and this has been the destiny of some of the pluvial lakes of southwestern United States, as we saw in Chapter 13.

Man's effect on lakes

Man's role in the history of lakes thus far has been one of speeding the aging process, hurrying them on to a premature end. One such example is the Caspian Sea, which we may view as a saltwater lake. Man's use of water that would normally come to the Caspian is carrying that lake closer to extinction, as suggested in Figure 22.8. It is also evident that man contributes to the destruction of a lake by increasing the rates of erosion and hence the amount of sediments that streams can deliver to a lake. We have already seen that intensive land use by man can increase the production of sediments one to four orders of magnitude above the natural rate. This can only increase the rate at which the lake basin is filled.

Man also contributes to lake decay by speeding up a process called *eutrophication. Eutrophic,* from the Greek *eutrophos,* "well-nourished," applies to lakes having abundant supplies of nutrients that support a dense growth of plant and animal life. Characteristically the decay of this organic matter depletes lake waters of oxygen particularly during the warmer months. Lakes characteristically begin as *oligotrophic lakes* (from the Greek meaning "sparsely-nourished") are low in accumulated nutrients and hence in organic productivity and high in dissolved oxygen. The lake ages and as more nutrients can accumulate biologic activity increases as the lake passes toward eutrophic conditions through a transitional *mesotrophic* stage as shown in Figure 22.9.

Under natural conditions the process of eutrophication proceeds at varying rates. Man plays a role in eutrophication by supplying large amounts of nutrient material to the lakes and bringing about an explosive increase in the biologic activity. Nitrogen and phosphorus are the most important elements contributed by man to the eutrophication process. Generally municipal sewage is the chief source of nutrients, although rural and industrial sources are also important. Experience has shown us that in a few decades oligotrophic lakes can be converted to strongly eutrophic lakes by turning them into receptacles for municipal sewage and industrial and rural wastes.

It was long thought that the Great Lakes were so large that eutrophication would not become a major problem. This we now know was an error in judgment for man has pushed one of them, Lake Erie, over the brink into the eutrophic stage (see Figure 22.10). This has, moreover, been an expensive misjudgment. In 1968 the Federal Water Pollution Control Administration estimated that it would take $1.1 billion to construct proper sewage facilities in the lake basin to take care of the needs of 1970, and in 1990 another $1.6 billion for improvement to handle the needs of an increased population. In addition, adequate control of industrial pollution of Lake Erie in 1970 would cost an estimated $285 million. The other Great Lakes have yet to reach the condition of Lake Erie, but each of them shows to some extent the impact of changes induced by man.

22.3 *Ground water*

Water underground is subject to contamination of the same type as is surface water. We have noted in Section 12.3 that water in limestone is more susceptible to pollution than is that flowing more slowly in, for example, a sandstone aquifer. Underground water may be subject to thermal pollution as well as microbial or chemical pollution. In this instance the very process of conserving water may give it the added undesirable quality of increased temperature. Thus, water used for air conditioning is withdrawn from the ground at one temperature, used in cooling, and returned to the ground at a higher temperature.

Saltwater invasion

A particular type of pollution to which underground freshwater supplies are subject is the invasion of salt water. This may be salt water of an adjacent ocean environment or it may be salt water trapped in rocks lying beneath the freshwater aquifers.

Fresh water has a lower specific gravity than does salt water: Normal sea water has specific gravity of about 1.025, compared with 1.000 for fresh water. Therefore fresh water will float on top of salt water. If there is little or no subsurface flow this can lead to equilibrium in which a body of fresh water is buoyed on salt water. In a groundwater situation in which fresh water and salt water are juxtaposed we might have a situation such as that shown in Figure 22.11. Here an island (or it could as well be a peninsula) is shown underlain by a homogeneous aquifer that extends under the adjacent ocean. Fresh water falling on the land has built up a prism of fresh water in hydrostatic equilibrium with the salt water that surrounds it. The height of a column of

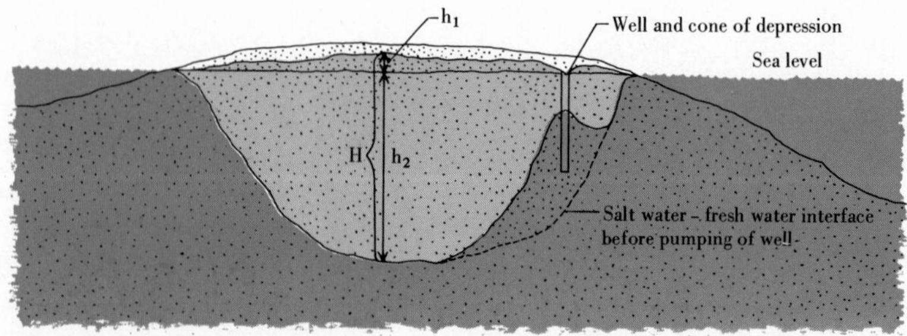

Figure 22.11 ***Cross section through an oceanic island (or a peninsula) underlain by homogeneous, permeable material. The fresh water forms a prism floating in hydrostatic balance with the neighboring salt water. Reduction of the groundwater table on the island will cause a change in the position of the saltwater–freshwater interface. See the text for discussion.***

fresh water is balanced by an equal mass of salt water. In the example given, the column of fresh water (H) is equal to h_1, the height of the water table above sea level, plus h_2, the thickness of the ground water below the sea level. A column of salt water equivalent in length to h_2 will balance a column of fresh water of length H or $h_1 + h_2$. Working this out we find that $h_2 = 40\ h_1$. This means that if the water table is 10 m above sea level then the freshwater–saltwater contact will be 400 m below sea level. It also means that as we lower the water table by whatever means we raise the elevation of the saltwater–freshwater interface at the rate of 40 m of rise for each meter of lowering of the water table. It is very possible, then, that a well which originally bottomed in fresh water could, by lowering the water table, turn into a saltwater well, as suggested in Figure 22.11.

Land subsidence

Excess pumping may deplete groundwater supplies and in certain places bring about saltwater invasion of wells. But another bothersome problem connected with heavy pumping of ground water is land subsidence. It expresses itself at the surface either as broad gentle depressions of the land or as catastrophic collapse and the formation of sinkholes.

Figure 22.12 **Principal areas in California where withdrawal of ground water has caused subsidence of the land surface.** (*After B. E. Lofgren, and R. L. Klausing, "Land Subsidence Due to Groundwater Withdrawal, Tulare–Wasco Area, California,"* U.S. Geol. Surv., *Professional Paper 437-B, Fig. 1, p. 2, 1969.*)

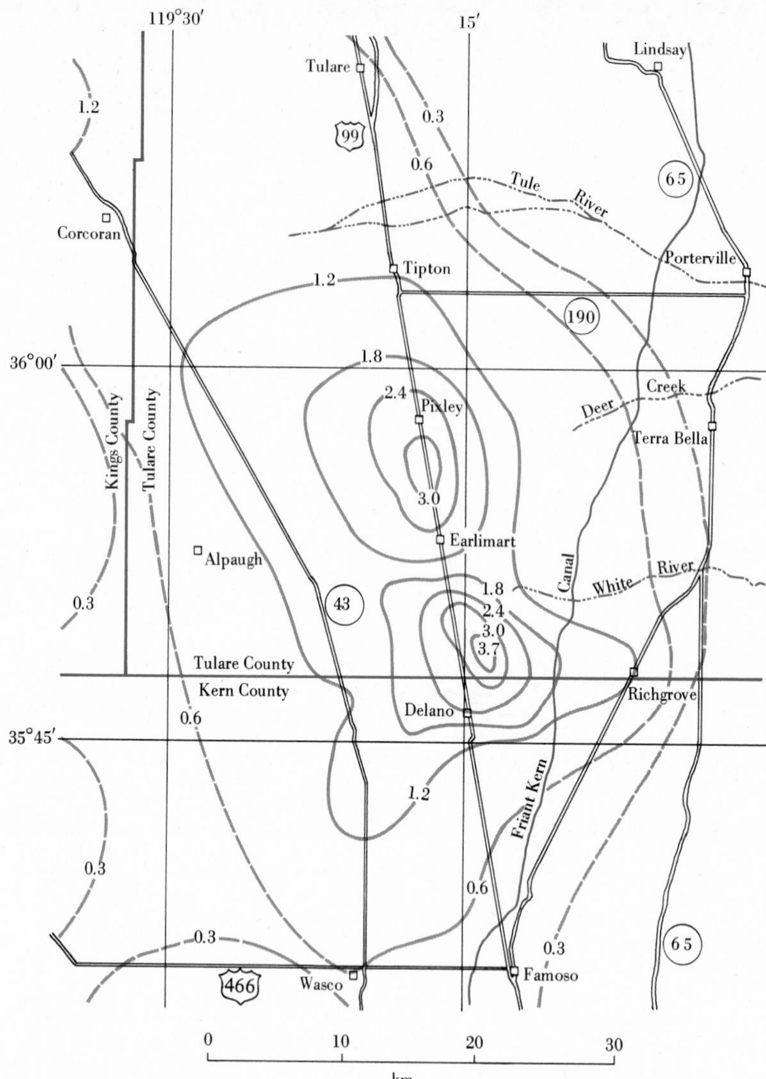

Figure 22.13 **Land subsidence in the Tulare–Wasco area, California, 1926–1962, due to withdrawal of ground water. Lines show equal subsidence in meters and are dashed where approximate.** (*After B. E. Lofgren and R. L. Klausing, "Land Subsidence Due to Groundwater Withdrawal, Tulare–Wasco Area, California,"* U.S. Geol. Surv., *Professional Paper 437-B, Fig. 50, p. 63, 1969.*)

Zones of broad subsidence have been created in California over large areas of the San Joaquin Valley and secondarily in the Santa Clara Valley south of Oakland and San Francisco (Figure 22.12). The phenomenon is related to the pumping of ground water and to the resulting compaction of sediments. The Tulare-Wasco area of the San Joaquin Valley is a good example of what happens. In 1905, intensive pumping for irrigation began in the area. By 1962, more than 20,000 km^2 of irrigable land had undergone subsidence of more than 0.3 m and as much as 4 m of subsidence had occurred in some places.

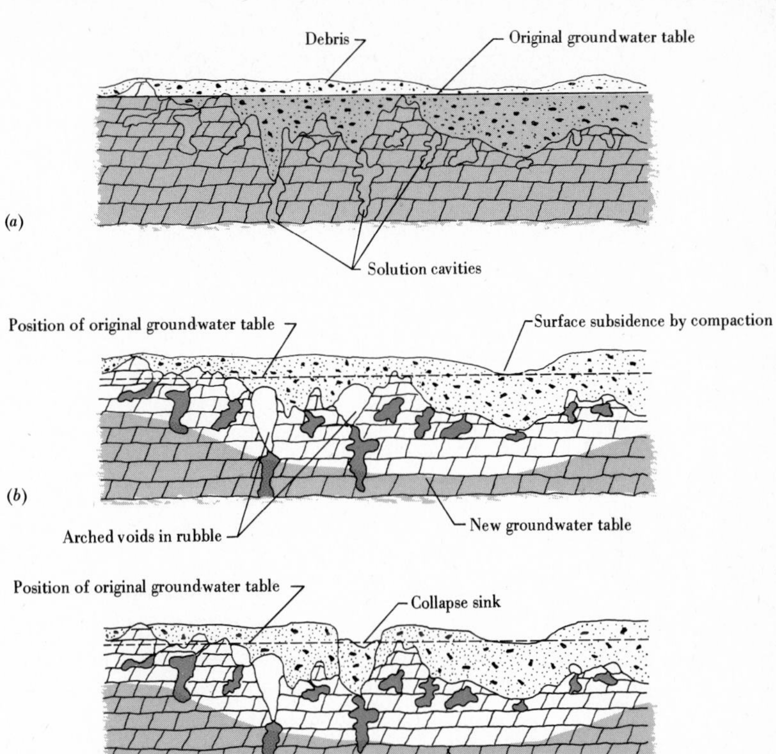

Figure 22.14 **Stages in the development of land subsidence and collapse in the Far West Rand mining district near Johannesburg, South Africa. Lowering of the groundwater table initiated the movements. See text for discussion.** (*After Richard M. Foose, "Sinkhole Formation by Groundwater Withdrawal, Far West Rand, South Africa,"* Science, *vol. 157, Fig. 3, p. 1047, 1967. Copyright 1967 by the American Association for the Advancement of Science.*)

Figure 22.13 shows subsidence in the area from 1926 to 1962. Subsidence was slow through this time and hardly noticeable to residents. But the casings of many wells have been damaged, the problems of surveying and construction have been complicated, and the operation of irrigation districts endangered.

Rapid subsidence resulting in sinkholes can take place with the lowering of the groundwater table, although it is not as common as the gentler subsidence described above. Richard Foose, of Amherst College, describes sinkhole formation resulting from the reduction of the groundwater table in a portion of the South African gold-mining district. Here the groundwater table was lowered at least 160 m as part of the mining program, and sinkholes up to 125 m wide and 50 m deep developed in an area underlain by a dolomite covered by a thick mantle of weathered material. Sinkholes develop over zones in which the surface of the dolomite is extremely irregular and characterized by pinnacles of unweathered rock separated by accumulations of debris. The process of sinkhole formation is promoted by the creation of large voids between pinnacles. This occurs in part by desiccation of the debris as the groundwater table is lowered, in part by washing of debris into openings in the dolomite at greater depths, and in part by sloughing off of debris from the roof of the upwardly enlarging void between pillars. At some point, the roof collapses and a sinkhole forms at the surface (see Figure 22.14).

22.4 Soils

Most arable soils are formed by weathering, which creates a zonation in the surficial material of the earth. The zones (soil horizons) thus formed enlarge downward with time, and concurrently surface material is slowly removed by normal erosive processes. The introduction of farming, grazing, lumbering, and construction increases the rate of erosion. In a few years or a few decades erosion is great enough to upset the balance between the downward growing soil zones and the material being removed at the surface. The rate of removal, along with man's turning and disruption of the soil, may in a few years or few decades remove all or large parts of the soil profile that has taken so long to form.

The loss of the soil through man-induced erosion can be compensated for in two ways. It may either be artificially regenerated by the introduction of nutrient-supplying chemical fertilizers, or land may be allowed to revert to its original state and given time to recover. In the latter case the recovery time will vary with the local situation, but several generations to hundreds of years must usually be allowed.

We are conditioned to think of topsoil as a "precious" material that should not be destroyed. But there are some situations in which one of the first things one wants to do is to get rid of the weathered material and get down to the less weathered earth materials. A moment of thought will tell us why this is so. Some soils have been long in the process of formation. Much of the nutrient material has been leached out of them and slow erosion at the surface has not been rapid enough to expose new unleached materials, which can provide the basis for plant growth. Under these conditions the removal of the relatively sterile upper portion and the exposure of the less weathered material still undepleted in the elements needed for plant growth may be beneficial.

In some instances the development of a heavy *B* horizon may create not only a low nutrient zone, but also a zone which renders a soil clayey and poorly drained.

Surface mining

Surface mining, the process of removing the overburden of soil and rock that covers a mineral deposit, has certain technical advantages over underground mining methods. But it also creates problems which include an instantaneous disruption or destruction of the soil and a rearrangement of the landscape.

Surface mining takes a number of forms. *Open-pit mining* is perhaps the most familiar and is represented by the gravel pit, the stone quarry, and the gigantic pits of some mines such as the vast excavations for low-grade copper in Arizona, Utah, and Nevada. *Strip mining* consists of moving to one side varying thicknesses of soil and rock until the mineral deposit is exposed for removal (see Figure 22.15). The process is most commonly used in the coal-

Figure 22.15 **Strip mining is an easy way to mine shallow mineral deposits but it disrupts the soil and rearranges the landscape, as shown in this operation.** (*Photograph by Tennessee Valley Authority.*)

mining industry. *Hydraulic mining* uses a strong jet of water to move deposits of sand and gravel that may contain or cover the sought-for ore. The ore is carried through concentrating equipment where the wanted mineral is separated, usually by utilizing differences in specific gravity. *Dredging* is yet another form of surface mining. In situations where sand and gravel are sought the products of dredging are all removed for consumption. In instances where higher-priced minerals are needed, they are separated from the dredgings and the debris that is discarded may be left as piles in the vicinity of the operation.

The primary effect of surface mining is a disturbance or, more usually, a destruction of the soil; the creation of gaping holes in the landscape; the formation of a jumbled artificial landscape unadjusted to normal surface processes; the disruption of the biota—both animal and plant; and the modification of water flow both on the surface and underground.

By 1965, a total of 12,800 km^2 of land had been disturbed by surface mining in the United States. This had risen by 1970 to an estimated 16,000 km^2. Of this total area, mining for coal accounted for 41 percent; for sand and gravel, 26 percent; for stone, 8 percent; for gold, 6 percent; for phosphate, 6 percent; for iron, 5 percent; for clay, 3 percent; and for all others, 5 percent.

Increasing attention is being paid to restoring land disrupted by surface mining. This is generally an expensive business, but it is most economically done if it is included as an integral part of the mining project rather than as a separate operation planned and carried out after the site has been abandoned by miners.

As of 1965, 34 percent of disturbed land had been reclaimed in one way or another. Of this percentage, 46 percent had been healed by nature, 40 percent had been voluntarily restored by industry, 11 percent had been restored by industry under law, and the remaining 3 percent had been restored by federal, state and local governments.

22.5 Solid wastes

We have referred to some of the problems raised by chemical and organic wastes that reach our streams and lakes. Man produces a very large amount of additional waste, which includes everything from paper clips to buses. Some of these wastes can be salvaged and reused, and many are. Many, however, must be disposed of and this raises an increasingly complex problem for densely populated areas.

New York City, for instance, dumps wastes from barges onto the continental shelf beyond New York Harbor at the rate of 2 kg/person per day. As such, New York City has become the major source of sediment in the area, far outstripping the contribution of the local rivers. By 1970 this dumping of wastes was already having a deleterious effect on the marine life in the area. Man, then, has become a major source of sediments. We found earlier that he had also become an important factor in increasing the rates at which the processes of erosion produce sediments. The figures given in Table 22.4 suggest the magnitude of man's contribution to sediments in the United States. At best they are approximations, but probably they are low. The total is staggering, particularly when we compare it to the 540 million metric tons of material estimated to have been carried annually by the streams of the United States before man influenced the rates of erosion.

22.6 Oceans

To anybody who has sailed the oceans or flown across them it is difficult to imagine that man could greatly affect such large bodies of water. In fact, except for local situations, the oceans have been fairly well immune to the attacks of man, largely because of their size. Let us look at some of the local effects of man on the marine environment.

Not unsurprisingly the near-shore marine environments near heavily popu-

Table 22.4 ***Major sources of unreclaimed solid wastes produced in the United States in 1970, including those estimated to result from man's acceleration of the rates of erosion***[a]

Source	*Metric tons* $\times 10^6$
Domestic, municipal, and industrial waste	325
Junked cars, trucks, and buses	10
Mining industry	1,300
Increased stream load due to man's activity	560
Total	2,195

[a] Estimates compiled from various sources.

lated areas suffer the most. We have already used as an example the dumping of New York City's solid wastes on the adjacent continental shelf.

Oil spills and blowouts

The *Torrey Canyon* tragedy of March 18, 1967, awakened the world to the problem of oil pollution at sea. The *Torrey Canyon,* a 295-m tanker carrying 117,000 tons of crude oil from Kuwait to English refineries, went aground on rocks off Land's End, at the southwestern tip of England. Over a two-week period the stranded hulk gave up its load of petroleum, bringing a black, oily contamination to the French and English beaches along the English Channel, and either directly or indirectly caused the death of untold numbers of birds and marine organisms. The blowout in 1969 of an oil well drilled in the Santa Barbara channel off the coast of California produced similar results and emphasized to all that the process of offshore drilling can produce ecological hazards.

What did we learn from these events? First, we recognized that we could expect more such disasters. Increased offshore drilling, increased size of tankers, and even the penetration of large tankers into the treacherous and icy Arctic waters tell us that we can expect more of the same in the future. Greater care will certainly reduce the number of incidents, but it seems unlikely that we will eliminate them entirely.

In addition, we have learned some techniques that are useful in handling an oil spill once it occurs. Petroleum spilled at sea moves as a coherent mass 5 to 60 cm thick and parallel with the wind direction at about 3.4 percent of the wind velocity. Therefore, knowing direction and velocity of the wind, we can predict direction of movement of an oil spill and its estimated time of arrival at a selected location.

Several techniques can be used to eliminate the oil before it fouls the shore. The oil can be surrounded by booms and barges and pumped from the sea's surface. In the *Torrey Canyon* disaster the French had some luck with this but were more successful in sprinkling the oil with a dust of chalk and sodium stearate. Oil adhered to the dust and then, because in combination with the dust the oil was heavier than water, it sank to the sea bottom. Over 23,000 tons of oil were eliminated from the surface in this way, but their effect on the sea bottom is not known. The use of detergents to disperse the oil so that it may be more quickly oxidized or become more susceptible to degradation by bacteria is not usually successful. The British used a great volume of detergent to disperse the *Torrey Canyon* petroleum at sea and even more on the beaches. The only real result after the application of nearly 10 million liters of detergent was the loss of marine life, particularly in the intertidal marine zone.

Natural processes aid in the elimination of the petroleum. Evaporation is effective and, in general, about 25 percent of an oil spill will evaporate in a few days. After three months, 85 percent will have volatilized. In addition, photochemical reactions, degradation by bacteria, and adhesion to sedimentary particles in suspension in the water all tend to get rid of the pollution.

Figure 22.16 ***A saltwater wedge may develop in some estuarine situations. The position of the wedge moves as the supply of fresh water varies relative to the supply of salt water.***

Estuaries

Estuaries are long, relatively narrow incursions of the marine environment into a land mass. In reality they are valleys turned into a near-shore marine environment by invasion of the ocean waters as sea level rises in relation to the land. Estuaries are probably much more common today than they were during most of geologic time. This is because during the Pleistocene, valley erosion of the exposed continental margins occurred during the low-water stages attending glaciations, and flooding of these valleys took place during periods of deglaciation and rising of the sea level.

The estuary has become an attractive geographic feature for man. It has provided protected anchorages and shorelines for port facilities. It gives penetration for marine shipping well inland beyond the open ocean, and has provided suitable environments for various facets of the fishing industry.

The estuary is an area where fresh water from the land meets the salt water from the ocean. The ways in which these two types of water circulate and are related to each other play an important role in the use of the estuary.

As shown in Figure 22.16, the fresh water fed into the estuary by the river at its head flows above the heavier salt water, which assumes a tapering form known as the *saltwater wedge.* The position of the wedge must change with the changing supplies of salt and fresh waters. Thus, during periods of drought and low river flow the saltwater wedge pushes farther and farther upstream, bringing the marine environment up the estuary with it. Conversely, in periods of heavy stream flow the saltwater wedge is pushed seaward. This has importance to the location of water supplies along the upper reaches of an estuary. It is possible that a freshwater intake may, in periods of low stream flow, find itself pumping brackish or salt water instead of fresh.

22.7 *The atmosphere*

The envelope of air surrounding our earth sustains life and plays a critical role in previously discussed geologic processes of weathering, oceanic circulation, stream flow, glacier formation and dissipation, and movement of solid materials by wind. Here let us examine some of the effects that man is having on the atmosphere.

Pollution

Weather, as we experience it, occurs in the lower 10 to 15 km of the atmosphere, a zone called the *troposphere* from the Greek *tropo,* "turning" or "mixing," and *sphere.* Within this zone there is a general decrease in temperature with increasing elevation until the *stratosphere* is reached, at which point

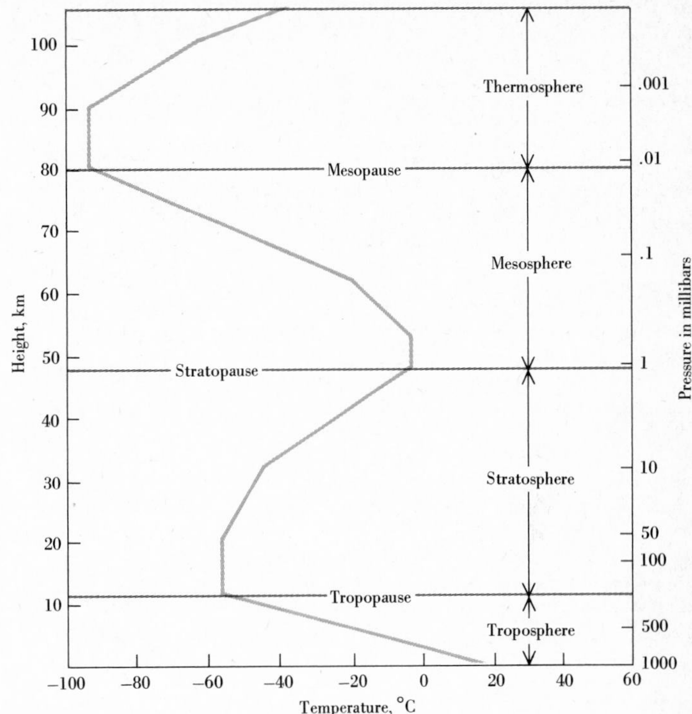

Figure 22.17 ***Thermal layers in the atmosphere.***

temperatures increase. This zone is overlain by the *mesosphere* and this in turn by the *thermosphere*, as shown in Figure 22.17.

Over the years, the use of coal, oil, and gas as energy sources has put into the atmosphere, particularly the troposphere, increasing amounts of the combustion products of these materials (see Table 22.5). Pollution of the atmosphere becomes noticeable, even dangerous, as these products become concentrated in areas of heavy population. Concentration of air pollutants occur in such areas not just because their rate of production is high there, but also because of certain characteristics of the lower troposphere. The meteorologic conditions that increase pollutant concentration are those which create stable air rather than unstable. Atmospheric stability allows an increasing supply of

Table 22.5 ***Air pollutant emissions, millions of short tons per year, 1965***[a]

	Carbon monoxide	*Sulfur dioxides*	*Hydrocarbons*	*Nitrogen oxides*	*Particles*	*Totals*	*% Totals*
Automobiles	66	1	12	6	1	86	60
Industry	2	9	4	2	6	23	17
Electric power plants	1	12	1	3	3	20	14
Space heating	2	3	1	1	1	8	6
Refuse disposal	1	1	1	1	1	5	3
Total	72	26	19	13	12	142	100

[a] From *The Sources of Air Pollution and Their Control*, Public Health Service Publication No. 1548, 1966.

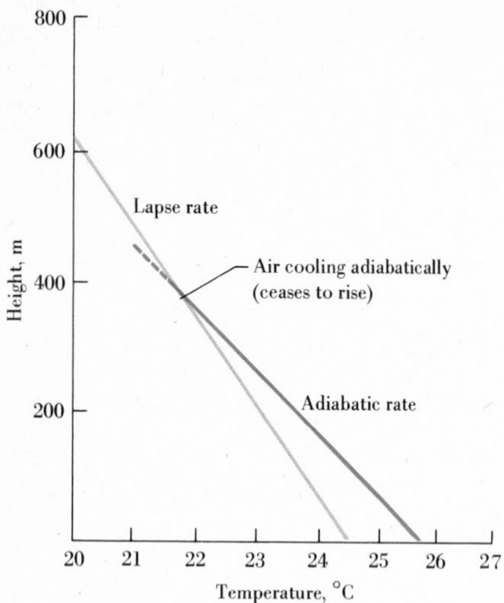

Figure 22.18 ***Assume that a body of air at the earth's surface is warmed to 25.5°C. Being warm, it rises. As it rises it expands and thus cools at its adiabatic rate. The lapse rate is shown here as less than the adiabatic rate. Therefore when the air rises to 400 m its temperature and the temperature of the air through which it has been rising are equal. Stability is attained.***

pollutants to be concentrated near their source of production, where they persist until unstable conditions remove them.

To understand atmospheric stability and how it affects air pollution we will find it useful to understand two characteristics of the atmosphere. One of these is the *adiabatic rate* of temperature change and the other is the *temperature lapse rate.*

When a body of air rises it expands, and as it expands it cools. Conversely, as air descends it compresses and warms. This change of temperature within the body of air takes place without gain or loss of heat to the surrounding air through which it moves. This is called an adiabatic change. For air that is saturated with moisture the rate is about 6°C/km, and for air that is unsaturated with respect to moisture it is 10°C/km. The lapse rate—the actual change in temperature of air with altitude—varies with time and place, but 6.5°C/km is an average value.

Turning to Figure 22.18, we see what might happen in an unstable atmospheric situation. In the example given the air close to the ground is warmed by terrestrial radiation and the lapse rate is high. The air at the ground surface, being warm, is lighter and therefore more buoyant than the air above. As a result it begins to rise and to cool at its adiabatic rate. It will continue to rise as long as its adiabatic rate of cooling does not lower its temperature below that of the surrounding air, whose temperature decreases with increasing elevation according to its lapse rate. Once the rising air cools to the same temperature as that of the surrounding air, movement stops because it has the same density and therefore the same mass and buoyancy as the surrounding air. Further upward motion of the adiabatically cooling air ceases at this level because any further upward motion would produce adiabatic cooling, which would make the air heavier than the surrounding air. Being heavier, it cannot rise.

Figure 22.19 shows another example. Here the lapse rate is shown with a *temperature inversion* at a distance above the ground. This merely means that the air temperature at ground level is lower than that at some elevation above the ground, and that as we go upward air becomes somewhat warmer. At some point this increase changes (or inverts) and the air becomes cooler with increased elevation. In this situation air at the ground surface cannot rise because its adiabatic rate is decreasing at a greater rate than is the lapse rate of the surrounding air. Therefore air at the ground surface is heavier, less buoyant than air above, and cannot move upward. If we had some mechanism of forcing the surface air upward it would become still cooler and heavier and would settle downward toward the ground once we removed our hypothetical mechanism.

There are a number of ways in which inversions are created in the atmosphere, but here we need only to note that whenever inversions occur the air becomes stable and can be trapped close to the earth's surface. This means that any pollutants put into the stable air will remain and can only be moved

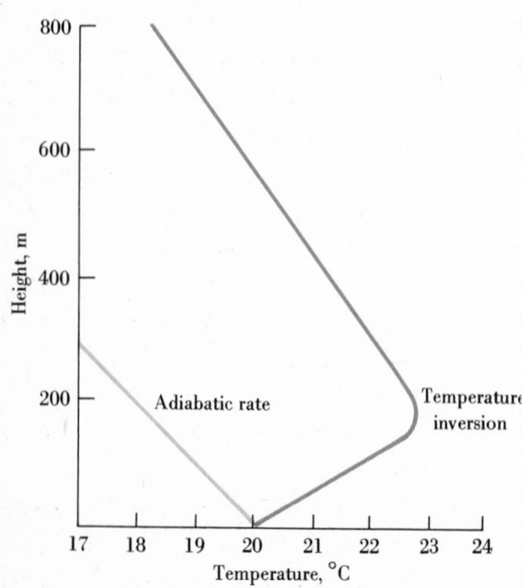

Figure 22.19 ***The temperature inversion diagrammed here causes stable air conditions. See the text for discussion.***

Table 22.6 ***Number of particles per cubic centimeter of air for various environments***[a]

Environment	*Particles/cm*^{3}
Urban	15,700
Suburban or town	3,700
Open countryside	1,000
Mountainous country	
Less than 1,000 m	600
1,000 to 2,000 m	200
Over 2,000 m	100
Ocean	100

[a]Modified from Department of Health, Education, and Welfare.

with the onset of unstable air. The result is the familiar zone of smoke, haze, and smog that blankets heavily industrialized and populated areas. At best, the resulting contamination is a nuisance; at its worst, it is a threat to health and to life itself.

SOLID PARTICLES In addition to gaseous pollutants there is a large increase of solid particles that is associated with population centers. The airborne dirt of urban areas is well-known. The magnitude of this concentration of particulate matter in urban air as compared with nonurban areas is suggested in Table 22.6.

CARBON DIOXIDE Carbon dioxide is a minor constituent of the atmosphere and amounts to approximately 300 ppm (parts per million) of the total composition of air. The combustion of oil, gas, and coal for industry, space heating, and internal combustion engines puts great quantities of CO_2 into the atmosphere each year. From 1850 to 1970 the percent of increase of CO_2 in the atmosphere from combustion sources has been variously estimated but it appears to approximate 10 percent. Inasmuch as we are burning fossil fuels at an ever-increasing rate it follows that the rate of increase should continue to accelerate in the years to come.

The question is what effect this increase may have on our climate. In general, because CO_2 absorbs heat radiated from the earth and the atmosphere, the prediction is that increasing CO_2 will increase worldwide temperatures. One estimate is that a doubling of atmospheric CO_2 will increase earth temperature by 2.36°C. If we assume that a 25 percent increase in CO_2 will occur in the next few decades, then an increase in temperature of about 0.7°C could be forecast. Other estimates suggest that such an increase in CO_2 would mean an increase of 1.5°C in global temperature. Regardless of the size, a change in temperature can be expected to trigger other changes, such as in cloudiness or moisture content of the air, although the magnitude and even the direction of such changes are still difficult, if not impossible, to predict.

URBANIZATION AND CLIMATE Not only has urbanization and its attendant activities changed the quality of the atmosphere but it has also changed the local climatic patterns. We are all familiar with the weather forecasts telling us to expect slightly cooler temperatures in the suburbs than in the city.

This difference is real and is caused by the agglomeration of streets, buildings, and parking areas, and extensive elimination of vegetative cover that characterize our cities. In addition, heat supplied by industrial and space heating in urban districts in temperature areas may, during winter months, equal or exceed the heat received from natural radiation. Therefore, temperatures in cities are higher than in adjacent open country.

Precipitation can be affected by cities as well. William C. Ackermann, chief of the Illinois State Water Survey, estimates that in and around major urban areas precipitation may increase from 5 to 30 percent over neighboring rural areas. Thus an examination of precipitation records from mid-Chicago and mid-St. Louis stations show that rainfall is up to 8 percent heavier than in nearby rural areas. Another study has demonstrated that heat and air pollution in the Chicago–Gary area has affected the weather 48 km downwind in LaPorte, Indiana. LaPorte reported 31 percent more precipitation, 38 percent more thunderstorms, and 24 percent more days of hail than recorded at surrounding stations.

It is reasonable to surmise that future studies will show that major urban areas play significant roles in the modification of our climate.

22.8 *Earthquakes*

In Chapter 16 we examined the general nature of earthquakes. Here we ask to what extent we can predict earthquakes and whether man may be able to control or at least influence earthquakes.

In Chapter 16, we found that earthquakes occur in well-defined, narrow belts that are the loci of present-day crustal instability. It is clear that, knowing the distribution of earthquake activity in the present and immediate past, we can predict with confidence that these same zones will most likely be the areas affected by future tremors. In other words we can expect more earthquake activity in such areas as California, southern Alaska, or Japan than in northern Wisconsin, Florida, or France. More difficult, but very important, are specific questions such as: "When will a quake occur?" "What magnitude will it have?" "How much damage will it do?" "Can we do anything to control a future quake?" We are only at the very beginning of the search for answers to these questions, as the following observations will indicate.

We also saw in Chapter 16 that during any one period of time the number of small earthquakes is very large when compared with large-magnitude quakes. Thus, worldwide, we observe that each year there are only two quakes of 7.7 magnitude or greater, whereas there are an estimated 100,000 quakes of magnitude 2.5 to 3. Estimates of the frequency of quakes of varying magnitude can be compiled for specific areas, as is shown for California in Table 22.7. This compilation is based on observations of the occurrence of past quakes; the record is not long enough or complete enough to be precise. Certainly for the great quakes (magnitude 8.0 and above) several centuries of future data are needed before their frequency of occurrence can be established with any statistical confidence. In addition, a frequency prediction tells us, of course, only that an event has a certain chance of occurring during a given interval of time. It does not tell us exactly when it will occur within that time interval, but we do get some idea of the chances that it will occur.

Table 22.7 gives some other predictions for California. It suggests that, given the present distribution of population in the state, only a certain number of quakes out of those that occur will produce significant damage.

Potential damage from earthquakes can be looked at in another way. Table 22.8 lists four modern U.S. earthquakes in order of increasing magnitude. The approximate loss in terms of dollars is listed for each of these quakes. What would have been the dollar losses had the quakes all occurred in fully urbanized areas? The estimates are shown in the final column. The figures

Table 22.7 **Frequency of earthquakes in California**[a]

Range	*Quakes/1,000 yr*[b]	*Radius of major damage, km*[c]	*Damaging quakes/100 yr*[d]
5.0–5.9	1,000	3 (local)	30
6.0–6.9	100	15	15
7.0–7.4	12	50	10
7.5–7.9	6	80	6
8.0–8.5	3	250	3

[a] Data, all of which are approximate, courtesy of Robert Phinney.
[b] The number of quakes is rounded; the probable variability is 20 to 30 percent. The estimate of great quakes (magnitude 8.0 and over) is very tentative; several centuries of data would be needed before a significant figure could be given.
[c] Area in which building damage becomes general. Corresponds to shaking intensities of VII to VIII and greater.
[d] The expected number of earthquakes per century that might cause appreciable damage in an urban area, based on distribution of population in 1970.

are approximations at best, but they clearly indicate that, had the Alaskan earthquake of 1964 occurred in an urban area, the monetary loss would have been astronomical, the entire national economy would have been affected, and damage would have been well beyond the ability of the insurance industry to underwrite. These figures, of course, cannot depict the human suffering and general damage to the social fabric that must necessarily accompany an event of this type and size.

We have found that man can induce small earthquakes in at least three ways—namely, by the injection of fluids into deep wells, by underground nuclear explosions, and by filling of reservoirs. At the Rocky Mountain Arsenal in Colorado, the U.S. Army Corps of Engineers in 1962 began to dispose of dangerous waste fluids by injecting them into a well some 4,000 m deep. Shortly after the start of the program, small earthquakes began to occur in the vicinity of the arsenal. In 1966, after nearly 700 million liters of waste fluids had been pumped into the well, the injection program was stopped, but the earthquakes continued. In 1968, a test program of pumping out the waste fluid was begun in an attempt to see whether their removal would affect

Table 22.8 **Effect of magnitude on losses resulting from four modern U.S. earthquakes**[a]

			Losses, \$ × 10^6	
Magnitude	*Earthquake*	*Location*	*Actual*	*Possible in fully urban area*[b]
5.3	Daly City, 1957	Urban	1	1
7.1	Imperial Valley, 1940	Agricultural, with towns	5	75
7.7	Kern County, 1952	Rural, with 5% urban	50	1,000
8.3	Alaska, 1964	Wilderness, with scattered centers, incl. Anchorage	311	25,000

[a] Courtesy of Robert Phinney.
[b] A rough extrapolation of losses for an equivalent earthquake whose damaged region is fully urbanized.

the occurrence of quakes. Four pumping tests were run and, although there was no appreciable change in seismic activity during pumping, there was an increase in activity after cessation of two of the pumping tests.

Detailed studies of seismic records of areas in which underground nuclear explosions have taken place show that there is a significant rise in earthquake activity immediately after the explosions. The amount of increase is directly related to the size of the blast.

The weight of water filling reservoirs may cause earthquakes. An example comes from the Zambezi River in Zambia. Here a hydroelectric dam, built in 1958, impounded the waters of the Zambezi to form Lake Kariba in the Kariba Gorge. A seismograph net has since recorded thousands of quakes, the largest with a magnitude of 5.8.

The ability of man to effect earthquake activity suggests the possibility that he might also be able to use artificial means to relieve earth stresses in small increments before they build to dangerous levels. Whether a program of planned nuclear explosions along fault zones with the purpose of releasing stress in small amounts is feasible is not known. The dangers, however, are obvious.

22.9 *Volcanoes*

In Chapter 4 we found that modern volcanoes have a certain distribution in the world, and we later noted that this distribution also coincides with earthquake activity. We can predict, then, that volcanoes and their activity will probably be confined to the belts where historic and present activity, both volcanic and seismic, have been concentrated.

We also found in Chapter 4 that individual volcanoes have some pattern of eruptive activity. But our data on volcanic activity are not so complete as they are for seismic events, in part because the seismic events are much more numerous. Nevertheless, for some volcanoes our records are good enough and the monitoring system is sensitive enough to predict eruptions. Kilauea in Hawaii is an example. Before an eruption occurs, this volcano goes through a period of swelling related to the rise of the magma toward the surface. Measurement of this swelling by a series of delicate tiltmeters warns of impending activity.

Outline

Man affects his physical environment in many ways.

Streams develop a high, short-period flood peak through *urbanization* and a low, long-extended flood peak through *flood control projects.*

Water quality is affected by solids and dissolved material, and by thermal changes.

Lakes are geologically shortlived. Man can hasten the death of a lake and push it into a *eutrophic* stage by addition of excess nutrients from industrial, rural, and municipal wastes.

Ground water may be contaminated by *saltwater invasion* and by *chemical, biological,* and *thermal* changes. Excess pumping can lead to *collapse* and *subsidence* of the ground surface.

Soils may be destroyed by agriculture, by surface mining, and by construction activity.

Solid wastes, produced as a result of man's activity, in the United States exceed the total amount of material carried annually to the oceans by the nation's rivers.

Oceans, large as they are, can be affected by man through, for example, *dumping of solids, oil spills,* and *blowouts.*

Estuaries are particularly easily influenced by man's presence.

Atmosphere can be polluted by man's activity, particularly when *temperature inversions* are present to trap pollutants close to the ground.

Solid particles are two orders of magnitude more abundant in urban air than in sea or mountain air.

The *carbon dioxide* content continues to increase due to burning of fossil fuels, but the long-range effects are unknown. Locally, man affects climate by atmospheric modification.

Earthquakes can be induced by man, but cannot yet be controlled. A major quake in an urbanized area would be a national disaster.

Supplementary readings

ADABASHEV, I: *Global Engineering.* Progress Publishers, Moscow, 1966. If you are concerned by the plans of the capitalist establishment for changing the physical environment you will find little solace in comrade Adabashev's suggestions.

AMERICAN CHEMICAL SOCIETY: *Cleaning our Environment: the Chemical Basis for Action,* American Chemical Society, Washington, D.C., 1969. A well-documented study covering air, water, solid wastes, and pesticides, compiled from the work of a committee of experts.

CLARK, JOHN R.: "Thermal Pollution and Aquatic Life," *Sci. Am.,* vol. 220, pp. 19–27, 1969. An easy-to-read report on the effect of thermal pollution on fish and other aquatic organisms.

COMMONER, BARRY: *Science and Survival,* The Viking Press, New York, 1967. A distinguished biologist looks at some of the vast forces unleashed by man and at what some of their long-range effects on the environment may be.

EISELEY, LOREN: "Man: The Lethal Factor," *Am. Scientist,* vol. 51, pp. 71–83, 1963. An anthropologist examines man's ability to affect the environment and other species.

FLAWN, PETER: *Environmental Geology.* Harper and Row, New York, 1970. A textbook dealing with the application of geologic knowledge to conservation, land-use planning and resource management.

LAUFF, GEORGE H. (ed.): *Estuaries,* American Association for the Advancement of Science, Washington, D.C., 1967. An extensive symposium on various aspects of estuaries, including some examples of how man modifies them.

LEOPOLD, LUNA B.: "Hydrology for Urban Planning—A Guidebook on the Hydrologic Effects of Urban Land Use," *U.S. Geol. Surv.*, Circ. 554, 1968. An investigation into what happens to streams as a result of increasing urbanization.

OSBORN, FAIRFIELD: *Our Plundered Planet*, Little, Brown & Company, Boston, 1948. A classic book on man's awakening (or reawakening) awareness of his place in the natural scheme of things.

PAKISER, L. C., and others: "Earthquake Prediction and Control," *Science*, vol. 166, pp. 1467–1474, 1969. A good review of the subject, which also provides references to the more important literature.

PRESIDENT'S SCIENCE ADVISORY COMMITTEE: *Restoring the Quality of Our Environment*, Government Printing Office, Washington, D.C., 1965. A good factual summary of the several facets of the problem.

STEINBRUGGE, KARL V.: *Earthquake Hazard in the San Francisco Bay Area: A Continuing Problem in Public Policy*, Institute of Governmental Studies, University of California, Berkeley, 1968. An authoritative and understandable treatment of California's sword of Damocles.

U.S. DEPARTMENT OF THE INTERIOR: *Lake Erie Report*, Federal Water Pollution Control Administration, Great Lakes Region, 1968. A well-documented report of the question "Who killed Lake Erie?"

U.S. DEPARTMENT OF THE INTERIOR: *Surface Mining and Our Environment*, Government Printing Office, Washington, D.C., 1967. Lavishly illustrated and easy to read, yet containing a wealth of information.

23

The moon, earth craters, tektites

IN THIS CHAPTER, WE EXAMINE THE EARTH'S NEAREST NEIGHBOR, THE MOON, whose motions have been watched and studied for unknown centuries. Some evidence suggests that paleolithic man kept track of the waxing and waning of the moon some 15,000 years ago. Certainly the observations of the moon in ancient times were made with greater exactness than were the observations in other physical sciences. Thus by the second century B.C., the Greek astronomer–mathematician Hipparchus had discovered the eccentricity of the moon's orbit and determined its inclination to the ecliptic. Through the years the motions of the moon have been determined with greater and greater precision, and refinements continue to the present. In this chapter, however, we are concerned not so much with lunar motion but rather with the nature of the moon. What is it made of? What processes go on at its surface? What processes go on beneath it? What is its history?

The first man to see the moon with other than the naked eye was the Italian Galileo Galilei (1564–1642). A Dutch lens grinder had just constructed the first telescope and Galileo, imitating it, built his own instrument. Early in 1609, Galileo, then a mathematics professor at the University of Padua, turned his telescope on, among other celestial objects, the moon. What he saw is published in a treatise *Sidereus Nuncius* (*The Starry Messenger*), one of the great landmarks of scientific discovery (Figure 23.1). For the first time man found that, as Galileo wrote, the "moon is not robed in a smooth and polished surface, but is rough and uneven" (Figure 23.2). In the last $3\frac{1}{2}$ centuries, better telescopes, the camera, space craft and their instruments, and direct observation by men have given us a very good idea of what the moon is like and how it got that way (Figure 23.3). Certainly we do not know everything about the moon, but then we do not know everything about the earth, either, and we have been on it a lot longer than we have been on the moon.

23.1 *Measure of the moon*

The moon revolves around the earth once every 29.53 days and rotates on its axis at the same average rate, a coincidence that means the moon always presents the same side to our view from earth. The mean distance from the center point of the moon to that of the earth is about 384,400 km, but the variation between the maximum and minimum distance approximates 44,800 km. The moon's equatorial diameter is just over 3,450 km, which is less than 0.28 that of the earth. Its mass is 0.012 of the earth. The overall specific gravity of the moon is 3.34 and the pull of gravity is 0.165 that of gravity on earth.

Atmosphere is lacking and temperatures range from over 100°C to perhaps as high as 130°C at lunar noon to less than −150°C at lunar midnight. The 2-week lunar day is long enough so that the changes in temperature are fairly slow, although during a lunar eclipse they are extremely rapid and changes of more than 140°C in less than 1 hr have been determined. Even though the surface undergoes extreme temperature variations, it has been

Figure 23.1 ***In 1610, Galileo published* The Starry Messenger, *which contained the results of the first telescopic examination of the heavens. The title page is shown here.*** *(Photograph by Willard Starks.)*

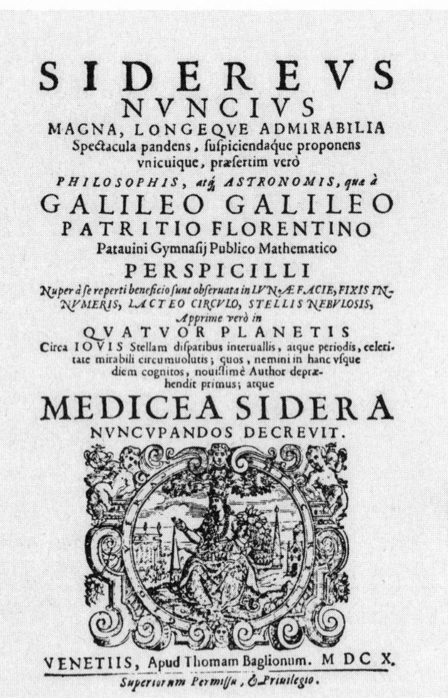

SIDEREVS
NVNCIVS
MAGNA, LONGEQVE ADMIRABILIA
Spectacula pandens, suspiciendaque proponens
vnicuique, præsertim verò
PHILOSOPHIS, atq; ASTRONOMIS, quæ à
GALILEO GALILEO
PATRITIO FLORENTINO
Patauini Gymnasij Publico Mathematico
PERSPICILLI
Nuper à se reperti beneficio sunt obseruata in LVNÆ FACIE, FIXIS INNVMERIS, LACTEO CIRCVLO, STELLIS NEBVLOSIS,
Apprime verò in
QVATVOR PLANETIS
Circa IOVIS Stellam disparibus interuallis, atque periodis, celeritate mirabili circumuolutis; quos, nemini in hanc vsque diem cognitos, nouissimè Author depræhendit primus; atque
MEDICEA SIDERA
NVNCVPANDOS DECREVIT.

VENETIIS, Apud Thomam Baglionum. M DC X.
Superiorum Permissu, & Priuilegio.

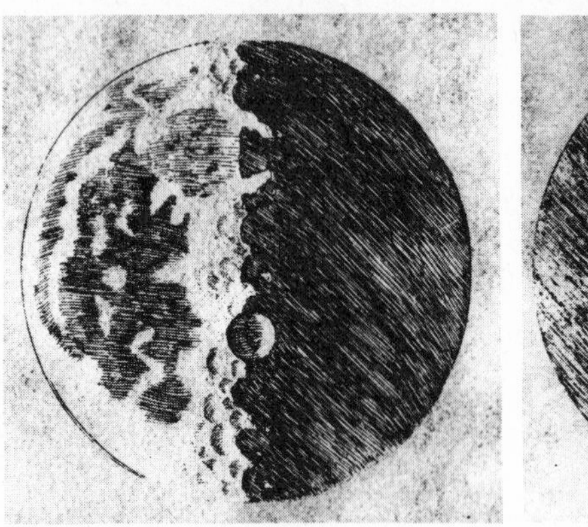

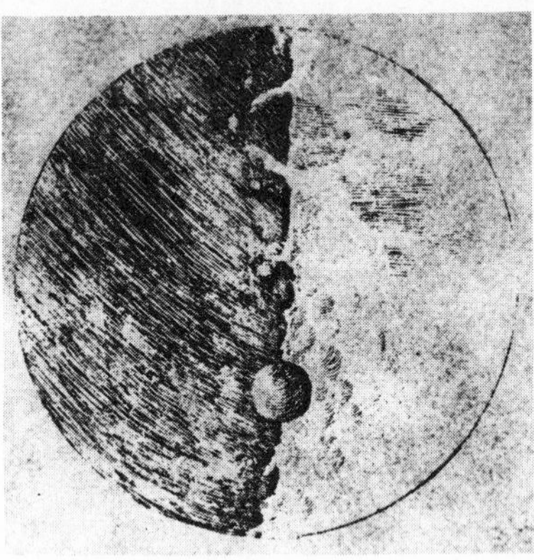

Figure 23.2 ***These reproductions from* The Starry Messenger *record part of what Galileo saw of the moon through his telescope. North is to the top, the last quarter is to the left, and the first quarter is to the right. Compare with Figure 23.3.*** *(Photographs by Willard Starks.)*

Figure 23.3 ***The moon as seen through a present-day optical telescope. North is to the top, the last quarter is to the left, and the first quarter is to the right.*** *(Photograph by Lick Observatory.)*

calculated that at a few meters below the surface the temperature is fairly constant and hovers around −25°C. At about a meter in depth the daily variations at the surface are damped out.

23.2 Composition of the moon

The mean density of the moon is 3.3, which is close to the density of the earth's upper mantle. Whether this density means a similar composition for the moon is a question still to be answered. However, because the mean density of the earth is about 5.5, it is clear that its mean composition must be different than that of the moon.

Before either manned or unmanned space probes began providing information about the moon, reflectivity studies of the moon's surface suggested that the surface rocks there could well be of basaltic composition. This deduction received support from instrumental data returned by the soft landings of 1967 and 1968. The rocks brought back from the moon by the *Apollo 11* and *12* missions provided our first close look at actual samples of the moon. The earlier speculations about the composition were in part confirmed for, indeed, the great bulk of the samples had a basaltic composition.

The samples from *Apollo 11* and *12* proved to be (1) crystalline rocks including both a fine-grained, vesicular basalt and a fine- to medium-grained gabbro; (2) breccias; and (3) lunar soil.

Crystalline rocks

The basalts and gabbros of Tranquillity Base contain the familiar calcium-rich plagioclase, pyroxene, and olivine of their earthly cousins. In contrast to the typical rock of basaltic composition on earth, however, the moon rocks contain a large amount of the mineral ilmenite, an oxide of titanium and iron. The overall chemistry of these lunar rocks is somewhat different also. Thus they contain very much more titanium and are lower in silica, alumina, and sodium than basalts and gabbros on earth. The densities of the various lunar samples from the *Apollo 11* mission vary somewhat around 3.3.

The site of *Apollo 12*'s landing provided more examples of igneous rocks. In chemical composition they have proved generally similar to those from the *Apollo 11* site, although the titanium content of the *Apollo 12* rocks is somewhat lower.

These crystalline rocks all formed from a silicate melt—hence we call them igneous. Some of these are aphanitic and vesicular, suggesting that they were derived from a lava flow. Others are medium-grained, with single crystals

Figure 23.4 ***A lunar gabbro as seen in a thin section magnified under the microscope. Minerals present are pyroxene, calcium plagioclase, and ilmenite. Section shown is about 875 μm wide.*** *(Photograph by U.S. Geological Survey.)*

reaching a maximum size of 3 mm, suggesting slower cooling beneath the lunar surface (see Figure 23.4).

Breccias

Numerous examples of breccia have been found on the moon's surface. Those collected on *Apollo 11* and *12* were made up predominately of small fragments of gabbro and basalt bound in a matrix of lunar soil (Figure 23.5). It is theorized that the materials were consolidated into a coherent rock by compaction arising from meteoritic impact.

Figure 23.5 **A breccia from the moon as seen in a thin section magnified under the microscope. The small fragments of minerals and crystalline rock are set in a fine-grained matrix. Section shown is about 875 μm wide.** (*Photograph by U.S. Geological Survey.*)

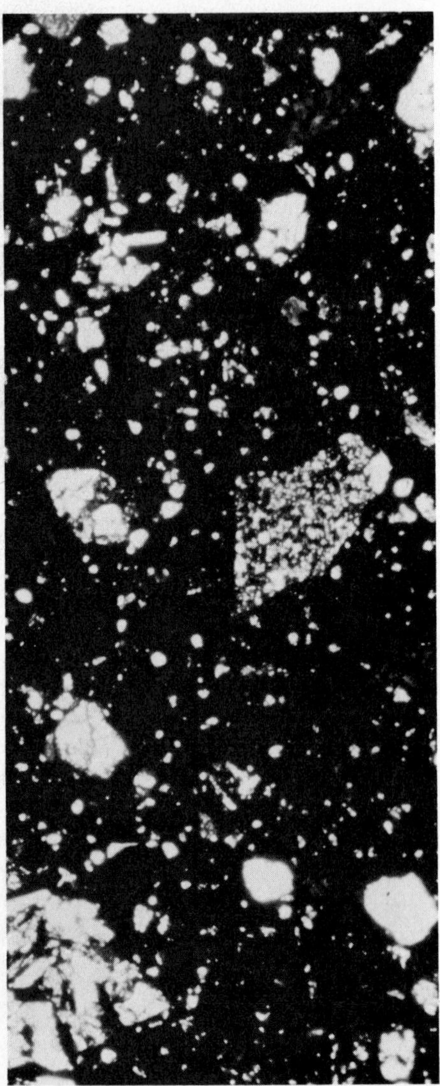

Lunar Soil

The soil from the *Apollo 11* site is made up of a mixture of crystalline fragments, small glassy objects, and a very small amount of meteoritic material.

The glass objects found are generally smaller than 0.2 mm. Some are colorless and others range in color from brown to red to green to yellow. Their shapes are variable, some being angular and others spheroidal, ellipsoidal, dumbbell-shaped, and teardrop-shaped. Their chemical composition is similar to the basaltic and gabbroic rocks of the moon.

The crystalline fragments found in the soil are largely basaltic in composition. But in addition there are fragments of a rock called *anorthosite,* and on the sample from the *Apollo 11* landing site these made up between 3 and 4 percent of the lunar soil. Anorthosite is made up of 90 percent or more of the plagioclase mineral anorthite, the other 10 percent being pyroxene and some olivine. As the percentage of anorthite falls, the rock becomes more and more gabbroic as the pyroxene and olivine increase proportionately. The rock then becomes known as a gabbroic anorthosite or anorthositic gabbro. The densities are around 2.8 to 2.9. All indications are that the maria are underlain by basaltic rocks. Therefore these fragments of anorthositic rock are foreign to the maria and it is reasoned that the fragments were blown to their present location by the explosions caused by meteoritic impact or volcanic activity.

As is to be expected, there are variations in the soil material from one place to another. Thus, although the material found at the *Apollo 11* site is very similar to that found at the *Apollo 12* site, the latter is somewhat lighter in tone, contains a larger percentage of olivine, and includes two or three samples that can be compared with volcanic ash.

Variation in lunar composition

These anorthositic fragments lead us to the question of what amount of differentiation we will find in the composition of the moon, either at depth or across its surface. The question cannot yet be resolved, but there are some clues. For instance, data collected by orbiting lunar satellites show that the large circular maria are areas of positive gravity anomalies. This points to the concentration of heavy masses below the surface. These are called *mascons* (from *mass* + *concentration*), and it is calculated that they comprise about 10^{-5} of the mass of the entire moon. It is variously suggested that these are due to upward movement of heavy material, to the remains of large meteorites, or to both. Another line of evidence concerns the highlands of the moon. Data gathered by the *Surveyor* soft-landing program suggested that the maria material was higher in iron and titanium than the material of the lunar highlands. This suggests a lower density for highland material than for maria material.

In summary, then, we can say that there is some evidence for compositional variation in the moon. Taking the evidence of the mascons, *Surveyor* data, lunar soil, and lunar topography, it may be argued that the lunar highlands are made of lighter-weight material (anorthosite?) and that the maria are underlain by heavier material (basalt). We could speculate further that the lunar highlands stand high because they are light in weight and are isostatic with the denser material of the maria. If so, then the earth's continents and ocean basins may offer an analogy to the moon's gross topography.

23.3 *Processes on the moon*

Much of our attention in this book has been directed to the processes that change the earth. We found that internal forces of volcanism and tectonism are countered by the surface process of water, wind, ice, and weathering, coupled with gravity. When we turn to the moon and examine the processes at work, we step into a scene at once familiar and unfamiliar when compared with that of our earthly experience.

The most fundamental difference is the absence of a lunar atmosphere. There is no air, nor are there swirling clouds which distribute moisture. There is, in fact, no surface moisture, although, as we shall see, there may be water or ice somewhere beneath the surface. Without an atmosphere, there is no wind. We trace this lack of atmosphere and hydrosphere to the moon's small mass and hence its low gravitational attraction. This means that the escape velocity, the velocity which a particle must achieve to escape the lunar gravitational field, is a low 2.38 km/sec. This is only a little over one-fifth of that on earth, which is 11.2 km/sec, and lower than the escape velocity of gases and most liquids. Therefore, whatever gases and liquids may escape from the lunar interior to the surface leak off into space, as helium does from the earth. Thus, although the earth can retain nitrogen, carbon dioxide, and water at its surface and other planets retain ammonia, methane, helium, and probably even hydrogen, the moon is much too small to keep any of these elements and compounds. Without water and an atmosphere, life as we know it cannot exist, much less evolve. It is not strange that the lunar surface should look different from that of the earth. And yet weathering does go on, movement does take place, and the moon's face does change, although differently and at a slower pace than on the earth.

Figure 23.6 ***This cobble on the moon's surface is about 12 cm long. The low ridge across its center suggests differential weathering, allowing the less resistant material of most of the rock to weather more rapidly than a thin band of resistant material.*** (*Photograph by National Aeronautics and Space Administration,* Surveyor I.)

Weathering

In Chapter 6 we defined terrestrial weathering as a change in material at or near the earth's surface, in response to atmosphere, water, and living matter. Obviously this definition is inapplicable in the lunar environment. The lunar surface material does change, however, and we can ascribe the changes to some sort of weathering processes.

Here are some examples of changes which we can attribute to lunar weathering. Figure 23.6 shows a cobble-sized fragment of moon rock. Note the ridge that cuts across the crest of the rock. It is reasonable to assume that this is a zone of somewhat more resistant material that has been left standing above the main part of the cobble, which has weathered more rapidly.

One of the first observations man made on the moon was that many of the rocks that litter the surface are rounded on their upper, exposed portions but angular where in contact with the surface or partially buried by it. The conclusion that some sort of weathering process is affecting the exposed parts

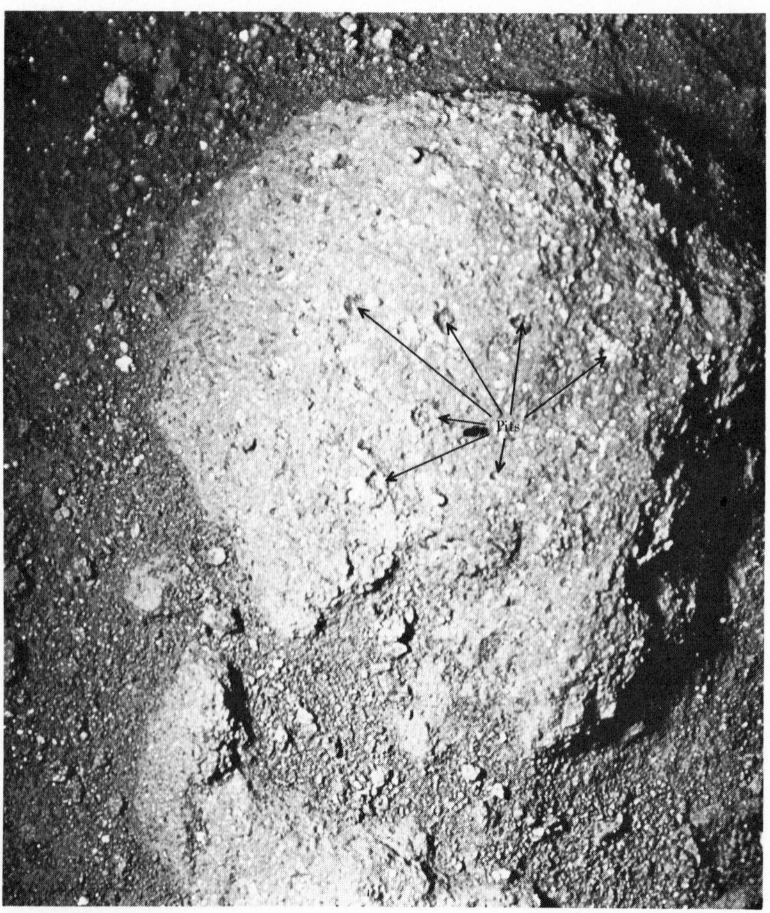

Figure 23.7 ***This stone at Tranquillity Base has a rounded upper surface and is set in finer material, part of which may have come from the breakdown of the stone itself. Arrows indicate location of some of the larger pits on the stone. Many of the bright spots represent reflections of pit-related glass. Reflections from the ground surface come from small fragments of glass in the granular lunar soil. Diameter of the rock is about 7 cm.*** (*Photograph by National Aeronautics and Space Administration,* Apollo 11.)

of the rocks is inescapable. Now that we have observed moon rocks at first hand we can see that they are marked by small pits ranging down to 20 μm or less in diameter. Those less than 4 mm in diameter are glass-lined and in some instances rimmed with a glassy material. These features suggest a high-energy impact of particles with the lunar material. In those instances in which a rock fragment displays both rounded and angular sides, the pits occur more frequently on the rounded surfaces. The surface of the rocks tends to be somewhat lighter in tone than the interior, a characteristic apparently related to a thin zone of microfractures or glass attributed to impact of particles on the rock (see Figure 23.7).

Many of the younger craters of the moon are called ray craters because of the associated rays of material that spread outward as light-colored streaks created at the time of crater formation. The youngness of these craters (Tycho, Kepler, and Copernicus are good examples) can be demonstrated because they and their associated features are superimposed on most other lunar features. The older craters, however, lack these rays and the consensus is that they were originally present but have been obscured or destroyed by some type of weathering. The variation in sharpness of adjacent craters can be cited as a third example of weathering. Thus Figure 23.8 is a photograph of craters just south of the equator on the lunar far side. The small- and intermediate-sized

Figure 23.8 ***Many of the small and intermediate-sized craters in this vertical photograph on the far side of the moon have sharp outlines. They are superposed on other, older craters of similar size and are identifiable by subdued outline. The longer weathering of the older craters is thought to account for the difference in crater freshness. Area shown is about 28 km across.*** (*Photograph by National Aeronautics and Space Administration,* Apollo 8.)

Figure 23.9 ***The foot of Neil Armstrong, first man on the moon, makes a clear imprint in the "soil" at Tranquillity Base. The soil has a granular texture, is slightly cohesive, and is made of discrete particles, many of which may represent the mechanical breakdown of larger rock masses by some processes of weathering. Depth of penetration of boot indicates that the soil here has a bearing strength between about 3 to 7 $\times$ 10^{-4} kg/cm^2.*** (*Photograph by National Aeronautics and Space Administration,* Apollo 11.)

craters show sharp, well-defined outlines in the low sun. They are superposed on a terrain pocked by craters of similar size but with much softer, more subdued outlines. Because of this superposition, we reason that the sharply outlined craters are younger than the softly outlined ones. We can then argue that the softened shapes of the older craters result from a longer period of lunar weathering than for the younger craters. A final example is provided by the material encountered on much of the moon's surface thus far. It is a finely divided granular material, slightly cohesive but still of discrete particles (Figure 23.9). These individual particles could very well represent, at least in part, the mechanical breakdown of larger, more coherent pieces.

Now let us look at some of the ways in which weathering might take place on the moon.

TEMPERATURE CHANGES The very great changes in temperature to which lunar materials are subjected may cause them to expand and contract enough so that they weaken and break. Probably the sudden changes that take place during lunar eclipses are the most effective, although the slower temperature changes of the lunar day may, given time, be effective as well. Such action might either break pebble-sized fragments or spall off small bits from the surface of larger fragments.

IMPACT BY METEORITES We have examples on earth of what happens when a meteorite of large size crashes into rock. First, the explosive impact creates a crater that is considerably larger than the meteorite. The rock is in part pulverized, and badly shattered and broken. The rocks at the lip of the crater are bent upward and turned outward and over on themselves, and the material blasted from the crater area can be thrown for long distances. The meteorite, then, can break material into smaller and smaller pieces, truly a process of mechanical weathering. On the moon, also, we can predict that large meteorites have broken down coherent rock into smaller pieces and spread the products of impact across the lunar surface. The moon, in addition, is subject to a type of meteoritic bombardment that the earth is spared. Small meteorites do not survive the trip through the earth's atmosphere, being either melted or vaporized because of the heat generated by friction with the air. The moon, however, is not protected by an atmospheric envelope and undergoes a fairly continuous rain of small meteorites. The small pits already described on surfaces of lunar rocks are in part due to the impact of small meteorites. The process tends to break down rock material into smaller and smaller particles and to churn the uppermost portion of the rubble mantle that appears to shroud most of the moon's surface. Eugene M. Shoemaker of the California Institute of Technology estimates that these small meteorites—micrometeorites as he terms them—produce a complete turnover of the surface material to a depth of 1 mm every 100 years.

FROST ACTION We see no water, ice, or clouds on the moon. How then do we invoke frost action as a weathering process? Admittedly we do not yet know whether the process does take place, but here is why some entertain the possibility.

Recall that the temperature a few meters below the surface is an estimated constant $-25°C$. Furthermore, as we shall see later, there is every indication that igneous activity occurs or has occurred on the moon. Igneous activity must produce volatile gases and fluids. Water released by crystalization of igneous material will migrate toward the lunar surface and eventually reach a level of freezing temperature. Permanently frozen ground must result. The upper limit of this zone is the level at which temperatures bring ice to the melting point. This level must fluctuate to some extent and it is in this zone of melting and freezing that frost action can be expected. Another spot where freezing and thawing could occur is adjacent to zones of volcanic activity, where volcanic heat creates a zone of melting of the ground ice, a zone which must vary with the varying supply of volcanic heat.

COSMIC RADIATION The lack of an atmosphere exposes the lunar surface to direct attack by cosmic radiation, a process from which the earth is largely shielded. This radiation, made up chiefly of protons and secondarily of alpha particles (the nuclei of hydrogen and helium atoms, respectively), is known to cause damage to the crystal lattice of minerals and to have additional effects which, over long periods of time, could lead to the physical breakdown of minerals. Cosmic radiation may be in part responsible for the obliteration of the rays associated with the younger craters.

TIDES We usually associate tides with our oceans. In addition, however, we know that the solid earth also reacts to the tidal pull of the moon and sun. We can measure these earth tides and find that at a maximum they amount to about 10 cm or less. The pull of the earth and sun on the moon is up to 80 times as strong as that exerted on the earth by sun and moon. The tides in lunar rock, therefore, must be correspondingly greater than in the earth rocks. This constant kneading of the moon must contribute to the weakening and breaking down of bodies of lunar rock.

RATES OF EROSION Studies of materials collected at Tranquillity Base have led to the conclusion that the rate of erosion of rocks on the lunar surface is about 10^{-7} cm/year, which contrasts with the average rate on earth (estimated at about 2.5×10^{-3} cm/year).

Transportation

With no wind and no water, we must resort to other processes to move material on the moon's surface.

MASS MOVEMENT The movement of earth material under the influence of gravity was considered in Chapter 10. On the moon, as well, gravity plays a continuing role in moving material from higher to lower locations. Movements of particularly large masses characterize the interior margins of crater rims, especially large ones (see Figures 23.10 and 23.11). Here, blocks of lunar rock material form great steplike features, each block being a slump similar

Figure 23.10 ***This oblique photograph is of a portion of the crater Copernicus. Beyond the central peaks (over 300 m high) are the slumped blocks along the inner wall of the crater. The crater lip is about 3,000 m above the crater floor and approximately 45 km beyond the central peaks. Compare with the vertical photograph in Figure 23.11.*** (*Photograph by National Aeronautics and Space Administration*, Orbiter II.)

to those described from the earth in Chapter 10 (see Figures 10.3 and 10.4).

Another example of gravity's action on surface material has been discovered in photography by lunar probe, particularly by the *Orbiter* missions. Boulders of varying size have left identifiable tracks hundreds of meters long as they bounced and rolled down lunar slopes.

To what extent slow, unspectacular downslope movements operate on the moon's surface is not yet known. On earth, where abundant moisture is present to facilitate movement, long-term creep of surficial material is important. Creep on lunar material remains to be demonstrated, as do the gravity movements of small individual particles, although it seems reasonable to expect both.

When considering gravity, one must not forget the possible presence of a zone of permanently frozen ground at some distance below the surface. Just above the top of this zone, we postulate a zone of moisture which might be alternately frozen and melted. This zone may well play an important role in the downslope motion of material.

Figure 23.11 ***A vertical view which includes a portion of the slump blocks along the inner wall of Copernicus shown in Figure 23.10. Some of the treads on the stepped slump blocks are very flat, probably the result of lava lakes filling the basins behind the blocks as they rotated downward and outward from the main wall. Area shown is about 50 km across.*** (*Photograph by National Aeronautics and Space Administration,* Orbiter IV.)

IMPACT AND VOLCANISM Impacting meteorites and explosive volcanism play their roles in moving material across the lunar surface. Particles thrown free of the moon's surface move farther than do particles ejected with the same force on the earth's surface. This is true for three reasons. First, there is no friction with an atmosphere to slow a particle moving across the moon's surface. Second, gravity is less than that on the earth. Third, the small radius of the moon means a greater angle of surface curvature than on the earth and thus a greater range for ejected particles. The rays of the crater Tycho, for instance, probably represent ejected material, and they extend for distances of 1,500 km from the crater.

IGNEOUS ACTIVITY Early in this chapter we noted that igneous rocks do in fact exist on the moon. Actually it is a good guess that a very large percentage of the near-surface rocks of the moon are igneous. The real question is not whether rocks on the moon crystallize from a silicate melt, but what the origin of the melt might be. We have two obvious choices of origin. First, the melt may be generated by internal heat of the moon in much the same way that magmas are generated on earth. Second, the melt may result from the impact of meteorites on the moon. On a very small scale the glassy film associated with the pits on moon rocks reflects this impact energy. Increase the size of the impacting object to that which can create a depression the size of one of the moon's maria and one may be able to create a magma by impact.

Let us consider some of the evidence for igneous activity on the moon similar to that which we know on earth. Later, in Section 23.4, we will turn to the evidence for large impact structures and, by extension, for the creation of melts by impact.

As early as 1587 an English observer of the moon reported a bright spot halfway between the points of the new moon. Since then many similar sightings have been recorded and are well enough authenticated to qualify as reality. Do these observations represent volcanic activity on the moon? Authorities are agreed that they do not represent glowing lava flows but that there is a strong possibility that they represent volcanic gases escaping to the surface. Substantiation that these observations reflect internal heat of the moon escaping to the surface can be found in photographs taken by *Ranger IX* before it crashed into the floor of the crater Alphonsus (see Figure 23.12). Previously, in 1958, the Russian Nikolai Kozyrev had reported a bright reddish cloudlike phenomenon in the same crater. In the photographs relayed back by *Ranger IX*, one can see small craters with darkened haloes at several spots on the floor of Alphonsus. Some are associated with the trenchlike depressions called *rilles*, which furrow the main crater floor. There are numerous examples of strings of craters on the moon, often aligned along rilles (see Figure 23.15). They bear close resemblance to the maar volcanic craters of the earth. These are found in a number of volcanic areas on our globe, one of the most important being the Eifel district of Germany, where they were first described and named. Usually less than 1 km in diameter, they lack a volcanic cone and display

Figure 23.12 **The crater Alphonsus as photographed from Ranger IX at a height of 425 km. Note the slightly darkened zones and related small craters on the floor of Alphonsus. They represent volcanic activity. The crater is about 100 km across.** *(Photograph by National Aeronautics and Space Administration.)*

at best only a thin layer of ejecta. They are thought to be a result of an explosive process deriving its energy from water superheated by rising magma. This water, confined by the overlying rock, eventually passes over to steam and breaks through the overlying seal. If some of the lunar craters are, in fact, counterparts of the maars, their presence gives added reason for believing that the moon has a supply of water stored as ice somewhere beneath the surface.

Some craters on the moon may actually be small volcanic cones, but volcanoes similar to our familiar Mt. Vesuvius or Fujiyama seem to be lacking. The bulk of the craters on the moon are thought by most students, although not by all, to be caused by impact.

Most scientists feel that the large, dark maria, the "seas" of the moon, are filled largely by lava flows. They envisage them as accumulations not unlike the flood basalts described in Chapter 4.

Regardless of whether the lunar rocks crystallized from a melt created by impact or by the internal heat of the moon, it is clear from a variety of laboratory experiments that the temperature of crystallization of the lunar rocks lies between 1210° and 1060°C. Final crystallization of the last remaining interstitial liquids have continued at temperatures down to 850°C. By comparison, lavas erupting on earth are a little above the melting point of material and range between about 1200° and 900°C. Depending on local conditions, they may still flow at temperatures down to less than 800°C.

TECTONISM The absence of long, linear mountain masses on the moon similar to those of the Andean, Himalayan, and Rocky Mountain systems seems to rule out tectonic activity comparable to that on the earth. Despite this, certain features suggest movement of the moon's crust. For instance, the well-known Straight Wall in the western side of the Mare Nubium is approximately 270 m high and shows every indication of having been formed by faulting. The many rilles of the lunar surface also indicate some stress in the lunar crust and are comparable to some of the grabens and rift valleys on earth. Some believe that there may be deep-seated convection currents which move blocks of the lunar crust. It is generally thought, however, that these structural features are due to some cause other than tectonism. Possible sources of stress include impact, volcanic eruptions, drainage away of molten lava beneath the surface, and melting of subsurface ice.

The *Apollo 11* and *12* missions conducted seismic experiments to measure directly any movement within the moon. These indicated that seismic activity on the moon is very slight indeed and certainly is in no way comparable to the extensive seismic activity recorded on earth.

23.4 *Features of the moon*

A full moon on a clear night shows its earth-turned face to be divided into dark and light areas. The dark areas are the *maria,* which lie well below the light-toned *highlands* or *terrae.* This difference in brightness is caused by a difference in reflectivity of the two different zones of the lunar surface, the highlands being rougher than the maria. Relief varies rapidly and the differences between the rims and bottoms of the larger craters ranges from 2 to over 5,000 m. Maximum relief on the moon is between 10,000 and 11,000 m, the lowest spots being in the Mare Imbrium, Mare Nectaris, and Mare Nubium.

We are used to seeing the dark zones of the maria, which cover about one-third of the visible moon. It came as a surprise, then, when the first pictures of the far side of the moon (taken by the Russians) showed maria to be almost completely lacking on that side.

Maria

The Italian astronomer Giovanni Riccioli named most of the moon's maria back in 1651. Even as Galileo had before him, he felt that these great dark areas were actual seas of water. Today we know they are not, and it is most generally believed that they are floored with the products of volcanic activity, chiefly lava flows.

Looking at a map of the moon, one can distinguish two main types of maria. First there are the large (hundreds of kilometers across) circular maria such as Imbrium, Serenitas, Crisium, Nectaris, Cognitum, and, part way around to the far side, Orientale. Then there is a group of irregular maria such as Frigoris, Fecunditatis, Tranquillitatis, and the great Oceanus Procellarum.

The circular maria are outlined by discontinuous arcs of mountainous masses that stand hundreds and in many places thousands of meters above the maria floors. The Appenine Mountains on the southwestern margin of the Mare Imbrium, for instance, reach an elevation of about 4,300 m above the neighboring mare floor. The maria basins are flooded with dark material, usually interpreted as lava flows. Earthbound geologists have mapped many of these flows and see their frontal margins as 20 to 100 m in height. These maria fillings are heavily pocked with later craters.

Mare Orientale looks like a great bull's-eye set in a series of concentric rings (Figure 23.13). It does not take much imagination to see the troughs and ridges which form these rings as great waves moving out from a common center. In fact, this is the interpretation that has been offered by some who see these features as "frozen" waves created by the explosion that formed the main mare. Similar, though less well displayed, shock wave features can be seen around Mare Nectaris and most of the other circular mare.

We have already noted that igneous rocks appear to fill the basins of the maria, and have also raised the question of what the source of the igneous melts on the moon could be. We can repeat here the observation that the lavas which are thought to floor the maria may be the result of very large impact events and not the product of heat originating in the subsurface.

The irregular maria also appear to be lava-filled basins. In a very general way, they lie outside of the circular maria. For instance, the continuous stretch of Mare Frigoris–Oceanus Procellarum surrounds Mare Nubium on three sides. On the west Nubium is contiguous to Serenitatis and smaller irregular maria

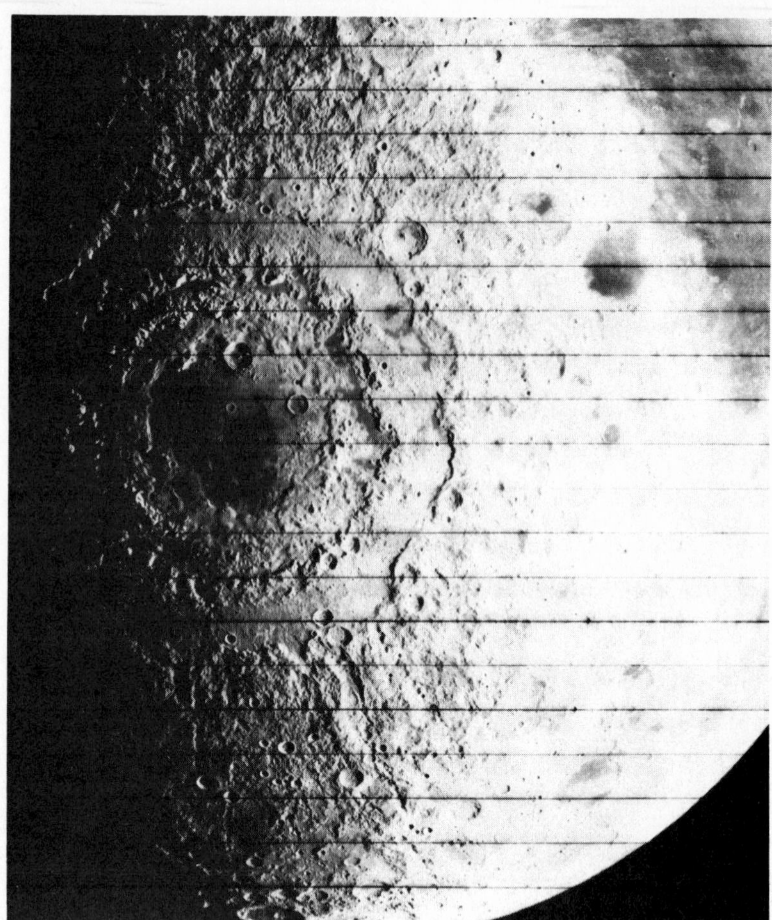

Figure 23.13 **Mare Orientale is partially visible from the earth on the extreme southwestern side of the moon. Here in an Orbiter IV picture we see the entire mare. The outer circular scarp measures nearly 1,000 km in diameter. The concentric rings suggest waves generated by an energy source at the center of the "bull's-eye."** (*Photograph by National Aeronautics and Space Administration.*)

more or less complete the ring around these two. In many places the surfaces of the irregular maria are continuous through the broken and missing walls of the circular maria, the most conspicuous example being the connection between Nubium and Procellarum.

Craters

A crater, more than any other feature, is the trademark of the moon, and the circular maria we have just described are craters now partially filled. Here we shall look at the familiar but smaller craters. The largest of these smaller craters on the moon's near side is the crater Bailly, nearly 300 km across. Most of the craters we see on the moon, however, are a few tens of kilometers in diameter.

From analogy with topographic features on earth, we know that craterlike features can form in a number of ways. These include (1) craters formed by impact of meteorites; (2) volcanic craters that may be explosive or due to caldera collapse; and (3) collapse of roof-rock into voids created by the solution of sedimentary deposits such as limestone or halite in the underground. Other depressions, not necessarily with true crater forms, are the result of wind action of the melting of blocks of stagnant glacier ice. On the moon, it is apparent that impact and volcanism are the chief processes of crater formations. Sinkholes, glacial kettle holes, and blowouts are ruled out because of the lack of an atmosphere on the moon and hence the lack of moisture, sedimentary rocks, processes of solution, and the action of wind. It is possible, however, that melting of ground ice may produce depressions.

We now know that there is a range in the size of moon craters from hundreds of kilometers in diameter to pits a few micrometers across. We can refer to these craters as *primary craters*, formed by the original impact of meteorites or by volcanic activity, and as *secondary* or *satellite craters*, made by the falling fragments ejected from a primary crater.

Copernicus, a well-known, fairly large primary crater, shows many of the features characteristic of moon craters. It measures about 90 km in diameter and over 3,300 m in depth. The relatively level floor has a raised center portion, a feature that is more pronounced as a central peak in somewhat smaller craters. The walls facing into the crater are terraced or stepped and represent the slumping of large slices of the crater walls into the central depression. The rim of the crater is about 1,000 m high and slopes away from the crest at the crater's edge, to feather out at the edge of an irregular zone up to 150 km wide and concentric to the crater.

The material that forms the rim is zoned into three rings. The interior zone is very hummocky, with ridges concentric to the crater. Outward beyond this, the material is less hummocky and the topography is one of branching ridges more or less radial to the central crater. The outermost zone is still less hummocky, but the radial ridges of the intermediate zone still persist in subdued form. This zone is pocked by many small satellite craters thought to have been formed by fragments thrown out of the primary crater. Superimposed on the rim material and spreading far beyond it to a distance 500 km from the crater are light-colored, discontinuous streaks, the *rays* of Copernicus (see Figure 23.14).

Copernicus is interpreted by most observers as an impact crater. The central rise on the floor of the crater is thought to result from upward adjustment following impact. The rim is thought to have been pushed upward and outward as a result of the impact explosion, and the hummocky ridges of the interior zone of the rim material are thought to stem from the same cause. Outward from this zone, the areas of radial ridges are interpreted as material ejected from the crater and arranged in ridges subparallel to the direction of blast. The ray material is believed to be finely crushed matter thrown out from the primary crater and, to a lesser extent, similar material kicked up during formation of the satellite craters. Copernicus is one of a number of *ray craters* on the moon. Most craters, however, lack rays. Their absence is attributed to the greater age, hence longer weathering, of the nonray craters.

Some craters show only a low lip for a rim. In general, these are old craters all but buried by material ejected from younger craters or by volcanic material, including lava flows, that filled the maria.

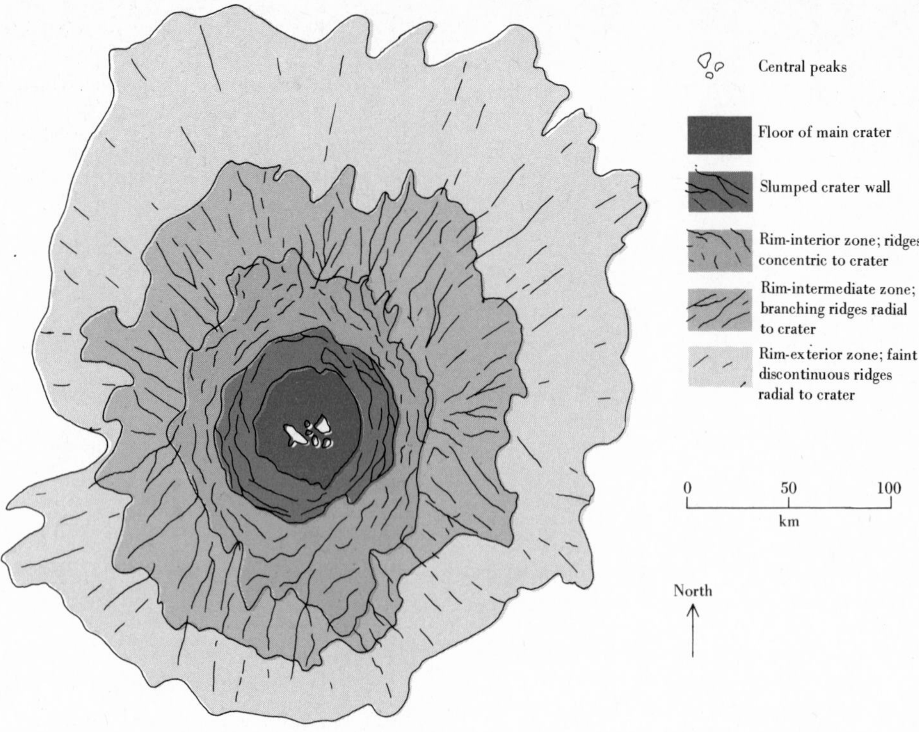

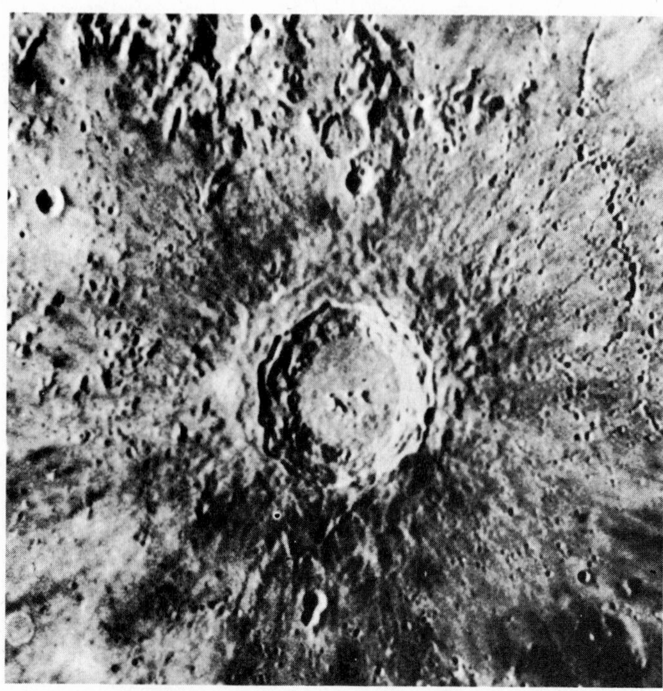

Figure 23.14 ***(a) Map of Copernicus showing the extent of the major zones of the central crater. (b) Telescopic view of Copernicus. The rays and smaller craters can be seen in the photograph.*** (*Photograph by Mt. Wilson Observatory. Map based on H. H. Schmitt et al., "Geologic Map of the Copernicus Quadrangle of the Moon,"* U.S. Geological Survey Atlas of the Moon, *Map I-515, LAC-58, 1967.*)

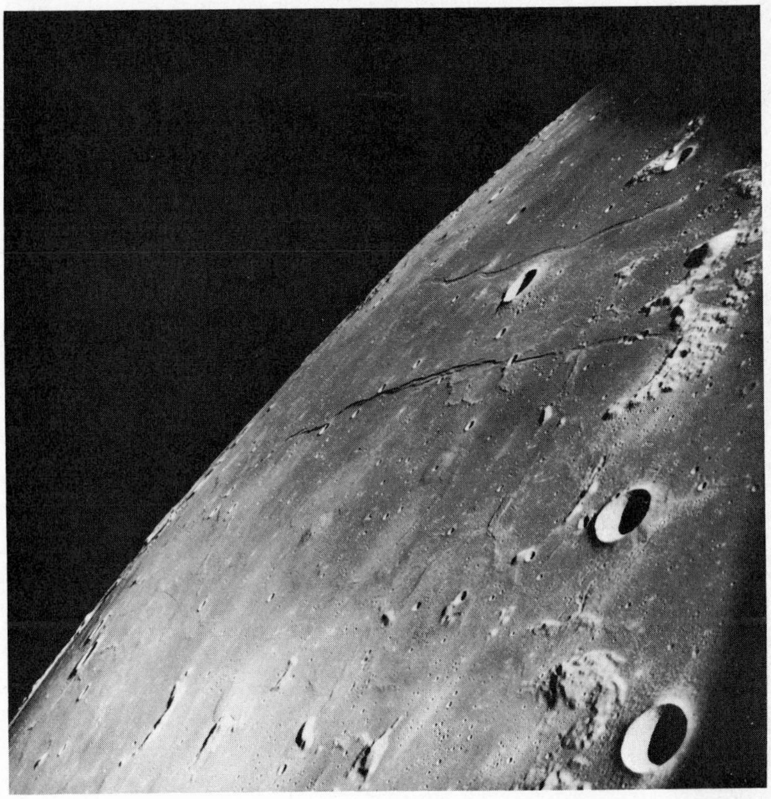

Figure 23.15 ***Mare Tranquillitatis from* Apollo 8. *The crater Cauchy lies between Cauchy rille and the closer Cauchy scarp. The scarp is a fault but turns into a rille toward the right side of the picture. A conical feature halfway between Cauchy rille and Cauchy scarp has a volcanic form and is superposed on the hilly country near the right side of the picture. Low domal features on the observer's side of Cauchy scarp may be due to laccolithic updomings. The great bulk of the material of the mare is thought to be* volcanic.** *(Photograph by National Aeronautics and Space Administration.)*

There are a number of small craters 1 to 5 km in diameter which can be assigned with some certainty to volcanic action. One of these types, the maar crater, has already been described. Such craters tend to be located in well-defined chains and may have a very low rim or none at all. A rimless type of crater is found on the crests of low, moundlike hills in the mare regions. The hills, which are 100 to 200 m high and up to 20 km in diameter, are interpreted as either low volcanic cones or as laccolithiclike uplifts (Figure 23.15). The craters are thought to be due to collapse of the caldera type described in Chapter 4.

Rilles, wrinkle ridges, faults, and fractures

At various places, the moon's surface is creased by long valleys called *rilles* (from the German for "furrow"). There are various forms but they are from 10 to 150 km in length, from 0.5 to 4 km wide, and from 50 to 500 m deep, although depths of up to 1,000 m are reported.

Most rilles are related to the maria or to smaller, filled craters. There are basically two forms. One is straight-walled, and when it changes direction, it does so at a well-defined angle. The second is sinuous and, in many instances, gives the impression of a meandering valley (see Figures 23.16 and 23.17).

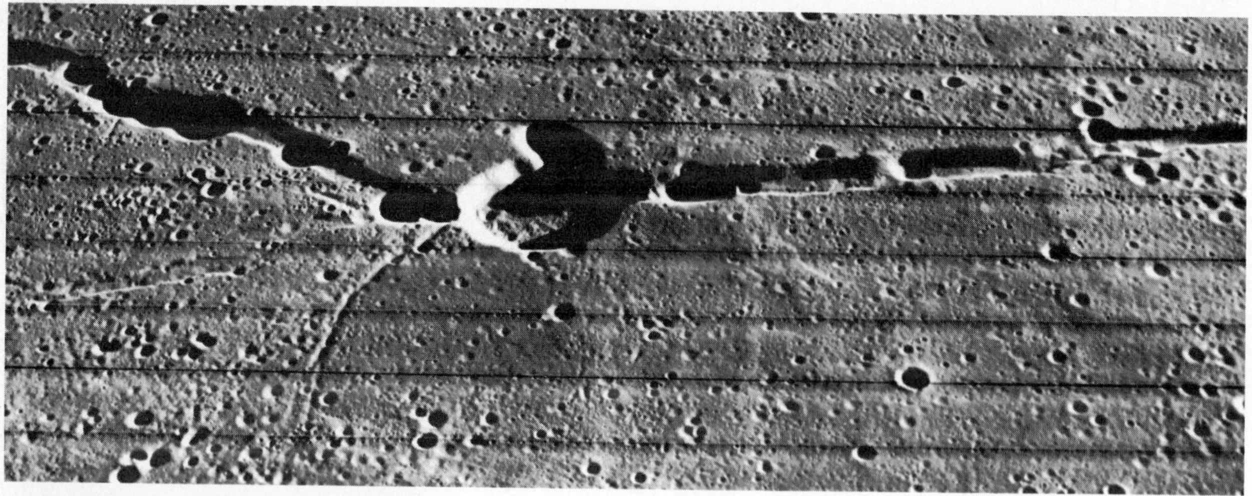

Figure 23.16 **Rilles lead off from the crater Hyginus. The two more prominent rilles are marked by several smaller craters. Hyginus is about 10 km across and the rille craters average about 2 km in diameter.** (*Photograph by National Aeronautics and Space Administration,* Orbiter III.)

The straight-walled types are generally thought to be downdropped strips of mare material. It is not clear whether the motion is one of collapse or whether it is caused by crustal stresses. Strings or chains of maarlike craters are located along a number of these rilles.

The sinuous rilles are particularly hard to interpret. The meandering outline suggests flowing water, a suggestion strengthened in the case of some rilles because smaller meandering channels are found in the rille floors. The similarity in form with river-formed features on earth has caused some to argue that water, in fact, caused these lunar features. The difficulty of obtaining an adequate supply of surface water has made this suggestion unacceptable to most investigators. Another possibility, however, is that volcanically warmed water may trickle in a meandering course through the subsurface, melting ground ice and creating a tunnel which later collapses to produce the feature we now see. Collapse could also result from the subsurface drainage of lava back into the magma chamber. The tunnel thus created might then collapse to produce the rille we now see.

Smaller rilles are found on the walls of some craters. They are different from those just described and display a border defined by levees. Very likely they represent streams of lava that flowed down into the crater, leaving levees along the cooling margins of the flows while the more molten centers drained craterward.

Figure 23.17 **The Alpine valley creases the mountains on the northeast side of Mare Imbrium. Down its center winds a sinuous, discontinuous rille. In the upper right, toward the Mare Imbrium, are other sinuous rilles. See the text for discussion.** (*Photograph by National Aeronautics and Space Administration,* Orbiter V.)

Across the floors of most maria and flooded craters, one sees long, low ridges comparable in scale to the large rilles. Some suggest that these are anticlinal ridges formed by laterally directed compressional stresses. Others regard them as uplifts caused by volcanic intrusions.

We have already mentioned the existence of faults on the moon. These and prominent fracture systems are particularly prevalent in the older rocks. Many seem to be related to the stresses developed during the formation of the maria, those radiating from Mare Imbrium being among the most striking. Others may be related to a moonwide fracture system.

At Tranquillity Base, sets of small intersecting grooves have been found in the lunar soil. These are a centimeter or less in depth, and from 3 to 300 cm long, 0.5 to 3 cm wide, and 3 to 5 cm apart. One interpretation of these features is that they are produced by the settling of fine-grained lunar soil into fractures in the underlying bedrock as it is joggled by seismic or impact events or by tidal motion.

23.5 *Lunar history*

We are not yet sure how the moon formed and came to be a satellite of the earth. Most workers currently think that the moon's origin is related in some way to the earth's origin. In this view, the moon and earth formed more or less simultaneously and adjacent to each other, building up material derived from the original dust cloud from which the planetary system is thought to have evolved. Others have suggested that the earth once had rings, such as those of the planet Saturn, and that these rings eventually became the moon. Another suggestion made originally by Sir George Darwin, second son of Sir Charles Darwin, is that the moon was torn from the earth at some past time. A fourth theory states that the moon once traveled in its own orbit around the sun but has since been captured by the earth and assumed its present orbit. About all we can conclude here is that the state of knowledge about the very early history of the moon is in an unsatisfactory state.

Even though we cannot yet determine the origin and earliest history of the moon to the satisfaction of most students of that body, we do have a fairly good idea of a great deal of the moon's history. Much of this story has been worked out by geologists of the U.S. Geological Survey working from photographs taken through earth-based telescopes and cameras carried in space probes. In fact, the U.S. Geological Survey is well into an ambitious program of geologic mapping, which is producing geologic maps of the moon at a scale of 1:1,000,000. Rock units are being mapped on the basis of relative age, surface characteristics, and genetic types. Ages are established by cross-cutting relationships and superposition, stratigraphic principles discussed in Chapter 9 on geologic time (Figure 23.18). As we acquire samples directly from the moon, radioactive dating has begun to provide a picture of actual ages of the several lunar events now established by the techniques of relative age determination.

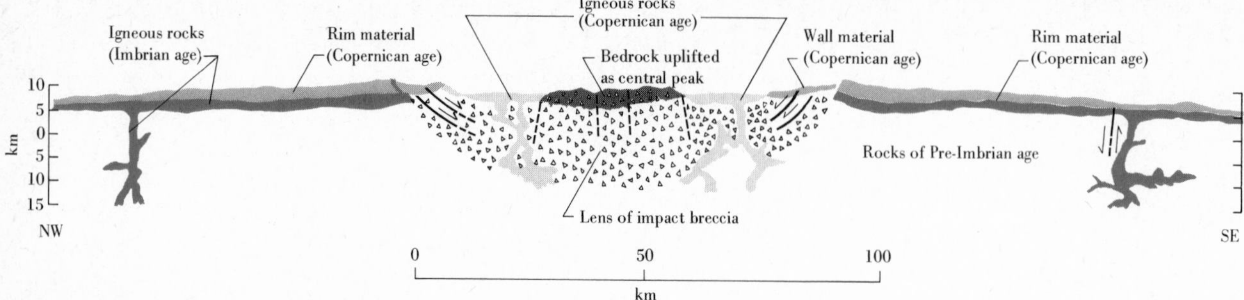

Figure 23.18 **Geologic cross section through the crater Theophilus. The distribution of intrusive rocks of Copernican and Imbrian age is hypothetical. Faults are shown as solid lines where definite and as dashed lines where covered or where the sense of movement is not known. See Table 23.1 and the text for discussion of age designations.** (*Modified from Daniel J. Milton, "Geologic Map of the Theophilus Quadrangle of the Moon,"* U.S. Geological Survey Atlas of the Moon, *Map I-546, LAC-78, 1968.*)

Table 23.1 outlines the scheme devised to designate time during lunar history. The system is slightly modified from a sequence first worked out in the area of the crater Copernicus. We can expect further refinements as the amount of information increases. In the meantime, we can characterize the lunar periods as follows.

PREIMBRIAN TIME Preimbrian time includes the earliest part of lunar history. We believe that during this period most of the material that makes up the terrae came into being. We think that toward the latter part of Preimbrian time the moon was subjected to the shattering explosions that formed the large basins we know as the maria. Preimbrian materials have a complex history and on them is impressed a lunar landscape of ridges, valleys, and craters. Rocks of this age date from the beginning of lunar and earthly history, over 4.6 billion years ago.

IMBRIAN PERIOD The Imbrian period begins with the creation of the Mare Imbrium and the concurrent ejection of a broad blanket of material from the basin area. Included also in this period are events which created most of the mare material, filling not only Mare Imbrium but the older maria as well. The general name of Procellarian (after the Oceanus Procellarum) is given to this material. Most of this infilling is interpreted as extrusive igneous material. At present, most geologists see this material as representing a long time of extensive volcanic activity, which included both fissure flows and fragmental ejecta.

ERATOSTHENIAN PERIOD The crater Eratosthenes, northwest of Copernicus, gives its name to the Eratosthenian period. The crater and its ejecta are superimposed on older material of Imbrian age. The period includes the formation of the crater Eratosthenes, its related ejecta, many of the moon's prominent craters, and the material blown from them. All these craters lack any visible rays.

COPERNICAN PERIOD Named for the crater Copernicus, the Copernican period comprises the time of formation of the ray craters of the moon and the ejecta blankets connected with crater formation.

Table 23.1 ***Time units of lunar history***

Time units
Present time
Copernican period
Eratosthenian period
Imbrian period
Preimbrian time
Beginning of lunar history

Absolute age determinations

By early 1970, radioactive age determinations had been carried out on samples from both the Mare Tranquillitatis and the Oceanus Procellarum. The basaltic rocks from the Mare Tranquillitatis crystallized about 3.7×10^9 years ago. This is indeed old by earthly standards and suggests that the moon's surface at that spot has been changed but little during that time. This date falls within the Imbrian period mentioned above.

In addition, at least one sample from Tranquillity Base is still older. A sample analyzed by Professor Gerald J. Wasserburg, of the California Institute of Technology, crystallized about 4.6×10^9 years, close to the presumed time of earth origin. It has been suggested that we are dealing here with materials dating from the formation of the oldest lunar crust and even the solar system.

Preliminary measurements of the age of crystalline basaltic rocks collected by *Apollo 12* in the Oceanus Procellarum indicate a time of crystallization of about 2.5×10^9 years ago. These rocks also were formed during the Imbrian period; thus the period appears to have covered a long segment of time.

Analysis of the nuclides produced by cosmic ray activity indicates that the material at Tranquillity Base has been within a few centimeters of the surface for the last 10 million years and within 1 or 2 meters for at least the last 500 million years.

23.6 *Impact craters on the earth*

There are several craters on earth whose origin we can assign with confidence to the impact of meteorites during the immediate geological past. There are about 16 such craters known and there are probably another 4 or 5 yet to be found. These craters, termed *astroblemes* (from the Greek word for "star" and the Middle English for "disfigurement") by the U.S. earth scientist R. S. Dietz, are concentrated in the arid and semiarid regions of the earth. This distribution is not surprising for these are the areas where erosion is slowest and the chance of their survival is greatest.

The best known of these is probably Meteor crater in Arizona, also known as Barringer crater, after an early student of the crater, and as Coon Butte crater (Figure 23.19). The crater measures about 1,200 m in diameter, is 150 m deep, and has a rim which averages 55 m above the surrounding country. The rim slopes away from a crest near the crater's edge.

In addition to its shape, other features point to an origin by impact. No igneous rocks are associated with the crater to suggest that it was once a volcano. The material scattered around the outside of the hole appears to have been "splashed" out as if some great force blew it from the crater site. Furthermore, the sedimentary layers at the rim have been turned upward and

Figure 23.19 **Low-angle photograph of Meteor crater, Arizona. The profile of the rim is seen. The stream valley is the Canyon Diablo.** (*Photograph by American Museum of Natural History.*)

in places older material lies on top of younger. Meteoritic fragments of small size have been found scattered in the immediate area. A jumbled mass of rock material characterizes the steep interior walls of the crater and the rim area. Drilling into the bottom of the crater shows the rock to be shattered and broken to depths of over 300 m before firm, unaltered rock is encountered.

The absence of a large meteoritic body at some depth beneath the floor is not an argument against the meteoritic impact. When a large meteorite falls, it does not bury itself beneath the crater it forms; instead, it explodes and vaporizes, leaving the crater scar and a few remaining fragments of meteoritic material.

What are some additional lines of evidence that might indicate impact? We might expect that, in addition to the rock shattered by the explosion, we would also find that some of the rock had been finely pulverized and sometimes fused. At the Arizona crater, we find that the local Coconino sandstone has in places been ground into a very fine-grained material of angular shape. In other places, the sandstone has a very porous texture, in part because of fusing of the quartz grains to a glass. More diagnostic than these features, however, are some minerals which form only under extremely high pressures, as well as some rock structures, called *shatter cones,* that so far seem unique to high-pressure environments.

Coesite and stishovite

Coesite and *stishovite* are high-pressure forms of quartz and may also indicate high temperatures. Their atomic structure varies from that of quartz and their density is higher. Coesite, for instance, has a density of 2.92 and stishovite of 4.34, as compared with the 2.65 of ordinary quartz. Both of these minerals were created in the laboratory before they were found in the field.

Coesite was first created in 1953 by Loring Coes, Jr., in the laboratories of the Norton Company, Worcester, Massachusetts. The mineral, formed under pressures of 35 kilobars (kbar) and in the 500° to 800°C temperature range, was named for Dr. Coes the following year by Robert Sosman of Rutgers University.

The story relating this laboratory creation to impact craters is an interesting one. Back in 1907, George Merrill studied the strongly shattered grains from Meteor crater in Arizona and noted a peculiar material in which the shattered grains seemed to be set. He identified it as silica glass or as opal. Half a century later, Eugene M. Shoemaker, then a graduate student at Princeton University, was studying the similarities between nuclear explosions and Meteor crater. Samples collected by him from the crater arrived some years later at the laboratories of the U.S. Geological Survey in Washington. There, E. C. T. Chao began a detailed examination of the material and found the same sort of thing that Merrill had in 1907. Chao, however, was suspicious of the original identification as glass or opal and made an X-ray study of the material. It turned out to be the first identified natural occurrence of coesite.

Since this discovery in 1960, coesite has been reported from the Wabar craters in Arabia, from the Ashanti crater in Ghana, and from the nuclear crater at Teapot Ess in the United States, to name just three of a number of localities.

Stishovite, discovered in laboratory work by the Russian mineralogist S. M. Stishov and his colleague, S. V. Popova, in 1961, is now also known from impact craters. In fact, coesite and stishovite are not known to exist in nature except from the high-pressure environments associated with very large natural or artificial explosions.

Shatter cones

As their name implies, shatter cones have a conical shape. They average in height from 6 to 12 cm, although many are smaller and some have been reported up to 2 m in height. The angle of the apex is about 90°. The conical surfaces are marked by fine striations that radiate from the apex toward the base (Figure 23.21). They are found most commonly in limestone or dolomite but are reported from shale, sandstone, quartzite, granite, and chert as well.

Figure 23.20 ***The New Quebec crater on the Ungava peninsula of northern Quebec measures over 3 km across. It is the result of the impact of a meteorite. The smaller lakes were formed by glacier ice which had disappeared before the creation of the crater. South is to the top.*** *(Photograph by Canadian Government Mines and Technical Surveys.)*

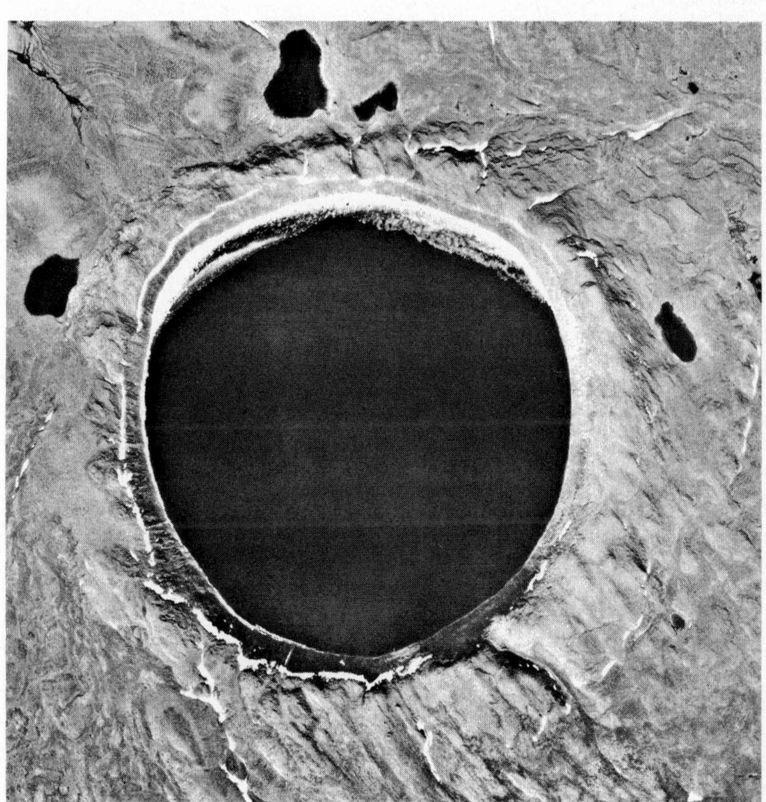

Figure 23.21 **Shatter cones in dolomite from the Well Creek Basin, Tennessee. Height is about 25 cm.** (*Photograph by Robert Dietz.*)

Of particular interest is the occurrence of shatter cones. They are associated with known impact craters. Furthermore, they are associated with large features called *cryptovolcanic structures,* a term referring to the fact that the rock involved appears to have been disrupted by volcanic activity but that there are, cryptically, no volcanic rocks associated with the structures. When a field orientation can be determined, the cones point upward or horizontally toward the center of the crater or the cryptovolcanic structure.

The deduction, first drawn by R. S. Dietz, is that the shatter cones reflect an explosion of large magnitude and that the apex of a cone points in the direction from which the energy or shock wave moves. The striations along the walls could be caused by the movement of conical slices of rock forced to move against each other by the violence of the impact.

Confirmation of the impact origin of shatter cones has come from laboratory experiments carried on jointly by scientists of the U.S. Geological Survey and the National Aeronautics and Space Administration. They subjected a small block of dolomite to the high-speed impact of a small aluminum sphere. The projectile hit the rock at the speed of 5,620 m/sec. The impact produced a crater 5 cm wide and 1.5 cm deep. In the crater bottom were three miniature, but easily recognized, shatter cones with striated surfaces. The cones measured between 2 and 7.5 mm in diameter at their bases, and their heights were comparable. The axes of the cones were approximately parallel to the path of the projectile and the apices of the cones pointed in the direction from which the projectile had moved.

23.7 *Tektites*

Nearly 200 years ago in the valley of the Moldau River of western Czechoslovakia the first of the small, enigmatic objects called tektites was found. The name *tektite* is derived from the Greek for "molten." And molten they once were, cooling quickly to form a glass that resembles the obsidian with which we are familiar from volcanic regions on the earth. You will remember that in Section 23.2 we described some glassy objects found in the lunar soil. Indeed, for many years prior to man's arrival on the moon tektites were thought by many to have been splashed from the moon's surface and been captured by the earth's gravitational field. Today it is generally thought that the earth's tektites did not originate on the moon, for reasons which we shall see.

Most tektites are small, 1 cm or so in diameter, although some are over 10 cm in dimensions. They are black, except for some found in the Moldau valley, which are a dark green. An estimated 1 million or more tektites have so far been discovered. They occur in fields or areas, the most extensive of which is Australasia. Here they are called *australites* and are reported from Australia, Tasmania, Indochina, Java, Borneo, the Philippines, and from the sediments in the neighboring seas. The next largest known field is in Czechoslovakia, where the tektites are called *moldavites.* Another field occurs in the Ivory Coast of west Africa and the adjacent Atlantic Ocean. In the United States, tektites have been found in Georgia, in Texas (where they are called *bediasites*) and a single tektite is reported from Martha's Vineyard, Massachusetts (see Figure 23.22).

The tektites are of different ages. Those from Australasia are Pleistocene and are about 700,000 years old. This date is arrived at in part by the radiometric dating of the tektites and in part by their discovery in deep-sea sediments just above the boundary between the Matuyama and Brunhes polarity epochs, a boundary which has this age. Tektites from the Ivory Coast are 1.1 million years of age. Those from Czechoslovakia are about 15 million years old and those from the United States are thought to be about 35 million years old.

The shape of tektites is variable, but one of the most common shapes is a disk resembling a button. Their surface is usually dull, pitted, and sometimes grooved. Internally glass shows a distorted, flow structure. A peculiar feature of the australites is that they show two stages of heating and cooling. An original glass has been remelted in part, presumably by passage through the earth's atmosphere.

The composition is largely silica (70 to 80 percent) and Al_2O_3 (11 to 15 percent). Iron, magnesium, calcium, potassium, sodium, titanium, and manganese make up the balance of all tektites. They contain virtually no water. Some tektites have voids which are essentially gas-free, i.e., vacuums. This suggests that their formation has been at very high temperatures or in an environment of virtually no atmosphere.

The origin of tektites is still unresolved. Some believe them to have been

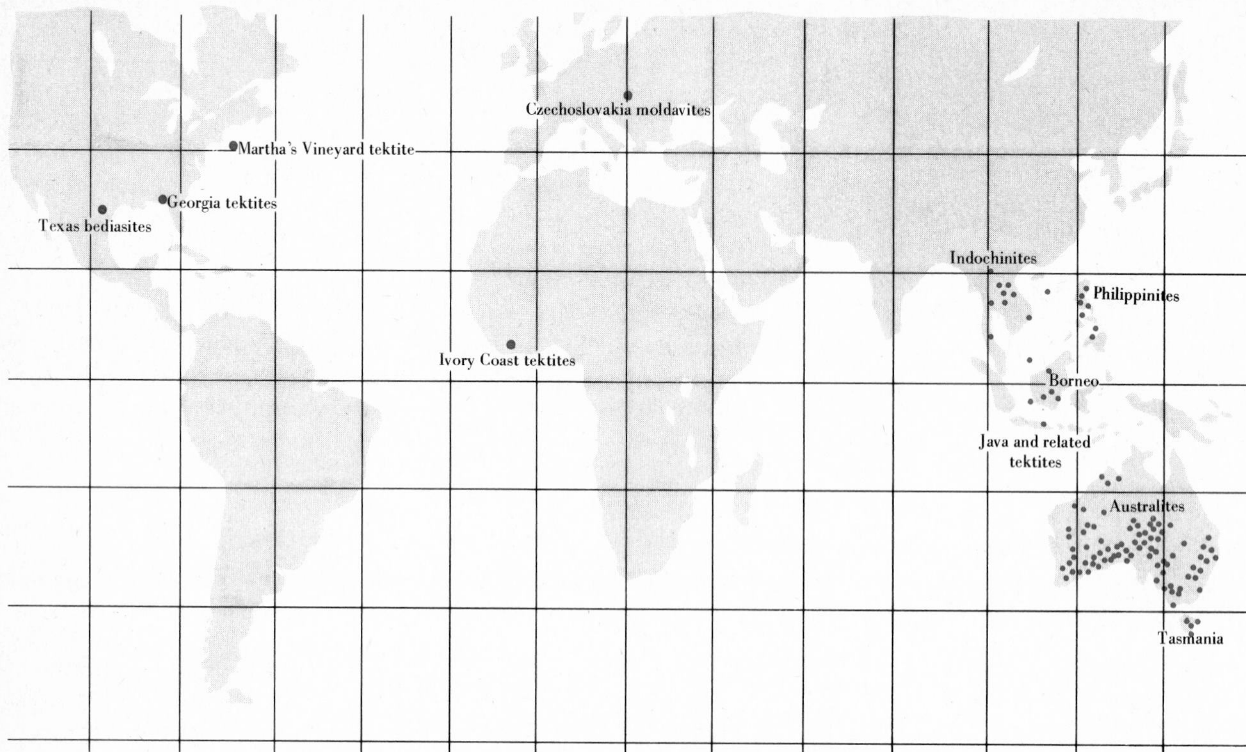

Figure 23.22 **Distribution of tektites.** (*After Brian Mason,* Meteorites, *p. 203, John Wiley & Sons, Inc., New York, 1962.*)

formed by the impact of meteorites with the earth's surface. In truth, their composition is not too far from some terrestrial rocks, both sedimentary and igneous. Others see them as fragments from outer space, although none have been observed to have fallen during a meteor shower, and their composition is not compatible with what we know about material from beyond the earth–moon system. Most students of the problem have discarded the idea of the moon as a source for tektites because the composition of the tektites is so different from what we now know of the composition of the moon. Silica content of lunar rocks is generally much lower, and the iron, titanium, and manganese higher. One possible exception is reported from the samples returned by *Apollo 12.* In this instance the glass approaches the composition of the Java tektites.

Outline

The *moon* revolves around the earth every 29.53 days and rotates at the same rate, always keeping the same face toward the earth. With a diameter of about 3,450 km and a density of 3.34 g/cm^2, it has a mass of 0.012 that of the earth and a gravity of 0.165 that of the earth. Temperatures range from $-150°C$ to more than 130°C.

Composition of the moon includes large amounts of basaltic material at the surface. Anorthosite may form the highlands.

Processes on the moon are at the same time similar to and different from those of the earth.

Weathering, because of lack of an atmosphere, is restricted to the effect of *gravity*, *temperature changes*, *impact by meteorites*, possibly *frost action*, *cosmic radiation*, and *tides*.

Transportation of materials is effected by *mass movement*, *impact*, and *volcanism*.

Igneous activity is expressed by small, explosive volcanoes, lava flows, and intrusions.

Features of the moon include the low-lying *maria* and the *highlands* or *terrae*.

Craters are the single most characteristic features. They include *primary* and *secondary* or *satellite* craters. They are formed by volcanic activity and by impact.

Rilles are long, straight-walled or sinuous trenches.

Wrinkle ridges occur on the surfaces of maria.

Fractures and *faults* are found over the entire lunar surface but more commonly in the highlands.

Origin of the moon is not yet understood. Later lunar history can be established, however, by application of the principles of superposition and cross-cutting relationships.

Impact craters on earth are caused by meteorites and show features similar to certain lunar craters. In addition, detailed studies have shown that high-pressure forms of quartz and shatter cones are associated with the earth's craters.

Tektites are small, glassy objects irregularly distributed on the earth's surface. Their origin is in dispute.

Supplementary readings

BALDWIN, RALPH B.: *The Measure of the Moon*, University of Chicago Press, Chicago, 1963. A very good general book on the moon.

FIELDER, GILBERT: *Lunar Geology*, Butterworth Press, London, 1965. The geology of the moon, with emphasis on structural features.

LOWMAN, PAUL D.: *Lunar Panorama*, Wletflugbild, Feldmeilen/Zurich. 1969. A handsomely illustrated discussion of lunar features.

LUNAR SAMPLE PRELIMINARY EXAMINATION TEAM: "Preliminary Examination of Lunar Samples from Apollo XI," *Science*, vol. 165, pp. 1211–1227, 1969. A team of 51 experts cooperate to produce the first scientific report on the rocks brought back to earth by the first men to walk on the moon.

LUNAR SAMPLE PRELIMINARY EXAMINATION TEAM: "Preliminary Examination of Lunar Samples from Apollo XII," *Science*, vol. 167, pp. 1325–1339, 1970. The first official report on the *Apollo 12* samples.

MASON, BRIAN: *Meteorites*, John Wiley & Sons, Inc., New York, 1962. A first-rate discussion of meteorites with a chapter on tektites.

Science, vol. 167, no. 3918, January 30, 1970. The entire issue is devoted to the scientific reports of the more than 500 scientists from nine countries who worked on the samples brought from the moon to the earth by the *Apollo 11* mission.

24

The earth's place in space

SPACE EXTENDS IN ALL DIRECTIONS. IT IS THAT IN WHICH ALL PHYSICAL THINGS are ordered and related. It provides a place for everything.

Looking out into space at night from the earth, we see many points of light. Most of these are *stars* that stay in the same positions relative to each other, but among them are a few that wander about from night to night. The latter are *planets.* And scattered about the sky are hazy patches of light that a telescope does not resolve as sharp points as it does the stars. These are *nebulae* (see Figure 24.1). The complete assemblage of all we can see and describe out there is called the *universe.*

All the stars we can see constitute the Milky Way and belong to a family of about 100 billion. This is our *galaxy.* It has been named the Milky Way Galaxy. The nebulae are other galaxies. It has been estimated that there are millions of these distributed through space.

In our galaxy, some stars appear to be grouped together. These are called *constellations.* About 90 constellations have been described. They were named centuries ago after mythical personages, animals, or inanimate objects they were fancied to resemble. They are still used today as convenient means of describing the locations of portions of the sky.

This chapter discusses the earth's place in space, and some of the ideas that have been put forward to explain its origin.

24.1 Our solar system

We are able to exist physically because of heat and light from our personal star, the sun. As we observe it from earth, the sun rises in the east to bring daylight and sets in the west to bring night. This phenomenon led to the assumption that the sun moves around a stationary earth. Likewise, the stars appear to rise in the east and set in the west. Therefore, for a long segment of human history, the earth was believed to be the center of the universe, about which everything else revolved.

This belief, supported by even noted Greek scientists who were dominated by the teachings of Aristotle, was widely held until about 3 centuries ago. One Greek, Aristarchus, had the temerity in the third century B.C. to suggest that the planets, including the earth, circle about the sun, and that the earth rotates on its axis, giving us night and day. But he failed to convince most of the people of his time, and his explanation for the movements of the planets around the sun was not to be firmly established until 1543, by another man—Copernicus. Still another century passed, however, before there was universal acceptance of Aristarchus' idea.

We now know that the sun is the center of our physical existence. Everything within the sun's gravitational control constitutes what we call our *solar system* (see Figure 24.2). The solar system includes nine planets, 31 satellites or "moons," thousands of asteroids, scores of comets, and uncounted millions of meteors. All these revolve about the sun, forming a system 16 billion km in diameter.

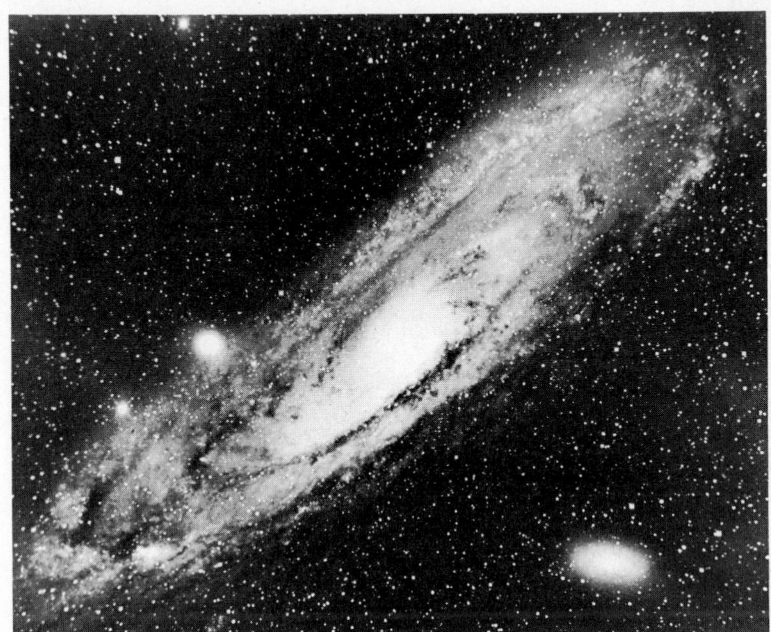

Figure 24.1 **The Great Nebula in Andromeda. It is a spiral nebula, similar in shape to our galaxy.**

The sun

The sun is a star. Like all other stars, it is a hot self-luminous ball of gas. Located 150 million km from the earth, it has a diameter of about 1,391,000 km and a mass 332,000 times that of the earth. It is the source of the solar system's light, heat, and charged particles. Its gravitation holds the system together.

Sixty-six elements have been recognized in the spectrum of the sun but they are in highly heated and excited atomic states and do not resemble their forms on earth. The bulk of the sun, however, is composed of two elements. Its volume is 81.76 percent hydrogen and 18.17 percent helium. All other elements total only 0.07 percent.

When the sun is at the zenith on a clear day, its light has a luminosity at the earth's surface of 10,000 candles.[1] The sun's light represents radiant energy upon which terrestrial life depends for existence.

The sun's radiant energy is believed to be generated by the conversion of matter in atomic reactions that form helium nuclei from hydrogen nuclei, in a process called *the carbon cycle*. This takes place in the sun's deep interior.

When the nucleus of hydrogen ^1_1H collides with the nucleus of carbon $^{12}_6C$ and joins it to form the nucleus of nitrogen $^{13}_7N$, radiant energy is emitted. After five more steps, the process terminates with the $^{12}_6C$ back in its original state but accompanied by a newly assembled helium nucleus that was formed

[1]The *candle*, unit of intensity, is defined as one-sixtieth of the intensity of 1 cm^2 of a black-body radiator operated at the temperature of freezing platinum.

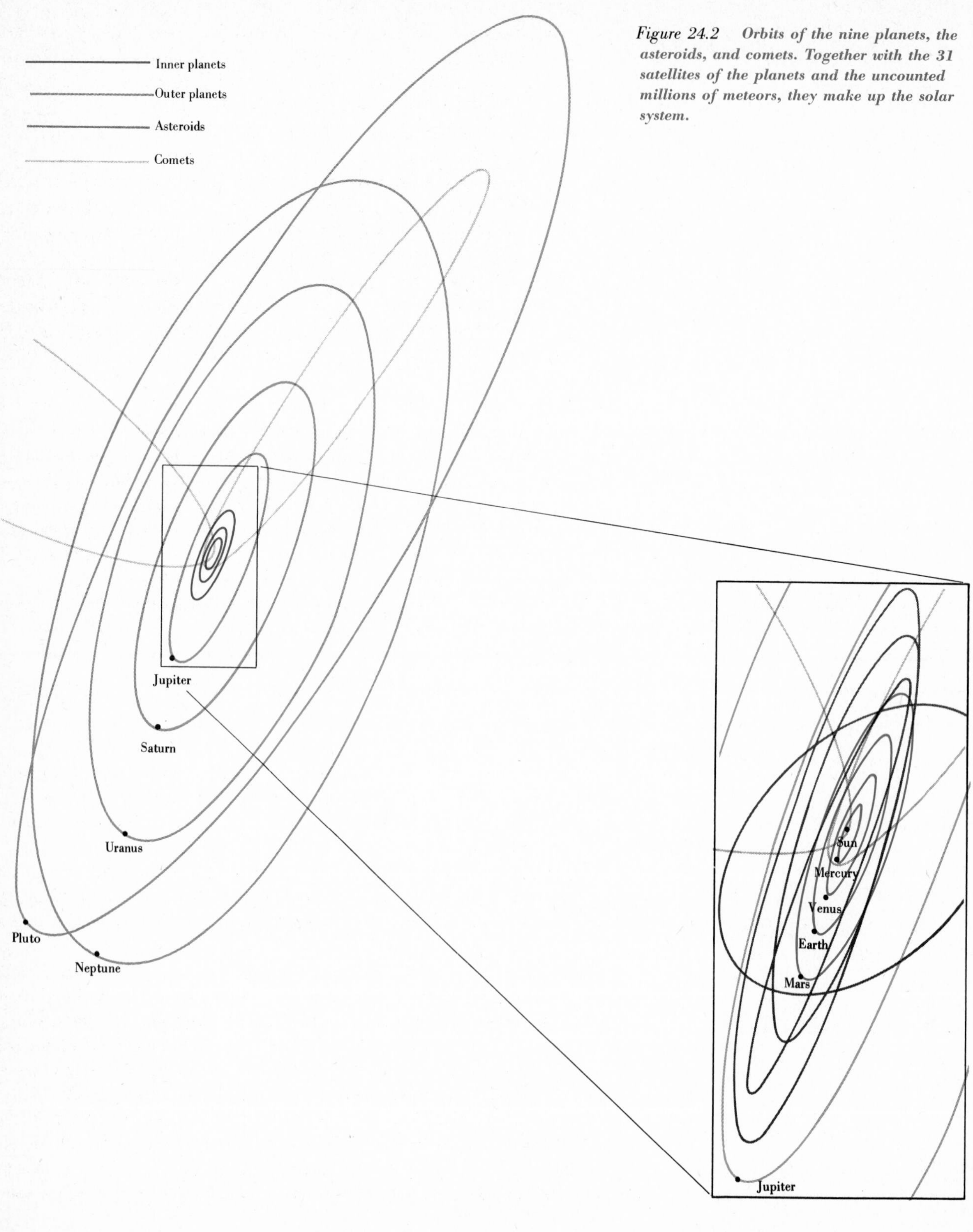

Figure 24.2 **Orbits of the nine planets, the asteroids, and comets. Together with the 31 satellites of the planets and the uncounted millions of meteors, they make up the solar system.**

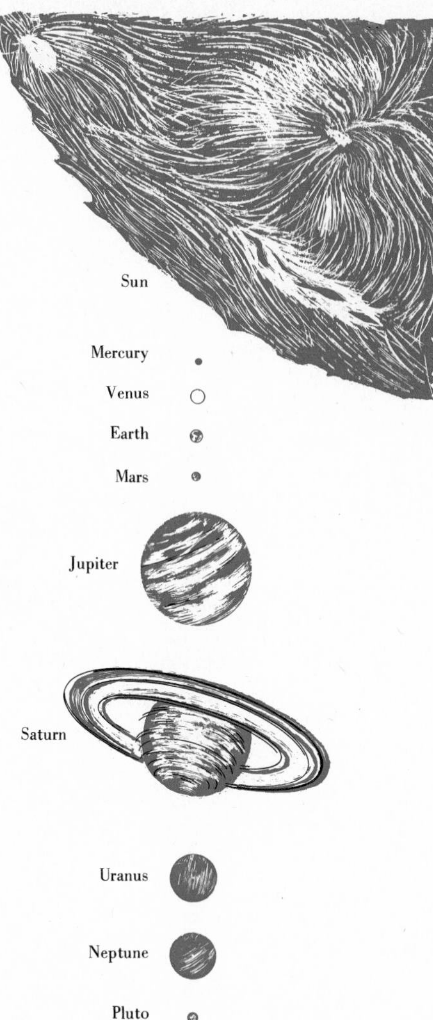

Figure 24.3 **Relative sizes of the planets.**

along the way and by the energy converted from mass during the synthesis of helium (see Chapter 2). The carbon has acted simply as a *catalyst;* it takes part in the reaction but emerges unchanged, so it is used over and over again. The hydrogen has been converted to helium. In effect, the sun is "burning" 4 million tons of hydrogen per second, producing an "ash" of helium. Even at this rate, the sun can keep on for 30 billion years.

These atomic reactions take place at temperatures of several million degrees and are automatically controlled. If the temperature increases, the process operates too rapidly and expands the gas of the sun. This expansion causes cooling, which lowers the rate of energy production.

The sun's nuclear reactions create lethal gamma rays which travel towards the sun's surface and change to X rays and ultraviolet rays on their journey as they collide with other atoms. The continuous output of solar energy has been calculated at 38×10^{16} w (watts).

Sunspots are eruptions of hot gas seen on the sun's surface and causing magnetic storms here on earth. They begin at about 20° or 30° north or south of the equator and move toward it. When they are within 8° of the equator, they disappear. Sunspots occur in groups with each individual one being short-lived, rarely lasting more than 50 days. They range in size from a few hundred to over 80,000 km in diameter. Sunspots wax and wane in 11-year cycles. They seem to be a part of a larger cycle in which the sun's magnetic field reverses itself every 22 years. The last reversal occurred in 1957–1958, when sunspots were at their greatest intensity.

There are other eruptions from the sun's surface which are related to magnetic discharges. The largest blasts of hot ionized gases—plasma—create the solar flares. These propel particles all the way to earth.

Planets and satellites

The name *planet* was given to certain celestial bodies that appear to wander about the sky, in contrast to the seemingly fixed stars. It came from the Greek *planetes,* meaning "wandering." Our earth is a planet, one of nine that circle the sun.

There is a systematic uniformity in the motion of the planets. All travel around the sun from west to east in nearly the same plane, the *ecliptic.* And they revolve on their respective axes in a "forward" sense, or from west to east, with the exception of Venus and Uranus.

In order of distance from the sun, the first four planets are Mercury, Venus, the earth, and Mars. These planets are about the same size and fairly dense, as though they were made of iron and stone. They are called the *terrestrial planets* because of their similarity to the earth. Next in order of distance from the sun are Jupiter, Saturn, Uranus, Neptune, and Pluto. The first four of these are of relatively large size and low density (see Figure 24.3). Little is known about Pluto, whose discovery was announced on March 12, 1930, but it is more like the terrestrial planets than the others. A uniform pattern of spacing outward from the sun is broken between Mars and Jupiter. In the "gap" are the thousands of asteroids ranging in size from 2 km to about

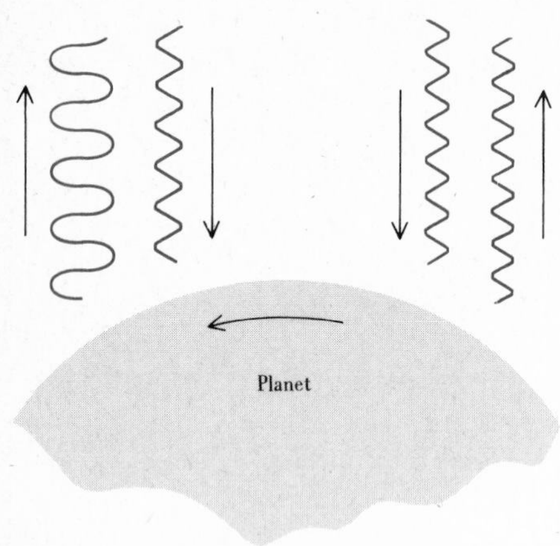

Figure 24.4 **Rotation determined by Doppler effect. Waves reflected from the center of a planet's disk show no change of frequency. Waves reflected from a receding limb show a shift to lower frequency.**

770 km in diameter. Some 1,500 have been catalogued to date. The asteroids occupy an orbit believed to have been either that of a planet that exploded or of matter that never completed the planet-forming process.

The planets emit radio waves. These penetrate the planetary atmospheres and cloud covers and carry signals to the earth, where they are picked up by radio telescopes at wavelengths between 1 cm and 15 m.

Also, radar signals transmitted from the earth and bounced back for reception on the radio telescopes have been used to gather information about planetary surfaces.

MERCURY The planet closest to the sun is Mercury. In its orbit around the sun, it sometimes comes as close to the earth as 77 million km. It is the smallest planet, not much larger than our moon. Its equatorial diameter is approximately 4,800 km compared to the moon's 3,450 km. Its density is practically the same as that of the earth and its mass and volume are about one-twentieth of that of the earth.

Although it is the fourth brightest of the planets, it has been difficult to observe because of its closeness to the sun. It was long thought to rotate on its axis once in 88 days, the time in which it completes a circuit of the sun, and, accordingly, always to present the same side to the sun. However, in 1965 evidence was obtained by radar recorded on a radio telescope in Arecibo, Puerto Rico, that Mercury actually rotates on its axis once in 59 days. Doppler differences between center and edge reflections provided the measurements (see Figure 24.4). With this discovery, we now know that the sun rises on Mercury every 170 earth days.

From the intensity of its radio emissions, it has been estimated that the surface temperature of Mercury gets up to 415°C, hot enough to melt tin or lead, at "high noon." From reflected radar signals, it has been found that the dark side of Mercury has a temperature of roughly 21°C, a great deal higher than the −263°C it was long believed to be.

Mercury's surface shows some intriguing dark spots which suggest that it may have a topography similar to that of the moon. It has a very thin atmosphere, with an approximate density about one-thousandth that of the earth's atmosphere. If it once had a thicker atmosphere, this has been driven off by the heat of the sun.

VENUS Venus is exceeded only by the sun and the moon in brightness. It is the planet that swings closest to the earth in its orbit around the sun, where it travels 107 million km in 224.7 earth days (Figure 24.5). Its density of 5.25 is close to the earth's 5.5. From the properties of its atmosphere and the height of its cloud deck, its size has been estimated as 12,042 km in diameter, compared to the earth's 12,682 km. It has no satellites and its surface is completely hidden from visual observation by a thick, bright, slightly yellowish unbroken cloud cover which prevents optical telescopes from obtaining much information about it. It receives twice as much solar radiation as the earth, but the brightness of its cloud cover has led to the conclusion that a possible 70 percent of this is reflected back into space.

With the development of high-altitude stratospheric balloon astronomy, instruments were sent high enough to get a spectrum of traces of H_2O above Venus' clouds. It is suspected that carbon dioxide vapor above the clouds is enormous compared with that in the earth's atmosphere. It may equal 1.6 km equivalent thickness at sea level pressure compared to 2 m for the earth's atmosphere. The cloud deck starts approximately 60 km above Venus' surface. Infrared measurements of temperature of the top of the clouds consistently show $-40°C$. The atmosphere between the cloud deck and the surface is believed to be 30 times as dense as that of the earth (see Figure 24.6). It has been suggested that the clouds' outer layer is composed of fine ice particles and the lower cloud layer of a mixture of carbon dioxide and water vapor. In 1960, it was determined by radar that Venus rotates in a retrograde sense at the slow rate of once in 243 earth days.

Figure 24.5 ***Phases of Venus as seen through a telescope. It appears full and small when on the far side of the sun. As it nears the earth, it becomes larger and a thin crescent. Then, almost between the earth and the sun, it shows a fuzzy partial halo as sunlight is scattered by its atmosphere.***

On December 14, 1962, the U.S. spacecraft *Mariner II* approached within 34,560 km of Venus and radioed scientific data 59 million km back to the earth. From these data, scientists were able to confirm their earlier conclusions that the surface of Venus must be dry and far too hot for the existence of any organic life (see Figure 24.7).

In October 1967, the U.S.S.R. probed the atmosphere and landed a spacecraft, *Venus 4*, on the planet's surface. It radioed back measures of temperature, pressure, and composition. Maximum temperature, which is probably the surface temperature, was 280°C. The surface pressure was 12 to 22 atm and atmospheric composition was 90 to 95 percent carbon dioxide, 0.4 percent water, less than 7 percent nitrogen, and less than 1.6 percent water plus oxygen. Thirty-four hours later, the U.S. *Mariner V* at a distance of 4,000 km detected a carbon dioxide concentration of 72 to 87 percent and found no evidence of oxygen. Also, the magnetic field of Venus was about 0.003 that of the earth.

Radio thermal measurements have shown that the temperature does not change much between the night and day sides of the planet. The total variation is about 15 to 40°C.

The surface of Venus is believed to be composed of loose dust. It is also extremely hot and dry. By analyzing the pattern of radar rays reflected off its surface, scientists have discovered that it is not smooth (see Figure 24.8). There are a couple of regions of high relief which are called "mountain ranges." One trends north–south over a distance of approximately 3,800 km and is several hundred kilometers across. Another range of similar size runs east and west.

A great unsolved mystery is how it is possible for Venus and the earth, so nearly the same in size, mass, and distance from the sun, to be so different.

THE EARTH The earth is a sphere 12,682 km in diameter. It revolves about the sun in a plane called the *ecliptic* and rotates about an axis the direction of which in space remains fixed and inclined relative to the ecliptic so that the equator makes an angle of 23.5° with it. The time required for one revolution around the sun is defined as 1 year and the time required for one complete rotation on its axis is 1 day. The annual revolution about the sun, combined with the inclination of its axis of rotation, produces the seasons. When the northern hemisphere is tipped toward the sun, that zone of the earth

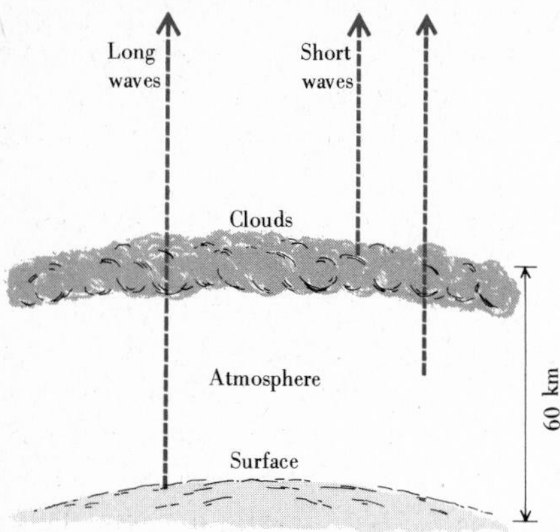

Figure 24.6 **Model of Venus' atmosphere and clouds, in part based on data from U.S. spacecraft Mariner II during its 1962 flyby.**

Figure 24.7 **Possible effect of Venus' clouds and atmosphere in trapping heat. The planet's surface receives short, visible light waves from the sun during the day, absorbs them and heats up, and then radiates long, infrared waves. Some of these escape, but enough are trapped to maintain the surface temperature at a high level. The same principle operates in a florist's greenhouse. In a similar way, our atmosphere operates to stabilize the earth's surface at about 10° C above what it would be without it.**

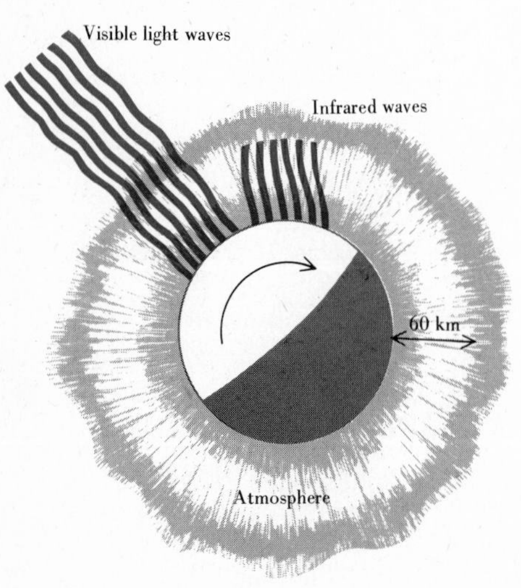

gets more heat and has summer. On the opposite side of the orbit, the northern hemisphere "leans" away from the sun and has winter (see Figure 24.9).

The earth is not a perfect sphere. The "solid" earth adjusts itself to the centrifugal force created by its rotation by bulging at the equator so that its radius there is 21.5 km greater than that at the poles. And measurements by artificial satellites have revealed two 30-m "dimples" in the northern hemisphere, which give the earth a slightly pear-shaped outline.

The sun provides the earth with a heat range sufficient to maintain its life forms, but life is possible only because this solar energy is controlled by the atmosphere. It keeps noon temperatures from rising too high and night temperatures from sinking too low, and it shields the surface from lethal quantities of ultraviolet and charged-particle radiation, as well as X-ray and gamma-ray effects (see Figure 24.10). It is doubtful that life in any form could exist without this shielding. The atmosphere also protects the earth's surface from thousands of millions of meteors each day. Most of these are no larger than a fair-sized speck of dust, much smaller than an average grain of sand, and weigh less than one-thousandth of a gram. But such a particle traveling 70 km/sec would strike with the same energy as a direct discharge from a 45-caliber pistol fired at point-blank range and would be dangerous to a person. In the atmosphere, these particles are vaporized by friction. Occasionally, some of the more massive meteors have been able to penetrate to the surface of the earth and cause damage. The Great Siberian meteor of 1908, on reaching the earth, exploded so violently that trees were laid flat out to 50 km from the area of impact. On the average, only one person is killed by a meteorite every 100 years, but if a large one like that of 1908 should hit a modern city, it would be more destructive than an atomic bomb blast.

Early man-made satellites took measurements that led to the discovery of the Van Allen belts (see Figure 2.23). These begin about 1,000 km above the earth's surface and are classed as parts of the earth's atmosphere. They consist mostly of very energetic ionized particles, mostly electrons and nuclei of hydrogen, trapped by the earth's magnetic field. They are dangerous to living organisms not protected by shielding equivalent to about 1.3 cm of lead.

The density of our atmosphere decreases with height. Half the air is contained in the first 6 km, half the remainder in the next 6 km, and so on. At almost 100 km above the surface, density is one-millionth the surface value. At this height, short-wave radio signals are reflected.

MARS The diameter of Mars is a little more than one-half that of the earth and its mass is about one-tenth. It travels in an elliptical orbit around the sun at a mean distance approximately 50 percent greater than that of the earth. It receives less than half as much sunlight and is therefore colder. Its axis is tipped at approximately the same angle as that of the earth, giving each hemisphere a summer and winter. Because it takes 687 earth days to complete its orbit around the sun, its year is a little less than two of our years, and its seasons are twice as long.

Mars has a thin hazy atmosphere thick enough to give it a climate and weather but not thick enough to hide its surface features. It has white polar "ice caps," which form in winter and disappear in summer. These suggest the possible

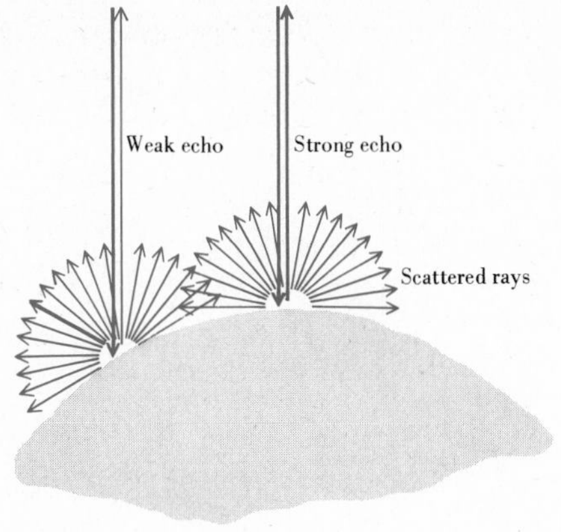

(*a*)

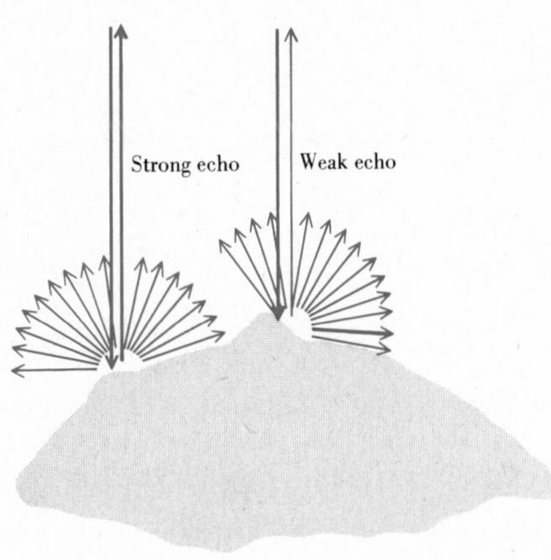

(*b*)

Figure 24.8 ***Mapping by reflected waves. A smooth surface (a) returns reflections that weaken progressively from the center of the disk. A rough surface (b) gives irregular variations in reflected signal strength.***

presence of water, but for all we can tell they may be no more than a few centimeters thick. Possibly the polar caps are carbon dioxide snow, mixed with a bit of water frost, and the white clouds and mists may also be mostly carbon dioxide.

Its surface colors change with the seasons. In the spring, its ice caps become smaller and dark areas spread from the polar regions toward the equator at about 2 km/hr. The southern hemisphere ice cap disappears entirely, but the northern cap does not totally disappear. The spreading dark areas of spring are reversed in the fall and it has been suggested that they are caused by organisms that hibernate below the surface in winter and move toward the surface in the summer. Others have suggested that the changes are due to changing wind patterns, with wind sweeping dust off the plateaus into the lowlands in spring and the reverse in the fall.

It has been suggested that some of the many pronounced markings on Mars' surface were sculptured by running water. At the turn of the twentieth century, man-made canals were thought to exist, but astronomers now generally reject this interpretation. Clouds frequently form, and in 1956, E. C. Slipher of Lowell Observatory observed an apparent physical reaction related to the clouds, a darkening of the soil when the clouds left an area. Because the planet's dry atmosphere prevents rainfall, rain as we know it could not have caused this change, but some kind of moisture transfer from cloud to soil is regarded as possible.

Extensive and persistent dust storms occur on Mars. There are only two substances that show the unusual light-scattering properties exhibited by Mars' surface, the bog-iron ores limonite and goethite, which are oxides of iron. Below a light dry dust of these rusty materials, the ground is probably solid and permanently frozen.

The range of uncertainty in much of our information about Mars has left room for speculation. It may be, for example, that there is plant life resembling desert vegetation on the earth—probably mostly mosses and lichens. Also, though little oxygen and water remain on the planet, they may have been abundant in the distant past. One hypothesis postulates that much of the oxygen was used up by combining with iron in the rocks. And great quantities of water may be held in chemical combination with rocks, as well as in the form of permafrost.

In 1965, the U.S. spacecraft *Mariner 4* traveled 500 million km and came within 10,000 km of the surface of Mars. It took pictures and relayed them along with some other information back to the earth. The pictures showed circular craters similar to those on the moon. They ranged in size from 5 to 120 km in diameter. Some appear to have been eroded. The agent of erosion is not known. Water is not present on Mars in significant quantity today, so if they were eroded by running water, it would have had to be in the distant past. The erosion agent may have been the wind.

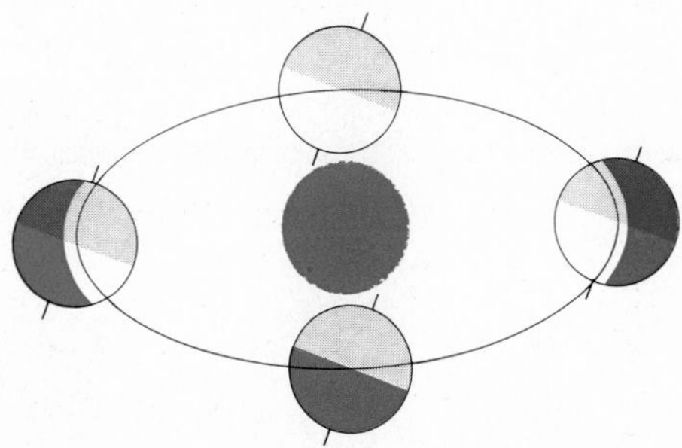

Figure 24.9 ***The seasons on a planet result from its axis of rotation's being tilted with reference to the plane in which it orbits the sun. This gives each hemisphere more sunlight for half a year.***

Figure 24.10 **The earth as seen from space by the Verner E. Suomi camera. ATS-1 photograph, December 11, 1966.** (*Courtesy V. E. Suomi and Goddard Space Flight Center.*)

Mariner 4 did not detect a magnetic field around Mars. It also confirmed that Mars' atmosphere is thin, but found it thinner than had been supposed. It is about as dense as our atmosphere is at a height of 32 km. Atmospheric pressure at its surface is 1 percent of that of the earth. This thin atmosphere does not trap and hold much of the sun's heat. The temperature at its equator at noon may be 20 to 30°C, dropping to −85°C at night. It is colder at the poles.

No oxygen has been found in Mars' atmosphere, which is composed primarily of carbon dioxide.

Radar astronomy seems to indicate that Mars has a variety of topographic features. Extensive highlands seem to rise as much as 20 km above low-lying deserts.

Mars has two small satellites orbiting it. Deimos, 5 km in diameter, orbits at a distance of 24,000 km. Phobos, 8 km in diameter, orbits at 9,300 km.

On July 31, 1969, *Mariner 6*, carrying two television cameras, flew by Mars and recorded 74 pictures. On August 5, 1969, *Mariner 7* passed Mars, acquiring 93 far-encounter and 33 near-encounter pictures[2] (see Figure 24.11).

Before the *Mariner 4* expedition, Mars was thought to be like the earth; after *Mariner 4*, it seemed to be like the moon; through *Mariners 6 and 7*, it was found to have its own distinctive features, unknown elsewhere in the solar system.

[2]Robert B. Leighton, Norman H. Horowitz, Bruce C. Murray, Robert P. Sharp, Alan G. Herriman, Andrew T. Young, Bradford A. Smith, Merton E. Davies, and Conway B. Leovy, "Mariner 6 Television Pictures: First Report," *Science*, vol. 165, pp. 684–690, August 15, 1969; "Mariner 7 Television Pictures: First Report," vol. 165, pp. 788–795, August 22, 1969.

Within one picture frame covering 625,000 km there were 156 craters ranging from 3 to 240 km in diameter. There is evidence of some weathering and transportational processes more effective on Mars than on the moon. Many Martian craters have flatter floors, and fewer central peaks, with no obvious secondary craters and rays. Mars has many "ghost" craters but has neither sinuous rilles like the moon nor earthlike features such as mountain ranges, tectonic basins, stream-cut surfaces, dune fields, playa flats, or other arid-region features.

The first impression of Mars given by the pictures is that the surface is generally visible without obscuring clouds or haze, with the exception of the polar regions and a few areas where afternoon "clouds" appear.

Mariners 6 and 7 confirmed the earlier evidence of a moonlike cratered appearance, but they added terrain data suggestive of more active and more recent surface processes than had been evident before. They showed three distinctive terrains not correlated in any simple manner with light and dark markings observed from the earth: (1) cratered terrains, (2) chaotic terrains, and (3) featureless terrains.

Terrains dominated by craters are known to be widespread in the southern hemisphere and as far north as latitude 20°. Our knowledge of cratered terrains in the northern hemisphere is less complete, partly because of poor photographic coverage and unfavorable sun angles. Craters are of two types, large and flat-bottomed or small and bowl-shaped. Diameters of the flat-bottomed craters range from a few kilometers to a few hundred kilometers, with estimated diameter-to-depth ratios of the order of 100 to 1. Some of the smaller, bowl-shaped craters appear to have interior slopes steeper than 20°.

Chaotic terrains emerge abruptly from relatively smooth cratered surfaces as irregularly shaped apparently lower areas of jumbled ridges and depressions 1 to 3 km wide and 2 to 10 km long.

Featureless terrains are typified by the largest such area thus far identified, the floor of the bright circular "desert" Hellas, centered at about latitude 40°S, longitude 290° on the International Astronomical Union's map of Mars. No area of comparable size and smoothness is known on the moon.

At the turn of the twentieth century, an intriguing suggestion was made by astronomer Percival Lowell that certain narrow markings on Mars were in fact artificial waterways or canals. Unfortunately for romanticists, it is now agreed that these do not constitute a network and they are not waterways, artificial or natural.

Inferred processes The absence of earthlike tectonic features suggests that the crust of Mars has not been subjected to the kinds of internal and external forces that have shaped the earth's surface. A related inference is that Mars never had an earthlike atmosphere.

It has been deduced that many if not most heavily cratered areas are primordial and have never been subjected to erosion by water. However, erosion of some kind, blanketing, and other surface processes must have been operating almost up to the present in the areas of chaotic and featureless terrains.

The chaotic terrain gives a general impression of collapse structures caused by large-scale withdrawal or alteration of underlying layers.

A preponderance of the observations indicates that the seasonal white polar

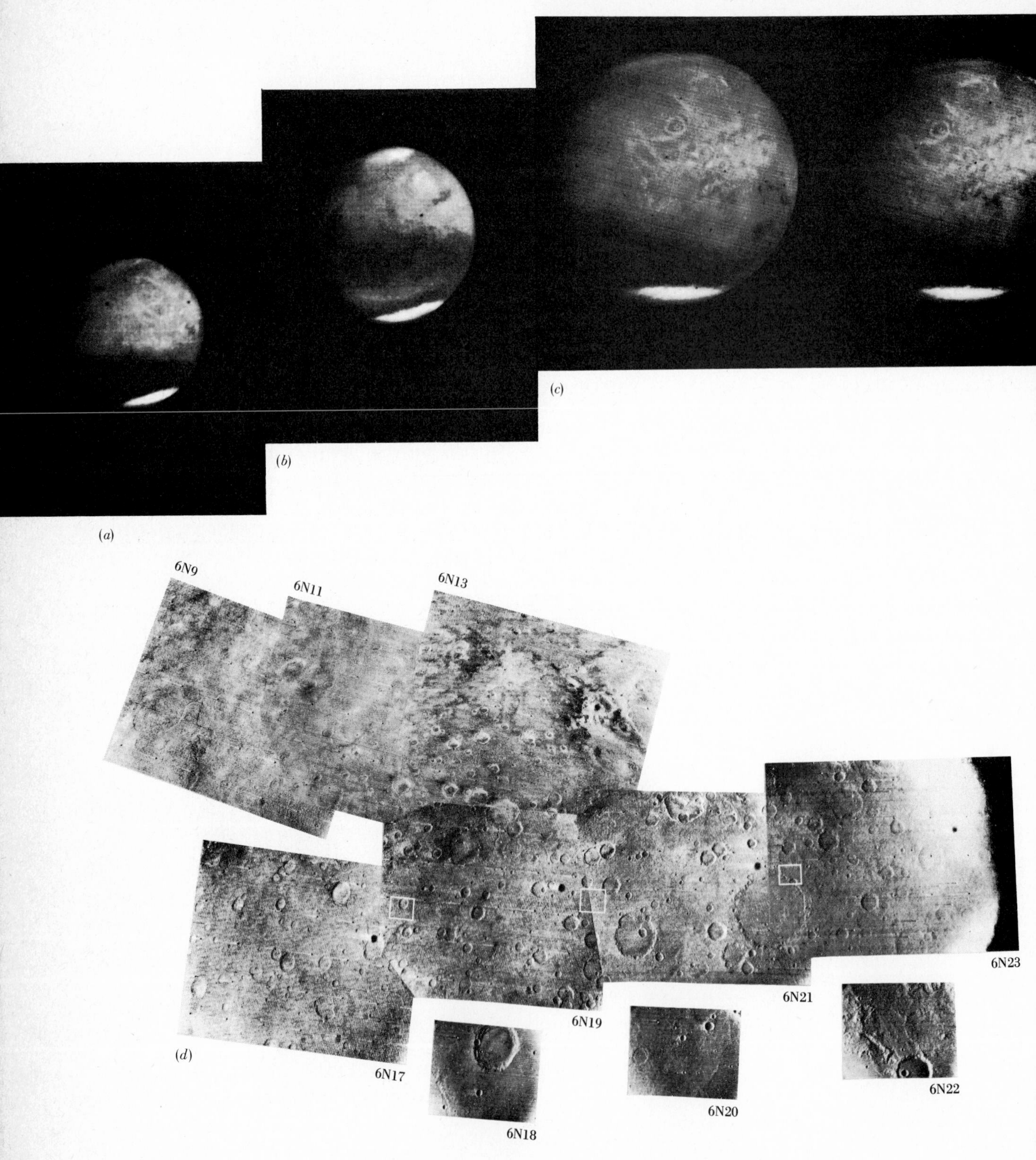
(a)
(b)
(c)
(d)
6N9
6N11
6N13
6N17
6N18
6N19
6N20
6N21
6N22
6N23

Figure 24.11 **Mars from far encounter to close approach as seen from Mariners 6 and 7. (a) Mariner 7 picture 7F47, from 1,014,870 km, shows the south polar cap and the crater Nix Olympica. (b) Mariner 7 picture 7F57, from 861,850 km, shows both polar caps (north at top). The bright ring at the upper right is in the area known as Elysium and appears to be a crater at least 320 km in diameter. (c) Mariner 7 stereo pair of picture 7F73 (on right), from 471,750 km, and 7F74, from 452,100 km. Taken 47 min apart. The prominent bright ring is the crater Nix Olympica. Mars' north is at the top when the picture is tipped 12° clockwise. (d) Mariner 6 mosaic P10332 from seven wide-angle pictures taken during close passage on July 30, 1969. In 6N22, the larger crater is about 24 km across and is on the edge of one shown in 6N21, which is 240 km in diameter. Craters as small as 300 m in diameter can be seen in the small, high-resolution pictures.** *(All photographs courtesy of NASA, by Jet Propulsion Laboratory of the California Institute of Technology.)*

"ice caps" are CO_2 snow with insignificant amounts of H_2O interspersed.

If there is any life on Mars, it would probably be microbial. The most serious limiting factor for life on Mars is scarcity of water, and the evidence to date is that there is little, if any, likelihood of finding life on Mars, or that there was a significant aqueous phase in the planet's history.

JUPITER, SATURN, URANUS, AND NEPTUNE These rapidly rotating planets differ so greatly from the terrestrial planets Mercury, Venus, the earth, Mars, and Pluto that they hardly seem to belong to the same system. Their densities range from Saturn's 0.7 (on a scale where water equals 1.0) through 1.33 for Jupiter, 1.53 for Uranus, to 2.41 for Neptune, in contrast to the terrestrial planets' 4 to 5.5. They have thick atmospheres of hydrogen, helium, methane, and in some cases ammonia. Their planetary surfaces are not clearly distinct from their atmospheres in the sense that they are for the earth and the other planets.

Jupiter Jupiter is the largest planet in our solar system. Its diameter of 142,400 km is over 11 times that of the earth. Its mass is 300 times and volume 1,000 times as great as that of the earth. All this enormity rotates at a tremendous speed to complete the day in 9 hr and 55 min.

Jupiter orbits the sun once in 12 years. At an average distance of 778 million km from the sun, it receives only one-twenty-seventh of the sunlight that the earth does. Recent evidence shows that it is emitting more heat than it is receiving, so it must be generating some of its own.

Jupiter has 12 satellites, more than any other planet (see Figure 24.12). Most of these travel in circular orbits. The inner 8 satellites orbit in the same direction that the planet rotates, but the outer 4 go in the reverse direction. The innermost satellite is approximately 112 km in diameter and orbits at 179,000 km. The third satellite is as large as the planet Mercury and orbits at 1 million km. The outer 4 satellites are less than 16 km in diameter and are roughly 22 million km from Jupiter.

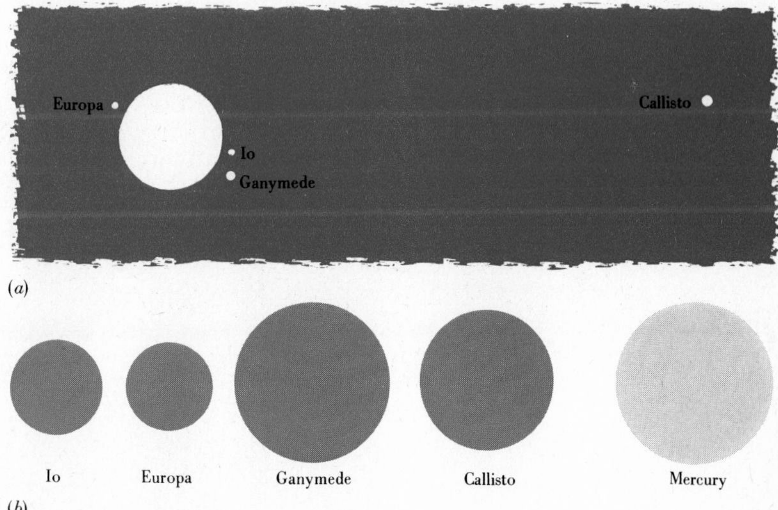

Figure 24.12 **Jupiter has 12 satellites. The four brightest are called the Galilean satellites, after their discoverer, and are named Io, Europa, Ganymede, and Callisto. (a) Callisto is nearly 2 million km away from Jupiter. The other three Galilean satellites have orbits from 420,000 to 1 million km away. (b) Ganymede's 5,000-km diameter exceeds that of the planet Mercury by 145 km. Callisto's diameter is 4,800 km; that of Io, 3,200; and that of Europa, 2,900.**

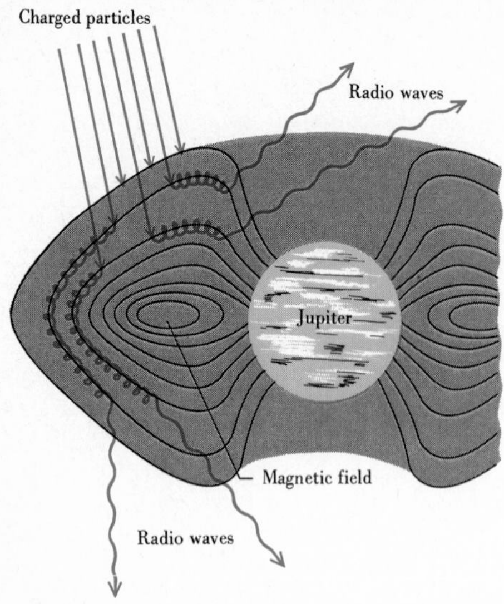

Figure 24.13 **Jupiter has a gigantic magnetic field around it which traps charged particles that emit radio waves and make it one of the strongest radio sources in the sky.**

Jupiter has several interesting features. It has alternating light and dark bands running parallel to its equator and covering most of its surface. These move around the planet at quite high speeds. They may be aligned by powerful trade winds. An unusual feature, first observed in 1878, is the Great Red Spot. It is 48,000 km long, 12,800 km wide, and slowly moves in relation to its immediate surroundings, although never out of Jupiter's south tropical zone. It drifts within the zone as though it were not fixed to a solid surface. It often changes color and sometimes may disappear from view for years. It has not been explained.

Jupiter contains twice as much matter as all the other planets combined, yet its average density is only 1.33 times that of water, which makes it about one-fourth as dense as the earth. To have this density, it is believed that Jupiter is composed primarily of hydrogen and helium. Computation has led to the conclusion that the ratio density at the center/density of upper levels is greater for Jupiter than it is for the earth. To explain this density distribution, it has been suggested that at the internal pressures which exceed 1 million earth atmospheres, the hydrogen atoms collapse. In this new atomic state they behave like a metal.

Jupiter's atmosphere is composed primarily of hydrogen and helium with some ammonia and methane. Some water vapor has been detected. The highest clouds are chiefly ammonia snowflakes at $-125°C$. It is believed to be thousands of kilometers thick, but we have no proof of this as Jupiter has no discernible inner surface. It is believed that hydrogen undergoes a gradual transition from gas to liquid to solid as pressure increases.

Jupiter has a powerful magnetic field encircling it, shaped like our magnetosphere (see Figure 24.13). It traps electrically charged particles. From this, it is believed that Jupiter must have a magnetic core.

Saturn Saturn is the most picturesque of the planets (see Figure 24.14). Its faint bandings are similar to Jupiter's, though more regular. Girdling it above its equator is a set of rings unique among the planets (see Figure 24.15). These rings are composed of individual sand-sized particles of matter, possibly coated with ice, each fragment moving in its own orbit about the planet. Each

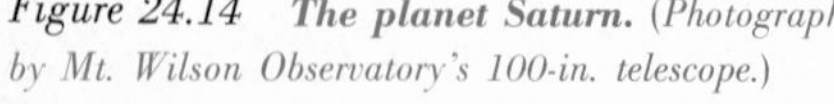
Figure 24.14 **The planet Saturn.** (*Photograph by Mt. Wilson Observatory's 100-in. telescope.*)

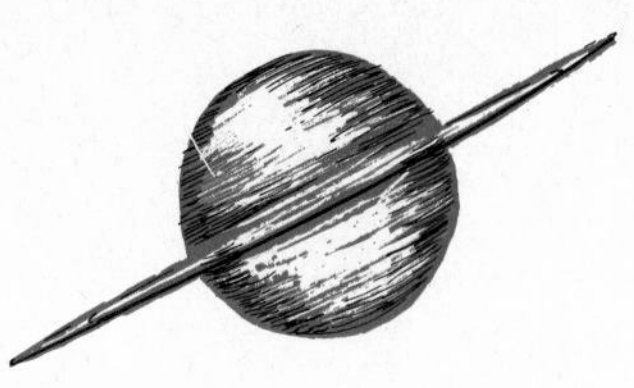

(a)

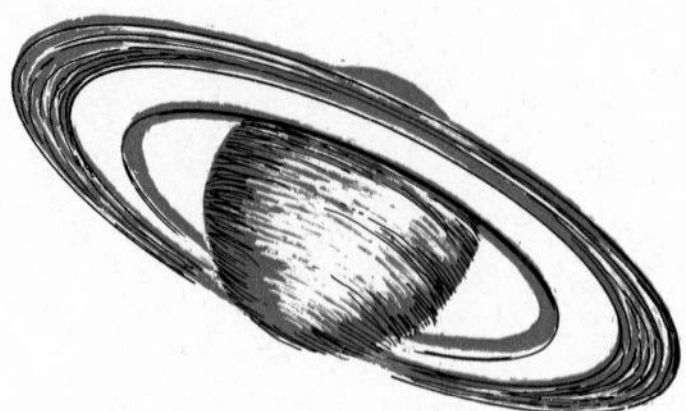

(b)

(c)

Figure 24.15 **Saturn's ring system (a) as seen edge-on from the earth in 1951; (b) as seen in 1960; and (c) as seen almost edge-on again in 1967.**

ring may be no more than a few centimeters thick, although the outermost has a diameter of 274,000 km.

Saturn is a planet of very low density, seven-tenths that of water. No oxygen has been detected in its atmosphere. It is composed mostly of hydrogen and helium, and its upper atmosphere has less ammonia and more methane than that of Jupiter. Its total mass is one-third that of Jupiter, and its gases are not as compressed as are those of Jupiter. It is 120,000 km in diameter and probably has a center of solid hydrogen. The top of its atmosphere is $-170°C$. It rotates in about 10 hr and takes 29 years to orbit the sun at an average distance of 1,427 million km.

It has 10 satellites. The tenth was discovered on December 15, 1966. The sixth is the only satellite in the solar system with an atmosphere. This atmosphere consists of about one-half as much methane as the atmospheres of Jupiter and Saturn.

Saturn seems to lack a magnetosphere, though it could be that it has one and that it has not been detected due to the rings.

Uranus and Neptune Their great distance from the earth makes it difficult to observe Uranus and Neptune. Uranus' average distance from the sun is 4,480 million km. It has a diameter of 48,000 km, rotates on its axis every 10 hr and 49 min, and orbits the sun every 84 years. It is mainly composed of hydrogen and helium, and large amounts of methane have been detected. Faint, slow-moving wind bands have been observed. These are probably frozen ammonia and have a temperature of about $-185°C$. Uranus has five satellites. Its plane of rotation is tipped 98° to the plane of revolution about the sun, so that its rotation is retrograde.

Neptune's average distance from the sun is 3,480 million km. Its diameter is 45,000 km. It rotates in 15 hr and orbits the sun in 165 years. It probably has a composition similar to that of Uranus, but no ammonia has been detected. No bands of clouds have been reported. Its outer atmospheric temperature is probably $-195°C$. Neptune has two satellites.

PLUTO Pluto, the most distant planet yet discovered, is a dwarf among giants in the outer reaches of the solar system. Determination of its equatorial diameter as in the vicinity of 5,700 km has been questioned, so the density is not known—except that it is probably significantly greater than that of the giant outer planets. Its mass has been calculated as approximately eight-tenths that of the earth. The temperature is probably lower than $-200°C$, only a few degrees from absolute zero. Most common gases would liquefy or freeze out on its surface. It has even been suggested that Pluto is not a true planet at all but a lost satellite from Neptune because it crosses the orbit of Neptune in its travel around the sun every 248 years at a speed of nearly 18,000 km/hr. Its eccentric elliptical orbit takes it to 4.5 billion km from the sun at the closest point and 7.4 billion km from it at the farthest point (see Table 24.1).

Table 24.1 ***The planets***

Planet	*Average distance from sun, km × 10^3*	*Rotation period, earth units*	*Length of year, earth units*	*Equatorial diameter, km*	*Mass (earth = 1)*	*Density (water = 1)*	*Satellites*	*Equivalent weight of 67.5 kg man*	*Gases identified in atmosphere*
Mercury	58,000	59 days	88.0 days	4,840	0.05	5.5	None	20	None
Venus	108,000	243 days	224.7 days	12,042	0.81	5.25	None	54	CO_2, H_2O (?), HCl, HF
Earth	149,598	23.9 hr	365.3 days	12,682	1.00	5.52	1	67.5	Many
Mars	229,000	24.6 hr	687.0 days	6,720	0.11	3.96	2	20	CO_2, H_2O
Jupiter	780,000	9.8 hr	11.9 years	141,920	317.9	1.33	12	175.5	CH_4, NH_3, H_2, He (?)
Saturn	1,431,000	10.2–10.6 hr	29.5 years	120,000	95.2	0.68	10	74	CH_4, NH_3, H_2, He (?)
Uranus	2,880,000	10.8 hr	84.0 years	47,360	14.5	1.60	5	61	CH_4, H_2
Neptune	4,510,000	15.7 hr	164.8 years	44,160	17.4	2.30	2	101	CH_4, H_2
Pluto	5,950,000	6.4 days	249.9 years	5,760	?	?	0	?	None

24.2 *Our galaxy*

All the stars visible without the aid of a telescope and most of those visible with a telescope are in our Milky Way Galaxy. This group of at least 100 billion stars seems to be distributed about a center in the vicinity of the constellation Sagittarius, the Archer, which is visible from northern latitudes as it hangs low in the southeastern sky in summer. Our galaxy is shaped like a discus of truly Olympian proportions: about 20,000 light-years thick at the center, thinner at the edges, and 100,000 light-years in diameter. Our solar system is located in the principal plane of the galaxy, about 30,000 light-years from the center, a position that makes it difficult for us to work out the details of star distribution. It seems probable, however, that from a distance of 1.5 million light-years the galaxy looks not unlike one of the nearest of our neighboring galaxies (see Figure 24.16). Such external galaxies are sometimes

Figure 24.16 ***A distant galaxy showing a discus shape similar to our galaxy viewed edgewise.*** (*Photograph by Mt. Wilson Observatory.*)

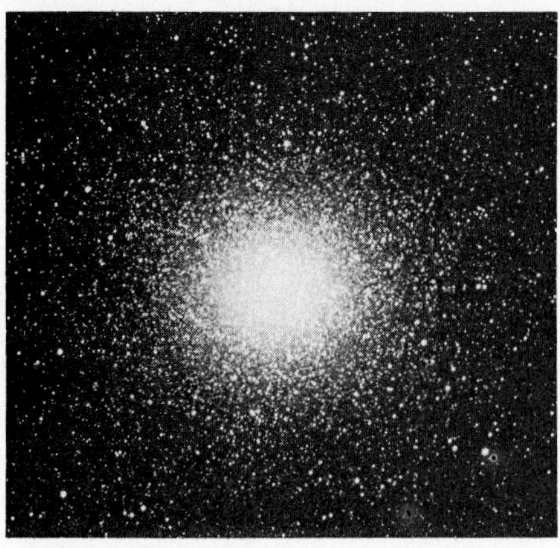

Figure 24.17 ***A globular cluster. At 22,000 light-years, this is one of the clusters nearest to us.*** *(Photograph by Harvard University's Boyden Station, Bloemfontein, South Africa.)*

called nebulae because of their filmy, nebulous appearance in the telescopes through which they were first seen. Those known to have a spiral structure are called *spiral nebulae.* Our Milky Way Galaxy is a spiral nebula.

Distributed fairly uniformly around the edge of our galaxy are about 100 globular clusters of stars (see Figure 24.17). Each cluster contains from 50,000 to 100,000 stars within a diameter of about 100 light-years. Because these stars are about 1 light-year apart, they are more closely crowded together than in any other region.

All the stars in our galaxy move around a galactic center. Our solar system travels at a speed of about 240 km/sec and requires 200 million years to complete a circuit.

Our sun is a very ordinary star compared to others in our galaxy. Some stars are so large that they would encompass the orbit of the sun and the earth. They are 5 times as hot as the sun, 10,000 times as bright, and 400 times its mass. Others are diminutive in comparison, having only one-fifth of the sun's mass, one-ten-thousandth of its brilliance, and one-third of its heat.

Matter in interstellar space

Distributed irregularly in interstellar space throughout the universe are myriads of tiny pieces of dust and gas. The dust and gas in the neighborhood of the sun, it has been estimated, account for half the total matter of our entire solar system. These pieces of matter are larger than molecules and atoms but so small that they could not be seen without the aid of the most powerful microscopes. They appear to be dusty grains of silica, or even frozen gases, or some other common substances converted to an unknown form by the low temperature, close to absolute zero, that prevails in interstellar space.[3]

Light that has passed through this interstellar matter in certain directions becomes polarized—that is, it ends up vibrating in only one plane, in contrast to the random pattern of vibration common to most light. This polarizing effect is interpreted as evidence that the pieces are not spherical but are rather elongated bodies aligned either by a gigantic magnetic field or by "winds" from moving gas.

The presence of interstellar matter is strikingly demonstrated in areas like the Horsehead Nebula in Orion, where some parts of the sky are blacked out by the material, whereas other areas glow with light that it reflects (see Figure 24.18).

It is now believed that most of the common elements found on the earth can be found in interstellar space. The density of this material is very low—about what we would get by pulverizing an ordinary marble and then spreading the dust uniformly throughout the volume of a sphere 1,600 km in diameter. Yet its presence is significant. There is too much of it to have been blown out of occasional exploding stars or to have escaped from others. Theorists believe this matter to be left over from the creation of the universe—pieces of the very stuff from which the stars and planets were compounded.

Figure 24.18 ***Horsehead Nebula in Orion, where some parts of the sky are blacked out while other areas glow with light reflected by interstellar matter.*** *(Photographed in red light by 200-in. telescope, Mt. Wilson and Palomar Observatories.)*

[3]Leo Goldberg and Lawrence H. Aller, *Atoms, Stars, and Nebulae,* McGraw-Hill-Blakiston, New York, 1943.

24.3 *Origin of our solar system*

Many ideas have been put forth on the origin of the planets. Some 2 centuries ago, the Comte de Buffon proposed that the planets were formed when a large mass collided with the sun and caused blobs of matter to be thrown out. These blobs became the planets. To calculate when this event took place, Buffon performed timed experiments on balls of metals and of rock. After heating these to white heat, he measured the rate at which they cooled. Then he computed how long it would have taken a ball as large as the earth to cool. His answer: 74,832 years. But Buffon overlooked some important factors: His samples were not fair representatives of the complex earth, and he neglected the fact that size governs rate of cooling.

A later idea on the origin of our solar system combined suggestions from three of science's great men—Kant, Laplace, and Helmholtz. The story begins with a very large, nebulous star, not very hot, because its matter was so spread out that it filled all the space included in the orbits of the planets then known. This original "sun" was assigned a diameter of 3.2 billion km. It was rotating. As it lost energy by radiating heat into space, it shrank; and as it shrank, it rotated faster and faster, as required by the law of conservation of angular momentum. As speed at its equator increased, centrifugal force there mounted. Finally this force became greater than the counterbalancing gravitative force, and some of the sun's matter was thrown off. This caused a minor collapse to make up for the lost matter. The ejected matter contracted into a sphere, forming a planet. The sun became stable until, losing heat through radiation, it began to shrink again and speed up, which led to throwing off more matter—another planet. This cycle continued until all the planets were formed. The remaining matter then collapsed into the present sun, which has not, up to now, lost enough heat to make it rotate fast enough to throw off any more planets.

Part of this hypothesis was sound, because a star can develop its heat by collapsing, or contracting. It happens that our sun does not. A sequence of events like these might produce a solar system, but it did not produce ours. By this hypothesis, the planets should all rotate in the plane of the sun's equator. They miss by about 6°. This is not much, but it is important. Also, this hypothesis would leave the sun with 98 percent of the solar system's angular momentum.[4] It actually has about 2 percent.

Failure of this hypothesis to explain some features of our solar system threw speculation back toward the Buffon idea. Instead of a massive body having collided with the sun, these later hypotheses proposed that such a body passed close by. The Chamberlain–Moulton planetesimal hypothesis blamed an invading star for tearing some matter from the sun. It proposed that our planets were built up from this separated matter. They started small in size and grew to present dimensions by sweeping up the detached matter as they circled

[4]The product of *mass* times *radius of orbit* times *velocity*.

the sun. In the Jeans–Jeffries tidal hypothesis, the planets condensed from gaseous filaments torn from both the sun and a passing star. These two hypothesis had to be modified before long, after it was shown that matter torn from the sun and the passing star would fall back into the parent bodies because it would not receive sufficient angular momentum from the pull to go into orbit. It was then postulated that the passing star was in fact a double star and that therefore three masses were involved. One of these was destroyed because of the pull of the other two stars on it. The debris from this destroyed star supplied the matter for the planets. Later calculations showed that stellar matter could be supplied under such conditions and could go into orbit around the sun but that it would not coagulate into planets. Science was making progress toward proving that our solar system could not be formed.

Early generations limited their speculation to origins of the planets. The origin of the sun itself was left to unexplained processes that occurred at an earlier date.

By the middle of the twentieth century, hypotheses that come under a general heading of *dust-cloud hypotheses* were sprouting and dying in close succession. Many included explanation of the origin of the sun as well as the planets. One begins with interstellar matter widely distributed throughout space. Light pressure from the stars is supposed to have caused this matter to pack together to form a cloud about 9,600 billion km in diameter. At about that size, the cloud began to collapse, very slowly at first, under the force of its own gravity. But having reached a diameter of about 5,900 million km, it swiftly completed the collapse in a few hundred years. The increased pressure in the contracting cloud greatly pushed up the temperature and in the last white-hot phase of its collapse, the sun began to radiate as a star. The planets and satellites were derived from minor dust streams in the original cloud before there had been much contraction.

Two important features of our solar system remained unexplained by dust-cloud hypotheses. They do not explain the spacing of the planets from the sun, and they do not explain the high angular momentum of the planets.

Dust-cloud hypotheses have been modified by using the properties of electric currents and magnetic fields to explain the angular momentum of the sun and planets. The physical laws governing behavior of the hot gases involved are thereby changed. Ordinarily, gases respond directly to the pull of gravity, rotation, and pressure. But in a magnetic field maintained by electric current—that is, a *magnetohydrodynamic field*—the ionized gas has a wiry strength that resists these forces. In 1960, Fred Hoyle proposed that magnetohydrodynamics governed the behavior of the original matter in the dust clouds, composed of rapidly rotating ionized gases. Through these gases ran lines of magnetohydrodynamic force similar in some respects to long, elastic threads tied to the gases. The gas in the outer portions of the cloud was revolving more slowly than the center, so the threads tended to wind around and around and stretch. This increased the angular momentum of the outer portion, which became the planets and slowed down the center, which became the sun. The amounts of slowing down of the sun and speeding up of the planets necessary to this hypotheses, however, are still not sufficient to explain the actual distribution of angular momentum in our solar system.

In the midtwentieth century, as radioactive dating methods (see Chapter 9) were applied to more and more of the earth's rocks, evidence accumulated that some of the oldest were formed over 3 billion years ago. Estimates of the age of the earth itself rose to around 5 or 6 billion years. At the same time, astronomers were considering hypotheses that the entire "known" universe is expanding from a volume of limited extent, from which it evolved 4 to 5 billion years ago. The elements composing the embryonic universe were formed within 5 min from a primordial plasma of photons, that is, light. At an early stage of the evolution from that point, "matter was dispersed throughout the universe in the form of its simplest elemental compound, hydrogen."[5] This condensed and under conditions of intense heat supplied by the gravitational potential energy underwent a series of nuclear reactions by means of which atoms of the heavier, more complex elements were built up. Nuclear reactions have been taking place in the hot interiors of stars, including our sun, ever since they came into existence.

Whatever the processes that have resulted in the formation of the universe, they seem to have produced an overwhelming preponderance of double stars, which are dynamically unlikely to be able to hold families of planets. Possibly 1 star in 100 is a single star, like our sun, and thus capable of having planets. And in our galaxy of 10^{11} stars, 1 in 1 million, or possibly even 1 in 1,000, may have a planet like the earth under similar conditions of heat supply. This statistical fact means that in our galaxy alone there may be 1 million to 1 billion planets on which "human" beings might exist. And there are millions of galaxies in the universe.

Outline

Our solar system consists of planets, satellites, asteroids, comets, and meteors held together by the sun's gravitation.
- The *sun* is a star that is the source of the solar system's light, heat, and charged particles.
- *Planets and satellites* are celestial bodies that appear to wander about the sky.
  - *Mercury* is the smallest planet and closest to the sun.
  - *Venus* is the most nearly like the earth in size and mass.
  - The *earth* is 12,800 km in diameter.
    - The *moon* is our nearest neighbor in space and our natural satellite.
  - *Mars* is crater-pocked, cold, and without oxygen.
  - *Jupiter* is the largest planet and has 12 satellites.
  - *Saturn* is girdled by a unique set of rings.
  - *Uranus* and *Neptune* are composed mainly of hydrogen and helium.
  - *Pluto*, the most distant planet, is more like the inner planets than its nearer neighbors.

Our galaxy is the Milky Way Galaxy with 100 billion stars and includes all we can see without the aid of a telescope.

[5]L. H. Ahrens, *Distribution of the Elements in our Planet*, McGraw-Hill Book Company, Inc., New York, 1965.

Matter in interstellar space is distributed irregularly throughout the universe as tiny pieces of dust and gas.

Origin of the solar system has been the subject of speculations ranging from collisions of stars to condensing of great quantities of interstellar matter.

Supplementary readings

BERGAMINI, DAVID, and THE EDITORS OF LIFE: *The Universe,* Time, Inc., New York, 1962. A technically competent presentation of well-chosen topics superbly illustrated. Its use of photography and color printing give any reader today a better comprehension than many professional astronomers were able to get a generation ago.

SAGAN, CARL, JONATHAN NORTON LEONARD, and THE EDITORS OF LIFE: *Planets,* Time, Inc., New York, 1966. An eminent authority on astronomy, Gerard P. Kuiper, Director of the Lunar and Planetary Laboratory, University of Arizona, says, "The text of this book is lucid, the discourse concise and intelligent."

WHIPPLE, FRED L.: *Earth, Moon, and Planets,* 3rd ed., Harvard University Press, Cambridge, Mass., 1968. This basic memoir, first published in 1941, has been fully revised to incorporate exciting expansions of our knowledge by the space program.

Appendix A

The elements

Table A.1 Alphabetical list of the elements

Element	Symbol	Atomic number	Element	Symbol	Atomic number
Actinium	Ac	89	Molybdenum	Mo	42
Aluminum	Al	13	Neodymium	Nd	60
Americium	Am	95	Neon	Ne	10
Antimony	Sb	51	Neptunium	Np	93
Argon	A	18	Nickel	Ni	28
Arsenic	As	33	Niobium	Nb	
Astatine	At	85	(or Columbium)	(Cb)	41
Barium	Ba	56	Nitrogen	N	7
Berkelium	Bk	97	Nobelium	No	102
Beryllium	Be	4	Osmium	Os	76
Bismuth	Bi	83	Oxygen	O	8
Boron	B	5	Palladium	Pd	46
Bromine	Br	35	Phosphorus	P	15
Cadmium	Cd	48	Platinum	Pt	78
Calcium	Ca	20	Plutonium	Pu	94
Californium	Cf	98	Polonium	Po	84
Carbon	C	6	Potassium	K	19
Cerium	Ce	58	Praseodymium	Pr	59
Cesium	Cs	55	Promethium	Pm	61
Chlorine	Cl	17	Protactinium	Pa	91
Chromium	Cr	24	Radium	Ra	88
Cobalt	Co	27	Radon	Rn	86
Columbium	Cb		Rhenium	Re	75
(or Niobium)	(Nb)	41	Rhodium	Rh	45
Copper	Cu	29	Rubidium	Rb	37
Curium	Cm	96	Ruthenium	Ru	44
Dysprosium	Dy	66	Samarium	Sm	62
Einsteinium	En	99	Scandium	Sc	21
Erbium	Er	68	Selenium	Se	34
Europium	Eu	63	Silicon	Si	14
Fermium	Fm	100	Silver	Ag	47
Fluorine	F	9	Sodium	Na	11
Francium	Fr	87	Strontium	Sr	38
Gadolinium	Gd	64	Sulfur	S	16
Gallium	Ga	31	Tantalum	Ta	73
Germanium	Ge	32	Technetium	Tc	43
Gold	Au	79	Tellurium	Te	52
Hafnium	Hf	72	Terbium	Tb	65
Helium	He	2	Thallium	Tl	81
Holmium	Ho	67	Thorium	Th	90
Hydrogen	H	1	Thulium	Tm	69
Indium	In	49	Tin	Sn	50
Iodine	I	53	Titanium	Ti	22
Iridium	Ir	77	Tungsten		
Iron	Fe	26	(or Wolfram)	W	74
Krypton	Kr	36	Uranium	U	92
Lanthanum	La	57	Vanadium	V	23
Lawrencium	Lw	103	Wolfram		
Lead	Pb	82	(or Tungsten)	W	74
Lithium	Li	3	Xenon	Xe	54
Lutetium	Lu	71	Ytterbium	Yb	70
Magnesium	Mg	12	Yttrium	Y	39
Manganese	Mn	25	Zinc	Zn	30
Mendelevium	Me	101	Zirconium	Zr	40
Mercury	Hg	80			

Table A.2 ***Electronic configuration of elements 1 through 30***

Atomic number[a]	Name	Symbol	Number of electrons in shell 1	shell 2		shell 3			shell 4				Mass number of stable isotopes[b]	Parts/10^6 in earth's crust[c]
			s	*s*	*p*	*s*	*p*	*d*	*s*	*p*	*d*	*f*		
1	Hydrogen	H	1										1, 2	1,400
2	Helium	He	2 (inert gas)										4, 3	0.003
3	Lithium	Li	2	1									7, 6	65
4	Beryllium	Be	2	2									9	<0.001
5	*Boron*	*B*	*2*	*2*	*1*								*11, 10*	*3*
6	Carbon	C	2	2	2								12, 13	320
7	Nitrogen	N	2	2	3								14, 15	46
8	Oxygen	O	2	2	4								16, 18	466,000
9	Fluorine	F	2	2	5								19	300
10	Neon	Ne	2	2	6 (inert gas)								20, 22, 21	<0.001
11	Sodium	Na	2	2	6	1							23	28,300
12	Magnesium	Mg	2	2	6	2							24, 25, 26	20,900
13	Aluminum	Al	2	2	6	2	1						27	81,300
14	*Silicon*	*Si*	*2*	*2*	*6*	*2*	*2*						*28, 29, 30*	*277,200*
15	Phosphorus	P	2	2	6	2	3						31	1,180
16	Sulfur	S	2	2	6	2	4						32, 34, 33, 36	520
17	Chlorine	Cl	2	2	6	2	5						35, 37	314
18	Argon	A	2	2	6	2	6 (inert gas)						40, 36, 38	0.04
19	Potassium	K	2	2	6	2	6		1				39, 41	25,900
20	Calcium	Ca	2	2	6	2	6		2				40, 42, 43, 44, 46, 48	36,300
21	Scandium	Sc	2	2	6	2	6	1	2				45	5
22	Titanium	Ti	2	2	6	2	6	2	2				48, 46, 47, 49, 50	4,400
23	Vanadium	V	2	2	6	2	6	3	2				51	150
24	Chromium	Cr	2	2	6	2	6	5	1				52, 53, 50, 54	200
25	Manganese	Mn	2	2	6	2	6	5	2				55	1,000
26	Iron	Fe	2	2	6	2	6	6	2				56, 54, 57, 58	50,000
27	Cobalt	Co	2	2	6	2	6	7	2				59	23
28	Nickel	Ni	2	2	6	2	6	8	2				58, 60, 62, 61, 64	80
29	Copper	Cu	2	2	6	2	6	10	1				63, 65	70
30	Zinc	Zn	2	2	6	2	6	10	2				64, 66, 68, 67, 70	132

Key: Metals *Metalloids* Nonmetals

[a] Atomic number = number of protons.

[b] Mass number = protons + neutrons. Isotopes are listed in the order of their abundance.

[c] Elements 1–30 together constitute 99.6 percent of the earth's crust.

Table A.3 ***Electronic configuration of elements 31 through 103***

Number of electrons in

Atomic number	Name	Symbol	shell 1	shell 2		shell 3			shell 4				shell 5				shell 6				shell 7			
			s	*s*	*p*	*s*	*p*	*d*	*s*	*p*	*d*	*f*	*s*	*p*	*d*	*f*	*s*	*p*	*d*	*f*	*s*	*p*	*d*	*f*
31	Gallium	Ga	2	2	6	2	6	10	2	1														
32	*Germanium*	*Ge*	*2*	*2*	*6*	*2*	*6*	*10*	*2*	*2*														
33	*Arsenic*	*As*	*2*	*2*	*6*	*2*	*6*	*10*	*2*	*3*														
34	*Selenium*	*Se*	*2*	*2*	*6*	*2*	*6*	*10*	*2*	*4*														
35	Bromine	Br	2	2	6	2	6	10	2	5														
36	Krypton	Kr	2	2	6	2	6	10	2	6	(inert gas)													
37	Rubidium	Rb	2	2	6	2	6	10	2	6			1											
38	Strontium	Sr	2	2	6	2	6	10	2	6			2											
39	Yttrium	Y	2	2	6	2	6	10	2	6	1		2											
40	Zirconium	Zr	2	2	6	2	6	10	2	6	2		2											
41	Niobium (Columbium)	Nb (Cb)	2	2	6	2	6	10	2	6	4		1											
42	Molybdenum	Mo	2	2	6	2	6	10	2	6	5		1											
43	Technetium	Tc	2	2	6	2	6	10	2	6	6		1											
44	Ruthenium	Ru	2	2	6	2	6	10	2	6	7		1											
45	Rhodium	Rh	2	2	6	2	6	10	2	6	8		1											
46	Palladium	Pd	2	2	6	2	6	10	2	6	10													
47	Silver	Ag	2	2	6	2	6	10	2	6	10		1											
48	Cadmium	Cd	2	2	6	2	6	10	2	6	10		2											
49	Indium	In	2	2	6	2	6	10	2	6	10		2	1										
50	Tin	Sn	2	2	6	2	6	10	2	6	10		2	2										
51	*Antimony*	*Sb*	*2*	*2*	*6*	*2*	*6*	*10*	*2*	*6*	*10*		*2*	*3*										
52	*Tellurium*	*Te*	*2*	*2*	*6*	*2*	*6*	*10*	*2*	*6*	*10*		*2*	*4*										
53	Iodine	I	2	2	6	2	6	10	2	6	10		2	5										
54	Xenon	Xe	2	2	6	2	6	10	2	6	10		2	6	(inert gas)									
55	Cesium	Cs	2	2	6	2	6	10	2	6	10		2	6			1							
56	Barium	Ba	2	2	6	2	6	10	2	6	10		2	6			2							
57	Lanthanum	La	2	2	6	2	6	10	2	6	10		2	6	1		2							
58	Cerium	Ce	2	2	6	2	6	10	2	6	10	1	2	6	1		2							
59	Praseodymium	Pr	2	2	6	2	6	10	2	6	10	2	2	6	1		2							
60	Neodymium	Nd	2	2	6	2	6	10	2	6	10	3	2	6	1		2							
61	Promethium	Pm	2	2	6	2	6	10	2	6	10	4	2	6	1		2							
62	Samarium	Sm	2	2	6	2	6	10	2	6	10	5	2	6	1		2							
63	Europium	Eu	2	2	6	2	6	10	2	6	10	6	2	6	1		2							
64	Gadolinium	Gd	2	2	6	2	6	10	2	6	10	7	2	6	1		2							
65	Terbium	Tb	2	2	6	2	6	10	2	6	10	8	2	6	1		2							
66	Dysprosium	Dy	2	2	6	2	6	10	2	6	10	9	2	6	1		2							
67	Holmium	Ho	2	2	6	2	6	10	2	6	10	10	2	6	1		2							
68	Erbium	Er	2	2	6	2	6	10	2	6	10	11	2	6	1		2							
69	Thulium	Tm	2	2	6	2	6	10	2	6	10	12	2	6	1		2							
70	Ytterbium	Yb	2	2	6	2	6	10	2	6	10	13	2	6	1		2							
71	Lutetium	Lu	2	2	6	2	6	10	2	6	10	14	2	6	1		2							
72	Hafnium	Hf	2	2	6	2	6	10	2	6	10	14	2	6	2		2							
73	Tantalum	Ta	2	2	6	2	6	10	2	6	10	14	2	6	3		2							

Table A.3 ***Electronic configuration of elements 31 through 103 (cont.)***

Atomic number	Name	Symbol	*Number of electrons in* shell 1	shell 2		shell 3			shell 4				shell 5				shell 6				shell 7			
			s	*s*	*p*	*s*	*p*	*d*	*s*	*p*	*d*	*f*	*s*	*p*	*d*	*f*	*s*	*p*	*d*	*f*	*s*	*p*	*d*	*f*
74	Wolfram (Tungsten)	W	2	2	6	2	6	10	2	6	10	14	2	6	4		2							
75	Rhenium	Re	2	2	6	2	6	10	2	6	10	14	2	6	5		2							
76	Osmium	Os	2	2	6	2	6	10	2	6	10	14	2	6	6		2							
77	Iridium	Ir	2	2	6	2	6	10	2	6	10	14	2	6	7		2							
78	Platinum	Pt	2	2	6	2	6	10	2	6	10	14	2	6	8		2							
79	Gold	Au	2	2	6	2	6	10	2	6	10	14	2	6	10		1							
80	Mercury	Hg	2	2	6	2	6	10	2	6	10	14	2	6	10		2							
81	Thallium	Tl	2	2	6	2	6	10	2	6	10	14	2	6	10		2	1						
82	Lead	Pb	2	2	6	2	6	10	2	6	10	14	2	6	10		2	2						
83	*Bismuth*	*Bi*	*2*	*2*	*6*	*2*	*6*	*10*	*2*	*6*	*10*	*14*	*2*	*6*	*10*		*2*	*3*						
84	*Polonium*	*Po*	*2*	*2*	*6*	*2*	*6*	*10*	*2*	*6*	*10*	*14*	*2*	*6*	*10*		*2*	*4*						
85	Astatine	At	2	2	6	2	6	10	2	6	10	14	2	6	10		2	5						
86	Radon	Rn	2	2	6	2	6	10	2	6	10	14	2	6	10		2	6 (inert gas)						
87	Francium	Fr	2	2	6	2	6	10	2	6	10	14	2	6	10		2	6			1			
88	Radium	Ra	2	2	6	2	6	10	2	6	10	14	2	6	10		2	6			2			
89	Actinium	Ac	2	2	6	2	6	10	2	6	10	14	2	6	10		2	6	1		2			
90	Thorium	Th	2	2	6	2	6	10	2	6	10	14	2	6	10	1	2	6	1		2			
91	Protactinium	Pa	2	2	6	2	6	10	2	6	10	14	2	6	10	2	2	6	1		2			
92	Uranium	U	2	2	6	2	6	10	2	6	10	14	2	6	10	3	2	6	1		2			
93	Neptunium	Np	2	2	6	2	6	10	2	6	10	14	2	6	10	4	2	6	1		2			
94	Plutonium	Pu	2	2	6	2	6	10	2	6	10	14	2	6	10	5	2	6	1		2			
95	Americium	Am	2	2	6	2	6	10	2	6	10	14	2	6	10	6	2	6	1		2			
96	Curium	Cm	2	2	6	2	6	10	2	6	10	14	2	6	10	7	2	6	1		2			
97	Berkelium	Bk	2	2	6	2	6	10	2	6	10	14	2	6	10	8	2	6	1		2			
98	Californium	Cf	2	2	6	2	6	10	2	6	10	14	2	6	10	9	2	6	1		2			
99	Einsteinium	En	2	2	6	2	6	10	2	6	10	14	2	6	10	10	2	6	1		2			
100	Fermium	Fm	2	2	6	2	6	10	2	6	10	14	2	6	10	11	2	6	1		2			
101	Mendelevium	Me	2	2	6	2	6	10	2	6	10	14	2	6	10	12	2	6	1		2			
102	Nobelium	No	2	2	6	2	6	10	2	6	10	14	2	6	10	13	2	6	1		2			
103	Lawrencium	Lw	2	2	6	2	6	10	2	6	10	14	2	6	10	14	2	6	1		2			

Key: Metals *Metalloids* Nonmetals

Appendix B

Powers of ten

WE NEED A SPECIAL VOCABULARY TO DESCRIBE THE SIZE OF THINGS RANGING FROM invisible atoms to the vast reaches of space around us. Such a vocabulary is supplied by the *powers of ten.*

$10^0 = 1$	1 with decimal point moved zero places
$10^1 = 10$	1 with decimal point moved 1 place to right
$10^2 = 100$	1 with decimal point moved 2 places to right
$10^3 = 1{,}000$	1 with decimal point moved 3 places to right
$10^4 = 10{,}000$	1 with decimal point moved 4 places to right
$10^5 = 100{,}000$	1 with decimal point moved 5 places to right
$10^6 = 1{,}000{,}000$	1 with decimal point moved 6 places to right
	etc.

In other words, the exponent of 10 indicates the number of places the decimal point is moved to the right (or, to put it another way, the number of zeros following 1).

For numbers smaller than 1, negative exponent is used.

$10^{-1} = 0.1$	1 with decimal point moved 1 place to left
$10^{-2} = 0.01$	1 with decimal point moved 2 places to left
$10^{-3} = 0.001$	1 with decimal point moved 3 places to left
$10^{-4} = 0.0001$	1 with decimal point moved 4 places to left
$10^{-5} = 0.00001$	1 with decimal point moved 5 places to left
$10^{-6} = 0.000001$	1 with decimal point moved 6 places to left
	etc.

In other words, the negative exponent of 10 indicates the number of places the decimal point is moved to the left of the number 1.

In comparing the sizes of things expressed in powers of 10, we need to recall certain laws of exponents taught in school algebra:

MULTIPLICATION Add exponents. Examples:

$$10^3 \cdot 10^3 = 10^{3+3} = 10^6$$

or

$$1{,}000 \cdot 1{,}000 = 1{,}000{,}000$$

$$10^{21} \cdot 10^6 = 10^{27}$$

(That is, 10^{27} is 10^6 or 1 million times as large as 10^{21}.)

DIVISION Subtract exponents. Examples:

$$10^6/10^3 = 10^{6-3} = 10^3$$

or

$$1{,}000{,}000/1{,}000 = 1{,}000$$

$$10^{27}/10^{21} = 10^{27-21} = 10^6$$

Table B.1, on page 616, shows some familiar distances and sizes expressed in centimeters at various powers of ten.

Table B.1 ***Some distances and sizes expressed in powers of ten***

	cm
1 light-year	10^{18}
	10^{17}
	10^{16}
	10^{15}
	10^{14}
Distance earth to sun (1.5×10^{13} cm)	10^{13}
	10^{12}
	10^{11}
	10^{10}
Diameter of earth (1.3×10^{9} cm)	10^{9}
	10^{8}
	10^{7}
Distant view	10^{6}
	10^{5}
Lengths of radio waves	10^{4}
	10^{3}
Meter stick	10^{2}
Width of hand	10^{1}
Width of pencil (1 cm)	10^{0}
	10^{-1}
Thickness of sheet of paper	10^{-2}
	10^{-3}
	10^{-4}
Wavelength of visible light	10^{-5}
Diameter of some molecules	10^{-6}
X rays	10^{-7}
Diameter of atom (1 A)	10^{-8}
	10^{-9}
	10^{-10}
	10^{-11}
Diameter of atomic nucleus	10^{-12}
	10^{-13}

Appendix C

Minerals

MANY OF THE MOST COMMON MINERALS MAY BE IDENTIFIED IN HAND SPECIMENS BY their physical properties. Among the characteristics useful for this purpose are (1) hardness, (2) specific gravity, (3) streak (sometimes color), (4) shape (that is, crystal form, cleavage, and fracture), and (5) response to light as indicated by luster and transparency.

Hardness

The hardness of a mineral is determined by scratching the smooth surface of one mineral with the edge of another. In making a hardness test, be sure that the mineral being tested is actually scratched. Sometimes particles simply rub off the specimen, suggesting that it has been scratched, even though it has not been.

Ten common minerals have been arranged in the Mohs scale of relative hardness, shown in Table C.1. Each of these minerals will scratch all those lower in number on the scale and will be scratched by all those higher. In other words, this is a *relative scale.* In terms of absolute hardness, the steps are nearly, though not quite, uniform up to 9. Number 7 is 7 times as hard as 1, and number 9 is 9 times as hard as 1. But number 10 is about 40 times as hard as 1.

Luster

Luster is the way a mineral looks in reflected light. There are several kinds of luster.

Metallic The luster of metals
Adamantine The luster of diamonds
Vitreous The luster of a broken edge of glass
Resinous The luster of yellow resin
Pearly The luster of pearl
Silky The luster of silk

Fracture

Many minerals that do not exhibit cleavage (see Chapter 3) do break, or fracture, in a distinctive manner. Some of the types of fracture are

Conchoidal Along smooth, curved surfaces like the surface of a shell (*conch*); commonly observed in glass and quartz
Fibrous or *splintery* Along surfaces roughened by splinters or fibers
Uneven or *irregular* Along rough, irregular surfaces
Hackly Along a jagged, irregular surface with sharp edges

Magnetism

Minerals which, in their natural state, will be attracted to a magnet are said to be *magnetic.* Magnetite, Fe_3O_4, and pyrrhotite, $Fe_{1-x}S$, with x between

Table C.1 *Mohs scale of hardness*

Scale	Mineral	Test
1	Talc	(*Softest*)
2	Gypsum	
$2\frac{1}{2}$		Fingernail
3	Calcite	Copper coin
4	Fluorite	
5	Apatite	
$5\frac{1}{2}$–6		Knife blade or plate glass
6	Orthoclase	
$6\frac{1}{2}$–7		Steel file
7	Quartz	
8	Topaz	
9	Corundum	
10	Diamond	(*Hardest*)

0 and 0.2, are the only common magnetic minerals, although many others containing iron are drawn to a powerful enough electromagnet.

Pyroelectricity

Pyroelectricity is the simultaneous development of positive and negative charges of electricity on different parts of the same crystal under the proper conditions of temperature change. Quartz is a good example. If it is heated to about 100°C, it will, on cooling, develop positive electric charges at three alternate prismatic edges and negative charges at the three other edges.

Piezoelectricity

Piezoelectricity is an electric charge developed in a crystallized body by pressure. Quartz is probably the most important piezoelectric mineral, for an extremely slight pressure parallel to its "electric axis" can be detected by the electric charge set up. It is used in specially oriented plates in radio equipment and in sonic sounders.

Fluorescence and phosphorescence

Minerals which become luminescent during exposure to ultraviolet light, X rays, or cathode rays are *fluorescent.* If the luminescence continues after the exciting rays are shut off, the mineral is said to be *phosphorescent.*

Fusibility

Minerals can be divided into those fusible and those infusible in a blowpipe flame. Seven minerals showing different degrees of fusibility have been used as a scale to which fusible minerals can be referred. They are listed in Table C.2.

Tenacity

A mineral's cohesiveness as shown by its resistance to breaking, crushing, bending, or tearing is known as its *tenacity.* Various kinds of tenacity in minerals include the following:

Brittle Breaks or powders easily
Malleable Can be hammered into thin sheets
Sectile Can be cut into thin shavings with a knife
Ductile Can be drawn into wire
Flexible Bends but does not return to its original shape when pressure is removed

Table C.2 **Scale of fusibility**

No.	*Mineral*	*Approximate fusing point, °C*	*Remarks*
1	Stibnite	525	Easily fusible in candle flame
2	Chalcopyrite	800	A small fragment fuses easily in Bunsen burner flame
3	Garnet (almandite)	1050	Infusible in Bunsen flame but easily fusible in blowpipe flame
4	Actinolite	1200	A sharp-pointed splinter fuses with little difficulty in blowpipe flame
5	Orthoclase	1300	Edges of fragments rounded with difficulty in blowpipe flame
6	Bronzite	1400	Only fine ends of splinters are rounded in blowpipe flame
7	Quartz	1470	Infusible in blowpipe flame

Elastic After being bent, will resume its original position upon release of pressure

Solubility

Concentrated hydrochloric acid, HCl, diluted with three parts of water, is commonly used for the solution of minerals being tested. Many other wet reagents are used for special tests to help identify minerals.

Silicates

More than 90 percent of rock-forming minerals are silicates, with structures based on the SiO_4 tetrahedron. Four important classes are listed in Table C.3.

Table C.3 **Classes of silicates**

Class[a]	*Arrangement of SiO_4 tetrahedra*	*Examples*
Nesosilicates	Isolated	Olivine
Inosilicates	Chains	
	Single	Augite
	Double	Hornblende
Phyllosilicates	Sheets	Talc; micas
Tectosilicates	Frameworks	Quartz; feldspars

[a] *Neso,* "island"; *ino,* "chain" or "thread"; *phyllo,* "sheet"; *tecto,* "framework."

Table C.4 **Plagioclase feldspars**

Species	Percent albite	Percent anorthite
Albite $Na(AlSi_3O_8)$	100–90	0–10
Oligoclase	90–70	10–30
Andesine	70–50	30–50
Labradorite	50–30	50–70
Bytownite	30–10	70–90
Anorthite $Ca(Al_2Si_2O_8)$	10–0	90–100

Plagioclase feldspars

The plagioclase feldspars, also called the soda–lime feldspars, form a complete solid-solution series from pure albite to pure anorthite. Calcium substitutes for sodium in all proportions, with accompanying substitution of aluminum for silicon. The series is divided into the six arbitrary species names listed in Table C.4.

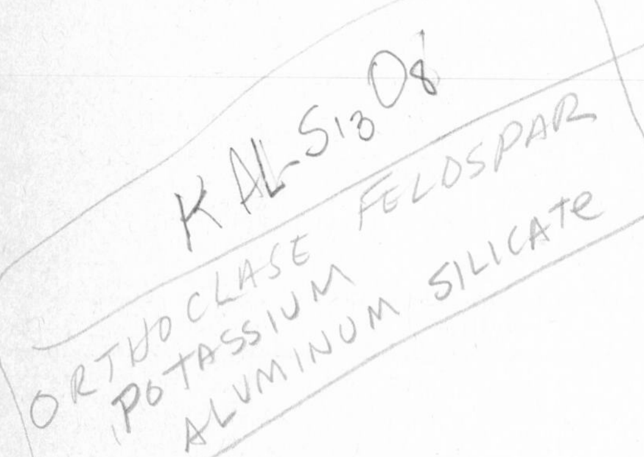

Pyroxenes and amphiboles

The pyroxene family of minerals and the amphibole family of minerals are inosilicates that parallel each other. The amphiboles contain OH. The pyroxenes crystallize at higher temperatures than their amphibole analogues. The two families are listed in Table C.5; for an explanation of cleavage differences between them, see Figure 3.10.

Table C.5 **Ions in common pyroxenes and amphiboles**

Ions			
X	*Y*	*Pyroxenes*[a]	*Amphiboles*[b]
Mg	Mg	Enstatite	Anthophyllite
Ca	Mg	Diopside	Tremolite
Li	Al	Spodumene	
Na	Al	Jadeite	Glaucophane
Na	Fe‴	Aegirite	Arfvedsonite
Ca, Na	Mg, Fe, Mn, Al, Fe‴, Ti	Augite	Hornblende

[a] Basic structure, single chain, SiO_3; formula, $XY(Si_2O_6)$.
[b] Basic structure, double chains, Si_4O_{11}; formula, $X_{0-7}Y_{7-14}Z_{16}O_{44}(OH)_4$.

Table C.6 ***Minerals arranged according to specific gravity***

Specific gravity	*Mineral*	*Specific gravity*	*Mineral*	*Specific gravity*	*Mineral*
1.9–2.2	Opal	3.0–3.3	Actinolite	3.6–4.0	Limonite
2.0–3.0	Bauxite	3.0–3.3	Tremolite	3.65–3.75	Staurolite
2.16	Halite	3.15–3.2	Apatite	3.77	Azurite
2.2–2.65	Serpentine	3.15–3.2	Spodumene	3.85	Siderite
2.3	Graphite	3.16	Andalusite	3.9–4.1	Sphalerite
2.32	Gypsum	3.18	Fluorite	4.0	Carnotite
2.57	Orthoclase	3.2	Hornblende	4.02	Corundum
2.6	Kaolinite	3.2–3.4	Augite	4.1–4.3	Chalcopyrite
2.6–2.9	Chlorite	3.2–3.5	Enstatite	4.25	Almandite
2.62	Albite	3.23	Sillimanite	4.52–4.62	Stibnite
2.65	Quartz	3.27–3.37	Olivine	4.6	Chromite
2.7–2.8	Talc	3.3–3.5	Jadeite	4.68	Zircon
2.72	Calcite	3.3–4.37	Goethite	5.02	Pyrite
2.76	Anorthite	3.35–3.45	Epidote	5.06–5.08	Bornite
2.76–3.1	Muscovite	3.4–3.55	Aegirite	5.18	Magnetite
2.8–2.9	Wollastonite	3.4–3.6	Topaz	5.26	Hematite
2.8–3.2	Biotite	3.45	Arfvedsonite	5.5–5.8	Chalcocite
2.85	Dolomite	3.5	Diamond	6.8–7.1	Cassiterite
2.85–3.2	Anthophyllite	3.5–4.1	Spinel	7.4–7.6	Galena
2.89–2.98	Anhydrite	3.5–4.3	Garnet	9.0–9.7	Uraninite
3.0–3.25	Tourmaline	3.56–3.66	Kyanite		

Table C.7 ***Minerals arranged according to hardness***

Hardness	*Mineral*	*Hardness*	*Mineral*	*Hardness*	*Mineral*
1	Talc	5	Apatite	$6–6\frac{1}{2}$	Pyrite
1–2	Graphite	5	Kyanite (along crystal)	6–7	Cassiterite
1–3	Bauxite	$5–5\frac{1}{2}$	Goethite	6–7	Epidote
2	Gypsum	$5–5\frac{1}{2}$	Limonite	6–7	Sillimanite
2	Stibnite	$5–5\frac{1}{2}$	Wollastonite	$6\frac{1}{2}–7$	Jadeite
$2–2\frac{1}{2}$	Chlorite	5–6	Actinolite	$6\frac{1}{2}–7$	Olivine
$2–2\frac{1}{2}$	Kaolinite	5–6	Augite	$6\frac{1}{2}–7$	Spodumene
$2–2\frac{1}{2}$	Muscovite	5–6	Diopside	$6\frac{1}{2}–7\frac{1}{2}$	Almandite
2–5	Serpentine	5–6	Hornblende	$6\frac{1}{2}–7\frac{1}{2}$	Garnet
$2\frac{1}{2}$	Galena	5–6	Opal	7	Kyanite (across crystal)
$2\frac{1}{2}$	Halite	5–6	Tremolite		
$2\frac{1}{2}–3$	Biotite	$5\frac{1}{2}$	Chromite	7	Quartz
$2\frac{1}{2}–3$	Chalcocite	$5\frac{1}{2}$	Enstatite	$7–7\frac{1}{2}$	Staurolite
3	Bornite	$5\frac{1}{2}$	Uraninite	$7–7\frac{1}{2}$	Tourmaline
3	Calcite	$5\frac{1}{2}–6$	Anthophyllite	$7\frac{1}{2}$	Andalusite
$3–3\frac{1}{2}$	Anhydrite	$5\frac{1}{2}–6\frac{1}{2}$	Hematite	$7\frac{1}{2}$	Zircon
$3\frac{1}{2}–4$	Chalcopyrite	6	Albite	8	Spinel
$3\frac{1}{2}–4$	Dolomite	6	Anorthite	8	Topaz
$3\frac{1}{2}–4$	Siderite	6	Arfvedsonite	9	Corundum
$3\frac{1}{2}–4$	Sphalerite	6	Magnetite	10	Diamond
4	Azurite	6	Orthoclase		
4	Fluorite	$6–6\frac{1}{2}$	Aegirite		

Table C.8 **The common minerals**

Mineral	*Chemical composition and name*	*Specific gravity*	*Streak*	*Hardness*	*Cleavage or fracture*	*Luster*
Actinolite (an asbestos; an amphibole)	$Ca_2(Mg, Fe)_5Si_8O_{22}(OH)_2$ Calcium iron silicate (tremolite with more than 2% iron)	3.0–3.3	Colorless	5–6	See *Amphibole*	Vitreous
Aegirite (a pyroxene)	$NaFe'''(Si_2O_6)$	3.40–3.55		6–6½	Imperfect cleavage at 87° and 93°	Vitreous
Albite	$Na(AlSi_3O_8)$ Sodic plagioclase feldspar	2.62	Colorless	6	Good in 2 directions at 93°34′	Vitreous to pearly
Almandite	$Fe_3Al_2(SiO_4)_3$	4.25	White	6½–7½		Vitreous to resinous
Amphibole family	See *Anthopyllite, Tremolite, Afrvedsonite, Hornblende*				Perfect prismatic at 56° and 124°, often yielding a splintery surface	
Andalusite	Al_2SiO_5 Aluminum silicate	3.16	Colorless	7½	Not prominent	Vitreous
Andesine	A plagioclase feldspar 50–70% albite					
Anhydrite	$CaSO_4$ Anhydrous calcium sulfate	2.89–2.98	Colorless	3–3½	3 directions at right angles to form rectangular blocks	Vitreous; pearly
Ankerite	Dolomite in which ferrous iron replaces more than half the magnesium					
Anorthite	$Ca(Al_2Si_2O_8)$ Calcic plagioclase feldspar	2.76	Colorless	6	Good in 2 directions at 94°12′	Vitreous to pearly
Anthophyllite (an amphibole)	$(Mg, Fe)_7(Si_8O_{22})(OH)_2$	2.85–3.2		5½–6	See *Amphibole*	Vitreous
Apatite	$Ca_5(F, Cl)(PO_4)_3$ Calcium fluophosphate	3.15–3.2	White	5	Poor cleavage, one direction; conchoidal fracture	Glassy
Arfvedsonite (a sodium-rich amphibole)	$Na_3Mg_4Al(Si_8O_{22})\cdot(OH, F)_2$	3.45		6	See *Amphibole*	Vitreous

Color	*Transparency*	*Form*	*Other properties*
White to light green	Transparent to translucent	Slender crystals, usually fibrous	A common ferromagnesian metamorphic mineral; a common component of green schists
Brown or green	Translucent	Slender prismatic crystals	Rare rock former, chiefly in rocks rich in soda and poor in silica
Colorless, white, gray	Transparent to translucent	Tabular crystals; striations caused by twinning	Opalescent variety, *moonstone*
Deep red	Transparent to translucent	12- or 24-sided; massive or granular	A garnet used to define one of the zones of middle-grade metamorphism; striking in schists
			A group of silicates with tetrahedra in double chains; hornblende is the most important; contrast with *pyroxene*
Flesh-red, reddish brown, olive-green	Transparent to translucent	Usually in coarse, nearly square prisms; cross section may show black cross	Found in schists formed by the middle-grade metamorphism of aluminous shales and slates; the variety *chiastolite* has carbonaceous inclusions in the pattern of a cross
			As grains in igneous rock; the chief feldspar in andesite lavas of the Andes Mountains
White; may have faint gray, blue, or red tinge	Transparent to translucent	Commonly in massive fine aggregates not showing cleavage; crystals rare	Found in limestones and in beds associated with salt deposits; heavier than calcite, harder than gypsum
			Formed in low-grade regional metamorphism from the conversion of calcite or dolomite or both between 80 and 120°C
Colorless, white, gray, green, yellow, red	Transparent to translucent	Striations caused by twinning; lathlike or platy grains	Occurs in many igneous rocks
Gray to various shades of green and brown		Lamellar or fibrous	From metamorphism of olivine
Green, brown, red	Translucent to transparent	Massive, granular	Widely disseminated as an accessory mineral in all types of rocks; unimportant source of fertilizer; a transparent variety is a gem, but too soft for general use
Deep green to black	Translucent	Long prismatic crystals	Rock former in rocks poor in silica

Table C.8 *The common minerals (cont.)*

Mineral	*Chemical composition and name*	*Specific gravity*	*Streak*	*Hardness*	*Cleavage or fracture*	*Luster*
Asbestos	See *Actinolite, Chrysotile, Serpentine*					
Augite (a pyroxene)	$Ca(Mg, Fe, Al)(Al, Si_2O_6)$ Ferromagnesian silicate	3.2–3.4	Greenish gray	5–6	Perfect prismatic along 2 planes at nearly right angles to each other, often yielding a splintery surface	Vitreous
Azurite	$Cu_3(CO_3)_2(OH)_2$ Blue copper carbonate	3.77	Pale blue	4	Fibrous	Vitreous to dull, earthy
Bauxite	Hydrous aluminum oxides of indefinite composition; not a mineral	2–3	Colorless	1–3	Uneven fracture	Dull to earthy
Biotite (black mica)	$K(Mg, Fe)_3AlSi_3O_{10}(OH)_2$ Ferromagnesian silicate	2.8–3.2	Colorless	$2\frac{1}{2}$–3	Perfect in one direction into thin, elastic, transparent, smoky sheets	Pearly, glassy
Bornite (peacock ore; purple copper ore)	Cu_5FeS_4 Copper iron sulfide	5.06–5.08	Grayish black	3	Uneven fracture	Metallic
Bronzite (a pyroxene)	Enstatite with 5–13% FeO					
Bytownite	A plagioclase feldspar 10–30% albite					
Calcite	$CaCO_3$ Calcium carbonate	2.72	Colorless	3	Perfect in 3 directions at 75° to form unique rhombohedral fragments	Vitreous
Carnotite	$K_2(UO_2)_2(VO_4)_2$ Potassium uranyl vanadate	4		Very soft	Uneven fracture	Earthy
Cassiterite (tin stone)	SnO_2 Tin oxide	6.8–7.1	White to light brown	6–7	Conchoidal fracture	Adamantine to submetallic and dull
Chalcocite (copper glance)	Cu_2S Copper sulfide	5.5–5.8	Grayish black	$2\frac{1}{2}$–3	Conchoidal fracture	Metallic

Color	Transparency	Form	Other properties
			General term for certain fibrous minerals with similar physical characteristics though different composition; chrisotile is the most common
Dark green to black	Translucent only on thin edges	Short, stubby crystals with 4- or 8-sided cross section; often in granular crystalline masses	An important igneous rock-forming mineral found chiefly in simatic rocks
Intense azure blue	Opaque	Crystals complex in habit and distorted; sometimes in radiating spherical groups	An ore of copper; a gem mineral; effervesces with HCl
Yellow, brown, gray, white	Opaque	In rounded grains; or earthy, claylike masses	An ore of aluminum; produced under subtropical to tropical climatic conditions by prolonged weathering of aluminum-bearing rocks; a component of *laterites;* clay odor when wet
Black, brown, dark green	Transparent, translucent	Usually in irregular foliated masses; crystals rare	Constructed around tetrahedral sheets; a common and important rock-forming mineral in both igneous and metamorphic rocks
Brownish bronze on fresh fracture; tarnishes to variegated purple and blue, then black	Opaque	Usually massive; rarely in rough cubic crystals	An important ore of copper
			Primarily as grains in igneous rocks
Usually white or colorless; may be tinted gray, red, green, blue, yellow	Transparent to opaque	Usually in crystals or coarse to fine granular aggregates; also compact, earthy; crystals extremely varied—over 300 different forms	A very common rock mineral, occurring in masses as limestone and marble; effervesces freely in cold dilute hydrochloric acid
Brilliant canary yellow	Opaque	Earthy powder	An ore of vanadium and uranium
Brown or black; rarely yellow or white	Translucent; rarely transparent	Commonly massive granular	The principal ore of tin
Shiny lead-gray; tarnishes to dull black	Opaque	Commonly aphanitic and massive; crystals rare; small, tabular with hexagonal outline	One of the most important ore minerals of copper; occurs principally as a result of secondary sulfide enrichment

Table C.8 ***The common minerals (cont.)***

Mineral	*Chemical composition and name*	*Specific gravity*	*Streak*	*Hardness*	*Cleavage or fracture*	*Luster*
Chalco-pyrite (copper pyrites; yellow copper ore; fool's gold)	$CuFeS_2$ Copper iron sulfide	4.1–4.3	Greenish black; also greenish powder in groove when scratched	$3\frac{1}{2}$–4	Uneven fracture	Metallic
Chlorite	$(Mg, Fe)_5(Al, Fe''')_2$-$Si_3O_{10}(OH)_8$ Hydrous ferromagnesian aluminum silicate	2.6–2.9	Colorless	2–$2\frac{1}{2}$	Perfect in 1 direction like micas, but into inelastic flakes	Vitreous to pearly
Chromite	$FeCr_2O_4$ Iron chromium oxide	4.6	Dark brown	$5\frac{1}{2}$	Uneven fracture	Metallic to sub-metallic or pitchy
Chrysotile (serpentine asbestos)	See *Serpentine*					
Clay	See *Kaolinite, Illite, Montmorillonite*					
Corundum (ruby, sapphire)	Al_2O_3 Aluminum oxide	4.02	Colorless	9	Basal or rhombohedral parting	Adamantine to vitreous
Diamond	C	3.5	Colorless	10	Octahedral cleavage	Adamantine; greasy
Diopside (a pyroxene)	$CaMg(Si_2O_6)$	3.2		5–6	Poor prismatic	Vitreous
Dolomite	$CaMg(CO_3)_2$ Calcium magnesium carbonate	2.85	Colorless	$3\frac{1}{2}$–4	Perfect in 3 directions at 73°45′	Vitreous or pearly
Emery	Black granular corundum intimately mixed with magnetite, hematite, or iron spinel					
Enstatite (a pyroxene)	$Mg_2(Si_2O_6)$ Magnesium inosilicate	3.2–3.5		$5\frac{1}{2}$	Good at 87° and 93°	Vitreous

Color	Transparency	Form	Other properties
Brass-yellow; tarnishes to bronze or iridescence, but more slowly than bornite or chalcocite	Opaque	Usually massive	An ore of copper; distinguished from pyrite by being softer than steel, distinguished from gold by being brittle; like pyrite, known as fool's gold
Green of various shades	Transparent to translucent	Foliated massive, or in aggregates of minute scales	A common metamorphic mineral characteristic of low-grade metamorphism
Iron-black to brownish black	Subtranslucent	Massive, granular to compact	The only ore of chromium; a common constituent of peridotites and serpentines derived from them; one of the first minerals to crystallize from a cooling magma
Brown, pink, or blue; may be white, gray, green, ruby-red, sapphire-blue	Transparent to translucent	Barrel-shaped crystals; sometimes deep horizontal striations; coarse or fine granular	Common as an accessory mineral in metamorphic rocks such as marble, mica schist, gneiss; occurs in gem form as *ruby* and *sapphire;* the abrasive emery is black granular corundum mixed with magnetite, hematite, or the magnesian aluminum oxide *spinel*
Colorless or pale yellow; may be red, orange, green, blue, black	Transparent	Octahedral crystals, flattened, elongated, with curved faces	Gem and abrasive; 95% of natural diamond production is from South Africa; abrasive diamonds have been made in commercial quantities in the laboratory in the United States
White to light green	Transparent to translucent	Prismatic crystals; also granular massive	Contact metamorphis mineral in crystalline limestones
Pink, flesh; may be white, gray, green, brown, black	Transparent to opaque	Rhombohedral crystals with curved faces; granular cleavable masses, or aphanitic compact	Occurs chiefly in rock masses of dolomitic limestone and marble, or as the principal constituent of the rock named for it; distinguished from limestone by its less vigorous action with cold hydrochloric acid (the powder dissolves with effervescence, large pieces in hot acid)
Grayish, yellowish, or greenish white to olive-green and brown	Translucent	Usually massive	Common in pyroxenites, peridotites, gabbros, and basalts; also in both stony and metallic meteorites

Table C.8 ***The common minerals (cont.)***

Mineral	*Chemical composition and name*	*Specific gravity*	*Streak*	*Hardness*	*Cleavage or fracture*	*Luster*
Epidote	$Ca_2(Al, Fe)Al_2O(SiO_4)$-$(Si_2O_7)(OH)$ Hydrous calcium aluminum iron silicate	3.35–3.45	Colorless	6–7	Good in 1 direction	Vitreous
Fayalite (an olivine)	$Fe_2(SiO_4)$ See *Olivine*					
Feldspars	Aluminosilicates	2.55–2.75		6	Good in 2 directions at or near 90°	
Fluorite	CaF_2 Calcium fluoride	3.18	Colorless	4	Good in 4 directions parallel to the faces of an octahedron	Vitreous
Forsterite (an olivine)	$Mg_2(SiO_4)$ See *Olivine*					
Galena	PbS Lead sulfide	7.4–7.6	Lead-gray	$2\frac{1}{2}$	Good in 3 directions parallel to the faces of a cube	Metallic
Garnet	$R''_3R'''_2(SiO_4)_3$ R″ may be Ca, Mg, Fe, or Mn. R‴ may be Al, Fe, Ti, or Cr.	3.5–4.3	Colorless	$6\frac{1}{2}$–$7\frac{1}{2}$	Uneven fracture	Vitreous to resinous
Glaucophane (a sodium-rich amphibole)	Variety of arfvedsonite					
Goethite (bog-iron ore)	$HFeO_2$	3.3–4.37	Yellowish brown	5–$5\frac{1}{2}$	Perfect 010	Adamantine to dull
Graphite (plumbago; black lead)	C Carbon	2.3	Black	1–2	Good in one direction; folia flexible but not elastic	Metallic or earthy
Gypsum	$CaSO_4 \cdot 2H_2O$ Hydrous calcium sulfate	2.32	Colorless	2	Good cleavage in one direction yielding flexible but inelastic flakes; fibrous fracture in another direction; conchoidal fracture in a third direction	Vitreous, pearly, silky
Halite (rock salt; common salt)	NaCl Sodium chloride	2.16	Colorless	$2\frac{1}{2}$	Perfect cubic cleavage	Glassy to dull

Color	*Transparency*	*Form*	*Other properties*
Pistachio-green, yellowish to blackish green	Transparent to translucent	Prismatic crystals striated parallel to length; usually coarse to fine granular; also fibrous	A metamorphic mineral often associated with chlorite; derived from metamorphism of impure limestone; characteristic of contact metamorphic zones in limestone
			The most common igneous rock-forming group of minerals; weather to clay minerals
Variable; light green, yellow, bluish green, purple, etc.	Transparent to translucent	Well-formed interlocking cubes; also massive, coarse or fine grains	A common, widely distributed mineral in dolomites and limestone; an accessory mineral in igneous rocks; used as a flux in making steel; some varieties fluoresce
Lead-gray	Opaque	Cube-shaped crystals; also in granular masses	The principal ore of lead; so commonly associated with silver that it is also an ore of silver
Red, brown, yellow, white, green, black	Transparent to translucent	Usually in 12- or 24-sided crystals; also massive granular, coarse or fine	Common and widely distributed, particularly in metamorphic rocks; brownish red variety *almandite*, q.v.
Yellowish brown to dark brown	Subtranslucent	Massive, in radiating fibrous aggregates; foliated	An ore of iron; one of the commonest minerals formed under oxidizing conditions as a weathering product of iron-bearing minerals
Black to steel-gray	Opaque	Foliated or scaly masses common; may be radiated or granular	Feels greasy; common in metamorphic rocks such as marble, schists, and gneisses
Colorless, white, gray; with impurities, yellow, red, brown	Transparent to translucent	Crystals prismatic, tabular, diamond-shaped; also in granular, fibrous, or earthy masses	A common mineral widely distributed in sedimentary rocks, often as thick beds; *satin spar* is a fibrous gypsum with silky luster; *selenite* is a variety which yields broad, colorless, transparent folia; *alabaster* is a fine-grained massive variety
Colorless or white; impure: yellow, red, blue, purple	Transparent to translucent	Cubic crystals; massive granular	Salty taste; permits ready passage of heat rays (i.e., diathermanous); a very common mineral in sedimentary rocks; interstratified in rocks of all ages to form a true rock mass

Table C.8 ***The common minerals (cont.)***

Mineral	*Chemical composition and name*	*Specific gravity*	*Streak*	*Hardness*	*Cleavage or fracture*	*Luster*
Hematite	Fe_2O_3 Iron oxide	5.26	Light to dark Indian-red; blackens on heating	$5\frac{1}{2}$–$6\frac{1}{2}$	Uneven fracture	Metallic
Hornblende (an amphibole)	Complex ferromagnesian silicate of Ca, Na, Mg, Ti, and Al	3.2	Colorless	5–6	Perfect prismatic at 56° and 124°	Vitreous; fibrous variety often silky
Hypersthene	Enstatite with greater than 13% FeO					
Illite (clay)						
Jadeite (a pyroxene)	$NaAl(Si_2O_6)$	3.3–3.5		$6\frac{1}{2}$–7	87° and 93°	Vitreous
Kaolinite (clay)	$Al_2Si_2O_5(OH)_4$ Hydrous aluminum silicate	2.6	Colorless	2–$2\frac{1}{2}$	None	Dull earthy
Kyanite	Al_2SiO_5 Aluminum silicate	3.56–3.66	Colorless	5 along, 7 across crystals	Good in one direction	Vitreous to pearly
Labradorite	A plagioclase feldspar 30–50% albite					
Limonite (brown hematite; bog-iron ore; rust)	Hydrous iron oxides; not a mineral	3.6–4	Yellow-brown	5–$5\frac{1}{2}$ (finely divided, apparent H as low as 1)	None	Vitreous
Magnetite	Fe_3O_4 Iron oxide	5.18	Black	6	Some octahedral parting	Metallic
Mica	See *Biotite; Muscovite*					
Montmorillonite (clay)	Hydrous aluminum silicate					

Color	*Transparency*	*Form*	*Other properties*
Reddish brown to black	Opaque	Crystals tabular; botryoidal; micaceous and foliated; massive	The most important ore of iron; red earthy variety known as *red ocher;* botryoidal form known as *kidney ore;* micaceous form *specular;* widely distributed in rocks of all types and ages
Dark green to black	Translucent on thin edges	Long, prismatic crystals; fibrous; coarse- to fine-grained masses	Distinguished from augite by cleavage; a common rock-forming mineral which occurs in both igneous and metamorphic rocks
			A general term for clay minerals that resemble micas; the chief constituent in many shales
Apple green, emerald green, white		Fibrous in compact massive aggregates	Occurs in large masses in serpentine by metamorphism of a nepheline–albite rock
White	Opaque	Claylike masses	Usually unctuous and plastic; other clay minerals similar in composition and physical properties, but different in atomic structure, are *illite* and *montmorillonite;* derived from the weathering of the feldspars
Blue; may be white, gray, green, streaked	Transparent to translucent	In bladed aggregates	Characteristic of middle-grade metamorphism; compare with andalusite, which has the same composition and is formed under similar conditions, but has a different crystal habit; contrast with sillimanite, which has the same composition but different crystal habit and forms at highest metamorphic temperatures
			Widespread as a rock mineral; the only important constituent in large masses of rocks called anorthosite
Dark brown to black	Opaque	Amorphous; mammillary to stalactitic masses; concretionary, nodular, earthy	Always of secondary origin from alteration or solution of iron minerals; mixed with fine clay, it is a pigment, *yellow ocher*
Iron-black	Opaque	Usually massive granular, granular or aphanitic	Strongly magnetic; may act as a natural magnet, known as *lodestone;* an important ore of iron; found in black sands on the seashore; mixed with corundum, it is a component of *emery*
			Unique capacity for absorbing water and expanding

Table C.8 The common minerals (cont.)

Mineral	*Chemical composition and name*	*Specific gravity*	*Streak*	*Hardness*	*Cleavage or fracture*	*Luster*
Muscovite (white mica; potassium mica; common mica)	$KAl_3Si_3O_{10}(OH)_2$ Nonferromagnesian silicate	2.76–3.1	Colorless	$2–2\frac{1}{2}$	Good cleavage in 1 direction, giving thin, very flexible and elastic folia	Vitreous, silky, pearly
Oligoclase	A plagioclase feldspar 70–90% albite					
Olivine (peridot)	$(Mg, Fe)_2SiO_4$ Ferromagnesian silicate	3.27–3.37	Pale green, white	$6\frac{1}{2}–7$	Conchoidal fracture	Vitreous
Opal	$SiO_2 \cdot nH_2O$ A mineraloid	1.9–2.2		5–6	Conchoidal fracture	Vitreous, resinous
Orthoclase	$K(AlSi_3O_8)$ Potassium feldspar	2.57	White	6	Good in 2 directions at or near 90°	Vitreous
Plagioclase	Soda-lime feldspar					
Pyrite (iron pyrites; fool's gold)	FeS_2 Iron sulfide	5.02	Greenish or brownish black	$6–6\frac{1}{2}$	Uneven fracture	Metallic
Pyroxene family	Inosilicates: See *Enstatite, Diopside, Spodumene, Jadeite, Aegirite, Augite*					
Quartz (silica)	SiO_2 Silicon oxide but structurally a silicate, with tetrahedra sharing oxygens in 3 dimensions	2.65	Colorless	7	Conchoidal fracture	Vitreous, greasy, splendent

Color	Transparency	Form	Other properties
Thin: colorless; thick: light yellow, brown, green, red	Thin: transparent; thick: translucent	Mostly in thin flakes	Widespread and very common rock-forming mineral; characteristic of sialic rocks; also very common in metamorphic rocks such as gneiss and schist; the principal component of some mica schists; sometimes used for stove doors, lanterns, etc., as transparent *isinglass;* used chiefly as an insulating material
			Found in various localities in Norway, with inclusions of hematite, which give it a golden shimmer and sparkle; this is called *aventurine* oligoclase, or *sunstone*
Olive to grayish green, brown	Transparent to translucent	Usually in embedded grains or granular masses	A common rock-forming mineral found primarily in simatic rocks; the principal component of peridotite; actually, a series grading from *forsterite* to *fayalite;* the most common olivines are richer in magnesium than in iron; the clear green variety *peridot* is sometimes used as a gem
Colorless; white; pale yellow, red, brown, green, gray, blue; opalescent	Transparent to translucent	Amorphous; massive; often botryoidal, stalactitic	Many varieties; lines and fills cavities in igneous and sedimentary rocks, where it was deposited by hot waters
White, gray, flesh pink	Translucent to opaque	Prismatic crystals; most abundantly in rocks as formless grains	Characteristic of sialic rocks
			A continuous series varying in composition from pure albite to pure anorthite; important rock-forming minerals; characteristic of simatic rocks
Brass-yellow	Opaque	Cubic crystals with striated faces; also massive	The most common of the sulfides; used as a source of sulfur in the manufacture of sulfuric acid; distinguished from chalcopyrite by its paler color and greater hardness; from gold by its brittleness and hardness
			A group of silicates with tetrahedra in single chains; augite is the most important; contrast with amphibole
Colorless or white when pure; any color from impurities	Transparent to translucent	Prismatic crystals with faces striated at right angles to long dimension; also massive forms of great variety	An important constituent of sialic rocks; coarsely crystalline varieties: *rock crystal, amethyst* (purple), *rose quartz, smoky quartz, citrine* (yellow), *milky quartz, cat's eye;* cryptocrystalline varieties: *chalcedony, carnelian* (red chalcedony), *chrysoprase* (apple-green chalcedony), *heliotrope* or *bloodstone* (green chalcedony with small red spots), *agate* (alternating layers of chalcedony and opal); granular varieties: *flint* (dull to dark brown), *chert* (like flint but lighter in color), *jasper* (red from hematite inclusions), *prase* (like jasper, but dull green)

Table C.8 ***The common minerals (cont.)***

Mineral	*Chemical composition and name*	*Specific gravity*	*Streak*	*Hardness*	*Cleavage or fracture*	*Luster*
Serpentine	$Mg_3Si_2O_5(OH)_4$ Hydrous magnesium silicate	2.2–2.65	Colorless	2–5	Conchoidal fracture	Greasy, waxy, or silky
Siderite (spathic iron; chalybite)	$FeCO_3$ Iron carbonate	3.85	Colorless	$3\frac{1}{2}$–4	Perfect rhombohedral cleavage	Vitreous
Sillimanite (fibrolite)	Al_2SiO_5 Aluminum silicate	3.23	Colorless	6–7	Good cleavage in 1 direction	Vitreous
Sphalerite (zinc blende; black jack)	ZnS Zinc sulfide	3.9–4.1	White to yellow and brown	$3\frac{1}{2}$–4	Perfect cleavage in 6 directions at 120°	Resinous
Spinel	$MgAl_2O_4$	3.5–4.1	White	8		Vitreous
Spinel group	AB_2O_4					

A	B = Al		B = Fe		B = Cr	
Mg	$MgAl_2O_4$	Spinel	$MgFe_2O_4$	Magnesio-ferrite	$MgCr_2O_4$	Magnesio-chromite
Fe	$FeAl_2O_4$	Hercynite	$FeFe_2O_4$	Magnetite	$FeCr_2O_4$	Chromite
Zn	$ZnAl_2O_4$	Gahnite	$ZnFe_2O_4$	Franklinite		
Mn	$MnAl_2O_4$	Galaxite	$MnFe_2O_4$	Jacobsite		

Mineral	*Chemical composition and name*	*Specific gravity*	*Streak*	*Hardness*	*Cleavage or fracture*	*Luster*
Spondumene (a pyroxene)	$LiAl(Si_2O_6)$ Lithium aluminum inosilicate	3.15–3.20		$6\frac{1}{2}$–7	Perfect at 87° and 93°	Vitreous
Staurolite	$Fe''Al_5Si_2O_{12}(OH)$ Iron aluminum silicate	3.65–3.75	Colorless	7–$7\frac{1}{2}$	Not prominent	Fresh: resinous, vitreous; altered: dull to earthy
Stibnite	Sb_2S_3 Antimony trisulfide	4.52–4.62	Lead-gray to black	2	Perfect 010	Metallic
Taconite	Not a mineral					
Talc (soapstone; steatite)	$Mg_3Si_4O_{10}(OH)_2$ Hydrous magnesium silicate	2.7–2.8	White	1	Good cleavage in 1 direction; thin folia, flexible but inelastic	Pearly to greasy

Color	*Transparency*	*Form*	*Other properties*
Variegated shades of green	Translucent	Platy or fibrous	Platy variety, *antigorite;* fibrous variety, *chrysotile,* an asbestos; an alteration product of magnesium silicates such as olivine, augite, and hornblende; common and widely distributed
Light to dark brown	Transparent to translucent	Granular, compact, earthy	An ore of iron; an accessory mineral in taconite
Brown, pale green, white	Transparent to translucent	Long, slender crystals without distinct terminations; often in parallel groups; frequently fibrous	Relatively rare, but important as a mineral characteristic of high-grade metamorphism; contrast with andalusite and kyanite, which have the same composition but form under conditions of middle-grade metamorphism
Pure: white, green; with iron: yellow to brown and black; red	Transparent to translucent	Usually massive; crystals many-sided, distorted	A common mineral; the most important ore of zinc; the red variety is called *ruby zinc;* streak lighter than corresponding mineral color
White, red, lavender, blue, green, brown, black	Usually translucent; may be clear and transparent	Octahedral crystals	A common metamorphic mineral imbedded in crystalline limestone, gneisses, and serpentine; when transparent and finely colored, it is a gem; the red is spinel ruby or *balas ruby;* some are blue
White, gray, pink, yellow, green	Transparent to translucent	Prismatic crystals; coarse, some large	A source of lithium which improves lubricating properties of greases; some gem varieties
Red-brown to brownish black	Translucent	Usually in crystals, prismatic, twinned to form a cross; rarely massive	A common accessory mineral in schists and slates; characteristic of middle-grade metamorphism; associated with garnet, kyanite, sillimanite, tourmaline
Lead-gray to black	Opaque	Slender prismatic habit; often in radiating groups	The chief ore of antimony, which is used in various alloys
			Unleached iron formation in the Lake Superior district, consists of chert (see *Quartz*) with hematite, magnetite, siderite, and hydrous iron silicates; an ore of iron
Gray, white, silver-white, apple-green	Translucent	Foliated, massive	Of secondary origin, formed by the alteration of magnesium silicates such as olivine, augite, and hornblende; most characteristically found in metamorphic rocks

Table C.8 **The common minerals (cont.)**

Mineral	*Chemical composition and name*	*Specific gravity*	*Streak*	*Hardness*	*Cleavage or fracture*	*Luster*
Topaz	$Al_2SiO_4(F, OH)_2$ Aluminum fluosilicate	3.4–3.6	Colorless	8	Good in 1 direction	Vitreous
Tourmaline	Complex silicate of boron and aluminum, with sodium, calcium, fluorine, iron, lithium, or magnesium	3–3.25	Colorless	7–7½	Not prominent; black variety fractures like coal	Vitreous to resinous
Tremolite (an amphibole)	$Ca_2Mg_5(Si_8O_{22})(OH)_2$	3.0–3.3		5–6	Cleavage at 56°	Vitreous
Uraninite (pitchblende)	Complex oxide of uranium with small amounts of Pb, Ra, Th, Y, N, He, and A	9–9.7	Brownish black	5½	Not prominent	Submetallic, pitchy
Wollastonite	$CaSiO_3$ Calcium silicate	2.8–2.9	Colorless	5–5½	Good cleavage in 2 directions at 84° and 96°	Vitreous or pearly on cleavage surfaces
Zircon	$Zr(SiO_4)$ Zirconium nesosilicate	4.68	Uncolored	7½		Adamantine

Source: Cornelius Hurlbut, Jr., *Dana's Manual of Mineralogy*, 17th ed., John Wiley & Sons, Inc., New York, 1961.

Color	*Transparency*	*Form*	*Other properties*
Straw-yellow, wine-yellow, pink, bluish, greenish	Transparent to translucent	Usually in prismatic crystals, often with striations in direction of greatest length	Represents 8 on Mohs scale of hardness; a gem stone
Varied: black, brown; red, pink, green, blue, yellow	Translucent	Usually in crystals; common: with cross section of spherical triangle	Gem stone; an accessory mineral in pegmatites, also in metamorphic rocks such as gneisses, schists, marbles
White to light green	Transparent to translucent	Often bladed or in radiating columnar aggregate	Frequently in impure, crystalline dolomitic limestones where it formed on recrystallization during metamorphism; also in talc schists
Black	Opaque	Usually massive and botryoidal (i.e., like a bunch of grapes)	An ore of uranium and radium; the mineral in which helium and radium were first discovered
Colorless, white or gray	Translucent	Commonly massive, fibrous, or compact	A common contact metamorphic mineral in limestones
Brown; also gray, green, red, and colorless	Translucent; sometimes transparent	Tetragonal prism and dipyramid	Transparent variety is a gem stone; a source of zirconium metal which is used in the construction of nuclear reactors

Appendix D

The earth and other planets

Table D.1 **Composition of sea water at 35 parts per thousand salinity**[a]

Element	μg/liter	Element	μg/liter
Hydrogen	1.10×10^8	Molybdenum	10
Helium	0.0072	Ruthenium	—
Lithium	170	Rhodium	—
Beryllium	0.0006	Palladium	—
Boron	4,450	Silver	0.28
Carbon (inorganic)	28,000	Cadmium	0.11
(dissolved organic)	2,000	Indium	—
Nitrogen (dissolved N_2)	15,500	Tin	0.81
(as NO_3^-, NO_2^-, NH_4^+)	670	Antimony	0.33
Oxygen (dissolved O_2)	6,000	Tellurium	—
(as H_2O)	8.83×10^8	Iodine	64
Fluorine	1300	Xenon	0.047
Neon	0.120	Cesium	0.30
Sodium	1.08×10^7	Barium	21
Magnesium	1.29×10^6	Lanthanum	0.0029
Aluminum	1	Cerium	0.0013
Silicon	2900	Praesodymium	0.00064
Phosphorus	88	Neodymium	0.0023
Sulfur	9.04×10^5	Samarium	0.00042
Chlorine	1.94×10^7	Europium	0.000114
Argon	450	Gadolinium	0.0006
Potassium	3.92×10^5	Terbium	0.0009
Calcium	4.11×10^5	Dysprosium	0.00073
Scandium	<0.004	Holmium	0.00022
Titanium	1	Erbium	0.00061
Vanadium	1.9	Thulium	0.00013
Chromium	0.2	Ytterbium	0.00052
Manganese	1.9	Lutetium	0.00012
Iron	3.4	Hafnium	<0.008
Cobalt	0.39	Tantalum	<0.0025
Nickel	6.6	Tungsten	<0.001
Copper	23	Rhenium	—
Zinc	11	Osmium	—
Gallium	0.03	Iridium	—
Germanium	0.06	Platinum	—
Arsenic	2.6	Gold	0.011
Selenium	0.090	Mercury	0.15
Bromine	6.73×10^4	Thallium	—
Krypton	0.21	Lead	0.03
Rubidium	120	Bismuth	0.02
Strontium	8,100	Radium	1×10^{-13}
Yttrium	0.003	Thorium	0.0015
Zirconium	0.026	Protactinium	2×10^{-10}
Niobium	0.015	Uranium	3.3

[a] Adapted from Karl K. Turekian, *Oceans*, Table 6-1, p. 92, Prentice-Hall, Inc., Englewood Cliffs, N.J., 1968.

Table D.2 **Composition of river waters of the world**[a]

Substance	*Parts/10^6*
HCO_3	58.4
Ca	15
SiO_2	13.1
SO_4	11.2
Cl	7.8
Na	6.3
Mg	4.1
K	2.3
NO_3	1
Fe	0.67

[a] From Daniel A. Livingstone, "Data of Geochemistry," *U.S. Geol. Surv. Professional Paper*, no. 440G, p. G-41, Table 81, 1963.

Table D.3 **Average composition of the lithosphere**[a]

	Average igneous rock	*Average shale*	*Average sandstone*	*Average limestone*	*Weighted average lithosphere*[b]
SiO_2	59.12%	58.11%	78.31%	5.19%	59.07%
TiO_2	1.05	0.65	0.25	0.06	1.03
Al_2O_3	15.34	15.40	4.76	0.81	15.22
Fe_2O_3	3.08	4.02	1.08 }	0.54 (Fe₂O₃ + FeO)	{ 3.10
FeO	3.80	2.45	0.30 }		{ 3.71
MgO	3.49	2.44	1.16	7.89	3.45
CaO	5.08	3.10	5.50	42.57	5.10
Na_2O	3.84	1.30	0.45	0.05	3.71
K_2O	3.13	3.24	1.32	0.33	3.11
H_2O	1.15	4.99	1.63	0.77	1.30
CO_2	0.10	2.63	5.04	41.54	0.35
ZrO_2	0.04	—	—	—	0.04
P_2O_5	0.30	0.17	0.08	0.04	0.30
Cl	0.05	—	Tr[c]	0.02	0.05
F	0.03	—	—	—	0.03
SO_3	—	0.65	0.07	0.05	—
S	0.05	—	—	0.09	0.06
$(Ce, Y)_2O_3$	0.02	—	—	—	0.02
Cr_2O_3	0.06	—	—	—	0.05
V_2O_3	0.03	—	—	—	0.03
MnO	0.12	Tr	Tr	0.05	0.11
NiO	0.03	—	—	—	0.03
BaO	0.05	0.05	0.05	0.00	0.05
SrO	0.02	0.00	0.00	0.00	0.02
Li_2O	0.01	Tr	Tr	Tr	0.01
Cu	0.01	—	—	—	0.01
C	0.00	0.80	—	—	0.04
Total	100.00	100.00	100.00	100.00	100.00

[a] After F. W. Clarke and H. S. Washington, "The Composition of the Earth's Crust," *U.S. Geol. Surv. Professional Paper*, no. 127, p. 32, 1924.
[b] Weighted average: igneous rock, 95%; shale, 4%; sandstone, 0.75%; limestone, 0.25%.
[c] Trace.

Table D.4 Earth areas

	× 10^6 km^2
Total area	510
Land (29.22% of total)	149
Oceans and seas (70.78% of total)	361
Glacier ice	15.6
Continental shelves	28.4

Table D.5 Earth size

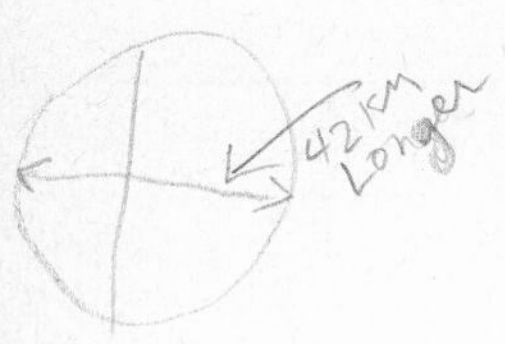

Equatorial radius	6,378 km
Polar radius	6,357 km
Mean radius[a]	6,371 km
Polar circumference	40,009 km
Equatorial circumference	40,077 km
Ellipticity, (equatorial radius − polar radius)/equatorial radius = 1/297	

[a] "Mean radius" is the term used by geophysicists to designate the radius of a sphere of equal volume.

Table D.6 Earth volume, density, and mass

	Average thickness or radius, km	*Volume, × 10^6 km^3*	*Mean density, g/cm^3*	*Mass, × 10^{24} g*
Total earth	6,371	1,083,230	5.52	5,976
Oceans and seas	3.8	1,370	1.03	1.41
Glaciers	1.6	25	0.9	0.023
Crust of continents	35	6,210	2.8	17.39
Crust of oceans	8	2,660	2.9	7.71
Mantle	2,883	899,000	4.5	4,068
Core	3,471	175,500	10.71	1,881

Table D.7 Some planetary data[a]

	Mean distance to sun, astr. units[b]	*Synodic period[c]*	*Eccentricity of orbit*	*Inclination of orbit to ecliptic*	*Orbital velocity,[d] km/sec*	*Volume, earth = 1*
Mercury	0.387	115.88 da	0.206	7.0°	47.7	0.056
Venus	0.723	583.92 da	0.007	3.4°	34.9	0.857
Earth	1	—	0.017	0°	29.6	1
Moon	1	29.53 da	0.05	5.1°	1.02	0.0203
Mars	1.524	779.9 da	0.093	1.8°	24	0.15
Jupiter	5.203	11.86 yr	0.048	1.3°	13	1,320
Saturn	9.54	1.035 yr	0.056	2.5°	9.6	769
Uranus	19.18	1.012 yr	0.047	0.8°	6.7	50
Neptune	30.07	1.006 yr	0.009	1.8°	5.4	42
Pluto	39.67	1.004 yr	0.247	17.2°	4.8	0.1
Sun	—	—	—	—	—	1,304,000

[a] Adapted from Fred L. Whipple, *Earth, Moon, and Planets*, 3rd ed., Harvard University Press, Cambridge, Mass., 1968.
[b] The astronomical unit is the mean distance from the earth to the sun, 149,598,000 km.
[c] Synodic period is the time of one revolution with respect to the sun as seen from the earth.
[d] Orbital velocity is the mean value.

Table D.8 The 25 most abundant elements in the earth's crust[a]

Rank	Element	Weight percent
1	Oxygen	46.60
2	Silicon	27.72
3	Aluminum	8.13
4	Iron	5.00
5	Calcium	3.63
6	Sodium	2.83
7	Potassium	2.59
8	Magnesium	2.09
9	Titanium	0.44
10	Hydrogen	0.14
Subtotal		99.17
11	Phosphorous	0.118
12	Manganese	0.100
13	Fluorine	0.070
14	Sulfur	0.052
15	Strontium	0.045
16	Barium	0.040
17	Carbon	0.032
18	Chlorine	0.020
19	Chromium	0.020
20	Zirconium	0.016
21	Rubidium	0.012
22	Vanadium	0.011
23	Nickel	0.008
24	Zinc	0.007
25	Nitrogen	0.005
Total		99.726

[a] Data from Brian Mason, *Principles of Geochemistry*, 3rd ed., Table 3.3, John Wiley & Sons, Inc., New York, 1966.

Mass,[e] *earth = 1*	*Surface gravity,*[f] *earth = 1*	*Velocity of escape,*[g] *km/sec*	*Albedo*[h]	*Density, water = 1*	*Period of rotation*	*Number of satellites*	*Equatorial diameter, km*
0.055	0.38	4.2	0.059	5.5	59 da	0	4,840
0.816	0.89	10.3	0.85	5.25	243 da	0	12,042
1	1	11.2	0.35	5.52	23 hr 56 min	1	12,682
0.01229	0.165	2.36	0.07	3.34	27.3 da	0	3,458
0.107	0.38	4.96	0.15	3.96	24.6 hr	2	6,720
317.9	2.6	59.2	0.58	1.33	9.8 hr	12	141,920
95.2	1.1	35.2	0.57	0.68	10.2–10.6 da	10?	120,000
14.5	0.96	22.4	0.8	1.6	10.8 hr	5	47,360
17.4	1.5	24.0	0.71	2.3	15.7 hr	2	44,160
?	?	?	0.15?	?	153 hr	0	5,760
332,950	—	613	—	1.41	25 da	—	1,384,000

[e] The earth weights 5.976×10^{21} metric tons.
[f] The weight of a given object on the earth when multiplied by the quantity in the table becomes the weight of the object at the surface of the planet.
[g] An object shot away from the equator with this velocity would escape forever into space (neglecting friction with an atmosphere).
[h] Albedo is the ratio of the total amount of light reflected by the planet to the light incident on it.

Appendix E

Tables of measurement

THE FUNDAMENTAL UNIT OF LENGTH IS THE METER (m), ORIGINALLY DEFINED AS 10^{-7} of the distance from the equator to the North Pole, but now defined as the distance between two marks inscribed on the standard meter bar in Paris.

1 kilometer (km) = 10^3 m   1 millimeter (mm) = 10^{-3} m
1 centimeter (cm) = 10^{-2} m   1 micrometer (μm) = 10^{-6} m

The official definition of an inch is based on the length of a meter: 1 inch (in.) = 2.54×10^{-2} m.

Temperature is measured in degrees Centigrade (°C), a scale in which the interval between the freezing point and the boiling point of water is divided into 100°, with 0° representing the freezing point and 100° the boiling point. One degree Fahrenheit (1°F) = $\frac{5}{9}$°C. One measure is converted to the other in this manner: If temperature is in degrees Fahrenheit, subtract 32° from that number; five-ninths of the difference is the temperature in degrees Centigrade. If temperature is in degrees Centigrade, multiply that number by 1.8; add 32° to the result, and the sum is the temperature in degrees Fahrenheit.

Tables E.1 through E.4 convert various metric and English measures.

Table E.1 **Conversion of mass (*M*) and weight (mlt^{-2})[a]**

	Grams, g	*Kilograms, kg*	*Pounds, lb*	*Ounces, oz*
g	1	1,000	453.6	28.35
kg	0.001	1	0.4536	2.835×10^{-2}
lb	2.205×10^{-3}	2.205	1	6.25×10^{-2}
oz	3.527×10^{-2}	35.27	16	1

[a] Multiply units in the column heads by the figures in the table to convert them to units in the stub. The mass conversion factors apply to *gravitational* units of force that have the same names. The dimensions of these units when used as gravitational units of force are mlt^{-2}.

Table E.2 **Conversion of volumes[a]**

	Cubic meters, m^3	*Cubic yards, yd^3*	*Cubic centimeters, cm^3*	*Cubic inches, in.^{3}*	*Cubic feet, ft^3*
m^3	1	0.7646	10^{-6}	1.639×10^{-5}	2.832×10^{-2}
yd^3	1.308	1	1.308×10^{-6}	2.143×10^{-5}	3.704×10^{-2}
cm^3	10^6	7.646×10^5	1	16.39	2.832×10^4
in.^{3}	6.102×10^4	46,656	6.102×10^{-2}	1	1,728
ft^3	35.31	27	3.531×10^{-5}	5.787×10^{-4}	1

[a] Multiply units in the column heads by the figures in the table to convert them to units in the stub.

Table E.3 **Conversion of lengths**[a]

	Meters, m	*Yards, yd*	*Centimeters, cm*	*Inches, in.*	*Feet, ft*	*Kilometers, km*	*Miles, mi*
m	1	0.9144	10^{-2}	2.54×10^{-2}	0.3048	10^{3}	1,609
yd	1.094	1	1.094×10^{-2}	2.778×10^{-2}	0.3333	1,094	1,760
cm	100	91.44	1	2.54	30.48	10^{5}	1.609×10^{5}
in.	39.37	36	0.3937	1	12	3.939×10^{4}	6.336×10^{4}
ft	3.281	3	3.281×10^{-2}	8.333×10^{-2}	1	3,281	5,280
km	10^{-3}	0.144×10^{-4}	10^{-5}	2.54×10^{-5}	3.048×10^{-4}	1	1.609
mi	6.214×10^{-4}	5.682×10^{-4}	6.214×10^{-6}	1.578×10^{-5}	1.894×10^{-4}	0.6214	1

[a] Multiply units in the column heads by the figures in the table to convert them to units in the stub.

Table E.4 **Conversion of areas**[a]

	Acres	*Square inches, in.²*	*Square centimeters, cm²*	*Square feet, ft²*	*Square meters, m²*	*Square miles, mi²*	*Square kilometers, km²*
Acres	1	6.27×10^{-6}	2.471×10^{-8}	3.296×10^{-5}	2.471×10^{-4}	640	247.1
$in.^2$	6,272,640	1	0.1550	144	1,550	4.015×10^{9}	1.550×10^{9}
cm^2	4.047×10^{7}	6.452	1	929	10^{4}	2.59×10^{10}	10^{10}
ft^2	4.356×10^{4}	6.944×10^{-3}	1.076×10^{-3}	1	10.76	2.788×10^{7}	1.076×10^{7}
m^2	4,047	6.452×10^{-4}	10^{-4}	9.290×10^{-2}	1	2.590×10^{6}	10^{6}
mi^2	1.562×10^{-3}	4.25×10^{-9}	3.861×10^{-11}	3.587×10^{-8}	3.861×10^{-7}	1	0.3861
km^2	4.047×10^{-3}	6.452×10^{-10}	10^{-10}		10^{-6}	2.590	1

[a] Multiply units in the column heads by the figures in the table to convert them to units in the stub.

Appendix F

Topographic and geologic maps

"TOPOGRAPHY" REFERS TO THE SHAPE OF THE PHYSICAL FEATURES OF THE LAND. A *topographic map* is the representation of the position, relation, size, and shape of the physical features of an area. In addition to hills, valleys, rivers, and mountains, most topographic maps also show the culture of a region—that is, roads, towns, houses, political boundaries, and similar features.

Topographic maps

Topographic maps are used in the laboratory for the observation and analysis of the effects of the several geological processes that are constantly changing the face of the earth.

Definitions

Relief of an area is the difference in elevation between the tops of hills and the bottoms of valleys.

Height is the vertical difference in elevation between an object and its immediate surroundings.

Elevation or *altitude* is the vertical distance between a given point and the datum plane.

Datum plane is the reference surface from which all altitudes on a map are measured. This is usually mean sea level.

Bench mark is a point of known elevation and position, which is usually indicated on a map by the letters B.M., with the altitude given, on American maps, to the nearest foot.

Contour line is a map line connecting points representing places on the earth's surface that have the same elevation. It thus locates the intersection with the earth's surface of a plane at any arbitrary elevation parallel to the datum plane. Contours represent the vertical or third dimension on a map, which has only two dimensions. They show the size and shape of physical features such as hills and valleys. A hachured contour line indicates a depression. It resembles an ordinary contour line except for the hachures or short dashes on one side pointing toward the center of the depression.

Contour interval is the difference in elevation represented by adjacent contour lines.

Scale of a map is the ratio of the distance between two points on the ground and the same two points on the map. It may be expressed in three ways:

Fractional scale If two points are exactly 1 mile apart, they may be represented on the map as being separated by some fraction of that distance, say 1 in. In this instance, the scale is 1 in. to the mile. There are 63,360 in. in 1 mile, so this scale can be expressed as the fraction or ratio 1:63,360. Actually, many topographic maps of the U.S. Geological Survey have a scale of 1:62,500. Many recent maps have a scale of 1:31,250 and others of 1:24,000.

Graphic scale This scale is a line printed on the map and divided into units that are equivalent to some distance, such as 1 mile or 1 km.

Verbal scale This is an expression in common speech, such as "an inch to a mile," "two miles to the inch," or "four centimeters to the kilometer."

Conventional symbols

An explanation of the symbols used on topographic maps is printed on the back of each topographic sheet, along the margin or, for newer maps, on a separate legend sheet. In general, culture (works of man) is shown in black. All water features, such as streams, swamps, and glaciers, are shown in blue. Relief is shown by contours in brown. Red may be used to indicate main highways, and green overprints may be used to designate areas of woods, orchards, vineyards, or scrub.

The U.S. Geological Survey distributes free of charge (apply to The Director, Geological Survey, Washington 25, D.C.) a single sheet entitled "Topographic Maps" that includes an illustrated summary of topographic map symbols.

Locating points

Any particular point or area may be located in several ways on a topographic map. The three most commonly used are

Figure F.1 *Subdivision by township and range. See the text for discussion.*

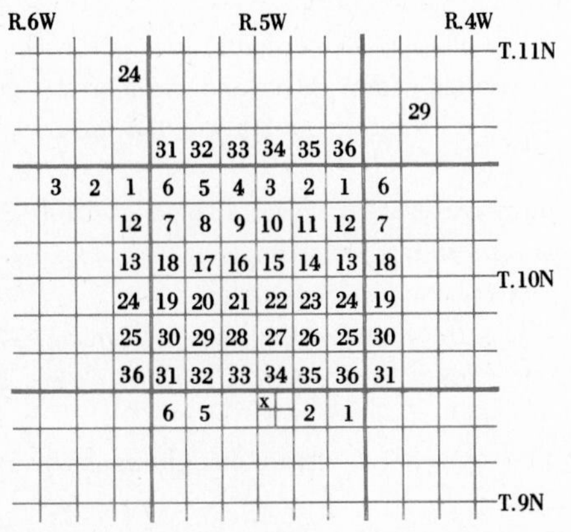

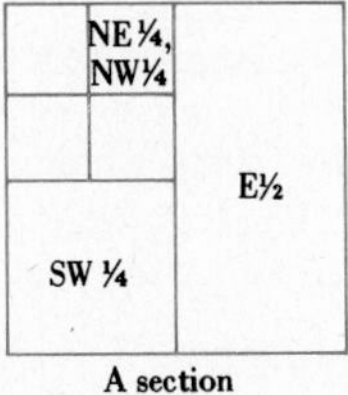

In relation to prominent features A point may be referred to as being so many kilometers in a given direction from a city, mountain, river mouth, lake, or other easily located feature on the map.

By latitude and longitude Topographic maps of the U.S. Geological Survey are bounded on the north and south by parallels of latitude and on the east and west by meridians of longitude. These intersecting lines form the grid into which the earth has been divided. Latitude is measured north and south from the equator, and longitude is measured east and west from the prime meridian that passes through Greenwich, England. Thus maps in the United States are within north latitude and west longitude.

By township and range The greater part of the United States has been subdivided by a system of land survey in which a square 6 miles on a side forms the basic unit, called a township. Not included in this system are all the states along the eastern seaboard (with the exception of Florida), West Virginia, Kentucky, Tennessee, Texas, and parts of Ohio. Townships are laid off north and south from a base line and east and west from a principal meridian. Each township is divided into 36 sections, usually 1 mile on a side. Each section may be further subdivided into half-sections, quarter-sections, or sixteenth-sections. Thus in Figure F.1 the point X can be located as in the northeast quarter of the northwest quarter of section 3, township 9 north, range 5 west. This is abbreviated as NE$\frac{1}{4}$ NW$\frac{1}{4}$ Sec 3, T9N, R5W, or NE NW Sec 3–9N–5W.

Contour sketching

Many contour maps are now made from aerial photographs. Before this can be done, however, the position and location of a number of reference points, or bench marks, must be determined in the field. If the topographic map is surveyed in the field rather than from aerial photographs, the topographer first determines the location and elevation of bench marks and a large number of other points that are selected for their critical position. Such points may be along streams, on hilltops, on the lowest point in a saddle between hills, or at places where there is a significant change in slope. On the basis of these points, contours may be sketched through points of equal elevation. Preferably, the contours are sketched in the field in order to include minor irregularities that are visible to the topographer.

Because contours are not ordinary lines, certain requirements must be met in drawing them to satisfy the definition of contour lines. These are listed below.

1 All points on one contour line have the same elevation.

2 Contours separate all points of higher elevation than the contour from all points of lower elevation.

3 The elevation represented by a contour line is always a simple multiple of the contour interval. Every contour line that is a multiple of 5 times the contour interval is heavier than the others. (Exception: 25-ft contours, in which every multiple of 4 times the interval is heavier.)

4 Contours never cross or intersect one another.

5 A vertical cliff is represented by coincident contours.

6 Every contour closes on itself either within or beyond the limits of the map. In the latter case, the contours will end at the edge of the map.

7 Contour lines never split.

8 Uniformly spaced contour lines represent a uniform slope.

9 Closely spaced contour lines represent a steep slope.

10 Contour lines spaced far apart represent a gentle slope.

11 A contour line that closes within the limits of the map indicates a hill.

12 A hachured contour line represents a depression. The short dashes or hachures point into the depression.

13 Contour lines curve up a valley but cross a stream at right angles to its course.

14 Maximum ridge and minimum valley contours always go in pairs. That is, no single lower contour can lie between two higher ones, and vice versa.

Topographic profiles

A topographic profile is a cross section of the earth's surface along a given line. The upper line of this section is irregular and shows the shape of the land along the line of profile or section.

Profiles are most easily constructed with graph paper. A horizontal scale, usually the map scale, is chosen. Then a vertical scale sufficient to bring out the features of the surface is chosen. The vertical scale is usually several times larger than the horizontal—that is, it is exaggerated. The steps in the construction of a profile are as follows:

1 Select a base (one of the horizontal lines on the graph paper). This may be sea level or any other convenient datum level.
2 On the graph paper, number each fourth or fifth line above the base, according to the vertical scale chosen.
3 Place the graph paper along the line of profile.
4 With the vertically ruled lines as guides, plot the elevation of each contour line that crosses the line of profile.
5 If great accuracy is not important, plot only every heavy contour and the tops and bottoms of hills and the bottoms of valleys.
6 Connect the points.
7 Label necessary points along the profile.
8 Give the vertical and horizontal scales.
9 State the vertical exaggeration.
10 Title the profile.

VERTICAL EXAGGERATION The profile represents both vertical and horizontal dimensions. These dimensions are not usually on the same scale, because the vertical needs to be greater than the horizontal to give a clear presentation of changes in level. Thus, if the vertical scale is 500 ft to the inch and the horizontal scale is 5,280 ft to the inch—or say 1:62,500—the vertical exaggeration is about 10 times, written 10×. This is obtained by dividing the horizontal scale by the vertical scale. Note that both horizontal and vertical scales must be expressed in the same unit (commonly feet to the inch) before dividing.

Geologic maps

Geologic maps show the distribution of earth materials on the surface. In addition, they indicate the relative age of these materials and suggest their arrangement beneath the surface.

Definitions

FORMATION The units depicted on a geologic map are usually referred to as formations. We define a formation as a rock unit with upper and lower boundaries that can be recognized easily in the field and that is large enough to be shown on the map. A formation receives a distinctive designation made up of two parts. The first part is geographic and refers to the place or general area where the formation is first described. The second refers to the nature of the rock. Thus *Trenton limestone* is a formation composed dominantly of limestone and is named after Trenton Falls in central New York State, where it was first formally described. *Wausau granite* designates a body of granite in the Wausau, Wisconsin, area. If the lithology is so variable that no single lithologic distinction is appropriate, the word *formation* may be used. For instance, the *Raritan formation* is named for the area of the Raritan River and Raritan Bay in New Jersey, and its lithology includes both sand and clay.

DIP AND STRIKE The dip and strike of a rock layer refers to its orientation in relation to a horizontal plane. In Chapter 16, we found that the dip is the acute angle that a tilted rock layer makes with an imaginary plane. We also found that the strike is the compass direction of a line formed by the intersection of the dipping surface with an imaginary horizontal plane. The direction of strike is always at right angles to the direction of dip. The dip-and-strike symbol used on a geologic map is in the form of a topheavy T. The cross bar represents the direction of the strike of the bed. The short upright represents the direction of the dip of the bed. This sometimes, but not always, has an arrow pointing in the direction of dip. Very often the angle of dip is indicated alongside the symbol.

Example:

⊤ 30	Strike E–W; dip 30°S
↗ 25	Strike N, 45°E; dip 25°SE

Note: In this example, the top of the page is considered to be north.

CONTACT A contact is the plane separating two rock units. It is shown on the geologic map as a line that is the intersection of the plane between the rock units and the surface of the ground.

OUTCROP An outcrop is an exposure of rock material that crops out at the surface through the cover of soil and weathered material. In areas of abundant rainfall, soil and vegetation obscure the underlying rock material and only a small fraction of 1 percent of the surface may be in outcrop. In dry climates

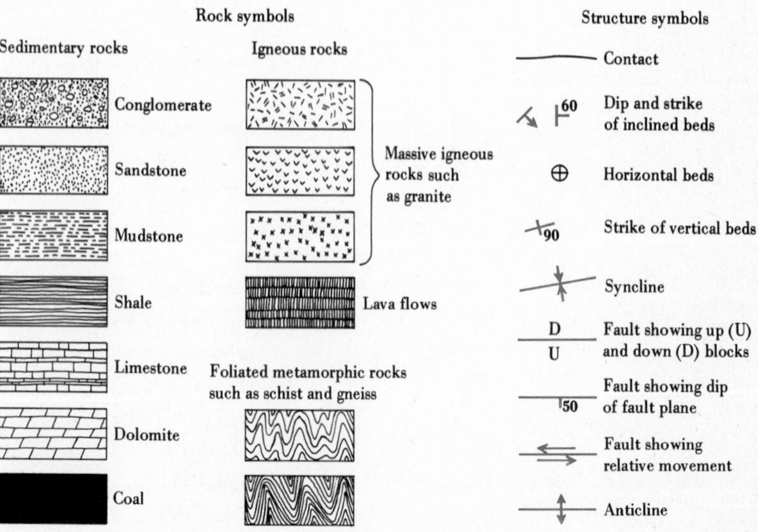

Figure F.2 Symbols commonly used on geologic maps.

Table F.1 ***Letter symbols and colors commonly used to designate units in the geologic column***

Period	*Symbol*	*Color*
Pleistocene	Q	Yellow and gray
Pliocene	Tpl	Yellow-ocher
Miocene	Tm	Yellow-ocher
Oligocene	To	Yellow-ocher
Eocene	Te	Yellow-ocher
Paleocene	Tp	Yellow-ocher
Cretaceous	K	Olive-green
Jurassic	J	Blue-green
Triassic	T_R	Light peacock-blue or bluish gray-green
Permian	P	Blue
Pennsylvanian	Cp	Blue
Mississippian	Cm	Blue
Devonian	D	Gray-purple
Silurian	S	Purple
Ordovician	O	Red-purple
Cambrian	€	Brick red
Precambrian	P€	Terra-cotta and gray-brown

where soils are shallow or absent and the plant cover is discontinuous, bedrock usually crops out much more widely.

LEGEND AND SYMBOLS A legend is an explanation of the various symbols used on the map. There is no universally accepted set of standard symbols, but some that are more widely used are given in Figure F.2. In addition to the graphic symbols in Figure F.2, letter symbols are sometimes used to designate rock units. Such a symbol contains a letter or letters referring to the geologic column, followed by a letter or letters referring to the specific name of the rock unit. Thus in the symbol Ot, the O stands for Ordovician and the t for the Trenton limestone of central New York State. The letters or abbreviations generally used for the geologic column are given in Table F.1.

Sometimes different colors are used to indicate different rock systems. There is no standardized color scheme, but many of the geologic maps of the U.S. Geological Survey use the colors given in Table F.1, combined with varying patterns, for systems of sedimentary rocks. No specific colors are designated for igneous rocks, but when colors are used, they are usually purer and more brilliant than those used for sedimentary rocks.

Construction of a geologic map

The basic idea of geologic mapping is simple. We are interested first in showing the distribution of the rocks at the earth's surface. Theoretically, all we need to do is plot the occurrence of the different rocks on a base map, and then we have a geologic map. Unfortunately, the process is not quite this simple.

In most areas the bedrock is more or less obscured in one way or another,

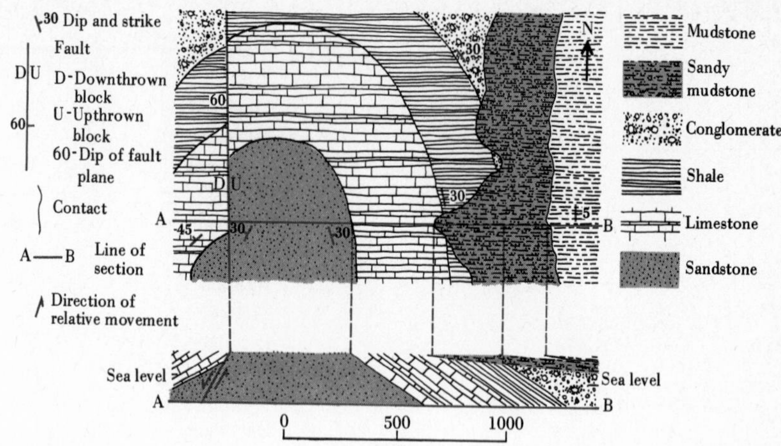

Figure F.3 ***Construction of a geologic cross section from a geologic map.***

and only a small amount of outcrop is available for observation, study, and sampling. From the few exposures available, the geologist must extrapolate the general distribution of rock types. In this extrapolation, his field data are obviously of prime importance. But he will also be guided by changes in soil, vegetation, and landscape, as well as by patterns that can be detected on aerial photographs. Furthermore, he may be aided by laboratory examination of field samples and by the records of both deep and shallow wells. The geologist may also have available to him geophysical data that help determine the nature of obscured bedrock. Eventually, when he has marshaled as many data as possible, he draws the boundaries delineating the various rock types.

In addition to the distribution of rock types, the geologist is also concerned with depicting, as accurately as he can, the ages of the various rocks and their arrangement beneath the surface. These goals, also, will be realized in part through direct observations in the field and in part through other lines of evidence. The preparation of an accurate, meaningful, geologic map demands experience, patience, and judgment.

Geologic cross sections

A geologic map tells us something of how rocks are arranged in the underground. Often, to show these relations more clearly, we find it convenient to draw geologic cross sections. Such a section is really a diagram showing a side view of a block of the earth's crust as it would look if we could lift it up to view. We have used cross sections in many illustrations throughout this book.

A geologic cross section is drawn, insofar as possible, at right angles to the general strike of the rocks. The general manner in which a geologic cross section is projected from a geologic map is shown in Figure F.3. If the projection is made onto a topographic profile in which the vertical scale has been exaggerated, then the angle of the dip of the rocks should be exaggerated accordingly.

Glossary

THE DEFINITIONS IN THIS GLOSSARY HAVE BEEN COORDINATED WITH THE "Glossary of Geology and Related Sciences," 2nd ed., published by the American Geological Institute, 2101 Constitution Avenue, Washington, D.C. 20037.

A Abbreviation for *angstrom*, a unit of length equal to 10^{-8} cm.

AA LAVA Lava whose surface is covered with a random mass of angular jagged blocks. High silica content characteristic.

ABLATION As applied to glacier ice, the process by which ice below the snow line is wasted by evaporation and melting.

ABRASION Erosion of rock material by friction of solid particles moved by water, ice, wind, or gravity.

ABSOLUTE TIME Geologic time measured in terms of years. Compare with *relative time.*

ACIDIC LAVA Composed of 70 percent or more of silica.

ACTINOLITE A metamorphic ferromagnesian mineral. An asbestos.

ADIABATIC RATE In a body of air moving upward or downward, the change in temperature that occurs without gain or loss of temperature to the air through which it moves.

AFTERSHOCK An earthquake that follows a larger earthquake and originates at or near the focus of the larger earthquake. Generally, major shallow earthquakes are followed by many aftershocks. These decrease in number as time goes on, but may continue for many days or even months.

AGATE A variety of chalcedony with alternating layers of chalcedony and opal.

A-HORIZON The soil zone immediately below the surface, from which soluble material and fine-grained particles have been moved downward by water seeping into the soil. Varying amounts of organic matter give the A-horizon a color ranging from gray to black.

AIRY HYPOTHESIS Explains isostasy by assuming the earth's crust has the same density everywhere and that differences in elevation result from differences in the thickness of the outer layer.

ALBITE The feldspar in which the diagnostic positive ion is Na^+. Sodic feldspar, $Na(AlSi_3O_8)$. One of the plagioclase feldspars.

ALLUVIAL FAN The land counterpart of a delta. An assemblage of sediments marking the place where a stream moves from a steep gradient to a flatter gradient and suddenly loses its transporting power. Typical of arid and semiarid climates, but not confined to them.

ALMANDITE A deep red garnet of iron and aluminum formed during regional metamorphism.

ALPHA DECAY Radioactive decay that takes place by the loss of an alpha particle from the nucleus. Mass of the element decreases by four and atomic number decreases by two.

ALPHA PARTICLE A helium atom lacking electrons and therefore having a double positive charge.

AMORPHOUS A state of matter in which there is no orderly arrangement of atoms.

AMPHIBOLE GROUP Ferromagnesian silicates with a double chain of silicon-oxygen tetrahedra. Common example: hornblende. Contrast with *pyroxene group.*

AMPHIBOLITE A faintly foliated metamorphic rock developed during the regional metamorphism of simatic rocks. Composed mainly of hornblende and plagioclase feldspars.

AMPHIBOLITE FACIES An assemblage of minerals formed at moderate to high pressures between 450°C and 700°C during regional metamorphism.

ANDALUSITE A silicate of aluminum built around independent tetrahedra (Al_2SiO_5). Characteristic of middle-grade metamorphism. Compare with *kyanite,* which has the same composition and forms under similar conditions, but which has a different crystal habit. Contrast with *sillimanite,* which has the same composition but different crystal habit and forms at highest metamorphic temperatures.

ANDESITE A fine-grained igneous rock with no quartz or orthoclase, composed of about 75 percent plagioclase feldspars and the balance ferromagnesian silicates. Important as lavas, possibly derived by fractional crystallization from basaltic magma. Widely characteristic of mountain-making processes around the borders of the Pacific Ocean. Confined to continental sectors.

ANDESITE LINE A map line designating the petrographic boundary of the Pacific Ocean. Extrusive rocks on the Pacific side of the line are basaltic and on the other side andesitic.

ANGSTROM A unit of length equal to one hundred-millionth of a centimeter (10^{-8} cm). Abbreviation A.

ANGULAR MOMENTUM A vector quantity, the product of mass times radius of orbit times velocity. The energy of motion of the solar system.

ANGULAR UNCONFORMITY An unconformity in which the older strata dip at a different angle from that of the younger strata.

ANHYDRITE The mineral calcium sulfate, $CaSO_4$, which is gypsum without water.

ANORTHITE The feldspar in which the diagnostic positive element is Ca^{2+}. Calcic feldspar, $Ca(Al_2Si_2O_8)$. One of the plagioclase feldspars.

ANORTHOSITE A plutonic igneous rock composed of 90 percent or more of the feldspar mineral anorthite. Pyroxene and some olivine usually make up the balance of the rock.

ANTARCTIC BOTTOM WATER Sea water that sinks to the ocean floor off Antarctica and flows equatorward beneath the North Atlantic Deep Water.

ANTARCTIC INTERMEDIATE WATER Sea water that sinks at about 50°S and flows northward above the North Atlantic Deep Water.

ANTECEDENT STREAM A stream that maintains after uplift the same course it originally followed prior to uplift.

ANTHRACITE Metamorphosed bituminous coal of about 95 to 98 percent carbon.

ANTICLINAL THEORY The theory that water, petroleum, and natural gas accumulate in up-arched strata in the order named (water lowest), provided the structure contains reservoir rocks in proper relation to source beds and capped by an impervious barrier.

ANTICLINE A configuration of folded, stratified rocks in which the rocks dip in two directions away from a crest, as the principal rafters of a common gable roof dip away from the ridgepole. The reverse of a *syncline.* The "ridgepole" or crest is called the axis.

APHANITIC TEXTURE Individual minerals are present, but in particles so small that they cannot be identified without the aid of a microscope.

AQUIFER A permeable material through which ground water moves.

ARÊTE A narrow, saw-toothed ridge formed by cirques developing from opposite sides into the ridge.

ARKOSE A detrital sedimentary rock formed by the cementation of individual grains of sand size and predominantly composed of quartz and feldspar. Derived from the disintegration of granite.

ARROYO Flat-floored, vertically walled channel of an intermittent stream typical of semiarid climates. Often applied to such features of southwestern United States. Synonymous with *wadi* and *wash.*

ARTESIAN WATER Water that is under pressure when tapped by a well and is able to rise above the level at which it is first encountered. It may or may not flow out at ground level.

ASBESTOS A general term applied to certain fibrous minerals that display similar physical characteristics although they differ in composition. Some asbestos has fibers long enough to be spun into fabrics with great resistance to heat, such as those used for automobile brake linings. Types with shorter fibers are compressed into insulating boards, shingles, etc. The most common asbestos mineral (95 percent of U.S. production) is chrysotile, a variety of serpentine, a metamorphic mineral.

ASH Volcanic fragments consisting of sharply angular glass particles, smaller than cinders.

ASPHALT A brown to black solid or semisolid bituminous substance. Occurs in nature but is also obtained as a residue from the refining of certain hydrocarbons (then known as "artificial asphalt").

ASTEROID Small bodies in an orbit believed to have been either that of a planet that exploded, or of matter that never completed the planet-forming process.

ASTROGEOLOGY The geology of celestial bodies other than the earth. Generally restricted in meaning to the geology of the extraterrestrial bodies of our planetary system.

ASYMMETRIC FOLD A fold in which one limb dips more steeply than the other.

ATLANTIC RIDGE Belt of mountains under the Atlantic Ocean extending from Iceland to Antarctica and situated nearly midway between the continents.

ATOLL A ring of low coral islands arranged around a central lagoon.

ATOM A combination of protons, neutrons, and electrons, of which 103 kinds are now known.

ATOMIC ENERGY Energy associated with the nucleus of an atom. It is released when the nucleus is split.

ATOMIC MASS The nucleus of an atom contains 99.95 percent of its mass. The total number of protons and neutrons in the nucleus is called the *mass number.*

ATOMIC NUMBER The number of positive charges on the nucleus of an atom; the number of protons in the nucleus.

ATOMIC REACTOR A huge apparatus in which a radioactive core heats water under pressure and passes it to a heat exchanger.

ATOMIC SIZE The radius of an atom (average distance from the center to the outermost electron of the neutral atom). Commonly expressed in angstroms.

AUGITE A rock-forming ferromagnesian silicate mineral built around single chains of silicon-oxygen tetrahedra.

AUREOLE A zone in which contact metamorphism has taken place.

AUSTRALITES See *tektites*.

AXIAL PLANE A plane through a rock fold that includes the axis and divides the fold as symmetrically as possible.

AXIS The ridge, or place of sharpest folding, of an anticline or syncline.

B Symbol for *bulk modulus*.

BACKSET BEDS Inclined layers of sand developed on the gentler dune slope to the windward. These beds may constitute a large part of the total volume of a dune, especially if there is enough vegetation to trap most of the sand before it can cross over to the slip face.

BACK SWAMP Marshy area of a flood plain at some distance from and lower than the banks of a river confined by natural levees.

BANKFULL STAGE The stage of flow at which a stream fills its channel up to the level of its bank. The recurrence interval is on the average once every 1.5 to 2 years.

BARCHAN A crescent-shaped dune with wings or horns pointing downwind. Has a gentle windward slope and steep lee slope inside the horns. About 30 m in height and 300 m wide from horn to horn. Moves with the wind at about 15 m per year across a flat, hard surface where a limited supply of sand is available.

BARRIER ISLAND A low, sandy island near the shore and parallel to it, on a gently sloping offshore bottom.

BARRIER REEF A reef that is separated from a landmass by a lagoon of varying width and depth opening to the sea through passes in the reef.

BASAL PLANE The lower boundary of a zone of concentric folding.

BASAL SLIP The movement of an entire glacier over the underlying ground surface.

BASALT A fine-grained igneous rock dominated by dark-colored minerals, consisting of plagioclase feldspars (over 50 percent) and ferromagnesian silicates. Basalts and andesites represent about 98 percent of all extrusive rocks.

BASE FLOW That portion of stream flow attributable to ground-water flow.

BASE LEVEL For a *stream*, a level below which it cannot erode. There may be temporary base levels along a stream's course, such as those established by lakes, or resistant layers of rock. Ultimate base level for a stream is sea level. For a *region*, a plane extending inland from sea level sloping gently upward from the sea. Erosion of the land progresses toward this plane, but seldom, if ever, quite reaches it.

BASEMENT COMPLEX Undifferentiated rocks underlying the oldest identifiable rocks in any region. Usually sialic, crystalline, metamorphosed. Often, but not necessarily, Precambrian.

BASIC LAVA Lava composed of less than 50 percent silica.

BATHOLITH A discordant pluton that increases in size downward, has no determinable floor, and shows an area of surface exposure exceeding $100 km^2$.

BAUXITE A mineraloid that is the chief ore of commercial aluminum. A mixture of hydrous aluminum oxides.

BAY BARRIER A sandy beach, built up across the mouth of a bay, so that the bay is no longer connected to the main body of water.

BEDDING (1) A collective term used to signify the existence of beds or layers in sedimentary rocks. (2) Sometimes synonymous with *bedding plane*.

BEDDING PLANE Surface separating layers of sedimentary rocks. Each bedding plane marks the termination of one deposit and the beginning of another of different character, such as the surface separating a sand bed from a shale layer. A rock tends to separate or break readily along bedding planes.

BEDIASITES See *tektites*.

BED LOAD Material in movement along a stream bottom, or, if wind is the moving agency, along the surface. Contrast with material carried in suspension or solution.

BEHEADED STREAM The lower section of a stream that has lost its upper portion through *stream piracy*.

BELT OF SOIL MOISTURE Subdivision of zone of aeration. Belt from which water may be used by plants or withdrawn by soil evaporation. Some of the water passes down into the intermediate belt, where it may be held by molecular attraction against the influence of gravity.

BENCH MARK See Appendix F.

BENIOFF ZONE A seismic zone dipping beneath a continental margin and having a deep-sea trench as its surface expression.

BERGSCHRUND The gap or crevasse between glacier ice and the headwall of a cirque.

BERMS In the terminology of coastlines, berms are stormbuilt beach features that resemble small terraces; on their seaward edges are low ridges built up by storm waves.

BETA DECAY Radioactive decay that takes place by the loss of a beta particle (an electron) from one of the neutrons in the nucleus. The mass of the element remains the same but the atomic number increases by one.

B-HORIZON The soil zone of accumulation that lies below the A-horizon. Here is deposited some of the material that has moved downward from the A-horizon.

BINDING ENERGY The amount of energy that must be supplied to break an atomic nucleus into its component fundamental particles. It is equivalent to the mass that disappears when fundamental particles combine to form a nucleus.

BIOCHEMICAL ROCK A sedimentary rock made up of deposits resulting directly or indirectly from the life processes of organisms.

BIOTITE "Black mica," ranging in color from dark brown to green. A rock-forming ferromagnesian silicate mineral with its tetrahedra arranged in sheets.

BITUMINOUS COAL Soft coal, containing about 80 percent carbon and 10 percent oxygen.

BLOCKS Pyroclastic debris consisting of coarse, angular pieces of the cone or masses broken away from rock that blocks the vent.

BLOWOUT A basin, scooped out of soft, unconsolidated deposits by the process of deflation. Ranges from a few meters to several kilometers in diameter.

BODY WAVE Push-pull or shake earthquake wave that travels through the body of a medium, as distinguished from waves that travel along a free surface.

BOMBS Pyroclastic debris in the form of rounded masses that congeal from magma as it travels through the air.

BORNITE A mineral, CU_5FeS_4. An important ore of copper.

BOTTOMSET BED Layer of fine sediment deposited in a body of standing water beyond the advancing edge of a growing delta. The delta eventually builds up on top of the bottomset beds.

BOULDER SIZE A volume greater than that of a sphere with a diameter of 256 mm.

BOULDER TRAIN A series of glacier erratics from the same bedrock source, usually with some property that permits easy identification. Arranged across the country in the shape of a fan with the apex at the source and widening in the direction of glacier movement.

BOWEN'S REACTION SERIES A series of minerals for which any early-formed phase tends to react with the melt that remains, to yield a new mineral further along in the series. Thus early-formed crystals of olivine react with remaining liquids to form augite crystals; these in turn may further react with the liquid then remaining to form hornblende. See also *continuous reaction series* and *discontinuous reation series.*

BRAIDED STREAM A complex tangle of converging and diverging stream channels separated by sand bars or islands. Characteristic of flood plains where the amount of debris is large in relation to the discharge.

BRECCIA A clastic rock made up of angular fragments of such size that an appreciable percentage of the volume of the rock consists of particles of granule size or larger.

BRITTLE A property of material whereby its strength in tension is greatly different from its strength in compression.

BROADLY UPWARPED MOUNTAINS Mountainous relief developed by broad upwarping of the earth's crust.

BROWN CLAY An extremely fine-grained deposit characteristic of some deep ocean basins, particularly those of the Pacific.

BULK MODULUS The number that expresses a material's resistance to elastic changes in volume. For example, the number of pounds per square inch necessary to cause a specified change in volume. Represented by the symbol B.

BURIAL METAMORPHISM Changes resulting from pressures and temperatures in rocks buried to depths of several kilometers.

CALCIC FELDSPAR Anorthite, $Ca(Al_2Si_2O_8)$.

CALCITE A mineral composed of calcium carbonate, $CaCO_3$.

CALDERA A roughly circular, steep-sided volcanic basin with a diameter at least three or four times its depth. Commonly at the summit of a volcano. Contrast with *crater.*

CALICHE A whitish accumulation of calcium carbonate in the soil profile.

CALVING As applied to glacier ice, the process by which a glacier that terminates in a body of water breaks away in large blocks. Such blocks form the icebergs of polar seas.

CAPACITY The amount of material that a transporting agency such as a stream, a glacier, or the wind can carry under a particular set of conditions.

CAPILLARY FRINGE Belt above zone of saturation in which underground water is lifted against gravity by surface tension in passages of capillary size.

CAPILLARY SIZE "Hairlike," or very small, such as tubes from 0.0025 to 0.25 cm. in diameter.

CARBOHYDRATE A compound of carbon, hydrogen, and oxygen. Carbohydrates are the chief products of the life process in plants.

CARBONATE MINERAL Mineral formed by the conbination of the complex ion $(CO_3)^{2-}$ with a positive ion. Common example: calcite, $CaCO_3$.

CARBON CYCLE The process in the sun's deep interior by which radiant energy is generated in the formation of helium from hydrogen.

CARBON-14 Radioactive isotope of carbon $^{14}_{6}C$, which has a half-life of 5,730 years. Used to date events back to about 50,000 years ago.

CARBON-RATIO A number obtained by dividing the amount of fixed carbon in coal by the sum of fixed carbon and volatile matter, and multiplying by 100. This is the same as the percentage of fixed carbon, assuming no moisture or ash.

CARLSBERG RIDGE A belt of mountains under the Indian Ocean extending from the Gulf of Aden south of Arabia to its juncture with the Kerguelen–Gaussberg Ridge near the Maldive Islands southwest of India.

CASSITERITE A mineral; tin dioxide, SnO_2. Ore of tin with a specific gravity of 7. Nearly 75 percent of the world's tin production is from placer deposits, mostly in the form of cassiterite.

CAVITATION A process of erosion in a stream channel caused by sudden collapse of vapor bubbles against the channel wall.

CELLULOSE The most abundant carbohydrate, $C_6H_{10}O_5$, with a chain structure like that of the paraffin hydrocarbons. With lignin, it forms an important constituent of plant material, from which coal is formed.

CEMENTATION The process by which a binding agent is precipitated in the spaces between the individual particles of an unconsolidated deposit. The most common cementing agents are calcite, dolomite, and quartz. Others include iron oxide, opal, chalcedony, anhydrite, and pyrite.

CEMENT ROCK A clayey limestone used in the manufacture of hydraulic cement. Contains lime, silica, and alumina in varying proportions.

CENTRAL VENT An opening in the earth's crust, roughly circular, from which magmatic products are extruded. A volcano is an accumulation of material around a central vent.

CHALCEDONY A general name applied to fibrous cryptocrystalline silica with a waxy luster. Deposited from aqueous solutions and frequently found lining or filling cavities in rocks. *Agate* is a variety with alternating layers of chalcedony and opal.

CHALCOCITE A mineral, copper sulfide, Cu_2S, sometimes called *copper glance.* One of the most important ore minerals of copper.

CHALCOPYRITE A mineral, a sulfide of copper and iron, $CuFeS_2$. Sometimes called copper pyrite or yellow copper ore.

CHALK A variety of limestone made up in part of biochemically derived calcite, in the form of the skeletons or skeletal fragments of microscopic oceanic plants and animals that are mixed with very fine-grained calcite deposits of either biochemical or inorganic chemical origin.

CHEMICAL ENERGY Energy released or absorbed when atoms form compounds. Generally becomes available when atoms have lost or gained electrons, and often appears in the form of heat.

CHEMICALLY ACTIVE FLUID Hydrothermal solutions and pore liquids that become involved in metamorphism.

CHEMICAL PROCESS One in which atoms of elements combine to form compounds, or compounds combine to form other compounds.

CHEMICAL ROCK In the terminology of sedimentary rocks, a chemical rock is composed chiefly of material deposited by chemical precipitation, either organic or inorganic. Compare with *detrital sedimentary rock.* Chemical sedimentary rocks may have either a clastic or nonclastic (usually crystalline) texture.

CHEMICAL WEATHERING The weathering of rock material by chemical processes that transform the original material into new chemical combinations. Thus chemical weathering of orthoclase produces clay, some silica, and a soluble salt of potassium.

CHERT Granular cryptocrystalline silica similar to flint but usually light in color. Occurs as a compact massive rock, or as nodules.

CHLORITE A family of tetrahedral sheet silicates of iron, magnesium, and aluminum, characteristic of low-grade metamorphism. It has green color, with cleavage like that of mica, except that small scales of chlorite are not elastic, whereas those of mica are.

C-HORIZON The soil zone that contains partially disintegrated and decomposed parent material. It lies directly under the B-horizon and grades downward into unweathered material.

CHROMITE A mineral. An oxide of iron and chromium, $FeCr_2O_4$, the only ore of commercial chromium. It is one of the first minerals to crystallize from a magma and is concentrated within the magma.

CHRYSOTILE A metamorphic mineral; an asbestos, the fibrous variety of serpentine. A silicate of magnesium, with tetrahedra arranged in sheets.

CHUTE OR CHUTE CUTOFF As applied to stream flow, the term "chute" refers to a new route taken by a stream when its main flow is diverted to the inside of a bend, along a trough between low ridges formed by deposition on the inside of the bend where water velocities were reduced. Compare with *neck cutoff.*

CINDERS Volcanic fragments that are small, slaglike, solidified pieces of magma 0.5 to 2.5 cm across.

CINDER CONE Structure built exclusively or predominantly of pyroclastic ejecta dominated by cinders. Parasitic to a major volcano, it seldom exceeds 500 m in height. Slopes up 30° to 40°. Example: Parícutin.

CIRQUE A steep-walled hollow in a mountainside at high elevation, formed by ice-plucking and frost action, and shaped like a half-bowl or half-amphitheater. Serves as principal gathering ground for the ice of a valley glacier.

CLASTIC TEXTURE Texture shown by sedimentary rocks formed from deposits of mineral and rock fragments.

CLAY MINERALS Finely crystalline, hydrous silicates that form as a result of the weathering of such silicate minerals as feldspar, pyroxene, and amphibole. The most common clay minerals belong to the kaolinite, montmorillonite, and illite groups.

CLAY SIZE A volume less than that of a sphere with a diameter of $\frac{1}{256}$ mm (0.004 mm).

CLEAVAGE (1) *Mineral cleavage:* A property possessed by many minerals of breaking in certain preferred directions along smooth plane surfaces. The planes of cleavage are governed by the atomic pattern, and represent directions in which atomic bonds are relatively weak. (2) *Rock cleavage:* A property possessed by certain rocks of breaking with relative ease along parallel planes or nearly parallel surfaces. Rock cleavage is designated as *slaty, phyllitic, schistose,* and *gneissic.*

COAL A sedimentary rock composed of combustile matter derived from the partial decomposition and alteration of cellulose and lignin of plant materials.

COBBLE SIZE A volume greater than that of a sphere with a diameter of 64 mm and less than that of a sphere with a diameter of 256 mm.

COESITE High-pressure form of quartz, with density of 2.92. Associated with impact craters and cryptovolcanic structures.

COL A pass through a mountain ridge. Created by the enlargement of two cirques on opposite sides of the ridge until their headwalls meet and are broken down.

COLD GLACIER One in which no surface melting occurs during the summer months and whose temperature is always below freezing.

COLLOIDAL SIZE Between 0.2 μm and 1 μm (0.0002 mm to 0.001 mm).

COLOR The sensation resulting from stimulation of the retina of the eye by light waves of certain lengths.

COLUMN A column or post of dripstone joining the floor and roof of a cave; the result of joining of a stalactite and a stalagmite.

COLUMNAR JOINTING A pattern of jointing that blocks out columns of rock. Characteristic of tabular basalt flows or sills.

COMPACTION Reduction in pore space between individual grains as a result of pressure of overlying sediments or pressures resulting from earth movement.

COMPETENCE The maximum size of particle that a transporting agency, such as a stream, a glacier, or the wind, can move.

COMPOSITE VOLCANIC CONE Composed of interbedded lava flows and pyroclastic material, and characterized by slopes of close to 30° at the summit, reducing progressively to 5° near the base. Example: Mayon.

COMPOUND A combination of the atoms or ions of different elements. The mechanism by which they are combined is called a *bond.*

COMPRESSION A squeezing stress that tends to decrease the volume of a material.

CONCENTRIC FOLDING An elastic bending of an originally horizontal sheet with all internal movements parallel to a basal plane (the lower boundary of the fold).

CONCHOIDAL FRACTURE A mineral's habit of fracturing to produce curved surfaces like the interior of a shell (*conch*). Typical of glass and quartz.

CONCORDANT PLUTON An intrusive igneous body with contacts parallel to the layering or foliation surfaces of the rocks into which it was intruded.

CONCRETION An accumulation of mineral matter that forms around a center or axis of deposition after a sedimentary deposit has been laid down. Cementation consolidates the deposit as a whole, but the concretion is a body within the host rock that represents a local concentration of cementing material. The enclosing rock is less firmly cemented than the concretion. Commonly spheroidal or disk-shaped and composed of such cementing agents as calcite, dolomite, iron oxide, or silica.

CONE OF DEPRESSION A dimple in the water table, which forms as water is pumped from a well.

CONE SHEET A dike that is part of a concentric set that dips inward, like an inverted cone.

CONGLOMERATE A detrital sedimentary rock made up of more or less rounded fragments of such size that an appreciable percentage of the volume of the rock consists of particles of granule size or larger.

CONNATE WATER Water that was trapped in a sedimentary deposit at the time the deposit was laid down.

CONSEQUENT STREAM A stream following a course that is a direct consequence of the original slope of the surface on which it developed.

CONSTELLATION Apparent groupings of stars named for their resemblance to animals or mythical personages.

CONTACT METAMORPHISM Metamorphism at or very near the contact between magma and rock during intrusion.

CONTINENTAL DRIFT Hypothesis that an original single continent, sometimes referred to as *Pangaea,* split into several pieces that "drifted" laterally to form the present-day continents. Assumes mountains folded by pressures from drifting blocks.

CONTINENTAL GLACIER An ice sheet that obscures mountains and plains of a large section of a continent. Existing continental glaciers are on Greenland and Antarctica.

CONTINENTAL RISE In some places the base of the continental slope is marked by the somewhat gentler continental rise, which leads downward to the deep ocean floor.

CONTINENTAL SHELF Shallow, gradually sloping zone extending from the sea margin to a depth at which there is a marked or rather steep descent into the depths of the ocean down the continental slope. The seaward boundary of the shelf averages about 130 m in depth but may be either more or less than this.

CONTINENTAL SLOPE Portion of the ocean floor extending downward from the seaward edge of the continental shelves. In some places, such as south of the Aleutian Islands, they descend directly to the ocean deeps. In other places, such as off eastern North America, they grade into the somewhat gentler continental rises, which in turn lead to the deep ocean floors.

CONTINUOUS REACTION SERIES That branch of Bowen's reaction series comprising the plagioclase feldspars, in which reaction of early-formed crystals with later liquids takes place continuously—that is, without abrupt phase changes.

CONTOUR INTERVAL See Appendix F.

CONTOUR LINE See Appendix F.

CONVECTION A mechanism by which material moves because its density is different from that of surrounding material. The density differences are frequently brought about by heating.

CONVECTION CELL A pair of *convection currents* adjacent to each other.

CONVECTION CURRENT A closed circulation of material sometimes developed during convection. Convection currents normally develop in pairs; each pair is called a *convection cell.*

CONVECTION CURRENT HYPOTHESIS Assumption that mountains folded and scraped off down-plunging currents.

COQUINA A coarse-grained, porous, friable variety of clastic limestone made up chiefly of fragments of shells.

CORDILLERA A broad belt of mountain ranges.

CORE The innermost zone of the earth, which is surrounded by the mantle.

CORE DRILLING Drilling with a hollow bit and barrel, which cut out and recover a solid core of the rock penetrated.

CORIOLIS EFFECT The tendency of any moving body, on or starting from the surface of the earth, to continue in the direction in which the earth's rotation propels it. The direction in which the body moves because of this tendency, combined with the direction in which it is aimed, determines the ultimate course of the body relative to the earth's surface. In the northern hemisphere, the coriolis effect causes a moving body to veer or try to veer to the right of its direction of forward motion; in the southern hemisphere, to the left. The magnitude of the effect is proportional to the velocity of a body's motion. This effect causes cyclonic storm-wind circulation to be counterclockwise in the northern hemisphere and clockwise in the southern hemisphere, and determines the final course of ocean currents relative to trade winds.

CORRELATION The process of establishing the contemporaneity of rocks or events in one area with other rocks or events in another area.

COVALENT BOND A bond in which atoms combine by the sharing of their electrons.

CRATER A roughly circular, steep-sided volcanic basin with a diameter less than three times its depth. Commonly at the summit of a volcano. Contrast with *caldera.* Applied also to depressions caused by meteorites either by direct impact or by explosive impact.

CREEP As applied to soils and surficial material, slow downward movement of a plastic type. As applied to elastic solids, slow permanent yielding to stresses that are less than the yield point if applied for a short time only.

CREVASSE (1) A deep crevice or fissure in glacier ice. (2) A breach in a natural levee.

CROSS-BEDDING See *inclined bedding.*

CROSSCUTTING RELATIONSHIPS, LAW OF A rock is younger than any rock across which it cuts.

CRUST The outermost zone of the earth. Composed of solid rock 35 to 50 km thick under continents, with sialic and simatic sections. It is thinner and simatic under permanent ocean basins and rests on the mantle; may be covered by sediments.

CRYPTOCRYSTALLINE A state of matter in which there is actually an orderly arrangement of atoms characteristic of crystals, but in which the units are so small (that is, the material is so fine-grained) that the crystalline nature cannot be determined with the aid of an ordinary microscope.

CRYPTOVOLCANIC STRUCTURE Geologic structure in which the rocks appear to have been disrupted by volcanic activity, but in which there are hidden no volcanic rocks.

CRYSTAL A solid with orderly atomic arrangement, which may or may not develop external faces that give it crystal form.

CRYSTAL FORM The geometrical form taken by a mineral, giving an external expression to the orderly internal arrangement of atoms.

CRYSTALLINE STRUCTURE The orderly arrangement of atoms in a crystal. Also called crystal structure.

CRYSTALLIZATION The process through which crystals separate from a fluid, viscous, or dispersed state.

CURIE TEMPERATURE The temperature above which ordinarily magnetic material loses its magnetism. On cooling below this temperature it regains its magnetism. Example: iron loses its magnetism above 760°C and regains it as it cools below this temperature. This is its curie temperature.

CURRENT BEDDING See *inclined bedding.*

CURRENT RIPPLE MARKS Ripple marks, asymmetric in form, formed by air or water moving more or less continuously in one direction.

CUTOFF See *chute cutoff; neck cutoff.*

CYCLE OF EROSION A qualitative description of river valleys and regions passing through the stages of youth, maturity, and old age with respect to the amount of erosion that has occurred.

DATUM PLANE See Appendix F.

DEBRIS SLIDE A small, rapid movement of largely unconsolidated material that slides or rolls downward to produce an irregular topography.

DECOMPOSITION Synonymous with *chemical weathering.*

DEEP FOCUS Earthquake focus deeper than 300 km. The greatest depth of focus known is 700 km.

DEEP-SEA TRENCHES See *island arc deeps.*

DEFLATION The erosive process in which the wind carries off unconsolidated material.

DEFORMATION OF ROCKS Any change in the original shape or volume of rock masses. Produced by mountain-building forces. Folding, faulting, and plastic flow are common modes of rock deformation.

DEHYDRATION A factor in some progressive metamorphism.

DELTA A plain underlain by an assemblage of sediments that accumulate where a stream flows into a body of standing water where its velocity and transporting power are suddenly reduced. Originally so named because many deltas are roughly triangular in plan, like the Greek letter *delta* (Δ), with the apex pointing upstream.

DENDRITIC PATTERN An arrangement of stream courses that, on a map or viewed from the air, resemble the branching habit of certain trees, such as the oaks or maples.

DENSITY A measure of the concentration of matter, expressed as the mass per unit volume. (Mass equals weight divided by acceleration of gravity.)

DENSITY CURRENT A current due to differences in the density of sea water from place to place caused by changes in temperature and variations in salinity or the amount of material held in suspension.

DEPOSITIONAL REMANENT MAGNETISM Magnetism resulting from the tendency of magnetic particles such as magnetite to orient themselves in the earth's magnetic field as they are deposited. Their orientation is maintained as the soft sediments are lithified and thus records the earth's field when the particles were laid down. Abbreviation DRM.

DESICCATION Loss of water from pore spaces of sediments through compaction or through evaporation caused by exposure to air.

DETRITAL SEDIMENTARY ROCKS Rocks formed from accumulations of minerals and rocks derived either from erosion of previously existing rock or from the weathered products of these rocks.

DIABASE A rock of basaltic composition, essentially labradorite and pyroxene, characterized by ophitic texture.

DIABASIC See *ophitic.*

DIAMOND A mineral composed of the element carbon; the hardest substance known. Used as a gem and industry in cutting tools.

DIATOMACEOUS OOZE A siliceous deep-sea ooze made up of the cell walls of one-celled marine algae known as diatoms.

DIFFERENTIAL EROSION The process by which different rock masses or different parts of the same rock erode at different rates.

DIFFERENTIAL WEATHERING The process by which different rock masses or different parts of the same rock weather at different rates.

DIKE A tabular discordant pluton.

DIKE SWARM Group of approximately parallel dikes.

DIORITE A coarse-grained igneous rock with the composition of andesite (no quartz or orthoclase), composed of about 75 percent plagioclase feldspars and the balance ferromagnesian silicates.

DIP (1) The acute angle that a rock surface makes with a horizontal plane. The direction of the dip is always perpendicular to the strike. (2) See *magnetic declination.*

DIPOLE Any object that is oppositely charged at two points. Most commonly refers to a molecule that has concentrations of positive or negative charge at two different points.

DIPOLE MAGNETIC FIELD The portion of the earth's magnetic field that can best be described by a dipole passing through the earth's center and inclined to the earth's axis of rotation. See also *nondipole field* and *external magnetic field.*

DIP POLE See *magnetic pole.*

DIP-SLIP FAULT Fault in which displacement is in the direction of the fault's dip.

DISCHARGE With reference to stream flow, the quantity of water that passes a given point in unit time. Usually measured in cubic feet per second (abbreviated cfs).

DISCONFORMITY An unconformity in which the beds on opposite sides are parallel.

DISCONTINUITY Within the earth's interior, sudden or rapid changes with depth in one or more of the physical properties of the materials constituting the earth, as evidenced by seismic data.

DISCONTINUOUS REACTION SERIES That branch of Bowen's reaction series including the minerals olivine, augite, hornblende, and biotite, for which each change in the series represents an abrupt phase change.

DISCORDANT PLUTON An intrusive igneous body with boundaries that cut across surfaces of layering or foliation in the rocks into which it has been intruded.

DISINTEGRATION Synonymous with *mechanical weathering.*

DISTANCE MEASURE (FOR EARTHQUAKES) A great circle arc between two points, expressed in kilometers, miles, or degrees of angle, subtended at the earth's center by that arc.

DISTRIBUTARY CHANNEL OR STREAM A river branch that flows away from a main stream and does not rejoin it. Characteristic of deltas and alluvial fans.

DIVIDE Line separating two drainage basins.

DOLOMITE A mineral composed of the carbonate of calcium and magnesium, $CaMg(CO_3)_2$. Also used as a rock name for formations composed largely of the mineral dolomite.

DOME An anticlinal fold without a clearly developed linearity of crest, so that the beds involved dip in all directions from a central area, like an inverted but usually distorted cup. The reverse of a *basin.*

DRAINAGE BASIN The area from which a given stream and its tributaries receive their water.

DRIFT Any material laid down directly by ice, or deposited in lakes, oceans, or streams as a result of glacial activity. Unstratified glacial drift is called *till* and forms *moraines.* Stratified glacial drift forms *outwash plains, eskers, kames,* and *varves.*

DRIPSTONE Calcium carbonate deposited from solution by underground water entering a cave in the zone of aeration. Sometimes called *travertine.*

DRM See *depositional remanent magnetism.*

DRUMLIN A smooth, streamlined hill composed of till. Its long axis is oriented in the direction of ice movement. The blunt nose points upstream and a gentler slope tails off downstream with reference to the ice movement. In height, drumlins range from about 8 to 60 m, with the average somewhat less than 30 m. Most drumlins are between 0.5 and 1 km in length, and the length is commonly several times the width. Diagnostic characteristics are the shape and the composition of unstratified glacial drift, in contrast to kames, which are of stratified glacial drift and random shapes.

DUCTILE A property of material when its strength in tension is not greatly different from its strength in compression.

DUNE A mound or ridge of sand piled by wind.

DUST-CLOUD HYPOTHESES Hypotheses that the solar system was formed from the condensation of interstellar dust clouds.

DUST SIZE A volume less than that of a sphere with a diameter of .06 mm; used in reference to particles carried in suspension by wind.

EARTHFLOW A combination of slump and mudflow.

EARTH WAVES Mechanism for transmitting energy from an earthquake focus.

ECLIPTIC The apparent path of the sun in the heavens; the plane of the planets' orbit.

ELASTIC DEFORMATION A nonpermanent deformation, after which the body returns to its original shape or volume when the deforming force is removed.

ELASTIC ENERGY The energy stored within a solid during elastic deformation, and released during elastic rebound.

ELASTICITY A property of materials that defines the extent to which they resist small deformations, from which they recover completely when the deforming force is removed. Elasticity = stress/strain.

ELASTIC LIMIT The maximum stress that produces only elastic deformation.

ELASTIC REBOUND The recovery of elastic strain when a material breaks or when the deforming force is removed.

ELASTIC SOLID A solid that yields to applied force by changing shape or volume, or both, but returns to its original condition when the force is removed. The amount of yield is proportional to the force.

ELECTRICAL ENERGY The energy of moving electrons.

ELECTRIC CHARGE A property of matter resulting from an imbalance between the number of protons and the number of electrons in a given piece of matter. The electron has a negative charge, the proton a positive charge. Like charges repel each other, unlike charges attract each other.

ELECTRIC CURRENT A flow of electrons.

ELECTRIC TURBINE GENERATOR Apparatus that uses steam from a heat exchanger to drive a turbine and generate electricity.

ELECTRON A fundamental particle of matter, the most elementary unit of negative electrical charge. Its mass is 0.00055 amu.

ELECTRON CAPTURE DECAY Radioactive decay that takes place as an orbital electron is captured by a proton in the nucleus. The mass of the element remains constant but the atomic number decreases by one.

ELECTRON SHELL An imaginary spherical surface representing all possible paths of electrons with the same average distance from a nucleus and with approximately the same energy.

ELEMENT A unique combination of protons, neutrons, and electrons that cannot be broken down by ordinary chemical methods. The fundamental properties of an element are determined by its number of protons. Each element is assigned a number that corresponds to its number of protons. Combinations containing from 1 through 108 protons are now known.

END MORAINE A ridge or belt of till marking the farthest advance of a glacier. Sometimes called *terminal moraine*.

ENERGY The capacity for producing motion. Energy holds matter together. It can become mass, or can be derived from mass. It takes such forms as kinetic, potential, heat, chemical, electrical, and atomic energy; one form of energy can be changed to another.

ENERGY LEVEL The distance from an atomic nucleus at which electrons orbit. May be thought of as a shell surrounding the nucleus.

ENTRENCHED MEANDER A meander cut into underlying bedrock when regional uplift allows the originally meandering stream to resume downward cutting.

EPICENTER An area on the surface directly above the focus of an earthquake.

EPIDOTE A silicate of aluminum, calcium, and iron characteristic of low-grade metamorphism and associated with chlorite and albite in the greenschist facies. Built around independent tetrahedra.

EPIDOTE-AMPHIBOLITE FACIES An assemblage of minerals formed between 250°C and 450°C during regional metamorphism.

ERG A unit of energy expressing the capacity for doing work. Equal to the energy expended when a force of 1 dyne acts through a distance of 1 cm.

EROSION Movement of material from one place to another on the earth's surface. Agents of movement include water, ice, wind, and gravity.

EROSIONAL FLOOD PLAIN A flood plain that has been created by the lateral erosion and the gradual retreat of the valley walls.

ERRATIC In the terminology of glaciation, a stone or boulder carried by ice to a place where it rests on or near bedrock of different composition.

ESCAPE VELOCITY The minimum velocity an object must have to escape from a gravitational field. For the moon this is about 2.38 km/sec. and for the earth, about 11.2 km/sec.

ESKER A widening ridge of stratified glacial drift, steep-sided, 3 to 15 m in height, and from a fraction of a mile to over 160 km in length.

EUSTATIC CHANGE OF SEA LEVEL A change in sea level produced entirely by an increase or a decrease in the amount of water in the oceans, hence of worldwide proportions.

EUTROPHIC In the aging of a lake, pertains to an old-age lake and indicates a high supply of nutrients supporting a high biologic productivity.

EVAPORATION The process by which a liquid becomes a vapor at a temperature below its boiling point.

EVAPORITE A rock composed of minerals that have been precipitated from solutions concentrated by the evaporation of solvents. Examples: rock salt, gypsum, anhydrite.

EXFOLIATION The process by which plates of rock are stripped from a larger rock mass by physical forces.

EXFOLIATION DOME A large, rounded domal feature produced in homogeneous coarse-grained igneous rocks and sometimes in conglomerates by the process of exfoliation.

EXTERNAL MAGNETIC FIELD A component of the earth's field originating from activity above the earth's surface. Small when compared with the dipole and nondipole components of the field, which originate beneath the surface.

EXTRUSIVE ROCK A rock that has solidified from a mass of magma that poured or was blown out upon the earth's surface.

FACIES An assemblage of mineral, rock, or fossil features reflecting the environment in which a rock was formed. See *sedimentary facies, metamorphic facies.*

FAULT A surface of rock rupture along which there has been differential movement.

FAULT-BLOCK MOUNTAIN A mountain bounded by one or more faults.

FELDSPARS Silicate minerals composed of silicon-oxygen and aluminum-oxygen tetrahedra linked together in three-dimensional networks with positive ions fitted into the interstices of the negatively charged framework of tetrahedra. Classed as aluminosilicates. When the positive ion is K^+, the mineral is orthoclase; when it is Na^+, the mineral is albite; when it is Ca^{2+}, the mineral is anorthite.

FELSITE A general term for light-colored, fine-grained igneous rocks.

FERROMAGNESIAN SILICATE A silicate in which the positive ions are dominated by iron, magnesium, or both.

FIBROUS FRACTURE A mineral's habit of breaking into splinters or fibers.

FIERY CLOUD (NUÉE ARDENTE) An avalanche of incandescent pyroclastic debris mixed with steam and other gases, heavier than air, and projected down a volcano's side.

FIORD A glacially deepened valley that is now flooded by the sea to form a long, narrow, steep-walled inlet.

FIRN Granular ice formed by the recrystallization of snow. Intermediate between snow and glacier ice. Sometimes called *névé.*

FISSILITY A property of splitting along closely spaced planes more or less parallel to the bedding. Its presence distinguishes shale from mudstone.

FISSURE ERUPTION Extrusion of lava from a fissure in the earth's crust.

FLASHY STREAM A stream with a high flood peak of short duration, which may be caused by urbanization.

FLINT Granular cryptocrystalline silica, usually dull and dark. Often occurs as lumps or nodules in calcareous rocks, such as the Cretaceous chalk beds of southern England.

FLOOD BASALT Basalt poured out from fissures in floods that tend to form great plateaus. Sometimes called *plateau basalt.*

FLOOD PLAIN Area bordering a stream, over which water spreads in time of flood.

FLOOD PLAIN OF AGGRADATION A flood plain formed by the building up of the valley floor by sedimentation.

FLOTATION A process that begins the concentration of ore minerals from gangue.

FLOW Deformation that occurs in rocks when they are in the plastic state.

FLUID Material that offers little or no resistance to forces tending to change its shape.

FLUX A substance or mixture that promotes fusion.

FOCUS The source of a given set of earthquake waves.

FOLD A bend, flexure, or wrinkle in rock produced when the rock was in a plastic state.

FOLD MOUNTAINS Mountains consisting primarily of elevated, folded sedimentary rocks.

FOLIATION A layering in some rocks caused by parallel alignment of minerals. A textural feature of some metamorphic rocks. Produces rock cleavage.

FOOTWALL One of the blocks of rock involved in fault movement. The one that would be under the feet of a person standing in a tunnel along or across the fault. Opposite the hanging wall.

FORE DUNE A dune immediately back of the shoreline of an ocean or large lake.

FORESET BEDS Inclined layers of sediment deposited on the advancing edge of a growing delta or along the lee slope of an advancing sand dune.

FORESHOCK A relatively small earthquake that precedes a larger earthquake by a few days or weeks and originates at or near the focus of the larger earthquake.

FOSSIL Evidence of past life, such as the bones of a dinosaur, the shell of an ancient clam, the footprint of a long-extinct animal, or the impression of a leaf in a rock.

FOSSIL FUELS Organic remains used to produce heat or power by combustion. Include petroleum, natural gas, and coal.

FRACTIONAL DISTILLATION The recovery, one or more at a time, of fractions of a complex liquid, each of which has a different density.

FRACTIONATION A process whereby crystals that formed early from a magma have time to settle appreciably before the temperature drops much further. They are effectively removed from the environment in which they formed.

FRACTURE As a mineral characteristic, the way in which a mineral breaks when it does not have cleavage. May be conchoidal (shell-shaped), fibrous, hackly, or uneven.

FRACTURE CLEAVAGE A system of joints spaced a fraction of a centimeter apart.

FRINGING REEF A reef attached directly to a landmass.

FRONT In connection with concepts of granitization, the limit to which diffusing ions of a given type are carried. The *simatic front,* for example, is the limit to which diffusing ions carried the calcium, iron, and magnesium that they removed from the rocks in their paths. The *granitic front* is the limit to which diffusing ions deposited granitic elements.

FROST ACTION Process of mechanical weathering caused by repeated cycles of freezing and thawing. Expansion of water during the freezing cycle provides the energy for the process.

FROST HEAVING The heaving of unconsolidated deposits as lenses of ice grow below the surface by acquiring capillary water from below.

FUMAROLE Vent for volcanic steam and gases.

FUNDAMENTAL PARTICLES Protons, neutrons, and electrons, which combine to form atoms. Each particle is defined in terms of its *mass* and its *electric charge.*

G Symbol for *rigidity modulus.*

GABBRO A coarse-grained igneous rock with the composition of basalt.

GALAXY A family of stars grouped in space. The earth belongs to the Milky Way galaxy, which contains about 100 billion stars.

GALENA A mineral; lead sulfide, PbS. The principal ore of lead.

GARNET A family of silicates of iron, magnesium, aluminum, calcium, manganese, and chromium, which are built around independent tetrahedra and appear commonly as distinctive twelve-sided, fully developed crystals. Characteristic of metamorphic rocks. Generally cannot be distinguished from one another without chemical analysis.

GAS (1) A state of matter that has neither independent shape nor volume, can be compressed readily, and tends to expand indefinitely. (2) In geology, the word "gas" is sometimes used to refer to *natural gas,* the gaseous hydrocarbons that occur in rocks, dominated by methane. Compare with use of the word "oil" to refer to *petroleum.*

GEODE A roughly spherical, hollow or partially hollow accumulation of mineral matter from a few centimeters to nearly half a meter in diameter. An outer layer of chalcedony is lined with crystals that project inward toward the hollow center. The crystals, often perfectly formed, are usually quartz, although calcite and dolomite are also found and, more rarely, other minerals. Geodes are most commonly found in limestone and more rarely in shale.

GEOGRAPHIC POLES The points on the earth's surface marked by the ends of the earth's axis of rotation.

GEOLOGIC COLUMN A chronologic arrangement of rock units in columnar form with the oldest units at the bottom and the youngest at the top.

GEOLOGIC TIME-SCALE A chronologic sequence of units of earth time.

GEOLOGY An organized body of knowledge about the earth. It includes both *physical geology* and *historical geology*.

GEOMAGNETIC POLES The dipole best approximating the earth's observed field is one inclined 11.5° from the axis of rotation. The points at which the ends of this imaginary magnetic axis intersect the earth's surface are known as the geomagnetic poles. They should not be confused with the magnetic, dip poles, or virtual geomagnetic poles.

GEOPHYSICAL PROSPECTING Mapping rock structures by methods of experimental physics. Includes measuring magnetic fields, the force of gravity, electrical properties, seismic wave paths and velocities, radioactivity, and heat flow.

GEOPHYSICS The physics of the earth.

GEOSYNCLINE Literally, an "earth syncline." The term now refers, however, to a basin in which thousands of meters of sediments have accumulated, with accompanying progressive sinking of the basin floor explained only in part by the load of sediments. Common usage of the term includes both the accumulated sediments themselves and the geometrical form of the basin in which they are deposited. All folded mountain ranges were built from geosynclines, but not all geosynclines have become mountain ranges.

GEOTHERMAL FIELD Area where wells are drilled in order to obtain elements contained in solution in hot brines, and to tap heat energy.

GEYSER A special type of thermal spring which intermittently ejects its water with considerable force.

GLACIER A mass of ice, formed by the recrystallization of snow, that flows forward, or has flowed at some time in the past, under the influence of gravity. By convention we exclude icebergs from this definition even though they are large fragments broken from the seaward end of glaciers.

GLACIER ICE A unique form of ice developed by the compression and recrystallization of snow, and consisting of interlocking crystals.

GLASS A form of matter that exhibits the properties of a solid but has the atomic arrangements, or lack of order, of a liquid.

GLOBIGERINA OOZE A deep-sea calcareous ooze in which limy shells of minute one-celled animals called *Globigerina* abound.

GLOSSOPTERIS FLORA A late Paleozoic assemblage of fossil plants named for the seed-fern *Glossopteris*, one of the plants in the flora. Widespread in South America, South Africa, Australia, India, and Antarctica.

GNEISS Metamorphic rock with gneissic cleavage. Commonly formed by the metamorphism of granite.

GNEISSIC CLEAVAGE Rock cleavage in which the surfaces of easy breaking, if developed at all, are from a few hundredths of a millimeter to a centimeter or more apart.

GOETHITE Hydrous iron oxide, FeO(OH).

GONDWANALAND Hypothetical continent thought to have broken up in the Mesozoic. The resulting fragments are postulated to form present-day South America, Africa, Australia, India, and Antarctica.

GRABEN An elongated, trenchlike structural form bounded by parallel normal faults created when the block that forms the trench floor moved downward relative to the blocks that form the sides.

GRADATION Leveling of the land. This is constantly being brought about by the forces of gravity and such agents of erosion as surface and underground water, and wind, glacier ice, and waves.

GRADE A term used to designate the extent to which metamorphism has advanced. Found in such combinations as high-grade or low-grade metamorphism. Compare with *rank*.

GRADED BEDDING The type of bedding shown by a sedimentary deposit when particles become progressively finer from bottom to top.

GRADIENT Slope of a stream bed.

GRANITE A coarse-grained igneous rock dominated by light-colored minerals, consisting of about 50 percent orthoclase, 25 percent quartz, and the balance of plagioclase feldspars and ferromagnesian silicates. Granites and granodiorites comprise 95 percent of all intrusive rocks.

GRANITIZATION A special type of metasomatism by which solutions of magmatic origin move through solid rocks, change ions with them, and convert them into rocks which achieve granitic character without having passed through a magmatic stage.

GRANODIORITE A coarse-grained igneous rock intermediate in composition between granite and diorite.

GRANULAR TEXTURE Composed of mineral grains large enough to be seen by the unaided eye.

GRAPHIC STRUCTURE An intimate intergrowth of potassic feldspar and quartz with the long axes of quartz crystals lining up parallel to a feldspar axis. The quartz part is dark and the feldspar is light in color, so the pattern suggests Egyptian hieroglyphs. Commonly found in pegmatites.

GRAPHITE A mineral composed entirely of carbon. "Black lead." Very soft because of its crystalline structure; diamond, in contrast, has the same composition but is the hardest substance known.

GRAVITATIONAL THEORY (OIL AND GAS ACCUMULATION) If oil, gas, and water are present in a reservoir bed, the oil and gas, being lighter than water, rise to the top, with gas uppermost.

GRAVITY ANOMALY Difference between observed value of gravity and computed value.

GRAVITY FAULT A fault in which the hanging wall appears to have moved downward relative to the footwall. Also called *normal fault.*

GRAVITY METER An instrument for measuring the force of gravity. Also called gravimeter.

GRAVITY PROSPECTING Mapping the force of gravity at different places to determine differences in specific gravity of rock masses, and, through this, the distribution of masses of different specific gravity. Done with a gravity meter (gravimeter).

GRAYWACKE A variety of sandstone generally characterized by its hardness, dark color, and angular grains of quartz, feldspar, and small rock fragments set in a matrix of clay-sized particles. Also called *lithic sandstone.*

GREAT RED SPOT Unexplained feature on Jupiter, 48,000 km long and 12,800 km wide. It slowly drifts, often changes color, and sometimes disappears from view for a number of years.

GREENSCHIST A schist characterized by green color. The product of regional metamorphism of simatic rocks. The green color is imparted by the mineral chlorite.

GREENSCHIST FACIES An assemblage of minerals formed between 150°C and 250°C during regional metamorphism.

GROUNDMASS The finely crystalline or glassy portion of a porphyry.

GROUND MORAINE Till deposited from a glacier as a veneer over the landscape and forming a gently rolling surface.

GROUND WATER Underground water within the zone of saturation.

GROUND-WATER TABLE The upper surface of the zone of saturation for underground water. It is an irregular surface with a slope or shape determined by the quantity of ground water and the permeability of the earth materials. In general, it is highest beneath hills and lowest beneath valleys. Also referred to as *water table.*

GUYOT A flat-topped *seamount* rising from the floor of the ocean like a volcano but planed off on top and covered by appreciable depth of water. Synonymous with *tablemount.*

GYPSUM Hydrous calcium sulfate, $CaSO_4 \cdot 2H_2O$. A soft, common mineral in sedimentary rocks, where it sometimes occurs in thick beds interstratified with limestones and shales. Sometimes occurs as a layer under a bed of rock salt since it is one of the first minerals to crystallize on the evaporation of sea water. Alabaster is a fine-grained massive variety of gypsum.

H Symbol for mineral hardness.

HACKLY FRACTURE A mineral's habit of breaking to produce jagged, irregular surfaces with sharp edges.

HALF-LIFE Time needed for one half of the nuclei in a sample of a radioactive element to decay.

HALITE A mineral; rock salt, or common salt, NaCL. Occurs widely disseminated, or in extensive beds and irregular masses, precipitated from sea water and interstratified with rocks of other types as a true sedimentary rock.

HANGING VALLEY A valley that has a greater elevation than the valley to which it is tributary, at the point of their junction. Often (but not always) created by a deepening of the main valley by a glacier. The hanging valley itself may or may not be glaciated.

HANGING WALL One of the blocks involved in fault movement. The one that would be hanging overhead for a person standing in a tunnel along or across the fault. Opposite the footwall.

HARDNESS A mineral's resistance to scratching on a smooth surface. The Mohs scale of relative hardness consists of ten minerals. Each of these will scratch all those below it in the scale and will be scratched by all those above it: (1) talc, (2) gypsum, (3) calcite, (4) fluorite, (5) apatite, (6) orthoclase, (7) quartz, (8) topaz, (9) corundum, (10) diamond.

HEAD Difference in elevation between intake and discharge points for a liquid. In geology, most commonly of interest in connection with the movement of underground water.

HEAT ENERGY A special manifestation of kinetic energy in atoms. The temperature of a substance depends on the average kinetic energy of its component particles. When heat is added to a substance, the average kinetic energy increases.

HEAT EXCHANGER A unit in atomic power generation that uses water heated under pressure by the atomic reactor to form steam from water in another system and to drive a turbine for generation of electricity.

HEAT FLOW The product of the thermal gradient and the thermal conductivity of earth materials. Its average over the whole earth is $1.2 \pm 0.15\ \mu\ cal/cm^2/sec$.

HEIGHT See Appendix F.

HEMATITE Iron oxide, Fe_2O_3. The principal ore mineral for about 90 percent of the commercial iron produced in the United States. Characteristic red color when powdered.

HINGE FAULT Fault in which displacement dies out perceptibly along the strike and ends at a definite point.

HISTORICAL GEOLOGY The branch of geology that deals with the history of the earth, including a record of life on the earth as well as physical changes in the earth itself.

HORN A spire of bedrock left where cirques have eaten into a mountain from more than two sides around a central area. Example: Matterhorn of the Swiss Alps.

HORNBLENDE A rock-forming ferromagnesian silicate mineral with double chains of silicon-oxygen tetrahedra. An amphibole.

HORNFELS Dense, granular metamorphic rock. Since this term is commonly applied to the metamorphic equivalent of any fine-grained rock, its composition is variable.

HORNFELS FACIES An assemblage of minerals formed at temperatures greater than 700°C during contact metamorphism.

HORST An elongated block bounded by parallel normal faults in such a way that it stands above the blocks on both sides.

HOT SPRING A *thermal spring* that brings hot water to the surface.

Water temperature usually 6.5°C or more above mean air temperature.

HYDRATION The process by which water combines chemically with other molecules.

HYDRAULIC GRADIENT Head of underground water divided by the distance of travel between two points. If the head is 10 m for two points 100 m apart, the hydraulic gradient is .1 or 10 percent. When head and distance of flow are the same, the hydraulic gradient is 100 percent.

HYDRAULIC MINING Use of a strong jet of water to move deposits of sand and gravel from their original site to separating equipment, where the sought-for mineral is extracted.

HYDROCARBON A compound of hydrogen and carbon that burns in air to form water and oxides of carbon. There are many hydrocarbons. The simplest, methane, is the chief component of natural gas. Petroleum is a complex mixture of hydrocarbons.

HYDROELECTRIC POWER Conversion of energy to electricity by the free fall of water. This method supplies about 25 percent of the world's electrical energy.

HYDROLOGIC CYCLE The general pattern of movement of water from the sea by evaporation to the atmosphere, by precipitation onto the land, and by movement back to the sea again under the influence of gravity.

HYDROTHERMAL SOLUTION A hot, watery solution that usually emanates from a magma in the late stages of cooling. Frequently contains and deposits in economically workable concentrations minor elements that, because of incommensurate ionic radii or electronic charges, have not been able to fit into the atomic structures of the common minerals of igneous rocks.

HYSTERESIS Retardation of recovery from elastic deformation after the stress is removed.

ICECAP A localized *ice sheet.*

ICE SHEET A broad moundlike mass of glacier ice of considerable extent that has a tendency to spread radially under its own weight. Localized ice sheets are sometimes called *icecaps.*

IGNEOUS ROCK An aggregate of interlocking silicate minerals formed by the cooling and solidification of magma.

ILLITE A clay mineral family of hydrous aluminous silicates. Structure is similar to that of montmorillonite, but aluminum is substituted for 10 to 15 percent of the silicon, which destroys montmorillonite's property of expanding with the addition of water because weak bonds are replaced by strong potassium ion links. Structurally, illite is intermediate between montmorillonite and muscovite. Montmorillonite converts to illite in sediments, while illite converts to muscovite under conditions of low-grade metamorphism. Illite is the commonest clay mineral in clayey rocks and recent marine sediments, and is present in many soils.

ILMENITE Iron titanium oxide. Accounts for much of the unique abundance of titanium on the moon.

INCLINED BEDDING Bedding laid down at an angle to the horizontal. Also referred to as *cross-bedding* and *current bedding.*

INDEX MINERALS Chlorite, low-grade metamorphism; almandite, middle-grade metamorphism; sillimanite, high-grade metamorphism.

INDUCED MAGNETISM In the terminology of rock magnetism, one of the components of the rock's natural remanent magnetism. It is parallel to the earth's present field and results from it.

INFILTRATION The soaking into the ground of water on the surface.

INERTIA MEMBER The central element of a seismograph, consisting of a weight suspended by a wire or spring so that it acts like a pendulum free to move in only one plane.

INOSILICATE Mineral with crystal structure containing SiO_4 tetrahedra in single or double chains.

INTENSITY (OF AN EARTHQUAKE) A measure of the effects of earthquake waves on man, structures, and the earth's surface at a particular place. Contrast with *magnitude,* which is a measure of the total energy released by an earthquake.

INTERMEDIATE BELT Subdivision of zone of aeration. The belt that lies between the belt of soil moisture and the capillary fringe.

INTERMEDIATE LAVA Lava composed of 60 to 65 percent silica.

INTERMITTENT STREAM A stream that carries water only part of the time.

INTRUSIVE ROCK A rock that solidified from a mass of magma that invaded the earth's crust but did not reach the surface.

INVERSION See *temperature inversion.*

ION An electrically unbalanced form of an atom, or group of atoms, produced by the gain or loss of electrons.

IONIC BOND A bond in which ions are held together by the electrical attraction of opposite charges.

IONIC RADIUS The average distance from the center to the outermost electron of an ion. Commonly expressed in angstroms.

ISLAND ARC DEEPS Arcuate trenches bordering some of the continents. Some reach depths of 9,000 m or more below the surface of the sea. Also called deep-sea trenches or trenches.

ISOCLINAL FOLDING Beds on both limbs are nearly parallel, whether the fold is upright, overturned, or recumbent.

ISOSEISMIC LINE A line connecting all points on the surface of the earth where the intensity of shaking produced by earthquake waves is the same.

ISOSTASY The ideal condition of balance that would be attained by earth materials of differing densities if gravity were the only force governing their heights relative to each other.

ISOTOPE Alternative form of an element produced by variations in the number of neutrons in the nucleus.

JASPER Granular cryptocrystalline silica usually colored red by hematite inclusions.

JET OR SHOOTING FLOW A type of flow, related to turbulent flow, occurring when a stream reaches high velocity along a sharply inclined stretch, or over a waterfall, and the water moves in plunging, jet-like surges.

JOINT A break in a rock mass where there has been no relative movement of rock on opposite sides of the break.

JOINT SYSTEM A combination of intersecting joint sets, often at approximately right angles.

JUVENILE WATER Water brought to the surface or added to underground supplies from magma.

KAME A steep-sided hill of stratified glacial drift. Distinguished from a drumlin by lack of unique shape and by stratification.

KAME TERRACE Stratified glacial drift deposited between a wasting glacier and an adjacent valley wall. When the ice melts, this material stands as a terrace along the valley wall.

KAOLINITE A clay mineral, a hydrous aluminous silicate. $Al_4Si_4O_{10}(OH)_8$. Structure consists of one sheet of silicon-oxygen tetrahedra each sharing three oxygens to give a ratio of Si_4O_{10}, linked with one sheet of aluminum and hydroxyl. The composition of pure kaolinite does not vary as it does for the other clay minerals, montmorillonite and illite, in which ready addition or substitution of ions takes place.

KARST TOPOGRAPHY Irregular topography characterized by sinkholes, streamless valleys, and streams that disappear underground, all developed by the action of surface and underground water in soluble rock such as limestone.

KERGUELEN–GAUSSBERG RIDGE A belt of mountains under the Indian Ocean between India and Antarctica.

KETTLE A depression in the ground surface formed by the melting of a block of ice buried or partially buried by glacial drift, either outwash or till.

KINETIC ENERGY Energy of movement. The amount possessed by an object or particle depends on its mass and speed.

KYANITE A silicate mineral characteristic of the temperatures of middle-grade metamorphism. Al_2SiO_5 in bladed blue crystals is softer than a knife along the crystal. Its crystalline structure is based on independent tetrahedra. Compare with *andalusite,* which has the same composition and forms under similar conditions, but has a different crystal habit. Contrast with *sillimanite,* which has the same composition but different crystal habit, and forms at highest metamorphic temperature.

L Symbol for earthquake surface waves.

LACCOLITH A concordant pluton that has domed up the strata into which it was intruded.

LAG TIME On a hydrograph of a stream, the time interval between the center of mass of the precipitation and the center of mass of the resulting flood.

LAMINAR FLOW Mechanism by which a fluid such as water moves slowly along a smooth channel, or through a tube with smooth walls, with fluid particles following straight-line paths parallel to the channel or walls. Contrast with *turbulent flow.*

LANDSLIDE A general term for relatively rapid mass movement, such as slump, rock slide, debris slide, mudflow, and earthflow.

LAPILLI Pyroclastic debris in pieces about the size of walnuts.

LAPSE RATE The actual rate of change in the temperature of the atmosphere with elevation.

LARGE WAVES Earthquake surface waves.

LATENT HEAT OF FUSION The number of calories per unit volume that must be added to a material at the melting point to complete the process of melting. These calories do not raise the temperature.

LATERAL MORAINE A ridge of till along the edge of a valley glacier. Composed largely of material that fell to the glacier from valley walls.

LATERITE Tropical soil rich in hydroxides of aluminum and iron formed under conditions of good drainage.

LAVA Magma that has poured out onto the surface of the earth, or rock that has solidified from such magma.

LEFT-LATERAL FAULT Strike-slip fault where the ground opposite you appears to have moved toward the left when you stand facing it.

LEVEE (NATURAL) Bank of sand and silt built by a river during floods, where suspended load is deposited in greatest quantity close to the river. The process of developing natural levees tends to raise river banks above the level of the surrounding flood plains. A break in a natural levee is sometimes called a crevasse.

LIGNITE A low-grade coal with about 70 percent carbon and 20 percent oxygen. Intermediate between peat and bituminous coal.

LIMB One of the two parts of an anticline or syncline, on either side of the axis.

LIMESTONE A sedimentary rock composed largely of the mineral calcite, $CaCO_3$, which has been formed by either organic or inorganic processes. Most limestones have a clastic texture, but nonclastic, particularly crystalline, textures are common. The carbonate rocks, limestone and dolomite, constitute an estimated 12 to 22 percent of the sedimentary rocks exposed above sea level.

LIMONITE Iron oxide with no fixed composition or atomic structure. A mineraloid; always of secondary origin and not a true mineral. Is encountered as ordinary rust, or the coloring material of yellow clays and soils.

LIQUID A state of matter that flows readily so that the mass assumes the form of its container, but retains its independent volume.

LITHIC SANDSTONE See *graywacke.*

LITHIFICATION The process by which unconsolidated rock-forming materials are converted into a consolidated or coherent state.

LOAD The amount of material that a transporting agency, such as a stream, a glacier, or the wind, is actually carrying at a given time.

LOESS An unconsolidated, unstratified aggregation of small, angular mineral fragments, usually buff in color. Generally believed to be wind-deposited. Characteristically able to stand on very steep to vertical slopes.

LONGITUDINAL DUNE A long ridge of sand oriented in the general direction of wind movement. A small one is less than 3 m in height and 60 m in length. Very large ones are called seif dunes.

LONGITUDINAL WAVE A push-pull wave.

LOPOLITH Tabular concordant pluton shaped like the bowl of a spoon, with both reef and floor sagging downward.

MAAR CRATER A volcanic crater with little or no cone. Explosive in origin.

MAGMA A naturally occurring silicate melt, which may contain suspended silicate crystals or dissolved gases, or both. These conditions may be met in general by a mixture containing as much as 65 percent crystals, but no more than 11 percent of dissolved gases.

MAGNETIC DECLINATION The angle of divergence between a geographic meridian and a magnetic meridian. It is measured in degrees east and west of geographic north.

MAGNETIC INCLINATION The angle that the magnetic needle makes with the surface of the earth. Also called dip of the magnetic needle.

MAGNETIC POLE The north magnetic pole is the point on the earth's surface where the north-seeking end of a magnetic needle free to move in space points directly down. At the south magnetic pole the same needle points directly up. These poles are also known as *dip poles.*

MAGNETIC REVERSAL A shift of 180° in the earth's magnetic field such that a north-seeking needle of a magnetic compass would point to the south rather than to the north magnetic pole.

MAGNETITE A mineral; iron oxide, Fe_3O_4. Black, strongly magnetic. An important ore of iron.

MAGNETOSPHERE A region from 1,000 to 64,000 km above the earth, where the magnetic field traps electrically charged particles from the sun and from space. First believed to consist of two bands, the Van Allen belts.

MAGNITUDE (OF AN EARTHQUAKE) A measure of the total energy released by an earthquake. Contrast with *intensity,* which is a measure of the effects of earthquake waves at a particular place.

MANTLE The intermediate zone of the earth. Surrounded by the crust, it rests on the core at a depth of about 2,900 km.

MANTLE BOILING HYPOTHESIS Explains continuous up-and-down history of mountain making and rock deformation as turbulence produced by boiling of the mantle under a relatively rigid crust.

MARBLE Metamorphic rock of granular texture, no rock cleavage, and composed of calcite or dolomite or both.

MARIA The dark-toned "seas" of the moon. They mark the topographically low areas of the moon.

MARL A calcareous clay, or intimate mixture of clay and particles of calcite or dolomite, usually fragments of shells.

MARSH GAS Methane, CH_4 the simplest paraffin hydrocarbon. The predominant component of natural gas.

MASCONS Concentrations of mass located beneath the surfaces of the lunar maria.

MASS A number that measures the quantity of matter. It is obtained on the earth's surface by dividing the weight of a body by the acceleration due to gravity.

MASSIVE PLUTON Any pluton that is not tabular in shape.

MASS MOVEMENT Surface movement of earth materials induced by gravity.

MASS NUMBER Number of protons and neutrons in the nucleus of an atom.

MASS UNIT One-twelfth the mass of the carbon atom. Approximately the mass of the hydrogen atom.

MATTER Anything that occupies space. Usually defined by describing its states and properties: solid, liquid, or gaseous; possesses mass, inertia, color, density, melting point, hardness, crystal form, mechanical strength, or chemical properties. Composed of atoms.

MEANDER (1) *n.*, A turn or sharp bend in a stream's course. (2)*v.i.*, To turn, or bend sharply. Applied to stream courses in geological usage.

MEANDER BELT The zone along a valley floor that encloses a meandering river.

MECHANICAL WEATHERING The process by which rock is broken down into smaller and smaller fragments as the result of energy developed by physical forces. Also known as *disintegration.*

MEDIAL MORAINE A ridge of till formed by the junction of two lateral moraines when two valley glaciers join to form a single ice stream.

MEDITERRANEAN WATER Water that flows outward from the Mediterranean Sea into the Atlantic Ocean through the Straits of Gibraltar.

MERCALLI INTENSITY SCALE A scale to evaluate the intensity of earthquake shaking on the basis of effects at a given place.

MESOSPHERE An atmospheric layer, about 25 km thick, which lies above the stratosphere and in which the temperature declines with increasing elevation.

MESOTROPHIC In the aging of a lake, a stage intermediate between oligotrophic and eutrophic.

METAL A substance that is fusible and opaque, is a good conductor of electricity, and has a characteristic luster. Examples: gold, silver, aluminum. Seventy-seven of the elements are metals.

METALLOID An element that has some metallic and some nonmetallic characteristics. There are nine metalloids. See Appendix A.

METAMORPHIC FACIES An assemblage of minerals that reached equilibrium during metamorphism under a specific range of temperature.

METAMORPHIC ROCK "Changed-form rock." Any rock that has been changed in texture or composition by heat, pressure, or chemically active fluids after its original formation.

METAMORPHIC ZONE An area subjected to metamorphism and characterized by a certain metamorphic facies that formed during the process.

METAMORPHISM A process whereby rocks undergo physical or chemical changes, or both, to achieve equilibrium with conditions other than those under which they were originally formed. Weathering is arbitrarily excluded from the meaning of the term.

The agents of metamorphism are heat, pressure, and chemically active fluids.

METASOMATISM A process whereby rocks are altered when volatiles exchange ions with them.

METEOR A transient celestial body that enters the earth's atmosphere with great speed and becomes incandescent from heat generated by resistance of the air.

METEORIC WATER Ground water derived primarily from precipitation.

METEORITE A stony or metallic body that has fallen to the earth from outer space.

METHANE The simplest paraffin hydrocarbon, CH_4. The principal constituent of natural gas. Sometimes called marsh gas.

MICAS A group of silicate minerals characterized by perfect sheet or scale cleavage resulting from their atomic pattern, in which silicon-oxygen tetrahedra are linked in sheets. Biotite is the ferromagnesian black mica. Muscovite is the potassic white mica.

MICROSEISM A small shaking. Specifically limited in technical usage to earth waves generated by sources other than earthquakes, and most frequently to waves with periods of from 1 sec to about 9 sec, from sources associated with atmospheric storms.

MIDATLANTIC RIDGE Belt of mountains under the Atlantic Ocean from Iceland to Antarctica, located nearly midway between the continents.

MIGMATITE A mixed rock produced by an intimate interfingering of magma and an invaded rock.

MINERAL A naturally occurring solid element or compound, exclusive of biologically formed carbon components. It has a definite composition, or range of composition, and an orderly internal arrangement of atoms (crystalline structure), which gives it unique physical and chemical properties, including a tendency to assume certain geometrical forms known as *crystals.*

MINERAL DEPOSIT An occurrence of one or more minerals in such concentration and form as to make possible removal and processing for use at a profit.

MINERALOID A substance that does not yield a definite chemical formula, and shows no sign of crystallinity. Bauxite, limonite, and opal are examples.

MINING Removing material from the earth.

MODULUS OF ELASTICITY The slope of the graph line relating stress to strain in elastic deformation.

MOHOROVICIC DISCONTINUITY (MOHO) Base of the crust marked by abrupt increases in velocities of earth waves.

MOLDAVITES See *tektites.*

MOLECULE The smallest unit of a compound which displays the properties of that compound.

MONADNOCK A hill left as a residual of erosion, standing above the level of a peneplain.

MONEL METAL Steel containing 68 percent nickel.

MONOCLINE A double flexure connecting strata at one level with the same strata at another level.

MONTMORILLONITE A clay mineral family, a hydrous aluminous silicate with a structural sandwich of one ionic sheet of aluminum and hydroxyl between two (Si_4O_{10}) sheets. These sandwiches are piled on each other with water between them, and with nothing but weak bonds to hold them together. As a result, additional water can enter the lattice readily. This causes the mineral to swell appreciably and further weakens the attraction between structural sandwiches. Consequently, a lump of montmorillonite in a bucket of water slumps rapidly into a loose, incoherent mass. Compare with the other clay minerals, kaolinite and illite.

MOON A natural satellite.

MORAINE A general term applied to certain landforms composed of till.

MOUNTAIN Any part of a landmass that projects conspicuously above its surroundings.

MOUNTAIN CHAIN A series or group of connected mountains having a well-defined trend or direction.

MOUNTAIN GLACIER Synonymous with *alpine glacier.*

MOUNTAIN RANGE A series of more or less parallel ridges, all of which were formed within a single geosyncline or on its borders.

MOUNTAIN STRUCTURE Structure produced by the deformation of rocks.

MUDCRACKS Cracks caused by the shrinkage of a drying deposit of silt or clay under surface conditions.

MUDFLOW Flow of a well-mixed mass of rock, earth, and water that behaves like a fluid and flows down slopes with a consistency similar to that of newly mixed concrete.

MUDSTONE Fine-grained, detrital sedimentary rock made up of silt and clay-sized particles. Distinguished from shale by lack of fissility.

MUSCOVITE "White mica." A nonferromagnesian rock-forming silicate mineral with its tetrahedra arranged in sheets. Sometimes called potassic mica.

NATIVE STATE State in which an element occurs uncombined in nature. Usually applied to the metals, as in native copper, native gold, etc.

NATURAL GAS Gaseous hydrocarbons that occur in rocks. Dominated by methane.

NATURAL REMANENT MAGNETISM The magnetism of a rock. May or may not coincide with present magnetic field of the earth. Abbreviation NRM.

NATURAL RESOURCES Energy and materials made available by geological processes.

NEAP TIDE Tide produced when sun, earth, and moon define a right angle.

NEBULA A hazy patch of light in the sky that a telescope does not resolve as a sharp point, as it does the stars.

NECK CUTOFF The breakthrough of a river across the narrow neck separating two meanders, where downstream migration of one has been slowed and the next meander upstream has overtaken it. Compare with *chute cutoff.*

NEGATIVE CHARGE A condition resulting from a surplus of electrons.

NESOSILICATE Mineral with crystal structure containing SiO_4 tetrahedra arranged as isolated units.

NEUTRON A proton and an electron combined and behaving like a fundamental particle of matter. Electrically neutral, with a mass of 1.00896 amu. If isolated, it decays to form a proton and an electron.

NÉVÉ Granular ice formed by the recrystallization of snow. Intermediate between snow and glacier ice. Sometimes called *firn.*

NICHROME A steel alloy with 35 to 85 percent nickel.

NICKEL STEEL Steel containing 2.5 to 3.5 percent nickel.

NIVATION Erosion beneath and around the edges of a snowbank.

NODULE An irregular, knobby-surfaced body of mineral that differs in composition from the rock in which it is formed. Silica in the form of chert or flint is the major component of nodules. They are commonly found in limestone and dolomite.

NONCLASTIC TEXTURE Applied to those sedimentary rocks in which the rock-forming grains are interlocked. Most sedimentary rocks with nonclastic texture are crystalline.

NONCONFORMITY An unconformity in which the older rocks are of intrusive igneous origin.

NONDIPOLE MAGNETIC FIELD That portion of the earth's magnetic field remaining after the dipole field and the external field are removed.

NONFERROMAGNESIANS Silicate minerals that do not contain iron or magnesium.

NONMETAL An element that does not exhibit metallic luster, conductivity, or other features of metals. Seventeen of the elements are nonmetals.

NORMAL FAULT A fault in which the hanging wall appears to have moved downward relative to the footwall. Opposite of a thrust fault. Also called *gravity fault.*

NORTH ATLANTIC DEEP WATER Sea water in the Arctic that sinks in the North Atlantic and drifts southward as far as 60° south of the equator.

NRM See *natural remanent magnetism.*

NUCLEUS The protons and neutrons constituting the central part of an atom.

NUÉE ARDENTE "Hot cloud." A French term (*pl. nuées ardentes*) applied to a highly heated mass of gas-charged lava ejected more or less horizontally from a vent or pocket at the summit of a volcano, onto an outer slope down which it moves swiftly, however slight the incline, because of its extreme mobility.

OBLIQUE SLIP FAULT Fault with components of relative displacement along both strike and slip.

OBSIDIAN Glassy equivalent of granite.

OIL In geology, refers to petroleum.

OIL SHALE Shale containing such a proportion of hydrocarbons as to be capable of yielding petroleum on slow distillation.

OLIGOTROPHIC In the aging of a lake, pertains to a youthful lake and indicates water low in accumulated nutrients and high in dissolved oxygen.

OLIVINE A rock-forming ferromagnesian silicate mineral that crystallizes early from a magma and weathers readily at the earth's surface. Its crystal structure is based on isolated SiO_4 ions and positive ions of iron or magnesium, or both. General formula: $(Mg,Fe)_2SiO_4$.

OÖLITES Spheroidal grains of sand size, usually composed of calcium carbonate, $CaCO_3$, and thought to have originated by inorganic precipitation. Some limestones are made up largely of oölites.

OOZE Deep-sea deposit consisting of 30 percent or more by volume of the hard parts of very small, sometimes microscopic, organisms. If a particular organism is dominant, its name is used as modifier, as in *globigerina* ooze, or *radiolarian* ooze.

OPAL Amorphous silica, with varying amounts of water. A mineral gel.

OPEN-PIT MINING Surface mining represented by sand and gravel pits, stone quarries, and the copper mines of some of the western states.

OPHITIC A rock texture in which lath-shaped plagioclase crystals are enclosed wholly or in part in later-formed augite, as commonly occurs in diabase.

ORDER OF CRYSTALLIZATION The chronological sequence in which crystallization of the various minerals of an assemblage takes place.

ORE DEPOSIT Metallic minerals in concentrations that can be worked at a profit.

OROGENY Process by which mountain structures develop.

ORTHOCLASE The feldspar in which K^+ is the diagnostic positive ion; $K(AlSi_3O_8)$.

ORTHOQUARTZITE A sandstone composed completely, or almost completely, of quartz grains. *Quartzose sandstone* is a synonym.

OUTWASH Material carried from a glacier by meltwater, and laid down in stratified deposits.

OUTWASH PLAIN Flat or gently sloping surface underlain by outwash.

OVERBANK DEPOSITS The sediments (usually clay, silt, and fine sand) deposited on a flood plain by a river overflowing its banks.

OVERTURNED FOLD A fold in which at least one limb is overturned—that is, has rotated through more than 90°.

OXBOW An abandoned meander, caused by a neck cutoff.

OXBOW LAKE An abandoned meander isolated from the main stream channel by deposition, and filled with water.

OXIDE MINERAL A mineral formed by the direct union of an element with oxygen. Examples: ice, corundum, hematite, magnetite, cassiterite.

P Symbol for earthquake primary waves.

PAHOEHOE LAVA Lava whose surface is smooth and billowy, frequently molded into forms that resemble huge coils of rope. Characteristic of basic lavas.

PAIRED TERRACES Terraces that face each other across a stream at the same elevation.

PALEOMAGNETISM The study of the earth's magnetic field as it has existed during geologic time.

PANGAEA A hypothetical continent from which all others are postulated to have originated through a process of fragmentation and drifting.

PARABOLIC DUNE A dune with a long, scoop-shaped form that, when perfectly developed, exhibits a parabolic shape in plan, with the horns pointing upwind. Contrast *barchan*, in which the horns point downwind. Characteristically covered with sparse vegetation, and often found in coastal belts.

PARTICLES, FUNDAMENTAL See *fundamental particles*.

PATER NOSTER LAKES A chain of lakes resembling a string of beads along a glaciated valley where ice-plucking and gouging have scooped out a series of basins.

PEAT Partially reduced plant or wood material containing approximately 60 percent carbon and 30 percent oxygen. An intermediate material in the process of coal formation.

PEBBLE SIZE A volume greater than that of a sphere with a diameter of 4 mm and less than that of a sphere of 64 mm.

PEDALFER A soil characterized by the accumulation of iron salts or iron and aluminum salts in the B-horizon. Varieties of pedalfers include red and yellow soils of the southeastern United States, and podsols of the northeastern quarter of the United States.

PEDIMENT Broad, smooth erosional surface developed at the expense of a highland mass in an arid climate. Underlain by beveled rock, which is covered by a veneer of gravel and rock debris. The final stage of a cycle of erosion in a dry climate.

PEDOCAL A soil characterized by an accumulation of calcium carbonate in its profile. Characteristic of low rainfall. Varieties include black and chestnut soils of the northern Plains states, and the red and gray desert soils of the drier western states.

PEDOLOGY The science that treats of soils—their origin, character, and utilization.

PEGMATITE A small pluton of exceptionally coarse texture, with crystals up to 12 m in length, commonly formed at the margin of a batholith and characterized by graphic structure. Nearly 90 percent of all pegmatites are simple pegmatites of quartz, orthoclase, and unimportant percentages of micas. The others are extremely rare ferromagnesian pegmatites, and complex pegmatites. Complex pegmatites have as their major components the sialic minerals of simple pegmatites, but they also contain a variety of rare minerals.

PELAGIC DEPOSIT Material formed in the deep ocean and deposited there. Example: ooze.

PENDULUM Inertia member so suspended that, after it is displaced a restoring force will return it to the starting position. If displaced, then released, it oscillates, completing one to-and-fro swing in a time called its period.

PENEPLAIN An extensive, nearly flat surface developed by subaerial erosion, and close to base level, toward which the streams of the region are reducing it. Orignally defined as forming in a humid climate.

PERCHED WATER TABLE The top of a zone of saturation that bottoms on an impermeable horizon above the level of the general water table in the area. Is generally near the surface, and frequently supplies a hillside spring.

PERIDOTITE A coarse-grained igneous rock dominated by dark-colored minerals, consisting of about 75 percent ferromagnesian silicates and the balance plagioclase feldspars.

PERIOD For an oscillating system, the length of time required to complete one oscillation.

PERMEABILITY For a rock or an earth material, the ability to transmit fluids. Permeability for underground water is sometimes expressed numerically as the number of gallons per day that will flow through a cross section of 1 ft^2 at 60°F, under a hydraulic gradient of 100 percent. Permeability is equal to velocity of flow divided by hydraulic gradient.

PETROLEUM A complex mixture of hydrocarbons, accumulated in rocks, and dominated by paraffins and cycloparaffins. Crude petroleums are classified as *paraffin-base* if the residue left after volatile components have been removed consists principally of a mixture of paraffin hydrocarbons; as *asphalt-base* if the residue is primarily cycloparaffins.

PHASE (IN PHYSICAL CHEMISTRY) A homogeneous, physically distinct portion of matter in a system that is not homogeneous, as in the three phases ice, water, and aqueous vapor.

PHASE (IN SEISMOLOGY) A group of waves of one type.

PHENOCRYST A crystal significantly larger than the crystals of surrounding minerals.

PHOSPHATE ROCK A sedimentary rock containing calcium phosphate.

PHOTOSYNTHESIS The process by which carbohydrates are compounded from carbon dioxide and water in the presence of sunlight and chlorophyll.

PHYLLITE A clayey metamorphic rock with rock cleavage intermediate between slate and schist. Commonly formed by the regional metamorphism of shale or tuff. Micas characteristically impart a pronounced sheen to rock cleavage surfaces. Has phyllitic cleavage.

PHYLLITIC CLEAVAGE Rock cleavage in which flakes are produced that are barely visible to the unaided eye. It is coarser than slaty cleavage, finer than schistose cleavage.

PHYLLOSILICATE Mineral with crystal structure containing SiO_4 tetrahedra arranged as sheets.

PHYSICAL GEOLOGY The branch of geology that deals with the nature and properties of material composing the earth, distribution of materials throughout the globe, the processes by which they are formed, altered, transported, and distorted, and the nature and development of landscape.

PIEDMONT GLACIER A glacier formed by the coalescence of valley glaciers and spreading over plains at the foot of the mountains from which the valley glaciers came.

PIRATE STREAM One of two streams in adjacent valleys that has been able to deepen its valley more rapidly than the other, that extended its valley headward until it breached the divide between them, and that captured the upper portion of the neighboring stream.

PLAGIOCLASE FELDSPARS Albite and anorthite.

PLAIN Region of low relief composed of nearly horizontal sedimentary rocks.

PLANET A heavenly body that changes its position from night to night with respect to the background of stars.

PLANETOLOGY An organized body of knowledge about the planetary system.

PLASTIC DEFORMATION Permanent change in shape or volume that does not involve failure by rupture, and that, once started, continues without increase in the deforming force.

PLASTIC SOLID A solid that undergoes deformation continuously and indefinitely after the stress applied to it passes a critical point.

PLATEAU BASALT Basalt poured out from fissures in floods that tend to form great plateaus. Sometimes called *flood basalt.*

PLAYA The flat-floored center of an undrained desert basin.

PLAYA LAKE A temporary lake formed in a playa.

PLEOCHROIC HALO Minute, concentric spherical zones of darkening or coloring that form around inclusions of radioactive minerals in biotite, chlorite, and a few other minerals. About 0.075 mm in diameter.

PLUNGE The acute angle that the axis of a folded rock mass makes with a horizontal plane.

PLUTON A body of igneous rock that is formed beneath the surface of the earth by consolidation from magma. Sometimes extended to include bodies formed beneath the surface of the earth by the metasomatic replacement of older rock.

PLUTONIC IGNEOUS ROCK A rock formed by slow crystallization, which yields coarse texture. Once believed to be typical of crystallization at great depth, but that is not a necessary condition.

PLUVIAL LAKE A lake formed during a pluvial period.

PLUVIAL PERIOD A period of increased rainfall and decreased evaporation, which prevailed in nonglaciated areas during the time of ice advance elsewhere.

PODSOL An ashy gray or gray-brown soil of the pedalfer group. This highly bleached soil, low in iron and lime, is formed under moist and cool conditions.

POINT BARS Accumulations of sand and gravel deposited in the slack waters on the inside of bends of a winding or meandering river.

POISSON'S RATIO Ratio of change of diameter per unit diameter to change of length per unit length in elastic stretching or compression of a cylindrical specimen.

POLAR COMPOUND A compound, such as water, with a molecule that behaves like a small bar magnet with a positive charge on one end and a negative charge on the other.

POLARITY EPOCH An interval of time during which the earth's magnetic field has been oriented dominantly in either a normal or a reverse direction. An epoch may be marked by shorter intervals of the opposite sign, called *polarity events.*

POLARITY EVENT See *polarity epoch.*

POLAR WANDERING OR MIGRATION A movement of the position of the magnetic pole during past time in relation to its present position.

POROSITY The percentage of open space or interstices in a rock or other earth material. Compare with *permeability.*

PORPHYRITIC A textural term for igneous rocks in which larger crystals, called phenocrysts, are set in a finer groundmass, which may be crystalline or glassy or both.

PORPHYRY An igneous rock containing conspicuous phenocrysts in a fine-grained or glassy groundmass.

PORTLAND CEMENT A hydraulic cement consisting of compounds of silica, lime, and alumina.

POSITIVE CHARGE A condition resulting from a deficiency of electrons.

POTASSIC FELDSPAR Orthoclase, $K(AlSi_3O_8)$.

POTENTIAL ENERGY Stored energy waiting to be used. The energy that a piece of matter possesses because of its position or because of the arrangement of its parts.

POTHOLE A hole ground in the solid rock of a stream channel by sands, gravels, and boulders caught in an eddy of turbulent flow and swirled for a long time over one spot.

PRAIRIE SOILS Transitional soils between pedalfers and pedocals.

PRATT HYPOTHESIS Explains isostasy by assuming that all portions of the crust have the same total mass above a certain elevation, called the level of compensation. Higher sections would have proportionately lower density.

PRECIPITATION The discharge of water, in the form of rain, snow, hail, sleet, fog, or dew, on a land or water surface. Also, the process of separating mineral constituents from a solution by evaporation (halite, anhydrite) or from magma to form igneous rocks.

PRESSURE Force per unit area.

PRIMARY WAVE Earthquake body waves that travel fastest and advance by a push-pull mechanism. Also known as longitudinal, compressional, or P-waves.

PROTON A fundamental particle of matter with a positive electrical charge of 1 unit (equal in amount but opposite in effect to the charge of an electron), and with a mass of 1.00758 amu.

PROTORE The original rock, too poor in mineral values to constitute an ore, from which desired elements have been leached and redeposited as an ore. The process of leaching and redeposition of desired elements is sometimes called supergene enrichment, or secondary sulfide enrichment.

PTEROPOD OOZE A calcareous deep-sea ooze dominated by the remains of minute molluscs of the group *Pteropoda.*

PUMICE Pieces of magma up to several centimeters across that have trapped bubbles of steam or other gases as they were thrown out. Sometimes they have enough bouyancy to float on water.

PUSH-PULL WAVE A wave that advances by alternate compression and rarefaction of a medium, causing a particle in its path to move forward and backward along the direction of the wave's advance. In connection with waves in the earth, also known as *primary wave*, compressional wave, longitudinal wave, or P-wave.

PYRITE A sulfide mineral. Iron sulfide, FeS_2.

PYROCLASTIC DEBRIS Fragments blown out by explosive volcanic eruptions and subsequently deposited on the ground. Include ash, cinders, lapilli, blocks, bombs, and pumice.

PYROXENE GROUP Ferromagnesian silicates with a single chain of silicon-oxygen tetrahedra. Common example: augite. Compare with *amphibole group* (example: hornblende), which has a double chain of tetrahedra.

PYRRHOTITE A mineral; iron sulfide. So commonly associated with nickel minerals that it has been called "the world's greatest nickel ore."

QUARTZ A silicate mineral, SiO_2, composed exclusively of silicon-oxygen tetrahedra with all oxygens joined together in a three-dimensional network. Crystal form is a six-sided prism tapering at the end, with the prism faces striated transversely. An important rock-forming mineral.

QUARTZITE Metamorphic rock commonly formed by the metamorphism of sandstone and composed of quartz. Has no rock cleavage. Breaks through sand grains as contrasted to sandstone, which breaks around the grains.

RADAR Ultra-high-frequency electromagnetic radiation.

RADIAL DRAINAGE An arrangement of stream courses in which the streams radiate outward in all directions from a central zone.

RADIANT ENERGY Electromagnetic waves traveling as wave motion.

RADIOACTIVITY The spontaneous breakdown of an atomic nucleus, with emission of radiant energy.

RADIOLARIAN OOZE A siliceous deep-sea ooze dominated by the delicate and complex hard parts of minute marine protozoa called *Radiolaria*.

RAIN WASH The water from rain after it has fallen on the ground and before it has been concentrated in definite stream channels.

RANGE (MOUNTAIN) An elongated series of mountain peaks considered to be a part of one connected unit, such as the Appalachian Range or the Sierra Nevada Range.

RANK A term used to designate the extent to which metamorphism has advanced. Compare with *grade*. Rank is more commonly employed in designating the stage of metamorphism of coal.

RAY CRATERS Those lunar craters that are marked by rays. They are young on the lunar time scale.

RAYS The light-toned streaks that spread outward from such lunar craters as Tycho, Kepler, and Copernicus.

REACTION SERIES See *Bowen's reaction series*.

RECESSIONAL MORAINE A ridge or belt of till marking a period of moraine formation, probably in a period of temporary stability or a slight re-advance, during the general wastage of a glacier and recession of its front.

RECORDER The part of a seismograph that makes a record of ground motion.

RECTANGULAR PATTERN An arrangement of stream course in which the tributaries flow into the larger streams at angles approaching 90°.

RECUMBENT FOLD A fold in which the axial plane is more or less horizontal.

REFLECTION SEISMIC PROSPECTING Uses reflected waves and places seismographs at distances only a fraction of the depths investigated.

REFRACTION SEISMIC PROSPECTING Uses travel times of refracted waves and spreads seismographs over lines roughly four times the depth being investigated.

REFRACTORY A mineral or compound that resists the action of heat and of chemical reagents.

REGIONAL METAMORPHISM Metamorphism occurring over tens or scores of kilometers.

REJUVENATION A change in conditions of erosion that causes a stream to begin more active erosion and a new cycle.

RELATIVE TIME Dating of events by means of their place in a chronologic order of occurrence rather than in terms of years. Compare with *absolute time*.

RELIEF See Appendix F.

REVERSE FAULT A fault in which the hanging wall appears to have moved upward relative to the footwall. Also called *thrust fault*. Contrast with *normal* or *gravity fault*.

RHYOLITE A fine-grained igneous rock with the composition of granite.

RIFT ZONE A system of fractures in the earth's crust. Often associated with extrusion of lava.

RIGHT-LATERAL FAULT Strike-slip fault in which the ground opposite you appears to have moved toward the right when you stand facing it.

RIGIDITY Resistance to elastic shear.

RIGIDITY MODULUS A measure that expresses a material's rigidity. Represented by the symbol G.

RILL Miniature stream channel which forms along the axis of a broad, shallow trough carrying sheet wash or sheet flow.

RILLES Trench-like depressions on the moon's surface. Some are straight-walled, others sinuous.

RING DIKE An arcuate (rarely circular) dike with steep dip.

RIPPLE MARKS Small waves produced in unconsolidated material by wind or water. See *ripple marks of oscillation; current ripple marks*.

RIPPLE MARKS OF OSCILLATION Ripple marks formed by oscillating movement of water such as may be found along a sea coast outside the surf zone. They are symmetrical, with sharp or slightly rounded ridges separated by more gently rounded troughs.

ROCHE MOUTONNÉE A sheep-shaped knob of rock (*pl. roches moutonnées*) that has been rounded by the action of glacier ice. Usually only a few feet in height, length, and breadth. A gentle slope faces upstream with reference to the ice movement. A steeper slope attributed to plucking action of the ice represents the downstream side.

ROCK An aggregate of minerals of different kinds in varying proportions.

ROCK CYCLE A concept of the sequences through which earth materials may pass when subjected to geological processes.

ROCK FLOUR Finely divided rock material pulverized by a glacier and carried by streams fed by melting ice.

ROCK-FORMING SILICATE MINERALS Minerals built around a framework of silicon-oxygen tetrahedra. Olivine, augite, hornblende, biotite, muscovite, orthoclase, albite, anorthite, quartz.

ROCK GLACIER A tongue of rock waste found in the valleys of certain mountainous regions. Characteristically lobate and marked by a series of arcuate, rounded ridges that give it the aspect of having flowed as a viscous mass.

ROCK MELT A liquid solution of rock-forming mineral ions.

ROCK SALT Halite, or common salt, NaCl.

ROCK SLIDE Sudden and rapid slide of bedrock along planes of weakness.

ROSSI-FOREL SCALE A scale for rating earthquake intensities. Devised in 1878 by de Rossi of Italy and Forel of Switzerland.

RUNOFF Water that flows off the land.

RUPTURE A breaking apart or state of being broken apart.

S Symbol for secondary wave.

SALT In geology this term usually refers to halite, or rock salt, NaCl, particularly in such combinations as salt water, and salt dome.

SALTATION Mechanism by which a particle moves by jumping from one point to another.

SALT DOME A mass of NaCl generally of roughly cylindrical shape and with a diameter of about 2 km near the top. These masses have been pushed through surrounding sediments into their present positions. Reservoir rocks above and alongside salt domes sometimes trap oil and gas.

SALTWATER WEDGE A body of water, found in some estuaries, which thins toward the head of the estuary and is overridden by fresh water from the land.

SAND Clastic particles of sand size, commonly but not always composed of the mineral quartz.

SAND SIZE A volume greater than that of a sphere with a diameter of 0.0625 mm and less than that of a sphere with a diameter of 2 mm.

SANDSTONE A detrital sedimentary rock formed by the cementation of individual grains of sand size and commonly composed of the mineral quartz. Sandstones constitute an estimated 12 to 28 percent of sedimentary rocks.

SAPROPEL An aquatic ooze or sludge that is rich in organic matter. Believed to be the source material for petroleum and natural gas.

SATELLITE CRATER A crater formed by the impact of a fragment ejected during the creation of a primary crater. Also called *secondary crater.*

SCALE See Appendix F.

SCHIST A metamorphic rock dominated by fibrous or platy minerals. Has schistose cleavage and is a product of regional metamorphism.

SCHISTOSE CLEAVAGE Rock cleavage in which grains and flakes are clearly visible and cleavage surfaces are rougher than in slaty or phyllitic cleavage.

SEA ARCH The roof of a cave cut by the sea through a headland.

SEA CAVE A cave formed by the erosive action of sea water.

SEA-FLOOR SPREADING The concept that the ocean floors spread laterally from the crests of the main ocean ridges. As material moves laterally from the ridge new material is thought to replace it along the ridge crest by welling upward from the mantle.

SEAMOUNT An isolated, steep-sloped peak rising from the deep ocean floor but submerged beneath the ocean surface. Most have sharp peaks, but some have flat tops and are called *guyots* or *tablemounts.* Seamounts are probably volcanic in origin.

SECONDARY CRATER See *satellite crater.*

SECONDARY WAVE An earthquake body wave slower than the primary wave. A *shear, shake,* or *S-wave.*

SECULAR VARIATION OF THE MAGNETIC FIELD A change in inclination, declination, or intensity of the earth's magnetic field. Detectable only from long historical records.

SEDIMENTARY FACIES An accumulation of deposits that exhibits specific characteristics and grades laterally into other sedimentary accumulations, which were formed at the same time but which exhibit different characteristics.

SEDIMENTARY ROCK Rock formed from accumulations of sediment, which may consist of rock fragments of various sizes, the remains or products of animals or plants, the product of chemical action or of evaporation, or mixtures of these. *Stratification* is the single most characteristic feature of sedimentary rocks, which cover about 75 percent of the land area of the world.

SEDIMENTATION The process by which mineral and organic matter are laid down.

SEIF DUNE A very large longitudinal dune. As high as 100 meters and as long as 100 kilometers.

SEISMIC PROSPECTING A method of determining the nature and structure of buried rock formations by generating waves in the ground (commonly by small charges of explosive) and measuring the length of time these waves require to travel different paths.

SEISMIC SEAWAVE A large wave in the ocean generated at the time of an earthquake. Popularly, but incorrectly, known as a *tidal wave.* Sometimes called a *tsunami.*

SEISMOGRAM The record obtained on a seismograph.

SEISMOGRAPH An instrument for recording vibrations, most commonly employed for recording earth vibrations during earthquakes.

SEISMOLOGY The scientific study of earthquakes and other earth vibrations.

SERPENTINE A silicate of magnesium common among metamorphic minerals. Occurs in two crystal habits, one platy, known as antigorite, the other fibrous, known as chrysotile. Chrysotile is an asbestos. The name "serpentine" comes from mottled shades of green on massive varieties, suggestive of the markings of a serpent.

S.G. Symbol for specific gravity.

SHAKE WAVE Wave that advances by causing particles in its path to move from side to side or up and down at right angles to the direction of the wave's advance, a shake motion. Also called *shear wave* or *secondary wave.*

SHALE A fine-grained detrital, sedimentary rock made up of silt- and clay-sized particles. Contains clay minerals as well as particles of quartz, feldspar, calcite, dolomite, and other minerals. Distinguished from mudstone by presence of fissility.

SHATTER CONE A conical fracture form in rock caused by the high energies associated with volcanic explosions or meteoritic impacts. Apices point toward points of explosion or impact.

SHEAR Change of shape without change of volume.

SHEAR MODULUS See *rigidity modulus.*

SHEAR WAVE Wave that advances by shearing displacements (which change the shape without changing the volume) of a medium. This causes particles in its path to move from side to side or up and down at right angles to the direction of the wave's advance. Also called *shake wave* or *secondary wave.*

SHEET FLOW See *sheet wash.*

SHEETING Joints that are essentially parallel to the ground surface. They are more closely spaced near the surface and become progressively farther apart with depth. Particularly well developed in granitic rocks, but sometimes in other massive rocks as well.

SHEET WASH The water accumulating on a slope in a thin sheet of water. May begin to concentrate in rills. Also called *sheet flow.*

SHIELD Nucleus of Precambrian rocks around which a continent has grown.

SHIELD VOLCANO A volcano built up almost entirely of lava, with slopes seldom as great as 10° at the summit and 2° at the base. Examples: the five volcanoes on the island of Hawaii.

SIAL A term coined from the symbols for silicon and aluminum. Designates the composite of rocks dominated by granites, granodiorites, and their allies and derivatives, which underlie continental areas of the globe. Specific gravity is considered to be about 2.7.

SIALIC ROCK An igneous rock composed predominantly of silicon and aluminum. The term is constructed from chemical symbols for silicon and aluminum. Average specific gravity is about 2.7.

SIDERITE A mineral; iron carbonate, $FeCO_3$. An ore of iron.

SILICATE MINERALS Minerals with crystal structure containing SiO_4 tetrahedra arranged as isolated units (nesosilicates), single or double chains (inosilicates), sheets (phyllosilicates), or three-dimensional frameworks (tectosilicates).

SILICON-OXYGEN TETRAHEDRON A complex ion composed of a silicon ion surrounded by four oxygen ions. It has a negative charge of 4 units and is represented by the symbol $(SiO_4)^{4-}$. It is the diagnostic unit of silicate minerals, and makes up the central building unit of nearly 90 percent of the materials of the earth's crust.

SILLIMANITE A silicate mineral, Al_2SiO_5, characteristic of highest metamorphic temperatures and pressures. Occurs in long slender crystals, brown, green, white. Its crystalline structure is based on independent tetrahedra. Contrast with *kyanite* and *andalusite,* which have the same composition but different crystal habits, and form at lower temperatures.

SILT SIZE A volume greater than that of a sphere with a diameter of 0.0039 mm and less than that of a sphere with a diameter of 0.0625 mm.

SIMA A term coined from the chemical symbols for silicon and magnesium. Designates a worldwide shell of dark, heavy rocks. The sima is believed to be the outermost rock layer under deep, permanent ocean basins, such as the mid-Pacific. Originally, the sima was considered basaltic in composition, with a specific gravity of about 3.0. It has been suggested also, however, that it may be peridotitic in composition, with a specific gravity of about 3.3.

SIMATIC ROCK An igneous rock composed predominantly of ferromagnesian minerals. Average specific gravity is 3.0 to 3.3.

SINK See *sinkhole.*

SINKHOLE Depression in the surface of the ground caused by the collapse of the roof over a solution cavern.

SLATE A fine-grained metamorphic rock with well-developed slaty cleavage. Formed by the low-grade regional metamorphism of shale.

SLATY CLEAVAGE Rock cleavage in which ease of breaking occurs along planes separated by microscopic distances.

SLIP-FACE The steep face on the lee side of a dune.

SLOPE FAILURE See *slump.*

SLUMP The downward and outward movement of rock or unconsolidated material as a unit or as a series of units. Also called *slope failure.*

SNOWFIELD A stretch of perennial snow existing in an area where winter snowfall exceeds the amount of snow that melts away during the summer.

SNOWLINE The lower limit of perennial snow.

SOAPSTONE See *talc.*

SODIC FELDSPAR Albite, $Na(AlSi_3O_8)$.

SOIL The superficial material that forms at the earth's surface as a result of organic and inorganic processes. Soil varies with climate, plant and animal life, time, slope of the land, and parent material.

SOIL HORIZON A layer of soil approximately parallel to the land surface with observable characteristics that have been produced through the operation of soil-building processes.

SOLAR CONSTANT The average rate at which radiant energy is received by the earth from the sun. It is equal to a little less than 2 $cal./cm^2$ on a plane perpendicular to the sun's rays at the outer edge of the atmosphere, when the earth is at a mean distance from the sun.

SOLAR SYSTEM The sun with a group of celestial bodies held by the sun's gravitational attraction and revolving around it.

SOLID Matter with a definite shape and volume and some fundamental strength. May be crystalline, glassy, or amorphous.

SOLIFLUCTION Mass movement of soil affected by alternate freezing and thawing. Characteristic of saturated soils in high latitudes.

SPACE LATTICE In the crystalline structure of a mineral, a three-dimensional array of points representing the pattern of locations of identical atoms or groups of atoms which constitute a mineral's *unit cell.* There are 230 pattern types.

SPECIFIC GRAVITY The ratio between the weight of a given volume of a material and the weight of an equal volume of water at 4°C.

SPECIFIC HEAT The amount of heat necessary to raise the temperature of 1 g of any material 1°C.

SPHALERITE A mineral; zinc sulfide, ZnS. Nearly always contains iron, (Zn,Fe)S. The principal ore of zinc. Also known as Zinc Blende or Black Jack.

SPHEROIDAL WEATHERING The spalling off of concentric shells from rock masses of various sizes as a result of pressures built up during chemical weathering.

SPIT A sandy bar built by currents into a bay from a promontory.

SPRING A place where the water table crops out at the surface of the ground and where water flows out more or less continuously.

SPRING TIDE Produced when sun, earth, and moon are in line.

STACK A small island that stands as an isolated, steep-sided rock mass just off the end of a promontory. Has been isolated from the land by erosion and by weathering concentrated just behind the end of a headland.

STALACTITE Icicle-shaped accumulation of dripstone hanging from a cave roof.

STALAGMITE Post of dripstone growing upward from the floor of a cave.

STAR A heavenly body that seems to stay in the same position relative to other heavenly bodies.

STAUROLITE A silicate mineral characteristic of middle-grade metamorphism. Its crystalline structure is based on independent tetrahedra with iron and aluminum. It has a unique crystal habit that makes it striking and easy to recognize: six-sided prisms intersecting at 90° to form a cross, or at 60°to form an X.

STISHOVITE High pressure form of quartz. Density, 4.34. Associated with impact craters and cryptovolcanic structures.

STOCK A discordant pluton that increases in size downward, has no determinable floor, and shows an area of surface exposure less than 100 km^2. Compare with *batholith.*

STOPING A mechanism by which batholiths have moved into the crust by the breaking off and foundering of blocks of rock surrounding the magma chamber.

STRAIN Change of dimensions of matter in response to stress. Commonly, unit strain, such as change in length per unit length (total lengthening divided by original length), change in width per unit width, change in volume per unit volume. Contrast with *stress.*

STRATIFICATION The structure produced by the deposition of sediments in layers or beds.

STRATIGRAPHIC TRAP A structure that traps petroleum or natural gas because of variation in permeability of the reservoir rock, or the termination of an inclined reservoir formation on the up-dip side.

STRATOSPHERE An atmospheric zone about 30 km thick, which lies above the troposphere and in which temperature generally increases upward.

STREAK The color of the fine powder of a mineral. May be different from the color of a hand specimen. Usually determined by rubbing the mineral on a piece of unglazed porcelain (hardness about 7), known as a streak plate, which is, of course, useless for minerals of greater hardness.

STREAM CAPTURE See *stream piracy.*

STREAM ORDER The hierarchy in which segments of a stream system are arranged.

STREAM PIRACY The process whereby a stream rapidly eroding headward cuts into the divide separating it from another drainage basin, and provides an outlet for a section of a stream in the adjoining valley. The lower portion of the partially diverted stream is called a *beheaded stream.*

STREAM TERRACE A surface representing remnants of a stream's channel or flood plain when the stream was flowing at a higher level. Subsequent downward cutting by the stream leaves remnants of the old channel or flood plain standing as a terrace above the present level of the stream.

STRENGTH The stress at which rupture occurs or plastic deformation begins.

STRESS Force applied to material that tends to change the material's dimensions. commonly, unit stress, or total force divided by the area over which it is applied. Contrast with *strain.*

STRIATION A scratch or small channel gouged by glacial action. Bedrock, pebbles, and boulders may show striations produced when rocks trapped by the ice were ground against bedrock or other rocks. Striations along a bedrock surface are oriented in the direction of ice flow across that surface.

STRIATIONS In minerals, parallel thread-like lines or narrow bands on the face of a mineral; reflect the internal atomic arrangement.

STRIKE The direction of the line formed by intersection of a rock surface with a horizontal plane. The strike is always perpendicular to the direction of the dip.

STRIKE-SLIP FAULT A fault in which movement is almost in the direction of the fault's strike.

Strip mining Surface mining in which soil and rock covering the sought-for commodity are moved to one side. Some coal mining is pursued in this manner.

STRUCTURAL RELIEF The difference in elevation of parts of a deformed stratigraphic horizon.

STRUCTURE The attitudes of deformed masses of rock.

SUBLIMATION The process by which solid material passes into the gaseous state without first becoming a liquid.

SUBSEQUENT STREAM A tributary stream flowing along beds of less erosional resistance, and parallel to beds of greater resistance. Its course is determined subsequent to the uplift that brought the more resistant beds within its sphere of erosion.

SUBSURFACE WATER Water below the surface of the ground. Also referred to as *underground water,* and *subterranean water.*

SUBTERRANEAN WATER Water below the surface of the ground. Also referred to as *underground water* and *subsurface water.*

SULFATE MINERAL Mineral formed by the combination of the complex ion $(SO_4)^{2-}$ with a positive ion. Common example: gypsum, $CaSO_4 \cdot 2H_2O$.

SULFIDE MINERAL Mineral formed by the direct union of an element with sulfur. Examples: argentite, chalcocite, galena, sphalerite, pyrite, and cinnabar.

SUN A hot, self-luminous ball of gas; a star.

SUPERHEAT Heat added to a substance after melting is complete.

SUPERIMPOSED STREAM A stream whose present course was established on young rocks burying an old surface. With uplift, this course was maintained as the stream cut down through the young rocks to and into the old surface.

SUPERPOSITION, LAW OF If a series of sedimentary rocks has not been overturned, the topmost layer is always the youngest and the lowermost is always the oldest.

SURFACE WAVE Wave that travels along the free surface of a medium. Earthquake surface waves are sometimes represented by the symbol L.

SURGE As applied to glaciers, a rapid, and sometimes catastrophic, advance of the ice.

SUSPENDED WATER Underground water held in the zone of aeration by molecular attraction exerted on the water by the rock and earth materials and by the attraction exerted by the water particles on one another.

SUSPENSION Process by which material is buoyed up in air or water and moved from one place to another without making contact with the surface while in transit. Contrasts with *traction.*

SYMMETRICAL FOLD A fold in which the axial plane is essentially vertical. The limbs dip at similar angles.

SYNCLINE A configuration of folded stratified rocks in which the rocks dip downward from opposite directions to come together in a trough. The reverse of an *anticline.*

TABLEMOUNT See *guyot.*

TABULAR A shape with large area relative to thickness.

TACONITE Unleached iron formation of the Lake Superior District. Consists of chert with hematite, magnetite, siderite, and hydrous iron silicates. An ore of iron. It averages 25 percent iron, but natural leaching turns it into an ore with 50 to 60 percent iron.

TALC A silicate of magnesium common among metamorphic minerals. Its crystalline structure is based on tetrahedra arranged in sheets. Greasy and extremely soft. Sometimes known as *soapstone.*

TALUS A slope established by an accumulation of rock fragments at the foot of a cliff or ridge. The rock fragments that form the talus may be rock waste, sliderock, or pieces broken by frost action. Actually, however, the term "talus" is widely used to mean the rock debris itself.

TARN A lake formed in the bottom of a cirque after glacier ice has disappeared.

TECTONIC CHANGE OF SEA LEVEL A change in sea level produced by land movement.

TECTOSILICATE Mineral with crystal structure containing SiO_4 tetrahedra arranged in three-dimensional frameworks.

TEKTITES Small glassy objects averaging 1 cm in diameter and often occurring in "fields" over wide areas of the earth's surface. Thought by some to be of meteoric origin or to have come from moon's surface. Known locally by various names including *australites, moldavites,* and *bediasites.*

TEMPERATURE A measure of the activity of atoms; degree of heat.

TEMPERATURE INVERSION In the atmosphere, a temperature inversion occurs when air becomes warmer with increasing altitude and then inverts and becomes cooler with a further increase in altitude.

TEMPORARY BASE LEVEL A base level that is not permanent, such as that formed by a lake.

TENSION Stretching stress that tends to increase the volume of a material.

TERMINAL MORAINE A ridge or belt of till marking the farthest advance of a glacier. Sometimes called *end moraine.*

TERMINAL VELOCITY The constant rate of fall eventually attained by a grain or a body when the acceleration caused by the influence of gravity is balanced by the resistance of the fluid through which the grain falls or the air through which a body falls.

TERRACE A nearly level surface, relatively narrow, bordering a stream or body of water, and terminating in a steep bank. Commonly the term is modified to indicate origin, as in *stream* terrace and *wave-cut* terrace.

TERRAE The light-toned highlands of the moon.

TERRIGENOUS DEPOSIT Material derived from above sea level and deposited in deep ocean. Example: volcanic ash.

TETRAHEDRON A four-sided solid (*pl., tetrahedra*). Used commonly in describing silicate minerals as a shortened reference to the silicon-oxygen tetrahedron.

TEXTURE The general physical appearance of a rock, as shown by the size, shape, and arrangement of the particles that make up the rock.

THERMAL CONTRACTION HYPOTHESIS Assumes that the earth's loss of heat results in contraction and compression of the crust as it adjusts to the shrinking interior. This causes mountains to fold.

THERMAL GRADIENT In the earth, the rate at which temperature increases with depth below the surface. A general average seems to be around 30°C increase per kilometer of depth.

THERMAL POLLUTION Increase in the normal temperatures of natural waters caused by the intervention of human activities.

THERMAL SPRING A spring that brings warm or hot water to the surface. Sometimes called *warm spring*, or *hot spring*. Temperature is usually 6.5°C or more above the mean air temperature.

THERMO REMANENT MAGNETISM Magnetism acquired by an igneous rock as it cools below the curie temperatures of magnetic minerals in the rock. Abbreviation, TRM.

THERMOSPHERE An atmospheric zone above the mesosphere in which temperature increases with increasing elevation.

THIN SECTION A slice of rock ground so thin as to be translucent.

THRUST FAULT A fault in which the hanging wall appears to have moved upward relative to the footwall. Also called *reverse fault*. Opposite of *gravity* or *normal fault*.

TIDAL CURRENT A water current generated by the tide-producing forces of the sun and the moon.

TIDAL INLET Waterway from open water into a lagoon.

TIDAL WAVE Popular but incorrect designation for *tsunami*.

TIDE Alternate rising and falling of the surface of the ocean, other bodies of water, or the earth itself, in response to forces resulting from motion of the earth, moon, and sun relative to each other.

TILL Unstratified and unsorted glacial drift deposited directly by glacier ice.

TILLITE Rock formed by the lithification of till.

TIME-DISTANCE GRAPH A graph of travel time against distance.

TOMBOLO A sand bar connecting an island to the mainland, or joining two islands.

TOPOGRAPHIC DESERTS Deserts deficient in rainfall either because they are located far from the oceans toward the center of continents, or because they are cut off from rain-bearing winds by high mountains.

TOPOGRAPHY The shape and physical features of the land.

TOPSET BED Layer of sediment constituting the surface of a delta. Usually nearly horizontal, and covers the edges of inclined foreset beds.

TOURMALINE A silicate mineral of boron and aluminum, with sodium, calcium, fluorine, iron, lithium, or magnesium. Formed at high temperatures and pressures through the agency of fluids carrying boron and fluorine. Particularly associated with pegmatites.

TOWNSHIP AND RANGE See Appendix F.

TRACTION The process of carrying material along the bottom of a stream. Traction includes movement by saltation, rolling, or sliding.

TRANSCURRENT FAULT A strike-slip fault.

TRANSDUCER A device that picks up relative motion between the mass of a seismograph and the ground and converts this into a form that can be recorded.

TRANSFORM FAULT A point at which strike-slip displacements stop and another structural feature, such as a ridge, develops.

TRANSITION ELEMENT An element in a series in which an inner shell is being filled with electrons after an outer shell has been started. All transition elements are metallic in the free state.

TRANSPIRATION The process by which water vapor escapes from a living plant and enters the atmosphere.

TRANSVERSE DUNE A dune formed in areas of scanty vegetation and in which sand has moved in a ridge at right angles to the wind. It exhibits the gentle windward slope and the steep leeward slope characteristic of other dunes.

TRANSVERSE WAVE Shear or shake wave.

TRAP ROCK A popular synonym for basalt.

TRAVEL TIME Total elapsed time for a wave to travel from its source to a designated point.

TRAVERTINE A form of calcium carbonate, $CaCO_3$, which forms stalactites, stalagmites, and other deposits in limestone caves, or as incrustations around the mouths of hot and cold calcareous springs. Sometimes known as *tufa*, or *dripstone*.

TRELLIS PATTERN A roughly rectilinear arrangement of stream courses in a pattern reminiscent of a garden trellis, developed in a region where rocks of differing resistance to erosion have been folded, beveled, and uplifted.

TRENCHES See *island arc deeps*.

TRM See *thermo remanent magnetism*.

TROPICAL DESERTS Deserts lying between 5° to 30° north and south of the equator.

TROPOSPHERE The lower 10 to 15 km of the atmosphere, in which the world's "weather" occurs. Temperature generally decreases upward.

TRUNCATED SPUR The beveled end of a divide between two tributary valleys where they join a main valley that has been glaciated. The glacier of the main valley has worn off the end of the divide.

TSUNAMI A large wave in the ocean generated at the time of an earthquake. Popularly, but incorrectly, known as a *tidal wave*. Sometimes called *seismic seawave*.

TUFA Calcium carbonate, $CaCO_3$, formed in stalactites, stalagmites, and other deposits in limestone caves, as incrustations around the mouths of hot and cold calcareous springs, or along streams carrying large amounts of calcium carbonate in solution. Sometimes known as *travertine*, or *dripstone*.

TUFF Rock consolidated from volcanic ash.

TUNDRA A stretch of arctic swampland developed on top of permanently frozen ground. Extensive tundra regions have developed in parts of North America, Europe, and Asia.

TURBIDITES Sedimentary deposits that have settled out of turbid water carrying particles of widely varying grade size. Characteristically display *graded bedding*.

TURBIDITY CURRENT A current in which a limited volume of turbid or muddy water moves relative to surrounding water because of its greater density.

TURBULENT FLOW Mechanism by which a fluid such as water moves over or past a rough surface. Fluid not in contact with the irregular boundary outruns that which is slowed by friction or deflected by the uneven surface. Fluid particles move in a series of eddies or whirls. Most stream flow is turbulent, and turbulent flow is important in both erosion and transportation. Contrast with *laminar flow*.

ULTIMATE BASE LEVEL Sea level, the lowest possible base level for a stream.

UNCONFORMITY A buried erosion surface separating two rock masses, the older of which was exposed to erosion for a long interval of time before deposition of the younger. If, in the process, the older rocks were deformed and were not horizontal at the time of subsequent deposition, the surface of separation is an *angular unconformity*. If the older rocks remained essentially horizontal during erosion, the surface separating them from the younger rocks is called a *disconformity*. An unconformity that develops between massive igneous rocks that are exposed to erosion and then covered by sedimentary rocks is called a *nonconformity*.

UNDERGROUND WATER Water below the surface of the ground. Also referred to as *subsurface water*, and *subterranean water*.

UNEVEN FRACTURE A mineral's habit of breaking along rough, irregular surfaces.

UNIFORMITARIANISM The concept that the present is the key to the past. This means that the processes now operating to modify the earth's surface have also operated in the geologic past, that there is a uniformity of processes past and present.

UNIT CELL In the crystalline structure of a mineral, a parallelepiped enclosing an atom or group of atoms arbitrarily selected so that the mineral's structure is represented by periodic repetition of this unit in a *space lattice*.

UNIVERSE The complete assemblage of everything that exists in space.

UNPAIRED TERRACE A terrace formed when an eroding stream, swinging back and forth across a valley, encounters resistant rock beneath the unconsolidated alluvium and is deflected, leaving behind a single terrace with no corresponding terrace on the other side of the stream.

VALLEY GLACIER A glacier confined to a stream valley. Usually fed from a cirque. Sometimes called *alpine glacier* or *mountain glacier*.

VALLEY TRAIN Gently sloping plain underlain by glacial outwash and confined by valley walls.

VAN ALLEN BELTS Zones in the earth's atmosphere above 1,000 km, where electrons and very energetic ionized nuclei of atoms, mostly hydrogen, are trapped by the earth's magnetic field. See *magnetosphere*.

VARVE A pair of thin sedimentary beds, one coarse and one fine. This couplet of beds has been interpreted as representing a cycle of one year, or an interval of thaw followed by an interval of freezing in lakes fringing a glacier.

VENTIFACT A pebble, cobble, or boulder that has had its shape or surface modified by wind-driven sand.

VERTICAL EXAGGERATION See Appendix F.

VESICLE A small cavity in an aphanitic or glassy igneous rock, formed by the expansion of a bubble of gas or steam during the solidification of the rock.

VIRTUAL GEOMAGNETIC POLE For any one locality, the pole consistent with the magnetic field as measured at that locality. The term refers to magnetic-field direction of a single point, in contrast to "geometric pole," which refers to the best fit of a geocentric dipole for the entire earth's field. Most paleomagnetic readings are expressed as virtual geomagnetic poles.

VISCOSITY An internal property of rocks that offers resistance to flow. The ratio of deforming force to rate at which changes in shape are produced.

VOLATILE COMPONENTS Materials in a magma, such as water, carbon dioxide, and certain acids, whose vapor pressures are high enough to cause them to become concentrated in any gaseous phase that forms.

VOLCANIC ASH A dust-sized pyroclastic particle: its volume is equal to or less than that of a sphere with a diameter of 0.06 mm.

VOLCANIC BLOCK An angular mass of newly congealed magma blown out in an eruption. Contrast with *volcanic bomb*.

VOLCANIC BOMB A rounded mass of newly congealed magma blown out in an eruption. Contrast with *volcanic block*.

VOLCANIC BRECCIA Rock formed from relatively large blocks of congealed lava embedded in a mass of ash.

VOLCANIC DUST Pyroclastic detritus consisting of particles of dust size.

VOLCANIC EARTHQUAKES Earthquakes caused by movements of magma or explosions of gases during volcanic activity.

VOLCANIC ERUPTION The explosive or quiet emission of lava, pyroclastics, or volcanic gases at the earth's surface, usually from a volcano but rarely from fissures.

VOLCANIC MOUNTAINS Mountains built up from the extrusion of lava and pyroclastic debris.

VOLCANIC NECK The solidified material filling a vent or pipe of a dead volcano.

VOLCANIC TREMOR Continuous shaking of the ground associated with a certain phase of a volcanic eruption.

VOLCANO A landform developed by the accumulation of magmatic products near a central vent.

VUG A small unfilled cavity in a rock, usually lined with a crystalline layer of different composition from the surrounding rock.

WADI See *arroyo*.

WARM GLACIER One which reaches the melting temperature throughout its thickness during the summer season.

WARM SPRING A spring that brings warm water to the surface. A *thermal spring*. Temperature is usually 6.5°C or more above mean air temperature.

WARP Large section of a continent composed of horizontal strata which have been gently bent upward or downward.

WASH See *arroyo*.

WATER GAP The gap cut through a resistant ridge by a superimposed or antecedent stream.

WATER TABLE The upper surface of the zone of saturation for underground water. It is an irregular surface with a slope or shape determined by the quantity of ground water and the permeability of the earth materials. In general, it is highest beneath hills and lowest beneath valleys.

WAVE A configuration of matter that transmits energy from one point to another.

WEATHERING The response of materials that were once in equilibrium within the earth's crust to new conditions at or near contact with water, air, or living matter.

WIND GAP The general term for an abandoned water gap.

WRINKLE RIDGES Ridges found on surfaces of lunar maria and flooded craters. May be caused by uplift due to volcanism or to compression.

YARDANG A sharp-edged ridge between two troughs or furrows excavated by wind action.

YARDANG TROUGH A trough excavated by wind action, between two yardangs.

YAZOO-TYPE RIVER A tributary that is unable to enter its main stream because of natural levees along the main stream. The Yazoo-type river flows along the back-swamp zone parallel to the main stream.

YIELD POINT The maximum stress that a solid can withstand without undergoing permanent deformation, either by plastic flow or by rupture.

ZONE OF AERATION A zone immediately below the surface of the ground, in which the openings are partially filled with air, and partially with water trapped by molecular attraction. Subdivided into (*a*) belt of soil moisture, (*b*) intermediate belt, and (*c*) capillary fringe.

ZONE OF SATURATION Underground region within which all openings are filled with water. The top of the zone of saturation is called the *water table.* The water contained within the zone of saturation is called *ground water.*

ZONES OF REGIONAL METAMORPHISM High grade, 700°C; middle grade, 400°C; low grade, 150°C.

Index

QE
501
.L35
1971
Leet, L. DOn
Physical geology
107152
Learning Resources
Brevard Community College
Cocoa, Florida

200,000,000 YEARS BENEATH THE SEA

Peter Briggs - HOLT, RINEHART & WINSTON 1971

THE RESTLESS EARTH - Nigel Calder
VIKING Press - 1972.

The near side of the moon from Apollo 11. (NASA.)

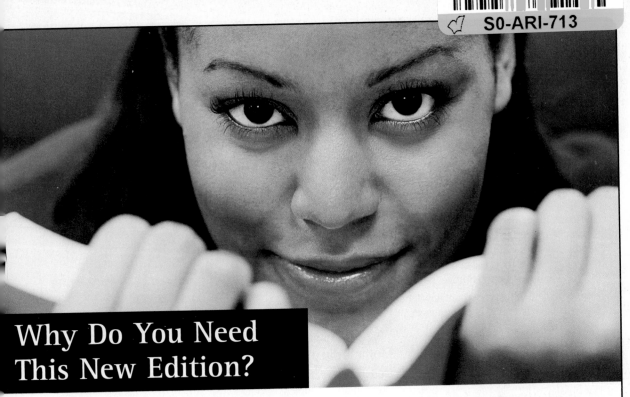

Why Do You Need This New Edition?

If you're wondering why you should buy this edition of *The Master Reader*, here are 6 good reasons!

1. **Review each chapter on-the-go with our new *Chapter Review Cards*.** Easy to tear out and take with you wherever you need to study—on the bus, at your job, in the library—*Chapter Review Cards* boil each chapter down to the fundamentals, making them the perfect tool to check your comprehension or prepare for an exam.

2. **Understand what you need to learn with *Learning Outcomes* at the beginning of each chapter.** New Learning Outcomes begin each chapter to help you focus on the key skills you need to learn to understand and apply to become a master reader.

3. **Write about what you read with new *What Do You Think?* writing prompts.** How should bullying be handled in schools? What is the solution to the obesity epidemic? In short, what do *you* think about today's issues? Throughout the book, new *What Do You Think?* questions challenge you to respond to today's issues as explored in our new, longer reading selections at the end of each chapter and in Part Two: Additional Readings.

4. **Explore new topics in our many new readings.** New topics include building good communication skills, developing good study skills, participating in your community, managing your finances, and many more! With over 25 percent of the readings and accompanying activities new to this edition, you will find topics that help you build new skills to master not only reading, but your life.

5. **Master finding the main idea and its supporting details.** The table of contents has been rearranged to teach these related topics in order—making it easier for you to learn these fundamental skills.

6. **Take your reading online with our new *Connect to MyReadingLab* features.** New *Connect to MyReadingLab* sections provide directions for accessing activities for additional practice on Pearson's MyReadingLab program.

Third Edition

The Master Reader

D. J. Henry
Daytona State College

Longman
Boston Columbus Indianapolis New York San Francisco Upper Saddle River
Amsterdam Cape Town Dubai London Madrid Milan Munich Paris Montreal Toronto
Delhi Mexico City Sao Paulo Sydney Hong Kong Seoul Singapore Taipei Tokyo

Editor in Chief: Eric Stano
Senior Acquisitions Editor: Kate Edwards
Editorial Assistant: Lindsey Allen
Associate Development Editor: Erin Reilly
Senior Supplements Editor: Donna Campion
Senior Media Producer: Stefanie Liebman
Marketing Manager: Tom DeMarco
Production Manager: Ellen MacElree
Project Coordination, Text Design, and Electronic Page Makeup: Nesbitt Graphics, Inc.
Cover Designer/Manager: Wendy Ann Fredericks
Cover Photo: Masterfile Corporation. All rights reserved.
Photo Researcher: Rona Tuccillo
Senior Manufacturing Buyer: Dennis J. Para
Printer and Binder: R.R.Donnelley Crawfordsville, IN
Cover Printer: Lehigh-Phoenix Color/Hagerstown

For permission to use copyrighted material, grateful acknowledgment is made to the copyright holders on pp. 715–719, which are hereby made part of this copyright page.

Library of Congress Cataloging-in-Publication Data

Henry, D. J. (Dorothy Jean)
The master reader / D.J. Henry. -- 3rd ed.
p. cm.
Includes bibliographical references and index.
ISBN 978-0-205-78086-0 (alk. paper)
1. Reading (Higher education) 2. Reading comprehension. 3. Critical thinking. I. Title.
LB2395.3.H49 2010
428.4071'1--dc22

2010043078

7 8 9 10—DOC—14 13

ISBN-10: 0-205-78086-5 (student ed.)

ISBN-13: 978-0-205-78086-0 (student ed.)

ISBN-10: 0-205-83519-8 (instructor's ed.)

ISBN-13: 978-0-205-83519-5 (instructor's ed.)

Longman
is an imprint of

www.pearsonhighered.com

Brief Contents

Detailed Contents

Preface

Dear Colleagues:

One of my personal heroes is Oprah Winfrey. She embodies so much of what is wonderful about the human spirit—with her tenacity, generosity, awe-inspiring work ethic, and wisdom, she has influenced our world for the good in profound ways. She is one of our premier teachers. I shall always be grateful for her devotion to reading and for her active work to inspire others to love the printed word. Her devotion to literacy underscores two ideals in our profession: reading empowers an individual life, and our work as instructors is of great and urgent importance. Many of our students come to us needing to reinforce the basic skills that make effective reading and clear thinking possible. Too often they struggle with text structure and feel uncertain about their comprehension. However, with solid instruction and guided practice, these students can discover the power and pleasure of reading. *The Master Reader,* Third Edition, has been designed to address these challenges.

New to This Edition

The following changes have been made to *The Master Reader,* Third Edition, to help students become master readers and critical thinkers.

- **New order for the Main Ideas and Supporting Details Coverage.** Based on extensive feedback from instructors across the country, the order of presentation has been rearranged to bring the main ideas and supporting details coverage together, closer to the beginning of the book. The table of contents now progresses from Main ideas to Implied Main Ideas, to Supporting Details—more in line with the way instructors nationwide present these topics. Of course, each chapter is still self-contained, allowing instructors to easily teach these important topics in any order they choose.
- **New Chapter Review Cards.** New chapter review cards will make studying more accessible and efficient by distilling chapter content down to the fundamentals, helping students to quickly master the basics, to review their understanding on the go, or to prepare for upcoming exams. Because

they are made of durable cardstock, students can keep these Review Cards for years to come and pull them out whenever they need a quick review.

- **New Learning Outcomes.** Each chapter now opens with learning outcomes keyed to Bloom's taxonomy. Learning outcomes help students to understand why they are learning the material and help them to set goals for their learning.
- **New "Connect to MyReadingLab" features.** New *Connect to MyReadingLab* sections offer specific activities tied to chapter content for use as lab activities or additional out-of-class practice. Never search for an appropriate online activity again; *Connect to MyReadingLab* provides easy-to-follow click paths and descriptions of specific MyReadingLab activities and how they fit with chapter content.
- **More attention to the connection between reading and writing.** Part Two now opens with a 6-Step Reading/Writing Action Plan that encourages students to see reading as a chance to enter a conversation with the writer and helps students to see how the reading process and the writing process work together to help them participate in this conversation.
- **New "What Do You Think?" writing prompts.** Found at the end of longer reading selections throughout the book, our new "What Do You Think?" writing prompts challenge students to respond to the issues explored.
- **New Reading Level Indications in the Annotated Instructor's Edition.** The reading level of all selections within our Review and Mastery Tests, Additional Readings, and Combined Skills Tests are now indicated in the Annotated Instructor's Edition (levels are not indicated in the student edition).
- **New Longer Reading Selections.** In addition to seven new longer readings in Part Two, over 25 percent of the reading selections and accompanying pedagogy throughout the text have been revised, giving students new reading material that is lively, up to date, and thought-provoking.
- **New Design.** In addition to appearing more modern and mature, the new design visually clarifies the text's different features to help students navigate and find the content they are looking for with greater ease.

Guiding Principles

The Master Reader, Third Edition, was written to develop in students the essential abilities that will enable them to become master readers and critical thinkers.

Practice and Feedback

Aristotle said, "What we have to learn to do, we learn by doing." We all know that the best way *to learn* is *to do*. Thus one of the primary aims of this text is to give students opportunity after opportunity to practice, practice, practice!

Every concept introduced in the book is accompanied by an **explanation** of the concept, an **example** with an explanation of the example, and one or more **practice** exercises. Each chapter also contains **brief skill applications**, **four review tests**, and **four mastery tests**. Furthermore, a durable, removable chapter review card is included for each chapter.

High-Interest Reading Selections

According to French poet, dramatist, and novelist, Victor Hugo, "To learn to read is to light a fire; every syllable that is spelled out is a spark." For developmental students we can fan the sparks by encouraging an enthusiasm for reading. For many, this enthusiasm can be stimulated by reading material that offers high-interest topics written in a fast-paced style. Every effort has been made to create reading passages in examples, reviews, and tests that students will find lively and engaging. Topics are taken from issues arising in popular culture and in current textbooks; some examples are gangs, movies, weight loss, sports figures, depression, interpersonal relationships, drug use, nutrition, inspirational and success stories, role models, stress management, and exercise—all written in active language using a variety of sentence patterns.

Integration of the Reading Process and Reading Skills

Master readers blend individual reading skills into a reading process such as SQ3R. Before reading, master readers skim for new or key vocabulary or main ideas. They create study questions and make connections to their prior knowledge. During reading, they check their comprehension. For example, they annotate the text. They notice thought patterns and the relationship between ideas. They read for the answers to the questions they created before reading. After reading, master readers use outlines, concept maps, and summaries to review what they have read and deepen their understanding. Students are taught to integrate each skill into a reading process in Part One.

In Chapter 1, "A Reading System for Master Readers," students are introduced to SQ3R. In every other Part One chapter students actively apply SQ3R reading strategies in "Before Reading About" and "After Reading About" activities. "Before Reading About" activities are pre-reading exercises that appear at the beginning of each chapter. These activities guide the student to review important concepts studied in earlier chapters, build on prior knowledge, and preview upcoming material. "After Reading About" activities are review activities that appear after the review tests in each chapter. These activities guide students to reflect on their achievements and assume responsibility for learning.

Comprehensive Approach

An ancient Chinese proverb states, "Skill comes from practice." *The Master Reader,* Third Edition, invites skill building by offering several levels of learning. First, students are given an abundance of practice. They are able to focus on individual reading skills through a chapter-by-chapter workbook approach. In each chapter of Part One, Review Tests 3 and 4 offer a multiparagraph passage with items on all the skills taught up to that point. In addition, Chapter 1, "A Reading System for Master Readers," teaches students how to apply their reading skills to the reading process before, during, and after reading by using SQ3R. Students also learn to apply all the skills in combination in Part Two, "Additional Readings," and Part Three, "Combined-Skills Tests."

Textbook Structure

To help students become master readers and critical thinkers, *The Master Reader,* Third Edition, introduces the most important basic reading skills in Part One and then provides sections of additional readings in Part Two, and combined-skills tests in Part Three.

Part One, Becoming a Master Reader

Essential reading skills are introduced sequentially in Part One. Each chapter focuses the student's attention on a particular skill.

- Chapter 1, "A Reading System for Master Readers," guides students through the reading process. Stages of the SQ3R process are explained thoroughly, with ample opportunities for practice, review, and mastery. The aim is to show students how to apply the skills they acquire in each of the chapters before, during, and after reading.
- Chapter 2, "Vocabulary Skills," fosters vocabulary acquisition during reading by using a mnemonic technique: SAGE stands for **S**ynonyms, **A**ntonyms, **G**eneral context, and **E**xample. The chapter also develops language skills by demonstrating how to determine word meanings from prefixes, roots, and suffixes.

- Chapter 3, "Stated Main Ideas," offers both verbal and visual strategies to enable students to see the building-block relationship among topics, main ideas, and supporting details and explains strategies to identify main ideas along with extensive practice in doing so. In addition, this chapter teaches students to identify the central idea of multiparagraph passages.
- Chapter 4, "Implied Main Ideas and Implied Central Ideas," furthers students' understanding about the central idea of longer passages and the main idea by explaining unstated main ideas and unstated central ideas. The chapter offers extensive practice.
- Chapter 5, "Supporting Details," identifies the differences between major and minor details.
- Chapter 6, "Outlines and Concept Maps," reinforces the skills of locating main ideas and identifying major and minor supporting details. The chapter teaches students the structure of a text by offering instruction and practice in the applications of outlines and concept maps.
- Chapter 7, "Transitions and Thought Patterns," introduces the fundamental thought patterns and the words that signal those patterns. Students are given numerous opportunities to practice identifying the signal words and their relationships to the thought patterns they establish. The chapter includes the time order, space order, listing, and classification patterns.
- Chapter 8, "More Thought Patterns," introduces more complex thought patterns and the words that signal those patterns. Just as in Chapter 7, students are given extensive practice opportunities. Chapter 8 introduces the comparison-and-contrast, cause-and-effect, generalization-and-example, and definition patterns.
- Chapter 9, "Fact and Opinion," explains the differences between fact and opinion and develops the higher-level thinking skills that enable students to separate fact from opinion through extensive practice.
- Chapter 10, "Tone and Purpose," continues the students' study of the importance of word choice and the author's purpose. Detailed instruction and extensive practice develop the students' ability to determine whether the author's purpose is to entertain, to inform, or to persuade.
- Chapter 11, "Inferences," carefully addresses the advanced skill of making inferences by dividing the necessary mental processes into units of activity. Students are taught the basic skills necessary to evaluate an author's purpose and choice of words.
- Chapter 12, "The Basics of Argument," teaches the fundamental logical thought process used to examine the author's claim and supports.

Students learn to recognize the author's claim and to evaluate supports as adequate and relevant.

- Chapter 13, "Advanced Argument: Persuasive Techniques," offers extensive explanations and practice of several common biased arguments that use logical fallacies and propaganda techniques. The logical fallacies include personal attack, straw man, begging the question, either-or, false comparison, and false cause. The propaganda techniques covered are name-calling, testimonials, bandwagon, plain folks, card stacking, transfer, and glittering generalities.

Part Two, Additional Readings

Part Two is a collection of ten reading selections followed by skills questions designed to give students real reading opportunities and the ability to gauge their growth. This part begins with a key discussion about the relationship between reading and writing and includes a few pointers on basic writing skills. The readings, which range from magazine articles to book excerpts, were chosen based on each selection's likelihood to engage, encourage, and motivate readers. Each selection is followed by skills questions so that students can practice again all the skills taught in Part One. The skills questions are followed by discussion and writing topics that encourage students to practice making connections among listening, speaking, reading, and writing.

Part Three, Combined-Skills Tests

Part Three is a set of ten reading passages and combined-skills tests. The purpose of this section is to offer the student more opportunities to apply reading skills and strategies comprehensively and to become more familiar with a standardized testing format. Increasing familiarity will help prepare them for the exit exams, standardized reading tests, and future course content quizzes, tests, and exams.

Chapter Features

Each chapter in Part One has several important features that help students become master readers.

Learning Outcomes: New to this edition, each chapter opens with learning outcomes to help students preview and assess their progress as they master chapter content.

"Before Reading About . . .": "Before Reading About . . ." activities appear at the beginning of Chapters 2–13 in Part One. These activities are pre-reading exercises based on SQ3R: they review important concepts studied in earlier chapters, build on prior knowledge, and preview the chapter. The purpose of "Before Reading About . . ." is to actively teach students to develop a reading process that applies individual reading skills as they study.

"After Reading About . . .": "After Reading About . . ." activities appear after Review Test 4 in Chapters 2–13 of Part One. Based on SQ3R, "After Reading About . . ." activities teach students to reflect on their achievements and assume responsibility for their own learning. These activities check students' comprehension of the skill taught in the chapter.

Instruction, example, explanation, and practice. The chapter skill is broken down into components, and each component is introduced and explained. Instruction is followed by an example, an explanation of the example, and a practice. Each of these components has its own instruction, example and explanation, and practice exercises.

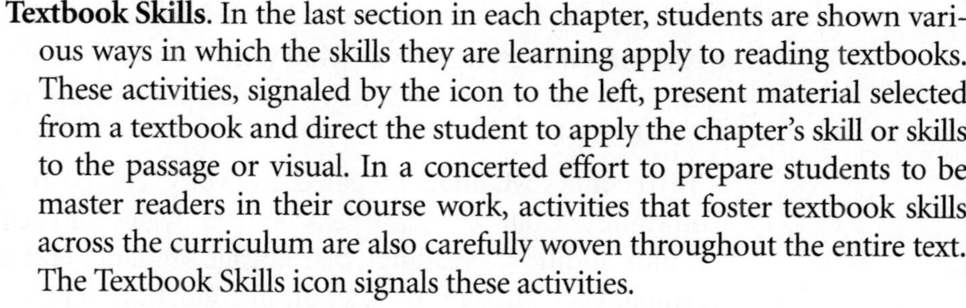
Textbook Skills. In the last section in each chapter, students are shown various ways in which the skills they are learning apply to reading textbooks. These activities, signaled by the icon to the left, present material selected from a textbook and direct the student to apply the chapter's skill or skills to the passage or visual. In a concerted effort to prepare students to be master readers in their course work, activities that foster textbook skills across the curriculum are also carefully woven throughout the entire text. The Textbook Skills icon signals these activities.

Visual Vocabulary. The influence of technology and the media on reading is evident in the widespread use of graphics in newspapers, magazines, and textbooks. Throughout *The Master Reader*, Third Edition, visual vocabulary is presented as part of the reading process, and students interact with these visuals by completing captions or answering skill-based questions. The aim is to teach students to value photos, graphs, illustrations, and maps as important sources of information.

Applications. Brief application exercises give students a chance to apply each component of the reading skill as a strategy.

Review Tests. Each chapter has four Review Tests. Review Tests 1 and 2 are designed to give ample opportunity for practice on the specific skill taught in the chapter; Review Tests 3 and 4 offer a multiparagraph passage with combined-skills questions based on all the skills taught up to and including that particular chapter. Review Tests 3 and 4 also give "What Do You Think?" writing prompts so that teachers have the opportunity to guide students as they develop critical thinking skills.

Mastery Tests. Each chapter includes four Mastery Tests. Most of the Mastery Tests are based on excerpts from science, history, psychology, social science, and literature textbooks.

Chapter Review Cards. New to this edition, a chapter review card is included for each chapter. The Chapter Review serves as a comprehension check for the reading concepts being taught.

The Longman Teaching and Learning Package

The Master Reader is supported by a series of innovative supplements. Ask your Pearson sales representative for a copy, or download the content at **www.pearsonhighered.com/irc**. Your sales representative will provide you with the username and password to access these materials.

The **Annotated Instructor's Edition** (AIE) is a replica of the student text, with all answers included. Ask your Longman sales representative for ISBN 0-205-83519-8.

The **Instructor's Manual,** prepared by Mary Dubbé of Thomas Nelson Community College, features teaching strategies for each textbook chapter, plus additional readings that engage students and encourage active learning. Each chapter includes an introduction, reproducible handouts, study-strategy cards, and a ten-item quiz. A supplemental section provides a sample syllabus, readability calculations for each reading in *The Master Reader*, Third Edition, nine book quizzes to encourage independent reading and the creation of book groups, sample THEA and Florida State Exit Exams, and a scaffolded book review form. ISBN 0-205-83518-X.

The **Lab Manual,** prepared by Mary Dubbé of Thomas Nelson Community College, is available as a separate student workbook and provides a collection of 65 activities that give additional practice, enrichment, and an assessment for the skills presented in *The Master Reader*, Third Edition. The activities for each chapter include practice exercises, one review test, and two mastery tests that mirror the design of *The Master Reader*, Third Edition. The lab manual is available packaged with *The Master Reader*, Third Edition, for an additional cost. ISBN 0-205-83520-1.

MyReadingLab is a website specifically created for developmental students that provides diagnostics, practice tests, and reporting on student reading skills and reading levels.

Acknowledgments

As I worked on the third edition of this reading series, I felt an overwhelming sense of gratitude and humility for the opportunity to serve the learning community as a textbook author. I would like to thank the entire Longman team for their dedication to providing the best possible materials to foster literacy. To every person, from the editorial team to the representatives in the field, all demonstrate a passion for students, teachers, and learning. It is a joy to be part of such a team. Special thanks are due to the following: Kate Edwards, Acquisitions Editor, and Erin Reilly, Development Editor, for their guidance and support; Kathy Smith with Nesbitt Graphics, Inc. for her tireless devotion to excellence; and Ellen MacElree and the entire production team for their work ethic and gracious attitudes. I would also like to thank Mary Dubbé for authoring the Lab Manual and the Instructor's Manual that supplement this reading series.

For nearly twenty-five years, I worked with the most amazing group of faculty from across the state of Florida as an item-writer, reviewer, or scorer of state-wide assessment exams for student learning and professional certification. The work that we accomplished together continues to inform me as a teacher, writer, and consultant. I owe a debt of gratitude to this group who sacrificed much for the good of our students.

I would also like to acknowledge two of my colleagues at Daytona State College: Dustin Weeks, Librarian, and Sandra Offiah-Hawkins, reading professor. As Tennyson extols in "Ulysses," these are the "souls that have toiled, and wrought, and thought with me." Their influence and support have made me a better person, teacher, and writer.

Finally, I would like to gratefully recognize the invaluable insights provided by the following colleagues and reviewers. I deeply appreciate their investment of time and energy: Tomekia Cooper, Albany Technical College; Jan Eveler, El Paso Community College; Annie Gonzalez, Laredo Community College; Yolanda Nieves, Wilbur Wright College; Margarita Sanchez, McLennan Community College; Michael Vensel, Miami Dade College; and Shari Waldrop, Navarro College.

D. J. Henry
Daytona Beach, Florida

PART ONE

Becoming a Master Reader

A Reading System for Master Readers

1

LEARNING OUTCOMES

After studying this chapter, you should be able to do the following:

1. Define prior knowledge.
2. Discuss the three phases of the reading process.
3. Illustrate SQ3R.
4. Describe your reading process.
5. Assess your comprehension of prior knowledge and the reading process.
6. Evaluate the importance of prior knowledge and SQ3R.
7. Activate prior knowledge and apply SQ3R to your reading process.
8. Develop textbook skills: Use SQ3R to master textbook reading.

Many people think that reading involves simply passing our eyes over the words in the order that they appear on the page. However, reading is an active process during which you draw information from the text to create meaning. When you understand what you've read, you've achieved **comprehension** of the material.

Comprehension is an understanding of what has been read.

Once we understand the **reading process**, we can follow specific steps and apply strategies that will make us master readers. The most important aspect of being a master reader is being an *active reader*.

Active reading means that you ask questions, find answers, and react to the author's ideas. For example, an active reader often marks, or annotates, key ideas by underlining the text or writing notes in the margin. (For more information about annotating a text, see page 594.) In addition, an active reader often checks comprehension by summarizing, or briefly restating the author's major points. (For more information about summarizing, see pages 595–597.) The activities in this chapter are designed to give you the skills you need to become an active reader.

Prior Knowledge

We all have learned a large body of information throughout a lifetime of experience. This body of information is called **prior knowledge**.

Knowledge is gained from experience and stored in memory. For example, a small child hears her father repeat the word *no* frequently as he takes away dangerous objects, removes her from unsafe situations, or wants her to stop certain behaviors. The child quickly learns and will remember the meaning of *no*.

> **Prior knowledge** is the large body of information that is learned throughout a lifetime of experience.

Prior Knowledge

Before Reading

Call up prior knowledge:

What do I already know?

During Reading

Check new information against prior knowledge:

Does this make sense based on what I know?

After Reading

Check for growth or change in knowledge base:

What did I learn?

Prior Knowledge Strategies in the Reading Process

Body of knowledge learned throughout a lifetime of experience

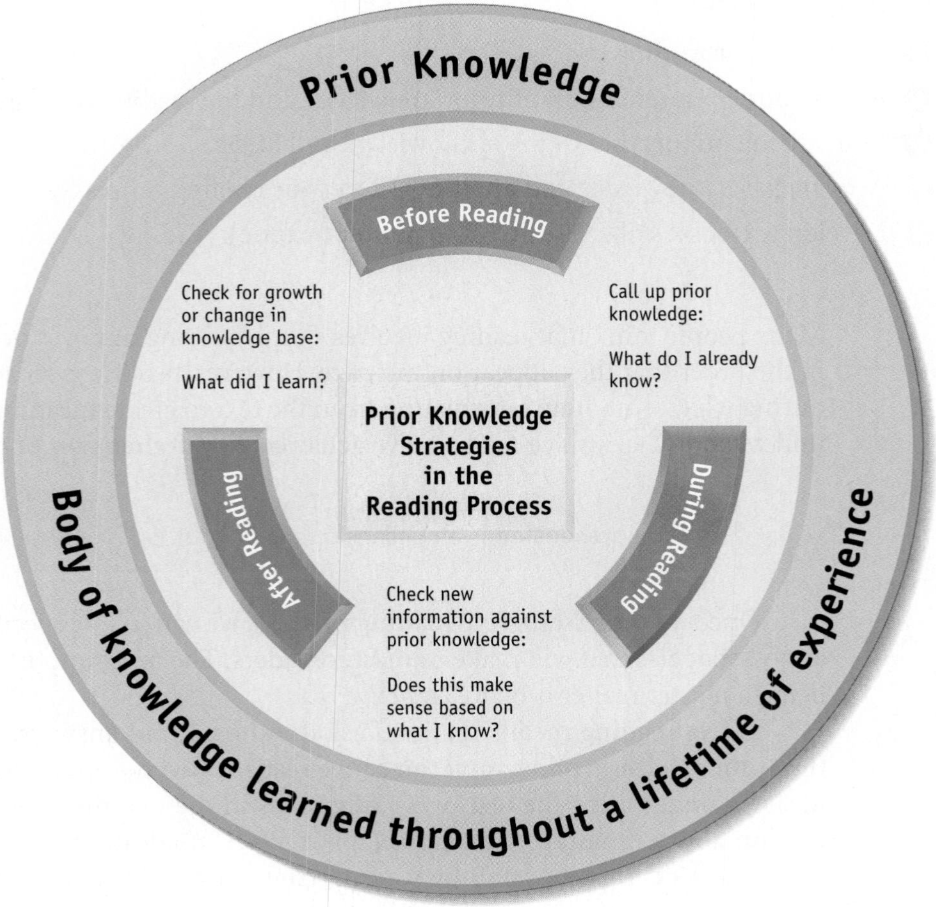

EXAMPLE Read the following passage taken from a college science textbook. In the space provided, list any topics from the passage about which you already have prior knowledge.

The Lunar Surface

[1]When Galileo first pointed his telescope toward the Moon, he saw two different types of terrains—dark lowlands and brighter, highly cratered highlands. [2]Because the dark regions resembled seas on Earth, they were called maria (*mar* = sea, singular, mare). [3]Today, we know that the maria are not oceans but, instead, are flat plains that resulted from immense outpourings of fluid basaltic lavas. [4]By contrast, the light-colored areas resemble Earth's continents, so the first observers dubbed them terrae (Latin for "land"). [5]These areas are now generally referred to as lunar highlands, because they are elevated several kilometers above the maria. [6]Together, the arrangement of terrae and maria result in the well-known "face of the moon."

—From *Foundations of Earth Science,* 5th Edition, p. 409 by Frederick K. Lutgens, Edward J. Tarbuck, and Dennis Tasa. Copyright © 2008 by Pearson Education, Inc. Printed and Electronically reproduced by permission of Pearson Education, Inc., Upper Saddle River, NJ.

____

____

____

EXPLANATION If you know something about the terrain of the moon, basaltic lava, or the actual length of a kilometer, then this passage makes more sense to you than it does to someone who does not possess such prior knowledge. However, even if you do not know much about those topics, you may have helpful prior knowledge about some of the other ideas in the passage. For example, most of us have seen and perhaps wondered about the "face of the moon." Our prior knowledge that comes from observing the moon with our own eyes connects with new information in the passage.

The more prior knowledge we have about a topic, the more likely we are to understand that topic. This is why master readers build their knowledge base by reading often!

PRACTICE 1

Read the passage from a college health textbook. Then, answer the questions that follow it.

From Everyday Problems to Emotionally Crippling Behavior

[1]Neuroses are the most common type of mental problem. [2]Although neuroses can cause emotional suffering, neurotic individuals usually can carry out day-to-day activities. [3]Anxiety and phobia are examples of

neuroses; they are generally viewed as cognitive distortions or unsatisfactory ways of reacting to life situations.

[4]Anxiety is often brought on by an imagined fear of impending danger. [5]Americans experience anxiety more than any other mental health problem. [6]Most anxiety is normal and may play an important role in anticipating situations. [7]For example, the fear of being attacked on a deserted street at night may cause a person to take appropriate actions to avoid such a possibility. [8]You have probably experienced anxiety before taking a final exam. [9]Some of the symptoms you may have had are sweating, dry mouth, heavy breathing, and insomnia. [10]This type of anxiety is in response to a realistic situation.

[11]However, some people suffer anxiety over a long period of time and without any apparent cause. [12]A person who has a phobia, for example, has an unreasonable fear of some object or situation. [13]Simple phobias include fear of heights or fear of bees and usually do not interfere with daily activities. [14]Some phobias, however, are more severe and can cause people to lead constricted lives. [15]Because of their irrational fears, some people do not leave the house. [16]Anxiety becomes a serious mental health problem when individuals suffering from it are so emotionally crippled that they cannot continue at school, hold a job, or otherwise lead a satisfying and productive life.

—From *Healthstyles: Decisions for Living* Well, 2nd ed., pp. 58–59 by B. E. Pruitt and Jane J. Stein.

1. What did you already know about neuroses such as anxiety and phobias? That is, what was your prior knowledge? ______________________________

____

2. When you think of anxiety and phobias, what do you think of? Describe ideas and experiences that come to mind. ______________________________

3. Was this an easy passage to understand? How does your prior knowledge affect your understanding of this passage? ______________________________

4. List any parts of the passage you had no prior knowledge of: ____________

____

The Reading Process

Triggering prior knowledge is a reading skill that you as an active reader can turn into a reading strategy by using it in your reading process. In each of the remaining chapters in Part One, you will be encouraged to begin thinking about how you can connect the reading skills that you are learning to the reading process. Master readers use reading skills as comprehension strategies throughout the reading process.

Master readers break reading into a three-step process. Each step uses its own thinking activities.

1. Before reading, look over, or *preview*, the material. (Previewing brings up prior knowledge.) Ask questions about the material you are about to read.
2. During reading, *test* your understanding of the material.
3. After reading, *review and react* to what you have learned.

One well-known way to apply this reading process is called SQ3R.

SQ3R stands for Survey, Question, Read, Recite, and Review.

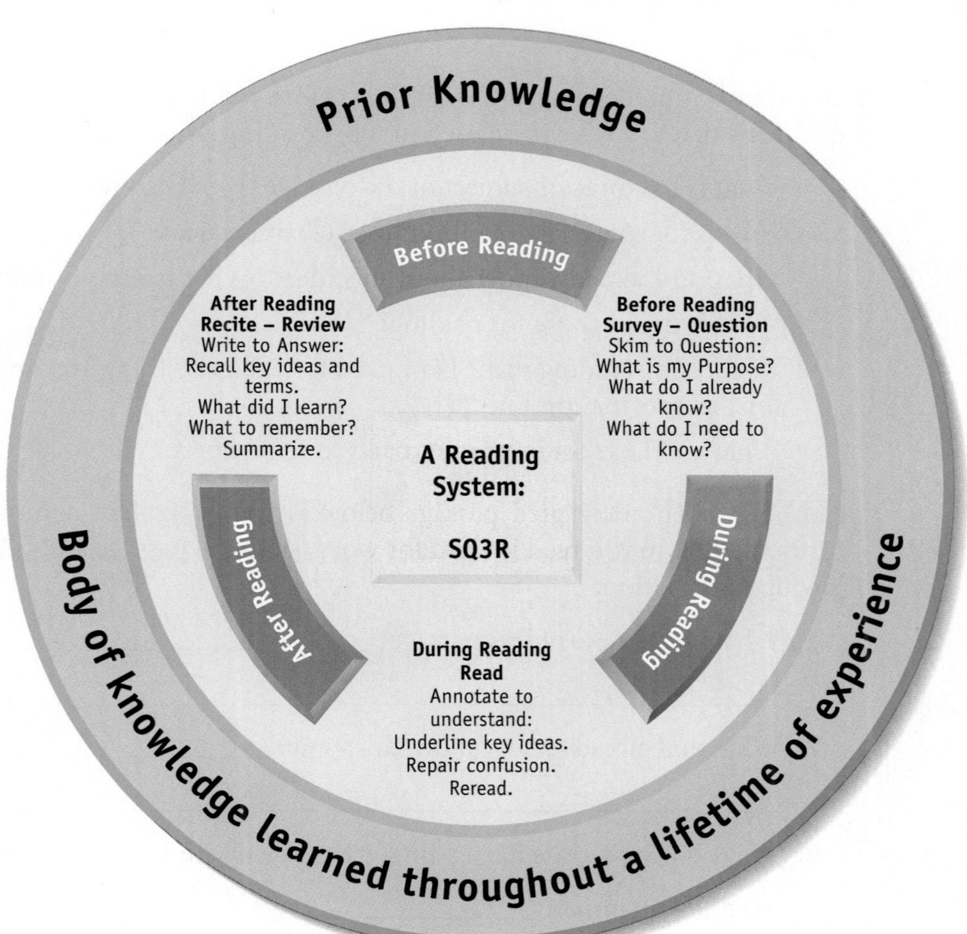

SQ3R activates prior knowledge and offers strategies for each phase of the reading process. The following graphic illustrates the phases of the reading process through SQ3R. Master readers repeat or move among phases as needed to repair comprehension.

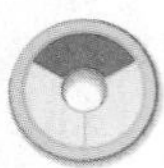

Before Reading: Survey and Question

A. Survey

Quickly look over, or **skim**, the reading passage for clues about how it is organized and what it is going to present or teach you.

To skim effectively, look at *italic* and **bold** type and take note of titles, the introduction, and headings. Also look at pictures and graphs. Finally, read the first paragraph, summaries, and questions. Each of these clues provides important information.

B. Question

To aid in comprehension, ask questions before you read. The list of prereading questions that follows can be used in most reading situations.

- What is the topic or subject of the passage? (See Chapter 3.)
- How is the material organized? (See Chapters 6 and 7.)
- What do I already know about this idea? (What is my prior knowledge?)
- What is my purpose for reading?
- What is my reading plan? Do I need to read everything, or can I just skim for the information I need?
- What are the most important parts to remember?

EXAMPLE The excerpted passage below is from a college communications textbook. Before you read it word for word, skim the passage and fill in the following information:

1. What is the topic of this passage? ______________________________

____

2. What do I already know about this topic? ______________________________

____

3. What is my purpose for reading? That is, why am I reading this? What do I need to remember? ______________________________

____

4. What ideas in the passage are in *italic* and/or **bold** type? ______________

____

____

____

Textbook Skills

Supply in the Marketplace of Ideas

[1]The *marketplace of ideas* assumes that all ideas should be discussed in the open. [2]The more ideas, the better. [3]Only when ideas (even those thought radical) gain access to the *public forum* can social change occur. [4]Women fought for almost a century before gaining the right to vote through a constitutional amendment adopted in 1920. [5]Their ideas were considered radical by many men—and women.

[6]The marketplace of ideas involves social discourse, and *social discourse* can lead to *policy formation*. [7]For example, the state of Oregon adopted strict recycling standards for its population by the early 1970s, ten to twenty years before most other states did the same. [8]Oregon's marketplace of ideas considered the recycling issue important, and the discussion in that marketplace led to policy changes that other states eventually adopted.

[9]The *supply* for the marketplace of ideas comes from *all sources of information*, including individuals, newspapers, movies, television, recordings, and radio. [10]Consumers, journalists, lobbyists, public-relations personnel, and government officials determine, both as individuals and as groups, what content has use in the marketplace of ideas.

—Folkerts & Lacy, *The Media in Your Life*, 2nd ed., pp. 43–44.

EXPLANATION

1. The title of the passage, "Supply in the Marketplace of Ideas," gives us a clue to the topic of the passage. By quickly looking at the terms in *italic* and **bold** type, you can see that this passage is about the marketplace of ideas.

2. The answer to this question on prior knowledge will vary for each of you. Some of you may be excellent shoppers and thus already have some experience and understanding about the marketplace and the complexity of supply and demand. If so, then you have prior knowledge that can be transferred to the "marketplace of ideas."

3. Wording of your purpose for reading will vary. A sample answer: I need to know what the "marketplace of ideas" means and about its source or supply.
4. The words in *italic* and **bold** type are *marketplace of ideas, public forum, social discourse, policy formation, supply, all sources of information,* and the title.

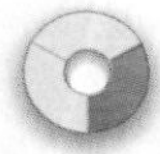

During Reading: Read and Annotate

After you have surveyed and asked questions about the text, it's time to read the entire passage. Use the following helpful practices while reading.

Read

As you read, think about the importance of the information by continuing to ask questions. For example:

- Does this new information agree with what I already knew?
- Do I need to change my mind about what I thought I knew?
- What is the significance of this information? Do I need to remember this?

In addition to asking questions while you read, acknowledge and resolve any confusion as it occurs.

- Create questions based on the headings, subheadings, and words in *italics* and **bold** type.
- Reread the parts you don't understand.
- Reread whenever your mind drifts during reading.
- Read ahead to see if an idea becomes clearer.
- Determine the meaning of words from their context.
- Look up new or difficult words.
- Think about ideas even when they differ from your own.

Annotate

Make the material your own. Make sure you understand it by repeating the information.

- Create a picture in your mind or on paper.
- Restate the ideas in your own words.
- Mark your text by underlining, circling, or highlighting topics, key terms, and main ideas. (See pages 595–597 for more about how to annotate.)

- Write out answers to the questions you created based on the headings, subheadings, and highlighted words.
- Write a summary of the passage or section.

EXAMPLE

A. Before you read the following excerpt from a college science textbook, survey the passage and answer the following questions.

1. What is the topic of the passage? ______________________________

2. What do I already know about the topic of this passage? What is my prior knowledge? ______________________________

3. What is important about this passage? What do I need to remember?

____

____

Textbook Skills

4. What are the words in **bold** type (which will help me remember what I need to know)? ______________________________

B. Once you have surveyed the information, read the passage. During reading, check your understanding by writing answers to the questions based on the ideas in **bold** type.

5. What new or difficult words do I need to look up?

____

____

6. What are circadian rhythms?

____

____

7. What is an example of circadian rhythms?

____

____

____

Biological Rhythms

[1]Many different types of animals show **circadian rhythms**—patterns of activity that approximate the length of a day. [2]For example, fruit flies (*Dacus tryoni*) always mate at about the same time in the evening; they continue to show this cyclical pattern of mating even when external cues indicating the time of day are not available. [3]This pattern enables the flies to mate under cover of darkness, when the danger of predation is reduced.

[4]Another common cycle in the animal world is a **circannual rhythm**, a cycle lasting approximately one year. [5]A circannual pattern makes it possible for an animal to engage in a behavior during the appropriate season. [6]For example, the golden-mantled ground squirrel (*Citellus lateralis*) enters into hibernation

8. What is a circannual rhythm?

____

____

____

____

9. What is an example of a circannual rhythm?

____

____

____

10. What do you need to reread to understand?

____

____

approximately the same time every year, during the late fall. [7]Like the fruit flies, the ground squirrel maintains its rhythm when kept in a laboratory, isolated from external factors that could indicate the season. [8]For example, ground squirrels kept in rooms with constant temperature and unchanging light-dark cycles showed a consistent hibernation schedule over a 4-year period.

—Maier, *Comparative Animal Behavior,* p. 431.

EXPLANATION Compare your answers to the ones below. Keep in mind that your wording and examples may be different.

A. Before Reading: Survey and Question

1. The topic of the passage is biological rhythms.
2. Answers to this question will vary.
3. Answers to this question will vary. A possible answer is: The two types of biological rhythms and examples of each type.
4. The terms in **bold** type are: circadian rhythms and circannual rhythm.

B. During Reading: Read and Annotate

5. Answers to this question will vary.
6. Circadian rhythms are patterns of activity that approximate the length of a day.
7. Fruit flies (*Dacus tryoni*) always mate about the same time in the evening.
8. A circannual rhythm is a cycle of behavior lasting approximately one year that allows an animal to engage in a behavior during the appropriate season.

9. The golden-mantled ground squirrel hibernates every year in the late fall.
10. Answers to this question will vary.

After Reading: Recite and Review

Recite

As part of your review, take time to think and write about what you have read.

- Connect new information to your prior knowledge about the topic.
- Form opinions about the material and the author.
- Notice changes in your opinions based on the new information.
- Write about what you have read.

Review

Once you have read the entire selection, go back over the material to review it.

- Summarize the most important parts (for more information about how to summarize, see pages 595–597).
- Return to and answer the questions raised by headings and subheadings.
- Review new words and their meanings based on the way they were used in the passage.

PRACTICE 2

Now that you have learned about each of the three phases of the reading process, practice putting all three together. Think before, during, and after reading. Apply SQ3R to the following passage. Remember the steps:

- **Survey**: Look over the whole passage.
- **Question**: Ask questions about the content. Predict how the new information fits together with or differs from what you already know about the topic.
- **Read**: Continue to question, look up new words, reread, and create pictures in your mind.
- **Recite**: Take notes; write out questions and answers, definitions of words, and new information.
- **Review**: Think about what you have read and written. Use journals to capture your opinions and feelings about what you have read.

A. **Before Reading: Survey and Question.** Skim the passage taken from a college history textbook, and then answer the following questions.

1. What is the topic of this passage? ______________________________

____

____

____

2. What do I already know about this information? ______________

____

____

____

3. What do I need to remember? ______________________________

____

____

____

4. What ideas are in **bold** type? ______________________________

____

____

____

5–6. *Before you go on:* Use the terms you listed in item 2 to create questions. Write the questions in the blank boxes labeled 5 and 6. Write your answers to these questions in the boxes during reading.

B. During Reading: Read and Recite. As you read, answer the questions you created from the ideas in **bold** type.

5. ______________________________

____

____

____

____

____

____

Separation of Powers and Checks and Balances

[1]The Founders believed that unlimited power was corrupting and that the concentration of power was dangerous. [2]James Madison wrote, "Ambition must be made to counteract ambition." [3]Articles I, II, and III of the Constitution created three separate branches within the national government: legislative, executive, and judicial. [4]This **separation of powers**

6. ______________________

____

____

____

____

was designed to place internal controls on governmental power. [5]Power is not only apportioned among three branches of government, but, perhaps, more important, each branch is given important **checks and balances** over the actions of others. [6]According to Madison, "The constant aim is to divide and arrange the several offices in such a manner as that each may be a check on the other." [7]No bill can become a law without the approval of both the House and the Senate. [8]The president shares legislative power through the power to sign or to veto laws of Congress, although Congress may override a presidential veto with a two-thirds vote in each house. [9]The president may also suggest legislation, "give to the Congress Information of the State of the Union, and recommend to their Consideration such Measures as he shall judge necessary and expedient." [10]The president may also convene special sessions of Congress.

[11]However, the president's power of appointment is shared by the Senate, which confirms cabinet and ambassadorial appointments. [12]The president must also secure the advice and consent of the Senate for any treaty. [13]The president must execute the laws, but it is Congress that provides the money to do so. [14]The president and the rest of the executive branch may not spend money that has not been appropriated by Congress. [15]Congress must also authorize the creation of executive departments and agencies. [16]Finally, Congress may impeach and remove the president from office for "Treason, Bribery, or other High Crimes and Misdemeanors."

—Adapted from *Politics in America,* 7th edition, pp. 72–73 by Thomas R. Dye. Copyright © 2007 by Pearson Education, Inc. Reprinted by permission of Pearson Education, Inc., Glenview, IL.

C. After Reading: Review

7. Why is the separation of power important? ______________________

____

____

8. Identify a time when the system of checks and balances has been noticeable.

____

____

____

Reading a Textbook: Using SQ3R to Master Textbook Reading

Textbook Skills

You can use the SQ3R reading system to increase your comprehension and retention of information from all types of reading materials, from newspapers to novels. As a college student, you will find SQ3R to be a particularly helpful study method when trying to master information in your textbooks. Most textbooks are designed with special features to help readers understand and learn the vast amount of information within them. Master readers actively use these features as they preview, read, and review textbook material. The following chart lists and explains several features common to most textbooks.

Textbook Features	
BOOK FEATURES	
Table of Contents	The table of contents, located in the front of the book, is a list of the chapters in the order in which they appear, along with their corresponding page numbers.
Index	The index, located in the back of the book, is an alphabetical list of all the specific topics discussed in the textbook, along with the precise page numbers that deal with each topic.
Glossary	A glossary is an alphabetical list of specialized words and their meanings. Glossaries can be located in a number of places in a textbook, including at the back of the book, in the chapter's preview material, in the margins of the pages where the words are first used, or at the end of the chapter as review material.
Preface	The preface is a type of introduction that discusses the textbook's overall purpose, format, and special features.
Appendices	Appendices are sections of additional material designed to supplement or support your learning; they appear at the back of many textbooks.
CHAPTER FEATURES	
Introductions and Previews	Each chapter usually begins with a brief overview of the chapter's contents. Sometimes, this section also includes key questions or objectives to keep in mind as you read.
Headings and Subheadings	Textbook authors divide complex information into smaller sections to organize the ideas. Each section is labeled with a heading. Some sections are divided into subgroups and are given subheadings.
Information Boxes	Information boxes are used to highlight important information; they frequently contain key concepts, definitions, activities, real-life connections, or summaries.

Summaries	A summary is a brief section at the end of a chapter that restates the chapter's main idea and major supporting details.
Review Questions	At the end of a section, the author may supply a set of questions designed to test your comprehension.
Typographical Features	Authors frequently use **bold** and *italic* type to draw attention to important ideas and terms.
Graphics	Graphs, tables, diagrams, maps, photographs, and other graphic aids support the information explained in the paragraphs. Graphics make the information more visually interesting and accessible to visual learners.

Master readers actively use textbook features throughout the process of reading an assignment. For example, a master reader sets up a reading session or a series of sessions to complete a reading assignment. Each reading session is broken into the three reading phases. The time spent in each phase of the process varies based on the reading assignment.

Before Reading: A master reader may begin a reading session with five to ten minutes of before-reading activities such as skimming, noting new or special terms, creating questions, and setting a purpose for the reading session.

During Reading: Then, the master reader reads for a set amount of time, such as thirty minutes. The duration of a reading session depends on many factors such as length or difficulty of the material, time-management issues, and reader interest. For example, the difficulty of the text may require several shorter sessions of reading smaller chunks of the assignment. Master readers avoid marathon sessions that lead to fatigue and loss of concentration. During reading, the master reader stays focused by highlighting the text, taking notes, and repairing any confusion when it occurs.

After Reading: The master reader spends five to fifteen minutes completing notes or writing a summary with information that fulfills the purpose for reading (which was established before reading).

EXAMPLE The following study plan outlines how one student plans to apply SQ3R to the process of reading textbooks. Test your understanding of the steps you can take to use SQ3R to understand textbook material. Fill in the blanks with information that best completes each idea. Answer the question that follows the chart.

Before Reading [5–10 minutes]— (1) ________ and (2) ________

- Preview textbook: Skim preface, table of contents, and locate index.
- Preview a chapter: Read chapter **(3)** ________. Preview material, including information boxes, graphics, and end-of-chapter questions.

- Turn chapter **(4)** ________, **(5)** __________, and terms in **bold** or *italic* type into questions.
- Preview **(6)** ________________ questions.

During Reading [around 30 minutes]— (7) _______

- Annotate to understand; underline **(8)** __________.
- Note definitions/examples of terms in **(9)** ________ or **(10)** ________ type.
- Repair **(11)** ________; check glossaries and indexes for more information.
- Reread or read ahead for clarification. Study graphics to understand key concepts.

After Reading [5–15 minutes]—(12) ________ **and (13)** ________

- Use notes to test recall. Use two-column notes. Consider the following headings: Topics and Details, Terms and Definitions, or Questions and Answers. Create flashcards for important terms.
- Answer **(14)** ____________ at the end of the chapter.
- Write a **(15)** __________ that paraphrases the chapter.

How can SQ3R help me read a textbook? ________________________

____

____

EXPLANATION Compare your answers to the ones given here.

Before Reading (1) *Survey* and (2) *Question*. Several ways to preview the textbook include the following: read the chapter (3) *introduction*, turn chapter (4) *headings* and (5) *subheadings* into questions, and preview (6) *end-of-chapter review* questions.

During Reading (7) *Read*; underline (8) *key ideas* and note the definitions and examples of terms in (9) *bold* or (10) *italic* type. Also repair (11) *confusion*.

After Reading (12) *Review* and (13) *Recite*. Answer (14) *review questions* and write a (15) *summary* using your own words.

How can SQ3R help me read a textbook? Although answers will vary, in general, SQ3R breaks the reading assignment into three phases of previewing, reading, and reviewing information. By asking questions, recording ideas, and reciting information, I am more likely to understand and remember information.

PRACTICE 3

The following selection is a section of a chapter from a geography textbook. Use SQ3R to comprehend the passage. Skim the passage and answer the "Before Reading" questions. During reading, underline key concepts and definitions. After reading, complete the three-column notes with information from the passage.

Before Reading—Survey and Question:

1. What do I need to know? ______________________________

____

____

2. What is my purpose? ______________________________

____

Rock Formation

[1]Although by human standards Earth's surface moves very slowly—by at most a few centimeters per year—this movement produces Earth's great diversity of rocks. [2]As Earth's crust moves, its materials are eroded and deposited, heated and cooled, buried and exposed.

Types of rocks [3]Rocks can be grouped into three basic categories that reflect how they form:

[4]**Igneous rocks** are formed when molten crustal material cools and solidifies. [5]The name derives from the Greek word for fire, which is the same root as for the English word ignite. [6]Examples of igneous rocks are basalt, which is common in volcanic areas, including much of the ocean floor, and granite, which is common in continental areas.

[7]**Sedimentary rocks** result when rocks eroded from higher elevations (mountains, hills, plains) accumulate at lower elevations (like swamps and ocean bottoms). [8]When subjected to high pressure and the presence of cementing materials to bind their grains together, rocks like sandstone, shale, conglomerate, and limestone are formed.

[9]**Metamorphic rocks** are created when rocks are exposed to great pressure and heat, altering them into more compact, crystalline rocks. [10]In Greek the name means "to change form." [11]Examples include marble (which metamorphosed from limestone) and slate (which metamorphosed from shale).

Minerals [12]Minerals are natural substances that comprise rocks. [13]Each type of mineral has specific chemical and crystalline properties. [14]Earth's rocks are diverse in part because the crust contains thousands of minerals. [15]The density of rocks depends on the kinds of materials they contain. [16]Denser rocks are dominated by compounds of silicon, magnesium, and iron minerals; they are called **sima** (for *si*licon-*ma*gnesium). [17]Less dense rocks are dominated by compounds of silicon and aluminum minerals; they are called **sial** (for *si*licon-*al*uminum).

[18]Denser sima rocks make up much of the oceanic crust. [19]Less dense sial rocks make up much of the continental crust. [20]The lower density and greater thickness of sial rocks cause the continents to have higher surface elevations than the oceanic crust, just as a less dense dry log will float higher in water than a denser wet one.

[21]The formation and distribution of many minerals is caused by the movements of Earth's crust. [22]Vast areas of the continental crust, known as **shields**, have not been significantly eroded or changed for millions of years.

[23]Shield areas often contain rich concentrations of minerals, such as metal ores and fossil fuels. [24]Shields are located in the core of large continents such as Africa, Asia, and North America. [25]Many of the world's mining districts exist where these continental shields are exposed at the surface.

—From *Introduction to Geography: People, Places, and Environment,* 4th ed., pp. 101–103 by Edward Bergman and William H. Renwick. Copyright © 2008 by Pearson Education, Inc. Printed and Electronically reproduced by permission of Pearson Education, Inc., Upper Saddle River, NJ.

After Reading: Review and Recite

Complete the following notes with information from the passage. Then, answer the summary question.

Chapter 3 Rock Formations **Textbook Notes pages 102–103**

Term	Definition	Examples
3. ____________	are formed when molten crustal material cools and solidifies	basalt, granite
4. ____________ ____________	**5.** ______________________ ______ higher elevations and collect at lower elevations	sandstone, shale conglomerate, limestone

6. ______________ ______________	are formed when rocks are exposed to great pressure and heat, altering them into more compact, crystalline rocks	**7.** ______________
Minerals	are natural elements that comprise rocks	silicon, magnesium, aluminum, metal ores, fossil fuels
sima	**8.** are denser rocks dominated by compounds of ______, __________ and iron minerals	make up much of oceanic crust
9. ____	are less dense rocks dominated by compounds of silicon and aluminum minerals	make up much of continental crust
shields	vast areas of continental crust, contain rich concentrations of minerals	core of large continents such as Africa, Asia, North America

Summary: What role do minerals play in rock formation and what are the three types of rocks?

10. ____

____

____

____

VISUAL VOCABULARY

Granite, ______________ rock, has become a popular building material for kitchen counter tops.

a. an igneous
b. a sedimentary
c. a metamorphic

APPLICATIONS

Application 1: Before, During, and After Reading

This activity is designed to give you an opportunity to apply the entire reading process to a short selection from a college health textbook. Survey the passage and answer the Before Reading question. Then, read the passage and answer the questions.

Textbook Skills

Types of Weight-Training Programs

[1]Weight-training programs can be divided into three general categories classified by the type of muscle contraction involved: isotonic, isometric, and isokinetic.

Isotonic Programs [2]Isotonic programs, like isotonic contractions, involve the concept of contracting a muscle against a moveable load (usually a free weight or weights mounted by cables or chains to form a weight machine). [3]Isotonic programs are very popular and are the most common type of weight-training program in use today.

Isometric Programs [4]An isometric strength-training program is based on the concept of contracting a muscle at a fixed angle against an immovable object, using an isometric or static contraction. [5]Interest in strength training increased dramatically during the 1950s with the finding that maximal strength could be increased by contracting a muscle for 6 seconds at two-thirds of maximal tension once per day for 5 days per week! [6]Although subsequent studies suggested that these claims were exaggerated, it is generally agreed that isometric training can increase muscular strength and endurance.

[7]Two important aspects of isometric training make it different from isotonic training. [8]First, in isometric training, the development of strength and endurance is specific to the joint angle at which the muscle group is trained. [9]Therefore, if isometric techniques are used, isometric contractions at several different joint angles are needed to gain strength and endurance throughout a full range of motion. [10]In contrast, because isotonic contractions generally involve the full range of joint motion, strength is developed over the full movement pattern. [11]Second, the static nature of isometric muscle contractions can lead to breath holding (called a **valsalva maneuver**), which can reduce blood flow to the brain and cause dizziness and fainting. [12]In an individual at high risk for coronary disease, the maneuver could be extremely dangerous and should always be avoided. [13]Remember: Continue to breathe during any type of isometric or isotonic contraction!

Isokinetic Programs [14]Isokinetic contractions are isotonic contractions performed at a constant speed (*isokinetic* refers to constant speed of movement). [15]Isokinetic training is a relatively new strength training method, so limited research exists to describe its strength benefits compared with those of isometric and isotonic programs. [16]Isokinetic exercises require the use of machines that govern the speed of movement during muscle contraction. [17]The first isokinetic machines available were very expensive and were used primarily in clinical settings for injury rehabilitation. [18]Recently, less expensive machines use a piston device (much like a shock absorber on a car) to limit the speed of movement throughout the range of the exercise. [19]Today, these machines are found in fitness centers across the United States.

—From *Total Fitness and Wellness,* 4th ed., pp. 131–133 by Scott K. Powers, Stephen L. Dodd and Virginia J. Noland. Copyright © 2006 by Pearson Education, Inc. Printed and Electronically reproduced by permission of Pearson Education, Inc., Upper Saddle River, NJ.

Before Reading

1. What is your prior knowledge about strength training? ______________

____

During and After Reading

2. What does an isotonic program involve? ______________________

____

3. What does an isometric program involve? ______________________

____

____

4. What is a *valsalva maneuver* and what are its effects? ______________

____

____

5. What does an isokinetic program involve? ______________________

____

After Reading

6. Write a one- or two-sentence summary. ______________________

____

____

____

____

____

7. Write a paragraph that states your personal views about what you have read. ____

____

____

VISUAL VOCABULARY

Jonathan Horton of the United States performs an ______________ muscle contraction as he competes in the rings event during the men's team gymnastics final at the 2008 Olympics in Beijing China.

a. isotonic
b. isometric
c. isokinetic

Textbook Skills

Application 2: Previewing a Textbook

Preview a textbook by completing the following textbook and chapter survey forms. Consider previewing *The Master Reader* or a textbook from another course you are taking.

Textbook Survey Form

Academic subject: ______________________________

Textbook: ______________________________

Author(s): ____________________

Most recent copyright date: ____________________

This textbook contains the following (where appropriate, write in the blank the number of versions of the feature):

____ Table of Contents	____ Index	____ Vocabulary Preview
____ Preface	____ Appendices	____ Additional Readings

Number of units or parts: ____ Number of Chapters: ____

Average length of chapters: ________________________

Other features: ____________________

How will I use the textbook features in my reading process? ______________

____

____

____

Chapter Survey Form

This chapter contains the following (where appropriate, write in the blank the number of times an item appears in the chapter):

Chapter Number and Title: ______________________________

Number of pages:

____ Introduction	____ Preview list of key ideas	____ Preview list of learning objectives
____ Headings	____ Subheadings	____ Information Boxes
____ Summaries	____ Review Questions	____ **Bold** or *italic* type
____ Photographs	____ Charts, tables, graphs	____ Illustrations, drawings

Other special features: ____________________

How will I use the chapter features in my reading process? ____________________

____

How many flash cards are needed for key terms and ideas? ____________________

Before reading, I will ____________________

During reading, I will ____________________

____

After reading I will ____________________

____

REVIEW TEST 1

Score (number correct) ______ × 10 = ______%

Before, During, and After Reading

A. Before you read, survey the following passage from a college health textbook, and then answer the questions listed here.

1. What is the topic of this passage? ____________________

____

2. What are the ideas in **bold** type? ____________________

____

3. What do I already know about this idea? ____________________

____

4. What do I need to remember? ____________________

____

Textbook Skills

B. Read the passage. As you read, answer the questions in the box.

5. What does **obsessiveness** mean?

____

____

____

Eating Disorders

[1]On occasion, over one-third of all Americans fit the description of obesity and diet obsessiveness. [2]For an increasing number of people, particularly young women, this obsessive relationship with food develops into an eating disorder.

6. What are the traits of anorexia nervosa?

____

____

____

____

____

7. What are the traits of bulimia nervosa?

____

____

____

____

____

8. What are the traits of binge eating disorder?

____

____

____

____

____

____

[3]**Anorexia nervosa** is a persistent, chronic eating disorder characterized by deliberate food restriction and severe, life-threatening weight loss. [4]Anorexia involves self-starvation motivated by an intense fear of gaining weight along with an extremely distorted body image. [5]Usually people with anorexia achieve their weight loss through initial reduction in total food intake, particularly of high-calorie foods. [6]What they do eat, they often purge through vomiting or using laxatives.

[7]**Bulimia nervosa** often involves binging and purging. [8]People with bulimia binge and then take inappropriate measures, such as secret vomiting, to lose the calories they have just acquired. [9]People with bulimia are obsessed with their bodies, weight gain, and how they appear to others. [10]Unlike those with anorexia, people with bulimia are often "hidden" from the public eye because their weight may vary only slightly or fall within normal range.

[11]**Binge eating disorder** (BED) also involves frequent bouts of binge eating. [12]People who suffer with BED binge like their bulimic counterparts, but they do not take excessive measures to lose the weight that they gain. [13]Often they are clinically obese, and they tend to binge much more often than the typically obese person who may consume too many calories but spaces his or her eating over a more normal daily eating pattern.

—Adapted from *Health: The Basics,* 7th ed., pp. 299–301 by Rebecca J. Donatelle. Copyright © 2007 by Pearson Education, Inc. Printed and Electronically reproduced by permission of Pearson Education, Inc., Upper Saddle River, NJ.

After Reading

9. What are the differences among the three types of eating disorders?

____

____

10. What is the main cause of eating disorders? ______________________

____

REVIEW TEST 2

Score (number correct) ______ × 10 = ______%

SQ3R

Read the following two passages from a chapter in a college geography textbook. The first passage is the chapter's preview; the second passage is the chapter's summary. Use the appropriate steps in SQ3R to read the material and complete the activities.

Before Reading: Survey and Question

1. What is the topic? ______________________________

2. How many subheadings are in the chapter? ______________________________

Chapter 9 Earth's Resources and Environmental Protection

A Look Ahead (Preview)

What is a Natural Resource?

[1]A natural resource is something that is useful to people. [2]Usefulness is determined by a mix of cultural, technological, and economic factors in addition to the properties of a given resource. [3]We often substitute one resource for another as our needs change.

Mineral and Energy Resources

[4]Mineral resources include metals and nonmetals. [5]As mineral resources are depleted, they are usually replaced with substitute materials. [6]Landfills accumulate used materials and may one day be a source of materials for reuse. [7]One principal energy resource comes from three fossil fuels—oil, coal, and gas. [8]As fossil fuels are depleted, they are likely to be replaced with nuclear and renewable energy.

Air and Water Resources

[9]Pollution results when a substance is discharged into the air or water faster than it can be dispersed or removed by natural processes. [10]Recycling and pollution prevention are growing in importance as methods for solving pollution problems.

Chapter Summary

[1]A *natural resource* is an element of the physical environment that is useful to people. [2]Cultural values determine how resources are used.

[3]Technological factors limit our use of some resources by determining the particular applications to which certain materials can be put. [4]Economic factors such as resource prices and levels of affluence influence whether a resource is used, and how much. [5]Renewable natural resources include air, water, soil, plants, and animals. [6]Nonrenewable resources include fossil fuels and nonenergy minerals. [7]Most resources are substitutable to some degree, so that if one resource is less available or more expensive another resource is available to take its place.

[8]We depend on a great many different materials in our daily lives. [9]Wealthy countries use large quantities of resources, causing depletion of some *mineral resources*. [10]Mineral wastes and other materials accumulate in landfills. [11]Recycling can help conserve landfill space as well as reducing resource use.

[12]Over time, society has changed its use of *energy resources* from wood to coal, oil and gas. [13]At present the world is dependent on fossil fuels for energy. [14]The United States has abundant coal resources. [15]Although the United States is a major oil producer it must import most its oil from other countries. [16]Growing worldwide demand has caused large increases in energy prices. [17]As fossil fuels are depleted, we will need to use energy more efficiently and develop other sources of energy. [18]Nuclear power, renewable electricity, and energy conservation are promising new sources of energy.

[19]*Air pollution* is a concentration of trace substances at a greater level than occurs in average air. [20]Acid deposition and pollution of urban areas are particularly harmful forms of contemporary air pollution. [21]Acid deposition is a regional problem that is most acute in areas that burn large amounts of coal, or have large numbers of automobiles and fossil-fuel fired power plants. [22]Air pollution is particularly severe in urban areas where there is a large concentration of pollution sources. [23]*Water pollution* results from both point and nonpoint sources. [24]Industrial facilities and municipal sewage plants are important point sources while agricultural and urban runoff are significant non point sources. [25]Pollution prevention is a promising approach to reducing water pollution.

[26]Forests are an example of a resource with many different uses, and conflicts over which uses are most important often arise. [27]Among the important uses of forests are timber products such as lumber and paper, recreation, biodiversity preservation, and carbon storage. [28]Sustained yield management is a strategy that attempts to balance the productive use of a resource while not depleting its supply.

—From *Introduction to Geography: People, Places, and Environment,* 4th ed., pp. 352–353, 388–389 by Edward Bergman and William H. Renwick. Copyright © 2008 by Pearson Education, Inc. Printed and Electronically reproduced by permission of Pearson Education, Inc., Upper Saddle River, NJ.

After Reading: Complete the following SQ3R Study Guide with information from the passages.

SQ3R Study Guide: *Earth's Resources and Environmental Protection*	
Natural resource	**3.** ________________________ ________________________
Renewable resources	**4.** ________________________
5. ____________ resources	**6.** ________________________
Energy sources	**7.** ________________________
New energy sources	**8.** ________________________ ________________________
Air pollution	**9.** ________________________ ________________________
Water pollution	**10.** ________________________ ________________________

VISUAL VOCABULARY

Steam rises from bore holes at Nesjavellir Geothermal Power Plant in Iceland. Geothermal power is a ____________ energy source.

a. nonrenewable
b. renewable

REVIEW TEST 3

Score (number correct) ______ × 25 = ______%

Reading Textbook Selections

Textbook Skills

Before Reading: Survey the following passage from a college education textbook. Study the words in the Vocabulary Preview; then skim the passage, noting terms in **bold** or *italic* type. Answer the Before Reading questions that follow the passage. Then, read the passage and answer the After Reading questions that follow.

Vocabulary Preview

preferences (sentence 9): favored choices
solitary (sentence 11): private, alone, independent
distinction (sentence 19): difference, trait
dimension (sentence chart): aspect, feature, trait
facets (sentence 21): aspect, feature, trait
cognitive (sentence 21): mental
spatial (sentence 21): physical space

Learning Preferences

[1]Since the late 1970s, a great deal has been written about differences in students' learning preferences. [2]Learning preferences are often called *learning styles* in these writings, but I believe preferences is a more accurate label because the "styles" are determined by your preferences for particular learning environments—for example, where, when, with whom, or with what lighting, food, or music you like to study. [3]I like to study and write during large blocks of time—all day, if I don't have classes. [4]I usually make some kind of commitment or deadline every week so that I have to work in long stretches to finish the work before that deadline. [5]Then I take a day off. [6]When I plan or think, I have to see my thinking in writing. [7]I have a colleague who draws diagrams of relationships when she listens to a speaker or plans a paper. [8]You may be similar or very different, but we all may work effectively. [9]But are these **preferences** important for learning?

[10]Some proponents of learning styles believe that students learn more when they study in their preferred setting and manner. [11]And there is evidence that very bright students need less structure and prefer quiet, **solitary** learning. [12]But most educational psychologists are skeptical about the value of learning preferences. [13]"The reason researchers roll their eyes at learning styles research is the utter failure to find that assessing children's learning styles and matching to instructional methods has any effect on their learning" (Stahl, 2002, p. 99).

[14]Students, especially younger ones, may not be the best judges of how they should learn. [15]Sometimes, students, particularly those who have difficulty, prefer what is easy and comfortable; real learning can be hard and uncomfortable. [16]Sometimes, students prefer to learn in a certain way because they have no alternatives; it is the only way they know how to approach the task. [17]These students may benefit from developing new—and perhaps more effective—ways to learn. [18]One final consideration: Many of the learning styles advocates imply that the differences in the learner are what matter.

Visual/Verbal Distinctions

[19]There is one learning styles **distinction** that has research support. [20]Richard Mayer has been studying the distinction between visual and verbal learners, with a focus on learning from computer-based multimedia. [21]He is finding that there is a visualizer-verbalizer dimension and that it has three **facets:** ***cognitive spatial*** *ability* (low or high), *cognitive style* (visualizer vs. verbalizer), and *learning preference* (verbal learner vs. visual learner), as shown in the following table.

Three Facets of the Visualizer-Verbalizer Dimension

There are three **dimensions** to visual versus verbal learning: ability, style, and preference. Individuals can be high or low on any or all of these dimensions.

Facet	Types of Learners	Definition
Cognitive Ability	High spatial ability	Good abilities to create, remember, and manipulate images and spatial information
	Low spatial ability	Poor abilities to create, remember, and manipulate images and spatial information
Cognitive Style	Visualizer	Thinks using images and visual information
	Verbalizer	Thinks using words and verbal information
Learning Preference	Visual learner	Prefers instruction using pictures
	Verbal learner	Prefers instruction using words

—From "Types of Learners Table" and "Three Facets of the Visualizer-Verbalizer Dimension" by Richard E. Mayer from *Journal of Educational Psychology*, 2003.

[22]The picture is more complex than simply categorizing a student as either a visual or a verbal learner. [23]Students might have preferences for learning with pictures, but their low spatial ability could make using pictures to learn less effective. [24]These differences can be reliably measured, but research has not identified the effects of teaching to these styles.

—From *Educational Psychology,* 10th ed., pp. 124–127 by Anita E. Woolfolk. Copyright © 2007 by Pearson Education, Inc. Reproduced by permission of Pearson Education, Inc., Boston, MA.

Before Reading

1. What do I already know about learning styles? ______________________

____

____

2. What do I need to learn? ______________________

____

After Reading

3. According to the chart, the three dimensions to visual versus verbal learning are ______________, ______________, and ______________.

4. Which of the terms—verbalizer or visualizer—describes you as a learner? Explain why. ______________________

____

WHAT DO YOU THINK?

What do you think is the best environment for learning? What kind of support and teaching do you need, based on your learning preference? Assume you have been asked to become a peer counselor in the learning center of your college or university. As part of the training program for peer counselors, you are required to go through some of the same activities required of students who seek academic support. You have been asked to prepare a report about your understanding and use of learning preferences. Based on the chart in the passage, identify and describe the ways in which you or someone you know is a visualizer-verbalizer. Be sure to give examples based on your experiences or observations of others.

REVIEW TEST 4

Score (number correct) ______ × 10 = ______%

Textbook Skills

Before Reading: Survey the following passage from the college textbook *Criminal Justice Today*. Study the words in the Vocabulary Preview. Then, skim the passage and answer the Before Reading questions that follow the passage. Next, read the passage and answer the After Reading questions that follow.

Vocabulary Preview

ingestible (sentence 2): able to be absorbed into the body
social convention (sentence 6): rule, principle, standard, custom
inherent (sentence 6): basic, natural, inbuilt
legitimate (sentence 25): legal, lawful
psychoactive (sentence 29): drugs or medications having a significant effect on mood or behavior
anesthetic (sentence 31): substance that dulls pain
advocated (sentence 33): supported, promoted
strictures (sentence 35): limits, restrictions, severe criticism

What Is a Drug?

[1]Before we begin any comprehensive discussion of drugs, we must first grapple with the concept of what a drug is. [2]In common usage, a *drug* may be any **ingestible** substance that has a noticeable effect on the mind or body. [3]Drugs may enter the body via injection, inhalation, swallowing, or even direct absorption through the skin or mucous membranes. [4]Some drugs, like penicillin and tranquilizers, are useful in medical treatment, while others, like heroin and cocaine, are attractive only to *recreational drug* users or to those who are addicted to them.

[5]In determining which substances should be called "drugs," it is important to recognize the role that social definitions of any phenomenon play in our understanding of it. [6]Hence, what Americans today consider to be a drug depends more on **social convention** or agreed-on

definitions than it does on any **inherent** property of the substance itself. [7]The history of marijuana provides a case in point. [8]Before the early twentieth century, marijuana was freely available in the United States. [9]Although alcohol was the recreational drug of choice at the time, marijuana found a following among some artists and musicians. [10]Marijuana was also occasionally used for medical purposes to "calm the nerves" and to treat hysteria. [11]Howard Becker, in his classic study of the early Federal Bureau of Narcotics (forerunner of the Drug Enforcement Administration, or DEA), demonstrates how federal agencies worked to outlaw marijuana in order to increase their power. [12]Federally funded publications voiced calls for laws against the substance. [13]And movies like *Reefer Madness* led the drive toward classifying marijuana as a dangerous drug. [14]The 1939 Marijuana Tax Act was the result, and marijuana has been thought of as a drug worthy of federal and local enforcement efforts ever since.

[15]Both the law and social convention make strong distinctions between drugs that are socially acceptable and those that are not. [16]Some substances with profound effects on the mind and body are not even thought of as drugs. [17]Gasoline fumes, chemical vapors of many kinds, perfumes, certain vitamins, sugar-rich foods, and toxic chemicals may all have profound effects. [18]Even so, most people do not think of these substances as drugs. [19]And they are rarely regulated by the criminal law.

[20]Recent social awareness has reclassified alcohol, caffeine, and nicotine as "drugs." [21]However, before the 1960s it is doubtful that most Americans would have applied that word to these three substances. [22]Even today, alcohol, caffeine, and nicotine are readily available throughout the country, with only minimal controls on their manufacture and distribution. [23]As a result, these three drugs continue to enjoy favored status in both our law and culture. [24]Nonetheless, alcohol abuse and addiction are commonplace in American society, and anyone who has tried to quit smoking knows the power that nicotine can wield.

[25]Occupying a middle ground on the continuum between acceptability and illegality are substances that have a **legitimate** medical use and are usually available only with a prescription. [26]Antibiotics, diet pills, and, in particular, tranquilizers, stimulants, and mood-altering chemicals (like the popular drug Prozac) are culturally acceptable but typically can be attained legally only with a physician's prescription.

[27]The majority of Americans clearly recognize these substances as drugs, albeit useful ones.

[28]Powerful drugs, those with the ability to produce substantially altered states of consciousness and with a high potential for addiction, occupy the forefront in social and legal condemnation. [29]Among them are ***psychoactive** substances* like heroin, peyote, mescaline, LSD, and cocaine. [30]Even here, however, legitimate uses for such drugs may exist. [31]Cocaine is used in the treatment of certain medical conditions and can be applied as a topical **anesthetic** during medical interventions. [32]LSD has been employed experimentally to investigate the nature of human consciousness, and peyote and mescaline may be used legally by members of the Native American Church in religious services. [33]Even heroin has been **advocated** as beneficial in relieving the suffering associated with some forms of terminal illness. [34]Hence, answers to the question of "What is a drug?" depend to a large extent on the social definitions and conventions operating at a given time and in a given place. [35]Some of the clearest definitional statements relating to controlled substances can be found in the law, although informal **strictures** and definitions guide much of everyday drug use.

—From *Criminal Justice Today: An Introductory Text for the 21st Century*, 10th ed., p. 578 by Frank J. Schmalleger. Copyright © 2009 by Pearson Education, Inc. Printed and Electronically reproduced by permission of Pearson Education, Inc., Upper Saddle River, NJ.

Before Reading

1. What is the topic? ______________________________

2. What is the purpose for reading? ______________________________

____

After Reading

Complete the following chart with information from the passage

The Continuum of Acceptability and Illegality of Drugs

Socially Acceptable	Middle Ground	Illegal
3. ________	**6.** prescription drugs, diet pills, tranquilizers, stimulants,	**8.** ________
4. ________		**9.** ________
5. ________	**7.** ________	**10.** ________

WHAT DO YOU THINK?

Would you be in favor of "reclassifying" any drugs that are currently considered illegal? Would you be in favor of reclassifying any drugs that are currently considered legal? Which ones and why? Have the government efforts to curtail use of illegal drugs been effective? Why or why not? Assume you are interested in a career in criminal justice and you are serving as a volunteer with your local police department. One of your duties is to help educate the public about important local issues. Your assignment today is to speak at a middle school about drugs. Based on the passage you just read, what do you want to share with the students? Write an essay in which you discuss the drug issue with this group of youth.

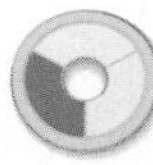

After Reading About a System for Master Readers

Before you move on to the Mastery Tests on a reading system for master readers, take time to reflect on your learning and performance by answering the following questions. Write your answers in your notebook.

- How has my knowledge base or prior knowledge changed about a reading system such as SQ3R?
- Based on my studies, how do I think I will perform on the Mastery Test(s)? Why do I think my scores will be above average, average, or below average?
- Would I recommend this chapter to other students who want to learn more about a reading system such as SQ3R. Why or why not?

Test your understanding of what you have learned about a Reading System for Master Readers by completing the Chapter 1 Review Card in the insert near the end of your text.

CONNECT TO PEARSON myreadinglab

To check your progress in meeting Chapter 1's learning outcomes, log in to www.myreadinglab.com and try the following activities.

- The "Memorization and Concentration" section of MyReadingLab ties the use of prior knowledge to your ability to focus on and remember what you have read. To access this resource, click on the "Study Plan" tab. Then click on "Memorization and Concentration." Under the heading "Review Materials," choose option #3 "Model: Concentration and Memorization."

- The "Active Reading Strategies" section of MyReadingLab offers an overview, model, slide show, practices, and tests about the reading process. To access this resource, go to MyReadingLab.com. Click on the "Study Plan" tab. Then choose "Active Reading Strategies" from the menu.
- The "Reading Textbooks" section of MyReadingLab offers an overview, model, slide show, practices, and tests about the reading process and textbooks. For example, to learn about how to survey a textbook, click on "Reading Textbooks." Then, under the heading "Review Materials," choose option #2 "Model: Reading Textbooks." To learn more about applying SQ3R to textbook reading, choose option #3 "Model: SQ3R."
- To measure your mastery of the content of this chapter, complete the tests in the "Reading Textbooks" section and click on Gradebook to find your results.

MASTERY TEST 1

Name ______________________ Section ______________

Date __________ **Score** (number correct) ______ × 20 = ______%

Textbook Skills

Using SQ3R, read the following passage from a college economics textbook.

A Nation on the Move

[1]World War II had a greater impact than the Depression on the future of American life. [2]While American soldiers and sailors fought abroad, the nation underwent sweeping social and economic changes at home.

[3]The war led to a vast migration of the American population. [4]Young men left their homes for training camps and then for service overseas. [5]Defense workers and their families, some nine million people in all, moved to work in the new booming shipyards, munitions factories, and aircraft plants. [6]Rural areas lost population while coastal regions, especially along the Pacific and the Gulf of Mexico, drew millions of people. [7]The location of army camps in the South and West created boom conditions in the future Sunbelt, as did the concentration of aircraft factories and shipyards in this region. [8]California had the greatest gains, adding nearly two million to its population in less than five years.

[9]This movement of people caused severe social problems. [10]Housing was in short supply. [11]Migrating workers crowded into house trailers and boardinghouses, bringing unexpected windfalls to landlords. [12]In one boomtown, a reporter described an old Victorian house that had five bedrooms on the second floor. [13]"Three of them," he wrote, "held two cots apiece, the two others held three cots." [14]But the owner revealed that "the third floor is where we pick up the velvet. . . . We rent to workers in different shifts . . . three shifts a day . . . seven bucks a week apiece."

[15]Family life suffered under these crowded living conditions. [16]An increase in the number of marriages, as young couples searched for something to hang on to in the midst of wartime turmoil, was offset by a rising divorce rate. [17]The baby boom that would peak in the 1950s began during the war and brought its own set of problems. [18]Only a few publicly funded day-care centers were available, and working mothers worried about their "latchkey children."

[19]Despite these problems, women found the war a time of economic opportunity. [20]The demand for workers led to a dramatic rise in women's employment, from fourteen million working women in 1940 to nineteen million by 1945. [21]Most of the new women workers were married and many were middle-aged, thus broadening the composition of the female workforce, which in the past had been composed primarily of young single women. [22]Women entered industries once viewed as exclusively male; by the end of the war, they worked alongside men tending blast furnaces in steel mills and welding hulls in shipyards. [23]Women enjoyed the hefty weekly paychecks, which rose by

50 percent from 1941 to 1943, and they took pride in their contributions to the war effort.

—From *The American Story,* 1st ed. pp. 886, 888, 890 by Robert A. Divine, George M. Fredrickson, R. Hal Williams and T. H. Breen. Copyright © 2002 by Pearson Education, Inc. Reprinted by permission of Pearson Education, Inc., Upper Saddle River, NJ.

1. What is the topic of this passage? ________________________________

____

2–5. Complete the concept map that follows with information from the passage.

A Nation on the Move

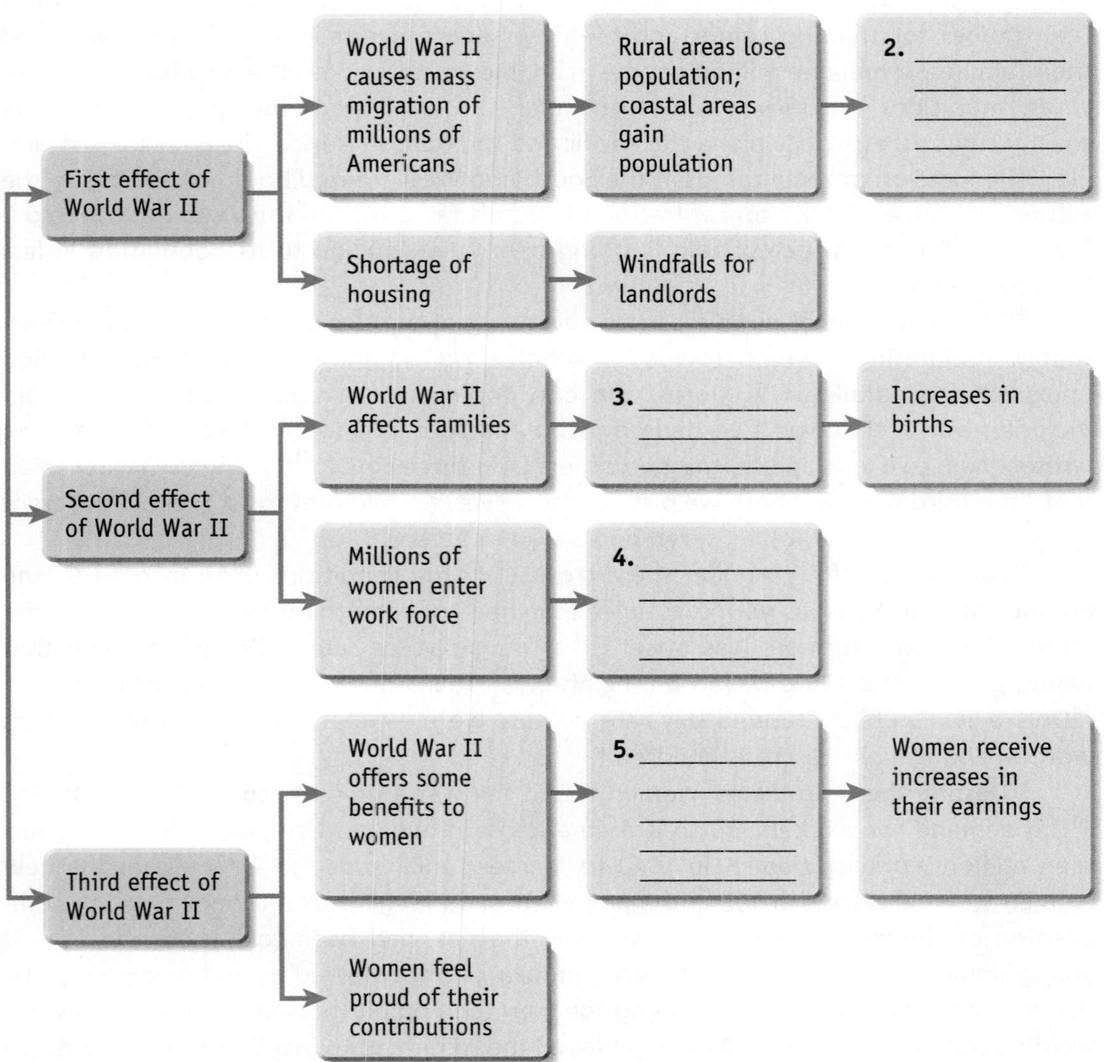

MASTERY TEST 2

Name ________________ Section ________

Date ________ **Score** (number correct) ______ × 20 = ______%

Using SQ3R, read the following passage taken from a college literature textbook.

Textbook Skills

Saying and Suggesting

[1]To write so clearly that they might bring "all things as near the mathematical plainness" as possible—that was the goal of scientists according to Bishop Thomas Sprat, who lived in the seventeenth century. [2]Such an effort would seem bound to fail, because words, unlike numbers, are ambiguous indicators. [3]Although it may have troubled Bishop Sprat, the tendency of a word to have multiplicity of meaning rather than mathematical plainness opens broad avenues to poetry.

[4]Every word has at least one **denotation**: a meaning as defined in a dictionary. [5]But the English language has many a common word with so many denotations that a reader may need to think twice to see what it means in a specific context. [6]The noun *field*, for instance, can denote a piece of ground, a sports arena, the scene of a battle, part of a flag, a profession, and a number system in mathematics. [7]Further, the word can be used as a verb ("he *fielded* a grounder") or an adjective (*field trip, field glasses*).

[8]A word also has **connotations**: overtones or suggestions of additional meaning that it gains from all the contexts in which we have met it in the past. [9]The word *skeleton*, according to a dictionary, denotes "the bony framework of a human being or other vertebrate animal, which supports the flesh and protects the organs." [10]But by its associations, the word can rouse thoughts of war, of disease and death, or (possibly) of one's plans to go to medical school. [11]Think, too, of the difference between "Old Doc Jones" and "Abner P. Jones, M.D." [12]In the mind's eye, the former appears in his shirtsleeves; the latter has a gold nameplate on his door. [13]That some words denote the same thing but have sharply different connotations is pointed out in this anonymous Victorian jingle:

> [14]Here's a little ditty that you really ought to know:
> Horses "sweat" and men "perspire," but ladies only "glow."

—From *Literature: An Introduction to Fiction, Poetry, and Drama*, 4th Compact Edition by X. J. Kennedy and Dana Gioia. Copyright © 2005 by Pearson Education, Inc. Reprinted by permission of Pearson Education, Inc., Glenview, IL.

1. What is the topic of this passage? ______________________________

____

____

Term	Definition	Example
2. ______________	**3.** ______________ ______________	*Field* can denote a piece of ground, a sports arena, the scene of a battle, part of a flag, a profession, and a number system in mathematics.
Connotation	**4.** ______________ ______________ ______________	**5.** ______________ ______________ ______________

MASTERY TEST 3

Name ______________________ Section ______________

Date __________ **Score** (number correct) ______ × 25 = ______%

Textbook Skills

Using SQ3R, read the following passage from a college psychology textbook.

1. What are the traits of nonassertive behavior?

____

____

____

____

____

2. What are the traits of aggressive behavior?

____

____

____

____

3. What are the traits of assertive behavior?

____

____

Nonassertiveness, Aggressiveness, and Assertiveness

[1]The nature of assertive communication can be better understood by distinguishing it from nonassertiveness and aggressiveness.

[2]**Nonassertiveness** refers to a lack of assertiveness in certain types of or in all communication situations. [3]People who are nonassertive fail to assert their rights. [4]In many instances, these people do what others tell them to do—parents, employers, and the like—without questioning and without concern for what is best for them. [5]They operate with a "You win, I lose" philosophy. [6]Nonassertive people often ask permission from others to do what is their perfect right. [7]Social situations create anxiety for these individuals, and their self-esteem is generally low.

[8]**Aggressiveness** is the other extreme. [9]Aggressive people operate with an "I win, you lose" philosophy; they care little for what the other person wants and focus only on their own needs. [10]Some people communicate aggressively only under certain conditions or in certain situations (for example, after being taken advantage of over a long period of time) while others communicate aggressively in all or at least most situations. [11]Aggressive communicators think little of the opinions, values, or beliefs of others and yet are extremely sensitive to others' criticism of their own behaviors. [12]Consequently, they frequently get into arguments with others.

[13]**Assertive behavior**—behavior that enables you to act in your own best interests without denying or infringing upon the rights of others—is the generally desired alternative to nonassertiveness or aggressiveness. [14]Assertive people operate with an "I win, you win" philosophy; they assume that both people can gain something from an interpersonal

4. What is the purpose of this passage?

____

____

____

____

____

____

interaction. [15]Assertive people are willing to assert their own rights. [16]Unlike their aggressive counterparts, however, they don't hurt others in the process. [17]Assertive people speak their minds and welcome others' doing likewise.

—From *The Interpersonal Communication Book,* 10th ed., p. 149 by Joseph A. DeVito. Copyright © 2004 by Pearson Education, Inc. Reproduced by permission of Pearson Education, Inc., Boston, MA.

VISUAL VOCABULARY

Road rage is an example of which type of behavior?

____

a. assertive
b. nonassertive
c. aggressive

MASTERY TEST 4

Name ______________________ Section ______________

Date __________ **Score** (number correct) ______ × 25 = ______%

Read the following passage from a college science textbook. As you read, answer the questions in the box.

1. What is a food chain?

____

____

____

____

2. What are basal species? Give an example.

____

____

____

3. What are intermediate species? Give an example.

____

____

____

____

4. What are top predators? Give an example.

____

____

____

____

Food Webs Describe Species Interactions

[1]Perhaps the most fundamental process in nature is the acquisition of food, or the energy and nutrients required for assimilation. [2]Ecologists studying the structure of communities often focus on the feeding relationships of the various species, or the manner in which species interact in the process of acquiring food.

[3]An abstract representation of feeding relationships within a community is the food chain. [4]A **food chain** is a descriptive diagram: a series of arrows, each pointing from one species to another. [5]The diagram represents the flow of food energy from prey (the consumed) to predator (the consumer). [6]For example, grasshoppers eat grass; clay-colored sparrows eat grasshoppers; and marsh hawks prey on the sparrows. [7]We write this relationship as follows:

grass ⟶ grasshopper ⟶ sparrow ⟶ hawk

[8]Feeding relationships in nature, however, are not simple, straight-line food chains. [9]Rather, they involve numerous food chains meshed into a complex food web with links leading from primary producers through an array of consumers. [10]Such food webs are highly interwoven. [11]Links represent a wide variety of species interactions.

[12]The species in a food web are distinguished by whether they are basal species, intermediate species,

or top predators. [13]**Basal species** feed on no other species but are fed upon by others. [14]**Intermediate species** feed on other species, and they are prey of other species. [15]Top **predators** are not subjected to predators and prey on intermediate and basal species.

—Adapted from *Elements of Ecology,* 6th ed., pp. 340–341 by Thomas M. Smith and Robert Leo Smith. Copyright © 2006 by Pearson Education, Inc. Reprinted by permission of Pearson Education, Inc., Glenview, IL.

VISUAL VOCABULARY

This photograph illustrates the three levels in the food chain: The grass is the ______________. The zebra are members of the ______________. And the lion is the ______________.

a. top predator.
b. basal species.
c. intermediate species.

Vocabulary Skills

2

LEARNING OUTCOMES

After studying this chapter, you should be able to do the following:

1. Define vocabulary.
2. Classify context clues using SAGE.
3. Describe how to comprehend new words as you read.
4. Assess your comprehension of context clues.
5. Analyze the meaning of a word based on word parts.
6. Use a glossary.
7. Evaluate the importance of what you have learned about vocabulary skills.
8. Apply what you have learned about vocabulary skills to your reading process.

Before Reading About Vocabulary Skills

Chapter 1 taught you the importance of surveying material before you begin reading by skimming the information for **bold** or *italic* type. Throughout this textbook, key ideas are emphasized in bold or italic print where they appear in the passage; often they are also set apart visually in a box that gives the definition or examples of the term. Skim the chapter for key ideas in boxes that will help you understand vocabulary skills. Refer to these boxes and create at least six questions that you can answer as you read the chapter. Write your questions in the following spaces (record the page number for the key term in each question):

____? (page ______)

____? (page ______)

____? (page ______)

____? (page ______)

____? (page ______)

____? (page ______)

Compare the questions you created with the following questions. Then, write the ones that seem most helpful in your notebook, leaving enough space after each question to record the answers as you read and study the chapter.

What is vocabulary? (page 48) What is a context clue? (page 49) What are the signal words for synonyms? (page 49) What are the signal words for antonyms? (page 51) What are the signal words for examples? (page 54) What is a glossary? (page 55) What are roots, prefixes, and suffixes? (page 59) How will knowing about these vocabulary skills help me develop my vocabulary? (pages 48 and 59)

Words Are Building Blocks

Words are the building blocks of meaning. Have you ever watched a child with a set of building blocks such as Legos? Hundreds of separate pieces can be joined together to create buildings, planes, cars, or even spaceships. Words are like that, too. A word is the smallest unit of thought. Words properly joined create meaning.

Vocabulary is all the words used or understood by a person.

How many words do you have in your **vocabulary**? If you are like most people, by the time you are 18 years old, you know about 60,000 words. During your college studies, you will most likely learn an additional 20,000 words. Each subject you study will have its own set of words. There are several ways to study vocabulary.

Context Clues: A SAGE Approach

Master readers interact with new words in a number of ways. One way is to use **context clues**. The meaning of a word is shaped by its context. The word *context* means "surroundings." The meaning of a word is shaped by the words surrounding it—its context. Master readers use context clues to learn new words.

> A **context clue** is the information that surrounds a new word, used to understand its meaning.

There are four types of context clues:

- **S**ynonyms
- **A**ntonyms
- **G**eneral context
- **E**xamples

Notice that, put together, the first letter of each context clue spells the word SAGE. The word *sage* means "wise." Using context clues is a wise—a SAGE—reading strategy.

Synonyms

A **synonym** is a word that has the same or nearly the same meaning as another word. Many times, an author will place a synonym near a new or difficult word as a context clue to the word's meaning. Usually, a synonym is set off with a pair of commas, a pair of dashes, or a pair of parentheses before and after it.

Synonym Signal Words	
or	that is

EXAMPLES Each of the following sentences has a key word in **bold** type. In each sentence, underline the signal words and then circle the synonym for the word in **bold**.

1. To ensure personal safety, be **cognizant,** or aware, of your surroundings.
2. Many crimes are committed against women who are **oblivious,** that is, uninformed and ignorant, about how to protect themselves.

VISUAL VOCABULARY

Self-defense training makes women ______________ of tactics they can use to ensure their safety.

a. fearful
b. cognizant
c. skilled

EXPLANATIONS

1. The signal word *or* clues the reader that the synonym for *cognizant* is *aware.*
2. The signal words *that is* clue the reader that the synonyms for *oblivious* are *uninformed* and *ignorant.*

PRACTICE 1

Each of the following sentences contains a word that is a synonym for the word in **bold** type. Underline the signal words and circle the synonym in each sentence.

1. Idris is known for his **discreet**, or tactful, manner of handling personnel issues.
2. **Indigent**, that is, poverty-stricken, people do not have access to health care.
3. Rocky's poetry has a **lyrical**, or musical, quality.

Antonyms

An **antonym** is a word that has the opposite meaning of another word. Antonyms help you see the shade of a word's meaning by showing you what the original word is *not*. The following contrast words often act as signals that an antonym is being used.

Antonym Signal Words

but	not	unlike
however	on the other hand	yet
in contrast		

Sometimes antonyms can be found next to the new word. In those cases, commas, dashes, or parentheses set them off. At other times, antonyms are placed in other parts of the sentence to emphasize the contrast between the ideas.

EXAMPLES In each sentence, underline the signal words and circle the antonym for the word in **bold** type. In the blank, write the letter of the word that best defines the word in **bold.**

______ **1.** Marcel **facilitated** the study group's progress with his thoughtful questions; in contrast, Randy hindered their ability to concentrate with his inappropriate jokes.

a. assisted
b. impeded
c. deepened
d. recorded

______ **2.** After purchasing a painting by the famous artist Monet, Charlene discovered the piece was a **facsimile**, not an original.

a. innovation
b. copy
c. rarity
d. spare

EXPLANATIONS

1. The signal words *in contrast* clue the reader to the antonym hindered. The best definition of the word *facilitated* is (a) *assisted.*

2. The signal word *not* clues the reader to the antonym *original.* The best definition of the word *facsimile* is (b) *copy.*

PRACTICE 2

In each sentence, underline the signal words and circle the antonym for the word in **bold** type. In the blank, write the letter of the word that best defines the word in **bold.**

______ **1.** The explanation the defense offered for the crime seems **credible.** It is not at all far-fetched, as the prosecution accuses.

a. mature
b. absurd
c. trustworthy
d. crazy

_____ **2.** The candidate gave a long, **discursive** speech when instead she should have made the speech short and direct.

a. entertaining
b. boring
c. important
d. rambling

_____ **3.** Before cosmetic surgery, Rory's face was **craggy** with age lines, but now after surgery, his complexion is smooth again.

a. sagging
b. damaging
c. rough
d. ugly

General Context

Often you will find that the author has not provided a synonym clue or an antonym clue. In that case, you will have to rely on the general context of the passage to figure out the meaning of the unfamiliar word. This requires you to read the entire sentence, or to read ahead for a few sentences, for information that will help you understand the new word.

Information about the word can be included in the passage in several ways. Sometimes a definition of the word may be provided. Vivid word pictures or descriptions of a situation can provide a sense of the word's meaning. Sometimes you may need to figure out the meaning of an unknown word by using logic and reasoning skills.

EXAMPLES In the blank, write the letter of the word that best defines the word in **bold** type.

Textbook Skills

_____ **1.** **Chronically** ill people may never regain the full level of health they experienced before the onset of their illnesses, and they may face a continuing loss of function and the constant threat of ever more serious medical problems as their illness progresses.

—From *Healthstyles: Decisions for Living Well*, 2nd ed., p. 344 by B. E. Pruitt and Jane J. Stein. Copyright © 1999 by Pearson Education, Inc. Printed and Electronically reproduced by permission of Pearson Education, Inc., Upper Saddle River, NJ.

a. slightly
b. permanently
c. briefly
d. effectively

_____ **2.** Nitrogen is one of the most **essential** nutrients for life, and it increases the fertility of soil and water.

—From *Elements of Ecology*, 4th ed., p. 349 by Thomas M. Smith and Robert Leo Smith. Copyright © 2000 by Pearson Education, Inc. Reprinted by permission of Pearson Education, Inc., Glenview, IL.

a. minor
b. optional
c. necessary
d. possible

EXPLANATIONS

1. The best meaning of the word *chronically* is (b) *permanently*. Clues from the sentence are the words and phrases *never regain*, *continuing loss*, *constant*, *ever more serious*, and *progresses*.
2. The best meaning of the word *essential* is (c) *necessary*. A clue word is *most*, which describes the word *essential*. The way the word *essential* is used in the sentence also provides a clue, for it describes *nutrients*. Even if the meaning of *nutrients* is not clear, the reader knows from the rest of the sentence that nutrients improve soil and water (by increasing its fertility). Thus the reader can conclude that the *most essential nutrients* are necessary.

PRACTICE 3

Each of the following sentences has a word in **bold** type. In the blank, write the letter of the word that best defines the word in **bold**.

Textbook Skills

_____ 1. Parental restrictiveness and harsh criticism are also associated with a greater likelihood of a teenager smoking. **Conversely**, affection, emotional support, and participation in meaningful conversations at home more often result in the teenager not smoking.

—From *HealthStyles: Decisions for Living Well*, 2nd ed., p. 181 by B. E. Pruitt and Jane J. Stein. Copyright © 1999 by Pearson Education, Inc. Printed and Electronically reproduced by permission of Pearson Education, Inc., Upper Saddle River, NJ.

a. additionally
b. in contrast
c. surprisingly
d. as a result

_____ 2. Fear **galvanized** my entire body, and I ran faster than I have ever run before, my heart thumping wildly.

a. froze
b. broke
c. stimulated
d. strained

_____ 3. The statement "Bill is **at least** 21 years old" can also be expressed or translated into the following mathematical language: $b > 21$.

a. less than or equal to
b. greater than or equal to
c. less than
d. equal to

Examples

Many times an author will show the meaning of a new or difficult word by providing an example. Signal words indicate that an example is coming.

Example Signal Words				
consists of	for example	for instance	including	such as

Colons and dashes can also indicate examples.

EXAMPLE Using example clues, choose the correct meaning of the words in **bold** type.

_____ **1.** Some people believe that the pesticides used on foods cause serious physical **impairments** such as weakened kidneys and a more fragile immune system.

a. improvements c. injuries
b. laws d. pairings

_____ **2.** Some students find **collaborative** learning helps them understand and retain information; for example, Nicole, Vejay, and Chad meet every Tuesday and Thursday in the library to compare notes and help each other prepare for tests.

a. additional c. independent
b. intense d. shared

EXPLANATIONS

1. The best meaning of the word *impairments* is (c) *injuries.*

2. The best meaning of the word *collaborative* is (d) *shared.*

PRACTICE 4

Using example clues, choose the correct meaning of the word in **bold** type.

Textbook Skills

_____ **1.** Your **habitat** could be the country you live in, your state and city of residence, or the location of your home. Depending on your activity, such as eating, it could be your kitchen. The same is true for other organisms.

—From *Elements of Ecology,* 4th ed., p. 16 by Thomas M. Smith and Robert Leo Smith. Copyright © 2000 by Pearson Education, Inc. Reprinted by permission of Pearson Education, Inc., Glenview, IL.

Textbook Skills

a. lifestyle c. territory
b. work d. reward

_____ **2.** **Figurative** language (also known as a figure of speech) departs from the literal meanings of words, usually by comparing very different ideas or objects; for instance, "thorny problems" compares the pain of a problem or situation to the pricking pain of a thorn in the flesh.

—Adapted from *The Little, Brown Handbook,* 9th ed., p. 560 by H. Ramsey Fowler and Jane E. Aaron. Copyright © 2004 by Pearson Education, Inc. Reprinted by permission of Pearson Education, Inc., Glenview, IL.

a. symbolic
b. factual
c. old-fashioned
d. basic

Textbook Skills

Reading a Textbook: Using a Glossary

Each subject or content area, such as science, mathematics, or English, has its own specialized vocabulary. As you learned in Chapter 1 (p. 16), some textbooks provide an extra section in the back of the book called a *glossary* that alphabetically lists all the specialized terms with their definitions as they were used throughout the textbook. Other textbooks may provide short glossaries within each chapter; in these cases, the glossaries may appear in the margins or in highlighted boxes, listing the words in the order that they appear on the page. The meanings given in a glossary are limited to the way in which the word or term is used in that content area.

> A **glossary** is a list of selected terms with their definitions as used in a specific area of study.

Glossaries provide excellent opportunities to use strategies before and after reading. Before reading, skim the section for specialized terms (usually these words are in **bold** or *italic* print). Checking the words and their meanings triggers prior knowledge or establishes meaning that will deepen your comprehension. In addition, you can create vocabulary review lists using glossary terms by paraphrasing or restating the definition in your own words. These vocabulary lists can be used after reading to review and test your recall of the material.

Textbook Skills

EXAMPLE The following selection is from a college psychology textbook. Before reading, use the glossary to complete your vocabulary review list. Then, read the passage. After reading, answer the questions.

Physiological Stress Reactions

[1]How would you respond if you arrived at a class and discovered that you were about to have a pop quiz? [2]You would probably agree that

this would cause you some stress, but what does that mean for your body's reactions? [3]Many of the physiological responses we described for emotional situations are also relevant to day-to-day instances of stress. [4]Such transient states of arousal, with typically clear onset and offset patterns, are examples of **acute stress.** [5]**Chronic stress,** on the other hand, is a state of enduring arousal, continuing over time, in which demands are perceived as greater than the inner and outer resources available for dealing with them. [6]An example of chronic stress might be a continuous frustration with your inability to find time to do all the things you want to do.

Emergency Reactions to Acute Threats

[7]In the 1920s, Walter Cannon outlined the first scientific description of the way animals and humans respond to danger. [8]He found that a sequence of activity is triggered in the nerves and glands to prepare the body either to defend itself and struggle or to run away to safety. [9]Cannon called this dual stress response the **fight-or-flight response.** [10]At the center of this stress response is the *hypothalamus,* which is involved in a variety of emotional responses. [11]The hypothalamus has sometimes been referred to as the *stress center* because of its twin functions in emergencies: (1) it controls the autonomic nervous system (ANS) and (2) it activates the pituitary gland.

—From *Psychology and Life,* 16th ed., p. 6 by Richard J. Gerrig and Philip G. Zimbardo. Copyright © 2002 by Pearson Education, Inc. Reproduced by permission of Pearson Education, Inc., Boston, MA.

Glossary

Acute stress A transient state of arousal with typically clear onset and offset patterns.

Chronic stress A continuous state of arousal in which an individual perceives demands as greater than the inner and outer resources available for dealing with them.

Fight-or-flight response A sequence of internal activities triggered when an organism is faced with a threat; prepares the body for combat and struggle or for running away to safety; recent evidence suggests that the response is characteristic only of males.

Before Reading

1. The process that readies the body to deal with danger and conflict is known as _______________.
2. _______________ is a temporary state that arises from day-to-day instances of stress.
3. Ongoing stress that challenges available resources is known as _______________.

After Reading

_____ **4.** The hypothalamus is also known as the
a. stress center.
b. autonomic nervous system.
c. the pituitary gland.

_____ **5.** Physiological stress reactions primarily involve
a. the mind.
b. the body.
c. the emotions.

EXPLANATION

1. The process that readies the body to deal with danger and conflict is known as *fight-or-flight*. Note that the paraphrase (restatement) of the definition for fight-or-flight draws on information given in the glossary.
2. *Acute stress* is a temporary state that arises from day-to-day instances of stress.
3. Ongoing stress that challenges available resources is known as *chronic stress*.
4. The hypothalamus is also known as the (a) stress center.
5. Physiological stress reactions primarily involve (b) the body.

PRACTICE 5

Textbook Skills

The following selection is from a college psychology textbook. Before reading, use the glossary to complete your vocabulary review list. Then, read the passage. After reading, answer the questions.

Glossary

Psychoactive drugs Chemicals that affect mental processes and behavior by temporarily changing conscious awareness of reality.

Dependence and Addiction

[1]**Psychoactive drugs** are chemicals that affect mental processes and behavior by temporarily changing conscious awareness. [2]Once in the brain, they attach themselves to synaptic receptors, blocking or stimulating certain reactions. [3]By doing so, they profoundly alter the brain's communication system, affecting perception, memory, mood, and behavior. [4]However, continued use of a given drug creates **tolerance**—greater dosages are required to achieve the same effect. . . . [5]Hand in

hand with tolerance is **physiological dependence,** a process in which the body becomes adjusted to and dependent on the substance, in part because neurotransmitters are depleted by the frequent presence of the drug. [6]The tragic outcome of tolerance and dependence is **addiction.** [7]A person who is addicted requires the drug in his or her body and suffers painful withdrawal symptoms (shakes, sweats, nausea, and, in the case of alcohol withdrawal, even death) if the drug is not present.

[8]When an individual finds the use of a drug so desirable or pleasurable that a *craving* develops, with or without addiction, the condition is known as **psychological dependence.** [9]Psychological dependence can occur with any drug. [10]The result of drug dependence is that a person's lifestyle comes to revolve around drug use so wholly that his or her capacity to function is limited or impaired. [11]In addition, the expense involved in maintaining a drug habit of daily—and increasing—amounts often drives an addict to robbery, assault, prostitution, or drug peddling. [12]One of the gravest dangers currently facing addicts is the threat of getting AIDS by sharing hypodermic needles—intravenous drug users can unknowingly share bodily fluids with those who have this deadly immune deficiency disease.

[13]Teenagers who use illicit drugs to relieve emotional distress and to cope with daily stressors suffer long-term negative consequences.

—From *Psychology and Life,* 16th ed., p. 174 by Richard J. Gerrig and Philip G. Zimbardo.

Glossary

Tolerance A situation that occurs with continued use of a drug in which an individual requires greater dosages to achieve the same effect.

Physiological dependence The process by which the body becomes adjusted to and dependent on a drug.

Addiction A condition in which the body requires a drug in order to function without physical and psychological reactions to its absence; often the outcome of tolerance and dependence.

Psychological dependence The psychological need or craving for a drug.

Before Reading

1. ______________ The body requires a drug and suffers painful withdrawal symptoms if the drug is not present.
2. ______________ Chemicals that affect an individual's thought processes and actions by temporarily altering the person's awareness of reality.
3. ____________________ The mental and emotional need or craving for a drug.
4. ____________________ The body's need for a drug.
5. ______________ Greater amounts of the drug are needed for the same effect.

After Reading

Complete the following summary. Fill in the blanks with the appropriate word from the glossary.

Continued use of **(6)** ________________ leads to several serious outcomes. First, psychoactive drugs alter a person's sense of reality, "affecting perception, memory, mood, and behavior." Ongoing use of the drug causes **(7)** ________________, a state in which the body needs more of the drug to produce the same effect. As the body develops tolerance, the body also develops **(8)** ________________ upon the drug. Tragically, the result of tolerance and dependence is **(9)** ________________. In addition, a person can suffer from **(10)** ________________, when a "craving" for a drug "develops with or without addiction."

Word Parts

Just as ideas are made up of words, words also are made up of smaller parts. *Word parts* can help you learn vocabulary more easily and quickly. In addition, knowing the meaning of the parts of words helps you understand a new word when you see it in context.

Many words are divided into the following three parts: *roots, prefixes*, and *suffixes.*

Root The basic or main part of a word. Prefixes and suffixes are added to roots to make a new word.
Example: *derm* means "skin."

Prefix A group of letters with a specific meaning added to the beginning of a word (root) to make a new word.
Example: the ***hypo*** in ***hypo****dermic* means "under."

Suffix A group of letters with a specific meaning added to the end of a word (root) to make a new word.
Example: the ***ic*** in *hypoderm****ic*** means "of, like, related to, being."

Master readers understand how the three word parts join together to make additional words.

Roots

The **root** is the basic or main part of a word. Many times a root combined with other word parts will create a whole family of closely related words. Even when the root word is joined with other word parts to form new words, the meaning of the root does not change. Knowing the commonly used roots will help you master many new and difficult words. A list of common roots is available on MyReadingLab.

EXAMPLES Study the following word parts. Using the meaning of the root *clamere* and the context of each sentence, put each word into the sentence that best fits its meaning. Use each word once.

Prefix:	*re-*	(again, back)
Root:	*clamere, claim, clam*	(call out, shout)
Suffix:	*-ant*	(one who does)

claimant	reclaim

1. The city may ask to ______________ the land it had leased to the shoe factory.
2. A ______________ is a person who files a loss with an insurance company.

EXPLANATIONS Both words contain the root *clamere* ("call out," "shout"), and each word uses the meaning differently. The additional word parts—the prefix and the suffix—created the different meanings.

1. The word *reclaim* combines the prefix *re-*, which means "again" or "back" with the root *claim*, which means "call out." *Reclaim* means "call back." The city may want to reclaim or call back control of the land it had leased.
2. The word *claimant* combines the root *claim*, which means "call out" with the suffix *-ant*, which means "one who does." A claimant is "one who calls out" for the insurance company to pay for the loss.

PRACTICE 6

Study the following word parts. Using the meaning of the root *dexter* and context clues, put each word into the sentence that best fits its meaning. Use each word once. Slight changes in spelling may be necessary.

Prefix:	*ambi-*	(both)
Root:	*dexter*	(on the right, skillful)
Suffix:	*-ous*	(of, like, related to, being)
	-ity	(a quality)

ambidextrous	dexterity	dexterous

1. Jolene is _______________; she can write and pitch a ball with either her right or her left hand.

2. Miguel, a _______________ mechanic, can rapidly rebuild an engine.

3. An Olympic gymnast must possess extraordinary _______________.

Prefixes

A **prefix** is a group of letters with a specific meaning added to the beginning of a word or root to make a new word. Although the basic meaning of a root is not changed, a prefix changes the meaning of the word as a whole.

The importance of prefixes can be seen in the family of words that comes from the root *struct*, which means "build." Look over the following examples of prefixes and their meanings. Note the change in the meaning of the whole word based on the meaning of the prefix.

Prefix	Meaning	Root	Meaning	Example
con-	with, together	*struct*	build	*construct* (to build)
de-	down away, reverse			*destruct* (to destroy)
in-	in, into			*instruct* (to teach)

EXAMPLES Using the meanings of the prefixes, root, and context clues, put each word into the sentence that best fits its meaning. Use each word once.

Prefix	Meaning	Root	Meaning
dis-	apart, in different directions	*sect*	cut
inter-	among, between, in the midst		

dissect	intersect

1. Some ethicists believe that biology students should learn to ________ using a computer program instead of real animals, such as frogs or cats.

2. My house sits on the corner where the roads New Trail and Rio Pinar

________.

EXPLANATIONS Compare your answers to the explanations below.

1. Some ethicists believe that biology students should learn how to cut apart, dissect, animals such as frogs and cats by using a computer program.

2. My house sits on the corner of two roads, New Trail and Rio Pinar, where they intersect, cross one another.

PRACTICE 7

Study the meaning of each of the following prefixes and roots.

Prefix	Meaning	Root	Meaning
circum-	around	*spect*	to look
intro-	inside, to the inside	*stance*	stand, position
retro-	backward		

Create at least three words by joining these prefixes and roots.

1. ________________________________

2. ________________________________

3. ________________________________

Suffixes

A **suffix** is a group of letters with a specific meaning added to the end of a word or root to make a new word. Although the basic meaning of a root does not change, a suffix can change the type of word and the way a word is used. (A list of common suffixes is available on MyReadingLab.) Look at the following set of examples.

Root	Meaning	Suffix	Meaning	Word
psych	mind	*-ical*	quality	*psychological*
		-ist	person	*psychologist*
		-ology	study of	*psychology*

Note that sometimes a word will use more than one suffix. For example, the word *psychologist* combines the root *psych* ("mind") with two suffixes: *-ology* ("study of") and *-ist* ("person"). Thus *psychologist* means "a person who studies the mind." The word *psychological* combines the root *psych* ("mind") with two suffixes: *-ology* ("study of") and *-ical* ("quality"). Thus the word *psychological* means "the study of the qualities of the mind." Note also that the *y* in the suffix *-ology* is dropped because the additional suffixes begin with an *i*, which takes the place of the *y*.

EXAMPLES Using the meanings of the root and the suffixes, and the context clues, put each of the words in the box below into the sentence that best fits its meaning. Use each word once.

Root	Meaning	Suffix	Meaning
err	wander, stray, deviate	*-ancy*	quality or state of
		-ant	one who, that which
		-or	a condition

errant	errors

1. News reporters are expected to avoid ________________ in their reports.
2. Acting the part of the ________________ son, Nathan often rebelled against his parents' wishes.

EXPLANATIONS

1. News reporters are expected to avoid *errors*, that is, wandering or deviating from the facts that serve as a basis for the report.
2. Acting the part of the *errant* son, Nathan often strayed or deviated from his parents' wishes.

PRACTICE 8

Using the meanings of the roots and the suffixes, and the context clues, put each word in the box into the sentence that best fits its meaning. Use each word once. Slight changes in spelling may be necessary.

Root	Meaning	Suffix	Meaning
icon	image, symbol	*-graph*	writing
		-ic	of, like, related to, being
		-ology	study of, science

iconic	iconography	iconology

1. Ted's interest in ______________ has taken him to Egypt to visit the pyramids.
2. Lady Gaga's musical abilities and controversial behaviors have turned her into an ______________ pop figure.
3. Ancient Egyptians practiced ______________ by creating a writing system using pictures for concepts instead of letters for sounds; their writing system is known as hieroglyphics.

Textbook Skills

Reading a Textbook: Discipline-Specific Vocabulary

During your college career, you will take classes in a variety of different course areas, or disciplines. Each discipline has special words that it uses to describe the concepts of its area of study. In other words, each discipline has its own vocabulary. This is part of the reason why reading a biology textbook may seem like a very different experience than reading an economics textbook. Courses in different disciplines may use the same words, but the words sometimes take on new or different meanings within the context of each course area. For example, when a biology textbook talks about a "consumer," it is referring to any organism that eats either plants or animals for food; but when an economics textbook talks about a "consumer," it is referring to any individual who uses a product or service. Although the word is used similarly, the different definitions are a result of their discipline-specific applications.

To help you incorporate these new words into your vocabulary, textbooks usually provide context clues. Word parts can also be an essential tool for deciphering discipline-specific vocabulary. Many academic disciplines, especially in the sciences, use a language that relies heavily on Latin roots, suffixes, and prefixes.

EXAMPLE Read the following text from the college textbook *Life on Earth.* Then, answer the questions, using context clues and the information about word parts as needed.

Textbook Skills

Energy Enters Community Through Photosynthesis

[1]During photosynthesis, pigments such as chlorophyll absorb specific wavelengths of sunlight. [2]This solar energy is then used in reactions that store energy in chemical bonds, producing sugar and other high-energy molecules. [3]Photosynthetic organisms, from mighty oak trees to single-celled diatoms in the ocean, are called autotrophs or producers, because they produce food for themselves using nonliving nutrients and sunlight. [4]In doing so, they directly or indirectly produce food for nearly all other forms of life as well. [5]Organisms that cannot photosynthesize, called heterotrophs or consumers, must acquire energy and many of their nutrients prepackaged in the molecules that comprise the bodies of other organisms.

—From *Biology: Life on Earth,* 8th ed., p. 561 by Teresa Audesirk, Gerald Audesirk, and Bruce E. Byers. Copyright © 2008 by Pearson Education, Inc. Printed and Electronically reproduced by permission of Pearson Education, Inc., Upper Saddle River, NJ.

Prefix	Meaning	Root	Meaning	Suffix	Meaning
auto-	self	*tithenai*	to do		
dia-	two	*tome*	segment	*-s*	more than one
eco-	habitat, environment	*troph*	to nourish	*-sis*	action, production
hetere-	other, different			*-ic*	pertaining to
photo-	light				
syn-	together				

_____ **1.** What is the best meaning of the word **photosynthesis**?

a. high-energy molecules
b. organisms that use light to produce food
c. solar energy
d. the action of using light to produce food

_____ **2.** What is the best meaning of **diatoms**?

a. multi-cell organisms
b. single-cell organisms with two-sided cell walls
c. sea creatures
d. mineral deposits

_____ **3.** What is the best meaning of **ecosystem**?

a. environment
b. security
c. nutritional
d. plants

_____ **4.** What is the best meaning of **trophic**?

a. environment
b. light levels
c. feeding level
d. energy supply

5. How would you define **autotrophs** and **heterotrophs**? Explain how you came to your definitions. ______________________________

____

____

____

EXPLANATION

1. Word parts and the general sense of the passage reveal the meaning of *photosynthesis*. The prefix *photo* means light; the prefix *syn* means "together"; the root *tithenai* means "to do," and the suffix *sis* indicates that the action of producing something is occurring. In addition, the author offers a defining phrase "use solar energy to make food." When combined, these word parts mean (d) "the action of using light to produce food."
2. Word parts and the general sense of the passage reveal the meaning of *diatoms*: the prefix *dia* means "two"; the root *tome* means "segment," and the suffix *s* indicates "more than one." The author states that diatoms are single-celled organisms, so the word parts indicate that the single cell must have two parts. Thus, the best meaning of *diatoms* is (b) "single-cell organisms with two-sided cell walls. "
3. Word parts reveal the meaning of *ecosystem*. The prefix *eco* means "habitat or environment." Thus (a) "environment" is the correct answer.
4. Word parts reveal the meaning of *trophic*. The root *troph* means "to nourish," and the suffix *ic* means "pertaining to." Thus (c) "feeding level" is the correct answer.
5. Although the wording of your answers may vary, you should have come to the following conclusion: "Autotrophs are organisms that can produce their own food." The suffix *auto* means "self" and the root *troph* means "to nourish." "Heterotrophs are organisms that are fed by others." The prefix *hetere* means "other," and the root *troph* means "to nourish."

PRACTICE 9

Textbook Skills

Study the following text from the college textbook *Biology: Life on Earth*. Then, complete the activity "Key Term Review," using context clues and the information given about word parts, as needed.

Leaves and Chloroplasts Are Adaptations for Photosynthesis

[1]The leaves of most land plants are only a few cells thick; their structure is elegantly adapted to the demands of photosynthesis. [2]The flattened shape of leaves exposes a large surface area to the sun, and their thinness ensures that sunlight can penetrate them to reach the light-trapping

chloroplasts inside. [3]Both the upper and lower surfaces of a leaf consist of a layer of transparent cells, the **epidermis**. [4]The outer surface of both epidermal layers is covered by the **cuticle**, a waxy, waterproof covering that reduces the evaporation of water from the leaf.

[5]A leaf obtains CO_2 for photosynthesis from the air; adjustable pores in the epidermis, called **stomata** (singular stoma; Greek for "mouth"), open and close at appropriate times to admit air carrying CO_2.

[6]Inside the leaf are layers of cells collectively called **mesophyll**. [7]The mesophyll cells contain most of a leaf's chloroplasts, and consequently photosynthesis occurs principally in these cells. [8]**Vascular bundles** supply water and minerals to the mesophyll cells and carry the sugars they produce to other parts of the plant.

[9]A single mesophyll cell can have from 40 to 200 chloroplasts, which are sufficiently small that 2000 of them lined up would just span your thumbnail. . . . [10]Chloroplasts are organelles that consist of a double outer membrane enclosing a semifluid medium, the stroma. [11]Embedded in the stroma are disk-shaped, interconnected membranous sacs called **thylakoids**. [12]The chemical reactions of photosynthesis that depend on light (light-dependent reactions) occur within the membranes of the thylakoids, while the photosynthetic reactions that can continue for a time in darkness (light-independent reactions) occur in the surrounding stroma.

—From *Biology: Life on Earth*, 8th ed., pp. 118–119 by Teresa Audesirk, Gerald Audesirk, and Bruce E. Byers. Copyright © 2008 by Pearson Education, Inc. Printed and Electronically reproduced by permission of Pearson Education, Inc., Upper Saddle River, NJ.

Prefix	Meaning	Root	Meaning	Suffix	Meaning
epi-	on	*cutis*	skin	*-ar*	of, like, related to
meso-	middle	*dermi*	skin	*-le*	of, like, related to
		phyllon	leaf	*-oid*	resembling
		thylakos	sack		
		vas	vessels		

Key Term Review Match the following terms to their definitions:

_____ **1.** cuticle — a. outer cell layer of plant; outer layer of skin

_____ **2.** epidermis — b. sacs in which photosynthesis takes place

_____ **3.** mesophyll — c. veins, fluid carrying vessels

_____ **4.** thylakoids — d. waxy or fatty covering on the exposed surfaces of plants

_____ **5.** vascular bundles — e. middle of leaf

Complete the Summary Fill in the blanks with terms in the Key Term Review.

A plant is made for the demands of photosynthesis. The waxy **(6)** ______________ covers the broad, thin **(7)** ______________ of the leaf. This flat expanse of the leaf epidermis takes in sunlight and its pores (called stomata) open and close to absorb CO_2. **(8)** ______________ carry water and CO_2 to the **(9)** ______________ cells, which contain chloroplasts; inside the chloroplasts are the **(10)** ______________, where photosynthesis occurs.

Reading a Textbook: Visual Vocabulary

Textbook Skills

Textbooks often make information clearer by giving a visual image such as a graph, chart, or photograph. Take time to study the visual images and captions in this section to figure out how each one ties in to the information given in words.

EXAMPLE Study the following image and its caption. Then answer the questions that follow. If necessary, review the passage "Leaves and Chloroplasts Are Adaptations for Photosynthesis" (page 66).

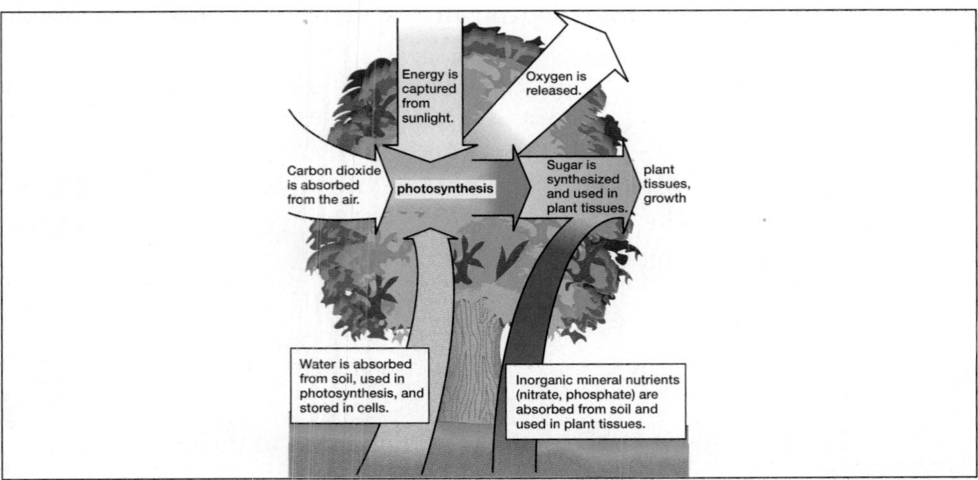

Figure 28-2 Primary productivity

______________ organisms, which capture solar energy and acquire inorganic nutrients from their environment, ultimately provide essentially all of the energy and most of the nutrients for organisms in higher trophic levels.

_____ Which term best completes the caption?

a. Ecosystem
b. Trophic
c. Photosynthetic
d. Heterotrophic

2. How does this image add to your understanding of the passage?

____

____

EXPLANATION

1. Photosynthetic best completes the caption.
2. This image illustrates how a photosynthetic organism captures solar energy and provides nutrients in the ecosystem.

PRACTICE 10

Study the following image and its caption. Then answer the questions that follow. If necessary, review the passage "Leaves and Chloroplasts Are Adaptations for Photosynthesis" (page 66).

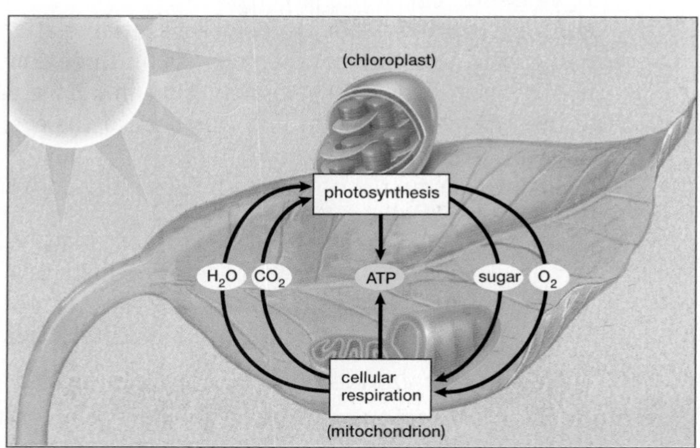

Interconnections between Photosynthesis and Cellular Respiration

Chloroplasts in green plants use the energy of sunlight to synthesize the high-energy carbon compounds such as glucose from low-energy molecules of water (H_2O) and carbon dioxide (CO_2). Plants themselves, and other organisms that eat plants or one another, extract energy from these organic molecules by cellular **respiration**, yielding water and carbon dioxide once again. This energy in turn drives all the reactions of life.

—From *Biology: Life on Earth,* 8th ed., Figure 28-2 from page 561 by Teresa Audesirk, Gerald Audesirk, and Bruce E. Byers. Copyright © 2008 by Pearson Education, Inc. Printed and Electronically reproduced by permission of Pearson Education, Inc., Upper Saddle River, NJ.

_____ **1.** The best definition of **respiration** in this context is the
a. energy-producing process in cells.
b. process of making glucose.
c. breathing process in animals.
d. death of a cell.

2. How does this graphic add to your understanding of the passage? _____

_____

_____

APPLICATIONS

Application 1: Context Clues and Glossary

The following passage from the college textbook *Psychology and Life* provides context clues and a glossary. Study the glossary. Read the passage, and circle the synonyms given for three of the words in **bold** type. Then, answer the questions.

Glossary

Convergence coming together
Depth perception ability to perceive spatial relationships, especially distances between objects, in three dimensions
Foveae a small cuplike depression or pit in a bone or organ
Retina the membrane that lines the back of the eye; receives the image created by the eye's lens
Retinal disparity the difference between the horizontal positions of matching images in two eyes

Textbook Skills

Depth Perception: Binocular Cues

[1]Have you ever wondered why you are **binocular**, or two-eyed, instead of **monocular**, or one-eyed? [2]The second eye is more than just a spare—it provides some of the best, most compelling information about depth. [3]The two sources of binocular depth information are *retinal disparity* and *convergence*. [4](The retina is the membrane that lines the back of the eye which receives the image created by the eye's lens.)

[5]Because the eyes are about 2 to 3 inches apart horizontally, they receive slightly different views of the world. [6]To convince yourself of this, try the following experiment. [7]First, close your left eye and use the right one to line up your two index fingers with some small object in the distance, holding one at arm's length and the other about a foot in front of your

face. [8]Now keeping your fingers stationary, close your right eye and open the left one while continuing to fixate on the distant object. [9]What happened to the position of your two fingers? [10]The second eye does not see them lined up with the distant object because it gets a slightly different view.

[11]This displacement between the horizontal positions of corresponding or matching images in your two eyes is called retinal disparity. [12]It provides depth information because the amount of **disparity**, or difference, depends on the relative distance of objects from you.

[13]Other binocular information about depth comes from **convergence**. [14]The two eyes turn inward to some extent whenever they are fixated on an object. [15]When the object is very close—a few inches in front of your face—the eyes must turn toward each other quite a bit for the same image to fall on both foveae.

—From *Psychology and Life*, 16th ed., Figure 5.21 (p. 137) and text adapted from pp. 136–137 by Richard J. Gerrig and Philip G. Zimbardo. Copyright © 2002 by Pearson Education, Inc. Reproduced by permission of Pearson Education, Inc., Boston, MA.

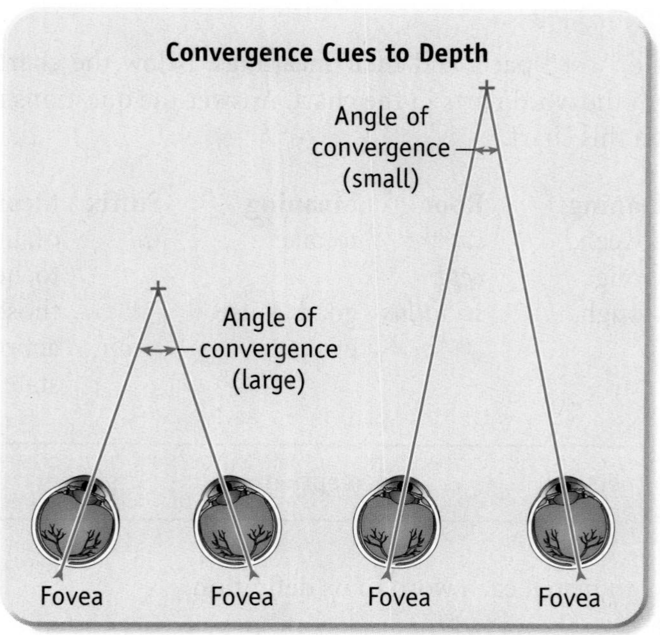

◄ When an object is close to you, your eyes must converge more than when an object is at a greater distance.

_____ **1.** What kind of context clue does the word **monocular** provide for the word **binocular**?

a. synonym
b. antonym
c. general context
d. example

_____ **2.** What is the best meaning of the word **convergence**?
a. merging
b. departure
c. reunion
d. fixated

_____ **3.** What context clue did you use to determine the meaning of **convergence**?
a. synonym
b. antonym
c. general context
d. example

_____ **4.** In the context of this reading selection, the words **disparity** and **convergence** are
a. antonyms.
b. synonyms.

Application 2: Word Parts

Study the chart of word parts and their meanings. Below the chart are three words made from the word parts in the chart. Answer the questions in sections A and B based on this chart.

Prefix	Meaning	Root	Meaning	Suffix	Meaning
inter-	between, among	*capere/ cept*	to take	*-al*	of, like, related to, being
per-	through	*stimulus*	goad, rouse, excite	*-i*	those that
				-tion	action or state of

interpersonal	perception	stimuli

A. Fill in the blanks to match each word to its definition.

1. _______________ describes that which is between or among people.

2. _______________ are those conditions that motivate or provoke a response.

3. _______________ is one's understanding, awareness, or grasp.

B. (4–6.) Using word meanings and context clues, put each word into the sentence that best fits its meaning. Use each word once.

interpersonal	perception	stimuli

Textbook Skills

The Stages of Perception

[1]_____________ is the process by which you become aware of objects, events, and, especially, people through your senses: sight, smell, taste, touch, and hearing. [2]Perception is an active, not a passive, process. [3]Your perceptions result from what exists in the outside world and from your own experiences, desires, needs and wants, loves and hatreds. [4]Among the reasons why perception is so important in _____________ communication is that it influences your communication choices. [5]The messages you send and listen to will depend on how you see the world, on how you size up specific situations, on what you think of the people with whom you interact.

[6]Interpersonal perception is a continuous series of processes that blend into one another. [7]For convenience of discussion we can separate them into five stages: (1) you sense, you pick up some kind of stimulation; (2) you organize the _____________ in some way; (3) you interpret and evaluate what you perceive; (4) you store it in memory; and (5) you retrieve it when needed.

—From *Essentials of Human Communication*, 4th ed., p. 55 by Joseph A. DeVito. Copyright © 2002 by Pearson Education, Inc. Reproduced by permission of Pearson Education, Inc., Boston, MA.

REVIEW TEST 1

Score (number correct) ______ × 10 = ______%

Context Clues

Read the following passage from the college textbook *Foundations of Earth Science*. Answer the questions that follow.

Textbook Skills

Earth's Internal Structure

[1]Earth's interior has three basic divisions—the iron rich **core**; the thin primitive crust; and Earth's largest layer, called the **mantle**, which is located between the core and the crust.

Earth's Crust [2]The crust, Earth's relatively thin, rocky outer skin, is of two types—**continental** crust and oceanic crust. [3]Both share the word crust, but the similarity ends there. [4]The oceanic crust is roughly 7 kilometers (4 miles) thick and composed of the dark igneous rock basalt. [5]By contrast, the continental crust averages 35 to 40 kilometers (22 to 25 miles) thick but may exceed 70 kilometers (40 miles) in some mountainous regions such as the Rockies and Himalayas. [6]Unlike the oceanic crust, which has a relatively **homogeneous** chemical composition, the continental crust consists of many rock types. [7]Although the upper crust has an average composition of a granitic rock called **granodiorite**, it varies considerably from place to place.

[8]Continental rocks have an average density of about 2.7 grams per cubic centimeter, and some are 4 billion years old. [9]The rocks of the oceanic crust are younger (180 million years or less) and denser (about 3.0 grams per cubic centimeter) than continental rocks.

Earth's Mantle [10]More than 82 percent of Earth's volume is contained in the mantle, a solid, rocky shell that extends to a depth of about 2900 kilometers (1800 miles). [11]The boundary between the crust and mantle represents a marked change in chemical composition. [12]The dominant rock type in the uppermost mantle is **peridotite**, which is richer in the metals magnesium and iron than the minerals found in either the continental or oceanic crust.

[13]The upper mantle extends from the crust-mantle boundary down to a depth of about 660 kilometers (410 miles). [14]The upper mantle can be divided into three different parts. [15]The top portion of the upper mantle is part of the stiff **lithosphere**, and beneath that is the weaker asthenosphere. [16]The bottom part of the upper mantle is called the transition zone.

[17]The lithosphere (sphere of rock) consists of the entire crust and uppermost mantle and forms Earth's relatively cool, rigid outer shell. [18]Averaging about 100 kilometers (62 miles) in thickness, the lithosphere is more than 250 kilometers (155 miles) thick below the oldest portions of the continents (Figure 6.22). [19]Beneath this stiff layer to a depth of about 350 kilometers (217 miles) lies a soft, comparatively weak layer known as the asthenosphere ("weak sphere"). [20]The top portion of the asthenosphere has a temperature/pressure regime that results in a small amount of melting. [21]Within this very weak zone, the lithosphere is mechanically detached from

the layer below. [22]The result is that the lithosphere is able to move independently of the asthenosphere, a fact we will consider in the next chapter.

[23]It is important to emphasize that the strength of various Earth materials is a function of both their composition and the temperature and pressure of their environment. [24]The entire lithosphere does not behave like a brittle solid similar to rocks found on the surface. [25]Rather, the rocks of the lithosphere get progressively hotter and weaker (more easily deformed) with increasing depth. [26]At the depth of the uppermost asthenosphere, the rocks are close enough to their melting temperature that they are very easily deformed, and some melting may actually occur. [27]Thus, the upper most asthenosphere is weak because it is near its melting point, just as hot wax is weaker than cold wax.

[28]From 660 kilometers (410 miles) deep to the top of the core, at a depth of 2900 kilometers (1800 miles), is the lower mantle. [29]Because of an increase in pressure (caused by the weight of the rock above), the mantle gradually strengthens with depth. [30]Despite their strength however, the rocks within the lower mantle are very hot and capable of very gradual flow.

Earth's Core [31]The composition of the core is thought to be an iron-nickel alloy with minor amounts of oxygen, silicon, and sulfur-elements that readily form compounds with iron. [32]At the extreme pressure found in the core, this iron-rich material has an average density of nearly 11 grams per cubic centimeter and approaches 14 times the density of water at Earth's center.

[33]The core is divided into two regions that exhibit very different mechanical strengths. [34]The outer core is a liquid layer 2270 kilometers (1410 miles) thick. [35]It is the movement of metallic iron within this zone that generates Earth's magnetic field. [36]The inner core is a sphere with a radius of 1216 kilometers (754 miles). [37]Despite its higher temperature, the iron in the inner core is solid due to the immense pressures that exist in the center of the planet.

—From *Foundations of Earth Science,* 5th ed., pp. 176–177 by Frederick K. Lutgens, Edward J. Tarbuck, and Dennis Tasa. Copyright © 2008 by Pearson Education, Inc. Printed and Electronically reproduced by permission of Pearson Education, Inc., Upper Saddle River, NJ.

_____ **1.** What is the best meaning of the word **continental** in sentence 2?

a. of or related to land
b. of or related to colonies
c. of or related to oceans
d. of or related to Europe

_____ **2.** Identify the context clue used for **continental** in sentence 2.

a. synonym
b. antonym
c. general sense of the passage
d. example

_____ **3.** What is the best meaning of **homogeneous** in sentence 6?
a. different than
b. liquid
c. of the same kind
d. solid

_____ **4.** Identify the context clue used for **homogeneous** in sentence 6.
a. synonym
b. antonym
c. general sense of the passage
d. example

Match the following glossary words to their definitions

_____ **5.** asthenosphere
_____ **6.** core
_____ **7.** granodiorite
_____ **8.** mantle
_____ **9.** lithosphere
_____ **10.** peridotite

a. a solid rocky shell, extends about 1800 miles below crust
b. dominant rock in uppermost mantle; made of magnesium and iron
c. innermost layer of Earth; mostly of iron-nickel alloy
d. middle, weaker layer of the upper mantle
e. a granitic rock that composes the upper crust
f. rigid outer layer of Earth; includes the crust and upper mantle

REVIEW TEST 2

Score (number correct) _____ × 10 = _____%

Context Clues

Choose the best definition for each word in **bold** type. Then, identify the context clue used in each passage. A clue may be used more than once.

Context Clues

Synonym **A**ntonym **G**eneral context **E**xample

_____ **1.** Jerry and Marie were **incredulous** when they heard that they had purchased the sole winning lottery ticket; they were skeptical that they could have won the entire $34 million jackpot.

a. excited
b. secretive
c. disbelieving
d. hysterical

2. Context clue: ______________________________

_____ **3.** The Underground Railroad was a route made up of a series of **clandestine** safe houses for African Americans fleeing the injustice of slavery in the South.

a. exposed
b. secret
c. dangerous
d. poor

4. Context clue: ______________________________

_____ **5.** Some people who are infected with West Nile virus develop mild flulike symptoms that disappear within a few days. However, in rare cases, West Nile virus leads to more **virulent** infections that could cause long-term disability or even death.

a. weak
b. hidden
c. contagious
d. dangerous

6. Context clue: ______________________________

_____ **7.** The play *The Taming of the Shrew* by William Shakespeare is known for its **trenchant,** or biting, wit about the battle of the sexes.

a. subtle
b. cutting
c. bland
d. rude

8. Context clue: ______________________________

_____ **9.** Often, humans are driven by their **temporal** needs for safety, food, housing, and other possessions and measure their success by how well these needs are met.

a. long-term
b. sordid
c. spiritual
d. worldly

10. Context clue: ______________________________

REVIEW TEST 3

Score (number correct) _____ × 10 = _____%

Vocabulary Skills

Before Reading: Study the glossary and survey the passage from a college sociology textbook. Locate the terms from the glossary in the passage; circle the

Textbook Skills

terms and underline the definitions as they are worded in context. Then, read the passage and respond to the After Reading questions and activities.

Glossary: Vocabulary Preview

megalopolis: an urban area consisting of at least two metropolises and their many suburbs

megacity: a city of 10 million or more residents

metropolitan statistical area: a central city and the urbanized counties adjacent to it

metropolis: a central city surrounded by smaller cities and their suburbs

urbanization: the process by which an increasing proportion of the population live in cities and have growing influence on the culture

The Process of Urbanization

[1]Although cities are not new to the world scene, urbanization is. [2]**Urbanization** refers to masses of people moving to cities, and these cities having a growing influence on society. [3]Urbanization is taking place all over the world. [4]In 1800, only 3 percent of the world's population lived in cities (Hauser and Schnore 1965). [5]Then in 2007, for the first time in history, more people lived in cities than in rural areas. [6]Urbanization is uneven across the globe. [7]For the industrialized world, it is 77 percent, and for the Least Industrialized Nations, it is 41 percent (Haub 2006; Robb 2007). [8]Without the Industrial Revolution, this remarkable growth could not have taken place, for an extensive infrastructure is needed to support hundreds of thousands and even millions of people in a relatively small area.

[9]To understand the city's attraction, we need to consider the "pulls" of urban life. [10]Because of its exquisite division of labor, the city offers incredible variety—music ranging from rap and salsa to country and classical, shops that feature imported delicacies from around the world and those that sell special foods for vegetarians and diabetics. [11]Cities also offer anonymity, which so many find refreshing in light of the tighter controls of village and small-town life. [12]And, of course, the city offers work.

[13]Some cities have grown so large and have so much influence over a region that the term city is no longer adequate to describe them. [14]The term **metropolis** is used instead. [15]This term refers to a central city surrounded by smaller cities and their suburbs. [16]They are linked by transportation and communication and connected economically, and sometimes politically, through county boards and regional governing bodies. [17]St. Louis is an example.

[18]Although this name, St. Louis, properly refers to a city of 340,000 people in Missouri, it also refers to another 2 million people

who live in more than a hundred separate towns in both Missouri and Illinois. [19]Altogether, the region is known as the "St. Louis or Bi-State Area." [20]Although these towns are independent politically, they form an economic unit. [21]They are linked by work (many people in the smaller towns work in St. Louis or are served by industries from St. Louis), by communications (they share the same area newspaper and radio and television stations), and by transportation (they use the same interstate highways, the Bi-State Bus system, and international airport). [22]As symbolic interactionists would note, shared symbols (the Arch, the Mississippi River, Busch Brewery, the Cardinals, the Rams, the Blues—both the hockey team and the music) provide the residents a common identity.

[23]Most of the towns run into one another, and if you were to drive through this metropolis, you would not know that you were leaving one town and entering another—unless you had lived there for some time and were aware of the fierce small town identifications and rivalries that coexist within this overarching identity.

[24]Some metropolises have grown so large and influential that the term **megalopolis** is used to describe them. [25]This term refers to an overlapping area consisting of at least two metropolises and their many suburbs. [26]Of the twenty or so megalopolises in the United States, the three largest are the Eastern seaboard running from Maine to Virginia, the area in Florida between Miami, Orlando, and Tampa, and California's coastal area between San Francisco and San Diego. [27]The California megalopolis extends into Mexico and includes Tijuana and its suburbs.

[28]This process of urban areas turning into a metropolis, and a metropolis developing into a megalopolis, occurs worldwide. [29]When a city's population hits 10 million, it is called a **megacity**. [30]In 1950, New York City was the only megacity in the world. [31]Today there are 19. [32]The figure entitled "How Many Millions of People Live in the World's Largest Megacities?" shows the world's 10 largest megacities. [33]Note that most megacities are located in the Least Industrialized Nations.

—From *Sociology: A Down-to-Earth Approach,* 9th ed., p. 606 by James M. Henslin. Copyright © 2009 by James Henslin. Reproduced by permission of Pearson Education, Inc., Boston, MA.

After Reading

Use the word parts and terms in the answer box to complete the word web based on information from the glossary and the passage.

Answer Box: city mega- megalopolis metropolis -polis- large megacity metro- mother urbanization

Prefixes:	Meaning
(1) ______	**(2)** ______
(3) ______	**(4)** ______
sub-	part

Root: Meaning	
(5) ______	**(6)** ______
-urbānus-	

Suffixes:	Meaning
-ation	action, process; noun
-an	of; related to adjective

Words Derived from Parts

Flow Chart of **(7)** ____________

Urban areas ⟶ **(8)** ________ ⟶ **(9)** ________ ⟶ **(10)** ________

VISUAL VOCABULARY

Fill in the blank with a word from the glossary that completes the caption.

The U.S. Census Bureau divides the country into 274 ____________ ____________ (MSAs). Each MSA consists of a central city of at least 50,000 people and the urbanized areas linked to it. About three of five Americans live in just fifty or so MSAs.

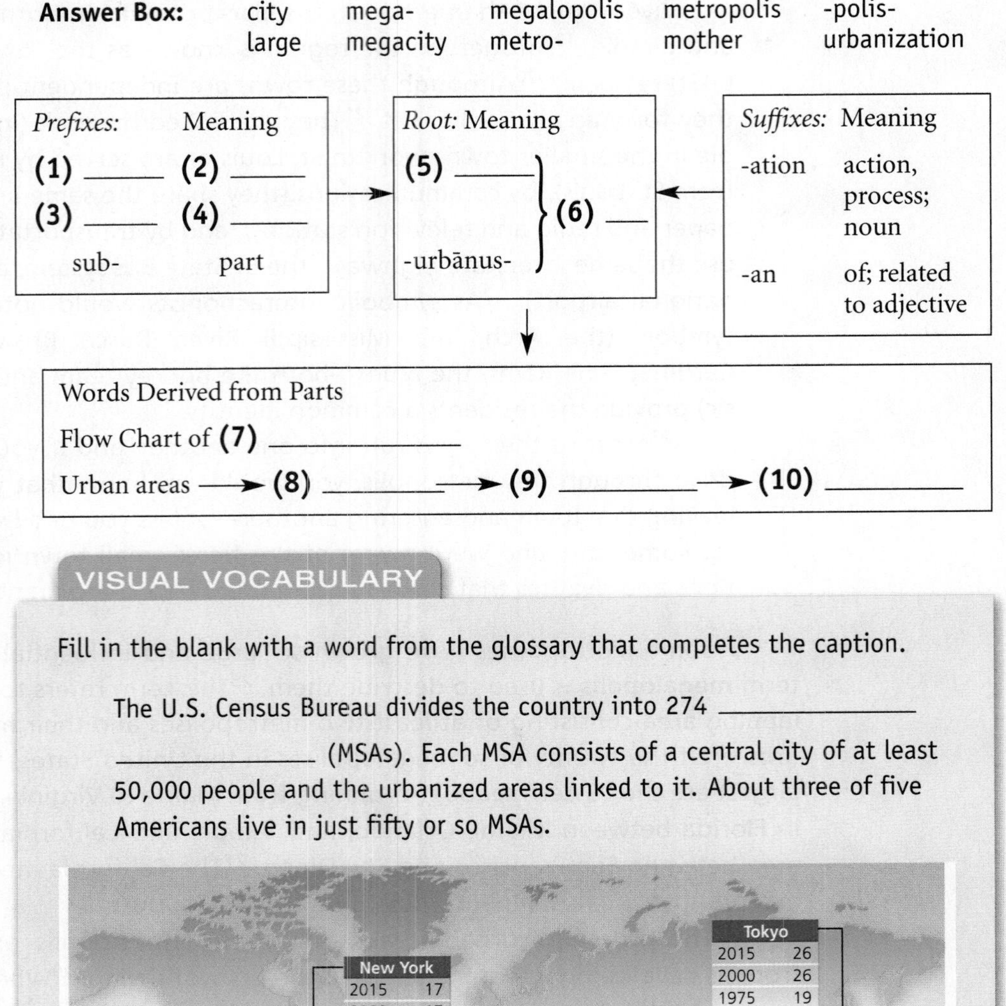

—From *Sociology: A Down-to-Earth Approach,* 9th ed., Figure 20.10 by James M. Henslin. Copyright © 2008 by James Henslin. Reproduced by permission of Pearson Education, Inc., Boston, MA.

WHAT DO YOU THINK?

What type of community do you live in—rural, suburban, or urban? What is the closest metropolis, megalopolis, and megacity near you? Reread the description of St. Louis. How has urbanization affected your area? Assume you are running for a local public office, such as a city commissioner or a mayor. Write an article for local publication that describes your vision for how the city should grow or control growth. Describe your community in terms of its urbanization, identify some strengths in your community, and identify one problem you would address.

REVIEW TEST 4

Score (number correct) ______ × 10 = ______%

Vocabulary Skills

Before Reading: Survey the following passage from the college textbook *World Civilizations: The Global Experience.* Study the words in the Vocabulary Preview; then skim the passage, noting the words in bold print. Answer the Before Reading questions that follow the passage. Then, read the passage. Finally, after reading, check the answers you gave before reading to make sure they are accurate. Respond to What Do You Think? to check your comprehension after reading.

Vocabulary Preview

Republic (sentence 2): State
Muslim (sentence 3): a believer of Islam
hold (sentence 21): the interior of a ship below deck; the cargo deck of a ship
singular (sentence 22): individual

Africa and the Africans in the Age of the Atlantic Slave Trade

[1]Sometimes a single, extraordinary life can represent the forces and patterns of a whole historical era. [2]Born in the early 19th century, Mahommah Gardo Baquaqua was a young man from the trading town of Djougou in what is now the Benin **Republic** in west Africa. [3]A **Muslim**, he could speak Arabic, Hausa, and a number of other languages, as was common among the trading peoples from which he came. [4]At a young age, Baquaqua was captured and **enslaved** during a war with a neighboring

African state; after gaining his freedom, he was enslaved again, and around 1845 he was sold into the Atlantic slave trade.

[5]Baquaqua was taken first to northeastern Brazil and from there was purchased by a ship captain from Rio de Janeiro. [6]After a number of voyages along the Brazilian coast, his ship eventually sailed for New York. [7]After a failed attempt to use the American courts to gain his freedom (a strategy that many slaves attempted), Baquaqua fled to Boston with the help of local **abolitionists.** [8]In that city he was **befriended** by antislavery Baptist missionaries. [9]With them, he sailed for Haiti. [10]Eventually he learned French and English and studied at a college in upstate New York in order to prepare for missionary work in his native Africa. [11]His life was not easy. [12]Eventually, because of racial incidents, he moved to Canada.

[13]Although Baquaqua had left the Baptist college, he did not abandon his desire to return to Africa, and he continued to seek ways to make that voyage. [14]In 1854, in an attempt to get the money he needed to realize his dream, he published his **autobiography,** [15]*An Interesting Narrative: Biography of Mahommah G. Baquaqua*. [16]In it he was able to provide his personal observations on his experiences. [17]On the slave ship, he reported:

> [18]O the loathsomeness and filth of that horrible place will never be **effaced** from my memory; nay as long as my memory holds her seat in this **distracted** brain, will I remember that. [19]My heart, even at this day, sickens at the thought of it. [20]Let those humane individuals, who are in favor of slavery, only allow themselves to take the slave's position in the noisome hold of a slave ship, just for one trip from Africa to America, and without going into the horrors of slavery further than this, if they do not come out thorough-going abolitionists, then I have no more to say in favor of abolition.

[21]The only place worse than the **hold** of a slave ship, said Baquaqua, was the place to which slave owners would be condemned in the next life. [22]We do not know whether Baquaqua finally returned to the land of his birth, but the life of this African, while **singular** in many aspects, represents the stories of millions of Africans in the age of the slave trade, and these make up an important part of world history.

—From *World Civilizations: The Global Experience, Combined Volume, Atlas Edition,* 5th ed., pp. 542–543 by Peter N. Stearns, Michael B. Adas, Stuart B. Schwartz, and Marc Jason Gilbert.

Before Reading

A. Use context clues to state in your own words the definition of the following terms. Indicate the context clue you used.

1. In sentence 4, what does the word **enslaved** mean? ____________

2. Identify the context clue used for the word **enslaved** in sentence 4.

____

3. In sentence 7, what does the word **abolitionists** mean? ____________

____

4. Identify the context clue used for the word **abolitionists** in sentences 7 and 20.

____

5. In sentence 8, what does the word **befriended** mean? ____________

____

6. Identify the context clue used for the word **befriended** in sentence 8. ____

____

7. In sentence 18, what does the word **distracted** mean? ____________

____

8. Identify the context clue used for the word **distracted** in sentence 18. ____

____

B. Study the following chart of word parts. Use context clues and the word parts to answer the questions that follow.

Prefix	Meaning	Root	Meaning	Suffix	Meaning
auto-	self	*bio*	life	*-ed*	completed action
ef-	out	*fac*	face	*-y*	quality, state
		graph	write		

______ **9.** What does the word **effaced** in sentence 18 mean?

a. carved
b. wiped out
c. seen
d. endured

10. What does the word **autobiography** in sentence 14 mean? ______

______

WHAT DO YOU THINK?

The passage opens with the statement "Sometimes a single, extraordinary life can represent the forces and patterns of a whole historical era." Do you agree that Mahommah Gardo Baquaqua represents the experiences of millions of Africans in the age of slave trade? Why or why not? How does an autobiography or a biography add to our understanding of history? In what ways does your life or the life of someone you know represent an important aspect of our current era? Assume you have been invited to participate in a community-wide project to record aspects of current life for future generations. Your community has decided to create a time capsule in which items and writings from local people will be enclosed. This time capsule will be opened and publically shared by officials in the year 2060. Write an autobiographical or biographical essay to include in the time capsule.

After Reading About Vocabulary Skills

Before you move on to the Mastery Tests on vocabulary skills for master readers, take time to reflect on your learning and performance by answering the following questions. Write your answers in your notebook.

- How has my knowledge base or prior knowledge about vocabulary and glossary skills changed?
- Based on my studies, how do I think I will perform on the Mastery Test(s)? Why do I think my scores will be above average, average, or below average?
- Would I recommend this chapter to other students who want to learn more about vocabulary skills? Why or why not?

Test your understanding of what you have learned about Vocabulary Skills for Master Readers by completing the Chapter 2 Review Card in the insert near the end of your text.

CONNECT TO PEARSON myreadinglab

To check your progress in meeting Chapter 2's learning outcomes, log in to **www.myreadinglab.com** and try the following activities.

- The "Vocabulary" section of MyReadingLab offers several tools that can help to accomplish this goal, including, the dictionary, vocabulary in context, synonyms and antonyms, and prefixes and suffixes. To access this resource, click on the "Study Plan" tab. Then click on "Vocabulary." Then click on the following links as needed:
  - "Overview"
  - "Model"
  - "Word Elements I: Root Words, Prefixes, and Suffixes (Flash Animation)"
  - "Compound Words (Flash Animation)"
  - "Word Elements II: Root Words, Prefixes, and Suffixes (Flash Animation)"
  - "Word Origins (Flash Animation)"
  - "Practice"
  - "Tests"
- The "Other Resources" section of MyReadingLab offers an overview of vocabulary and dictionary skills, along with an audio dictionary, activities, exercises and quizzes, spelling activities, and flashcards. To access resources for these skills, go to the "Other Resources" on the Home page of MyReadingLab. Click on the link labeled "Vocabulary Website." Under "select your level," choose the link labeled "Intermediate." Click on the following as needed: "Dictionary," "Context Clues," "Synonyms," "Antonyms," "Prefixes," "Roots," "Suffixes," and "Flashcards."
- To measure your mastery of the content of this chapter, complete the tests in the "Vocabulary" section and click on Gradebook to find your results.

MASTERY TEST 1

Name _______________ Section _______________

Date _________ **Score** (number correct) ______ × 10 = ______%

Read the following passage from the college textbook *Psychology and Life*. Answer the questions that follow.

Textbook Skills

Sleep Apnea

[1]Sleep apnea is a sleep disorder that causes the person to stop breathing while asleep. [2]When this happens, the blood's oxygen level drops and emergency hormones are **secreted**, which then causes the sleeper to awaken and begin breathing again. [3]Most people have a few apnea **episodes** a night. [4]However, someone with sleep apnea disorder can have hundreds of such cycles every night. [5]Sometimes apnea episodes frighten the sleeper. [6]But often they are so brief that the sleeper fails to credit **accumulating** or mounting sleepiness to them.

[7]Consider, for example, the case of a famous psychologist who, because of undetected sleep apnea, could not stay awake during research meetings and lectures. [8]When his wife made him aware of his disturbing nighttime behavior, he went to a sleep disorder clinic. [9]The treatment he received **reinvigorated** or revived his career (Zimbardo, personal communication, 1991). [10]In other cases, people have lost their jobs, friends, and even spouses.

[11]Sleep apnea is also frequent in premature infants. [12]These babies sometimes need physical **stimulation** to start breathing again. [13]Because of their immature respiratory system, these infants must remain attached to monitors in intensive care nurseries as long as the problem continues.

—From *Psychology and Life*, 16th ed., p. 165 by Richard J. Gerrig and Philip G. Zimbardo. Copyright © 2002 by Pearson Education, Inc. Reproduced by permission of Pearson Education, Inc., Boston, MA.

_____ **1.** What is the best meaning of the word **secreted** in sentence 2?

a. hidden
b. produced
c. suppressed
d. disturbed

_____ **2.** Identify the context clue used for **secreted** in sentence 2.

a. synonym
b. antonym
c. general context
d. example

_____ **3.** What is the best meaning of the word **episodes** in sentence 3?

a. scares
b. events
c. reasons
d. treatments

_____ **4.** Identify the context clue used for **episodes** in sentence 3.
a. synonym
b. antonym
c. general context
d. example

_____ **5.** What is the best meaning of the word **accumulating** in sentence 6?
a. lessening
b. embarrassing
c. increasing
d. confusing

_____ **6.** Identify the context clue used for **accumulating** in sentence 6.
a. synonym
b. antonym
c. general context
d. example

_____ **7.** What is the best meaning of the word **reinvigorated** in sentence 9?
a. repressed
b. ruined again
c. cured
d. strengthened again

_____ **8.** Identify the context clue used for **reinvigorated** in sentence 9.
a. synonym
b. antonym
c. general context
d. example

_____ **9.** What is the best meaning of the word **stimulation** in sentence 12?
a. encouragement
b. hindrance
c. pain
d. relief

_____ **10.** Identify the context clue used for **stimulation** in sentence 12.
a. synonym
b. antonym
c. general context
d. example

MASTERY TEST 2

Name ______________________ Section ______________

Date __________ **Score** (number correct) ______ × 10 = ______%

Read the following passage adapted from the college humanities textbook *The Creative Impulse: An Introduction to the Arts.* Using context clues, write the definition for each word in **bold** type. Choose definitions from the box. You will not use all the definitions. Then, identify the context clue you used to determine the meaning of each word.

Textbook Skills

Cave Painting

[1]The Cave of Lascaux (lahs-KOH) in France lies slightly over a mile from the little town of Montignac, in the valley of the Vézère (vay-Zair) River. [2]The cave itself was discovered in 1940 by a group of children who, while investigating a tree uprooted by a storm, scrambled down a **fissure** (crevice) into a world undisturbed for thousands of years. [3]The cave was sealed in 1963 to protect it from **atmospheric** damage, and visitors now see Lascaux II, an exact replica, which is sited in a quarry 600 feet away.

[4]Perhaps a **sanctuary** for the performance of sacred rites and ceremonies, the Main Hall, or Hall of the Bulls, elicits a sense of power and grandeur. [5]The thundering herd moves below a sky formed by the rolling contours of the stone ceiling of the cave, sweeping our eyes forward as we travel into the cave itself. [6]At the entrance of the main hall, the 8-foot "unicorn" begins a larger-than-lifesize **montage** of bulls, horses, and deer, which are up to 12 feet tall. [7]Their shapes intermingle with one another, and their colors radiate warmth and power. [8]These magnificent creatures remind us that their creators were capable technicians, who, with artistic skills at least equal to our own, were able to capture the essence, that is heart, beneath the visible surface of their world. [9]The paintings in the Main Hall were created over a long period of time and by a succession of artists. [10]Yet their **cumulative** or combined effect in this 30- by 100-foot domed gallery is that of a single work, carefully composed for maximum dramatic and communicative impact.

—From *The Creative Impulse: An Introduction to the Arts,* 8th ed., pp. 38–39 by Dennis J. Sporre. Copyright © 2009 by Pearson Education, Inc. Reprinted by permission of Pearson Education, Inc., Upper Saddle River, NJ.

ancient	composite	environmental	spirit
collective	crack	shelter	sweeping

Context Clues

Synonym
Antonym
General context
Example

1. In sentence 2, **fissure** means ______________________.
2. Context clue used for **fissure**: ______________________
3. In sentence 3, **atmospheric** means ______________________.
4. Context clue used for **atmospheric**: ______________________
5. In sentence 4, **sanctuary** means ______________________.
6. Context clue used for **sanctuary**: ______________________
7. In sentence 6, **montage** means ______________________.
8. Context clue used for **montage**: ______________________
9. In sentence 10, **cumulative** means ______________________.
10. Context clue used for **cumulative**: ______________________

VISUAL VOCABULARY

This painting in the Cave of Lascaux, France, is a ____________ of larger-than-lifesize animals.

MASTERY TEST 3

Name ____________ Section ____________

Date ________ **Score** (number correct) ________ × 20 = ________%

Read the following passage from the college textbook *Cultural Anthropology*. Answer the questions that follow the passage.

Textbook Skills

Intermarriage

[1]Fears of **ethnic** divide in the United States are being challenged by the rate at which couples of different races and ethnicities are marrying one another.

[2]Since 1970, the number of **interracial** marriages in the United States increased more than ten-fold, from less than 1 percent to more than 5 percent of the estimated 57 million couples recorded in the 2000 census. [3]This reflects both the population growth and an increased tendency to marry across racial lines. [4]At the same time, there have been dramatic improvements in the American attitudes toward race. [5]A Gallup poll conducted at the end of 2003 shows that 66 percent of White respondents said they would accept a child or grandchild marrying someone of a different race.

[6]Based on the 2000 census, interracial marriage has increased across most racial groups and is generating a growing population of **multiracial** Americans.

- [7]The typical interracial couple is a White person with a non-White spouse. [8]Intermarriage between two people from minority racial groups is relatively infrequent.
- [9]About one-fourth of Hispanic couples are inter-Hispanic, a rate that has been fairly stable since 1980.
- [10]Whites and Blacks have the lowest intermarriage rates while American Indians, Hawaiians, and multiple-race individuals have the highest.
- [11]Black men are more likely to **intermarry** than Black women, while Asian women are more likely to intermarry than Asian men. [12]Men and women from other racial groups are equally likely to intermarry.
- [13]Younger and better-educated Americans are more likely to intermarry than older and less-educated Americans.

[14]Intermarriage and the resulting mixed-race children they produce are gradually blurring the racial boundaries that have long divided the nation.

[15]Some **demographers** note that race could eventually lose much of its meaning in the United States, much as ethnicity has lost its meaning

among many Anglo Americans. [16]Interracial tolerance is increasing as the nation's Hispanic and Asian populations continue to grow. [17]Moreover, many new immigrants come from countries with mixed-race traditions, which may make them more open to interracial marriage.

—From *Cultural Anthropology,* 7th ed., p. 220 by Marvin Harris and Orna Johnson.

Use context clues and the word parts in the chart below to state in your own words the definition of the following words from the passage.

Prefix	Meaning	Root	Meaning	Suffix	Meaning
im-	into	*demos*	people	*-al, ic*	related to, state
inter-	between, among	*ethnos*	race, nation	*-er*	one who is
multi-	many	*graph*	writing	*-y*	quality, trait
		marier	marry		
		migrare	move		
		razza	race		

1. In sentence 1, what does the word **ethnic** mean? ____________________

____

2. In sentences 2, 7, 16, and 17, what does the word **interracial** mean? ________

____

3. In sentence 6, what does the word **multiracial** mean? ______________

____

4. In sentences 11, 12, and 13, what does the word **intermarry** mean? ________

____

5. In sentence 15, what does the word **demographers** mean? ____________

____

MASTERY TEST 4

Name ______________________ Section ______________

Date __________ Score (number correct) ______ × 10 = ______%

A. Using context clues, select the letter of the best meaning for each word in **bold** type. Then, identify the context clue you used.

______ **1.** To obtain a driver's license, a person should offer **verification**, or evidence, of good eyesight, fast reflexes, and reasoning skills.

a. proof
b. test
c. opinion
d. denial

______ **2.** The context clue used for the word *verification* in sentence 1 is called

a. a synonym.
b. an antonym.
c. the general context.
d. an example.

______ **3.** For dramatic impact, artists often favor the use of **complementary** colors such as orange and blue, which share no common colors because orange is made with yellow and red.

a. bold
b. related
c. contrasting
d. mysterious

______ **4.** The context clue used for the word *complementary* in sentence 3 is called

a. a synonym.
b. an antonym.
c. the general context.
d. an example.

B. Using context clues, write the definition for each word in **bold** type. Choose definitions from the box. Use each definition once.

distorted	doubts	imperfect	practical	pride	reasonable

5. Ancient tragedy teaches that all humans are **fallible**, or faulty, in motives, thoughts, or deeds.

Fallible means ______________________________________.

6. Ancient tragic heroes lose power, respect, and even their lives due to **hubris;** for example, Oedipus arrogantly believes he can outwit the gods and control his own life.

Hubris means ______________________________________.

7. Even though he still had **qualms** that he would fail, Angelo scored a 95 on the test, which was the highest grade in the class.

Qualms means ______________________________.

8. Due to the serious loss of life in manned space flights, some experts think that only unmanned missions should be thought of as **feasible** risks.

Feasible means ______________________________.

9. The loud evening songs of the cicada may sound **garbled** to the human ear, yet to the insects, they are clear messages.

Garbled means ______________________________.

10. A personal finance course teaches **pragmatic** steps to managing money such as how to balance a bank book, use credit cards, and evaluate financial risks.

Pragmatic means ______________________________.

VISUAL VOCABULARY

Vincent van Gogh used ______________ colors by placing brilliant orange flowers against a bright blue background.

Stated Main Ideas

3

LEARNING OUTCOMES

After studying this chapter you should be able to do the following:

1. Identify the following: traits of a main idea, the topic of a paragraph, a topic sentence.
2. Distinguish the movement of ideas from general and specific.
3. Define the following terms: *central idea* and *thesis statement.*
4. Locate the stated main idea in a paragraph or longer passage.
5. Assess your comprehension of locating stated main ideas.
6. Evaluate the importance of locating stated main ideas.

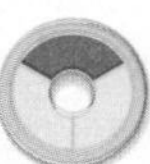

Before Reading About Stated Main Ideas

Effective use of the reading process relies on developing questions about the material that will guide you as you read. Using the chapter learning outcomes above, create at least five questions that you can answer as you study the chapter. Write your questions in the following spaces:

____? (page ______)

____? (page ______)

____? (page ______)

____? (page ______)

____? (page ______)

Compare the questions you created based on the chapter learning outcomes with the following questions. Then, write the ones that seem the most helpful in your notebook, leaving enough space after each question to record the answers as you read and study the chapter.

What are the traits of a main idea? (p. 96) What is the difference between a topic and a topic sentence? (p. 101) How is the flow of ideas related to the placement of topic sentences? (p. 105) What is the central idea? (p. 113) What is the difference between the central idea and a topic sentence? (p. 113)

The Traits of a Main Idea

A **main idea** is the author's controlling point about the topic. It usually includes the topic and the author's attitude or opinion about the topic, or the author's approach to the topic.

To identify the main idea, ask yourself two questions:

- Who or what is the paragraph about? The answer is the *topic*. The topic can be stated in just a few words.
- What is the author's controlling point about the topic? The answer is the *main idea*. The main idea is stated in one sentence.

Consider these questions as you read the following paragraph from the college textbook *Essentials of Human Communication*.

Textbook Skills

Communication Context

[1]The context of communication influences what you say and how you say it. [2]You communicate differently depending on the context you're in. [3]The communication context consists of at least four aspects. [4]The *physical context* refers to the tangible environment, the room, park, or auditorium; you don't talk the same way at a noisy football game as you do at a quiet funeral. [5]The *cultural context* refers to lifestyles, beliefs, values, behavior, and communication of a group; it is the rules of a group of people for considering something is right or wrong. [6]The *social-psychological context* refers to the status relationships among speakers, the formality of the situation; you don't talk the same way in the cafeteria as you would at a formal dinner at your boss's house. [7]The *temporal context* refers to the position in which a message fits into a sequence of events; you don't talk the same way after someone tells of the death of a close relative as you do after someone tells of winning the lottery.

—From *Essentials of Human Communication*, 4th ed., p. 7 by Joseph A. DeVito.

- *Who or what is the paragraph about*? The topic of the paragraph is "the communication context."
- *What is the author's controlling point about the topic*? The controlling point is that it "consists of at least four aspects." Putting topic and controlling point together, the main idea is "The communication context consists of at least four parts."

To better understand the traits of a main idea, compare a passage to a well-planned house of ideas. The *topic* or general subject matter is the roof. The roof covers all the rooms of the house. The *main idea* is the frame of the house, and the supporting details are the different rooms. The following diagram shows the relationship of the ideas:

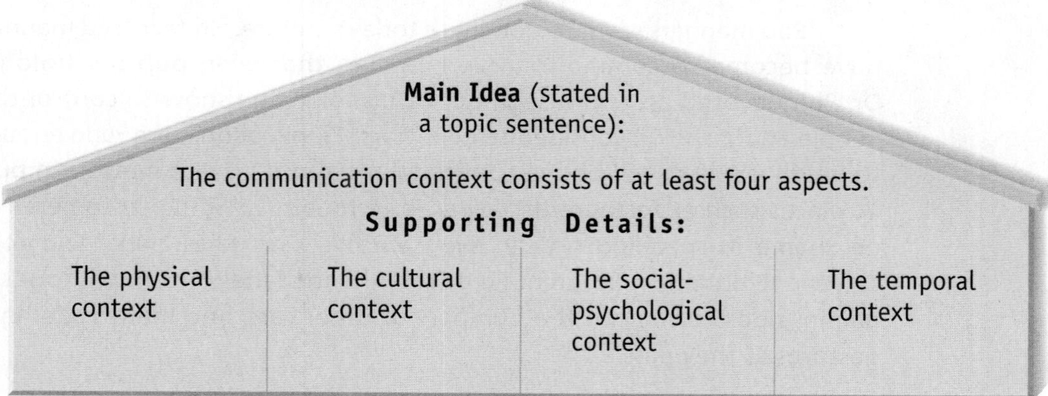
Topic: the communication context
Main Idea (stated in a topic sentence):
The communication context consists of at least four aspects.
Supporting Details:
The physical context
The cultural context
The social-psychological context
The temporal context

Each of the supporting details explains one aspect of the communication process.

Identifying the Topic of a Paragraph

When you ask the question "Who or what is the paragraph about?" you must be sure that your answer is neither too general nor too specific. A general subject needs specific ideas to support or explain it. However, no single paragraph can discuss all the specific ideas linked to a general idea. So an author narrows the general subject to a topic that needs a specific set of ideas to support it. For example, the very general subject "humor" can be narrowed to "stand-up comics." And the specific details related to stand-up comics might include different comics, ranging from Chris Rock to Ellen DeGeneres to Dane Cook.

However, a piece of writing dealing with the general topic "humor" will include a very different set of specific ideas than the narrower topic of "stand-up comics." The more general category of humor might include jokes and television sitcoms, for example.

Often an author shows the relationship between the topic and the specific details by repeating the topic throughout the paragraph as new pieces of information are introduced. To identify the topic, a master reader often skims the material for this recurring idea. Skimming for the topic allows you to grasp the relationship among a general subject, the topic, and specific details.

EXAMPLE Skim the following paragraph. Circle the topic as it recurs throughout the paragraph. Answer the question that follows.

A Question of Manners

[1]Bad manners seem to dominate today's culture. [2]In fact, bad manners have become so common in everyday life that even pop psychologist Dr. Phil devoted an entire segment of his television show to confronting people about their inconsiderate behaviors. [3]Some people are rude enough to loudly conduct cell phone conversations about private matters in public places such as restaurants, stores, and theatres. [4]Some customers feel no shame in speaking rudely, even abusively, to sales clerks and food servers. [5]Frequently, automobile drivers change lanes or turn without signaling, ride too near to the bumpers of other cars, and make rude hand gestures as they pass.

_____ Which of the following best states the topic?

a. manners
b. bad manners
c. confronting people about their inconsiderate behaviors

EXPLANATION "Manners" is too general, for it could cover good manners, business manners, or eating manners. "Confronting people about their inconsiderate behaviors" is too specific. This idea is a supporting detail, just one piece of evidence for the claim that bad manners dominate today's culture.

The topic of this paragraph is (b), "bad manners." You should have circled the following phrases: "Bad manners" (sentences 1 and 2), "inconsiderate behaviors" (sentence 2), "rudely" (sentence 4), and "rude" (sentences 3 and 5). Notice how the author uses the synonyms "rude" and "rudely" to restate the topic "bad manners."

PRACTICE 1

Skim each of the following paragraphs and circle the topic as it recurs throughout the paragraph. Then, identify the idea that correctly states the topic. (Hint: one idea is too general to be the topic; another idea is too specific.)

_______ **1.** [1]Reflexology is a form of natural healing based on the belief that reflexes in the hands, feet, and ears correspond to specific parts, glands, and organs of the body. [2]By applying pressure on these reflexes, reflexology aids the natural function of the related body areas by relieving tension and improving circulation. [3]Ancient humans stimulated reflexes naturally by working and building with their hands and walking barefoot over rough ground. [4]Modern humans have lost much of nature's way of maintaining a balanced and healthy equilibrium. [5]Reflexology aids natural health and vitality in several ways.

a. glands and organs
b. natural forms of healing
c. reflexology

VISUAL VOCABULARY

A ______________ uses a method of massage that eases tension through the use of finger pressure, especially to the feet.

a. reflex
b. reflexologist
c. reflexology

Textbook Skills

_______ **2.** [1]Just as we must critically evaluate information we read, we must also use critical thinking skills as we view images. [2]Every day we are bombarded with images—pictures on billboards, commercials on television, graphs and charts in newspapers and textbooks, to name just a few examples. [3]Most images slide by without our noticing them, or so we think. [4]But images, sometimes even more than text, can influence us. [5]Their creators have purposes, some worthy, some not. [6]And understanding those purposes requires that we think critically. [7]The methods of viewing images critically are the same as those we use for reading text critically. [8]For example, we preview, analyze, interpret, and (often) evaluate images.

—From *The Little, Brown Compact Handbook,* 4th ed., p. 64 by Jane E. Aaron. Copyright © 2001 by Pearson Education, Inc. Reprinted by permission of Pearson Education, Inc., Glenview, IL.

a. viewing images critically
b. graphs and charts
c. images

Identifying a Topic Sentence

Most paragraphs have three parts:

- A topic (the general idea or subject)
- A main idea (the controlling point the author is making about the topic, often stated in a topic sentence)
- Supporting details (the specific ideas to support the main idea)

Think again of the house of ideas that a writer builds. Remember, the main idea *frames* the specific ideas. Think of all the different rooms in a house: the kitchen, bedroom, bathroom, living room. Each room is a different part of the house. The frame determines the space for each room and the flow of traffic between rooms. Similarly, the main idea determines how much detail is given and how each detail flows into the next. The main idea of a paragraph is usually stated in a single sentence called the **topic sentence**. The topic sentence—the stated main idea—is unique in two ways.

First, the topic sentence contains two types of information: the topic and the author's controlling point, which restricts or qualifies the topic. At times, the controlling point may be expressed as the author's opinion using biased words. (For more information on biased words see Chapter 9, "Fact and Opinion.") For example, in the topic sentence "Bad manners seem to dominate today's culture," the biased words "bad," "seem," and "dominate" limit and control the general topic "manners."

Other times, the controlling point may express the author's thought pattern, the way in which the thoughts are going to be organized. (For more information on words that indicate thought patterns, see Chapters 7 and 8.) For example, the topic sentence, "Just as we must critically evaluate information we read, we must also use critical thinking skills as we view images" uses the phrase "just as" to reveal that the author will control the topic by comparing our responses to text and images.

Often, an author will use both biased words and a thought pattern to qualify or limit the topic. For example, the topic sentence, "Reflexology aids natural health and vitality in several ways," combines the biased word "aids" and the phrase "several ways" to indicate a list of positive examples and explanations has been provided.

These qualifiers—words that convey the author's bias or thought pattern—helped you correctly identify the topic in the previous section. An important difference between the topic and the topic sentence is that the topic sentence states the author's main idea in a complete sentence.

A **topic sentence** is a single sentence that states the topic and words that qualify the topic by revealing the author's opinion or approach to the topic.

The second unique trait of the topic sentence is its scope: the topic sentence is a general statement that all the other sentences in the paragraph explain or support. A topic sentence states an author's opinion or thought process, which must be explained further with specific supporting details. For example, in the paragraph about the communication context, the topic and the author's controlling point about the topic are stated in the first sentence. Each of the other sentences in the paragraph states and describes the four aspects of communication context:

Topic — Author's attitude — Author's thought pattern

The *communication context* consists of *at least four aspects.*

First aspect	Second aspect	Third aspect	Fourth aspect
The physical context	The cultural context	The social-psychological context	The temporal context

Supporting details are specific ideas that *develop, explain,* or *support* the main idea.

The supporting details of a paragraph are framed by the main idea, and all work together to explain or support the author's view of the topic. As a master reader, you will see that every paragraph has a topic, a main idea, and supporting details. It is much easier to tell the difference between these three parts of a passage once you understand how each part works. A topic, as the general subject of the paragraph, can be expressed in a word or phrase. The main idea contains both the topic and the author's controlling point about the topic and can be stated in one sentence called the topic sentence. The supporting details are all the sentences that state reasons and explanations for the main idea. To locate the topic sentence of a paragraph ask yourself two questions:

- Which sentence contains qualifiers that reveal the author's controlling point—that is, the author's attitude about the topic or approach to the topic?
- Do all the specific details in the passage support this statement?

EXAMPLES

A. The following group of ideas contains a topic, a main idea, and two supporting details. Circle the topic and underline the author's controlling point. Then, answer the questions.

a. A tsunami (tsoo-nah-me) can be the most disastrous of all ocean waves.

b. Tsunamis are powerful sea waves caused by seismic activity such as volcano eruptions, landslides, or earthquakes on the sea floor.

c. These waves, reaching heights over 100 feet and traveling as rapidly as 500 miles per hour, killed tens of thousands of people in the past century.

_____ **1.** Which of the following best states the topic?

a. a tsunami
b. seismic activity
c. an ocean wave

_____ **2.** Which sentence is the stated main idea?

B. Read the following paragraph. Circle the topic and underline the author's controlling point. Then, answer the two questions that follow it.

Drugs in Our Lives

Textbook Skills

[1]Today, more than ever, drugs affect our daily lives. [2]It is difficult to pick up a newspaper or to watch television without finding a report or program that concerns drug use or some issue associated with it. [3]In your personal life, you have had to confront the reality of drugs around you. [4]In school, you have probably been taught the risks involved in drug use,

and very likely you have had to contend with pressure to share a drug experience with your friends or the possibility of drugs being sold to you. [5]High school students on a Monday morning may boast about how much beer they consumed at a keg party over the weekend. [6]For some students, alcohol consumption begins in junior high school or earlier. [7]Experimentation with marijuana and mind-altering pills of all sorts seems commonplace. [8] Tobacco consumption among young people, despite the fact that it is illegal for those under eighteen years old to purchase tobacco products, remains a major societal problem. [9]Whether we like it or not, the decision to use drugs of all types and forms has become one of life's choices in American society and in communities around the world.

—Levinthal, *Drugs, Behavior, and Modern Society*, 3rd ed., p. 2.

1. What is the topic of the paragraph? ______________________

2. Which sentence states the author's main idea? ______________________

EXPLANATIONS

A. The topic is (a) "a tsunami." The first item (a) is the main idea. Note that the main idea is stated as a topic sentence, and contains both the topic "tsunami," and the author's controlling point about the topic, "most disastrous of all ocean waves." The second and third items give two details that support the main idea, by describing the disastrous traits of a tsunami.

B. Every idea in the paragraph has to do with drug usage. Sentence 1 states the main idea as a topic sentence: "Today, more than ever, drugs affect our daily lives." Notice that a topic sentence states the topic, and it states the author's controlling point about the topic. All the other sentences give reasons or evidence that "drugs affect our daily lives."

PRACTICE 2

Textbook Skills

The following groups of ideas come from the college textbook *Biology: Life on Earth*. Each group contains a main idea and two supporting details. In each group, first identify the topic; then, identify the statement main idea. (Hint: circle the topic and underline the author's controlling point in each group.)

Group 1

a. Living things are defined by distinct characteristics.
b. Living things are composed of cells that have a complex organized structure.

c. Living things acquire and use materials and energy from their environment and convert them into different forms.

—From *Biology: Life on Earth,* 8th ed., p. 11 by Teresa Audesirk, Gerald Audesirk, and Bruce E. Byers.

_____ **1.** Which of the following best states the topic?

a. cells
b. traits of living things
c. materials and energy
d. living things

_____ **2.** Which sentence is the stated main idea?

Group 2

a. All cells contain genes, units of heredity that provide the information needed to control the life of the cell and the small structures called organelles that are specialized to carry out specific functions such as moving the cell, obtaining energy, or synthesizing large molecules.
b. Even a single cell has an elaborate internal structure.
c. Cells are always surrounded by a thin plasma membrane that encloses the cytoplasm (organelles and the fluid surrounding them) and separates the cell from the outside world.

—From *Biology: Life on Earth,* 8th ed., p. 11 by Teresa Audesirk, Gerald Audesirk, and Bruce E. Byers.

_____ **3.** Which of the following best states the topic?

a. cells
b. organelles
c. hereditary information
d. the internal structure of cells

_____ **4.** Which sentence is the stated main idea?

Read the following paragraph from the college textbook *Biology: Life on Earth.* Circle the topic and underline the author's controlling point. Then, answer the questions that follow.

How Do Scientists Categorize the Diversity of Life?

[1]Although all living things share the general characteristics discussed earlier, evolution has produced an amazing variety of life-forms. [2]Organisms can be grouped into three major categories, called domains: Bacteria, Archaea, and Eukarya. [3]This classification reflects fundamental differences among the cell types that compose these organisms. [4]Members of both the Bacteria and the Archaea usually consist of single, simple cells. [5]Members of the Eukarya have bodies composed of one or more

highly complex cells. [6]This domain includes three major subdivisions or kingdoms: the Fungi, Plantae, and Animalia, as well as a diverse collection of mostly single-celled organisms collectively known as "protists." [7]There are exceptions to any simple set of criteria used to characterize the domains and kingdoms, but three characteristics are particularly useful: cell type, the number of cells in each organism, and how it acquires energy.

—From *Biology: Life on Earth*, 8th ed., pp. 14–15 by Teresa Audesirk, Gerald Audesirk, and Bruce E. Byers. Copyright © 2008 by Pearson Education, Inc. Printed and Electronically reproduced by permission of Pearson Education, Inc., Upper Saddle River, NJ.

5. What is the topic of the paragraph? ______________________

6. What is the topic sentence stating the main idea? ______________________

VISUAL VOCABULARY

Trichophyton rubrum fungus This fungus infects tissues such as skin, nails, and hair in humans and other animals. This organism is classified as part of the ______________ domain.

a. Bacteria
b. Archaea
c. Eukarya

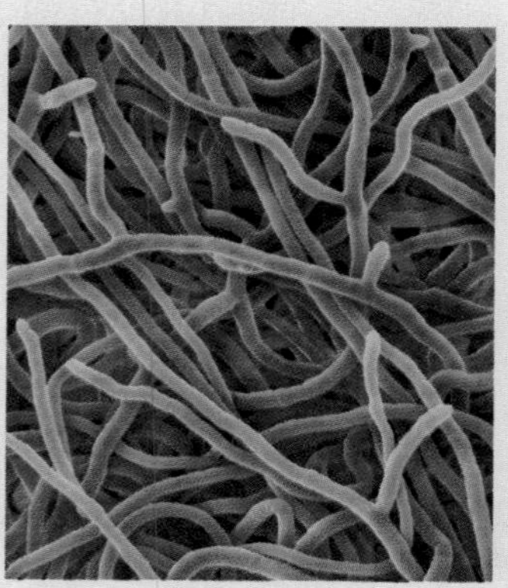

The Flow of Ideas and Placement of Topic Sentences

So far, many of the paragraphs you have worked on in this textbook have placed the topic sentence/main idea as the first sentence in the paragraph. However, not all paragraphs put the main idea first. In fact, a topic sentence can be placed at the **beginning** of a paragraph, **within** a paragraph, at the **end** of a paragraph, or at both the beginning and the end of a paragraph. The placement of the topic sentence controls the flow of ideas. In a sense, when a writer builds a house of ideas, the floor plan—the flow of ideas—changes based on where in the paragraph the topic sentence is located. One of the first things a master reader looks for is the location of the topic sentence.

Topic Sentence at the Beginning of a Paragraph

A topic sentence that begins a paragraph signals a move from general ideas to specific ideas. This flow from general to specific, in which an author begins with a general statement and moves to specific reasons and supports, is also known as deductive reasoning. Articles in encyclopedias and news stories in magazines and newspapers typically use the deductive flow of ideas. The following chart shows this flow from general to specific ideas.

Main idea: Topic sentence
Supporting detail
Supporting detail
Supporting detail
Supporting detail

EXAMPLE Read the paragraph, and identify its topic sentence. Remember to ask, "Does this sentence cover all the ideas in the passage?"

Textbook Skills

Cancer

[1]Cancer is a far-reaching malady. [2]Statistics tell us that it affects three of four families. [3]About a third of all Americans living today will contract cancer in their lifetime. [4]And cancer kills more children between ages three and fourteen years than any other disease. [5]Someone dies from cancer in the United States approximately every minute.

—Adapted from *Mind/Body Health: The Effects of Attitudes, Emotions,* and Relationships, 2nd ed., p. 13 by Keith J. Karren, Brent Q. Hafen, Lee Smith, and Kathryn J. Frandsen. Copyright © 2002 by Pearson Education, Inc. Reprinted by permission of Pearson Education, Inc., Upper Saddle River, NJ.

Topic sentence: ______

EXPLANATION The topic sentence of this paragraph is sentence 1: "Cancer is a far-reaching malady." All the other sentences explain the extent of cancer's effects.

Topic Sentence Within a Paragraph

Topic sentences within a paragraph can be near the beginning or in the middle of the paragraph.

Near the Beginning

A paragraph does not always start with the topic sentence. Instead, it may begin with a general overview of the topic. These introductory sentences are used to get the reader interested in the topic. They also lead the reader to the topic sentence. Sometimes introductory sentences tell how the ideas in one paragraph tie in to the ideas of earlier paragraphs. At other times, the introductory sentences give background information about the topic.

The flow of ideas remains deductive as it moves from general ideas (the introduction) and main idea (topic sentence) to specific ideas (supporting details). Human interest stories and editorials in magazines and newspapers, as well as academic papers, often rely on this flow of ideas. The following diagram shows this flow from general to specific ideas:

Introductory sentence(s)
Main idea: Topic sentence
Supporting detail
Supporting detail
Supporting detail

EXAMPLE Read the following paragraph, and identify its topic sentence. Remember to ask, "Does this sentence cover all the ideas in the passage?"

Textbook Skills

Understanding and Meeting Guest Needs

[1]The first step in delivering high-quality service is to learn and fully understand what customers want in a particular tourism service. [2]Tourism managers can uncover specific needs and expectations of customers in a number of ways. [3]First, marketing research can be used to gather information from potential and existing customers. [4]Many companies regularly survey members of their target market to better understand the changing needs and desires of segments they hope to serve. [5]For example, when PepsiCo acquired Taco Bell, management conducted a study of fast-food customers (any fast-food customer, not simply customers who liked Mexican food). [6]From this survey, PepsiCo concluded that fast-food customers had expectations about four things, which can be remembered by the acronym FACT. [7]Customers wanted their fast food really *Fast;* they expected their orders to be *Accurately* delivered; they wanted the premises to be *Clean;* and they expected foods and beverages to be served at appropriate *Temperatures.* [8]With this knowledge, top management

redesigned the entire Taco Bell system to better deliver these expected qualities.

—From *Tourism: The Business of Travel*, 4th ed., pp. 69–70 by Roy A. Cook, Laura J. Yale and Joseph J. Marqua. Copyright © 2010 by Pearson Education, Inc. Printed and Electronically reproduced by permission of Pearson Education, Inc., Upper Saddle River, NJ.

Topic sentence: ______

EXPLANATION Sentence 2 is the topic sentence of this paragraph. Sentence 1 offers background information about the general topic "meeting guest needs." The purpose of sentence 1 is to get the reader's attention and tie the upcoming ideas with previously discussed ideas. Sentences 3 through 8 offer supporting details that explain and illustrate how a company can "meet guest needs."

In the Middle

At times, an author begins a paragraph with a few attention-grabbing details. These details are placed first to stir the reader's interest in the topic. The flow of ideas no longer follows the deductive pattern of thinking because the material now moves from specific ideas (supporting details) to a general idea (the topic sentence) to specific ideas (additional supporting details). Creative essays and special interest stories that strive to excite reader interest often employ this approach. Often television news stories begin with shocking details to hook the viewer and prevent channel surfing. The following diagram shows this flow of ideas:

Supporting detail
Supporting detail
Main idea: Topic sentence
Supporting detail
Supporting detail

EXAMPLE Read the following paragraph, and identify its topic sentence. Remember to ask, "Does this sentence cover all the ideas in the passage?"

Textbook Skills

Adolescence

[1]John and Jorge are two 16-year-old neighbors and friends of different ethnic backgrounds. [2]At the same time, they are so much alike. [3]They both wear adult-size clothes, both have a shadow of a mustache on their

upper lip, both play computer games for many hours a day, both contemplate getting a summer job, and both think of attending a local college in two years. [4]As adolescents, they both have reached sexual maturity but have not yet taken on the rights and responsibilities of the adult status. [5]Adolescence is viewed not only as a developmental stage but also as a cultural phenomenon. [6]For instance, extended schooling in many developed countries stretches the period from childhood to adulthood. [7]On the contrary, many non-industrialized cultures encourage their members to take on adult roles as early as possible. [8]Thus, the adolescent stage becomes almost indistinguishable. [9]In some countries, such as Brazil, many children begin to work full time and take care of other family members as early as 12 and sometimes even earlier. [10]In other societies such as India, a girl can marry in her early teens and move to her husband's home to accept the roles of wife and mother.

—Shiraev & Levy, *Cross-Cultural Psychology: Critical Thinking and Contemporary Applications,* 3rd ed., pp. 235–236.

Topic sentence: ______

EXPLANATION Sentences 1 through 4 offer introductory examples that capture the readers' interest, set up the topic, and give concrete examples of the developmental traits of adolescents. Sentence 5 is the topic sentence. Sentences 6 through 10 illustrate that adolescence is a cultural phenomenon.

Topic Sentence at the End of a Paragraph

Sometimes an author waits until the end of the paragraph to state the topic sentence and main idea. This approach can be very effective, for it allows the details to build up to the main idea. The pattern is sometimes called climactic order.

The flow of ideas is known as inductive as the author's thoughts move from specific (supporting details) to general (the topic sentence). Inductive reasoning is often used in math and science to generate hypotheses and theories, and to discover relationships between details. In addition, inductive reasoning is often used in argument (for more about argument, see Chapters 12 and 13). Politicians and advertisers use this approach to convince people to agree with their ideas or to buy their products. If a politician begins with a general statement such as "Taxes must be raised," the audience members may strongly disagree. If, however, the politician begins with the details and leads up to the main idea, people are more likely to agree. For example, people are more likely to agree that roads need to be repaired, even if it means raising taxes. Inductive reasoning is the process of arriving at a general understanding based on specific details. The following diagram shows the ideas moving from specific to general.

Supporting detail
Supporting detail
Supporting detail
Supporting detail
Main idea: Topic sentence

EXAMPLE Read the following paragraph, and identify its topic sentence. Remember to ask, "Does this sentence cover all the ideas in the passage?"

Roofies and Rape

[1]Charlene sat perched on a bar stool as she sipped her drink and chatted flirtatiously with the cute stranger who had introduced himself as Roger. [2]After a few minutes of conversation, she excused herself for a quick trip to the restroom. [3]While she was gone, Roger dropped a little white pill into her glass and stirred until it dissolved. [4]When she returned, she finished her drink. [5]About ten minutes later, Charlene began to feel dizzy, disoriented, and nauseated; she felt simultaneously too hot and too cold. [6]Roger asked her if she felt all right and offered to drive her home. [7]Because she had difficulty speaking and moving, Roger easily guided her out of the bar and into his car. [8]Shortly thereafter, she passed out. [9]Roger took her to a remote wooded area, raped her, and left her there. [10]The next morning, she had no memory of what had happened. [11]Charlene became one of the thousands of victims of Rohypnol, also known as "roofies." [12]Rohypnol is used as a "date rape" drug, which causes a victim to pass out so she cannot resist or recall what happened to her.

Topic sentence: ______

EXPLANATION Sentences 1 through 11 tell the story of one woman's experience with Rohypnol. Sentence 12 states the main idea. Starting the passage with the details of the incident draws readers into the situation and heightens their interest. Ending the passage with the main idea is very powerful.

Topic Sentence at the Beginning and the End of a Paragraph

A paragraph may start and end by stating one main idea in two different sentences. Even though these two sentences state the same idea, they usually word the idea in different ways. A topic sentence presents the main idea at the beginning of the paragraph. Then, at the end of the paragraph, the main idea is stated again, this time using different words. This flow of ideas is based on the age-old advice given to writers to "tell the reader what you are going to say; say it; then tell the reader what you said." Many essays written by college students rely on this presentation of ideas. The following diagram shows this flow of ideas:

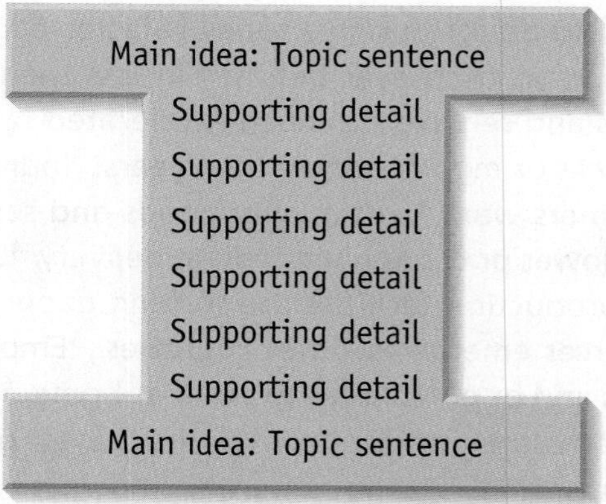

EXAMPLE Read the following paragraph, and identify its topic sentences. Remember to ask, "Do these sentences cover all the ideas in the passage?"

Recreational Therapy

[1]Recreational therapy benefits the physical and mental health of many patients. [2]Recreational therapy includes music, art, drama, and dance. [3]Patients who benefit from recreational therapy include those who suffer from substance abuse, eating disorders, violent abuse, homelessness, autism, and complications due to aging. [4]This type of therapy improves a person's physical, mental, social, and emotional abilities. [5]Recreational therapy promotes good health and well-being, which leads to a higher quality of life. [6]This therapy also reduces the need for extended healthcare services by nurturing independent living skills. [7]Recreational therapy is an excellent method to improve or maintain physical and mental health.

Topic sentences: _____

EXPLANATION Sentences 1 and 7 both state the main idea of the passage: Recreational therapy benefits patients. Notice how the wording changes at the end of the passage. Repeating the main idea makes the point much stronger and more likely to be remembered.

PRACTICE 3

Textbook Skills

Read the following series of paragraphs taken from the same section of the college textbook *Business*. Identify the topic sentence of each paragraph. Remember to ask, "Which sentence covers all the details in each paragraph."

The Business Environment

[1]No doubt business today is faster paced, more complex, and more demanding than ever before. [2]In less than a decade, the hunt for new goods and services has been accelerated by product life cycles measured in weeks or months rather than years. [3]Individual consumers and business customers want high-quality goods and services—often customized and with lower prices and immediate delivery. [4]Sales offices, service providers, and production facilities are shifting geographically as new markets and resources emerge in other countries. [5]Employees want flexible working hours and opportunities to work at home. [6]Stockholder expectations also add pressure for productivity increases, growth in market share, and larger profits. [7]At the same time, however, a more vocal public demands greater honesty, fair competition, and respect for the environment.

—From *Business*, 8th ed., p. 44 by Ricky W. Griffin and Ronald J. Ebert. Copyright © 2006 by Pearson Education, Inc. Printed and Electronically reproduced by permission of Pearson Education, Inc., Upper Saddle River, NJ.

Topic sentence(s): ______

Redrawing Corporate Boundaries

[1]Successful companies are responding to these challenges in new, often unprecedented ways. [2]To stay competitive, they are redrawing traditional organizational boundaries. [3]Today, firms join together with other companies, even with competitors, to develop new goods and services. [4]Some of these relationships are permanent. [5]But others are temporary alliances formed on short notice so that, working together, partners can produce and deliver products with shorter lead times than either firm could manage alone. [6]Increasingly, the most successful firms are getting leaner by focusing on their core competencies. [7]Core competencies are the skills and resources with which they compete best and create the most value for owners. [8]They *outsource* noncore business processes. [9]They pay suppliers and distributors to perform these secondary processes

and thereby increase their reliance on suppliers. [10]These new business models call for unprecedented coordination—not only among internal activities, but also among customers, suppliers, and strategic partners.

—From *Business*, 8th ed., pp. 44–45 by Ricky W. Griffin and Ronald J. Ebert. Copyright © 2006 by Pearson Education, Inc. Printed and Electronically reproduced by permission of Pearson Education, Inc., Upper Saddle River, NJ.

Topic sentence(s): ______

Outsourcing

[1]Outsourcing is the strategy of paying suppliers and distributors to perform certain business processes or to provide needed materials or services. [2]For example, the cafeteria at a large bank may be important to employees and some customers. [3]But running it is not the bank's main line of business and expertise. [4]Bankers need to focus on money management and financial services, not foodservice operations. [5]That's why most banks outsource cafeteria operations to foodservice management companies whose main line of business includes cafeterias. [6]The result, ideally, is more attention to banking by bankers, better food service for cafeteria customers, and formation of a new supplier-client relationship (foodservice company/bank). [7]Firms today outsource numerous activities. [8]These include payroll, employee training, and research and development. [9]Outsourcing is an increasingly popular strategy because it helps firms focus on their core activities and avoid getting sidetracked onto secondary activities.

—From *Business*, 8th ed., pp. 45–46 by Ricky W. Griffin and Ronald J. Ebert. Copyright © 2006 by Pearson Education, Inc. Printed and Electronically reproduced by permission of Pearson Education, Inc., Upper Saddle River, NJ.

Topic sentence(s): ______

The Central Idea and the Thesis Statement

Just as a single paragraph has a main idea, longer passages made up of two or more paragraphs also have a main idea. You encounter these longer passages in articles, essays, and textbooks. In longer passages, the main idea is called the **central idea**. Often the author will state the central idea in a single sentence called the **thesis statement**.

The **central idea** is the main idea of a passage made up of two or more paragraphs.

The **thesis statement** is a sentence that states the topic and the author's controlling point about the topic for a passage of two or more paragraphs.

You find the central idea of longer passages the same way you locate the main idea or topic sentence of a paragraph. The thesis statement is the one sentence that is general enough to include all the ideas in the passage.

EXAMPLE Read the following passage from a college health textbook, and identify the thesis statement, which states the central idea.

Textbook Skills

Sunlight and Depression

[1]Depression can strike anyone, and it has been known to affect people of all ages and from all walks of life. [2]We do know that certain people are at increased risk for depression. [3]One of the most significant factors is sunlight. [4]Recent experiments at the National Institute of Mental Health have concluded that the ability to deal with stress—and thus to avoid depression—can be significantly influenced by the amount of sunlight received each day. [5]Data from the studies indicate that a number of people seem better able to cope with stress, change, and challenge during the spring, summer, and early autumn months. [6]As winter approaches and the days grow shorter and darker, many persons become lethargic, anxious, and depressed.

[7]Investigators at the National Institute of Mental Health noticed that during the winter, people tend to slow down, gain weight, and sleep more. [8]Many also seem to struggle more with depression in the winter months, when there is less light. [9]Some even develop serious problems more than just a mild case of the blues. [10]Some become incapacitated. [11]Some even become suicidal.

[12]Believing that light played a role, investigators exposed persons predisposed to suffer from "light deprivation" to strong artificial, broad-spectrum light for at least five hours a day. [13]For those exposed to light, symptoms of depression and distress were significantly reduced or even completely eliminated. [14]The therapeutic light is broad-spectrum white light; yellow incandescent bulbs do not work. [15]Since those early studies, the daily regimen has been much simplified and has produced much the same benefits. [16]The usual regimen today is exposure to bright lights for 30 to 60 minutes in the early morning.

—From *Mind/Body Health: The Effects of Attitudes, Emotions, and Relationships*, 2nd ed., pp. 254–255 by Keith J. Karren, Brent Q. Hafen, Lee Smith, and Kathryn J. Frandsen.

Thesis statement: ______

EXPLANATION The first three sentences hook the reader's interest in the topic "sunlight and depression." Sentence 4 is the central idea of the passage.

Note that sentence 4 includes the topic "sunlight and depression" and the author's controlling point about the topic: "the ability to deal with stress—and thus to avoid depression—can be significantly influenced by the amount of sunlight received each day." Sentences 5 through 16 are supporting details that list the observations that investigators made about the role of light in depression.

PRACTICE 4

Read the following passage from a college history textbook. Identify the thesis statement, which states the central idea.

Textbook Skills

God Kings

[1]Divine kingship was the cornerstone of Egyptian life. [2]Initially, the king was the incarnation of Horus, a sky and falcon god; later, the king was identified with the sun god Ra, as well as with Osiris, the god of the dead. [3]As divine incarnation, the king was obliged above all to care for his people. [4]It was he who ensured the annual flooding of the Nile, which brought water to the parched land. [5]His commands preserved *maat*, the ideal state of the universe and society, a condition of harmony and justice. [6]In the poetry of the Old Kingdom, the king was the divine herdsman, while the people were the cattle of god:

[7]Well tended are men, the cattle of god.

[8]He made heaven and earth according to their desire and repelled the demon of the waters . . .

[9]He made for the rulers (even) in the egg, a supporter to support the back of the disabled.

[10]Unlike the rulers in Mesopotamia, the kings of the Old Kingdom were not warriors but divine administrators. [11]Protected by the Sahara, Egypt had few external enemies and no standing army. [12]A vast bureaucracy of literate court officials and provincial administrators assisted the god-king. [13]They wielded wide authority as religious leaders, judicial officers, and when necessary, military leaders. [14]A host of subordinate overseers, scribes, metalworkers, stonemasons, artisans, and tax collectors rounded out the royal administration. [15]At the local level, governors administered provinces called *nomes*, the basic unit of Egyptian local government.

—Adapted from Kishlansky, Geary, & O'Brian, *Civilization in the West*, 4th ed., pp. 18–19.

Central idea: ______

Reading a Textbook: Topics, Main Ideas, and Central Ideas

Textbooks identify topics in the title of each chapter. An excellent study strategy is to read a textbook's table of contents, including all of the chapters' titles, which list the general topics covered in the textbook. In addition to providing topics in chapter titles, textbooks also identify topics within each chapter. Other publications, such as newspapers and magazines, also use titles and headings to point out topics.

Textbook authors often state the topic of a passage or paragraph in a heading. For example, titles of graphs often help readers identify the main idea of the graph by stating the topic. Identifying the topic in a heading makes it easier to identify the central idea and supporting details.

EXAMPLE Read the following passage from a psychology textbook. Then, answer the questions that follow it.

Workplace Violence: Aggression on the Job

[1]City of Industry, California—A postal worker walked up to his boss, pulled a gun from a paper bag and shot him dead, the latest incident in an alarming increase in workplace violence. (*Los Angeles Times*, July 18, 1995)

[2]Portland, Oregon—A man accused of shooting two people and taking four others hostage in an office tower appeared in court on Friday. . . . [3]Police initially said Rancor intended to shoot female office workers for having him fired from his job—but investigators said Friday that Rancor had problems with authority in general. (Associated Press, 1996)

[4]Reports of events such as these have appeared with alarming frequency in recent years. [5]They appear to reflect a rising tide of violence in the workplace. [6]In fact, more than eight hundred people are murdered at work each year in the United States (National Institute for Occupational Safety and Health, 1993). [7]While these statistics seem to suggest that workplaces are becoming truly dangerous locations where disgruntled employees frequently attack or even shoot one another, two facts should be carefully noted. [8]First, a large majority of violence occurring in work settings is performed by outsiders who do not work there but who enter a workplace to commit robbery or other crimes. [9]Second, surveys indicate

that threats of physical harm or actual harm in work settings are actually quite rare. [10]In fact, for most occupations, the chances of being killed at work (by outsiders and coworkers combined) are something like 1 in 450,000.

[11]In sum, growing evidence suggests that while workplace violence is certainly an important topic worthy of careful study, it is relatively rare, and violence in work settings is, in fact, only the dramatic tip of the much larger problem of workplace aggression. [12]**Workplace aggression** is any form of behavior through which individuals seek to harm others in their workplace (Baron & Neuman, 1996; Neuman & Baron, 1998).

—Text and graph from *Social Psychology with Research Navigator*, 10th ed., pp. 460–461 and Figure 11.15 (p. 461) by Robert A. Baron and Donn Byrne. Copyright © 2004 by Allyn & Bacon, Inc. Reproduced by permission of Pearson Education, Inc., Boston, MA.

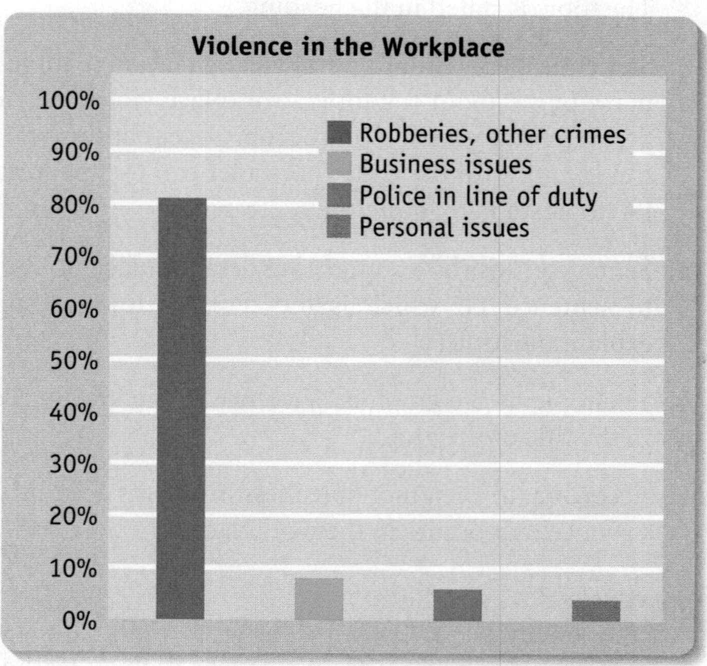

1. The topic of the paragraph is ______________________.

______ **2.** Sentence 11 is
a. the central idea. b. a supporting detail.

______ **3.** Sentence 10 is
a. the central idea. b. a supporting detail.

4. The topic of the graph is ______________________________.

5. ______ According to the graph, what percentage of workplace violence is tied to business problems?
a. about 80%
b. about 10%
c. a bit more than 5%
d. a bit less than 5%

EXPLANATION

1. The topic of the passage is "Workplace Violence: Aggression on the Job." The topic is stated in the heading.

2. Sentence 11 is (a) the central idea; it states the topic and the author's primary point about the topic: workplace violence is just part of the larger problem of aggression on the job. Note that this passage includes the central idea toward the end. The reader can use the heading, which identifies the topic, to help locate the central idea.

3. Sentence 10 is (b) a supporting detail. In fact, all the sentences leading up to sentence 11, which is the central idea, are supporting details that explain the topic.

4. The topic of the graph is "Violence in the Workplace." The topic is stated in the title of the graph.

5. According to the graph, business problems account for about 10% of the violence that occurs in the workplace.

PRACTICE 5

Read the following passage from a college biology textbook. Then, answer the questions that follow it.

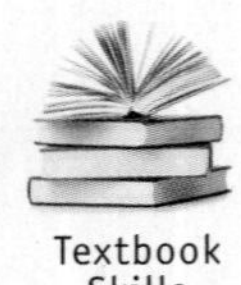
Textbook Skills

The Scientific Method Is the Basis of Scientific Inquiry

[1]A biologist's job is to answer questions about life. [2]What causes cancer? [3]Why are frog populations shrinking worldwide? [4]What happens

when a sperm and egg meet? [5]How does HIV cause AIDS? [6]When did the earliest mammals appear? [7]How do bees fly? [8]The list of questions is both endless and endlessly fascinating.

[9]Anyone can ask an interesting question about life. [10]A biologist, however, is distinguished by the manner in which he or she goes about finding answers. [11]A biologist is a scientist and accepts only answers that are supported by evidence, and then only by a certain kind of evidence. [12]Scientific evidence consists of observations or measurements that are easily shared with others and that can be repeated by anyone who has the appropriate tools. [13]The process by which this kind of evidence is gathered is known as the **scientific method.**

[14]The scientific method proceeds step-by-step. [15]It begins when someone makes an **observation** of an interesting pattern or phenomenon. [16]The observation, in turn, stimulates the observer to ask a **question** about what was observed. [17]Then, after a period of contemplation (that perhaps also includes reflecting on the scientific work of others who have considered related questions), the person proposes an answer to the question, an explanation for the observation. [18]This proposed explanation is a **hypothesis.** [19]A good hypothesis leads to a **prediction,** typically expressed in "if . . . then" language. [20]The prediction is tested with further observations or with experiments. [21]These experiments produce results that either support or refute the hypothesis, and a **conclusion** is drawn about it. [22]A single experiment is never an adequate basis for a conclusion; the experiment must be repeated not only by the original experimenter but also by others.

[23]You may find it easier to visualize the steps of the scientific method through an example from everyday life. [24]Imagine this scenario: Late to an appointment, you rush to your car and make the *observation* that it won't start. [25]This observation leads directly to a *question:* Why won't the car start? [26]You quickly form a *hypothesis:* The battery is dead. [27]Your hypothesis leads in turn to an if-then *prediction:* If the battery is dead, then a new battery will cause the car to start. [28]Next, you design an *experiment* to test your prediction: You replace your battery with the battery from your roommate's new car and try to start your car again. [29]Your car starts immediately, and you reach the *conclusion* that your dead battery hypothesis is correct.

—From *Life on Earth*, 5th ed., Figure 1-3 (p. 363) and pp. 3–4 by Teresa Audesirk, Gerald Audesirk and Bruce E. Byers. Copyright © 2009 by Pearson Education, Inc. Reprinted by permission of Pearson Education, Inc., Upper Saddle River, NJ.

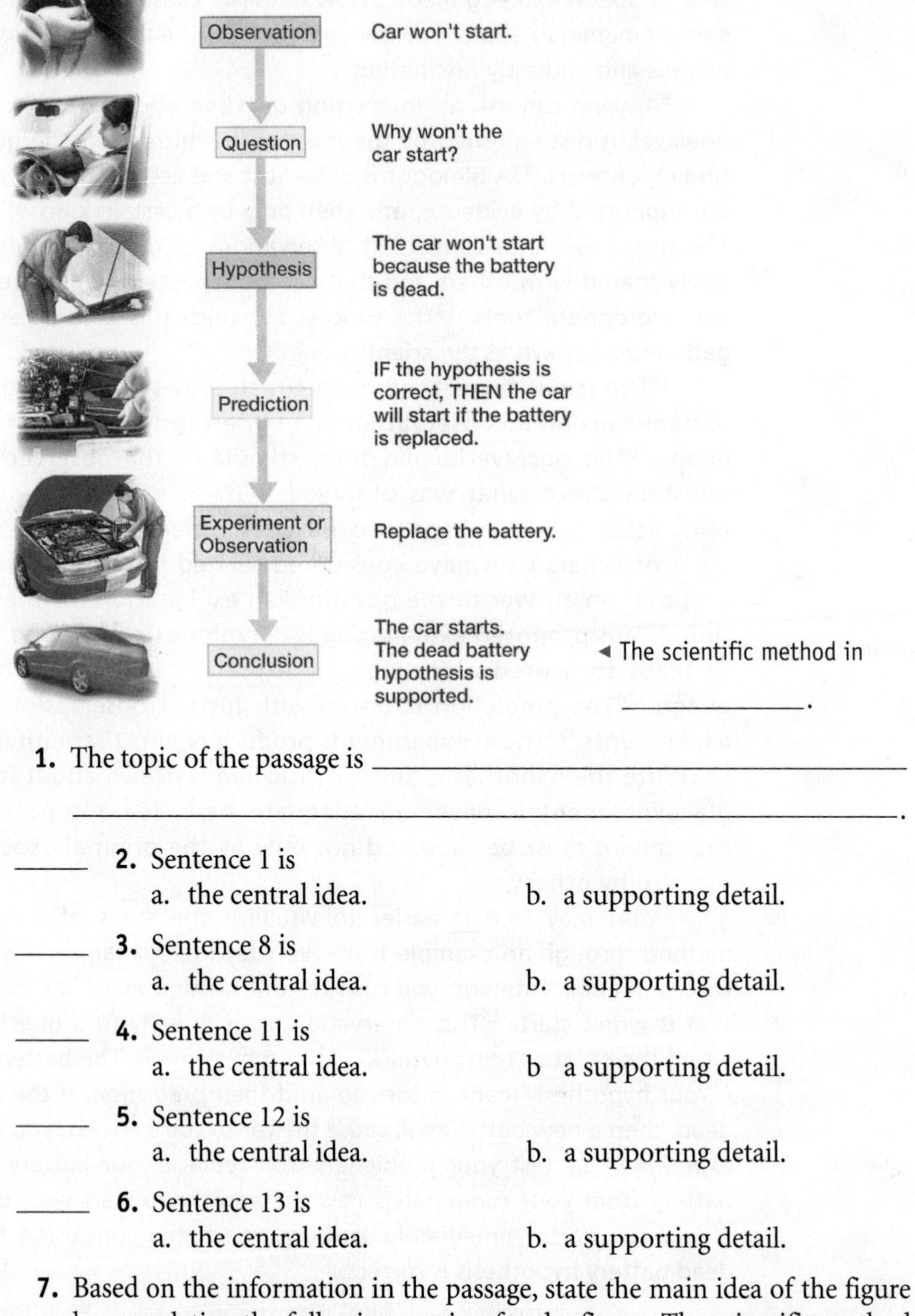
Observation
Car won't start.
Question
Why won't the car start?
Hypothesis
The car won't start because the battery is dead.
Prediction
IF the hypothesis is correct, THEN the car will start if the battery is replaced.
Experiment or Observation
Replace the battery.
Conclusion
The car starts. The dead battery hypothesis is supported.

◄ The scientific method in ____________.

1. The topic of the passage is ____.

_____ **2.** Sentence 1 is
a. the central idea. b. a supporting detail.

_____ **3.** Sentence 8 is
a. the central idea. b. a supporting detail.

_____ **4.** Sentence 11 is
a. the central idea. b a supporting detail.

_____ **5.** Sentence 12 is
a. the central idea. b. a supporting detail.

_____ **6.** Sentence 13 is
a. the central idea. b. a supporting detail.

7. Based on the information in the passage, state the main idea of the figure by completing the following caption for the figure: The scientific method in ____.

APPLICATIONS

Application 1: Identifying Topics

Read the following information that was posted on a government website. Circle the key terms that identify the topic as they recur throughout the paragraph. Then, answer the question that follows.

USAID Disaster Assistance Response Team: Haiti Earthquake

[1]The U.S. Agency for International Development (USAID) is an independent federal agency. [2] USAID provides economic development and humanitarian assistance around the world in support of the foreign policy goals of the United States. [3]Within USAID, a select group of staff is prepared to respond to disasters abroad, no matter when or where they occur. [4]This group is known as the Disaster Assistance Response Team (DART). [5]They are experienced professionals, dedicated to saving lives, easing suffering, and reducing the social and economic impact of disasters. [6]Within 24 hours of the 7.0 magnitude earthquake that rocked Haiti on January 12, 2010, the DART arrived in Haiti. [7]They immediately set to work. [8]DART conducted urban search and rescue (USAR) operations. [9]They organized logistics, established critical telecommunications support for the team and for the Government of Haiti. [10]In addition, they provided information updates on the unfolding situation. [11]The DART coordinated the delivery of more than 40 planeloads of relief. [12]This included more than 111,000 water containers, nearly 75,000 hygiene kits, more than 10,000 rolls of plastic sheeting, and 5,000 kitchen sets. [13]In addition, the six USAID-funded USAR teams saved the lives of 47 people trapped in collapsed structures. [14]Within six weeks, USAID programmed more than $400 million to address immediate food, water, health, and shelter needs for earthquake-affected populations.

—Adapted from "SUCCESS STORY: USAID Disaster Assistance Response Team: Haiti Earthquake." USAID 2 March 2010. http://www.usaid.gov/helphaiti/.

What is the topic of the paragraph? ______________________

____

Application 2: The Topic and the Main Idea

Read the following paragraph. Then, answer the questions that follow it.

Computer Worms, Viruses, and Bugs

[1]Worms, viruses, and bugs have long been loathed by many in the natural world as annoying and even dangerous pests. [2]Likewise, the artificial world of computers constantly battles the infestation of worms, viruses, and bugs. [3]A computer virus is a programming code that is deliberately designed as a sneak attack on other people's computers. [4]A virus may be designed to deliver a harmless, silly message or to strike a destructive blow to files on hard drives. [5]Computer viruses depend on the aid of the unwitting computer user. [6]For example, viruses are usually spread by copying or transferring a file. [7]A computer worm is a program deliberately designed to reproduce itself on a network of computers. [8]Unlike a virus, a computer worm does not need the aid of the user to wreak havoc. [9]In contrast, a computer bug is an unintended flaw in a computer's program. [10]Often, a bug occurs because the rules of the programming language were broken or misunderstood. [11]Or, sometimes, a step or detail in the programming process was skipped.

1. What is the topic of the paragraph? ________________

_____ **2.** Which sentence states the author's main idea?

a. sentence 1
b. sentence 2
c. sentence 4

Application 3: Topics, Main Ideas, and Supporting Details

Each of the following groups of ideas has a main idea and two supporting details. In each group, first identify the topic. Then, identify the stated main idea. (Hint: circle the topic and underline the author's controlling point in each group.)

Group 1

a. A geyser is a powerful eruption of a hot spring into a column of hot water or steam.
b. For example, the geyser Old Faithful erupts into a 60-foot tower of 350-degree water.
c. Old Faithful erupts about every 30 minutes around the clock, and each eruption lasts approximately three or four minutes.

_____ **1.** Which of the following best states the topic?

a. Old Faithful
b. geysers
c. hot springs

_____ **2.** Which sentence is the stated main idea?

Group 2

a. The Emotional Competence Inventory is a testing tool that measures emotional intelligence.
b. For example, the abilities to manage negative feelings and control impulses are signs of high emotional intelligence.
c. Emotional intelligence is knowing what your feelings are and making good decisions based on wise use of your emotions.

_____ **3.** Which of the following best states the topic?
a. emotional intelligence
b. The Emotional Competence Inventory
c. feelings

_____ **4.** Which sentence is the stated main idea?

Application 4: Location of the Topic Sentence

Read the following paragraph from a college finance textbook and identify its topic sentence or sentences. Remember to ask, "Does this sentence (Do these sentences) cover all the ideas in the passage?"

Textbook Skills

Disadvantages of Using Credit

[1]There can be a high cost to using credit. [2]If you borrow too much money, you may have difficulty making your credit card payments. [3]It is easier to obtain credit than to pay it back. [4]And having a credit line can tempt you to make impulse purchases that you cannot afford. [5]Eighty-three percent of all students have at least one credit card, and almost a third have four or more. [6]The average credit card balance for 21 percent of undergraduates is between $3,000 and $7,000. [7]Many students make minimum payments on their credit cards while in school with the expectation that they will be able to pay off their balance once they graduate and are working full-time. [8]Yet the accumulating interest fees catch many by surprise, and the debt can quickly become difficult to manage. [9]Today's graduating students have an average of $24,567 in combined education loan and credit card balances. [10]If you are unable to repay the credit you receive, you may not be able to obtain credit again or will have to pay a very high interest rate to obtain it. [11]Your ability to save money will also be reduced if you have large credit payments. [12]If your spending and credit card payments exceed your net cash flow, you limit your ability to increase your personal wealth. [13]Credit can be costly.

—Adapted from *Personal Finance*, 3rd ed., pp. 174–175 by Jeff Madura. Copyright © 2007 by Pearson Education, Inc. Reproduced by permission of Pearson Education, Inc., Boston, MA.

Topic sentence(s): _____

REVIEW TEST 1

Score (number correct) _____ × 20 = _____ %

Topics, Main Ideas, and Supporting Details

Textbook Skills

A. Each of the following groups of ideas includes one topic, one main idea, and two supporting details. In each group, first identify the topic. Then, identify the stated main idea. (Hint: circle the topic and underline the author's controlling point in each group.)

Group 1

a. People use several methods to avoid conflict.
b. In addition, quick acceptance of a suggested solution avoids conflict.
c. For example, some stop themselves from raising controversial aspects of an issue.

—Adapted from Folger, Peale, & Stutman, *Working Through Conflict*, 4th ed., p. 24.

_____ **1.** Which of the following best states the topic?
a. avoiding controversial issues
b. quick acceptance of a solution
c. avoiding conflict

_____ **2.** Which sentence best states the main idea?

Group 2

a. The difference between sun and shade plants has practical importance.
b. Look in a nursery catalog and you will find plants keyed by symbols to indicate their adaptation to full sun, partial shade, and full shade.
c. Foresters and landscape designers base a part of their management plans on the shade tolerance or intolerance of plants.

—Adapted from Smith & Smith, *Elements of Ecology*, 4th ed., p. 48.

_____ **3.** Which of the following best states the topic?
a. plants
b. sun and shade plants
c. landscape designers

_____ **4.** Which sentence best states the main idea?

B. Read the paragraph from a college education textbook. Then, answer the questions that follow it.

Classroom Tests

[1]Teacher-made tests are a traditional and indispensable part of evaluation. [2]They serve several functions, including, of course, measuring achievement. [3]The items on a test provide the teacher with the opportunity to select for the student those aspects of the unit of study that are most important and thus reinforce learning of the material one last time during the study of the unit. [4]Furthermore, students' attention is entirely devoted to the test for the duration of the time, resulting in some of the most intense learning that occurs in a classroom. [5]This is a strong rationale for writing powerful tests.

—Wilen, Ishler, Hutchinson, & Kindsvatter, *Dynamics of Effective Teaching*, 4th ed., p. 354.

_____ **5.** Which sentence states the main idea of the paragraph?

a. sentence 1
b. sentence 2
c. sentence 3
d. sentence 5

REVIEW TEST 2

Score (number correct) _____ × 10 = _____%

The following passage is the summary of Chapter 4, "Biogeochemical Cycles in the Biosphere," in the college textbook *Geography: People, Places, and Environment.* Read the passage. Then, answer the questions that follow.

Textbook Skills

Biogeochemical Cycles in the Biosphere: Chapter Summary

[1]Climate is the dominant control on local environmental processes, primarily through its influence on water availability and movement. [2]Knowledge of the water budget is essential to understanding both physical and biological processes because the water budget is a critical regulator of ecosystem activity. [3]This influence extends to human use of the landscape, through such issues as determining the natural vegetation cover or agricultural potential.

[4]Carbon is the basis of life on Earth; it cycles through the atmosphere, oceans, biosphere, and soil. [5]Photosynthesis transfers carbon from the atmosphere to the biosphere, and respiration returns it to the atmosphere. [6]Plant growth is most prolific in warm, humid climates where water and energy are plentiful. [7]Plant growth is less in dry or cool climates. [8]Large amounts of carbon are exchanged between the atmosphere and the oceans, and the oceans are a major sink for carbon dioxide that is added to the atmosphere by fossil-fuel combustion.

[9]The outermost layer of Earth's surface in land areas is soil, which is a mixture of mineral and organic matter formed by physical, chemical, and biological processes. [10]Climate is a major regulator of soil development, through its control on water movement. [11]Plant and animal activity in the soil produces organic matter and mixes the upper soil layers. [12]In humid regions, water moves downward through the soil, carrying dissolved substances lower in the profile or removing them altogether. [13]In arid regions, these substances are not so easily removed. [14]Soil characteristics reflect these processes, and there is close correspondence between the world climate map and the world soil map.

[15]Photosynthesis is the basis of food chains and ecological systems. [16]Plants compete for water, sunlight, and nutrients, and plants that are best adapted to compete for limiting factors will dominate a given environment. [17]Such adaptations help explain the world distribution of major vegetation types. [18]Disturbances modify ecological communities, and succession following disturbance is a fundamental process of landscape change. [19]Very large portions of the world's land surface have been modified by human activity, and humans are major players in most of the world's ecosystems.

[20]The world vegetation map closely mirrors the world climate map. [21]Ecologically diverse and complex forests occupy humid environments, storing most nutrients in their biomass. [22]In arid and semiarid regions, sparse vegetation is adapted to moisture stress. [23]Forests adapted to winter cold are in humid midlatitude climates, developing as broadleaf forests in warmer areas and coniferous forests in subarctic latitudes. [24]In high-latitude climates, cold-tolerant short vegetation occupies areas that have a mild summer season. [25]Vegetation is absent in ice-bound polar climates.

—From *Introduction to Geography: People, Places, and Environment*, 4th ed., p. 161 by Edward Bergman and William H. Renwick. Copyright © 2008 by Pearson Education, Inc. Printed and Electronically reproduced by permission of Pearson Education, Inc., Upper Saddle River, NJ.

_____ **1.** Which of the following best states the topic of the first paragraph (sentences 1–3)?

a. climate
b. water budget
c. human use of landscape
d. agricultural potential

2. The topic sentence of the first paragraph (sentences 1–3) is _____.

_____ **3.** Which of the following best states the topic of the second paragraph (sentences 4–8)?

a. carbon
b. plant growth
c. the oceans
d. the atmosphere

4. The topic sentence of the second paragraph (sentences 4–8) is _____.

_____ **5.** Which of the following best states the topic of the third paragraph (sentences 9–13)?

a. the Earth's surface
b. climate
c. soil
d. plant and animal activity

6. The topic sentence of the third paragraph (sentences 9–14) is _____.

_____ **7.** Which of the following best states the topic of the fourth paragraph (sentences 15–19)?

a. photosynthesis
b. food chains
c. adaptations
d. ecosystems

8. The topic sentence of the fourth paragraph (sentences 15–19) is _____.

_____ **9.** Which of the following best states the topic of the fifth paragraph (sentences 20–25)?

a. vegetation
b. forests
c. world climate map
d. climates

10. The topic sentence of the fifth paragraph (sentences 20–25) is _____.

REVIEW TEST 3

Score (number correct) _____ × 10 = _____%

Topics and Main Ideas

Textbook Skills

Before Reading: Survey the following passage adapted from the college textbook *Psychology and Life.* Skim the passage, noting the words in bold print. Answer the Before Reading questions that follow the passage. Then, read the passage. Next, answer the After Reading questions. Use the discussion and writing topics as activities to do after reading.

Vocabulary Preview

incredibly (1): extremely

mania (3): craze, obsession

Depression: A Mood Disorder

[1]There have almost certainly been times in your life when you would have described yourself as terribly depressed or **incredibly** happy. [2]For some people, however, extremes in mood disrupt normal life experiences. [3]A mood disorder is an emotional disturbance, such as a severe depression or depression **alternating** with **mania.** [4]Researchers estimate that about 19 percent of adults have suffered from mood disorders.

psychopathy (6): mental disorder
clinical (11): medically shown or proven by medical testing

[5]One common type of mood disorder is the major depressive disorder.

[6]Depression has been characterized as the "common cold of **psychopathy**" for two reasons. [7]First, it occurs frequently, and, second, almost everyone has experienced elements of the full-scale disorder at some time in their life. [8]Everyone has, at one time or another, experienced grief after the loss of a loved one or felt sad or upset when failing to achieve a desired goal. [9]These sad feelings are only one symptom experienced by people suffering from a major depressive disorder.

[10]People suffering with depression differ in terms of the **severity** and duration of their symptoms. [11]Many people struggle with **clinical** depression for only several weeks at one point in their lives. [12]Others suffer depression **episodically** or chronically for many years. [13]Estimates of the prevalence of mood disorders reveal that about 21 percent of females and 13 percent of males suffer a major depression at some time in their lives.

[14]Depression takes an enormous toll on those who suffer, their families, and society. [15]One European study found that people with **chronic** depression spend a fifth of their entire adult lives in the hospital. [16]Twenty percent of sufferers are totally disabled by their symptoms and do not ever work again. [17]In the United States, depression accounts for the majority of all mental hospital admissions. [18]Still, many experts believe that the disorder is under-diagnosed and under-treated. [19]Fewer than half of those who struggle with major depressive disorder receive any professional help.

—From *Psychology and Life,* 16th ed., pp. 482–484 by Richard J. Gerrig and Philip G. Zimbardo.

Before Reading

Vocabulary in Context

_____ **1.** The word **alternating** in sentence 3 means
a. remaining.
b. taking turns.
c. shutting out.
d. encouraging.

_____ **2.** The word **severity** in sentence 10 means
a. gentleness.
b. length.
c. visibility.
d. harshness.

_____ **3.** The word **episodically** in sentence 12 means
a. continually.
b. deeply.
c. regularly.
d. rarely.

_____ **4.** The word **chronic** in sentence 15 means
a. constant.
b. mild.
c. treatable.
d. occasional.

Topics and Main Ideas

_____ **5.** What is the topic of the passage?
a. grief
b. major depressive disorder
c. normal life
d. mental hospital admissions

After Reading

Main Ideas and Supporting Details

_____ **6.** Which sentence states the central idea of the passage?
a. sentence 1
b. sentence 5
c. sentence 10
d. sentence 14

_____ **7.** What is the topic of the third paragraph (sentences 10–13)?
a. people suffering from depression
b. differences in the severity and duration of the symptoms of depression
c. the number of males and females suffering from depression

_____ **8.** Which sentence states the main idea of the third paragraph?
a. sentence 10
b. sentence 11
c. sentence 12
d. sentence 13

9–10. Label each of the following two sentences from the fourth paragraph (sentences 14–19). Write **A** if it states the main idea or **B** if it supplies a supporting detail.

_____ **9.** Depression takes an enormous toll on those who suffer, their families, and society.

_____ **10.** Fewer than half of those who struggle with major depressive disorder receive any professional help.

WHAT DO YOU THINK?

What do you think are the major causes of depression? In what ways do you think depression affects a person's ability to work or learn? How do you think most people cope with depression? Assume you are taking a college course in psychology and your professor assigned the class to read the passage "Depression: A Mood Disorder." Your professor has also assigned you to begin the class discussion of the passage with a written response to the following prompt: "Predict or guess the difference between the terms **episodic depression** and **chronic depression**." Use details from the passage and from a dictionary to define each term. Can you think of possible examples to illustrate the differences? If so, include them in your written response.

REVIEW TEST 4

Score (number correct) ______ × 10 = ______%

Topics and Main Ideas

Before Reading: Survey the following passage adapted from the college textbook *Cross-Cultural Psychology*. Skim the passage, noting the words in bold print. Answer the Before Reading questions that follow the passage. Then, read the passage. Next, answer the After Reading questions. Use the discussion and writing topics as activities to do after reading.

Textbook Skills

Vocabulary Preview

psychological (1): mental, emotional

formulated (2): made, created

assumptions (3): beliefs, ideas

phenomenon (3): event, occurrence

Cross-Cultural Sensitivity

[1]One type of **psychological** knowledge represents a collection of *popular beliefs,* often called ***folk theories.*** [2]This knowledge is a type of "everyday **psychology**" that is **formulated** by the people and for the people. [3]These popular beliefs are shared **assumptions** about certain aspects of human psychological **phenomena.** [4]These assumptions vary from being general, such as the belief in the ability of dreams to predict the future, to quite specific, that a particular item of clothing will bring good luck. [5]We can increase our cross-cultural sensitivity by understanding the psychological power of popular beliefs.

[6]When Jeff, an exchange student from Oregon, was invited to a birthday party, he was thrilled. [7]This was the first party he would attend in Russia, and he knew how well

Russians mastered the art of celebration. [8]The day of the birthday, he dug out a nice souvenir from his suitcase, then caught a taxicab and decided to stop by a flower market to buy a nice bouquet—he was invited by a female student and he thought flowers would be a nice addition to the souvenir he brought from Portland. [9]He could not anticipate that the flowers would cause so much anxiety and frustration an hour later. [10]He bought a dozen roses—a nice gesture according to U.S. standards. [11]But when he presented flowers to the host, he noticed how visibly upset she became when she put the flowers into a vase. [12]He even saw her crying in the kitchen. [13]A couple of friends were trying to comfort her. [14]Jeff began to wonder if his behavior had been the cause of the young woman's crying. [15]What he learned, as he later said, was one of the strangest experiences in his life. [16]He said that the young woman was extremely upset because he brought an even number of flowers. [17]**Coincidentally,** she had recently survived a deadly illness and was extremely sensitive to the issue of death and dying. [18]Apparently, Russians bring an even number of flowers to funerals, memorial services in church, and cemeteries. [19]An odd number of flowers is designed for dates, weddings, and other happy celebrations. [20]Apparently the flowers—the number of them, in fact—that Jeff brought to the party became a disturbing signal that brought the woman's traumatic experience back to her memory.

[21]In general, Russians will not react in the same dramatic way if you bring an even number of flowers to their celebrations. [22]However, you will notice that one flower—out of the dozen or half-a-dozen you bring—disappears from the vase. [23]Fears, phobias, and superstitions are at times rooted in folk customs and practices.

—Adapted from Shiraev & Levy, *Cross-Cultural Psychology: Critical Thinking and Contemporary Applications*, 3rd ed., pp. 11, 264.

Before Reading

Vocabulary in Context

Use your own words to define each of the following terms. Identify the context clue you used to make meaning of the word.

1. The word **theories** in sentence 1 means ______________________.

2. Context clue used for **theories:** ______________________

3. The word **coincidentally** in sentence 17 means ________________.

4. Context clue used for **coincidentally:** ________________________

Topics and Main Ideas

5. What is the topic of the passage? ________________________

After Reading

Main Ideas and Supporting Details

6. Which sentence states the central idea of the passage? _____

_____ **7.** What is the topic of the second paragraph (sentences 6–20)?
a. the young woman's near death experience
b. Jeff's strange experience
c. a birthday party

8–10. Label each of the following sentences from the third paragraph (sentences 21–23). Write **A** if it states the main idea or **B** if it supplies a supporting detail.

_____ **8.** In general, Russians will not react in the same dramatic way if you bring an even number of flowers to their celebrations.

_____ **9.** However, you will notice that one flower—out of the dozen or half-a-dozen you bring—disappears from the vase.

_____ **10.** Fears, phobias, and superstitions are at times rooted in folk customs and practices.

WHAT DO YOU THINK?

Was Jeff wrong to bring flowers to the birthday party? Why or why not? Why was the young woman so upset? Why didn't she tell Jeff why she was upset? How would you define "cross-cultural sensitivity"? Assume you are a mentor to the young woman who became upset with Jeff's gesture of giving a dozen roses. Write a letter to her in which you give her advice and comfort.

After Reading About Stated Main Ideas

Before you move on to the Mastery Tests on stated main ideas, take time to reflect on your learning and performance by answering the following questions. Write your answers in your notebook.

- How has my knowledge base or prior knowledge about stated main ideas changed?
- Based on my studies, how do I think I will perform on the Mastery Test(s)? Why do I think my scores will be above average, average, or below average?
- Would I recommend this chapter to other students who want to learn more about stated main ideas? Why or why not?

Test your understanding of what you have learned about Vocabulary Skills for Master Readers by completing the Chapter 3 Review Card in the insert near the end of your text.

CONNECT TO PEARSON myreadinglab

To check your progress in meeting Chapter 3's learning outcomes, log in to **www.myreadinglab.com** and try the following activities.

- The "Main Idea section of MyReadingLab provides review materials, practice activities, and tests about topics and main ideas. To access this resource, click on the "Study Plan" tab. Then click on "Main Idea." Then click on the following links as needed: "Overview," "Model," "Practice," and "Tests."
- To measure your mastery of the content of this chapter, complete the tests in the "Main Idea" section and click on Gradebook to find your results.

MASTERY TEST 1

Name ______________________ Section ______________

Date __________ **Score** (number correct) ______ × 25 = ______%

A. Skim each of the following paragraphs and circle the topic as it recurs throughout the paragraph. Then, identify the idea that correctly states the topic. (Hint: one idea is too general to be the topic; another idea is too specific.)

Textbook Skills

Animals

[1]It is difficult to devise a concise definition of the term "animal." [2]No single feature fully characterizes animals, so the group is defined by a list of characteristics. [3]None of these characteristics is unique to animals, but together they distinguish animals from members of other kingdoms:

- [4]Animals are multicellular.
- [5]Animals obtain their energy by consuming the bodies of other organisms.
- [6]Animals typically reproduce sexually. [7]Although animal species exhibit a tremendous diversity of reproductive styles, most are capable of sexual reproduction.
- [8]Animal cells lack a cell wall.
- [9]Animals are motile (able to move about) during some stage of their lives. [10]Even the relatively stationary sponges have a free-swimming larval stage (a juvenile form).
- [11]Most animals are able to respond rapidly to external stimuli as a result of the activity of nerve cells, muscle tissue, or both.

—From *Biology: Life on Earth*, 8th ed., p.301, by Teresa Audesirk, Gerald Audesirk, and Bruce E. Byers. Copyright © 2008 by Pearson Education, Inc. Printed and Electronically reproduced by permission of Pearson Education, Inc., Upper Saddle River, NJ.

______ **1.**

a. animals
b. animal kingdoms
c. characteristics of animals

VISUAL VOCABULARY

Sponges, such as the Erect Rope sponge and the Encrusting sponge, are often described as the most primitive of ____________.

a. life forms.
b. creatures.
c. animals.

B. The following group of ideas is from a college psychology textbook. Read the ideas, and answer the questions that follow.

Textbook Skills

a. Implicit memories are of three major types.
b. An implicit memory is a memory that cannot be voluntarily called to mind but still influences behavior or thinking in certain ways.
c. One type of implicit memory is habits, which are well-learned responses that are carried out without conscious thought.

—From Stephen M. Kosslyn & Robin S. Rosenberg, *Psychology: The Brain, The Person, The World*, p. 206. Published by Allyn and Bacon, Boston, MA. Copyright © 2002 by Pearson Education. Reprinted by permission of the publisher.

______ **2.** Statement (a) is
a. the main idea. b. a supporting detail.

______ **3.** Statement (b) is
a. the main idea. b. a supporting detail.

______ **4.** Statement (c) is
a. the main idea. b. a supporting detail.

MASTERY TEST 2

Name ________________ Section ________

Date __________ **Score** (number correct) ______ × 33.3 = ______%

Identify the topic sentence in each of the following paragraphs.

A. Paragraph from a college accounting textbook

Textbook Skills

Estate Planning: Living Wills and Power of Attorney

[1]A living will is a simple legal document in which individuals state their preferences if they become mentally or physically disabled. [2]For example, many individuals have a living will that expresses their desire not to be placed on life support if they become terminally ill. [3]In this case, a living will also has financial implications because an estate could be charged with large medical bills resulting from life support. [4]In this way, those who do not want to be kept alive by life support can ensure that their estate is used in the way that they prefer. [5]A power of attorney is a legal document granting a person the power to make decisions for you in the event that you are incapacitated. [6]For example, you may name a family member or a close friend to make your investment and housing decisions if you become ill. [7]A durable power of attorney for health care is a legal document granting a person the power to make specific health care decisions for you. [8]While a living will states many of your preferences, a situation may arise that is not covered by your living will. [9]A durable power of attorney for health care means that the necessary decisions will be made by someone who knows your preferences, rather than by a health care facility. [10]Estate planning involves these decisions about a living will and power of attorney.

—Adapted from *Personal Finance*, 3rd ed., p. 543 by Jeff Madura. Copyright © 2007 by Pearson Education, Inc. Reproduced by permission of Pearson Education, Inc., Boston, MA.

Topic sentence(s): ______

B. Paragraph from a college health textbook

Textbook Skills

Caffeine Addiction

[1]As the effects of caffeine wear off, users may feel let down. [2]They may feel mentally or physically depressed, exhausted, and weak. [3]To offset this, people commonly choose to drink another cup of coffee. [4]Habitual

use leads to tolerance and psychological dependence. [5]Until the mid-1970s, caffeine was not medically recognized as addictive. [6]Chronic caffeine use and the behaviors linked to it was called "coffee nerves." [7]This syndrome is now recognized as caffeine intoxication, or **caffeinism**. [8]Symptoms of caffeinism include chronic insomnia, jitters, irritability, nervousness, anxiety, and involuntary muscle twitches. [9]Withdrawing the caffeine may make the effects worse and lead to severe headaches. [10](Some physicians ask their patients to take a simple test for caffeine addiction: don't consume anything containing caffeine, and if you get a severe headache within four hours, you are addicted.) [11]Caffeine meets the requirements for addiction, which are tolerance, psychological dependence, and withdrawal symptoms; thus, it can be classified as addictive.

—Adapted from *Health: The Basics*, 5th ed., p. 215 by Rebecca J. Donatelle. Copyright © 2003 by Pearson Education, Inc. Printed and Electronically reproduced by permission of Pearson Education, Inc. Upper Saddle River, NJ.

Topic sentence(s): ______

C. Paragraph from a college psychology textbook

Touch and Skin Senses

Textbook Skills

[1]The skin is a remarkably versatile organ. [2]In addition to protecting you against surface injury, holding in body fluids, and helping control body temperature, it contains nerve endings that produce sensations of pressure, warmth, and cold. [3]These sensations are called cutaneous senses (skin senses). [4]Because you receive so much sensory information through your skin, many different types of receptor cells operate close to the surface of the body. [5]Two examples are the *Meissner corpuscles* and *Merkel disks*. [6]The Meissner corpuscles respond best when something rubs against the skin, and the Merkel disks are most active when a small object exerts steady pressure against the skin.

—From *Psychology and Life*, 16th ed., pp. 108–109 by Richard J. Gerrig and Philip G. Zimbardo. Copyright © 2002 by Pearson Education, Inc. Reproduced by permission of Pearson Education, Inc., Boston, MA.

Topic sentence(s): ______

MASTERY TEST 3

Name ______________________ Section ______________

Date __________ **Score** (number correct) ______ × 25 = ______%

Read the passage below. Then, answer the questions that follow it.

Textbook Skills

The Theater of Sophocles

[1]For the citizens of Athens, Greece, in the fifth century B.C., theater was both a religious and a civic occasion. [2]Seated in the open air in a hillside amphitheater, as many as 17,000 spectators could watch a performance that must have somewhat resembled an opera or musical. [3]Plays were presented only twice a year on religious festivals. [4]These festivals were related to Dionysius, the god of wine and crops; in January there was the Lenaea, the festival of the winepress, when plays, especially comedies, were performed. [5]But the major theatrical event of the year came in March at the Great Dionysia. [6]This was a citywide celebration that included sacrifices, prize ceremonies, and spectacular processions as well as three days of drama.

[7]Each day at dawn a different author presented a trilogy of tragic plays. [8]A trilogy was made up of three related dramas that depicted an important mythic or legendary event. [9]Each intense tragic trilogy was followed by a satyr play, an obscene parody of a mythic story. [10]In a satyr play the chorus dressed as satyrs, unruly followers of Dionysius who were half goat or horse and half human.

[11]The Greeks loved competition and believed it fostered excellence; even theater was a competitive event—not unlike the Olympic games. [12]A panel of five judges voted each year at the Great Dionysia for the best dramatic presentation, and a substantial cash prize was given to the winning poet-playwright (all plays were written in verse). [13]Any aspiring writer who has ever lost a literary contest may be comforted to know that Sophocles, who triumphed in the competition twenty-four times, seems not to have won the annual prize for *Oedipus the King*. [14]Although this play ultimately proved to be the most celebrated Greek tragedy ever written, it lost the award to a revival of Aeschylus, a playwright who had recently died.

—Adapted from *Literature: An Introduction to Fiction, Poetry, and Drama,* 8th ed., p. 1375 by X. J. Kennedy and Dana Gioia. Copyright © 2002 by Pearson Education, Inc. Reprinted by permission of Pearson Education, Inc., Glenview, IL.

1. What is the topic of the passage? ______________________________

_____ **2.** Which sentence (or sentences) states the central idea of the passage?

a. sentence 1
b. sentence 3
c. sentence 10
d. sentences 1 and 14

_____ **3.** Which sentence (or sentences) state(s) the main idea of the second paragraph?

a. sentence 7
b. sentence 8
c. sentence 9
d. sentence 10

_____ **4.** Which sentence (or sentences) state(s) the main idea of the third paragraph?

a. sentences 11 and 14
b. sentence 11
c. sentence 12
d. sentence 14

VISUAL VOCABULARY

This ancient _______________ is the Theatre of Dionysus, built between 342–326 B.C., Athens, Greece.

a. competition
b. amphitheater
c. region

MASTERY TEST 4

Name ______________________ Section ______________

Date __________ **Score:** (number correct) ______ × 10 = ______%

Read the following passage from a college psychology textbook. Answer the questions that follow.

Textbook Skills

Biological Approaches to Motivation

[1]Perhaps you have heard the term instinct used to explain why spiders spin webs or birds fly south in the winter. [2]An instinct is a fixed behavior pattern that is characteristic of every member of a species and is assumed to be genetically programmed. [3]Thus, instincts represent one kind of biological motivation. [4]Psychologists generally agree that no true instincts motivate human behavior. [5]However, most also agree that biological forces underlie some human behaviors.

[6]One biological approach to motivation, drive-reduction theory, was popularized by Clark Hull (1943). [7]According to Hull, all living organisms have certain biological needs that must be met if they are to survive. [8]A need gives rise to an internal state of tension called a drive, and the person or organism is motivated to reduce it. [9]For example, when you are deprived of food or go too long without water, your biological need causes a state of tension in this case, the hunger or thirst drive. [10]You become motivated to seek food or water to reduce the drive and satisfy your biological need.

[11]Drive-reduction theory is derived largely from the biological concept of homeostasis—the tendency of the body to maintain a balanced, internal state to ensure physical survival. [12]Body temperature, blood sugar level, water balance, blood oxygen level—in short, everything required for physical existence must be maintained in a state of equilibrium, or balance. [13]When such a state is disturbed, a drive is created to restore the balance.

[14]Drive-reduction theory assumes that humans are always motivated to reduce tension. [15]Other theorists argue just the opposite, that humans are sometimes motivated to increase tension. [16]These theorists use the term arousal to refer to a person's state of alertness and mental and physical activation. [17]Arousal levels can range from no arousal (when a person is comatose), to moderate arousal (when pursuing normal day-to-day activities), to high arousal (when excited and highly stimulated). [18]Arousal theory states that people are motivated to maintain an optimal level of arousal. [19]If arousal is less than the optimal level, we do something to stimulate it; if arousal exceeds the optimal level, we seek to reduce the stimulation.

[20]When arousal is too low, stimulus motives—such as curiosity and the motives to explore, to manipulate objects, and to play—cause humans and other animals to increase stimulation. [21]Think about sitting in an airport or at a bus stop or any other place where people are waiting. [22]How many people do you see playing games on their cellphones or personal digital assistants (PDAs)? [23]Waiting is boring; in other words, it provides no sources of arousal. [24]Thus, people turn to electronic games to raise their level of arousal.

—From *Mastering the World of Psychology,* 3rd ed., pp. 290–291 by Samuel E. Wood, Ellen Green Wood, and Denise Boyd.

_____ **1.** Which of the following best states the topic of the first paragraph (sentences 1–5)?
a. instinct
b. motivation
c. human behavior
d. psychologists

2. The topic sentence of the first paragraph (sentences 1–5) is _____.

_____ **3.** Which of the following best states the topic of the second paragraph (sentences 6–10)?
a. Clark Hull
b. needs
c. drive-reduction theory
d. drives

4. The topic sentence of the second paragraph (sentences 6–10) is _____.

_____ **5.** Which of the following best states the topic of the third paragraph (sentences 11–13)?
a. drive-reduction theory
b. homeostasis
c. motivation
d. physical survival

6. The topic sentence of the third paragraph (sentences 11–13) is _____.

_____ **7.** Which of the following best states the topic of the fourth paragraph (sentences 14–19)?
a. drive-reduction theory
b. tension
c. stimulation
d. arousal theory

8. The topic sentence of the fourth paragraph (sentences 14–19) is _____.

_____ **9.** Which of the following best states the topic of the fifth paragraph (sentences 20–24)?
a. stimulus motives
b. arousal
c. boredom
d. games

10. The topic sentence of the fifth paragraph (sentences 20–24) is _____.

Implied Main Ideas and Implied Central Ideas

4

LEARNING OUTCOMES

After studying this chapter you should be able to do the following:

1. Define the term *implied main idea.*
2. Analyze supporting details to identify and state a topic of a passage.
3. Determine an implied main idea of a passage based on the topic, supporting details, and thought patterns of a passage.
4. Create a topic sentence that states the implied main idea of a paragraph.
5. Determine the implied central idea of a longer passage.
6. Create a thesis statement that states the implied central idea of a passage.
7. Evaluate the importance of stating implied main ideas.

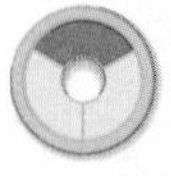

Before Reading About Implied Main Ideas and Implied Central Ideas

Take a moment to study the chapter learning outcomes. Underline key words that refer to ideas you have already studied in previous chapters. Each of these key words represents a great deal of knowledge upon which you will build as you learn about implied main ideas and implied central ideas. These key terms have been listed below. In the given spaces, write what you already know about each one.

- Main ideas: ______________________________

  ____

- Supporting details: ______________________________

  ____

  ____

  ____

- A summary: ______________________________
- Central ideas: ______________________________

____

Compare what you wrote to the following paragraph, which summarizes this vital prior knowledge:

> Main ideas are stated in a topic sentence. A topic sentence includes a topic and the author's controlling point. Supporting details explain the main idea. There are two types of supporting details. Major supporting details directly explain the topic sentence (or thesis statement), and minor supporting details explain the major supporting details. A summary condenses a paragraph or passage to its main idea. Central ideas are the main ideas of longer passages. Stated central ideas are called thesis statements.

Recopy the list of key words in your notebook; leave several blank lines between each idea. As you work through this chapter, record how you apply each idea in the list to the new information you learn about implied main ideas and implied central ideas.

What Is an Implied Main Idea?

As you learned in Chapter 3, sometimes authors state the main idea of a paragraph in a topic sentence. However, other paragraphs do not include a stated main idea. Even though the main idea is not stated in a single sentence, the paragraph still has a main idea. In these cases, the details clearly suggest or imply the author's main idea.

> An **implied main idea** is a main idea that is not stated directly but is strongly suggested by the supporting details in the passage.

When the main idea is not stated, you must figure out the author's controlling point about a topic. One approach is to study the facts, examples, descriptions, and explanations given—the supporting details. Another approach is to identify the author's thought pattern. A master reader often uses both approaches. Learning how to develop a main idea based on the supporting details and thought patterns will help you develop several skills. You will learn how to study information, value the meaning of supporting details, recognize the relationships among ideas, and use your own words to express an implied main idea.

Many different types of reading material use implied main ideas. For example, you will often need to formulate the implied main idea when you read literature. Short stories, novels, poems, and plays rely heavily on vivid details to suggest the author's point. The following short story is taken from a college literature textbook. Read the story, asking yourself, "What is the main idea?"

The Appointment in Samarra

W. Somerset Maugham

[1]*Death speaks*: There was a merchant in Baghdad who sent his servant to market to buy provisions and in a little while the servant came back, white and trembling, and said, Master, just now when I was in the marketplace, I was jostled by a woman in the crowd and when I turned I saw it was Death that jostled me. [2]She looked at me and made a threatening gesture; now, lend me your horse, and I will ride away from this city and avoid my fate. [3]I will go to Samarra and there Death will not find me. [4]The merchant lent him his horse, and the servant mounted it, and he dug his spurs in its flanks and as fast as the horse could gallop he went. [5]Then the merchant went down to the marketplace and he saw me standing in the crowd and he came to me and said, Why did you make a threatening gesture to my servant when you saw him this morning? [6]That was not a threatening gesture I said, it was only a start of surprise. [7]I was astonished to see him in Baghdad, for I had an appointment with him tonight in Samarra.

Did you notice that every sentence in this paragraph is a supporting detail? No single sentence covers all the other ideas. To figure out the implied main idea, ask the following questions:

Questions for Finding the Implied Main Idea

- What is the topic, or subject, of the paragraph?
- What are the major supporting details?
- Based on the details about the topic, what point or main idea is the author trying to get across?

Apply these three questions to the Somerset Maugham story by writing your responses to each question in the following blanks.

1. What is the topic of the story? ____________________

____

In this case, the title of the story tells us that the topic is an appointment, and each detail in the story leads to the understanding that the appointment was to be with Death.

2. What are the major supporting details?

____

____

____

____

____

____

____

____

____

____

3. What is the main idea the author is trying to get across? ____________________

____

____

In order to formulate a main idea statement for this story, you had to consider each of the details within it. First, the author has made Death a humanlike character. Next, when the servant has an encounter with Death in the marketplace, he misunderstands her gesture and in fear runs from her. Third, consider the difference between the ways the merchant and his servant interact with Death. The servant fears and misunderstands Death and runs away; the merchant faces Death and seeks to understand Death's intentions. Interestingly, the servant's fear of Death and his decision to run from Death drives him to meet Death, his death, at the appointed time and place.

This story is an excellent example of the different shades of meaning an author can imply through the use of supporting details. Implying the main idea allows the reader to use creative and critical thinking skills to make meaning of the piece.

Asking and answering these questions allows you to think about the impact of each detail and how the details fit together to create the author's most important point. Searching for an implied main idea is like a treasure hunt. You must carefully read the clues provided by the author. The following examples and practices are designed to strengthen this important skill.

Using Supporting Details and Thought Patterns to Find Implied Main Ideas

Remember that the main idea of a paragraph is like a frame of a house. Just as a frame includes all the rooms, a main idea must cover all the details in a paragraph. Therefore, the implied main idea will be general enough to cover all the details, but it will not be so broad that it becomes an overgeneralization or a sweeping statement that suggests details not given; nor can it be so narrow that some of the given details are not covered. Instead, the implied main idea must cover *all* the details given.

Having the skill of identifying a stated main idea will also help you grasp the implied main idea. You learned in Chapter 3 that the stated main idea (the topic sentence) has two parts. A main idea is made up of the topic and the author's controlling point about the topic. One trait of the controlling point is the author's opinion or bias. A second trait is the author's thought pattern. Consider, for example, the topic sentence "Older people benefit from volunteer work for several reasons." "Older people" and "volunteer work" make up the topic. "Benefit" states the opinion, and "several reasons" states the thought pattern. When you read material that implies the main idea, you should mentally create a topic sentence based on the details in the material.

Textbook Skills

EXAMPLE Read the following list of supporting details from a criminal justice textbook. Circle the topic as it recurs throughout the list of details. Underline transition words and double underline the biased (opinion) words. Then, choose the statement that best expresses the author's controlling point about the topic.

- Lay witnesses are non-expert witnesses who may be called to testify in court for either the prosecution or the defense.

- Lay witness may be eyewitnesses who saw the crime being committed or who came upon the crime scene shortly after the crime had occurred.
- Another type of lay witness is the character witness.
- A character witness frequently provides information about the personality, family life, business acumen, and so on of the defendant.
- Of course the victim may also be a lay witness, providing detailed and sometimes lengthy testimony about the defendant and the crime.

_____ Which statement best expresses the implied main idea?

a. Character witnesses offer important information on behalf of the defendant.
b. Witnesses are important to both the prosecution and the defense.
c. Several types of lay witnesses, or non-expert witnesses, may be called to testify for the prosecution or the defense.
d. Eyewitnesses and victims frequently provide details about the defendant and the crime.

EXPLANATION The best statement of the implied main idea is (c) Several types of lay witnesses, or non-expert witnesses, may be called to testify for the prosecution or the defense. Option (a) is too general, and options (b) and (d) state specific details. The terms *lay witness, witness, eyewitness, character witness,* and *victim* state the various forms of the topic and should have been circled. The words *another type, frequently, also,* and *sometimes* are transition words that should have been underlined. The phrase *another type* suggests that this list of details describes types of lay witnesses. The word *may* is a biased word that suggests a possibility rather than a factual event and, therefore, should be double underlined.

PRACTICE 1

Read the following groups of supporting details. Circle the topic as it recurs throughout the list of details. Underline transition words to help you locate the major details. Also, double underline the biased words to determine the author's opinion. Then, select the sentence that best expresses the implied main idea.

Group 1

- Many people volunteer within their communities because they want to make a difference.
- Others volunteer so that they can develop new skills.
- One reason some volunteer is to explore career paths.
- In addition, volunteering is a good way for young people to have fun working with friends.
- Finally, volunteers feel good about themselves.

_____ **1.** Which statement best expresses the implied main idea?
a. Volunteer work is a good way to build a résumé.
b. Volunteering provides an excellent opportunity to meet new people.
c. People engage in volunteer work for several reasons.
d. Volunteer work is very rewarding.

Group 2

Textbook Skills

- Ancient Roman entertainment included public fights between gladiators.
- Many gladiators were volunteers who willingly gave up their rights and property and pledged not only to suffer intensely but to die fighting.
- Romans valued a concept called *virtus* (honor above life).
- *Virtus* was a specifically male trait and was most often applied to a gladiator who had fought well or continued to struggle even after defeat.
- If a gladiator displayed *virtus* but was clearly losing, he could raise one finger in a plea for mercy.
- Usually, Romans granted *missio* or mercy for gladiators who displayed virtus so that they could live to fight another day.

—Adapted from *The Creative Impulse: An Introduction to the Arts,* 8th ed., p. 133, by Dennis J. Sporre. Copyright © 2009 by Pearson Education, Inc. Reprinted by permission of Pearson Education, Inc., Upper Saddle River, NJ.

_____ **2.** Which statement best expresses the implied main idea?
a. Ancient Romans enjoyed violent entertainment.
b. Ancient Romans valued honor above life.
c. Ancient Romans were often merciful.
d. Ancient Romans were virtuous.

Group 3

- A person who is supporting a family will normally incur more expenses than a single person without dependents.
- As people get older, they tend to spend more money on expensive houses, cars, and vacations.
- In addition, people's consumption behavior varies greatly.
- For example, at one extreme are people who spend their entire paycheck within a few days of receiving it; at the other extreme are "big savers" who reduce their spending and focus on saving for the future.

—Adapted from *Personal Finance,* 2nd ed., p. 32 by Jeff Madura. Copyright © 2004 by Pearson Education, Inc. Reproduced by permission of Pearson Education, Inc., Boston, MA.

______ **3.** Which statement best expresses the implied main idea?

a. Some people are better at managing their money than others.
b. The size of one's family affects cash outflows.
c. Several factors affect cash outflows.
d. People spend more money as they age.

Group 4

- Nomadic Native American tribes followed migratory prey and the seasonal changes of plant life.
- Nomadic Native American tribes needed housing structures that were easily erected and dismantled.
- For example, tipis were easily erected and dismantled.
- Agricultural Native American tribes, on the other hand, could enjoy more permanent dwellings, like pueblos.

______ **4.** Which statement best expresses the implied main idea?

a. Native American tribes were constantly on the move.
b. The lifestyle of a Native American tribe influenced the types of dwellings they inhabited.
c. Native American tribes were excellent architects.
d. Native American tribes lived in substandard housing.

Finding the Implied Main Ideas of Paragraphs

So far, you have learned to recognize the implied main idea by studying the specific details in a group of sentences. In this next step, the sentences will form a paragraph, but the skill of recognizing the implied main idea is exactly the same.

The implied main idea of paragraphs must not be too broad or too narrow, so study the supporting details and look for thought patterns and opinions that suggest the main idea.

EXAMPLE Read the following paragraph. Circle the topic as it recurs throughout the paragraph. Underline transition words to help you locate the major details. Also, double underline biased words to determine the author's opinion. Then, select the sentence that best expresses the implied main idea.

Textbook Skills

The Faces of Fetal Alcohol Syndrome

[1]"The guilt is tremendous . . . I did it again and again . . . I don't know how to tell them. [2]It was something I could have prevented." [3]Debbie has had seven children. [4]One of her older daughters, Corey, has been diagnosed with fetal alcohol syndrome (FAS), the most serious type of alcohol damage. [5]At age three she was hyperactive and talked like a one-year-old. [6]Doctors believe that Debbie's youngest child Sabrina is almost certainly a victim as well. [7]Her face bears the characteristic features of FAS, including small eyes, a short nose, and a small head. [8]At 7 months, she was weak, began having seizures, and could not eat solid food because she was unable to close her mouth around the spoon. [9]As this pregnant mother repeatedly got drunk, so did her developing children.

—From *Life on Earth*, 5th ed., p. 486 by Teresa Audesirk, Gerald Audesirk, and Bruce E. Byers. Copyright © 2009 by Pearson Education, Inc. Printed and Electronically reproduced by permission of Pearson Education, Inc., Upper Saddle River, NJ..

_____ Which sentence best states the implied main idea?

a. The effects of Fetal Alcohol Syndrome can be avoided.
b. Pregnant women can consume alcohol in moderate amounts.
c. Pregnant mothers who frequently consume alcohol cause preventable physical and mental damage to their developing children.

EXPLANATION Implied main ideas require that the reader be actively involved in the meaning-making process. The author states the damage caused to developing children by their mothers' prenatal consumption of alcohol in the following expression: "most serious type of alcohol damage." The author reinforces the point by describing the physical and mental traits of FAS in sentences 5 and 7. Item (c) is the best statement of the implied main idea of this paragraph. Item (a) is too narrow, and item (b) jumps to the wrong conclusion based on the evidence in the paragraph. Nowhere in the paragraph does the author imply that occasional or moderate consumption of alcohol is safe or leads to FAS.

PRACTICE 2

Read the following paragraphs. In each paragraph, circle the topic as it recurs throughout the paragraph. Underline transition words to help you locate the major details. Also, double underline the biased words to determine the author's opinion. Then, select the sentence that best expresses the implied main idea.

Textbook Skills

Opposition to Capital Punishment

[1]The first recorded effort to eliminate the death penalty occurred at the home of Benjamin Franklin in 1787. [2]At a meeting there on March 9 of that year, Dr. Benjamin Rush, a signer of the Declaration of Independence and a leading medical pioneer, read a paper against capital punishment to a small but influential audience. [3]Although his immediate efforts came to nothing, his arguments laid the groundwork for many debates that followed. [4]Michigan, widely regarded as the first abolitionist state, joined the Union in 1837 without a death penalty. [5]A number of other states, including Alaska, Hawaii, Massachusetts, Minnesota, West Virginia, and Wisconsin, have since spurned death as a possible sanction for criminal acts. [6]However, the death penalty remains a sentencing option in 38 of the states and in all federal jurisdictions.

—Adapted from *Criminal Justice Today: An Introductory Text for the 21st Century*, 10th ed., p. 409 by Frank J. Schmalleger. Copyright © 2009 by Pearson Education, Inc. Printed and Electronically reproduced by permission of Pearson Education, Inc., Upper Saddle River, NJ.

______ **1.** Which sentence best states the implied main idea?

a. Many oppose capital punishment.
b. Capital punishment should not be a viable sanction for criminal acts.
c. Dr. Benjamin Rush was the first to publically oppose capital punishment.
d. Attempts have been made to abolish capital punishment since the founding of the United States.

[1]The *personal* aspects of emotional intelligence are a set of components that include awareness and management of our own emotions. [2]People who are able to monitor their feelings as they arise are less likely to be ruled by them. [3]However, managing emotions does not mean suppressing them; nor does it mean giving free rein to every feeling. [4]Instead, effective management of emotions involves expressing them appropriately. [5]Emotion management also involves engaging in activities that cheer us up, soothe our hurts, or reassure us when we feel anxious. [6]The *interpersonal*

aspects of emotional intelligence make up the second set of components. [7]Empathy, or sensitivity to others' feelings, is one such component. [8]One key indicator of empathy is the ability to read others' nonverbal behavior—the gestures, vocal inflections, tones of voice, and facial expressions of others. [9]Another of the interpersonal components is the capacity to manage relationships. [10]However, it is related to both the personal aspects of emotional intelligence and to empathy. [11]In other words, to effectively manage the emotional give-and-take involved in social relationships, we have to be able to manage our own feelings and be sensitive to those of others.

—Adapted from *Mastering the World of Psychology*, 3rd ed., p. 237 by Samuel E. Wood, Ellen Green Wood, and Denise Boyd. Copyright © 2008 by Pearson Education, Inc. Reproduced by permission of Pearson Education, Inc., Boston, MA.

_____ **2.** Which sentence best states the implied main idea?
a. Emotional intelligence includes two sets of components.
b. Emotional intelligence is made up of personal aspects.
c. Emotional intelligence is made up of interpersonal aspects.
d. Emotional intelligence is the ability to manage emotions effectively.

Creating a Summary from the Supporting Details

You have developed the skill of figuring out implied main ideas, which will serve you well throughout college. The next step is to state the implied main idea in your own words. When you are reading and studying, you must be able to summarize the most important details in a one-sentence statement; in other words, you must create a topic sentence.

To formulate this one-sentence summary, find the topic, determine the author's opinion by examining the biased words, and use the thought pattern to locate the major details. Then, combine these ideas in a single sentence. The summary sentence includes the topic and the author's controlling point, just like a topic sentence. The statement you come up with must not be too narrow, for it must cover all the details given. On the other hand, it must not be too broad or go beyond the supporting details.

Remember that a main idea is always written as a complete sentence.

EXAMPLE Read the following paragraph. Circle the topic as it recurs throughout the paragraph. Underline words that reveal thought patterns and bias to discover the controlling point. Then, write a sentence that best states the implied main idea. Remember: not too narrow, not too broad—find that perfect fit!

The Cost of Owning a Horse

[1]People invest in horses for a variety of reasons, including riding for pleasure or therapy, competing in shows, farming, breeding, and racing. [2]The first cost incurred is, of course, the purchase of the horse. [3]Depending on the pedigree, health, and prior training of the horse, prices vary from a few hundred to several thousand dollars. [4]The ongoing cost of boarding the horse must be considered as well. [5]Boarding costs vary based on location and the type of services. [6]For example, some facilities offer just a pasture while others offer full-service stall boarding and daily exercise. [7]A third major expense involved in owning a horse is feed. [8]Horses are usually fed a grain mix, grass hay, and salt and minerals. [9]Next, health care costs include dewormings four times a year and a variety of vaccinations. [10]In addition, medical attention associated with breeding and emergencies can cause sizeable increases in expenses. [11]Farrier costs must also be considered; this expense includes at a minimum trimming the hooves every couple of months, but may require, depending on the type of activity and the owner's taste, shoeing the horse or resetting the horse's shoes. [12]Another cost that must be taken into account is the cost of stall bedding. [13]The dollar amount varies based on different bedding sources and quantity used. [14]Finally, the costs for equipment most often include items needed for grooming, feeding, and cleaning such as brushes, buckets, shovels, and forks; in addition, tack must be bought and kept in good repair.

Implied main idea: ____

____

VISUAL VOCABULARY

A synonym for *farrier* is

______________.

a. veterinarian.
b. groomer.
c. blacksmith.

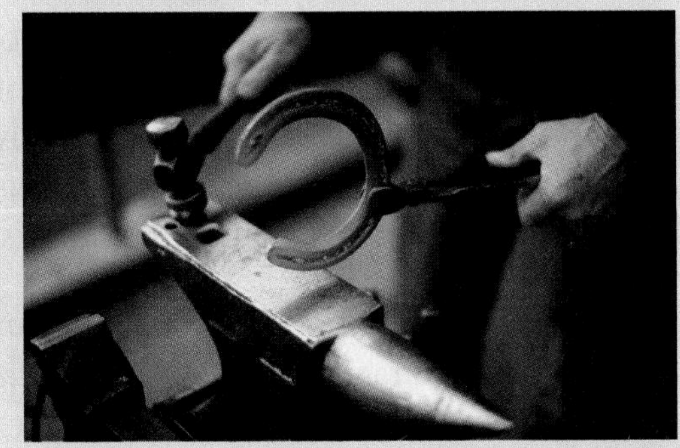

▸ A farrier making a horse shoe.

EXPLANATION The following sentence is one way to state the main idea of the paragraph: "Many expenses are involved in owning a horse." Keep in mind that using your own words to formulate an implied main idea means that the wording of answers will vary.

PRACTICE 3

Read the following paragraphs. Annotate the text. Then, write a sentence that states the implied main idea.

From *Narrative of the Life of Frederick Douglass, an American Slave. Written by Himself*

[1]I was born in Tuckahoe, near Hillsborough, and about twelve miles from Easton, in Talbot county, Maryland. [2]I have no accurate knowledge of my age, never having seen any authentic record containing it. [3]By far the larger part of the slaves know as little of their ages as horses know of theirs, and it is the wish of most masters within my knowledge to keep their slaves thus ignorant. [4]I do not remember to have ever met a slave who could tell of his birthday. [5]They seldom come nearer to it than planting-time, harvest-time, cherry-time, spring-time, or fall-time. [6]A want of information concerning my own was a source of unhappiness to me even during childhood. [7]The white children could tell their ages. [8]I could not tell why I ought to be deprived of the same privilege. [9]I was not allowed to make any inquiries of my master concerning it. [10]He deemed all such inquiries on the part of a slave improper and impertinent, and evidence of a restless spirit. [11]The nearest estimate I can give makes me now between twenty-seven and twenty-eight years of age. [12]I come to this, from hearing my master say, some time during 1835, I was about seventeen years old.

—From Douglass, Frederick, *Narrative of the Life of Frederick Douglass, an American Slave. Written by Himself,* 1845. pp. 1–2.

1. Implied main idea: ______________________________

____

From *A Personal Journal of an American Woman*

[1]For eons and in many societies, it seems, most women have been fastened to the traditional roles dictated by biology and culture. [2]However, women such as Oprah Winfrey, Diane Sawyer, Toni Morrison, Sally Ride,

and Sandra Day O'Connor prove to me that I am free. [3]As so many American freedoms, my freedom—the freedom of my American sisters—was bequeathed to me—to us—by warrior women who fought for the right to vote, the right to education, the right to fair wages, the right to birth control. [4]Women such as Harriet Tubman, Elizabeth Cady Stanton, Lucretia Mott, and Margaret Sanger possessed the riches of vision, courage, and fortitude, and they invested their wealth in our lives long before we were born. [5]And now as I live in the 21st century, I am free to be me: educator, author, entrepreneur, wife, and mother.

2. Implied main idea: ______________________________

____

____

The Implied Central Idea

Just as a single paragraph can have an implied main idea, longer passages made up of two or more paragraphs can also have an implied main idea. As you learned in Chapter 3, the stated main idea or central idea of these longer passages is called the *thesis statement*. When the main idea of several paragraphs is implied, it is called the **implied central idea**. You use the same skills to formulate the implied central idea of a longer passage that you use to formulate the implied main idea of a paragraph.

> **The implied central idea** is the main idea suggested by the details of a passage made up of two or more paragraphs.

Annotating the text is a helpful tool in determining the implied central idea. Just as you did to grasp the implied main idea for paragraphs, circle the topic. Underline the signal words for thought patterns. Remember, transition words introduce supporting details. An author often pairs a transition word with a major supporting detail. Consider the following examples: *the first reason, a second cause, the final effect, another similarity, an additional difference*, and so on. When you see phrases such as these, your one-sentence summary may include the following kinds of phrases: *several effects, a few differences*, and so on.

A longer passage often contains paragraphs with stated main ideas. The stated main idea of a paragraph is a one-sentence summary of that paragraph and can be used as part of your summary of the implied central idea.

EXAMPLE Read the following passage from a college humanities textbook. Annotate the text. Then, select the sentence that summarizes its implied central idea.

Textbook Skills

Hesiod

[1]Born at the end of the eighth century B.C.E., Hesiod (HEE-see-uhd) was the son of an emigrant farmer from Asia Minor. [2]He depicts this hard life in his poem *Works and Days*. [3]Hesiod's lesson is that men must work, and he sketches the work that occupies a peasant throughout the year, recounting how an industrious farmer toils and how work brings prosperity, in contrast to the ruin brought about by idleness. [4]He captures the changing seasons and countryside with vivid word pictures. [5]He also describes life at sea, and concludes with a catechism detailing how people should deal with each other. [6]His style is similar to Homer's, but his subject matter is quite different, for rather than telling of heroes and gods, Hesiod dwells on the mundane and mortal.

[7]Hesiod's most famous work, however, is the *Theogony*, in which he traces the mythological history of the Greek gods. [8]He describes the rise of the earth out of chaos, the overthrow of the Titans by Zeus, and the emergence of each god and goddess. [9]Scholars attempting to understand Greek art and religion have found this work invaluable as a source of information.

—From The Creative *Impulse: An Introduction to the Arts,* 8th ed., p. 70 by Dennis J. Sporre. Copyright © 2009 by Pearson Education, Inc. Reprinted by permission of Pearson Education, Inc., Upper Saddle River, NJ.

Implied central idea: ____

____

____

EXPLANATION Hesiod, a Greek writer, is the topic of this passage. The topic sentences of each paragraph tell the reader the two major points the author is using to limit the topic. In the first paragraph, sentence 2 is the topic sentence, and it focuses on the subject matter in one of Hesiod's writings: "hard life in his poem *Works and Days*." The additional details make clear that the poem is about everyday Greek life. In the second paragraph, sentence 7 is the topic sentence, and it focuses on another type of subject that Hesiod wrote about in a different work: "*Theogony*, in which he traces the mythological history of the Greek gods." By combining these two main ideas, a master reader creates a one-sentence summary of the passage.

Though answers will vary, one possible answer is "Born in the eighth century B.C.E., the Greek writer Hesiod wrote about both the struggles of everyday life and the history of the Greek gods." This sentence covers all the major supporting details but does not go beyond the information given in the passage.

PRACTICE 4

Textbook Skills

Read the following passage from a college sociology textbook. Write a sentence that states the implied central idea.

[1]Prostitution is illegal because it is a social problem that has a corrosive effect on society—it has widespread implications that go beyond prostitutes and their customers. [2]Prostitution keeps pimps and madams (who are constantly on the lookout for new employees) in business and can corrupt police officers who are bribed to look the other way. [3]Most importantly, it can feed into larger, more harmful organized crimes, such as the drug trade.

[4]Any neighborhood with the presence of prostitution will be hurt. [5]Property values will decrease substantially. [6]Not only will the prostitution activity be unwanted, but it will also attract undesirable people to the neighborhood. [7]Residents who have children will relocate, and prospective home buyers will look elsewhere.

[8]Prostitution can rapidly spread disease and is an enormous health risk. [9]Because of the many sexual partners that a prostitute encounters on a regular basis, one prostitute who carries an STD could infect dozens of people. [10]If the disease goes undetected and those clients have other sexual partners, more could be infected as well.

[11]Prostitution is immoral and undermines an accepted sense of decency. [12]The selling of sex and the sex acts that are performed by prostitutes violate long-held beliefs about intimacy and sexuality. [13]Prostitution ruins marriages, tears families apart, and leads to corruption in the home and community.

[14]Women are often treated in a manner in which they are objectified, and prostitution only serves to continue—and even enhance this mind-set. [15]The profession itself is degrading to women, and because of the stigma attached to prostitution, women are often seen and treated as second class citizens.

Implied central idea: ____

____

Reading a Textbook: How to Read a Map

Textbook Skills

Many academic courses, particularly in the social sciences, rely on maps to communicate important geographical, political, social, and economic information. Maps provide information about what occurs in time and space. Not only do maps show the exact locations of particular places, but they also document the movements of civilizations, people, ideas, and trade. To analyze a map, follow these basic steps:

- **Read the title and caption to identify the main topic of the map.** Is the map illustrating a political idea, physical features, or a particular theme? Read the title and caption before you study the map to focus your attention on the cartographer's purpose for the map.
- **Activate your prior knowledge.** What do you already know about the topic, the region, and the time period? How does this map fit in with what you already know? What do you need to know to understand the map?
- **Discern the orientation of the map.** The orientation, the direction in which the map is pointed, is usually shown by a compass. Most current maps place north at the top of the map.
- **Identify the date of the map.** Because our world (and our understanding of our world) is ever-changing politically and socially, older maps may not be relevant. In addition, a map may be illustrating a historical perspective.
- **Understand and apply the legend.** The legend, or key, which shows the meaning of the colors, symbols, and other markings on the map, is crucial to understanding what the map represents.
- **Understand and apply the scale.** The scale tells you the relationship between the distance on the map and the corresponding distance on the ground. For example, one inch may represent 100 miles. Understanding the scale enables the reader to visualize the size of and distance between places.

EXAMPLE The following map is from the textbook *World Civilizations: The Global Experience.* Read the map and answer the questions. Then, state the implied main idea of the map in one sentence.

European Population Density, c.1600

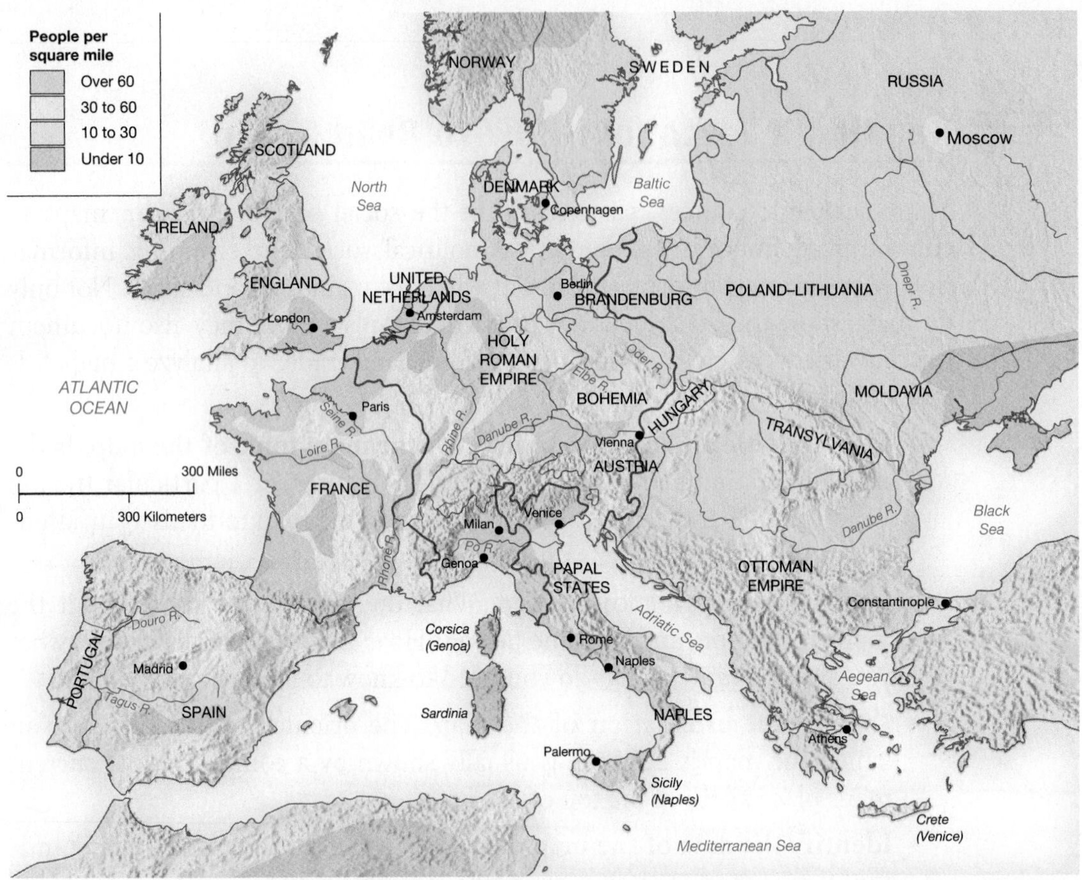

—From *World Civilizations: The Global Experience, Combined Volume, Atlas Edition*, 5th ed, Map 22.2 (p. 490) by Peter N. Stearns, Michael B. Adas, Stuart B. Schwartz, and Marc Jason Gilbert. Copyright © 2008 by Pearson Education, Inc. Printed and Electronically reproduced by permission of Pearson Education, Inc., Upper Saddle River, NJ.

1. What is the title and date of the map? ______________________________

____

2. According to the legend, what do the colors represent? ______________

____

3. Name three densely populated regions of Europe. ____________________

____

4. Why might most of the most heavily populated areas be located alongside rivers, oceans, or seas? ______________________________

____

Implied main idea of the map: ______________________________

____

EXPLANATION Compare your answers to the following.

(1) The date of the map is included in its title "European Population Density, c. 1600." Based on the date, the reader understands that this map offers a historical view of the region. (2) The legend indicates four levels of "people per square mile." (3) The map indicates more than three regions that are densely populated, so answers may vary, but should include three of the following: Ireland, England, France, United Netherlands, Holy Roman Empire, and Papal States. (4) Answers may vary, but prior knowledge supports the idea that during this time water provided the major means of transportation of people and trade, so the following statement is an appropriate answer: "Waterways make trade easier, and people settle along trade routes." Although your interpretation of the main idea of the map may vary, based on the title, the legend, and the answers to the preceding questions, the implied main idea of the map is as follows: "In 1600, European population density was greater alongside rivers, oceans, and seas." Quite often, to properly analyze a map, a master reader uses prior knowledge (see page 8 in this textbook for more information about this term).

PRACTICE 5

Study the map from the textbook *Introduction to Geography: People, Places, and Environment.* Answer the questions, and write a sentence that states the implied main idea of the map.

1. What prior knowledge is needed to understand the implied main idea of the map? ______________________________

2. According to the caption, what is the topic of the map? ______________________________

____

3. According to the legend, what is the unit of measurement? ______________

____

4. What is the implied main idea of the map? ______________

____

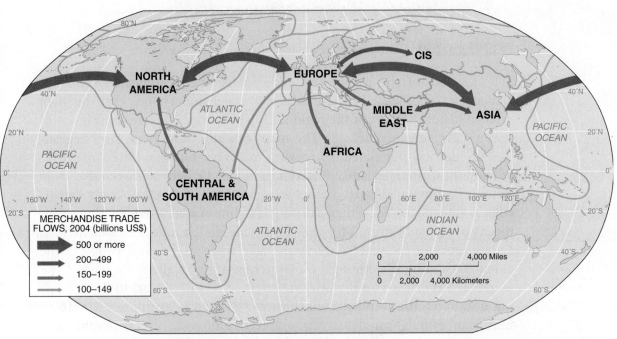

"Figure 12.32 World Merchandize trade by Region"

—From *Introduction to Geography: People, Places, and Environment,* 4th ed., Figure 12.32 by Edward Bergman and William H. Renwick. Copyright © 2008 by Pearson Education, Inc. Reprinted by permission of Pearson Education, Inc., Upper Saddle River, NJ.

A Final Note About Experience and Perspective

As you have worked through this chapter, hopefully you have had some lively discussions about the possible answers for activities that asked you to state implied main ideas. Often a set of details will suggest many things to many people. Determining main ideas requires that the reader bring personal understandings and experience to the task. Thus people with different perspectives may disagree about what the details suggest. Another complex aspect of determining main ideas is that authors may choose to give a collection of details because the idea suggested is difficult to sum up in one sentence. The author intends several meanings to coexist. The important point to remember is that the main idea you formulate should be strongly supported by the details in the paragraph or longer passage.

APPLICATIONS

Application 1: Creating a Summary from Supporting Details

A. Read each group of supporting details. Annotate the text by circling the topic and underlining key words of support and transition. In the spaces provided, use your own words to write the implied main idea for each group.

Supporting details from a college health textbook:

Textbook Skills

- For men, excessive fat is concentrated in the waist, abdomen, and upper body.
- For women, excessive fat is concentrated in the hips and thighs.
- Excessive fat concentrated in the waist, abdomen, and upper body is less healthy than fat located in the hips and thighs.
- Experts are not sure what accounts for these differences in fat distribution between men and women.

—Adapted from McGuigan, *Encyclopedia of Stress,* 1999, p. 77.

Implied main idea: ______________________________

____

____

B. Read the following two paragraphs from a college history textbook. Annotate the text. In the spaces provided, write a sentence that states the implied main idea for each paragraph.

Education for Democracy

Textbook Skills

[1]During the mid-1800s, except on the edge of the frontier and in the South, most youngsters between the ages of 5 and 10 attended school for at least a couple of months of the year. [2]These schools, however, were privately run and charged fees. [3]Attendance was not required and fell off sharply once children learned to read and do their sums well enough to get along in day-to-day life. [4]The teachers were usually young men waiting for something better to do. [5]All this changed with the rise of the common school movement. [6]At the heart of the movement was the belief, widely expressed in the first days of the Republic, that a government based on democratic rule must provide the means, as Jefferson put it, to

"diffuse knowledge throughout the mass of the people." [7]This meant free tax-supported schools, which all children were expected to attend. [8]It also came to mean that such an educational system should be administered on a statewide basis and that teaching should become a profession that required formal training.

—Adapted from *The American Nation: A History of the United States to 1877: Volume 1,* 10th ed., pp. 316–317 by John A. Garraty and Mark A. Carnes. Copyright © 2000 by Pearson Education, Inc. Printed and Electronically reproduced by permission of Pearson Education, Inc., Upper Saddle River, NJ.

1. Implied main idea: ______________________________

____

____

____

Textbook Skills

The Blue and the Gray

[1]At the beginning of the Civil War, more than 20 million people lived in the northern states (excluding Kentucky and Missouri, where opinion was divided). [2]In contrast, only 9 million people lived in the South, and of those about 3.5 million were slaves whom the whites hesitated to trust with arms. [3]In addition, the North's economic capacity to wage war was even more impressive. [4]It was manufacturing nine times as much as the Confederacy (including 97 percent of the nation's firearms) and had a far larger and more efficient system than the South. [5]Northern control of the merchant marine and the navy made possible a blockade of the Confederacy. [6]A blockade was a potent threat to a region so dependent on foreign markets.

—Adapted from *The American Nation: A History of the United States to 1877: Volume 1,* 10th ed., p. 405 by John A. Garraty and Mark A. Carnes. Copyright © 2000 by Pearson Education, Inc. Printed and Electronically reproduced by permission of Pearson Education, Inc., Upper Saddle River, NJ.

2. Implied main idea: ______________________________

____

Application 2: Implied Central Idea

Read each of the following passages from a college science textbook. Annotate the text. Then, answer the questions that follow each excerpt.

Gambling

[1]Compulsive—or addictive—behavior often fits a positive feedback model. [2]In this model there is repeated uncontrolled acting out of the

behavior pattern until the abuser's system goes out of control. [3]For example, a cocaine addict may ingest a substance until death, an alcoholic may similarly experience blackout, and a gambler may run out of money or collateral. [4]Compulsive gamblers usually continue gambling regardless of whether they are winning or losing. [5]However, lack of control is more serious when they are losing, perhaps because they are obsessed with the notion that a big win is just one bet away. [6]Some gamblers typically believe that the laws of probability, which are certain to lead to losses, do not apply to them. [7]But even if they understand that they will eventually lose, their control is less than that of non-compulsive gamblers.

[8]Compulsive gamblers often destroy their families as they ignore family needs and lose money to gambling. [9]The compulsion to make up for their losses often leads them to run up large debts and then even to steal. [10]The compulsive gambler is also often self-destructive. [11]For example, one-third of the compulsive gamblers sampled in one study had serious weight problems, 42 percent were drug abusers, 38 percent had cardiovascular problems, and 50 percent had violated laws.

—Adapted from *Encyclopedia of Stress,* 1st ed., by F. J. McGuigan. Copyright © 1999 by Pearson Education, Inc. Printed and Electronically reproduced by permission of Pearson Education, Inc., Upper Saddle River, NJ.

_____ **1.** Which sentence is the best statement of the implied main idea?

a. Compulsive gamblers are self-destructive.
b. Many people are compulsive gamblers.
c. Compulsive gamblers lack self-control.
d. Compulsive gambling is a serious disorder that harms the gambler and his family.

Insomnia

[1]The term **insomnia** typically means "sleeplessness" or "inability to sleep." [2]More precisely, common patterns are (1) resistance to falling asleep, (2) difficulty in remaining asleep, (3) poor sleep quality, and (4) waking up too early.

[3]Common problems related to chronic insomnia are fatigue, irritability, inability to concentrate, poor short-term memory, drowsiness, and abuse of stimulants. [4]Short-term insomnia occurs in most people who are under great distress or emotional upheaval. [5]Women tend to be more affected than men. [6]The frequency of both increases with age. [7]Chronic insomnia can begin at any age. [8]However, most patients experience ongoing difficulty prior to age 40.

[9]Physical factors that may affect sleep include heart disease, high blood pressure, "heartburn" associated with a hernia, and chronic

breathing problems. [10]However, most cases lack any obvious physical basis. [11]Stressful experiences that create unresolved excess tension are the main factors that cause insomnia. [12]Psychological factors such as anxiety and anger can keep a person "on alert." [13]And a highly active imagination and the reviewing of the day's activities can hinder one's ability to fall asleep.

—Adapted from *Encyclopedia of Stress,* 1st ed., p. 119 by F. J. McGuigan. Copyright © 1999 by Pearson Education, Inc. Printed and Electronically reproduced by permission of Pearson Education, Inc., Upper Saddle River, NJ.

2. Implied central idea: ______________________________

____

Application 3

The following passage is taken from a chapter review of a college psychology textbook. This portion of the review covers the first major section of the chapter. This major section has four subdivisions as indicated. Complete the study outline that follows the passage. First, determine the implied main idea of each subdivision. Then, determine the implied central idea of the entire section.

Textbook Skills

Explaining Motivation

A. How do the three components of motivation work together to influence behavior?

Activation is the component of motivation in which an individual takes the first steps toward a goal. Persistence is the component of motivation that enables a person to continue to work toward the goal even when he or she encounters obstacles. The intensity component of motivation refers to the energy and attention a person must employ to reach a goal.

B. What is the difference between intrinsic and extrinsic motivation?

With intrinsic motivation, an act is performed because it is satisfying or pleasurable in and of itself. With extrinsic motivation, an act is performed to bring a reward or to avert an undesirable sequence.

C. How do drive-reduction and arousal theory explain motivation?

Drive-reduction theory suggests that a biological need creates an unpleasant state of emotional arousal that impels the organism to engage in behavior that will reduce the arousal level. Arousal theory suggests that the aim of motivation is to maintain an optimal level of arousal.

D. According to Maslow, how do individuals attain self-actualization?

Maslow claimed that individuals must satisfy physiologic, safety needs before addressing needs for love, esteem, and self-actualization.

—From *Mastering the World of Psychology*, 3rd ed., p. 314 by Samuel E. Wood, Ellen Green Wood, and Denise Boyd. Copyright © 2008 by Pearson Education, Inc. Reproduced by permission of Pearson Education, Inc., Boston, MA.

1. Implied Central Idea: __________________________________

A. Implied main idea: __________________________________

B. Implied main idea: __________________________________

C. Implied main idea: __________________________________

D. Implied main idea: __________________________________

REVIEW TEST 1

Score (number correct) ______ × 25 = ______%

Implied Main Ideas

A. Read each group of supporting details. Annotate the text. Then, choose the sentence that best expresses the implied main idea for each group.

1. Supporting details:

- **Education.** About 1 in 4 adults with an advanced degree engage in a high level of overall physical activity, compared with 1 in 7 of those with less than a high school diploma.
- **Income.** Adults with incomes below the poverty level are three times as likely to be physically inactive as adults in the highest income group.

- **Marital Status.** Married women are more likely than never-married women to engage in a high level of overall physical activity.
- **Geography**. Adults in the South are more likely to be physically inactive than adults in any other region.
- This information is based on approximately 32,000 interviews with adults ages 18 and over, regardless of employment status, from the National Health Interview Survey, conducted by CDC's National Center for Health Statistics (NCHS).

—"HHS Issues New Report on Americans' Overall Physical Activity Levels," National Center for Health Statistics, 14 May 2003, Centers for Disease Control and Prevention. 2 Dec 2003 http://www.cdc.gov/nchs/releases/03news/physicalactivity.htm

_____ Which statement best expresses the implied main idea?

a. According to the National Center for Health Statistics, several factors affect the levels of physical activity of Americans.
b. People who live in the South are lazier than those who live in any other region.
c. Married women are in better physical condition than single women.
d. Exercise is an important aspect of a healthy lifestyle.

2. Supporting details:

- Graffiti consists of inscriptions, slogans and drawings scratched, scribbled or painted on a wall or other public or private surface.
- Graffiti is most often thought of as an eyesore.
- Graffiti generates fear of neighborhood crime and instability.
- It is costly, destructive, and lowers property values.
- Graffiti sends a message that people of the community are not concerned about the appearance of their neighborhoods.
- It is also against the law!

—Adapted from "What Is Graffiti?" Graffiti Control Program, The City of San Diego. 2 Dec. 2003 http://www.sannet.gov/graffiti/whatis.shtml

_____ Which sentence best states the implied main idea?

a. Graffiti should be considered as a right to free speech.
b. Graffiti is not art; it is vandalism.
c. Graffiti lowers the value of property.
d. Graffiti is illegal.

B. Read the following paragraph. Annotate the text. In the space provided, write the letter of the best statement of the implied main idea.

Building Healthy Bones

[1]The main mineral in bones is calcium, one of whose functions is to add strength and stiffness to bones, which they need to support the body. [2]To lengthen long bones during growth, the body builds a scaffold of protein and fills this scaffold in with calcium-rich mineral. [3]From the ages of 11 through 24, an individual needs about 1,200 milligrams (mg) of calcium each day. [4]Bone also needs vitamin D, to move calcium from the intestine to the bloodstream and into bone. [5]Vitamin D can be obtained from short, normal day-to-day exposure of arms and legs to sun and from foods fortified with the vitamin. [6]Also needed are vitamin A, vitamin C, magnesium and zinc, as well as protein for the growing bone scaffold. [7]Many foods provide these nutrients. [8]Growing bone is especially sensitive to the impact of weight and pull of muscle during exercise, and responds by building stronger, denser bones. [9]That's why it's especially important to be physically active on a regular basis. [10]Bone building activities include sports and exercise such as football, basketball, baseball, jogging, dancing, jumping rope, inline skating, skateboarding, bicycling, ballet, hiking, skiing, karate, swimming, rowing a canoe, bowling, and weight-training.

—Adapted from Farely, Dixie. "Bone Builders: Support Your Bones with Healthy Habits." *FDA Consumer Magazine*. September–October 1997. U. S. Food and Drug Administration. 19 March 2004. http://www.fda.gov/fdac/features/1997/697_bone.html

______ **3.** The best statement of the implied main idea is

a. Healthy diet leads to healthy bones.

b. Calcium is necessary to building healthy bones.

c. One must eat a well-balanced diet and exercise regularly to build healthy bones.

d. Many people develop weak bones due to poor diet and lack of exercise.

C. Study the following map from the textbook *Cultural Anthropology*. Write a sentence that states the implied main idea of the map.

4. Implied main idea: ______________________

____

____

____

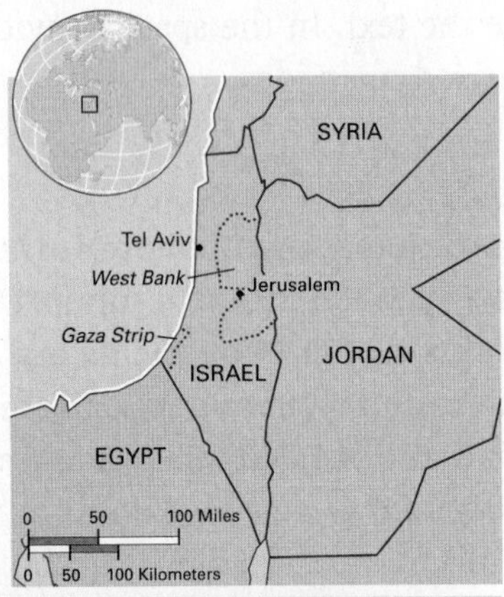

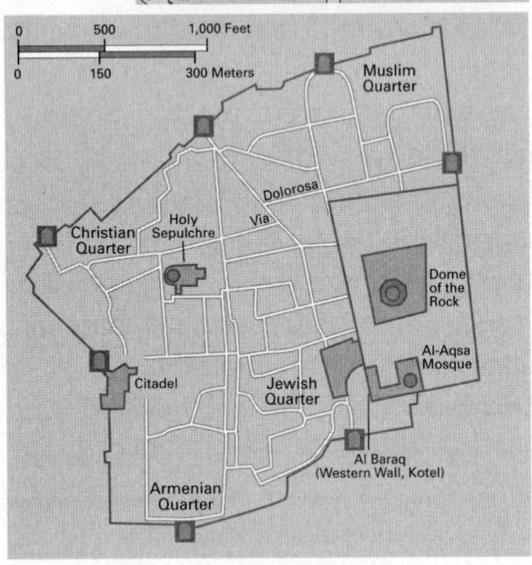

◄ Sacred Sites in the Old City of Jerusalem, Israel. Jerusalem is the holiest city of Judaism, the third holiest city of Islam, and holy to some Christian denominations. The section called the Old City is surrounded by walls that have been built, razed, relocated, and rebuilt over several hundred years. The Old City contains four quarters: Armenian, Christian, Jewish, and Muslim, and many sacred sites such as the Kotel and the via Dolorosa.

REVIEW TEST 2

Score (number correct) ______ × 25 = ______ %

Read the following passage from an American history college textbook. Answer the questions that follow the passage.

Textbook Skills

Taking Sides: Constitutional Principles

[1]What is the best way of understanding the Constitution? [2]Does it embody universal values, or should it be understood in terms of contemporary society? [3]Is there a difference between the founding conceptions of equality and those of today? [4]Does a natural rights understanding assume more individual freedom?

Overview

[5]When the Constitution was ratified in June 1788, the United States had a population of roughly 3.9 million, with an urban population of less than 500,000 citizens (approximately 11 percent). [6]The overwhelmingly dominant religion was Protestantism, and the right to vote was held principally by those who owned property. [7]In 2005, the United States had a population of nearly 300 million, with an urban population of roughly 225 million (72.5 percent). [8]How is it that the Constitution can incorporate the differing social and political views of a multicultural and diverse nation?

[9]Some scholars argue that the Constitution has been a successful document. [10]Its success is based upon the principles of natural law and natural rights—principles holding that all human beings are created equal and endowed with certain inalienable rights. [11]These principles do not change over time. [12]And political institutions can be created to reflect natural equality and human dignity.

[13]Others argue that the Constitution is a flexible instrument created to adapt to social, historical, and political change. [14]This view—positivism—holds that constitutions and laws should reflect prevailing social convention and thought, and it is in this way that the Constitution has been able to be interpreted to allow for equality and social justice. [15]Just what does the Constitution mean and how will this question determine the near future of American history?

Supporting a Natural Rights Interpretation of the Constitution

[16]**Natural rights theory assumes a higher moral law.** [17]The founders were correct in their supposition that it is through liberty and justice that individuals can realize their potential and approach happiness, and the Constitution was created to embody these values. [18]These values do not change over time.

[19]**A natural rights interpretation assumes the use of reason.** [20]Alexander Hamilton argues in the Federalist Papers that the Constitution represents "good government" created by "reflection and choice." [21]The founders used reflection and reason to create a new form of government based on the natural rights principle that all political power is derived from the people exercising their right to create government and to live under laws of their own choosing.

[22]**Natural rights theory embodies the principle of political equality.** [23]The Constitution should be interpreted as incorporating the principle found in the Declaration of Independence that "all men are created

equal" and should have equal political rights. [24]This allows the rich and the poor, the highly educated and the ignorant, the secular and the religious, and the interested and the apathetic to have a say and a share in government.

Against a Natural Rights Interpretation of the Constitution

[25]**The founders simply used the prevailing philosophies of their times.** [26]There is no way to determine if natural rights theory is true. [27]The founders lived in a certain moment in history and they had no way of knowing what the future held in the way of new philosophies and science of government. [28]For example, they did not consider that government could be used for social purposes, such as ensuring social welfare through government policy.

[29]**The Constitution must be interpreted in light of advances in technology and social organization.** [30]The United States of 2008 is a different nation than the America of 1788. [31]It is highly unlikely that the founders could envision the complex evolution of human society and technology—how could they consider freedom of speech issues and the Internet? [32]To apply constitutional law to Internet speech issues necessarily means interpreting the Constitution in a way undreamed of by the founders.

[33]**Natural rights theory as understood by the founders leads to inequality.** [34]For example, the Declaration of Independence declares all men are equal, yet it allowed for slavery and unregulated free markets. [35]The Constitution must be interpreted with a view to new understandings of social and political equality.

_____ **1.** Which sentence best states the implied main idea of paragraph 2 (sentences 5–8)?

a. The United States was founded on Protestant religious values.
b. The Constitution is based on universal values that do not change.
c. The United States has changed dramatically in size and culture since the Constitution was ratified.
d. The Constitution is outdated.

_____ **2.** Which sentence best states the implied central idea of paragraphs 5 through 7 (sentences 16–24)?

a. Natural rights theory is based on the moral law of values that do not change.

b. Natural rights theory is the foundation of the Constitution.
c. The Constitution is a well-thought out document.
d. The Constitution states "all men are created equal."

_____ **3.** Which sentence best states the implied central idea of paragraphs 8 through 10 (sentences 25–35)?
a. Technology and social organization have outpaced natural rights.
b. Natural rights is a theory of the founders who were limited by their own culture and ideas.
c. The natural rights theory did not prevent slavery.
d. Natural rights theory is an outdated interpretation of the Constitution for three reasons.

_____ **4.** Which sentence best states the implied central idea of the entire passage?
a. Positivism views the Constitution as a flexible document.
b. The Constitution is based on universal values of natural rights and natural law that do not change.
c. The meaning of the Constitution is a matter of debate.
d. Scholars debate the meaning of the Constitution based on the conflicting theories of natural rights and positivism.

REVIEW TEST 3

Score (number correct) _____ × 10 = _____%

Implied Main Ideas

Before you read, skim the following passage. Answer the Before Reading questions. Read the passage and annotate the text. Then, answer the After Reading questions.

Vocabulary Preview

ornamental (5): decorative, attractive

pigments (10): colors

spectrum (12): range

synthesize (17): manufacture

foliage (35): plant life

The Awe in Autumn: Why Leaves Change Their Color

[1]Did you know that the state of Maine is the most heavily forested state in the union? [2]It's true. [3]Over 90 percent of the state's land mass is devoted to growing trees. [4]That's 17 million acres! [5]In addition, we have a wide variety of trees, with over 50 native species, and many more **ornamental** trees in various neighborhoods. [6]With such an abundance of trees, it is no wonder that we have such a breathtaking season.

[7]Every autumn, magic **transformations** begin as trees begin to prepare for winter. [8]In certain regions, such as our

own, the shedding of leaves is preceded by a spectacular color show. [9]Formerly green leaves turn to brilliant shades of yellow, orange, and red. [10]These color changes are the result of transformations in leaf **pigments**.

[11]All during the spring and summer, leaves serve as factories where most of the foods necessary for the trees' growth are manufactured. [12]This food-making process takes place inside the leaf in numerous cells that contain the pigment chlorophyll, which actually masks the true color of the leaf by absorbing all other light colors and reflecting only the green light **spectrum**—thus the leaves appear to our eyes as green.

[13]The energy of the light absorbed by chlorophyll is converted into chemical energy stored in carbohydrates (sugars and starches). [14]By the plant absorbing water and nutrients through its roots, and absorbing carbon dioxide from the air, the chemical process of photosynthesis takes place. [15]This chemical change drives the biochemical reactions that cause plants to grow, flower, and produce seed. [16]Chlorophyll is not a very stable compound; bright sunlight causes it to **decompose**. [17]To maintain the amount of chlorophyll in their leaves, plants continuously **synthesize** it. [18]The synthesis of chlorophyll in plants requires sunlight and warm temperatures. [19]During the summer months, chlorophyll is continuously broken down and regenerated in the leaves of the trees.

[20]The shortening of the days and the cooler temperatures at night trigger several changes within the tree. [21]One of these changes is the growth of a corky membrane between the branch and the leaf stem. [22]This membrane interferes with the flow of nutrients into the leaf. [23]As this nutrient flow is interrupted, the production of chlorophyll in the leaf declines, and the green color begins to fade.

[24]The different chemical components that are always present inside the structure of the individual leaf will determine its fleeting brilliant color in the fall. [25]But weather also plays a factor in determining color. [26]The degree of color may vary from tree to tree. [27]For example, leaves directly exposed to the sun may turn red, while those on the shady side of the same tree may appear yellow or orange. [28]Also, the colors on the same tree may vary from year to year, depending upon the combination of weather conditions.

[29]According to the tree experts at several universities, including North Carolina State, the U.S. Department of Agriculture, and the U.S. Forestry Service, the most vivid colors appear after a warm, dry summer

and early autumn rains, which will help prevent early leaf fall. [30]Long periods of wet weather in late fall will produce a rather drab coloration. [31]Droughts favor the chemical formation of anthocyanin, which is the chemical that is most responsible for the brilliant reds and purples. [32]Drought conditions, especially in late summer, also favor red pigment formation due to the reduction of nitrate absorption through the tree's roots.

[33]The best colors seem to be displayed when relatively warm and sunny days are followed by cold nights through early October. [34]Light frost will enhance the colors, but a hard, killing frost will hasten the actual fall of the leaves from the trees. [35]The right combination of temperature, rainfall, and sunshine can prolong the autumn **foliage** brilliance by as much as two weeks.

—"The Awe in Autumn: Why Leaves Change Their Color," *Maine-ly Weather,* Fall/Winter 2002–2003, National Weather Service, NOAA. 2 Dec. 2003 http://www.erh.noaa.gov/er/car/WCM/Winter0203/page6.html

Before Reading

Vocabulary in Context

______ **1.** The best definition of the word **transformations** in sentence 7 is

a. mysteries.
b. changes.
c. shows.
d. seasons.

______ **2.** The best meaning of the word **decompose** in sentence 16 is

a. stabilize.
b. grow.
c. decay.
d. stink.

Thought Patterns

______ **3.** What is the thought pattern suggested by the title of the passage?

a. cause and effect
b. classification
c. spatial order
d. comparison

Stated Main Ideas

______ **4.** Which sentence is the topic sentence in the fifth paragraph (sentences 20–23)?

a. sentence 20
b. sentence 21
c. sentence 22
d. sentence 23

After Reading

Supporting Details

______ **5.** According to tree experts from universities and government agencies,
a. autumn is a breathtaking season.
b. trees go through magic transformations.
c. the most vivid colors appear after a warm, dry summer.
d. carbohydrates are sugars and starches.

______ **6.** Based on the passage, chlorophyll
a. causes the color of the leaves to turn brilliant colors.
b. is a stable compound.
c. is a pigment which masks the true color of the leaf.
d. is a corky membrane between the branch and the leaf stem.

Concept Maps and Charts

7. Complete the concept map with information from the passage.

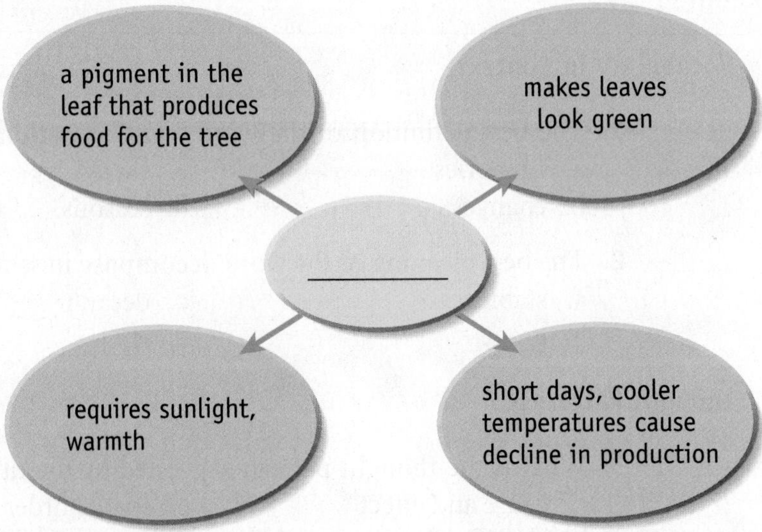

Implied Main Ideas and Implied Central Idea

______ **8.** Which sentence best states the implied main idea of the third, fourth, and fifth paragraphs (sentences 11–23)?
a. The production of chlorophyll affects the color of the leaves on a tree.

b. Chlorophyll produces food for the tree.
c. When chlorophyll production declines, leaves lose their green color.
d. Cooler weather slows down the production of chlorophyll.

_____ **9.** Which sentence best states the implied main idea of the eighth paragraph (sentences 33–35)?
a. A hard frost will cause autumn leaves to fall from trees early in the season.
b. Warm and sunny days produce brilliant colors.
c. Climate plays a significant role in the brilliance of the autumn colors.
d. Many factors affect the brilliance of the autumn colors.

_____ **10.** Which sentence best states the implied central idea of the reading passage?
a. Autumn is a beautiful season.
b. Chlorophyll is a major component in the process of leaves turning colors.
c. Several factors work together to produce the awe-inspiring colors of fall.
d. Weather is a major factor that affects the quality of the color of autumn leaves.

WHAT DO YOU THINK?

Do you think the change of seasons affects humans physically? Emotionally? Why or why not? Assume you are a real estate salesperson. Your firm is planning a national advertising campaign to persuade people to move to your region to buy a home or commercial property. Write a couple of paragraphs that describe the appealing aspects of a particular season or all the seasons in your area.

REVIEW TEST 4

Score (number correct) _____ × 10 = _____%

Implied Main Ideas

Before you read, skim the following passage. Answer the Before Reading questions. Read the passage and annotate the text. Then, answer the After Reading questions.

Vocabulary Preview

scavenged (5): hunted, searched

migrating (6): traveling, moving

manipulate (15): use, operate

mechanisms (22): means, methods

hierarchies (23): levels of power or importance

The Food-Producing Revolution

[1]For the first thousands of millennia of their existence, modern humans, known as *Homo sapiens sapiens* ("most intelligent people"), did not produce food. [2]Between 200,000 and 100,000 years ago, *Homo sapiens sapiens* first appeared in Africa and began to spread to other continents. [3]Scientists refer to this stage of human history as the Paleolithic Age, or Old Stone Age, because people made tools by cracking rocks and using their sharp edges to cut and chop. [4]*Homo sapiens sapien's* use of tools demonstrated adaptation to new environments and practical needs. [5]They **scavenged** for wild food and became shrewd observers of the natural environment. [6]They followed **migrating** herds of animals, hunting with increasing efficiency as their weapons improved. [7]They also created beautiful works of art by carving bone and painting on cave walls. [8]By 45,000 years ago, these humans had reached most of Earth's **habitable** regions, except for Australia, the islands of the South Pacific, and North and South America.

[9]The end of the last Ice Age about 15,000 years ago ushered in an era of momentous change: the food-producing revolution. [10]As the Earth's climate became warmer, causing changes in vegetation, humans began to interact with the natural environment in new ways. [11]The warmer climate allowed cereal grasses to spread quickly over large areas; hunter-gatherers learned to collect these wild grains and grind them up for food. [12]Some groups of hunter-gatherers settled in semi-permanent camps near rivers and wetlands, where wild grains grew. [13]When people learned that the seeds of wild grasses could be transplanted and grown in new areas, the domestication of plants was under way (see Map).

[14]At the same time that people discovered the benefits of planting seeds, they also began domesticating pigs, sheep, goats, and cattle, which eventually replaced wild game as the main source of meat. [15]**Domestication** occurs when humans **manipulate** the breeding of animals in order to serve their own purposes—for example, making wool (lacking on wild sheep), laying extra eggs (not done by undomesticated chickens), and producing extra milk (wild cows produce only enough milk for their offspring). [16]The first signs of goat domestication occurred about 8900 B.C.E. in the Zagros Mountains in Southwest Asia. Pigs, which adapt very well to human settlements because they eat garbage, were first domesticated around 7000 B.C.E. [17]By around 6500 B.C.E. domesticated cattle, goats, and sheep had become widespread.

[18]Farming and herding required hard work. [19]Even simple agricultural methods could produce about fifty times more food than hunting and gathering. [20]Thanks to the increased food supply, more newborns survived past infancy. [21]Populations expanded, and so did human settlements. [22]With the mastery of food production, human societies developed the **mechanisms** not only to feed themselves, but also to produce a surplus, which could then be traded for other resources. [23]Such economic activity allowed for economic specialization and fostered the growth of social, political, and religious **hierarchies**.

—Adapted from *The West: Encounters & Transformation, Atlas Edition,* Combined Edition, 2nd ed., Map 1.1, pp. 13–14, 27 by Brian Levack, Edward Muir, Meredith Veldman and Michael Maas. Copyright © 2008 by Pearson Education, Inc. Printed and Electronically reproduced by permission of Pearson Education, Inc., Upper Saddle River, NJ.

The Beginnings of Food Production

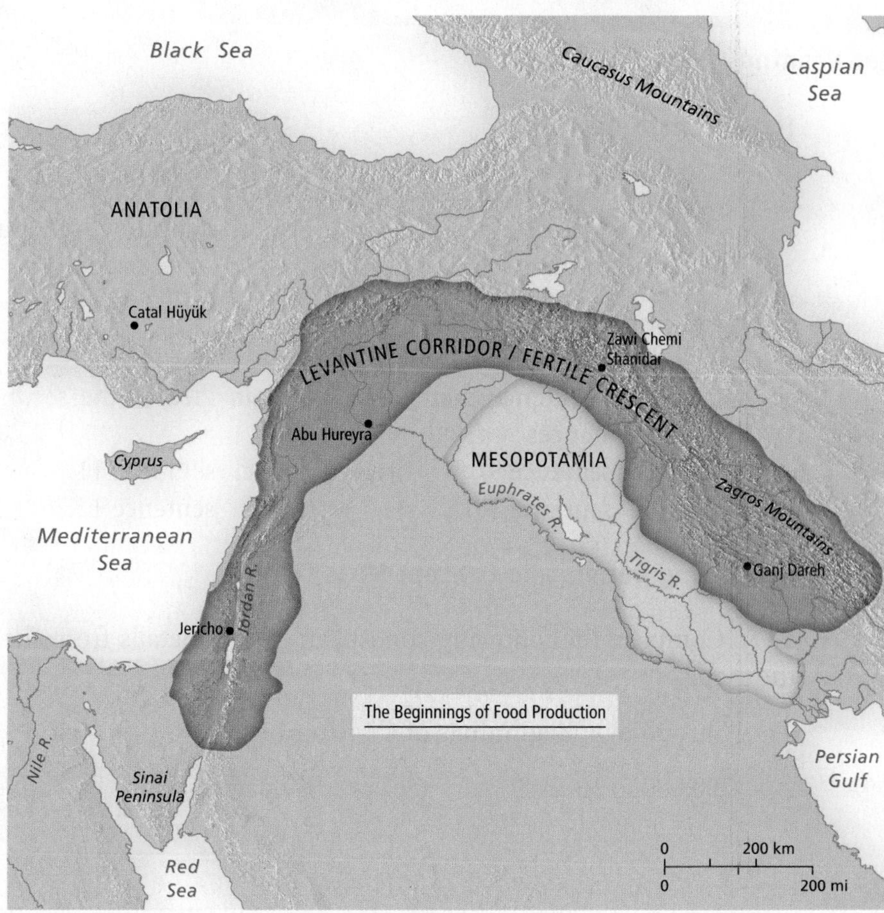

Before Reading

Vocabulary in Context

1–2. Study the following chart of word parts and the context in which the words are used. Then, use your own words to define the terms.

Root	Meaning	Suffix	Meaning
habitare	possess	*-able*	quality, degree
domus	house	*-ation*	action, process

1. In sentence 8, **habitable** means ______________________________

____

2. In sentence 15, **domestication** means ______________________________

____

After Reading

Thought Patterns

_____ **3.** The overall thought pattern used in this passage is
a. time order.
b. comparison and contrast.
c. definition and example.
d. space order.

Stated Main Ideas

_____ **4.** The sentence that states the main idea of the second paragraph (sentences 9–13) is
a. sentence 9.
b. sentence 10.
c. sentence 11.
d. sentence 12.

Supporting Details and Concept Maps

5–7. Complete the following concept map with details from the passage and the map.

Chronology: Beginnings of Civilization

200,000 to 100,000 years ago	Modern humans first appear in Africa.
45,000 years ago	**5.** ______________________________

15,000 years ago	Ice Age ends.
	6. ____________________
8900 B.C.E.	First signs of goat domestication occurred.
7000 B.C.E.	Pigs were domesticated.
6500 B.C.E.	**7.** ____________________ ____________________

Implied Main Ideas and Implied Central Ideas

_____ **8.** Which sentence best states the implied main idea of the fourth paragraph (sentences 18–23)?
a. Domesticating plants and animals was a difficult process.
b. The benefits of farming and herding were worth the effort.
c. Farming and herding increased trade.
d. Farming and herding lengthened the life expectancy of infants.

_____ **9.** Which sentence best states the implied central idea of the passage?
a. Food production requires great effort.
b. Domestication of plants and animals occurred about 15,000 years ago.
c. Increased food supplies lengthened the life of humans.
d. Food production made civilization possible.

Reading a Map

10. Study the map. State in one sentence the implied main idea of the map.

____

____

WHAT DO YOU THINK?

With the advancement of technology, humans now have the ability to feed the entire world. Why do you think, then, that famine and hunger are still a global problem? Which regions of the world are most affected by famine and starvation? What steps can we take as a country or as individuals to offer relief? Assume you are a concerned citizen who wants to help a particular group of people who are suffering from hunger. Write a letter to the editor of your local newspaper that calls on the public to help this group of people. Explain the problem and ask for action.

After Reading About Implied Main Ideas and Implied Central Ideas

Before you move on to the Mastery Tests on implied main ideas and implied central ideas, take time to reflect on your learning and performance by answering the following questions. Write your answers in your notebook.

- How has my knowledge base or prior knowledge about implied main ideas changed?
- Based on my studies, how do I think I will perform on the Mastery Test(s)? Why do I think my scores will be above average, average, or below average?
- Would I recommend this chapter to other students who want to learn more about implied main ideas? Why or why not?

Test your understanding of what you have learned about Implied Main Ideas for Master Readers by completing the Chapter 4 Review Card in the insert near the end of your text.

CONNECT TO PEARSON myreadinglab

To check your progress in meeting Chapter 4's learning outcomes, log in to **www.myreadinglab.com** and try the following activities.

- The "Main Idea" section of MyReadingLab provides review materials, practice activities, and tests about topics and main ideas. To access this resource, click on the "Study Plan" tab. Then click on "Main Idea." Then click on the following links as needed: "Overview," "Model," "Practice," and "Tests."
- To measure your mastery of the content in this chapter, complete the tests in the "Main Idea" section and click on Gradebook to find your results.

MASTERY TEST 1

Name ______________________ Section ______________

Date ____________ **Score** (number correct) ________ × 50 = ________%

A. Study the map and its caption from a college history textbook. Then, write a sentence that states the implied main idea of the map.

Textbook Skills

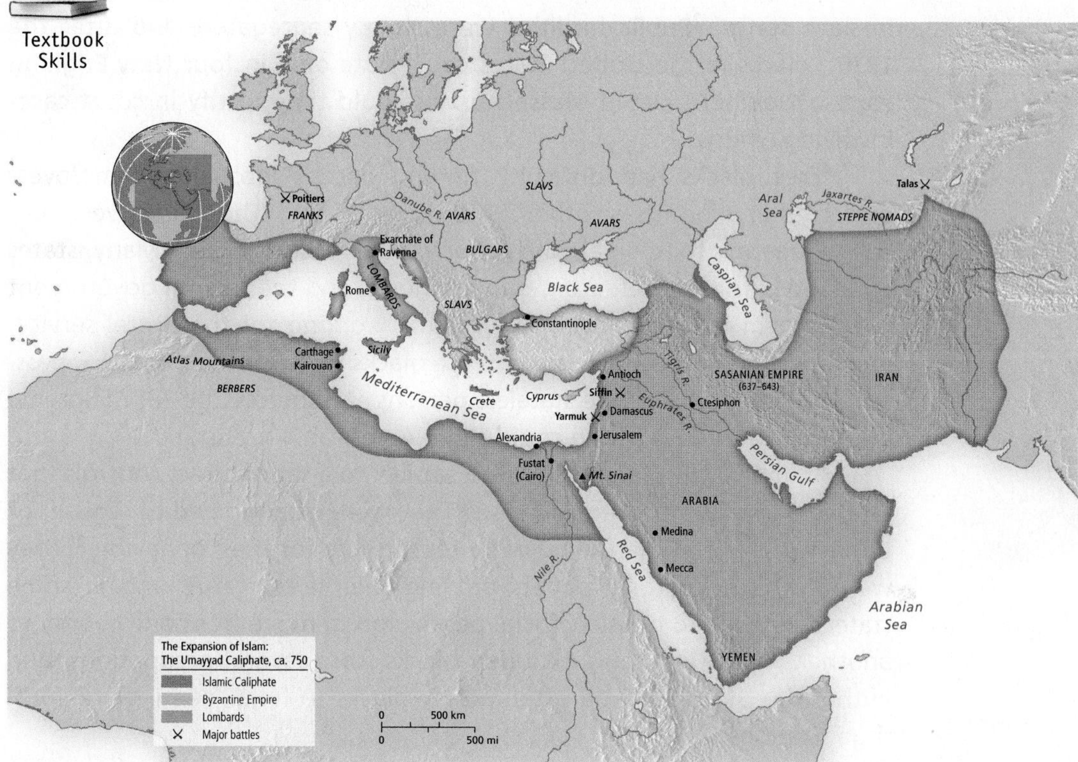

▲ Islam originated in the early seventh century among the inhabitants of the Arabian peninsula. Through conquest and expansion, Muslims created a single Islamic Empire stretching from Spain to central Asia by 750.

—Adapted from *The West: Encounters & Transformations, Atlas Edition, Combined Edition,* 2nd ed., Map 7.3, pp. 219, 221 by Brian Levack, Edward Muir, Meredith Veldman and Michael Maas. Copyright © 2008 by Pearson Education, Inc. Printed and Electronically reproduced by permission of Pearson Education, Inc., Upper Saddle River, NJ.

1. Implied main idea of map: ____

____

____

B. Determine the central idea stated in a thesis sentence for the following passage from a college history textbook. Annotate the text. Using your own words, write the central idea in the space provided.

Textbook Skills

The Struggles of Free Blacks

[1]In addition to the 4 million blacks in bondage, there were approximately 500,000 free African Americans in 1860, about half of them living in slave states. [2]Public facilities were strictly segregated, and after the 1830s, blacks in the United States could vote only in four New England states. [3]Nowhere but in Massachusetts could they testify in court cases involving whites.

[4]Free blacks had difficulty finding decent jobs; most employers preferred immigrants or other whites over blacks. [5]The blacks were usually relegated to menial and poorly paid occupations. [6]Many states excluded blacks entirely from public schools, and the federal government barred them from serving in the militia, working for the postal service, and laying claim to public lands. [7]Free blacks were even denied U.S. passports. [8]In effect, they were stateless persons even before the 1857 Supreme Court ruling that no Negro could claim American citizenship.

[9]In the South, free blacks were subject to a set of direct controls that tended to make them semi-slaves. [10]They were often forced to register or to have white guardians who were responsible for their behavior. [11]They were required to carry papers proving their free status, and in some states, they had to obtain official permission to move from one county to another. [12]Licensing laws excluded blacks from several occupations, and authorities often blocked attempts by blacks to hold meetings or form organizations.

—Adapted from *The American Story,* 1st ed., pp. 427–428 by Robert A. Divine, George M. Fredrickson, R. Hal Williams and T. H. Breen.

2. Central idea stated in a thesis sentence: ______________________________

____

____

MASTERY TEST 2

Name ______________________ Section ____________

Date ________ Score ______ (number correct) × 25 = ______%

Read the following textbook passages. Annotate the texts. Then, either write the letter of the best statement of the implied main idea or use your own words to state the implied main idea.

Interpersonal Relationships and the Media

[1]Whether for good or ill, millions of people tune in to the ever-present talk show to learn about friendship, romantic, family, and workplace relationships. [2]Talk shows often present themselves as educational, as therapy, and as authoritative about how you, your parents, and your friends should develop and maintain relationships. [3]Especially in the speech that closes many of the shows, talk show hosts with little to no professional training give out interpersonal advice. [4]For instance, they often give communication advice with no basis in scientific research. [5]They foster myths about communication. [6]Perhaps the most common myth is the idea that the more communication a couple has the better the relationship will be.

[7]Newspaper and magazine columnists offer advice without knowing even a scrap of information about who you are. [8]They do not know what difficulties you may be facing, nor do they know what you really want and need. [9]And astrology columns and psychic hotlines give you advice that has no basis in research.

—From *The Interpersonal Communication Book,* 10th ed., p. 274 by Joseph A. DeVito. Copyright © 2004 by Pearson Education, Inc. Reproduced by permission of Pearson Education, Inc., Boston, MA.

______ **1.** Which sentence best states the implied central idea?

a. The media are currently teaching millions of people ideas about interpersonal relationships that are not based on research.

b. Television talk shows foster myths about communication.

c. The more communication a couple has the better their relationship will be.

d. Newspaper and magazine columnists offer advice to millions of people.

Writing and Speaking

[1]Writing and speechmaking have much in common: both require careful thought about your subject, purpose, and audience. [2]Thus the

mental and physical activities that go into the writing process can also help you prepare and deliver a successful speech.

[3]Despite many similarities, however, writing for readers is not the same as speaking to listeners. [4]Whereas a reader can go back and reread a written message, a listener cannot stop a speech to rehear a section. [5]Several studies have reported that immediately after hearing a short talk, most listeners cannot recall half of what was said.

[6]Effective speakers adapt to their audience's listening ability by restating and repeating their ideas. [7]They use simple words, short sentences, personal pronouns, and informal language. [8]In formal writing, these strategies might seem out of place and too informal. [9]But in speaking, they improve listeners' comprehension.

—Adapted from *The Little, Brown Handbook,* 9th ed., p. 905 by H. Ramsey Fowler and Jane E. Aaron. Copyright © 2004 by Pearson Education, Inc. Reprinted by permission of Pearson Education, Inc., Glenview, IL.

2. Implied central idea of paragraphs 1 and 2: ______________________

____

____

3. Implied main idea of paragraph 3: ______________________

____

Textbook Skills

Advertisements' Purpose

[1]*Business ads* try to influence people's attitudes and behaviors toward the product and services a business sells. [2]Most of these ads try to persuade consumers to buy something. [3]But sometimes a business tries to improve its image through advertising.

[4]*Public service ads* promote behaviors and attitudes that are beneficial to society and its members. [5]These ads may be either national or local and usually are the product of donated labor and media time or space.

[6]*Political ads* aim to persuade voters to elect a candidate to political office or to influence the public on legislative matters. [7]These advertisements are run at the local, state, and national levels.

—Adapted from Folkerts & Lacy, *The Media in Your Life,* 2nd ed., p. 428.

4. Implied central idea: ______________________

____

____

MASTERY TEST 3

Name ______________________ Section ____________

Date __________ **Score** (number correct) ______ × 25 = ______%

Read the following selections. Annotate the texts. In the spaces provided, use your own words to state the implied main idea of each one.

A poem

Dark house, by which once more I stand (1850)

by Alfred Lord Tennyson

Dark house, by which once more I stand
    Here in the long unlovely street,
    Doors, where my heart was used to beat
So quickly, waiting for a hand,

A hand that can be clasped no more—
    Behold me, for I cannot sleep,
    And like a guilty thing I creep
At earliest morning to the door.

He is not here; but far away
    The noise of life begins again,
    And ghastly through the drizzling rain
On the bald street breaks the blank day.

—Tennyson, Alfred Lord. *In Memoriam* (London: E Moxon, 1850).

1. Implied main idea: ____

____

Paragraph from a college science textbook

Textbook Skills

Lives Up in Smoke

[1]Mark Twain once said "Quitting smoking is easy, I've done it a thousand times." [2]Researchers have found that, like cocaine and heroin, nicotine activates the brain's reward center. [3]The brain adjusts by becoming less sensitive, requiring larger quantities of nicotine to experience the same rewarding effect, and causing the reward center to feel understimulated when nicotine is withdrawn. [4]Withdrawal symptoms include nicotine craving, depression, anxiety, irritability, difficulty concentrating, headaches, and disturbed sleep. [5]Although at least 70% of smokers would like to quit, only about 2.5% of smokers are successful each year.

—From *Life on Earth,* 5th ed., p. 390 by Teresa Audesirk, Gerald Audesirk, and Bruce E. Byers. Copyright © 2009 by Pearson Education, Inc. Printed and Electronically reproduced by permission of Pearson Education, Inc., Upper Saddle River, NJ.

2. Implied main idea: ____

____

Paragraph from a college history textbook

Textbook Skills

Abraham Lincoln's Humble Beginnings

[1]Born to poor, illiterate parents on the Kentucky frontier in 1809, Abraham Lincoln received a few months of formal schooling in Indiana after the family moved there in 1816. [2]But mostly, he educated himself. [3]He read and reread a few treasured books by firelight. [4]In 1831, when the family moved to Illinois, he left home to make a living for himself in the struggling settlement of New Salem. [5]After failing as a merchant, he found a path to success in law and politics.

—Adapted from *The American Story,* 1st ed., p. 468 by Robert A. Divine, George M. Fredrickson, R. Hal Williams and T. H. Breen.

3. Implied main idea: ____

____

____

Visual graphic or concept map from a college health textbook

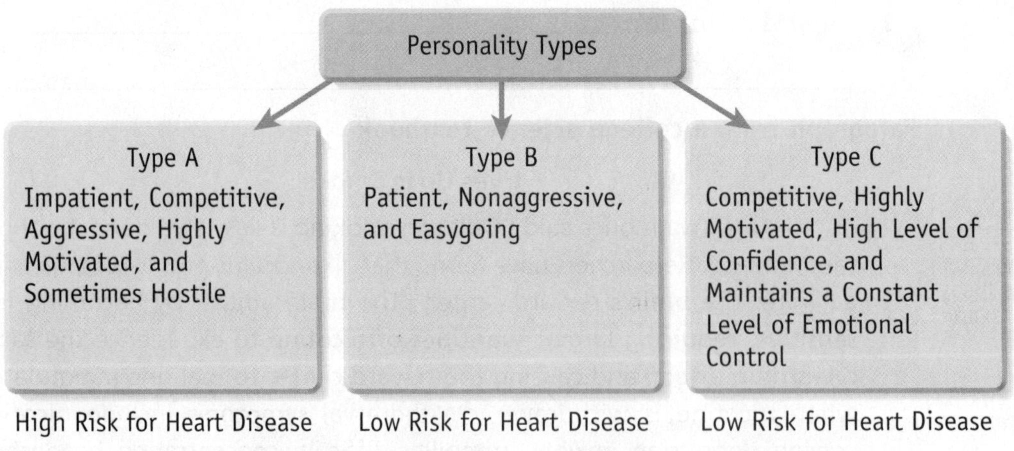

4. Implied main idea: ____

____

MASTERY TEST 4

Name __________________ Section __________

Date __________ Score (number correct) ______ × 25 = ______%

Textbook Skills

Read the passage and answer the questions that follow.

History of Drug Abuse

[1]Drug use is nothing new to the human race. [2]As long as fruit has fermented and been consumed, our species has experienced the effects of drugs, substances that have psychological or physical effects. [3]Intentional fermentation—the process used to make alcohol—dates back to the Stone Age. [4]Beer jugs from the Neolithic period serve as evidence that alcoholic beverages were produced as early as 10,000 B.C.E. [5]Narcotics, drugs that are considered illegal today, also date back thousands of years.

Drug Use in Ancient Times [6]Citizens of Sumer, the first true civilization, used opium for medical and recreational purposes as early as 3000 B.C.E.; early tablets reveal that Sumerians referred to the substance as "Gil Hul" or "joy plant." [7]Ancient texts such as the Torah describe the early origins of winemaking in the Middle East, and the Ebers Papyrus, a medical document from ancient Egypt, outlines the many medicinal uses of opium.

Drug Use in the 18th Century [8]Although opium was originally used as a medicine, it was also one of the first drugs to be abused for recreational purposes. [9]Opium addiction became a serious social problem in China during the 18th century. [10]Initially, it was used to sedate patients and relieve tension amid pain. [11]However, by 1729, opium use became so rampant that the emperor outlawed the cultivation, sale, and use of the drug.

Early Drug Use in the United States [12]During the Civil War in the United States, both opium and its derivative morphine were regularly given to soldiers to relieve the pain of battle wounds. [13]But dependence on the drugs lasted long after the wounds were healed, leading the addiction to become known as "army disease." [14]In 1898, Bayer, the German company known for producing aspirin, developed heroin. [15]Ironically, this drug was marketed as a non-addictive substitute for morphine, as well as a cough suppressant. [16]Heroin was in fact stronger than morphine, and by 1900, approximately 1 percent of the U.S. population was addicted to some form of opiates.

[17]In the early 1900s, harmful and addictive drugs were readily available to consumers. [18]Drugstores were free to sell virtually anything they wished as

long as a person had a prescription, and many over-the-counter drugs contained heroin, morphine, and other opiates. [19]Even Coca-Cola—a seemingly harmless product—contained cocaine until the company switched to caffeine in 1906. [20]However, around the same time, temperance movements for alcohol and drugs began to arise. [21]The first federal law to restrict the sale, manufacturing, and distribution of drugs was the Harrison Narcotic Act of 1914.

—From *Think Social Problems*, p. 142 by John D. Carl.

_____ **1.** Which sentence best states the implied main idea of paragraphs 1 and 2 (sentences 1–7)?
a. Narcotics have been in use for thousands of years.
b. The earliest drug to be produced and used was beer.
c. Drug use first developed for medical and religious reasons.
d. Humans have used drugs since the beginning of civilization.

_____ **2.** Which sentence best states the implied central idea of paragraph 3 (sentences 8–11)?
a. Opium was not intended to be used as a recreational drug.
b. Outlawing a drug does not stop its abuse.
c. Opium was first used as a recreational drug in the 18th century.
d. The Chinese were the first to outlaw the use of opium.

_____ **3.** Which sentence best states the implied central idea of paragraphs 4 and 5 (sentences 12–21)?
a. Drug use was prevalent during the Civil War.
b. Drug abuse in the United States is a historical problem.
c. In the past, the United States encouraged the use of harmful and addictive drugs.
d. The social problem of drug use is a significant current issue in the United States.

_____ **4.** Which sentence best states the implied central idea of the entire passage?
a. The United States struggles to address the social problems of drug use.
b. People have always used drugs.
c. The social problems caused by drug use began in ancient times and continue throughout history.
d. Drug abuse cannot be eliminated.

Supporting Details

5

LEARNING OUTCOMES

After studying this chapter you should be able to do the following:

1. Define *major supporting details, minor supporting details*, and *summary*.
2. Create questions to locate supporting details.
3. Distinguish between major and minor supporting details.
4. Complete a simple chart that outlines the topic, main idea, and supporting details of a passage.
5. Annotate a passage to create a summary.
6. Create a summary of a passage.
7. Evaluate the importance of supporting details.

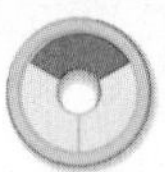

Before Reading About Supporting Details

In Chapter 3, you learned several important ideas that will help you as you work through this chapter. Use the following questions to call up your prior knowledge about supporting details.

What is a main idea? (page 96) ______________________________

____

____

What are the three parts of most paragraphs? (page 100) ______________,

______________, and ______________________________.

Define supporting details. (page 101) ______________________________

____

What are the different possible locations of topic sentences? (pages 105–111.)

____

____

What is a central idea? (page 113.) ______________________________________

____

Questions for Locating Supporting Details

To locate supporting details, a master reader turns the main idea into a question by asking one of the following reporter's questions: *who, what, when, where, why,* or *how.* The answer to this question will yield a specific set of supporting details. For example, the question *why* is often answered by listing and explaining reasons or causes. The question *how* is answered by explaining a process. The answer to the question *when* is based on time order. An author strives to answer some or all of these questions with the details in the paragraph. You may want to try out several of the reporter's questions as you turn the main idea into a question. Experiment to discover which question is best answered by the details.

> **Supporting details** explain, develop, and illustrate the main idea.

Take, for example, the subject "youth joining gangs." For this topic an author might choose to write about ways parents can prevent their child from becoming involved in gangs. The main idea of such a paragraph may read as follows:

Main idea: To prevent their children from joining gangs, parents must take the following two steps.

Using the word *how* turns the main idea into this question: "How can parents prevent their children from joining gangs?" Read the following paragraph for the answers to this question.

Parental Love:
The First Line of Defense in the War Against Gangs

[1]To prevent their children from joining gangs, parents must take the following two steps. [2]First, and most important, parents should

spend time with their children. [3]One-on-one time with a parent who is giving undivided attention builds trust, establishes communication, and fosters self-esteem. [4]Likewise, family trips to parks, libraries, museums, or the beach create a sense of belonging so that children are not tempted to seek acceptance elsewhere. [5]Second, parents should set and abide by reasonable rules. [6]Establishing boundaries that ensure safety and teach respect for authority is an expression of love. [7]Parents who expect children to value an education, abide by a curfew, and choose friends wisely are thwarting the chances that their children will join a gang.

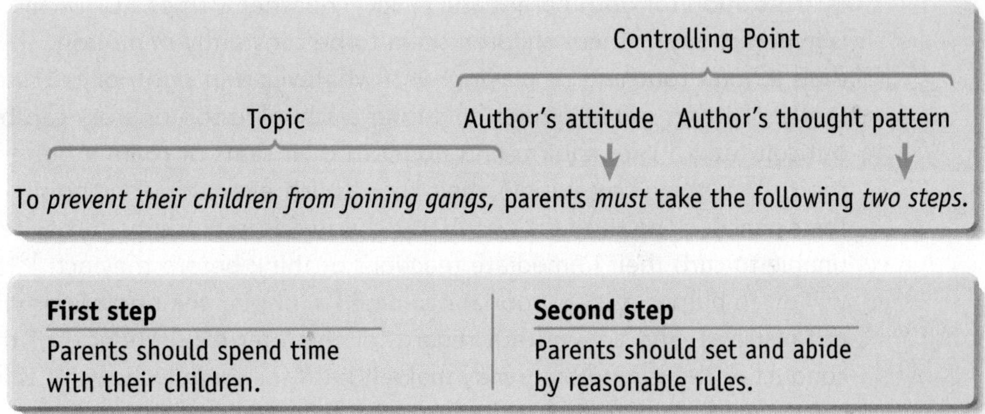
Controlling Point
Topic
Author's attitude
Author's thought pattern
To *prevent their children from joining gangs,* parents *must* take the following *two steps.*
First step
Parents should spend time with their children.
Second step
Parents should set and abide by reasonable rules.

The supporting details for this main idea answer the question "how?" by listing the two steps that are likely to prevent a child from joining a gang. Then, the paragraph discusses why these steps may help.

Note the relationship between the author's controlling point and the supporting details.

Notice that the details about situations directly explain the main idea. However, additional supporting details were given in the paragraph that are not listed. Each of the main supporting details needed further explanation. This paragraph shows us that there are two kinds of supporting details: details that explain the main idea and details that explain other details.

EXAMPLE Read the paragraph. Turn the topic sentence into a question using one of the reporter's questions (Who? What? When? Where? Why? How?). Write the question in the space. Fill in the chart with the answers to the questions you have created.

The Traits of Attention Deficit Hyperactivity Disorder

[1]Attention Deficit Hyperactivity Disorder (ADHD) has three basic traits. [2]One trait of ADHD children is **inattention.** [3]They have a hard time keeping their minds on any one thing and may get bored with a task after only a few minutes. [4]If they are doing something they really enjoy, they have no trouble paying attention. [5]But focusing deliberate, conscious attention to organizing and completing a task or learning something new is difficult. [6]Homework is particularly hard for these children. [7]They will forget to write down an assignment, or leave it at school. [8]They will forget to bring a book home, or bring the wrong one. [9]The homework, if finally finished, is full of errors and erasures. [10]Homework is often accompanied by frustration for both parent and child. [11]Another trait of ADHD children is **hyperactivity.** [12]These children seem to be constantly in motion. [13]They dash around touching or playing with whatever is in sight, or talk incessantly. [14]Sitting still at dinner or during a school lesson or story can be a difficult task. [15]They squirm and fidget in their seats or roam around the room. [16]Or they may wiggle their feet, touch everything, or noisily tap their pencil. [17]The third trait of ADHD children is **impulsivity.** [18]They seem unable to curb their immediate reactions or think before they act. [19]They will often blurt out inappropriate comments, display their emotions without restraint, and act without regard for the later consequences of their conduct. [20]Their impulsivity may make it hard for them to wait for things they want or to take their turn in games. [21]They may grab a toy from another child or hit when they're upset. [22]Even as teenagers or adults, they may impulsively choose to do things that have an immediate but small payoff rather than engage in activities that may take more effort yet provide much greater but delayed rewards.

—Adapted from "Attention Deficit Hyperactivity Disorder."
National Institute of Mental Health. 9 April 2004. 13 Dec 2007
http://www.nimh.nih.gov/publicat/adhd.cfm#symptoms

Question based on topic sentence: ____________________

Topic sentence: ____________________

First trait	**Second trait**	**Third trait**
(1) __________	(2) __________	(3) __________

EXPLANATION Using the reporter's question *What?*, you should have turned the topic sentence into the following question: *What are the three basic traits of*

ADHD? The answer to this question yields the supporting details. Compare your answers to the following: (1) Inattention, (2) Hyperactivity, and (3) Impulsivity.

PRACTICE 1

Read the paragraph. Turn the topic sentence into a question using one of the reporter's questions (Who? What? When? Where? Why? How?). Write the question in the space. Fill in the chart with the answers to the question you have created.

Ozone and Your Health

[1]Ozone is a gas that occurs both in the Earth's upper atmosphere and at ground level. [2]Ozone can be helpful or harmful, depending on where it is found. [3]The helpful ozone occurs naturally in the Earth's upper atmosphere—10 to 30 miles above the Earth's surface. [4]At this level, it shields us from the sun's harmful ultraviolet rays. [5]The harmful ozone is found in the Earth's lower atmosphere, near ground level. [6]Harmful ozone is formed when pollutants emitted by cars, power plants, industrial boilers, refineries, chemical plants, and other sources react chemically in the presence of sunlight. [7]Ozone at the ground level can harm your health. [8]First, ozone can irritate your respiratory system, causing you to start coughing, feel an irritation in your throat or experience an uncomfortable sensation in your chest. [9]Second, ozone can reduce lung function and make it more difficult for you to breathe as deeply and vigorously as you normally would. [10]When this happens, you may notice that breathing starts to feel uncomfortable. [11]If you are exercising or working outdoors, you may notice that you are taking more rapid and shallow breaths than normal. [12]Third, ozone can aggravate asthma. [13]When ozone levels are high, more people with asthma have attacks that require a doctor's attention or the use of additional medication. [14]One reason this happens is that ozone makes people more sensitive to allergens, which are the most common triggers for asthma attacks. [15]Also, asthmatics are more severely affected by the reduced lung function and irritation that ozone causes in the respiratory system. [16]Fourth, ozone can inflame and damage cells that line your lungs. [17]Within a few days, the damaged cells are replaced and the old cells are shed—much in the way your skin peels after a sunburn. [18]Fifth, ozone may aggravate chronic lung diseases such as emphysema and bronchitis and reduce the immune system's ability to fight off bacterial infections in the respiratory system. [19]Sixth, ozone may cause permanent lung damage. [20]Repeated

short-term ozone damage to children's developing lungs may lead to reduced lung function in adulthood. [21]In adults, ozone exposure may accelerate the natural decline in lung function that occurs as part of the normal aging process.

—Adapted from "Ozone and Your Health." *AIRNow.* The Environmental Protection Agency. 15 Dec 2004. 13 Dec 2007 http://www.airnow.gov/index.cfm?action=static.brochure

Question based on topic sentence: ______________________________

____

Topic sentence: ______________________________

Harmful Effects:

(1) ________ **(2)** ________ **(3)** ________ **(4)** ________ **(5)** ________ **(6)** ________

Major and Minor Details

A supporting detail will always be one of two types:

> A **major detail** directly explains, develops, or illustrates the *main idea.*
> A **minor detail** explains, develops, or illustrates a *major detail.*

A **major detail** is directly tied to the main idea. Without the major details, the author's main idea would not be clear because the major details are the principal points the author is making about the topic.

In contrast, a **minor detail** explains a major detail. The minor details could be left out, and the main idea would still be clear. Thus minor details are not as important as major details. Minor details are used to add interest and to give further descriptions, examples, testimonies, analysis, illustrations, and reasons for the major details. To better understand the flow of ideas, study the following diagram:

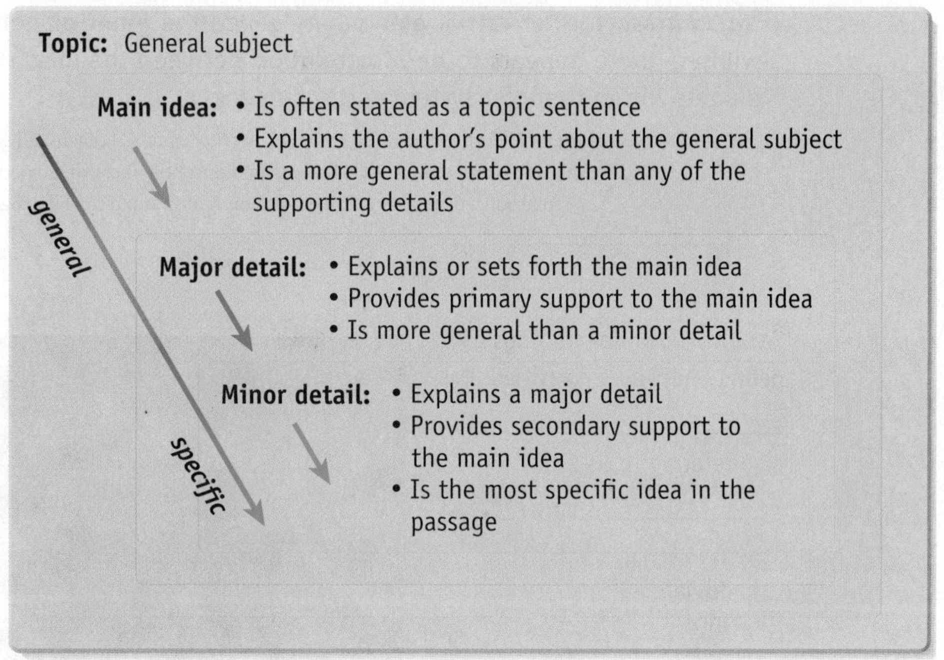

Topic: General subject
Main idea: • Is often stated as a topic sentence
• Explains the author's point about the general subject
• Is a more general statement than any of the supporting details
Major detail: • Explains or sets forth the main idea
• Provides primary support to the main idea
• Is more general than a minor detail
Minor detail: • Explains a major detail
• Provides secondary support to the main idea
• Is the most specific idea in the passage
general
specific

As ideas move from general to specific details, the author often uses signal words to introduce a new detail. These signal words—such as *first, second, next, in addition*, or *finally*—can help you identify major and minor details.

EXAMPLE See if you can tell the difference between major and minor details. Read the following paragraph. Then, complete the exercise that follows.

Terrestrial and Jovian Planets

[1]Planets fall quite nicely into two groups: the terrestrial (Earthlike) planets (Mercury, Venus, Earth, and Mars), and the Jovian (Jupiter-like) planets (Jupiter, Saturn, Uranus, and Neptune). [2]Pluto was recently demoted to a dwarf planet—a new class of solar system objects that have an orbit around the Sun but share their space with other celestial bodies. [3]The most obvious difference between the terrestrial and the Jovian planets is their size. [4]The largest terrestrial planets (Earth and Venus) have diameters only one quarter as great as the diameter of the smallest Jovian planet (Neptune). [5]Also, their masses are only 1/17 as great as Neptune's. [6]Hence, the Jovian planets are often called giants. [7]In addition to their size, planets are distinguished based on their location. [8]Because of their relative locations, the four Jovian planets are also referred to as the **outer planets**.

[9]In contrast, the terrestrial planets are called the **inner planets**. [10]As we will see, there appears to be a correlation between the location of these planets within the solar system and their sizes.

—From *Foundations of Earth Science,* 5th ed., p. 407 by Frederick K. Lutgens, Edward J. Tarbuck, and Dennis Tasa. Copyright © 2008 by Pearson Education, Inc. Printed and Electronically reproduced by permission of Pearson Education, Inc., Upper Saddle River, NJ.

VISUAL VOCABULARY

Saturn, known for its rings of debris and ice, is classified as a ______________ planet and also referred to as an ______________ planet.

a. Terrestrial
b. Jovian
c. inner
d. outer

Complete the following outline of the paragraph by supplying the appropriate major and minor details.

Stated main idea: Planets fall quite nicely into two groups: the terrestrial (Earthlike) planets (Mercury, Venus, Earth, and Mars), and the Jovian (Jupiter-like) planets (Jupiter, Saturn, Uranus, and Neptune).

A. ____

____

1. The largest terrestrial planets (Earth and Venus) have diameters only one quarter as great as the diameter of the smallest Jovian planet (Neptune).
2. Also, their masses are only 1/17 as great as Neptune's.
3. Hence, the Jovian planets are often called giants.

B. ____

1. ____

____

2. In contrast, the terrestrial planets are called the inner planets.
3. As we will see, there appears to be a correlation between the location of these planets within the solar system and their sizes.

EXPLANATION This paragraph provides two major supporting details that help explain the main idea: sentences 3 and 7 state these two major supporting details that describe the classification of planets based on size and location. Sentences 4, 5, and 6 are minor details that explain the differences between the two types of planets based on size. Sentences 8 and 9 are minor details; they explain the differences between the types of planets based on location. Sentence 10 is a minor detail; it suggests a relationship between the size and location of a planet. Interestingly, sentence 2 is a detail that does not directly relate to the two types of planets discussed in the paragraph. The author may have felt this detail was necessary, however, since Pluto, long thought of as a planet, has just recently been reclassified and is no longer considered a planet at all. Thus, sentence 2 acknowledges that some readers have prior knowledge about Pluto's former status as a planet. Sentence 2 preempts reader confusion and readjusts prior knowledge to current scientific standards.

PRACTICE 2

Read the following paragraph. Then, identify the major and minor details by completing the outline that follows it.

Weight Loss in Our Pets

[1]Weight management is an important aspect of ensuring a long, healthy, and happy life for our pets. [2]As a caring pet owner, you need to be aware of the various reasons pets lose weight and when weight loss signals trouble. [3]Of course, at times, some weight loss is expected. [4]For example, dropping a pound or so may be perfectly normal, particularly if your pet has put on a few extra pounds over the years. [5]In addition, some pets lose weight during those hot summer months when soaring temperatures can curb an appetite. [6]Likewise, some pets lose weight during the frigid winter months as they burn more calories to keep warm. [7]And female pets who are nursing mothers may lose weight as they burn calories to produce milk. [8]Another major cause of weight loss is the environment. [9]Even healthy pets that are boarded may lose weight because of the stress of a different environment. [10]What you feed your pet is a key aspect of the animal's environment; diet plays a vital role in weight

management because poor quality, spoiled, or inedible food can result in undernourishment and weight loss. [11]Finally, many underlying medical conditions are often signaled by sudden or dramatic weight loss. [12]A few examples of conditions to be on guard against are dental disease, gastrointestinal disorders (which may include parasites), diabetes, liver or kidney disease, heart disease, and cancer. [13]Any one of these conditions can interfere with eating and the absorption of nutrients.

Stated main idea: As a caring pet owner, you need to be aware of the various reasons pets lose weight and when weight loss signals trouble.

A. ______________________________

1. For example, dropping a pound or so may be perfectly normal, particularly if your pet has put on a few extra pounds over the years.
2. In addition, some pets lose weight during those hot summer months when soaring temperatures can curb an appetite.
3. Likewise, some pets lose weight during the frigid winter months as they burn more calories to keep warm.
4. ______________________________

   ____

B. ______________________________

1. ______________________________

   ____

2. What you feed your pet is a key aspect of the animal's environment; diet plays a vital role in weight management because poor quality, spoiled, or inedible food can result in undernourishment and weight loss.

C. ______________________________

____

1. A few examples of conditions to be on guard against are dental disease, gastrointestinal disorders (which may include parasites), diabetes, liver or kidney disease, heart disease, and cancer.
2. Any one of these conditions can interfere with eating and the absorption of nutrients.

Creating a Summary from Annotations

Reading for main ideas and major supporting details is an excellent study technique. Writing down main ideas and major supporting details in a summary after you read is an effective strategy that will deepen your understanding and provide you with study notes for review and reflection. (Chapter 1 presented a number of reading strategies.) A **summary** condenses a paragraph or passage down to only its primary points by restating the main idea, major supporting details, and important examples. Often, you will want to paraphrase or restate the ideas in your own words; other times, you may need to use the exact language of the text to ensure accuracy. For example, scientific or medical terms have precise meanings that must be memorized; thus your summaries of these types of ideas would include the original language of the text.

> A **summary** is a brief, clear restatement of the most important points of a paragraph or passage.

Different lengths of text require different lengths of summaries. For example, a paragraph can be summarized in one sentence or a few sentences. A passage of several paragraphs can be reduced to one paragraph, and a much longer selection such as a chapter in a textbook may require a summary of a page or two in length.

To help create a summary after reading, you can **annotate**, or mark, your text *during reading*. For example, as you read, circle the main idea and underline the major supporting details and important examples. To learn more about annotating a text see pages 594–595 in Part Two.

Textbook Skills

EXAMPLE Read the following paragraph from a college accounting textbook. Annotate the text by circling the main idea and underlining the major supporting details and the important examples. Then, complete the summary by filling in the blanks with information from the passage.

Income Method and Life Insurance

[1]The **income method** is a general formula for determining how much life insurance you should maintain based solely on your income. [2]This method normally specifies the life insurance amount as a multiple of your income, such as 10 times your annual income. [3]For example, if you have

an annual income of $40,000, this formula would suggest that you need $400,000 in life insurance. [4]This method is very easy to use. [5]The disadvantage is that it does not consider your age and your household situation. [6]Thus, it does not make a distinction between a household that has no children and one with four children. [7]The household with four children will likely need more life insurance because its expenses will be higher.

—Adapted from *Personal Finance*, 3rd ed., p. 372 by Jeff Madura. Copyright © 2007 by Pearson Education, Inc. Reproduced by permission of Pearson Education, Inc., Boston, MA.

Summary

The **income method** ____________________________________

____

____

____

For example, ____

____

____

EXPLANATION The main idea of the paragraph is "The income method is a general formula for determining how much life insurance you should maintain based solely on your income." Including important details in your summary clarifies the way the necessary amount of insurance is calculated. The disadvantages of this formula are minor details that you do not need to include since the main idea states that the formula is based solely on income.

Compare your summary to the following: The **income method** is a general formula for determining how much life insurance you should maintain based on a multiple of your income, such as 10 times your annual income. For example, for an annual income of $40,000, you probably need $400,000 in life insurance.

PRACTICE 3

Read the following passage from a college textbook about tourism. Annotate the text by circling the central idea and underlining the major supporting details and important examples. Then, compose a summary of the passage.

Textbook Skills

[1]For decades, tourism researchers have grouped tourist motivations as push or pull factors. [2]The notion is that travelers are both "pushed" to travel by personality traits or individual needs and wants, and "pulled" to travel by appealing attributes of travel destinations. [3]Traditionally, the push motivations have been thought useful for explaining the desire for travel while the pull motivations have been thought useful for explaining the actual destination choice.

[4] This "theory" of travel motivation highlights the fact that tourists are pushed (motivated) to travel by many factors simultaneously, and destinations pull (attract) visitors with a combination of resources. [5]For instance, a tourist generates the desire to escape from his mundane day-to-day routine and seeks a destination that seems to offer the "ticket" to that escape. [6]Research has shown that push and pull factors are matched by travelers. [7]For example, studies have found a large percentage of travelers are motivated to travel by a desire to be pampered, comfortable, and entertained. [8]Destinations that generate the most "pull" for this group of travelers are cities and beach resorts.

[9]Several "push" factors have been identified and researched as personality traits (such as novelty seeking). [10]An additional and particularly appropriate personality trait theory that relates to tourism is optimal arousal theory. [11]Briefly, the core of this theory is that each of us has some optimal level of arousal at which we feel most comfortable. [12]For some, that level is quite low, leading to a relaxed, slower-paced lifestyle, whereas for many, the optimal arousal level is very high, driving individuals constantly to seek new and challenging activities. [13]A person who is stressed out by work may desire to reduce arousal by seeking a quiet seaside resort to spend some quiet time with a loved one. [14]Another who is bored by the routine of his job and life may instead decide to travel to Europe and test his mettle on the ski slopes of the Alps.

—From *Tourism: The Business of Travel*, 4th ed, pp. 34–35 by Roy A. Cook, Laura J. Yale and Joseph J. Marqua. Copyright © 2010 by Pearson Education, Inc. Printed and Electronically reproduced by permission of Pearson Education, Inc., Upper Saddle River, NJ.

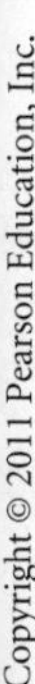

Creating a Summary: Implied Main Ideas

At times, an author may choose to imply a main idea instead of directly stating it. As you learned in Chapter 4, you can use supporting details to create a topic sentence or thesis statement when the main idea or central point is implied. Annotating your text will also help you create a summary for passages with an implied main idea.

First, identify the topic of the passage. Underline recurring words or phrases. Locate each heading or major supporting detail in the passage. (Remember, minor details explain or support major details. Thus, when creating a summary, you can ignore these minor details.) Assign a number or letter to each of the headings or major details you identified. Next, for each piece of information you have marked, ask the question "What controlling point or opinion about the topic does this detail reveal?" Often, a main heading can be turned into a question that will help you determine the implied main idea. Then, write a brief answer to each question in the margin next to the detail you marked. Next, after you finish reading, create a topic sentence or thesis statement for the passage. Finally, use only a few brief sentences to create the entire summary.

EXAMPLE Read the following passage taken from a college psychology textbook. As you read, complete the following steps. Then, create a summary of the passage in the space provided after the passage.

Step 1. Annotate the text: Underline the recurring key terms or phrases and label the major supporting details with a number or letter.

Step 2. Turn the main heading into a question to determine the implied main idea. If needed, turn major details into questions that reveal the author's controlling point.

Step 3. Answer each question in your own words.

Step 4. Create a thesis statement based on the main heading and/or supporting details.

Step 5. Write a summary that combines the thesis statement and the major supporting details into one sentence or a few brief sentences.

Textbook Skills

Three Ways to Measure Memory

[1]We have all had the embarrassing experience of being unable to recall the name of a person whom we are sure that we have already met. [2]This happens to everyone because recognition is an easier memory task than recall. [3]In 1 **recall**, a person must produce required information

simply by searching memory. [4]Trying to remember someone's name, the items on a shopping list, or the words of a speech or a poem is a recall task. [5]A recall task may be made a little easier if cues are provided to jog memory. [6]A **retrieval cue** is any stimulus or bit of information that aids in retrieving a particular memory. [7]Think about how you might respond to these two test questions:

[8]What are the four basic memory processes?

[9]The four processes involved in memory are e_____, s_____, c_____, and r_____.

[10]Both questions require you to recall information. [11]However, most students would find the second question easier to answer because it includes four retrieval cues.

2 [12]**Recognition** is exactly what the name implies. [13]A person simply recognizes something as familiar—a face, a name, a taste, a melody. [14]Multiple-choice, matching, and true false questions are examples of test items based on recognition. [15]The main difference between recall and recognition is that a recognition task does not require you to supply the information but only to recognize it when you see it. [16]The correct answer is included along with other items in a recognition question.

[17]There is another, more sensitive way to measure memory. [18]With the 3 **relearning method**, retention is expressed as the percentage of time saved when material is relearned relative to the time required to learn the material originally. [19]Suppose it took you 40 minutes to memorize a list of words, and one month later you were tested on those words, using recall or recognition. [20]If you could not recall or recognize a single word, would this mean that you had absolutely no memory of anything on the list? [21]Or could it mean that the recall and recognition tasks were not sensitive enough to measure what little information you may have stored? [22]How could a researcher measure such a remnant of former learning? [23]Using the relearning method, a researcher could time how long it would take you to relearn the list of words. [24]If it took 20 minutes to relearn the list, this would represent a 50% savings over the original learning time of 40 minutes. [25]The percentage of time saved—the savings score—reflects how much material remains in long-term memory.

[26]College students demonstrate the relearning method each semester when they study for comprehensive final exams. [27]Relearning material for a final exam takes less time than it took to learn the material originally.

—Adapted from *Mastering the World of Psychology,* 3rd ed., pp. 184–185 by Samuel E. Wood, Ellen Green Wood and Denise Boyd. Copyright © 2008 by Pearson Education, Inc. Reproduced by permission of Pearson Education, Inc., Boston, MA.

Summary: ____

____

____

____

____

____

EXPLANATION Think about the following question based on the title of the passage: "What are the three ways to measure memory?" To answer this question, you should have located and numbered the major supporting details, which also state the central idea of the passage. The author used bold print to highlight the three major details that answer the question. By using your own words to combine the thesis statement and the major supporting details, you create your summary. Compare your summary to the following: The three ways to measure memory include recall, recognition, and the relearning method. Recall is the task of producing information by searching memory. Often a retrieval cue, a bit of information or stimulus, aids in the recall of information from long-term memory. Recognition is the task of identifying material as familiar. Finally, the relearning method measures the percentage of time saved when material is relearned compared to the time needed to learn the material originally.

PRACTICE 4

Read the following passage from a college textbook about human anatomy. Then, create a summary of the passage in the space provided after the passage.

Textbook Skills

Muscle Tissue

[1]Muscle tissues are highly specialized to contract, or shorten, to produce movement.

[2]**Skeletal muscle** tissue is packaged by connective tissue sheets into organs called skeletal muscles, which are attached to the skeleton. [3]These muscles, which can be controlled voluntarily (or consciously), form the flesh of the body, the so-called muscular system. [4]When the skeletal muscles contract, they pull on bones or skin. [5]The result of their action is gross body movements—or changes in our facial expressions. [6]The cells of skeletal muscle are long, cylindrical, multinucleate, and they have obvious striations (stripes). [7]Because skeletal muscle cells are elongated to provide a long axis for contraction, they are often called muscle fibers.

[8]**Cardiac muscle** is found only in the heart. [9]As it contracts, the heart acts as a pump and propels blood through the blood vessels. [10]Like skeletal muscle, cardiac muscle has striations, but cardiac cells are uninucleate, relatively short, branching cells that fit tightly together (like clasped fingers) at junctions called intercalated disks. [11]These intercalated disks contain gap junctions that allow ions to pass freely from cell to cell, resulting in rapid conduction of the exciting electrical impulse across the heart. [12]Cardiac muscle is under involuntary control, which means that we cannot consciously control the activity of the heart. [13](There are, however, rare individuals who claim they have such an ability.)

[14]**Smooth muscle**, or visceral muscle, is so called because no striations are visible. [15]The individual cells have a single nucleus and are spindle-shaped (pointed at each end). [16]Smooth muscle is found in the walls of hollow organs such as the stomach, uterus, and blood vessels. [17]As smooth muscle in its walls contracts, the cavity of an organ alternately becomes smaller (constricts on smooth muscle contraction) or enlarges (dilates on smooth muscle relaxation) so that substances are propelled through the organ along a specific pathway. [18]Smooth muscle contracts much more slowly than the other two muscle types. [19]Peristalsis (per″ĭstal'sis), a wavelike motion that keeps food moving through the small intestine, is typical of its activity.

—From *Essentials of Human Anatomy and Physiology,* 9th ed., pp. 66, 98 by Elaine N. Marieb. Copyright © 2009 by Pearson Education, Inc. Reprinted by permission of Pearson Education, Inc., Glenview, IL.

Summary: ____

____

____

____

____

Textbook Skills

Reading a Textbook: Chapter-end Questions

Textbooks often provide questions at the end of a chapter or section to help you identify and remember the most important points. These questions typically ask you to summarize the main idea and major supporting details. The chapter-end questions may also ask you to note one or two minor supporting details as examples. Annotating your text by circling the main idea and underlining major supporting details and examples as you read will help you answer chapter-end questions and summarize what you are learning.

EXAMPLE Read the following paragraph and chapter-end question about the paragraph from a college communications textbook. Circle the topic sentence and underline the major supporting details and the important examples. Then, answer the question.

Textbook Skills

Types of Media

[1]The mass media fall broadly into two categories or groups. [2]These two groups have certain elements in common, yet they possess different physical traits. [3]The first group, **print**, includes newspapers, magazines, newsletters, and books. [4]Their words create images in the mind as well as convey information. [5]The second group is **electronics and film**. [6]This group includes radio, recordings, television, still and motion pictures, and video. [7]These media send their messages through visual and audio impact on the senses, sometimes with great emotional power.

Chapter-end Question: What are the two basic categories of media? Name at least two media in each category.

—Adapted from Agee, Ault, & Emery, *Introduction to Mass Communication*, 12th ed., pp. 6, 20.

Answer:

____

____.

EXPLANATION By answering this question, you are restating the main idea and the major supporting details. In the second part of the question, the textbook author asks you to name two examples. All of this information can be stated in one sentence: "The two basic categories of media are (1) print, which includes magazines and books, and (2) electronics and film, which includes radio and television."

PRACTICE 5

Read the following paragraph and chapter-end question about the paragraph from a college communications textbook. Circle the topic sentence and underline the major supporting details. Then, answer the question.

Textbook Skills

Friendship

[1]Friendship is an interpersonal relationship between two persons that is mutually productive and is characterized by mutual positive regard.

[2]Several types of friendships exist. [3]One type of friendship is known as reciprocity, characterized by loyalty, self-sacrifice, mutual affection, and generosity. [4]A second type is based on receptivity, characterized by a comfortable and positive imbalance in the giving and receiving of rewards; each person's needs are satisfied by the exchange. [5]Another type of friendship identified as association is a transitory relationship, more like a friendly relationship than a true friendship. [6]Friendships meet our needs and give us a variety of values, among which are the values of utility, affirmation, ego support, stimulation, and security. [7]Interestingly, women share more and are more intimate with same-sex friends than are men. [8]Men's friendships are often built around shared activities rather than shared intimacies. [9]By understanding the nature and function of friendship, the types of friendships, and the ways in which friendships differ between men and women, we can improve the effectiveness of our interpersonal communication with our friends.

Chapter-end Question: What is friendship, its types and purposes, and the ways it differs between men and women?

—From *The Interpersonal Communication Book*, 11th ed., pp. 282–283 by Joseph A. DeVito. Copyright © 2007 by Pearson Education, Inc. Reproduced by permission of Pearson Education, Inc., Boston, MA.

Answer:

____

____

____

____

APPLICATIONS

Application 1: Main Ideas, Major Supporting Details, and Minor Supporting Details

Read the following paragraph, and then, answer the questions.

Acne

[1]Acne is a skin condition characterized by whiteheads, blackheads, and inflamed red pimples. [2]The condition occurs when tiny holes on the surface of the skin, called pores, become plugged. [3]Each pore is an

opening to a canal called a follicle, which contains a hair and an oil gland. [4]Normally, the oil glands help keep the skin lubricated and help remove old skin cells. [5]When glands produce too much oil, the pores can become blocked, accumulating dirt, debris, and bacteria. [6]The blockage or plug is often called a comedone. [7]The top of the plug may be white (whitehead) or dark (blackhead). [8]If the comedone ruptures, the material inside, including oil and bacteria, can spread to the surrounding area and cause an inflammatory reaction. [9]The inflammation usually takes the form of pimples or "zits." [10]If the inflammation is deep in your skin, the pimples may enlarge to form firm, painful cysts. [11]Acne commonly appears on the face and shoulders, but may also occur on the trunk, arms, legs, and buttocks.

—Adapted from "Acne," *Medical Encyclopedia*, 10 Oct. 2003, MedlinePlus, U.S. Library of Medicine. National Institutes of Health. 10 Oct 2007 http://www.nlm.nih.gov/medlineplus/ency/article/000873.htm

_____ **1.** Sentence 1 is a
a. main idea.
b. major supporting detail.
c. minor supporting detail.

_____ **2.** Sentence 4 is a
a. main idea.
b. major supporting detail.
c. minor supporting detail.

_____ **3.** Sentence 9 is a
a. main idea.
b. major supporting detail.
c. minor supporting detail.

Application 2: Using the Main Idea and Supporting Details to Summarize

Read the following paragraph taken from a government Web site. Annotate the paragraph by circling the topic sentence and underlining the major supporting details. Then, write a one or two sentence summary of the information.

The Rights and Responsibilities of Citizenship

[1]Citizens of the United States have the right and the responsibility to participate in their government. [2]This process of self-government insures that power will always remain where it belongs—with the people. [3]The most important right citizens have is the right to vote. [4]By voting, the people have a voice in the government. [5]Officials can be voted in or out of office. [6]Every person's vote counts the same as another person's vote.

[7]It is important for all citizens to vote in every election to make sure that the democratic, representative system of government is maintained. [8]Before voting in an election, each citizen should be well informed about the issues and candidates. [9]A citizen of the United States also has the responsibility to serve on a jury if called to do so by the government. [10]If an individual is chosen for jury duty, he or she must stop work and attend the trial as long as he or she is needed. [11]The members of the jury need to decide the case in as fair a way as they can. [12]In addition, every person is responsible for obeying the laws of the community, state and country in which he or she lives, and all persons living in the United States are expected to pay the income taxes and other taxes honestly and on time. [13]Another duty of citizenship is to serve in the armed forces as needed. [14]During times of war, any man who is physically able can be called upon to fight for the United States; in peaceful times, there can be a draft, or men and women can enlist voluntarily.

—Adapted from "Responsibilities of Citizenship," *Ben's Guide to U.S. Government for Kids*, 1 Dec. 1999, U.S. Government Printing Office. 18 Nov 2007 http://bensguide.gpo.gov/6-8/citizenship/responsibilities.html

Summary: ____

____

____

____.

Textbook Skills

Application 3: Using Supporting Details to Summarize a Passage with an Implied Central Idea

Read the following information from a college textbook about marriage and families. Annotate the passage and write a summary of the passage.

Characteristics of Successful Stepfamilies

[1]Overcoming the challenges associated with stepfamily life is not an easy task, and many families fall apart under the pressure. [2]Approximately 60 percent of remarriages end in divorce. [3]Despite the risks, many stepfamilies do function successfully. [4]Consider the following information

about successful stepfamilies provided by marriage and family therapists Emily Visher and John Visher:

- (1) [5]They allow losses to be mourned. [6]Adults grieve for the loss of a previous relationship and allow children to share feelings of anger, fear, and guilt about their parents' divorce.
- (2) [7]They have realistic expectations. [8]Adults and children come to realize that their stepfamily will be different from their first family and that expectations of instant love and adjustment are not realistic. [9]They understand that relationships develop over time.
- (3) [10]Adult couples have a strong relationship. [11]Remarried adults find time away from the children to nurture their relationship.
- (4) [12]They establish family traditions. [13]Successful remarried families develop their own rituals and traditions to help them bond as a family.
- (5) [14]They develop step-relationships. [15]Stepparents gradually take on a disciplinary role, allowing the biological parent to perform primary parenting duties until a solid relationship with the stepparent is formed.
- (6) [16]They cooperate with the absent parent. [17]Stepfamilies are able to collaborate over child care arrangements, enabling children to transfer between households without being caught in the crossfire between parents and stepparents.

Summary: ______________________________

____

____

____

____

REVIEW TEST 1

Score (number correct) ______ × 20 = ______%

Main Ideas and Supporting Details

Read the paragraph, and then, answer the questions.

A Primer on Gasoline Prices

[1]The cost to produce and deliver gasoline to consumers includes the cost of crude oil to refiners, taxes, refinery processing costs, marketing and distribution costs, and finally the retail station costs. [2]In 2005 the price of crude oil averaged $50.23 per barrel, and crude oil accounted for about 53 percent of the cost of a gallon of regular grade gasoline. [3]In comparison, the average price for crude oil in 2004 was $36.98 per barrel, and it comprised 47 percent of the cost of a gallon of regular gasoline. [4]Federal, state, and local taxes are a large component of the retail price of gasoline. [5]Taxes (not including county and local taxes) account for approximately 19 percent of the cost of a gallon of gasoline. [6]Within this national average, federal excise taxes are 18.4 cents per gallon and state excise taxes average about 21 cents per gallon. [7]Also, eleven states levy additional state sales and other taxes. [8]Additional local county and city taxes can have a significant impact on the price of gasoline. [9]Refining costs and profits comprise about 19 percent of the retail price of gasoline. [10]Distribution, marketing and retail dealer costs and profits combined make up 9 percent of the cost of a gallon of gasoline. [11]From the refinery, most gasoline is shipped first by pipeline to terminals near consuming areas, then loaded into trucks for delivery to individual stations. [12]Some retail outlets are owned and operated by refiners, while others are independent businesses that purchase gasoline for resale to the public. [13]The price on the pump reflects the retailer's purchase cost for the product, the other costs of operating the service station, and local market conditions, such as the desirability of the location and the marketing strategy of the owner.

—Adapted from "A Primer on Gasoline Prices." May 2006. Energy Information Administration, U.S. Department of Energy 1 Oct. 2007 http://www.eia.doe.gov/pub/oil_gas/petroleum/analysis_publications/primer_on_gasoline_prices/html/petbro.html.

_____ **1.** Sentence 1 is
a. a main idea.
b. a major supporting detail.
c. a minor supporting detail.

_____ **2.** Sentence 2 is
a. a main idea.
b. a major supporting detail.
c. a minor supporting detail.

_____ **3.** Sentence 9 is
a. a main idea.
b. a major supporting detail.
c. a minor supporting detail.

______ **4.** Sentence 10 is
a. a main idea.
b. a major supporting detail.
c. a minor supporting detail.

______ **5.** Sentence 13 is
a. a main idea.
b. a major supporting detail.
c. a minor supporting detail.

VISUAL VOCABULARY

Complete the graphic based on the information in the passage. Fill in the blanks with the appropriate labels that identify what we pay for when we buy a gallon of gasoline.

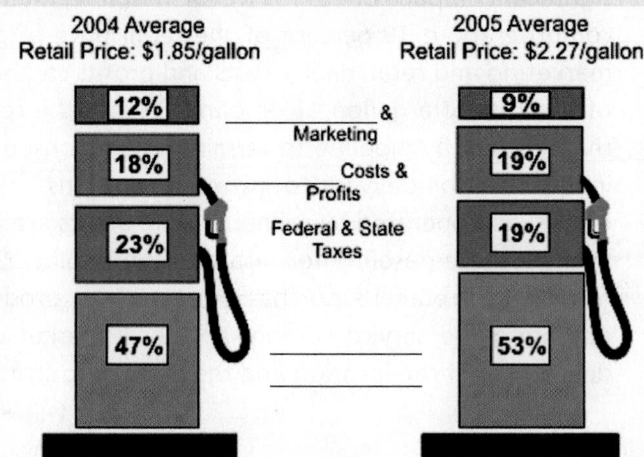

Source: A Primer on Gasoline Prices. Energy Information Administration, U.S. Department of Energy, May 2006

REVIEW TEST 2

Score (number correct) ______ × 10 = ______%

Main Ideas, Supporting Details, and Summarizing

Read the following passage from a college geography textbook. Answer the questions that follow.

Textbook Skills

The Differences between Folk Culture and Popular Culture

[1]The term **folk culture** refers to a culture that preserves traditions. [2]Folk groups are often bound by a distinctive religion, national background, or language, and are conservative and resistant to change. [3]Most folk-culture groups are rural, and relative isolation helps these groups maintain their integrity. [4]Folk-culture groups, however, also include urban neighborhoods of immigrants struggling to preserve their native cultures in their new homes. [5]Folk culture suggests that any culture identified by the term is a lingering remnant of something that is embattled by the tide of modern change.

[6]Folk geographic studies in the United States range from studies of folk songs, folk foods, folk medicine, and folklore to objects of folk material as diverse as locally produced pottery, clothing, tombstones, farm fencing, and even knives and guns. [7]In North America, the Amish provide an example of a folk culture. [8]The Amish are notable because they wear plain clothing and shun modern education and technology.

[9]**Popular culture**, by contrast, is the culture of people who embrace innovation and conform to changing norms. [10]Popular culture may originate anywhere. [11]And it tends to diffuse rapidly, especially wherever people have time, money, and inclination to indulge in it.

[12]Popular material culture usually means mass culture. [13]For example, items such as clothing, processed food, CDs, and household goods are mass produced for mass distribution. [14]Whereas folk culture is often produced or done by people at-large (folk singing and dancing, cooking, costumes, woodcarving, etc.), popular culture, by contrast, is usually produced by corporations and purchased.

—Adapted from *Introduction to Geography: People, Places, and Environment,* 4th ed., pp. 221–223 by Edward Bergman and William H. Renwick. Copyright © 2008 by Pearson Education, Inc. Printed and Electronically reproduced by permission of Pearson Education, Inc., Upper Saddle River, NJ.

_____ **1.** Sentence 1 states _____ of the first paragraph (sentences 1–5).
a. the main idea
b. a major supporting detail
c. a minor supporting detail

_____ **2.** Sentence 2 states _____ of the first paragraph (sentences 1–5).
a. the main idea
b. a major supporting detail
c. a minor supporting detail

_____ **3.** Sentence 7 states _____ of the second paragraph (sentences 6–8).
a. the main idea
b. a major supporting detail
c. a minor supporting detail

______ **4.** Sentence 9 states ______ of the third paragraph (sentences 9–11).
a. the main idea
b. a major supporting detail
c. a minor supporting detail

______ **5.** Sentence 13 states ______ of the fourth paragraph (sentences 12–14).
a. the main idea
b. a major supporting detail
c. a minor supporting detail

Complete the following summary with information from the passage.

(**6**) ____________ differs from (**7**) ____________. On the one hand folk culture refers to a culture that (**8**) ______________________________. On the other hand, popular culture refers to the culture that (**9**) ________________________________. Folk culture is produced by local people in rural regions while popular culture is mass produced, mass distributed, and sold to the masses by (**10**) ____________.

REVIEW TEST 3

Score (number correct) ______ × 10 = ______%

Main Ideas, Supporting Details, and Summarizing

Textbook Skills

Before you read the following passage from a college sociology textbook, skim the material and answer the Before Reading questions. Read the passage. Then, answer the After Reading questions.

Vocabulary Preview

incompatible (1): mismatched, opposed

fragile (10): weak, unstable, breakable

perceptions (16): views, insights, understandings

interpersonal (19): between people

relative (20): in relation to

Conflict Defined

[1]Conflict is the interaction of interdependent people who perceive **incompatible** goals and interference from each other in achieving those goals. [2]This definition has the advantage of providing a much clearer focus than definitions that view conflict simply as disagreement, as competition, or as the presence of opposing interests.

[3]The most important feature of conflict is that it is based in interaction. [4]Conflicts are formed and sustained by the behaviors of the parties involved and their verbal and nonverbal reactions to one another. [5]Conflict interaction takes many forms, and each form presents special problems and requires special handling. [6]The most familiar type of conflict interaction

is marked by shouting matches or open competition in which each party tries to defeat the other. [7]But conflicts may also be more subtle. [8]People may react to conflict by suppressing it. [9]A husband and wife may communicate in ways that allow them to avoid confrontation. [10]Either they are afraid the conflict may damage a **fragile** relationship, or they convince themselves that the issue "isn't worth fighting over." [11]This response is as much a part of the conflict process as fights and shouting matches.

[12]People in conflict perceive that they have incompatible goals or interests; they also believe that others are a barrier to achieving their goals. [13]The key word here is *perceive*. [14]Regardless of whether goals are actually incompatible, if the parties believe them to be incompatible, then conditions are ripe for conflict. [15]Regardless of whether one employee really stands in the way of a co-worker's promotion, if the co-worker interprets the employee's behavior as interfering with his promotion, then a conflict is likely to occur. [16]Communication is important; it is the key to shaping and maintaining the **perceptions** that guide conflict behavior.

[17]Communication problems can be an important cause of conflict. [18]Conflict may result from misunderstandings that occur when people have different communication styles. [19]For example, Tannen argues that men and women have different approaches to **interpersonal** communication. [20]Men are mostly task-oriented; they are concerned with establishing their position **relative** to others in conversations. [21]In contrast, women use conversations to build relationships and establish connections with others. [22]As a result, men and women may interpret the same act in very different ways. [23]When a man makes a demand during a conflict, he might mean to signal that he is strong and has a definite position. [24]A woman hearing this demand is likely to focus more on its **implications** for their relationship. [25]Thus she may interpret it as a signal that it will be very difficult to deal with this man. [26]As a result, the woman may become more competitive toward her male partner than she would have had she seen his demand in the man's terms. [27]According to Tannen, stylistic differences of this sort create communication barriers that make misunderstanding—and the conflict that results from it—**inevitable** in male-female relationships. [28]Similar communication problems can occur across almost any social divide, such as those between people of different cultures, ages, educational backgrounds, and socioeconomic classes.

—Adapted from Folger, Poole, & Stutman, *Working Through Conflict*, 4th ed., pp. 5–6.

Before Reading

Vocabulary in Context

_____ **1.** In sentence 24 of the passage, the word **implications** means

a. conflicts. c. accusations.
b. meanings. d. levels.

_____ **2.** In sentence 27 of the passage, the word **inevitable** means

a. delayed. c. hidden.
b. long term. d. unavoidable.

After Reading

Main Ideas

_____ **3.** Which sentence states the central idea of the passage?

a. sentence 1 c. sentence 3
b. sentence 2 d. sentence 4

_____ **4.** Which sentence is the topic sentence of the fourth paragraph (sentences 17–28)?

a. sentence 17 c. sentence 20
b. sentence 18 d. sentence 27

Supporting Details

_____ **5.** Sentence 7 is

a. a major supporting detail. b. a minor supporting detail.

_____ **6.** Sentence 9 is

a. a major supporting detail. b. a minor supporting detail.

_____ **7.** Sentence 15 is

a. a major supporting detail. b. a minor supporting detail.

8–10. Complete the summary with information from the passage.

Conflict is the ____________ of interdependent people who perceive incompatible goals and interference from each other in achieving those goals. Perceptions based on ____________ reactions, nonverbal reactions, and differing interpersonal ____________ styles are key sources of conflict.

WHAT DO YOU THINK?

Do you agree with Tannen's description of the different approaches to communication used by men and women? Why or why not? What are some other interactions between people that often lead to conflict? How can communication add to or ease conflict? Assume you are a supervisor of a sales staff at a local retail store, and two employees under your supervision are in conflict with each other. Assume you have met with them both privately and individually to understand and resolve the conflict. Write a memo to be sent to both employees and placed in their personnel files. Describe the conflict and suggest steps that you expect both to take to resolve the issue.

REVIEW TEST 4

Score (number correct) ______ × 10 = ______%

Main Ideas and Supporting Details

Textbook Skills

Before Reading: Survey the following passage adapted from the college textbook *World Civilizations: The Global Experience.* Skim the passage, noting the words in bold print. Answer the Before Reading questions that follow the passage. Then, read the passage. Next, answer the After Reading questions. Use the discussion and writing topics as activities to do after reading.

Vocabulary Preview

periodically (1): every so often
militants (3): violent protestors
authoritarian (5): controlling
alliance (5): partnership
justifiable (8): acceptable, valid, proper
fundamentalist (13): a person who strictly follows a set of basic ideas or principles

A Very Present Problem

[1]American interests had **periodically** been the targets of terrorist attacks since the 1960s. [2]Hijacking of airplanes and other moves frequently expressed hostility to U.S. policies. [3]But the massive attacks on the World Trade Center and the Pentagon by Islamic **militants** on September 11, 2001, created a new level of threat. [4]The attacks reflected concern about specific U.S. policies in the Middle East. [5]These concerns included support for **authoritarian** governments, the **alliance** with Israel, and the stationing of troops on "sacred ground" in Saudi Arabia. [6]The terrorists were also hostile to wider U.S. power, or as they termed it, arrogance. [7]Their response, hijacking airliners to crash into buildings that symbolized American financial and military might, killed about 3,000 people. [8]The terrorists regarded this as **justifiable** action against a nation they could not hope to fight by conventional means.

Vocabulary Preview

harbored (13): protected

Al Qaeda (13): a radical Sunni Muslim organization dedicated to the elimination of a Western presence in Arab countries

erroneous (17): wrong, incorrect

ensuing (20): resulting, following

[9]The attacks clearly altered U.S. policy and focused the administration on a war against terrorism. [10]"War on Terror" was the new catchphrase. [11]And a number of measures were taken, including heightened screening of international visitors. [12]The problem dominated American foreign policy. [13]A first response involved a military attack that successfully toppled the Islamic **fundamentalist** regime in Afghanistan that had **harbored** the **Al Qaeda** group behind the September 11 attacks. [14]World opinion largely supported this move. [15]The United States established new military bases near several possible centers of terrorist activity.

[16]In 2003 U.S. attention turned to Iraq, which was accused of amassing dangerous weaponry and aiding terrorists. [17]Evidence for these charges proved largely **erroneous.** [18]And world opinion turned heavily against the American move, with millions of demonstrators protesting the impending war in February 2003. [19]But the United States, joined by several **allies,** including Britain, invaded and quickly conquered the country. [20]The **ensuing** occupation, however, was extremely troubled, as the United States could not clearly restore order against a variety of **insurgents.** [21]The results of this war, in terms of Iraq's future, broader global reactions to the United States, and the flexibility of the American policy itself, are not yet clear.

—From *World Civilizations: The Global Experience, Combined Volume, Atlas Edition,* 5th ed., pp. 966–967 by Peter N. Stearns, Michael B. Adas, Stuart B. Schwartz, and Marc Jason Gilbert.

Before Reading

Vocabulary in Context

1–2. Use your own words to define the following term. Identify the context clue you used to make meaning of the word.

The word **allies** in sentence 19 means ______________________

Context clue used for **allies** ______________________

3. Study the following chart of word parts.

Prefix	Meaning	Root	Meaning	Suffix	Meaning
in-	in, into, on	*surge*	rise up	*-ent*	one who, one that

Use context clues and the word parts to state the definition of the word **insurgent** in sentence 20: ______________________________

Topic

4. What is the topic of the passage? ______________________________

After Reading

Main Ideas and Supporting Details

Complete the following study notes with information from the passage.

Thesis Statement: The attacks clearly altered U.S. policy and focused the administration on a war against terrorism.

Major Supports	**Minor Supports**
American interests had been targets of terrorist attacks since **(5)** _____	Hijacking of airplanes and other moves frequently expressed hostility to U.S. policies.
	6. ______________________________
7. Attacks reflected ______________________________	U.S. "arrogance" revealed by the following: Support of authoritarian governments; the alliance with Israel;
	8. ______________________________
9. Attacks altered U.S. policy and focused on ______________________________	"War on Terror" is a new catchphrase. Military attack topples Islamic regime that harbored Al Qaeda in Afghanistan.
	U.S. established military bases near centers of terrorist activities.

10. ______________________________

____

Ensuing occupation could not restore order against insurgents.

WHAT DO YOU THINK?

September 11, 2001 stands as a significant event in United States history. Did you or someone you know witness the event or the news coverage of the events surrounding 9/11? Do you think many people still think about 9/11? Should September 11, 2001 be officially recognized as a day to remember? Why or why not? Assume your college or university holds an annual day of observance every September 11. Also assume you are a member of the Student Government Association, and you have been asked to speak at this year's college-wide memorial service. A copy of your speech will also be printed in the student newspaper.

After Reading About Supporting Details

Before you move on to the Mastery Tests on supporting details, take time to reflect on your learning and performance by answering the following questions. Write your answers in your notebook.

- How has my knowledge base or prior knowledge about supporting details changed?
- Based on my studies, how do I think I will perform on the Mastery Test(s)? Why do I think my scores will be above average, average, or below average?
- Would I recommend this chapter to other students who want to learn more about supporting details? Why or why not?

Test your understanding of what you have learned about Supporting Details for Master Readers by completing the Chapter 5 Review Card in the insert near the end of your text.

CONNECT TO PEARSON myreadinglab

To check your progress in meeting Chapter 5's learning outcomes, log in to **www.myreadinglab.com** and try the following activities.

- The "Supporting Details" section of MyReadingLab offers more information about supporting details. You will find an overview, model, review materials, practice activities, and tests. To access this resource, click on the "Study Plan" tab. Then click on "Supporting Details." Then click on the following links as needed: "Overview," "Model," "Practice," and "Tests."

- The "Outlining and Summarizing" section of MyReadingLab gives an overview about memory and active reading. This section also provides a model for summarizing. You will find practice activities and tests. To access this resource, click on the "Study Plan" tab. Then click on "Outlining and Summarizing." Then click on the following links as needed: "Overview," "Model: Summarizing," "Practice," and "Tests."
- To measure your mastery of the content in this chapter, complete the tests in the "Supporting Details" section and click on Gradebook to find your results.

MASTERY TEST 1

Name ______________________ Section ____________

Date __________ **Score** (number correct) ______ × 10 = ______%

Read the passage from a college communications textbook. Then, complete the summary that follows it.

Textbook Skills

The Context of Communication

[1]Communication always takes place within a context that influences the form and the content of communication. [2]At times this context is so natural that you ignore it, like street noise. [3]At other times the context stands out, and the ways in which it affects the flow of your communications are obvious. [4]Think, for example, of the different ways you'd talk at a funeral, in a quiet restaurant, and at a rock concert.

[5]The context of communication has at least four dimensions. [6]The room, workplace, or park in which communication takes place—the concrete environment—is the *physical dimension.* [7]When you communicate face-to-face you're both in the same physical environment. [8]In computer-based communication, you may be in very different environments. [9]One of you may be on a beach in San Juan while another is in a Wall Street office. [10]The *cultural dimension* is passed from one generation to another and refers to the rules, the beliefs, and the attitudes of the people communicating. [11]For example, in some cultures it's considered polite to talk to strangers. [12]In others it's something to be avoided. [13]In some cultures, direct eye contact between child and adult shows directness and honesty; in others it shows defiance and arrogance. [14]The *social-psychological dimension* includes several aspects. [15]For example, it refers to the status relationships among the participants, such as who is the employer and who is the employee or who is the salesperson and who is the store owner. [16]The formality or informality, friendliness or hostility, and cooperativeness or competitiveness of the interaction are also part of this dimension. [17]The *temporal* or *time dimension* refers to where a particular message fits into the sequence of communication events. [18]For example, if you tell a joke about sickness immediately after your friend tells you she is sick, the joke will be seen differently than the same joke told as one of a series of similar jokes to your friends in the locker room of the gym.

—From *Messages: Building Interpersonal Communication Skills,* 4th ed., pp. 16–17 by Joseph A. DeVito. Copyright © 1999 by Pearson Education, Inc. Reproduced by permission of Pearson Education, Inc., Boston, MA.

_____ **1.** Sentence 4 states a
a. main idea.
b. major supporting detail.
c. minor supporting detail.

_____ **2.** Sentence 5 states a
a. main idea.
b. major supporting detail.
c. minor supporting detail.

_____ **3.** Sentence 6 states a
a. main idea.
b. major supporting detail.
c. minor supporting detail.

_____ **4.** Sentence 9 states a
a. main idea.
b. major supporting detail.
c. minor supporting detail.

_____ **5.** Sentence 17 states a
a. main idea.
b. major supporting detail.
c. minor supporting detail.

6–10. Complete the summary with information from the paragraph.

The context of communication has at least **(6)** _______________ dimensions. The physical dimension is the **(7)** _______________ environment, such as communicating via computers from different physical locations. The **(8)** _______________ dimension is passed from one generation to another and refers to attitudes, beliefs, and cultural rules. The social-psychological dimension includes the **(9)** _______________ of the people communicating and the tone and purpose of the communication. The **(10)** _______________ or time dimension refers to the sequence of the communication, such as the proper time to tell a joke.

MASTERY TEST 2

Name ______________________ Section ______________

Date __________ **Score** (number correct) ______ × 10 = ______%

Read this passage from a college communications textbook. Then, answer the questions that follow it.

Textbook Skills

Guidelines for Communicating Emotions

[1]Communicating your emotions and responding appropriately to the emotional expressions of others are as important as they are difficult. [2]There are, however, a variety of suggestions that will make these tasks easier and more effective. [3]Expressing emotions effectively begins with understanding your emotions and then deciding if and how you wish to express them.

[4]Your first task is to understand the emotions you're feeling. [5]For example, consider how you would feel if your best friend just got the promotion that you wanted or if your brother, a police officer, was shot while breaking up a street riot. [6]Think about your emotions as objectively as possible. [7]Think about both the bodily reactions you'd be experiencing as well as the interpretations and evaluations you'd be giving to these reactions. [8]Further, identify the prior conditions (in as specific terms as possible) that may be influencing your feelings. [9]Try to answer the question, "Why am I feeling this way?" or "What happened to lead me to feel as I do?"

[10]Your second task is to decide if, in fact, you want to express your emotions. [11]It will not always be possible to stop to think about whether you wish to express your emotions—at times you may respond almost automatically. [12]More often than not, however, you will have the time and the chance to ask yourself whether you wish to express your emotions. [13]When you do have this chance, remember that it isn't always necessary or wise to express every feeling you have. [14]Remember, too, the irreversibility of communication; once you communicate something, you cannot take it back. [15]Therefore, consider carefully the arguments for and against expressing your emotions for each decision you have to make.

—From *Messages: Building Interpersonal Communication Skills, 4th ed.*, p. 184 by Joseph A. DeVito. Copyright © 1999 by Pearson Education, Inc. Reproduced by permission of Pearson Education, Inc., Boston, MA.

______ **1.** Sentence 4 is

a. a main idea.
b. a major supporting detail.
c. a minor supporting detail.

______ **2.** Sentence 5 is

a. a main idea.
b. a major supporting detail.
c. a minor supporting detail.

_____ **3.** Sentence 7 is
a. a main idea.
b. a major supporting detail.
c. a minor supporting detail.

_____ **4.** Sentence 10 is
a. a main idea.
b. a major supporting detail.
c. a minor supporting detail.

_____ **5.** Sentence 15 is
a. a main idea.
b. a major supporting detail.
c. a minor supporting detail.

Read the following passage from a museum's website. Then, complete the summary. (**Hint:** As you read, annotate the information.)

Mexican Genius: Artist Diego Rivera

[1]Diego María Rivera (1886–1957) is one of the most prominent Mexican artists of the twentieth century. [2]He gained international acclaim as a leader of the Mexican mural movement that sought to bring art to the masses through large-scale works on public walls. [3]In his murals of the 1920s and 1930s, Rivera developed a new, modern imagery to express Mexican national identity. [4]His style featured stylized representations of the working classes and indigenous cultures and supported revolutionary ideals. [5]During his time abroad, Rivera drew upon the radical innovations of cubism, inaugurated a few years earlier by Pablo Picasso and Georges Braque. [6]Rivera adopted their dramatic fracturing of form; he used multiple perspective points and flattened the picture plane. [7]He also borrowed favorite cubist motifs, such as liqueur bottles, musical instruments, and painted wood grain. [8]Yet Rivera's cubism is formally and thematically distinctive. [9]Characterized by brighter colors and a larger scale than many early cubist pictures, his work also features highly textured surfaces executed in a variety of techniques.

—"Introduction" from *The Cubist Paintings of Diego Maria Rivera.* http://www.nga.gov/exhibitions/2004/rivera/intro.shtm. Reproduced by permission of National Gallery of Art, Washington, D.C.

6–10. Complete the summary with information from the paragraph.

Diego María Rivera (1886–1957), a prominent **(6)** _______________ of the twentieth century, is celebrated for his large-scale **(7)** _______________ on public walls that brought **(8)** _______________. Influenced by **(9)** _______________ and Braque, Rivera adopted **(10)** _______________ motifs and techniques. However, in contrast to early cubist artists, Rivera used brighter colors, highly textured surfaces, and a larger scale to express Mexican national identity.

MASTERY TEST 3

Name ______________________ Section ______________

Date __________ **Score** (number correct) ______ × 20 = ______%

Read this passage from a college history textbook. Then, answer the questions that follow it.

Textbook Skills

The Jazz Age

[1]"The music from the trumpet at the Negro's lips is honey mixed with liquid fire." [2]With these words, Langston Hughes described the new music called jazz. [3]But among black intellectuals, Hughes's opinion was in the minority. [4]Many black Americans distanced themselves from the sensuous music that brought white attention to black culture. [5]Jazz music's improvisational style, they argued, only reinforced the stereotype that black people were impetuous and lacked intellectual discipline. [6]Other black people criticized this "folk art" because it was born in brothels and performed in speakeasies—illegal bars that opened during the decade of Prohibition, when the government outlawed alcohol. [7]But jazz was not just a musical innovation. [8]At a time when all other social institutions in the United States were segregated, a few jazz clubs brought together music lovers of all races.

[9]Some historians consider jazz the only truly American art form. [10]Rather than deriving from another culture, it was born in the United States from a blend of European and African percussion, horn, and piano melodies. [11]Many jazz artists were trained in the disciplined formality of classical piano. [12]Most jazz was feel-good music, embracing a bewildering array of forms—from spirituals and the mournful tones of the blues to the aggressive percussion of marching band rhythms, from the asymmetry of ragtime to the staid patterns of European tradition. [13]Jazz offered deep, throaty voices, and provocative brass, woodwind, and percussion sounds. [14]Building on written scores, band members improvised during performances. [15]Thus jazz was always in motion, always transforming something familiar into something new. [16]Hughes felt that jazz embodied a cultural richness that opened new doors for black Americans. [17]Through jazz, Hughes commented, "Josephine Baker goes to Paris, Robeson to London, Jean Toomer to a Quaker Meeting House, Garvey to Atlanta Federal Penitentiary . . . and Duke Ellington to fame and fortune."

—Carson, Lapsansky-Werner, & Nash, *The Struggle for Freedom: A History of African Americans, Volume II*, p. 363.

______ **1.** Sentence 1 is
a. a main idea.
b. a major supporting detail.
c. a minor supporting detail.

______ **2.** Sentence 3 is
a. a main idea.
b. a major supporting detail.
c. a minor supporting detail.

______ **3.** Sentence 10 is
a. a main idea.
b. a major supporting detail.
c. a minor supporting detail.

______ **4.** Sentence 13 is
a. a main idea.
b. a major supporting detail.
c. a minor supporting detail.

______ **5.** Sentence 16 is
a. a main idea.
b. a major supporting detail.
c. a minor supporting detail.

VISUAL VOCABULARY

Jazz percussionist Hamid Drake performs at the 2008 New Orleans Jazz and Heritage Festival.

A percussionist ____________.

a. plays the drums.
b. makes music by striking instruments.
c. is a set of drums.

MASTERY TEST 4

Name ______________________ Section ______________

Date __________ **Score** (number correct) ______ × 20 = ______%

Read the following passage from a college history textbook. Answer the questions that follow.

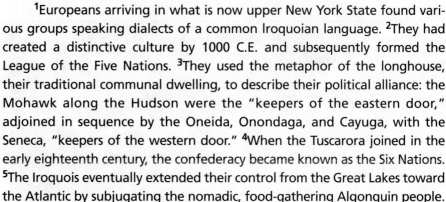

Textbook Skills

The Iroquois of the Northeast Woodlands

[1]Europeans arriving in what is now upper New York State found various groups speaking dialects of a common Iroquoian language. [2]They had created a distinctive culture by 1000 C.E. and subsequently formed the League of the Five Nations. [3]They used the metaphor of the longhouse, their traditional communal dwelling, to describe their political alliance: the Mohawk along the Hudson were the "keepers of the eastern door," adjoined in sequence by the Oneida, Onondaga, and Cayuga, with the Seneca, "keepers of the western door." [4]When the Tuscarora joined in the early eighteenth century, the confederacy became known as the Six Nations. [5]The Iroquois eventually extended their control from the Great Lakes toward the Atlantic by subjugating the nomadic, food-gathering Algonquin people.

[6]The Iroquois had the advantage of being agriculturists with permanent villages. [7]Some of these had several hundred residents and extensive fields where maize, beans, squash, and tobacco were grown. [8]Fish traps were built across streams, and smokehouses preserved joints of game. [9]Related families lived in the longhouses, long rectangular buildings protected by high wooden palisades. [10]Women played a notable part: they owned the homes and gardens, and, since descent was matrilineal, chose the leaders. [11]If the men chosen did not provide good leadership, they could be replaced.

—From *Civilizations Past & Present, Combined Volume,* 12th ed., p. 357 by Robert R. Edgar, Neil J. Hackett, George F. Jewsbury, Barbara S. Molony, Matthew Gordon. Copyright © 2008 by Pearson Education, Inc. Reprinted by permission of Pearson Education, Inc., Upper Saddle River, NJ.

______ **1.** Sentence 1 states ____ of paragraph 1 (sentences 1–5).
a. the main idea
b. a major supporting detail
c. a minor supporting detail

______ **2.** Sentence 2 states ____ of paragraph 1 (sentences 1–5).
a. the main idea
b. a major supporting detail
c. a minor supporting detail

______ **3.** Sentence 5 states ____ of paragraph 1 (sentences 1–5).
a. the main idea
b. a major supporting detail
c. a minor supporting detail

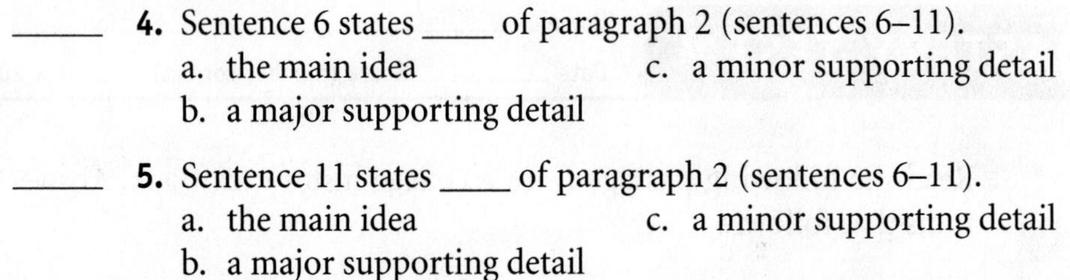

______ **4.** Sentence 6 states ____ of paragraph 2 (sentences 6–11).
a. the main idea
b. a major supporting detail
c. a minor supporting detail

______ **5.** Sentence 11 states ____ of paragraph 2 (sentences 6–11).
a. the main idea
b. a major supporting detail
c. a minor supporting detail

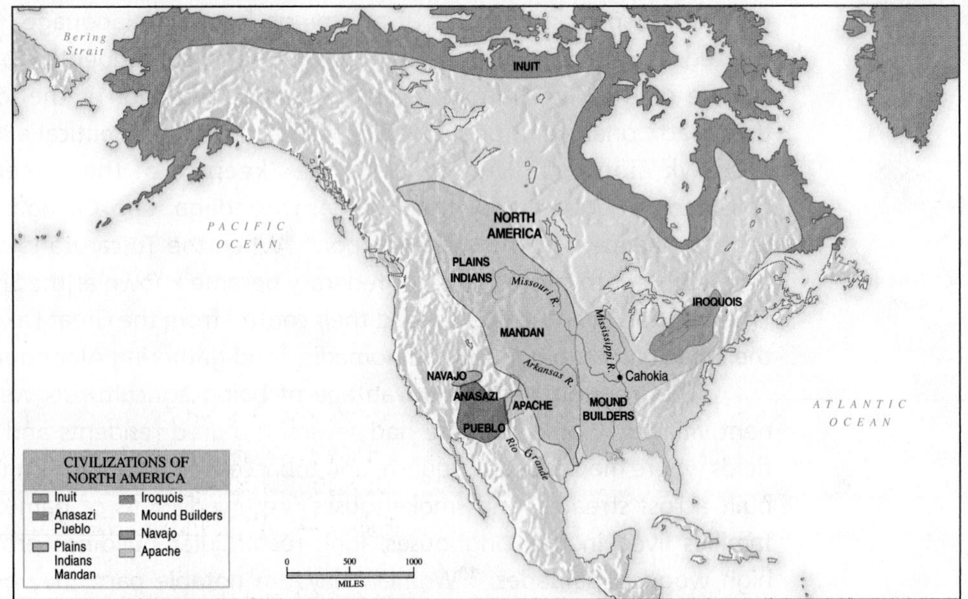

In the vastness of the North American continent, the varied environmental challenges led to the development of hundreds of different Indian tribes—more than 252 alone in the present-day state of California.

—From *Civilizations Past & Present, Combined Volume,* 12th ed., p. 357 by Robert R. Edgar, Neil J. Hackett, George F. Jewsbury, Barbara S. Molony, Matthew Gordon.

According to the legend, which group of Native Americans developed in the closest location to the Iroquois? ______

a. Inuit
b. Plains Indians
c. Navajo
d. Mound Builders

Outlines and Concept Maps

6

LEARNING OUTCOMES

After studying this chapter you should be able to do the following:

1. Define the terms *outline* and *concept map*.
2. Create an outline.
3. Create a concept map.
4. Evaluate the importance of outlines and concept maps.

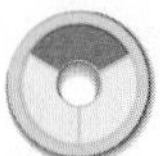

Before Reading About Outlines and Concept Maps

In Chapter 5, you learned several important ideas that will help you use outlines and concept maps effectively. To review, reread the diagram about the flow of ideas on page 197 in Chapter 5. Next, skim this chapter for key ideas in boxes about outlines, concept maps, and the table of contents in a textbook. Refer to the diagrams and boxes and create at least three questions that you can answer as you read the chapter. Write your questions in the following spaces (record the page number for the key term in each question):

____

____? (page ______)

____

____? (pages ____________)

____? (page ______)

Compare the questions you created with those that follow. Then, write the ones that seem most helpful in your notebook, leaving enough space after each question to record the answers as you read and study the chapter.

How does an outline show the relationships among the main idea, major supporting details, and minor supporting details? Where are main ideas used in an outline, concept map, and table of contents? Where are major supporting details used in an outline, concept map, and table of contents? Where are minor supporting details used in an outline, concept map, and table of contents? What is the difference between a formal outline and an informal outline?

Outlines

An outline shows how information in a paragraph or passage moves from a general idea to specific supporting details; thus it helps you make sense of the ways ideas relate to one another. A master reader uses an outline to see the main idea, major supporting details, and minor supporting details.

> An **outline** shows the relationships among the main idea, major supporting details, and minor supporting details.

Outlines can be formal or informal. A **formal** or **traditional outline** uses Roman numerals to indicate the main idea, capital letters to indicate the major details, and Arabic numbers to indicate minor details. It can be composed of full sentences or of topics or sentence parts that begin with a capital letter. A formal outline is particularly useful for studying complex reading material. Sometimes, you may choose to use an **informal outline** and record only the main ideas and the major supporting details. Because these outlines are informal, they may vary according to each student's notetaking style. Elements may or may not be capitalized. Also, one person might label the main idea with the number 1 and the major supporting details with letters *a*, *b*, *c*, *d*, and so on. Another person might not label the main idea at all and might label each major supporting detail with either letters or numbers.

EXAMPLE Read the following paragraph from a college finance textbook. Fill in the details to complete the outline. Then, answer the questions that follow it.

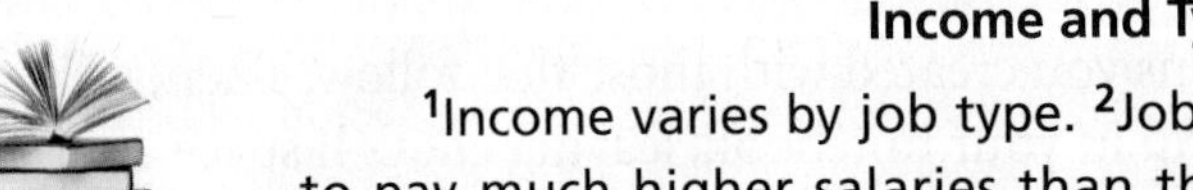

Textbook Skills

Income and Types of Jobs

[1]Income varies by job type. [2]Jobs that require specialized skills tend to pay much higher salaries than those that require skills that can be obtained very quickly and easily. [3]The income level associated with specific skills is also affected by the demand for those skills. [4]For example, the demand for people with a nursing license has been very high in recent years, so hospitals have been forced to pay high salaries to outbid other

hospitals for nurses. [5]Conversely, the demand for people with a history or English literature degree is low because more students major in these areas than there are jobs.

—From *Personal Finance*, 3rd ed., p. 27 by Jeff Madura. Copyright © 2007 by Pearson Education, Inc. Reproduced by permission of Pearson Education, Inc., Boston, MA.

Outline

I. (Main idea) ______________________________

A. Jobs that require specialized skills tend to pay much higher salaries than those that require skills that can be obtained very quickly and easily.

B. ______________________________

____

1. ______________________________

____

____

____

2. Conversely, the demand for people with a history or English literature degree is low because more students major in these areas than there are jobs.

Questions

_______ **1.** Sentences 4 and 5 are

a. major supporting details.

b. minor supporting details.

_______ **2.** The outline used in this activity is an example of

a. an informal outline.

b. a formal outline.

EXPLANATION The main idea of this paragraph is stated in sentence 1: "Income varies by job type." The author suggests that a list of supporting details will be presented and explained with the phrase "varies by job type." The second major detail, stated in sentence 3, is signaled with the word *also*. Sentences 4 and 5 are minor supporting details, introduced with the words *for example* and *conversely.* This outline is an example of a formal outline that includes the main idea, the major supporting details, and the minor supporting details of one paragraph.

An informal outline of the information looks like the following:

Stated main idea: Income varies by job type.

1. Jobs that require specialized skills tend to pay much higher salaries than those that require skills that can be obtained very quickly and easily.
2. The income level associated with specific skills is also affected by the demand for those skills.

Notice how this informal outline of the main idea and the major supporting details—without the minor supporting details—condenses the material into a summary of the author's primary points.

PRACTICE 1

Read the following paragraph from a college communications textbook. Then, answer the questions that follow it.

Textbook Skills

Active Listening

[1]Active listening serves several important functions. [2]First, it helps you as a listener to check your understanding of what the speaker said and, more important, what he or she meant. [3]Reflecting perceived meanings back to the speaker gives the speaker an opportunity to offer clarification and correct any misunderstandings. [4]Second, through active listening you let the speaker know that you acknowledge and accept his or her feelings. [5]An active learner accepts and identifies feelings with statements such as "You sound angry and upset." [6]Third, active listening stimulates the speaker to explore feelings and thoughts. [7]Elaborating on feelings helps you deal with them by talking them through with responses such as "I am upset and afraid, but I am not angry."

—From *The Interpersonal Communication Book,* 10th ed., p. 126 by Joseph A. DeVito.

1–5. Complete the following outline.

I. (Stated main idea) ______________________________

A. It clarifies and corrects.

1. ______________________________

____

2. ____

____

B. It offers acceptance.

1. Second, through active listening you let the speaker know that you acknowledge and accept his or her feelings.

2. ____

____

C. It deals with feelings.

1. Third, active listening stimulates the speaker to explore feelings and thoughts.

2. ____

____

VISUAL VOCABULARY

The best synonym for *empathy* is ______________.

a. understanding.
b. pity.
c. sorrow.

► Active listening communicates empathy for the feelings of another.

Concept Maps

An outline is one way to see the details that support a main idea. Another way to see details is through the use of a concept map. A **concept map** is a diagram that shows the flow of ideas from the main idea to the supporting details. Think of what you already know about a map. Someone can tell you how to get somewhere, but it is much easier to understand the directions if you can see how each road connects to the others by studying a map. Likewise, a concept map shows how ideas connect to one another.

A **concept map** is a diagram that shows the flow of ideas from the main idea to the supporting details.

To make a concept map, a master reader places the main idea in a box or circle as a heading and then places the major supporting details in boxes or circles beneath the main idea. Often arrows or lines are used to show the flow of ideas.

EXAMPLE Read the following passage. Then, complete the concept map by filling in the major and minor supporting details from the paragraph.

The Dangers of Texting While Driving

[1]Putting the brakes on the distracted driving epidemic will require both dedication and creative thinking, and the FCC is committed to doing its part to address this growing crisis.

Distracted Driving Is Dangerous

[2]The popularity of mobile devices has had some unintended and even dangerous consequences. [3]We now know that mobile communications is linked to a significant increase in distracted driving, resulting in injury and loss of life. [4]The National Highway Traffic Safety Administration reported in 2008 that driver distraction was the cause of 16 percent of all fatal crashes—5,800 people killed—and 21 percent of crashes resulting in an injury—515,000 people wounded. [5]According to AAA, nearly 50 percent of teens admit to texting while driving. [6]Distracted driving endangers life and property and the current levels of injury and loss are unacceptable.

What You Can Do

[7]*Give Clear Instructions* [8]Give teen drivers simple, clear instructions not to use their wireless devices while driving. [9]According to CTIA, the easiest way to say it is: "On the road, off the phone." [10]Before new drivers get their licenses, discuss the fact that taking their eyes off the road—even for a few seconds—could cost someone injury or even death.

[11]*Lead by Example* [12]Children learn from their parent's behavior. No one should text and drive. [13]Be an example for your children and if you need to text or talk on the phone, pull over to a safe place.

[14]*Become Informed and Be Active* [15]Review the information in our Clearinghouse and the literature on the web sites mentioned above. [16]Set rules for yourself and your household regarding distracted driving. [17]Tell family, friends and organizations to which you belong about the importance of driving without distractions. [18]Take information to your children's schools and ask that it be shared with students and parents.

—Federal Communications Commission. "FCC Consumer Advisory: The Dangers of Texting while Driving." Consumer & Government Affairs Bureau. 17 March 2010. http://www.fcc.gov/cgb/consumerfacts/drivingandtexting.html.

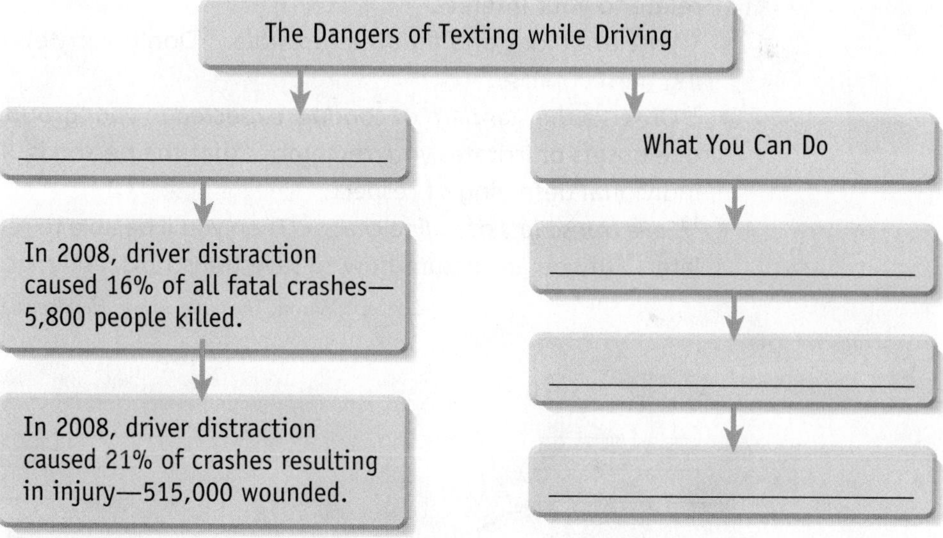

EXPLANATION The author of this passage used bold and italic print to emphasize the important ideas in the passage. The title of the passage states the topic and the author's controlling point. The two subheadings in bold print state the two major supporting details. The details in italic print highlight important minor details—action steps citizens can take in response to the dangers of texting while driving.

PRACTICE 2

Read the following passage from a college English handbook. Then, fill in the concept map with the central idea and major supporting details from the reading.

Using Online Chat

[1]You may be familiar with chat conversations from using instant messaging with friends and family. [2]In academic settings, chat will likely occur with coursework such as *WebCT* or *Blackboard*. [3]Collaborating via chat discussions will be more productive if you follow these tips:

- [4]*Use the chat space for brainstorming topics and exchanging impressions.* [5]The pace of online chat rarely allows lengthy consideration and articulation of messages.
- [6]*Focus on a thread or common topic.* [7]Online chat can be the electronic equivalent of a busy hallway outside a classroom, with different conversations occurring in the same space. [8]If you have trouble keeping up with all the messages, concentrate on the ones that relate to your interest.
- [9]*Write as quickly and fluidly as possible.* [10]Don't worry about producing perfect prose.
- [11]*Observe the standard of conduct expected in your group.* [12]If someone upsets or irritates you, remember that the person is, like you, an individual deserving of respect.
- [13]*Save transcripts of online chats.* [14]Then you'll be able to refer to them later. [15](If you are unsure how to save transcripts, ask your instructor.)

—Fowler & Aaron, *The Little, Brown Handbook*, 9th ed., p. 233.

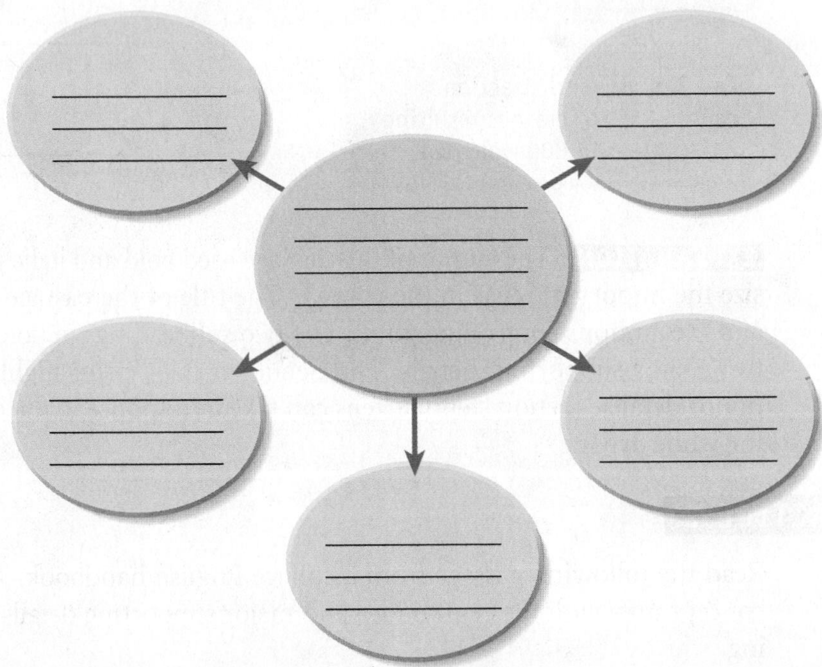

Textbook Skills

Reading a Textbook: The Table of Contents in a Textbook

The table of contents is a special kind of outline that is based on topics and subtopics. A **topic** is the *general subject,* so a **subtopic** is a *smaller part* of the topic. The general subject of the textbook is stated in the textbook's title. For example, the title *Health in America: A Multicultural Perspective* tells us that the book is about health concerns from the view of different cultures. Textbooks divide the general subject into smaller sections or subtopics. These subtopics form the chapters of the textbook. Because a large amount of information is found in each chapter, chapters are further divided into smaller parts or subtopics, and each subtopic is labeled with a heading. The table of contents lists the general subjects and subtopics of each chapter. Many textbooks provide a brief table of contents that divides the textbook into sections and lists the chapter titles for each section. There also may be a separate detailed table of contents that lists the subtopics for each chapter. A master reader examines the table of contents of a textbook to understand how the author has organized the information and to determine where specific information can be found.

EXAMPLE Survey, or look over, the following brief table of contents. Then, answer the questions.

1. What is the general topic of this textbook? ______________________

____

2. How many parts did the author use to divide the general topic? ________

3. How many chapters are in each part? ________ What is the approximate length of each chapter? ________ pages.

4. Write a one-sentence summary using the topic and subtopics for Part Two.

____

____

EXPLANATION The general topic of this textbook is stated in its title: *Messages: Building Interpersonal Communication Skills.* The author divides the general topic into three parts: "Messages about the Self and Others," "Messages: Spoken and Unspoken," and "Messages in Context." Each part is divided into

Messages: Building Interpersonal Communication Skills, 6th ed.

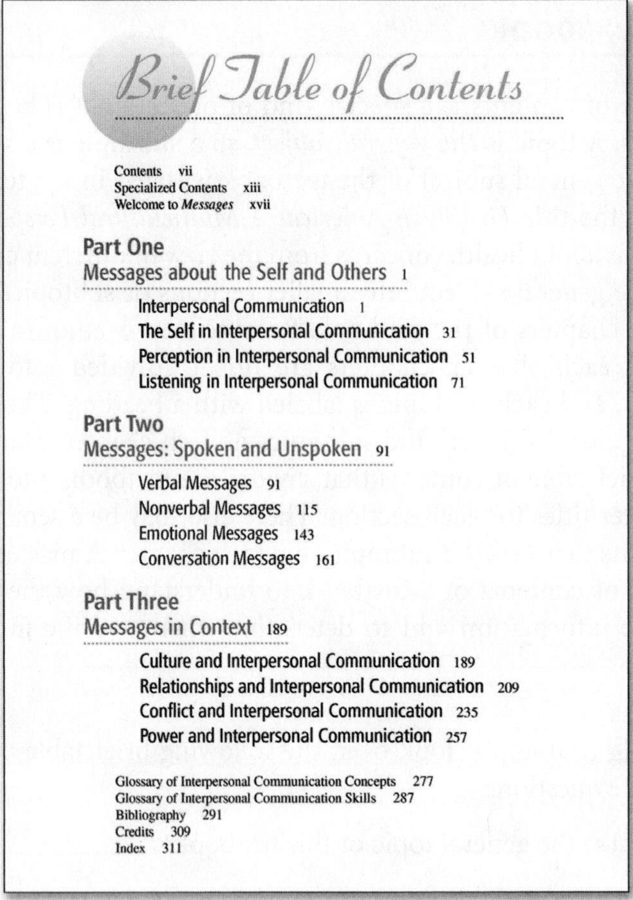

Brief Table of Contents

—From *Messages: Building Interpersonal Communication Skills,* 6th ed, pp. v–vi by Joseph A. DeVito. Boston: Allyn & Bacon, 2005.

four chapters. Knowing the length of each chapter helps you set aside the proper amount of time needed to read and study. In this textbook, each chapter is about twenty to thirty pages in length. One way to get a general sense of the ideas in a chapter or part is by writing a summary. Compare your summary to the following: Spoken and unspoken messages include verbal messages, nonverbal messages, emotional messages, and conversation messages.

PRACTICE 3

Study the following detailed table of contents for Chapter 1 of *Messages: Building Interpersonal Communication Skills,* 6th ed.

1. What is the topic of the chapter? ______________________________

2. How many subtopics are listed for the section "Principles of Interpersonal Communication"? ______________________

3. On what page does the discussion about competence in interpersonal communication begin? ______________________

4. What are the major supporting details of this chapter? ______________________

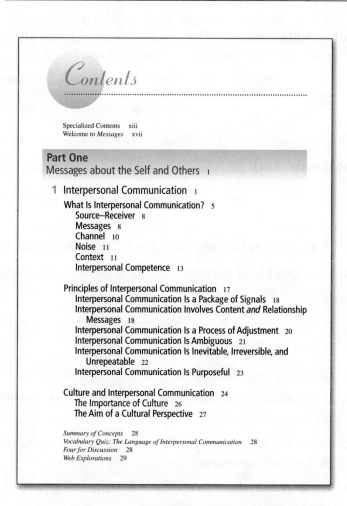

____

____

____

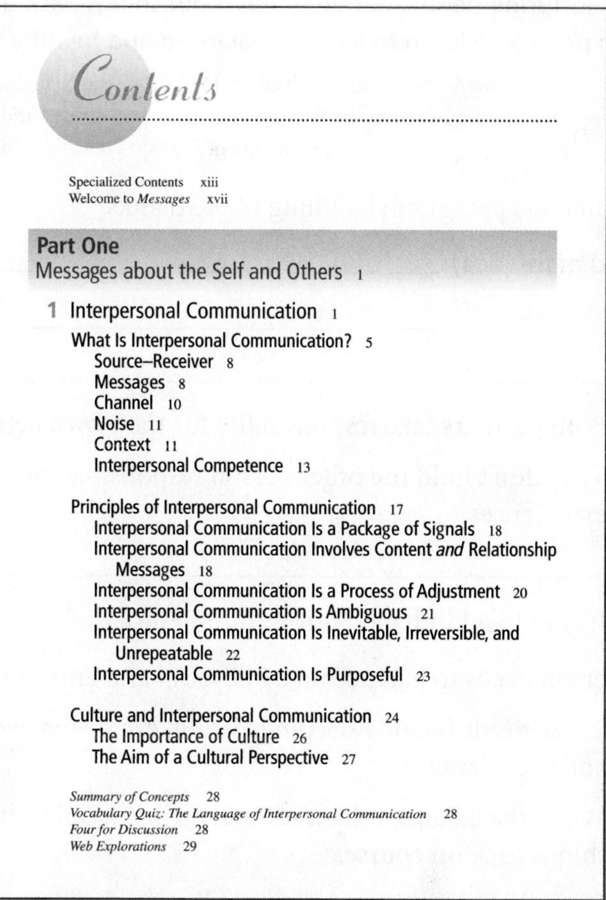

Contents

—From *Messages: Building Interpersonal Communication Skills,* 6th ed, p. vii by Joseph A. DeVito. Boston: Allyn & Bacon, 2005.

APPLICATIONS

Application 1: Major Supporting Details and Outlines

Read the following paragraph from a college health textbook.

Textbook Skills

Balanced Living

[1]Accountability and self-nurturance are very important to a good relationship. [2]**Accountability** means that both partners take responsibility for their own actions. [3]They don't hold the other person responsible for positive or negative experiences. [4]**Self-nurturance** goes hand in hand with accountability. [5]In order to make good choices in life, a person needs to balance many physical and emotional needs. [6]These needs include sleeping, eating, exercising, working, relaxing, and socializing. [7]When the balance is lost, self-nurturing people patiently try to put things back on course. [8]It is a lifelong process to learn to live in a balanced and healthy way.

—From *Access to Health,* 7th ed., p. 144 by Rebecca J. Donatelle and Lorraine G. Davis.

1–3. Outline the paragraph by filling in the blanks.

I. (Stated main idea) ______________________________

____

A. ______________________________

1. Both partners take responsibility for their own actions.
2. They don't hold the other person responsible for positive or negative experiences.

B. ______________________________

1. It goes hand in hand with accountability.

C. A person needs to balance many physical and emotional needs.

1. These needs include sleeping, eating, exercising, working, relaxing, and socializing.
2. When the balance is lost, self-nurturing people patiently try to put things back on course.
3. It is a lifelong process to learn to live in a balanced and healthy way.

______ **4.** The outline above is an example of

a. an informal outline.

b. a formal outline.

______ **5.** Sentence 6 is a
a. major supporting detail.
b. minor supporting detail.

VISUAL VOCABULARY

The best antonym for *nurturance* is ______________.

a. neglect.
b. challenge.
c. loving care.

► Self-nurturance means taking time to exercise and have fun.

Application 2: Major Details, Minor Supporting Details, and Outlines

Read the following paragraph from a college sociology textbook.

Textbook Skills

Can Deviance Really Be Functional for Society?

[1]When we think of deviance, its dysfunctions are likely to come to mind. [2]Functionalists, in contrast, are as likely to stress the functions of deviance as they are to emphasize its dysfunctions. [3]Most of us are upset by deviance, especially crime, and assume that society would be better off without it. [4]The classic functionalist theorist Emile Durkheim, however, came to a surprising conclusion. [5]Deviance, he said, including crime, is functional for society, for it contributes to the social order. [6]Its three main functions include the following. [7]*Deviance clarifies moral boundaries and affirms norms.* [8]A group's ideas about how people should act and think mark its *moral boundaries.* [9]Deviant acts challenge those boundaries. [10]To call a member into account is to say, in effect, "You broke an important rule, and we cannot tolerate that." [11]To punish deviants affirms the group's norms and clarifies what it means to be a member of the group. [12]*Deviance promotes social unity.* [13]To affirm the group's moral boundaries by punishing deviants fosters a "we" feeling among the group's members. [14]In saying, "You can't get by with that," the group collectively

affirms the rightness of its own ways. [15]*Deviance promotes social change.* [16]Groups do not always agree on what to do with people who push beyond their accepted ways of doing things. [17]Some group members may even approve of the rule-breaking behavior. [18]Boundary violations that gain enough support become new, acceptable behaviors. [19]Thus, deviance may force a group to rethink and redefine its moral boundaries, helping groups—and whole societies—to change their customary ways.

—From *Essentials of Sociology: A Down-to-Earth Approach,* 7th ed., p. 148 by James M. Henslin.

1–5. Complete the outline of the paragraph by filling in the blanks.

Stated main idea: ______________________________

____

A. ______________________________

1. A group's ideas about how people should act and think mark its moral boundaries. Deviant acts challenge those boundaries.

2. ______________________________

____

3. To punish deviants affirms the group's norms and clarifies what it means to be a member of the group.

B. ______________________________

1. To affirm the group's moral boundaries by punishing deviants fosters a "we" feeling among the group's members.

2. In saying, "You can't get by with that," the group collectively affirms the rightness of its own ways.

C. ______________________________

1. Groups do not always agree on what to do with people who push beyond their accepted ways of doing things.

2. Some group members may even approve of the rule-breaking behavior.

3. Boundary violations that gain enough support become new, acceptable behaviors.

4. Thus, deviance may force a group to rethink and redefine its moral boundaries, helping groups—and whole societies—to change their customary ways.

Application 3: Concept Maps and Signal Words

Read the following paragraph. Then, complete the concept map with the missing information from the paragraph.

A Caste System

[1]A **caste system** is a type of social stratification in which status is determined by birth and is life-long. [2]Someone who is born into a low-status group will always have low status, no matter how much that person may accomplish in life. [3]Achieved status cannot change an individual's place in this system. [4]India provides the best example of a caste system. [5]Based not on race but on religion, India's caste system has existed for nearly 3,000 years. [6]India's four main castes are subdivided into thousands of subcastes or jati. [7]Each subcaste specializes in a particular occupation. [8]For example, one jati washes clothes, another sharpens knives, and yet another repairs shoes. [9]Of the four main castes, the highest caste, the **Brahman**, is made up of priests and teachers. [10]Next is the **Kshatriya** caste of rulers and soldiers; followed by the **Vaishya** of merchants and traders while the lowest caste of the four is the **Shudra** comprised of peasants and laborers. [11]The lowest group in India's caste system is the **Dalit**, who make up India's "untouchables." [12]If a Dalit touches someone of a higher caste, that person becomes unclean. [13]Although the Indian government formally abolished the caste system in 1949, centuries-old practices cannot be eliminated so easily, and the caste system remains a part of everyday life in India.

—Adapted from *Essentials of Sociology: A Down-to-Earth Approach*, 7th ed., p. 171 by James M. Henslin. Copyright © 2007 by Pearson Education, Inc. Reproduced by permission of Pearson Education, Inc., Boston, MA.

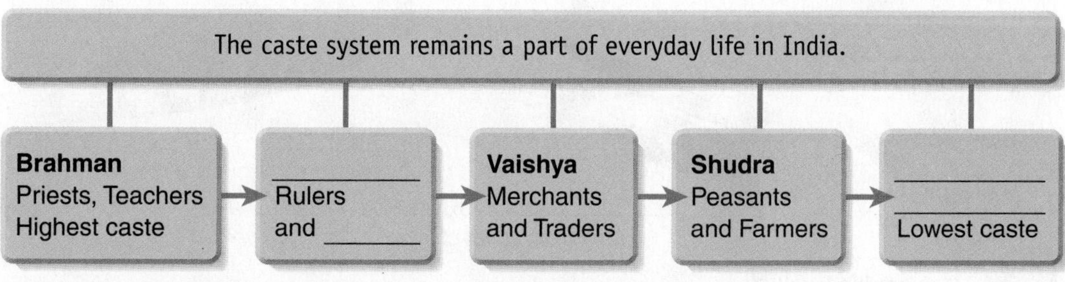

Application 4: Summary

Read the following paragraph from a college biology textbook. Complete the summary.

Textbook Skills

Energy Passes from One Trophic Level to Another

[1]Energy flows through communities from producers through several levels of consumers. [2]Each category of organism is called a **trophic level.** [3]The producers are the first trophic level, generally obtaining their energy directly from sunlight. [4]Consumers occupy several trophic levels. [5]Some consumers feed directly on producers, the most abundant living energy source in any ecosystem. [6]These **primary consumers,** or **herbivores** (plant eaters), ranging from grasshoppers to giraffes, form the second trophic level. [7]The third trophic level consists of **secondary consumers** or **carnivores** (meat eaters), predators such as the spider, eagle, and wolf that feed mainly on primary consumers. [8]Some carnivores occasionally eat other carnivores. [9]When doing so, they occupy the fourth trophic level, **tertiary consumers.**

—From *Life on Earth,* 5th ed., p. 578 and Figure 29–4a (p. 578) by Teresa Audesirk, Gerald Audesirk, and Bruce E. Byers.

Complete the illustration with information from the passage. Then, complete the summary that follows.

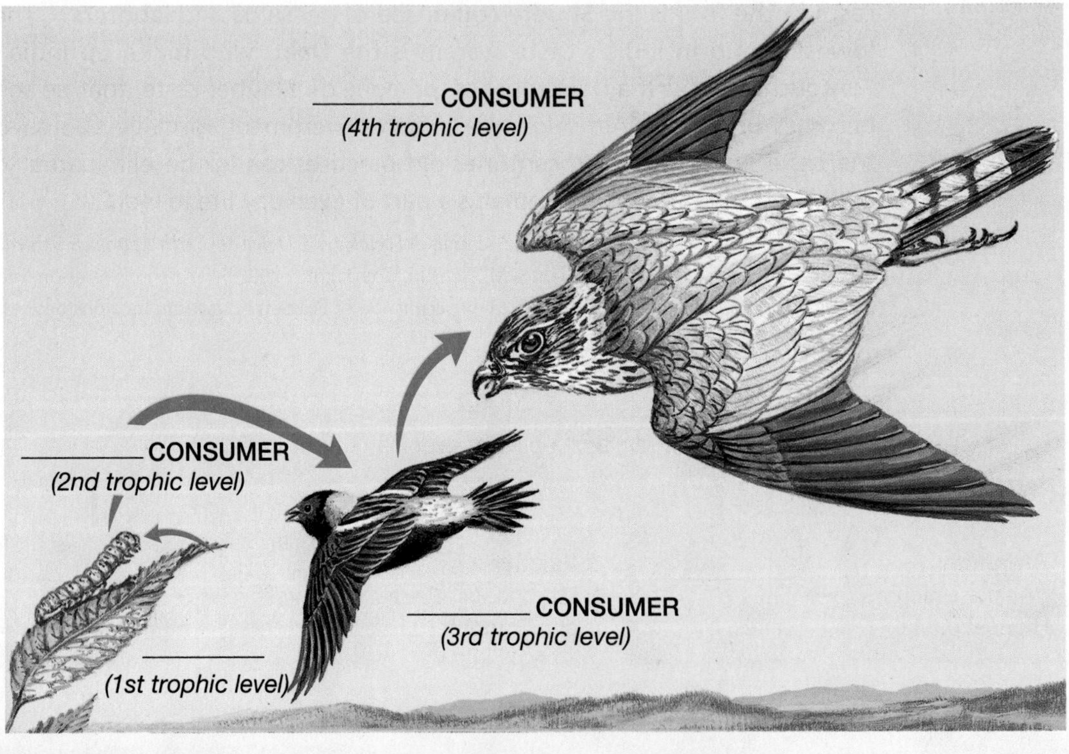

Summary

Energy flows through communities from producers through several levels of consumers. Each category of organism is called a __________ level. Producers, or plants, which get their energy directly from the __________, make up the first trophic level. The second trophic level is made up of primary consumers called __________ or plant eaters; the third trophic level consists of secondary consumers called carnivores or __________. The __________ trophic level includes tertiary consumers which consist of carnivores that eat other __________.

REVIEW TEST 1

Score (number correct) ______ × 10 = ______%

Main Ideas, Major and Minor Supporting Details, and Outlines

Read the following passage from a college business textbook. Then, complete the outline with the major and minor details from the passage.

Textbook Skills

Consumer Rights

[1]Much of the current interest in business responsibility toward customers can be traced to the rise of consumerism-social activism dedicated to protecting the rights of consumers in their dealings with businesses. [2]The first formal declaration of consumer rights protection came in the early 1960s when President John F. Kennedy identified four basic consumer rights. [3]Since that time, general agreement on two additional rights has also emerged; these rights are also backed by numerous federal and state laws. Legally, consumers have six basic rights:

- [4]*Consumers have a right to safe products.* [5]Businesses can't knowingly sell products that they suspect of being defective. [6]For example, a central legal argument in the recent problems involving Firestone tires was whether or not company officials knew in advance that the firm was selling defective tires.
- [7]*Consumers have a right to be informed about all relevant aspects of a product.* [8]For example, apparel manufacturers are now required to provide full disclosure on all fabrics used (cotton, silk, polyester, and so forth) and instructions for care (dry-clean, machine wash, hand wash).

- [9]*Consumers have a right to be heard.* [10]Labels on most products sold today have either a telephone number or address through which customers can file complaints or make inquiries.
- [11]*Consumers have a right to choose what they buy.* [12]Customers getting auto-repair service are allowed to know and make choices about pricing and warranties on new versus used parts. [13]Similarly, with the consent of their doctors, people have the right to choose between name-brand medications versus generic products that might be cheaper.
- [14]*Consumers have a right to be educated about purchases.* [15]All prescription drugs now come with detailed information regarding dosage, possible side effects, and potential interactions with other medications.
- [16]*Consumers have a right to courteous service.* [17]This right, of course, is hard to legislate. [18]But as consumers become increasingly knowledgeable, they're more willing to complain about bad service. [19]Consumer hotlines can also be used to voice service-related issues.

[20]American Home Products provides an instructive example of what can happen to a firm that violates one or more of these consumer rights. [21]For several years the firm aggressively marketed the drug Pondimin, a diet pill containing fenfluramine. [22]During its heyday, doctors were writing 18 million prescriptions a year for Pondimin and other medications containing fenfluramine. [23]The FDA subsequently discovered a link between the pills and heart-valve disease. [24]A class action lawsuit against the firm charged that the drug was unsafe and that users had not been provided with complete information about possible side effects. [25]American Home Products eventually agreed to pay $3.75 billion to individuals who had used the drug.

Stated central idea: **(1)** ______________________________

A. (2) ______________________________

1. (3) ______________________________

____

2. For example, a central legal argument in the recent problems involving Firestone tires was whether or not company officials knew in advance that the firm was selling defective tires.

B. (4) ______________________________

1. For example, apparel manufacturers are now required to provide full disclosure on all fabrics used (cotton, silk, polyester, and so forth).
2. Also required are instructions for care (dry-clean, machine wash, hand wash).

C. (5) ______________________________

1. Labels on most products sold today have either a telephone number or address.
2. Thus, customers can file complaints or make inquiries.

D. (6) ______________________________

1. Customers getting auto-repair service are allowed to know and make choices about pricing and warranties on new versus used parts.
2. (7) ______________________________

____

____

E. (8) ______________________________

1. All prescription drugs now come with detailed information.
2. Information may include dosage, possible side effects, and potential interactions with other medications.

F. (9) ______________________________

1. (10) ______________________________
2. But as consumers become increasingly knowledgeable, they're more willing to complain about bad service.
3. Consumer hotlines can also be used to voice service-related issues.

REVIEW TEST 2

Score (number correct) ______ × 25 = ______%

Supporting Details, Outlines, and Concept Maps

Read the following paragraph from a college ecology textbook. Complete the concept map with the main idea and the major supporting details from the paragraph.

Coral Reefs

Textbook Skills

[1]Lying in the warm, shallow waters about tropical islands and continental land masses are colorful rich oases, the coral reefs. [2]They are a unique buildup of dead skeletal material from a variety of organisms, including coral, certain types of algae, and mollusks. [3]Built only underwater at shallow depths, coral reefs need a stable foundation upon which to grow. [4]Such foundations are provided by shallow continental shelves and submerged volcanoes. [5]Coral reefs are of three types. [6](1) *Fringing reefs* grow seaward from the rocky shores of islands and continents. [7](2) *Barrier reefs* parallel shorelines of continents and islands and are separated from land by shallow lagoons. [8](3) *Atolls* are horseshoe-shaped rings of coral reefs and islands surrounding a lagoon, formed when a volcanic mountain subsided. [9]Such lagoons are about 40m deep, usually connect to the open sea by breaks in the reef, and may hold small islands of patch reefs. [10]Reefs build up to sea level.

—Adapted from Smith & Smith, *Elements of Ecology*, 4th ed. update, p. 504.

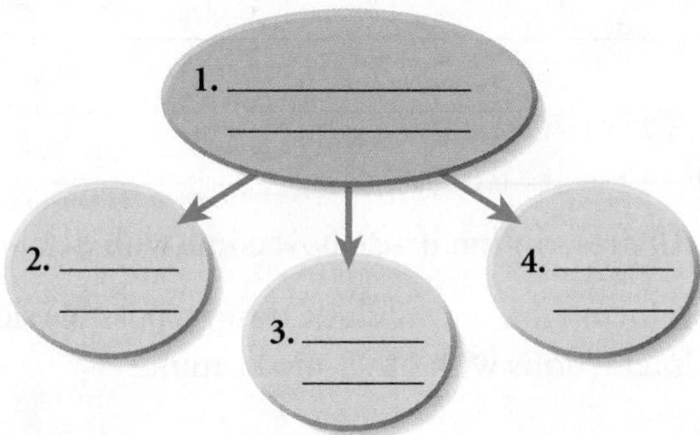

VISUAL VOCABULARY

Belize's Lighthouse Reef is ______________ that surrounds the Blue Hole, a sunken cave system that is a haven for marine life and scuba divers.

a. a fringe
b. a barrier
c. an atoll

REVIEW TEST 3

Score (number correct) ______ × 10 = ______%

Supporting Details and Outlines

Textbook Skills

Before you read the following passage from a college textbook about tourism, skim the material and answer the Before Reading questions. Read the passage. Then, answer the After Reading questions and activities.

The Art of Travel Writing

by William R. Gray

Vocabulary Preview

reportorial (1): having the traits of reporting news or facts

vanguard (4): the leading position in a movement, field, or trend

platoons (4): bodies of people with a common purpose or goal

atmospherics (10): the mood, tone, environment, climate

raucous (15): loud, noisy

[1]Over the past three decades or so, travel writing—like other nonfiction—has undergone a transformation from a more **reportorial** style to one that is propelled by the literary voice of the author. [2]This approach is personal and features the active participation of the writer in the subject matter—and thus is highly involving for readers, who feel they are being taken on a journey of discovery.

[3]Writers such as Edward Abbey, John McPhee, William Least Heat-Moon (particularly in *Blue Highways*), and Paul Theroux were in the **vanguard** of this approach to nonfiction writing, and today there are literally **platoons** of excellent writers following their lead. [4]And the great travel magazines overflow each issue with generally superb writing.

[5]In teaching my writing courses—including travel writing—and speaking at conferences, I try to instill the key elements that contribute to quality nonfiction writing, and all of these apply most emphatically to travel. [6]Focusing on these attributes helps elevate the quality of the writing, the personal involvement of the author, and draws readers in—and as writers we always need to keep our readers in the forefront of our minds.

- [7]Keen Observer: always seek the descriptive element, the penetrating detail that truly brings your subject to life and illuminates its uniqueness. [8]When interviewing someone, avoid the obvious descriptors such as color of hair and eyes; instead focus on timbre of voice, tilt of head, the gesture or mannerism that reveals character, the way light and shadow play across the face. [9]Likewise with place, use imagery—to evoke essential qualities; **atmospherics** and color are integral—the play of clouds, the tone of wind; use color dramatically—a cobalt sky; a sunset the color of apricots.

- [10]Active Participant: be both dynamically and intimately involved with your subject; follow every lead and participate in every activity appropriate to the place you are writing about. [11]Get up early, stay up late, and experience new things all day long.
- [12]Depth of Feeling: develop empathy for your subject; seek to experience it emotionally and do not hesitate to reveal something of yourself—how events and people affect you.
- [13]Openness to Experience: seek the **harrowing**, the exhilarating, the unusual. [14]Canoe through the crocodile-infested waters; climb the treacherous cliff; stay up all night with the **raucous** carnival celebrants. [15]In other words, be open to doing everything that might contribute to your understanding of the place you are writing about. [16]Be spontaneous—one of these activities may form the basis of the lead to your story.
- [17]Desire to Seek Knowledge and Understanding: strive to know your subject deeply; it's the way it is today because of the **confluence** of history, geology, exploration, warfare, cultural traditions, racial configuration, tourism, and a dozen other factors. [18]Understanding these elements will give perspective to your writing and help make it insightful and profound.
- [19]Desire to Seek a High Level in the Craft of Writing: use a strong first person point of view and a clear literary style to elevate and distinguish your writing. [20]Begin with an effective, original lead and end meaningfully. [21]Always be aware of **diction**, sentence variety, and quality of transitions. [22]Write so creatively and imaginatively that you leave your readers impatiently waiting to devour your next piece of writing.

—From *Tourism: The Business of Travel*, 4th ed., pp. 121–122 by Roy A. Cook, Laura J. Yale and Joseph J. Marqua.

Before Reading

Vocabulary in Context

_____ **1.** In sentence 13 of the passage, the word **harrowing** means

a. fun. c. frightening.
b. pleasant. d. fulfilling.

_____ **2.** In sentence 17 of the passage, the word **confluence** means

a. meeting. c. departure.
b. multitude. d. being.

_____ **3.** In sentence 21 of the passage, the word **diction** means
a. speech.
b. wording.
c. accent.
d. organization.

After Reading

Main Ideas and Supporting Details

4–10. Complete the following outline with the main idea and major supporting details of the passage.

Stated central idea: (4) ______________________________

- **(5)** ______________________________
- **(6)** ______________________________
- **(7)** ______________________________
- **(8)** ______________________________
- **(9)** Desire to Seek Knowledge and Understanding
- **(10)** Desire to Seek a High Level in the Craft of Writing

WHAT DO YOU THINK?

Do you have a favorite travel destination? If you could go anywhere in the world, where would you go and why? What place do you think everyone should visit at least once in her or his lifetime? Assume you are a travel writer and an online travel blog has asked you to write an article about a place of your choosing. Consider the following possibilities: your home town; a tourist attraction such as Disney World or Universal Studios; a state or national park. As you write and revise, use the elements of effective travel writing suggested by William R. Gray.

REVIEW TEST 4

Score (number correct) _____ × 10 = _____%

Supporting Details and Concept Maps

Before Reading: Survey the following passage adapted from the college textbook *The Interpersonal Communication Book.* Skim the passage, noting the words in bold print. Answer the Before Reading questions that follow the passage. Then, read the passage. Next, answer the After Reading questions. Use the discussion and writing topics as activities to do after reading.

Textbook Skills

Vocabulary Preview

implications (4): meanings
counter (5): opposite
subsequent (8): following, later
alibi (16): explanation, excuse
assurance (17): promise, guarantee

Repairing Conversational Problems: The Excuse

[1]At times you may say the wrong thing; then, because you can't erase the message (communication really is irreversible), you may try to account for it. [2]Perhaps the most common method for doing so is the excuse. [3]You learn early in life that when you do something that others will view negatively, an excuse is in order to justify your performance. [4]***Excuses***, central to all forms of communication and interaction, are "explanations or actions that lessen the negative **implications** of an actor's performance, thereby maintaining a positive image for oneself and others" (Snyder, 1984; Snyder, Higgins, & Stucky, 1983).

[5]Excuses seem especially in order when you say or are accused of saying something that runs **counter** to what is expected, sanctioned, or considered "right" by the people with whom you're talking. [6]Ideally, the excuse lessens the negative impact of the message.

Some Motives for Excuse Making

[7]The major motive for excuse making seems to be to maintain your self-esteem, to project a positive image to yourself and to others. [8]Excuses also represent an effort to reduce stress: You may feel that if you can offer an excuse—especially a good one that is accepted by those around you—it will reduce the negative reaction and the **subsequent** stress that accompanies a poor performance.

[9]Excuses also may enable you to maintain effective interpersonal relationships even after some negative behavior. [10]For example, after criticizing a friend's behavior and observing the negative reaction to your criticism, you might offer an excuse such as, "Please forgive me; I'm really exhausted. [11]I'm just not thinking straight." [12]Excuses enable you to place your messages—even your possible failures—in a more favorable light.

Types of Excuses

[13]Think of the recent excuses you have used or heard. [14]Did they fall into any of these three classes (Snyder, 1984)?

- [15]*I didn't do it:* Here you deny that you have done what you're being accused of. [16]You may then bring up an **alibi** to prove you couldn't have done it, or perhaps you may accuse another person of doing what you're being blamed for ("I never said that" or "I wasn't even near the place when it happened"). [17]These "I didn't do it" types are the worst excuses, because they fail to acknowledge responsibility and offer no **assurance** that this failure will not happen again.
- [18]*It wasn't so bad:* Here you admit to doing it but claim the offense was not really so bad or perhaps that there was justification for the behavior ("I only padded the expense account, and even then only modestly" or "Sure, I hit him, but he was asking for it").
- [19]*Yes, but:* Here you claim that **extenuating circumstances** accounted for the behavior; for example, that you weren't in control of yourself at the time or that you didn't intend to do what you did ("It was the liquor talking" or "I never intended to hurt him; I was actually trying to help").

—Adapted from *The Interpersonal Communication Book*, 11th ed., pp. 210–211 by Joseph A. DeVito. Copyright © 2007 by Pearson Education, Inc. Reproduced by permission of Pearson Education, Inc., Boston, MA.

Before Reading

Vocabulary in Context

Use your own words to define the following terms. Identify the context clue you used to make meaning of the word.

1. The word **counter** in sentence 5 means ______________________

2. Context clue used for **counter** ______________________

____

3. The phrase **extenuating circumstances** in sentence 19 means __________

____

4. Context clue used for **extenuating circumstances** ______________

____

After Reading

Main Ideas and Supporting Details

Complete the following graphic organizer with information from the passage.

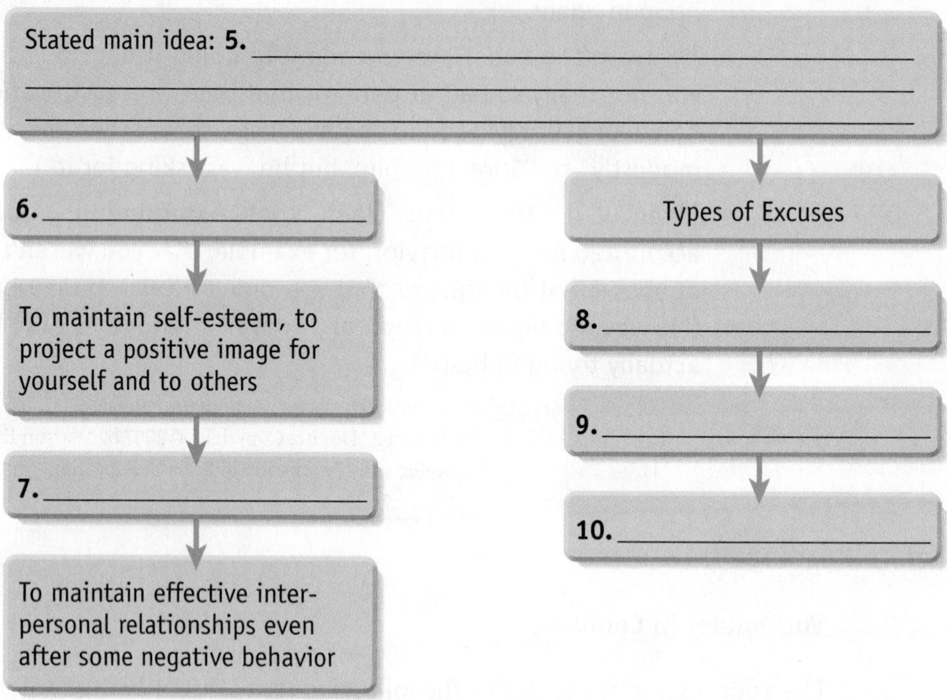

WHAT DO YOU THINK?

What kinds of excuses do you hear in informal groups on your campus? Do excuses on the job differ from excuses heard at school? Why or why not? Assume you are a manager of a local retail store. Tardiness to work is a growing problem among a significant number of employees in your department. At a manager's meeting, you discover that managers in other departments are experiencing the same problem. Write an article for your company's newsletter to educate workers about excuse making. Explain possible motivations, types of excuses, and the possible impact of excuse making on their chances for promotion. Also suggest a solution to avoid the need for excuses.

After Reading About Outlines and Concepts

Before you move on to the Mastery Tests on outlines and concept maps, take time to reflect on your learning and performance by answering the following questions. Write your answers in your notebook.

- How has my knowledge base or prior knowledge about outlines and concept maps changed?
- Based on my studies, how do I think I will perform on the Mastery Test(s)? Why do I think my scores will be above average, average, or below average?
- Would I recommend this chapter to other students who want to learn more about outlines and concept maps? Why or why not?

Test your understanding of what you have learned about Outlines and Concept Maps for Master Readers by completing the Chapter 6 Review Card in the insert near the end of your text.

CONNECT TO PEARSON myreadinglab

To check your progress in meeting Chapter 6's learning outcomes, log in to **www.myreadinglab.com** and try the following activities.

- The "Outlining and Summarizing" section of MyReadingLab gives an overview about memory and active reading. This section also provides a model for outlining and a model for mapping. You will also find practice activities and tests. To access this resource, click on the "Study Plan" tab. Then click on "Outlining and Summarizing." Then click on the following links as needed: "Overview," "Model: Outlining," "Model: Mapping," "Practice," and "Tests."
- To measure your mastery of the content of this chapter complete the tests in the "Outlining and Summarizing" section and click on Gradebook to find your results.

MASTERY TEST 1

Name ______________________ Section ____________
Date __________ **Score** (number correct) _____ × 10 = _____%

A. Read the following passage from a college communications textbook.

Textbook Skills

Networking Relationships

[1]**Networking** is a process of using other people to try to help you solve your problems, or at least offer insights that bear on your problem—for example, how to publish your manuscript, where to look for low-cost auto insurance, how to find an affordable apartment, or how to empty your cache. [2]Networking comes in two forms: informal and formal. [3]*Informal* networking is what we do every day when we find ourselves in a new situation or unable to answer questions. [4]Thus, for example, if you're new at school, you might ask someone in your class where to eat or shop for new clothes, or who's the best teacher. [5]In the same way, if you enter a new work environment, you might ask more experienced workers how to perform certain tasks or whom you should avoid or approach when you have questions. [6]*Formal* networking is the same thing except that it's much more systematic and strategic. [7]It's the establishment of connections with people who can help you—answer your questions, get you a job, help you get promoted, help you relocate, or accomplish any task you want to accomplish.

—Adapted from *The Interpersonal Communication Book,* 11th ed., pp. 305–306 by Joseph A. DeVito. Copyright © 2007 by Pearson Education, Inc. Reproduced by permission of Pearson Education, Inc., Boston, MA.

1–2. Complete the summary with information from the paragraph.

Networking is (1) ____

____—for example, (**2**)

____

____

_____ **3.** Sentence 1 is a
a. main idea.
b. major supporting detail.
c. minor supporting detail.

_____ **4.** Sentence 4 is a
a. main idea.
b. major supporting detail.
c. minor supporting detail.

5. What signal phrase introduces the second minor supporting detail for informal networking? ______________________

B. Read the following paragraph from a college government textbook.

Textbook Skills

News by Trial Balloon

[1]Those who make the news depend on the media to spread certain information and ideas to the general public. [2]Sometimes they feed stories to reporters in the form of **trial balloons**: information leaked to see what the political reaction will be. [3]For example, a few days prior to President Clinton's admission that he had an "inappropriate relationship" with Monica Lewinsky, top aides to the president leaked the story to Richard Berke of the *New York Times*. [4]The timing of the leak was obvious; the story appeared just before Clinton had to decide how to testify before Kenneth Starr's grand jury. [5]When the public reacted that it was about time he admitted this relationship, it was probably easier for him to do so—at least politically.

—Edwards, Wattenberg, & Lineberry, *Government in America*, 5th ed., Brief Version, p. 171.

6–7. Complete the summary with information from the paragraph.

(6) ______________________

__________—for example, **(7)** ______________________

____

____

_____ **8.** In the paragraph, sentence 3 is
a. a main idea.
b. a major supporting detail.
c. a minor supporting detail.

_____ **9.** In the paragraph, sentence 4 is
a. a main idea.
b. a major supporting detail.
c. a minor supporting detail.

10. What signal word introduces the find minor supporting detail? __________

MASTERY TEST 2

Name ______________ Section ______________

Date __________ **Score** (number correct) ______ × 10 = ______%

Textbook Skills

Read the following paragraph from a college psychology textbook. Then, answer the questions that follow it.

What Is Substance Abuse?

[1]What is the difference between drug use and substance abuse? [2]The impact on people's lives is one way to gauge use versus abuse. [3]People who experience negative effects as a result of their drug use cross the line into substance abuse. [4]A **substance abuser** is a person who overuses and relies on drugs to deal with everyday life. [5]Most substance abusers use alcohol or tobacco, but the whole range of psychoactive drugs and combinations of these drugs presents possibilities for substance abuse. [6]A person is a substance abuser if all three of the following statements apply:

- [7]The person has used the substance for at least a month.
- [8]The use has caused legal, personal, social, or vocational problems.
- [9]The person repeatedly uses the substance even in situations when doing so is hazardous, such as when driving a car.

[10]Missing work or school, spending too much money on drugs, damaging personal relationships, and getting into legal trouble are all signs of substance abuse. [11]When substance abusers try to decrease or quit their drug use, they may experience withdrawal symptoms. [12]**Withdrawal symptoms** are the reactions experienced when a substance abuser stops using a drug with dependence properties. [13]For example, an alcoholic may experience the "shakes" as he or she attempts to quit drinking alcohol.

—Adapted from Lefton & Brannon, *Psychology*, 8th ed., pp. 215–216.

______ **1.** Which sentence states the main idea of the passage?

a. sentence 1
b. sentence 2
c. sentence 3
d. sentence 8

______ **2.** In general, the major details of this passage are

a. facts that show the number of people who are substance abusers.
b. ways to avoid substance abuse.
c. reasons for substance abuse.
d. definitions of substance abuse, substance abuser, and withdrawal symptoms.

______ **3.** In the passage, sentence 4 is

a. a main idea.
b. a major supporting detail.
c. a minor supporting detail.

______ **4.** In the passage, sentence 7 is

a. a main idea.
b. a major supporting detail.
c. a minor supporting detail.

5–10. Complete the informal outline with the key terms, definitions, and examples from the paragraph.

Main idea: Drug abuse can be measured by the impact on the lives of users.

A. ______________: Negative effects as a result of drug use

Example: Missing work or school, spending too much money on drugs, damaging personal relationships, and getting into legal trouble are all signs of substance abuse.

B. Substance abuser: ______________________________
____

Example: ______________________________
____
____

C. ______________: ______________________________
____

Example: ______________________________
____

VISUAL VOCABULARY

Which word or phrase best states the meaning of *psychoactive*?

____

a. mind-altering
b. satisfying
c. addictive

▸ Caffeine is one example of a commonly used *psychoactive* drug.

MASTERY TEST 3

Name ______________________ Section ______________

Date __________ **Score** (number correct) ______ × 10 = ______%

Read the following passage from a college humanities textbook.

Textbook Skills

The Difference Between Problems and Issues

[1]To be able to think clearly and effectively, you must be able to understand and recognize problems and issues. [2]Problems and issues differ in some respects. [3]A problem is a situation that we regard as unacceptable. [4]In contrast, an issue is a matter about which intelligent, informed people disagree to some extent. [5]Solving problems, therefore, means deciding what action will change the situation for the best. [6]On the other hand, resolving issues means deciding what belief or viewpoint is the most reasonable.

[7]Whenever you are uncertain whether to treat a particular challenge as a problem or an issue, apply this test: Ask whether the matter involved tends to arouse partisan feelings and to divide informed, intelligent people. [8]If it does not, treat it as a problem. [9]If it does, treat it as an issue. [10]Here are some sample problems: a student trying to study in a noisy dormitory, a child frightened by the prospect of going to the hospital, a businesswoman dealing with sexual harassment. [11]Here are some sample issues: a public school teacher leading students in prayer in a public school classroom, a member of Congress proposing a cut in Social Security benefits for the elderly, an anthropologist stating that human beings are violent by nature.

—Adapted from Ruggiero, *The Art of Thinking*, 7th ed., p. 109.

______ **1.** Sentence 2 is a
a. main idea.
b. major supporting detail.
c. minor supporting detail.

______ **2.** Sentence 3 is a
a. main idea.
b. major supporting detail.
c. minor supporting detail.

______ **3.** Sentence 4 is a
a. main idea.
b. major supporting detail.
c. minor supporting detail.

______ **4.** Sentence 8 is a
a. main idea.
b. major supporting detail.
c. minor supporting detail.

______ **5.** Sentence 11 is a
a. main idea.
b. major supporting detail.
c. minor supporting detail.

6–10. Complete the concept map (a contrast chart) with information from the passage.

Differences between:	**6.** ____________	**Issues**
Definitions	an unacceptable situation	**7.** ____________
Goals	positive actions	**8.** ____________
Traits	**9.** ____________	arouse feelings and divide people
Examples	**10.** ____________	a public school teacher leading students in prayer in a public classroom, a member of Congress proposing a cut in Social Security benefits for the elderly, an anthropologist stating that human beings are violent by nature

VISUAL VOCABULARY

This photograph entitled "Pro-Life Demonstrators Face Pro-Choice Demonstrators" illustrates a significant ____________ in the United States.

a. problem
b. issue

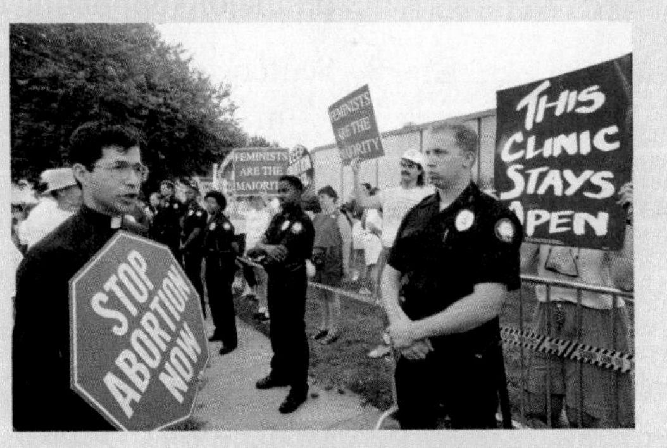

MASTERY TEST 4

Name ______________________ Section ______________

Date __________ **Score** (number correct) ______ × 20 = ______%

Read the following passage from a college textbook. Complete the concept map with information from the passage.

Popular Areas of Small-Business Enterprise

[1]Not surprisingly, small businesses are more common in some industries than in others. [2]The major small-business industry groups are *services, construction, finance and insurance, wholesaling,* and *transportation and manufacturing*. [3]Each industry differs in its needs for employees, money, materials, and machines. [4]But as a general rule, the more resources required, the harder it is to start a business and the less likely an industry is dominated by small firms.

[5]Remember, too, that small is a relative term. [6]The criteria (number of employees and total annual sales) differ from industry to industry and are often meaningful only when compared with truly large businesses. [7]The following discussion focuses on U.S. businesses across industry groups employing fewer than 20 employees.

Services [8]Small-business services range from marriage counseling to computer software, from management consulting to professional dog walking. [9]Partly because they require few resources, service providers are the fastest-growing segment of small business. [10]A retailer, for example, sells products made by other firms directly to consumers. [11]Usually, people who start small retail businesses favor specialty shops—say, big men's clothing or gourmet coffees—that let them focus limited resources on narrow market segments.

Construction [12]About 10 percent of businesses with fewer than 20 employees are involved in construction. [13]Because many construction jobs are small local projects, local firms are often ideal contractors.

Finance and Insurance [14]Financial and insurance firms also account for about 10 percent of all firms with fewer than 20 employees. [15]Most of these businesses are affiliates of or agents for larger national firms.

Wholesaling [16]Small-business owners often do well in wholesaling. [17]About 8 percent of businesses with fewer than 20 employees are wholesalers. [18]Wholesalers buy products from manufacturers or other producers and sell them to retailers. [19]They usually purchase goods in bulk and store them in quantities at locations convenient for retailers. [20]For a given volume of business, therefore, they need fewer employees than manufacturers, retailers, or service providers.

Transportation and Manufacturing [21]Some small firms—about 5 percent of all companies with fewer than 20 employees—do well in transportation and related businesses. [22]These include taxi and limousine companies, charter airplane services, and tour operators. [23]More than any other industry, manufacturing lends itself to big business. [24]But this doesn't mean that no small businesses do well in manufacturing; rather, about 5 percent of firms with fewer than 20 employees are involved in manufacturing. [25]Indeed, small manufacturers sometimes outperform big ones in such innovation-driven industries as electronics, toys, and computer software.

—Adapted from *Business,* 8th ed., pp. 92–93 by Ricky W. Griffin and Ronald J. Ebert. Copyright © 2006 by Pearson Education, Inc. Printed and Electronically reproduced by permission of Pearson Education, Inc., Upper Saddle River, NJ.

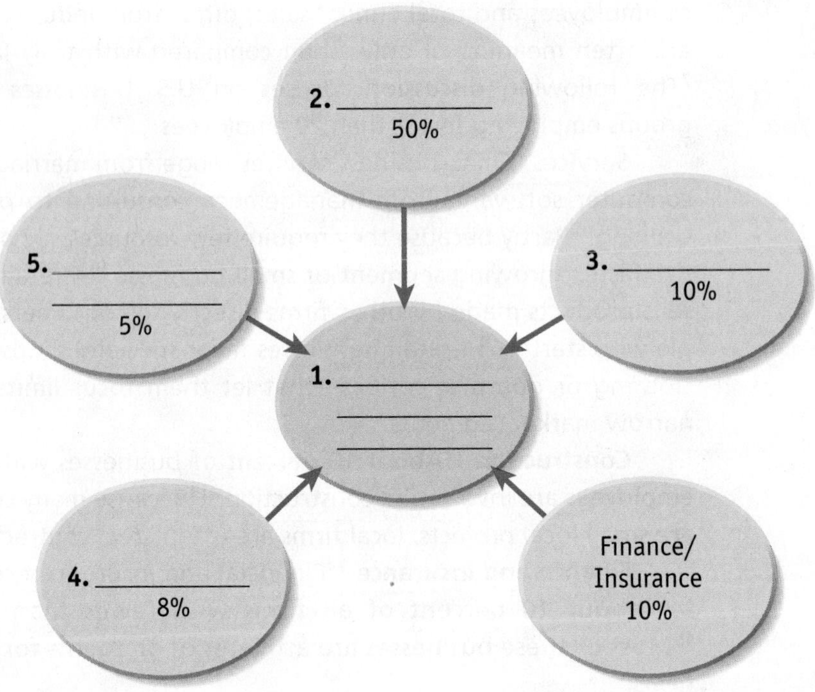

Transitions and Thought Patterns

7

LEARNING OUTCOMES

After studying this chapter you should be able to do the following:

1. Define the terms *transitions* and *thought patterns.*
2. Determine the relationships of ideas within a sentence.
3. Determine the relationships of ideas between sentences.
4. Recognize the following thought patterns and their signal words: *time order, space order, listing, classification.*
5. Determine the thought pattern used to organize a passage.
6. Evaluate the importance of transitions and thought patterns.

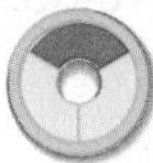

Before Reading About Transitions and Thought Patterns

Using the reporter's questions (Who? What? When? Where? Why? and How?), refer to the chapter learning outcomes and create at least three questions that you can answer as you study the chapter. Write your questions in the following spaces:

____

____

____

Now take a few minutes to skim the chapter for ideas and terms that you have studied in previous chapters. List those ideas in the following spaces:

____

____

Compare the questions you created based on the chapter learning outcomes with the following questions. Then, write the ones that seem the most helpful in your notebook, leaving enough space after each question to record the answers as you read and study the chapter.

What are transitions? What are thought patterns? What is the relationship between transitions and thought patterns? How do thought patterns use transition words?

On page 274, the terms *main idea* and *supporting details* are discussed in relationship to transitions and thought patterns. Consider the following study questions based on these ideas: How can transitions help me understand the author's main idea? How can transitions help me create an outline?

Transition Words: Relationships Within a Sentence

Read the following set of ideas. Which word makes the relationship between the ideas clear?

Marcus was asked to leave the movie __________ he was talking loudly.

a. however b. because c. finally

The word that makes the relationship between ideas within the sentence clear is (b) *because.* The first part of the sentence states what happened to Marcus. The second part of the sentence explains why he was asked to leave the theatre. The word *because* best signals the cause and effect relationship between Marcus's behavior and the result of his behavior.

Transitions are key words and phrases that signal the logical relationships both within a sentence and between sentences. **Transitions** help you make sense of an author's idea in two basic ways. First, transitions join ideas within a sentence. Second, transitions establish **thought patterns** so readers can understand the logical flow of ideas between sentences.

Transitions are words and phrases that signal thought patterns by showing the logical relationships both within a sentence and between sentences.
A **thought pattern** is established by using transitions to show the logical relationship between ideas in a paragraph or passage.

Some transition words have similar meanings. *For example, also, too,* and *furthermore* all signal the relationship of addition or listing. Sometimes a single word can serve as two different types of transitions, depending on how it is used. For example, the word *since* can reveal time order, or it can signal a cause. Notice the difference in the following two sentences.

Since I began studying every day, my understanding and my grades have improved dramatically.

Since you have the flu, please stay home from work to avoid spreading the disease.

EXAMPLE Complete the following ideas with a transition that shows the relationship within each sentence. Fill in each blank with a word from the box. Use each word once.

adjacent	another	during	one type

1. In the ancient Aztec culture, priests offered cacao seeds and chocolate drinks to their gods ____________ sacred ceremonies.
2. One way to shop for a vehicle is to visit local dealerships; ____________ way is to use online resources such as AutoTrader.com.
3. Fear of failure is ____________ of attitude that drives academic cheating.
4. The registrar's office is ____________ to the financial aid office.

EXPLANATION

1. The word *during* indicates *when* the Aztec priests offered the cacao seeds and chocolate drinks to their gods.
2. *Another* signals the *addition* of the second idea in this sentence.
3. The phrase *one type* suggests that other types of thinking also motivate academic cheating.
4. The word *adjacent* signals that the registrar's office is located *next to* the financial aid office.

PRACTICE 1

Complete the following ideas with a transition that shows the relationship within each sentence. Fill in each blank with a word from the box. Use each word once.

furthermore	in the foreground	one category	subsequently

1. An American citizen is entitled to a fair and speedy trial whether _____________ found guilty or innocent.
2. Walking—a safe, efficient, and popular method of attaining and maintaining physical fitness—strengthens the heart and lungs; _____________, it builds bones and tones muscles.
3. In Monet's 1877 painting *Arrival of a Train*, the smoke of the trains _____________ of the painting blurs the cathedral, which is barely visible past the smoke.
4. The white lie is only _____________ of the myriad types of lies that do much harm.

Master readers look for transition words, study their meaning in context, and use them as keys to unlock the author's thought pattern.

Thought Patterns: Relationships Between Sentences

Not only do transitions reveal the relationships among ideas within a sentence, they also show the relationship of ideas between sentences.

Walking strengthens the heart and lungs. It *also* builds and tones muscles.

EXAMPLE Complete each of the following ideas with a transition that makes the relationship between the sentences clear. Fill in each blank with a word from the box. Use each word once.

as a result	finally	for example	next

1. According to the National Weather Service, frostbite is the damage caused to body tissue due to freezing. _________, frostbite causes a loss of feeling and white or pale coloring of the affected areas.

2. Your family should have an emergency or disaster kit on hand. __________, you should have water, non-perishable food, and first aid supplies.

3. Four test-taking tips may prove helpful if you experience test anxiety. First, keep a positive attitude. __________, to gain confidence and build up points, answer the easy questions before you answer tougher ones.

4. Third, don't let the hard questions stump you; circle them, move on, and come back to them later if you can. __________, take time to proofread for careless errors.

VISUAL VOCABULARY

The root **dermis** means __________.

a. degree.
b. skin.
c. layer of skin or tissue.

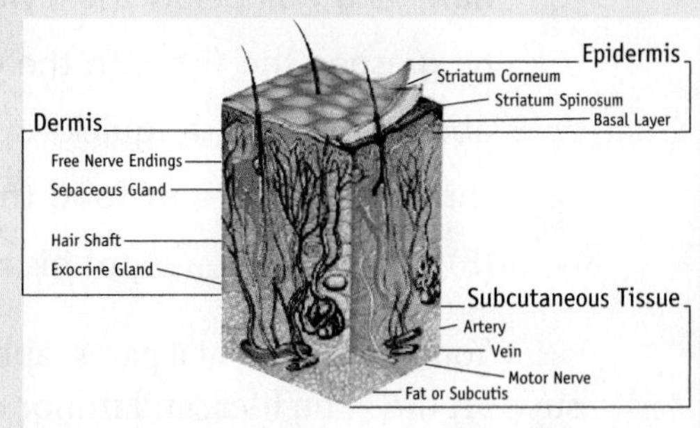

► **Cross-Section of the Three Layers of Human Skin**
First-degree frostbite injuries involve the epidermis, and fourth-degree injuries involve the epidermis, dermis, and subcutaneous tissue.

—Adapted from *Occupational Dermatoses*, Page 3, Slide 14, 21 Apr. 2001; National Institute for Occupational Safety and Health Photolibrary, CDC. http://www.cdc.gov/niosh/ocderm3.html

EXPLANATION

1. The transition *as a result* indicates that the second sentence offers specific symptoms of frostbite.

2. The transition phrase *for example* logically introduces example items to include in a family disaster kit.

3–4. Items 3 and 4 list four test-taking tips. In item 3, the transition *next* logically introduces the second tip, and in item 4, the transition *finally* introduces the last of the four test-taking tips.

PRACTICE 2

Complete the following ideas with transitions. Fill in each blank with a word from the box. Use each word once.

immediately	meanwhile	next	on	out

Andre began choking. **(1)** _______________ Daine wrapped his arms around Andre's waist. **(2)** _______________, Daine made a fist with his right hand. He then placed the thumb side of his fist **(3)** _______________ the middle of Andre's abdomen just above his navel and well below the lower tip of his breastbone. **(4)** _______________, he grasped the fist with the other hand. He pressed his fist into Andre's abdomen with a quick upward thrust. After several quick, hard thrusts, the piece of food that had been blocking Andre's airway flew **(5)** _______________ of his mouth.

You will recall that a paragraph is made up of a group of ideas. Major details support the main idea, and minor details support the major details. Transitions make the relationship between these three levels of ideas clear, smooth, and easy to follow.

Before beginning to write, an author must ask, "What thought pattern best expresses these ideas?" or "How should these ideas be organized so that the reader can follow and understand my point?" A **thought pattern** (also called a **pattern of organization**) allows the author to arrange the supporting details in a clear and smooth flow by using transition words.

Thought patterns (or **patterns of organization**) are signaled by using transitions to show the logical relationship between ideas in a paragraph, passage, or textbook chapter.

As you learned in Chapter 3, a main idea is made up of a topic and the author's controlling point about the topic. One way an author controls the topic is by using a specific thought pattern. Read the following paragraph. Identify the topic sentence by circling the topic and underlining the controlling point.

Interpersonal Perception

Textbook Skills

[1]Perception takes place in three stages that flow into one another and overlap. [2]At the first stage, you sense the stimuli. [3]At the second phase, you organize the stimulations. [4]The third stage in the process is interpretation-evaluation. [5]These stages are not separate; they're continuous and blend into and overlap one another.

—Adapted from *Messages: Building Interpersonal Communication Skills,* 4th ed., p. 66 by Joseph A. DeVito. Copyright © 1999 by Pearson Education, Inc. Reproduced by permission of Pearson Education, Inc., Boston, MA.

The topic is "perception" and the controlling point is the phrase "three stages that flow into one another and overlap." The phrase "that flow into one another and overlap" states the author's opinion. The words "three stages" state the author's thought pattern. The author's controlling point limits the supporting details to listing and describing the three stages of perception. The transition phrases *first stage, second phase,* and *third stage* signal each of the supporting details. Authors often introduce supporting details with transition words based on the controlling point. Creating an outline using transition words is an excellent way to grasp an author's thought pattern.

EXAMPLE Read the following paragraph. Complete the informal outline, then answer the question.

Types of Coping Styles

Textbook Skills

[1]Each of us can learn to control stress by adopting coping strategies that are consistent with our lifestyles. [2]According to Lazarus and Folkman (1984; Folkman and Lazarus, 1991), there are two types of coping responses. [3]The first type of coping style, **problem-focused coping,** is directed toward the source of the stress. [4]For example, if the stress is job-related, a person might try to change conditions at the job site or take courses to acquire skills that would enable him or her to obtain a different job. [5]The second type of coping style, **emotion-focused coping,** is directed toward a person's own personal reaction to a stressor. [6]For example, a person might try to relax and forget about the problem or find solace in the company of friends.

—Adapted from Carlson & Buskist, *Psychology: The Science of Behavior*, 5th ed., p. 552.

Topic sentence: ______________________________

A. ______________________________

For example, if the stress is job-related, a person might try to change conditions at the job site or take courses to acquire skills that would enable him or her to obtain a different job.

B. ____

____

For example, a person might try to relax and forget about the problem or find solace in the company of friends.

_____ What is the author's thought pattern?
a. time order b. classification

EXPLANATION Compare your outline to the following:

Topic sentence: There are two types of coping responses.

A. Problem-focused coping is directed toward the source of the stress.

For example, if the stress is job-related, a person might try to change conditions at the job site or take courses to acquire skills that would enable him or her to obtain a different job.

B. Emotion-focused coping is directed toward a person's own personal reaction to a stressor.

For example, a person might try to relax and forget about the problem or find solace in the company of friends.

The topic is "coping responses." The thought pattern is signaled by the phrase "two types." The transitions clearly carry out the thought pattern by introducing the major supporting details (the two types of coping responses) with the phrases "first type" and "second type." In addition, the author provides minor details in the form of examples, and each example is introduced with the phrase "for example." In this paragraph, these transitions establish the (b) classification thought pattern, which is discussed later in the chapter.

PRACTICE 3

Read the following paragraph. Then, complete the informal outline.

Textbook Skills

Biodiversity Loss and Species Extinction

[1]Biodiversity at all levels is being lost to human impact, most irretrievably in the extinction of species. [2]Once vanished, a species can never

return. [3]**Extinction** occurs when the last member of a species dies and the species ceases to exist, as apparently was the case with Monteverde's golden toad. [4]The disappearance of a particular population from a given area, but not the entire species globally, is referred to as **extirpation**. [5]For example, the tiger has been extirpated from most of its historic range, but it is not yet extinct. [6]However, extirpation is an erosive process that can, over time, lead to extinction.

—From *Essential Environment: The Science Behind the Stories,* 3rd ed., p. 165 by Jay H. Withgott and Scott R. Brennan. Copyright © 2009 by Pearson Education, Inc. Reprinted by permission of Pearson Education, Inc., Upper Saddle River, NJ.

Topic sentence: ______________________________

____

a. ______________________________

____

b. ______________________________

____

1. ______________________________

____

2. ______________________________

____

In this chapter, we discuss four common thought patterns and the transition words and phrases used to signal each:

- The time order pattern
- The space order pattern
- The listing pattern
- The classification pattern

Some additional common thought patterns are covered in Chapter 8.

The Time Order Pattern

The **time order** thought pattern generally shows a chain of events. The actions or events are listed in the order in which they occur. This is called *chronological* *order*. Two types of chronological order are narration and process. An author uses narration to describe a chain of points such as a significant event in history or a story. The second type of chronological order is process. Process is used to give directions to a task in order, like in steps, stages, or directions.

Narration: A Chain of Events

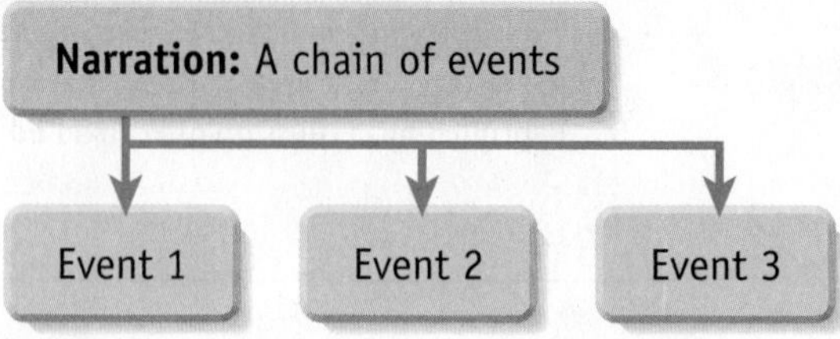

Transitions of **time** signal that the writer is describing when things occurred and in what order. The writer presents an event and then shows when each of the additional details or events flowed from the first event. Thus the details follow a logical order based on time.

Stretching your muscles *before* they are warmed up may cause injury.

Notice that this sentence warns about the relationship between stretching muscles, warming up muscles, and possible injury to muscles. The transition word tells us the order in which certain actions should occur and why. Muscles should be warmed up *before* they are stretched to avoid injury.

Transitions Used in the Time Order Pattern for Events or Stages				
after	during	last	often	then
afterward	eventually	later	previously	ultimately
as	finally	meanwhile	second	until
before	first	next	since	when
currently	immediately	now	soon	while

EXAMPLE Determine the logical order of the following seven sentences from a college history textbook. Write **1** in front of the sentence that should come first, **2** by the sentence that should come second, **3** by the sentence that should come third, and so on. (**Hint:** Circle the time transition words.)

The Beringia Land Bridge

_____ Eventually, this process dramatically lowered ocean levels.

_____ Ultimately, a land bridge emerged in the area of the Bering Strait.

_____ Today, 56 miles of ocean separate Siberia from Alaska.

_____ At times this land bridge between Asia and America, *Beringia*, may have been 1,000 miles wide.

_____ In turn, these ice caps spread over vast reaches of land.

_____ Year after year water being drawn from the oceans formed into mighty ice caps.

_____ The world was a much colder place 75,000 years ago; a great ice age, known as the Wisconsin glaciation, had begun.

—Adapted from Martin et al., *America and Its People*, 3rd ed., p. 5.

EXPLANATION Compare your answers to the sentences arranged in the order used by the author of a college history textbook. The transitions are in **bold** type.

The Beringia Land Bridge

Textbook Skills

[1]The world was a much colder place 75,000 years ago; a great ice age, known as the Wisconsin glaciation, had begun. [2]**Year after year** water being drawn from the oceans formed into mighty ice caps. [3]**In turn,** these ice caps spread over vast reaches of land. [4]**Eventually,** this process dramatically lowered ocean levels. [5]**Ultimately,** a land bridge emerged in the area of the Bering Strait. [6]**At times** this land bridge between Asia and America, *Beringia*, may have been 1,000 miles wide. [7]**Today,** 56 miles of ocean separate Siberia from Alaska.

—Adapted from *America and Its Peoples, Volume 1: A Mosaic in the Making,* 3rd ed., p. 5 by James Kirby Martin, Randy J. Roberts, Steven Mintz, Linda O. McMurry and James H. Jones. Copyright © 1997 by Pearson Education, Inc. Printed and Electronically reproduced by permission of Pearson Education, Inc., Upper Saddle River, NJ.

PRACTICE 4

Determine the logical order for the following six sentences. Write **1** by the sentence that should come first, **2** by the sentence that should come second, **3** by the sentence that should come third, and so on. (**Hint:** Circle the time transition words.)

Textbook Skills

Initiating

_____ During the first stage of a relationship, we make conscious and unconscious judgments about others.

_____ Although we are cautious at this stage, we have usually sized up the other person within 15 seconds.

_____ Immediately, he stereotyped and classified available women according to his own personal preferences.

_____ Finally, he began a conversation by asking, "Would you like a drink?" "Are you waiting for someone?" "You look like you need company," or "It was nice of you to save me this seat."

_____ For example, David entered a singles' bar and scanned the room for prospective dancing partners.

_____ After narrowing the field to two women, David planned his approach.

—Adapted from Barker and Gaut, *Communication*, 8th ed., p. 131.

Process: Steps, Stages, or Directions

The process thought pattern for steps, stages, or directions shows actions that can be repeated at any time with similar results. This pattern is used to provide steps or give directions for completing a task.

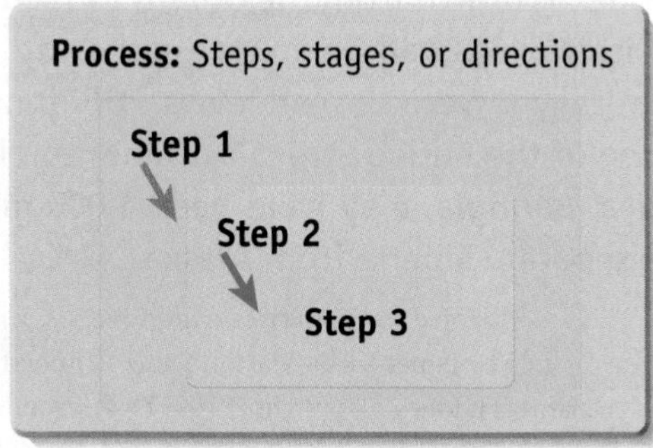

Read the following topic sentences. Circle the words that signal process time order.

1. Follow six simple steps to lose weight.
2. Photosynthesis is the process by which plants harness light energy from the sun.
3. Effective communication occurs through the cycle of sending and receiving messages.

Sentence 1 uses the word *steps* to introduce directions for the reader to follow. Sentence 2 uses the word *process* to indicate that these are stages plants undergo in photosynthesis. Sentence 3 signals a pattern of giving and receiving messages in communication using the word *cycle*. In paragraphs that developed these topic sentences, transitions of time order would likely signal the supporting details.

Transitions Used in the Time Order Pattern for Process				
after	during	later	previously	ultimately
afterward	eventually	meanwhile	second	until
as	finally	next	since	when
before	first	now	soon	while
currently	last	often	then	

EXAMPLE The following passage from the government website Small Business Administration uses the time order process to offer advice to people who are starting up their own small businesses. Read the passage and list the major steps to time management for small business owners. (**Hint:** Circle the time order transition words.)

Making Time

[1]The first step in learning how to manage your time is to develop a general work schedule. [2]Your work schedule should include time for yourself as well as time for the maintenance of your business.

[3]After you've defined the major elements of your workload, the next step is to prioritize them by identifying critical deadlines, routine maintenance items, and fun/relaxation time. [4]Answering questions like "How much time do I have to make this decision, finish this task, or contact this person?" will help you to start identifying what needs to be done immediately versus what can wait. [5]Setting priorities depends on deadlines, how many people you must call to get the information you need, and whether you can delegate or get assistance from others. [6]If you are involved in group projects, reserve additional time for communication and problem-solving.

[7]Once you have identified your priorities, look at all of your options for achieving them. [8]Evaluate and move forward with the ones you feel are the most useful for you. [9]The only time to consider changing approaches mid-task is when you know the change will save time. [10]If you are in doubt, it is usually best to consider staying in the direction you started.

[11]By setting up your work schedule and identifying your priorities, you have already started down the road to more effective time management.

—U.S. Small Business Association. "Manage Your Business." 18 March 2010.

VISUAL VOCABULARY

A successful business person must **prioritize** goals. The best synonym for prioritize is ______________.

a. complete.
b. rank.
c. identify.

I. ____________________

____

II. ____________________

____

____

III. ____________________

____

EXPLANATION Compare your answers to the following:

I. The first step in learning how to manage your time is to develop a general work schedule.

II. After you've defined the major elements of your workload, the next step is to prioritize them by identifying critical deadlines, routine maintenance items, and fun/relaxation time.

III. Once you have identified your priorities, look at all of your options for achieving them.

PRACTICE 5

The following paragraph uses the time order pattern for steps or directions to organize its ideas. Complete the list of steps that follows it by giving the missing details in their proper order. (**Hint:** Circle the time order transition words.)

Seirta's System to Collect and Organize Information

[1]During her first week of attending her composition, psychology, and history classes, Seirta noted that each course required class discussion and papers based on current events and how those events relate to the information studied in each course. [2]Seirta liked the idea of connecting class work to real life; thus she devised and carried out the following plan. [3]First, she skimmed the two newspapers delivered to her home, looking for headlines that tied in to the topics discussed that week in her classes (one paper was the local newspaper; the other was *USA Today*). [4]As she skimmed, if a piece seemed relevant, she placed a large check with a red pencil next to that headline. [5]Later, after all her family members had read the papers, she cut out each article she had checked. [6]While doing so, she was also careful to record the following information directly on the cut-out article: the name of the newspaper, the date of the newspaper, and the section and page number(s) where the article appeared. [7]Next, she carefully read each article and underlined relevant details with a yellow highlighter. [8]Often she quickly wrote a paragraph explaining how the article related to a particular concept covered in class. [9]She stapled her response to the article. [10]Then she placed the article in a file folder labeled with the topic of the article. [11]For example, one file folder was labeled "Women in Politics," another was labeled "Gang Violence," and a third was labeled "Stress." [12]As the semester progressed, she added more articles to each folder or created new folders for new topics. [13]Seirta deliberately worked quickly so the entire process took less than 30 minutes. [14]Every day, Seirta carried her folders of information with her to class and referred to them during class discussion. [15]When the time came to write an essay, Seirta used much of the information she had collected in her file.

Seirta's System to Collect and Organize Information

Step 1: She skimmed the two newspapers delivered to her home, looking for headlines that tied in to the topics discussed that week in her classes.

Step 2: ______________________________

____

Step 3: ______________________________

____

Step 4: ______________________________

____

____

____

Step 5: Next, she carefully read each article and underlined relevant details with a yellow highlighter.

Step 6: Often she quickly wrote a paragraph explaining how the article related to a particular concept covered in class.

Step 7: She stapled her response to the article.

Step 8: ______________________________

____

Step 9: As the semester progressed, she added more articles to each folder or created new folders for new topics.

Step 10: ______________________________

____

Step 11: When the time came to write an essay, Seirta used much of the information she had collected in her file.

The Space Order Pattern

The **space order pattern** allows authors to describe a person, place, or thing based on its location or the way it is arranged in space. In the space order pattern, also known as spatial order, the writer often uses descriptive details to help readers create vivid mental pictures of the subject being described. An author may choose to describe an object from top to bottom, from bottom to top, from right to left, from left to right, from near to far, from far to near, from inside to outside, or from outside to inside.

Space Order: Descriptive Details

Descriptive detail 1 → Descriptive detail 2 → Descriptive detail 3

Transition words of **space order** signal that the details follow a logical order based on two elements: (1) how the object, place, or person is arranged in space, and (2) the starting point from which the author chooses to begin the description.

Transition Words Used in the Space Order Pattern						
above	at the side	below	center	front	middle	there
across	at the top	beneath	close to	here	nearby	under
adjacent	back	beside	down	in	next to	underneath
around	backup	beyond	far away	inside	outside	within
at the bottom	behind	by	farther	left	right	

EXAMPLE Determine a logical order for the following six sentences. Write **1** by the sentence that should come first, **2** in front of the sentence that should come second, **3** in front of the sentence that should come third, **4** in front of the sentence that should come fourth, **5** in front of the sentence that should come fifth, and **6** in front of the sentence that should come last. (**Hint:** Circle the space order transition words.)

_____ In a tropical rain forest, vertical stratification, or layering, provides several discernible strata or layers of plant growth.

_____ At the upper canopy level, the treetops range from 100 to 160 feet, out of which emergents may soar to 200 feet.

_____ The middle layer contains another level of treetops that are covered with vines and air plants.

_____ The lowest and the thinnest layer is the forest floor of seedlings, shoots, and herbaceous plants.

_____ Between the middle layer and the forest floor is the shrub understory; thus very little light passes through the dense middle layer to the forest floor.

_____ This upper canopy is uneven, allowing sunlight to filter through treetops to the middle layer.

EXPLANATION Compare your answers to the sentences arranged in the proper order in the paragraph below. The transition words are in **bold** print.

[1]**In** a tropical rain forest, vertical stratification, or layering, provides several discernible strata or layers of plant growth. [2]**At the upper** canopy level, the treetops range from 100 to 160 feet, **out of** which emergents may soar to 200 feet. [3]This **upper** canopy is uneven, allowing sunlight to filter through treetops to the middle layer. [4]The **middle** layer contains another level of treetops that are covered with vines and air plants. [5]**Between the middle layer and the forest floor** is the shrub **understory**; thus very little light passes through the dense **middle** layer to the forest **floor**. [6]The **lowest** and the thinnest layer is the forest **floor** of seedlings, shoots, and herbaceous plants.

PRACTICE 6

Textbook Skills

Determine a logical order for the following five sentences. Write **1** in front of the sentence that should come first, **2** in front of the sentence that should come second, **3** in front of the sentence that should come third, **4** in front of the sentence that should come fourth, and **5** in front of the sentence that should come last. (**Hint:** Circle the space order transition words.)

The Anatomy of a Generalized Cell

_____ In turn, the cytoplasm is enclosed by the plasma membrane.

_____ The plasma membrane forms the outer cell boundary.

VISUAL VOCABULARY

Label the three regions of a generalized cell.

a. cytoplasm
b. nucleus
c. plasma membrane

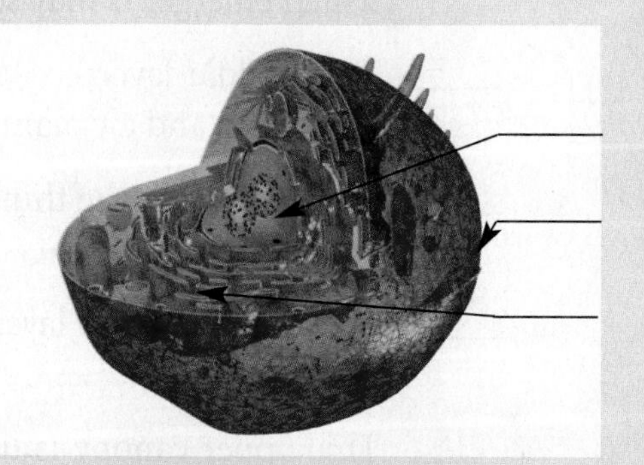

______ The nucleus is surrounded by the semifluid cytoplasm.

______ The nucleus is usually located near the center of the cell.

______ In general, all cells have three main regions; the *nucleus* (nu′kle-us), *cytoplasm* (si′to-plazm″), and a plasma membrane.

—From *Essentials of Human Anatomy and Physiology*, 9th ed., p. 66 and figure by Elaine N. Marieb. Copyright © 2009 by Pearson Education, Inc. Reprinted by permission of Pearson Education, Inc., Glenview, IL.

The Listing Pattern

Often authors want to present an orderly series or set of reasons, details, or points. These details are listed in an order that the author has chosen. Changing the order of the details does not change their meaning. Transitions of addition, such as *and, also,* and *furthermore,* are generally used to indicate a *listing pattern.*

Listing Pattern
Idea 1
Idea 2
Idea 3

Transitions of **addition** signal that the writer is adding to an earlier thought. The writer presents an idea and then *adds* other ideas to deepen or clarify the first idea. Thus, transitions of addition are used to establish the listing pattern.

Addition Transitions Used in the Listing Pattern				
also	final	for one thing	last of all	second
and	finally	furthermore	moreover	third
another	first	in addition	next	
besides	first of all	last	one	

EXAMPLE Refer to the box of addition transitions used in the listing pattern. Complete the following paragraph with transitions that show the appropriate relationship between sentences.

Benefits of Weightlifting

Weightlifting is an exercise that offers many benefits. **(1)** __________, increased muscle mass causes the body to burn fat, and lowered body fat reduces several health risks. **(2)** ______________, weightlifting builds bone density and thus reduces the risk of osteoporosis (brittle bones). **(3)** ______________, weightlifting increases strength, independence, and self-confidence.

EXPLANATION Compare your answers to the following:

Benefits of Weightlifting

[1]Weightlifting is an exercise that offers many benefits. [2]**First**, increased muscle mass causes the body to burn fat, and lowered body fat reduces several health risks. [3]**In addition**, weightlifting builds bone density and thus reduces the risk of osteoporosis (brittle bones). [4]**Finally,** weightlifting increases strength, independence, and self-confidence.

The paragraph on the benefits of weightlifting begins with a general idea that is then followed by three major supporting details. Each detail requires a transition to show addition.

PRACTICE 7

The following paragraph adapted from a college health textbook uses the listing thought pattern. Finish the outline that follows it by listing the major supporting details in their proper order. (**Hint:** Circle the addition transition words.)

Life Style Assessment

Textbook Skills

[1]Life style assessment is an important life skill. [2]First, life style assessment raises "awareness of areas that increase risk of disease, injury, and possibly premature death." [3]For example, examining specific patterns of behavior such as eating and sleeping allows us to be sure we are acting in healthy ways. [4]In addition, life style assessment increases the likelihood of making more positive life style choices in areas that have been unhealthy. [5]For example, smokers who think seriously about how many cigarettes they smoke in a day, how much money they spend weekly on smoking, and whether they have a chronic smoker's cough will understand the costs and risks of smoking better. [6]Once the risks become

real, our behaviors are easier to change. [7]Besides assessing eating, sleeping, and patterns of addiction, life style assessment looks at other patterns of behavior related to stress, relationships, physical fitness, and so on. [8]The final point to remember is that you have control over your life style choices.

—Adapted from Powers & Dodd, *Behavior Change Log Book*, 3rd ed., pp. 1–3.

Life Style Assessment

Life style assessment is an important life skill.

A. ______________________________

____

B. ______________________________

____

C. ______________________________

____

____

D. ______________________________

____

The Classification Pattern

Authors use the **classification pattern** to sort ideas into smaller groups and describe the traits of each group. Each smaller group, called a *subgroup*, is based on shared traits or characteristics. The author lists each subgroup and describes its traits.

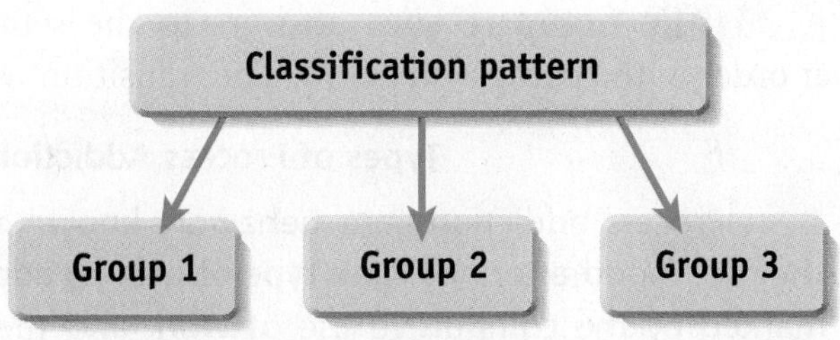

Because groups and subgroups are listed, transitions of addition are also used in this thought pattern. These transitions are coupled with words that indicate classes or groups. Examples of classification signal words are *first type*, *second kind*, *another group*, *order*, and *traits*.

Transitions Used in the Classification Pattern	
another (group, kind, type)	first (group, category, kind, type)
characteristics	second (group, class, kind, type)

EXAMPLE Determine a logical order for the following sentences. Write **1** in front of the sentence that should come first, **2** in front of the sentence that should come second, **3** in front of the sentence that should come third, and **4** in front of the sentence that should come last. (**Hint:** Circle the classification signal words.)

Textbook Skills

Types of Process Addictions

______ Another type of process addiction is money addiction, which includes gambling, spending, and borrowing; money addicts develop tolerance and also experience withdrawal symptoms such as depression, anxiety, and anger.

______ One type of process addiction is work addiction, marked by the compulsive use of work and the work persona to fulfill needs of intimacy, power, and success.

______ A third type of this kind of addiction is exercise addiction; addictive exercisers abuse exercise the same way that alcoholics abuse alcohol and face effects similar to those found in other addictions: alienation of family and friends, injuries from overdoing it, and a craving for more.

______ Process addictions are behaviors known to be addictive because they are mood-altering.

—Adapted from *Access to Health*, 7th ed., pp. 317–320 by Rebecca J. Donatelle and Lorraine G. Davis. Copyright © 2002 by Pearson Education, Inc. Printed and Electronically reproduced by permission of Pearson Education, Inc., Upper Saddle River, NJ.

EXPLANATION Compare your answers to the sentences arranged in their proper order in the paragraph below. The transition words are in **bold** type.

Types of Process Addictions

[1]Process addictions are behaviors known to be addictive because they are mood-altering. [2]**One type** of process addiction is work addiction, marked by the compulsive use of work and the work persona to fulfill

needs of intimacy, power, and success. [3]**Another type** of process addiction is money addiction, which includes gambling, spending, and borrowing; money addicts develop tolerance and also experience withdrawal symptoms such as depression, anxiety, and anger. [4]A **third type** of this kind of addiction is exercise addiction; addictive exercisers abuse exercise the same way that alcoholics abuse alcohol and face effects similar to those found in other addictions: alienation of family and friends, injuries from overdoing it, and a craving for more.

In this paragraph, transitions of addition work with the classification signal words. In this case, *one*, *another*, and *third* convey the order of the types listed.

PRACTICE 8

The following passage uses the classification pattern of organization. Fill in the outline that follows by giving the missing major details in their proper order. (**Hint:** Circle the classification transition words.)

Types of Volcanoes

[1]Geologists generally group volcanoes into four main kinds—cinder cones, composite volcanoes, shield volcanoes, and lava domes. [2]The first type, the cinder cone volcanoes, are the simplest type of volcano. [3]They are built from particles and blobs of congealed lava ejected from a single vent. [4]As the gas-charged lava is blown violently into the air, it breaks into small fragments that solidify and fall as *cinders* around the vent to form a circular or oval cone. [5]Most cinder cones have a bowl-shaped *crater* at the summit and rarely rise more than a thousand feet or so above their surroundings.

[6]The second type of volcano is the *composite* volcano. [7]The essential feature of a composite volcano is a conduit system through which magma from a reservoir deep in the Earth's crust rises to the surface. [8]The volcano is built up by the accumulation of material erupted through the conduit and increases in size as lava, cinders, ash, and so on, are added to its slopes. [9]Most composite volcanoes have a crater at the summit which contains a central vent or a clustered group of vents. [10]Lavas either flow through breaks in the crater wall or issue from fissures on the flanks of the cone. [11]Lava, solidified within the fissures, forms dikes that act as ribs, which greatly strengthen the cone.

[12]Shield volcanoes, the third type of volcano, are built almost entirely of fluid lava flows. [13]Flow after flow pours out in all directions from a central

summit vent, or group of vents, building a broad, gently sloping cone of flat, domical shape, with a profile much like that of a warrior's shield. [14]They are built up slowly by the accretion of thousands of highly fluid lava flows called basalt lava that spread widely over great distances, and then cool as thin, gently dipping sheets. [15]Lavas also commonly erupt from vents along fractures (rift zones) that develop on the flanks of the cone.

[16]The fourth type, volcanic or lava domes, are formed by relatively small, bulbous masses of lava too viscous to flow any great distance; consequently, on extrusion, the lava piles over and around its vent. [17]A dome grows largely by expansion from within. [18]As it grows, its outer surface cools and hardens, then shatters, spilling loose fragments down its sides. [19]Some domes form craggy knobs or spines over the volcanic vent, whereas others form short, steep-sided lava flows known as "coulees." [20]Volcanic domes commonly occur within the craters or on the flanks of large composite volcanoes.

—Adapted from Tilling, "Volcanoes," 6 Feb. 1997, U.S. Geological Survey. 22 Dec. 2003 http://pubs.usgs.gov/gip/volc/types.html

Types of Volcanoes

I. ______________________________

II. ______________________________

III. ______________________________

IV. ______________________________

VISUAL VOCABULARY

Which type of volcano is this Hawaiian volcano?

____

a. cinder cone
b. shield
c. composite
d. lava dome

Reading a Textbook: Thought Patterns in Textbooks

Textbook Skills

Textbook authors often use transitions to make relationships between ideas clear and easy to understand. However, often an author will use more than one type of transition. For example, classification combines words that indicate addition and types. Sometimes addition and time words are used in the same paragraph or passage for a specific purpose. Furthermore, authors may mix thought patterns in the same paragraph or passage. Finally, be aware that relationships between ideas still exist even when transition words are not explicitly stated. The master reader looks for the author's primary thought pattern.

EXAMPLE Read the following paragraphs from college textbooks. Circle the transitions or signal words used in each paragraph. Then, identify the primary thought pattern used in the paragraph.

A. Reinforcement

[1]A positive reinforcement is a reward that is given to increase the likelihood that a behavior change will occur. [2]Most positive reinforcers can be classified under five headings. [3]The first type is consumable reinforcers, which are delicious edibles, such as candy, cookies, or gourmet meals. [4]Second, activity reinforcers are opportunities to do something enjoyable, such as to watch TV, go on a vacation, or go swimming. [5]Third, manipulative reinforcers are incentives, such as getting a lower rent in exchange for mowing the lawn or the promise of a better grade for doing an extra-credit project. [6]A fourth type of positive reinforcer is the possessional reinforcers, which are tangible rewards, such as a new TV or a sports car. [7]The final type consists of social reinforcers that express signs of appreciation, approval, or love, such as loving looks, affectionate hugs, and praise.

—Adapted from *Health: The Basics*, 5th ed., p. 21 by Rebecca J. Donatelle. Copyright © 2003 by Pearson Education, Inc. Printed and Electronically reproduced by permission of Pearson Education, Inc. Upper Saddle River, NJ.

_____ The primary thought pattern of the paragraph is

a. time order. b. classification.

B. Interpersonal Perception

[1]Interpersonal perception is a continuous series of processes that blend into one another. [2]For convenience of discussion we can separate them into five stages. [3]During the first stage, you sense, you pick up some

kind of stimulation. [4]Next, you organize the stimuli in some way. [5]Third, you interpret and evaluate what you perceive. [6]Then, you store it in memory, and finally, in stage five, you retrieve it when needed.

—Adapted from *The Interpersonal Communication Book,* 10th ed., p. 91 by Joseph A. DeVito. Copyright © 2004 by Pearson Education, Inc. Reproduced by permission of Pearson Education, Inc., Boston, MA.

_____ The primary thought pattern of the paragraph is

a. time order. b. listing.

EXPLANATION Compare your answers to the following answers.

Passage A, "Reinforcement," is organized by (b) classification.

Passage B, "Interpersonal Perception," is organized by (a) time order.

PRACTICE 9

Read the following paragraph taken from a college communications textbook. Circle the transitions or signal words used in the passage. Then, identify the primary thought pattern used in the paragraph.

Textbook Skills

Skill Acquisition

[1]You may develop communication apprehensions largely because you see yourself as having inadequate skills. [2]So you logically fear failing. [3]The following suggestions will enable you to acquire the skills necessary to master your fears. [4]First, prepare and practice. [5]The more preparation and practice you put into something, the more comfortable you feel with it, and, thus, the less anxious you feel. [6]Second, focus on success. [7]Think positively. [8]Concentrate your energies on doing the very best job you can in whatever situation you are in. [9]Visualize yourself succeeding, and you stand a good chance of doing just that. [10]Third, familiarize yourself with the situation. [11]The more familiar you are with the situation, the better. [12]The reason is simple: When you're familiar with the situation and with what will be expected of you, you're better able to predict what will happen.

—Adapted from *The Interpersonal Communication Book,* 10th ed., p. 87 by Joseph A. DeVito. Copyright © 2004 by Pearson Education, Inc. Reproduced by permission of Pearson Education, Inc., Boston, MA.

_____ The primary thought pattern used in the paragraph is

a. time order. b. listing.

APPLICATIONS

Application 1: Identifying Transitions

Fill in each blank with one of the words from the box below. Use each word once.

during	eventually	now	within

(1) _______________ the last ice age (about 10,000 years ago), the Sahara was largely savanna and supported animals like giraffes, lions, and elephants, which are now found in East Africa, plus wandering bands of hunters. With the warming and drying trend of the past 10,000 years, the savanna gave way to desert grassland, which supported herds of cattle. The herders and the cattle **(2)** _______________ disappeared as the desert reclaimed the land, from about 8,000 years ago to the present.

In Roman times, however, the region just east of the Atlas Mountains was still able to support irrigated fields of wheat, and the area which is **(3)** _______________ stark desert was known as the bread basket of the Roman Empire.

Today, an estimated 2 million people live **(4)** _______________ the Sahara Desert. About two-thirds of these people are concentrated in oases, and the remaining one-third are nomadic people who travel throughout the desert.

—Adapted from "People," *Sahara*, PBS. org, December 29, 2003, http://www.pbs.org/sahara/people/people.htm Reprinted by permission of Telenova Productions.

Application 2: Identifying Thought Patterns

Textbook Skills

Identify the thought pattern suggested by each of the following topic sentences taken from college textbooks.

_____ **1.** It is convenient to divide up conversation into chunks or stages and view each stage as a choice as to what you'll say and how you'll say it.

—Adapted from *The Interpersonal Communication Book,* 10th ed., p. 214 by Joseph A. DeVito. Copyright © 2004 by Pearson Education, Inc. Reproduced by permission of Pearson Education, Inc., Boston, MA.

a. time order b. space order

_____ **2.** The use of genetically modified organisms in agriculture is controversial for two major reasons.

—From *Life on Earth,* 5th ed., p. 207 by Teresa Audesirk, Gerald Audesirk, and Bruce E. Byers. Copyright © 2009 by Pearson Education, Inc. Printed and Electronically reproduced by permission of Pearson Education, Inc., Upper Saddle River, NJ.

a. time order b. listing

_____ **3.** The most popular types of insurance are term insurance, whole life insurance, and universal life insurance.

—From *Personal Finance,* 2nd ed., p. 345 by Jeff Madura. Copyright © 2004 by Pearson Education, Inc. Reproduced by permission of Pearson Education, Inc., Boston, MA.

a. time order b. classification

_____ **4.** The civilization that would become the Roman Empire arose at the same time as that of ancient Greece.

—From *The Creative Impulse: An Introduction to the Arts,* 8th ed., p. 113 by Dennis J. Sporre. Copyright © 2009 by Pearson Education, Inc. Reprinted by permission of Pearson Education, Inc., Upper Saddle River, NJ.

a. time order b. listing

_____ **5.** English builds all sentences on five basic sentence patterns.

—Fowler and Aaron, *The Little, Brown Handbook,* 9th ed., p. 243.

a. space order b. classification

REVIEW TEST 1

Score (number correct) _____ × 10 = _____%

Transitions and Thought Patterns

Fill in each blank with a transition from the box. Use each transition once. Then, answer the questions that follow the paragraph.

already	and	in addition	when
also	another	one	

Textbook Skills

Barriers to Memory

It is likely that you will remember things that support your position **(1)** _______________ forget or distort information that contradicts your current beliefs. **(2)** _______________ reason, of course, may be selective attention; you focus on things you want to hear or things you expect to hear. Distortion can **(3)** _______________ be explained, in part, through schema theory. Ideas that fit into your existing framework are stored more readily than bits of information that don't make sense in light of what you **(4)** _______________ know. **(5)** _______________, threatening or unpleasant experiences are often blocked, or repressed, while positive images remain vivid. **(6)** _______________ reason why you forget is that the information you are trying to store in long-term memory becomes mixed with information already in your memory system. **(7)** _______________ these two information sets become confused, there is a "backward" impact of new learning on the material stored earlier.

—Adapted from *Listening: Attitudes, Principles, and Skills*, 2nd ed. pp. 155–156 by Judi Brownell. Copyright © 2002 by Pearson Education, Inc. Reproduced by permission of Pearson Education, Inc., Boston, MA.

_____ **8.** The relationship signaled by the word **already** is
a. time order. b. classification.

_____ **9.** The relationship signaled by the transition word **one** is
a. addition. b. space order.

_____ **10.** The overall thought pattern of the paragraph is
a. listing. b. time order.

REVIEW TEST 2

Score (number correct) ______ × 10 = ______%

Transitions

Select a transition word for each of the blanks. Then, identify the type of transition you chose.

A. The thief had bought thousands of dollars worth of goods with Aimee's credit card ______________ she even knew it was missing.

______ **1.** The best transition word for sentence A is
a. meanwhile.
b. after.
c. before.

______ **2.** The relationship between the ideas in sentence A is one of
a. classification.
b. time order.

B. Every cell has a plasma membrane which forms its ______________ border and sets the cell off from its fluid environment.

______ **3.** The best transition word for the sentence above is
a. another.
b. outside.
c. one.

______ **4.** The relationship between the ideas above is one of
a. space order.
b. addition.

C. Everything inside a cell ______________ the plasma membrane and the nucleus is called the cytoplasm.

______ **5.** The best transition word for the sentence above is
a. between.
b. before.
c. since.

______ **6.** The relationship between the ideas above is one of
a. listing.
b. space order.

D. Poetry can be broken into two very broad groups. The first type, dramatic poetry, tells a story using characters, setting, and conflict. The ______________, lyric poetry, directly expresses the emotions of the poet.

______ **7.** The best transition word or phrase for the sentences in section D is
a. additional part.
b. second type.
c. later time.

______ **8.** The relationship between the ideas in the sentences in section D is one of

a. classification. b. time order.

E. [1]For producing electricity, hydropower has two clear advantages over fossil fuels. [2]First, it is renewable; as long as precipitation falls from the sky and fills rivers and reservoirs, we can use water to turn turbines. [3]The ________ advantage of hydropower over fossil fuels is its cleanliness. [4]No carbon compounds are burned in the production of hydropower, so no carbon dioxide or other pollutants are emitted into the atmosphere. [5]Of course, fossil fuels are used in constructing and maintaining dams. [6]Moreover, recent evidence indicates that reservoirs release the greenhouse gas methane as a result of anaerobic decay in deep water. [7]Overall, hydropower accounts for only a fraction of the greenhouse gas emissions typical of fossil fuel combustion.

—From *Essential Environment: The Science Behind the Stories,* 3rd ed., p. 363 by Jay H. Withgott and Scott R. Brennan. Copyright © 2009 by Pearson Education, Inc. Reprinted by permission of Pearson Education, Inc., Upper Saddle River, NJ.

______ **9.** The best transition word for the blank in sentence 3 is

a. later.
b. addition.
c. second.

______ **10.** The main thought pattern of the paragraph is

a. time order. b. listing.

VISUAL VOCABULARY

The Grand Coulee Dam in Washington State, the largest electric power-producing facility in the United States, is an example of hydropower. The best meaning of hydro is

______________.

a. electric.
b. pressure.
c. wind.
d. water.

REVIEW TEST 3

Score (number correct) _____ × 5 = _____ %

Transitions and Thought Patterns

Textbook Skills

The following passage appears in a college psychology textbook. Before you read, skim the passage and answer the Before Reading questions. Read the passage. Then, answer the After Reading questions.

Categorizing Personality by Types

[1]We are always categorizing people according to distinguishing features. [2]These include college class, academic major, sex, and race. [3]Some personality theorists also group people into distinct, nonoverlapping categories that are called personality **types**. [4]Personality types are all-or-none **phenomena**, not matters of degree: If a person is assigned to one type, he or she could not belong to any other type within that system. [5]Many people like to use personality types in everyday life because they help simplify the complex process of understanding other people.

[6]One of the earliest type theories was originated in the 5th century B.C. by Hippocrates, the Greek physician who gave medicine the Hippocratic Oath. [7]He theorized that the body contained four basic fluids, or **humors**, each associated with a particular temperament, a pattern of emotions and behaviors. [8]In the 2nd century A.D., a later Greek physician, Galen, suggested that an individual's personality depended on which humor was predominant in his or her body. [9]Galen paired Hippocrates's body humors with personality temperaments according to the following scheme:

- [10]**Blood**. Sanguine temperament: cheerful and active
- [11]**Phlegm. Phlegmatic** temperament: apathetic and sluggish
- [12]**Black bile**. Melancholy temperament: sad and brooding
- [13]**Yellow bile**. Choleric temperament: irritable and excitable

[14]The theory proposed by Galen was believed for centuries, up through the Middle Ages, although it has not held up to modern scrutiny.

[15]In modern times, William Sheldon (1898–1977) originated a type theory that related physique to temperament. [16]Sheldon (1942) assigned people to three categories based on their body builds: *endomorphic* (fat, soft, round), *mesomorphic* (muscular, rectangular, strong), or *ectomorphic* (thin, long, fragile). [17]Sheldon believed that endomorphs are relaxed, fond of eating, and sociable. [18]Mesomorphs are physical people, filled with energy, courage, and assertive tendencies. [19]Ectomorphs are brainy, artistic, and introverted; they would think about life, rather than consuming it or acting

on it. [20]For a period of time, Sheldon's theory was sufficiently influential that nude "posture" photographs were taken of thousands of students at U.S. colleges like Yale and Wellesley to allow researchers to study the relationships between body type and life factors. [21]However, like Hippocrates's much earlier theory, Sheldon's notion of body types has proven to be of very little value in predicting an individual's behavior (Tyler, 1965).

[22]More recently, Frank Sulloway (1996) has proposed a contemporary type theory based on birth order. [23]Are you the firstborn child (or only child) in your family, or are you a laterborn child? [24]Because you can take on only one of these birth positions, Sulloway's theory fits the criteria for being a type theory. [25](For people with unusual family constellations—for example, a very large age gap between two children—Sulloway still provides ways of categorizing individuals.) [26]Sulloway makes birth-order predictions based on Darwin's idea that organisms diversify to find niches in which they will survive. [27]According to Sulloway, firstborns have a ready-made niche: They immediately command their parents' love and attention; they seek to maintain that initial attachment by identifying and complying with their parents. [28]By contrast, laterborn children need to find a different niche—one in which they don't so clearly follow their parents' example. [29]As a consequence, Sulloway characterizes laterborns as "born to rebel": "they seek to excel in those domains where older siblings have not already established superiority. [30]Laterborns typically cultivate openness to experience—a useful strategy for anyone who wishes to find a novel and successful niche in life" (Sulloway, 1996, p. 353).

—From *Psychology and Life,* 19th ed., pp. 407–408 by Richard J. Gerrig and Philip G. Zimbardo.

Before Reading

Vocabulary in Context

______ **1.** The word **phenomena** in sentence 4 means
a. spectacles. c. facts.
b. experiences. d. features.

______ **2.** The best meaning of the word **humors** in sentence 7 is
a. funny qualities. c. effects of body fluids.
b. emotions. d. behaviors.

______ **3.** The best synonym for the word **phlegmatic** in sentence 10 is
a. unmotivated. c. calm.
b. sickly. d. determined.

After Reading

Concept Maps and Graphic Organizers

4–15. Complete the following two-column notes with information from the passage.

Categorizing Personality by Types	
Theorist	**Theories of Personality Type**
4. ____________________	**5.** ____________________ cheerful and active **6.** Phlegm. Phlegmatic temperament: ____________________ Black bile. Melancholy temperament: sad and brooding **7.** ____________________ ____________________
8. ____________________	**9.** ____________ (fat, soft, round) **10.** Mesomorphic ____________________ **11.** ____________________
12. ____________________	**13.** ________ command parents' love and attention; identify and comply with parents **14.** ________ are born to rebel; they seek to excel in those domains where older siblings have not already established superiority

Central Ideas

_____ **15.** Which sentence states the central idea of the essay?

a. sentence 1
b. sentence 2
c. sentence 3
d. sentence 5

Supporting Details

_____ **16.** Sentence 20 states a
a. a main idea.
b. major supporting detail.
c. minor supporting detail.

Transitions and Thought Patterns

_____ **17.** What thought pattern does the word **types** in sentence 3 signal?
a time order
b. classification
c. space order

_____ **18.** What is the relationship of ideas within sentence 16?
a. listing
b. time order
c. space order

_____ **19.** What is the relationship of ideas between sentence 7 and sentence 8?
a. time order
b. classification
c. listing

_____ **20.** Based on the title, the main thought pattern of the passage is
a. time order.
b. space order.
c. classification.

WHAT DO YOU THINK?

Do you know people whom you would label as particular "types"? What particular "type" are you? Does "type" include all there is to know about the person? Assume you are taking a college level psychology class, and your professor has given you the following assignment: Choose a fictional character from a television show, movie, book, graphic novel, etc. Analyze the character using each of the personality types described in the passage. Describe the moods and behaviors of the character to support your use of each label.

REVIEW TEST 4

Score (number correct) _____ × 10 = _____ %

Transitions and Thought Patterns

Textbook Skills

Before Reading: Survey the following passage from the college textbook *Life on Earth.* Skim the passage, noting the words in **bold** print. Answer the Before Reading questions that follow the passage. Then, read the passage. Next, answer the After Reading questions. Use the discussion and writing topics as after reading activities.

Vocabulary Preview

mammalian (3): related to mammals, animals that have a spine and milk glands

mechanoreceptor (3): a nerve ending that responds to a mechanical stimulus (as in a change of pressure)

auditory (12): hearing

cerebellum (12): located at the back of the brain, the control center for muscle tone, balance, and coordination of movement

pharynx (14): a hollow tube that begins behind the nose and ends at the top of the windpipe

membrane (17): a thin barrier that surrounds a cell or parts of a cell

receptors (20): receivers

basilar (21): located at the base of a structure

cilia (22): short, hairlike projections from the surface of certain cells. The movement of cilia aids the movement of cells and fluids

The Perception of Sound Is a Specialized Type of Mechanoreception

[1]Sound is produced by any vibrating object—a drum, vocal cords, or the speaker of your CD player. [2]These vibrations, or sound waves, are transmitted through the air and intercepted by our ears, which convert them to signals that our brains interpret as the direction, pitch, and loudness of sound. [3]The **mammalian** ear consists of a variety of structures (for example, the outer ear and eardrum) that transmit vibrations to specialized **mechanoreceptor** cells deep in the inner ear.

The Ear Converts Sound Waves into Electrical Signals

[4]The ear of humans and most other vertebrates consists of three parts: the outer, middle, and inner ear. [5]The *outer ear* consists of the *pinna* and *auditory canal*. [6]The pinna is the flap of skin-covered cartilage attached to the surface of the head. [7]The pinna collects sound waves and modifies them in various ways. [8]Humans and other fairly large animals determine sound direction by differences in *when* sound arrives at the two ears and in *how loud* it is in each ear. [9]The shape and mobility of the pinna further contribute to sound localization. [10]Bats have probably the most precise sound localization in the animal kingdom. [11]Insect-eating bats emit extremely high-pitched shrieks that reflect off moths and other insect prey. [12]Large ears, a highly developed **auditory** cortex, and an enormous **cerebellum** (for precise control of flying) combine to allow the bats to intercept their flying prey in pitch darkness.

[13]The air-filled auditory canal conducts the sound waves to the *middle ear*, consisting of the *tympanic membrane*, or *eardrum;* three tiny bones called the *hammer, anvil,* and *stirrup;* and the auditory tube (also called the *Eustachian tube*). [14]The auditory tube connects the middle ear to the **pharynx** and equalizes the air pressure between the middle ear and the atmosphere.

[15]Sound waves traveling down the auditory canal vibrate the tympanic membrane, which in turn vibrates the hammer, the anvil, and the stirrup. [16]These bones transmit vibrations to

embedded (23): fixed in, set in

gelatinous (23): a jelly-like texture, quality, or consistency

potentials (25): stored energies

axons (26): large extensions of nerve cells, reaching from the cell body to the ends of other nerve cells or muscles

the *inner ear,* which contains the spiral-shaped *cochlea.* [17]The stirrup bone transmits vibrations to the fluid within the cochlea by vibrating a **membrane** on the cochlea called the *oval window.* [18]The *round window* is a second membrane that allows fluid within the cochlea to shift back and forth as the stirrup bone vibrates the oval window.

Sound Is Converted into Electrical Signals in the Cochlea

[19]The cochlea, in cross section, consists of three fluid-filled canals. [20]The central canal houses the **receptors** and the supporting structures that activate them in response to sound vibrations. [21]The floor of this central canal is the ***basilar*** *membrane*, on top of which sit mechanoreceptors called *hair cells.* [22]Hair cells have small cell bodies topped by hairlike projections that resemble stiff **cilia**. [23]Some of these hairs are **embedded** in a **gelatinous** structure called the *tectorial membrane*.

[24]The oval window passes vibrations from the bones of the middle ear to the fluid in the cochlea, which in turn vibrates the basilar membrane relative to the tectorial membrane. [25]This movement bends the hairs of the hair cells, producing receptor **potentials**. [26]The hair cells ______ release **neurotransmitters** onto neurons whose **axons** form the auditory nerve. [27]Action potentials triggered in these axons travel to auditory processing centers within the brain.

[28]The structures of the inner ear allow us to perceive *loudness* (the magnitude of sound vibrations) and *pitch* (the frequency of sound vibrations). [29]Soft sounds cause small vibrations, which bend the hairs only slightly and result in a small receptor potential and a low rate of action potentials in axons of the auditory nerve. [30]Loud sounds cause large vibrations, which cause greater bending of the hairs, a larger receptor potential, and a high rate of action potentials in the axons of the auditory nerve. [31]Loud sounds sustained for a long time can damage the hair cells, resulting in hearing loss, a fate suffered by many rock musicians and their fans. [32]In fact, many sounds in our everyday environment have the potential to damage hearing, especially if exposure to them is prolonged.

[33]The perception of pitch is a little more complex. [34]The basilar membrane resembles a harp in shape and stiffness: narrow and stiff at the end near the oval window but wider and more flexible near the tip of the cochlea. [35]In a harp, the short, tight strings produce high notes and the long, looser strings produce low notes. [36]In the basilar membrane, the **progressive** change in shape and stiffness causes each portion to

vibrate most strongly to a particular frequency of sound: high notes near the oval window and low notes near the tip of the cochlea. [37]The brain interprets signals from hair cells near the oval window as high-pitched sound, whereas signals from hair cells located progressively closer to the tip of the cochlea are interpreted as increasingly lower in pitch. [38]Young people with undamaged cochleas can detect sounds from about 30 vibrations per second (very low pitched) to about 20,000 vibrations per second (very high pitched).

—From *Life on Earth*, 5th ed., pp. 476–477 by Teresa Audesirk, Gerald Audesirk and Bruce E. Byers. Copyright © 2009 by Pearson Education, Inc. Reprinted by permission of Pearson Education, Inc., Upper Saddle River, NJ.

Before Reading

Vocabulary

1–2. Study the following chart of word parts. Read the sentences in which the words appear for context clues. Then, use your own words to define the terms.

Prefix	Meaning	Root	Meaning	Suffix	Meaning
pro-	before	*gress*	to step	*-er*	doer
trans-	across	*mitt*	to send	*-ive*	of, relating to, the quality or nature of
		neuron	nerve cell		

1. In sentence 26, **neurotransmitters** are ________________________

____.

2. In sentence 36, **progressive** means ________________________

____.

After Reading

3–5. Central Idea and Supporting Details

Complete the following study notes with information from the passage and the illustration of the ear in the Visual Vocabulary activity.

Thesis Statement: (3) ________________________

____.

Questions Based on Major Supporting Details	Answers Based on Minor Supporting Details
What does the outer ear consist of?	the flap-like pinna and the auditory canal
What does the middle ear consist of?	the tympanic membrane (eardrum); three bones called the hammer, **(4)** __________, and stirrup; and the auditory tube (Eustachian tube)
What does the inner ear consist of?	the **(5)** __________ (three fluid-filled chambers); the basilar membrane; hair cells, and the auditory nerve

Transitions and Thought Patterns

_____ **6.** The main thought pattern for the fifth paragraph (sentences 19–23) is

a. time order.
b. space order.
c. listing.

VISUAL VOCABULARY

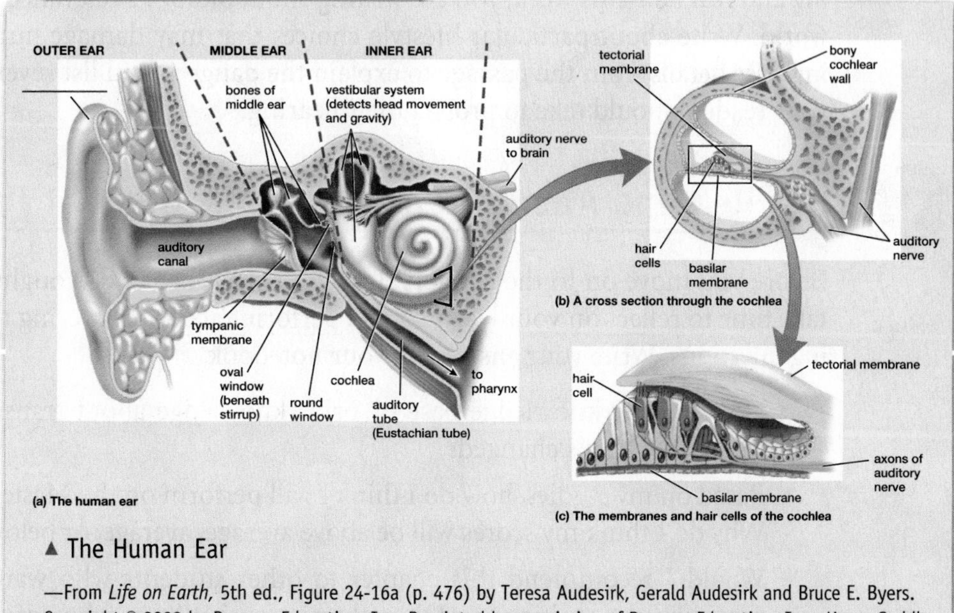

▲ The Human Ear

—From *Life on Earth,* 5th ed., Figure 24-16a (p. 476) by Teresa Audesirk, Gerald Audesirk and Bruce E. Byers. Copyright © 2009 by Pearson Education, Inc. Reprinted by permission of Pearson Education, Inc., Upper Saddle River, NJ.

_____ **7.** What is the relationship of ideas within sentence 4?
a. time order
b. space order
c. classifications

_____ **8.** What is the main relationship of ideas between sentences 13 and 14?
a. space order
b. time order
c. listing

_____ **9.** Which is the best transition word(s) for the blank in sentence 26?
a. then
b. after
c. in addition

_____ **10.** What is the relationship of ideas in sentence 34?
a. time order
b. space order
c. classification

WHAT DO YOU THINK?

How could the study of biology affect our lifestyle choices? For example, what lifestyle choices would or should the following people make after reading this passage: construction worker, factory worker, music lovers who use ear buds or attend concerts, airport baggage carriers. Assume you are writing a series of articles for your college newspaper about the relationship between classroom learning and real life. This week, you are writing about biology's relevance to the real world. Write about particular lifestyle choices that may damage human hearing. Use details from the passage to explain the dangers, and list several actions your readers should take to protect their hearing.

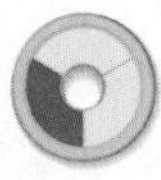

After Reading About Transitions and Thought Patterns

Before you move on to the Mastery Tests on transitions and thought patterns, take time to reflect on your learning and performance by answering the following questions. Write your answers in your notebook.

- How has my knowledge base or prior knowledge about transitions and thought patterns changed?
- Based on my studies, how do I think I will perform on the Mastery Test(s)? Why do I think my scores will be above average, average, or below average?
- Would I recommend this chapter to other students who want to learn more about transitions and thought patterns? Why or why not?

Test your understanding of what you have learned about Transitions and Thought Patterns for Master Readers by completing the Chapter 7 Review Card in the insert near the end of your text.

CONNECT TO PEARSON myreadinglab

To check your progress in meeting Chapter 7's learning outcomes, log in to **www.myreadinglab.com** and try the following exercises.

- The "Patterns of Organization" section of MyReadingLab gives additional information about transitions and patterns of organization. The section provides a model, practices, activities, and tests. To access this resource, click on the "Study Plan" tab. Then click on "Patterns of Organization." Then click on the following links as needed: "Overview," "Model," "Signal Words (Flash Animation)," "Other Patterns of Organization (Flash Animation)," "Practice," and "Tests."
- To measure your mastery of the content of this chapter, complete the tests in the "Transitions" section and click on Gradebook to find your results.

MASTERY TEST 1

Name __________ Section __________

Date __________ **Score** (number correct) ______ × 10 = ______%

A. The following information is from a college science textbook. Determine a logical order for the ideas to create a paragraph that makes sense. Indicate the proper order of ideas by writing **1, 2, 3, 4,** and **5** in the spaces provided.

Textbook Skills

______ Prey have evolved a number of defense mechanisms.

______ A fourth and more subtle defense is the timing of reproduction so that most of the offspring are produced in a short period of time.

______ One defense mechanism is cryptic coloration, which includes colors, patterns, shapes, and postures that allow prey to blend into the background.

______ A third defense mechanism employs physical means; for example, clams, armadillos, turtles, and numerous beetles all withdraw into their armor coat or shell when danger approaches.

______ Another defense is behavioral, such as alarm calls and distraction displays.

—Adapted from Smith and Smith, *Elements of Ecology*, updated 4th ed., pp. 198–199.

______ **6.** What overall thought pattern is shown in the paragraph?

a. time order
b. listing

VISUAL VOCABULARY

The best synonym for the phrase **cryptic coloration** is __________.

a. camouflage.
b. hidden.
c. colorful.

► An excellent example of cryptic coloration is the Peppered Moth.

B. Read the following paragraph from a college communications textbook. Fill in each blank (**7–9**) with the correct transition from the box. Use each transition once.

also	in addition	last

Textbook Skills

Persuasive Proof

In a persuasive speech, your support is proof, and your proof is material that offers evidence, argument, and motivational appeal. **(7)** _______________, your support establishes your credibility and reputation. You can persuade your audience with several types of support. First, logical support is built on specific examples and general concepts. Logical supports **(8)** _______________ come from comparisons and contrasts, from causes and effects, and from signs. Second, motivational support appeals to the emotions of your audience. These supports appeal to their desire for status, financial gain, or increased self-esteem. The **(9)** _______________ type, credibility appeals, are built on your own personal reputation or trustworthiness. This type of appeal rests on your skill, high moral character, and personal charm.

—Adapted from *Essentials of Human Communication*, 4th ed., p. 308 by Joseph A. DeVito. Copyright © 2002 by Pearson Education, Inc. Reproduced by permission of Pearson Education, Inc., Boston, MA.

______ **10.** What thought pattern is used in the paragraph?
a. classification
b. time order

MASTERY TEST 2

Name ______________________ Section ______________

Date __________ **Score** (number correct) ______ × 10 = ______%

A. Read the following paragraph from a college biology textbook.

Textbook Skills

How Do Biologists Study Life?

[1]Biology is the science of life. [2]Knowledge in biology is acquired through the scientific method. [3]First, an observation is made, which leads to a question. [4]Then a hypothesis is formulated that suggests a possible answer to the question. [5]The hypothesis is used to predict the outcome of further observations or experiments. [6]A conclusion is then drawn about the hypothesis. [7]Conclusions are based only on results that can be shared, verified, and repeated. [8]A scientific theory is a general explanation of natural phenomena, developed through extensive and reproducible experiments and observations.

—From *Life on Earth,* 5th ed., p. 13 by Teresa Audesirk, Gerald Audesirk, and Bruce E. Byers. Copyright © 2009 by Pearson Education, Inc. Printed and Electronically reproduced by permission of Pearson Education, Inc., Upper Saddle River, NJ.

_____ **1.** The relationship between sentence 3 and sentence 4 is based on
a. time. b. addition.

_____ **2.** The overall thought pattern that organizes the paragraph is
a. classification. b. time order.

B. Read the following information from a college mathematics textbook.

Textbook Skills

Types of Real Numbers

[1]The set of real numbers is the set of all numbers corresponding to points on the number line. [2]In addition, real numbers consist of several levels of relationships among various kinds of numbers. [3]First, real numbers are divided into two subsets: rational numbers and irrational numbers. [4]Rational numbers are next classified either as rational numbers that are not integers (such as 2/3, −4/5, and 8.2) or as integers. [5]Three types of integers exist: positive integers (1, 2, 3 . . .), zero (0), and negative integers (−1, −2, −3 . . .). [6]Every rational number has a point on the number line. [7]However, there are points on the line for which there are no rational numbers. [8]These points correspond to the second subset of real numbers—irrational numbers. [9]One example is the number π, which is used to

find the area and the circumference of a circle. [10] ________ example of an irrational number is the square root of 2, named $\sqrt{2}$.

—Adapted from *Introductory and Intermediate Algebra: A Combined Approach,* 2nd ed., pp. 14–17 by Marvin L. Bittinger and Judith A. Beecher. Copyright © 2003 by Pearson Education, Inc. Reproduced by permission of Pearson Education, Inc., Boston, MA.

_____ **3.** Which transition word best fits the blank in sentence 10?
a. First
b. Third
c. Another

_____ **4.** What is the thought pattern used in the paragraph?
a. classification
b. time order

Complete the concept map with the information from the passage.

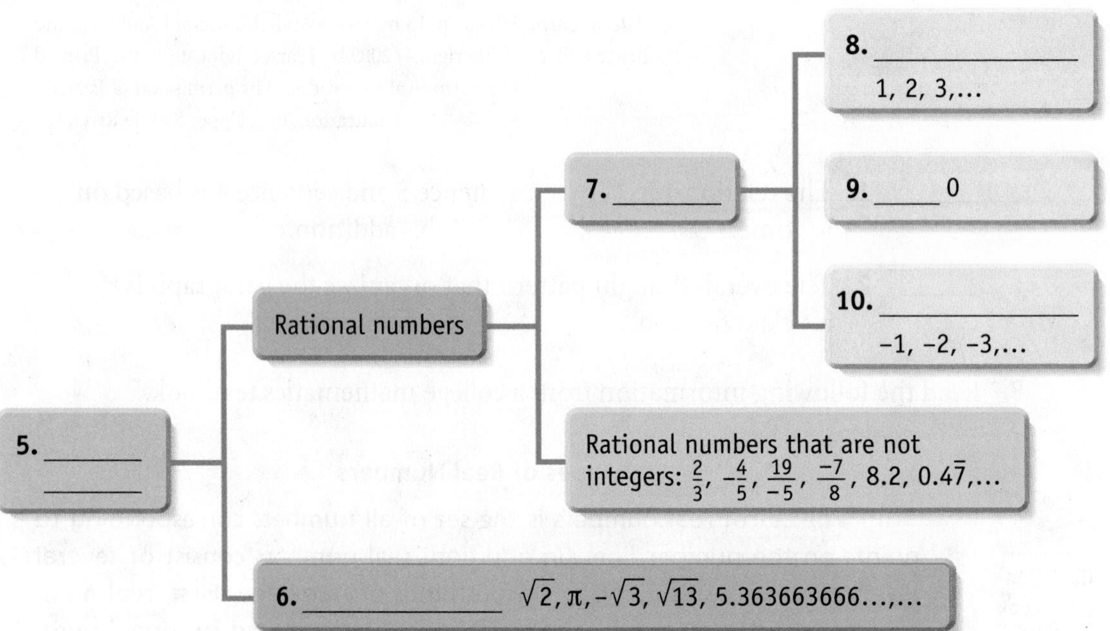

—Adapted from *Introductory and Intermediate Algebra: A Combined Approach,* 2nd ed., p. 17 by Marvin L. Bittinger and Judith A. Beecher. Copyright © 2003 by Pearson Education, Inc. Reproduced by permission of Pearson Education, Inc., Boston, MA.

MASTERY TEST 3

Name ______________________ Section ______________

Date __________ **Score** (number correct) ________ × 10 = ________%

A. Read the following poem by Elizabeth Barrett Browning, originally poem XLIII in *Sonnets from the Portuguese.*

How Do I Love Thee? Let Me Count the Ways

How do I love thee? Let me count the ways.
I love thee to the depth and breadth and height
My soul can reach, when feeling out of sight
For the ends of Being and ideal Grace.
I love thee to the level of every day's
Most quiet need, by sun and candle-light.
I love thee freely, as men strive for Right.
I love thee purely, as they turn from Praise.
I love thee with the passion put to use
In my old griefs, and with my childhood's faith.
I love thee with a love I seemed to lose
With my lost saints—I love thee with the breath,
Smiles, tears, of all my life!—and, if God choose,
I shall but love thee better ________ death.

______ **1.** What is the relationship between the ideas in line 2?
a. listing
b. time

______ **2.** Which transition word best fits the blank in line 14?
a. after
b. when
c. through

______ **3.** In this paragraph, the author
a. explains step by step how to fall in love.
b. gives a list of descriptions to express the quality of her love.

B. Read the following passage from a college government textbook.

Interest Groups

Textbook Skills

[1]Interest groups are everywhere in the American political system. [2]Political scientists loosely categorize interest groups into clusters. [3]An examination of the four distinct types of interest groups gives a good picture of much of the American interest group system.

[4]One type of group focuses on economic interests. [5]All economic interests are ultimately concerned with wages, prices, and profits. [6]In the American economy, the government does not determine these directly. [7]More commonly the government affects economic interests through regulations and taxes, among other things. [8]Business executives, factory workers, and farmers seek to affect government's impact on their livelihoods.

[9]Another type of interest group is made up of the environmentalists. [10]Environmental groups have promoted pollution-control policies, wilderness protection, and population control. [11]They have fought against strip mining, supersonic aircraft, the Alaskan oil pipeline, offshore oil drilling, and nuclear power plants.

[12]The next kind of interest group fights for equality interests. [13]The Fourteenth Amendment assures equal protection under the law. [14]American history, though, shows that this is easier said than done. [15]Interest groups representing minorities and women have made equal rights their main policy goal.

[16]The fourth type of interest group deals with consumers and public interests. [17]Today, over 2,000 organized groups are championing various causes or ideas in the public interest. [18]If products are made safer by the lobbying of consumer protection groups, it is not the members of such groups alone that benefit; everyone should be better off. [19]In addition to consumer groups, other groups speaking for those who cannot speak for themselves seek to protect children, animals, and the mentally ill.

—Adapted from Edwards, Wattenberg, & Lineberry, *Government in America*, Brief 5th ed., pp. 253–258.

______ **4.** The overall thought pattern for the passage is

a. time order. b. classification.

5. The second major supporting detail is signaled by the transition word(s)

______________________________.

6–10. Complete the outline with information from the paragraph.

Main idea stated in a topic sentence: ______________________________

____

____

Major supporting details:

a. ______________________________

b. ______________________________

c. ______________________________

d. ______________________________

MASTERY TEST 4

Name ______________________ Section ______________

Date ______________ Score (number correct) ______ × 10 = ______%

Textbook Skills

Read the following passage from a college world history textbook. Answer the questions that follow.

Out of the Mud: Farming and Herding after the Ice Age

[1]We can divide environments suited to early agriculture into three broad types: swampy wetlands, uplands, and alluvial plains, where flooding rivers or lakes renew the topsoil. [2](Cleared woodlands and irrigated drylands are also suitable for agriculture, but as far as we know, farming never originated in these environments. [3]Rather, outsiders brought it to these areas from someplace else.) [4]Each of the three types developed with peculiar characteristics and specialized crops. [5]It is worth looking at each in turn.

Swampland [6]Swamp is no longer much in demand for farming. [7]Nowadays, in the Western world, if we want to turn bog into farmland we drain it. [8]But it had advantages early on. [9]Swamp soil is rich, moist, and easy to work with simple technology. [10]At least one staple grows well in waterlogged land—rice. [11]We still do not know where or when rice was first cultivated, or even whether any of these wetland varieties preceded the dryland rice that has gradually become more popular around the world. [12]Most evidence, however, suggests that people were producing rice at sites on the lower Ganges River in India and in parts of Southeast Asia some 8,000 years ago, and in paddies in the Yangtze River valley in China not long afterward.

Uplands [13]Like swamplands, regions of high altitude are not places that people today consider good for farming. [14]Farmers have usually left these regions to the herdsmen and native upland creatures, such as sheep, goats, yaks, and llamas. [15]There are three reasons for doing so. [16]First, as altitude increases, cold and the scorching effects of solar radiation in the thin atmosphere diminish the variety of viable plants. [17]Second, slopes are subject to erosion (although this has a secondary benefit because relatively rich soils collect in valley bottoms). [18]Finally, slopes in general are hard to work once you have come to rely on plows, but this does not stop people who do not use plows from farming them. [19]Nonetheless, in highlands suitable for plant foods, plant husbandry or mixed farming did develop.

Alluvial Plains [20]Although swamps and rain-fed highlands have produced spectacularly successful agriculture, farmers get the best help from nature in alluvial plains, flat lands where mud carried by overflowing rivers or lakes renews the soil. [21]If people can channel the floods to keep crops from being swept away on these

—Adapted from *The World: A History, Combined Volume,* 2nd ed., pp. 41–47 by Felipe Fernandez-Armesto. Copyright © 2010 by Pearson Education, Inc. Reprinted by permission of Pearson Education, Inc., Upper Saddle River, NJ.

plains, alluvium, made up of sediment and other organic matter, restores nutrients and compensates for lack of rain. [22]Alluvial soils in arid climates sustained some of the world's most productive economies until late in the second millennium B.C.E. [23]Wheat and barley grew in the black earth that lines Egypt's Nile, the floodplains of the lower Tigris and Euphrates rivers in what is now Iraq, and the Indus River in what is now Pakistan. [24]People first farmed millet on alluvial soils in a somewhat cooler, moister climate in China, in the crook of the Yellow River and the Guanzhong Basin around 7,000 years ago. [25]And in the warm, moist climate of Indochina in what is now Cambodia, three crops of rice a year could grow on soil that the annual counter flow of the Mekong River created. [26]The Mekong becomes so torrential that the delta—where the river enters the sea—cannot funnel its flow, and water is forced back upriver.

_____ **1.** The relationship of ideas within sentence 23 is
a. listing.
b. space order.
c. time order.

_____ **2.** The relationship of ideas between sentence 15 and sentence 16 is
a. addition.
b. time order.
c. classification.

_____ **3.** The thought pattern of the third paragraph (sentences 13–19) is
a. space order.
b. listing.
c. classification.

_____ **4.** The main thought pattern of the passage is
a. time order.
b. space order.
c. classification.

5–10. Complete the following study notes with information from the passage.

Central Idea (Thesis statement): (5) ______________________________
____
____

Type	Trait
I. **(6)** ______________	**(9)** ______________________
II. **(7)** ______________	High altitudes; harsh, thin atmosphere; subject to erosion; hard to plow; left to herdsmen
III. **(8)** ______________	Flood waters **(10)** ______________ soil; mud made up of sediment and organic matter; sustained economies; grew wheat, barley, millet, and rice

More Thought Patterns

8

LEARNING OUTCOMES

After studying this chapter you should be able to do the following:

1. Recognize the following relationships or thought patterns: *comparison, contrast, cause, effect, generalization-and-example, definition-and-example.*
2. Determine the relationships of ideas within a sentence.
3. Determine the relationships of ideas between sentences.
4. Determine the thought pattern used to organize a passage.
5. Evaluate the importance of transitions and thought patterns.

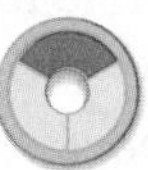

Before Reading About More Thought Patterns

In Chapter 7, you learned several important ideas that will help you as you work through this chapter. Use the following questions to call up your prior knowledge about transitions and thought patterns.

What are transitions? (Refer to page 270.) ______________________

____

____

What are thought patterns? (Refer to page 270.) ______________________

____

____

What is important to know about mixed thought patterns? Give an example from Chapter 7. (Refer to page 293.) ______________________

____

____

____

You have learned that transitions and thought patterns show the relationships of ideas within sentences as well as between sentences and paragraphs, and you studied four common types: time order, space order, listing, and classification. In this chapter, we will explore some other common thought patterns:

- The comparison-and-contrast patterns
- The cause-and-effect pattern
- The generalization-and-example pattern
- The definition-and-example pattern

The Comparison-and-Contrast Patterns

Many ideas become clearer when they are thought of in relation to one another. For example, comparing the prices of different phone plans makes us smarter shoppers. Likewise, noting the difference between mature and immature behavior helps us grow. The comparison-and-contrast patterns enable us to see these relationships. This section discusses both comparison and contrast, starting with comparison. The discussion then turns to the important and effective comparison-and-contrast pattern, in which these two basic ways of organizing ideas are combined when writing an explanation, a description, or an analysis.

Comparison

Comparison points out the ways in which two or more ideas are alike. Sample signal words are listed in the box.

Words and Phrases of Comparison				
a kind of	comparable	in like manner	likewise	same
alike	equally	in the same way	matching	similar
as	in a similar fashion	just as	near to	similarity
as well as	in character with	like	resemble	similarly

When comparison is used to organize an entire paragraph, the pattern looks like this.

Comparison Pattern		
Idea 1		**Idea 2**
Idea 1	*like*	Idea 2
Idea 1	*like*	Idea 2
Idea 1	*like*	Idea 2

Words and phrases of comparison state the relationship of ideas within and between sentences.

EXAMPLE Fill in each blank with a word or phrase that shows comparison.

1. Earning a promotion is _______________ running a marathon.
2. An aspiring employee must work to achieve _______________ a competitive athlete trains to win.
3. A raise in rank and salary is _______________ trophy.

EXPLANATION Each of the example sentences compares two topics to make a point: *earning a promotion* and *running a marathon*. Note that the relationship within each sentence helps to establish a comparison pattern for a three-sentence paragraph as illustrated in the following chart. Compare your choice of words of comparison with the ones given in the chart.

Comparison Pattern: Promotion and Marathon		
Idea 1		**Idea 2**
earning a promotion	(1) *like*	running a marathon
aspiring employee works	(2) *in the same way*	competitive athlete trains
raise	(3) *a kind of*	trophy

PRACTICE 1

Determine the logical order for the following four sentences. Write **1** by the sentence that should come first, **2** by the sentence that should come second, **3** by the sentence that should come third, and **4** by the sentence that should come last. Then, use the information to fill in the chart.

_____ The careers of Jennifer Lopez and Beyoncé Knowles are similar in several respects.

_____ An overall similarity between these two mega-stars is their financial success; both are extraordinarily rich with growing incomes.

_____ Both women also enjoy success as singers, record producers, actors, and fashion icons.

_____ Both Lopez and Knowles are known world-wide by their single names: J-Lo and Beyoncé.

Similarities between Jennifer Lopez and Beyoncé Knowles	
Jennifer Lopez	**Beyoncé Knowles**
5. ______________________________	**6.** ______________________________
______________________________	______________________________
7. ______________________________	**8.** ______________________________
______________________________	______________________________
9. ______________________________	**10.** ______________________________

Contrast

Contrast points out the ways in which two or more ideas are different. Sample signal words are listed in the box below.

Words and Phrases of Contrast				
all	counter to	differently	instead	than
although	despite	even though	nevertheless	to the contrary
as opposed to	differ	however	on the contrary	unlike
at the same time	difference	in contrast	on the one hand	while
but	different	in spite of	on the other hand	yet
conversely	different from	incompatible with	still	

When contrast is used to organize an entire paragraph, the pattern looks like this.

Contrast Pattern		
Idea 1		**Idea 2**
Idea 1	*differs from*	Idea 2
Idea 1	*differs from*	Idea 2
Idea 1	*differs from*	Idea 2

Words and phrases of contrast state the relationship of ideas within and between sentences.

EXAMPLE Fill in each blank with a word or phrase that shows contrast.

Brain development **(1)** _______________ between girls and boys. Girls develop the right side of the brain faster **(2)** _______________ boys. **(3)** _______________, boys develop the left side faster **(4)** _______________ girls. So girls may talk and read earlier **(5)** _______________ boys **(6)** _______________ boys may figure out how to build block towers and solve puzzles earlier **(7)** _______________ girls. Girls and boys also **(8)** _______________ in what they talk about. **(9)** _______________, girls tend to talk about other people; secrets in order to bond friendships; and school, wishes, and needs. **(10)** _______________, boys talk about things and activities such as what they are doing and who is best at the activity.

EXPLANATION

Every sentence in the paragraph contrasts two topics to make a point. Note that the relationships within and between sentences establish a contrast pattern for the paragraph as illustrated in the following chart. Compare your choice of words of contrast with the ones given in the chart.

Contrast Pattern: Difference between Girls and Boys in Brain Development

Idea 1		**Idea 2**
Brain development of girls	(1) *differs*	**Brain development of boys**
Right side develops faster (2) *than*	(3) *in contrast*	Left side develops faster (4) *than*
Talk and read earlier (5) *than*	(6) *while*	Build block towers and solve puzzles earlier (7) *than*
What they talk about	(8) *differ*	What they talk about
People, secrets, school, wishes, needs	(9) *on the one hand*	Other people, secrets, school, wishes, needs
	(10) *on the other hand*	Things and activities such as what they are doing and who is best

PRACTICE 2

Determine a logical order for the following five sentences. Write **1** by the sentence that should come first, **2** in front of the sentence that should come second, **3** by the sentence that should come third, and so on. Then, use the information to fill in the chart.

Textbook Skills

Cultural Differences and Touch

_____ Thus, on the one hand, southern Europeans may view the Japanese as cold, distant, and uninvolved.

_____ On the other hand, the Japanese may view the southern Europeans as pushy, aggressive, and too intimate.

_____ Members of a contact culture, such as southern Europeans, maintain close distances and touch each other in conversation, face each other directly, and maintain eye contact.

_____ In contrast, members of a noncontact culture, such as the Japanese, maintain greater distances in their interactions, rarely touch each other, avoid facing each other, and have much less direct eye contact.

_____ Some cultures are contact cultures, as opposed to others that are noncontact cultures.

—From *The Interpersonal Communication Book*, 10th ed., p. 193. by Joseph A. DeVito.

Cultural Differences and Touch	
Contact Cultures	**Noncontact Cultures**
1. Southern European	1. ______________________
2. maintain close distances and touch each other in conversation, face each other directly, and maintain eye contact	2. ______________________ ______________________ ______________________
3. ______________________ ______________________ ______________________	3. The Japanese may view the southern Europeans as pushy, aggressive, and too intimate.

Comparison and Contrast

The **comparison-and-contrast pattern** shows how two things are similar and how they are different.

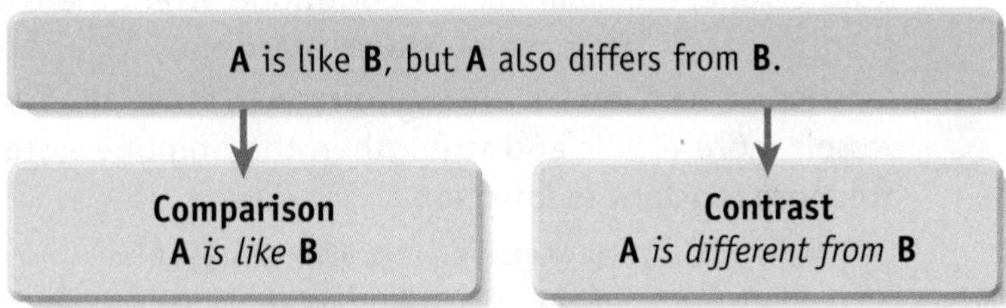

The following short paragraph offers an example of a comparison-and-contrast between two types of weight training. Read the paragraph and circle the comparison and contrast words and phrases.

> [1]Although exercising with free weights is similar to exercising with machine weights, important differences exist between the two methods. [2]Both methods offer resistance training, and both improve posture, build bones and muscles, and improve one's overall well-being. [3]However, free weights and machine weights differ in cost and convenience. [4]On the one hand, free weights are inexpensive and portable. [5]On the other hand, machines, such as Bowflex and Nautilus, costs hundreds of dollars and require a fixed location.

Did you circle the following nine signal words and phrases: *although, similar, differences, both, both, however, differ, on the one hand,* and *on the other hand.* These signal words and phrases state the relationship within and between sentences and establish the paragraph's thought pattern.

EXAMPLE Read the following paragraph from a college sociology textbook. Circle the comparison-and-contrast words; then, answer the questions that follow the paragraph.

Textbook Skills

Men and Women and Hormones

> [1]Scientists don't know why women and men differ but believe that hormones provide part of the explanation. [2]All males and females share three sex hormones. [3]Estrogen is dominant in females and is produced by the ovaries. [4]Progesterone is present in high levels during pregnancy and

is also secreted by the ovaries. [5]Testosterone is dominant in males, where it is produced by the testes. [6]All of these hormones are produced in very small quantities in both sexes before puberty.

[7]After puberty, different levels of these hormones in females and males produce different changes in bodily processes. [8]For example, testosterone, the main male sex hormone, strengthens muscles but threatens the heart. [9]Thus males are at twice the risk of heart disease as are females. [10]The main female sex hormones, especially estrogen, make blood vessels more elastic and strengthen the immune system. [11]Thus females are more resistant to infection.

—Adapted from *Marriages and Families: Changes, Choices, and Constraints*, 4th ed., p. 75 by Nijole V. Benokraitis. Copyright © 2002 by Pearson Education, Inc. Reproduced by permission of Pearson Education, Inc., Boston, MA.

1. What two ideas are being compared and contrasted? ____________________

____

2. List four different comparison-and-contrast words or phrases in the paragraph. ____________________

____

EXPLANATION (1) The paragraph compares and contrasts men's hormones and women's hormones. (2) You were correct to choose any four of these comparison-and-contrast words: *differ, but, share, both, all, different, as, more.*

PRACTICE 3

This paragraph from a college sociology textbook uses comparison and contrast. Read the paragraph and then, answer the questions that follow.

Textbook Skills

[1]The *authoritative* parent and the *authoritarian* parent may seem similar in purpose but differ in style. [2]The similar spelling of the terms *authoritative* and *authoritarian* requires that you read these terms carefully. [3]The first part of both words, which comes from the word authority, suggests that these parents are willing to "take charge" of their children. [4]However, they take charge in quite different ways. [5]On the one hand, authoritative parents are warm (accepting) and firm (controlling). [6]They use reason to gain compliance. [7]They explain rules and encourage verbal give-and-take with their children. [8]They are flexible in setting limits and

are responsive to their children's needs. [9]They encourage independent thinking and are accepting of opposing points of view. [10]During a conflict, an authoritative parent might ask a child, "What do you think we should do?" [11]In contrast, authoritarian parents value obedience above all. [12]They limit their child's freedom by imposing many rules, which they strictly enforce. [13]They favor punishment and forceful measures and mainly use power to gain compliance. [14]They value order and tradition and do not compromise. [15]They do not encourage verbal give-and-take with their children. [16]During a conflict, this parent is likely to assert power with statements such as "Because I said so."

—Adapted from Jaffe, *Understanding Parenting,* 2nd ed., pp. 158, 160.

1. What two ideas are being discussed? ______________________________

and ______________________________

2. How are authoritative and authoritarian parents similar? ______________

____

3. Complete the following chart with information from the passage:

The *authoritative* parent and the *authoritarian* parent may seem similar in purpose but differ in style.

____

Both **authoritative** and **authoritarian** parents "take charge" of a child.

Contrast	
Authoritative Parents	**Authoritarian Parents**
encourage independent thinking use reason to gain compliance are flexible ______________ ______________	value obedience above all ______________ ______________ are inflexible say, "Because I said so."

The Cause-and-Effect Pattern

Sometimes an author talks about *why* something happened or *what* results came from an event. A **cause** states why something happens; an **effect** states a result or outcome. Sample signal words are listed in the box.

Cause-and-Effect Words			
accordingly	by reason	leads to	so
as a result	consequently	on account of	thereby
because	due to	results in	therefore
because of	if . . . then	since	thus

Here are some examples:

Because Zahira is the oldest of eight siblings, she has developed many leadership skills.

Due to a collision involving 34 vehicles, all lanes of Interstate 4 have been shut down for hours.

Each of the example sentences has two topics: one topic causes or has an effect on the second topic. The cause or effect is the author's main point. For example, the two topics in the first sentence are Zahira's birth order and her leadership skills. The stated main idea is that being the oldest of eight siblings is the cause of her leadership skills. The cause is introduced by the word *because*. In the second sentence the two topics are the collision of 34 vehicles and the closing of Interstate 4; the author focuses on the cause by using the signal phrase *due to*.

Note that cause and effect has a strong connection to time, and many of the transitions for this pattern have a time element. Although many cause-and-effect transition words have similar meanings and may be interchangeable, authors carefully choose the transition that best fits the context.

EXAMPLE Read the following paragraph. Fill in each blank (**1–5**) with a cause-and-effect word or phrase from the box. Each is used only once.

because	due to	leads to	result	therefore

The State of Knowledge about Climate Science

Scientists know with virtual certainty several causes and effects of climate change. The composition of Earth's atmosphere is changing (**1**) _______________ of human activities. Increasing levels of greenhouse gases like carbon dioxide (CO_2) in the atmosphere since pre-industrial times are well-documented and understood. The atmospheric buildup of CO_2 and other greenhouse gases is largely the (**2**) _______________ of human activities such as the burning of fossil fuels. An "unequivocal" warming trend of about 1.0 to 1.7°F occurred from 1906–2005. Warming occurred in both the Northern and Southern Hemispheres, and over the oceans. The major greenhouse gases emitted **(3)** ______________ human activities remain in the atmosphere for periods ranging from decades to centuries. It is **(4)** ______________ virtually certain that atmospheric concentrations of greenhouse gases will continue to rise over the next few decades. Increasing greenhouse gas concentrations **(5)** ______________ the warming of the planet.

—U. S. Environmental Protection Agency. "Climate Change: State of Knowledge." 28 Sept. 2009. http://www.epa.gov/climatechange/science/stateofknowledge.html.

VISUAL VOCABULARY

The best meaning of radiation is ______________.

a. gases.
b. atmosphere.
c. heat.

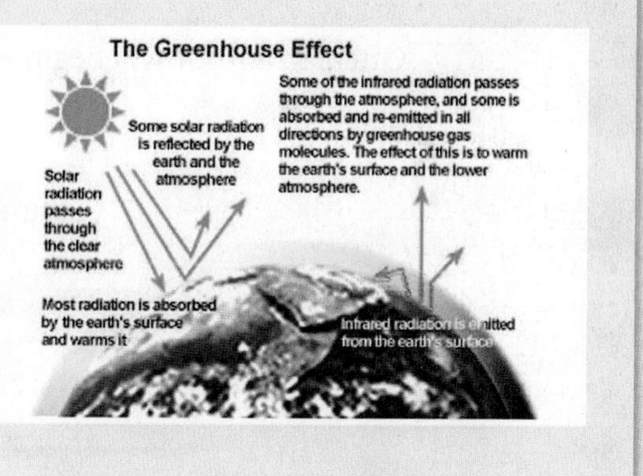

The Greenhouse Effect
Some of the infrared radiation passes through the atmosphere, and some is absorbed and re-emitted in all directions by greenhouse gas molecules. The effect of this is to warm the earth's surface and the lower atmosphere.
Some solar radiation is reflected by the earth and the atmosphere
Solar radiation passes through the clear atmosphere
Most radiation is absorbed by the earth's surface and warms it

EXPLANATION Compare your answers to the following: (1) because, (2) result, (3) due to, (4) therefore, (5) leads to.

Common Cause-and-Effect Patterns

The writer using cause and effect introduces an idea or event, then provides supporting details to show how that idea *results in* or *leads to* another idea. Many times, the second idea comes about because of the first idea. Thus the first idea is the cause, and the following ideas are the effects.

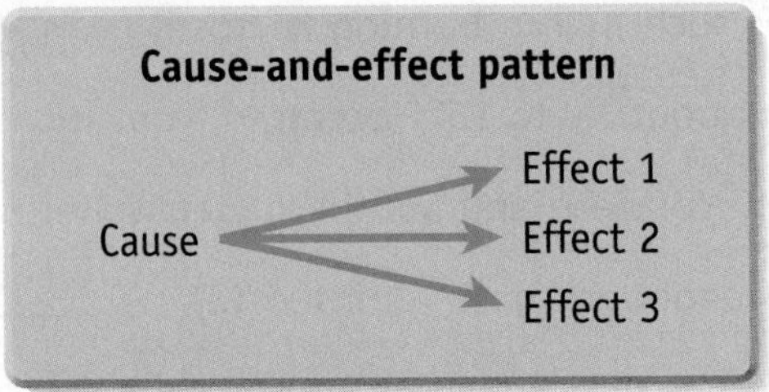

For example, read the following topic sentence:

Music therapy leads to the improvement of a person's physical and psychological well-being.

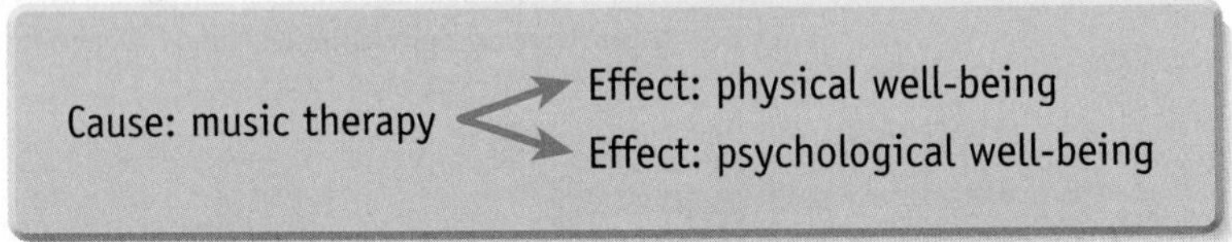

Often an author will begin with an effect and then give the causes.

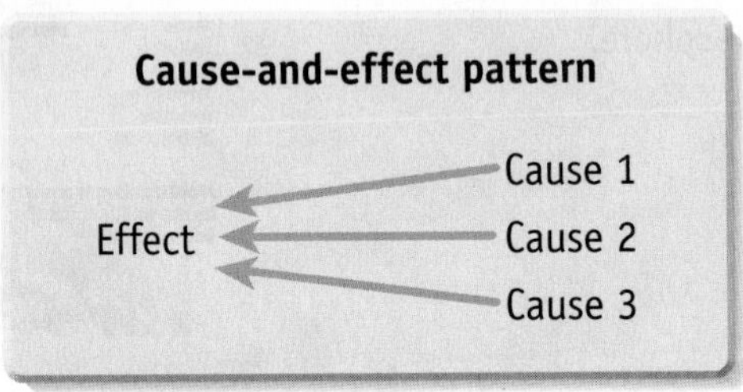

For example, read the following topic sentence.

Some people turn to music therapy due to a loss of physical or psychological well-being.

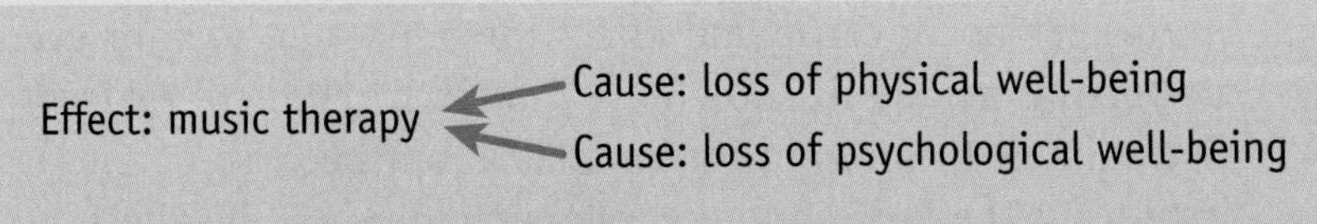

Sometimes the author may wish to emphasize a chain reaction.

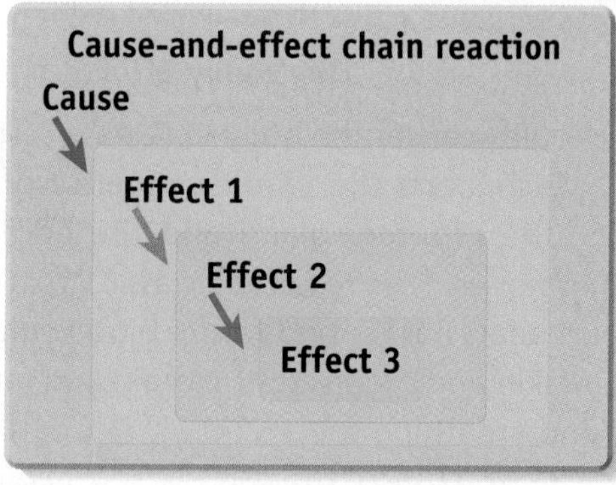

For example, read the following topic sentence.

Music therapy reduces stress and anxiety, which leads to positive emotional states and can ultimately result in feelings of control and empowerment.

Cause-and-effect chain reaction

Cause: music therapy → Effect: reduces stress and anxiety → Effect: positive emotional states → Effect: feelings of control and empowerment

EXAMPLE Determine a logical order for the following three sentences from a college health textbook. Write **1** by the sentence that should come first, **2** in front of the sentence that should come second, and **3** by the sentence that should come last.

Textbook Skills

_____ First, CRH instructs the pituitary gland and the adrenal glands to secrete special stress hormones.

_____ Hostility causes the body to release corticotrophin-releasing-hormone (CRH), which starts the whole sequence of stress hormones.

_____ The result is a classic stress response: blood pressure rises, the heart beats harder and faster, blood volume is increased, blood moves from the skin and organs to the brain and muscles, the liver releases stored sugar, and breathing speeds up.

—From *Mind/Body Health: The Effects of Attitudes, Emotions, and Relationships,* 2nd ed., p. 218 by Keith J. Karren, Brent Q. Hafen, Lee Smith, and Kathryn J. Frandsen. Copyright © 2002 by Pearson Education, Inc. Reprinted by permission of Pearson Education, Inc., Upper Saddle River, NJ.

EXPLANATION Here are the sentences arranged in their proper order. The transition and signal words are in bold type.

The Bodily Effects of Hostility

[1]Hostility **causes** the body to release corticotrophin-releasing-hormone (CRH), which starts the whole sequence of stress hormones. [2]**First**, CRH instructs the pituitary gland and the adrenal glands to secrete special stress hormones. [3]The **result** is a classic stress response: blood pressure rises, the heart beats harder and faster, blood volume is increased, blood moves from the skin and organs to the brain and muscles, the liver releases stored sugar, and breathing speeds up.

—From *Mind/Body Health: The Effects of Attitudes, Emotions, and Relationships,* 2nd ed., p. 218 by Keith J. Karren, Brent Q. Hafen, Lee Smith, and Kathryn J. Frandsen. Copyright © 2002 by Pearson Education, Inc. Reprinted by permission of Pearson Education, Inc., Upper Saddle River, NJ.

In this paragraph, the cause-and-effect signal words are *causes* and *result.* The listing word *first* indicates the order of the cause-and-effect discussion. The addition word *and* is used to list the bodily effects of hostility. This paragraph actually uses two thought patterns to make the point—listing, as well as cause and effect. Even though two patterns are used, the cause-and-effect pattern is the primary organizing pattern.

PRACTICE 4

This excerpt from a college health textbook uses the cause-and-effect thought pattern to organize information. Read the paragraph, circle the cause-and-effect signal words in it, and complete the concept map that follows.

Causes of Loneliness

Textbook Skills

[1]According to researchers with the Department of Health and Human Services, loneliness has several basic causes. [2]First, many lonely people have distinctive social traits that make it difficult for them to form and keep relationships. [3]They may be shy or lack self esteem and are unable to be assertive. [4]In addition, certain situations can lead to loneliness. [5]Time, distance, and money may hinder a person from becoming involved in meaningful relationships. [6]Cultural values may also lead to loneliness; American culture encourages us to be independent and mobile. [7]The nature of relationships is another cause of loneliness. [8]Lonely people have fewer social contacts and relationships than people who are not lonely; thus they

spend less time with other people and are likely to spend time with people they are not close to rather than with good friends. [9]Finally, countless events in life (many beyond our control) can make us feel lonely, rejected, alone, or inadequate. [10]Examples of the most frequent events that cause loneliness are the death of a spouse, divorce, or a geographical move.

—Adapted from *Mind/Body Health: The Effects of Attitudes, Emotions, and Relationships*, 2nd ed., pp. 346–348 by Keith J. Karren, Brent Q. Hafen, Lee Smith, and Kathryn J. Frandsen. Copyright © 2002 by Pearson Education, Inc. Reprinted by permission of Pearson Education, Inc., Upper Saddle River, NJ.

Effect: loneliness ← **1.** Cause: ______________

2. Cause: ______________

3. Cause: ______________

4. Cause: ______________

5. Cause: ______________

The Generalization-and-Example Pattern

Curiosity can quickly lead to danger: Rayanne, a curious five year old, found a hairpin on the floor and stuck it in an electrical outlet to see what would happen.

Some people may read this sentence and think that the author's focus is on Rayanne. However, the author's general point is about curiosity, not Rayanne. In the first part of the sentence the author has made a generalization: "Curiosity can quickly lead to danger." Adding an **example word** or phrase makes it clear that the author is using Rayanne as only one example of the general dangers of curiosity.

Read the sentence about Rayanne again. Note how the use of the signal word makes the relationship between ideas clear.

> Curiosity can quickly lead to danger; *for instance*, Rayanne, a curious five year old, found a hairpin on the floor and stuck it in an electrical outlet to see what would happen.

In the generalization-and-example thought pattern, the author makes a general statement and then offers an example or a series of examples to clarify the generalization.

The Generalization-and-Example Pattern

Statement of a general idea
  Example
  Example

Example words signal that a writer is giving an instance of a general idea.

Words and Phrases That Introduce Examples			
an illustration	for instance	once	to illustrate
for example	including	such as	typically

EXAMPLE Read each of the following items and fill in the blanks with an appropriate example word or phrase.

1. History is full of people who overcame significant challenges and went on to achieve success; _______________, Abraham Lincoln lost at least six major elections for various political offices before becoming the sixteenth president of the United States.

2. Women can now earn some of the highest salaries. _______________, in 2009, Forbes reported that Tyra Banks earned $30 million.

3. Television programs _______________ *Sesame Street* and *Blue's Clues* have helped educate millions of children.

EXPLANATION Many words and phrases that introduce examples are easily interchanged. Compare your choices with the following: (1) *for example,* (2) *For instance*, and (3) *such as*. Notice that in the first two of these three examples, the phrases *for example* and *for instance* are similar in meaning, signaling that at least one example will be offered. The use of the transition phrase *such as* in the third example, however, tends to signal a list. Even though transition words or phrases have similar meanings, authors carefully choose transitions based on style and meaning.

PRACTICE 5

Determine a logical order for the following three sentences. Write a **1** by the sentence that should come first, a **2** in front of the sentence that should come second, and **3** by the sentence that should come last.

_____ The $250,000–$450,000 million loss in crops that occurred in 2010 further exemplifies the damage of such a Florida freeze.

_____ Sunny, warm Florida can experience crop-damaging cold spells.

_____ In one instance, the temperature in agricultural areas during January 2010 dipped as low as 17 degrees.

The Definition-and-Example Pattern

Textbooks are full of new words and special terms. Even if the word is common, it can take on a special meaning in a specific course. To help students understand the ideas, authors often include a definition of the new or special term. Then, to make sure the meaning of the word or concept is clear, the author will also give examples.

Euphemisms are rephrasings of harsh terms; they attempt to avoid offending or to skirt an unpleasant issue. For instance, the Federal Reserve Board is fond of calling a bad market "a market imbalance." A bereaved person might rather hear, "I was sorry to hear of your grandmother's passing" than "I was sorry to hear your grandmother died."

—Faigley, *The Penguin Handbook,* 2003, p. 506.

In the paragraph above, the term *euphemisms* is defined first. Then, the author gives two examples to make the term clear to the reader.

The Definition Pattern

Term and definition
  Example
  Example

- The **definition** explains the meaning of new, difficult, or special terms. Definitions include words like *is, are,* and *means*: "Euphemisms *are* rephrasings of harsh terms; they attempt to avoid offending or to skirt an unpleasant issue."
- The **examples** follow a definition to show how the word is used or applied in the content course. Examples are signaled by words like *for example* and *such as:* "For instance, the Federal Reserve Board is fond of calling a bad market 'a market imbalance.'"

EXAMPLE Determine a logical order for the following three sentences. Write a **1** by the sentence that should come first, a **2** in front of the sentence that should come second, and a **3** by the sentence that should come last. Then, read the explanation.

_____ They also help clarify and intensify your verbal messages.

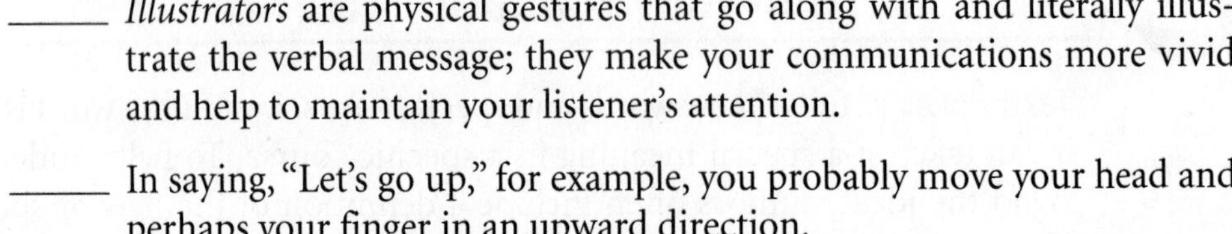

_____ *Illustrators* are physical gestures that go along with and literally illustrate the verbal message; they make your communications more vivid and help to maintain your listener's attention.

_____ In saying, "Let's go up," for example, you probably move your head and perhaps your finger in an upward direction.

—From *The Interpersonal Communication Book,* 10th ed., p. 182 by Joseph A. DeVito.

EXPLANATION The sentences have been arranged in the proper order in the paragraph below. The definition, example, and transition words are in **bold** type.

Illustrators, One Type of Body Gesture

Illustrators **are** physical gestures that go along with and literally illustrate the verbal message; they make your communications more vivid and help to maintain your listener's attention. They **also** help clarify and intensify your verbal messages. In saying, "Let's go up," **for example,** you probably move your head and perhaps your finger in an upward direction.

This sequence of ideas begins by introducing the term *illustrators.* The term is linked to its definition with the verb *are.* The author signals two additional traits of illustrators with the addition word *also*. The sentence that contains *for example* logically follows the definition and additional traits of the term *illustrators.*

PRACTICE 6

Read the paragraph. Finish the definition concept table that follows it by adding the missing details in the proper order.

Affect Displays, A Type of Gesture

[1]Affect displays are the movements of the face that convey emotional meaning—the expressions that show anger and fear, happiness and surprise, eagerness and fatigue. [2]They're the facial muscles that give you away when you try to present a false image and that lead people to say "You look angry. [3]What's wrong?" [4]We can, however, consciously control affect displays, as actors do when they play a role. [5]Affect displays

may be unintentional (as when they give you away) or intentional (as when you want to show anger, love, or surprise).

—From *The Interpersonal Communication Book,* 10th ed., p. 182 by Joseph A. DeVito. Copyright © 2004 by Pearson Education, Inc. Reproduced by permission of Pearson Education, Inc., Boston, MA.

Term: ______________________________

Definition: ______________________________

Example: ______________________________
____

Example: ______________________________

Example: ______________________________

VISUAL VOCABULARY

Provide captions for the two photographs that correctly label the types of gestures represented by each of the images.

____

____

Reading a Textbook: Thought Patterns and Textbooks

Textbook Skills

Textbook authors rely heavily on the use of transitions and thought patterns to make information clear and easier to understand.

EXAMPLE The following topic sentences have been taken from college textbooks. Identify the *primary* thought pattern that each sentence suggests.

_____ **1.** After 1660, the Virginia economy steadily declined for several reasons.

a. cause and effect
b. comparison and contrast
c. definition and example

_____ **2.** Domain-specific knowledge is knowledge that pertains to a particular task or subject; for example, knowing that the shortstop plays between second and third base is specific to the domain of baseball.

a. cause and effect
b. comparison and contrast
c. definition and example

EXPLANATION Topic sentence 1, taken from a history textbook, uses (a) cause and effect, signaled by the phrase *for several reasons*. Topic sentence 2, from an education textbook, is organized according to (c) definition and example, using the verb *is* in the definition and the phrase *for example* to introduce the example.

PRACTICE 7

The following topic sentences have been taken from college textbooks (in mathematics, social science, health, and literature). Identify the primary thought pattern that each sentence suggests.

_____ **1.** A stereotype, a fixed impression about a group, may influence your views of individual group members because you may see individuals only as members of a group rather than as unique individuals.

a. cause and effect
b. comparison and contrast
c. definition and example

_____ **2.** Note that the expressions $x + y$ and $y + x$ have the same values no matter what the variables stand for; they are equivalent.

a. cause and effect
b. comparison and contrast
c. generalization and example

_____ **3.** Excessive time pressure, excessive responsibility, lack of support, or excessive expectations of yourself and those around you may result in overload.
a. cause and effect
b. comparison and contrast
c. definition and example

_____ **4.** Poets have long been fond of retelling myths, narrowly defined as traditional stories about the deeds of immortal beings; myths tell us stories of gods or heroes—their battles, their lives, their loves, and often their suffering—all on a scale of magnificence larger than our life.
a. cause and effect
b. comparison and contrast
c. definition and example

APPLICATIONS

Application 1: Identifying Thought Patterns

Read the following paragraph and identify the thought pattern used in its organization. Fill in the concept map that follows with the missing details.

Yoga or Pilates?

[1]Pilates and Yoga seem similar in many ways. [2]They both strengthen the muscles of the arms, legs, abdomen, and back. [3]Both demand concentration, rely on proper breathing techniques, and result in better posture. [4]However, Yoga and Pilates differ in their origins, purposes, methods, and outcomes. [5]The differences between these activities are grounded in their origin. [6]Yoga originated around 5,000 years ago in India as an Eastern philosophy. [7]Yoga, which means "discipline," refers to a method of meditation to gain release from the bondage of Karma (or fate) and achieve rebirth. [8]In contrast, Pilates is an exercise system developed 80 years ago by Joseph Pilates, a Western athlete from Germany, to rehabilitate injured soldiers. [9]The difference in their origins is also seen in the different purposes of the activities. [10]While Yoga is a holistic spiritual discipline, based on meditation to create a life of balance and composure, Pilates is aimed at physical fitness that develops coordination between muscles to stabilize the body. [11]Yoga emphasizes the union of body, mind, and spirit; Pilates strengthens the core and lengthens the spine. [12]Given these core differences in origin and purpose, it makes sense that their methods vary as well. [13]Many styles of Yoga exist, yet all are performed on a mat and may use props such as blocks and straps to foster fluid movement into and out of poses that stretch and strengthen the body. [14]In contrast, Pilates can be performed in two basic ways: on a mat as callisthenic exercises or on

special machines for movement against resistance. [15]Although both activities increase flexibility and improve posture, they do differ in some outcomes. [16]Yoga lubricates the joints and massages the internal glands and organs, as well as detoxifies and tones the body. [17]On the other hand, Pilates develops a flat abdomen and strong back.

1. What is the main thought pattern? ______________________________

Pilates or Yoga?

	Yoga	Pilates
Origin	2. ______________________	80-year-old exercise system developed by Joseph Pilates, a Western athlete from Germany
Purpose	a holistic spiritual discipline aimed at creating a life of meditation, balance and composure a series of poses that emphasizes the union of the body, mind, and spirit	3. ______________________ an exercise system that strengthens the core of the body and lengthens the spine
Method	4. ______________________ employs fluid movement into and out of poses that stretch and strengthen the body	Pilates can be performed on a mat as callisthenic exercises or on special machines for movement against resistance.
Effects	increases flexibility, improves posture, lubrication of joints, massages the internal glands and organs of the body, detoxifies and tones the body	5. ______________________

Application 2: Using Several Thought Patterns in One Passage

Read the following passage from a college social science textbook. Answer the questions that follow.

Textbook Skills

Comparable Worth

[1]**Comparable worth** is an idea that calls for equal pay for males and females doing work that requires the same skill, effort, level of responsibility, and working conditions. [2]Supporters of this idea argue that jobs can be measured in terms of needed education, skills, experience, mental demands, and working conditions. [3]The basic worth of the job can also be assessed based on its importance to society. [4]Assigning point values to these and other traits, several communities found that women were making much lower salaries than men, even though their jobs had high points. [5]For example, a legal secretary was paid $375 a month less than a carpenter, but both received the same number of job evaluation points. [6]Since 1984, a few states (for example, Minnesota, South Dakota, New Mexico, and Iowa) have raised women's wages after conducting comparable worth studies.

—Adapted from *Marriages and Families: Changes, Choices, and Constraints,* 4th ed., p. 485 by Nijole V. Benokraitis. Copyright © 2002 by Pearson Education, Inc. Reproduced by permission of Pearson Education, Inc., Boston, MA.

1. What is the main organizing thought pattern? ______________

2. What is the relationship of ideas *within* sentence 4? ______________

3. What is the relationship *between* sentence 4 and sentence 5? ______________

4. What is the relationship of ideas *within* sentence 5? ______________

____

VISUAL VOCABULARY

Comparable worth calls for

______________ pay for men and women based on skill, effort, responsibility, and working conditions.

Which term best completes the idea?

a. discriminating
b. selective
c. equivalent

REVIEW TEST 1

Score (number correct) ______ × 10 = ______%

Transition Words and Thought Patterns

A. Circle the signal words used in these sentences from a college psychology text. Then, identify the organizational pattern in each, as follows:

a. cause and effect
b. comparison or contrast or both
c. definition and example

Textbook Skills

______ **1.** Social negotiation is an aspect of the learning process that relies on collaboration with others and respect for different points of view. For example, during class discussions, while students talk and listen to each other, they learn to establish and defend their own positions and respect the opinions of others.

—Adapted from *Educational Psychology*, 8th ed., p. 335, 369–370 by Anita E. Woolfolk. Copyright © 2001 by Pearson Education, Inc. Reproduced by permission of Pearson Education, Inc., Boston, MA.

______ **2.** Higher mental processes develop through social negotiation and interaction, so collaboration in learning is valued.

—Adapted from Woolfolk, p. 335.

______ **3.** A reward is an attractive object or event supplied as a result of a particular behavior. In contrast, an incentive is an object or event that encourages or discourages the behavior. The promise of an A may be an incentive for a student to work hard at studying, while actually receiving the A is her reward.

—Adapted from Woolfolk, pp. 369–370.

______ **4.** If we are consistently reinforced for certain behaviors, then we may develop habits or tendencies to act in certain ways.

—Adapted from Woolfolk, p. 370.

B. Read the following passage from a college health textbook. Then, answer the questions that follow.

Beliefs and Attitudes

Textbook Skills

[1]Even if you know why you should change a specific behavior, your beliefs and attitudes about the value of your actions will significantly affect what you do. [2]We often assume that _ rational people realize there is a risk in what they are doing, then they will act to reduce that risk. [3]But this is not necessarily true. [4]Consider, for example, the number of physicians

and other health professionals who smoke, fail to manage stress, consume high-fat diets, and act in other unhealthy ways. [5]They surely know better, but their "knowing" is disconnected from their "doing." [6]Why is this so? [7]Two strong influences on behaviors are beliefs and attitudes.

[8]A **belief** is an appraisal of the relationship between some object, action, or idea and some attribute of that object, action, or idea (_____________, smoking is expensive, dirty, causes cancer—or it is relaxing). [9]Beliefs may develop from direct experience (such as whether you have trouble breathing after smoking for several years). [10]Or they may develop from secondhand experience or knowledge conveyed by other people (for example, if you see your grandfather die of lung cancer after he smoked for years). [11]Although most of us have a general idea about what makes up a belief, we may be a bit unsure about what makes up an attitude. [12]We often hear or say comments such as "He's got a rotten attitude" or "She needs an attitude adjustment." [13]However, we may still be unable to define attitude. [14]**Attitude** is a relatively stable set of beliefs, feelings, and behaviors in relation to something or someone.

—Adapted from *Access to Health*, 7th ed., pp. 19–20 by Rebecca J. Donatelle and Lorraine G. Davis. Copyright © 2002 by Pearson Education, Inc. Printed and Electronically reproduced by permission of Pearson Education, Inc., Upper Saddle River, NJ.

5. Fill in the blank in sentence 2 with a transition word or phrase that makes sense.

_____ **6.** The relationship of ideas within sentence 2 is one of
a. generalization and example.
b. comparison and contrast.
c. cause and effect.

_____ **7.** The primary thought pattern of organization for the first paragraph is
a. generalization and example.
b. comparison and contrast.
c. cause and effect.

8. Fill in the blank in sentence 8 with a transition word or phrase that makes sense.

_____ **9.** The relationship of ideas between sentence 12 and sentence 13 is one of
a. definition and example.
b. comparison and contrast.
c. cause and effect.

_____ **10.** The primary thought pattern for the second paragraph is
a. definition and example.
b. comparison or contrast or both.
c. cause and effect.

REVIEW TEST 2

Score (number correct) ______ × 20 = ______%

Read the passage and complete the concept map with information from the passage.

Textbook Skills

Sleep Deprivation

¹What is the longest you have ever stayed awake? ²Most people have missed no more than a few consecutive nights of sleep, perhaps studying for final exams. ³If you have ever missed two or three nights of sleep, you may remember having had difficulty concentrating, lapses in attention, and general irritability. ⁴Research indicates that even the rather small amount of sleep deprivation associated with delaying your bedtime on weekends leads to decreases in cognitive performance and increases in negative mood on Monday morning. ⁵Thus, the familiar phenomenon of the "Monday morning blues" may be the result of staying up late on Friday and Saturday nights. ⁶Moreover, sleep loss is known to impair learning in both children and adults.

⁷How does a lack of sleep affect the brain? ⁸The effects of sleep deprivation go beyond simply feeling tired. ⁹In fact, research has shown that failing to get enough sleep affects your ability to learn. ¹⁰So, if you stay up all night to study for a test, you may actually be engaging in a somewhat self-defeating behavior.

—From *Mastering the World of Psychology*, 3rd ed., p. 123 by Samuel E. Wood, Ellen Green Wood, and Denise Boyd.

1–5. Complete the following cause-and-effect concept map with information from the passage.

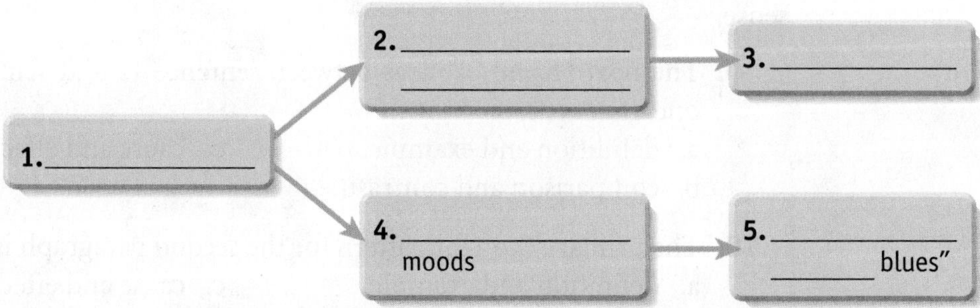

REVIEW TEST 3

Score (number correct) ______ × 10 = ______%

Transitions and Thought Patterns

Textbook Skills

Before you read the following essay from the college textbook *Psychology and Life*, skim the passage and answer the Before Reading questions. Read the essay. Then, answer the After Reading questions.

Vocabulary Preview

affiliate (7): associate, connect with

aesthetic (12): artistic

hierarchy (1, 13): ladder, series of levels

actualization (14): realization, fulfillment

transcendence (16): the state of going beyond usual limits

A Hierarchy of Needs

[1]Humanist psychologist Abraham Maslow (1970) formulated the theory that basic motives form a **hierarchy** of needs. [2]In Maslow's view, the needs are arranged in a sequence from primitive to advanced, and the needs at each level must be satisfied before the next level can be reached. [3]At the bottom (primitive level) of this hierarchy are the basic biological needs, such as hunger and thirst. [4]They must be met before any other needs can begin to operate. [5]When *biological needs* are pressing, other needs are put on hold and are unlikely to influence your actions. [6]When they are reasonably well satisfied, the needs at the next level—*safety needs*—motivate you. [7]When you are no longer concerned about danger, you become motivated by *attachment needs*—needs to belong, to **affiliate** with others, to love, and to be loved. [8]If you are well fed and safe and if you feel a sense of belonging, you move up to *esteem needs*—to like oneself, to see oneself as **competent** and effective, and to do what is necessary to earn the esteem of others.

[9]Humans are thinking beings, with complex brains that demand the stimulation of thought. [10]You are motivated by strong ***cognitive** needs* to know your past, to comprehend the puzzles of current existence, and to predict the future. [11]It is the force of these needs that enables scientists to spend their lives discovering new knowledge. [12]At the next level of Maslow's hierarchy comes the human desire for beauty and order, in the form of ***aesthetic** needs* that give rise to the creative side of humanity.

[13]At the top of the **hierarchy** are people who are nourished, safe, loved and loving, secure, thinking, and creating. [14]These people have moved beyond basic human needs in the quest for the fullest development of their abilities, or *self-**actualization***. [15]A self-actualizing person has many positive traits such as being self-aware, self-accepting, socially responsive, creative, and open to change. [16]*Needs for **transcendence*** may lead to higher states of

awareness and a cosmic vision of one's part in the universe. [17]Very few people move beyond the self to achieve union with spiritual forces.

[18]Maslow's theory is a particularly upbeat view of human motivation. [19]At the core of the theory is the need for each individual to grow and actualize his or her highest potential. [20]Can we maintain such an unfailingly positive view? [21]The data suggests that we cannot. [22]Alongside the needs Maslow recognized, we find that people express power, dominance, and aggression. [23]You also know from your own experience that Maslow's strict hierarchy breaks down. [24]You're likely, for example, to have ignored hunger on occasion to follow higher-level needs. [25]Even with these limitations, however, Maslow's scheme may enable you to bring some order to different aspects of your motivational experiences.

—From *Psychology and Life*, 16th ed., pp. 388–389 by Richard J. Gerrig and Philip G. Zimbardo.

VISUAL VOCABULARY

Which level of Maslow's hierarchy of needs is represented by the picture?

____

Before Reading

Vocabulary in Context

_____ **1.** The word **competent** in sentence 8 means
- a. prideful.
- b. happy.
- c. capable.
- d. clear.

_____ **2.** The best synonym for **cognitive** in sentence 10 is
- a. emotional.
- b. thinking.
- c. demanding.
- d. planning.

After Reading

Concept Maps

3–4. Finish the concept map that follows with information from the passage.

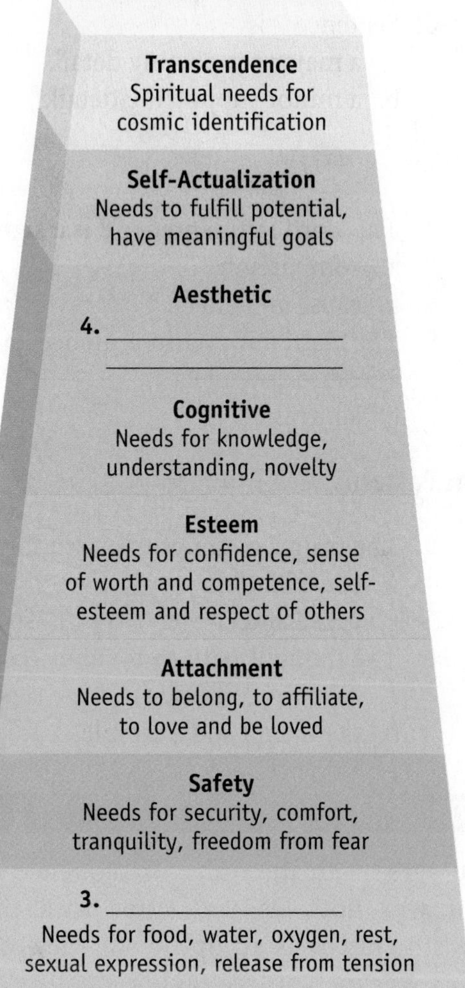

▲ **Maslow's Hierarchy of Needs**
According to Maslow, needs at the lower level of the hierarchy dominate an individual's motivation as long as they are unsatisfied. Once these needs are adequately met, the higher needs occupy the individual's attention.

—From *Psychology and Life*, 16th ed., Figure 12.7 (p. 390) by Richard J. Gerrig and Philip G. Zimbardo. Copyright © 2002 by Pearson Education, Inc. Reproduced by permission of Pearson Education, Inc., Boston, MA.

Central Idea and Main Idea

______ **5.** Which sentence states the central idea of the passage?
a. sentence 1
b. sentence 2
c. sentence 17
d. sentence 24

Supporting Details

______ **6.** Sentence 7 is
a. a major supporting detail.
b. a minor supporting detail.

Transitions

______ **7.** The word *if* in sentence 8 is a signal word that shows
a. comparison.
b. cause and effect.
c. time order.

______ **8.** The word *however* in sentence 25 signals
a. cause and effect.
b. contrast.
c. addition.

Thought Patterns

______ **9.** The primary thought pattern for the entire passage is
a. classification.
b. cause and effect.
c. comparison and contrast.

______ **10.** The thought pattern for the second paragraph (sentences 9–12) is
a. cause and effect.
b. definition and example.
c. listing.

WHAT DO YOU THINK?

Do you agree with the order or sequence of needs in Maslow's hierarchy? Why or why not? Do you agree with the textbook author's view that Maslow's hierarchy is an optimistic view of human development? Why or why not? Assume you are taking a college course in psychology, and your professor has assigned a weekly essay response to your textbook reading. Choose one of the following topics and write a few paragraphs in response to Maslow's hierarchy: Describe your motivations or the motivations of someone you know based on Maslow's hierarchy of needs. Describe a public figure who has reached self-actualization. Use generalizations, definitions, and examples in your response.

REVIEW TEST 4

Score (number correct) ______ × 10 = ______%

Transitions and Thought Patterns

Before Reading: Survey the following passage adapted from the college textbook *The Interpersonal Communication Book.* Skim the passage, noting the words in bold print. Answer the Before Reading questions that follow the passage. Then, read the passage. Next, answer the After Reading questions. Use the discussion and writing topics as activities to do after reading.

Gossip

Vocabulary Preview

solidifies (6): strengthens

camaraderie (6): friendship

exhortation (11): urging, advice, warning

ethical (17): moral, principled, right

[1]Gossip is social talk that involves making evaluations about persons who are not present during the conversation; it generally occurs when two people talk about a third party (Eder & Enke, 1991). [2]As you obviously know, a large part of your conversation at work and in social situations is spent gossiping (Lachnit, 2001; Waddington, 2004; Carey, 2005). [3]In fact, one study estimates that approximately two-thirds of people's conversation time is devoted to social topics, and that most of these topics can be considered gossip (Dunbar, 2004). [4]Gossiping seems universal among all cultures (Laing, 1993), and among some it's a commonly accepted ritual (Hall, 1993).

[5]Lots of reasons have been suggested for the popularity and persistence of gossip. [6]One reason often given is that gossip bonds people together and **solidifies** their relationship; it creates a sense of **camaraderie** (Greengard, 2001; Hafen, 2004). [7]At the same time, of course, it helps to create an in-group (those doing the gossiping) and an out-group (those being gossiped about). [8]Gossip also serves a persuasive function in teaching people the cultural rules of their society. [9]That is, when you gossip about the wrong things that so-and-so did, you're in effect identifying the rules that should be followed and perhaps even the consequences that follow when the rules are broken. [10]Gossip enables you to learn what is and what is not acceptable behavior (Baumeister, Zhang, & Vohs, 2004). [11]Within an organization, gossip helps to regulate organization behavior: Gossip enables workers to learn who the organizational heroes are, what they did, and how they were rewarded—and carries an implicit **exhortation** to do likewise. [12]And, of course, negative gossip enables workers to learn who broke the rules and what punishments resulted from such rule-breaking-again,

with an accompanying implicit **admonition** to avoid such behaviors (Hafen, 2004).

[13]People often engage in gossip for some kind of reward; for example, to hear more gossip, gain social status or control, have fun, cement social bonds, or make social comparisons (Rosnow, 1977; Miller & Wilcox, 1986; Leaper & Holliday, 1995; Wert & Salovey, 2004). [14]Research is not consistent on the consequences of gossip for the person gossiping. [15]One research study argues that gossiping leads others to see you more negatively, regardless of whether your gossip is positive or negative and whether you're sharing this gossip with strangers or friends (Turner, Mazur, Wendel, & Winslow, 2003). [16]Another study finds that positive gossip leads to acceptance by your peers and greater friendship intimacy (Cristina, 2001).

[17]As you might expect, gossiping often has **ethical** implications, and in many instances gossip would be considered unethical. [18]Some such instances: when gossip is used to unfairly hurt another person, when you know it's not true, when no one has the right to such personal information, or when you've promised secrecy (Bok, 1983).

—From *The Interpersonal Communication Book,* 11th ed., pp. 212–213 by Joseph A. DeVito.

Before Reading

Vocabulary in Context

1. In sentence 12, **admonition** means ______________________

2. Context clue used for **admonition** ______________________

____

After Reading

3–6. Central Idea, Supporting Details, and Outlines

Complete the following outline with information from the passage.

Thesis Statement: (3) ______________________

____

____

I. Gossip

A. Definition: (**4**) ______________________________

____

1. Two-thirds of conversation time is considered gossip.

2. Gossip is a universal ritual among all cultures.

B. Reasons for gossip

1. Gossip bonds people and solidifies relationships.

2. Gossip creates an in-group and an out-group.

3. (**5**) ______________________________

C. (**6**) ______________________________

1. Gossiping may lead others to see you negatively.

2. Gossiping may lead to acceptance and intimacy.

Transitions and Thought Patterns

______ **7.** The main thought pattern for the first paragraph (sentences 1–4) is
a. comparison.
b. cause and effect.
c. definition and example.

______ **8.** The main thought pattern for the second paragraph (sentences 5–12) is
a. comparison.
b. cause and effect.
c. definition and example.

______ **9.** What is the relationship of ideas between sentences 8 and 9?
a. cause and effect
b. contrast
c. generalization and example

______ **10.** What is the relationship of ideas within sentence 13?
a. definition and example
b. generalization and example
c. contrast

WHAT DO YOU THINK?

Why do people gossip? When is gossip unethical or harmful? Can gossip ever be beneficial? How or why? Have you or someone you know been the subject of gossip? How did being the topic of gossip make you or that person feel? What role does gossip play in online social networks like Facebook and Twitter? Assume you have taken the role of a mentor to a younger person through your public service work with a youth organization such as the Boys Club, the Girls Club, Boys Scouts, Girl Scouts, or the YMCA. Write an article to post on the organization's local Web page about the causes and effects of gossip—particularly in the context of online social networks.

After Reading About More Thought Patterns

Before you move on to the Mastery Tests on thought patterns, take time to reflect on your learning and performance by answering the following questions. Write your answers in your notebook.

- How has my knowledge base or prior knowledge about more thought patterns changed?
- Based on my studies, how do I think I will perform on the Mastery Test(s)? Why do I think my scores will be above average, average, or below average?
- Would I recommend this chapter to other students who want to learn about thought patterns? Why or why not?

Test your understanding of what you have learned about Thought Patterns for Master Readers by completing the Chapter 8 Review Card in the insert near the end of your text.

CONNECT TO PEARSON myreadinglab

To check your progress in meeting Chapter 8's learning outcomes, log in to **www.myreadinglab.com** and try the following exercises.

- The "Patterns of Organization" section of MyReadingLab gives additional information about transitions and patterns of organization. The section provides a model, practices, activities, and tests. To access this resource, click on the "Study Plan" tab. Then click on "Patterns of Organization." Then click on the following links as needed: "Overview," "Model," "Signal Words (Flash Animation)," "Other Patterns of Organization (Flash Animation)," "Practice," and "Tests."
- To measure your mastery of the content of this chapter, complete the tests in the "Patterns of Organization" section and click on Gradebook to find your results.

MASTERY TEST 1

Name ______________________ Section ______________

Date __________ Score (number correct) ________ × 10 = ________ %

Read the following paragraph from a college social science textbook. Fill in the blanks (**1–5**) with transition words from the box. Then, answer the questions following the paragraph.

differences	effect	more	same	than

Textbook Skills

Differences in the Effects of Drugs

Some differences in drug effects may be related to an interaction between the drug itself and the specific traits of the person taking the drug. One characteristic is a person's weight. In general, a heavier person will require a greater amount of a drug **(1)** ______________ a lighter person to receive the same drug **(2)** ______________. It is for this reason that drug dosages are expressed as a ratio of drug amount to weight. Another characteristic is gender. Even if a man and a woman are exactly the **(3)** ______________ weight, differences in drug effects can still result on the basis of gender **(4)** ______________ in body composition and sex hormones. Women have, on average, a higher proportion of fat, due to a greater fat-to-muscle ratio, and a lower proportion of water than men. When we look at the effects of alcohol consumption in terms of gender, we find that the lower water content in women makes them feel **(5)** ______________ intoxicated than men, even if the same amount of alcohol is consumed.

—Adapted from Levinthal, *Drugs, Behaviors, and Modern Society*, 3rd ed., pp. 58–59.

______ **6.** The transition used to fill blank (1) signals

a. cause and effect.

b. generalization and example.

c. comparison and contrast.

_____ **7.** The transition used to fill blank (2) signals
a. cause and effect.
b. generalization and example.
c. comparison and contrast.

_____ **8.** The transition used to fill blank (3) signals
a. cause and effect.
b. definition and example.
c. comparison and contrast.

_____ **9.** The transition used to fill blank (4) signals
a. cause and effect.
b. generalization and example.
c. comparison and contrast.

_____ **10.** The transition used to fill blank (5) signals
a. cause and effect.
b. definition.
c. comparison and contrast.

MASTERY TEST 2

Name ______________ Section ______________
Date ________ **Score** (number correct) ________ × 5 = ______ %

A. Write the numbers **1** to **9** in the spaces provided to identify the correct order of the ideas. Then, answer the question that follows the list.

Textbook Skills

Physical Effects of Modern Life

_____ One of the most pervasive problems is shortened muscles, tendons, and ligaments.

_____ For example, continual sitting at work and leisure results in shortened muscles, tendons, and ligaments and eventually results in restricted movements.

_____ However, our daily activities have changed drastically.

_____ Our modern living patterns are directly responsible for a host of health problems.

_____ First, our basic body structure and function have not changed since the days of prehistoric humans.

_____ It is not unusual for adults in their mid-twenties to be so restricted that touching their toes while keeping the knees straight is impossible or painful.

_____ In addition to restricted movement, chronic sitting affects the spine.

_____ The result will be a sitting-type posture when one is standing or walking; this posture produces pain and tension because body parts are not positioned naturally.

_____ The spine will condition itself to the demand of chronic sitting and lose its ability to remain erect.

—Adapted from Girdano, Everly, & Dusek, *Controlling Stress and Tension*, 6th ed., p. 248.

_____ **10.** The primary thought pattern is
a. generalization and example.
b. comparison and contrast.
c. cause and effect.

B. Read the following paragraph from a college social science textbook. Then, answer the questions and complete the concept map.

Insomnia

[1]The term *insomnia* typically means "sleeplessness" or the "inability to sleep." [2]Insomnia has several causes and effects. [3]Common effects of insomnia include fatigue, irritability, poor concentration, poor short-term memory, drowsiness, and abuse of stimulants. [4]Several factors that contribute to insomnia are physical, such as heart disease, high blood pressure, "heartburn," and chronic breathing problems. [5]However, mental factors such as tension caused by anxiety and anger are the main causes of insomnia.

—Adapted from *Encyclopedia of Stress*, 1st ed., p. 119 by F. J. McGuigan. Copyright © 1999 by Pearson Education, Inc. Printed and Electronically reproduced by permission of Pearson Education, Inc., Upper Saddle River, NJ.

______ **11.** The relationship between sentence 3 and sentence 4 is

a. cause and effect.
b. addition.
c. contrast.

Causes and Effects of Insomnia

Main idea (stated in the topic sentence): Insomnia has several causes and effects.

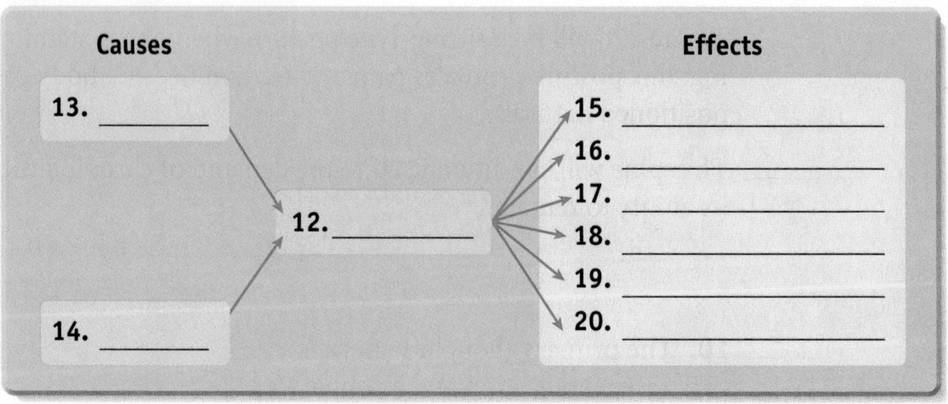

MASTERY TEST 3

Name ______________ Section ______________

Date ________ **Score** (number correct) ______ × 10 = ______ %

Read the following paragraphs from college textbooks. Then, answer the questions and complete the outline and the concept table.

Road Rage

[1]"Road rage" is the term used to describe the attitude of extremely angry, hostile, and aggressive drivers. [2]Several factors are linked to road rage. [3]First, road congestion has become worse in the past 10 years due to an increase in both the number of cars and total miles driven, without an expansion in the amount of road space available. [4]Road congestion adds to the amount of time a driver spends on the road and frequently causes delay as well as late or missed appointments. [5]This is an added stressor for people whose schedules are already overloaded due to work and family demands. [6]Thus, poor time management skills may contribute to road rage. [7]Finally, a lack of anger control often leads to road rage. [8]Demands on time increase the tendency to overreact to what otherwise might be considered slight inconveniences. [9]Thus, the angry, stressed driver is likely to see the presence of others on the road as "unreasonable" behavior and an excuse to disregard legal and safety rules so that she or he can get past the "morons."

—Adapted from *Encyclopedia of Stress*, 1st ed., pp. 185–186 by F. J. McGuigan. Copyright © 1999 by Pearson Education, Inc. Printed and Electronically reproduced by permission of Pearson Education, Inc., Upper Saddle River, NJ.

______ **1.** The thought pattern suggested by sentence 1 is
   a. definition and example.
   b. comparison and contrast.
   c. cause and effect.

______ **2.** The primary thought pattern
   a. discusses the similarities between normal drivers and "morons."
   b. explains some of the causes of road rage.
   c. compares individual responses to road rage.

3–5. **Road Rage**

Main idea (stated in the topic sentence): Several factors are linked to road rage.

A. ______________________________

B. ______________________________

C. ______________________________

Textbook Skills

Semantic versus Episodic Memory

[1]In your long-term memory (LTM), you store two different types of information: semantic and episodic memories. [2]Semantic memories are memories of the meanings of words (a pine is an evergreen tree with long needles), concepts (heat moves from a warmer object to a cooler one), and general facts about the world (the original 13 colonies were established by the British). [3]For the most part, you don't remember when, where, or how you learned this information. [4]In contrast, episodic memories are memories of events that are linked to a time, place, and situation (when, where, and how). [5]In other words, episodic memories provide a context. [6]The meaning of the word *memory* is no doubt firmly implanted in your semantic memory; whereas the time and place you first began to read this book are most likely in your episodic memory. [7]Brain scanning studies have given evidence that semantic and episodic memories are distinct. [8]The frontal lobe of the brain, for instance, plays a key role in looking up stored information. [9]However, when we recall semantic memories, the left frontal lobe tends to be more active than the right. [10]The reverse is true when we recall episodic memories.

—Adapted from Kosslyn & Rosenberg, *Fundamentals of Psychology*, p. 163.

______ **6.** The relationship between sentences 7 and 8 is
a. generalization and example.
b. cause and effect.
c. comparison and contrast.

Semantic versus Episodic Memory

(7) Main idea (Stated in the topic sentence): ______________________________

____

	Semantic Memory	Episodic Memory
First difference	(8) ______________ ______________	Memories of events that are linked to time, place, and situation
Second difference	(9) ______________ ______________ ______________	Episodic memories are learned in context.
Third difference	The left frontal lobe is more active than the right frontal lobe when we recall semantic memories.	(10) ______________ ______________ ______________ ______________

MASTERY TEST 4

Name ______________________ Section ______________
Date __________ **Score** (number correct) ______ × 20 = ______ %

Textbook Skills

Read the following passage from a college science textbook. After reading, complete the concept map with information from the passage.

Human Use and Impact: Oceans

[1]Our species has a long history of interacting with the oceans. [2]We have traveled across their waters, clustered our settlements along their coastlines, and been fascinated by their beauty, power, and vastness. [3]We have also left our mark upon them by exploiting oceans for their resources and polluting them with our waste.

Marine pollution threatens resources [4]People have long made the oceans a sink for waste and pollution. [5]Oil, plastic, industrial chemicals, sewage, excess nutrients, and abandoned fishing gear all eventually make their way into the oceans.

Nets and plastic debris [6]Plastic bags and bottles, fishing nets, gloves, fishing line, buckets, floats, abandoned cargo, and much else that people transport on the sea or deposit into it can harm marine organisms. [7]Because most plastic is not biodegradable, it can drift for decades before washing up on beaches. [8]Marine mammals, seabirds, fish, and sea turtles may mistake floating plastic debris for food and can die as a result of ingesting material they cannot digest or expel. [9]Fishing nets that are lost or intentionally discarded may continue snaring animals for decades. [10]We can all help minimize this type of harm by reducing our use of plastics and by picking up trash on beaches.

Oil pollution [11]Major oil spills—such as the one that occurred in 1989 when the Exxon Valdez struck a reef in Prince William Sound, Alaska, and spilled 42 million L (11 million gal) of crude oil—make headlines and cause serious environmental problems. [12]Yet the majority of oil pollution in the oceans accumulates from innumerable widely spread small sources, including leakage from small boats and runoff from human activities on land. [13]Moreover, the amount of petroleum spilled into the oceans in recent years is equaled by the amount that seeps up from naturally occurring seafloor deposits. [14]Nonetheless, minimizing the amount of oil we release into coastal waters is important. [15]Petroleum can physically coat and kill marine organisms and, when ingested, can poison them. [16]Oil spills can devastate fisheries and the local economies dependent on them.

Excess nutrients [17]Pollution from fertilizer runoff or other nutrient inputs can create dead zones in coastal marine ecosystems, as with the Gulf of Mexico. [18]Excessive nutrient concentrations also may give rise to population explosions among several species of marine algae that produce powerful toxins that attack the nervous systems of vertebrates. [19]Blooms of these algae are known as harmful algal blooms. [20]Some algal species produce reddish pigments that discolor surface waters, and blooms of these species are nicknamed red tides. [21]Harmful algal blooms can cause illness and death among zooplankton, birds, fish, marine mammals, and people as their toxins are passed up the food chain. [22]They also cause economic loss for communities dependent on fishing or beach tourism.

[23]As severe as the impacts of all these types of marine pollution can be, however, most marine scientists concur that the more worrisome dilemma is overharvesting.

—Adapted from *Essential Environment: The Science behind the Stories*, 3rd ed., pp. 275–276 by Jay H. Withgott and Scott R. Brennan.

1. ________ →	**Marine Pollution** →	**Effects**
Trash	**2.** ________________	poisons, snares, and kills marine life
Oil spills		
3. ________________ ________________	Oil pollution	poisons, coats, kills marine life
Runoff from land		
Natural seafloor deposits		
		creates dead zones;
Fertilizer runoff	**4.** ________________	population explosion of toxic algae and algal blooms
		5. ________________ ________________

Fact and Opinion

9

LEARNING OUTCOMES

After studying this chapter you should be able to do the following:

1. Define the following terms: *fact* and *opinion*.
2. Ask questions to identify facts.
3. Analyze biased words to identify opinions.
4. Analyze supposed "facts."
5. Distinguish between fact and opinion.
6. Evaluate the importance of facts and opinions.
7. Interpret fact(s) and opinion(s) to improve comprehension.

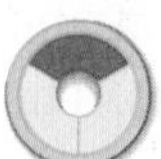

Before Reading About Fact and Opinion

You are most likely already familiar with the commonly used words *fact* and *opinion,* and you probably already have an idea about what each one means. Take a moment to clarify your current understanding about fact and opinion by writing a definition for each one in the spaces below:

Fact: ____

____

Opinion: ____

____

____

As you work through this chapter, compare what you already know about fact and opinion to new information that you learn using the following method:

On a blank page in your notebook draw a line down the middle of the page to form two columns. Label one side "Fact" and the other "Opinion." Below each heading, copy the definition you wrote for each one. As you work through the chapter, record new information you learn about facts and opinions in their corresponding columns.

What Is the Difference Between Fact and Opinion?

Fact: The average American child will witness 12,000 violent acts on television each year, amounting to about 200,000 violent acts by age 18.

Opinion: American television programming is too violent.

Master readers must sort fact from opinion to properly understand and evaluate the information they are reading.

A **fact** is a specific detail that is true based on objective proof. A fact is discovered.

An **opinion** is an interpretation, value judgment, or belief that cannot be proved or disproved. An opinion is created.

Objective proof can be physical evidence, an eyewitness account, or the result of an accepted scientific method.

Most people's points of view and beliefs are based on a blend of fact and opinion. Striving to remain objective, many authors rely mainly on facts. The main purpose of these authors is to inform. For example, textbooks, news articles, and medical research rely on facts. In contrast, editorials, advertisements, and fiction often mix fact and opinion. The main purpose of these types of writing is to persuade or entertain.

Separating fact from opinion requires you to think critically because opinion is often presented as fact. The following clues will help you separate fact from opinion.

Fact	Opinion
Is objective Is discovered States reality Can be verified Is presented with unbiased words	Is subjective Is created Interprets reality Cannot be verified Is presented with biased words
Example of a fact Spinach is a source of iron.	*Example of an opinion* Spinach tastes awful.

A fact is a specific, objective, and verifiable detail; in contrast, an opinion is a biased, personal view created from feelings and beliefs.

EXAMPLE Read the following statements, and mark each one **F** if it states a fact or **O** if it expresses an opinion.

_____ **1.** The most beautiful spot on earth is a serene lake nestled in the foothills of the Appalachian Mountains.

_____ **2.** Tippah Lake is located just outside of the small town of Ripley, Mississippi, in the foothills of the Appalachian Mountains.

EXPLANATION The first sentence expresses an opinion in the form of a personal interpretation about what is beautiful. The second sentence is a statement of fact that gives the location of the lake.

PRACTICE 1

Read the following statements and mark each one **F** if it states a fact or **O** if it expresses an opinion.

_____ **1.** Living in a large city is detrimental to one's psychological well-being.

_____ **2.** New York City offers an array of cultural experiences including Broadway theatre, historic sites such as Ellis Island and the Statue of Liberty, and many museums such as the Children's Museum of Manhattan.

Ask Questions to Identify Facts

To test whether a statement is a fact, ask these three questions:

- Can the statement be proved or demonstrated to be true?
- Can the statement be observed in practice or operation?
- Can the statement be verified by witnesses, manuscripts, or documents?

If the answer to any of these questions is no, the statement is not a fact. Instead, it is an opinion. Keep in mind, however, that many statements blend both fact and opinion.

EXAMPLE Read the following statements and mark each one **F** if it states a fact or **O** if it expresses an opinion.

_____ **1.** In the 1700s women sewed lead weights into the bottoms of their bathing gowns to prevent the gowns from floating up and exposing their legs.

_____ **2.** Bikinis reveal too much skin and are indecent.

_____ **3.** Arnold Schwarzenegger, governor of California, was born in Austria.

_____ **4.** Arnold Schwarzenegger's distinguished film career and admirable devotion to public service make him a unique American hero.

EXPLANATION Compare your answers to the ones below.

1. F: This statement can be easily verified by doing research.

2. O: This is a statement of personal opinion. Words and phrases such as *too much* and *indecent* are judgments that vary from person to person.

3. F: These facts are common knowledge and easily verified.

4. O: Several words in this statement, such as *distinguished, admirable,* and *unique,* reveal that it is a statement of opinion based on personal views.

PRACTICE 2

Read the following statements, and mark each one **F** if it states a fact or **O** if it expresses an opinion or **F/O** if it expresses both a fact and an opinion.

_____ **1.** Ninety-five percent of the world's consumers live outside the United States.

_____ **2.** During the creation of his novel *Carrie*, Stephen King threw the novel into the trash, and his wife dug it out and convinced him to finish it.

_____ **3.** Low-carbohydrate diets such as Atkins and South Beach are ineffective and dangerous.

_____ **4.** Balto was a dog who, in the winter of 1925, led a dogsled team through a blizzard to get medicine to Nome, Alaska, because of a fatal diphtheria epidemic; a statue stands in his honor in Central Park, New York, at 67th Street and Fifth Avenue, just inside the Park.

_____ **5.** Cats can be irritating pets because they are independent and unaffectionate.

_____ **6.** Newspapers are the most reliable source for news.

_____ **7.** Danica Sue Patrick is an American auto racing driver, currently competing in the IndyCar Series, the ARCA Racing Series, and the NASCAR Nationwide Series.

_____ **8.** Danica Patrick is also a beautiful woman who has appeared on the cover of *Sports Illustrated.*

_____ **9.** Regular dental flossing reduces gum disease.

_____ **10.** People who live in the Northeast part of the United States are much more open-minded and liberal than people who live in the South.

Note Biased Words to Identify Opinions

Be on the lookout for biased words. **Biased words** express opinions, value judgments, and interpretations. They are often loaded with emotion. The box contains a small sample of these kinds of words.

Biased Words					
amazing	best	favorite	great	miserable	stupid
awful	better	frightful	greatest	more	ugly
bad	disgusting	fun	handsome	most	unbelievable
beautiful	exciting	good	horrible	smart	very

Realize that a sentence can include both facts and opinions. The part of the sentence that includes a biased word may be an opinion about another part of the sentence that is a fact.

EXAMPLE Read the following sentences. Underline the biased words.

1. John Coltrane was a jazz musician whose quartet produced hit albums that remain the most innovative and expressive jazz music of our time.

2. Even though marijuana use is widespread in America, it should not be legalized.

EXPLANATION In the first sentence, the first part "John Coltrane was a jazz musician whose quartet produced hit albums" is a fact that can be proved. The second part of the sentence, however, expresses an opinion about his music using the biased words *most innovative* and *expressive.* In the second sentence, the widespread use of marijuana is well documented, but the idea that "it should not be legalized" is an opinion.

PRACTICE 3

Read the following two sentences. Underline the biased words.

1. The works of Joan Didion, one of the most brilliant writers of this generation, should be required reading in a liberal arts education.
2. Writing is a satisfying process that can lead to alarming self-discoveries, comforting insights, and much needed wisdom.

Note Qualifiers to Identify Opinions

Be on the lookout for words that qualify an idea. A qualifier may express an absolute, unwavering opinion using words like *always* or *never*. Other times a qualifier expresses an opinion in the form of a command as in *must*, or the desirability of an action with a word like *should.* Qualifiers may indicate different degrees of doubt with words such as *seems* or *might*. The box contains a few examples of these kinds of words.

Words That Qualify Ideas

all	could	likely	never	possibly, possible	sometimes
always	every	may	often	probably, probable	think
appear	has/have to	might	only	seem	usually
believe	it is believed	must	ought to	should	

Remember that a sentence can include both fact and opinion. Authors use qualifiers to express opinions about facts.

EXAMPLE Read the following sentences. Underline the qualifiers.

1. Dissatisfaction with work often spills over into relationships; most people are not able to separate work problems from their relationship lives.
2. In his book *Messages: Building Interpersonal Communication Skills*, Joseph A. DeVito states, "Commitment may take many forms; it may be an engagement or a marriage; it may be a commitment to help the person or be with the person, or a commitment to reveal your deepest secrets." (1999, p. 264)

EXPLANATION

1. The qualifiers in this sentence are *often* and *most*.
2. The qualifiers in this sentence are *may* and *many*.

PRACTICE 4

Read the following sentences. Underline the qualifiers.

1. Oral hygiene is the only way to avoid possible gum disease; you must brush your teeth three times a day and floss at least once daily.
2. All citizens should save some portion of their earnings for their probable retirement.

Think Carefully About Supposed "Facts"

Beware of **false facts,** or statements presented as facts that are actually untrue. At times, an author may mislead the reader with a false impression of the facts. Political and commercial advertisements often present facts out of context, exaggerate the facts, or give only some of the facts. For example, Governor People boasts that under his leadership, his state has attained the highest literacy rates ever achieved. Although his state's literacy rates did reach their highest historical levels, the actual rise was slight, and his state's literacy levels remained the lowest in the country. The governor misled people by leaving out important facts.

Sometimes an author deliberately presents false information. Journalist Jayson Blair concocted a host of false facts in many of the articles he wrote for the *New York Times*. For example, in his five articles about the capture and rescue of Private Jessica Lynch, Blair fabricated many facts. First, he wrote his stories as if he had conducted face-to-face interviews, yet he never traveled to meet the family, and he conducted all of his interviews by telephone. Then, in a March 27, 2003, article, he described Lynch's father as being "choked up as he stood on his porch here overlooking the tobacco fields and cattle pastures." The statement was false: He never witnessed such a scene, and the Lynch home does not overlook tobacco fields or cattle pastures. Blair's use of false facts ruined his career and damaged the reputation of the prestigious *New York Times*.

Most often, however, false facts are mere errors. Read the following two examples of false facts:

1. Thomas Alva Edison invented the first light bulb.
2. The common cold is most commonly spread by coughing and sneezing.

Research quickly proves the first statement to be a false fact; light bulbs were being used 50 years before Edison patented his version of the light bulb. The second statement is a false fact because studies show that a cold is more likely to be spread by physical contact such as shaking hands or sharing a phone than by inhaling particles in the air. Often some prior knowledge of the topic is needed to identify false facts. The more you read, the more masterful you will become at evaluating facts as true or false.

False facts can be used to mislead, persuade, or entertain. For example, read the following claims about a nutritional supplement that allegedly prevents a wide range of illnesses including colds, flus, cancer, irritable bowel syndrome, *Candida Albicans,* and heart disease.

BodyWise

[1]Every day your body is under attack and the results can be deadly. [2]The best thing you can do is to bolster your immune system so it can defend itself from infection and foreign invaders. [3]We believe BodyWise International's all natural AG-Immune formula is the most powerful product ever created to promote a healthy immune system . . . and a healthier future. [4]And while it's [sic] primary ingredient, Ai/E10™, has proven to be clinically effective for use with both immune and autoimmune disorders, you don't have to be immune system challenged to enjoy the amazing benefits of this exciting new supplement. [5]Think of it as a prudent preventative measure that promotes optimal health by increasing the activity of natural killer (NK) cells—your body's last line of defense against illness and disease. [6]AG-Immune has already changed the lives of thousands of people across North America. [7]Imagine what it could do for you: [8]"One day I happened to notice that, on the cover of The Ultimate Nutrient booklet, *candida* was listed as one of the areas Ai/E10 was clinically successful in treating. [9]I began taking two a day and literally—within 36 hours—the pain began to dissipate. [10]After only three weeks, my symptoms were completely gone! [11]I have been on BodyWise for almost eight years and at the risk of sounding over-dramatic, I have to say, the products have definitely changed my quality of life."

—Adapted from *United States of America, Plaintiff, v. Body Wise International, Inc.,* and *Jesse A. Stoff, M.D., Defendants,* United States District Court, Central District of California, Southern Division. Civil Action No.: SACV-05-43 (DOC) (Anx) 5 Nov. 2007. http://www.ftc.gov/os/caselist/bodywise/050118compbodywise.pdf

The Federal Trade Commission sued and won a settlement against the company for false advertisement. The company could not support its claims to prevent or improve disease symptoms with scientific evidence. Note the biased words the author used to describe the effectiveness of the supplement: *attack, deadly, best, bolster, most powerful, healthier, amazing benefits, exciting, prudent,* and *optimal,* to name a few.

VISUAL VOCABULARY

The best synonym for "clinically" is ______________.

a. accidentally.
b. scientifically.
c. humanly.

▸ Ai/E10™ has proven to be clinically effective for use with both immune and autoimmune disorders.

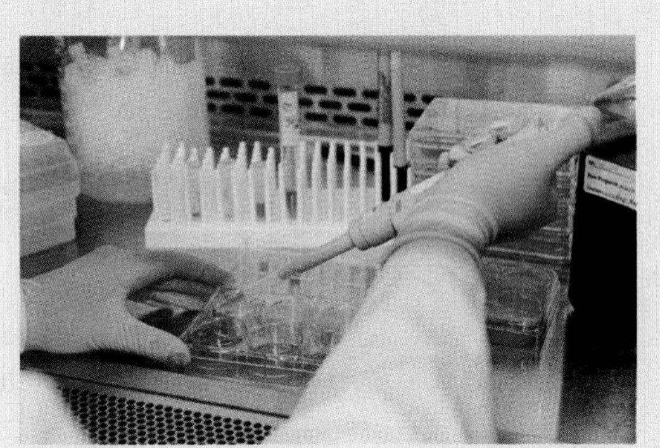

In addition to thinking carefully about false facts, beware of opinions worded to sound like facts. Remember that facts are specific details that can be researched and verified as true. However, opinions may be introduced with phrases like *in truth, the truth of the matter,* or *in fact.* Read the following two statements:

1. In truth, reproductive cloning is expensive and highly inefficient.

2. More than 90% of cloning attempts fail to produce viable offspring, and cloned animals tend to have much weaker immune function and higher rates of infection, tumor growth, and other disorders; in fact, reproductive cloning is expensive and highly inefficient.

—Adapted from "Cloning Fact Sheet," 19 Nov. 2003, *Human Genome Program,* U.S. Department of Energy. 29 Jan. 2004, http://www.ornl.gov/sci/techresources/Human_Genome/elsi/cloning.shtml#risks

The first statement is a general opinion that uses the biased words *expensive, highly,* and *inefficient.* The second statement is a blend of fact and opinion.

It begins with the facts, but then uses the phrase *in fact* to introduce the opinion.

EXAMPLES Read the following statements, and mark each one as follows:

F if it states a fact

O if it states an opinion

F/O if it combines fact and opinion

_____ **1.** Public funding for art education, at both national and local levels, has declined in recent years.

_____ **2.** Diet Pepsi tastes better than Diet Coke.

_____ **3.** Microwave popcorn is a nutritious, low-calorie, and tasty snack food.

_____ **4.** T'ai Chi and Yoga include relaxation and breathing techniques that improve mental and physical health.

_____ **5.** With humankind's ability to produce synthetic fibers that can clothe us with beauty and comfort, humans should refrain from murdering animals so their hides, coats, or furs can be used as clothing.

EXPLANATIONS

1. This is a statement of fact that can be researched.
2. This is a statement of opinion that includes the biased word *better*. Taste preferences are a matter of personal opinion.
3. This is a blend of fact and opinion. Popcorn is a low-calorie snack food, and its nutritional value can be verified; however, the value word *tasty* states an opinion.
4. This is a statement of fact that can be verified through research.
5. This is a statement that blends fact and opinion. The biased words *beauty*, *should*, and *murdering* are used.

EXAMPLE Look at the following cartoon. Then, write one fact and one opinion about the cartoon.

© Jeff Parker 1/14/04 *Florida Today* and Cagle Cartoons, Inc.

Fact: ______________________________

____

Opinion: ______________________________

____

EXPLANATION Compare your answers to the ones below.

Fact: Many Americans ignore the Surgeon General's warning against smoking. The percentage of Americans who smoke has declined since 1964.

Opinion: The Surgeon General's warning against smoking has had a significant effect on American smoking behaviors. People who smoke are foolish.

PRACTICE 5

A. Read the following statements and mark each one as follows:

F if it states a fact

O if it expresses an opinion

F/O if it combines fact and opinion

_____ **1.** I believe capital punishment is a crime against humanity.

_____ **2.** Executives of large corporations receive outrageously large salaries that unfairly reward them for successes created by lower paid rank-and-file workers.

______ **3.** *E. coli*, influenza, tuberculosis, measles, mumps, and smallpox are just a few of the diseases that are preventable through vaccinations.

______ **4.** More than 8 million Africans are infected with tuberculosis, with nearly 3 million dying each year; wealthier and more developed countries, such as the United States, should help fund widespread vaccination programs in Africa.

______ **5.** Astrology offers information that is effective in making important life decisions.

B. Read the following short reviews of destinations, restaurants, movies, and plays. Mark each one as follows:

F if it states a fact

O if it expresses an opinion

F/O if it combines fact and opinion

______ **6.** The 14th Street Bar and Grill in Boulder, Colorado, offers excellent food, average service, and a spectacular view of the Rocky Mountains.

______ **7.** Drago's Seafood in Metairie, Louisiana, offers Creole cuisine ranging from $21 to $24, with lobster and seafood as the main entrees.

______ **8.** Judson College, the nation's sixth-oldest women's college, offers a challenging and high-quality liberal arts education.

______ **9.** One bat can eat as many as one thousand insect pests, such as mosquitoes, in an hour.

______ **10.** *TMZ TV*, a television magazine show that delves into the private lives of celebrities, stands as an embarrassing example of the public's interest in gossip and the misfortune of other people.

Read Critically: Evaluate Details as Fact or Opinion in Context

Because the printed word seems to give authority to an idea, many of us accept what we read as fact. Yet much of what is published is actually opinion. Master readers question what they read. Reading critically is noting the use of fact and opinion in the context of a paragraph or passage, the author, and the type of source in which the passage is printed.

Evaluate the Context of the Passage

EXAMPLE Read the passage, and identify each sentence as follows:

F if it states a fact

O if it expresses an opinion

F/O if it combines fact and opinion

Vermicomposting

[1]Composting is an excellent way to recycle organic materials like food scraps and yard clippings. [2]Vermicomposting uses red worms to decompose organic material and create fertilizer for gardens. [3]In addition, it provides an easy, clean, odor-free, and environmentally correct way to get rid of garbage. [4]The worms are able to eat over half their body weight in one day, and their waste material, known as castings, is full of nutrients and microbes that promote plant growth. [5]Everyone should use vermicomposting.

1. _____ **2.** _____ **3.** _____ **4.** _____ **5.** _____

EXPLANATION Compare your answers to the ones below.

1. **F/O:** Composting is a way to recycle organic materials like food scraps and yard clippings; however, the word *excellent* is a biased word.
2. **F:** This statement can be verified.
3. **F/O:** Vermicomposting is a way to get rid of garbage; however, the words *easy, clean*, and *environmentally correct* are biased words or phrases.
4. **F:** This statement can be checked and verified as true.
5. **O:** This statement expresses an opinion with the words *everyone* and *should.*

PRACTICE 6

Read the passage, and identify each sentence as follows:

F if it states a fact

O if it expresses an opinion

F/O if it combines fact and opinion

What Makes a Home Energy Star?

[1]ENERGY STAR is the government-backed symbol for energy efficiency; homes that earn the ENERGY STAR are significantly more energy efficient than standard homes. [2]ENERGY STAR builders achieve this high efficiency for you by selecting from a variety of features such as the following:

Tight Construction and Ducts

[3]Advanced techniques for sealing holes and cracks in the home's "envelope" and in heating and cooling ducts help reduce drafts, moisture, dust, pollen, pests, and noise. [4]A tightly sealed home improves comfort and indoor air quality while lowering utility and maintenance costs.

Effective Insulation Systems

[5]Properly installed and inspected insulation in floors, walls, and attics ensures even temperatures throughout the house, while using less energy. [6]The result is lower utility costs and a quieter, more comfortable home.

Efficient Heating and Cooling Equipment

[7]An energy-efficient, properly installed heating and cooling system uses less energy to operate, which reduces your utility bills. [8]This system can also be quieter, reduce indoor humidity, and improve the overall comfort of your home.

High Performance Windows

[9]Energy-efficient windows employ advanced technologies, such as protective coatings and improved frame assemblies, to help keep heat in during the winter and out during the summer. [10]These windows also block damaging ultraviolet sunlight that can discolor carpets and furnishings.

—Adapted from U. S. Environmental Protection Agency. *Energy Star® Qualified New Homes*. EPA 430-F-09-053 March 2009. http://www.energystar.gov/ia/partners/downloads/consumer_brochure.pdf.

1. _____ **2.** _____ **3.** _____ **4.** _____ **5.** _____

6. _____ **7.** _____ **8.** _____ **9.** _____ **10.** _____

VISUAL VOCABULARY

Energy construction materials, including insulation, replaces a tornado destroyed home.
The best definition of **insulation** is material that ________ heat, electricity, or sound.

a. isolates
b. enhances
c. blocks

Evaluate the Context of the Author

Even though opinions can't be proved true like facts can, many opinions are still sound and valuable. To judge the accuracy of the opinion, you must consider the source: the author of the opinion. Authors offer two types of valid opinions: informed opinions and expert opinions.

> An author develops an **informed opinion** by gathering and analyzing evidence.
> An author develops an **expert opinion** through much training and extensive knowledge in a given field.

EXAMPLE Read the topic and study the list of authors who have written their opinions about the topic. Identify each person as **IO** if he or she is more likely to offer an informed opinion and **EO** if he or she is more likely to offer an expert opinion.

How to Lose Weight Safely and Effectively

______ **1.** Michael F. Roizen, MD, and Mehmet C. Oz, MD, authors of *You: On a Diet.*

______ **2.** An advice columnist such as Ann Landers or Dear Abby responding to a reader's question.

_____ **3.** Martha Stewart giving cooking tips on her nationally syndicated talk show.

_____ **4.** A student essay based on research.

EXPLANATION

1. Drs. Roizen and Oz are considered experts in the field of weight loss. One way to identify an expert opinion is to note if the person giving the opinion holds an advanced degree or title or has published articles or books about the topic being discussed. Both Roizen and Oz have the education and the achievement of being successful authors about this topic.
2. Advice columnists offer informed opinions on a wide range of topics. They often cite experts in their advice.
3. Martha Stewart has had extensive experience cooking and writing cookbooks and can provide an expert opinion about food preparation.
4. A student offers an informed opinion based on research.

Evaluate the Context of the Source

Often people turn to factual sources to find the factual details needed to form informed opinions and expert opinions. A medical dictionary, an English handbook, and a world atlas are a few excellent examples of factual sources.

EXAMPLE Read the following passage from a college psychology textbook. Answer the questions that follow.

Textbook Skills

Interpreting Dreams

[1]You may have wondered whether dreams, especially those that frighten us or that recur, have hidden meanings. [2]Sigmund Freud believed that dreams function to satisfy unconscious sexual and aggressive desires. [3]Because such wishes are unacceptable to the dreamer, they have to be disguised and therefore appear in dreams in symbolic forms. [4]Freud claimed that objects such as sticks, umbrellas, tree trunks, and guns symbolize the male sex organ; objects such as chests, cupboards, and boxes represent the female sex organ. [5]Freud differentiated between the **manifest content** of a dream—the content of the dream as recalled by the

dreamer—and the **latent content**—or the underlying meaning of the dream—which he considered more significant.

[6]Beginning in the 1950s, psychologists began to move away from the Freudian interpretation of dreams. [7]For example, Hall (1953) proposed a **cognitive theory of dreaming**. [8]He suggested that dreaming is simply thinking while asleep. [9]Advocates of Hall's approach argued for a greater focus on the manifest content. [10]The actual dream itself is seen as an expression of a broad range of the dreamer's concerns rather than as an expression of sexual impulses (Webb, 1975).

[11]Well-known sleep researcher J. Allan Hobson (1988) rejects the notion that nature would equip humans with the capability of having dreams that would require a specialist to interpret. [12]Hobson and McCarley (1977) advanced the **activation-synthesis hypothesis of dreaming**. [13]This hypothesis suggests that dreams are simply the brain's attempt to make sense of the random firing of brain cells during REM-sleep. [14]Just as people try to make sense of input from the environment during their waking hours, they try to find meaning in the conglomeration of sensations and memories that are generated internally by this random firing of brain cells. [15]Hobson (1989) believes that dreams also have psychological significance, because the meaning a person imposes on the random mental activity reflects that person's experiences, remote memories, associations, drives, and fears.

—From *Mastering the World of Psychology*, 3rd ed., p. 127 by Samuel E. Wood, Ellen Green Wood, and Denise Boyd. Copyright © 2008 by Pearson Education, Inc. Reproduced by permission of Pearson Education, Inc., Boston, MA.

_____ **1.** Sentence 1 states
a. a fact.
b. an opinion.
c. a fact and an opinion.

_____ **2.** Sentence 3 states
a. a fact.
b. an opinion.
c. a fact and an opinion.

_____ **3.** Sentence 6 states a
a. a fact.
b. an opinion.
c. a fact and an opinion.

_____ **4.** Overall, the theories of Freud, Hall, Hobson, and McCarley offer
a. informed opinions.
b. expert opinions.
c. factual details.

EXPLANATION Compare your answers to the following:

1. Sentence 1 states (b) an opinion. Many readers may have never "wondered" about dreams. This statement cannot be verified or proven.
2. Sentence 3 states the expert (b) opinion of Sigmund Freud—which is a partial explanation of his theory of dream interpretation.
3. Sentence 6 states (a) fact—an event in the history of dream interpretation.
4. Overall, the theories of Freud, Hall, Hobson, and McCarley offer (b) expert opinions. Scientific theories are plausible explanations for a fact or occurrence that can be observed. Thus, theories are not themselves facts, but reasonable explanations by experts supported by facts. As seen in this passage, conflicting theories about an occurrence can exist.

PRACTICE 7

Read the passage, and then, answer the questions that follow it.

Phobias

[1]All of us have experienced that gnawing feeling of fear in the pit of our stomach. [2]Perhaps we feared failing a test, losing a job, or disappointing a loved one, and each of these circumstances caused us to experience legitimate fear. [3]In contrast, a significant number of us experience fear for no known or logical reason. [4]According to the National Institute of Mental Health, millions of Americans, as many as one out of ten, suffer from phobias. [5]F. J. McGuigan, in his book *Encyclopedia of Stress,* defines a phobia as "an irrational, obsessive, and intense fear that is focused on a specific circumstance, idea, or thing." [6]Experts discuss phobias in terms of three subgroups: agoraphobia, social phobia, and specific phobia. [7]Agoraphobia is defined by *Merriam-Webster's Collegiate Dictionary* as the "abnormal fear of being helpless in an embarrassing or unescapable situation that is characterized especially by the avoidance of open or public places." [8]For example, McGuigan states that people who suffer agoraphobia tend to avoid crowds, tunnels, bridges, public transportation, and elevators. [9]In the textbook *Psychology and Life,* authors Richard Gerrig and Philip Zimbardo write that social phobia occurs as a person anticipates a public situation in which he or she will be observed by others. [10]People who suffer from social phobias fear public embarrassment. [11]"Finally, specific phobias occur in response to a specific object or situation." [12]For example, some people have an irrational fear of snakes, insects, or heights.

_____ **1.** Sentence 1 is
a. a fact. b. an opinion.

_____ **2.** In sentence 4, the National Institute of Mental Health is offered as
a. a factual resource. b. an informed opinion.

_____ **3.** In sentence 5, the *Encyclopedia of Stress* offers
a. a factual resource. b. an opinion.

_____ **4.** In sentence 7, *Merriam-Webster's Collegiate Dictionary* acts as
a. an expert opinion. b. a factual resource.

_____ **5.** In sentence 9, Gerrig and Zimbardo's definition is
a. an informed opinion. b. an expert opinion.

_____ **6.** Overall, this paragraph is
a. a factual resource. c. an expert opinion.
b. an informed opinion.

VISUAL VOCABULARY

1. Identify the meaning of the word parts:

agora means ____________ ____________

phobia means ____________

2. What type of phobia is agoraphobia? ____________ ____________

▲ Maya overcame agoraphobia and now enjoys city life.

Textbook Skills

Reading a Textbook: The Use of Graphics, Fact, and Opinion in a Textbook

Most textbook authors are careful to present only ideas based on observation, research, and expert opinion.

In addition, textbook authors often use pictures, drawings, or graphics to make the relationship between the main idea and supporting details clear. These graphics, most of which are based on factual data and expert opinions, require careful analysis to discern fact from opinion as you interpret them.

EXAMPLE Read the following passage and graphic illustration from a college nutrition textbook. Identify the sentences in the passage and the statements about the graphic as follows:

F if it states a fact

O if it states an opinion

F/O if it combines fact and opinion

[1]The Best Way to Meet Your Nutrient Needs Is with a Well-Balanced Diet

[2]Many foods provide a variety of nutrients. [3]For example, low-fat milk is high in carbohydrates and protein and provides a small amount of fat; milk is also a good source of the vitamins A, D, and riboflavin, as well as the minerals potassium and calcium, and is approximately 90 percent water by weight. [4]Whereas milk contains a substantial variety of all six classes of nutrients, as a single food item it doesn't have to provide all nutrients in order to be good for you. [5]Rather, a well-balanced diet composed of a variety of foods can provide you with all of these important nutrients (see Figure 1). [6]Also the delicious texture and aroma of foods coupled with the social interaction of meals are lost when you pop a pill to meet your nutrient needs; that said, some individuals should take a supplement if food alone can't meet their needs.

—From *Nutrition and You*, 1st ed., p. 12 by Joan Salge Blake. Copyright © 2008 by Pearson Education, Inc. Printed and Electronically reproduced by permission of Pearson Education, Inc., Upper Saddle River, NJ.

1. ______ **2.** ______ **3.** ______ **4.** ______ **5.** ______ **6.** ______

______ **7.** According to Figure 1, popcorn popped in oil is empty in nutritional value.

______ **8.** Broccoli is a low-fat, high-fiber food.

______ **9.** Chicken is probably more nutritional without the skin.

______ **10.** Broccoli contains more nutritional value than popcorn.

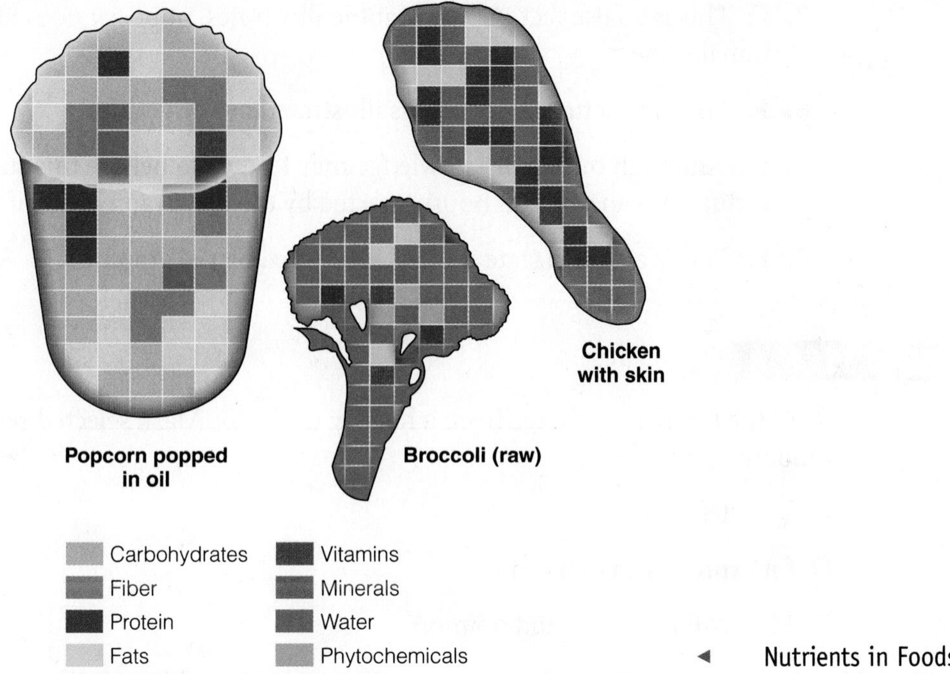

◄ Nutrients in Foods

—From *Nutrition and You*, 1st ed., Figure 1.2 (p. 12) by Joan Salge Blake. Copyright © 2008 by Pearson Education, Inc. Printed and Electronically reproduced by permission of Pearson Education, Inc., Upper Saddle River, NJ.

EXPLANATION Compare your answers to the ones that follow:

1. **O:** The word *Best* states a value judgment, which is based on expert opinion and facts. The last sentence actually contradicts this opinion by stating that some people cannot meet their nutritional needs through food alone.
2. **F:** This sentence states a commonly known and accepted fact.
3. **F/O:** Although the information in this sentence can be verified, the author includes the value word *good*. Some might say that milk is an *excellent* or *adequate* source of these nutrients.
4. **F/O:** Again, the author includes value words such as *substantial* and *good*. Again, these value judgments are based on facts and expert opinions, but these words are vague, subjective, and arguable in their meaning.
5. **F:** This sentence states a commonly known and accepted fact supported by the graphic illustration.
6. **O:** This sentence offers the author's opinion as indicated by the subjective tone words *delicious, pop a pill,* and *should*.

7. **F:** This is a false fact. As the graphic illustrates, popcorn does have nutritional value.

8. **F:** This is a factual statement as illustrated by the graphic.

9. **O:** Although our prior knowledge may lead us to believe this statement is factual, it is an opinion not supported by the paragraph or the illustration.

10. **F:** This is a factual statement as illustrated by the graphic.

PRACTICE 8

A. Read the following passage from a history textbook. Mark selected sentences as follows:

F if it states a fact

O if it expresses an opinion

F/O if it combines fact and opinion

Textbook Skills

Women and World War II

[1]World War II had a dramatic impact on women. [2]Easily the most visible change involved the sudden appearance of large numbers of women in uniform. [3]The military organized women into auxiliary units with special uniforms, their own officers, and amazingly, equal pay. [4]By 1945, more than 250,000 women had joined the Women's Army Corps (WAC), the Army Nurses Corps, the Women Accepted for Voluntary Emergency Service (WAVES), the Navy Nurses Corps, the Marines, and the Coast Guard. [5]Most women who joined the armed services either filled traditional women's roles, such as nursing, or replaced men in noncombat jobs.

[6]Women also substituted for men on the home front. [7]For the first time in history, married working women outnumbered single working women, as 6.3 million women entered the workforce during the war. [8]The war challenged the conventional image of female behavior, as "Rosie the Riveter" became the popular symbol of women who abandoned traditional female occupations to work in defense industries.

[9]Women paid a price for their economic independence, though. [10]Outside employment did not free wives from domestic duties. [11]The same women who put in full days in offices and factories went home to cook, clean, shop, and care for their children. [12]They had not one job, but two, and the only way they could fill both was to sacrifice relaxation, recreation, and sleep. [13]Outside employment also raised the problem of

child care. [14]A few industries, such as Kaiser Steel, offered day-care facilities, but most women had to make their own informal arrangements.

—Adapted from *America and Its Peoples, Volume II: A Mosaic in the Making,* 3rd ed., pp. 875–876 by James Kirby Martin, Randy J. Roberts, Steven Mintz, Linda O. McMurry and James H. Jones. Copyright © 1997 by Pearson Education, Inc. Printed and Electronically reproduced by permission of Pearson Education, Inc., Upper Saddle River, NJ.

______ **1.** Sentence 1

______ **2.** Sentence 2

______ **3.** Sentence 3

______ **4.** Sentence 4

______ **5.** Sentence 8

______ **6.** Sentence 9

______ **7.** Sentence 12

B. Study the image of "Rosie the Riveter." Then, identify each sentence, based on the picture, as follows:

F if it states a fact

O if it expresses an opinion

F/O if it combines fact and opinion

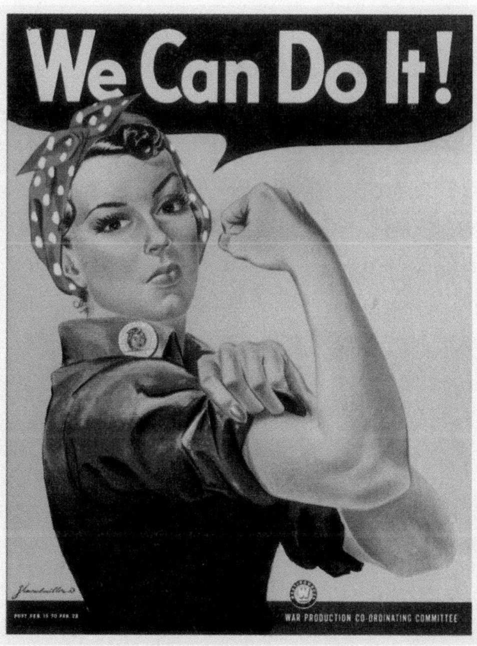

—"We Can Do It!" poster by J. Howard Miller. *Produced by Westinghouse for the War Production Co-Ordinating Committee, NARA Still Picture Branch (NWDNS-179-WP-1563)*, National Archives and Records Administration. © Courtesy of National Archives, photo war & conflict #798_we_can_do_it.

______ **8.** Some women gave up their traditional feminine dress to enter the workforce.

_____ **9.** "Rosie the Riveter" has the look of a strong, competent woman dressed in overalls and bandanna.

_____ **10.** "Rosie the Riveter" is the ideal patriotic woman.

APPLICATIONS

Textbook Skills

Application 1: Fact and Opinion

The following graph comes from a health textbook. Study the graph, and label each statement, based on the graph, as follows:

F if it states a fact

O if it expresses an opinion

F/O if it combines fact and opinion

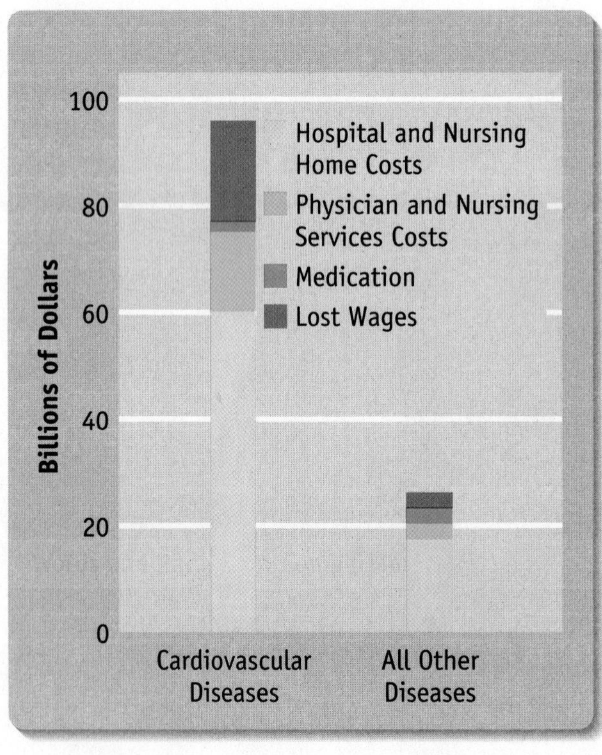

◄ Annual Economic Costs of Cardiovascular Diseases and All Other Diseases in the United States

—From *Total Fitness and Wellness Student Textbook Component,* 3rd ed. by Scott K. Powers and Stephen L. Dodd. Copyright © 2003 by Pearson Education, Inc. Printed and Electronically reproduced by permission of Pearson Education, Inc., Upper Saddle River, NJ.

_____ **1.** The economic costs of cardiovascular diseases are alarmingly higher than the costs for all other diseases in the United States.

_____ **2.** The yearly costs of cardiovascular diseases run more than $90 billion.

_____ **3.** In contrast to all other diseases, more dollars are lost in wages to cardiovascular diseases.

_____ **4.** Developing a national strategy to reduce the risk of cardiovascular disease should be a major health priority.

Application 2: Opinions, Biased Words, and Qualifiers

Read the following statements. Underline any biased words or qualifiers that are used. Then, mark each statement as follows:

F if it states a fact

O if it expresses an opinion

F/O if it combines fact and opinion

Textbook Skills

_____ **1.** Hybrid electric vehicles (HEVs) combine the internal combustion engine of a conventional vehicle with the battery and electric motor of an electric vehicle; the proven result is twice the fuel economy of conventional vehicles.

_____ **2.** The practical benefits of HEVs include not only improved fuel economy but also lower emissions, making HEVs the best cars on the market.

_____ **3.** By combining gasoline with electric power, hybrids have the same or greater range than traditional combustion engines.

_____ **4.** The 2008 Toyota Pruis gets between 43–56 miles per gallon.

_____ **5.** HEVs should be the only cars on the market.

—Adapted from United States Environmental Protection Agency. "Hybrid Vehicles." 5 Nov. 2007 http://www.fueleconomy.gov/

Application 3: Informed and Expert Opinions

Read the list of ten sources from which information can be obtained. Label each source as follows:

IO if it offers an informed opinion

EO if it offers an expert opinion

FS if it is a factual source

______ **1.** The National Archives and Records Administration (NARA), an independent federal agency that preserves our nation's history by overseeing the management of all federal records

______ **2.** A physical education instructor's recommendation for a knee injury

______ **3.** A physician's recommendation for a knee injury

______ **4.** A family member who has researched features and prices for several brands of dishwashers

______ **5.** A medical dictionary

______ **6.** *The Little, Brown Handbook,* a textbook used in college writing courses

______ **7.** A college course syllabus

______ **8.** About communication: Deborah Tannen, Professor of Linguistics, Georgetown University; author of several books and dozens of articles about communication

______ **9.** Lance Armstrong, six-time winner of the Tour de France, on the sport of cycling

______ **10.** The Web site *howstuffworks,* a commercial online resource for explanations about how things work

REVIEW TEST 1

Score (number correct) ______ × 20 = ______%

Read the following article from a college health textbook. Answer the questions that follow.

Textbook Skills

Do Restaurants and Food Marketers Encourage Overeating?

[1]When you go out to your local restaurant, do you think your dinner looks the same as one your grandmother might have ordered 50 years ago? [2]Would you be surprised to learn that today's serving portions are significantly larger than those of past decades? [3]Today's popular restaurant foods dwarf their earlier counterparts. [4]A 25 ounce prime-rib dinner served at one local steak chain, for example, contains nearly 3,000 calories and 150 grams of fat in the meat alone. [5]Add a baked potato with

sour cream or butter, a salad loaded with creamy dressing, and fresh bread with real butter, and the meal may surpass the 5,000-calorie mark and ring in at close to 300 grams of fat. In other words, it exceeds what most adults should eat in 2 days!

[6]And that is just the beginning. [7]Soft drinks, once commonly served in 12-ounce sizes, now come in "big gulps" and 1-liter bottles. [8]Cinnamon buns used to be the size of a dinner roll; now one chain sells them in giant, butter-laden, 700-calorie portions.

[9]What accounts for the increased portion sizes of today? [10]Restaurant owners might say that they are only giving customers what they want. [11]While there may be some merit to this claim, it's also true that bigger portions can justify higher prices, which help increase an owner's bottom line.

[12]A quick glance at the fattening of Americans provides growing evidence of a significant health problem. [13]According to Donna Skoda, a dietitian and chair of the Ohio State University Extension Service, "People are eating a ton of extra calories. [14]For the first time in history, more people are overweight in America than are underweight. [15]Ironically, although the U.S. fat intake has dropped in the past 20 years from an average of 0 to 33 percent of calories, the daily calorie intake has risen from 1,852 calories per day to over 2,000 per day. [16]In theory, this translates into a weight gain of 15 pounds a year." [17]Skoda and others say that the main reason Americans are gaining weight is that people no longer recognize a normal serving size. [18]The National Heart, Lung, and Blood Institute has developed a pair of "Portion Distortion" quizzes that show how today's portions compare with those of 20 years ago. [19]Test yourself online at http://hin.nhlbi.nih.gov/portion to see whether you can guess the differences between today's meals and those previously considered normal. [20]Just one example is the difference between an average cheeseburger 20 years ago and the typical cheeseburger of today. [21]According to the "Portion Distortion" quiz, today's cheeseburger has 590 calories—257 more calories than the cheeseburger of 20 years ago!

—Adapted from *Access to Health*, 11th ed., Green Edition, p. 292 by Rebecca Donatelle.

______ **1.** Sentence 3 states

a. a fact.
b. an opinion.
c. a fact and an opinion.

_____ **2.** Sentence 10 states
a. a fact.
b. an opinion.
c. a fact and an opinion.

_____ **3.** Sentence 11 states
a. a fact.
b. an opinion.
c. a fact and an opinion.

_____ **4.** Sentence 14 states
a. a fact.
b. an opinion.
c. a fact and an opinion.

_____ **5.** In the article, Donna Skoda offers
a. an informed opinion based on personal experience.
b. an expert opinion based on education and professional experience.
c. factual details only.

REVIEW TEST 2

Score (number correct) _____ × 10 = _____%

Fact and Opinion

Read the review, and answer the questions that follow it.

From the Pen of the Mighty King

[1]Stephen King, with his fantastical imagination, has built an impressive body of literary work worthy of recognition.[2]The scope of his writing includes short stories, novels, screenplays for movies and television series, nonfiction, illustrated novels, a comic book, and a children's cookbook. [3]However, he may be most widely known for those stories that explore the darkness of our fury and fears. [4]In fact, an editor for Barnes & Noble stated on the bookseller's website, "Stephen King proves once again why he is the reigning master of dark fiction" in their review of his novel *Dreamcatcher*. [5]His first novel, *Carrie*, the novel he allegedly threw in the trash, reveals the macabre side of his imagination. [6](Thank God his wife allegedly dug it out, flung it back and said, "finish it!") [7]*Carrie* is a dark, dark tale fueled by the ferocity of an adolescent experience;

the story encompasses outcasts, bullies, guilt, anger, sin, mayhem, death, justice, and fear. [8]In *Carrie*, King revealed his ability to make nightmares seem real. [9]In contrast, King also has the ability to probe into common human experiences with wit and tenderness, as in his short story "The Body," also made into a Hollywood movie, *Stand By Me*. [10]In this story, King's pen renders a compelling portrait of a young boy's end of innocence, his conflict with mortality, and his induction into adulthood. [11]Stephen King's body of work has made him the 2003 recipient of *The National Book Foundation Medal for Distinguished Contribution to American Letters*.

_____ **1.** Sentence 1 is a statement of
a. fact.
b. opinion.
c. a combination of fact and opinion.

_____ **2.** Sentence 2 is a statement of
a. fact.
b. opinion.
c. a combination of fact and opinion.

_____ **3.** Sentence 3 is a statement of
a. fact.
b. opinion.
c. a combination of fact and opinion.

_____ **4.** Sentence 5 is a statement of
a. fact.
b. opinion.
c. a combination of fact and opinion.

_____ **5.** Sentence 6 is a statement of
a. fact.
b. opinion.
c. a combination of fact and opinion.

_____ **6.** Sentence 7 is a statement of
a. fact.
b. opinion.
c. a combination of fact and opinion.

_____ **7.** Sentence 9 is a statement of
a. fact.
b. opinion.
c. a combination of fact and opinion.

_____ **8.** Sentence 10 is a statement of
a. fact.
b. opinion.
c. a combination of fact and opinion.

_____ 9. Sentence 11 is a statement of
a. fact.
b. opinion.
c. a combination of fact and opinion.

_____ 10. The editor from Barnes & Noble in sentence 4 offers
a. an informed opinion.
b. an expert opinion.
c. a fact.

REVIEW TEST 3

Score (number correct) _____ × 10 = _____%

Fact and Opinion

Textbook Skills

Before you read this passage from a psychology textbook, skim the passage and answer the Before Reading questions. Read the passage. Then, answer the After Reading questions.

Vocabulary Preview

sanctioning (2): permitting, authorizing
enterprise (7): venture, project
priorities (40): main concerns

Gambling and Sports

[1]Objections to the legalization of sports gambling appear to fall into two classes, moral and financial. [2]Many people do not like the idea of legally **sanctioning** a vice like gambling. [3]Part of the reason major sports organizations object is the justified fear that gambling would taint the image of professional sports. [4]Image is an important financial asset to sports teams and leagues. [5]A public suspicion that games might be "fixed" by crooked gambling interests could result in lowered gate receipts and a reduced TV audience (Prey, 1985, p. 210). [6]Curiously, gambling interests, even legal gambling interests, may oppose legalization of sports gambling (as did the Canadian provinces) to prevent a loss of their share of the gambling market. [7]For example, lawful off-track betting parlors could be expected to lose business to a newly legalized sports bookmaking **enterprise**.

[8]The argument that legal gambling might somehow threaten the integrity of sports appears to be wrong. [9]Over 95 percent of the sports betting in the United States, about $24 billion worth, is handled by illegal bookmakers (Prey, 1985, p. 196). [10]Legalization of sports gambling could be expected to draw money away from these underworld bookies toward aboveboard, legally regulated bookies. [11]It seems naive to believe that gamblers placing bets with legal bookmakers would be

likely to try to fix games while those already betting with illegal, possibly mob-controlled bookmakers would not. [12]Insofar as gambling threatens the integrity of sports, sports are already thoroughly threatened. [13]Legalization of sports gambling could only be expected to reduce this threat (Smith, 1990).

[14]Legal and illegal sports bookies serve as brokers of bets on highly visible sports events. [15]To bet on the outcome of a professional football game a person must stop by the bookie's office. [16]The office may be a bar, a magazine store, a gas station, or some other front business. [17]Good customers can place their bets by telephone, in effect taking a loan from the bookmaker. [18]To place a $100 wager, the bettor hands the bookie $110 or so. [19]The extra $10 is a commission paid to the bookmaker for his or her services. [20]This commission is only paid on losses. [21]Winners get their $110 back, plus $100 in winnings.

[22]The successful bookmaker must have several business skills. [23]The bookmaker makes a 5 percent commission on all bets handled if an equal amount is bet on each team. [24]To encourage equal wagering the bookie creates a line that requires bettors to predict not only the winning team but the point differential as well. [25]Thus, "Chicago over Minnesota by 9" means a Chicago backer wins only if Chicago wins by 9 points or more. [26]If not enough people bet on Minnesota and Chicago wins by 9 or more points, the bookie loses money. [27]It is important that the line attract an equal amount of wagering on the two teams. [28]A successful bookie has the knowledge to do this.

[29]A successful bookmaker may occasionally "take a bath" by accepting too many bets on the losing side of a game or, more correctly, a number of games, thereby having to pay out more money to winners than he or she collected from losers. [30]This temporary cash-flow problem cannot as a rule be quickly covered by a legitimate bank loan. [31]Herein lies one of the reasons bookmakers are likely to be linked with organized crime. [32]The best, and perhaps the only, source of a short-term, no-questions-asked loan is the local loan shark. [33]Loan sharking is a standard organized crime business that makes loans to "friends" at **exorbitant** (20 percent per week) interest, but this may seem a good deal to a bookie with temporary cash-flow problems. [34]Bookmakers may also occasionally require debt-collection services, which loan sharks can provide.

[35]A second tie-in with organized crime is protection from authorities. [36]Bookmaking operations, like any service business, must be accessible to the public if they are to make money. [37]This means remaining at one location with one telephone number for a long time. [38]If a would-be bettor can find a bookmaker, so can an arresting officer. [39]No illegal

bookmaking operation can succeed without somehow avoiding police interference. [40]Law-enforcement officials may choose to overlook a harmless, honest bookie in that no one is being hurt, otherwise law-abiding citizens are involved in placing bets, and police have more important **priorities** than to prosecute a bookmaker who may well be acquitted even if the case against the bookie is strong. [41]They may also refrain from interfering thanks to payoffs or favors to political leaders.

[42]These payoffs may be provided by an organized crime figure. [43]Few bookmakers are so successful that they can afford to pay off mayors and city council members, but an organized crime figure may assure protection of smalltime bookies in return for a share of the profits. [44]The natural tie-in with organized crime and government corruption provides the strongest argument for the legalization of gambling as well as some other vices (Barnes & Teeters, 1959).

—From *Sociology of Sport,* pp. 243–245 by John C. Phillips. Copyright © 1993 by Allyn & Bacon. Reproduced by permission of Pearson Education, Inc., Boston, MA.

Before Reading

Vocabulary in Context

_____ **1.** What does the word **exorbitant** mean in sentence 33?
a. excessive
b. reasonable
c. expected
d. low

After Reading

Main Idea and Implied Central Idea

_____ **2.** The main idea of paragraph 1 (sentences 1–7) is stated in
a. sentence 1.
b. sentence 2.
c. sentence 6.
d. sentence 7.

_____ **3.** Which of the following sentences best states the implied central idea of the passage?
a. Sports bookmaking is a billion dollar industry in the United States.
b. Sports gambling should be legalized to end the link to organized crime and government corruption.
c. Sports gambling, a billion dollar business, is a controversial issue for several reasons.

Supporting Details

_____ **4.** According to the author, lawful off-track betting parlors oppose the legalization of sports gambling because
a. gambling is a vice.
b. of the tie between sports gambling and organized crime.
c. they could lose money to newly legalized sports bookmaking businesses.
d. they fear damage to their image.

Transitions

_____ **5.** The relationship of ideas between sentence 6 and sentence 7 is one of
a. cause and effect.
b. time order.
c. contrast.
d. generalization and example.

_____ **6.** The relationship between the ideas within sentence 26 is one of
a. cause and effect.
b. time order.
c. contrast.
d. addition.

Thought Patterns

_____ **7.** The primary thought pattern that organizes paragraph 6 (sentences 35–41) is
a. cause and effect.
b. time order.
c. examples.
d. contrast.

Fact and Opinion

_____ **8.** Sentence 1 is
a. a fact.
b. an opinion.
c. both fact and opinion.

_____ **9.** Sentence 3 is
a. a fact.
b. an opinion.
c. a mixture of fact and opinion.

_____ **10.** Sentence 9 is
a. a fact.
b. an opinion.
c. a mixture of fact and opinion.

WHAT DO YOU THINK?

Is gambling a vice or just another form of entertainment? In what ways do you think cell phones and wireless computers have affected the bookmaking business? Assume that your city government wants to legalize gambling in your area. Write a letter to the editor of your local newspaper, which you also plan to mail to the mayor and city council members. In your letter, express your support of or opposition to legalized gambling in your area.

REVIEW TEST 4

Score (number correct) ______ × 10 = ______%

Fact and Opinion

Textbook Skills

Before Reading: Survey the following passage adapted from a college nutrition textbook. Skim the passage, noting the words in bold print. Answer the Before Reading questions that follow the passage. Then, read the passage. Next, answer the After Reading questions. Use the discussion and writing topics as activities to do after reading.

Vocabulary Preview

saturated (1): of or relating to an organic compound, especially a fatty acid, containing the maximum number of hydrogen atoms and only single bonds between the carbon atoms

unsaturated (2): of or relating to an organic compound with a double or triple bond that links two atoms, usually of carbon

What Is *Trans* Fat and Where Do You Find It?

[1]At one time, saturated fats from animal sources, like lard, and highly **saturated** tropical plant oils, like coconut and palm oils, were staples in home cooking and commercial food preparation. [2]These saturated fats work well in commercial food products because they provide a rich, flaky texture to baked goods and are more resistant to **rancidity** than the **unsaturated** fats found in oils. [3](The double bonds in unsaturated fats make them more susceptible to being damaged by oxygen, and thus, becoming rancid.) [4]Then, in the early twentieth century, a German chemist discovered the technique of **hydrogenation** of oils, which caused the unsaturated fatty acids in the oils to become more saturated. [5]*Trans fats* were born.

[6]The process of hydrogenation involves heating an oil and exposing it to **hydrogen** gas, which causes some of the double bonds in the unsaturated fatty acid to become saturated with hydrogen. [7]Typically, the hydrogens of a double bond are lined up in a *cis* (*cis* = same) **configuration**, that is,

hydrogenation (4): the adding of hydrogen to an unsaturated fatty acid to make it more saturated and solid at room-temperature

hydrogen (6): a colorless, tasteless, odorless, flammable gaseous substance that is the simplest and most abundant element

configuration (7): pattern, design

reformulated (13): changed, remade

cholesterol (17): a soft, waxy substance found among the lipids (fats) in the bloodstream and in the body's cells

they are all on the same side of the carbon chain in the fatty acid. [8]During hydrogenation, some hydrogens cross to the opposite side of the carbon chain, resulting in a *trans (trans* = cross) configuration. [9]The newly configured fatty acid is now a *trans fatty acid.*

[10]*Trans* fats provide a richer texture, a longer shelf life, and better resistance to **rancidity** than unsaturated fats, so food manufacturers use them in many commercially made food products. [11]The first partially hydrogenated shortening, Crisco, was made from cottonseed oil, and became available in 1911.

[12]*Trans* fats came into even more widespread commercial use when saturated fat fell out of favor in the 1980s. [13]Research had confirmed that saturated fat played a role in increased risk of heart disease, so food manufacturers **reformulated** many of their products to contain less saturated fat. [14]The easiest solution was to replace the saturated fat with *trans* fats. [15]Everything from cookies, cakes, and crackers to fried chips and doughnuts used *trans* fats to maintain their texture and shelf life. [16]*Trans* fats were also frequently used for frying at fast-food restaurants.

[17]We now know that *trans* fats are actually worse for heart health than saturated fat because they not only raise the LDL **cholesterol** levels, but they also lower HDL cholesterol in the body. [18]*Trans* fat currently provides an estimated 2.5 percent of the daily calories in the diets of adults in the United States. [19]Of this amount, about 25 percent of them are coming from naturally occurring *trans* fats that are found in meat and dairy foods. [20]We don't yet know if the naturally occurring *trans* fats have the same heart-unhealthy effects as do those that are created through hydrogenation. [21]*Trans* fats use should be kept as low as possible in the diet.

[22]The major sources of *trans* fats are commercially prepared baked goods, margarines, fried potatoes, snacks, shortenings, and salad dressings. [23]Whole grains, fruits, and vegetables don't contain any *trans* fats, so consuming a plant-based diet with minimal commercially prepared foods will go a long way toward preventing *trans* fat (and saturated fat) from overpowering your diet.

—Adapted from *Nutrition and You,* 1st ed., p. 147 by Joan Salge Blake. Copyright © 2008 by Pearson Education, Inc. Printed and Electronically reproduced by permission of Pearson Education, Inc., Upper Saddle River, NJ.

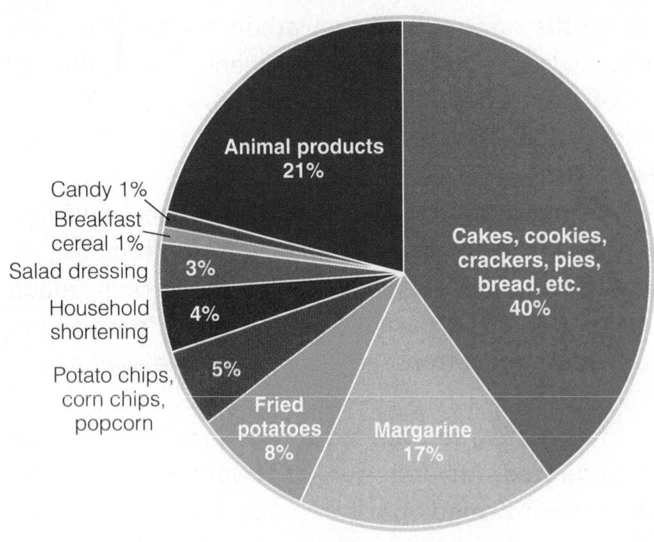

◄ **Major Food Sources of Trans Fat for U.S. Adults** Commercially made baked goods and snack items are the major contributors of trans fat in the diet.
—Source FDA, Questions and Answers About Trans Fat Nutrition Labeling, 2003.

Before Reading

Vocabulary in Context

Use context clues to state in your own words the definition of the following terms. Indicate the context clue you used.

1. In sentences 2 and 10 what does the word **rancidity** mean? ____________

____

2. Identify the context clue used for the word **rancidity** in sentences 2 and 10.

____

After Reading

Implied Central Idea

3. Using your own words, write a sentence that best states the implied main idea of the passage. ______________________________

____

____

Main Idea and Supporting Details

_____ **4.** In paragraph 4 (sentences 12–16), sentence 16 is a
a. main idea.
b. major supporting detail.
c. minor supporting detail.

Transitions

_____ **5.** The relationship of ideas between sentence 15 and sentence 16 is one of
a. cause and effect.
b. listing.
c. time order.
d. definition and example.

_____ **6.** The relationship of ideas within sentence 23 is one of
a. time order.
b. contrast.
c. cause and effect.
d. definition and example.

Thought Patterns

_____ **7.** The overall thought pattern for this passage is
a. time order.
b. definition and example.
c. space order.
d. comparison and contrast.

Fact and Opinion

_____ **8.** Sentence 4 is a statement of
a. fact.
b. opinion.
c. fact and opinion.

_____ **9.** Sentence 14 is a statement of
a. fact.
b. opinion.
c. fact and opinion.

_____ **10.** Read the caption for the graphic. The caption is a statement of
a. fact.
b. opinion.
c. fact and opinion.

WHAT DO YOU THINK?

Why are trans fats so harmful to human health? Assume you taking a college-level health course. Your professor has assigned weekly written responses to what you are learning through your studies. This week, you have decided to respond to what you have learned about a healthful diet and trans fats. You have narrowed your written response to two topics. Choose one of these topics and write several paragraphs using information from the passage about trans fats: (1) Evaluate your own diet and discuss ways in which your diet is healthful or needs to change to become healthful. (2) Persuade people who consume a diet high in trans fats to change to a diet balanced with grains, fruits, vegetables, and protein.

After Reading About Fact and Opinion

Before you move on to the Mastery Tests on fact and opinion, take time to reflect on your learning and performance by answering the following questions. Write your answers in your notebook.

- How has my knowledge base or prior knowledge about fact and opinion changed?
- Based on my studies, how do I think I will perform on the Mastery Test(s)? Why do I think my scores will be above average, average, or below average?
- Would I recommend this chapter to other students who want to learn about fact and opinion? Why or why not?

Test your understanding of what you have learned about fact and opinion by completing the Chapter 9 Review Card in the insert near the end of your text.

CONNECT TO PEARSON myreadinglab

To check your progress in meeting Chapter 9's learning outcomes log in to **www.myreadinglab.com** and try the following exercises.

- The "Critical Thinking" section of MyReadingLab gives additional information about fact and opinion. The section provides a model, practices, activities, and tests. To access this resource, click on the "Study Plan" tab. Then click on "Critical Thinking." Then click on the following links as needed: "Overview," "Model," "Critical Thinking: Facts and Opinions (Flash Animation)," "Practice," and "Tests."
- To measure your mastery of this chapter, complete the tests in the "Critical Thinking" section and click on Gradebook to find your results.

MASTERY TEST 1

Name ______________ Section ______________

Date ______________ **Score** (number correct) ______ × 10 = ______%

A. Read the following statements, and mark each one as follows:

F if it states a fact

O if it expresses an opinion

F/O if it combines fact and opinion

______ **1.** The barbell squat is the best exercise to sculpt beautiful, toned, firm legs.

______ **2.** The barbell squat develops the quadriceps muscles (the front upper leg), the hamstring muscles (back of upper leg), and the gluteal muscles (the buttocks).

______ **3.** In today's fast-paced, high-stress society, too many people eat to satisfy emotional needs.

______ **4.** Because nutritional supplements prevent nutritional deficiencies, everyone should take supplements.

______ **5.** Vitamin C is a water-soluble vitamin that may aid dieters by suppressing cortisol, a hormone that aids in fat storage and breaks down muscle tissue.

B. Read the following short reviews. Mark each one as follows:

F if it states a fact

O if it expresses an opinion

F/O if it combines fact and opinion

______ **6.** *Animal Farm* by George Orwell, Penguin, 1951.

Animal Farm is easily the most famous work of political allegory ever written. The animals take over the running of a farm, and everything is wonderful for a while—until the pigs get out of hand. It is a brilliant description of what happens when the revolution goes astray. Allegory is hard to do gracefully, but Orwell manages it superbly: while true appreciation of *Animal Farm* requires an understanding of the history of the Russian revolution, those without it will still get the point. And *Animal Farm* can even be appreciated as a story by children with no understanding of the political message at all!

—Yee, *Danny Yee's Book Reviews*, 10 Aug. 1992. 4 Feb. 2004 http://dannyreviews.com/s/romance.html

_____ **7.** *Beyond a Boundary* by C. L. R. James, Duke University Press, 1993.

Beyond a Boundary blends personal memoir, social history, and sports commentary. James's subject is cricket and its role in his own life and in the history of the West Indies and England. He begins with the place of cricket in his family history, in his childhood and schooling in Trinidad, and in the social stratification of West Indian society. He then recounts his personal experiences of some of the great West Indian cricketers, among them George John, Wilton St Hill, and above all Learie Constantine, who was a personal friend. In the three essays on Constantine, James also discusses league cricket and his own move to England and involvement with West Indian politics.

—Yee, *Danny Yee's Book Reviews*, 1998. 31 Jan. 2003 http://dannyreviews.com/h/Beyond_Boundary.html

_____ **8.** *Statistical Abstract of the United States, 2007.*

The *Statistical Abstract of the United States,* published since 1878, is the authoritative and comprehensive summary of statistics on the social, political, and economic organization of the United States. Sources of data include the Census Bureau, Bureau of Labor Statistics, Bureau of Economic Analysis, and many other Federal agencies and private organizations.

—United States. *The 2007 Statistical Abstract: The National Data Book.* U.S. Census Bureau. 5 Nov. 2007. http://www.census.gov/compendia/statab/

_____ **9.** The *Encyclopedia of Earth,* <http://www.eoearth.org/>.

The *Encyclopedia of Earth* is an electronic reference about the Earth, its natural environments, and their interaction with society. The *Encyclopedia* is a free, fully searchable collection of articles. Articles are written by scholars, professionals, educators, and experts who collaborate and review each other's work.

_____ **10.** Gladwell, Malcolm, *What the Dog Saw,* Little, Brown, and Co., 2009, $27.99.

Gladwell's latest book, *What the Dog Saw,* a package of his favorite articles from the *New Yorker* since he joined as a staff writer in 1996, is divided into three sections: The first deals with what he calls obsessives and minor geniuses; the second with flawed ways of thinking; the third on how we make predictions about people: will they make a good employee, are they capable of great works of art, or are they the local serial killer? Brought together, the pieces form a dazzling record of Gladwell's art, according to Ian Sample of the *Guardian.*

MASTERY TEST 2

Name ______________ Section ______________

Date ________ **Score** (number correct) ______ × 5 = ______%

A. Read the list of statements, and mark each one as follows:

F if it states a fact; **O** if it expresses an opinion; **F/O** if it combines fact and opinion

______ **1.** Capital punishment is immoral and cruel. It should be outlawed in a civilized society.

______ **2.** Any criminal, no matter how hardened, can be and should be rehabilitated to reenter society.

______ **3.** As of January 2007, 3,350 men and women were on death rows across the United States.

______ **4.** Capital punishment is the only form of justice for heinous crimes against humanity.

______ **5.** A majority of nations have ended capital punishment in law or practice.

B. Read the paragraph from a college history textbook. Identify the numbered sentences as follows:

F if it states a fact; **O** if it expresses an opinion; **F/O** if it combines fact and opinion

Textbook Skills

Settling the Far West

[6]During the 1840s thousands of pioneers headed westward toward California and Oregon. [7]In 1841, the first party of 69 pioneers left Missouri for California, led by an Ohio schoolteacher named John Bidwell. [8]The hardships the party endured were nearly unbearable. [9]They were forced to abandon their wagons and eat their pack animals, "half roasted, dripping with blood." [10]Over the next 25 years, 350,000 more made the trek along the overland trails.

—From *America and Its Peoples, Volume I: A Mosaic in the Making*, 3rd ed., pp. 423–424 by James Kirby Martin, Randy J. Roberts, Steven Mintz, Linda O. McMurry and James H. Jones. Copyright © 1997 by Pearson Education, Inc. Printed and Electronically reproduced by permission of Pearson Education, Inc., Upper Saddle River, NJ.

6. ______ **7.** ______ **8.** ______ **9.** ______ **10.** ______

C. Read the following passage from a college history textbook. Identify each numbered sentence as follows:

F if it states a fact; **O** if it expresses an opinion; **F/O** if it combines fact and opinion

Textbook Skills

Trailblazing

[11]In 1811 and 1812 fur trappers marked out the Oregon Trail, the longest and most famous pioneer route in American history. [12]This trail crossed about 2000 miles from Independence, Missouri, to the Columbia River country of Oregon. [13]During the 1840s, 12,000 pioneers traveled the Trail's entire length to Oregon.

[14]Travel on the Oregon Trail was a tremendous test of human endurance. [15]The journey by wagon train took six months. [16]Settlers encountered prairie fires, sudden blizzards, and impassable mountains. [17]Cholera and other diseases were common; food, water, and wood were scarce. [18]Only the stalwart dared brave the physical hardship of the westward trek.

—From *America and Its Peoples, Volume I: A Mosaic in the Making*, 3rd ed., p. 425 by James Kirby Martin, Randy J. Roberts, Steven Mintz, Linda O. McMurry and James H. Jones.

11. ______ **12.** ______ **13.** ______ **14.** ______ **15.** ______

16. ______ **17.** ______ **18.** ______

D. Study the picture and its caption. Mark each idea as follows:

F if it states a fact; **O** if it expresses an opinion; **F/O** if it combines fact and opinion

◂ Life along the westward trails was a tremendous test of human endurance.

______ **19.** Pioneers encountered arid desert, difficult mountain passes, dangerous rivers, and quicksand.

______ **20.** Still, despite the hardships of the experience, few emigrants ever regretted their decision to move west.

—Text and image adapted from *America and Its Peoples, Volume I: A Mosaic in the Making*, 3rd ed., p. 425 by James Kirby Martin, Randy J. Roberts, Steven Mintz, Linda O. McMurry and James H. Jones.

MASTERY TEST 3

Name ______________________ Section ______________

Date __________ **Score** (number correct) ______ × 10 = ______%

A. The following passage and the figure in section B were published together in a health textbook. Read the passage, and identify the numbered sentences as follows:

F if it states a fact

O if it expresses an opinion

F/O if it combines fact and opinion

Textbook Skills

Content and Information Regulation

[1]Direct content regulation emerged from government efforts to balance the free flow of information and ideas against the negative effects of media products. [2]Part of the news media's role, as H. L. Mencken said, "is to comfort the afflicted and afflict the comfortable." [3]Content regulation tries to reduce unjustified, unnecessary, and unreasonable harm to people from media content. [4]Such regulation can occur before or after distribution. [5]Some types of speech, such as political speech, are more protected than others, such as commercial speech.

—Folkerts & Lacy, *The Media in Your Life*, 2nd ed., p. 327.

1. ______ **2.** ______ **3.** ______ **4.** ______ **5.** ______

VISUAL VOCABULARY

The business conversation of this real estate agent is an example of ______________ speech.

a. political
b. commercial

B. Study the graph. Read the statements that are based on the figure. Then, identify each one as follows:

Textbook Skills

F if it states a fact

O if it expresses an opinion

F/O if it combines fact and opinion

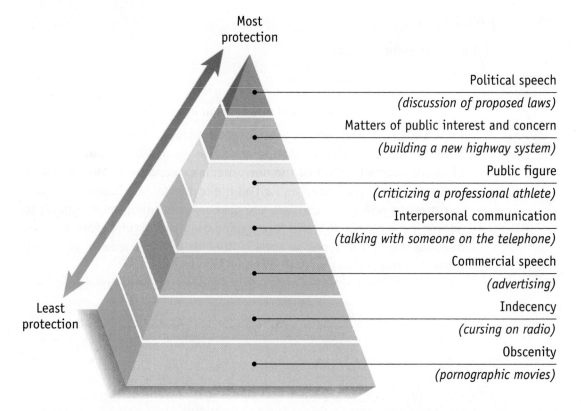

▲ **Levels of Protected Communication**

—From Folkerts & Lacy, *The Media in Your Life*, p. 327. Published by Allyn and Bacon. Copyright © 2001. "Levels of Protected Communication" by Todd F. Simon. Reprinted by permission of author.

______ **6.** Political speech includes discussions of proposed laws.

______ **7.** This chart outlines the levels of protected speech.

______ **8.** Talking with someone on a telephone has a degree of protection.

______ **9.** Obscenity should not be protected communication.

______ **10.** According to the graphic, political communication receives more protection than commercial communication; thus a politician's speech is a more important form of communication than is a commercial.

MASTERY TEST 4

Name ______________________ Section ____________

Date __________ **Score** (number correct) ______ × 25 = ______%

Read the following information posted on the website of the Smithsonian Institute. Answer the questions that follow.

Tattoos: The Ancient and Mysterious History

[1]Humans have marked their bodies with tattoos for thousands of years. [2]These permanent designs are sometimes plain, sometimes elaborate, always personal. [3]They have served as amulets, status symbols, declarations of love, signs of religious beliefs, adornments and even forms of punishment. [4]Joann Fletcher, research fellow in the department of archaeology at the University of York in Britain, describes the history of tattoos and their cultural significance to people around the world.

The earliest evidence of tattoos

[5]In terms of tattoos on actual bodies, the earliest known examples were for a long time Egyptian, and tattoos were present on several female mummies dated to c. 2000 B.C. [6]But the more recent discovery of the Iceman from the area of the Italian-Austrian border in 1991 and his tattoo patterns serves as an earlier example. [7]He was carbon-dated at around 5,200 years old.

Description of the tattoos on the Iceman and their significance

[8]According to Professor Don Brothwell of the University of York, one of the specialists who examined him, the distribution of the tattooed dots and small crosses on his lower spine and right knee and ankle joints correspond to areas of strain-induced degeneration. [9]This pattern suggests that the tattoos may have been applied to alleviate joint pain, and they were therefore essentially therapeutic. [10]This would also explain their somewhat "random" distribution in areas of the body. [11]The marks would not have been that easy to display had they been applied as a form of status marker.

Evidence that ancient Egyptians had tattoos

[12]There's certainly evidence that women had tattoos on their bodies and limbs from figurines from 4000-3500 B.C. [13]Occasional female figures are represented in tomb scenes around 1200 B.C. [14]And tattoos are evident in figurine form around 1300 B.C. All these had tattoos on their thighs. [15]Also small bronze implements identified as tattooing tools were discovered at the town site of Gurob in northern Egypt and dated 1450 B.C. [16]And then, of course, there are the

mummies with tattoos. [17]Included are the three women already mentioned and dated to c. 2000 B.C. [18]And several later examples of female mummies with these forms of permanent marks were found in Greco-Roman burials at Akhmim.

The function served by tattoos: Who and why

[19]Because this seemed to be an exclusively female practice in ancient Egypt, mummies found with tattoos were usually dismissed by the (male) excavators who seemed to assume the women were of "dubious status." [20]They were described in some cases as "dancing girls." [21]The female mummies had nevertheless been buried at Deir el-Bahari (opposite modern Luxor) in an area associated with royal and elite burials. [22]And we know that at least one of the women described as "probably a royal concubine" was actually a high-status priestess named Amunet, as revealed by her funerary inscriptions.

[23]And although it has long been assumed that such tattoos were the mark of prostitutes or were meant to protect the women against sexually transmitted diseases, I personally believe that the tattooing of ancient Egyptian women had a therapeutic role and functioned as a permanent form of amulet during the very difficult time of pregnancy and birth. [24]This is supported by the pattern of distribution, largely around the abdomen, on top of the thighs and the breasts. [25]This would also explain the specific types of designs, in particular the net-like distribution of dots applied over the abdomen. [26]During pregnancy, this specific pattern would expand in a protective fashion in the same way bead nets were placed over wrapped mummies to protect them and "keep everything in." [27]The placing of small figures of the household deity Bes at the tops of their thighs would again suggest the use of tattoos as a means of safeguarding the actual birth. [28]Bes was the protector of women in labor, and his position at the tops of the thighs is a suitable location. [29]This would ultimately explain tattoos as a purely female custom.

—Adapted from Lineberry, "Tattoos: The Ancient and Mysterious History." Smithsonian.com 1 Jan. 2007. 23 March 2010. http://www.smithsonianmag.com/history-archaeology/tattoo.html.

______ **1.** Sentence 1 states
a. a fact. b. an opinion. c. both fact and opinion.

______ **2.** Sentence 5 states
a. a fact. b. an opinion. c. both fact and opinion.

______ **3.** Sentence 23 states
a. a fact. b. an opinion. c. both fact and opinion.

______ **4.** Sentence 27 states
a. a fact. b. an opinion. c. both fact and opinion.

Tone and Purpose

10

LEARNING OUTCOMES

After studying this chapter you should be able to do the following:

1. Define the following terms: *tone* and *purpose*.
2. Understand how tone is established.
3. Identify subjective and objective tone words.
4. Determine the general purpose in the main idea.
5. Determine the primary purpose.
6. Evaluate a passage for the use of irony.
7. Use tone and purpose to improve comprehension.

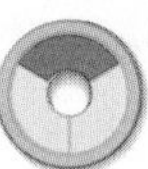

Before Reading About Tone and Purpose

Study the chapter learning outcomes and underline words that relate to ideas you have already studied. Did you underline the following terms: subjective, objective, and main idea? What you already know about these topics will help you learn about tone and purpose. Use the blanks that follow to write a short one- or two-sentence summary about each topic:

Subjective words: ____

____.

Objective words: ____

____.

Main idea: ____

____.

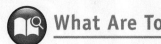

What Are Tone and Purpose?

Read the following two paragraphs. As you read, think about the difference in the tone and purpose of each one:

> Youth substance abuse can lead to many other problems, including the development of delinquent behavior, anti-social attitudes, and health-related issues. These problems not only affect the child, but can also influence the child's family, community, and ultimately society.
>
> —Executive Office of the President of the United States, "Juveniles and Drugs." Executive Office of the President. Office of National Drug Control Policy. 2 Nov. 2007. http://www.whitehousedrugpolicy.gov/drugfact/juveniles/index.html

> My name is Sheanne; I am 17 years old, and I am an alcoholic. I began drinking when I was 9 years old. My older cousins used to party pretty hard, and it seemed so cool, like they were having so much fun. They always got a kick out of giving me a beer on the sly, and I loved being part of the scene. By the time I was 11, I was guzzling hard booze. I fought with my family all the time, and school was a blur. Then last year, drunk as usual, I insisted on driving home from a late-night party. My twin sister, Shannon, rode with me. I don't remember what happened; I blacked out. Shannon died in that crash. I have been sober for 305 days. Every day is a struggle. Take my advice: Don't drink! If you do, don't drink and drive!
>
> —Sheanne's speech to a high school audience

The differences in the tone and purpose of these two paragraphs are obvious. The first paragraph was written by the government to inform the public about juveniles and drug abuse, using unbiased words and an objective, formal tone. The second paragraph approaches the same subject, drug abuse, with a different purpose—to persuade youth to avoid alcohol abuse. Sheanne conveys a painful personal experience using biased words and a subjective, informal tone.

Every text is created by an author who has a specific attitude toward the chosen topic and a specific reason for writing and sharing that attitude. The author's attitude is conveyed by the tone. **Tone** is the emotion or mood of the author's written voice. Understanding tone is closely related to understanding the author's reason for writing about the topic. This reason for writing is known as the author's **purpose.** Tone and purpose work together to convey the author's meaning.

Tone and purpose are greatly influenced by the audience the author is trying to reach. The audience for the first paragraph is members of the general public who need factual information; the objective presentation of facts best serves such a wide-ranging audience. The audience for the second example is teenagers who need to be persuaded; the informal, personal approach is much more likely to reach this audience. Tone and purpose are established with word choice. Master readers read to understand the author's tone and purpose. To identify tone and purpose, you need to build on several skills you have already studied: vocabulary, fact and opinion, and main ideas.

Tone is the author's attitude toward the topic.
Purpose is the reason the author writes about a topic.

Understand How Tone Is Established

The author's attitude is expressed by the tone of voice he or she assumes in the passage. An author carefully chooses words that will make an impact on the reader. Sometimes an author wants to appeal to reason by using an objective tone, and just gives facts and factual explanations. At other times, an author wants to appeal to emotions by using a subjective tone to stir the reader to feel deeply.

Study the following list of words that describe the characteristics of tone.

Characteristics of Tone Words

Objective Tone	**Shows no feelings for or against a topic**	**Subjective Tone**	**Shows favor for or against a topic**
unbiased	—remains impartial	biased	—makes it personal
neutral	—focuses on facts	emotional	—focuses on feelings
formal	—uses higher level words —avoids personal pronouns *I* and *you* —creates distance between writer and reader	informal	—uses conversational language —uses personal pronouns *I* and *you* —creates a connection between writer and reader
An objective tone is impartial, unbiased, neutral and most often formal.		A subjective tone is personal, biased, emotional, and often informal.	

For example, in an effort to share reliable information, textbooks strive for an objective tone, one that is matter-of-fact and neutral. The details given in an objective tone are likely to be facts. In contrast, sharing an author's personal world view through fiction and personal essays often calls for a subjective tone. A subjective tone uses words that describe feelings, judgments, or opinions. The details given in a subjective tone are likely to include experiences, senses, feelings, and thoughts.

EXAMPLES Look at the following list of quotations. Based on word choice, choose the tone word that best describes each statement.

______ **1.** "Permanent success cannot be achieved except by incessant intellectual labour, always inspired by the ideal."

Sarah Bernhardt (1845–1923), French actor

—From *The Art of the Theatre*, ch. 3 (1924); qtd. in *The Columbia World of Quotations*, 1996. *Bartleby.com*, 2001. 7 Nov. 2007 www.bartleby.com/66/

a. objective b. subjective

______ **2.** "It is . . . to my three children [that] I owe my very being. In attempting to fulfill my duty to them as a mother, I met the challenge of their helplessness, their innocence, their dependence. Despising cowardice in others, I wished to prove myself no coward. Believing in the good, the gentle, the beautiful things of life, I addressed myself to the sweet duty of keeping these attributes for my children's sake and my own. And in striving to provide a living for them, I found a success beyond my wildest dreams."

Alice Foote MacDougall (1867–1945), U.S. businesswoman

—From *The Autobiography of a Business Woman*, ch. 4 (1928); qtd. in *The Columbia World of Quotations*, 1996. *Bartleby.com*, 2001. 7 Nov. 2007 www.bartleby.com/66/

a. emotional b. neutral

______ **3.** "Success is a great deodorant. It takes away all your past smells."

Elizabeth Taylor, American film actor

—From ABC-TV interview April 6, 1977; qtd. in *The Columbia World of Quotations*, 1996. *Bartleby.com*, 2001. 7 Nov. 2007 www.bartleby.com/66/

a. formal b. informal

EXPLANATIONS

1. (b) subjective: Sarah Bernhardt establishes a subjective tone through the use of biased language with words such as *incessant*, *intellectual*, *labor*, *inspire*, or *ideal.*

2. (a) emotional: MacDougall expresses a passionate drive for success.

3. (b) informal: Taylor uses colloquial language (commonly spoken words) for a witty effect.

PRACTICE 1

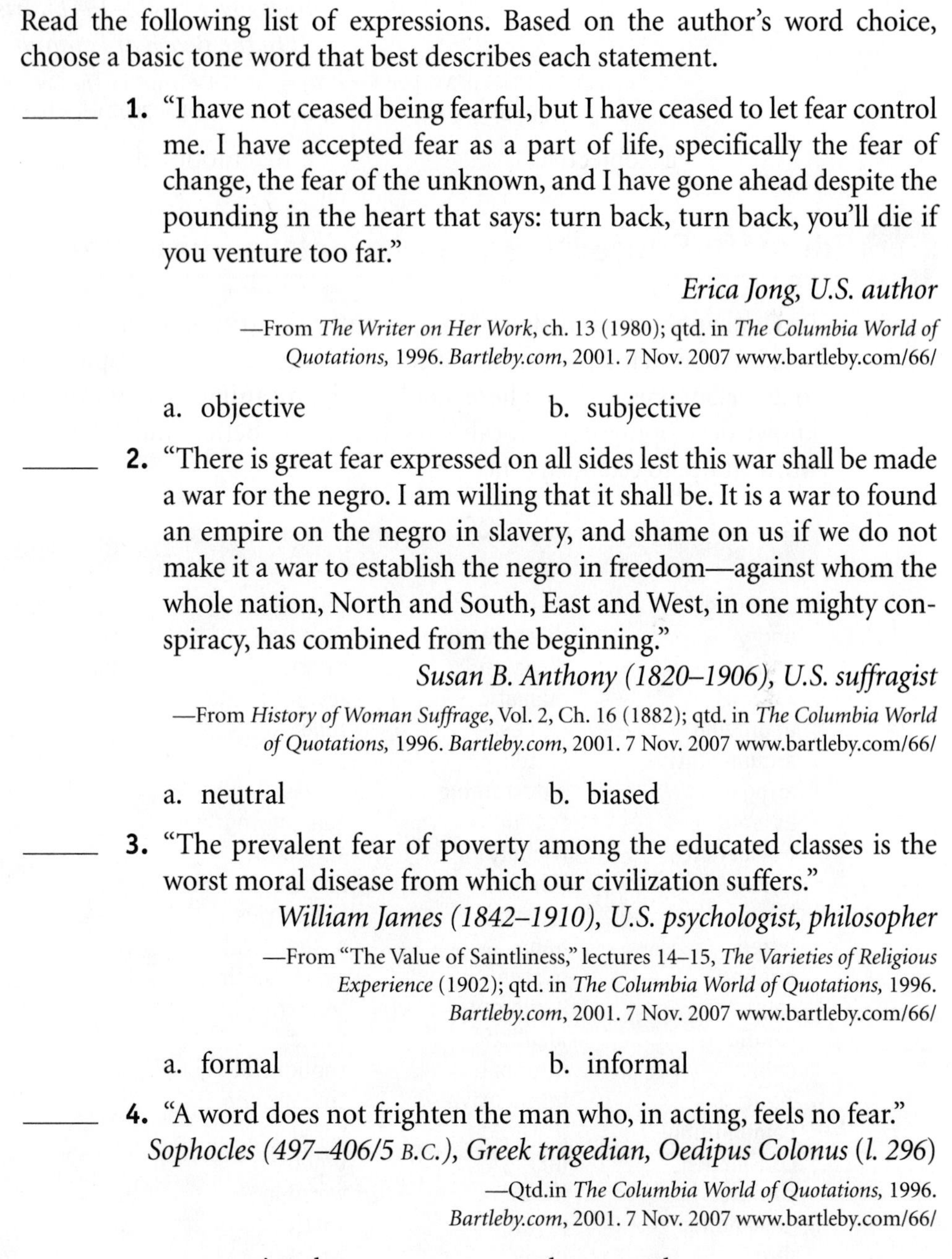

Read the following list of expressions. Based on the author's word choice, choose a basic tone word that best describes each statement.

_____ **1.** "I have not ceased being fearful, but I have ceased to let fear control me. I have accepted fear as a part of life, specifically the fear of change, the fear of the unknown, and I have gone ahead despite the pounding in the heart that says: turn back, turn back, you'll die if you venture too far."

Erica Jong, U.S. author

—From *The Writer on Her Work*, ch. 13 (1980); qtd. in *The Columbia World of Quotations*, 1996. *Bartleby.com*, 2001. 7 Nov. 2007 www.bartleby.com/66/

a. objective   b. subjective

_____ **2.** "There is great fear expressed on all sides lest this war shall be made a war for the negro. I am willing that it shall be. It is a war to found an empire on the negro in slavery, and shame on us if we do not make it a war to establish the negro in freedom—against whom the whole nation, North and South, East and West, in one mighty conspiracy, has combined from the beginning."

Susan B. Anthony (1820–1906), U.S. suffragist

—From *History of Woman Suffrage*, Vol. 2, Ch. 16 (1882); qtd. in *The Columbia World of Quotations*, 1996. *Bartleby.com*, 2001. 7 Nov. 2007 www.bartleby.com/66/

a. neutral   b. biased

_____ **3.** "The prevalent fear of poverty among the educated classes is the worst moral disease from which our civilization suffers."

William James (1842–1910), U.S. psychologist, philosopher

—From "The Value of Saintliness," lectures 14–15, *The Varieties of Religious Experience* (1902); qtd. in *The Columbia World of Quotations*, 1996. *Bartleby.com*, 2001. 7 Nov. 2007 www.bartleby.com/66/

a. formal   b. informal

_____ **4.** "A word does not frighten the man who, in acting, feels no fear."

Sophocles (497–406/5 B.C.), Greek tragedian, Oedipus Colonus (l. 296)

—Qtd.in *The Columbia World of Quotations*, 1996. *Bartleby.com*, 2001. 7 Nov. 2007 www.bartleby.com/66/

a. emotional   b. neutral

_____ 5. Around us fear, descending
Darkness of fear above
And in my heart how deep unending
Ache of love!

James Joyce (1882–1941), Irish writer,
"On the Beach at Fontana" (l. 9–12)

—From *Oxford Book of Modern Verse*, Yeats, ed. (1936); qtd. in *The Columbia World of Quotations*, 1996. *Bartleby.com*, 2001. 7 Nov. 2007 www.bartleby.com/66/

a. objective b. emotional

Identify Subjective and Objective Tone Words

Recognizing tone and describing an author's attitude deepens your comprehension and helps you become a master reader. A small sample of words used to describe tone are listed here. Look up the meanings of any words you do not know; developing your vocabulary helps you better understand an author's word choice to establish tone.

Subjective			Objective
admiring	disbelieving	persuasive	accurate
angry	discouraged	pleading	factual
annoyed	disdainful	poetic	impartial
anxious	dramatic	reverent	matter-of-fact
approving	earnest	rude	straightforward
argumentative	elated	sad	truthful
arrogant	entertaining	sarcastic	
assured	fearful	self-pitying	
belligerent	friendly	serious	
biting	funny	sincere	
bitter	gloomy	supportive	
bored	happy	suspenseful	
bubbly	hostile	sympathetic	
calm	humorous	tender	
candid	idealistic	tense	
cold	informal	thoughtful	
comic	informative	threatening	
complaining	irritated	timid	
confident	joking	urgent	
cynical	jovial	warning	
demanding	joyful	wistful	
direct	lively	wry	
disappointed	loving		

EXAMPLE Read the following quotations. Choose a word that best describes the tone of each statement.

cautionary	confident	critical	joyous	reverent

1. "Queenliness is an attitude that starts on the inside and works its way out. The way you hold your head up makes you a queen . . . I know who I am. I am confident. I know God. I can take care of myself. I share my life with others, and I love—I am worthy of the title Queen."

   Queen Latifah, actress, singer, rap artist

   —Source: From *Ladies First: Revelations of a Strong Woman* by Queen Latifah and Karen Hunter. Copyright 1999 by Queen Latifah, Inc.

   Tone: ______________________________

2. "Music puts me in touch with something beyond the intellect, something otherworldly, something sacred."

   Sting, musician

   —"Mystery and Religion of Music," Berklee College of Music, 1994.

   Tone: ______________________________

3. "After a century of striving, after a year of debate, after a historic vote, health care reform is no longer an unmet promise. It is the law of the land. It is the law of the land."

   President Barack Obama

   —Remarks by the President on Health Insurance Reform at the Department of the Interior, 23 March 2010 whitehouse.gov/the-press-office.

   Tone: ______________________________

4. "Americans wanted us to get at the root of this problem, which is cost. Instead, Democrats are spending trillions more on a system that already costs too much and forcing seniors, small business owners and middle class families to pay for it. You can call that a lot of things. You might even call it historic. But you can't call it reform."

   Senator Mitch McConnell, Republican-Kentucky, Minority Leader

   —Remarks in press release regarding health care bill, 23 March 2010 http://mcconnell.senate.gov/public/index.cfm?p=PressReleases&ContentRecord_id=3ae19c9a-060b-47e3-9558-60e262b02427.

   Tone: ______________________________

5. "As though a deep gulf were yawning below, As crossing thin ice, Take heed how ye go."

Confucius

—*The Sayings of Confucius.* Vol. XLIV, Part 1. *The Harvard Classics.* New York: P.F. Collier & Son, 1909–14; Bartleby.com, 2001. www.bartleby.com/44/1/. [24 March 2010].

Tone: ______________________________

VISUAL VOCABULARY

Use a tone word to describe the mood of President Obama and his team as they work to pass the health care reform bill in 2010.

____

____

EXPLANATION Compare your answers to the following: (1) confident, (2) reverent, (3) joyous, (4) critical, (5) cautionary.

PRACTICE 2

Read the following items. Based on the author's word choice, choose a word from the box that best describes the tone of each statement.

emotional	humble	matter-of-fact	reflective	respectful

1. **Harvard University, Cambridge, May 28, 1896**
   President Booker T. Washington

MY DEAR SIR: Harvard University desires to confer on you at the approaching Commencement an honorary degree; but it is our custom to confer degrees only on gentlemen who are present. Our Commencement occurs this year on June 24, and your presence would be desirable from about noon till about five o'clock in the afternoon. Would it be possible for you to be in Cambridge on that day?

Believe me, with great regard,
Very truly yours,
CHARLES W. ELIOT.

—From Washington, Booker, *Up from Slavery: An Autobiography,* ch. XVII, p. 4 (1901); *Bartleby.com,* 2000. 24 Feb. 04 www.bartleby.com/1004/

Tone: ______________________________

2. "This was a recognition that had never in the slightest manner entered into my mind, and it was hard for me to realize that I was to be honoured by a degree from the oldest and most renowned university in America."

Booker T. Washington, President of Tuskegee College

—From *Up from Slavery: An Autobiography,* ch. XVII, p. 5 (1901); *Bartleby.com,* 2000. 24 Feb. 04 www.bartleby.com/1004/

Tone: ______________________________

3. "As I sat upon my veranda, with this letter in my hand, tears came into my eyes. My whole former life—my life as a slave on the plantation, my work in the coal-mine, the times when I was without food and clothing, when I made my bed under a sidewalk, my struggles for an education, the trying days I had had at Tuskegee, days when I did not know where to turn for a dollar to continue the work there, the ostracism and sometimes oppression of my race,—all this passed before me and nearly overcame me."

Booker T. Washington, President of Tuskegee College

—From *Up from Slavery: An Autobiography,* ch. XVII, p. 5 (1901); *Bartleby.com,* 2000. 24 Feb. 04 www.bartleby.com/1004/

Tone: ______________________________

4. "As this was the first time that a New England university had conferred an honorary degree upon a Negro, it was the occasion of much newspaper comment throughout the country."

Booker T. Washington, President of Tuskegee College

—From *Up from Slavery: An Autobiography,* ch. XVII, p. 10 (1901); *Bartleby.com,* 2000. 24 Feb. 04 www.bartleby.com/1004/

Tone: ______________________________

5. "If my life in the past has meant anything in the lifting up of my people and the bringing about of better relations between your race and mine, I assure you from this day it will mean doubly more."

Booker T. Washington, President of Tuskegee College

—From *Up from Slavery: An Autobiography,* ch. XVII, p. 9 (1901); *Bartleby.com,* 2000. 24 Feb. 04 www.bartleby.com/1004/

Tone: ______________________________

Discover the General Purpose in the Main Idea

In Chapter 3, you learned that a main idea is made up of a topic and the author's controlling point. You identified the controlling point by looking for thought patterns and biased (tone) words. The next two sections will build on what you have learned. First, you will study the relationship between the general purpose and the author's main idea. Then, you will apply what you have learned to figure out an author's primary purpose.

Many reasons can motivate a writer. These can range from the need to take a stand on a hotly debated issue to the desire to entertain an audience with an amusing story. Basically, an author writes to share a main idea about a topic. An author's main idea, whether stated or implied, and the author's purpose are directly related. One of the following three general purposes will drive a main idea: to inform, to persuade, and to entertain.

- **To inform.** When a writer sets out to inform, he or she shares knowledge and information or offers instruction about a particular topic. A few tone words typically used to describe this purpose include *objective, matter-of-fact,* and *straightforward.* Authors use facts to explain or describe the main idea to readers. Most textbook passages are written to inform. The following topic sentences reflect the writer's desire to inform.

    1. A sensible weight management program combines a healthful diet and regular exercise.
    2. *Narrative of the Life of Frederick Douglass, an American Slave, Written by Himself* records Douglass's life as a slave in the United States.

In sentence 1, the topic is weight management, and the words that reveal the controlling point are *sensible, healthful,* and *regular.* The author uses a tone that is unbiased and objective, so the focus is on the information. In sentence 2, the topic is Frederick Douglass's autobiography, and the words that reveal the controlling point are *records, life,* and *slave.* Again, the author chooses words that are matter-of-fact and suggest that factual details will follow. Both topic sentences indicate that the author's purpose is to provide helpful information.

- **To persuade.** A writer who sets out to persuade tries to bring the reader into agreement with his or her view on the topic. A few of the tone words typically used to describe this purpose include *argumentative, persuasive, forceful, controversial, positive, supportive, negative,* and *critical.* Authors combine facts with emotional appeals to sway the reader to their point of view. Politicians and advertisers often write and speak to persuade. The following topic sentences reflect the writer's desire to persuade.

3. Resistance training is the best method of shaping the body.
4. *Narrative of the Life of Frederick Douglass, an American Slave, Written by Himself* should be required reading in American public high schools.

In sentence 3, the topic is resistance training. The word that reveals the author's controlling point is *best.* In sentence 4, the topic is Douglass's autobiography. The words that reveal the controlling point are *should be,* which are followed by a recommendation for action. The author is offering a debatable personal opinion about what should be required high school reading. In both of these sentences, the authors are out to convince others to agree with them.

- **To entertain.** A writer whose purpose is to entertain sets out to captivate or interest the audience. A few of the tone words typically used to describe this purpose include *amusing, entertaining, lively, humorous,* and *suspenseful.* To entertain, authors often use expressive language and creative thinking. Most readers are entertained by material that stirs an emotional reaction such as laughter, sympathy, or fear. Thus, authors engage readers creatively through vivid images, strong feelings, or sensory details (such as sights, sounds, tastes, textures, and smells). Both fiction and nonfiction writers seek to entertain. The following topic sentences reflect the writer's desire to entertain.

5. Leona had the fashion style of a cantaloupe.
6. Anton woke up to the cozy aromas of coffee, bacon, and freshly baked bread, just as he had every day of his twenty years, yet today promised to be very different. He just knew it.

You may have found identifying the topic and controlling point a little more challenging in these two sentences. Often, when writers entertain, they imply the main idea. And when they use an implied main idea, they rely much more heavily on tone words. Sentence 5 deals with the topic of Leona and her fashion style. In this sentence, the author focuses the topic with the phrase "style of a cantaloupe." The use of *cantaloupe* (a biased word when used in this context) offers a strong clue that the author's purpose is to entertain. The author seeks to amuse the reader with the contrast between style and cantaloupe. Surprising contrasts often amuse the reader. Authors also use other methods to entertain, including exaggerations, vivid details, and dramatic descriptions.

Sentence 6 deals with the topic of Anton's experience. In this case, the words that reveal the author's controlling point are *cozy, aromas, freshly baked,* which are pleasant sensory details, and *promised* and *different.* However, the main idea is not really about Anton's experience. The point seems to be about a change or hope for change in Anton's life.

These six sentences show that a topic can be approached in a variety of ways. The author chooses a topic and a purpose. The purpose shapes the focus of the main idea. The author carefully chooses tone words to express the main idea in light of the purpose. Each of these choices then controls the choices of supporting details and the thought pattern used to organize them.

EXAMPLES Read each of the following paragraphs. Annotate them for main idea and tone. Then, identify the author's purpose as follows:

I = to inform **P** = to persuade **E** = to entertain

______ **1.** [1]Darla shivered, pulled her jacket's fuzzy collar up to the tops of her ears, and tucked her chin into its warmth; her arms and legs felt heavy and difficult to move as she started across the street. [2]It was only November, and already she longed for the bright, long days of June and July. [3]She dreaded the looming blackness of January and February. [4]She did not mesh well with the rhythm of their abbreviated days and unending nights.

[5]Ah, she thought to herself, I am so incredibly tired. [6]She felt the familiar cravings for something sweet and crunchy; she mentally pictured the box of Cocoa Puffs sitting on the pantry shelf, and she longed to be home. [7]She didn't care that she was already beginning to show her dreaded winter weight, she was going to have a bowl, or maybe two, as soon as she could.

______ **2.** [1]Some people suffer from symptoms of depression during the winter months, with symptoms subsiding during the spring and summer months. [2]This may be a sign of Seasonal Affective Disorder (SAD).

[3]SAD is a mood disorder linked to two factors: episodes of depression and the seasonal variations of daylight.

[4]Some experts believe as seasons change, a shift in our "biological internal clocks" occurs, partly because of the changes in sunlight patterns. [5]This may cause our biological clocks to be out of "step" with our daily schedules. [6]The most difficult months for some SAD sufferers seem to be January and February. [7]In addition, younger persons and women seem to be at higher risk.

—Adapted from "Seasonal Affective Disorder," Mental Health America, Copyright 2010 Mental Health America, http://www.mentalhealthamerica.net/go/sad

______ **3.** [1]If you suffer from seasonal affective disorder (SAD), also known as winter depression, you should implement the following lifestyle as a remedy. [2]First, you need to spend at least thirty minutes outside every morning, even when it's very cloudy, for the effects of daylight are very beneficial. [3]You must eat a well-balanced diet that includes sufficient amounts of vitamins and minerals as recommended by the FDA. [4]You should exercise for 30 minutes a day, three times a week. [5]Proper diet and exercise give you more energy, even though your body is craving starchy and sweet foods. [6]Just as importantly, you must develop an active social life and become involved with regular activities. [7]Of course, you should also seek professional counseling if needed, during the winter months.

—Adapted from "What Is Seasonal Disorder?" The Cleveland Clinic Health Information Center. 22 Feb. 2004 http://www.clevelandclinic.org/health/health-info/docs/2300/2361.asp?index=9293

4. Study each of the photographs of people in public speaking situations. Label the purpose of each speaker with one of the following:

E = to entertain **I** = to inform **P** = to persuade

a. ______ b.  ______ c. ______

EXPLANATIONS

Passages 1 through 3 used the same topic: seasonal affective disorder (SAD). However, the purpose of each passage differed. Note how the difference in purpose affected the selection and presentation of supporting details.

1. The topic is Darla's experience with SAD. This topic is implied. In the first paragraph, the author uses vivid details, descriptions, and appeals to the senses *to entertain* the reader through an unfolding dramatic event.
2. In the second paragraph, the author's purpose is simply *to inform* the reader about the symptoms of seasonal affective disorder (SAD).
3. However, in the third paragraph, the author attempts *to persuade* the reader that the only solution to SAD is a change in lifestyle. The persuasive tone is established and carried through by using the verbs *should, need to,* and *must.*
4. a. P, to persuade b. I, to inform c. E, to entertain

PRACTICE 3

Read the following topic sentences. Label each according to its purpose:

I = to inform **P** = to persuade **E** = to entertain

_____ 1. Understanding the aging process and those who seem to defy it is vital to our future.

_____ 2. According to Elisabeth Kubler-Ross, people go through five stages of grief.

_____ 3. To experience the hospitality and delights of a true summer resort away from the daily stresses, take a trip back in time to Mackinac Island in Michigan.

_____ 4. The sun spilled its brilliance from above so that every leaf, blade of grass, and ripple of water on the lake shimmered with no regard for my great grief.

Figure Out the Primary Purpose

In addition to the three general purposes, authors often write to fulfill a more specific purpose. The following table offers several examples of specific purposes.

General and Specific Purposes

To Inform	To Entertain	To Persuade
to analyze	to amuse	to argue for
to clarify	to delight	to argue against
to discuss	to frighten	to criticize
to establish		to convince
to explain		to inspire (motivate a change)

Often a writer has two or more purposes in one piece of writing. Blending purposes adds interest and power to a piece of writing. Take, for example, the award-winning documentary *Sicko* by Michael Moore. This film attempts to inform and entertain, but its primary purpose is to argue. The film uses facts, personal bias, and humor to take a strong stand against the current system of health care in America. When an author has more than one purpose, only one purpose is in control overall. This controlling purpose is called the **primary purpose.**

You have studied several reading skills that will help you grasp the author's primary purpose. For example, the author's primary purpose is often suggested by the main idea, the thought pattern, and the tone of the passage. Read the following topic sentence. Identify the author's primary purpose by considering the main idea, thought pattern, and tone.

______ Spanking must be avoided as a form of discipline due to its long-term negative effects on the child.

a. to discuss the disadvantages of spanking
b. to argue against spanking as a means of discipline
c. to make fun of those who use spanking as a means of discipline

This topic sentence clearly states a main idea "against spanking" using the tone words *must* and *negative*. The details will be organized using the thought pattern "long-term effects." Based on the topic sentence, the author's primary purpose is (b) to argue against spanking as a means of discipline. Even when the main idea is implied, tone and thought patterns point to the author's primary purpose.

You should also take into account titles, headings, and prior knowledge about the author. For example, it's easy to see that Jay Leno's primary purpose is to entertain us with his book *If Roast Beef Could Fly*. The title is funny, and we know Jay Leno is a comedian. A master reader studies the general context of the passage to find out the author's primary purpose.

Primary purpose is the author's main reason for writing the passage.

EXAMPLE Read the following paragraphs. Identify the primary purpose of each.

1. **Coacoochee**

Coacoochee. A Seminole warrior, nicknamed "Wildcat" and "Shrieky Scream," led raids against Americans, including Captain Dummett and Sergeant Ormond on the Halifax River. Famous for escaping jail in St. Augustine [Florida], he was banished to Mexico where he died at 49 of smallpox. His mother said he was made of Florida sands.

—*Images of America: Ormond Beach,* published by The Ormond Beach Historical Trust, 1999, p. 15.

______ The main purpose of this paragraph is

a. to share historical information about Coacoochee, a Seminole Indian.
b. to amuse the reader with entertaining details about Coacoochee.
c. to convince readers that Coacoochee was a hero.

2. **Early Biological Thought Did Not Include the Concept of Evolution**

Textbook Skills

[1]Pre-Darwinian science, heavily influenced by theology, held that all organisms were created simultaneously by God and that each distinct life-form remained fixed and unchanging from the moment of its creation. [2]This explanation of how life's diversity arose was elegantly expressed by the ancient Greek philosophers especially Plato and Aristotle. [3]Plato (427–347 B.C.) proposed that each object on Earth was merely a temporary reflection of its divinely inspired "ideal form." [4]Plato's student Aristotle (384–322 B.C.) categorized all organisms into a linear hierarchy that he called the "ladder of Nature."

[5]These ideas formed the basis of the view that the form of each type of organism is permanently fixed. [6]This view reigned unchallenged for nearly 2,000 years. [7]By the eighteenth century, however, several lines of newly emerging evidence began to erode the dominance of this static view of creation.

—From *Life on Earth,* 5th ed., pp. 211–212 by Teresa Audesirk, Gerald Audesirk, and Bruce E. Byers. Copyright © 2009 by Pearson Education, Inc. Printed and Electronically reproduced by permission of Pearson Education, Inc., Upper Saddle River, NJ.

______ The main purpose of this passage is
a. to offer entertaining details about the idea of fixed species.
b. to convince the reader that the idea of fixed species is a proven fact.
c. to explain the origin of the idea of fixed species.

3. Should Children Pledge Allegiance?

Textbook Skills

[1]A government that is built upon respect for the dignity and the rights of every human being deserves the allegiance or loyalty of its citizens. [2]The United States is built upon respect for the dignity and the rights of every human being. [3]Therefore, the United States deserves the allegiance of its citizens. [4]In addition, an effort to develop in citizens an appreciation for such an allegiance is acceptable. [5]That effort, however, must not harm the dignity of the individual. [6]In addition, the effort must not cause the individual to act against his or her personal beliefs. [7]Sadly, requiring students to recite the pledge of allegiance in school does, in some cases, cause students to act against their personal beliefs. [8]And, in some cases, when students refuse to act against their beliefs in this way, they have been abused. [9]Therefore, requiring students to recite the pledge of allegiance is not acceptable.

—Adapted from *The Art of Thinking: A Guide to Critical and Creative Thought*, 7th ed., p. 200 by Vincent R. Ruggiero. Copyright © 2004 by Pearson Education, Inc. Reprinted by permission of Pearson Education, Inc., Glenview, IL.

______ The main purpose of this paragraph is
a. to argue against requiring students to recite the pledge of allegiance.
b. to entertain the reader with a clever view about requiring students to recite the pledge of allegiance.
c. to inform the reader about the pledge of allegiance.

EXPLANATION

1. The first paragraph is taken from a book produced by the historical society of the small coastal town of Ormond Beach, Florida. The purpose is (a), to share historical information about Coacoochee, a Seminole Indian.

2. The second paragraph comes from a college biology textbook. The main purpose is (c), to explain the origin of the idea of fixed species.

3. The third paragraph is taken from a college textbook on critical thinking. Its main purpose is (a), to argue against requiring students to recite the pledge of allegiance.

PRACTICE 4

A. Read the passage from a college finance textbook. Answer the questions that follow it.

Textbook Skills

Personal Financial Planning

[1]Where does it all go? [2]It seems like the last paycheck is gone before the next one comes in. [3]Money seems to burn a hole in your pocket, yet you don't believe that you are living extravagantly. [4]Last month you made a pledge to yourself to spend less than the month before. [5]Somehow, though, the weeks seem to go by, and you are again in the same position as you were last month. [6]Your money is gone. [7]Is there any way to plug the hole in your pocket?

[8]As with any campaign, the first step is to gather information. [9]How much income do you earn? [10]How much is left after withholding taxes? [11]What are your expenses? [12]For many people, the first obstacle is to correctly assess their true expenses. [13]Calculating your net income is easy; just look at your pay statement. [14]But expenses are sly little creatures. [15]Each one seems so harmless and worthwhile, but combined together they can be a pack of piranhas that quickly gobble up your modest income. [16]What can you do to gain control of your personal finances?

[17]The solution is simple, but not easy. [18]The solution is simple because others have blazed the path; however, your task is not easy, because it takes self-discipline and there may be no immediate reward. [19]The result is often like a diet: easy to get started, but hard to carry through.

[20]Your tools are the personal balance statement, the personal cash flow statement, and a budget. [21]These three personal financial statements show you where you are, predict where you will be after three months or a year, and help you control those pesky expenditures; the potential benefits are reduced spending, increased savings and investments, and peace of mind from knowing that you are in control.

_____ **1.** The purpose of sentence 8 in the second paragraph is
- a. to inform the reader that financial planning is similar to any operating plan.
- b. to state a point against extravagant spending.
- c. to entertain the reader with a vivid detail about planning.

_____ **2.** The main purpose of the second paragraph is
- a. to explain to the reader that everyone struggles with financial problems.

b. to convince the reader that personal financial planning is necessary.
c. to amuse the reader with a real-life story.

_____ **3.** The main purpose of the passage overall is
a. to explain and illustrate the process of personal financial planning.
b. to offer an amusing approach with real life details to the topic "personal financial planning."
c. to argue against extravagant spending.

Recognize Irony Used for Special Effects

Irony is a tone often used in both conversation and written text. An author uses **irony** when he or she says one thing but means something else. Irony is the contrast between what is stated and what is implied, or between actual events and expectations.

Irony is often used to entertain and enlighten. For example, in the novel *Huckleberry Finn* by Mark Twain, the boy Huckleberry Finn believes he has done something wrong when he helps his older friend Jim escape slavery. The ironic contrast lies between what Huckleberry Finn thinks is wrong and what really is wrong: slavery itself. Twain set up this ironic situation to reveal the shortcomings of society.

Irony is also used to persuade. In her essay "I Want a Wife," Judy Brady seems to be saying she wants a wife to take care of the children, do the household chores, and perform all the other countless duties expected of a wife in the mid-twentieth century. However, she doesn't really want a wife; she wants equality with men. As she describes the role of a wife as a submissive servant, she argues against the limitations that society has placed on women.

Due to the powerful effects of irony, authors use it in many types of writing. You may come across irony in fiction, essays, poetry, comedy routines, and cartoons. When authors use irony, they imply their main ideas and rely heavily on tone. Thus you need to understand two common types of irony so that you can see and enjoy their effects: verbal irony and situational irony.

> **Verbal irony** occurs when the author's words state one thing but imply the opposite.

- During a disagreement, your friend says, "I can't wait to hear your next great idea!"
- After eating large portions of a full-course meal, a diner says, "Good thing I wasn't hungry."
- After completing a challenging exam, a student says, "Well, that was easy."

Situational irony occurs when the events of a situation differ from what is expected.

- A popular singer suffers from severe stage fright.
- The person voted "most likely to succeed" in high school becomes a homeless drifter.
- A wealthy person who has never taken a vacation.

EXAMPLES Read the items, and identify the type of irony used in each.

_____ **1.** Referring to Katharine Hepburn, in a theater review of *The Lake*, Dorothy Parker wrote, "She runs the gamut of emotions from A to B."

—In *The Columbia World of Quotations*, 1996; *Bartleby.com*, 2001. 7 Nov. 2007 www.bartleby.com/66/

a. verbal irony
b. situational irony
c. no irony

_____ **2.** "Talkative people who wish to be loved are hated; when they desire to please, they bore; when they think they are admired, they are laughed at; they injure their friends, benefit their enemies, and ruin themselves."

—Plutarch, Greek essayist and biographer (46–119 A.C.E.)

—From *The International Dictionary of Thoughts*, Comp. Bradley, Daniels, and Jones, 1969, p. 707.

a. verbal irony
b. situational irony
c. no irony

_____ **3.** "C-O-P-D [Chronic Obstructive Pulmonary Disease] is the fourth leading cause of death in the United States. It is a serious lung disease that takes away the breath of 12 million Americans."

—Kenneth P. Moritsugu, Acting Surgeon General, U.S. Department of Health and Human Services

—From Remarks at 52nd International Respiratory Congress. 11 Nov. 2006. Washington, D.C. 7 Nov. 2007 http://www.surgeongeneral.gov/news/speeches/12112006.html.

a. verbal irony
b. situational irony
c. no irony

EXPLANATIONS Compare your answers with these:

1. (a) verbal irony: Dorothy Parker is known for her ironic wit. In this case, Parker's biting wit arises out of the contrast between the word *gamut* and the phrase "A to B." The word *gamut* means "scope" or "range," and sets up the expectation that Parker is going to imply that Ms. Hepburn displays an impressive breadth of emotion; however, she surprisingly limits Hepburn's range of emotion to the first two letters of the alphabet. What initially seems to be a compliment ends as a humorous insult.

2. (b) situational irony: Most people who are talkative are trying to make a connection or make a difference through the power of talk. Plutarch points out the irony of the situation when their excessive talk brings about the opposite of their hopes.

3. (c) no irony: The author directly states facts without implying any other meaning than the one stated.

PRACTICE 5

Read the items below and identify the type of irony used in each.

______ **1.** "I love working sixteen-hour days for minimum wage."
a. verbal irony
b. situational irony
c. no irony

______ **2.** A recent college graduate takes a job as an intern in a company as a stepping-stone into a salaried position with full benefits. However, the intern's supervisor feels threatened by the efficiency and skill of the intern and denies her the full-time position.
a. verbal irony
b. situational irony
c. no irony

______ **3.** "Notice—By Order of the Author. Persons attempting to find a motive in this narrative will be prosecuted; persons attempting to find a moral in it will be banished; persons attempting to find a plot in it will be shot."

—Mark Twain, *Huckleberry Finn* (1884).

a. verbal irony
b. situational irony
c. no irony

_____ **4.** A student studies for weeks for final exams and earns the highest grade point average in her classes.
a. verbal irony
b. situational irony
c. no irony

_____ **5.** On his way home from an Alcoholics Anonymous meeting, a reformed alcoholic is killed in an automobile accident by a drunk driver who walks away from the accident with only minor injuries.
a. verbal irony
b. situational irony
c. no irony

Reading a Textbook: Author's Tone and Purpose

Textbook Skills

Read the following passage from the textbook *Access to Health*. Then, answer the questions about the author's tone and purpose.

Societal Causes of Violence

[1]Although the underlying causes of violence and abuse are as varied as the individual crimes and people involved, several social, cultural, and individual factors seem to increase the likelihood of violent acts (see Figure 10.1). [2]Included among those factors most commonly listed are the following:

Poverty. [3]Low socioeconomic status and poor living conditions can create an environment of hopelessness, leaving one feeling trapped and seeing violence as the only way to obtain what is needed or wanted.

Unemployment. [4]It is a well-documented fact that when the economy goes sour, violent crime, suicide, assault, and other crimes increase.

Parental influence. [5]Violence is cyclical. [6]Children raised in environments in which shouting, slapping, hitting, and other forms of violence are commonplace are more apt to "act out" these behaviors as adults. [7]Horrifying reports in recent years have made this pattern impossible to ignore.

Cultural beliefs. [8]Cultures that objectify women and empower men to be tough and aggressive tend to have increased rates of violence in the home.

The media. [9]A daily dose of murder and mayhem can take a toll on even resistant minds.

Discrimination/oppression. [10]Whenever one group is oppressed by another, seeds of discontent are sown and hate crimes arise.

Religious differences. [11]Religious persecution has been a part of the human experience since the earliest times. [12]These battles between right and wrong, good and evil, and the attempt to impose beliefs on others have often led to violence.

Breakdown in the criminal justice system. [13]Overcrowded prisons, lenient sentences, early releases from prisons, and trial errors subtly encourage violence in a number of ways.

Stress. [14]People who suffer from inordinate amounts of stress or are in crisis are more apt to be highly reactive, striking out at others or acting irrationally.

—Text and graphic from *Access to Health,* 7th ed., pp. 94–95 and Figure 4.1 (p. 95) by Rebecca J. Donatelle and Lorraine G. Davis. Copyright © 2002 by Pearson Education, Inc. Printed and Electronically reproduced by permission of Pearson Education, Inc., Upper Saddle River, NJ.

_____ **1.** The author's primary purpose for this section of the textbook is
a. to inform.
b. to entertain.
c. to persuade.

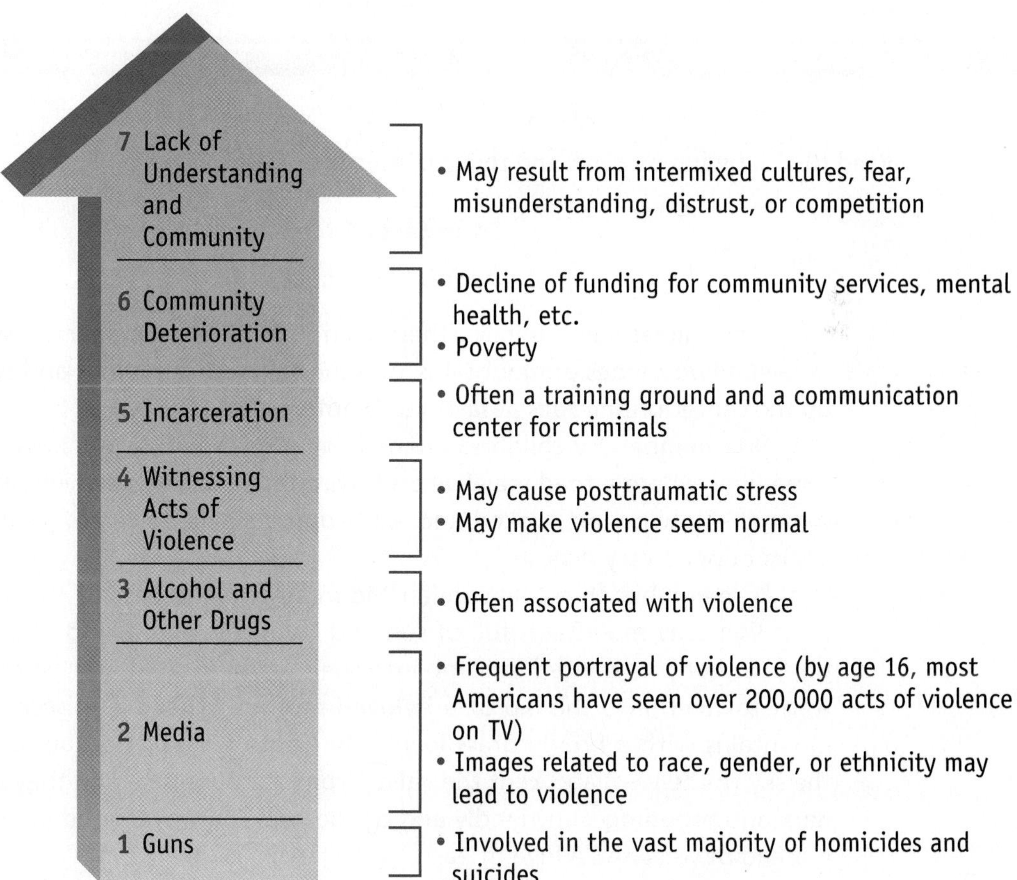

▲ **Figure 4.1** Correlates to Violence

_____ **2.** The overall tone of the text is
a. biased. b. objective.

_____ **3.** The purpose of Figure 10.1, "Correlates to Violence," is
a. to inform. c. to persuade.
b. to entertain.

_____ **4.** The tone of sentence 6 is
a. ironic. c. sarcastic.
b. angry. d. matter-of-fact.

_____ **5.** The tone of sentence 7 is
a. irritated. c. firm.
b. matter-of-fact. d. sarcastic.

APPLICATIONS

Application 1: Author's Tone and Purpose

Read the following passage, and then, answer the questions.

From *East of Eden*
by John Steinbeck

[1]The Salinas Valley is in Northern California. [2]It is a long narrow swale between two ranges of mountains, and the Salinas River winds and twists up the center until it falls at last into Monterey Bay.

[3]I remember my childhood names for grasses and secret flowers. [4]I remember where a toad may live and what time the birds awaken in the summer—how people looked and walked and smelled even. [5]The memory of odors is very rich.

[6]I remember that the Gabilan Mountains to the east of the valley were light gay mountains full of sun and loveliness and a kind of invitation, so that you wanted to climb into their warm foothills almost as you want to climb into the lap of a beloved mother. [7]They were beckoning mountains with a brown grass love. [8]The Santa Lucias stood up against the sky to the west and kept the valley from the open sea, and they were dark and brooding—unfriendly and dangerous. [9]I always found in myself a dread of west and a love of east.

— From *East of Eden* by John Steinbeck. New York: Penguin Group (USA) Inc., 1952, 1980.

_____ **1.** The tone of sentences 1 and 2 is
a. factual.
b. reflective.
c. cold.

_____ **2.** The tone of sentences 3 to 5 is
a. encouraging.
b. reflective.
c. sarcastic.

_____ **3.** The tone of sentences 6 and 7 is
a. matter-of-fact.
b. prideful.
c. admiring.

_____ **4.** The tone of sentence 8 is
a. sarcastic.
b. critical.
c. objective.

_____ **5.** The purpose of the first paragraph is
a. to inform the reader about certain aspects of the Salinas Valley.
b. to entertain the reader with little-known facts about the Salinas Valley.
c. to persuade the reader that the Salinas Valley is beautiful.

_____ **6.** The overall purpose of the passage is
a. to inform.
b. to entertain.
c. to persuade.

Application 2: Irony

Read the items below. Identify the type of irony used, if any.

_____ **1.** "The brain is a wonderful organ; it starts working the moment you get up, and does not stop until you get to the office."

—*Robert Frost*

—In Eldin, *Jokes & Quotes for Speeches*, 1989, p. 42.

a. verbal irony
b. situational irony
c. no irony

_____ **2.** "The drive down to see you was just wonderful, simply wonderful! How boring it would have been if we had simply zipped down here in the usual one-hour drive time. We were thrilled by the three-hour traffic jam caused by a minor accident that every driver on the road just had to slow down to inspect."

—*A son's response to his parents' inquiry about his drive to visit them*

a. verbal irony
b. situational irony
c. no irony

_____ **3.** One of the surprising parts of maturing as a woman is the way in which one grows to resemble one's mother, both in looks and ways.

a. verbal irony
b. situational irony
c. no irony

ZITS © 2010 Zits Partnership, King Features Syndicate

_____ **4.** What irony is used in the cartoon?

a. verbal irony
b. situational irony
c. no irony

REVIEW TEST 1

Score (number correct) _____ × 10 = _____ %

Author's Purpose

A. Read the topic sentences. Label each one according to its purpose as follows:

I = to inform **P** = to persuade **E** = to entertain

_____ **1.** A wedding should be viewed as one of the happiest of events, and we worked very hard to behave as if that were true for us, too—but more and more, the day felt like a collision of families in slow motion.

_____ **2.** Americans consume more calories per person than does any other group of people in the world.

_____ **3.** The Federal Drug Administration has a moral responsibility to the public that includes regulating alternative or herbal medicines.

_____ **4.** "The process by which money that you currently hold accumulates interest over time is referred to as compounding."

—From *Personal Finance,* 2nd ed., p. 61 by Jeff Madura. Copyright © 2004 by Pearson Education, Inc. Reproduced by permission of Pearson Education, Inc., Boston, MA.

_____ **5.** A dollar received today is worth more than a dollar received tomorrow because it can be saved and invested; the time value of money is one of the strongest forces on earth—so save and invest.

—Adapted from *Personal Finance,* 2nd ed., p. 59 by Jeff Madura. Copyright © 2004 by Pearson Education, Inc. Reproduced by permission of Pearson Education, Inc., Boston, MA.

_____ **6.** Adults with higher levels of education and income generally have more favorable health behaviors in terms of cigarette smoking, leisure-time physical activity, and body weight status; therefore, it is imperative that all people are afforded the opportunity for higher education.

_____ **7.** In 2000, a total of 2,403,351 deaths occurred in the United States; life expectancy at birth was 76.9 years, and the leading causes of death were as follows: heart disease, cancer, stroke, chronic lower respiratory diseases, accidents, diabetes, influenza and pneumonia, Alzheimer's disease, and kidney disease.

—Adapted from "National Vital Statistics Reports," 50: 15, CDC, 11 June 2003. 3 May 2008 http://www.cdc.gov/nchs/data/nvsr/nvsr50/nvsr50_15.pdf

_____ **8.** One way to achieve the status of expert on a subject is through firsthand experience; thus I think it honest and helpful to offer myself as an expert of the highest order on the subject of smoking cessation, having quit over a dozen times.

B. Read the following items. Identify the primary purpose of each.

Editorial Cartoon Published in a Newspaper

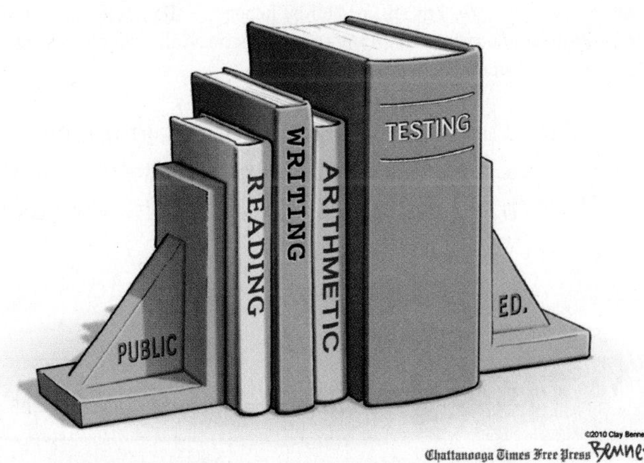

© 2010 Clay Bennett Chattanooga Times Free Press

______ **9.** The primary purpose of this cartoon is
a. to inform the reader about the quality of public education.
b. to entertain the reader by mocking the public education system.
c. to persuade the reader to oppose the emphasis on testing in public education.

Graphic from a Health Textbook

Textbook Skills

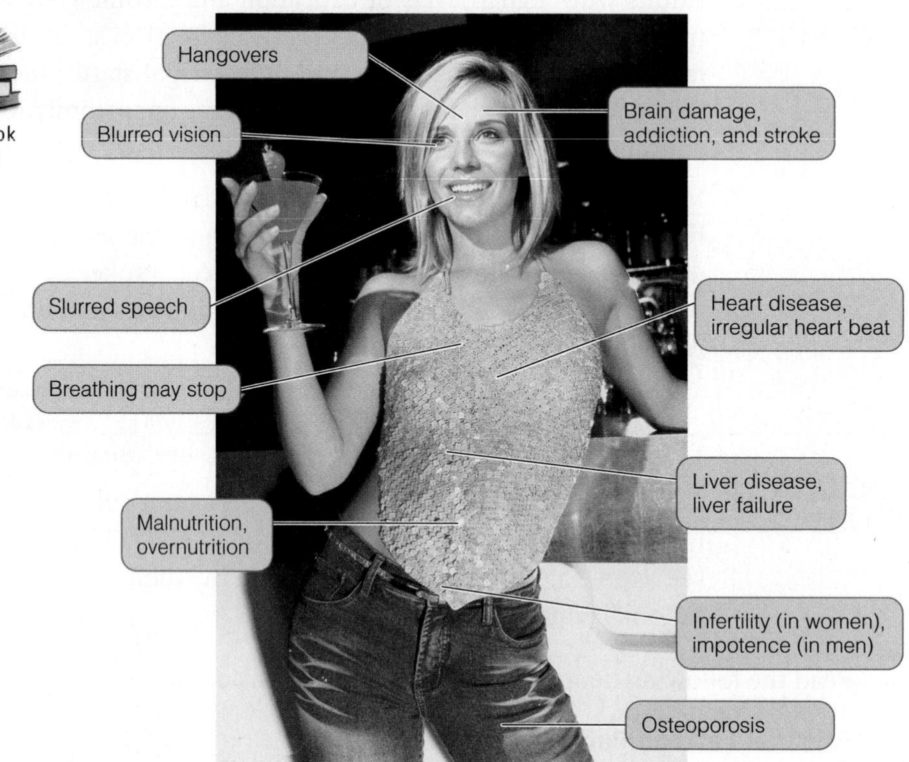

—From *Access to Health*, 7th ed., p. 380 by Rebecca J. Donatelle and Lorraine G. Davis. Copyright © 2002 by Pearson Education, Inc. Printed and Electronically reproduced by permission of Pearson Education, Inc., Upper Saddle River, NJ.

______ **10.** The primary purpose of this graphic is
a. to persuade readers to abstain from alcohol consumption.
b. to inform readers about the various negative effects of alcohol on the body.
c. to interest readers with an interesting illustration.

REVIEW TEST 2

Score (number correct) ______ × 25 = ______%

Read the following passage from a college history textbook. Answer the questions that follow.

Textbook Skills

Native Americans

[1]Native Americans are the first true Americans, and their status under U.S. law is unique. [2]Under the U.S. Constitution, Indian tribes are considered distinct governments. [3]This situation has affected Native Americans' treatment by the Supreme Court in contrast to other groups of ethnic minorities. [4]And, minority is a term that accurately describes American Indians. [5]It is estimated that there were as many as 10 million Indians in the New World at the time Europeans arrived in the 1400s. [6]The actual number of Indians is hotly contested. [7]Estimates vary from a high of 150–200 million to a low of 20–50 million throughout North and South America. [8]By 1900, the number of Indians in the continental United States had plummeted to less than 2 million. [9]Today, there are 2.8 million.

[10]Many commentators would agree that for years Congress and the courts manipulated Indian law to promote the westward expansion of the United States. [11]The Northwest Ordinance of 1787, passed by the Continental Congress, specified that "good faith should always be observed toward the Indians; their lands and property shall never be taken from them without their consent, and their property rights, and liberty, they shall never be invaded or disturbed, unless in just and lawful wars authorized by Congress." [12]These strictures were not followed. [13]Instead, over the years, "American Indian policy has been described as 'genocide-at-law' promoting both land acquisition and cultural extermination." [14]During the eighteenth and nineteenth centuries, the U.S. government isolated Indians on reservations as it confiscated their lands and denied them basic political rights. [15]Indian reservations were administered by the federal government. [16]And Native Americans often lived in squalid conditions.

[17]With passage of the Dawes Act in 1887, however, the government switched policies to promote assimilation over separation. [18]Each Indian family was given land within the reservation. [19]The rest was sold to whites, thus reducing Indian lands from about 140 million acres to about 47 million. [20]Moreover, to encourage Native Americans to assimilate, Indian children were sent to boarding schools off the reservation, and native languages and rituals were banned. [21]Native Americans didn't become U.S. citizens nor were they given the right to vote until 1924.

[22]At least in part because tribes were small and scattered (and the number of Indians declining), Native Americans formed no protest movement in reaction to these drastic policy changes. [23]It was not until the 1960s that Indians began to mobilize. [24]During this time, Indian activists,

many trained by the American Indian Law Center at the University of New Mexico, began to file hundreds of test cases in the federal courts involving tribal fishing rights, tribal land claims, and the taxation of tribal profits.

[25]Native Americans have won some very important victories concerning hunting, fishing, and land rights. [26]Native American tribes all over America have sued to reclaim lands they say were stolen from them by the United States, often more than 200 years ago. [27]Today, these land rights allow Native Americans to play host to a number of casinos across the country.

—Adapted from O'Connor and Sabato. *American Government: Continuity and Change*, 2008 ed., pp. 226–227.

_____ **1.** The overall purpose of the passage is to
a. inform the reader with an overview of the history of Native Americans.
b. amuse the reader with graphic details from the history of Native Americans.
c. persuade the reader to support the Native American cause.

_____ **2.** The overall tone of the passage is
a. informal. b. formal.

_____ **3.** The overall tone of the passage is
a. neutral. b. biased.

_____ **4.** Overall, the difficulties faced by the Native American reveals
a. verbal irony. c. no irony.
b. situational irony.

REVIEW TEST 3

Score (number correct) _____ × 10 = _____%

Author's Purpose and Tone

Textbook Skills

Before Reading: Survey the following passage adapted from the college textbook *Atlas of Slavery*. Skim the passage, noting the words in **bold** print. Answer the Before Reading questions that follow the passage. Then, read the passage. Next, answer the After Reading questions. Use the discussion and writing topics as activities to do after reading.

Vocabulary Preview

integral (2): essential
artifacts (3): manufactured objects
barter (4): exchange
dragooned (7): forced
legions (17): masses
retinue (19): servants, followers
ubiquitous (20): everywhere

Slavery in the Ancient World

[1]Slavery was commonplace in a host of ancient and traditional societies. [2]For example, it played a major, **integral** role in societies from ancient Egypt to the wider world of classical **antiquity**. [3]Indeed, many of the surviving **artifacts** and buildings of those societies, now awash with millions of tourists, from the Pyramids to the urban fabric of ancient Greece, were built by slaves. [4]In all those societies, slaves were recruited from outside: by warfare, by raids and kidnapping and by trade and **barter** in humanity. [5]There were slave markets for the transfer and supply of fresh slaves arriving from the edges of trade and military conquest. [6]Throughout the history of ancient Egypt, large numbers of slaves were acquired by military means (though the slave population also grew of its own accord). [7]The New Kingdom (1560–1070 B.C.) expanded throughout the present day Middle East and Sudan; many thousands of Nubians from the Sudan were **dragooned** as slaves, working in agriculture, in manufacturing and in the military. [8]Females inevitably found their way into domestic slavery. [9]Other African slaves came to Egypt from the coast of Somalia.

[10]Greek civilization was equally bound up with slavery, and trade thrived in enslaved peoples, especially in the classical period (450–330 B.C.). [11]Slaves could be found in large numbers in all the major city-states, where they were bought and sold as objects, with men used in heavy work, women normally in domestic chores. [12]Again, slaves were recruited by violent expeditions. [13]Curiously, the rise of slavery paralleled the rise of Greek democracy, for the slaves gave citizens the freedom to take part in their civic duties. [14]Although slavery changed over time, it became a defining characteristic of Greek civilization. [15]Slaves from the East passed through the Greek slave markets of Chios, Delos and Rhodes.

[16]Slavery in the Roman Empire is better known (because it was more recent and therefore better documented). [17]The successes of the Roman legions throughout that far-flung empire provided **legions** of slaves for the Roman heartlands, from the northern boundaries of the empire in Britain to the southern limits in Africa. [18]It has been calculated that the golden age of Rome was maintained by the importation of upwards of half a million slaves in one year. [19]Indeed, Roman armies returned in triumph to the imperial capital ahead of their **retinue** of conquered slaves from as far afield as Germany, Britain and Gaul. [20]Slaves were **ubiquitous:** as domestics, as labourers in Roman agriculture, in the hell of the mines and, of course, for entertainment in the Colosseum. [21]Few areas of

Roman life remained untouched by slaves. [22]Rome may not have been built in a day, but it was certainly built by slaves.

—Adapted from Walvin, James. *Atlas of Slavery*. pp. 16, 18. © James Walvin, 2006. Reprinted by permission of Pearson Education Limited.

Before Reading

Vocabulary in Context

1. In sentence 2, what does the word **antiquity** mean? ________________

____

2. Identify the context clue used for the word **antiquity** in sentence 2. ______

____

After Reading

Central Idea

______ **3.** Which sentence best states the central idea of the passage?
a. sentence 1
b. sentence 3
c. sentence 2
d. sentence 22

Supporting Details

______ **4.** One of the benefits of slavery in Greek civilization was
a. the construction of pyramids.
b. the rise of democracy.
c. entertainment in the Colosseum.
d. the establishment of city-states.

Transitions

______ **5.** The relationship of ideas within sentence 10 is
a. cause and effect.
b. space order.
c. comparison.
d. listing.

Thought Patterns

______ **6.** The overall thought pattern for the passage is
a. cause and effect.
b. space order.
c. time order.
d. definition and example.

Fact and Opinion

_____ **7.** Sentence 20, "Slaves were ubiquitous: as domestics, as labourers in Roman agriculture, in the hell of the mines and, of course, for entertainment in the Colosseum" is a statement of
a. fact.
b. opinion.
c. fact and opinion.

Tone and Purpose

_____ **8.** The tone of sentence 22 "Rome may not have been built in a day, but it was certainly built by slaves" is
a. objective.
b. sarcastic.
c. witty.

_____ **9.** The overall tone of the passage is
a. informative.
b. argumentative.
c. witty.

_____ **10.** The primary purpose of the passage is
a. to persuade the reader that slavery in ancient times was inhumane.
b. to engage the reader with interesting facts about slavery in ancient times.
c. to inform the reader about the importance of slavery in ancient times.

WHAT DO YOU THINK?

Why were slaves important to developing civilization in the ancient world? What contributions were made by slaves in the ancient world? Assume you are taking a college history course, and your professor requires written responses to what you are reading. Choose one of the following topics and write a three-paragraph essay: (1) Assume the role of an Egyptian, Greek, or Roman citizen who opposes slavery. Write a draft of a speech you plan to give during a public forum; in your speech explain your reasons for opposing slavery. Include descriptions of daily tasks, working conditions, and inhumane treatment of a slave in ancient times. (2) Explain several reasons why slavery existed in ancient Egypt, Greece, and Rome.

REVIEW TEST 4

Score (number correct) ______ × 10 = ______%

Author's Purpose and Tone

Textbook Skills

Before Reading: Survey the following passage adapted from the college textbook *The West: Encounters & Transformations.* Skim the passage, noting the words in **bold** print. Answer the Before Reading questions that follow the passage. Then, read the passage. Next, answer the After Reading questions. Use the discussion and writing topics as activities to do after reading.

Vocabulary Preview

SS (2): abbreviation for the German *word Schutzstaffel,* which translates as Protective Squadron, a personal guard unit for Hitler, responsible for the mass murder of Jews and others deemed undesirable

Einsatzgruppen (3): strike or task force, SS units given the task of murdering Jews and Communist Party members in the areas of the Soviet Union occupied by Germany during World War II

indoctrinated (8): instructed, taught

Roma (12): gypsies

conglomerate (15): corporation, business

ghettos (20): in certain European cities, a section to which Jews were formerly restricted

The Death Camps: Murder by Assembly Line

[1]On January 20, 1942, senior German officials met in a villa in Wannsee, outside Berlin, to finalize plans for killing every Jew in Europe. [2]**SS** lieutenant colonel Adolf Eichmann (1906–1962) listed the number of Jews in every country; even the Jewish populations in neutral countries such as Sweden and Ireland showed up on the target list. [3]The Wannsee Conference marked the beginning of a more systematic approach to murdering European Jews, one that built on the experience gained by the **Einsatzgruppen** in the Soviet war.

[4]To accomplish mass murder, the Einsatzgruppen had become killing machines. [5]By trial and error, they discovered the most efficient ways of identifying and rounding up Jews, shooting them quickly, and burying the bodies. [6]But the Einsatzgruppen actions also revealed the limits of conventional methods of killing. [7]Shooting took time, used up valuable ammunition, and required large numbers of men.

[8]Moreover, even the best-trained and carefully **indoctrinated** soldiers eventually cracked under the strain of shooting unarmed women and children at close range. [9]A systematic approach was needed, one that would utilize advanced killing technology and provide a comfortable distance between the killers and the killed. [10]This perceived need resulted in a key Nazi innovation: the death camp.

[11]The death camp was a specialized form of a concentration camp. [12]From 1933 on, Hitler's government had sentenced communists, Jehovah's Witnesses, the **Roma**, and anyone else defined as an enemy of the **regime** to forced labor in concentration camps. [13]After the war began, the concentration camp system

extermination (21): execution, death
culled (25): picked

expanded dramatically. [14]Scattered throughout Nazi-controlled Europe, concentration camps became an essential part of the Nazi war economy. [15]Some firms, such as the huge chemical **conglomerate** I. G. Farben, established factories inside or right next to camps, which provided vital supplies of forced labor. [16]All across Europe during the war, concentration camp inmates died in huge numbers from the brutal physical labor, torture, and diseases brought on by malnutrition and inadequate housing and sanitary facilities. [17]But it was only in Poland that the Nazis constructed death camps, specialized concentration camps with only one purpose—murder, primarily the murder of Jews.

[18]The death camps marked the final stage in a vast assembly line of murder. [19]In early 1942 the trains conveying victims to the death camps began to rumble across Europe.

[20]Jewish **ghettos** across Nazi-occupied Europe emptied as their inhabitants moved in batches to their deaths. [21]Individuals selected for **extermination** followed orders to gather at the railway station for deportation to "work camps" farther east. [22]They were then packed into cattle cars, more than 100 people per car, all standing up for the entire journey. [23]Deprived of food and water, with hardly any air, and no sanitary facilities, often for several days, many Jews died en route. [24]The survivors stumbled off the trains into a nightmare world. [25]At some camps, SS guards **culled** stronger Jews from each transport to be worked to death as slave laborers. [26]Most, however, walked straight from the transport trains into a reception room, where they were told to undress, and then herded into a "shower room"—actually a gas chamber. [27]Carbon monoxide gas or a pesticide called Zyklon-B killed the victims. [28]After the poison had done its work, Jewish slaves emptied the chamber and burned the bodies in vast crematoria, modeled after industrial bake ovens. [29]The Nazis thus constructed a vast machine of death.

—From *The West: Encounters & Transformations, Atlas Edition, Combined Edition*, 2nd edition., pp. 870–871 by Brian Levack, Edward Muir, Meredith Veldman and Michael Maas. Copyright © 2008 by Pearson Education, Inc. Printed and Electronically reproduced by permission of Pearson Education, Inc., Upper Saddle River, NJ.

Before Reading

Vocabulary in Context

1. In sentence 12, what does the word **regime** mean? ____________________

____

2. Identify the context clue used for the word **regime** in sentence 12. ______

____

After Reading

Central Idea

______ **3.** Which sentence best states the central idea of the passage?
a. sentence 1
b. sentence 3
c. sentence 4
d. sentence 29

Supporting Details

______ **4.** According to the passage, the death camps were constructed
a. across Europe.
b. only in Poland.
c. in Berlin.
d. in Jewish ghettos.

Transitions

______ **5.** The relationship between sentence 5 and sentence 6 is
a. time order.
b. contrast.
c. listing.
d. definition and example.

Thought Patterns

______ **6.** The thought pattern for paragraph 4 (sentences 11–17) is
a. listing.
b. time order.
c. contrast.
d. definition and example.

Fact and Opinion

______ **7.** Sentence 8 is a statement of
a. fact.
b. opinion.
c. fact and opinion.

Tone and Purpose

______ **8.** The tone of sentence 18 is
a. matter-of-fact.
b. condemning.
c. bitter.
d. understanding.

______ **9.** The overall tone of the passage is
a. horrified.
b. neutral.
c. condemning.
d. sorrowful.

_____ **10.** The primary purpose of the passage is
a. to persuade the reader about the brutality of the Nazi death camps.
b. to shock the reader with graphic details about the Nazi death camps.
c. to inform the reader about the Nazi commitment to exterminating Jews and others they deemed undesirable.

WHAT DO YOU THINK?

Do you think that most of the citizens of Germany knew about the death camps? Do you think you would have had the courage to defy the Nazi regime and help the Jews and other "enemies of the state"? Assume that a local chapter of the American Jewish League is sponsoring an essay contest with a $250 prize in honor of those who suffered and died under the Nazi regime and to educate the next generation about the consequences of prejudice. Choose one of the following two topics and write a 500-word essay:

1. The Nazi regime exhibited the extreme effects of prejudice and hatred during the early twentieth century. Today, in the early twenty-first century, we also see the extreme effects of prejudice and hatred of certain regimes in the world. Identify and discuss a current regime and its actions of terror that should be stopped.

2. Both before and during the Nazi regime many German citizens did not raise objections to the actions of terror that occurred under Hitler's rule; other German citizens tried to secretly help those who were targeted for prison or death. Write an essay that discusses several steps the average German citizen could have taken to take a stand against Hitler and his reign of terror.

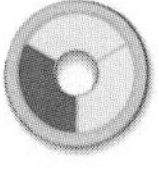

After Reading About Tone and Purpose

Before you move on to the Mastery Tests on tone and purpose, take time to reflect on your learning and performance by answering the following questions. Write your answers in your notebook.

- How has my knowledge base or prior knowledge about tone and purpose changed?
- Based on my studies, how do I think I will perform on the Mastery Test(s)? Why do I think my scores will be above average, average, or below average?

- Would I recommend this chapter to other students who want to learn about tone and purpose? Why or why not?

Test your understanding of what you have learned about tone and purpose by completing the Chapter 10 Review Card in the insert near the end of your text.

CONNECT TO PEARSON myreadinglab

To check your progress in meeting Chapter 10's learning outcomes, log in to **www.myreadinglab.com** and try the following exercises.

- The "Purpose and Tone" section of MyReadingLab provides an overview, model, practices, activities, and tests. To access this resource, click on the "Study Plan" tab. Then click on "Purpose and Tone." Then click on the following links as needed: "Overview," "Model," "Practice," and "Tests."
- To measure your mastery of this chapter, complete the tests in the "Purpose and Tone" section and click on Gradebook to find your results.

MASTERY TEST 1

Name ______________________ Section ______________

Date ____________ Score (number correct) ______ × 20 = ______%

Read the poem, and then, answer the questions that follow it.

Fear Not the Fall

by Billie Jean Young

Fear not the fall.
Better to fall from the strength
of the sound of one's voice
speaking truth to the people
than to spend a lifetime
mired in discontent,
groveling in the sty of certainty
of acceptance,
of pseudo-love,
of muteness,
and easy-to-be-around-ness.

Better to fall screaming—
arms flailing,
legs askew
clawing in your intensity
for the right to be—
than silently,
like a weakened snowbird,
sentenced to the boredom of earth
too sick to sing.

— From *Fear Not the Fall: Poems and a Two-Act Drama* by Billie Jean Young,
Reprinted with permission of NewSouth Books, Montgomery, AL. p. 112.

______ **1.** In the first stanza, the words *strength*, *sound*, *speaking*, and *truth* are best described by the tone word
- a. confident.
- b. humble.
- c. afraid.
- d. angry.

______ **2.** In the second stanza, the words *screaming*, *flailing*, *askew*, *clawing*, and *intensity* are best described by the tone word
- a. composed.
- b. defeated.
- c. hostile.
- d. sad.

______ **3.** The tone of the last three lines of the second stanza is

a. hateful.
b. cold.
c. solemn.
d. neutral.

______ **4.** The phrase "fear not the fall" is ironic because

a. the poem states "fear not" but implies the opposite.
b. people often warn against falling or failing, but this poem advises "fear not."
c. the poem suggests that failure is certain.

______ **5.** The overall purpose of this poem is

a. to persuade the reader to risk living honestly and avoid settling for security.
b. to inform the reader about the effect of fear and the need to directly face fear.
c. to entertain the reader with a personal experience about the fear of falling.

MASTERY TEST 2

Name ______________________ Section ______________

Date __________ **Score** (number correct) ______ × 20 = ______%

Read the following passage from a college literature textbook. Then, answer the questions.

Textbook Skills

Blues

[1]Among the many song forms to have shaped the way poetry is written in English, no recent form has been more influential than the blues. [2]Originally a type of folk music developed by black slaves in the South, blues songs have both a distinctive form and tone. [3]They traditionally consist of three-line stanzas in which the first two identical lines are followed by a concluding, rhyming third line.

[4]To dream of muddy water—trouble is knocking at your door.
[5]To dream of muddy water—trouble is knocking at your door.
[6]Your man is sure to leave you and never return no more.

[7]Early blues lyrics almost always spoke of some sadness, pain, or deprivation—often the loss of a loved one. [8]The melancholy tone of the lyrics, however, is not only world-weary but also world-wise. [9]The blues expound the hard-won wisdom of bitter life experience. [10]They frequently create their special mood through down-to-earth, even gritty, imagery drawn from everyday life. [11]Although blues reach back into the nineteenth century, they were not widely known outside African American communities before 1920 when the first commercial recordings appeared. [12]Their influence on both music and song from that time was rapid and extensive. [13]By 1930 James Weldon Johnson could declare, "It is from the blues that all that may be called American music derives its most distinctive characteristic." [14]Blues have not only become an enduring category of popular music; they have helped shape virtually all the major styles of contemporary pop—jazz, rap, rock, gospel, country, and of course, rhythm-and-blues.

[15]The style and structure of blues have also influenced modern poets. [16]Not only African American writers like Langston Hughes, Sterling A. Brown, Etheridge Knight, and Sonia Sanchez have written blues poems, but white poets as dissimilar as W. H. Auden, Elizabeth Bishop, Donald Justice, and Sandra McPherson have employed the form. [17]The classic touchstones of the blues, however, remain the early singers like Robert Johnson, Ma Rainey, Blind Lemon Jefferson, Charley Patton, and—perhaps preeminently—Bessie Smith, "the Empress of the Blues."

—Kennedy & Gioia, *Literature: An Introduction to Fiction, Poetry, and Drama,* 3rd Compact ed., pp. 586–587.

_____ **1.** Which term best describes the tone of sentence 1?
a. objective
b. subjective

_____ **2.** The tone of sentence 2 and sentence 3 is
a. objective.
b. subjective.

_____ **3.** The tone of the words *preeminently* and *Empress of the Blues* in sentence 17 is
a. mocking.
b. understated.
c. objective.
d. admiring.

_____ **4.** Which word describes the overall tone of the passage?
a. humorous
b. informal
c. academic
d. argumentative

_____ **5.** The overall purpose of this passage is
a. to inform the reader about the general importance, history, and characteristics of the blues.
b. to entertain the reader with lively details about the lives of blues singers and artists.
c. to persuade the reader that blues is the most important form of music and poetry in the English language.

MASTERY TEST 3

Name ______________________ Section ______________

Date __________ **Score** (number correct) ______ × 20 = ______%

Read the following passage from a college communications textbook. Then, answer the questions.

Textbook Skills

Advertisement's Purpose

[1]Although most advertisements promote a product or service, not all do. [2]Advertisements fall into three categories: business ads, public service ads, and political ads.

[3]**Business ads** try to influence people's attitudes and behaviors toward the product and services a business sells, toward the business itself, or toward an idea that the business supports. [4]Most of these ads try to persuade customers to buy something, but sometimes a business tries to improve its image through advertising. [5]This often occurs when a company has been involved in a highly publicized incident that might negatively affect its image. [6]Exxon had to rebuild its image after the *Valdez* accident in 1989, in which an Exxon tank spilled millions of gallons of oil in Prince William Sound and damaged the Alaska Coast.

[7]**Public service ads** promote behaviors and attitudes that are beneficial to society and its members. [8]These ads may be either national or local and usually are the product of donated labor and media time or space. [9]The Advertising Council produces the best known public service ads. [10]The council, which is supported by advertising agencies and the media, was formed during World War II to promote the war effort. [11]It now runs about twenty-five campaigns a year. [12]These campaigns must be in the public interest, timely, noncommercial, nonpartisan, nonsectarian, and nonpolitical.

[13]**Political ads** aim to persuade voters to elect a candidate to political office or to influence the public on legislative issues. [14]These advertisements run at the local, state, and national levels. [15]They incorporate most forms of media but use newspapers, radio, television, and direct mail most heavily. [16]During a presidential election year, more than $1 billion is spent on political ads.

—Folkerts & Lacy, *The Media in Your Life*, 2nd ed., pp. 428–429.

______ **1.** The tone of the first paragraph (sentences 1–2) is

a. neutral.
b. approving.
c. bitter.
d. bored.

_____ **2.** The purpose of sentence 6 is
a. to condemn Exxon for spilling millions of gallons of oil.
b. to keep the reader entertained with a vivid detail.
c. to illustrate the need for a business ad to improve a company's image.

_____ **3.** The tone of sentence 16 is
a. shocked.
b. shameful.
c. matter-of-fact.
d. abrupt.

_____ **4.** Which word best describes the overall tone of this passage?
a. unbiased
b. biased

_____ **5.** The overall purpose of this passage is
a. to inform the reader about different types of advertising.
b. to persuade the reader that advertising serves beneficial purposes.
c. to entertain the reader with interesting details about advertising.

VISUAL VOCABULARY

This advertisement is an example of a ________________ ________________

—Library of Congress, Prints & Photography Division, Yanker Poster Collection, POS6-U.S., no. 1384

MASTERY TEST 4

Name ______________________ Section ____________

Date __________ **Score** (number correct) ______ × 25 = ______%

Read the following passage from a college social science textbook. Then, answer the questions.

Textbook Skills

Earth Watch: Tangled Troubles—Logging and Bushmeat

[1]The bushmeat trade in Africa is a prime example of how threats to biodiversity interact and amplify one another. [2]Historically, rural Africans have supplemented their diet by hunting a variety of animals, collectively called "bushmeat." [3]Traditional subsistence hunting by small tribes using primitive weapons did not pose a serious threat to animal populations.

[4]Now, however, as logging roads penetrate deeper into rain forests, hunters follow, using shotguns and snares to kill any animal large enough to eat. [5]Communities that spring up along logging roads develop a culture of hunting and selling bushmeat, and become dependent on this new and profitable industry. [6]Logging trucks are sometimes used to carry the meat to urban markets. [7]The World Conservation Society estimates that the bushmeat harvest in equatorial Africa exceeds a million tons annually. [8]Many of the hunted animals play an important role in dispersing tree seeds. [9]Thus, loss of these animals reduces the ability of the logged forest to regenerate.

[10]Bushmeat hunters pay no attention to the sex, age, size, or rarity of the animal. [11]Therefore, many threatened species are declining rapidly. [12]For example, despite estimates that only about 2,000 to 3,000 pygmy hippos remain in the wild, pygmy hippo meat has been found in bushmeat markets, as has meat from elephants and rhinos.

[13]The profitability of bushmeat has helped overcome traditional African taboos against eating primates. [14]Although one-third of all primates (apes, monkeys, and lemurs) are threatened with extinction, in some bushmeat markets, 15% of the meat comes from primates. [15]In Cameroon, Africa, endangered gorillas are a favored target of poachers because of their large size. [16]Even endangered chimpanzees and bonobos, our closest relatives, end up in cooking pots. [17]Experts now believe that bushmeat hunting is an even greater threat than habitat loss for Africa's great apes, and that the combined threats of hunting and habitat loss make extinction in the wild a very real possibility for some of these magnificent, intelligent species.

[18]Recognizing the threats to wildlife, several central African countries are working to reduce illegal timber harvesting and wildlife poaching. [19]Together, they have established protected areas in the African rain forest of the Congo River Basin. [20]Although enormous logging enterprises continue along the borders of these reserves, protecting them is a crucial step toward preserving some of Africa's rich natural heritage.

—Adapted from *Life on Earth*, 5th ed., p. 631 by Teresa Audesirk, Gerald Audesirk, and Bruce E. Byers.

_____ **1.** The overall tone of the passage is
a. biased against bushmeat hunters.
b. biased for bushmeat hunters.
c. neutral about bushmeat hunters.

_____ **2.** The tone of sentence 16 is
a. angry.
b. ironic.
c. neutral.
d. mournful.

_____ **3.** In sentence 17, the words that describe the apes are
a. demeaning.
b. bitter.
c. admiring.
d. sorrowful.

_____ **4.** The overall purpose of the passage is to
a. inform the reader about the bushmeat trade in Africa.
b. entertain the reader with graphic details about the bushmeat trade in Africa.
c. persuade the reader that the bushmeat trade in Africa is a serious problem that must be addressed.

Inferences

11

LEARNING OUTCOMES

After studying this chapter you should be able to do the following:

1. Define the term *inference.*
2. Distinguish between a valid and an invalid inference.
3. Identify and apply the five steps for making a VALID inference.
4. Form valid inferences.
5. Define the following terms: *connotation, metaphor, personification, simile,* and *symbol.*
6. Infer meanings based on connotations, metaphors, personification, simile, and symbol.
7. Evaluate the importance of inferences.

Before Reading About Inferences

Predict what you need to learn based on the learning outcomes for this chapter by completing the following chart.

What I Already Know	What I Need to Learn
______________________	______________________.
______________________	______________________.
______________________	______________________.

Copy the following study outline in your notebook. Leave ample blank spaces between each topic. Use your own words to fill in the outline with information about each topic as you study about inferences.

Reading Skills Needed to Make VALID Inferences

I. **V**erify facts.

II. **A**ssess prior knowledge.

III. **L**earn from text.

  a. Context clues

  b. Thought patterns

  c. Implied main ideas

IV. **I**nvestigate bias.

V. **D**etect contradictions.

Inferences: Educated Guesses

Read the following paragraph.

> As Professor De Los Santos handed out the graded essays, Kassie's stomach rumbled with a familiar queasiness. She avoided looking around and focused with intensity on the opened textbook, not really seeing the pages that were before her. Aaron, who sat next to her, exclaimed with pleasure as he received his paper. "Unbelievable," he said as he turned to Kassie in surprise, "I thought I had blown this assignment." Kassie did not look up or respond to him, but continued staring at her book as she listened intently for her name. A girl with short, spiky hair across the room frowned after she received her paper and asked, "Dr. De Los Santos, may we rewrite this assignment?" The professor shook her head. Finally, Kassie heard her name, looked up, and took her essay. "Good job, Kassie," Dr. De Los Santos said as she handed her the paper, "Your work shows much improvement." Kassie smiled and carefully read each suggestion the teacher had written on her essay.

Which of the following statements might be true, based on the ideas in the passage?

______ Kassie is extremely nervous about her essay.

______ The girl with short, spiky hair is not satisfied with the grade she received.

______ Aaron did not put much effort into his essay.

______ Kassie's grade was "A" on her essay.

Did you choose the first two statements? Congratulations! You just made a set of educated guesses, or **inferences**. An author suggests or **implies** an idea, and the reader comes to a conclusion and makes an inference about what the author means.

In the paragraph about Kassie, the first two statements are firmly based on the information in the passage. However, the last two statements are not backed by the supporting details. The facts only suggest that Aaron thought he hadn't performed well on the assignment. He could have put a great deal of effort into the work; perhaps he underestimated his abilities. We just don't have enough information to make either assumption. Likewise, the passage does not suggest that Kassie made an "A," only that her work had improved.

What Is a Valid Inference?

People constantly draw conclusions about what they notice. We observe, gather information, and make inferences all the time.

> An **inference** or **conclusion** is an idea that is suggested by the facts or details in a passage.

For example, study the photo, and then answer the accompanying question.

VISUAL VOCABULARY

This business man is wearing a face mask because

____

____

____

If you wrote *of air pollution*, you made an inference based on the clues given by the visual details of the situation.

A **valid inference** is a rational judgment based on details and evidence. The ability to make a valid inference is a vital life skill. Making valid inferences aids us in our efforts to care for our families, succeed in our jobs, and even guard our health.

For example, doctors strive to make inferences about our health based on our symptoms. In the case of a heart attack, a doctor notes the observable symptoms such as pressure or pain in the center of the chest, shortness of breath, a cold sweat, nausea, or lightheadedness. To confirm the diagnosis, the doctor checks the patient's heart rate and blood pressure, takes a blood sample and performs a test called an electrocardiogram. If her educated guess (a guess based on evidence) is correct, she prescribes the best treatment she deems necessary.

A **valid inference** is a logical conclusion based on evidence.

EXAMPLE Read the following passage. Write **V** beside the three valid inferences. (**Hint:** valid inferences are firmly supported by the details in the passage.)

Pollen and Allergies

[1]For the past two weeks a heavy dusting of yellow pollen from the pine trees coated all the cars in the parking lot at Robert's job site. [2]The thick layering of pine pollen masked the fact that the oak and juniper trees were also pollinating. [3]In addition, every yard in Robert's neighborhood was covered with the purple flowering heads of Bermuda grass. [4]Interestingly, Robert linked the obvious pine pollen to allergy flare-ups, until he read a scientific report that stated that the heaviness of pine pollen causes it to fall to the ground rather than be carried by the wind like most allergy-causing pollen. [5]The same report stated that some of the most widely known plants that cause allergic reactions are Kentucky bluegrass, Johnson grass, Bermuda grass, redtop grass, orchard grass, and sweet vernal grass. [6]Trees that produce allergenic pollen include oak, ash, elm, hickory, pecan, box elder, and mountain cedar. [7]An allergic reaction includes frequent sneezing, clear and watery nasal discharge and congestion, itchy eyes, nose, and throat, and watery eyes. [8]For over a week, Robert has been experiencing these symptoms.

_____ **1.** Robert is experiencing an allergic reaction to pollen.

_____ **2.** Pine pollen is a major source of allergic reactions.

_____ **3.** Robert's yard consists of Bermuda grass.

_____ **4.** Robert's allergy attack is caused by the pollen from oak trees, juniper trees, or Bermuda grass.

_____ **5.** Kentucky bluegrass does not grow where Robert lives.

_____ **6.** Robert suffers with allergy attacks from pollen every year.

EXPLANATION Statements 1, 3, and 4 are firmly based on the information in the passage. It is reasonable to infer that Robert is experiencing an allergic reaction to pollen based on his symptoms and the fact that Robert's yard consists of Bermuda grass since "every" yard in his neighborhood does. It is also reasonable to infer that either one or some combination of the pollen from the oak trees, juniper trees, and Bermuda grass is causing his allergic reaction, based on the list of plants known to cause allergic reactions.

Statements 2, 5, and 6 are not based on the information in the passage. It is unreasonable to assume that pine pollen is a major source of allergic reactions because of information in the scientific report that explained that this pollen drops to the ground and is not carried by the wind and thus not breathed in by humans. It is also wrong to infer that Kentucky bluegrass does not grow where Robert lives simply because it is not present in the yards in his neighborhood or at his place of business. Finally, it is incorrect to infer that Robert experiences an allergic reaction to pollen every year. The passage only focuses on the last few weeks of Robert's life, so we do not have enough information to know whether he has suffered from these symptoms before.

PRACTICE 1

Each of the following items contains a short passage and three inferences. In each item, only one inference is valid. In the space provided, write the letter of the inference that is clearly supported by each passage.

_____ **1.** Jermaine and his office coworkers all arrived at work at the same time Monday morning. As usual, Jermaine unlocked the only entrance to the building, and then everyone unlocked their offices to begin their work routine. Immediately, Jermaine and his coworkers discovered that their computers and printers were missing from their offices. Each one reported that they had locked their offices with their computers and printers securely inside before they left on Friday. As they looked around the office, nothing indicated a forced entry. Jermaine called the police to report the theft of the computers and printers.

a. Jermaine stole the computers.

b. The office computers and printers were stolen by someone who had access to the office keys.

c. Jermaine and his coworkers were not surprised to find the computers and printers were missing.

_____ **2.** Alex and Mindy sit opposite each other in a booth at a diner. Mindy's face is streaked with tears. Alex's face is turned away as if in deep thought. His mouth is drawn into a thin frown, and his arms are folded across his chest. They do not speak to one another. As the waitress fills their coffee cups, Alex's frown turns into a smile, and he orders a large breakfast. Mindy shakes her head to indicate that she does not want anything to eat. As fresh tears begin to flow, she quickly ducks her head.

a. Alex and Mindy are having a fight.
b. Mindy is more upset than is Alex.
c. Alex is insensitive to Mindy's feelings.

_____ **3.** The patrol officer flashes his lights to pull over the sedan that has been weaving across the center line that divides the two-way traffic. After inspecting the driving license of Laurel, the young woman behind the wheel, he asks her to step out of the car, stand with her arms outstretched, and touch the tip of her nose with an index finger. She loses her balance as she tries to do so.

a. Laurel is drunk.
b. The police officer is harassing Laurel.
c. The police officer suspects that Laurel has been driving while under the influence of alcohol.

Making VALID Inferences and Avoiding Invalid Conclusions

Two of the most common pitfalls of making inferences are ignoring the facts and relying too much on personal opinions and bias. Often we are tempted to read too much into a passage because of our own prior experiences or beliefs. Of course, to make a valid inference, we must use clues based on logic and our experience. However, the most important resource must be the written text. As master readers, our main goal is to find out what the author is stating or implying. Master readers learn to use a VALID thinking process to avoid drawing false inferences or coming to invalid conclusions.

An **invalid conclusion** is a false inference that is not based on the details, or facts in the text or on reasonable thinking.

The VALID approach is made up of 5 steps:

Step 1. **V**erify and value the facts.

Step 2. **A**ssess prior knowledge.

Step 3. **L**earn from the text.

Step 4. **I**nvestigate for bias.

Step 5. **D**etect contradictions.

Step 1: Verify and Value the Facts

Develop a devotion to finding the facts. In Chapter 9, you learned to identify facts and to beware of false facts. You learned that authors may mix fact with opinion or use false information for their own purposes. Just as authors may make this kind of mistake, readers may, too. Readers may draw false inferences by mixing the author's facts with their own opinions or by misreading the facts. So it is important to find, verify, and stick to factual details. Once you have all the facts, only then can you begin to interpret the facts by making inferences.

EXAMPLE Read the following short passage. Then, write **V** next to the two valid inferences firmly supported by the facts.

Child Passenger Safety

[1]Motor vehicle injuries are the leading cause of death among children in the United States. [2]But many of these deaths can be prevented. [3]In the United States during 2005, 1,451 children ages 14 years and younger died as occupants in motor vehicle crashes, and approximately 203,000 were injured. [4]That's an average of 4 deaths and 556 injuries each day. [5]Of the children ages 0 to 14 years who were killed in motor vehicle crashes during 2005, nearly half were unrestrained.

[6]One out of four of all occupant deaths among children ages 0 to 14 years involve a drinking driver. [7]More than two-thirds of these fatally injured children were riding with a drinking driver. [8]Restraint use among young children often depends upon the driver's restraint use. [9]Almost 40% of children riding with unbelted drivers were themselves unrestrained. [10]Child restraint systems are often used incorrectly; one study found that 72% of nearly 3,500 observed child restraint systems were misused in a way that could be expected to increase a child's risk of injury during a crash. [11]All children under the age of 12 should ride in the back

seat. [12]Putting children in the back seat places children in the safest part of the vehicle in the event of a crash.

—Adapted from United States, Department of Human and Health Services. "Child Passenger Safety: Fact Sheet." *National Center for Injury Prevention and Control, Centers for Disease Control and Prevention.* 2 Feb 2010. http://www.cdc.gov/ncipc/factsheets/childpas.htm

_____ **1.** Most of the deaths and injuries related to motor vehicle crashes among children could be prevented.

_____ **2.** Placing children 14 years and younger in age-appropriate restraint systems could reduce serious and fatal injuries by half.

_____ **3.** Parents are the chief cause of deaths and injuries related to motor vehicle crashes among children.

EXPLANATION The first two statements are correct inferences based on the facts. However, there is no hint or clue that parents are the chief cause of these injuries and deaths. Children often ride in vehicles with people other than their parents. The third statement goes beyond the facts without any reason to do so.

VISUAL VOCABULARY

Fill in the blanks with words from the passage "Child Passenger Safety" that best complete the caption.

This child is _____________ restrained, _____________ the child's risk of injury during a crash.

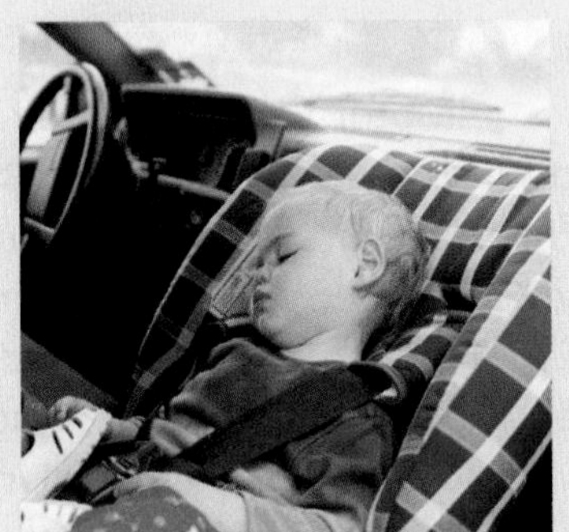

Step 2: Assess Prior Knowledge

Once you are sure of the facts, the next step is to draw on your prior knowledge. What you have already learned and experienced can help you make accurate inferences.

EXAMPLE 1 Read the following passage from a college science textbook. Identify the facts. Check those facts against your own experience and understanding. Write **V** next to the four inferences firmly supported by the facts in the passage.

Textbook Skills

The Flow of Energy and the Food Chain

[1]Energy stored by plants moves through the ecosystem in a series of steps of eating and being eaten, the **food chain**. [2]Food chains are descriptive diagrams—a series of arrows, each pointing from one species to another for which it is a source of food. [3]For example, grasshoppers eat grass, clay-colored sparrows eat the grasshoppers, and marsh hawks prey upon the sparrows. [4]We write this relationship as follows:

[5]grass ⟶ grasshopper ⟶ sparrow ⟶ marsh hawk

[6]However, the food chain is not linear. [7]Resources are shared, especially at the beginning of the food chain. [8]The same plant is food for a variety of animals and insects, and the same animal is food for several predators. [9]Thus food chains link to form a **food web**, the complexity of which varies within and among ecosystems.

—Adapted from *Elements of Ecology,* 4th ed., p. 329 by Thomas M. Smith and Robert Leo Smith. Copyright © 2000 by Pearson Education, Inc. Reprinted by permission of Pearson Education, Inc., Glenview, IL.

______ **1.** An *ecosystem* is a community of organisms and its environment functioning as a unit.

______ **2.** Grasshoppers eat only grass.

______ **3.** Plant life is the basis of the food chain used as an example.

______ **4.** The phrase *prey upon* in sentence 3 means "feed on."

______ **5.** The survival of the marsh hawk could be dependent upon the availability of grasshoppers.

______ **6.** Grass is always a basic element in a food chain or food web.

EXPLANATION Items 1, 3, 4, and 5 are sound inferences based on information in the passage. However, we have no evidence that grasshoppers eat only grass (item 2), nor that grass is always a basic element in a food chain or food web (item 6). In fact, sentence 9 states that ecosystems vary as does their complexity. For example, the food chain in a lake or ocean would not necessarily include grass.

EXAMPLE 2 Study the editorial cartoon. In the space provided, write an inference that is firmly supported by the details.

____

____

EXPLANATION Consider the following inference solidly based on the cartoon (and prior knowledge about fossil fuels and the environment): Global use of fossil fuels by industries is killing the planet.

Step 3: Learn from the Text

When you value and verify facts, you are learning from the text. A valid inference is always based on what is stated or implied by the details in the text; in contrast, an invalid inference goes beyond the evidence. Thus, to make a valid inference, you must learn to rely on the information in the text. Many of the skills you have studied in previous chapters work together to enable you to learn from the text. For example, context clues unlock the meaning of an author's use of vocabulary. Becoming aware of thought patterns teaches you to look for the logical relationship between ideas. Learning about stated and implied main ideas trains you to examine supporting details. (In fact, you use inference skills to find the implied main idea.) In addition, tone and purpose reveal the author's bias and intent. (Again, you often use inference skills to grasp the author's tone and purpose.) As you apply these skills to your reading process, you are carefully listening to what the author has to say. You are learning from the text. Only after you have learned from the text can you make a valid inference. The following examples show you how you learn from the text.

EXAMPLE Read the following statement. Answer the questions that follow.

Dr. Mehmet Oz, a consummate heart surgeon, deserves recognition as one of the most important figures in the 21st century because he's the most important and most accomplished celebrity doctor in history.

______ **1.** The best synonym for the word **consummate** is
a. excellent.
b. expensive.
c. demanding.

______ **2.** What is the primary relationship of ideas within the sentence?
a. addition
b. cause and effect
c. generalization and example

______ **3.** The author's tone is
a. sarcastic.
b. neutral.
c. flattering.

______ **4.** The author's purpose is to
a. inform.
b. entertain.
c. persuade.

______ **5.** Which of the following is a valid inference?
a. Dr. Oz only treats celebrities.
b. Dr. Oz has a large public following through the media.
c. Most people agree that Dr. Oz is "one of the most important and accomplished celebrity doctors in history."

PRACTICE 2

Read the following passages from college textbooks. Then, answer the questions that follow.

Textbook Skills

Alcohol, Cancer, and Heart Disease

[1]Heavy drinking is associated with an increased risk for several cancers—notably, cancer of the mouth, esophagus, pharynx, larynx, liver, and pancreas. [2]An analysis of multiple studies published in the mid-1990s found that having two alcoholic drinks per day (any type of alcohol) increased a woman's chances of developing breast cancer by nearly 25 percent. [3]The reasons for this are unclear. [4]However, researchers speculate that alcohol influences the metabolism of estrogen and that prolonged exposure to high levels of estrogen increases breast cancer risk, particularly for women on hormone replacement therapy (HRT). [5]The effect of

one drink per day on increased risk is a matter of debate. [6]Most experts feel that one drink per day does not increase risk. [7]But before you decide to toss out all your alcohol, you should know that there is increasing evidence that a glass of red wine seems to provide protection against heart disease.

—From *Access to Health*, 7th ed., p. 442 by Rebecca J. Donatelle and Lorraine G. Davis.

______ **1.** The best meaning of the word **speculate** in sentence 4 is
a. know.
b. study.
c. think.

______ **2.** What is the relationship of ideas between sentence 3 and sentence 4?
a. listing
b. cause and effect
c. contrast

______ **3.** What is the primary thought pattern of the paragraph?
a. contrast
b. cause and effect
c. time order

______ **4.** Which of the following best describes the author's tone and purpose?
a. to convince the reader to avoid consuming wine
b. to inform the general public about the possible link between alcohol and increased risk of cancer
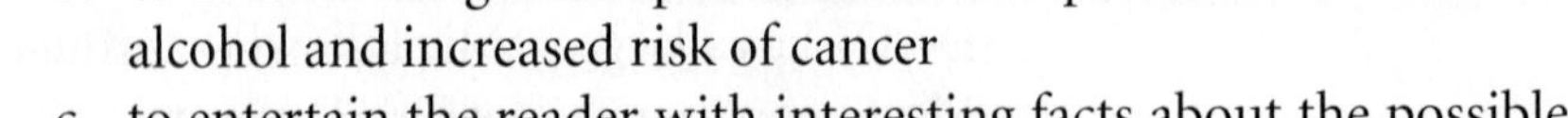
c. to entertain the reader with interesting facts about the possible effects of alcohol
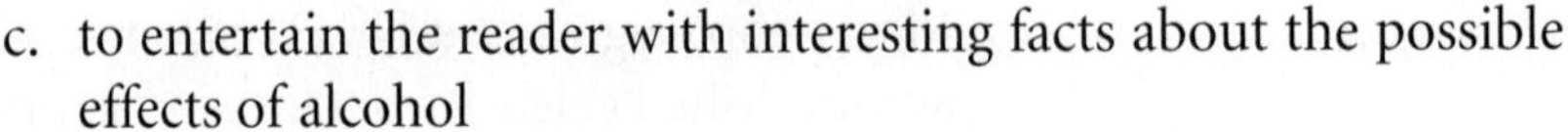

Textbook Skills

The Cracking Glass Ceiling

[1]What keeps women from breaking through the glass ceiling, the mostly invisible barrier that prevents women from reaching the executive suite? [2]Researchers have identified a "pipeline" that leads to the top: the marketing, sales, and production positions that directly affect the corporate bottom line. [3]Men, who dominate the executive suite, stereotype women as being less capable of leadership than they are. [4]Viewing women as good at "support," they steer women into human resources or public relations. [5]There, successful projects are not appreciated in the same way as those that bring corporate profits—and bonuses for their managers.

[6]Another reason the glass ceiling is so powerful is that women lack mentors—successful executives who take an interest in them and teach them the ropes. [7]Lack of a mentor is no trivial matter, for mentors can

provide opportunities to develop leadership skills that open the door to the executive suite.

[8]The glass ceiling is cracking, however. [9]A look at women who have broken through reveals highly motivated individuals with a fierce competitive spirit who are willing to give up sleep and recreation for the sake of career advancement. [10]They also learn to play by "men's rules," developing a style that makes men comfortable. [11]Most of these women also have supportive husbands who share household duties and adapt their careers to accommodate the needs of their executive wives. [12]In addition, women who began their careers 20 to 30 years ago are running many major divisions within the largest companies. [13]With this background, some of these women will emerge as the new top CEOs.

—From *Essentials of Sociology: A Down-to-Earth Approach,* 7th ed., pp. 271–272 by James M. Henslin. Copyright © 2007 by Pearson Education, Inc. Reproduced by permission of Pearson Education, Inc., Boston, MA.

_____ **1.** The best meaning of the word **mentor** in sentence 6 is
a. employer.
b. adviser.
c. men.

_____ **2.** What is the relationship of ideas between sentence 7 and sentence 8?
a. listing
b. cause and effect
c. contrast

_____ **3.** What is the primary thought pattern of the paragraph?
a. classification
b. cause and effect
c. contrast

_____ **4.** Which of the following statements best describes the author's tone and purpose?
a. to inform the reader about the causes of the glass ceiling
b. to persuade the reader to protest the glass ceiling
c. to entertain the reader with interesting facts about the glass ceiling

Step 4: Investigate for Bias

One of the most important steps in making a valid inference is confronting your biases. Each of us possesses strong personal views that influence the way we process information. Often our personal views are based on prior experiences.

For example, if we have had a negative prior experience with a used car salesperson, we may become suspicious and stereotype all used car salespeople as dishonest. Sometimes, our biases are based on the way in which we were raised. Some people register as Democrats or Republicans and vote for only Democratic or Republican candidates simply because their parents were dedicated supporters of one party. To make a valid inference, we must investigate our response to information for bias. Our bias can shape our reading of the author's meaning. To investigate for bias, note biased words and replace them with factual details as you form your conclusions.

EXAMPLE Reword each of the following statements to eliminate the use of biased words.

1. Simon Cowell is best known for his roles as a brutally honest judge on the silly TV talent shows *The X Factor* and *American Idol.*

___

___

2. Surprisingly, Simon Cowell won the prestigious International Emmy award because he completely reshaped 21st century television and music around the world.

___

___

EXPLANATION Compare your answers to the following:

1. Simon Cowell is known for his roles as judge on the TV talent shows *The X Factor* and *American Idol.* The following biased words were deleted: *best*, *brutally honest*, and *silly*.
2. Simon Cowell won the International Emmy award because he has influenced 21st century television and music around the world. The following biased words were deleted: *Surprisingly*, *prestigious*, *completely reshaped.*

PRACTICE 3

Study the photo and caption that appears in a college sociology textbook. Reword the caption to eliminate any bias.

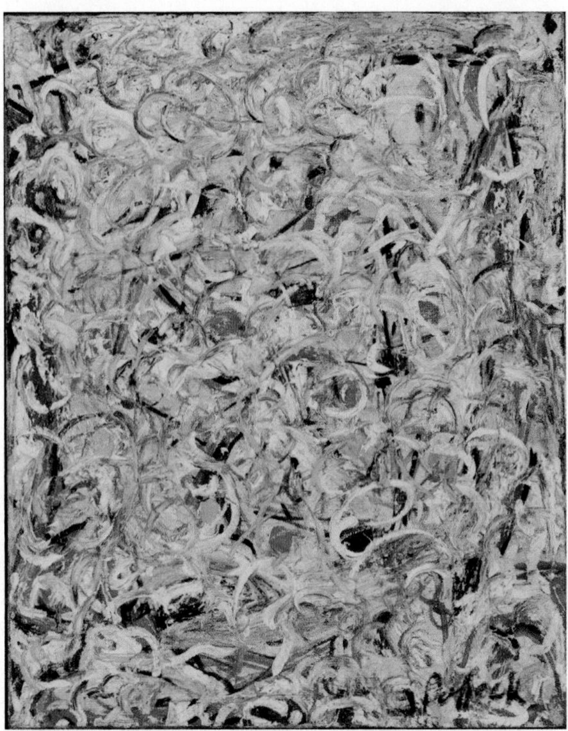

◄ **Jackson Pollock, *Shimmering Substance.* (1946)** Many of Pollock's postwar works—huge paintings that pulse with power—show an obsession with heat and light, surely no coincidence in the dawn of the nuclear age.

—From *The West: Encounters & Transformations, Atlas Edition, Combined Edition,* 2nd ed., p. 910 by Brian Levack, Edward Muir, Meredith Veldman and Michael Maas. Copyright © 2008 by Pearson Education, Inc. Printed and Electronically reproduced by permission of Pearson Education, Inc., Upper Saddle River, NJ.

Textbook Skills

____

____

Step 5: Detect Contradictions

Have you ever misjudged a situation or had a wrong first impression? For example, have you ever assumed a person was conceited or rude only to find out later that he or she was acutely shy? Many times, there may be a better explanation for a set of facts than the first one that comes to mind. The master reader hunts for the most reasonable explanation. The best way to do this is to consider other explanations that could logically contradict your first impression.

EXAMPLE 1 Read the following list of symptoms. Then, in the blank, write as many explanations for the symptoms as you can.

- Fatigue
- Muscle and joint pain
- Erythema migrans (bull's-eye skin rash)
- Swollen lymph nodes
- Chills and fever
- Headache

____

____

EXPLANATION Some people may think the symptoms describe the flu, but it is actually a list of symptoms for the first stage of Lyme disease. The symptom that clearly identifies Lyme disease is the bull's-eye rash; however, not everyone who has the disease has this symptom, and others may have this symptom but not have Lyme disease. Thus Lyme disease can be misdiagnosed in its earliest stage. A reader who does not think about other possible views can easily jump to a wrong conclusion. Master readers consider all the facts and all the possible explanations for those facts.

EXAMPLE 2 Study the graphic. In the space provided, write an inference that is firmly supported by the details.

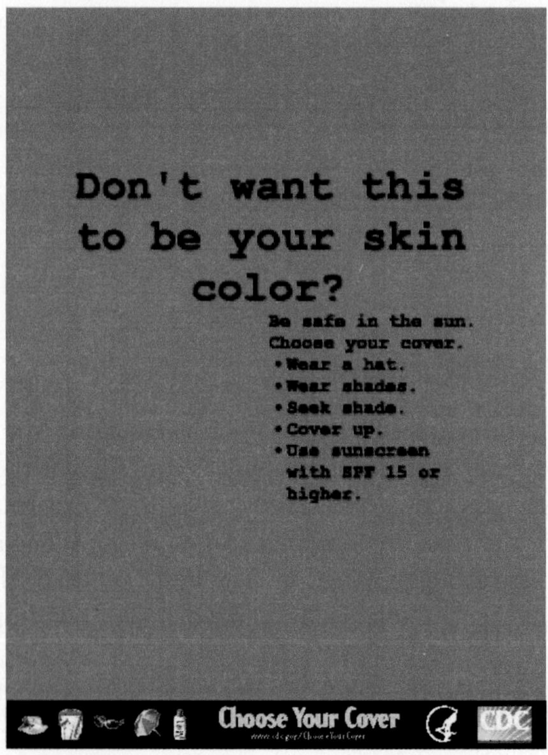

—"Don't Want This to Be Your Skin Color?" CDC http://www.cdc.gov/cancer/skin/pdf/poster-red.pdf

____

____

EXPLANATION This poster implies that protection against the sun's harmful UV rays is necessary. Consider the following statement of an inference firmly based on the information in the poster: "You should protect yourself from severe sunburn and sun damage by covering up with one or more of the following: a hat, clothing, sunscreen, and sunglasses."

PRACTICE 4

Read the following passage. Investigate the list of inferences that follow for bias. Underline biased words. Mark each inference as follows: **V** for a valid inference firmly supported by the facts, **I** for an invalid inference.

Marching for the Vote: The Woman Suffrage Parade of 1913

[1]On Monday, March 3, 1913, clad in a white cape astride a white horse, lawyer Inez Milholland led the great woman suffrage parade down Pennsylvania Avenue in the nation's capital. [2]Behind her stretched a long line with nine bands, four mounted brigades, three heralds, about twenty-four floats, and more than 5,000 marchers. [3]The procession began late, but all went well for the first few blocks. [4]Soon, however, the crowds, mostly men in town for the following day's inauguration of Woodrow Wilson, surged into the street making it almost impossible for the marchers to pass. [5]Occasionally only a single file could move forward. [6]Women were jeered, tripped, grabbed, shoved, and many heard "indecent epithets" and "barnyard conversation." [7]Instead of protecting the parade, the police "seemed to enjoy all the ribald jokes and laughter and participated in them." [8]One policeman explained that they should stay at home where they belonged. [9]The men in the procession heard shouts of "Henpecko" and "Where are your skirts?" [10]As one witness explained, "There was a sort of spirit of levity connected with the crowd. [11]They did not regard the affair very seriously."

—Adapted from Harvey, "Marching for the Vote: The Woman Suffrage Parade of 1913," *American Women,* 18 June 2003, Library of Congress. 12 Feb. 2004 http://memory.loc.gov/ammem/awhhtml/aw01e/aw01e.html#ack

_____ **1.** The word *suffrage* in sentence 1 means "right to vote."

_____ **2.** Nearly all of the 5,000 marchers were women.

_____ **3.** The police supported women's right to vote.

______ **4.** Men opposed a woman's right to vote.

______ **5.** Many men in the crowd opposed the women's right to vote.

Inferences in Creative Expression

As you have learned, nonfiction writing, such as in textbooks and news articles, directly states the author's point. Everything is done to make sure that the meanings are clear and unambiguous (not open to different interpretations). However, in many other types of writing, both fiction and nonfiction, authors use creative expression to suggest layers of meaning. Creative expressions are also known as literary devices. The following chart shows a few common literary devices, their meanings, and an example of each.

Creative Expression: Literary Devices

Literary Device	Meanings	Example
Connotations of words	The emotional meaning of words	My home is for sale.
Metaphor	A direct comparison	Lies are sinkholes.
Personification	Giving human traits to things that are not human	The sun woke slowly.
Simile	An indirect comparison	Lies are like sticky webs.
Symbol	Something that stands for or suggests something else	A skull and crossbones is a symbol for poison and death.

By using these devices, a writer creates a vivid mental picture in the reader's mind. When a creative expression is used, a reader must infer the point the writer is making from the effects of the images. The following paragraph is the introduction to an essay about anger. Notice its use of literary devices. After you read, write a one-sentence statement of the author's main idea for this paragraph.

A Growing Anger

[1]At first, Julio sat quietly shaking his head; then his forehead crumpled into a wadded frown, and his mouth flattened into a pencil-thin line. [2]As the speaker continued, Julio's heartbeat sounded like drums beating in his ears, and his entire body flushed as a searing heat rushed up from his toes to the roots of his hair. [3]His jaw tightened and his fists

clenched and unclenched as he constantly shifted his weight in his chair. [4]Finally, unable to contain himself any longer, he abruptly rose, jostling the table and tumbling the wine glasses. [5]His body trembled, and his voice shook as he exclaimed, "You have no idea what you are talking about." [6]Just as suddenly, he thudded back into his chair, fists still clenched, and stared without seeing at the white tablecloth bloodied with spilled wine.

____

____

This paragraph uses several creative expressions. *Forehead crumpled into a wadded frown* and *mouth flattened into a pencil-thin line* are metaphors. The author is showing the physical effects of anger by directly comparing Julio's frown to a crumpled wad and his mouth to a thin pencil line. The phrase *heartbeat sounded like drums* is a simile that indirectly compares the sound of Julio's heartbeat to a drum. The author chooses words that have rich connotations in the phrase *flushed as a searing heat rushed,* suggesting that Julio is becoming inflamed and passionate about what he is hearing. And finally, the author uses personification with the phrase *tablecloth bloodied with spilled wine.* The phrase suggests that the tablecloth is injured rather than merely soiled and that the wine bled instead of spilled. All of these details work to create a vivid mental image, and based on these details, the reader must come to the conclusion that Julio becomes increasingly angry. To understand plays, novels, short stories, and poems, the master reader must use inference skills.

EXAMPLE Read the following poem written by Langston Hughes. Answer the questions that follow.

I, Too

by Langston Hughes

I, too, sing America.

I am the darker brother.
They send me to eat in the kitchen
When company comes,
But I laugh,
And eat well,
And grow strong.

Tomorrow,
I'll be at the table
When company comes.
Nobody'll dare
Say to me,
"Eat in the kitchen,"
Then.

Besides,
They'll see how beautiful I am
And be ashamed–

I, too, am American.

—Hughes, Langston, "I, Too," from *The Collected Poems of Langston Hughes,* edited by Arnold Rampersad with David Roessel, Associate Editor, copyright © 1994 by The Estate of Langston Hughes. Used by permission of Alfred A. Knopf, a division of Random House, Inc.

1–3. Choose **three** inferences that are most firmly based on the information in the passage by writing a **V** next to each one.

_____ a. When the speaker of the poem refers to himself as "the darker brother," he is implying that he is an African American.

_____ b. The line "they send me to eat in the kitchen" refers to the historical unequal treatment of African Americans in America.

_____ c. The speaker of the poem is angry and resentful.

_____ d. The speaker of the poem possesses pride and hope.

_____ e. The speaker does not believe that African Americans will ever be appreciated in America.

_____ **4.** Based on the details of the poem, we can conclude that the implied main idea of the poem could be that
a. the speaker is celebrating being an African American.
b. the speaker is condemning America.
c. the speaker is celebrating America.

EXPLANATION

1–3. The valid inferences are (a), (b), and (d).

4. Based on the details of the poem, we can conclude that the main idea of the poem could be (a) the speaker is celebrating being an African American.

PRACTICE 5

Textbook Skills

Read the poem written by Robert Frost. Then, choose the inferences that are most logically based on the details in the poem.

The Road Not Taken

by Robert Frost

Two roads diverged in a yellow wood,
And sorry I could not travel both
And be one traveler, long I stood
And looked down one as far as I could
To where it bent in the undergrowth;

Then took the other, as just as fair,
And having perhaps the better claim,
Because it was grassy and wanted wear;
Though as for that the passing there
Had worn them really about the same,

And both that morning equally lay
In leaves no step had trodden black.
Oh, I kept the first for another day!
Yet knowing how way leads on to way,
I doubted if I should ever come back.

I shall be telling this with a sigh
Somewhere ages and ages hence:
Two roads diverged in a wood, and I—
I took the one less traveled by,
And that has made all the difference.

—From *Mountain Interval* by Robert Frost.
New York: Henry Holt, 1920.

_____ **1.** The best meaning of the word **diverged** in the first line is
a. parted.
b. ended.
c. merged.

_____ **2.** In the poem, the roads represent
a. travelers.
b. fears.
c. choices.

_____ **3.** The second verse describes the "other" road he took as
a. less traveled than the first road.
b. completely untraveled by others.
c. worn with the constant travel of others.

_____ **4.** In the first two lines of the third verse, the speaker implies that
a. others had traveled on both roads earlier that day.
b. no other person had traveled on either road that day.
c. he already regrets his decision to take the other road.

_____ **5.** The main idea of the poem is that the speaker
a. is an individualist.
b. has many regrets.
c. conforms to the leadership of others.

Reading a Textbook: Inferences and Visual Aids

Textbook Skills

Textbook authors often use pictures, photos, and graphs to imply an idea. These visuals are used to reinforce the information in that section of the texbook.

EXAMPLE Study this figure, taken from the textbook *Psychology and Life.* Then answer the question.

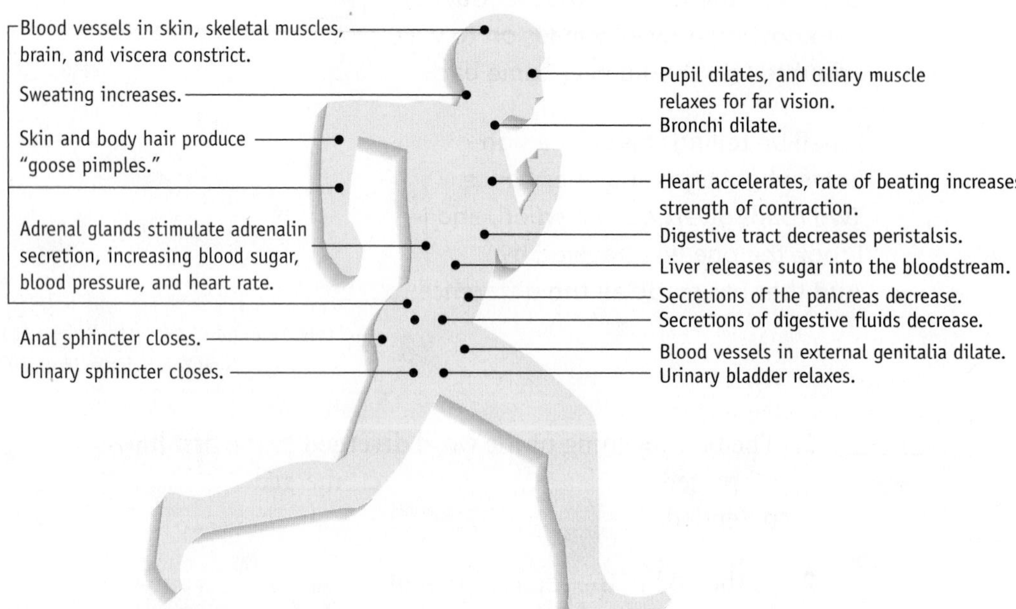
Blood vessels in skin, skeletal muscles, brain, and viscera constrict.
Sweating increases.
Skin and body hair produce "goose pimples."
Adrenal glands stimulate adrenalin secretion, increasing blood sugar, blood pressure, and heart rate.
Anal sphincter closes.
Urinary sphincter closes.
Pupil dilates, and ciliary muscle relaxes for far vision.
Bronchi dilate.
Heart accelerates, rate of beating increases strength of contraction.
Digestive tract decreases peristalsis.
Liver releases sugar into the bloodstream.
Secretions of the pancreas decrease.
Secretions of digestive fluids decrease.
Blood vessels in external genitalia dilate.
Urinary bladder relaxes.

▲ **The Body's Reaction to Stress**

—From *Psychology and Life,* 19th ed., Figure 13.7, (p. 407) by Richard J. Gerrig and Philip G. Zimbardo.

_____ **1.** What is the topic or title of the chapter?
   a. Jogging and Fitness
   b. Cardiovascular Health
   c. Emotion, Stress, and Health

2. Write a caption in a complete sentence that best states the main idea of the figure.

____

____

EXPLANATION

1. This figure appeared in a chapter with the title "Emotion, Stress, and Health," choice (c).

2. The figure shows at least 15 different ways in which stress affects the human body. Consider the following statement of the figure's main idea: Stress produces a wide range of physiological changes in the body.

PRACTICE 6

Study this figure, taken from the textbook *Life on Earth* and read the list of statements about the figure. Write a **V** in the blanks next to the four inferences that are most firmly based on the figure.

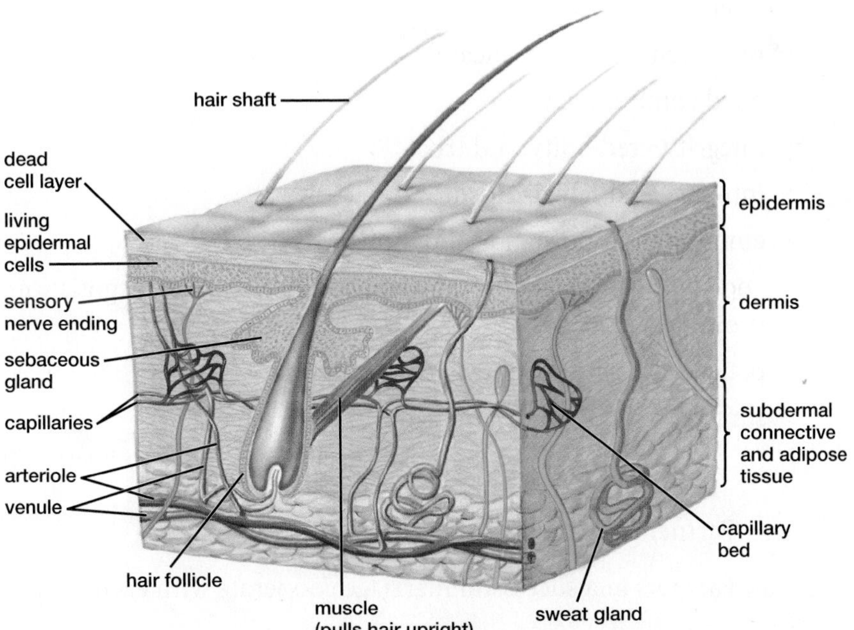

—From *Life on Earth,* 5th ed., Figure 19.9 (p. 363) by Teresa Audesirk, Gerald Audesirk and Bruce E. Byers. Copyright © 2009 by Pearson Education, Inc. Reprinted by permission of Pearson Education, Inc., Upper Saddle River, NJ.

_____ **1.** This figure appears in the chapter titled "The Organization of the Animal Body."

_____ **2.** This figure appears in the chapter titled "Animal Behavior."

_____ **3.** The epidermis is one thin layer of living cells.

_____ **4.** The epidermis is the outer layer of skin.

_____ **5.** Mammalian skin is made up of three layers.

_____ **6.** Skin contains muscle tissue.

_____ **7.** Human skin differs from mammalian skin.

_____ **8.** Skin is not an organ.

APPLICATIONS

Application 1: Making Inferences from a List

Read the following list of details. Write a **V** in the blank next to the three inferences that are most firmly based on the information.

Characteristics of Successful Marriages

- commitment
- interdependence
- reciprocity (desire to please)
- good communication
- integrity, credibility, and trust
- intimacy and self-disclosure
- emotional support and validation
- good conflict resolution skills, especially flexibility, negotiation, and compromise
- patience, sensitivity, and tolerance
- similar views of the relationship

—Adapted from Jaffe, *Understanding Parenting,* 2nd ed., p. 103.

_____ Partners in a successful marriage never disagree.

_____ Partners in a successful marriage cooperate with each other.

_____ Intimacy is enhanced by self-disclosure.

______ Flexibility and compromise are key elements in effectively resolving conflicts.

______ Successful marriages are difficult to achieve and less and less common.

______ The most important trait of a successful marriage is commitment.

Application 2: Making Inferences from a Visual

Textbook Skills

The following comic strip from *Frank and Earnest* appeared in a college sociology textbook in the chapter about socialization and gender. Read the cartoon; then write an inference about gender that is soundly based on the information in the comic strip.

—©2002 Thaves/Dist. by NEA, Inc. www.frankandernest.com <http://www.frankandernest.com>

____

Application 3: Making Inferences from Literature

Read the following excerpt from the well-known play *Hamlet,* written by William Shakespeare. Then, using inference skills, answer the questions.

Hamlet: To be, or not to be: that is the question:
Whether 'tis nobler in the mind to suffer
The slings and arrows of outrageous fortune,
Or to take arms against a sea of troubles,
And by opposing end them? To die: to sleep;
No more; and, by a sleep to say we end
The heart-ache and the thousand natural shocks
That flesh is heir to, 'tis a consummation
Devoutly to be wish'd. To die, to sleep;
To sleep: perchance to dream: ay, there's the rub;
For in that sleep of death what dreams may come
When we have shuffled off this mortal coil,

Must give us pause. There's the respect
That makes calamity of so long life;
For who would bear the whips and scorns of time,
The oppressor's wrong, the proud man's contumely,
The pangs of dispriz'd love, the law's delay,
The insolence of office, and the spurns
That patient merit of the unworthy takes,
When he himself might his quietus make
With a bare bodkin? who would fardels bear,
To grunt and sweat under a weary life,
But that the dread of something after death,
The undiscover'd country from whose bourn
No traveller returns, puzzles the will,
And makes us rather bear those ills we have
Than fly to others that we know not of?
Thus conscience does make cowards of us all;
And thus the native hue of resolution
Is sicklied o'er with the pale cast of thought,
And enterprises of great pith and moment
With this regard their currents turn awry,
And lose the name of action.

—From *Hamlet.* In *The Complete Works of William Shakespeare.* Edited by W. J. Craig. London: Oxford University Press, 1914.

1. What action is Hamlet considering?

____

____

2. To what does the metaphor "mortal coil" refer?

____

____

3. To what does the metaphor "The undiscover'd country from whose bourn No traveller returns" refer?

____

____

4. What stops Hamlet from taking the action he is considering?

____

REVIEW TEST 1

Score (number correct) ______ × 20 = ______ %

Making Inferences

Read the poem, and then answer the questions that follow.

My Heart Leaps up When I Behold

by William Wordsworth (1807)

My heart leaps up when I behold
  A rainbow in the sky:
So was it when my life began;
So is it now I am a man;
So be it when I shall grow old,
  Or let me die!
The Child is father of the Man;
  I could wish my days to be
Bound each to each by natural piety.

—Wordsworth, William, *The Complete Poetical Works.* London: Macmillan and Co., 1888; Bartleby.com, 1999. www.bartleby.com/145/. 28 March 2010.

_____ **1.** Based on the context of the poem, the best synonym for the word **piety** is

a. youthfulness.
b. courage.
c. goodness.
d. law.

_____ **2.** What emotion does the title of the poem suggest?

a. fear
b. confidence
c. doubt
d. joy

_____ **3.** The rainbow represents

a. nature.
b. mankind.
c. youthfulness.
d. old age.

_____ **4.** In the line "The Child is the father of the Man," the poet suggests

a. mankind is as innocent as a child.
b. the person we are as an adult is connected to the person we are as a child.
c. life passes by too quickly.

______ **5.** Overall, the poet expresses
a. faith in the goodness of human beings.
b. a life-long love of nature.
c. a fear of death.

REVIEW TEST 2

Score (number correct) ______ × 20 = ______%

Textbook Skills

Read the following passage from a college psychology textbook. Complete the activities that follow.

Episodic and Semantic Memories

[1]When we discussed the functions of memories earlier, we made a distinction between declarative and procedural memories. [2]We can define another dimension along which declarative memories differ with respect to the cues that are necessary to retrieve them from memory. [3]Canadian psychologist Endel Tulving (1972) first proposed the distinction between episodic and semantic types of declarative memories.

[4]**Episodic memories** preserve, individually, the specific events that you have personally experienced. [5]For example, memories of your happiest birthday or of your first kiss are stored in episodic memory. [6]To recover such memories, you need retrieval cues that specify something about the time at which the event occurred and something about the content of the events. [7]Depending on how the information has been encoded, you may or may not be able to produce a specific memory representation for an event. [8]For example, do you have any specific memories to differentiate the tenth time ago you brushed your teeth from the eleventh time ago?

[9]Everything you know, you began to acquire in some particular context. [10]However, there are large classes of information that, over time, you encounter in many different contexts. [11]These classes of information come to be available for retrieval without reference to their multiple times and places of experience. [12]These **semantic memories** are generic, categorical memories, such as the meanings of words and concepts. [13]For most people, facts like the formula $E = MC^2$ and the capital of France don't require retrieval cues that make reference to the episodes, the original learning contexts, in which the memory was acquired.

[14]Of course, this doesn't mean that your recall of semantic memories is foolproof. [15]You know perfectly well that you can forget many facts that have become dissociated from the contexts in which you learned them. [16]A

good strategy when you can't recover a semantic memory is to treat it like an episodic memory again. [17]By thinking to yourself, "I know I learned the names of the Roman emperors in my Western civilization course," you may be able to provide the extra retrieval cues that will shake loose a memory.

—From *Psychology and Life,* 19th ed., p. 206 by Richard J. Gerrig and Philip G. Zimbardo.

1–3. Write **V** for valid by the three inferences that are most firmly supported by the ideas in the passage.

_____ The term episodic refers to actions that occur in time.

_____ The term semantic refers to language and symbols.

_____ Episodic and semantic memories never work together to retrieve information.

_____ Most people would rely on episodic memories to retrieve the dates of the Civil War.

_____ There are at least four categories of long-term memories.

4–5. Complete the concept map with information from the passage.

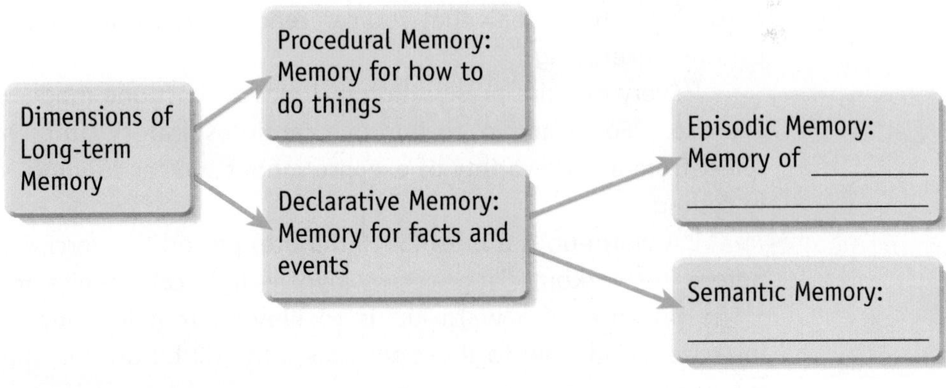

REVIEW TEST 3

Score (number correct) _____ × 10 = _____%

Making Inferences

Before you read the following passage from a college health textbook, skim the passage. Answer the Before Reading questions. Then, read the passage and answer the After Reading questions.

Vocabulary Preview

incentive (8): motivation, reason

cardiorespiratory (18): of or relating to the heart and respiratory system

calisthenics (20): exercises

elevate (21): raise

duration (23): length, period of time

Exercise Prescription

[1]Doctors often prescribe medications to treat certain diseases. [2]For every individual there is an appropriate dosage of medicine to cure an illness. [3]Similarly, for each individual, there is a correct dosage of exercise to best promote physical fitness. [4]An **exercise prescription** should be tailored to meet the needs of the individual. [5]It should include fitness goals, mode of exercise, a warm-up, a primary conditioning period, and a cool-down.

[6]Establishing short-term and long-term fitness goals is an important part of an exercise prescription. [7]Goals serve as motivation to start an exercise program. [8]Further, reaching your fitness goals improves self-esteem and provides the **incentive** to make a lifetime commitment to regular exercise. [9]A logical and common type of fitness goal is a performance goal. [10]Increased flexibility is an example of a performance goal. [11]You should set both long-term and short-term performance goals, and you should be tested to determine when you have achieved them. [12]In addition to performance goals, consider setting exercise **adherence** goals. [13]That is, set a goal to exercise a specific number of days per week. [14]Exercise adherence goals are important because fitness will improve only if you exercise regularly. [15]The most important rule in setting goals is that you must create realistic ones. [16]Set fitness goals that you can reach so you will be encouraged to continue exercising.

[17]Every exercise prescription includes at least one mode, or type, of exercise. [18]For example, to improve **cardiorespiratory** fitness, you could select from a wide variety of exercise modes such as running, swimming, or cycling.

[19]A warm-up is a brief (5–15 minute) period of exercise that occurs before the workout. [20]It generally involves light **calisthenics** or stretching. [21]The purpose of a warm-up is to **elevate** muscle temperature and increase blood flow to those muscles that will be used in the workout. [22]A warm-up can also reduce the strain on the heart and may reduce the risk of muscle and tendon injuries.

[23]The primary conditioning period includes the mode of exercise and the frequency, intensity, and **duration** of the workout. [24]The frequency of exercise is the number of times per week that you exercise. [25]In general, the suggested number is three to five times per week. [26]The intensity of exercise is the amount of stress or overload placed on the body during the exercise. [27]The method of measuring intensity

varies with the type of exercise. [28]For example, the heart rate is a standard means of tracking intensity during training to improve cardiorespiratory fitness. [29]In strength training, the number of repetitions that can be performed before muscle fatigue occurs is used. [30]In addition, flexibility is measured by the degree of tension or discomfort felt during the stretch.

[31]Another key **component** of the exercise prescription is the duration of exercise. [32]This is the amount of time spent in performing the primary workout. [33]Note that the duration of exercise does not include the warm-up or cool-down. [34]In general, research has shown that 20 to 30 minutes per exercise session is the least amount of time needed to improve physical fitness.

[35]The cool-down is a 5- to 15-minute period of low-intensity exercise that immediately follows the primary workout. [36]For instance, a period of slow walking might be used as a cool-down after a running workout. [37]A cool-down accomplishes several goals. [38]First, the cool-down allows blood to be returned from the muscles back toward the heart. [39]During exercise, large amounts of blood are pumped to the working muscles. [40]Once exercise ends, blood tends to remain in large blood vessels located around the muscles that were worked. [41]This is known as pooling. [42]Failure to redistribute pooled blood after exercise could result in you feeling lightheaded or even fainting. [43]Finally, some experts argue that post-exercise soreness may be reduced as a result of a cool-down.

—From *Total Fitness and Wellness Student Textbook Component,* 3rd ed., pp. 65–69 by Scott K. Powers and Stephen L. Dodd. Copyright © 2003 by Pearson Education, Inc. Printed and Electronically reproduced by permission of Pearson Education, Inc., Upper Saddle River, NJ.

Before Reading

Vocabulary in Context

_____ **1.** The word **adherence** in sentence 12 means
a. obedience. c. time.
b. esteem.

_____ **2.** The word **component** in sentence 31 means
a. benefit. c. factor.
b. goal.

Concept Maps

Finish the concept map with information from the passage.

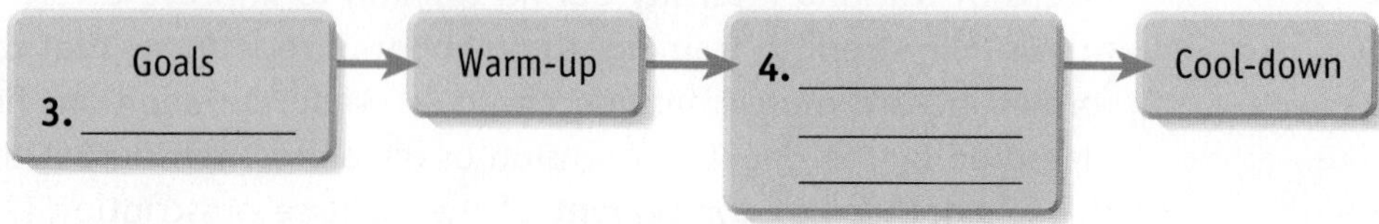

After Reading

Central Idea and Main Idea

_____ **5.** Which sentence best states the main idea of the passage?
a. sentence 3
b. sentence 5
c. sentence 6

Supporting Details

_____ **6.** Sentence 28 is a
a. major supporting detail.
b. minor supporting detail.

Transitions

_____ **7.** What thought pattern describes the relationship between sentences 35 and 36?
a. cause and effect
b. time order
c. definition and example

Thought Patterns

_____ **8.** The thought pattern used in the fifth paragraph (sentences 23–30) is
a. cause and effect.
b. time order.
c. definition and example.

Tone and Purpose

_____ **9.** The author's tone and purpose in the passage are
a. to inform with objective details.
b. to entertain with subjective details.
c. to persuade with discouraging details.

Inferences

10. Write **V** for valid by the inference that is most firmly based on the information in the passage.

_____ A performance goal that is set too high can become a block to motivation.

_____ Sudden and intense exercise helps increase muscle mass more than a slower build up.

_____ In order to experience a complete workout, each exercise period should be at least two hours.

WHAT DO YOU THINK?

Why is exercise so important for a healthy lifestyle? Why is frequency, intensity, and the duration of exercise so important? What are some of the barriers that keep people from committing to an exercise program? Assume you are a personal trainer and create a personal exercise prescription for one or both of the following two potential clients: (1) Shantel, age 45, no known medical problems or limitations, has a desk job, and is 30 pounds overweight; (2) Richard, age 17, no known medical problems or limitations, trying out for a position on his high school track team, is at a good weight, but is out of shape.

REVIEW TEST 4

Score (number correct) _____ × 10 = _____%

Making Inferences

Textbook Skills

Before Reading: Survey the following passage adapted from the college textbook *Essentials of Sociology*. Skim the passage, noting the words in **bold** print. Answer the Before Reading questions that follow the passage. Then, read the passage. Next, answer the After Reading questions. Use the discussion and writing topics as activities to do after reading.

Lara Croft, Tomb Raider: Changing Images of Women in the Mass Media

[1]The mass media reflect women's changing role in society. [2]Portrayals of women as passive, as subordinate, or as mere background objects remain, but a new image has broken through. [3]Although this new image

Vocabulary Preview

fundamental (3): basic
intricate (7): complex
socializes (8): teaches
essence (11): core, real meaning
vanquish (12): conquer, defeat
protagonist (14): hero
encapsulated (15): summarize
predicament (17): dilemma, problem
voluptuous (18): sensual pleasure due to fullness and beauty of form

exaggerates changes, it also illustrates a **fundamental** change in gender relations. [4]Lara Croft is an outstanding example of this change.

[5]Like books and magazines, video games are made available to a mass audience. [6]And with digital advances, they have crossed the line from what is traditionally thought of as games to something that more closely resembles interactive movies. [7]Costing an average of $10 million to produce and another $10 million to market, video games now have **intricate** subplots and use celebrity voices for the characters (Nussenbaum 2004). [8]Sociologically, what is significant is that the content of video games **socializes** their users. [9]As they play, gamers are exposed not only to action but also to ideas and images. [10]The gender images of video games communicate powerful messages, just as they do in other forms of the mass media.

[11]Lara Croft, an adventure-seeking archeologist and star of *Tomb Raider* and its many sequels, is the **essence** of the new gender image. [12]Lara is smart, strong, and able to utterly **vanquish** foes. [13]With both guns blazing, she is the cowboy of the twenty-first century, the term *cowboy* being purposefully chosen, as Lara breaks stereotypical gender roles and dominates what previously was the domain of men.

[14]She was the first female **protagonist** in a field of muscle-rippling, gun-toting **macho-caricatures** (Taylor 1999). [15]Yet the old remains powerfully **encapsulated** in the new. [16]As the photos on [the next] page make evident, Lara is a fantasy girl for young men of the digital generation.

[17]No matter her foe, no matter her **predicament**, Lara oozes sex. [18]Her form-fitting outfits, which flatter her **voluptuous** physique, reflect the mental images of the men who fashioned this digital character.

[19]Lara has caught young men's fancy to such an extent that they have bombarded corporate headquarters with questions about her personal life. [20]Lara is the star of two movies and a comic book. [21]There is even a Lara Croft candy bar.

[22]A sociologist who reviewed this text said, "It seems that for women to be defined as equal, we have to become symbolic males—warriors with breasts." [23]Why is gender change mostly one-way—females adopting traditional male characteristics? [24]To see why men get to keep their gender roles, these two questions should help:

[25]Who is moving into the traditional territory of the other? [26]Do people prefer to imitate power or powerlessness? [27]Finally, consider just

how far stereotypes have actually been left behind. [28]The ultimate goal of the video game, after foes are vanquished, is to see Lara in a nightie.

—From *Essentials of Sociology: A Down-to-Earth Approach,* 7th ed., p. 72. by James M. Henslin. Copyright © 2007 by Pearson Education, Inc. Reproduced by permission of Pearson Education, Inc., Boston, MA.

▲ The mass media not only reflect gender stereotypes but also they play a role in changing them. Sometimes they do both simultaneously. The images of Lara Croft not only reflect women's changing role in society, but also, by exaggerating the change, they mold new stereotypes.

Before Reading

Vocabulary in Context

1–2. Study the following chart of word parts. Then, read the passage for context clues to help you define the meaning of the word. Next, use your own words to define the terms. Finally, identify the context you used to determine the word's meaning.

Root	Meaning	Suffix	Meaning
macho	male	*-ure*	act, function
caricare	to load	*-s*	plural

1. In sentence 14, **macho-caricatures** are ______________________________

____.

2. What is the context clue used to determine the meaning of **macho-caricatures** in sentence 14? ______________________________

____

After Reading

Implied Central Idea

_____ **3.** Which sentence best states the main idea of the passage?
a. The mass media reflect women's changing role in society.
b. Lara Croft encapsulates the changing roles of women in society.
c. Lara Croft is a stereotype of feminine sexuality.
d. Lara Croft appeals mostly to men.

Supporting Details

_____ **4.** Sentence 7 is a
a. major supporting detail. b. minor supporting detail.

Transitions

_____ **5.** The relationship within sentence 5 is
a. cause and effect. c. comparison.
b. listing. d. definition and example.

Thought Patterns

_____ **6.** The main thought pattern for the second paragraph (sentences 5–10) is
a. listing. c. cause and effect.
b. space order. d. generalization and example.

Tone and Purpose

_____ **7.** The author's tone and purpose in the passage are
a. to inform the reader with objective details.
b. to entertain with subjective observations.
c. to persuade with graphic details.

Inferences

8–10. Write **V** for valid by the three inferences that are most firmly based on the information in the passage and the images of Lara Croft.

_____ Lara Croft has developed a strong following among male consumers of entertainment.

_____ Lara Croft does not appeal to women consumers of entertainment.

_____ Lara Croft video games and movies contain violence and sexuality.

_____ Lara Croft breaks completely free of female stereotypes.

_____ Lara Croft is an aggressive action figure and main character in a series of video games, movies, and a comic book.

WHAT DO YOU THINK?

Do you think females are seen as less powerful than men? Why or why not? Have perceptions about men and women changed in recent years? Do you think video games send powerful messages about power or gender? Why or why not? Assume you are a concerned parent and you have decided to write an article about the messages contained in movies and video games to post on the community blog sponsored by your local newspaper. Consider using one of the following topics for your article: (1) Discuss what one may learn about our culture by playing video games, or a particular video game; (2) Write a review of a movie or video game based on the message it may send about violence or gender. Take a stand in support of or in opposition to the movie or video game.

After Reading About Inferences

Before you move on to the Mastery Tests on inferences, take time to reflect on your learning and performance by answering the following questions. Write your answers in your notebook.

- How has my knowledge base or prior knowledge about inferences changed?
- Based on my studies, how do I think I will perform on the Mastery Test(s)? Why do I think my scores will be above average, average, or below average?
- Would I recommend this chapter to other students who want to learn more inferences? Why or why not?

Test your understanding of what you have learned about inferences by completing the Chapter 11 Review Card in the insert near the end of your text.

CONNECT TO myreadinglab

To check your progress in meeting Chapter 11's learning outcomes, log in to www.myreadinglab.com and try the following exercises.

- The "Inference" section of MyReadingLab gives additional information about inferences. The section provides an overview, model, practices, and tests. To access this resource, click on the "Study Plan" tab. Then click on "Inference." Then click on the following links as needed: "Overview," "Model," "Practice," and "Test."
- To measure your mastery of this chapter, complete the tests in the "Inference" section and click on Gradebook to find your results.

MASTERY TEST 1

Name ______________________ Section ____________

Date __________ **Score** (number correct) _____ × 20 = _____%

Read the following passage, adapted from a college biology textbook. Then, write **V** for valid by the five inferences most soundly based on the information.

Textbook Skills

Matters of the Heart

[1]Cardiovascular disease, which impairs the heart and blood vessels, is the leading cause of death in the United States, killing nearly 1 million Americans annually. [2]And no wonder. [3]Your heart must contract vigorously more than 2.5 billion times during your lifetime without once stopping to rest, forcing blood through vessels whose total length would encircle the globe twice. [4]Because these vessels may become constricted, weakened, or clogged, the cardiovascular system is a prime candidate for malfunction.

Atherosclerosis Obstructs Blood Vessels

[5]*Atherosclerosis* causes the walls of the large arteries to thicken and lose their elasticity. [6]This change is caused by deposits called *plaques,* which are composed of cholesterol and other fatty substances as well as calcium and fibrin. [7]Plaques are deposited within the wall of the artery between the smooth muscle and the tissue that lines the artery. [8]A plaque may rupture through the lining into the interior of the vessel, stimulating platelets to adhere to the vessel wall and initiate blood clots. [9]These clots further obstruct the artery and may completely block it. [10]Or a clot may break loose and clog a narrower artery "downstream." [11]Arterial clots are responsible for the most serious consequences of atherosclerosis: heart attacks and strokes.

[12]A *heart attack* occurs when one of the coronary arteries is blocked. [13](Coronary arteries supply the heart muscle itself.) [14]Deprived of nutrients and oxygen, a heart muscle whose blood supply is curtailed by a blocked artery dies rapidly and painfully. [15]Heart attacks are the major cause of death from atherosclerosis.

[16]About 1.1 million Americans suffer heart attacks each year, and about half a million people die from them. [17]But atherosclerosis also causes plaques and clots to form in arteries throughout the body. [18]If a clot or plaque obstructs an artery that supplies the brain, it can cause a stroke, in which brain function is lost in the area deprived of blood and its vital oxygen and nutrients.

Treatment of Atherosclerosis

[19]The exact cause of atherosclerosis is unclear, but it is promoted by high blood pressure, cigarette smoking, genetic predisposition, obesity, diabetes, lack of exercise, and high blood levels of a certain type of cholesterol bound to a carrier molecule called low-density lipoprotein (LDL). [20]If LDL-bound cholesterol levels are too high, cholesterol may be deposited in arterial walls. [21]In contrast, cholesterol bound to high-density lipoprotein (HDL) is metabolized or excreted and hence is often called "good" cholesterol. [22]Treatment of atherosclerosis includes the use of drugs or changes in diet and lifestyle to lower blood pressure and blood cholesterol levels.

—Text and Figure E20-1 (p. 382) below from *Life on Earth*, 5th ed., p. 382 by Teresa Audesirk, Gerald Audesirk, and Bruce E. Byers. Copyright © 2009 by Pearson Education, Inc. Printed and Electronically reproduced by permission of Pearson Education, Inc., Upper Saddle River, NJ.

______ A synonym for coronary artery is blood vessel.

______ Plaque is a sticky substance that hardens in the arteries.

______ Cardiovascular diseases include heart attacks and strokes.

______ Most heart attacks come on suddenly.

______ Cholesterol promotes a healthy heart.

______ The tendency to develop heart problems can be inherited.

______ Lifestyle choices have a significant effect on the health of the heart.

______ Most people are too busy to exercise.

VISUAL VOCABULARY

Fill in the blanks with words from the passage that best complete the caption.

______ clogs arteries: When the fibrous cap ruptures, a blood clot forms, obstructing the ______.

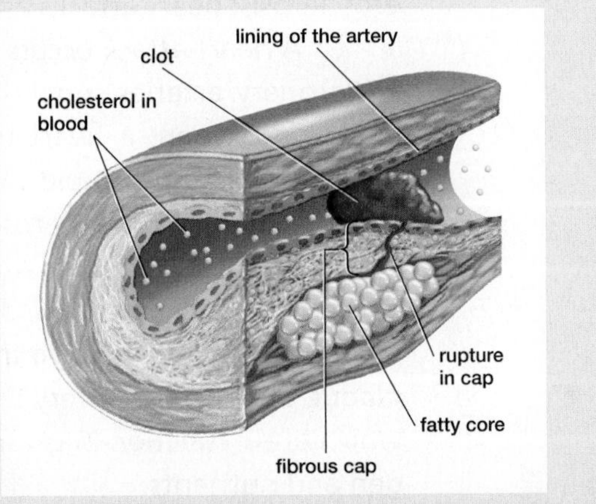

▸ **Figure E20-1**

MASTERY TEST 2

Name ____________________ Section ____________

Date __________ Score (number correct) ______ × 25 = ______%

Read the poem, and then answer the questions that follow.

Mr. Flood's Party

by Edwin Arlington Robinson

Old Eben Flood, climbing along one night
Over the hill between the town below
And the forsaken upland *hermitage — a dwelling where one lives alone
That held as much as he should ever know
On earth again of home, paused *warily. — cautiously
The road was his with not a native near;
And Eben, having leisure, said aloud,
For no man else in Tilbury Town to hear:

"Well, Mr. Flood, we have the harvest moon
Again, and we may not have many more;
The bird is on the wing, the poet says,
And you and I have said it here before.
Drink to the bird." He raised up to the light
The jug that he had gone so far to fill,
And answered huskily: "Well, Mr. Flood,
Since you propose it, I believe I will."

Alone, as if enduring to the end
A valiant armor of scarred hopes outworn,
He stood there in the middle of the road
Like Roland's ghost winding a silent horn.
Below him, in the town among the trees,
Where friends of other days had honored him,
A phantom salutation of the dead
Rang thinly till old Eben's eyes were dim.

Then, as a mother lays her sleeping child
Down tenderly, fearing it may awake,
He set the jug down slowly at his feet
With trembling care, knowing that most things break;
And only when assured that on firm earth
It stood, as the uncertain lives of men
Assuredly did not, he paced away,

And with his hand extended paused again:
"Well, Mr. Flood, we have not met like this
In a long time; and many a change has come
To both of us, I fear, since last it was
We had a drop together. Welcome home!"
*Convivially returning with himself, sociably
Again he raised the jug up to the light;
And with an acquiescent quaver said:
"Well, Mr. Flood, if you insist, I might.
"Only a very little, Mr. Flood—
For auld lang syne. No more, sir; that will do."
So, for the time, apparently it did,
And Eben evidently thought so too;
For soon amid the silver loneliness
Of night he lifted up his voice and sang,
Secure, with only two moons listening,
Until the whole harmonious landscape rang—

"For auld lang syne." The weary throat gave out,
The last word wavered; and the song being done,
He raised again the jug regretfully
And shook his head, and was again alone.
There was not much that was ahead of him,
And there was nothing in the town below—
Where strangers would have shut the many doors
That many friends had opened long ago.

—From E. A. Robinson, *Collected Poems*. New York, The Macmillan Company, 1921.

_____ **1.** The poem suggests that the narrator is
a. young and distrustful.
b. old and lonely.
c. middle-aged and frightened.

_____ **2.** In the second and fifth stanzas, Eben Flood is talking to
a. a friend.
b. God.
c. himself.

_____ **3.** In the second stanza, the narrator implies that Eben
a. went to town to buy liquor.
b. lives in town.
c. loves poetry and birds.

_____ **4.** Overall, the poet suggests that Eben Flood is
a. well known and well respected by the townspeople.
b. lonely because he has outlived those who knew him.
c. fearful of death.

MASTERY TEST 3

Name ______________________ Section ______________

Date __________ **Score** (number correct) ______ × 20 = ______%

Read the following passage taken from the novel *Forever Island* by Patrick Smith. Using inference skills, mark each of the statements that follow:

T for true

F for false

UK for unknown based on the information given

From *Forever Island*

[1]Charlie had heard his father and his grandfather speak of the time when the Seminole lived on the land far to the north, the rich rolling hills with fertile soil that grew corn and squash and pumpkin in abundance, and the game was plentiful and the streams as clear as an open sky, and then the white man came and took the land, and there was fighting and blood and death and hunger; then the Seminole moved south, giving way to the white man, settling on the land north of the big lake, but the white man came again and wanted the land, more fighting and more running and more hunger, and they moved south again. [2]Then the white man said they could have this land and be bothered no more; but he came again and the blood ran red, and the white man brought in dogs to track the Seminole like an animal. [3]Then they placed a bounty on the Seminole, fifty dollars for a man, twenty-five dollars for a woman, fifteen dollars for a child, bands of white hunters surrounding the chickees at night, capturing the men and women and children, binding their hands and feet with rope, throwing them into wagons like sacks of feed and hauling them to the fort in the north where they collected the bounty and returned to the swamp for another hunt; then the fighting and the running until the Seminole disappeared into the heart of the swamp and into the Sea of Grass and did not come out again until it was safe to do so, and some had not come out yet.

[4]Charlie himself had seen the white man come into this land and slaughter the egret for its feathers, shooting them only when nesting on the rookery, killing them by the hundreds of thousands and leaving the young either to die in the nest and be eaten by vultures or fall out of the nest and drown, and the water around the mangroves turning red with blood; and he had seen the white man come into this land and slaughter the alligator, shipping out their hides fifty thousand at a time to be made into wallets and shoes, and he had once seen a thousand alligators killed

in one pond in one day, but this time not for the hides, for the white men pulled out the alligator's teeth to be sold as watch fobs and left the bodies with their hides to rot in the blistering sun; and he had seen the white man come with his mules and his curses and his saws and his puffing trains and strip the land of the giant bald cypress, cutting them down like fields of sugar cane; and he had seen the white man wipe out the tree snails so that their shells could be sold as trinkets; and he had seen the white man dig the canals and drain the land and come closer and closer until he was now here again, once more telling the Seminole that he could not live on this land because the white man wanted it.

_____ **1.** Charlie and his family are Seminoles.

_____ **2.** Charlie and his family live in the Everglades in Florida.

_____ **3.** At first the white man tried to live in peace with the Seminole.

_____ **4.** The white man destroyed the natural environment because he was greedy.

_____ **5.** The conflict between the white man and the Seminoles was over long ago.

MASTERY TEST 4

Name ______________________ Section ____________

Date ____________ **Score** (number correct) ________ × 25 = ________%

Read the following passage from a college history textbook. Mark each of the statements that follow as **T** for **true**, **F** for **false**, or **UK** for **unknown** based on the information given.

Textbook Skills

Cuba

[1]Cuba's modern history has been dominated by its revolution of 1959. [2]Before the revolution, Cuba had been controlled by a former army sergeant, Fulgencio Batista, who had ruled on his own or through civilian presidents since 1934. [3]Cuba's economy largely depended on sugar exports and American investment. [4]And Cuba's capital, Havana, was a haven for American tourists (and gangsters), lured by the tropical climate and gambling casinos.

[5]Cuba's revolutionary movement was led by Fidel Castro (b. 1926). [6]The son of a well-to-do sugarcane farmer, Castro was introduced to revolutionary politics as a law student at the University of Havana. [7]In July 1953, he organized a disastrous attack on a garrison at Santiago. [8]After serving a short prison term, he fled to Mexico and plotted a return. [9]He and his rebel band of 80 sailed on the Granma in late 1956, but they were nearly all killed when they landed in Oriente province. [10]Castro and a small band of rebels escaped to the Sierra Maestra Mountains in southeast Cuba, from which they waged a guerrilla war. [11]With popular support for Castro's movement growing and Batista's National Guard collapsing, Batista abruptly fled after his annual party on New Year's Eve, 1958. [12]Castro's rebels marched unopposed into Havana.

[13]Although he billed himself as a nationalist reformer when he took power, Castro moved sharply to the left in 1960 and proclaimed the revolution socialist the following year. [14]His rhetoric became stridently anti-American. [15]Shortly after John F. Kennedy became the U.S. president in 1961, the United States supported an ill-fated invasion by Cuban exiles at the Bay of Pigs off the southern coast of Cuba. [16]By then, the Cuban government was forging close ties with the socialist bloc, especially the Soviet Union.

—From *Civilizations Past and Present, Combined Volume,* 12th ed., p. 1041 by Robert R. Edgar, Neil J. Hackett, George F. Jewsbury, Barbara S. Molony, Matthew Gordon. Copyright © 2008 by Pearson Education, Inc. Reprinted by permission of Pearson Education, Inc., Upper Saddle River, NJ.

[17]Castro's policies were dogmatically socialist. [18]He built up the Communist party and jailed thousands of opponents, including former comrades. [19]His government created a command economy in which the government controlled most sectors of the economy. [20]It seized American property and nationalized businesses. [21]It also addressed social problems and launched successful campaigns to eradicate illiteracy and to provide basic health care to the lower classes. [22]Educational and health standards rose appreciably, as did living conditions among the peasants, who constituted the great majority of the population. [23]The professional and middle classes, however, suffered losses in both living standards and personal liberties, and many hundreds of thousands fled to the United States.

_____ **1.** Most people in Cuba think of Fidel Castro as a hero who liberated Cuba.

_____ **2.** During the 1960s, Cuba and Russia became allies.

_____ **3.** Communism is a type of socialism.

_____ **4.** All social classes from the poor to the wealthy benefitted from the changes Castro brought to Cuba.

VISUAL VOCABULARY

Before Castro's revolution, Havana, Cuba, was a haven for American tourists. The best synonym for haven is

_____________.

a. trap.
b. retreat.

The Basics of Argument

12

LEARNING OUTCOMES

After studying this chapter you should be able to do the following:

1. Define the terms *argument, claim,* and *evidence.*
2. Identify the author's claim and evidence.
3. Determine whether or not the evidence is relevant.
4. Determine whether or not the evidence is adequate.
5. Analyze the argument for bias.
6. Apply inference skills to evaluate arguments for validity.
7. Evaluate the importance of the basics of argument.

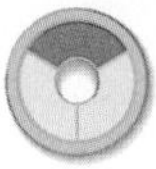

Before Reading About the Basics of Argument

Many of the same skills you learned to make valid inferences will help you master the basics of argument. Take a moment to review the five steps in the VALID approach to making sound inferences. Fill in the following blanks with each of the steps.

Step 1: ______________________________

Step 2: ______________________________

Step 3: ______________________________

Step 4: ______________________________

Step 5: ______________________________

Use your prior knowledge about valid inferences, other reading skills, and the learning outcomes to create at least three questions that you can answer as you study:

1. ____
____?

2. ____
____?

3. ____
____?

Compare the questions you created based on your prior knowledge and the chapter learning outcomes with the following questions. Then, write the ones that seem the most helpful in your notebook, leaving enough space between questions to record your answers as you read and study the chapter.

How will verifying and valuing the facts help me decide if supports in an argument are relevant? How will learning from the text help me decide if supports in an argument are adequate? How does an argument use bias? What is the relationship between main ideas and the author's claim? How does opinion affect an argument? What is the connection between tone, purpose, and the basics of argument?

What Is an Argument?

Have you noticed how many of us enjoy debating ideas and winning arguments? Many television shows such as the talk show *The View* and political programs such as *The Daily Show with Jon Stewart* or *Glenn Beck* thrive on conflict and debate. Likewise, talk radio devotes hours of air time to debating cultural and political issues on shows like *Rush Limbaugh* and *The Bill Press Show*. In everyday life, a couple may argue about spending priorities. Colleagues may argue how best to resolve a workplace dispute. One politician may argue in favor of corporate tax cuts while another politician argues against any kind of tax cut. Certain topics stir debate and argument. For many, sports contests can cause heated debate, and topics such as religion and evolution often evoke powerful debates.

Some people are so committed to their ideas that they become emotional, even angry. However, effective **argument** is reasoned: it is a process during which a claim is made and logical details are offered to support that claim.

An **argument** is made up of two types of statements:
1. The author's claim—the main point of the argument
2. The supports—the evidence or reasons that support the author's claim

The purpose of an argument is to persuade the reader that the claim is valid, that is, sound or reasonable. To decide if a claim is valid, you must analyze the argument in four basic steps.

1. Identify the author's claim and supports.
2. Decide whether the supports are relevant.
3. Decide whether the supports are adequate.
4. Check the argument for bias.

Step 1: Identify the Author's Claim and Supports

Read the following claim:

> Competitive sports provide several benefits to participants and should be funded by public funds.

The claim certainly states the speaker's point clearly, but it probably wouldn't inspire most of us to support or become involved in competitive sports. Instead, our first response to the claim is likely to be, "What are the benefits to participants of competitive sports?" We need reasons before we can decide whether or not we think the claim is valid.

Read the following statements:

1. Competitive sports teach participants physical, social, emotional, and psychological skills.
2. Competitive sports set standards against which participants can measure growth and ability.
3. Competitive sports allow participants to experience the connection between effort and success.

These three sentences offer the supports for the author's claim. We are now able to understand the basis of the argument, and we now have details about which we can agree or disagree.

Writers frequently make claims that they want us to accept as valid. To assess whether the claim is valid, a master reader first identifies the claim and the supports.

EXAMPLES

A. Read each of the following groups of ideas. Mark each statement **C** if it is an author's claim or **S** if it provides support for the claim.

1. ______ a. Staph is a particularly dangerous type of bacteria.
   ______ b. Staph may cause skin infections that look like pimples or boils, which may be red, swollen, painful, or have pus or other drainage.
   ______ c. Some strains of staph have developed antibiotic resistance, making a staph infection sometimes difficult to treat.
2. ______ a. Aerobic training such as walking, biking, and swimming strengthens the heart and lungs.
   ______ b. A comprehensive fitness program must combine aerobic and anaerobic training.
   ______ c. Anaerobic training, using free weights, weight machines, and resistance bands, builds muscles and bones.
3. ______ a. All Siamese cats are domesticated cats.
   ______ b. All domesticated cats are excellent pets.
   ______ c. Therefore, all Siamese cats are excellent pets.

B. Political cartoons offer arguments through the use of humor. The cartoonist has a claim to make and uses the situation, actions, and words in the cartoon as supporting details. Study the cartoon reprinted here. Then, write a claim based on the supports in the cartoon.

____

____

____

____

____

____

EXPLANATIONS

A. Compare your answers to these.

1. In this example the opinion is phrased *particularly dangerous* and then supported with the evidence. Thus (a) is the claim and (b) and (c) are the supports.

2. The use of the word *must* identifies (b) as the claim; the other sentences provide support.

3. The word *therefore* signals the conclusion or claim.

B. Several claims may be suggested by the details in the cartoon. One possible claim is "Gang life leads to death or prison."

PRACTICE 1

A. Read the following groups of ideas. Mark each statement **C** if it is an author's claim or **S** if it provides support for the claim.

1. ______ a. As many as 9 million Americans have their identities stolen each year.

______ b. Identity theft is a serious crime that affects millions of people each year.

2. ______ a. Exposure to media violence causes increases in aggression and violence.

______ b. By the time a typical child finishes elementary school, he or she will have seen around 8,000 murders and more than 100,000 other acts of violence on TV.

3. ______ a. The use of cell phones while driving contributes to 25% of all traffic crashes.

______ b. A federal law should be passed to ban talking on a cell phone while driving.

4. ______ a. Concealed handguns are not an effective form of self-defense.

______ b. A recent study found that someone carrying a gun for self-defense was 4.5 times more likely to be shot during an assault than a victim without a gun.

B. Study the following editorial cartoon. Write a claim based on the details of the cartoon.

© 2010 Mike Keefe, The Denver Post, and PoliticalCartoons.com <http://PoliticalCartoons.com/>

____

Step 2: Decide Whether the Supports Are Relevant

In Step 1, you learned to identify the author's claim and supports. The next step is to decide whether the supports are relevant to the claim. Remember, a claim, like any main idea, is made up of a topic and a controlling point. Irrelevant supports change the topic or ignore the controlling point. Relevant supports will answer the reporter's questions (*Who? What? When? Where? Why?* and *How?*). Use these questions to decide whether the supports for a claim are relevant.

For example, read the following argument a student makes about her grades. Identify the support that is irrelevant to her claim.

> [1]I deserve a higher grade than the one I received. [2]Even though I have missed some classes, I made sure to get all the lecture notes from classmates. [3]I turned in all my assignments on time, and my test scores average at the A-level. [4]It's not my fault I missed so many classes. [5]My mother became seriously ill, and she is the one who watches my children for me so I can come to school. [6]I also had car problems.

By turning this student's claim into a question, she and her teacher can test her ability to offer valid reasons: "How have I shown I have mastery of the course material?" Sentences 2 and 3 offer relevant examples of her mastery of the course material. However, sentences 4, 5, and 6 state irrelevant supports that shift the focus away from her grade. The argument is about her grade based on an appeal to pity or emotions.

When evaluating an argument, it is important to test each piece of supporting evidence to determine whether it is relevant.

EXAMPLES

A. Read the following lists of claims and supports. Mark each support **R** if it is relevant to the claim or **N** if it is not relevant to the claim.

1. Claim: All police should be equipped with Taser guns.
Supports:

_____ a. Taser guns were developed by two brothers who wanted their mother to own a weapon for self-defense with which she could feel comfortable.

_____ b. The Taser gun is a *nonlethal* stun gun that disables suspects for five seconds by shooting them with 50,000 volts of electricity.

_____ c. Human rights activists warn that Taser guns can be abused as instruments of torture, especially since they leave no significant or enduring mark on the skin.

_____ d. Police officers in Miami, Florida, and Seattle, Washington, have had no incidents or sharply lower incidents of fatal police shootings since they began using Taser guns.

_____ e. The electrical shock caused by Taser guns has no long-term negative effect.

B. Argument is also used in advertisements. It is important for you to be able to understand the points and supports of ads. Many times advertisers appeal to emotions, make false claims, or give supports that are not relevant because their main aim is to persuade you to buy their product. Study the following mock advertisement for Cabre, a fictitious brand of red wine. Mark each of the statements **R** if it is relevant to the claim or **N** if it is not relevant to the claim.

◀ Cabre! To a happy and long life. Recent research suggests that one glass of red wine daily may do more than reduce the risk of high levels of cholesterol. The grape skin and seeds appear to hold a natural cancer-fighting chemical.*

*Research conducted on animals only.

2. Claim: Moderate consumption of red wine may have some health benefits.
Supports:

_____ a. A beautiful woman drinks red wine.

_____ b. Red wine may reduce the risk of high levels of cholesterol.

_____ c. The grape skin and seeds appear to hold a natural cancer-fighting chemical.

_____ d. The woman is happy.

_____ e. Research conducted on animals only.

EXPLANATIONS

A. **1.** Items *b*, *d*, and *e* are relevant to the claim. Items *a* and *c* are not relevant. The reason the Taser guns were created is not important to their use by police officers; and a warning about the possibility of Taser guns being used as instruments of torture is not the issue.

B. **2.** Items *a*, *d*, and *e* are not relevant to the claim that moderate consumption of red wine may have some health benefits. Advertisers often use beautiful models who appear to be pleased with the product, but neither beauty nor personal preference of the model is relevant to health issues. Items *b* and *c* are relevant to the claim. However, interestingly, the smallest line of print in this mock advertisement discloses that the research only involved animals. Thus the "research" is not relevant to the claim. In fact, the disclaimer could lead a reasonable reader to discard all of the supports as not relevant.

PRACTICE 2

A. Read the following lists of claims and supports. Mark each support **R** if it is relevant to the claim or **N** if it is not relevant to the claim.

1. Claim: The need for organ and tissue donors is critical.
Supports:

_____ a. According to the Organ Procurement and Transplantation Network (OPTN), as of November 13, 2007, over 92,000 people were on a waiting list for an organ transplant.

_____ b. According to the OPTN, only 19,249 transplants took place between January and August of 2007.

_____ c. Each day 19 people on the waiting list die.

_____ d. The organs needed for transplant include heart, kidneys, pancreas, lungs, liver, and intestines; the tissues needed are cornea, skin, bone marrow, heart valves, and connective tissue.

_____ e. To become an organ donor, fill out and carry a donor card, indicate your intent to donate on your driver's license, and tell your family about your decision.

—Adapted from "Organ Donation," *FirstGov*. 13 Nov. 2007
http://www.organdonor.gov

2. Claim: Binge drinking is dangerous and irresponsible.
Supports:

_____ a. Binge drinking is commonly defined as drinking large quantities of alcohol in a short period of time: 5 drinks in a row for men; 4 drinks in a row for women.

_____ b. Many young people do not think it is important to have alcohol at a party.

_____ c. Eating a large meal before drinking slows down the effect of the alcohol.

_____ d. The intent and purpose of binge drinking is to become intoxicated and to lose control.

_____ e. Binge drinking is linked to vandalism, confrontations with police, injuries, alcohol poisoning, and death.

3. Claim: If expressions of religion are allowed in national government institutions, then prayer should be allowed in public schools.

Supports:

_____ a. Banning prayer in public schools leads to moral decline.

_____ b. Congress prays at the opening of each session.

_____ c. Federal officials, including the President of the United States, take their oaths of office upon a Bible.

_____ d. Moses and the Ten Commandments are prominently displayed in the Supreme Court building.

_____ e. A majority of Americans favor prayer in schools.

4. Claim: Volunteers perform many selfless acts.

Supports:

_____ a. Volunteers promote worthwhile causes and improve the lives of others.

_____ b. Volunteers boost their own sense of self-worth and well-being.

_____ c. Volunteers conserve funds for charities, nonprofits, and other community organizations by contributing their time.

_____ d. They share their skills and expertise.

_____ e. Volunteers improve their reputations and promote their careers.

B. Study the public service poster sponsored by the White House Office of National Drug Control Policy on the following page. Read the claim, and then mark each support **R** if it is relevant to the claim or **N** if it is not relevant to the claim.

Claim: Steer Clear of Pot.

Supports:

_____ a. Stoners laugh a lot, eat junk food, and don't bug anyone.

_____ b. It's dangerous to drive stoned.

_____ c. An estimated 38,000 high school seniors in the United States reported in 2001 that they had crashed while driving under the influence of marijuana.

_____ d. Everyone smokes pot.

_____ e. Research shows that smoking weed affects a driver's perception, coordination, and reaction time.

—United States Executive Office of the President. "Steer Clear of Pot." The White House Office of National Drug Control Policy. Free Resources. 13 Nov. 2007. http://www.freevibe.com/Drug_Facts/free-resources.asp

Now that you have practiced identifying relevant supports in a list format, you are ready to isolate relevant supports in reading passages. In a passage, the thesis statement states the author's claim. Each of the supporting details must be evaluated as relevant or irrelevant supports for the thesis statement.

EXAMPLE Read the following passage.

Shell Eggs from Farm to Table

[1]Eggs are among the most nutritious foods on earth and can be part of a healthy diet; however, they are perishable just like raw meat, poultry, and fish. [2]Unbroken, clean, fresh shell eggs may contain *Salmonella enteritidis* (SE) bacteria that can cause foodborne illness. [3]While the number of eggs affected is quite small, several cases of foodborne illness have surfaced in the last few years. [4]To be safe, eggs must be properly handled and cooked.

[5]Bacteria can be present on the exterior of a shell egg because the egg exits the hen's body through the same passageway as feces is excreted; thus, eggs are washed and sanitized at the processing plant. [6]Bacteria can be present inside an uncracked, whole egg. [7]This interior contamination may be due to bacteria within the hen's ovary or oviduct before the shell forms around the yolk and white. [8]*Salmonella enteritidis* doesn't make the hen sick. [9]It is also possible for eggs to become infected by *Salmonella enteritidis* fecal contamination through the pores of the shells after they have been laid.

[10]Researchers say that, if present, the *Salmonella enteritidis* is usually in the yolk or yellow; however, experts cannot rule out the bacteria's presence in egg whites. [11]As a result, everyone is advised against ingesting raw or undercooked egg yolks and whites or products containing raw or undercooked eggs.

—Adapted from "Egg Products Preparation," Food Safety and Inspection Service, U.S. Department of Agriculture, 7 March 2007. 4 May 2008. http://www.fsis.usda.gov/Fact_sheets/Focus_on_shell_Eggs/index.asp

1. Underline the thesis statement (the sentence that states the author's claim).

_____ **2.** Which sentence is *not* relevant to the author's point?

a. sentence 2 c. sentence 3
b. sentence 6 d. sentence 8

EXPLANATION

1. Sentence 4 is the central idea and states the author's claim.
2. The sentence that is *not* relevant to the author's point is *d*, sentence 8. The author's claim focuses on the health risks associated with eggs, not with the health of the hens laying the eggs.

PRACTICE 3

Read the following passages, and respond to the two questions following each of them.

The Street Outreach Program

[1]Today, in communities across the country, some young people are living on the streets, running from or being asked to leave homes characterized by abuse, neglect, or parental drug or alcohol abuse. [2]Once on the streets, such youth are at risk of being sexually exploited or abused by adults for pleasure or profit.

[3]To prevent the sexual abuse or exploitation of these young people by providing them with services that help them leave the streets, Congress established the Education and Prevention Services to Reduce Sexual Abuse of Runaway, Homeless, and Street Youth Program, through the Violence Against Women Act of the Violent Crime Control and Law Enforcement Act of 1994. [4]That program created Grants for the Prevention of Sexual Abuse and Exploitation (also known as the Street Outreach Program). [5]Naturally, not all at-risk youth can be reached through this program.

[6]The Street Outreach Program offers many services including the following: street-based education and outreach, access to emergency shelter, survival aid, individual assessments, treatment and counseling, prevention and education activities, information and referrals, crisis intervention, and follow-up support.

[7]The Education and Prevention Services to Reduce Sexual Abuse of Runaway, Homeless, and Street Youth Program should be supported with increased funding.

—Excerpted from "Family and Youth Services Bureau Street Outreach Program," Administration for Children and Families, 13 Nov. 2007 http://www.acf.hhs.gov/programs/fysb/streetout.htm

1. Underline the thesis statement (the sentence that states the author's claim).

_____ 2. Which sentence is *not* relevant to the author's point?

a. sentence 3
b. sentence 4
c. sentence 5
d. sentence 6

Preventing Identity Theft Is Your Responsibility!

[1]As with any crime, you can't guarantee that you will never be a victim, but you can take responsibility to minimize your risk. [2]To protect yourself against identity theft, you should take the following steps.

[3]First, don't give out personal information on the phone, through the mail, or over the Internet unless you've initiated the contact or are sure you know whom you're dealing with. [4]Identity thieves may pose as representatives of banks, Internet service providers (ISPs) and even government agencies to get you to reveal your SSN, mother's maiden name, account numbers, and other identifying information.

[5]Second, secure personal information in your home, especially if you have roommates, employ outside help, or are having service work done in your home. [6]Don't carry your SSN card; leave it in a secure place.

[7]Third, guard your mail and trash from theft. [8]Deposit outgoing mail in post office collection boxes or at your local post office rather than in an unsecured mailbox. [9]Promptly remove mail from your mailbox. [10]If you're planning to be away from home and can't pick up your mail, call the U.S. Postal Service to request a vacation hold. [11]You can have the amount of junk mail delivered to your home reduced by contacting Mail Preference Service, PO Box 643, Carmel, NY 10512.

[12]Finally, to thwart an identity thief who may pick through your trash or recycling bins to capture your personal information, tear or shred your charge receipts, copies of credit applications, insurance forms, physician statements, checks and bank statements, expired charge cards that you're discarding, and credit offers you get in the mail.

—Adapted from "Protecting Against Identity Theft," Federal Trade Commission, 2003. 13 Nov. 2007 http://www.consumer.gov/idtheft/protect_againstidt.html#5

3. Underline the thesis statement (the sentence that states the author's claim).

______ **4.** Which sentence is *not* relevant to the author's point?

a. sentence 8　　c. sentence 10
b. sentence 9　　d. sentence 11

Step 3: Decide Whether the Supports Are Adequate

In Step 1 you learned to identify the author's claim and supports. In Step 2 you learned to make sure the supports are relevant. In Step 3 you must decide whether the supports are adequate. A valid argument is based not only on a claim and relevant support but also on the amount and quality of the support given. That is, supports must give enough evidence for the author's claim to be convincing. Just as you used the reporter's questions to decide whether supports are relevant, you can use them to test whether supports are adequate. Supporting details fully explain the author's controlling point about a topic. Remember, those questions are *Who? What? When? Where? Why?* and *How?*

For example, you may argue, "A low-carbohydrate diet is a more healthful diet. I feel much better since I stopped eating so much bread and pasta." However, the reporter's question "Why?" reveals that the support is inadequate. The answer to "Why is a low-carbohydrate diet a more healthful diet?" should include expert opinions and facts, not just personal opinion. Often in the quest to support a claim, people oversimplify their reasons. Thus, they do not offer enough information to prove the claim. Instead of logical details, they may offer false causes, false comparisons, forced choices, or leave out facts that hurt the claim. You will learn more about inadequate argument in Chapter 13.

In Chapter 11, you studied how to avoid invalid conclusions and make valid inferences (see pages 462–463). The same thinking steps you use to make valid inferences help you identify valid claims: consider the facts, don't infer anything that is not there, and make sure nothing contradicts your conclusion.

EXAMPLE Read the list of supports.

- Increasing levels of greenhouse gases like carbon dioxide (CO_2) in the atmosphere since pre-industrial times are well documented and understood.
- The atmospheric buildup of CO_2 and other greenhouse gases is largely the result of human activities such as the burning of fossil fuels.
- An "unequivocal" warming trend of about 1.0 to 1.7°F occurred from 1906–2005. Warming occurred in both the Northern and Southern Hemispheres, and over the oceans.
- The major greenhouse gases emitted by human activities remain in the atmosphere for periods ranging from decades to centuries. It is therefore virtually certain that atmospheric concentrations of greenhouse gases will continue to rise over the next few decades.
- Increasing greenhouse gas concentrations tend to warm the planet.

—U.S. Environmental Protection Agency. "Climate Change: Science: State of Knowledge." 28 Sept. 2009. http://epa.gov/climatechange/science/stateofknowledge.html

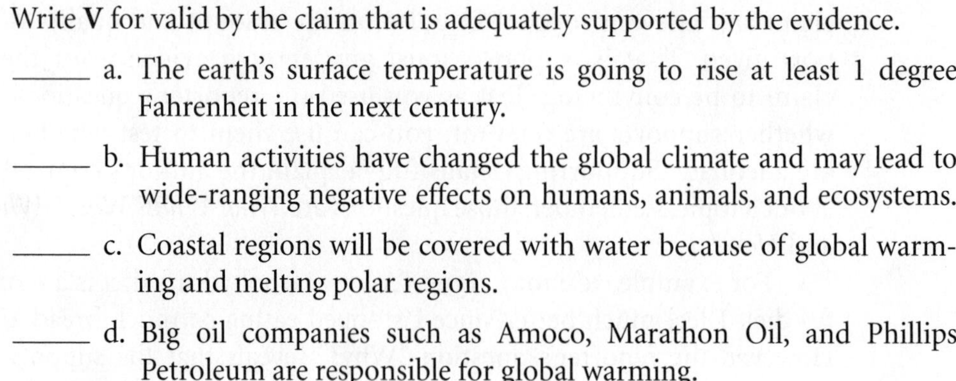

Write **V** for valid by the claim that is adequately supported by the evidence.

_____ a. The earth's surface temperature is going to rise at least 1 degree Fahrenheit in the next century.

_____ b. Human activities have changed the global climate and may lead to wide-ranging negative effects on humans, animals, and ecosystems.

_____ c. Coastal regions will be covered with water because of global warming and melting polar regions.

_____ d. Big oil companies such as Amoco, Marathon Oil, and Phillips Petroleum are responsible for global warming.

EXPLANATION Choices (a) and (c) jump to conclusions about the future effects of global warming, and choice (d) uses the evidence to jump to conclusions about the role of big oil companies in the cause of global warming. More evidence is needed to support any one of those claims. The only logical conclusion based on the evidence is (b).

PRACTICE 4

A. Read the list of supports taken from an article about using corn as an alternative energy. Then, write **V** next to the claim that is adequately supported by the evidence.

- Dwindling foreign oil, rising prices at the gas pump, and hype from politically well-connected U.S. agribusiness have combined to create a frenzied rush to convert food grains into ethanol fuel.
- Using corn or any other biomass for ethanol requires huge regions of fertile land, plus massive amounts of water and sunlight to maximize crop production.
- If the entire national corn crop were used to make ethanol, it would replace a mere 7% of U.S. oil consumption—far from making the U.S. independent of foreign oil.
- The environmental impacts of corn ethanol production include severe soil erosion of valuable food cropland; the heavy use of nitrogen fertilizers and pesticides that pollute rivers; and the required use of fossil fuels, releasing large quantities of carbon dioxide into the atmosphere, adding to global warming.
- More than 40 percent of the energy contained in one gallon of corn ethanol is expended to produce it, and that expended energy to make ethanol comes mostly from highly valuable oil and natural gas.

- Growing crops for fuel squanders land, water, and energy vital for human food production.

—Pimentel, David. "Corn Can't Save Us: Debunking the Biofuel Myth." *Kennebec Journal*. 28 March 2008. http://www.kjonline.com/archive/corn-can_t-save-us-debunking-the-biofuel-myth.html?searchterm=biofuel+myt.

_____ a. Ethanol made from corn and other biomasses is an acceptable alternative energy form.

_____ b. The United States is too dependent on foreign oil.

_____ c. Fuels made from biomasses can reduce the United States's dependency on foreign oil.

_____ d. The use of corn and other biofuels to solve our energy problem is an ethically, economically, and environmentally unworkable sham.

B. Study the following graph. Then, write **V** by the claim that is adequately supported by the evidence in the graph.

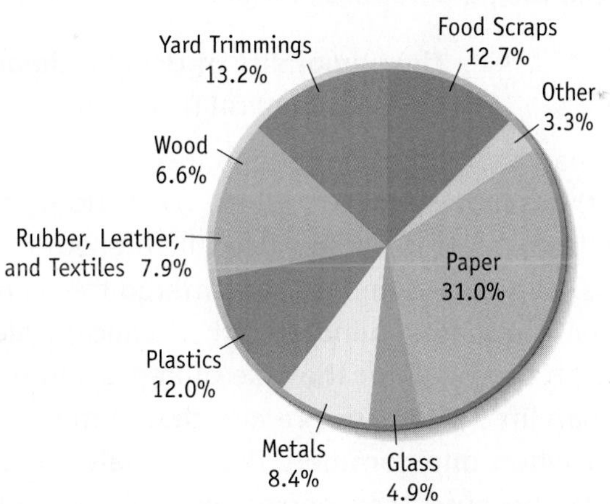

—U. S. Environmental Protection Agency. *Municipal Solid Waste Generation, Recycling, and Disposal in the United States: Facts and Figures for 2008.* http://www.epa.gov/osw/nonhaz/municipal/pubs/msw2008rpt.pdf

_____ a. The United States government has ineffective policies for recycling.

_____ b. The people of the United States generate too much trash.

_____ c. The United States needs more landfills.

_____ d. Recycling organic material will make the largest impact on waste management.

Step 4: Check the Argument for Bias

In Step 1, you learned to identify the author's claim and supports. In Step 2, you learned to make sure the supports are relevant to the claim. In Step 3, you learned to avoid false inferences and identify valid claims based on adequate supports. Again, the skills you use to make sound inferences help you determine whether an argument is valid. In Step 4, you must also check for the author's bias for or against the topic. Authors may use emotionally slanted language or biased words to present either a favorable or negative view of the topic under debate. In addition, authors may include only the details that favor the stances they have taken. A valid argument relies on objective, factual details. As you evaluate the argument for the author's bias, ask the following questions:

- Does the author provide mostly positive or negative supports?
- Does the author provide mostly factual details or rely on biased language?
- Does the writer include or omit opposing views?

EXAMPLE Read the paragraph. Then, answer the questions.

The Immorality of Using Animals in Medical Research

[1]Animals should not be used in medical research. [2]The lives of hundreds of thousands of mice, rabbits, birds, dogs, cats, horses, and pigs, among others, have been sacrificed in the quest to save human lives. [3]Hundreds of thousands more have suffered the pain and distress of being infected with diseases, blinded with chemicals, electrically shocked, or injected with drugs, all for the sake of improving the longevity and quality of human life. [4]The basic premise that humans are superior life forms falls short when one considers that animals are used in these horrific experiments because they possess enough physiological similarities to humans to make the research findings worthwhile for humans. [5]For example, much of what is known about the human immune system came from studies with mice, and understandings about the human cardiovascular system came from studies with dogs. [6]Even if humans believe themselves more worthy because they are self-aware creatures capable of reasoning, able to establish complex social orders, and gifted with written language, humans prove themselves to be mere animals when they inflict pain upon defenseless fellow creatures for personal gain, in an instinctive grapple

for "survival of the fittest." [7]Non-human animals are sentient beings, capable of intense suffering. [8]This fact alone makes the cost they pay too high—and human morals too low.

_____ **1.** Overall, the passage relies on
a. factual details.
b. emotionally slanted language.

_____ **2.** Which of the following sentences includes an opposing view?
a. sentence 5
b. sentence 6
c. sentence 7
d. sentence 8

_____ **3.** In this passage, the author expressed a biased attitude
a. in favor of using animals in medical research.
b. in favor of finding alternatives to using animals in medical research.
c. against the belief that humans are a superior life form.
d. against the belief that animals are capable of suffering.

EXPLANATION

1. Although the author uses emotionally slanted words occasionally, overall, the passage relies on (a) factual details that can be verified through research. Some of the biased words are *should not, sacrificed, horrific, mere, defenseless,* and *grapple*.

2. (b) Sentence 6 refers to the opposing view that humans are superior to other animals and thus their lives are more worthy of saving. The sentence even lists a few of the reasons the opposition holds this belief.

3. In this passage, the author expresses a bias (c) against the belief that humans are a superior life form by emphasizing the human willingness to inflict pain and suffering on other creatures.

PRACTICE 5

Read the paragraph. Then, answer the questions.

School Uniforms Benefit Students

[1]School uniforms should be mandatory in all public schools. [2]Uniforms promote school spirit and school values by establishing a group identity and making students feel as if they are part of a team. [3]School uniforms promote modesty and eliminate the distraction of sexually provocative

clothing. [4]They contribute to the sense that school is a place of order and work, as well as foster a respect for authority. [5]In such an environment, students are likely to approach learning more seriously. [6]School uniforms also minimize cliques based on socioeconomic differences; they increase attendance and reduce dropout rates. [7]In addition, school uniforms increase student safety in several ways. [8]Gang- and drug-related activity decline; students are less likely to bring weapons to school undetected; and teachers and administrators are more able to distinguish their students from outsiders. [9]For those who cannot afford uniforms or oppose uniforms on the basis of legitimate religious or philosophical beliefs, an "opt-out" policy should be in place so that students with the support of their families can apply for exemptions.

______ **1.** Overall, the passage offers supports that are
- a. mainly positive.
- b. mainly negative.

______ **2.** Which of the following sentences refers to an opposing view?
- a. sentence 6
- b. sentence 7
- c. sentence 8
- d. sentence 9

______ **3.** In this passage, the author expressed a biased attitude
- a. in favor of individual competition.
- b. in favor of freedom of expression.
- c. against the "opt-out" policy.
- d. against cliques based on socioeconomic differences.

Reading a Textbook: The Logic of Argument

Textbook Skills

Most of the subjects you will study in college rely on research by experts, and these experts may have differing views on the same topic. Often textbooks spell out these arguments. Sometimes textbook authors will present several experts' views. But occasionally, only one view will be given. In this case, be aware that there may be other sides to the story.

Textbook arguments are usually well developed with supports that are relevant and adequate. These supports may be studies, surveys, expert opinions, experiments, theories, examples, or reasons. Textbooks may also offer graphs, charts, and photos as supports. A master reader tests passages in textbooks for the logic of the arguments they present. The exercise that follows is designed to give you practice evaluating the logic of arguments in textbooks.

PRACTICE 6

Read the following passage from a college communications textbook and study the *Cathy* cartoon that accompanies it. Mark each statement in the passage **C** if it is an author's claim or **S** if it provides support for a claim. Then, write a sentence that states a logical conclusion that can be drawn from the *Cathy* cartoon about interpersonal communication.

Textbook Skills

Interpersonal Communication Involves Content and Relationship Messages

[1]Many conflicts arise because people misunderstand relationship messages and cannot clarify them. [2]Other problems arise when people fail to see the difference between content messages and relationship messages. [3]A good example occurred when my mother came to stay for a week at a summer place I had. [4]On the first day she swept the kitchen floor six times. [5]I had repeatedly told her that it did not need sweeping, that I would be tracking in dirt and mud from the outside. [6]She persisted in sweeping, however, saying that the floor was dirty. [7]On the content level, we were talking about the value of sweeping the kitchen floor. [8]On the relationship level, however, we were talking about something quite different. [9]We were each saying, "This is my house." [10]When I realized this, I stopped complaining about the relative usefulness of sweeping a floor that did not need sweeping. [11]Not surprisingly, she stopped sweeping.

Ignoring Relationship Considerations

[12]Examine the following interchange and note how relationship considerations are ignored:

Messages	*Comments*
[13]PAUL: I'm going bowling tomorrow. The guys at the plant are starting a team.	[14]He focuses on the content and ignores any relationship implications of the message.
[15]JUDY: Why can't we ever do anything together?	[16]She responds primarily on a relationship level and ignores the content implications of the message, and expresses her displeasure at being ignored in his decision.
[17]PAUL: We can do something together any time; tomorrow's the day they're organizing the team.	[18]Again, he focuses almost exclusively on the content.

[19]This example reflects research findings that show that men focus more on content messages; women focus more on relationship messages. [20]Once you recognize this gender difference, you can increase your sensitivity to the opposite sex.

—Text and cartoon from *Messages: Building Interpersonal Communication Skills,* 4th ed., p. 21. by Joseph A. DeVito. Copyright © 1999 by Pearson Education, Inc. Reproduced by permission of Pearson Education, Inc., Boston, MA.

_____ **1.** sentence 1
_____ **2.** sentence 2
_____ **3.** sentence 3
_____ **4.** sentence 4
_____ **5.** sentence 5
_____ **6.** sentence 6
_____ **7.** sentence 7
_____ **8.** sentence 8
_____ **9.** sentence 9
_____ **10.** sentence 10
_____ **11.** sentence 11
_____ **12.** sentence 12
_____ **13.** sentence 13
_____ **14.** sentence 14
_____ **15.** sentence 15
_____ **16.** sentence 16
_____ **17.** sentence 17
_____ **18.** sentence 18
_____ **19.** sentence 19
_____ **20.** sentence 20

____

____

____

Note: Sentences 13, 15, and 17 are examples of interpersonal communication; sentences 14, 16, and 18 are the textbook author's claims about the meanings (the interpretations) of each example.

APPLICATIONS

Application 1: Argument—Author's Claim and Supports

A. Read the following groups of ideas. Mark each statement as **C** if it's an author's claim or **S** if it provides support for the claim.

_____ a. Zoo staffers design activities that will occupy and challenge animals such as placing their food in stumps and forcing the animal to work to gain food.

_____ b. Animals housed in large zoos also receive high-quality veterinary care.

_____ c. Large American zoos today are concerned with the welfare of the animals in their exhibits.

_____ d. Today, animals are housed in large, spacious natural settings that try to recreate the animal's habitat of origin.

_____ e. In addition, zoo staffers work to provide opportunities for animals to fulfill natural functions, such as roaming and foraging.

B. Study the cartoon by Mike Lester from the *Rome News Tribune.* Then, write a sentence that states a claim or conclusion that can be asserted based on the details in the cartoon.

© 2010 Mike Lester/Cagle Cartoons

Claim: ____

____

Application 2: Argument: Relevant Supports and Author's Bias

Read the following editorial by Pamela G. Bailey, president and CEO of the Grocery Manufacturers Association, published in *USA Today*. Then, answer the questions.

Opposing View: Don't Blame Us

[1]As a mother of five, I know how hard it is to ensure kids are leading a healthy lifestyle. [2]Working parents have less time to prepare meals. [3]Schools are reducing PE. [4]The Internet and other distractions limit active playtime. [5]More spread out communities mean more driving and less walking and biking.

[6]All this contributes to the childhood obesity epidemic. [7]It will take robust solutions to address the crisis. [8]That's why it's so encouraging that first lady Michelle Obama's Let's Move initiative focuses on positive steps we all can take, rather than assigning blame.

[9]We recognize our role in fighting obesity. [10]In the past few years, we've changed more than 10,000 recipes to reduce calories, fat, sugar and salt and increase whole grains, fiber and other nutrients. [11]In recent weeks, we've seen announcements of sweeping reductions in salt and sugar and other healthy improvements.

[12]These measures will make a difference because they give parents choices that fit their lives—foods that are healthy, as well as convenient, tasty and affordable. [13]These recipe changes aren't the result of a government mandate. [14]They are the result of hundreds of millions of dollars in R&D and market research. [15]It is about businesses responding to consumers' changing preferences.

[16]A similar responsiveness is leading to marketing changes. [17]Today, the majority of advertising during children's programming showcases active lifestyles and healthy choices such as 100% fruit juices. [18]Labels are improving to provide consumers with clear and useful nutrition information. [19]And we are working with Congress and the White House to strengthen a law that would set science-based nutrition standards for foods in schools.

[20]America's food companies will keep making our products healthier, our marketing more responsible and our labeling more informative. [21]And in partnership with government and with consumers, we are committed to winning the battle over the childhood obesity crisis in our lifetimes.

—"Opposing View: Don't Blame Us" by Pamela G. Bailey from USA TODAY, March 31, 2010. Reprinted by permission of USA Today.

1. Underline the thesis statement that states the author's claim.

_____ **2.** Which sentence is not relevant to the author's point?

a. sentence 1
b. sentence 3
c. sentence 8
d. sentence 15

_____ **3.** In this article, the author expresses a biased attitude in favor of

a. physical education classes in public schools.
b. government mandates to monitor the grocery business.
c. the efforts of the grocery business to address childhood obesity.
d. assigning blame for childhood obesity.

Application 3: Argument: Adequate Supports

Study the graphic, which was published in *USA Today*. Then, choose the conclusion that is **not** adequately supported by evidence. (Indicate your choice with a check (✓) in the blank.)

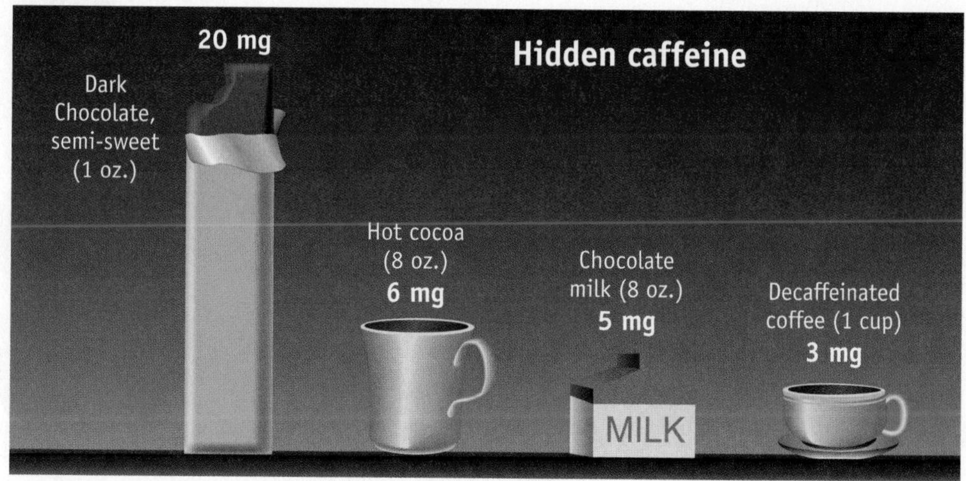

—"Hidden Caffeine" by Ying Lou and Gia Kereselidze from USA TODAY, March 3, 2004. Reprinted by permission of USA Today.

_____ a. Caffeine is found in coffee, chocolate, and cocoa.

_____ b. Among the sources of caffeine that are compared, dark chocolate and semi-sweet chocolate contain the highest levels of caffeine per ounce.

_____ c. The public prefers candy bars over cocoa, chocolate milk, or decaffeinated coffee.

_____ d. Decaffeinated coffee contains some caffeine.

REVIEW TEST 1

Score (number correct) _____ × 10 = _____%

Argument

A. Study the following mock advertisement. Then, study the claim and supports based on it. Mark each support **R** if it is relevant to the claim or **N** if it is not relevant to the claim. Then, answer the questions.

JUST SAY, "NO THANKS!"

- In 1999, 61% of adults in the United States were overweight or obese.
- Approximately 300,000 deaths each year in the United States may be attributable to obesity.
- Overweight and obesity are associated with heart disease, certain types of cancer, type 2 diabetes, stroke, arthritis, breathing problems, and psychological disorders, such as depression.
- Choose a diet that is low in saturated fat and cholesterol and moderate in total fat.
- Exercise three times a week for 30 minutes.

—Data from "Fact Sheet: What You Can Do," Overweight and Obesity. United States Office of the Surgeon General, 11 Jan. 07. 4 May 08. http://www.surgeongeneral.gov/topics/obesity/calltoaction/fact_whatcanyoudo.htm

Claim: The health risks associated with obesity can be reduced by diet choices.

Supports:

_____ **1.** In 1999, 61% of adults in the United States were overweight or obese.

_____ **2.** A meal consisting of a cheeseburger, French fries, and soda is high in saturated fat, total fat, and cholesterol.

_____ **3.** Overweight and obesity are associated with heart disease, certain types of cancer, type 2 diabetes, stroke, arthritis, breathing problems, and psychological disorders, such as depression.

_____ **4.** Exercise helps reduce obesity.

_____ **5.** Overall, the relevant supports are
a. statements of factual details.
b. statements using biased language.

B. Read the following group of ideas from a government website. Mark each statement **C** if it is an author's claim or **S** if it provides support for the claim.

_____ **6.** One small chocolate chip cookie (50 calories) is equivalent to walking briskly for 10 minutes.

_____ **7.** The difference between a large gourmet chocolate chip cookie and a small chocolate chip cookie could be about 40 minutes of raking leaves (200 calories).

_____ **8.** One hour of walking at a moderate pace (20 min/mile) uses about the same amount of energy that is in one jelly-filled doughnut (300 calories).

_____ **9.** A fast-food "meal" containing a double-patty cheeseburger, extra-large fries, and a 24 oz. soft drink is equal to running $2^1/_2$ hours at a 10 min/mile pace (1,500 calories).

_____ **10.** Food intake should be balanced with activity.

"Overweight and Obesity: What You Can Do," United States Office of the Surgeon General, 11 Jan. 07. 3 May 2008 http://www.surgeongeneral.gov/topics/obesity/calltoaction/fact_whatcanyoudo.htm

REVIEW TEST 2

Score (number correct) _____ × 20 = _____%

Argument

Read the following passage, and then, answer the questions.

Study Examines Spanking Among Minnesota Parents

[1]Parents can effectively replace physical punishment with different methods of discipline. [2]When parents use nurturing and teaching methods

instead of spanking, their children become less aggressive and violent. [3]Those are the findings of an 8-year study led by the University of Minnesota's Extension Service in Goodhue County. [4]Researchers followed 1,000 parents of children younger than 13 to learn about their attitudes toward spanking.

[5]"If you hit your children, it will be very difficult to teach them not to hit others because they have experienced it from the most important person in their lives," said retired sociologist Ron Pitzer, who led the study. [6]During the study, a public awareness and educational campaign called "Kids: Handle with Care" sent the message to Goodhue County residents that it is never okay to spank a child. [7]The message was circulated through newspaper articles, radio programs, restaurant table tent cards, grocery bags and carts, church programs, parade floats, and a county fair exhibit.

[8]The following results were reported among Goodhue County parents at the conclusion of the study:

- [9]The use of physical punishment dropped from 36 percent to 12 percent.
- [10]Parents who spanked their children reported a sizeable increase in their children's aggressiveness.
- [11]Parents who reduced physical punishment reported their children were less aggressive.
- [12]Parents who attended classes were better at setting limits and enforcing consequences, and they were more calm and nurturing. [13]Their children were more obedient, communicated more openly, had a better attitude, and were calmer.
- [14]Fathers matched or exceeded mothers in alternative, more positive discipline methods.

[15]The findings indicate that the county-wide educational effort was successful in helping make the decision to eliminate spanking, resulting in happier parents and happier children.

—Adapted from "Study Examines Spanking Among Minnesota Parents," Children's Bureau Express, U.S. Department of Human and Health Services, Apr. 2002. 3 May 2008 http://cbexpress.acf.hhs.gov/printer_friendly.cfm?issue_id=2002-04&prt_iss=1

_____ **1.** Which sentence states the author's claim?

a. sentence 1 c. sentence 4
b. sentence 3 d. sentence 5

Mark each of the following sentences (**2–5**) **C** if it is a claim or **S** if it provides support for the claim.

______ **2.** sentence 5

______ **3.** sentence 6

______ **4.** sentence 7

______ **5.** In this passage, the author expresses a biased attitude
a. in favor of spanking children as a form of punishment.
b. against the use of physical punishment as a form of discipline.

VISUAL VOCABULARY

An antonym for the word *nurturing* is ______________.

a. disengaged.
b. involved.
c. needed.

▸ A father who is nurturing has a positive impact on his child's development.

REVIEW TEST 3

Score (number correct) ______ × 10 = ______%

Argument

Before Reading: Skim the following editorial published by *USA Today*. Answer the Before Reading questions that follow the passage. Then, read the passage. Next, answer the After Reading questions. Use the discussion and writing topics as activities to do after reading.

Our View on Obesity: Hooked on Junk Food

[1]"Betcha can't eat just one" may be a Frito-Lay marketing slogan for potato chips, but, as any dieter can tell you, most junk food creates addict-like behavior—and not just in adults.

[2]Nearly a third of children are overweight or obese, triggering a backlash—most recently Michelle Obama's Let's Move initiative to make kids eat better and exercise.

Vocabulary Preview

deter (19): discourage, prevent

akin (21): similar, related

applicably (22): relevant, connected, affecting

enticing (29): tempting, inviting

[3]But it's unlikely that Obama's effort—or any child-specific approach—will succeed by itself. [4]The food industry has just gotten too good at doing what businesses are supposed to do: give customers what they want. [5]Years of research and sales results have taught it that layering fat, sugar and salt into foods will give consumers the ultimate taste sensation. [6]All three trigger such a pleasurable response in the brain that people will keep eating even after they're stuffed. [7]Better yet for sales, the brain then remembers that enjoyable sensation, launching a pattern of craving and overeating that becomes an **entrenched**—and health-toxic—habit.

[8]This cycle, and the way it evolved, was documented a year ago by David Kessler, a former commissioner of the Food and Drug Administration and former dean of Yale's medical school, in his ground-breaking book *The End of Overeating*.

[9]While Kessler avoids using the word "addiction," a new study, published in *Nature Neuroscience*, demonstrates it's exactly the kind of behavior that continual access to junk food causes. [10]While rats that had occasional access to junk food could resist eating if electroshocks were administered, rats that had constant access kept eating even while being electrocuted. [11]Does that sound like anyone you know?

[12]If there's anything American culture offers, it's continuous access to junk food. [13]The latest entry: Wendy's Baconator Triple burger, weighing in at a whopping 1330 calories. [14]Add supersized portions and supermarkets crammed with fatty, processed foods lacking nutritional value, and you have a problem. [15]Eating never stops. [16]The average child consumes nearly three snacks a day.

[17]The first lady's proposals can help at the edges. [18]Increased activity burns off calories, while healthier school lunches could reduce calories consumed. [19]More prominent calorie labeling might **deter** some people. [20]But those messages will still be competing with more potent signals from the brain.

[21]Kessler's solution, like Obama's, is to induce a broad-based shift in social attitudes, **akin** to the one that has cut smoking in half. [22]That evolution has had many drivers: warning labels, taxes (which had a particular impact on teens), public education and, less **applicably**, lawsuits and bans. [23]But everyone agrees when the tide turned. [24]As evidence of smoking's health effects became undeniable, the first surgeon general's warning was ordered onto cigarette packs in 1965. [25]Odd as it may seem now, the move set off an uproar.

[26]The push for healthier food has not attained the same critical mass, but the trend is strikingly similar to the early '60s. [27]Evidence linking obesity to illnesses is mounting, and an industry's attempt to induce addictive behavior is being dragged into public view. [28]Now Obama's move has brought the White House into the picture. [29]Given all the **enticing** bait that's so scientifically laid before us, the tide had better shift toward healthier food soon. [30]In a sense, we're all lab rats—only the shock treatment is missing.

—"Our View on Obesity: Hooked on Junk Food." from USA TODAY, March 31, 2010. Reprinted by permission of USA Today.

Before Reading

Vocabulary in Context

_____ **1.** Based on the context of the passage, what is the best meaning of the word **entrenched** (sentence 7)?
a. flexible
b. positive
c. fixed
d. negative

Tone and Purpose

_____ **2.** The overall tone of the passage is
a. objective.
b. biased.

_____ **3.** The primary purpose of the passage is to
a. inform.
b. entertain.
c. persuade.

After Reading

Central Idea and Main Idea

_____ **4.** Which sentence best states the implied central idea of the passage?
a. Childhood obesity cannot be overcome.
b. Americans are hooked on junk food.
c. The food industry, like the tobacco industry did with cigarettes, has made money by getting consumers addicted to junk food.
d. The continuous access to junk food has caused an addiction to junk food that will require a broad-based shift in social attitudes to overcome childhood obesity.

Supporting Details

______ **5.** In the sixth paragraph (sentences 12–16), sentence 15 states a
a. main idea.
b. major supporting detail.
c. minor supporting detail.

Transitions

______ **6.** The relationship within sentence 21 is
a. time order.
b. listing.
c. comparison.
d. cause and effect.

Thought Patterns

______ **7.** The overall thought pattern of the passage is
a. time order.
b. listing.
c. comparison.
d. cause and effect.

Fact and Opinion

______ **8.** Sentence 3 is a statement of
a. fact.
b. opinion.
c. fact and opinion.

Inferences

______ **9.** Based on the information in the passage, the reader can infer that
a. many Americans are addicted to junk food.
b. people choose to be obese.
c. electroshock treatment may be a useful tool to fight childhood obesity.
d. Michelle Obama's Let's Move initiative will have a profound impact on reducing childhood obesity.

Argument

______ **10.** Which sentence is not relevant to the author's claim?
a. sentence 1
b. sentence 8
c. sentence 24
d. sentence 30

WHAT DO YOU THINK?

Do you agree with the editorial's view that the food industry has attempted to "induce addictive behavior" for a profit? Why or why not? Do you agree with the author's comparison of the food industry to the tobacco industry? Should the government use warning labels, taxes, and bans to fight childhood obesity? Why or why not? If so, what foods should be banned or taxed? What kind of labels should be on foods? Assume you are taking a college health class and your professor has asked you to write about a health issue of current importance. Take a stand for or against a new law to fight obesity that includes banning and taxing certain foods. In your article, identify the foods that might be taxed or banned.

REVIEW TEST 4

Score (number correct) ______ × 10 = ______%

Argument

Textbook Skills

Skim the following passage adapted from a college textbook about world civilizations. Answer the Before Reading questions that follow the passage. Then, read the passage. Next, answer the After Reading questions.

Vocabulary Preview

migratory (4): moving, traveling

ecological (6): natural, environmental

imperialism (6): government or authority of an empire

immunities (7): protections

The Great Exchange

[1]The arrival of the Spaniards and the Portuguese in the Americas began one of the most extensive and profound changes in the history of humankind. [2]The New World, which had existed in isolation since the end of the last ice age, was now brought into continual contact with the Old World. [3]The peoples and cultures of Europe and Africa came to the Americas through voluntary or forced immigration. [4]Between 1500 and 1850, perhaps 10 to 15 million Africans and 5 million Europeans crossed the Atlantic and settled in the Americas as part of the great **migratory** movement. [5]Contact also initiated a broader biological and ecological exchange that changed the face of both the Old World and the New World—the way people lived, what they ate, and how they died, indeed how many people there were in different regions—as the animals, plants, and diseases of the two hemispheres were transferred.

[6]It was historian Alfred Crosby who first called this process the *Columbian exchange*, and he has pointed out its profound effects as the

first stage of the "**ecological imperialism**" that accompanied the expansion of the West. [7]Long separated from the populations of the Old World and lacking **immunities** to diseases such as measles and smallpox, populations throughout the Americas suffered disastrous losses after initial contact. [8]Throughout North America, contact with Europeans and Africans resulted in epidemics that devastated the **indigenous** populations.

[9]Disease may have also moved in the other direction. [10]Some authorities believe that syphilis had an American origin and was brought to Europe only after 1492. [11]In general, however, forms of life in the Old World—diseases, plants, and animals—were more complex than those in the Americas and thus displaced the New World varieties in open competition. [12]The diseases of Eurasia and Africa had a greater impact on America than American diseases had on the Old World.

[13]With animals also, the major exchange was from the Old World to the New World. [14]Native Americans had domesticated dogs, guinea pigs, some fowl, and llamas, but in general domesticated animals were far less important in the Americas than in the Old World. [15]In the first years of settlement in the Caribbean, the Spanish introduced horses, cattle, sheep, chickens, and domestic goats and pigs, all of which were considered essential for civilized life as the Iberians understood it. [16]Some of these animals thrived in the New World. [17]For example, a hundred head of cattle abandoned by the Spanish in the Rio de la Plata area in 1587 had become 100,000 head 20 years later.

[18]In the exchange of foods, the contribution of America probably outweighed that of Europe, however. [19]New World plants such as tomatoes, squash, sweet potatoes, types of beans, and peppers, became essential foods in Europe. [20]Tobacco and cacao, or chocolate, both American in origin, became widely distributed throughout the world.

[21]Even more important were the basic crops such as potato, maize, and manioc, all of which yielded more calories per acre than all the Old World grains, except rice. [22]The high yield of calories per acre of maize and potatoes had supported the high population densities of the American civilizations. [23]After the Columbian voyages, these foods began to produce similar effects in the rest of the world. [24]After 1750, the world population experienced a dramatic rise. [25]The reasons were many, but the contribution of the American foodstuff was a central one.

—Adapted from *World Civilizations: The Global Experience Combined Volume, Atlas Edition*, 5th ed., Atlas Ed., p. 526 by Peter N. Stearns, Michael B. Adas, Stuart B. Schwartz, and Marc Jason Gilbert.

Before Reading

Vocabulary in Context

Use context clues to state in your own words the definition of the following term. Indicate the context clue you used.

1. In sentence 8, what does the word **indigenous** mean?

____

2. Identify the context clue used for the word **indigenous** in sentence 8.

____

Tone and Purpose

______ **3.** What is the overall tone and purpose of the passage?

a. to inform the reader about the impact of the Columbian exchange
b. to persuade the reader that the Columbian exchange benefited America more than Europe
c. to engage the reader with little known facts about early American history

After Reading

Central Idea

4. Use your own words to state the central idea of the passage:

____

____

Main Idea and Supporting Details

______ **5.** Sentence 17 states a

a. main idea.
b. major supporting detail.
c. minor supporting detail.

Transitions

______ **6.** The relationship between sentences 8 and 9 is

a. cause and effect.
b. addition.
c. time order.
d. contrast.

Thought Patterns

_____ **7.** The overall thought pattern of the passage is
a. cause and effect.
b. listing.
c. time order.
d. contrast.

Fact and Opinion

_____ **8.** Sentence 1 is a statement of
a. fact.
b. opinion.
c. fact and opinion.

Inferences

_____ **9.** Based on the context of the information in the second paragraph (sentences 6–8), the term "ecological imperialism" in sentence 6 implies that
a. natural resources exchanged between America and Europe dramatically changed the natural environments of both regions.
b. Europe damaged the natural resources of America.
c. the immigration of Europeans brought diseases and caused epidemics that weakened the resistance of native peoples to European control.

Argument

_____ **10.** The following list of statements contains one claim and several supports from the third paragraph (sentences 9–12). Identify the support that is **not** relevant to the claim.
a. Disease may have also moved in the other direction (from America to the Old World).
b. Some authorities believe that syphilis had an American origin and was brought to Europe only after 1492.
c. The diseases of Eurasia and Africa had a greater impact on America than American diseases had on the Old World.

WHAT DO YOU THINK?

What is the significance of the Columbian exchange? Which region benefitted the most from the Columbian exchange? How might life in the United States be different today if the Columbian exchange had not occurred? Assume you are taking a college course in world civilizations. Your professor has assigned a writing assignment as a chapter test. Choose one of the following prompts and write a three-paragraph essay: (1) Define the Columbian exchange and discuss key examples of its effect on America and Western Europe; (2) Take a stand in support of or in opposition to the following claim by the Office of the United States Trade Representative: "American families benefit from trade and open markets every day. Trade delivers a greater choice of goods—everything from food and furniture to computers and cars—at lower prices." Offer examples to support your stance.

After Reading About the Basics of Argument

Before you move on to the Mastery Tests on the basics of argument, take time to reflect on your learning and performance by answering the following questions. Write your answers in your notebook.

- How has my knowledge base or prior knowledge about the basics of argument changed?
- Based on my studies, how do I think I will perform on the Mastery Test(s)? Why do I think my scores will be above average, average, or below average?
- Would I recommend this chapter to other students who want to learn more about the basics of argument? Why or why not?

Test your understanding of what you have learned about the basics of argument by completing the Chapter 12 Review Card in the insert near the end of your text.

CONNECT TO PEARSON myreadinglab

To check your progress in meeting Chapter 12's learning outcomes, log in to **www.myreadinglab.com** and try the following exercises.

- The "Critical Thinking" section of MyReadingLab gives additional information about the basics of argument. The section provides an overview, model, practices, and tests. To access this resource, click on the "Study Plan" tab. Then click on "Critical Thinking." Then click on the following links as needed: "Overview," "Model," "Critical Thinking: Facts and Opinions (Flash Animation)," "Practice," and "Test."

- The "Study Skills Website" section of MyReadingLab also gives additional information about the basics of argument. To access these resources, go to the "Other Sources" box on the home page of MyReadingLab. Click on "Study Skills Website." Then scroll down the page to the heading "Life Skills" and click on "Critical Thinking." Explore each of the links on the bar on the left side of the page.
- To measure your mastery of this chapter, complete the tests in the "Critical Thinking" section and click on Gradebook to find your results.

MASTERY TEST 1

Name ______________________ Section ______________

Date __________ **Score** (number correct) ______ × 20 = ______%

Read the following passage and examine the graphic in this excerpt from a college sociology textbook. Then, answer the questions.

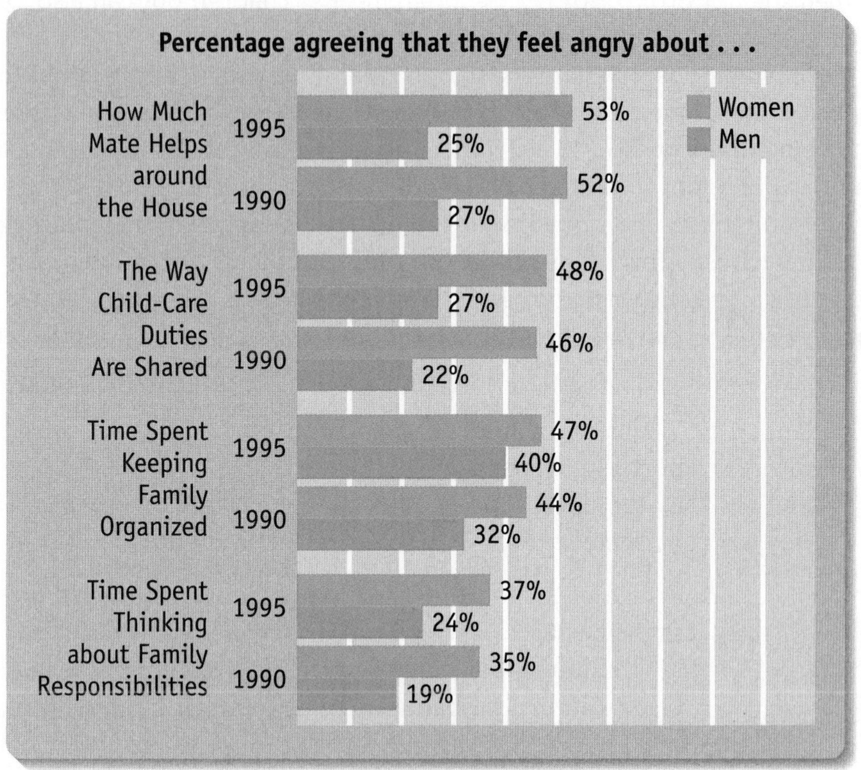

▲ **Sources of Resentment When Wives Are Employed**

—Figure adapted from the 1995 Virginia Slims Opinion Research Poll conducted by Roper Starch Worldwide, 1996, p. 82; text and adapted figure from *Marriages and Families: Changes, Choices, and Constraints,* 4th ed., pp. 91–92, Figure 4.3 (p. 91). by Nijole V. Benokraitis. Copyright © 2002 by Pearson Education, Inc. Reproduced by permission of Pearson Education, Inc., Boston, MA.

Textbook Skills

Gender Roles

[1]Traditional gender roles promote stability, continuity, and predictability. [2]Because each person knows what is expected of him or her, rights and responsibilities are clear. [3]Men and women do not have to argue over who does what: if the house is clean, she is a "good wife"; if the bills are paid, he is a "good provider." [4]If the costs and benefits of the relationship are fairly balanced and each partner is relatively happy, traditional gender roles can work well. [5]As long as both partners live up

to their role expectations, they are safe in assuming that they will take care of each other financially, emotionally, and sexually.

[6]Some women stay in traditional relationships because as long as they live up to the idealized role, they don't have to make autonomous decisions or assume responsibility when things go wrong. [7]An accommodating wife or mother can enjoy both power and prestige through her husband's accomplishments. [8]As a good mother, she not only controls and dominates her children but can also be proud of guiding and enriching their lives.

[9]But traditional roles are changing. [10]Many men have lost their jobs to downsizing and more women are working to pursue their own professional interests as well as to help out financially. [11]Black fathers, especially, are losing well-paying, low-skilled manufacturing jobs that are eliminated by automation and global job restructuring. [12]As a result, black and other ethnic fathers are now sharing the breadwinning role with their wives. [13]This has created more strain at home. [14]Although husbands of employed wives may be doing more of the household tasks than their fathers did, there is some evidence that men's resentment about the division of labor may be growing.

—Benoknaitis, *Marriages and Families*, 4th ed., 2002, pp. 91–92.

______ **1.** Which of the following sentences is a claim that is adequately supported by the details?

a. sentence 1
b. sentence 6
c. sentence 7
d. sentence 8

______ **2.** Which of the following is a logical conclusion based on the details in the passage?

a. Traditional wives are completely dependent upon their husbands.
b. Traditional husbands often help out with household chores.
c. Traditional wives take care of the home and children; traditional husbands provide income for the family.
d. Traditional gender roles are the ideal roles for men and women.

3–5. Based on the information in the graphic, place a check (✔) by the three statements that assert logical conclusions.

______ Women are much more likely than men to feel angry about how much their mate helps around the house.

______ More men lost their jobs as women entered the workforce.

______ Men's resentment rose between 1990 and 1995, even though women say they do the bulk of the household chores and child care.

______ Divorce rates increased from 1990 to 1995.

______ Both wives and their mates were more dissatisfied overall with their share of family responsibilities in 1995 than in 1990.

MASTERY TEST 2

Name ______________ Section ______________

Date ________ **Score** (number correct) ______ × 20 = ______%

Read the following passage from a college history textbook, and then answer the questions.

Textbook Skills

Is There a Right to Privacy?

[1]Nowhere does the Bill of Rights say that Americans have a right to privacy. [2]Clearly, however, the First Congress had the concept of privacy in mind when it created the first 10 amendments. [3]For example, freedom of religion implies the right to exercise private beliefs, and protections against "unreasonable searches and seizures" make persons secure in their homes. [4]In addition, private property cannot be seized without "due process of law." [5]In 1928, Justice Brandeis hailed privacy as "the right to be left alone—the most comprehensive of the rights and the most valued by civilized men."

[6]The idea that the Constitution guarantees a right to privacy was first expressed in a 1965 case involving the conviction of a doctor and family planning specialist for disseminating birth control devices in violation of a little-used Connecticut law. [7]The state reluctantly brought them to court. [8]They were convicted. [9]After wrestling with the privacy issue in *Griswold* v. *Connecticut*, seven Supreme Court justices finally decided that various portions of the Bill of Rights cast "penumbras" (or shadows), unstated liberties implied by the stated rights. [10]These protected a right to privacy, including a right to family planning between husband and wife. [11]Supporters of privacy rights argued that this ruling was a reasonable interpretation of the Fourth Amendment. [12]There were many critics of the ruling. [13]They claimed that the Supreme Court was inventing protections not specified by the Constitution.

[14]The most important application of the privacy rights, however, came not in the area of birth control but in the area of abortion. [15]The Supreme Court unleashed a constitutional firestorm in 1973 that has not yet abated.

—From *Government in America: People, Politics, and Policy,* 5th ed., p. 103 by George C. Edwards III, Martin P. Wattenberg and Robert L. Lineberry. Copyright © 2000 by Pearson Education, Inc. Reprinted by permission of Pearson Education, Inc., Glenview, IL.

______ **1.** The claim of the entire passage is stated in

a. sentence 1.
b. sentence 2.
c. sentence 3.
d. sentence 15.

_____ **2.** The claim of the second paragraph is stated in
a. sentence 6.
b. sentence 11.
c. sentence 9.
d. sentence 13.

_____ **3.** In the first paragraph, sentence 4 states
a. a detail that is relevant.
b. a detail that is not relevant.

_____ **4.** In the second paragraph, sentence 7 states
a. a detail that is relevant.
b. a detail that is not relevant.

_____ **5.** In the second paragraph, sentence 12 states
a. a detail that is relevant.
b. a detail that is not relevant.

VISUAL VOCABULARY

Which is the best synonym for *disseminated?*

____

a. blown
b. gathered
c. spread

► Upon maturation, the dandelion turns into the well-known puffball containing seeds that are disseminated by the wind.

MASTERY TEST 3

Name ______________________ Section ____________

Date ________ Score (number correct) ______ × 20 = ______%

Read the following passage from a college sociology textbook. Then, answer the questions.

Textbook Skills

The Impact of the Consumer Culture on the Growth of Suburbia

[1]Following World War II, a combination of factors had brought in a consumer culture, which has since come to dominate American culture. [2]The essence of the consumer culture is a focus on acquisition of material objects made possible by affluence, mobility, and leisure. [3]For instance, the automobile, mass produced and, thus, cheap enough to be available to a majority of people, led to the growth of the highway system on which the automobile could move freely and rapidly. [4]In turn, highways were especially responsible for the spread of suburbanization to the extent that it became a pervasive national phenomenon. [5]At the same time, other facets of social change became apparent and were reinforced by suburbanization.

[6]As the cities had done before them, the suburbs spawned their own lifestyle, revolving around the absence of the father for long portions of the day, the necessity for a private vehicle, and the chauffeuring of children to various activities. [7]In addition, trips to the shopping center or suburban mall had become the focus of recreation for suburbanites. [8]The booming economy permitted one family member—usually the husband and father—to be the sole breadwinner, leaving the wife and mother to do the chauffeuring and shopping chores.

[9]Originally, the suburbs attracted young married couples who were planning fairly large families—the baby boom parents of the years following World War II. [10]The postwar economy increased individual incomes and made mortgage money available, so that for the first time home ownership became a dream almost all could fulfill. [11]Most of the early suburbanites were between 25 and 45 years old, with children ranging in age from infancy to the teens. [12]They were predominantly white, middle-income, high school graduates, politically conservative, and morally "proper." [13]From these beginnings there developed the stereotype of the suburbs as "bedroom" communities of almost identical homes.

—John & Perry, *Contemporary Society: An Introduction to Social Science*, 10th ed., pp. 264–265.

_____ **1.** Sentence 1 is a statement that is
  a. a claim.
  b. a support for a claim.

_____ **2.** Sentence 3 is a statement that is
  a. a claim.
  b. a support for a claim.

_____ **3.** Sentence 5 is a statement that is
  a. a claim.
  b. a support for a claim.

_____ **4.** Sentence 6 is a statement that is
  a. a claim.
  b. a support for a claim.

_____ **5.** Sentence 11 is a statement that is
  a. a claim.
  b. a support for a claim.

MASTERY TEST 4

Name ______________________ Section ______________

Date __________ **Score** (number correct) ______ × 10 = ______%

Read the following passage from a college biology textbook. Then, answer the questions.

Textbook Skills

Natural Selection Tends to Preserve Genes That Help an Organism Survive and Reproduce

[1]On average, organisms that best meet the challenges of their environment will leave the most offspring; these offspring will inherit the genes that made their parents successful. [2]Thus, natural selection preserves genes that help organisms flourish in their environment. [3]To create a hypothetical example, a mutated gene that caused ancestral beavers to grow larger teeth allowed those with this mutation to chew down trees more efficiently, build bigger dams and lodges, and eat more bark than "ordinary" beavers could. [4]Because these big-toothed beavers obtained more food and better shelter, they were able to raise more offspring who inherited their parents' genes for larger teeth. [5]Over time, less-successful, smaller-toothed beavers became increasingly scarce; after many generations, all beavers had large teeth.

[6]Structures, physiological processes, or behaviors that aid in survival and reproduction in a particular environment are called adaptations. [7]Most of the features that we admire so much in our fellow life-forms, such as the long limbs of deer, the wings of eagles, and the mighty trunks of redwood trees, are adaptations molded by millions of years of natural selection acting on random mutations.

[8]Over millennia, the interplay of environment, genetic variation, and natural selection inevitably results in evolution: a change in the genetic makeup of species. [9]This change has been documented innumerable times both in laboratory settings and in the wild. [10]For example, antibiotics have acted as agents of natural selection on bacterial populations, causing the evolution of antibiotic-resistant forms. [11]Lawn mowers have caused changes in the genetic makeup of populations of dandelions, favoring those that produce flowers on very short stems. [12]Scientists have documented the spontaneous emergence of entirely new species of plants due to mutations that alter their chromosome number.

[13]What helps an organism survive today can become a liability tomorrow. [14]If environments change—for example, as global warming occurs—the genetic makeup that best adapts organisms to their environment will also change over time. [15]When random new mutations increase the fitness of an organism in the altered environment, these mutations will spread throughout the population. [16]Populations within a species that live in different environments will be subjected to different types of natural selection. [17]If the differences are

great enough and continue for long enough, they may eventually cause the populations to become sufficiently different from one another to prevent interbreeding—a new species will have evolved.

[18]If, however, favorable mutations do not occur, a changing environment may doom a species to extinction. [19]Dinosaurs are extinct not because they were failures—after all, they flourished for 100 million years—but because they could not adapt rapidly enough to changing conditions.

[20]Within particular habitats, diverse organisms have evolved complex interrelationships with one another and with their nonliving surroundings. [21]The diversity of species and the interactions that sustain them are encompassed by the term biodiversity. [22]In recent decades, the rate of environmental change has been drastically accelerated by human activities. [23]Many wild species are unable to adapt to this rapid change. [24]In habitats most affected by humans, many species are being driven to extinction.

—From *Biology: Life on Earth*, 8th ed., p. 10 by Teresa Audesirk, Gerald Audesirk, and Bruce E. Byers.

_____ **1.** Sentence 1 states a
a. claim. b. support.

_____ **2.** Sentence 2 states a
a. claim. b. support.

_____ **3.** Sentence 6 states a
a. claim. b. support.

_____ **4.** Sentence 7 states a
a. claim. b. support.

_____ **5.** Sentence 8 states a
a. claim. b. support.

_____ **6.** Sentence 9 states a
a. claim. b. support.

_____ **7.** Sentence 13 states a
a. claim. b. support.

_____ **8.** Sentence 17 states a
a. claim. b. support.

_____ **9.** Sentence 18 states a
a. claim. b. support.

_____ **10.** Sentence 19 states a
a. claim. b. support.

Advanced Argument: Persuasive Techniques

13

LEARNING OUTCOMES

After studying this chapter you should be able to do the following:

1. Define the terms *fallacy, propaganda, personal attack, straw man, begging the question, name-calling, testimonial, bandwagon, plain folks, either-or, false comparison, false cause, card stacking, transfer,* and *glittering generalities.*
2. Detect fallacies based on irrelevant arguments: *personal attack, straw man,* and *begging the question.*
3. Detect propaganda techniques based on irrelevant arguments; *name-calling, testimonials, bandwagon,* and *plain folks.*
4. Detect fallacies based on inadequate arguments: *either-or, false comparison,* and *false cause.*
5. Detect propaganda techniques based on inadequate arguments: *card stacking, transfer,* and *glittering generalities.*
6. Evaluate the importance of advanced arguments: persuasive techniques.

Before Reading About Advanced Argument

In this chapter, you will build on the concepts you studied in Chapter 12 about the basics of argument. Take a moment to review the four steps in analyzing an argument. Fill in the following blanks with each of the steps.

Step 1. ____________________

Step 2. ____________________

Step 3. ____________________

Step 4. ____________________

To help you master the material in the chapter, create a three-column chart in your notebook. In the left column copy the headings as in the example that follows. Leave enough room between each heading to fill in definitions and examples as you work through the chapter.

General definition: A fallacy is

Fallacy	**Definition**	**Example**
Personal attack		
Straw man		

General definition: Propaganda is

Propaganda technique	**Definition**	**Example**
Name-calling		
Testimonials		

Biased Arguments

Much of the information that we come in contact with on a daily basis is designed to influence our thoughts and behaviors. Advertisements, editorials, and political campaigns offer one-sided, biased information to sway public opinion.

This biased information is based on two types of reasoning: the use of **fallacies** in logical thought and the use of **propaganda**. A master reader identifies and understands the use of these persuasion techniques in biased arguments.

What Is a Fallacy in Logical Thought?

You have already studied logical thought in Chapter 12. Logical thought or argument is a process that includes an author's claim, relevant support, and a valid conclusion. A **fallacy** is an error in the process of logical thought. A fallacy leads to an invalid conclusion. You have also studied two general types of fallacies: irrelevant details and inadequate details. By its nature, a fallacy is not persuasive because it weakens an argument. However, fallacies are often used to convince readers to accept an author's claim. In fact, the word *fallacy* comes from a Latin word that means "to deceive" or "trick." You will learn more about irrelevant and inadequate arguments in the next sections of this chapter.

Fallacies are not to be confused with false facts. A fact, true or false, is stated without bias, and facts can be proven true or false by objective evidence. In contrast, a fallacy is an invalid inference or biased opinion about a fact or set of facts. Sometimes the word *fallacy* is used to refer to a false belief or the reasons for a false belief.

A **fallacy** is an error in logical thought.

EXAMPLE Read the following sets of ideas. Mark each statement as follows:

UB for unbiased statements
B for biased arguments

_____ **1.** Henry Ford invented the automobile.

_____ **2.** Jamal had been wearing his lucky shirt during every game in which he scored a winning point, and when he didn't wear his lucky shirt, he did not score. Now he wears his lucky shirt every game without fail.

_____ **3.** Joanne lies all the time because she is incapable of telling the truth.

_____ **4.** Based on eyewitness accounts and the statement of Marji Thompson, the police report stated that Thompson had thought the car was in drive when it was actually in reverse. When she stepped on the gas, she backed into the car parked behind her, damaging her back bumper and the front bumper and hood of the other car.

_____ **5.** Four-time winner of NASCAR's Nextel Cup Championship, Jeff Gordon set new standards in the sport during the 2007 season with 6 victories, 21 top-five finishes, a record-breaking 30 top-ten finishes, and a 7.3 average finishing position with one DNF (Did Not Finish). Even though he came in second place behind Jimmie Johnson for the title during the season's Chase, Gordon has more career wins than almost all other drivers; only Richard Petty with seven wins has more victories.

EXPLANATION

1. This statement is unbiased (UB), but it is not true. The automobile was not invented by Henry Ford. The automobile evolved from inventors' efforts worldwide; the modern automobile is the result of over 100,000 patents. Ford did invent an improved assembly line.
2. This is a biased argument (B) based on a fallacy in logical thought. By not considering other reasons for his success, Jamal has identified a false cause and made an invalid inference. You will learn more about the fallacy of false cause later in this chapter.
3. This is a biased argument (B) based on a fallacy in logical thought. The statement about Joanne uses circular thinking; the statement restates the claim as

its own proof. The phrase "lies all the time" is the claim, which is restated by the phrase "incapable of telling the truth." No factual evidence is given.

4. This is an unbiased statement (UB). Every detail can be verified through the police report and eyewitness accounts of the incident.

5. This is an unbiased statement (UB). This statement is factual and can be proven with eyewitness accounts and newspaper reports.

PRACTICE 1

Read the following sets of ideas. Mark each statement as follows:

UB for unbiased statements
B for biased arguments

______ **1.** The flea is the smallest insect.

______ **2.** Watching violent programs causes youth to become violent criminals.

______ **3.** Convicted felons should not have the right to vote because they have broken the law.

______ **4.** Cars that run on hydrogen are not dependent upon imported oil.

What Is Propaganda?

Propaganda is a means by which an idea is widely spread. The word *propaganda* comes from a Latin term that means to "propagate" or "spread." **Propaganda** is a biased argument that advances or damages a cause. Propaganda is often used in politics and advertising.

Read the following two descriptions of a mock sandwich called The Two-Fisted Bacon Burger. The first description is an advertisement for the sandwich. The second is the nutritional information for a meal that would include the sandwich and a small order of fries.

Advertisement: You're gonna love The Two-Fisted Bacon Burger! Two juicy 1/2 lb. patties fully loaded with crisp smokehouse bacon, mayo, ketchup, and thick-sliced cheddar cheese inside a fresh, soft bun. Nothing compares! Two handfuls of lip-smacking, mouth-watering pleasure. Get the combo! The Two-Fisted Bacon Burger with small fries is umm, umm, *delicious*!

Nutritional Facts:

	Calories	Calories from fat	Total fat (g)	Saturated fat (g)	Trans fat (g)	Cholesterol (mg)	Sodium (mg)	Carbs (g)	Fiber (g)	Sugar (g)	Protein (g)
The Two-Fisted Bacon Burger	1601	944	78.48	31.06	0	389	1398	18.8	.6	2	118.82
Small French Fries	271	130	14.5	3.4	3.7	0	165	31.9	3.4	0	4

The advertisement uses tone words that appeal to the senses of hunger—smell, taste, and vision—such as *two-fisted, juicy, crisp, smokehouse, thick-sliced, fresh, soft, lip-smacking, mouth-watering,* and *um, um, delicious*. These words are meant to tempt the customer into purchasing the sandwich. (For more about tone, see Chapter 10, "Tone and Purpose.") In contrast, the nutritional facts reveal objective data. These facts might keep someone who is concerned about a healthful diet from buying and eating this meal.

This advertisement is an example of **propaganda**. Propaganda uses fallacies to spread biased information.

Propaganda is an act of persuasion that systematically spreads biased information that is designed to support or oppose a person, product, cause, or organization.

Propaganda uses a variety of techniques that are based on **emotional appeal.** If you are not aware of these techniques, you may be misled by the way information is presented and come to invalid conclusions. Understanding propaganda techniques will enable you to separate factual information from emotional appeals so that you can come to valid conclusions.

Emotional appeal is the arousal of emotion to give meaning or power to an idea.

EXAMPLE Read the following sets of ideas. Mark each statement as follows:

UB for unbiased statements
B for biased arguments

_____ **1.** Studies show that green tea contains antioxidants and polyphenols, as well as a wide variety of vitamins and minerals.

_____ **2.** A vote for Senator Manness is a vote for the worker on the job, the teacher in the classroom, the nurse in the hospital, the ordinary person trying to make ends meet.

_____ **3.** Certain foods, such as chocolate, nuts, and soft drinks, cause acne.

_____ **4.** Mueller's dried pasta is America's favorite pasta.

_____ **5.** Zora Neale Hurston, author of *Their Eyes Were Watching God,* records the unique customs and speech of the rural black town of Eatonville, Florida, where she was born and raised.

EXPLANATION

1. This is an unbiased statement (UB). It is factual and can be proven with objective evidence and expert opinions.
2. This is a biased argument (B) using the emotional appeal of propaganda. The statement uses the "plain folks" appeal. Senator Manness is identified as an everyday person with the same values as everyday people. You will learn more about this propaganda technique later in this chapter.
3. This is an unbiased statement (UB); however, it is a false fact. Current research does not support this claim.
4. This is a biased argument (B) using the emotional appeal of the propaganda techniques of "transfer" and "bandwagon." You will learn more about these propaganda techniques later in this chapter.
5. This is an unbiased statement (UB). This statement offers factual information that can be verified by reading the novel and researching African American customs and speech.

PRACTICE 2

Read the following sets of ideas. Mark each statement as follows:

UB for unbiased statements
B for biased arguments

_____ **1.** Look like Sandra Bullock, Heidi Klum, or Jennifer Lopez—use Love Your Hair for silky, shimmering hair.

_____ **2.** Be a true patriot; buy Liberty Bonds to support our troops.

_____ **3.** People with apple-shaped figures are at greater risk for heart disease, stroke, diabetes, and some types of cancer.

_____ **4.** Research exists both for and against milk consumption; while many experts agree that milk contains important vitamins and minerals for healthy bones, such as Vitamin D and calcium, other experts say milk contributes to heart disease and obesity.

Irrelevant Arguments: Fallacies

Writing based in logical thought offers an author's claim and relevant supporting details, and it arrives at a valid conclusion. Fallacies and propaganda offer irrelevant arguments based on irrelevant details. Irrelevant details draw attention away from logical thought by ignoring the issue or changing the subject.

Personal Attack

Personal attack is the use of abusive remarks in place of evidence for a point or argument. Also known as an *ad hominem* attack, a personal attack attempts to discredit the point by discrediting the person making the point.

For example, Maurice Long, a recovering alcoholic, decides to run for mayor. His opponent focuses attention on Maurice's history of alcohol abuse and ignores his ideas about how to make the community stronger with statements like "Don't vote for a Long record of poor decisions," or "Don't give the keys to the city to the town drunk." However, Maurice's past struggle with alcoholism has nothing to do with his current ability to work as a public servant; making this argument a personal attack.

EXAMPLE Read this discussion between a husband and wife, and then underline two uses of the logical fallacy of *personal attack.*

Budget Woes

[1]"Glenn, I am more than concerned about the amount of money you are spending," Jean says in a concerned but polite voice.

[2]"Oh, really?" Glenn replies with surprise.

[3]"How can you be surprised? [4]We worked out this budget together, so you must know that you are spending more than we can afford," Jean retorts with exasperation.

[5]"I seem surprised because I have worked hard to stay within the budget, but the cost of gas at the pump has skyrocketed and is much more expensive than what we figured, and . . . " Glenn says defensively.

[6]Jean interrupts, "Stop! [7]The price of gas is not the issue; you are eating out every day at lunch while the pantry is full of groceries you purchased, with my credit card, by the way, so you could save money by taking your lunch. [8]You are playing golf twice a week, while you promised to cut back to playing only once a week."

[9]"Well, I only agreed to that budget because you are such a control freak and have to have everything your way," Glenn shouts, then continues in a controlled, measured voice. [10]"The budget is unreasonable; we make enough money to live comfortably. [11]I wish you wouldn't worry so much."

[12]"What! [13]Well, maybe I wouldn't get so 'freaked out' if you weren't such a manipulator." [14]Jean's voice trembles with emotion as she continues. [15]"You shouldn't have agreed to the budget if you didn't think it was realistic. [16]It's not fair to agree with me knowing you are going to do as you please; we have to be honest with each other."

[17]"Okay, okay, so let's sit down together later tonight and create a budget that we can both live with," Glenn says as he embraces his wife.

EXPLANATION Although much of this conversation is emotional, most of it is based on facts. However, sentences 9 and 13 sink into an exchange of personal attacks. Glenn accuses his wife of being a *control freak*, and she responds by calling him a *manipulator*.

Straw Man

A **straw man** is a weak argument substituted for a stronger one to make the argument easier to challenge. A straw man fallacy distorts, misrepresents, or falsifies an opponent's position. The name of the fallacy comes from the idea that it is easier to knock down a straw man than a real man who will fight back. The purpose of this kind of attack is to shift attention away from a strong argument to a weaker one that can be more easily overcome. Study the following example.

Governor Goodfeeling is campaigning against a movement by the populace to limit the size of public school classes to 15 students per teacher. His opponents run political advertisements that state the following:

"Governor Goodfeeling would rather invest in special interest tax breaks than our children's future. Governor Goodfeeling uses taxpayer money to fund her campaign against class-size reductions."

This passage doesn't mention Governor Goodfeeling's reason for opposing the limit on class size: the state does not have the additional $2.3 billion needed to implement such a major reform without a drastic tax increase. Her opponents used straw man fallacies to put forth views that do not address the governor's line of reasoning and are easier to attack than is her valid stand against smaller class sizes.

EXAMPLE Read the following paragraph, and then underline the logical fallacy of *straw man.*

[1]For the past five years, Senator Richy and his political party have consistently blocked action on the Patient's Bill of Rights while Americans have suffered mercilessly from HMO abuse and neglect. [2]For every day that Congress delays, medical treatment for tens of thousands of Americans is effectively denied or delayed by insurance companies. [3]Those, like Senator Richy, who stand against the Patients' Bill of Rights stand against the spirit of the original Bill of Rights and other fundamental freedoms, such as the freedom of speech, religion, and the freedom from unreasonable searches and seizures.

EXPLANATION The third sentence sets up a straw man, a misleading notion that those who oppose the Patients' Bill of Rights oppose basic constitutional rights. This approach effectively draws attention away from the actual reasons Senator Richy opposed this bill and denies the public an honest discussion of the facts involved in the issue.

Begging the Question

Begging the question restates the point of an argument as the support and conclusion. Also known as *circular reasoning*, begging the question assumes that an unproven or unsupported point is true. For example, the argument "Exercise is tiring because it is strenuous" begs the question. The point "Exercise is tiring" is assumed to be true because it is restated in the term *strenuous* without specific supports that give logical reasons or explanations. Compare the same idea stated without begging the question: "I find exercise such as weightlifting tiring because I have little endurance and strength."

EXAMPLE Read the following paragraph. Underline the irrelevant argument of *begging the question.*

It's the Law for a Reason

[1]Current laws that mandate the use of bicycle helmets must not be repealed. [2]Properly fitted helmets protect the brain against debilitating injuries. [3]According to the Centers for Disease Control and Prevention, more than 500,000 people in the United States are treated in emergency rooms, and more than 700 people die as a result of bicycle-related injuries annually. [4]In addition, if bicycle helmet laws are repealed, then these laws would cease to exist.

EXPLANATION Sentence 4 is a statement of begging the question. To say *these laws would cease to exist* is a restatement of the word *repeal*, which means "cancel." As a master reader, you want to know the reasons that explain how helmet laws have reduced the risks or incidents of injuries.

PRACTICE 3

Identify the fallacy in each of the following items. Write **A** if it begs the question, **B** if it constitutes a personal attack, or **C** if it is a straw man.

_____ **1.** Health care is a universal right because everyone deserves adequate health care.

_____ **2.** Sarah Clinton is a fatal cancer to the Republican party.

_____ **3.** Opponents of health reform are afraid of the facts themselves. Their attempt to drown out opposing views is simply un-American.

_____ **4.** "Nothing is not an option. You didn't send me to Washington to do nothing," President Obama, in defense of his $800 billion economic stimulus package in 2009.

Irrelevant Arguments: Propaganda Techniques

Name-Calling

Name-calling uses negative labels for a product, idea, or cause. The labels are made up of emotionally loaded words and suggest false or irrelevant details that cannot be verified. Name-calling is an expression of personal opinion. For example, a bill for censorship of obscene speech on television and radio may be labeled "anti-American" to generate opposition to the bill. The "anti-American" label suggests that any restriction placed on the right of public speech is *against basic American values,* for which the Revolutionary War was fought.

EXAMPLE Read the following paragraph. Underline the irrelevant details that use *name-calling.*

[1]State Attorney John Q. Private's appointment to the U.S. Court of Appeals for the Sixth Circuit should not be confirmed. [2]Private is an ideological zealot with a long history of undermining legal and constitutional protections for ordinary Americans. [3]He has diligently opposed reproductive rights, environmental protections, and the separation of church and state. [4]In addition, he has consistently showed favoritism for big tobacco and the gun lobby. [5]It is clear that Private is a right-wing conservative, a partisan who values corporate interests more than the interests of the people.

EXPLANATION The paragraph offers little factual evidence to support the assertion that John Q. Private should not be confirmed to the U.S. Court of Appeals. Phrases such as *undermining legal and constitutional protections* and *diligently opposed reproductive rights, environmental protections, and the separation of church and state* are emotionally laden accusations that remain unsupported with evidence. The author couples these biased generalizations with name-calling. Sentences 2 and 5 label John Q. Private an *ideological zealot,* a *right-wing conservative,* and a *partisan.*

Testimonials

Testimonials use irrelevant personal opinions to support a product, idea, or cause. Most often the testimonial is provided by a celebrity whose only qualification as a spokesperson is fame. For example, a famous actor promotes a certain brand of potato chips as his favorite, or a radio talk show host endorses a certain type of mattress.

EXAMPLE Read the following paragraph. Underline the irrelevant details that use a *testimonial.*

The Brilliance of a Smile

[1]SmileBrite offers the fastest and best whitening results of all the leading paint-on and strip whiteners. [2]SmileBrite's secret is in its method

of application and its secret combination of ingredients, which include peroxide. [3]Film star Julia Famous loves SmileBrite: "SmileBrite is safe and effective; I trust my smile to SmileBrite." [4]This special formula is gentle and will not damage the tooth's enamel, yet delivers tremendous results in just a few applications.

EXPLANATION Sentences 1, 2, and 4 offer details about SmileBrite that can be verified. However, sentence 3 uses the testimonial of a famous actress. Being a famous actress doesn't make the spokesperson an expert on the safety or effectiveness of a product. A dentist, dental hygienist, or scientific researcher could offer a relevant expert opinion.

PRACTICE 4

Identify the propaganda technique used in each of the following items. Write **A** if the sentence is an example of name-calling or **B** if it is a testimonial.

_____ **1.** Senator Fleming is a big-government socialist who works against the American way.

_____ **2.** "I'm winning at losing weight. Winning at losing on Weight Watchers. Weight Watchers—because it works!" Jennifer Hudson.

_____ **3.** "When my husband lost his job, we lost our health insurance. Then, my five year old was diagnosed with a brain tumor. When my husband finally got a new job several months later, the company's health insurance denied payment to treat our daughter, saying she had a pre-existing condition. We need health reform!" — Jennifer Cortez

_____ **4.** Jay Leno is a thief who stole *The Tonight Show* from David Letterman and Conan O'Brian.

Bandwagon

The **bandwagon** appeal uses or suggests the irrelevant detail that "everyone is doing it." This message plays on the natural desire of most individuals to conform to group norms for acceptance. The term *bandwagon* comes from the 19th-century use of a horse-drawn wagon that carried a musical band to lead circus parades and political rallies. To *jump on the bandwagon* meant to follow the crowd, usually out of excitement and emotion stirred up by the event rather than out of thoughtful reason or deep conviction.

EXAMPLE Read the following paragraph. Underline the irrelevant details that use the *bandwagon* appeal.

A Matter of Honor

[1]Tyrell hated to write. [2]He often struggled with finding a topic, and once he found a topic, he then struggled with finding the words to express his ideas. [3]He knew he was in danger of failing his composition course, and the stress made it even harder for him to think of an idea for his major research project. [4]While he was searching the Internet for ideas, he came across a Web site that sold research essays. [5]As he paid for an essay, downloaded, and printed it out, he thought to himself, "No big deal, everybody cheats at least once in their college career."

EXPLANATION In sentence 5, Tyrell uses the bandwagon fallacy to justify his cheating. Finding permission in "everybody cheats" is not valid reasoning. The emotional appeal is the relief that you are one of "everybody," not an individual who has decided to engage in dishonest and unethical behavior.

Plain Folks

The **plain folks** appeal uses irrelevant details to build trust based on commonly shared values. Many people distrust the wealthy and powerful, such as politicians and the heads of large corporations. Many assume that the wealthy and powerful cannot relate to the everyday concerns of plain people. Therefore, the person or organization in power puts forth an image to which everyday people can more easily relate. For example, a candidate may dress in simple clothes, pose for pictures doing everyday chores like shopping for groceries, or talk about his or her own humble beginnings to make a connection with "plain folks." These details strongly suggest that "you can trust me because I am just like you." The appeal is to the simple, everyday experience, and often the emphasis is on a practical or no-nonsense approach to life.

EXAMPLE Read the following paragraph. Underline the irrelevant details that appeal to *plain folks*.

For the Good of the Children

[1]Helen McCormick, a well-known multimillionaire, is running for the school board. [2]Every day she dresses in sensible shoes and a conservative business suit and goes door to door in neighborhoods populated with young families with school-age children. [3]As she visits with them, she

reminds them of her success and attributes it to hard work and a good education, even though she inherited her fortune and has never held a job. [4]Her newspaper ads run the slogan "I am committed to public education. [5]My own children attend public school; help me help our children." [6]On the day of the election, Helen stands on the corner of the busiest intersection holding a sign that reads, "Helen cares about our children."

EXPLANATION Candidate Helen is described as wearing clothes that many "plain folks" also wear: *sensible shoes and a conservative business suit.* In sentence 3, the candidate asserts ideals valued by many people: *hard work and a good education.* She emphasizes her common bond with voters by stating *My own children attend public school; help me help our children.* Helen sets a friendly tone with the use of her first name. In addition, Helen's willingness to go door to door and stand on the street corner indicates she is humble and working hard for every vote. All of these details suggest that this multimillionaire, Helen, is just one of the "plain folks."

PRACTICE 5

Label each of the following items according to the propaganda techniques they employ:

A. plain folks
B. bandwagon
C. testimonial
D. name-calling

_____ **1.** An owner of a local car dealership stars in a commercial for his business dressed in jeans and a flannel shirt with his sleeves rolled up.

_____ **2.** Nine out of ten customers express deep satisfaction with Relora, the natural way to ease tension and lose weight.

_____ **3.** Don't be a butt head; stop smoking today!

_____ **4.** "Proactive keeps your skin acne free. Proactive means clear skin," says Jessica Simpson.

Inadequate Arguments: Fallacies

In addition to offering relevant supporting details, logical thought relies on adequate supporting details. A valid conclusion must be based on sufficient support. Fallacies and propaganda offer inadequate arguments that lack details.

Inadequate arguments oversimplify the issue and do not give a person enough information to draw a proper conclusion.

Either-Or

Either-or assumes that only two sides of an issue exist. Also known as the *black-and-white fallacy*, either-or offers a false dilemma because more than two options are usually available. For example, the statement "If you don't vote for social security reform, you don't care about the elderly" uses the either-or fallacy. The statement assumes only one reason for not voting for social security reform—not caring about the elderly. Yet it may be that a person doesn't approve of the particular reform being considered, prefers another solution, or, perhaps, believes the current social security program is strong and serves the elderly as well as could be expected. Either-or leaves no room for the middle ground or other options.

EXAMPLE Read the following paragraph taken from President Bush's Address to the Joint Session of Congress and the American people on September 20, 2001. Then, underline the logical fallacy of *either-or*.

> [1]Our response involves far more than instant retaliation and isolated strikes. [2]Americans should not expect one battle, but a lengthy campaign, unlike any other we have ever seen. [3]It may include dramatic strikes, visible on TV, and covert operations, secret even in success. [4]We will starve terrorists of funding, turn them one against another, drive them from place to place, until there is no refuge or no rest. [5]And we will pursue nations that provide aid or safe haven to terrorism. [6]Every nation, in every region, now has a decision to make. [7]Either you are with us, or you are with the terrorists. [8]From this day forward, any nation that continues to harbor or support terrorism will be regarded by the United States as a hostile regime.

—George W. Bush, *The White House*, 20 Sept. 2001. 10 Apr. 2004. http:// www.whitehouse.gov/news/releases/2001/09/20010920-8.html

EXPLANATION President Bush's address is in direct response to the September 11, 2001, terrorist attacks on the United States. No one would dispute that the shocking, horrific loss that resulted from that attack demanded a strong response. However, Bush's use of the either-or fallacy in sentence 7 is undeniable. A reader may choose to assume that Bush is speaking only of terrorists and those who actively support their acts of terror. However, his use of this either-or statement allows his critics to infer that he equates those who may disagree with America's response to the situation with terrorists. In addition, his statement leaves no room for a country to take a neutral stance. This issue is difficult and emotional, which places an even greater responsibility on writers, speakers, listeners, and readers to analyze such statements and ideas for logical soundness.

False Comparison

False comparison assumes that two things are similar when they are not. This fallacy is also known as a *false analogy*. An analogy is a point-by-point comparison that is used to explain an unfamiliar concept by comparing it to a more familiar one.

For example, some people have compared the human heart to a plumbing system. The heart is thought of as a pump that sends blood throughout the body to vital organs, and the arteries and veins are compared to pipes that carry this nutrient-enriched fluid to its destination and take away waste material. However, the analogy breaks down when one considers all the differences between a mechanical plumbing system and the human anatomy. First, the heart is not totally responsible for transporting the blood throughout the body. For instance, arteries dilate and contract and move in a wavelike manner to regulate and move blood through the body. Also, the relationship between the pressure, volume, and rate of the heartbeat is very different from the workings of a mechanical pump. A false comparison occurs when the differences outweigh the similarities.

EXAMPLE Read the following paragraph. Underline the logical fallacy of *false comparison.*

> 1Here at Great Foods we are in business to win. 2We want to win customers and profits. 3We do what it takes to make our products the best and to satisfy our customers. 4We only hire the most talented and competitive workers, and we give them rigorous training in sales and service. 5Just like a football team, we at Great Foods are a committed team, playing hard to beat the competition and build a loyal following.

EXPLANATION This paragraph draws a false comparison between a business and a football team. While a business may compete against other businesses for customers, the analogy breaks down upon closer examination. For example, in football, one team always wins and the other clearly loses; in contrast, two similar businesses can thrive in close proximity to each other. Furthermore, for a business to make a profit, the customer may lose a bargain. In football, each player is assigned a distinct role; oftentimes in a business, an employee may fulfill a variety of roles. In football, the rules of the game are fixed and known to all who play. However, in business, the "rules" may vary based on who is in charge and the type of business, and customers may not always understand the rules ("the fine print") until after a purchase is made. This false comparison relies on superficial similarities and oversimplifies the complexities of running a successful business.

False Cause

False cause, also known as **Post Hoc,** assumes that because events occurred around the same time, they have a cause-and-effect relationship. For example, a black cat crossing your path, walking under a ladder, and breaking a mirror are said to cause bad luck. What are the other possible causes? The Post Hoc fallacy is the false assumption that because event B *follows* event A, event B *was caused by* event A. A master reader does not assume a cause without thinking about other possible causes.

EXAMPLE Read the following paragraph. Underline the logical fallacy of *false cause.*

> [1]Leanne, a single mother of two children, decided to continue her education and become a registered nurse. [2]The long hours of study required a drastic change in her lifestyle. [3]Before enrolling in college, her lifestyle had been moderately active, and, as a conscientious mother, she had prepared well-balanced, healthy meals. [4]However, once she began her college career, her lifestyle became much more sedentary due to the long hours of attending class and studying at home. [5]In addition, she found that she was eating on the go (although she usually ordered healthful meals such as salad and grilled chicken sandwiches). [6]Over the course of her four-year college career, she gained forty pounds. [7]Leanne thought she probably had a thyroid problem, since her mother had had one. [8]When she went to the doctor, she discovered she did not. [9]Her weight gain was caused by her lack of exercise and poor diet. [10]Leanne started exercising regularly and eating better. [11]Within six months, she had lost the extra weight.

EXPLANATION Leanne jumped to a false conclusion about the cause of her weight gain. She had attributed her weight gain to a medical condition, when the real cause was changes in her lifestyle.

PRACTICE 6

Identify the fallacy in each of the following items. Write **A** if the sentence states a false cause, **B** if it makes a false comparison, or **C** if it employs the either-or fallacy.

______ **1.** Gossip is just like murder; it destroys a life.

______ **2.** Every time I made spaghetti for dinner, my husband and I fought, so I don't fix spaghetti any more.

______ **3.** I failed the test because the teacher doesn't like me.

______ **4.** A father said to his son, "I will only pay for your education if you decide to become a doctor."

______ **5.** Either you give me a raise, or I am going to quit immediately.

______ **6.** Which logical fallacy does the World War II poster reprinted here use?

—© Courtesy of National Archives, photo NWDNS-188-pp. 42.

Inadequate Arguments: Propaganda Techniques

Card Stacking

Card stacking omits factual details in order to misrepresent a product, idea, or cause. Card stacking intentionally gives only part of the truth. For example, a commercial for a snack food labels the snack "low in fat," which suggests that it is healthier and lower in calories than a product that is not low in fat. However, the commercial does not mention that the snack is loaded with sugar and calories.

EXAMPLE Read the following list of details about the product Carblaster. Place a check beside the detail(s) that would be omitted by *card stacking*.

_____ **1.** Carblaster guarantees weight loss by blocking as much as 45 grams of carbohydrates from entering the body.

_____ **2.** Carblaster contains the miracle ingredient phaselous vulgaris, an extract from the northern white kidney bean.

_____ **3.** The extract hinders an enzyme in the body that breaks down carbohydrates into glucose.

_____ **4.** Side effects can include severe gastrointestinal distress, heartburn, excessive gas, and diarrhea.

EXPLANATION The detail that would be left out by a writer or speaker using the method of card-stacking is the last detail in the list: *Side effects can include severe gastrointestinal distress, heartburn, excessive gas, and diarrhea.* Consumers might choose to avoid the risk of such side effects and not buy the product.

Transfer

Transfer creates an association between a product, idea, or cause with a symbol or image that has positive or negative values. This technique carries the strong feelings we may have for one thing over to another thing.

Symbols stir strong emotions, opinions, or loyalties. For example, a cross represents the Christian faith; a flag represents a nation; and a beautiful woman or a handsome man represents acceptance, success, or sex appeal. Politicians and advertisers use symbols like these to win our support. For example, a political candidate may end a speech with a prayer or the phrase "God bless America," to suggest that God approves of the speech. Another example of transfer is the television spokesperson who wears a white lab coat and quotes studies about the health product she is advertising.

Transfers can also be negative. For example, skull and crossbones together serve as a symbol for death. Therefore, placing a skull and crossbones on a bottle transfers the dangers of death to the contents of the bottle.

EXAMPLE Read the following paragraph. Underline the irrelevant details that use *transfer.*

Senator Edith Public

[1]Senator Edith Public is running for reelection to represent California in the Senate and is the guest of honor at a lavish Hollywood-sponsored fundraising party, which is being covered by CNN. [2]She is introduced by Ron Massey, a highly popular film actor. [3]And as she speaks, she is surrounded by some of the best known, successful producers, directors, and actors in Hollywood. [4]Senator Public says, "Thank you, Ron, for the warm and enthusiastic support. [5]I know that all of you in this room are concerned about education, the environment, and national health care. [6]As you know, the polls indicate that millions of hardworking Americans across this country are also concerned about these very issues. [7]I promise to work hard to encourage funding for research for alternative sources of energy, to increase national funding for public education, and to pass a Patients' Bill of Rights.

EXPLANATION Senator Edith Public is hoping that the public's affection and respect for these individuals will transfer to her and motivate the fans to vote for her. Interestingly, in her speech, the senator uses the plain folks appeal in the phrase *hardworking Americans* and the bandwagon appeal in her allusion to the polls that show these issues to be popular with the public.

Glittering Generalities

Glittering generalities offer general positive statements that cannot be verified. A glittering generality is the opposite of name-calling. Often words of virtue and high ideals are used, and the details are inadequate to support the claim. For example, words like *truth, freedom, peace,* and *honor* suggest shining ideals and appeal to feelings of love, courage, and goodness.

EXAMPLE Read the following paragraph. Underline the irrelevant details that use *glittering generalities.*

People for Democracy

[1]People for Democracy is a nonprofit organization dedicated to actively preserving the rights of democracy, which are under dire attack by the

conservative fanatics who are currently in control of the government. [2]People for Democracy dedicate our time, energies, and talents to fight tirelessly and courageously for the rights and freedoms that make the United States a vibrant and diverse democracy. [3]We are pro-life, fighting for a woman's right to choose. [4]We advocate civil rights, fighting, like Lincoln, for the equal rights of all. [5]We defend constitutional liberties, ensuring separation of church and state. [6]We will not rest as long as radical demigods threaten our way of life. [7]Our fight is the fight for freedom and justice.

EXPLANATION Most of the glittering generalities used to describe this fictitious organization call to mind American virtues; even the name of the organization is chosen to communicate democratic ideals. The paragraph offers no substantive details but relies on glittering generalities, which include the following: sentence 1 uses *preserving the rights of democracy;* sentence 2 uses *fight tirelessly and courageously, rights and freedoms,* and *vibrant and diverse democracy;* sentence 4 uses *equal rights of all;* sentence 6 uses *our way of life;* and sentence 7 states *Our fight is the fight for freedom and justice.* Sentence 1 and 6 also use name-calling with the labels *conservative fanatics* and *radical demigods,* and sentence 4 uses the technique of transfer by mentioning President Lincoln, a revered martyr for all these values.

PRACTICE 7

Label each of the following items according to the propaganda techniques they employ:

A. transfer
B. glittering generality
C. card stacking

_____ **1.** A vote for Joan Willis is a vote for honesty, dedication, and fairness.

_____ **2.** Hayden Panettiere, a star of the popular television series *Heroes,* poses in a "Got Milk" advertisement with a white milk mustache and says, "You don't have to be a hero to feel invincible. That's why I drink milk."

_____ **3.** Dewdrop deodorant will increase your confidence and poise with its light and refreshing scent.

_____ **4.** An individual dressed in a white medical coat holds a box of pain reliever and says, "Need immediate relief? Take Pain-away. Guaranteed to relieve headache pain in minutes."

_____ **5.** Use Curl Right to turn frizzy hair into luscious, long-lasting curls. Curl Right's alcohol-free and vitamin-enriched formula restores your hair's natural moisture and vitality. Curl Right gives that hard-to-handle hair lasting volume and styling control.

Reading a Textbook: Examining Biased Arguments

Textbook Skills

Textbooks strive to present information in a factual, objective manner with relevant and adequate support, in keeping with their purpose to inform. However, textbook authors may choose to present biased arguments for your examination. As a master reader, you are expected to evaluate the nature of the biased argument and the author's purpose for including the biased argument.

EXAMPLE The following passage appears in a college mass communications textbook. As you read the passage, underline biased information. After you read, answer the questions.

Trash TV

[1]Television programs fell still lower in quality during the late 1980s and the 1990s with the popularity of "trash TV." [2]Network programs and syndication shows alike tried to shock audiences with bloody reconstructions of lurid crimes, emphasis on sexual deviations, and confrontations on racial, sexual, and family problems.

[3]Many of these programs were presented in news and feature formats that emphasized such "news" as exposés of sexual misconduct by doctors, detailed depictions of grisly murders, and friendly interviews with prostitutes. [4]News directors expressed anger, contending that the format debased their own legitimate work.

[5]The prime example of sensationalized TV was Geraldo Rivera, who hit perhaps his lowest level with a show on Satanism on NBC in which dismembered corpses and ritualistic child abuse were displayed.

[6]To the dismay of many television officials and critics, this program was the highest-rated two-hour documentary ever shown on network television. [7]Indeed, putting this show in the same category with such true documentary classics as CBS's "The Selling of the Pentagon" was a gross misnomer.

[8]In the 1990s, some signs emerged that the orgy of sleaze was diminishing as advertisers began to shun the more extreme programs under pressure from viewers' groups and Congressional inquiries. [9]Sleaze persisted, however, in such shows as "Cops" and its spin-off series including "Top Cops" and "Real-Life Cops" as well as "Studs," "The Love Connection," "Married . . . with Children," "X-Files," and "Tales from the Crypt."

[10]Violence predominated in many other shows not considered among the sleaze genre. [11]The National Coalition on Television Violence attacked two extremely violent series, "Freddy's Nightmare" and "Friday the 13th." [12]The former featured intense torture, rape, the meathouse slaughter of women, cannibalism, and the killing of parents. [13]Equally violent themes, including the serial killing of prostitutes, slow-motion murders, and satanic human sacrifice, characterized the "Friday the 13th" series. [14]Late-evening horror movies replaced the two series in the mid-1990s.

—Agee, Ault, & Emery. *Introduction to Mass Communications*, 12th ed., pp. 83–84.

_____ **1.** Overall, the tone of the passage is
- a. positive about "trash TV."
- b. negative about "trash TV."
- c. neutral toward "trash TV."

_____ **2.** The primary purpose of the passage is
- a. to encourage readers to condemn "trash TV."
- b. to inform the reader about the nature of "trash TV."
- c. to persuade readers to boycott "trash TV."

_____ **3.** In sentence 3, the word "*news*" illustrates the propaganda technique
- a. glittering generalities.
- b. testimonial.
- c. bandwagon.
- d. false cause.

_____ **4.** The words *deviations, sleaze,* and *orgy* are examples of
- a. begging the question.
- b. personal attacks.
- c. transfer.
- d. straw man.

_____ **5.** Sentence 6 illustrates the effect of the propaganda technique
- a. card stacking.
- b. bandwagon.
- c. plain folks.
- d. testimonial.

EXPLANATION The biased information includes the following words: *Trash, shock, bloody, lurid, deviations, confrontations, misconduct, grisly murders, anger, debased, legitimate, sensationalized, lowest level, Satanism, abuse, dismay, gross misnomer, orgy of sleaze, shun, extreme, sleaze, extremely violent, torture, rape, meathouse slaughter, cannibalism, killing, murders, satanic, sacrifice, horror.*

1. This list of biased words indicates the negative tone (b) toward "trash TV."
2. The primary purpose of the passage is (b) to inform the reader about the nature of "trash TV."

3. In sentence 3, the word "*news*" illustrates (a) glittering generalities. News is thought of as being objective and factual. Calling the content described as trash TV "news" gives a positive or glittering spin to this kind of programming, which is news only in the most general sense of the word.

4. The words *deviations*, *sleaze*, and *orgy* are examples of the use of transfer. These traits transfer a negative value to "trash TV."

5. Sentence 6 illustrates the effect of the (b) bandwagon technique. High ratings means millions of people jumped on the bandwagon and made trash TV popular.

PRACTICE 8

The following passage appears in a college history textbook. As you read the passage, underline biased words. After you read, answer the questions.

Textbook Skills

Demon Rum

[1]Reformers must interfere with the affairs of others. [2]Thus there is often something of the busy-body and arrogant meddler about them. [3]How they are regarded usually turns on the observer's own attitude toward their objectives. [4]What is to some an unjustified infringement on a person's private affairs is to others a necessary intervention for that person's own good and for the good of society. [5]Consider the temperance movement, the most widely supported and successful reform of the Age of Reform. [6]Americans in the 1820s consumed huge amounts of alcohol, more than ever before or since. [7]Not that the colonists had been teetotalers. [8]Liquor, mostly in the form of rum or hard apple cider, was cheap and everywhere available. [9]Taverns were an integral part of colonial society. [10]There were alcoholics in colonial America, but because neither political nor religious leaders considered drinking dangerous, there was no alcohol "problem." [11]Most doctors recommended the regular consumption of alcohol as healthy. [12]John Adams, certainly the soul of propriety, drank a tankard of hard cider every day for breakfast. [13]Dr. Benjamin Rush's *Inquiry into the Effects of Ardent Spirits* (1784), which questioned the medicinal benefits of alcohol, fell on deaf ears.

[14]However, alcohol consumption increased markedly in the early years of the new republic, thanks primarily to the availability of cheap corn and rye whiskey distilled in the new states of Kentucky and Tennessee. [15]In the 1820s the per capita consumption of hard liquor reached 5 gallons, well over twice what it is today. [16]Since small children and many grown

people did not drink that much, others obviously drank a great deal more. [17]Many women drank, if mostly at home. [18]Reports of carousing among 14-year-old college freshmen show that youngsters did too. [19]But the bulk of the heavy drinking occurred when men got together, at taverns or grogshops and at work. [20]Many prominent politicians, including Clay and Webster, were heavy consumers. [21]Webster is said to have kept several thousand bottles of wine, whiskey, and other alcoholic beverages in his cellar.

[22]Artisans and common laborers regarded their twice-daily "dram" of whiskey as part of their wages. [23]In workshops, masters were expected to halt production periodically to drink with their apprentices and journeymen. [24]Trips to the neighborhood grogshop also figured into the workaday routine. [25]In 1829 Secretary of War John Eaton estimated that three-quarters of the nation's laborers drank at least 4 ounces of distilled spirits a day.

[26]The foundation of the American Temperance Union in 1826 signaled the start of a national crusade against drunkenness. [27]Employing lectures, pamphlets, rallies, essay contests, and other techniques, the union set out to persuade people to "sign the pledge" not to drink liquor. [28]Primitive sociological studies of the effects of drunkenness added to the effectiveness of the campaign. [29]Reformers were able to show a high statistical correlation between alcohol consumption and crime.

[30]In 1840 an organization of reformed drunkards, the Washingtonians, set out to reclaim alcoholics. [31]One of the most effective Washingtonians was John B. Gough, rescued by the organization after seven years in the gutter. [32]"Crawl from the slimy ooze, ye drowned drunkards," Gough would shout, "and with suffocation's blue and livid lips speak out against the drink!"

[33]Revivalist ministers like Charles Grandison Finney argued that alcohol was one of the great barriers to conversion, which helps explain why Utica, a town of fewer than 13,000 residents in 1840, supported four separate temperance societies in that year. [34]Employers all over the country also signed on, declaring their businesses henceforward to be "cold-water" enterprises. [35]Soon the temperance movement claimed a million members.

—From *The American Nation: A History of the United States to 1877, Volume I,* 10th ed., pp. 287–288 by John A. Garraty and Mark A. Carnes.

_____ **1.** Overall the tone of the passage
   a. is positive about alcohol consumption (drinking).
   b. is negative about alcohol consumption (drinking).
   c. remains neutral toward alcohol consumption (drinking).

_____ **2.** The author's purpose is
a. to argue against alcohol consumption.
b. to inform the reader about the reason for the temperance movement.
c. to amuse the reader by sharing the history of alcohol consumption in the United States.

_____ **3.** Sentence 2 is an example of the propaganda technique
a. bandwagon. c. transfer.
b. testimonials. d. name-calling.

_____ **4.** Sentences 34–35 are an example of
a. bandwagon. c. transfer.
b. testimonials. d. name-calling.

APPLICATIONS

Application 1

Read the following mock advertisement. Label each sentence using one of the following letters (some answers may be used more than once):

a. unbiased statement e. transfer
b. bandwagon f. testimonial
c. plain folks g. glittering generality
d. false cause h. false comparison

Be Free with NicoFree

[1]Are you one of the 46 million smokers struggling to kick the habit? [2]Are you tired of seeing your hard-earned dollars and good health go up in smoke? [3]Take heart—NicoFree chewing gum guarantees instant relief from nicotine addiction. [4]Just like former Dallas Cowboys' quarterback Troy Aikman, you too can give up smoking for good. [5]NicoFree allows nicotine to be absorbed through the mucus membranes in the mouth when the gum is chewed. [6]First, chew one piece of NicoFree until you feel its refreshing, tingling sensation. [7]Then place the piece between your cheek and your gums or under your tongue. [8]NicoFree turns the switch to your addiction off. [9]NicoFree unplugs your cravings. [10]Rap star Under Dawg says, "Like Me, Be Free with NicoFree!"

_____ **1.** _____ **6.**

_____ **2.** _____ **7.**

_____ **3.** _____ **8.**

_____ **4.** _____ **9.**

_____ **5.** _____ **10.**

Application 2

Study the following photograph used in an anti-litter campaign.

_____ **1.** Which of the following propaganda techniques is used in the advertisement?

a. testimonial
b. plain folks
c. transfer

2. Write a caption for the photograph that uses a propaganda technique.

____

REVIEW TEST 1

Score (number correct) _____ × 20 = _____%

Biased Arguments

Read the following sets of ideas. Write **UB** if the statement is unbiased, or **B** if the idea is a biased argument.

_____ **1.** Many experts believe that human activities are causing temperatures around the world to increase.

_____ **2.** Rising temperatures, without doubt, will lead to rising sea levels, catastrophic natural disasters, agricultural loss, and widespread disease.

_____ **3.** The Kyoto Protocol, established by the United Nations, is an international agreement that aims to reduce carbon dioxide emissions and the presence of greenhouse gases caused by human activities.

_____ **4.** The effort to stop global warming by liberals like Al Gore and other rich celebrities are misguided and will damage the global economy.

_____ **5.** Action must be taken to halt global warming.

REVIEW TEST 2

Score (number correct) _____ × 5 = _____%

Biased Arguments: Fallacies in Logical Thought and Propaganda

A. Write the letter of the fallacy next to its definition.

a. begging the question	d. false cause
b. personal attack	e. false comparison
c. straw man	f. either-or

_____ **1.** In this fallacy, the original argument is replaced with a weaker version that is easier to challenge than the original argument.

_____ **2.** This fallacy assumes two things are similar when they are not.

_____ **3.** This fallacy assumes that because events occurred around or near the same time, they have a cause-and-effect relationship.

_____ **4.** This fallacy assumes that only two sides of an issue exist.

_____ **5.** This fallacy restates the point of an argument as the support and conclusion.

_____ **6.** This fallacy uses abusive remarks in place of evidence for a point or argument.

B. Write the letter of the propaganda technique next to its definition.

a. plain folks
b. bandwagon
c. testimonial
d. transfer
e. name-calling
f. glittering generality
g. card stacking

_____ **7.** This technique uses irrelevant personal opinions to support a product, idea, or cause.

_____ **8.** This technique uses or suggests the irrelevant detail that "everyone is doing it."

_____ **9.** This technique omits factual details in order to misrepresent a product, idea, or cause.

_____ **10.** This technique uses irrelevant details to build trust based on commonly shared values.

_____ **11.** This technique creates an association between a product, idea, or cause with a symbol or image that has positive or negative values.

_____ **12.** This technique uses negative labels for a product, idea, or cause.

_____ **13.** This technique offers general positive statements that cannot be verified.

C. Write the letter of the fallacy used in each of the following items.

_____ **14.** If you don't like the rules of the house, get out and live on your own.

a. begging the question
b. either-or
c. personal attack
d. straw man

_____ **15.** The teacher doesn't care if you pass or fail—he still gets paid.

a. false cause
b. either-or
c. begging the question
d. personal attack

_____ **16.** Studying for final exams is like running in a marathon.

a. false comparison
b. straw man
c. begging the question
d. false cause

_____ **17.** Every time you get into serious trouble, Sam and Maurice are nearby. You are not allowed to see either one of them again.

a. straw man
b. either-or
c. false cause
d. personal attack

D. Write the letter of the propaganda technique used in each of the following items.

_____ **18.** A vote for Social Security Reform is a vote for peace of mind.

a. bandwagon
b. glittering generalities
c. plain folks
d. transfer

_____ **19.** A decorated war hero and proven military leader, General Smithenhouser says, "I fought shoulder to shoulder with Senator Treat in the trenches of Viet Nam and in the mountains of Afghanistan. I tell you, you will find no greater man for President of the United States. I pledge my full support to his candidacy."

a. bandwagon
b. glittering generalities
c. plain folks
d. transfer

_____ **20.** If you long for fluffy, light biscuits just like the ones fresh from your grandmother's oven, then buy Country Biscuits.

a. bandwagon
b. glittering generalities
c. plain folks
d. transfer

REVIEW TEST 3

Score (number correct) _____ × 10 = _____%

Advanced Argument

Before Reading: Survey the following editorial from *Washington Post.* Skim the passage, noting the words in **bold** print and answer the Before Reading questions. Then, answer the After Reading questions and respond to the writing prompt.

Vocabulary Preview

statutory rape (9): sexual relations with a minor, someone who has not reached the legal age of consent

consensual (14): involving the agreement of all

assaultive (17): extremely aggressive or disposed to attack

prevailing (20): current, usual, popular

sentiment (20): feeling, response, opinion

Should We Be Criminalizing Bullies?

by Ruth Marcus

[1]My heart aches for the parents of Phoebe Prince, the 15-year-old Massachusetts high school student who committed suicide in January after being relentlessly bullied at school and online.

[2]My heart aches for her younger sister, who found Phoebe hanging in the stairwell of the family's home. [3]A scarf the sister had bought her as a Christmas gift was knotted around Phoebe's neck.

[4]My heart aches for Phoebe, who arrived from Ireland last fall only to endure months of abuse from classmates at South Hadley High School, the apparent result of Phoebe's brief fling with a popular football player.

[5]My heart aches, but I also question the wisdom of filing criminal charges against nine of Phoebe's former classmates, as happened last week. [6]Bullying should be taken seriously—by teachers, administrators, parents and, yes, fellow students. [7]I'm doubtful, though, that criminal prosecution is the best way to punish or prevent it.

[8]Nine students were charged, including three girls not named because they are juveniles. [9]Two boys, 17 and 18, were accused of **statutory rape**; the age of consent in Massachusetts is 16.

[10]One of the juveniles is charged with "assault by means of a dangerous weapon, to wit: a bottle, can or similar beverage container"—apparently throwing a soda can at Phoebe as she walked home from school the day she died. [11]The other charges include stalking, harassment, violation of civil rights and, my favorite, disturbance of a school assembly.

[12]If this sounds derisive, it's not because I doubt the seriousness of the conduct but because the specific counts underscore how clumsy a tool criminal law is to deal with such behavior. [13]Charging nine students is casting an awfully wide net.

[14]The statutory rape charges are especially troubling, assuming the sex was **consensual**. [15]Teenage boys engage in this conduct with teenage girls every day without being prosecuted. [16]That activity, however unwise, does not suddenly acquire criminal overtones because the girl involved killed herself.

[17]In announcing the charges, District Attorney Elizabeth Scheibel described "a nearly three-month campaign of verbally abusive, **assaultive**

behavior and threats of physical harm . . . relentless activity directed toward Phoebe, designed to humiliate her and to make it impossible for her to remain at school." [18]The bullying, Scheibel said, "far exceeded the limits of normal teenage relationship-related quarrels."

[19]How does she know—and do we routinely want prosecutors making these calls? [20]*Slate's* Emily Bazelon reported that among South Hadley students, "the **prevailing sentiment** was that, yes, Phoebe had been mistreated but not in some **unprecedented** way. [21]'A lot of it was normal girl drama,' one girl told me. [22]'If you want to label it bullying, then I've bullied girls and girls have bullied me. . . . [23]It was one of the worst things I've heard of some girls doing to another girl. [24]But it wouldn't have hurt most people that much.' "

[25]The criminalization of bullying risks a slippery slope down the age range. [26]In Waltham, Mass., in January an 11-year-old was charged with two counts of assault and battery with a dangerous weapon—using her foot and a locker door—and one count of assault with a dangerous weapon using scissors.

[27]The kids who bullied Phoebe Prince should be punished—suspended, expelled, required to attend counseling. [28]Still, to be a teenager is to do stupid things. [29]The teenage brain is a work in progress. [30]The prefrontal cortex, the part linked to impulse control, judgment and decision-making, is still maturing. [31]This is why all teenagers need adult supervision, from parents and teachers.

[32]And it is why not enough responsibility has been placed on those whose brains were fully developed: the school staff who apparently knew of the harassment and did not do enough to stop it. [33]As Scheibel reported, "The investigation has revealed that certain faculty, staff and administrators of the high school also were alerted to the harassment of Phoebe Prince before her death."

[34]The school says it did what it could when it knew. [35]To its credit, it had brought in an expert on bullying even before Phoebe's problems came to light. [36]Still, the consultant told *USA Today* that when she returned to the school after Phoebe's death, "I was told there was no visible sign these kids had faced consequences for what they'd done."

[37]As a legal matter, this is not a crime. [38]In a broader sense, it is nothing short of criminal.

—"Should We Be Criminalizing Bullies?" *The Washington Post,*

Before Reading

Vocabulary

_____ **1.** What is the best meaning of the word **unprecedented** in sentence 20?
a. normal
b. predictable
c. extraordinary
d. brutal

Tone and Purpose

_____ **2.** The author's tone and purpose is
a. to entertain readers with shocking details about bullying.
b. to inform readers with objective details about the consequences of bullying.
c. to persuade readers with facts and opinions against prosecuting bullies.

After Reading

Central Idea

_____ **3.** Which sentence best states the author's central idea?
a. sentence 1
b. sentence 7
c. sentence 6
d. sentence 38

Supporting Details

_____ **4.** According to the passage, how long had Phoebe Prince been bullied?
a. a few weeks
b. 15 months
c. 11 months
d. 3 months

Thought Patterns

_____ **5.** The thought pattern of paragraph 12 (sentences 27–31) is
a. cause and effect.
b. contrast.
c. time order.
d. classification.

Fact and Opinion

_____ **6.** Sentences 37 and 38 are statements of
a. fact. b. opinion. c. fact and opinion.

Inferences

______ **7.** Based on the details of the passage, which of the following is a valid inference?
a. Phoebe was not bullied, but only experienced normal teenage relationship-related quarrels.
b. The bullies should not face serious consequences because a teenager does stupid things.
c. Bullying is unavoidable and cannot be stopped by parents or educators.
d. School administrators, faculty, and staff have a duty to report and stop bullying at school.

Argument

______ **8.** Bullying is a form of
a. personal attack.
b. testimonial.
c. straw man.
d. false comparison.

______ **9.** In sentence 28, the author uses the fallacy of
a. false comparison.
b. personal attack.
c. bandwagon.
d. begging the question.

______ **10.** In sentences 37 and 38, the author uses the fallacy of
a. false cause.
b. personal attack.
c. bandwagon.
d. begging the question.

WHAT DO YOU THINK?

Have you or anyone you know been involved in bullying? Do you think bullying is a widespread problem or are the cases mentioned in the article isolated events? Do you agree with the author's claim that bullying is not a crime? Why or why not? Assume you are taking a college course in sociology, and your professor has assigned weekly writing assignments on current events. Write an essay in which you agree with or disagree with the claim that teenagers should not be legally prosecuted for bullying other teenagers. In your essay, discuss the social climate in which bullying occurs. Discuss both the causes and effects of bullying.

REVIEW TEST 4

Score (number correct) ______ × 10 = ______ %

Advanced Argument

Textbook Skills

Before Reading: Survey the following passage adapted from a college sociology textbook. Skim the passage, noting the words in **bold** print. Answer the Before Reading questions that follow the passage. Then, read the passage. Next, answer the After Reading questions. Use the discussion and writing topics as activities to do after reading.

Vocabulary Preview

rampant (3): uncontrolled

nomadic (5): wandering

ethnocentric (7): belief in the superiority of one's own ethnic group

immunity (15): resistance

accommodated (21): adjusted to

intervene (29): interfere

pacification (31): soothing, subduing

Native Americans

[1]"I don't go so far as to think that the only good Indians are dead Indians, but I believe nine out of ten are—and I shouldn't inquire too closely in the case of the tenth. [2]The most vicious cowboy has more moral principle than the average Indian."

—Teddy Roosevelt, 1886 President of the United States, 1901–1909

[3]This quote from Teddy Roosevelt provides insight into the **rampant** racism of earlier generations. [4]Yet, even today, thanks to countless grade B Westerns, some Americans view the original inhabitants of what became the United States as wild, uncivilized savages, a single group of people subdivided into separate tribes. [5]The European immigrants to the colonies, however, encountered diverse groups of people with a variety of cultures—from **nomadic** hunters and gatherers to people who lived in wooden houses in settled agricultural communities. [6]Altogether, they spoke over 700 languages. [7]Each group had its own norms and values—and the usual **ethnocentric** pride in its own culture. [8]Consider what happened in 1744, when the colonists of Virginia offered college scholarships for "savage" lads. [9]The Iroquois replied:

> [10]"Several of our young people were formerly brought up at the colleges of Northern Provinces. [11]They were instructed in all your sciences. [12]But when they came back to us, they were bad runners, ignorant of every means of living in the woods, unable to bear either cold or hunger, knew neither how to build a cabin, take a deer, or kill an enemy . . . [13]They were totally good for nothing."
>
> [14]They added, "If the English gentlemen would send a dozen or two of their children to Onondaga, the great Council would take

care of their education, bring them up in really what was the best manner and make men of them." (Nash 1974; in McLemore 1994)

[15]Native Americans, who numbered about 10 million, had no **immunity** to the diseases the Europeans brought with them. [16]With deaths due to disease—and warfare, a much lesser cause—their number was reduced to about one-twentieth its original size. [17]A hundred years ago, the Native American population reached a low point of a half million. [18]Native Americans, who now number about 2 million, speak 150 different languages. [19]Like Latinos and Asian Americans, they do not think of themselves as a single people who fit neatly within a single label.

[20]At first, relations between the European settlers and the Native Americans were by and large peaceful. [21]The Native Americans **accommodated** the strangers, as there was plenty of land for both the newcomers and themselves. [22]As Native Americans were pushed aside and wave after wave of settlers continued to arrive, however, Pontiac, an Ottawa chief, saw the future—and didn't like it. [23]He convinced several tribes to unite in an effort to push the Europeans into the sea. [24]He almost succeeded, but failed when the English were reinforced by fresh troops.

[25]A pattern of deception developed. [26]The U.S. government would make treaties to buy some of a tribe's land, with the promise to honor forever the tribe's right to what it had not sold. [27]European immigrants, who continued to pour into the United States, would disregard these boundaries. [28]The tribes would resist, with death tolls on both sides. [29]The U.S. government would then **intervene**—not to enforce the treaty but to force the tribe off its lands. [30]In its relentless drive westward, the U.S. government embarked on a policy of **genocide**. [31]It assigned the U.S. cavalry the task of "**pacification**," which translated into slaughtering Native Americans who "stood in the way" of this territorial expansion.

—From *Essentials of Sociology: A Down-to-Earth Approach*, 7th ed., pp. 247–248. by James M. Henslin.

Before Reading

Vocabulary in Context

Use context clues to state in your own words the definition of the following term. Indicate the context clue you used.

1. In sentence 30, what does the word **genocide** mean? ______________

____

2. Identify the context clue used for the word **genocide** in sentence 30.

____

Tone and Purpose

_____ 3. What is the overall tone and purpose of the passage?
  a. to inform the reader about the historical relationship between Native Americans, European settlers, and the U.S. government
  b. to persuade the reader that Native Americans have endured a pattern of racism that continues today
  c. to entertain the reader with surprising details about Native American history

After Reading

Central Idea

4. Use your own words to state the central idea of the passage: ______________

____

____

Main Idea and Supporting Details

_____ 5. Sentence 15 states a
  a. main idea.
  b. major supporting detail.
  c. minor supporting detail.

Transitions

_____ 6. The relationship between sentences 4 and 5 is
  a. cause and effect.
  b. addition.
  c. time order.
  d. contrast.

Thought Patterns

_____ 7. The overall thought pattern of the passage is
  a. cause and effect.
  b. listing.
  c. time order.
  d. contrast.

Fact and Opinion

_____ **8.** Sentences 1 and 2 are statements of
a. fact.
b. opinion.
c. fact and opinion.

Inferences

_____ **9.** Based on the information in sentence 31, the U.S. government's use of "pacification" implies that
a. the U.S. government tried to coexist peacefully with Native Americans.
b. the U.S. government used a glittering generality as a propaganda technique to describe its violent actions against Native Americans.
c. the Native Americans initiated violence and caused the actions taken against them by the U.S. government.

Argument

_____ **10.** Sentences 1 and 2 are examples of the logical fallacy
a. begging the question.
b. straw man.
c. false comparison.
d. personal attack.

WHAT DO YOU THINK?

How did the values of the Iroquois differ from those of the colonists? Based on this passage, which group seems more "savage" and why? How would you define ethnocentric pride? What examples of ethnocentric pride have you observed or learned about? Assume your community is going to celebrate the various ethnic groups that live in your area during the traditional Fourth of July celebration. You have been asked to represent your ethnic group by giving a short speech that explains your heritage. Write a draft of your speech that is at least three paragraphs long. In your speech, define your ethnic group, the core values of your culture, and a key tradition that reflects those values.

After Reading About Advanced Argument

Before you move on to the Mastery Tests on advanced argument, take time to reflect on your learning and performance by answering the following questions. Write your answers in your notebook.

- How has my knowledge base or prior knowledge about advanced argument, persuasive techniques changed?
- Based on my studies, how do I think I will perform on the Mastery Test(s)? Why do I think my scores will be above average, average, or below average?
- Would I recommend this chapter to other students who want to learn more about advanced argument: persuasive techniques? Why or why not?

Test your understanding of what you have learned about advanced argument: persuasive techniques by completing the Chapter 13 Review Card in the insert near the end of your text.

CONNECT TO PEARSON myreadinglab

To check your progress in meeting Chapter 13's learning outcomes, log in to **www.myreadinglab.com** and try the following exercises.

- The "Critical Thinking" section of MyReadingLab gives additional information about advanced argument, persuasive techniques. The section provides an overview, model, practices, and tests. To access this resource, click on the "Study Plan" tab. Then click on "Critical Thinking." Then click on the following links as needed: "Overview," "Model," "Critical Thinking: Facts and Opinions (Flash Animation)," "Practice," and "Test."
- The "Study Skills Website" section of MyReadingLab also gives additional information about advanced argument, persuasive techniques. To access these resources, go to the "Other Sources" box on the home page of MyReadingLab. Click on "Study Skills Website." Then scroll down the page to the heading "Life Skills" and click on "Critical Thinking." Explore each of the links on the bar on the left side of the page.
- To measure your mastery of this chapter, complete the tests in the "Critical Thinking" section and click on Gradebook to find your results.

1 Chapter Review

Summary of Key Concepts of a Reading System for Master Readers

LEARNING OUTCOME 1 2 3 5 8

Assess your comprehension of prior knowledge and the reading process.

- Comprehension is ____________________.
- Prior knowledge is ____________________
  ____________________.
- Use prior knowledge to ____________________.
  - Activate __________ by asking ____________________
    ____
  - Check __________ against prior knowledge by asking, __________
    ____
  - Check for ____________________ by asking,
    ____
- The reading process has three phases: ____________________
  ____
- SQ3R, an acronym for a reading process, stands for __________
  __________. SQ3R activates prior knowledge and offers strategies for each phase of the reading process:
  - Before Reading, ______: Skim ____________________
    __________. ______ : Ask ____________________
    ____
  - During Reading: __________. __________ key words and ideas. Repair confusion. Reread.

- After Reading: ______________: Answer questions such as ________
  ____
  ______________: Recall ________________________. Summarize.
- Textbook features help readers understand and learn the vast amount of information in a textbook. A few examples are as follows:
  - Table of Contents: ____
  - Index: ____
  - Glossary: ____
  - Preface: ____
    ____
  - Appendices: ____
  - Typographical features: _______________________________________
    ____
  - Graphics: ____
    ____

Test Your Comprehension of A Reading System for Master Readers

Respond to the following question and prompt.

LEARNING OUTCOME 1 2 3 4 5 6 7 8

Describe your reading process. How did you read before your studied this chapter? Will you change your reading process? If so, how? If not, why not? ____
____
____
____

2 Chapter Review

Summary of Key Concepts of Vocabulary Skills

LEARNING OUTCOME 1 4 Assess your comprehension of vocabulary skills.

- **Vocabulary** is ____.
- The acronym SAGE identifies four of the most common types of context clues: The four context clues and their definitions are as follows:

Context Clue	Definition
■ ________________	____ ____
■ ________________	____ ____
■ ________________	____ ____ ____
■ ________________	____ ____ ____

- The three basic word parts and their definitions are as follows:

Word Part	Definition
■ ________________	____
■ ________________	____ ____ ____
■ ________________	____ ____

- Each subject matter has its own ____________________.
- A glossary is ____________________ ____________________, often provided as an extra section in ____________________.

Test Your Comprehension of Vocabulary Skills

Respond to the following prompts.

LEARNING OUTCOME 2 3 4 6 8

Demonstrate your use of context clues. Use the headings below and create a chart based on the four types of context clues. Then, complete the chart with new words you have come across recently.

Type of Clue	New Word	Meaning of Word	Source Sentence of Word

LEARNING OUTCOME 3 5 6 7 8

Demonstrate your ability to decode the meaning of words using word parts. Use the blank word web below to create your own web of words linked by word parts. Use a family of words you have come across recently.

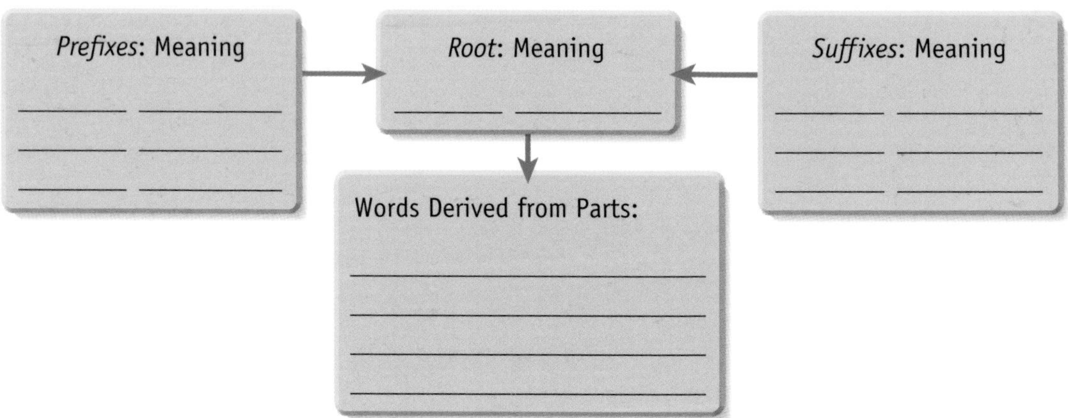

LEARNING OUTCOME 3 5 6 7 8

Create a glossary of terms for a subject you are currently studying; use the headings below to create your glossary. Then, complete the chart with new words you have come across recently.

Word	Definition	Source Sentence of Word

3 Chapter Review

Summary of Key Concepts of Stated Main Ideas

LEARNING OUTCOME 1 3 5

Assess your comprehension of stated main ideas. Complete the following two-column notes with information from the chapter.

Reading Concept	Definition
Topic Sentence	
Central Idea	
Thesis Statement	
Various Locations of the Stated Main Idea	
Deductive Thinking	
Inductive Thinking	

Test Your Comprehension of Stated Main Ideas

Respond to the following questions and prompts.

LEARNING OUTCOME 2 5

In your own words, what is the difference between deductive and inductive thinking? ____________________

____

____

____

LEARNING OUTCOME 1 3

In your own words, what is the difference between a topic sentence and a thesis statement? ____________________

____

____

____

LEARNING OUTCOME 2 4 5

Draw and label four graphs that show the possible locations of stated main ideas.

LEARNING OUTCOME 5 6

Identify and discuss the two most important ideas in this chapter that will help you improve your reading comprehension. ____________________

____

____

4 Chapter Review

Summary of Key Concepts of Implied Main Ideas

LEARNING OUTCOME 1

Assess your comprehension of implied main ideas. Complete the following two-column notes, set up as question-answers, with information from the chapter.

Implied Main Ideas	
What is an implied main idea?	
What three questions can help determine an implied main idea?	
What should be annotated in a passage to help determine a main idea?	
What is the implied central idea?	

Test Your Comprehension of Implied Main Ideas

Respond to the following questions and prompts.

LEARNING OUTCOME 2 3

How can the skills you use to identify the stated main idea help you determine the implied main idea? ______

LEARNING OUTCOME 7

Describe how you will use what you have learned about implied main ideas in your reading process to comprehend textbook material.

LEARNING OUTCOME 3 4 6

Study the following concept map. Then, write the implied main idea suggested by the details in the map.

Gardening

→ **Increases Property Value**

→ **Relieves Stress**

→ **Provides Useful Exercise**

Implied main idea: ______

LEARNING OUTCOME 5 7

Identify and discuss the two most important ideas in this chapter that will help you improve your reading comprehension. ______

5 Chapter Review

Summary of Key Concepts of Supporting Details

LEARNING OUTCOME 1 3

Assess your comprehension of supporting details. Fill in the blanks with information from the chapter.

- To locate supporting details in a passage, a master reader turns the ________________ into a ________________________________.
- A major supporting detail ________________________________ ____.
- A minor supporting detail ________________________________ ____.
- A ________________ is a ________________________________ ____.
- Often you will want to ________________, or restate, ideas in your own words.
- ________________ or marking your text ________________ reading will help you create a ________________________ after you read.
- To create a summary for a passage with a stated main idea, ____________ ____.
- To create a summary for a passage with an implied main idea, __________ ____.

Test Your Comprehension of Supporting Details

Respond to the following questions and prompts.

LEARNING OUTCOME 1 3

In your own words, how do major and minor supporting details differ? ____
____

LEARNING OUTCOME 2 4 5

In the space below, outline the steps for creating a summary for stated and implied main ideas. See pages 201–204.

LEARNING OUTCOME 5 6 7

Summarize the two most important ideas in this chapter that will help you improve your reading comprehension.

6 Chapter Review

Summary of Key Concepts of Outlines and Concept Maps

LEARNING OUTCOME 1 Assess your comprehension of outlines and concept maps. Complete the chart with information from the chapter.

Outlines and Concept Maps: Question-Answer Two-Column Notes	
What is the purpose of an outline?	______________________ ______________________ ______________________
Which signal words may an author use to introduce a main idea?	______________________ ______________________ ______________________ ______________________
Which signal words may an author use to introduce a supporting detail?	______________________ ______________________ ______________________ ______________________
How does a formal outline use Roman numerals and numbers?	A formal outline uses ____________ to indicate the ____________, ____________ to indicate the ____________, and ____________ to indicate the ____________.
What is a concept map?	A concept map is a ______________________ ______________________ ______________________

Test Your Comprehension of Outlines and Concept Maps

Respond to the following questions and prompts.

LEARNING OUTCOME 2 3 In the space below, create an outline and a concept map for the following terms from a college anatomy and physiology textbook.

Skeletal Muscle; Smooth Muscle; Muscle Types; Cardiac Muscle; Producing Movement; Maintaining Posture; Muscle Functions; Stabilizing Joints

____________ ________________________

____

____

____

____

____

____

____

____

____

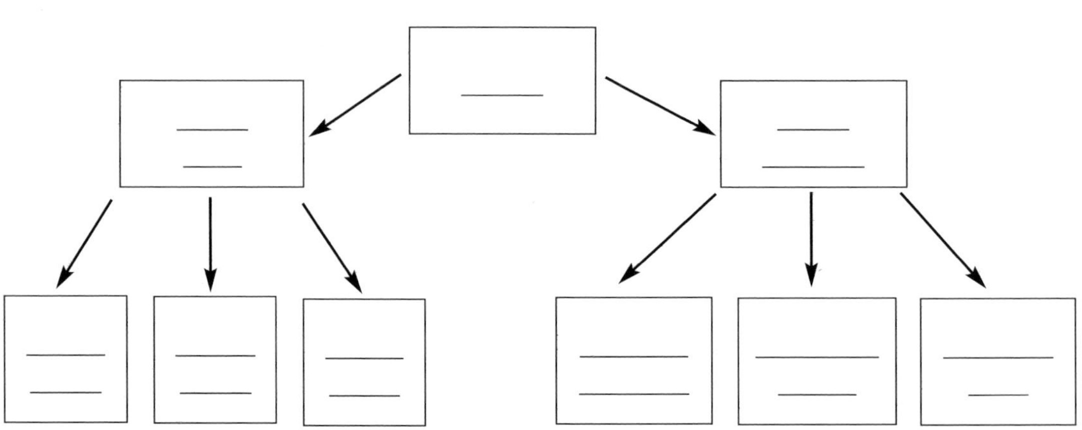

LEARNING OUTCOME 4 Summarize the two most important ideas in this chapter that will help you improve your reading comprehension. ________________

____

____

7 Chapter Review

Summary of Key Concepts of Transitions and Thought Patterns

LEARNING OUTCOME 1 4

Assess your comprehension of transitions and thought patterns. Complete the following two-column notes of terms and definitions with information from the chapter.

Term	Definition/Examples
Transitions	______________________________ ______________________________ ______________________________
Thought pattern	______________________________ ______________________________ ______________________________
Time order: Narration Time order: Process	______________________________ ______________________________ ______________________________
Listing pattern	______________________________ ______________________________
Examples of signal words for listing	______________________________
Classification pattern	______________________________ ______________________________
Examples of classification signal words	______________________________
Space order pattern	______________________________ ______________________________
Examples of words used to establish space order pattern	______________________________

Test Your Comprehension of Transitions and Thought Patterns

Respond to the following questions and prompts.

LEARNING OUTCOME 1

In your own words, what is a transition? ____________________

____

____

LEARNING OUTCOME 1

In your own words, what is a thought pattern? ____________________

____

LEARNING OUTCOME 2 3 5

In the space below, illustrate the thought pattern of the following ideas from a college human anatomy textbook by creating an outline or concept map.

Burns are classified according to their severity (depth) as first-, second-, or third-degree burns. In first-degree burns, only the epidermis is damaged. The area becomes red and swollen. First-degree burns are not usually serious and generally heal in two to three days without any special attention. Sunburn is usually a first-degree burn. Second-degree burns involve injury to the epidermis and the upper region of the dermis. The skin is red and painful, and *blisters* appear. Because sufficient numbers of epithelial cells are still present, regrowth can occur. First- and second-degree burns are referred to as partial-thickness burns. Third-degree burns destroy the entire thickness of the skin, so these burns are also called full-thickness burns. In third-degree burns, regeneration is not possible, and skin grafting must be done to cover the underlying exposed tissues.

—Adapted from Lutgens & Tarbuck. *Foundations of Earth Science*, 5th ed., pp. 4–6.

LEARNING OUTCOME 6

Summarize the two most important ideas in this chapter that will help you improve your reading comprehension. ____________________

____

____

8 Chapter Review

Summary of Key Concepts of More Thought Patterns

LEARNING OUTCOME 1 Assess your comprehension of thought patterns. Complete the following two-column notes of terms and definitions with information from the chapter.

Term	Definition/Examples
Comparison	______________________
Words and phrases of comparison	______________________
Contrast	______________________ ______________________
Words and phrases of contrast	______________________
Cause	______________________
Effect	______________________
Words and phrases of cause and effect	______________________
Generalization and example	______________________ ______________________ ______________________
Definition and example	______________________ ______________________ ______________________
Words and phrases of example	______________________

Test Your Comprehension of More Thought Patterns

Respond to the following questions and prompts.

LEARNING OUTCOME 1

In your own words, what is the difference between comparison and contrast? ______________________

LEARNING OUTCOME 1

In your own words, what is the difference between a cause and an effect? ______________________

LEARNING OUTCOME 1

In your own words, what are the similarities and differences between these two thought patterns: *generalization and example and definition and example?* ______________________

LEARNING OUTCOME 2 3 4

In the space below, create a concept map based on the information in the paragraph.

According to research, meditation can have positive emotional and physical effects. Regular meditation leads to increased control over one's emotions. Regular meditation also contributes to lower blood pressure and cholesterol levels.

LEARNING OUTCOME 5

Summarize the two most important ideas in this chapter that will help you improve your reading comprehension.

____

9 Chapter Review

Summary of Key Concepts of Fact and Opinion

LEARNING OUTCOME 1

Assess your comprehension of fact and opinion. Complete the following two-column notes of terms and definitions with information from the chapter.

Term	Definition
A fact	______________________ ______________________
An opinion	______________________ ______________________
Objective proof	______________________ ______________________
An informed opinion	______________________ ______________________
An expert opinion	______________________ ______________________
A fact	______________________
An opinion	______________________
Biased words	______________________ ______________________
A qualifier	______________________ ______________________

Test Your Comprehension of Fact and Opinion

Respond to the following questions and prompts.

LEARNING OUTCOME 1 5

In your own words, what is the difference between a fact and an opinion? ______________________________

____

LEARNING OUTCOME 2 3 4

In your own words, what is the difference between an informed opinion and an expert opinion? ______________________________

____

____

____

____

____

____

LEARNING OUTCOME 1 2 3 5

In your own words, describe how to distinguish between fact and opinion. ______________________________

____

____

____

____

____

LEARNING OUTCOME 7

Identify the following two statements as fact or opinion. Then explain the importance of knowing the difference between fact and opinion.

______ Most fast food is high in calories and fat and low in nutrition.

______ Fast food tastes great.

____

____

____

____

LEARNING OUTCOME 6

Summarize the two most important ideas in this chapter that will help you improve your reading comprehension. ______________________________

____

____

10 Chapter Review

Summary of Key Concepts of Tone and Purpose

LEARNING OUTCOME 1 3

Assess your comprehension of tone and purpose. Complete the following two-column notes of terms and definitions with information from the chapter.

Term	Definition
Tone	______________________
Objective tone words	______________________ ______________________
Subjective tone words	______________________
The author's purpose	______________________
The primary purpose	______________________ ______________________
The Purpose: To Inform	______________________
The Purpose: To Entertain	______________________
The Purpose: To Persuade	______________________ ______________________
Verbal irony	______________________ ______________________
Situational irony	______________________ ______________________

Test Your Comprehension of Tone and Purpose

Respond to the following questions and prompts.

LEARNING OUTCOME 2 4 5

In your own words, what is the relationship between tone and purpose? ____________

LEARNING OUTCOME 2

In your own words, what is the difference between verbal and situational irony? ____________

LEARNING OUTCOME 6 7

Use a checklist to help you determine the tone and purpose of passages. Select a passage to analyze; then complete the following checklist with information from the passage.

Title of Passage:			
Subjective Tone	**Yes**	**No**	**Examples (words or phrases)/Explanations**
Objective Tone			
Irony			
Verbal Irony			
Situational Irony			
Primary Purpose			
To Inform			
To Persuade			
To Entertain			

LEARNING OUTCOME 7

Summarize the two most important ideas in this chapter that will help you improve your reading comprehension. ____________

11 Chapter Review

Summary of Key Concepts about Inferences

LEARNING OUTCOME 1 3 5

Assess your comprehension of inferences. Complete the following two-column notes of terms and definitions with information from the chapter.

Term	Definition
Inference	________________
Valid inference	________________
Invalid conclusion	________________
The 5-step VALID Thinking Process	Step 1: ________ Step 2: ________ Step 3: ________ Step 4: ________ Step 5: ________

Test Your Comprehension of Inferences

Respond to the following questions and prompts.

LEARNING OUTCOME 1 2

In your own words, what is the difference between a valid and an invalid inference? ________________

____

____

LEARNING OUTCOME 4 5 6

Complete the following three-column notes about creative expression. Use your own examples.

Creative Expression: Literary Devices

Literary Device	Meaning	Example
Connotations of words	____________	____________
	____________	____________
	____________	____________
	____________	____________
Metaphor	____________	____________
	____________	____________
Personification	____________	____________
	____________	____________
	____________	____________
Simile	____________	____________
Symbol	____________	____________
	____________	____________

LEARNING OUTCOME 7

Describe how you will use what you have learned about inferences in your reading process to comprehend textbook material. __________

____

____

____

____

LEARNING OUTCOME 7

Summarize the two most important ideas in this chapter that will help you improve your reading comprehension. ______________

____

____

12 Chapter Review

Summary of Key Concepts about the Basics of Argument

LEARNING OUTCOME 1 Assess your comprehension of the basics of argument. Complete the question-answer study notes with information from the chapter.

What is an effective argument?	______________________
What two types of statements make up an argument?	(a) ______________________ (b) ______________________
What is an invalid conclusion?	______________________
What are the four steps to analyze an argument?	(1) ______________________ (2) ______________________ (3) ______________________ (4) ______________________
What is a claim made without providing adequate support for the claim?	______________________

Test Your Comprehension of The Basics of Argument

Respond to the following questions and prompts.

LEARNING OUTCOME 1 In your own words, explain the relationship between making an inference and analyzing an argument. ______________________

____

LEARNING OUTCOME 2 3 4 5 6 Create a valid argument. Assume you are going to create and distribute a brochure about distracted driving. You have found the following photograph to use in your brochure. Using details from the photo, write a claim and two supports that clearly support the claim.

Claim: ______

Support 1: ______

Support 2: ______

LEARNING OUTCOME 7 Describe how you will use what you have learned about the basics of argument in your reading process to comprehend written material in various sources such as textbooks or web sites. ______

LEARNING OUTCOME 6 Summarize the two most important ideas in this chapter that will help you improve your reading comprehension. ______

13 Chapter Review

Summary of Key Concepts about Advanced Argument: Persuasive Techniques

LEARNING OUTCOME 1 2 3 4 5

Assess your comprehension of advanced arguments and persuasive techniques. Complete the two-column notes with information from the chapter.

Fallacy	
Irrelevant details	
Inadequate details	
Propaganda	
Emotional appeal	
Personal Attack	
Straw Man	
Begging the question	
Name-calling	
Testimonials	
Bandwagon	

Plain Folks	__________
Either-or	__________
False comparison	__________
False cause	__________
Card stacking	__________
Transfer	__________
Glittering generalities	__________

Test Your Comprehension of Advanced Argument: Persuasive Techniques

Respond to the following questions and prompts.

LEARNING OUTCOME 1

In your own words, explain the relationship among the following terms: *fallacy*, *irrelevant details*, *emotional appeal*, *inadequate details*, and *propaganda*. __________

LEARNING OUTCOME 6

Summarize the two most important ideas in this chapter that will help you improve your reading comprehension. __________

MASTERY TEST 1

Name _______________ Section _______________

Date _______ **Score** (number correct) _______ × 10 = _______%

Write the letter of the fallacy used in each statement.

_____ **1.** If you cannot perform the duties listed on your job description, you'll have to find another job.

a. begging the question
b. either-or
c. personal attack
d. straw man

_____ **2.** The government doesn't care about you. You are just a student who doesn't pay taxes.

a. false cause
b. either-or
c. begging the question
d. personal attack

_____ **3.** Falling in love is like jumping off the high diving board.

a. false comparison
b. straw man
c. begging the question
d. false cause

_____ **4.** Whenever you eat ice cream, you get irritable. We're not having dessert anymore.

a. straw man
b. either-or
c. false cause
d. personal attack

_____ **5.** Most politicians are like alcoholics—drunk on power, in denial, and unwilling to stop their addiction to lying.

a. false comparison
b. begging the question
c. straw man
d. false cause

_____ **6.** *Parent to teenager:* "You can't be trusted because you have proven you are untrustworthy."

a. straw man
b. personal attack
c. begging the question
d. either-or

_____ **7.** My grandmother says that to cure a wart you should cut a potato in half, rub the wart with the cut potato, and then bury the potato at night during a full moon. The wart will go away within a week.

a. false cause
b. straw man
c. either-or
d. personal attack

_____ **8.** *Speaker 1:* I propose a solution that will address the concerns of both developers and environmentalists. In every new neighborhood, developers will set aside green areas that can be used as parks or environmental havens. Funds to offset the cost of these green areas could be raised by a nominal county tax and donations from the Green-Friendly Organization.

Speaker 2: My opponent's solution to everything is to raise taxes.

a. begging the question
b. personal attack
c. false comparison
d. straw man

_____ **9.** My Lexus is my favorite car of all those that I have ever owned because I like it so much.

a. straw man
b. personal attack
c. begging the question
d. either-or

_____ **10.** City commissioner Jerome Little is a political bully who is determined to protect the wealthy at the expense of the middle class.

a. personal attack
b. straw man
c. false cause
d. either-or

MASTERY TEST 2

Name ____________________ Section ____________

Date __________ **Score** (number correct) ______ × 20 = ______%

A. Identify the propaganda technique used in each statement. Some techniques are used more than once.

a. plain folks
b. bandwagon
c. testimonial
d. name-calling

_____ **1.** The founder of our company is just like you; she understands the pressures of everyday life, so she made sure that we offer affordable goods in the most efficient way possible to save you money and time.

_____ **2.** Don't delay! Buy a lot in this gated community before they are all gone. Lots are moving quickly! Don't miss out on this opportunity to live in one of the most sought-after neighborhoods.

_____ **3.** Of course Gerald Homey would vote for an increase in minimum wage; he is the puppet of labor unions.

B. Read each fictitious advertisement. Then, identify the detail from the list that was omitted from the ad for the purpose of card stacking.

_____ **4.** For Sale: This two-year-old female Yorkie terrier has been neutered, house-trained, and has had all the required vaccinations. She is bright, energetic, and loving. She weighs three pounds and has silver coloring.

a. The family is moving to a condominium where no pets are allowed.
b. The terrier is a thoroughbred with registered papers.
c. The terrier has a feisty personality with a history of biting.

_____ **5.** Ageless Solutions is a miracle supplement that turns back the aging process. With a secret blend of natural ingredients, including vitamin D, the recommended daily dosage helps the body rebuild bone, protecting you from osteoporosis and other bone diseases.

a. Ageless Solutions contains a complete daily dose of several necessary vitamins, including vitamin A, vitamin C, riboflavin, niacin, and folic acid.

b. Certain batches of Ageless Solutions have been recalled due to labeling errors that led to excessive dosages of vitamin D. Taking excessive amounts of this vitamin can cause weakness, fatigue, headaches, nausea, vomiting, diarrhea, mental status changes, and even coma in severe cases.

c. Ageless Solutions has been clinically tested with proven results.

MASTERY TEST 3

Name ______________________ Section ______________

Date __________ Score (number correct) ______ × 10 = ______%

Propaganda Techniques

A. Identify the propaganda technique used in each statement. A technique may be selected more than once or not at all.

a. plain folks

b. bandwagon

c. testimonial

d. transfer

e. name-calling

f. glittering generality

______ **1.** Sandra's Fruitcake is just like the one your mother made every Christmas. For the same fresh-baked, homemade taste, buy Sandra's Fruitcake.

______ **2.** A candidate says, "If you like what George Washington, Abraham Lincoln, and Martin Luther King, Jr. stood for, then you will cast your vote for me."

______ **3.** I would never recommend Professor Higgins; he is anti-American and subversive.

______ **4.** Please, Mother, I have to have a pair of low-riding jeans and a midriff shirt; it's what everybody is wearing.

______ **5.** Actress Heather McCoy says, "True Blond leaves my hair silky, shiny, and much more manageable."

______ **6.** The owner of a national fast-food chain dressed in a cotton short-sleeved shirt, sits at a table in one of his restaurants, surrounded by working-class people who are happily eating hamburgers and French fries. He says, "My wife and I started this business 30 years ago, and we are still here for you. Drop on in. We'll be glad to see you."

______ **7.** Famous race car driver Rick Ellington states, "I only use ValvoClean on the race track and on the road."

______ **8.** Supplies Limited. Only a few special edition Bag Babies are left. No more shipments expected before Christmas. Hurry before they are all gone.

_____ 9.

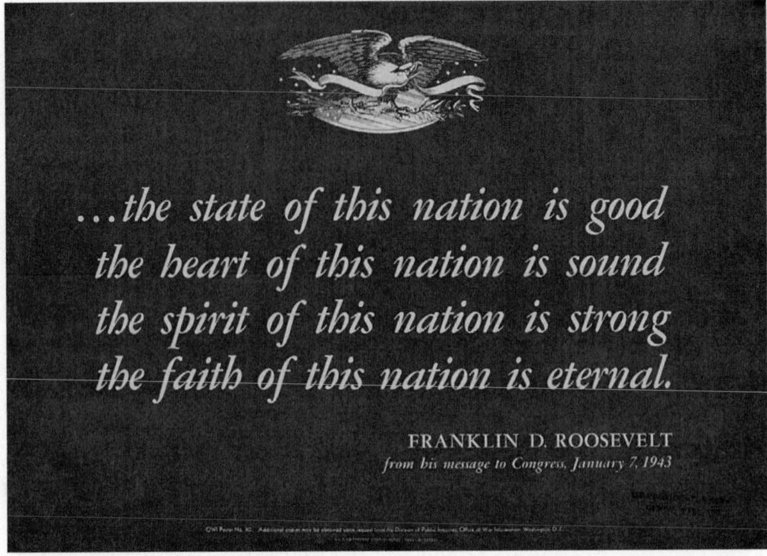

...the state of this nation is good
the heart of this nation is sound
the spirit of this nation is strong
the faith of this nation is eternal.
FRANKLIN D. ROOSEVELT
from his message to Congress, January 7, 1943

—From World War II Poster Collection, Northwestern University Library, 29 Oct. 2002. "The state of this nation is good," Franklin Roosevelt, message to Congress, 1943, U.S. Office of War Information. © Courtesy Northwestern University Library (www.library.northwestern.edu/govpub/collections/)

B. Read this fictitious advertisement. Then, identify the detail that was omitted for the purpose of card stacking.

_____ **10.** Fun Water will turn your backyard into a water park for your children and their friends. Simply hook your water hose to Fun Water and watch the fun begin. Fun Water throws a sheet of water down a brightly colored slide that gently slopes into a shallow pool. Your children will slip and slide their way to fun all summer long.

a. Fun Water is a highly popular, fast-selling product.

b. Fun Water doesn't waste water because it recycles and reuses water.

c. Several children have experienced minor injuries using Fun Water.

MASTERY TEST 4

Name ______________________ Section ______________

Date ____________ **Score** (number correct) ______ × 20 = ______%

Textbook Skills

Read the following passage from a college history textbook. Answer the questions that follow.

Moving Westward: Society and Politics in the "Age of the Common Man"

[1]Campaigning for political office in Tennessee in the 1820s was not an activity for the faint of heart. [2]Candidates competed against each other in squirrel hunts, the loser footing the bill for the barbecue that followed. [3]A round of speechmaking often was capped by several rounds of whiskey drinking enjoyed by candidates and supporters alike. [4]Into this boisterous arena stepped a man unrivaled as a campaigner. [5]David Crockett ran successfully for several offices, including local justice of the peace in 1818, state legislator in 1821 and 1823, and member of the U.S. House of Representatives in 1827, 1829, and 1833. [6]The plainspoken Crockett knew how to play to a crowd and rattle a rival. [7]He bragged about his skill as a bear hunter and ridiculed the fancy dress of his opponents.

[8]He condemned closed-door political caucuses (small groups of party insiders who hand-picked candidates) and praised grassroots democracy. [9]Crockett claimed he could out-shoot, out-drink, and out-debate anyone who opposed him. [10]If his opponent lied about him, why, then, he would lie about himself: "Yes fellow citizens, I can run faster, walk longer, leap higher, speak better, and tell more and bigger lies than my competitor, and all his friends, any day of his life." [11]Crockett's blend of political theater and folksy backwoods banter earned him the allegiance of voters like him—people who, though having little formal education, understood the challenges of carving a homestead out of the dense thickets of western Tennessee.

[12]Crockett's raucous brand of campaigning appealed to Westerners: European Americans living just west of the Appalachian Mountains. [13]His social betters might sniff that he was a rough, ignorant man—in the words of one Tennessee political insider, "more in his proper place, when hunting a Bear in the cane Brake, than he will be in the Capital." [14]But newspaper reporters and defeated opponents alike grew to respect his ability to champion ordinary farmers. [15]As a politician, Crockett spoke for debtors, squatters, and militia veterans of the Revolutionary War. [16]He scorned the well born in favor of those who could shoot down and skin a wolf.

[17]During the 1820s, European American settlers in the trans-Appalachian West transformed the style and substance of American politics. [18]Beginning with Kentucky in 1792, western states began to relax or abolish property requirements for adult male voters. [19]Even the English that Americans spoke changed. [20]New words introduced into the political vocabulary reflected the rough-hewn, woodsman quality of western electioneering: candidates hit the campaign trail, giving stump speeches along the way. [21]They supported their party's platform with its planks (positions on the issues). [22]As legislators, they voted for pork-barrel projects that would benefit their constituents at home. [23]Emphasizing his modest origins, David Crockett became widely known as Davy Crockett. [24](It is hard to imagine anyone calling the Sage of Monticello Tommy Jefferson.)

[25]These developments reflected the movement of European Americans westward. [26]In 1790, 100,000 Americans (not including Indians) lived west of the Appalachian Mountains; half a century later that number had increased to 7 million, or about four out of ten Americans.

—From *Created Equal: A Social and Political History of the United States, Combined Volume*, 2nd ed., pp. 355–356 by Jacqueline Jones, Peter H. Wood, Thomas Borstelmann, Elaine Tyler May, Vicki L. Ruiz.

_____ **1.** In sentence 4, the description of Davy Crockett uses
a. glittering generalities. c. personal attacks.
b. false comparisons. d. testimonials.

_____ **2.** Overall, Davy Crockett's political appeal was based on
a. personal attacks. c. plain folks.
b. transfer. d. card stacking.

_____ **3.** In sentence 7, Davy Crockett's boast in his skill as a bear hunter is an example of
a. false comparison. c. straw man.
b. bandwagon. d. testimonial.

_____ **4.** In sentence 13, the opponents of Davy Crockett used a _______ against him.
a. false comparison c. testimonial
b. bandwagon d. personal attack

_____ **5.** Sentence 22 is an example of the political use of
a. false comparison. c. straw man.
b. bandwagon. d. personal attack.

Additional Readings

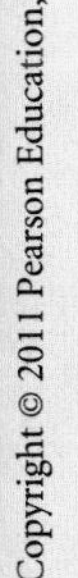

The Connection Between Reading and Writing

The link between reading and writing is vital and natural. Written language allows an exchange of ideas between a writer and a reader. Thus, writing and reading are two equal parts of the communication process. In fact, reading is a form of listening or receiving information. And writing is like speaking—the sending of information. So a master reader makes every effort to understand and respond to the ideas of the writer. Likewise, a master writer makes every effort to make ideas clear so the reader can understand and respond to those ideas. Most writers find that reading improves their writing. Reading builds prior knowledge and fuels ideas for writing.

Because of this close relationship between reading and writing, both share similar thinking steps in their processes. In Chapter 1, you learned that the reading process has three phases: Before Reading, During Reading, and After Reading. The writing process also has three phases that occur before, during, and after writing: Prewriting, Drafting, and Proofing. By coordinating these two sets of process, you can improve both your reading and your writing. For example, the following statements sum up one way to connect reading and writing:

Reading is a prewriting activity. Drafting is an after reading activity.

Once you think of reading as a prewriting activity, you become a responsive or active reader during the reading process. In fact, you can begin using your writing skills as you read by annotating the text.

Annotating a Text

The word *annotate* suggests that you "take notes" in your book. Writing notes in the margin of a page as you read keeps you focused and improves your comprehension. You can quickly note questions where they occur, move on in your reading, and later return to clarify answers. In addition, after reading, your annotations help you review and respond to the material. The following suggestions offer one way to annotate a text:

How to Annotate a Text

- Circle important terms.
- Underline definitions and meanings.
- Note key ideas with a star or a check.
- Place question marks above words that are unknown or confusing.
- Number the steps in a process or items in a list.
- Write summaries at the end of long sections.
- Write recall questions in the margin near their answers.
- Write key words and meanings in the margin.

EXAMPLE The passage from a college communications textbook on page 596 is marked up as an example of an annotated text. Read the passage. Study the annotations. Then work with a peer or in a small group and create a summary of the text based on the annotations. See below to review how to write a summary.

Writing a Summary

Writing a summary is an effective step in the reading and studying process.

A **summary** is a brief, clear restatement of a longer passage.

A summary includes only the passage's most important points. Often a summary is made up of the main idea and major supporting details. The length of a summary should reflect your study needs and the kind of passage you are trying to understand. For example, a paragraph might be summarized in a sentence or two, an article might be summarized in a paragraph, and a textbook chapter might be summarized in a page or two.

You can discover how well you understand a passage by writing a summary of it as an after reading activity. Use the annotations you make during reading to create your summary.

For example, read the following summary of "Recognize Culture Shock" from a college communications textbook. Underline the words and phrases that were annotated in the longer section on page 596:

> [1]Culture shock is the "psychological reaction you experience when you're in a culture very different from your own." [2]When you are in culture shock, you can't communicate effectively and don't know even simple customs. [3]For example, you don't know how to ask for a favor, how to order a meal, or how to dress. [4]Anthropologist Kalervo Oberg coined the term "culture shock" and identifies its four stages. [5]Stage one, "the honeymoon," is filled with excitement and adventure. [6]Stage two, "the crisis," brings confusion and frustration due to unfamiliar customs and surroundings. [7]Stage three, "the recovery," is the process of learning how to function. [8]And stage four, "the adjustment," brings independence and enjoyment. [9]As the process indicates, culture shock can be overcome.

This summary includes the author's main idea and the major supporting details. However, this summary also brings in a few minor supporting details. For example, sentence 3 gives examples of customs that a person would not know. Including these details makes the summary longer than may be necessary. The version on page 601 includes only the main idea and the major supporting details.

Recognize Culture Shock Culture shock refers to the psychological reaction you experience when you're in a culture very different from your own (Furnham & Bochner, 1986). Culture shock is normal; most people experience it when entering a new and different culture. Nevertheless, it can be unpleasant and frustrating. Part of this results from feelings of alienation, conspicuousness, and difference from everyone else. When you lack knowledge of the rules and customs of the new society, you cannot communicate effectively. You're apt to blunder frequently and seriously. In your culture shock you may not know basic things:

1. ■ how to ask someone for a favor or pay someone a compliment
2. ■ how to extend or accept an invitation for dinner
3. ■ how early or how late to arrive for an appointment
4. ■ how long you should stay when visiting someone
5. ■ how to distinguish seriousness from playfulness and politeness from indifference
6. ■ how to dress for an informal, formal, or business function
7. ■ how to order a meal in a restaurant or how to summon a waiter

7 basic communication tasks & simple customs you may not know

Anthropologist Kalervo Oberg (1960), who first used the term culture shock, notes that it occurs in stages. These stages are useful for examining many encounters with the new and the different. Going away to college, moving in together, or joining the military, for example, can also result in culture shock. In explaining culture shock, we use the example of moving away from home into your own apartment to illustrate its four stages.

Who first used the term "culture shock"?

1 **Stage One: The Honeymoon** At first you experience fascination, even enchantment, with the new culture and its people. You finally have your own apartment. You're your own boss. Finally, on your own! When in groups of people who are culturally different, this stage is characterized by cordiality and friendship in these early and superficial relationships. Many tourists remain at this stage because their stay in foreign countries is so brief.

How many stages are there?

2 **Stage Two: The Crisis** Here, the differences between your own culture and the new one create problems. No longer do you find dinner ready for you unless you do it yourself. Your clothes are not washed or ironed unless you do them yourself. Feelings of frustration and inadequacy come to the fore. This is the stage at which you experience the actual shock of the new culture. One study of foreign students coming from over 100 different countries studying in 11 different countries found that 25 percent of the students experienced depression (Klineberg & Hull, 1979).

What are the labels for each stage?

3 **Stage Three: The Recovery** During this period you gain the skills necessary to function effectively. You learn how to shop, cook, and plan a meal. You find a local laundry and figure you'll learn how to iron later. You learn the language and ways of the new culture. Your feelings of inadequacy subside.

4 **Stage Four: The Adjustment** At this final stage, you adjust to and come to enjoy the new culture and the new experiences. You may still experience periodic difficulties and strains, but on a whole, the experience is pleasant. Actually, you're now a pretty decent cook. You're even coming to enjoy it. You're making a good salary so why learn to iron?

Culture shock can be overcome!

[1]Culture shock is the "psychological reaction you experience when you're in a culture very different from your own." [2]When you are in culture shock, you can't communicate effectively and don't know even simple customs. [3]Anthropologist Kalervo Oberg coined the term "culture shock" and identifies its four stages: stage one is "the honeymoon"; stage two is "the crisis"; stage three is "the recovery"; and stage four is "the adjustment." [4]Culture shock can be overcome.

Remember, the length of the summary depends on your study needs as well as the length of the passage you are summarizing.

A Reading-Writing Plan of Action

Can you see how annotating a text lays the ground upon which you can build a written response? The steps you take during reading feed into the process of writing a response after reading.

Remember, reading and writing is a conversation between the writer and the reader. One writes; the other reads. But the conversation often doesn't end there. A reader's response to a piece of writing keeps the dialogue going. When you write a summary, your response is to restate the author's ideas. It's like saying to the author, "If I understood you, you said . . ." When you offer your own views about the author's ideas, you are answering the author's implied question, "What do you think?" In your reading and writing classes, your teacher often steps into the conversation. He or she stands in for the author and becomes the reader of your written response. In this case, your teacher evaluates both your reading and writing abilities. Your teacher checks your response for accuracy in comprehension of the author's message and development of your ability to write. The following chart illustrates this exchange of ideas.

The Conversation among Writers and Readers

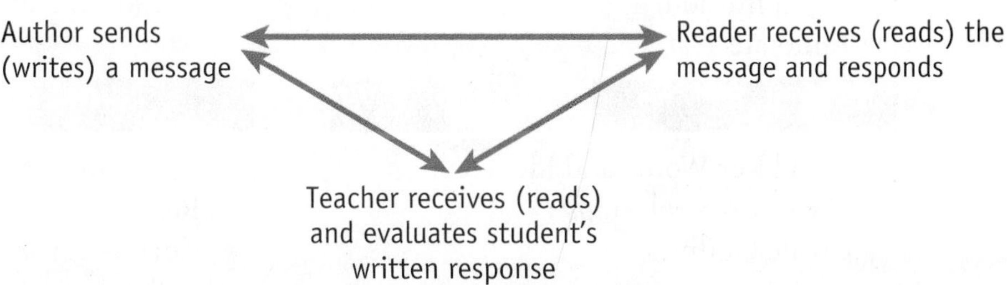
Author sends (writes) a message
Reader receives (reads) the message and responds
Teacher receives (reads) and evaluates student's written response

In each skill chapter of this textbook, the question "What Do You Think?" is posed after Review Tests 3 and 4. This question also appears after each reading selection in this section. The "What Do You Think?" writing assignments prompt you to respond to what you have read. This activity creates a writing situation and gives you a purpose for your written response. Just like a vocabulary word makes more sense in context than in a list, a writing assignment in context is more meaningful than an isolated topic or set of disconnected questions. The goal of "What Do You Think?" is to strengthen your connection between reading and writing. Because reading and writing are two distinct processes, it is helpful to have a guide that shows how to efficiently coordinate them. The following chart lays out a reading-writing plan of action. Note that the chart breaks the reading-writing process into a series of 6 steps. Keep in mind that any step can be repeated as needed. Also, you can move back and forth between steps as needed.

Study the 6-Step Reading-Writing Action Plan. Then work with a peer or small group of classmates and discuss the relationship between reading and writing, and how you will put this plan to use.

A 6-Step Reading-Writing Action Plan

Read		Write
1. Survey and Question	**BEFORE**	**4. Prewrite**
Call on Prior Knowledge Identify Topic Identify Key or New Words Identify Patterns of Organization Note Visual Aids Skim Introductions and Conclusions		Build Prior Knowledge* Gather Information* Read and Annotate* Brainstorm Ideas Choose Your Topic Generate Your Details Create a Concept Map Outline Ideas
2. Read	**DURING**	**5. Draft**
Monitor Comprehension Fix Confusion Annotate Text		Write Introduction, Body, and Conclusion
3. Review and Recite	**AFTER**	**6. Revise and Proofread**
Recall Key Words and Ideas Create Concept Maps Create Outlines Write a Summary Write a Response		Revise to Organize Revise for Exact Wording Correct Errors Fragments and Run-ons Spelling Punctuation

* Prewriting steps accomplished during reading

READING 1

Sex, Lies and Conversation: Why Is It So Hard for Men and Women to Talk to Each Other?

by Deborah Tannen

Deborah Tannen is University Professor of Linguistics in the Department of Linguistics at Georgetown University in Washington, D.C. Tannen, the author of nineteen published books and nearly a hundred articles and the recipient of five honorary doctorates, is an expert on the nature of human language. She lectures worldwide and is also a frequent guest on television and radio news and information shows. Tannen is perhaps best known to the general public, however, as the author of *You Just Don't Understand: Women and Men in Conversation.* This groundbreaking book about gender-based communication was on the *New York Times* best-seller list for nearly four years, including eight months as the number one best-selling book. It was also a best-seller in Brazil, Canada, England, Germany, Holland, and Hong Kong. *You Just Don't Understand* offers helpful insights about gender differences in communication style. This article, adapted from *You Just Don't Understand: Women and Men in Conversation,* appeared in *The Washington Post.*

Vocabulary Preview

anecdotes (paragraph 1): illustrations, stories
concurred (paragraph 1): agreed
crystallizes (paragraph 2): becomes well defined
havoc (paragraph 2): chaos, disorder
tangible (paragraph 4): concrete, real, touchable
inequities (paragraph 4): injustices
socialization (paragraph 8): training to gain skills necessary to function successfully in society
attuned (paragraph 12): adjusted, in tune with
analogous (paragraph 13): similar, comparable
alienated (paragraph 14): separated
requisite (paragraph 19): necessary, essential
paradox (paragraph 22): contradiction
ethnic (paragraph 26): cultural

1 I was addressing a small gathering in a suburban Virginia living room—a women's group that had invited men to join them. Throughout the evening, one man had been particularly talkative, frequently offering ideas and **anecdotes**, while his wife sat silently beside him on the couch. Toward the end of the evening, I commented that women frequently complain that their husbands don't talk to them. This man quickly **concurred**. He gestured toward his wife and said, "She's the talker in our family." The room burst into laughter; the man looked puzzled and hurt. "It's true," he explained. "When I come home from work I have nothing to say. If she didn't keep the conversation going, we'd spend the whole evening in silence."

2 This episode **crystallizes** the irony that although American men tend to talk more than women in public situations, they often talk less at home. And this pattern is wreaking **havoc** with marriage.

3 The pattern was observed by political scientist Andrew Hacker in the late '70s. Sociologist Catherine Kohler Riessman reports in her new book "Divorce Talk" that most of the women she interviewed—but only a few of the men—gave lack of communication as the reason for their divorces. Given the current divorce rate of nearly 50 percent, that amounts to millions of cases in the United States every year—a virtual epidemic of failed conversation.

4 In my own research, complaints from women about their husbands most often focused not on **tangible inequities** such as having given up the chance for a career to accompany a husband to his, or doing far more than their share of daily life-support work like cleaning, cooking, social arrangements and errands. Instead, they focused on communication: "He doesn't listen to me," "He doesn't talk to me." I found, as Hacker observed years before, that most wives want their husbands to be, first and foremost, conversational partners, but few husbands share this expectation of their wives.

In short, the image that best represents the current crisis is the stereotypical cartoon scene of a man sitting at the breakfast table with a newspaper held up in front of his face, while a woman glares at the back of it, wanting to talk. 5

Linguistic Battle of the Sexes

How can women and men have such different impressions of communication in marriage? Why the widespread imbalance in their interests and expectations? 6

In the April issue of *American Psychologist*, Stanford University's Eleanor Maccoby reports the results of her own and others' research showing that children's development is most influenced by the social structure of peer interactions. Boys and girls tend to play with children of their own gender, and their sex-separate groups have different organizational structures and interactive norms. 7

I believe these systematic differences in childhood **socialization** make talk between women and men like cross-cultural communication, heir to all the attraction and pitfalls of that enticing but difficult enterprise. My research on men's and women's conversations uncovered patterns similar to those described for children's groups. 8

For women, as for girls, intimacy is the fabric of relationships, and talk is the 9 thread from which it is woven. Little girls create and maintain friendships by exchanging secrets; similarly, women regard conversation as the cornerstone of friendship. So a woman expects her husband to be a new and improved version of a best friend. What is important is not the individual subjects that are discussed but the sense of closeness, of a life shared, that emerges when people tell their thoughts, feelings, and impressions.

10 Bonds between boys can be as intense as girls', but they are based less on talking, more on doing things together. Since they don't assume talk is the cement that binds a relationship, men don't know what kind of talk women want, and they don't miss it when it isn't there.

11 Boys' groups are larger, more inclusive, and more **hierarchical**, so boys must struggle to avoid the subordinate position in the group. This may play a role in women's complaints that men don't listen to them. Some men really don't like to listen, because being the listener makes them feel one-down, like a child listening to adults or an employee to a boss.

12 But often when women tell men, "You aren't listening," and the men protest, "I am," the men are right. The impression of not listening results from misalignments in the mechanics of conversation. The misalignment begins as soon as a man and a woman take physical positions. This became clear when I studied videotapes made by psychologist Bruce Dorval of children and adults talking to their same-sex best friends. I found that at every age, the girls and women faced each other directly, their eyes anchored on each other's faces. At every age, the boys and men sat at angles to each other and looked elsewhere in the room, periodically glancing at each other. They were obviously **attuned** to each other, often mirroring each other's movements. But the tendency of men to face away can give women the impression they aren't listening even when they are. A young woman in college was frustrated: Whenever she told her boyfriend she wanted to talk to him, he would lie down on the floor, close his eyes, and put his arm over his face. This signaled to her, "He's taking a nap." But he insisted he was listening extra hard. Normally, he looks around the room, so he is easily distracted. Lying down and covering his eyes helped him concentrate on what she was saying.

Analogous to the physical alignment 13
that women and men take in conversation is their topical alignment. The girls in my study tended to talk at length about one topic, but the boys tended to jump from topic to topic. The second-grade girls exchanged stories about people they knew. The second-grade boys teased, told jokes, noticed things in the room and talked about finding games to play. The sixth-grade girls talked about problems with a mutual friend. The sixth-grade boys talked about 55 different topics, none of which extended over more than a few turns.

Listening to Body Language

Switching topics is another habit that 14
gives women the impression men aren't listening, especially if they switch to a topic about themselves. But the evidence of the 10th-grade boys in my study indicates otherwise. The 10th-grade boys sprawled across their chairs with bodies parallel and eyes straight ahead, rarely

looking at each other. They looked as if they were riding in a car, staring out the windshield. But they were talking about their feelings. One boy was upset because a girl had told him he had a drinking problem, and the other was feeling **alienated** from all his friends.

15 Now, when a girl told a friend about a problem, the friend responded by asking probing questions and expressing agreement and understanding. But the boys dismissed each other's problems. Todd assured Richard that his drinking was "no big problem" because "sometimes you're funny when you're off your butt." And when Todd said he felt left out, Richard responded, "Why should you? You know more people than me."

16 Women perceive such responses as belittling and unsupportive. But the boys seemed satisfied with them. Whereas women reassure each other by implying, "You shouldn't feel bad because I've had similar experiences," men do so by implying, "You shouldn't feel bad because your problems aren't so bad."

17 There are even simpler reasons for women's impression that men don't listen. Linguist Lynette Hirschman found that women make more listener-noise, such as "mhm," "uhuh," and "yeah," to show "I'm with you." Men, she found, more often give silent attention. Women who expect a stream of listener noise interpret silent attention as no attention at all.

18 Women's conversational habits are as frustrating to men as men's are to women. Men who expect silent attention interpret a stream of listener noise as overreaction or impatience. Also, when women talk to each other in a close, comfortable setting, they often overlap, finish each other's sentences and anticipate what the other is about to say. This practice, which I call "participatory listenership," is often perceived by men as interruption, intrusion and lack of attention.

19 A parallel difference caused a man to complain about his wife, "She just wants to talk about her own point of view. If I show her another view, she gets mad at me." When most women talk to each other, they assume a conversationalist's job is to express agreement and support. But many men see their conversational duty as pointing out the other side of an argument. This is heard as disloyalty by women, and refusal to offer the **requisite** support. It is not that women don't want to see other points of view, but that they prefer them phrased as suggestions and inquiries rather than as direct challenges.

20 In his book "Fighting for Life," Walter Ong points out that men use "agonistic" or warlike, oppositional formats to do almost anything; thus discussion becomes debate, and conversation a competitive sport. In contrast, women see conversation as a ritual means of establishing rapport. If Jane tells a problem and June says she has a similar one, they walk away feeling closer to each other. But this attempt at establishing rapport can backfire when used with men. Men take too literally women's ritual "troubles talk," just as women mistake men's ritual challenges for real attack.

The Sounds of Silence

21 These differences begin to clarify why women and men have such different

expectations about communication in marriage. For women, talk creates intimacy. Marriage is an orgy of closeness: you can tell your feelings and thoughts, and still be loved. Their greatest fear is being pushed away. But men live in a hierarchical world, where talk maintains independence and status. They are on guard to protect themselves from being put down and pushed around.

22 This explains the **paradox** of the talkative man who said of his silent wife, "She's the talker." In the public setting of a guest lecture, he felt challenged to show his intelligence and display his understanding of the lecture. But at home, where he has nothing to prove and no one to defend against, he is free to remain silent. For his wife, being home means she is free from the worry that something she says might offend someone, or spark disagreement, or appear to be showing off; at home she is free to talk.

23 The communication problems that endanger marriage can't be fixed by mechanical engineering. They require a new conceptual framework about the role of talk in human relationships. Many of the psychological explanations that have become second nature may not be helpful, because they tend to blame either women (for not being assertive enough) or men (for not being in touch with their feelings). A sociolinguistic approach by which male-female conversation is seen as cross-cultural communication allows us to understand the problem and forge solutions without blaming either party.

24 Once the problem is understood, improvement comes naturally, as it did to the young woman and her boyfriend who seemed to go to sleep when she wanted to talk. Previously, she had accused him of not listening, and he had refused to change his behavior, since that would be admitting fault. But then she learned about and explained to him the differences in women's and men's habitual ways of aligning themselves in conversation. The next time she told him she wanted to talk, he began, as usual, by lying down and covering his eyes. When the familiar negative reaction bubbled up, she reassured herself that he really was listening. But then he sat up and looked at her. Thrilled, she asked why. He said, "You like me to look at you when we talk, so I'll try to do it." Once he saw their differences as cross-cultural rather than right and wrong, he independently altered his behavior.

Women who feel abandoned and de- 25
prived when their husbands won't listen to or report daily news may be happy to discover their husbands trying to adapt once they understand the place of small talk in women's relationships. But if their husbands don't adapt, the women may still be comforted that for men, this is not a failure of intimacy. Accepting the difference, the wives may look to their friends or family for that kind of talk. And husbands who can't provide it shouldn't feel their wives have made unreasonable demands. Some couples will still decide to divorce, but at least their decisions will be based on realistic expectations.

In these times of resurgent **ethnic** 26
conflicts, the world desperately needs cross-cultural understanding. Like charity, successful cross-cultural communication should begin at home.

VISUAL VOCABULARY

_______ Which phrase best expresses the meaning of *rapport* in the caption?

a. a bond
b. an idea
c. a commitment

▸ Women see conversation as a way to establish rapport.

Choose the best meaning of each word in **bold** type. Use context clues to make your choice.

Vocabulary in Context _______ **1.** The word **Linguistic** in the heading "Linguistic Battle of the Sexes" (above paragraph 6) relates to

a. gender.
b. language.
c. relationship.
d. education.

Vocabulary in Context _______ **2.** "Boys' groups are larger, more inclusive, and more **hierarchical**, so boys must struggle to avoid the subordinate position in the group." (paragraph 11)

a. ranked
b. brutal
c. logical
d. submissive

Central Idea and Main Ideas _______ **3.** Which sentence is the best statement of the central idea of the passage?

a. "Given the current divorce rate of nearly 50 percent, that amounts to millions of cases in the United States every year—a virtual epidemic of failed conversation." (paragraph 3)
b. "My research on men's and women's conversations uncovered patterns similar to those described for children's groups." (paragraph 8)
c. "A sociolinguistic approach by which male-female conversation is seen as cross-cultural communication allows us to understand the problem and forge solutions without blaming either party." (paragraph 23)
d. "Like charity, successful cross-cultural communication should begin at home." (paragraph 26)

Central Idea and Main Ideas ______ **4.** Which sentence is the best statement of the implied main idea of paragraphs 14 through 20?

a. Men and women have different views of communication in marriage.
b. Men switch topics more often than women do during a conversation.
c. Women and men differ in their use of body language during the communication process.
d. Women are more effective listeners than men.

Supporting Details ______ **5.** According to Catherine Kohler Riessman, which of the following reasons did most women give for their divorces?

a. giving up the chance for a career to follow a husband
b. doing far more than their share of daily life-support work like cleaning
c. sexual infidelity
d. lack of communication

Supporting Details ______ **6.** Men reassure each other by

a. making direct eye contact when they communicate.
b. making listener-noise such as "mhm," "uhuh," and "yeah" to show "I'm with you."
c. implying "you shouldn't feel bad because your problems aren't so bad."
d. practicing "participatory listenership."

Transitions ______ **7.** "Little girls create and maintain friendships by exchanging secrets; similarly, women regard conversation as the cornerstone of friendship. So a woman expects her husband to be a new and improved version of a best friend." (paragraph 9)

The relationship of ideas between these two sentences is

a. time order.
b. cause and effect.
c. comparison and contrast.

Transitions ______ **8.** "The misalignment begins as soon as a man and a woman take physical positions." (paragraph 12)

The relationship of ideas within this sentence is

a. time order.
b. cause and effect.
c. comparison and contrast.

Thought Patterns ______ **9.** The overall thought pattern used in this passage is

a. a discussion of the differences in the ways men and women communicate.

b. an argument against the various ways men use communication.
c. a narrative that illustrates the different problems men and women face as they attempt to communicate.
d. a step-by-step description of how to effectively communicate with the opposite sex.

Thought Patterns

_______ **10.** The main thought pattern used in paragraph 24 is
a. comparison.
b. definition and example.
c. time order.

Fact and Opinion

_______ **11.** Overall, the ideas in this passage are mainly based on
a. the opinions and experiences of ordinary men and women.
b. fictitious or hypothetical details.
c. the research and observation of the author and other experts in the field of linguistics and communication.

Fact and Opinion

_______ **12.** "Like charity, successful cross-cultural communication should begin at home." (paragraph 26)

This sentence is a statement of
a. fact.
b. opinion.
c. fact and opinion.

Tone and Purpose

_______ **13.** Which word best expresses the overall tone of the passage?
a. biased
b. condescending
c. balanced
d. emotional

Tone and Purpose

_______ **14.** The overall tone of paragraph 1 is
a. dismayed.
b. humorous.
c. scornful.
d. neutral.

Tone and Purpose

_______ **15.** The tone of paragraph 26 is
a. mocking.
b. objective.
c. hopeful.
d. persuasive.

Tone and Purpose

_______ **16.** The author's main purpose is
a. to persuade men and women to change their communication styles.
b. to entertain readers with amusing incidents of miscommunication between men and women.
c. to inform the reader about the differences in communication styles between men and women so that they can better understand one another.

Inferences ______ **17.** Based on the information in the article, generally men
a. do not listen as well as women.
b. are more competitive than women.
c. prefer intimacy over independence, unlike women.

Inferences ______ **18.** The author implies that generally women
a. are more comfortable and better at giving advice than are men.
b. are more interested in establishing networks of support than are men.
c. are more skilled at effective communication than are men.

Inferences ______ **19.** The details in paragraphs 21–26 imply that
a. men are just as emotional as women.
b. the hurt and tension between men and women are often based on misunderstanding the communication purpose of the other.
c. women need to communicate more than men.

Argument ______ **20.** Men live in a hierarchical world, where talk maintains independence and status.

Which statement does *not* support this claim?
a. Men are on guard to protect themselves from being put down and pushed around.
b. Men use "agnostic" or warlike, oppositional formats to do almost anything.
c. He said, "You like me to look at you when we talk, so I'll try to do it."

Mapping

Complete the chart with information from the passage. Wording may vary.

Communication Approach of Men	Communication Approach of Women
Establish bonds based on ______________	Establish bonds based on talking
Jump from topic to topic	Talk at length about one topic
Expect silent attention	Practice "participatory listenership"
"Agnostic" or warlike, oppositional formats	Ritual means of establishing rapport
Discussion becomes debate	Talk creates ______________

WHAT DO YOU THINK?

Do you think men and women differ in the way they communicate? Explain why or why not with examples you have observed. Given that men and women do differ in their approach to communication, how do these differences affect communication at work or school? Assume a friend of yours is having difficulty communicating with a boss or friend of the opposite sex. Write a letter to your friend in which you give advice based on what you learned from this article to help him or her communicate more effectively. Begin your letter with a summary of the author's main idea and major supporting details.

MASTER READER Scorecard

"Sex, Lies and Conversation"

Skill	Number Correct		Points		Total
Vocabulary					
Vocabulary in Context (2 items)	_______	×	4	=	_______
Comprehension					
Central Idea and Main Ideas (2 items)	_______	×	4	=	_______
Supporting Details (2 items)	_______	×	4	=	_______
Transitions (2 items)	_______	×	4	=	_______
Thought Patterns (2 items)	_______	×	4	=	_______
Fact and Opinion (2 items)	_______	×	4	=	_______
Tone and Purpose (4 items)	_______	×	4	=	_______
Inferences (3 items)	_______	×	4	=	_______
Argument (1 item)	_______	×	4	=	_______
Mapping (2 items)	_______	×	10	=	_______
			Comprehension Score		_______

READING 2

Teens, Nude Photos and the Law

by Dahlia Lithwick

Dahlia Lithwick is a contributing editor at *Newsweek* and senior editor at *Slate*, where she has written for the *Supreme Court Dispatches* and *Jurisprudence* columns since 1999. Her writing has also appeared in *The New Republic*, the *Washington Post*, and the *New York Times*, and she has also been a frequent guest on the NPR program *Day-to-Day*. Ms. Lithwick was awarded the Online News Association's award for online commentary in 2001. In the following article for *Newsweek* magazine, Lithwick questions whether the "police should be involved when tipsy teen girls e-mail their boyfriends naughty Valentine's Day pictures."

Vocabulary Preview

confiscated (paragraph 1): took away
iteration (paragraph 1): repetition, instance
narcissism (paragraph 1): self-admiration, egotism, conceit
venue (paragraph 3): setting, scene
purveyors (paragraph 5): suppliers, sellers, sources
paternalism (paragraph 5): telling people what is best, a style of governing or managing in which the desire to assist and protect reduces personal choice and responsibility
perpetuates (paragraph 6): spreads, carries on, brings about

---

1 Say you're a middle-school principal who **confiscated** a cell phone from a 14-year-old boy, only to discover it contains a nude photo of his 13-year-old girlfriend. Do you (a) call the boy's parents in despair; (b) call the girl's parents in despair; or (c) call the police? More and more, the answer is (d) all of the above. Which could result in criminal charges for both of your students, and their eventual designation as sex offenders. "Sexting" is the clever new name for the act of sending, receiving or forwarding naked photos via your cell phone, and I wasn't fully convinced that America was facing a sexting epidemic, as opposed to a journalists-writing-about-sexting epidemic, until I saw a new survey done by the National Campaign to Prevent Teen and Unplanned Pregnancy. One teenager in five reported having sent or posted naked photos of themselves. Whether all this reflects a new child-porn epidemic, or just a new **iteration** of the old teen **narcissism** epidemic, remains unclear.

Last month, three girls (ages 14 or 15) 2
in Greensburg, Pa., were charged with **disseminating** child pornography for sexting

their boyfriends. The boys who received the images were charged with possession. A teenager in Indiana faces felony obscenity charges for sending a picture of his genitals to female classmates. A 15-year-old girl in Ohio and a 14-year-old girl in Michigan were charged with felonies for sending nude images of themselves to classmates. Some of these teens have pleaded guilty to lesser charges. Others have not. If convicted, these young people may have to register as sex offenders, in some cases for a decade or two. Similar charges have been brought in cases reported in Alabama, Connecticut, Florida, New Jersey, New York, Pennsylvania, Texas, Utah and Wisconsin.

3 One quick clue that the criminal-justice system is probably not the best **venue** for addressing sexting? A survey of the charges brought in the cases reflects that—depending on the jurisdiction—prosecutors have charged the senders of smutty photos, the recipients of smutty photos, those who save the smutty photos and the hapless forwarders of smutty photos with the same crime: child pornography. Who is the victim here? Everybody and nobody.

4 There may be an argument for police intervention in cases that involve a genuine threat or cyberbullying, such as a recent Massachusetts incident in which the picture of a naked 14-year-old girl was allegedly sent to more than 100 cell phones, or a New York case involving a group of boys who turned a nude photo of a 15-year-old girl into crude animations and PowerPoint presentations. But ask yourself whether those cases are the same as the cases in which tipsy teen girls send their boyfriends naughty Valentine's Day pictures.

5 The argument for hammering every such case seems to be that sending naked pictures might have serious consequences, so let's charge these kids with felonies, which will surely have serious consequences. In the Pennsylvania case a police captain explained that the charges were brought because "it's very dangerous. Once it's on a cell phone, that cell phone can be put on the Internet where everyone in the world can get access to that juvenile picture." The argument that we must prosecute kids as the producers and **purveyors** of kiddie porn because they are too dumb to understand that their seemingly innocent acts can harm them goes beyond **paternalism**. Child-pornography laws intended to protect children should not be used to prosecute and then label children as sex offenders. We seem to forget that kids can be as tech-savvy as Bill Gates but as **gullible** as Bambi. Even in the age of the Internet, young people fail to appreciate that naked pictures want to roam free.

6 The real problem with criminalizing teen sexting as a form of child pornography is that the great majority of these kids are not predators. They think they're being brash and sexy. And while some of the reaction to sexting reflects legitimate concerns about children as sex objects, some **perpetuates** legal stereotypes and fallacies. A recent *New York Times* article quotes the Family Violence Prevention Fund, a nonprofit domestic-violence-awareness group, saying that the sending of nude pictures, even if done voluntarily, constitutes "digital dating violence." But do we truly believe that one in five teens is participating in an act of violence? Experts insist the sexting trend hurts

teen girls more than boys, fretting that they feel "pressured" to take and send naked photos. Paradoxically, the girls in the Pennsylvania case were charged with "manufacturing, disseminating or possessing child pornography" while the boys were merely charged with possession. If the girls are the real victims, why are we treating them more harshly than the boys?

7 Judging from the sexting prosecutions in Pennsylvania, Ohio and Indiana this year, it's clear that the criminal-justice system is too blunt an instrument to resolve a problem that reflects more about the volatile combination of teens and technology than about some national cybercrime spree. Parents need to remind their teens that a dumb moment can last a lifetime in cyberspace. But judges and prosecutors need to understand that a lifetime of cyberhumiliation shouldn't be grounds for a lifelong real criminal record.

—"Teens, Nude Photos and the Law: Ask Yourself: Should the Police Be Involved When Tipsy Teen Girls E-mail Their Boyfriends Naughty Valentine's Day Pictures?" by Dahlia Lithwick from *Newsweek*, February 14 Copyright © 2009 Newsweek, Inc. All rights reserved. Used by permission and protected by the Copyright Laws of the United States. The printing, copying, redistribution, or retransmission of the Material without express written permission is prohibited. www.newsweek.com.

Choose the best meaning of each word in **bold.** Use context clues to make your choice.

Vocabulary in Context ______ **1.** "Last month, three girls (ages 14 or 15) in Greensburg, Pa., were charged with **disseminating** child pornography for sexting their boyfriends." (paragraph 2)

a. creating
b. distributing
c. concealing
d. enjoying

Vocabulary in Context ______ **2.** "We seem to forget that kids can be as tech-savvy as Bill Gates but as **gullible** as Bambi." (paragraph 5)

a. dumb
b. clever
c. naive
d. playful

Central Idea ______ **3.** Which of the following sentences states the central idea of the passage?

a. "One teenager in five reported having sent or posted naked photos of themselves." (paragraph 1)
b. "Child-pornography laws intended to protect children should not be used to prosecute and then label children as sex offenders." (paragraph 5)
c. "The real problem with criminalizing teen sexting as a form of child pornography is that the great majority of these kids are not predators." (paragraph 6)

d. "Parents need to remind their teens that a dumb moment can last a lifetime in cyberspace." (paragraph 7)

Supporting Details _____ **4.** According to the passage, teens caught sexting risk being charged with

a. child pornography.
b. sexting.
c. prostitution.
d. dating violence.

Supporting Details _____ **5.** A recent *New York Times* article quotes the Family Violence Prevention Fund, a nonprofit domestic-violence-awareness group, saying that the sending of nude pictures, even if done voluntarily, constitutes

a. cybercrime.
b. cyberbulling.
c. prostitution.
d. digital dating violence.

Transitions _____ **6.** "Do you (a) call the boy's parents in despair; (b) call the girl's parents in despair; or (c) call the police?" (paragraph 1)

The relationship of ideas within this sentence is

a. listing.
b. classification.
c. comparison and contrast.
d. generalization and example.

Transitions _____ **7.** "Experts insist the sexting trend hurts teen girls more than boys, fretting that they feel 'pressured' to take and send naked photos. Paradoxically, the girls in the Pennsylvania case were charged with 'manufacturing, disseminating or possessing child pornography' while the boys were merely charged with possession." (paragraph 6)

The relationship of ideas between these sentences is

a. cause and effect.
b. classification.
c. comparison and contrast.
d. generalization and example.

Thought Patterns _____ **8.** The thought pattern used in paragraph 4 is

a. definition.
b. classification.
c. comparison and contrast.
d. space order.

Thought Patterns _____ **9.** The thought pattern for paragraph 5 is

a. cause and effect.
b. classification.
c. comparison and contrast.
d. generalization and example.

Fact and Opinion _____ **10.** "One teenager in five reported having sent or posted naked photos of themselves." (paragraph 1)

This sentence is a statement of

a. fact.
b. opinion.
c. fact and opinion.

Fact and Opinion ______ **11.** "Child-pornography laws intended to protect children should not be used to prosecute and then label children as sex offenders." (paragraph 5)

This sentence is a statement of
a. fact.
b. opinion.
c. fact and opinion.

Fact and Opinion ______ **12.** "A survey of the charges brought in the cases reflects that—depending on the jurisdiction—prosecutors have charged the senders of smutty photos, the recipients of smutty photos, those who save the smutty photos and the hapless forwarders of smutty photos with the same crime: child pornography." (paragraph 3)

This sentence is a statement of
a. fact.
b. opinion.
c. fact and opinion.

Tone ______ **13.** The overall tone of the author is
a. objective.
b. bitter.
c. optimistic.
d. skeptical.

Purpose ______ **14.** The overall purpose of the author is
a. to inform the public about the teen trend of sexting and its consequences.
b. to entertain the public with graphic details of teen sexting.
c. to persuade the public to decriminalize teen sexting.

Inferences ______ **15.** Based on the details in paragraph 1, we can infer that
a. most teens are involved in sexting.
b. sexting among teens is not a serious social problem.
c. only teens are involved in sexting.
d. most teens are not involved in sexting.

Inferences ______ **16.** Based on the details in paragraph 2, we can infer that
a. teens who are convicted of sexting could spend time in jail.
b. teens are not facing serious consequences when they are caught sexting.
c. teens who participate in sexting are sex offenders.
d. teen sexting is limited and occurs in only a few places.

Inferences ______ **17.** Based on the details in the passage, we can infer that
a. poor parenting is the cause of teen involvement in sexting.
b. teenagers are too immature to understand the consequences of sexting.

c. teen sexting should be punished as a serious crime.
d. teen sexting is a harmless rite of passage into adulthood.

Argument ______ **18.** Identify the persuasive technique used in the following sentence:

"The argument for hammering every such case seems to be that sending naked pictures might have serious consequences, so let's charge these kids with felonies, which will surely have serious consequences." (paragraph 5)

a. false cause c. begging the question
b. personal attack d. transfer

Argument ______ **19.** Identify the persuasive technique used in the following sentence:

"The argument that we must prosecute kids as the producers and purveyors of kiddie porn because they are too dumb to understand that their seemingly innocent acts can harm them goes beyond paternalism." (paragraph 5)

a. false analogy c. begging the question
b. personal attack d. transfer

Argument ______ **20.** Identify the persuasive technique used in the following sentence:

"We seem to forget that kids can be as tech-savvy as Bill Gates but as gullible as Bambi." (paragraph 5)

a. false analogy c. begging the question
b. personal attack d. transfer

Mapping

Complete the following pro-con chart for argumentation with information from the passage. Wording may vary.

Felony Charges for Teen Sexting	
Pro (for)	**Con (against)**
Teen sexting publishes (1) ______________.	Kids are not (4) ______________
It disseminates (2) ______________.	Punishment is too harsh: Conviction of
It raises concerns about children as sex objects.	a felony and (5) ______________ ______________
It constitutes (3) ______________	

WHAT DO YOU THINK?

Do you think teen sexting is a serious social problem? Why or why not? Why do you think one in five teens participate in sexting? How should society respond to teen sexting? Assume you are volunteer teen counselor at your local Boys and Girls Club, and it has come to your attention that several teens with whom you work are involved in sexting. In response, the Club has asked you to give a five-minute talk to the youth about the seriousness of teen sexting. Write a three-paragraph draft of a speech in which you warn teens about the dangers of sexting.

MASTER READER Scorecard

"Teens, Nude Photos and the Law"

Skill	**Number Correct**		**Points**		**Total**
Vocabulary					
Vocabulary in Context (2 items)	______	×	4	=	______
Comprehension					
Central Idea (1 item)	______	×	4	=	______
Supporting Details (2 items)	______	×	4	=	______
Transitions (2 items)	______	×	4	=	______
Thought Patterns (2 items)	______	×	4	=	______
Fact and Opinion (3 items)	______	×	4	=	______
Tone and Purpose (2 items)	______	×	4	=	______
Inferences (3 items)	______	×	4	=	______
Argument (3 items)	______	×	4	=	______
Mapping (5 items)	______	×	4	=	______
	Comprehension Score				______

READING 3

Binge Drinking, A Campus Killer

by Sabrina Rubin Erdely

Sabrina Rubin Erdely, an investigative journalist, graduated from the University of Pennsylvania in 1994. Currently a senior staff writer at *Philadelphia* magazine and a contributing writer for *Cosmopolitan*, Erdely's work has earned her a number of awards, including nomination for the National Magazine Award in public interest writing. In the article reprinted here, Erdely explores the nationwide chronic problem of binge drinking on college and university campuses.

Vocabulary Preview

indulgences (paragraph 1): pleasures, excesses
surmise (paragraph 11): conclude, speculate
gratification (paragraph 11): satisfaction, pleasure
concoctions (paragraph 15): mixture, blend, brew
defibrillation (paragraph 18): an electrical shock applied to restore the heart's rhythm
equivalent (paragraph 19): equal, corresponding
illicit (paragraph 38): illegal, illegitimate, banned

---

1 Pregame tailgating parties, post-exam celebrations and Friday happy hours—not to mention fraternity and sorority mixers—have long been a cornerstone of the collegiate experience. But on campuses across America, these **indulgences** have a more alarming side. For some of today's college students, binge drinking has become the norm.

2 This past February I headed to the University of Wisconsin-Madison, rated the No. 2 party school in the nation by the college guide *Princeton Review,* to see the party scene for myself. On Thursday night the weekend was already getting started. At a **raucous** off-campus gathering, 20-year-old Tracey Middler struggled to down her beer as fist-pumping onlookers yelled, "Chug! Chug! Chug!"

In the kitchen, sophomore Jeremy Budda drained his tenth beer. "I get real wasted on weekends," he explained. Nearby, a 19-year-old estimated, "I'll end up having 17, 18 beers." 3

Swept up in the revelry, these partiers aren't thinking about the alcohol-related tragedies that have been in the news. All they're thinking about now is the next party. The keg is just about empty. 4

5 As the 19-year-old announces loudly, these college students have just one objective: "to get drunk!"

6 The challenge to drink to the very limits of one's endurance has become a celebrated staple of college life. In one of the most extensive reports on college drinking thus far, a 1997 Harvard School of Public Health study found that 43 percent of college students admitted binge drinking in the preceding two weeks. (Defined as four drinks in a sitting for a woman and five for a man, a drinking binge is when one drinks enough to risk health and well-being.)

7 "That's about five million students," says Henry Wechsler, who co-authored the study. "And it's certainly a cause for concern. Most of these students don't realize they're engaging in risky behavior." University of Kansas Chancellor Robert Hemenway adds, "Every year we see students harmed because of their involvement with alcohol."

8 Indeed, when binge drinking came to the forefront last year with a rash of alcohol-related college deaths, the nation was stunned by the loss. There was Scott Krueger, the 18-year-old fraternity pledge at the Massachusetts Institute of Technology, who died of alcohol poisoning after downing the equivalent of 15 shots in an hour. There was Leslie Baltz, a University of Virginia senior, who died after she drank too much and fell down a flight of stairs. Lorraine Hanna, a freshman at Indiana University of Pennsylvania, was left alone to sleep off her night of New Year's Eve partying. Later that day her twin sister found her dead—with a blood-alcohol content (BAC) of 0.429 percent. (Driving with a BAC of 0.1 percent and above is illegal in all states.)

Experts estimate that excessive drinking is involved in thousands of student deaths a year. And the Harvard researchers found that there has been a dramatic change in why students drink: 39 percent drank "to get drunk" in 1993, but 52 percent had the same objective in 1997. 9

"What has changed is the across-the-board *acceptability* of intoxication," says Felix Savino, a psychologist at UW-Madison. "Many college students today see not just drinking but being *drunk* as their primary way of socializing." 10

The reasons for the shift are complex and not fully understood. But researchers **surmise** that it may have something to do with today's instant-**gratification** lifestyle—and young people tend to take it to the extreme. 11

In total, it is estimated that America's 12 million undergraduates drink the equivalent of six million gallons of beer a week. When that's combined with teenagers' need to drink secretly, it's no wonder many have a dangerous relationship with alcohol. 12

The biggest predictor of bingeing is fraternity or sorority membership. Sixty-five percent of members qualified as binge drinkers, according to the Harvard study. 13

August 25, 1997, was meant to be a night the new Sigma Alpha Epsilon pledges at Louisiana State University in Baton Rouge would never forget, and by 8 P.M. it was certainly shaping up that way. The revelry had begun earlier with a keg party. Then they went to a bar near campus, where pledges consumed massive quantities of alcohol. 14

Among the pledges were Donald Hunt, Jr., a 21-year-old freshman and Army veteran, and his roommate, Benjamin 15

Wynne, a 20-year-old sophomore. Friends since high school, the two gamely drank the alcoholic **concoctions** offered to them and everyone else.

16 Before long, many in the group began vomiting into trash cans. (Donald Hunt would later allege in a lawsuit that these "vomiting stations" were set up for that very purpose, something the defendants adamantly deny.) About 9:30, **incapacitated** pledges were taken back to sleep it off at the frat house.

17 The 911 call came around midnight. Paramedics were stunned at what they found: more than a dozen young men sprawled on the floor, on chairs, on couches, reeking of alcohol. The paramedics burst into action, shaking the pledges and shouting, "Hey! Can you hear me?" Four couldn't be roused, and of those, one had no vital signs: Benjamin Wynne was in cardiac arrest.

18 Checking to see that nothing was blocking Wynne's airway, the paramedics began CPR. Within minutes they'd inserted an oxygen tube into his lungs, hooked up an I.V., attached a cardiac monitor and begun shocking him with **defibrillation** paddles, trying to restart his heart.

19 Still not responding, Wynne was rushed by ambulance to Baton Rouge General Hospital. Lab work revealed that his blood-alcohol content was an astonishing 0.588 percent, nearly six times the legal driving limit for adults—the **equivalent** of taking about 21 shots in an hour.

20 Meanwhile, three other fraternity pledges were undergoing similar revival efforts. One was Donald Hunt. He would suffer severe alcohol poisoning and nearly die.

After working furiously on Wynne, the hospital team admitted defeat. He was pronounced dead of acute alcohol poisoning. 21

One simple fact people tend to lose sight of is that alcohol is a poison—often pleasurable, but a toxin nonetheless. And for a person with little experience processing this toxin, it can come as something of a physical shock. 22

In general, a bottle of beer has about the same alcohol content as a glass of wine or shot of liquor. And the body can remove only the equivalent of less than one drink hourly from the bloodstream. 23

Many students are not just experimenting once or twice. In the Harvard study, half of binge drinkers were "frequent binge drinkers," meaning they had binged three or more times in the previous two weeks. 24

It also is assumed by some that bingeing is a "guy thing," an activity that, like cigar smoking and watching televised sports, belongs in the realm of male bonding. Statistics, however, show that the number of heavy-drinking young women is significant. Henry Wechsler's Harvard study found that a hefty 48 percent of college men were binge drinkers, and women were right behind them at 39 percent. 25

Howard Somers had always been afraid of heights. Perhaps his fear was some sort of an omen. On an August day in 1997 he helped his 18-year-old daughter, Mindy, move into her dorm at Virginia Tech. As they unloaded her things in the eighth-floor room, Somers noted with unease the position of the window. It opened inward like an oven door, its lip about level with her bed. He mentioned it, but Mindy dismissed his concern with a smile. 26

27 "I have gone through more guilt than you can imagine," Somers says now quietly. "Things I wish I had said or done. But I never thought this would happen. Who would?"

28 Mindy Somers knew the dangers of alcohol and tried to stay aware of her limits. She'd planned not to overdo it that Friday night, since her mother was coming in that weekend to celebrate Mindy's 19th birthday on Sunday. But it was Halloween, the campus was alive with activity, and Mindy decided to stop in at several off-campus parties.

29 When she returned to her room at 3 A.M., she was wiped out enough to fall into bed fully clothed. Mindy's bed was pushed lengthwise against the long, low window. Her roommate and two other girls, who were on the floor, all slept too soundly to notice that sometime after 4 A.M. Mindy's bed was empty.

30 When the paperboy found her facedown on the grass at 6:45 A.M., he at first thought it was a Halloween prank. Police and EMTs swarmed to the scene in minutes. Somers was pronounced dead of massive chest and abdominal injuries. She had a blood-alcohol content of 0.21 percent, equal to her having drunk about five beers in one hour.

31 Police surmised that Mindy had tried to get out of bed during the night but, disoriented, had slipped out the window, falling 75 feet to her death. "It was a strange, tragic accident," Virginia Tech Police Chief Michael Jones says.

32 A terrible irony was that the week prior to Mindy's death had been Virginia Tech's annual Alcohol Awareness Week.

33 While binge drinking isn't always lethal, it does have other, wide-ranging effects. Academics is one realm where it takes a heavy toll.

During my trip to Wisconsin most students told me they didn't plan on attending classes the following day. "Nah, I almost never go to class on Friday. It's no big deal," answered Greg, a sophomore. According to a survey of university administrators, 38 percent of academic problems are alcohol-related, as are 29 percent of dropouts. 34

Perhaps because alcohol increases aggression and impairs judgment, it is also related to 25 percent of violent crimes and roughly 60 percent of vandalism on campus. According to one survey, 79 percent of students who had experienced unwanted sexual intercourse in the previous year said that they were under the influence of alcohol or other drugs at the time. "Some people believe that alcohol can provide an excuse for inappropriate behavior, including sexual aggression," says Jeanette Norris, a University of Washington researcher. Later on, those people can claim, "It wasn't me—it was the booze." 35

Faced with the many potential dangers, college campuses are scrambling for ways to reduce binge drinking. Many offer seminars on alcohol during freshman orientation. Over 50 schools provide alcohol-free living environments. At the University of Michigan's main campus in Ann Arbor, for instance, nearly 30 percent of undergrads living in university housing now choose to live in alcohol-free rooms. Nationwide several fraternities have announced that by the year 2000 their chapter houses will be alcohol-free. 36

After the University of Rhode Island topped the *Princeton Review* party list two 37 years in a row, administrators banned alcohol at all student events on campus; this year URI didn't even crack the top ten. Some campuses respond even more severely, unleashing campus raids and encouraging police busts.

38 Researchers debate, however, if such "zero-tolerance" policies are helpful or if they might actually result in more secret, off-campus drinking. Other academics wonder if dropping the drinking age to 18 would take away the **illicit** thrill of alcohol and lower the number of kids drinking wildly. Others feel this would just create more drinking-related fatalities.

39 Whatever it takes, changing student behavior won't be easy. "What you've got here are people who think they are having fun," Harvard's Henry Wechsler explains. "You can't change their behavior by preaching at them or by telling them they'll get hurt."

Around 2 A.M. at UW-Madison a hundred kids congregate at a downtown intersection in a nightly ritual. One girl is trying to pull her roommate up off the ground. "I'm not that drunk," the one on the ground insists. "I just can't stand up." 40

Two fights break out. A police car cruises by and the crowd thins, some heading to after-hours parties. Then maybe at 3 or 4 A.M. they'll go home to get some sleep, so they will be rested for when they start to drink again. Tomorrow night. 41

Choose the best meaning of each word in **bold** type. Use context clues to make your choice.

Vocabulary in Context ______ **1.** "At a **raucous** off-campus gathering, 20-year-old Tracey Middler struggled to down her beer as fist-pumping onlookers yelled, 'Chug! Chug! Chug!'" (paragraph 2)

a. fun
b. dangerous
c. illegal
d. wild

Vocabulary in Context ______ **2.** "About 9:30, **incapacitated** pledges were taken back to sleep it off at the frat house." (paragraph 16)

a. mindless
b. disabled
c. dangerous
d. young

Central Idea and Main Ideas ______ **3.** Which sentence is the best statement of the implied central idea of the article?

a. Binge drinking is a normal and widespread trend in American college life.
b. Parents, educators, and researchers are unable to stop the dangerous trend of college binge drinking.
c. College students indulge in binge drinking to have fun and assert adultlike independence.

d. The national problem of binge drinking among college students is an alarming and escalating threat to their physical and academic well-being that has college campuses scrambling for solutions.

Central Idea and Main Ideas ______ **4.** Which sentence states the main idea of paragraph 6?

a. "The challenge to drink to the very limits of one's endurance has become a celebrated staple of college life."
b. "In one of the most extensive reports on college drinking thus far, a 1997 Harvard School of Public Health study found that 43 percent of college students admitted to drinking in the preceding two weeks."
c. "Defined as four drinks in a sitting for a woman and five for a man, a drinking binge is when one drinks enough to risk health and well-being."

Supporting Details ______ **5.** Which university—"rated as the No. 2 party school in the nation"—did the author visit to research her article?

a. Harvard
b. Princeton
c. University of Wisconsin-Madison
d. University of Pennsylvania

Supporting Details ______ **6.** For a man, binge drinking is defined by downing how many drinks in one sitting?

a. 4
b. 5
c. 8
d. 12

Transitions ______ **7.** "On Thursday night the weekend was already getting started. At a raucous off-campus gathering, 20-year-old Tracey Middler struggled to down her beer as fist-pumping onlookers yelled, 'Chug! Chug! Chug!'" (paragraph 2)

What is the relationship of ideas between these two sentences?

a. time order
b. contrast
c. example

Transitions ______ **8.** "In general, a bottle of beer has about the same alcohol content as a glass of wine or shot of liquor." (paragraph 23)

Which term best describes the relationship of ideas within this sentence?

a. time order
b. example
c. comparison

Thought Patterns

______ **9.** The overall thought pattern used by the author to organize her article is
a. examples of binge drinking and its consequences.
b. an argument to eliminate alcohol from college and university campuses.
c. a discussion of the causes of binge drinking.
d. a narrative about binge drinking.

Thought Patterns

______ **10.** The thought pattern in paragraph 11 is
a. example.
b. cause and effect.
c. time order.

Fact and Opinion

______ **11.** Overall, the ideas in this passage
a. are based on research and statistics.
b. are based on research and the personal observations of the author.
c. objectively present the views of college students.
d. angrily describe the causes and effects of binge drinking.

Fact and Opinion

______ **12.** "Many college students today see not just drinking but being *drunk* as their primary way of socializing." (paragraph 10)

This sentence is a statement of
a. fact.
b. opinion.
c. fact and opinion.

Tone and Purpose

______ **13.** The overall tone of the passage is
a. sympathetic.
b. humorous.
c. objective.
d. outraged.

Tone and Purpose

______ **14.** The tone of paragraph 8 is
a. harsh.
b. regretful.
c. matter-of-fact.
d. disgusted.

Tone and Purpose

______ **15.** The tone of paragraph 27 is
a. angry.
b. puzzled.
c. humble.
d. remorseful.

Tone and Purpose

______ **16.** The author's main purpose is
a. to persuade parents, educators, and students to take action against binge drinking.
b. to shock readers with alarming details about binge drinking.
c. to inform readers about the depth and scope of the alarming national trend of binge drinking among college students.

Inferences ______ **17.** From paragraphs 8 and 9, we can conclude that

a. deaths from binge drinking are inevitable.
b. universities and colleges are responsible for the thousands of student deaths that occur each year.
c. thousands of lives could be saved each year if binge drinking could be controlled.
d. college students understand the risks associated with binge drinking.

Inferences ______ **18.** Based on the details about Mindy Somers (paragraphs 28–32), we can conclude that

a. Mindy engaged in binge drinking frequently.
b. Virginia Tech's annual Alcohol Awareness Week did not effectively educate Mindy about the dangers of binge drinking.
c. Mindy's father should have known about the risks related to her drinking habits.
d. binge drinking is always lethal.

Inferences ______ **19.** The details in paragraphs 2, 3, and 34 imply that

a. the weekend binge drinking cycle probably begins on Thursday.
b. most college students do not "get wasted" during the week.
c. most college students are more interested in partying than in studying.
d. males are more likely to binge and miss classes than are females.

Argument ______ **20.** The following list of ideas from paragraphs 9–11 contains a claim and supports for that claim. In the space, write the letter of the claim for the argument.

a. "The Harvard researchers found that there has been a dramatic change in why students drink: 39 percent drank 'to get drunk' in 1993, but 52 percent had the same objective in 1997."
b. "What has changed is the across-the-board *acceptability* of intoxication," says Felix Savino, a psychologist at UW-Madison.
c. "Researchers surmise that it may have something to do with today's instant-gratification life-style—and young people tend to take it to the extreme."
d. "Many college students today see not just drinking but being *drunk* as their primary way of socializing."

Mapping

Complete the concept map below. Fill in the blanks with the central idea and the missing supporting details from "Binge Drinking, A Campus Killer."

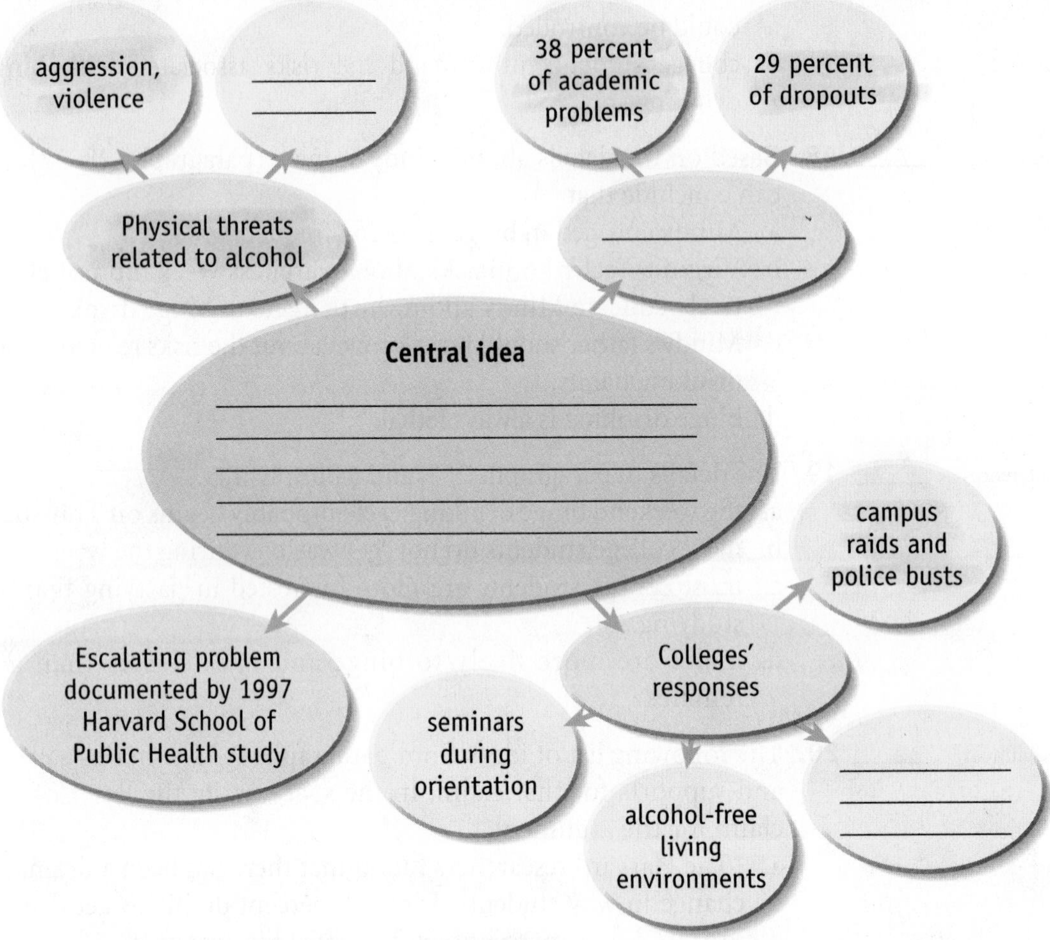

WHAT DO YOU THINK?

Do you think binge drinking is a problem at your college or university? What do you think is the best solution to binge drinking on college campuses? Assume you are a member of your Student Government Association (SGA). SGA has decided to raise awareness about this issue, so you have been asked to write an article for publication in the college and local newspapers. In your editorial, advise college students about the danger of binge drinking and suggest ways to avoid the dangers based on what you have learned from reading this article. Begin your editorial with a summary of the author's main ideas and major supporting details. Also identify and discuss some obvious and underlying reasons for student binge drinking.

MASTER READER Scorecard

"Binge Drinking, A Campus Killer"

Skill	Number Correct		Points		Total
Vocabulary					
Vocabulary in Context (2 items)	______	×	4	=	______
Comprehension					
Central Idea and Main Ideas (2 items)	______	×	4	=	______
Supporting Details (2 items)	______	×	4	=	______
Transitions (2 items)	______	×	4	=	______
Thought Patterns (2 items)	______	×	4	=	______
Fact and Opinion (2 items)	______	×	4	=	______
Tone and Purpose (4 items)	______	×	4	=	______
Inferences (3 items)	______	×	4	=	______
Argument (1 item)	______	×	4	=	______
Mapping (4 items)	______	×	5	=	______
			Comprehension Score		______

READING 4

Is Substance Abuse a Social Problem?

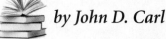

Textbook Skills

by John D. Carl

John D. Carl holds a Ph.D. in Sociology from the University of Oklahoma. His work history includes not only colleges but hospitals, schools, churches, and prisons. This passage is taken from his textbook *Think: Social Problems*. Carl's passion for sociology is evident as he expresses his goal for writing: "the goal of this book is rather simple: To teach students to view social problems critically and to use sociological thinking to help them do that." How would you evaluate the use of drugs in America?

Vocabulary Preview

plethora (paragraph 1): large or excessive amount or number
inebriation (paragraph 1): intoxicated, drunk
gratification (paragraph 9): satisfaction, fulfillment
socialization (paragraph 10): training or developing skills needed to function successfully in society

---

BASICS OF DRUG USE IN AMERICA

1 As you can imagine, drug use and abuse creates a **plethora** of social problems for society. These problems are not only because of the extra costs associated with trying to help people "kick the habit" but also due to the lost productivity, destroyed relationships, and personal injury that such abuse can cause. Of all the drugs available in the United States, tobacco and alcohol are the two most commonly used. Recent surveys have shown that 20.6 percent of Americans over 18 are current smokers. In addition, 23 percent of adults drink alcohol at a dangerous level, and even among youth, 91 percent of drinkers 12 to 14 years old admit to binge drinking. College students in particular fall into the binge-drinking scene, as many students view **inebriation** as a rite of passage. This attitude has created a serious issue for schools, as alcohol-related hospitalizations and deaths are regular occurrences on many college campuses. In fact, numerous colleges are now making it illegal for anyone to consume alcohol on campus, regardless of their age.

CHARACTERISTICS OF DRUG USERS

Age

2 If you think teenagers and college students make up the majority of "users," you're mistaken. It's true that almost 20 percent of

18- to 25-year olds frequently use illegal drugs, compared to 5.8 percent of those 26 and older. However, a 2007 survey on drug use showed that 42 percent of Americans over 26 have used marijuana, while only about 16 percent of young adults have tried this drug. Researchers attribute this pattern to the aging baby boomer population who grew up in a culture of drug use.

3 Even if the researchers are correct in saying that anti-drug ads have been counterproductive, teen drug use has been steadily decreasing since the turn of the millennium. In 1999, it was estimated that more than half of high school seniors used illicit drugs, mostly marijuana. Even pre-teen drug use was high, as one fourth of eighth graders admitted to having been drunk at least once, and 44 percent reported that they smoked cigarettes. Compare those numbers to the current ones that estimate 44 percent of all high school seniors admit to alcohol use, and 18 percent admit to marijuana use.

Race

4 Although drug abuse occurs among all races in the United States, it is more prevalent in some races than others. Asians currently show the lowest rate of drug use among persons 12 and older (3.6 percent); Native Americans and Alaska Natives have the highest percentage of drug use in the United States (13.7 percent).

Socioeconomic Status

5 How does drug use effect socioeconomic status? Certainly, there are many studies that show the connection between illegal drug use and dropping out of school, lower educational attainment, unemployment, and low rates of advancement in one's career. However, in terms of earnings, research suggests that lower incomes for drug users takes time to **manifest** itself. In other words, the effect of drug use has little impact on the individual when he or she is younger, but increases over time. This is in part due to the fact that early in their work lives, drug users tend to take jobs with little potential for growth. Over time, that choice keeps them from advancing, unlike their non-user peers.

PRESCRIPTION DRUG USE

As government and private organizations work to thwart current forms of drug use, Americans are finding new ways to get high. The most current trend in drug abuse is the misuse of prescription drugs. According to the Centers for Disease Control and Prevention, more than 1.8 billion prescription drugs were ordered or provided in 2006, the most frequently prescribed drugs being analgesics, or painkillers. Although these drugs are created and prescribed with the intention of healing or providing relief, they are sometimes abused rather than used appropriately. 6

The three types of prescription drugs that are most commonly abused are opiates (medically prescribed to treat pain), central nervous system depressants (prescribed to treat anxiety and sleep disorders), and stimulants (prescribed to treat disorders such as narcolepsy and attention deficit disorder). Although the use of these types of drugs can help an individual, abuse of these drugs can be extremely harmful to a user's body. 7

Amphetamines, such as Adderall or Ritalin, are some of the most widely abused 8 prescription drugs, especially among teens. Ritalin, a drug prescribed to individuals with attention deficit disorder, has been used without a prescription by as many as 1 in 10 teenagers. As you may already know, some of the most frequent abuse of amphetamines occurs on college campuses. Whereas some students might drink a can of Red Bull to stay alert for a long night of studying, others might turn to amphetamines. One student even admitted to me that he took Adderall to stay awake when he needed to pull all-nighters.

WHY DO WE USE DRUGS?

9 Our bodies have natural defenses against infection and disease; when we're in pain, our brains produce chemicals that dull the ache and make us feel happy. So, why do we need drugs? Some suggest that this is the wrong question. People have always used drugs and will likely continue to use them. Most of society's legal efforts to control drug use fail; the issue becomes how to control use while avoiding the harmful aspects of drug abuse. The causes of drug abuse are difficult to determine and often involve claims that abusers cannot delay **gratification** and have low self-control. However, many people do not agree that addiction is a sign of moral or psychological weakness. Treating addiction as a medical condition—as opposed to a moral failure—arose in the 20th century and resulted in a number of treatment programs and models. In general, these models hold two common beliefs: (1) Individuals have biological predispositions to addiction, and (2) these predispositions can be overcome through treatment.

SOCIALIZATION

Sociological theory and medical research are aware that determining why people use drugs is a complex issue. Researcher Denise Kandel suggests that there is an interaction between how a person is socialized and with whom they interact. Recall that we are socialized by a variety of individuals. Parents provide long-term values, and their use of drugs has the potential to influence their children. However, one's peers provide even more powerful **socialization**. Teens who get involved in drugs and/or alcohol generally have peers who are users. The selection of one's peer group provides the foundation for the likelihood of use by adolescents. This happens in part because they bond to users and learn to share their values and behaviors. Of course, we choose our friends not merely based on who we live near, but also by who interests us. In this way, people who get into drug-using groups in a sense choose their path, although, at the same time, they are being steered toward this path by their group. This is a dynamic process that usually starts with minor alcohol use, such as beer or wine, moves to cigarettes and hard alcohol, then to marijuana, and possibly to harder drugs. To put it simply, people can drift into drugs through a process of being socialized into their use. 10

Another part of drug socialization stems from the fact that the United States is increasingly becoming a medicalized society. Medicalization is the process by which we expand the use of medical terms and solutions to non-medical problems. Medical personnel often claim that certain social 11 components are diseases in need of treatment. Such an attitude expands the power of the medical community, but also increases the public's desire for medical solutions. This leads us to seek a pill to solve our problems. In 2006 alone, more than 1.8 billion prescription drugs were ordered or provided. Does that necessarily make things better? It's clear that what it does do is help socialize us into a mind-set whereby drug use is common.

12 At what point do we change from a culture of drug users to a culture of drug abusers? When so many people seem motivated to abuse substances and modify their behavior with drugs, it's easy to see how this problem can spread to the greater society. A student of mine who had an alcoholic husband put it this way: "Living with an addict is like living with an elephant in your house. It stinks up the place, makes lots of messes, and it's too big to simply act like it's not there because doing that will destroy your family, your children, and your sense of reality." The social problems of substance abuse are not just individual issues.

GO GLOBAL: THE MEXICAN DRUG WAR

13 The real Mexican-American War ended in 1848, but lately a new, covert war has begun. Like any other war, this conflict has resulted in casualties: 6,300 bodies in 2008 alone. This time, however, the Mexican and U.S. governments are working together, battling the rampant drug trafficking that goes on at the border.

14 It's not just drugs that officials are worried about. According to the Federal Bureau of Alcohol, Tobacco, Firearms and Explosives, roughly 90 percent of guns seized in raids of Mexican dealers are traced back to the United States. Drug traffickers cross the border on three-day shopping visas, purchase large quantities of assault weapons and ammunition from local merchants, then return home. It's estimated that nearly 2,000 firearms cross the border into Mexico daily.

15 Americans aren't just providing drug cartels with guns; they're also providing them with money. The majority of drugs cultivated in Mexico are sold to Americans, and Mexican drug dealers generate profits of $15 billion to $25 billion a year solely from the United States.

16 Mexico's drug war isn't only an issue south of the border. The U.S. Justice Department has stated that Mexican drug trafficking organizations represent the greatest organized crime threat to the United States. President Obama has been working with Mexican President Felipe Calderon to build a more aggressive offense against drug lords. "We are absolutely committed to working in partnership with Mexico to make sure that we are dealing with this **scourge** on both sides of the border," Obama said after meeting with Calderon. "You can't fight this war with just one hand. You can't have Mexico making an effort and the United States not making an effort."

17 So far, the President has sent Congress a war-spending request for $350 million to increase security along the United States-Mexico border. More efforts from the U.S. government are expected to be initiated as violence from the Mexican drug war continues to spread.

—Adapted from *Think Social Problems*, pp. 144–147 by John D. Carl. Copyright © 2011 by Pearson Education, Inc. Reproduced by permission of Pearson Education, Inc., Boston, MA.

Choose the best meaning of each word in **bold**. Use context clues to make your choice.

Vocabulary in Context ______ 1. "However, in terms of earnings, research suggests that lower incomes for drug users takes time to **manifest** itself." (paragraph 5)

a. list
b. reveal
c. clear
d. hide

Vocabulary in Context ______ 2. "'We are absolutely committed to working in partnership with Mexico to make sure that we are dealing with this **scourge** on both sides of the border,' Obama said . . ." (paragraph 16)

a. whip
b. law
c. misery
d. issue

Central Ideas and Main Ideas ______ 3. Which of the following sentences from paragraph 1 states the central idea of the passage?

a. "As you can imagine, drug use and abuse creates a plethora of social problems for society."
b. "These problems are not only because of the extra costs associated with trying to help people 'kick the habit' but also due to the lost productivity, destroyed relationships, and personal injury that such abuse can cause."
c. "Of all the drugs available in the United States, tobacco and alcohol are the two most commonly used."
d. "Recent surveys have shown that 20.6 percent of Americans over 18 are current smokers."

Central Ideas and Main Ideas ______ 4. Which of the following sentences states the central idea of paragraphs 6 through 8?

a. "As government and private organizations work to thwart current forms of drug use, Americans are finding new ways to get high." (paragraph 6)
b. "The most current trend in drug abuse is the misuse of prescription drugs." (paragraph 6)
c. "The three types of prescription drugs that are most commonly abused are opiates (medically prescribed to treat pain), central nervous system depressants (prescribed to treat anxiety and sleep disorders), and stimulants (prescribed to treat disorders such as narcolepsy and attention deficit disorder)." (paragraph 7)
d. "Amphetamines, such as Adderall or Ritalin, are some of the most widely abused prescription drugs, especially among teens." (paragraph 8)

Supporting Details ______ **5.** According to the passage, which of the following has the greatest influence an individual's drug use?

a. parents c. peers
b. doctors d. one's self

Supporting Details ______ **6.** According to the passage, how much money has President Obama requested from Congress to increase security along the United States-Mexican border?

a. $15 billion c. $1.8 billion
b. $25 billion d. $350 million

Transitions ______ **7.** "Although the use of these types of drugs can help an individual, abuse of these drugs can be extremely harmful to a user's body." (paragraph 7)

The relationship of ideas within this sentence is
a. addition. c. comparison and contrast.
b. cause and effect. d. generalization and example.

Transitions ______ **8.** "Recent surveys have shown that 20.6 percent of Americans over 18 are current smokers. In addition, 23 percent of adults drink alcohol at a dangerous level, and even among youth, 91 percent of drinkers 12 to 14 years old admit to binge drinking." (paragraph 1)

The relationship of ideas between these sentences is
a. cause and effect. c. comparison and contrast.
b. addition. d. generalization and example.

Thought Patterns ______ **9.** The overall thought pattern for paragraphs 2 through 5 is
a. cause and effect. c. comparison and contrast.
b. classification. d. definition and example.

Thought Patterns ______ **10.** The thought pattern for paragraph 4 is
a. cause and effect. c. comparison and contrast.
b. classification. d. generalization and example.

Thought Patterns ______ **11.** The thought pattern for paragraph 5 is
a. cause and effect. c. comparison and contrast.
b. classification. d. generalization and example.

Fact and Opinion ______ **12.** "Recent surveys have shown that 20.6 percent of Americans over 18 are current smokers." (paragraph 1)

This sentence is a statement of
a. fact. c. fact and opinion.
b. opinion.

Fact and Opinion ______ **13.** "Living with an addict is like living with an elephant in your house." (paragraph 12)

This sentence is a statement of
a. fact. c. fact and opinion.
b. opinion.

Fact and Opinion ______ **14.** "This leads us to seek a pill to solve our problems." (paragraph 11)

This sentence is a statement of
a. fact. c. fact and opinion.
b. opinion.

Fact and Opinion ______ **15.** "This time, however, the Mexican and U. S. governments are working together, battling the rampant drug trafficking that goes on at the border." (paragraph 13)

This sentence is a statement of
a. fact. c. fact and opinion.
b. opinion.

Tone ______ **16.** The overall tone of the passage is
a. informal. b. formal.

Tone ______ **17.** The overall tone of the author is
a. light-hearted. c. dismayed.
b. belittling. d. balanced.

Purpose ______ **18.** The author's purpose is to
a. inform the student about the sociological perspective of drug use in America.
b. entertain the student with interesting facts about drug use in America.
c. persuade the student to take a stand against drug abuse in America.

Inferences ______ **19.** Based on the details in paragraphs 7 and 8, we can infer that Adderall and Ritalin are
a. opiates. c. stimulants.
b. depressants. d. analgesics.

Argument ______ **20.** Identify the persuasive technique used in the following statements:

"Living with an addict is like living with an elephant in your house. It stinks up the place, makes lots of messes, and it's too big to simply act like it's not there because doing that will destroy your family, your children, and your sense of reality." (paragraph 12)

a. false analogy
b. personal attack
c. begging the question
d. transfer

Outlining

Complete the following formal outline with information from the passage.

Central Idea: ____

____

I. Basics of Drug Use in America

II. ______________________

  A. Age

  B. ____________

  C. ____________

III. Prescription Drug Use

IV. Why Do We Use Drugs?

V. Socialization

VI. ______________________

WHAT DO YOU THINK?

Did you learn any new information from this passage? If so, what did you learn? If you already knew all of the information in the passage, which parts of the passage are the most important and why? Do you think America is largely responsible for the drug trafficking between Mexico and the United States? Why or why not? Assume you are taking a college course in sociology, and your professor has handed out study topics for a unit review worth 25% of your final grade. When you and your study group divided the topics up among yourselves to answer and share with the group, you drew the following two topics: (1) Write a short essay that discusses and illustrates how a person is socialized into using drugs; (2) Give several reasons to support or oppose President Obama's request for money to increase security along the United States-Mexico border.

MASTER READER Scorecard

"Is Substance Abuse a Social Problem?"

Skill	Number Correct		Points		Total
Vocabulary					
Vocabulary in Context (2 items)	________	×	4	=	________
Comprehension					
Central Idea and Main Ideas (2 items)	________	×	4	=	________
Supporting Details (2 items)	________	×	4	=	________
Transitions (2 items)	________	×	4	=	________
Thought Patterns (3 items)	________	×	4	=	________
Fact and Opinion (4 items)	________	×	4	=	________
Tone and Purpose (3 items)	________	×	4	=	________
Inferences (1 item)	________	×	4	=	________
Argument (1 item)	________	×	4	=	________
Outlining (5 items)	________	×	4	=	________
			Comprehension Score		________

READING 5

Think You're Operating on Free Will? Think Again

by Eben Harrell

Do you believe in free will? Do you believe that you can control all the choices that you make? Do you believe that your choices determine your future? Why or why not? If not, what controls our decisions? Our genes, our society, supernatural powers? The following article by Eben Harrell, a reporter for *Time* magazine, points us to some interesting findings in recent scientific studies about our "will" to choose. Prior to joining the London bureau in 2007, Harrell worked at the *Aspen Times* in Colorado and the *Scotsman* in Edinburgh.

Vocabulary Preview

rational (paragraph 2): logical, reasonable, sound
subliminally (paragraph 4): without awareness
constructive (paragraph 5): positive, helpful
automaticity (paragraph 6): done unconsciously or from force of habit
skepticism (paragraph 6): disbelief, doubt
exceptionalism (paragraph 10): being different from the norm
undermines (paragraph 10): weakens, undercuts
id (paragraph 10): according to Sigmund Freud, the unconscious part of the mind, the source of inborn impulses and drives
voyagers (paragraph 11): travelers, explorers

---

1 Studies have found that upon entering an office, people behave more competitively when they see a sharp leather briefcase on the desk, they talk more softly when there is a picture of a library on the wall, and they keep their desk tidier when there is a vague scent of cleaning agent in the air. But none of them are consciously aware of the influence of their environment.

2 There may be few things more fundamental to human identity than the belief that people are **rational** individuals whose behavior is determined by conscious choices. But recently psychologists have compiled an impressive body of research that shows how

deeply our decisions and behavior are influenced by unconscious thought, and how greatly those thoughts are swayed by stimuli beyond our immediate comprehension.

3 In an intriguing review in the July 2 edition of the journal *Science*, published online Thursday, Ruud Custers and Henk Aarts of Utrecht University in the Netherlands lay out the mounting evidence of the power of what they term the "unconscious will." "People often act in order to realize desired outcomes, and they assume that consciousness drives that behavior. But the field now challenges the idea that there is only a **conscious** will. Our actions are very often initiated even though we are unaware of what we are seeking or why," Custers says.

4 It is not only that people's actions can be influenced by unconscious stimuli; our desires can be too. In one study cited by Custers and Aarts, students were presented with words on a screen related to puzzles—*crosswords*, *jigsaw piece*, etc. For some students, the screen also flashed an additional set of words so briefly that they could only be detected **subliminally**. The words were ones with positive associations, such as *beach*, *friend* or *home*. When the students were given a puzzle to complete, the students exposed unconsciously to positive words worked harder, for longer, and reported greater motivation to do puzzles than the control group.

5 The same priming technique has also been used to prompt people to drink more fluids after being subliminally exposed to drinking-related words, and to offer **constructive** feedback to other people after sitting in front of a screen that subliminally flashes the names of their loved ones or occupations associated with caring like *nurse*. In other words, we are often not even consciously aware of why we want what we want.

John Bargh of Yale University, who 6
10 years ago predicted many of the findings discussed by Custers and Aarts in a paper entitled "The Unbearable **Automaticity** of Being," called the *Science* paper a "landmark—nothing like this has been in *Science* before. It's a large step toward overcoming the **skepticism** surrounding this research." But Bargh says the field has actually moved beyond the use of subliminal techniques, and studies show that unconscious processes can even be influenced by stimuli within the realms of consciousness, often in unexpected ways. For instance, his own work has shown that people sitting in hard chairs are more likely to be more rigid in negotiating the sales price of a new car, they tend to judge others as more generous and caring after they hold a warm cup of coffee rather than a cold drink, and they evaluate job candidates as more serious when they review their résumés on a heavy clipboard rather than a light one.

"These are **stimuli** that people are 7
conscious of—you can feel the hard chair, the hot coffee—but were unaware that it influenced them. Our unconscious is active in many more ways than this review suggests," he says.

Custers says his work demonstrates 8
that subliminal-advertising techniques—which some countries have outlawed—can

be effective. But he says people concerned about being unconsciously manipulated "should be much more scared of commercials they can see, rather than those they can't see." Many soda commercials, he says by way of example, show the drink with positive-reward cues such as friends or beaches. "If you are exposed to these advertisements over and over again, it does create an association in your mind, and your unconscious is more likely to suddenly decide you want a Coke," he says.

9 But he also says that, at least when it comes to unhealthy food, policymakers are beginning to understand that "personal choice" may be a weak counter to heavy advertising. "We are starting to talk about 'toxic environments' with food and to understand how easy it is to mindlessly reach for a bag of potato chips. Removing such stimuli from the environment can be very effective," he says.

10 Both Custers and Bargh acknowledge that their research undermines a fundamental principle used to promote human **exceptionalism**—indeed, Bargh has in the past argued that his work **undermines** the existence of free will. But Custers also points out that his conclusions are not new: people have long sensed that they are influenced by forces beyond their immediate recognition—be it Greek gods or Freud's unruly **id**. What's more, the unconscious will is vital for daily functioning and probably evolved before consciousness as a handy survival mechanism—Bargh calls it "the evolutionary foundation upon which the scaffolding of consciousness is built." Life requires so many decisions, Bargh says, "that we would be swiftly overwhelmed if we did not have the automatic processes to deal with them."

For his part, Custers says that it is true that our conscious selves are sometimes **voyagers** on a vessel of which they have little control, but he does not see this as a cause for helplessness. "We have to trust that our unconscious sense of what we want and what is good for us is strong, and will lead us largely in the right direction." 11

—"Think You're Operating on Free Will? Think Again." by Eben Harrell from *Time Magazine*, July 2, 2010.

Choose the best meaning of each word in **bold**. Use context clues to make your choice.

Vocabulary in Context  ______ **1.** "But the field now challenges the idea that there is only a **conscious** will. Our actions are very often initiated even though we are unaware of what we are seeking or why," Custers says. (paragraph 3)

a. uncontrolled
b. deliberate
c. innocent
d. strong

Vocabulary in Context __d__ 2. "These are **stimuli** that people are conscious of—you can feel the hard chair, the hot coffee—but were unaware that it influenced them." (paragraph 7)

a. responses
b. supports
c. effects
d. causes

Central Idea __d__ 3. Which of the following sentences from the first two paragraphs states the central idea of the passage?

a. "Studies have found that upon entering an office, people behave more competitively when they see a sharp leather briefcase on the desk, they talk more softly when there is a picture of a library on the wall, and they keep their desk tidier when there is a vague scent of cleaning agent in the air."
b. "But none of them are consciously aware of the influence of their environment."
c. "There may be few things more fundamental to human identity than the belief that people are **rational** individuals whose behavior is determined by conscious choices."
d. "But recently psychologists have compiled an impressive body of research that shows how deeply our decisions and behavior are influenced by unconscious thought, and how greatly those thoughts are swayed by stimuli beyond our immediate comprehension."

Main Idea ______ 4. Which of the following sentences states the main idea of paragraph 4?

a. "It is not only that people's actions can be influenced by unconscious stimuli; our desires can be too."
b. "In one study cited by Custers and Aarts, students were presented with words on a screen related to puzzles—*crosswords*, *jigsaw piece*, etc."
c. "For some students, the screen also flashed an additional set of words so briefly that they could only be detected subliminally."
d. "When the students were given a puzzle to complete, the students exposed unconsciously to positive words worked harder, for longer, and reported greater motivation to do puzzles than the control group."

Main Idea ______ 5. Which of the following sentences states the main idea of paragraph 8?

a. "Custers says his work demonstrates that subliminal-advertising techniques—which some countries have outlawed—can be effective."

b. "But he says people concerned about being unconsciously manipulated 'should be much more scared of commercials they can see, rather than those they can't see.'"
c. "Many soda commercials, he says by way of example, show the drink with positive-reward cues such as friends or beaches."
d. "'If you are exposed to these advertisements over and over again, it does create an association in your mind, and your unconscious is more likely to suddenly decide you want a Coke,' he says."

Supporting Details ___C___ **6.** According to the passage, words such as *beach*, *friend*, or *home* were used subliminally
a. to prompt people to drink more fluids.
b. by John Bargh in the paper he wrote "The Unbearable Automaticity of Being."
c. in a scientific study involving students.
d. to prove the existence of free will.

Supporting Details ___b___ **7.** According to the passage, after holding a warm cup of coffee, people are more likely to
a. be rigid in negotiating the sales price of a new car.
b. judge others as more generous and caring.
c. evaluate job candidates as more serious.
d. be motivated to solve puzzles.

Transitions ___b___ **8.** "But recently psychologists have compiled an impressive body of research that shows how deeply our decisions and behavior are influenced by unconscious thought, and how greatly those thoughts are swayed by stimuli beyond our immediate comprehension." (paragraph 2)

The relationship of ideas within this sentence is
a. classification. c. comparison and contrast.
b. cause and effect. d. time order.

Transitions ___C___ **9.** "Our actions are very often initiated even though we are unaware of what we are seeking or why." (paragraph 3)

The relationship of ideas within this sentence is
a. classification. c. comparison and contrast.
b. cause and effect. d. time order.

Transitions ______ **10.** "'People often act in order to realize desired outcomes, and they assume that consciousness drives that behavior. But the field now challenges the idea that there is only a conscious will. Our actions are very often initiated even though we are unaware of what we are seeking or why,' Custers says." (paragraph 3)

The relationship of ideas between these sentences is

a. cause and effect. c. comparison and contrast.
b. classification. d. generalization and example.

Thought Patterns __b__ **11.** The thought pattern for paragraph 8 is

a. classification. c. comparison and contrast.
b. cause and effect. d. time order.

Thought Patterns ______ **12.** The overall thought pattern for the passage is

a. cause and effect. c. comparison and contrast.
b. classification. d. generalization and example

Fact and Opinion __b__ **13.** "There may be few things more fundamental to human identity than the belief that people are rational individuals whose behavior is determined by conscious choices." (paragraph 2)

This sentence is a statement of

a. fact. c. fact and opinion.
b. opinion.

Fact and Opinion __c__ **14.** "In an intriguing review in the July 2 edition of the journal *Science*, published online Thursday, Ruud Custers and Henk Aarts of Utrecht University in the Netherlands lay out the mounting evidence of the power of what they term the 'unconscious will.'" (paragraph 3)

This sentence is a statement of

a. fact. c. fact and opinion.
b. opinion.

Fact and Opinion __a__ **15.** "In one study cited by Custers and Aarts, students were presented with words on a screen related to puzzles—*crosswords, jigsaw piece,* etc." (paragraph 4)

This sentence is a statement of

a. fact. c. fact and opinion.
b. opinion.

Tone ___d___ **16.** The overall tone of the passage is

a. doubtful.
b. neutral.
c. ironic.
d. persuasive.

Purpose ___c___ **17.** The primary purpose of the passage is

a. to inform.
b. to entertain.
c. to persuade.

Inferences ______ **18.** Based on the details in the passage, we can infer that

a. people do not have the power to make a free choice in any situation, decision, or action.
b. people are not responsible for the decisions they make or the actions they take.
c. we are deeply influenced by our surroundings.
d. humans are puppets controlled by unseen forces.

Argument ______ **19.** Read the claim and supports taken from the passage. Then identify the detail that does not support the claim.

Claim: "It is not only that people's actions can be influenced by unconscious stimuli; our desires can be too." (paragraph 3)

a. "There may be few things more fundamental to human identity than the belief that people are rational individuals whose behavior is determined by conscious choices." (paragraph 2)
b. "But recently psychologists have compiled an impressive body of research that shows how deeply our decisions and behavior are influenced by unconscious thought, and how greatly those thoughts are swayed by stimuli beyond our immediate comprehension." (paragraph 2)
c. "'Our unconscious is active in many more ways than this review suggests,' he says." (paragraph 7)
d. "We are starting to talk about 'toxic environments' with food and to understand how easy it is to mindlessly reach for a bag of potato chips." (paragraph 9)

Argument ___d___ **20.** Identify the persuasive technique used by "subliminal-advertising techniques" as described in paragraph 8.

a. false analogy
b. personal attack
c. begging the question
d. transfer

Summarizing

Complete the following summary of the article "Think You're Operating on Free Will? Think Again" by Eben Harrell. Fill in the blanks with information from the article.

Recent research confronts the long-held belief in the ______________ of human beings. Studies now offer strong evidence that our decisions and behaviors are affected by ______________ beyond our awareness. According to researchers ______________, of Utrecht University in the Netherlands, our ______________ sways our actions and our desires. Their findings agree with the earlier research of John Bargh of Yale University. Bargh also says that the unconscious will likely evolved before the ______________ will as a tool for survival. The unconscious will helps us handle the endless number of decisions we face each day.

WHAT DO YOU THINK?

The article states that some countries have banned subliminal-advertising techniques. Do you think the United States should as well? What about banning the use of propaganda techniques such as transfer, the use of images like the beach or a group of friends, to sell products? Do you think that "'personal choice' may be a weak counter to heavy advertising"? Or do you think consumers are wise enough to resist the power of advertisements? Assume that a senator from your state is pushing a law that strictly controls the kind of advertisements that can be made in the United States. Write a letter to the senator that supports or opposes the ban. Use specific examples to support your stand.

MASTER READER Scorecard

"Think You're Operating on Free Will? Think Again"

Skill	**Number Correct**		**Points**		**Total**
Vocabulary					
Vocabulary in Context (2 items)	_______	×	4	=	_______
Comprehension					
Central Idea and Main Ideas (3 items)	_______	×	4	=	_______
Supporting Details (2 items)	_______	×	4	=	_______
Transitions (3 items)	_______	×	4	=	_______
Thought Patterns (2 items)	_______	×	4	=	_______
Fact and Opinion (3 items)	_______	×	4	=	_______
Tone and Purpose (2 items)	_______	×	4	=	_______
Inferences (1 item)	_______	×	4	=	_______
Argument (2 items)	_______	×	4	=	_______
Summarizing (5 items)	_______	×	4	=	_______
			Comprehension Score		_______

READING 6

The Day Language Came into My Life

by Helen Keller

Helen Keller (1880–1968) was born a hearing child whose bout with disease left her both blind and deaf at the age of 19 months. With the dedicated and loving help of her gifted teacher Anne Sullivan, Helen Keller overcame these barriers to lead an extraordinary life. She graduated with honors from Radcliffe, devoted her life to good works, and was awarded the Presidential Medal of Freedom by Lyndon Johnson in 1964. Her autobiography, *The Story of My Life* (1902), was the basis of a play and two movies. The following passage, an excerpt from her autobiography, takes the reader through the miraculous events that taught her the meaning of words.

Vocabulary Preview

immeasurable (paragraph 1): vast
dumb (paragraph 2): silent
languor (paragraph 2): weariness
succeeded (paragraph 2): followed
plummet (paragraph 3): ball of lead
sounding-line (paragraph 3): a line or wire weighted, usually with a plummet, to measure the depth of water
quiver (paragraph 8): tremble, shiver

---

1 The most important day I remember in all my life is the one on which my teacher, Anne Mansfield Sullivan, came to me. I am filled with wonder when I consider the **immeasurable** contrast between the two lives which it connects. It was the third of March 1887, three months before I was seven years old.

2 On the afternoon of that eventful day, I stood on the porch, **dumb,** expectant. I guessed vaguely from my mother's signs and from the hurrying to and fro in the house that something unusual was about to happen, so I went to the door and waited on the steps. The afternoon sun penetrated the mass of honeysuckle that covered the porch and fell on my upturned face. My fingers lingered almost unconsciously on the familiar leaves and blossoms which had just come forth to greet the sweet southern spring. I did not know what the future held of marvel or surprise for me. Anger and bitterness had preyed upon me continually for weeks and a deep **languor** had **succeeded** this passionate struggle.

3 Have you ever been at sea in a dense fog, when it seemed as if a tangible white

darkness shut you in, and the great ship, tense and anxious, groped her way toward the shore with **plummet** and **sounding-line**, and you waited with beating heart for something to happen? I was like that ship before my education began, only I was without compass or sounding-line and had no way of knowing how near the harbor was. "Light! Give me light!" was the wordless cry of my soul, and the light of love shone on me in that very hour.

4 I felt approaching footsteps. I stretched out my hand as I supposed to my mother. Someone took it, and I was caught up and held close in the arms of her who had come to reveal all things to me, and, more than all things else, to love me.

5 The morning after my teacher came she led me into her room and gave me a doll. The little blind children at the Perkins Institution had sent it and Laura Bridgman had dressed it; but I did not know this until afterward. When I had played with it a little while, Miss Sullivan slowly spelled into my hand the word "d-o-l-l." I was at once interested in this finger play and tried to imitate it. When I finally succeeded in making the letters correctly I was flushed with childish pleasure and pride. Running downstairs to my mother I held up my hand and made the letters for doll. I did not know that I was spelling a word or even that words existed; I was simply making my fingers go in monkeylike imitation. In the days that followed I learned to spell in this uncomprehending way a great many words, among them *pin*, *hat*, *cup* and a few verbs like *sit*, *stand* and *walk*. But my teacher had been with me several weeks before I understood that everything has a name.

One day, while I was playing with my 6
new doll, Miss Sullivan put my big rag doll into my lap also, spelled "d-o-l-l" and tried to make me understand that "d-o-l-l" applied to both. Earlier in the day we had had a tussle over the words "m-u-g" and "w-a-t-e-r." Miss Sullivan had tried to impress it upon me that "m-u-g" is mug and that "w-a-t-e-r" is water, but I persisted in **confounding** the two. In despair she had dropped the subject for the time, only to renew it at the first opportunity. I became impatient at her repeated attempts and, seizing the new doll, I dashed it upon the floor. I was keenly delighted when I felt the fragments of the broken doll at my feet. Neither sorrow nor regret followed my passionate outburst. I had not loved the doll. In the still, dark world in which I lived there was no strong sentiment or tenderness. I felt my teacher sweep the fragments to one side of the hearth, and I had a sense of satisfaction that the cause of my discomfort was removed. She brought me my hat, and I knew I was going out into the warm sunshine. This thought, if a wordless sensation may be called a thought, made me hop and skip with pleasure.

We walked down the path to the well- 7
house, attracted by the fragrance of the honeysuckle with which it was covered. Someone was drawing water and my teacher placed my hand under the spout. As the cool stream gushed over one hand she spelled into the other the word *water*, first slowly, then rapidly. I stood still, my whole attention fixed upon the motions of her fingers. Suddenly I felt a misty consciousness as of something forgotten—a thrill of returning thought; and somehow the

mystery of language was revealed to me. I knew then that "w-a-t-e-r" meant the wonderful cool something that was flowing over my hand. The living word awakened my soul, gave it light, hope, joy, set it free! There were barriers still, it is true, but barriers that could in time be swept away.

8 I left the well-house eager to learn. Everything had a name, and each name gave birth to a new thought. As we returned to the house every object which I touched seemed to **quiver** with life. That was because I saw everything with the strange, new sight that had come to me. On entering the door I remembered the doll I had broken. I felt my way to the hearth and picked up the pieces. I tried vainly to put them together. Then my eyes filled with tears; for I realized what I had done, and for the first time I felt **repentance** and sorrow.

I learned a great many new words that 9
day. I do not remember what they all were; but I do know that *mother*, *father*, *sister*, *teacher* were among them—words that were to make the world blossom for me, "like Aaron's rod, with flowers."* It would have been difficult to find a happier child than I was as I lay in my crib at the close of that eventful day and lived over the joys it had brought me, and for the first time longed for a new day to come.

* Aaron's rod is an allusion to a miracle recorded in the Old Testament wherein God caused a dead stick to sprout with flowers.

Choose the best meaning of each word in **bold**. Use context clues to make your choice.

Vocabulary in Context ______

1. "Miss Sullivan had tried to impress it upon me that 'm-u-g' is mug and that 'w-a-t-e-r' is water, but I persisted in **confounding** the two." (paragraph 6)

a. confusing
b. repeating
c. hating
d. remembering

Vocabulary in Context ______

2. "Then my eyes filled up with tears; for I realized what I had done and for the first time I felt **repentance** and sorrow." (paragraph 8)

a. hurt
b. resentment
c. remorse
d. relief

Central Idea and Main Ideas ______

3. Which sentence best states the central idea of the passage?

a. "The most important day I remember in all my life is the one on which my teacher, Anne Mansfield Sullivan, came to me."
b. "I am filled with wonder when I consider the immeasurable contrast between the two lives which it connects."

c. "I was like that ship before my education began, only I was without compass or sounding-line and had no way of knowing how near the harbor was."

d. "It would have been difficult to find a happier child than I was as I lay in my crib at the close of that eventful day and lived over the joys it had brought me, and for the first time longed for a new day to come."

Central Idea and Main Ideas __c__

4. Which sentence is the best statement of the implied main idea of paragraph 3?

a. Helen Keller was afraid of the sea.
b. Helen Keller felt unloved.
c. Helen Keller couldn't wait to begin learning.
d. Helen Keller understood the sea.

Supporting Details __C__

5. The morning after Miss Sullivan arrived, she gave Helen

a a hug.
b. a cup of water.
c. a doll from the little blind children at the Perkins Institution.
d. a dose of discipline.

Supporting Details __C__

6. Helen broke her doll because

a. she didn't care for it.
b. she was jealous of it.
c. she was angry at having to work with spelling words.
d. she was spoiled and mean.

Thought Patterns __u__

7. The main thought pattern for the overall passage is

a. time order.
b. cause and effect.
c. definition.
d. classification.

Thought Patterns __C__

8. The thought pattern for paragraph 3 is

a. cause and effect
b. classification.
c. comparison and contrast.

Transitions __a__

9. "I stood still, my whole attention fixed upon the motions of her fingers. Suddenly I felt a misty consciousness as of something forgotten—a thrill of returning thought; and somehow the mystery of language was revealed to me." (paragraph 7)

The relationship of ideas between these two sentences is

a. time order.
b. contrast.
c. example.

Transitions ______ **10.** "I felt my teacher sweep the fragments to one side of the hearth, and I had a sense of satisfaction that the cause of my discomfort was removed." (paragraph 6)

The relationship of ideas within this sentence is

a. time order.
b. example.
c. addition.

Fact and Opinion ______ **11.** "I learned a great many new words that day."

This sentence from paragraph 9 is a statement of

a. fact.
b. opinion.
c. fact and opinion.

Fact and Opinion ______ **12.** "I do not remember what they all were; but I do know that *mother, father, sister, teacher* were among them—words that were to make the world blossom for me, 'like Aaron's rod, with flowers.' "

This sentence from paragraph 9 is a statement of

a. fact.
b. opinion.
c. fact and opinion.

Tone and Purpose __a__ **13.** The overall tone of the passage is

a. reflective and joyful.
b. angry and bitter.
c. matter of fact.
d. calm and soothing.

Tone and Purpose ______ **14.** "Then my eyes filled up with tears; for I realized what I had done, and for the first time I felt repentance and sorrow."

The tone of this statement from paragraph 8 is

a. harsh.
b. hopeful.
c. regretful.
d. admiring.

Tone and Purpose ______ **15.** "As the cool stream gushed over one hand she spelled into the other the word *water*, first slowly, then rapidly. I stood still, my whole attention fixed upon the motions of her fingers. Suddenly I felt a misty consciousness as of something forgotten—a thrill of returning thought; and somehow the mystery of language was revealed to me."

The tone of these sentences from paragraph 7 is

a. excited.
b. confused.
c. pained.
d. humorous.

Tone and Purpose ______ **16.** The author's main purpose in "The Day Language Came into My Life" is

a. to persuade others about the joy of teaching the blind and deaf.
b. to entertain readers with a lighthearted story that has a happy ending.
c. to inform the reader about the significance of Anne Sullivan's impact on Helen Keller's life.

Inferences ______ **17.** From the passage, we can conclude that

a. the author was an intelligent and quick learner.
b. the author did not want to learn.
c. Helen Keller would never have learned to communicate without Anne Sullivan.
d. the author's family was ashamed of Helen's disabilities.

Inferences ______ **18.** Based on the details in paragraphs 5–7, we can conclude that

a. an important step in the learning of language is practicing skills.
b. Helen's stubbornness slowed down her learning process.
c. Anne Sullivan was a strict teacher.
d. Helen loved her doll.

Argument ______ **19.** "On entering the door I remembered the doll I had broken. I felt my way to the hearth and picked up the pieces. I tried vainly to put them together. Then my eyes filled with tears; for I realized what I had done, and for the first time I felt repentance and sorrow."

These details from paragraph 8 imply that

a. the doll meant more to Helen once she understood it had a name.
b. the doll can be fixed.
c. the doll was cheaply made and easily broken.
d. Helen had never had a doll before.

Argument ______ **20.** The following list of ideas contains a claim and the supports for that claim. In the given space, write the letter of the claim.

a. Anne Sullivan was a loving and patient teacher to Helen Keller.
b. Helen says that Sullivan "had come to reveal all things to me, and, more than all things else, to love me."
c. When Helen did not make progress with the words *mug* and *water*, Sullivan dropped the subject for a time, only to renew it at the next opportunity.
d. When Helen broke the doll, Sullivan swept the fragments out of the way and continued the lesson at the well-house.

Mapping

Complete the following time line with details for the passage.

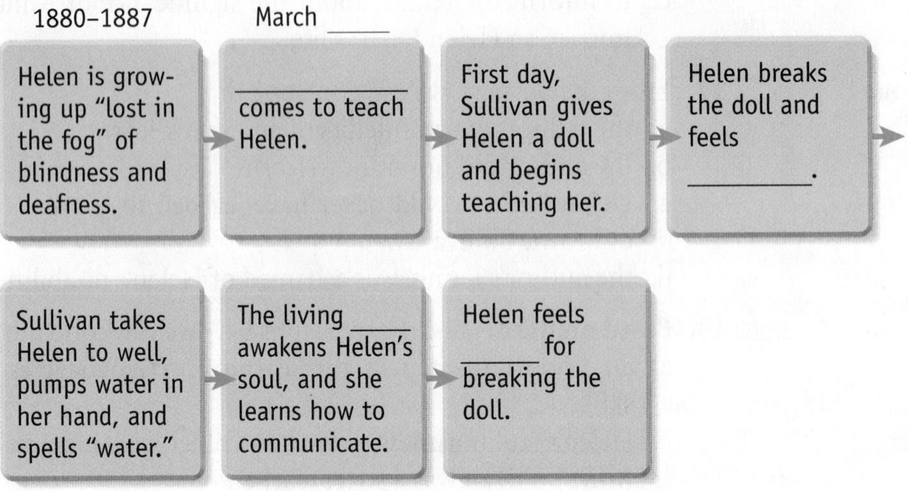

1880–1887
March ____
Helen is growing up "lost in the fog" of blindness and deafness.
________ comes to teach Helen.
First day, Sullivan gives Helen a doll and begins teaching her.
Helen breaks the doll and feels ________.
Sullivan takes Helen to well, pumps water in her hand, and spells "water."
The living ____ awakens Helen's soul, and she learns how to communicate.
Helen feels ______ for breaking the doll.

WHAT DO YOU THINK?

Like Helen Keller, many of us have overcome some obstacle in order to achieve an education. Assume you are writing a letter to a younger member of your family who is struggling academically. This person is seeking your advice about why it is important to attend school. Think back to a defining time in your education. What do you value most about your education, and who has helped you to pursue this goal?

MASTER READER Scorecard

"The Day Language Came into My Life"

Skill	Number Correct		Points		Total
Vocabulary					
Vocabulary in Context (2 items)	______	×	4	=	______
Comprehension					
Central Idea and Main Ideas (2 items)	______	×	4	=	______
Supporting Details (2 items)	______	×	4	=	______
Thought Patterns (2 items)	______	×	4	=	______
Transitions (2 items)	______	×	4	=	______
Fact and Opinion (2 items)	______	×	4	=	______
Tone and Purpose (4 items)	______	×	4	=	______
Inferences (2 items)	______	×	4	=	______
Argument (2 items)	______	×	4	=	______
Mapping (5 items)	______	×	4	=	______
	Comprehension Score				______

READING 7

Fannie Lou Hamer

by Maya Angelou

Poet, writer, performer, teacher, and director, Maya Angelou was raised in Stamps, Arkansas, then moved to San Francisco. In addition to her best-selling autobiographies, beginning with *I Know Why the Caged Bird Sings*, she has also written a cookbook, *Hallelujah! The Welcome Table*, and five poetry collections, including *I Shall Not Be Moved* and *Shaker, Why Don't You Sing?* The following passage appears in *Letter to My Daughter*, Angelou's first original collection in 10 years. In this collection of essays, she shares lessons based on the distilled knowledge of a lifetime well-lived. If you could choose a public figure to honor, whom would you choose? What lessons could be learned from this person's life?

Vocabulary Preview

imperative (paragraph 2): of vital importance, urgent
sequestered (paragraph 3): placed in isolation
laud (paragraph 5): praise
avert (paragraph 5): turn away
embolden (paragraph 11): encourage

---

1 "All of this on account we want to register, to become first-class citizens, and if the Freedom Democratic Party is not seated now, I question America, is this America, the land of the free and the home of the brave, where we have to sleep with our telephones off the hooks because our lives be threatened daily because we want to live as decent human beings, in America? Thank you."

—FANNIE LOU HAMER

2 It is important that we know that those words come from the lips of an African American woman. It is **imperative** that we know those words come from the heart of an American.

3 I believe that there lives a burning desire in the most **sequestered** private heart of every American, a desire to belong to a great country. I believe that every citizen wants to stand on the world stage and represent a noble country where the mighty do not always crush the weak and the dream of a democracy is not the sole possession of the strong.

4 We must hear the questions raised by Fannie Lou Hamer forty years ago. Every

American everywhere asks herself, himself, these questions Hamer asked:

5 What do I think of my country? What is there, which elevates my shoulders and stirs my blood when I hear the words, the United States of America: Do I praise my country enough? Do I **laud** my fellow citizens enough? What is there about my country that makes me hang my head and **avert** my eyes when I hear the words the United States of America, and what am I doing about it? Am I relating my disappointment to my leaders and to my fellow citizens, or am I like someone not involved, sitting high and looking low? As Americans, we should not be afraid to respond.

6 We have asked questions down a pyramid of years and given answers, which our children memorize, and which have become an integral part of the spoken American history. Patrick Henry remarked, "I know not what course others may take, but as for me, give me liberty or give me death."

7 George Moses Horton, the nineteenth century poet, born a slave, said, "Alas, and was I born for this, to wear this brutish chain? I must slash the handcuffs from my wrists and live a man again."

8 "The thought of only being a creature of the present and the past was troubling. I longed for a future too, with hope in it. The desire to be free, awakened my determination to act, to think, and to speak."

—FREDERICK DOUGLASS

9 The love of democracy motivated Harriet Tubman to seek and find not only her own freedom, but to make **innumerable** trips to the slave South to gain the liberty of many slaves and instill the idea into the hearts of thousands that freedom is possible.

10 Fannie Lou Hamer and the Mississippi Democratic Freedom Party were standing on the shoulders of history when they acted to unseat evil from its presumed safe perch on the backs of the American people. It is fitting to honor the memory of Fannie Lou Hamer and surviving members of the Mississippi Democratic Freedom Party. For their gifts to us, we say thank you.

11 The human heart is so delicate and sensitive that it always needs some **tangible** encouragement to prevent it from faltering in its labor. The human heart is so robust, so tough, that once encouraged it beats its rhythm with a loud unswerving insistency. One thing that encourages the heart is music. Throughout the ages we have created songs to grow on and to live by. We Americans have created music to **embolden** the hearts and inspire the spirit of people all over the world.

12 Fannie Lou Hamer knew that she was one woman and only one woman. However, she knew she was an American, and as an American she had a light to shine on the darkness of racism. It was a little light, but she aimed it directly at the gloom of ignorance.

13 Fannie Lou Hamer's favorite was a simple song that we all know. We Americans have sung it since childhood . . .

14 "This little light of mine, I'm going to let it shine, Let it shine,

15 Let it shine,

16 Let it shine.

—"Fannie Lou Hamer" from *Letter to My Daughter* by Maya Angelou, pp. 83–85. Copyright © 2008 by Maya Angelou. Used by permission of Random House, Inc.

Choose the best meaning of each word in **bold**. Use context clues to make your choice.

Vocabulary in Context ______ **1.** "The love of democracy motivated Harriet Tubman to seek and find not only her own freedom, but to make **innumerable** trips to the slave South to gain the liberty of many slaves and instill the idea into the hearts of thousands that freedom is possible." (paragraph 9)

a. few
b. specific
c. immense
d. countless

Vocabulary in Context ______ **2.** "The human heart is so delicate and sensitive that it always needs some **tangible** encouragement to prevent it from faltering in its labor." (paragraph 11)

a. obvious
b. important
c. subtle
d. indescribable

Central Idea ______ **3.** Which of the following sentences states the central idea of the passage?

a. "It is important to know that those words come from the lips of an African American woman." (paragraph 2)
b. "We must hear the questions raised by Fannie Lou Hamer forty years ago." (paragraph 4)
c. "It is fitting to honor the memory of Fannie Lou Hamer and surviving members of the Mississippi Democratic Freedom Party." (paragraph 10)
d. "However, she knew she was an American, and as an American she had a light to shine on the darkness of racism." (paragraph 12)

Main Idea ______ **4.** Which of the following sentences states the main idea of paragraph 11?

a. "The human heart is so delicate and sensitive that it always needs some tangible encouragement to prevent it from faltering in its labor."
b. "The human heart is so robust, so tough, that once encouraged it beats its rhythm with a loud unswerving insistency."
c. "One thing that encourages the heart is music."
d. "Throughout the ages we have created songs to grow on and to live by."

Supporting Details ______ **5.** According to the passage, who said "I must slash the handcuffs from my wrists . . ."?

a. Patrick Henry
b. George Moses Horton
c. Fannie Lou Hamer
d. Frederick Douglass

Supporting Details ______ **6.** According to the passage, Harriet Tubman was motivated by
a. fear of slavery.
b. the shoulders of history.
c. love of democracy.
d. the work of Fannie Lou Hamer.

Transitions ______ **7.** "It was a little light, but she aimed it directly at the gloom of ignorance." (paragraph 12)

The relationship of ideas within this sentence is
a. cause and effect.
b. listing.
c. comparison and contrast.
d. generalization and example.

Transitions ______ **8.** "Fannie Lou Hamer knew that she was one woman and only one woman. However, she knew she was an American, and as an American she had a light to shine on the darkness of racism." (paragraph 12)

The relationship of ideas between these sentences is
a. cause and effect. c. comparison and contrast.
b. listing. d. generalization and example.

Thought Patterns ______ **9.** "The human heart is so delicate and sensitive that it always needs some tangible encouragement to prevent it from faltering in its labor. The human heart is so robust, so tough, that once encouraged it beats its rhythm with a loud unswerving insistency." (paragraph 11)

The thought pattern established by these two sentences is
a. cause and effect. c. comparison and contrast.
b. listing. d. definition and example.

Thought Patterns ______ **10.** The overall thought pattern for paragraph 5 is
a. cause and effect. c. comparison and contrast.
b. listing. d. definition and example.

Fact and Opinion ______ **11.** "It is imperative that we know those words come from the heart of an American." (paragraph 2)

This sentence is a statement of
a. fact. c. fact and opinion.
b. opinion.

Fact and Opinion ______ **12.** "I believe that there lives a burning desire in the most sequestered private heart of every American, a desire to belong to a great country." (paragraph 3)

This sentence is a statement of
a. fact.
b. opinion.
c. fact and opinion.

Fact and Opinion ______ **13.** "We must hear the questions raised by Fannie Lou Hamer forty years ago." (paragraph 4)

This sentence is a statement of
a. fact.
b. opinion.
c. fact and opinion.

Tone ______ **14.** The overall tone of the author is
a. complaining.
b. balanced.
c. bitter.
d. inspirational.

Tone ______ **15.** The tone of the words of Fannie Lou Hamer in the first paragraph is
a. challenging.
b. accepting.
c. frightened.
d. respectful.

Purpose ______ **16.** The purpose of the author is
a. to inform the reader about the racism faced by African Americans such as Fannie Lou Hamer.
b. to entertain the reader with the words and deeds of highly regarded civil rights leaders.
c. to persuade the reader to appreciate and be inspired by Fannie Lou Hamer and the history she represents.

Inferences ______ **17.** Based on the details in the passage, we can infer that Fannie Lou Hamer was working to
a. gain freedom for slaves.
b. be officially represented in the political process.
c. be seated in a restaurant.
d. produce music to further the civil rights movement.

Inferences ______ **18.** Based on the details in the passage, we can infer that
a. there will always be injustice and oppression.
b. determined individuals can significantly impact society for the good.
c. freedom always comes through violence and loss of life.
d. citizens should love their country unconditionally.

Argument _____ **19.** Identify the persuasive technique used in the following sentence:

"I believe that every citizen wants to stand on the world stage and represent a noble country where the mighty do not always crush the weak and the dream of a democracy is not the sole possession of the strong." (paragraph 3)

a. false cause
b. personal attack
c. glittering generality
d. black and white fallacy

Argument _____ **20.** Read the claim and supports taken from the passage. Then identify the detail that does not support the claim.

". . . give me liberty or give me death." (paragraph 6)

a. false cause
b. personal attack
c. glittering generality
d. black and white fallacy

Summarizing

Complete the following summary of the passage:

(1) _______________ and surviving members of the Mississippi Democratic Freedom Party should be honored because they stood on the historical shoulders of (2) _______________, (3) _______________, (4) _______________, and (5) _______________ to shine a light on the gloom of ignorance and the darkness of racism.

WHAT DO YOU THINK?

Do you think it is important to pay tribute to people who have impacted our lives? Why or why not? Whom would you choose to honor for his or her impact on society? Assume you are taking a college course in sociology. Your professor sponsors a course webpage on which he posts samples of student essays about social issues. The current topic is as follows: After reading Maya Angelou's essay "Fannie Lou Hamer," write your own tribute (400–750 words) to a person who has had a significant impact on society. In your essay, identify the social problem and describe how this person made a difference.

MASTER READER Scorecard

"Fannie Lou Hamer"

Skill	Number Correct		Points		Total
Vocabulary					
Vocabulary in Context (2 items)	______	×	4	=	______
Comprehension					
Central Idea and Main Ideas (2 items)	______	×	4	=	______
Supporting Details (2 items)	______	×	4	=	______
Transitions (2 items)	______	×	4	=	______
Thought Patterns (2 items)	______	×	4	=	______
Fact and Opinion (3 items)	______	×	4	=	______
Tone and Purpose (3 items)	______	×	4	=	______
Inferences (2 items)	______	×	4	=	______
Argument (2 items)	______	×	4	=	______
Outlining (5 items)	______	×	4	=	______
				Comprehension Score	______

READING 8

The Truman Library Speech

by Kofi Annan

Kofi Annan delivered his final speech as United Nations Secretary General at the Truman Presidential Museum and Library in Independence, Missouri, in 2006. Throughout his speech, he discusses five lessons he learned as a world leader; the following excerpt is the final portion of his speech. What do you know about the United Nations? Does the world need an international organization to ensure peace and prosperity for all?

Vocabulary Preview

constrained (paragraph 3): controlled, limited
philanthropic (paragraph 4): charitable
Compact (paragraph 6): agreement
multilateral (paragraph 8): many-sided, involving many parties
bequeathed (paragraph 8): bestowed, handed down

---

1 My fourth lesson—closely related to the last one—is that governments must be accountable for their actions in the international arena, as well as in the domestic one.

2 Today the actions of one state can often have a decisive effect on the lives of people in other states. So does it not owe some account to those other states and their citizens, as well as to its own? I believe it does.

3 As things stand, accountability between states is highly skewed. Poor and weak states are easily held to account, because they need foreign assistance. But large and powerful states, whose actions have the greatest impact on others, can be **constrained** only by their own people, working through their domestic institutions.

4 That gives the people and institutions of such powerful states a special responsibility to take account of global views and interests, as well as national ones. And today they need to take into account also the views of what, in UN jargon, we call "non-state actors." I mean commercial corporations, charities and pressure groups, labor unions, **philanthropic** foundations, universities and think tanks—all the **myriad** forms in which people come together voluntarily to think about, or try to change, the world.

5 None of these should be allowed to substitute itself for the state, or for the democratic process by which citizens choose their governments and decide policy. But they all have the capacity to influence political

processes, on the international as well as the national level. States that try to ignore this are hiding their heads in the sand.

6 The fact is that states can no longer—if they ever could—confront global challenges alone. Increasingly, we need to enlist the help of these other actors, both in working out global strategies and in putting those strategies into action once agreed. It has been one of my guiding principles as Secretary General to get them to help achieve UN aims—for instance through the Global **Compact** with international business, which I initiated in 1999, or in the worldwide fight against polio, which I hope is now in its final chapter, thanks to a wonderful partnership between the UN family, the US Centers for Disease Control and—crucially—Rotary International.

7 So that is four lessons. Let me briefly remind you of them: First, we are all responsible for each other's security. Second, we can and must give everyone the chance to benefit from global prosperity. Third, both security and prosperity depend on human rights and the rule of law. Fourth, states must be accountable to each other, and to a broad range of non-state actors, in their international conduct.

8 My fifth and final lesson derives inescapably from those other four. We can only do all these things by working together through a **multilateral** system, and by making the best possible use of the unique instrument **bequeathed** to us by Harry Truman and his contemporaries, namely the United Nations.

9 In fact, it is only through multilateral institutions that states can hold each other to account. And that makes it very important to organize those institutions in a fair and democratic way, giving the poor and the weak some influence over the actions of the rich and the strong.

That applies particularly to the international financial institutions, such as the World Bank and the International Monetary Fund. Developing countries should have a stronger voice in these bodies, whose decisions can have almost a life-or-death impact on their fate. And it also applies to the UN Security Council, whose membership still reflects the reality of 1945, not of today's world. 10

That is why I have continued to press for Security Council reform. But reform involves two separate issues. One is that new members should be added, on a permanent or long-term basis, to give greater representation to parts of the world, which have limited voice today. The other, perhaps even more important, is that all Council members, and especially the major powers who are permanent members, must accept the special responsibility that comes with their privilege. The Security Council is not just another stage on which to act out national interests. It is the management committee, if you will, of our **fledgling** collective security system. 11

As President Truman said, "The responsibility of the great states is to serve and not dominate the peoples of the world." He showed what can be achieved when the US assumes that responsibility. And still today, none of our global institutions can accomplish much when the US remains aloof. But when it is fully engaged, the sky is the limit. 12

These five lessons can be summed up as five principles, which I believe are essential 13 for the future conduct of international relations: collective responsibility, global solidarity, the rule of law, mutual accountability, and multilateralism. Let me leave them with you, in solemn trust, as I hand over to a new Secretary General in three weeks' time.

14 My friends, we have achieved much since 1945, when the United Nations was established. But much remains to be done to put those five principles into practice.

15 Standing here, I am reminded of Winston Churchill's last visit to the White House, just before Truman left office in 1953. Churchill recalled their only previous meeting, at the Potsdam conference in 1945. "I must confess, sir," he said boldly, "I held you in very low regard then. I loathed your taking the place of Franklin Roosevelt." Then he paused for a moment, and continued: "I misjudged you badly. Since that time, you more than any other man, have saved Western civilization."

My friends, our challenge today is not to save Western civilization—or Eastern, for that matter. All civilization is at stake, and we can save it only if all peoples join together in the task. 16

You Americans did so much, in the last century, to build an effective multilateral system, with the United Nations at its heart. Do you need it less today, and does it need you less, than 60 years ago? 17

Surely not. More than ever today Americans, like the rest of humanity, need a functioning global system through which the world's peoples can face global challenges together. And in order to function, the system still cries out for far-sighted American leadership, in the Truman tradition. 18

I hope and pray that the American leaders of today, and tomorrow, will provide it. 19

—From "The Truman Library Speech" by Kofi Annan, December 11, 2006, http://www.un.org/News/ossg/sg/stories/statments_full.asp?statID=40. Reprinted by permission of the United Nations.

VISUAL VOCABULARY

The ______________ efforts of the United States supply donated food in Addis Abba, Ethiopia.

a. constrained
b. bequeathed
c. philanthropic

Choose the best meaning of each word in **bold.** Use context clues to make your choice.

Vocabulary in Context ______ **1.** "I mean commercial corporations, charities and pressure groups, labor unions, philanthropic foundations, universities and think tanks—all the **myriad** forms in which people come together voluntarily to think about, or try to change, the world." (paragraph 4)

a. real
b. little
c. limited
d. countless

Vocabulary in Context ______ **2.** "It is the management committee, if you will, of our **fledgling** collective security system." (paragraph 11)

a. developing
b. expert
c. weakening
d. soaring

Central Idea ______ **3.** Which of the following sentences states the central idea of paragraphs 1 through 6?

a. "My fourth lesson—closely related to the last one—is that governments must be accountable for their actions in the international arena, as well as in the domestic one." (paragraph 1)
b. "As things stand, accountability between states is highly skewed." (paragraph 3)
c. "But large and powerful states, whose actions have the greatest impact on others, can be constrained only by their own people, working through their domestic institutions." (paragraph 3)
d. "The fact is that states can no longer—if they ever could—confront global challenges alone." (paragraph 6)

Central Idea ______ **4.** Which of the following sentences states the central idea of paragraphs 8 through 12?

a. "My fifth and final lesson derives inescapably from those other four." (paragraph 8)
b. "We can only do all these things by working together through a multilateral system, and by making the best possible use of the unique instrument bequeathed to us by Harry Truman and his contemporaries, namely the United Nations." (paragraph 8)
c. "Developing countries should have a stronger voice in these bodies, whose decisions can have almost a life-or-death impact on their fate." (paragraph 10)
d. "And still today, none of our global institutions can accomplish much when the US remains aloof." (paragraph 12)

Supporting Details ______ **5.** According to the passage Rotary International partnered with the UN to
a. engage in worldwide commerce.
b. address worldwide poverty.
c. fight polio worldwide.
d. create multilateral institutions.

Supporting Details ______ **6.** According to passage, the United Nations was established
a. in 1945.
b. in 1953.
c. in 1999.
d. 60 years ago.

Transitions ______ **7.** "These five lessons can be summed up as five principles, which I believe are essential for the future conduct of international relations: collective responsibility, global solidarity, the rule of law, mutual accountability, and multilateralism." (paragraph 13)

The relationship of ideas within this sentence is
a. cause and effect.
b. time order.
c. comparison and contrast.
d. listing.

Transitions ______ **8.** "None of these should be allowed to substitute itself for the state, or for the democratic process by which citizens choose their governments and decide policy. But they all have the capacity to influence political processes, on the international as well as the national level." (paragraph 5)

The relationship of ideas between these sentences is
a. cause and effect.
b. classification.
c. comparison and contrast.
d. generalization and example.

Thought Patterns ______ **9.** The thought pattern for paragraph 7 is
a. cause and effect.
b. listing.
c. time order.
d. comparison and contrast.

Thought Patterns ______ **10.** The thought pattern for paragraph 11 is
a. cause and effect.
b. listing.
c. time order.
d. comparison and contrast.

Fact and Opinion ______ **11.** "In fact, it is only through multilateral institutions that states can hold each other to account." (paragraph 9)

This sentence is a statement of
a. fact.
b. opinion.
c. fact and opinion.

Fact and Opinion ______ **12.** "My friends, we have achieved much since 1945, when the United Nations was established." (paragraph 14)

This sentence is a statement of

a. fact.
b. opinion.
c. fact and opinion.

Fact and Opinion ______ **13.** "You Americans did so much, in the last century, to build an effective multilateral system, with the United Nations at its heart." (paragraph 17)

This sentence is a statement of

a. fact.
b. opinion.
c. fact and opinion.

Tone ______ **14.** The overall tone of the author is

a. optimistic.
b. pessimistic.
c. neutral.

Tone ______ **15.** The tone of paragraph 3 is

a. balanced.
b. timid.
c. emotional.
d. frank.

Purpose ______ **16.** The primary purpose of the author is

a. to inform the audience about the purpose of the United Nations.
b. to entertain the audience with personal experiences and opinions about the United Nations.
c. to inspire American support of the United Nations with lessons learned by serving as leader of the UN.

Inferences ______ **17.** Based on the details in the passage, we can infer that the United States is

a. a superpower state with worldwide influence.
b. not well respected by the international community.
c. unfair in its treatment of other nations.
d. opposed to the United Nations.

Inferences ______ **18.** Based on the details in the passage, we can infer that Kofi Annan calls for

a. one government, the United Nations, to rule the world.
b. the United Nations to grow in military strength.
c. a balance of worldwide power and responsibility through the United Nations.
d. financial aid for the United Nations.

Argument ______ **19.** Identify the persuasive technique used in the following sentence:

"All civilization is at stake, and we can save it only if all peoples join together in the task." (paragraph 16)

a. false analogy
b. either-or
c. personal attack.
d. transfer

Argument ______ **20.** Identify the persuasive technique used in the following sentence:

"And in order to function, the system still cries out for far-sighted American leadership, in the Truman tradition." (paragraph 18)

a. false analogy
b. either-or
c. personal attack
d. transfer

Mapping

Complete the following concept map with information from the passage.

Kofi Annan's Lessons and Principles: The United Nations

Five Lessons	Five Principles
First, we are all responsible for each other's security.	______________
Second, we can and must give everyone the chance to benefit from global prosperity.	______________
Third, both security and prosperity depend on human rights and the rule of law.	______________
Fourth, states must be accountable to each other, and to a broad range of non-state actors, in their international conduct.	______________
Fifth, we can only do all these things by working together through a multilateral system—The United Nations.	______________

WHAT DO YOU THINK?

Do you think powerful countries like the United States have a responsibility to other countries (such as Haiti, Cuba, Mexico, Israel, Iraq, Afghanistan, Great Britain to name a few)? Why or why not? Do you agree with Kofi Annan's opinions in this speech? Why or why not? Assume you eventually want to work with the United Nations, and have been surfing the Internet for information. Through your research, you learn about The World Bank's Essay Competition. The prizes range from $3,000 to $1,000 plus a trip to Sweden for the awards ceremony. The topic is *How can you tackle youth drug abuse through youth-led solutions?* Please answer all questions: (1) How does youth drug abuse affect you, your country, town, or local community? (2) What can you do, working together with your peers, to find a sustainable solution for youth drug abuse? (3) What national or international groups or organizations would you seek help from to reduce youth drug abuse?

MASTER READER Scorecard

From "The Truman Library Speech"

Skill	**Number Correct**		**Points**		**Total**
Vocabulary					
Vocabulary in Context (2 items)	______	×	4	=	______
Comprehension					
Central Idea (2 items)	______	×	4	=	______
Supporting Details (2 items)	______	×	4	=	______
Transitions (2 items)	______	×	4	=	______
Thought Patterns (2 items)	______	×	4	=	______
Fact and Opinion (3 items)	______	×	4	=	______
Tone and Purpose (3 items)	______	×	4	=	______
Inferences (2 items)	______	×	4	=	______
Argument (2 items)	______	×	4	=	______
Mapping (5 items)	______	×	4	=	______
		Comprehension Score			______

READING 9

The Price of Greatness

by Winston S. Churchill

Winston S. Churchill (1874–1965) was the acclaimed Prime Minister of the United Kingdom. At the outbreak of the Second World War, he was appointed First Lord of the Admiralty—a post he had held earlier from 1911 to 1915. In May, 1940, he became Prime Minister and Minister of Defence and remained in office until 1945. Known for his magnificent oratory, Churchill also won the Nobel Prize for Literature in 1953, and Queen Elizabeth II conferred upon him the dignity of Knighthood. In 1963, President Kennedy conferred on him the honorary citizenship of the United States. The following passage is an excerpt from the speech he gave in 1943 when he received an honorary degree at Harvard. Do you think the United States has a responsibility to the world?

Vocabulary Preview

remorselessly (paragraph 2): showing no pity or compassion; continuing without lessening in strength or intensity
indisputable (paragraph 4): beyond doubt, undeniable
prodigious (paragraph 5): extraordinary, impressive
Parliamentarians (paragraph 5): members of a parliament, a type of governing body
anarchy (paragraph 6): disorder, chaos, rebellion
conceptions (paragraph 8): broad understandings, ideas, theories
vigilant (paragraph 10): watchful, on guard
munitions (paragraph 11): weapons, arms
plenary (paragraph 12): comprehensive, complete, fully represented
Bismarck (paragraph 16): Prime Minister, of Prussia from 1862–1890, he oversaw the unification of Germany and designed the German Empire in 1871
amenity (paragraph 18): pleasantness, courtesy, advantage

---

1 Twice in my lifetime the long arm of destiny has reached across the oceans and involved the entire life and manhood of the United States in a deadly struggle.

2 There was no use in saying "We don't want it; we won't have it; our forebears left Europe to avoid these quarrels; we have founded a new world which has no contact

with the old." There was no use in that. The long arm reaches out **remorselessly**, and every one's existence, environment, and outlook undergo a swift and irresistible change. What is the explanation, Mr. President, of these strange facts, and what are the deep laws to which they respond? I will offer you one explanation—there are others, but one will suffice.

3 The price of greatness is responsibility. If the people of the United States had continued in a **mediocre** station, struggling with the wilderness, absorbed in their own affairs, and a factor of no consequence in the movement of the world, they might have remained forgotten and undisturbed beyond their protecting oceans: but one cannot rise to be in many ways the leading community in the civilized world without being involved in its problems, without being convulsed by its agonies and inspired by its causes.

4 If this has been proved in the past, as it has been, it will become **indisputable** in the future. The people of the United States cannot escape world responsibility. Although we live in a period so tumultuous that little can be predicted, we may be quite sure that this process will be intensified with every forward step the United States makes in wealth and in power. Not only are the responsibilities of this great Republic growing, but the world over which they range is itself contracting in relation to our powers of locomotion at a positively alarming rate.

5 We have learned to fly. What **prodigious** changes are involved in that new accomplishment! Man has parted company with his trusty friend the horse and has sailed into the azure with the eagles, eagles being represented by the infernal (loud laughter)—I mean internal—combustion engine. Where, then, are those broad oceans, those vast staring deserts? They are shrinking beneath our very eyes. Even elderly **Parliamentarians** like myself are forced to acquire a high degree of mobility.

But to the youth of America, as to the youth of all the Britains, I say "You cannot stop." There is no halting-place at this point. We have now reached a stage in the journey where there can be no pause. We must go on. It must be world **anarchy** or world order. 6

Throughout all this ordeal and struggle which is characteristic of our age, you will find in the British Commonwealth and Empire good **comrades** to whom you are united by other ties besides those of State policy and public need. To a large extent, they are the ties of blood and history. Naturally I, a child of both worlds, am conscious of these. 7

Law, language, literature—these are considerable factors. Common **conceptions** of what is right and decent, a marked regard for fair play, especially to the weak and poor, a stern sentiment of impartial justice, and above all the love of personal freedom, or as Kipling put it: 8

"Leave to live by no man's leave, underneath the law"—these are common conceptions on both sides of the ocean among the English-speaking peoples. We hold to these conceptions as strongly as you do. 9

We do not war primarily with races as such. Tyranny is our foe, whatever trappings or disguise it wears, whatever language it speaks, be it external or internal, we must forever be on our guard, ever mobilized, ever **vigilant,** always ready to spring 10 at its throat. In all this, we march together. Not only do we march and strive shoulder to shoulder at this moment under the fire of the enemy on the fields of war or in the air, but also in those realms of thought which are consecrated to the rights and the dignity of man.

11 At the present time we have in continual vigorous action the British and United States Combined Chiefs of Staff Committee, which works immediately under the President and myself as representative of the British War Cabinet. This committee, with its elaborate organization of Staff officers of every grade, disposes of all our resources and, in practice, uses British and American troops, ships, aircraft, and **munitions** just as if they were the resources of a single State or nation.

12 I would not say there are never divergences of view among these high professional authorities. It would be unnatural if there were not. That is why it is necessary to have a **plenary** meeting of principals every two or three months. All these men now know each other. They trust each other. They like each other, and most of them have been at work together for a long time. When they meet they thrash things out with great candor and plain, blunt speech, but after a few days the President and I find ourselves furnished with sincere and united advice.

13 This is a wonderful system. There was nothing like it in the last war. There never has been anything like it between two allies. It is reproduced in an even more tightly-knit form at General Eisenhower's headquarters in the Mediterranean, where everything is completely intermingled and soldiers are ordered into battle by the Supreme Commander or his deputy, General Alexander, without the slightest regard to whether they are British, American, or Canadian, but simply in accordance with the fighting need.

Now in my opinion it would be a most foolish and **improvident** act on the part of our two Governments, or either of them, to break up this smooth-running and immensely powerful machinery the moment the war is over. For our own safety, as well as for the security of the rest of the world, we are bound to keep it working and in running order after the war—probably for a good many years, not only until we have set up some world arrangement to keep the peace, but until we know that it is an arrangement which will really give us that protection we must have from danger and aggression, a protection we have already had to seek across two vast world wars. 14

I am not qualified, of course, to judge whether or not this would become a party question in the United States, and I would not presume to discuss that point. I am sure, however, that it will not be a party question in Great Britain. We must not let go of the securities we have found necessary to preserve our lives and liberties until we are quite sure we have something else to put in their place, which will give us an equally solid guarantee. 15

The great **Bismarck**—for there were once great men in Germany—is said to have observed towards the close of his life that the most potent factor in human society at the end of the nineteenth century was the fact that the British and American peoples spoke the same language. 16

That was a pregnant saying. Certainly it has enabled us to wage war together with 17 an intimacy and harmony never before achieved among allies.

18 This gift of a common tongue is a priceless inheritance, and it may well someday become the foundation of a common citizenship. I like to think of British and Americans moving about freely over each other's wide estates with hardly a sense of being foreigners to one another. But I do not see why we should not try to spread our common language even more widely throughout the globe and, without seeking selfish advantage over any, possess ourselves of this invaluable **amenity** and birthright.

Choose the best meaning of each word in **bold**. Use context clues to make your choice.

Vocabulary in Context ______ **1.** "The price of greatness is responsibility. If the people of the United States had continued in a **mediocre** station, struggling with the wilderness, absorbed in their own affairs, and a factor of no consequence in the movement of the world . . ." (paragraph 3)

a. unusual
b. commonplace
c. special
d. extreme

Vocabulary in Context ______ **2.** "Throughout all this ordeal and struggle which is characteristic of our age, you will find in the British Commonwealth and Empire good **comrades** to whom you are united by other ties besides those of State policy and public need." (paragraph 7)

a. enemies
b. soldiers
c. friends
d. ideas.

Central Idea ______ **3.** Which of the following sentences states the central idea of the passage?

a. "Twice in my lifetime the long arm of destiny has reached across the oceans and involved the entire life and manhood of the United States in a deadly struggle." (paragraph 1)
b. "The price of greatness is responsibility." (paragraph 3)
c. "It must be world anarchy or world order." (paragraph 6)
d. "This gift of a common tongue is a priceless inheritance, and it may well someday become the foundation of a common citizenship." (paragraph 18)

Supporting Details ______ **4.** According to the passage prodigious changes in the world have occurred because of
a. world wars.
b. law.
c. high mobility.
d. the establishment of the United States.

Supporting Details ______ **5.** According to Bismarck, what was the most potent fact in human society at the end of the nineteenth century?
a. free movement
b. the great men of Germany
c. the common values of Britain and the U. S.
d. the common tongue of Britain and the U. S.

Transitions ______ **6.** "We hold to these conceptions as strongly as you do." (paragraph 9)

The relationship of ideas within this sentence is
a. cause and effect.
b. time order.
c. comparison and contrast.
d. generalization and example.

Transitions ______ **7.** "There is no halting-place at this point. We have now reached a stage in the journey where there can be no pause." (paragraph 6)

The relationship of ideas between these sentences is
a. cause and effect.
b. time order.
c. comparison and contrast.
d. generalization and example.

Thought Patterns ______ **8.** The thought pattern for paragraph 3 is
a. cause and effect.
b. time order.
c. comparison and contrast.
d. definition and example.

Thought Patterns ______ **9.** The thought pattern for paragraph 4 is
a. cause and effect.
b. time order.
c. comparison and contrast.
d. definition and example.

Fact and Opinion ______ **10.** "Certainly it has enabled us to wage war together with an intimacy and harmony never before achieved among allies." (paragraph 17)

This sentence is a statement of
a. fact.
b. opinion.
c. fact and opinion.

Fact and Opinion ______ **11.** "At the present time we have in continual vigorous action the British and United States Combined Chiefs of Staff Committee, which works immediately under the President and myself as representative of the British War Cabinet." (paragraph 11)

This sentence is a statement of

a. fact.
b. opinion.
c. fact and opinion.

Fact and Opinion ______ **12.** "Now in my opinion it would be a most foolish and improvident act on the part of our two Governments, or either of them, to break up this smooth-running and immensely powerful machinery the moment the war is over." (paragraph 14)

This sentence is a statement of

a. fact.
b. opinion.
c. fact and opinion.

Tone ______ **13.** The overall tone of the author is

a. neutral.
b. critical.
c. arrogant.
d. approving.

Tone ______ **14.** The tone of the author in paragraph 15 is

a. humiliated.
b. arrogant.
c. confident.
d. timid.

Purpose ______ **15.** The primary purpose of the author is

a. to inform the audience of the accomplishments of Great Britain and the United States during the war.
b. to celebrate the shared values, accomplishments, and responsibilities of Great Britain and the United States during and after the war.
c. to persuade the audience to support the war efforts of Great Britain and the United States.

Inferences ______ **16.** Based on the details in paragraph 14, we can infer that **improvident** means

a. wise.
b. fateful.
c. important.
d. careless.

Inferences ______ **17.** Based on the details in the passage, we can conclude that Churchill believes that the United States is responsible for

a. World War II.
b. the security of the rest of the world after the war.
c. our powers of locomotion.
d. the gift of a common language.

Argument ______ **18.** Identify the persuasive technique used in the following sentence:

"Man has parted company with his trusty friend the horse and has sailed into the azure with the eagles." (paragraph 5)

a. plain folks
b. either-or
c. glittering generality
d. false cause

Argument ______ **19.** Identify the persuasive technique used in the following sentence:

"It must be world anarchy or world order." (paragraph 6)

a. false comparison
b. either-or
c. plain folks
d. transfer

Argument ______ **20.** Identify the persuasive technique used in the following sentence:

"Not only do we march and strive shoulder to shoulder at this moment under the fire of the enemy on the fields of war or in the air, but also in those realms of thought which are consecrated to the rights and the dignity of man." (paragraph 10)

a. glittering generality
b. either-or
c. false cause
d. false comparison

Summarizing

Complete the following summary with information from the passage.

According to Winston Churchill in his speech given when he received an honorary degree at Harvard, the people of the United States cannot escape world ______________. The responsibilities of the United States grow as its wealth and power increases, and the changes brought by ______________ shrinks the range of the world. The shared ______________ and the common ______________ of Great Britain and the United States built a smooth-running and immensely powerful machinery of war that should continue until ______________ ______________.

WHAT DO YOU THINK?

Do you agree with Churchill's claim that a "gift of a common tongue" [the English language] is a priceless inheritance, and it may well someday become the foundation of a common citizenship? Why or why not? Why would a common language help secure world peace and prosperity? Do you think the United States should create a law that makes English its official language? Or do you think the United States should encourage its citizens to learn more than one language? If so, which language should be learned? Assume that your local Board of Education is considering adding a foreign language as a requirement for graduation from high school. Write a letter to the board in favor of or in opposition to this proposal. If you support the proposal, identify the language you believe should be taught and why. If you oppose the proposal, explain why students should learn only English.

MASTER READER Scorecard

From "The Price of Greatness"

Skill	Number Correct		Points		Total
Vocabulary					
Vocabulary in Context (2 items)	______	×	4	=	______
Comprehension					
Central Idea (1 item)	______	×	4	=	______
Supporting Details (2 items)	______	×	4	=	______
Transitions (2 items)	______	×	4	=	______
Thought Patterns (2 items)	______	×	4	=	______
Fact and Opinion (3 items)	______	×	4	=	______
Tone and Purpose (3 items)	______	×	4	=	______
Inferences (2 items)	______	×	4	=	______
Argument (3 items)	______	×	4	=	______
Summarizing (5 items)	______	×	4	=	______
			Comprehension Score		______

READING 10

Real People in the "Age of the Common Man"

Textbook Skills

by Jacqueline Jones, Peter H. Wood, Thomas Borstelmann, Elaine Tyler May, and Vicki L. Ruiz

In the preface of the *Created Equal* textbook from which this passage comes, the authors state, "*Created Equal* tells the dramatic, evolving story of America in all its complexity—a story of a diverse people 'created equal' yet struggling to achieve equality." The following excerpt addresses the American ideal of equality and the concept of the "common man." What do you think the term "common man" meant during the 1800s? How would you define the "common man" in America today?

Vocabulary Preview

hierarchy (paragraph 1): formally ranked group, chain of command
suffrage (paragraph 1): right to vote, act of voting
egalitarian (paragraph 1): democratic, equal, free
denigrate (paragraph 3): belittle, scorn, degrade
assimilate (paragraph 8): integrate, blend in, conform
interlopers (paragraph 9): intruders, trespassers
deference (paragraph 17): respect, submission
subversive (paragraph 19): rebellious, defiant
atole (paragraph 25): hot sweet drink made from corn dough
vanguard (paragraph 30): front line, forerunner
advocate (paragraph 35): supporter, activist

---

1 In the early 1830s, a wealthy Frenchman named Alexis de Tocqueville visited the United States and wrote about the contradictions he saw. In his book *Democracy in America* (published in 1835), Tocqueville noted that the United States lacked the rigid **hierarchy** of class privilege that characterized European nations. With universal white manhood **suffrage**, white men could vote and run for office regardless of their class or religion. However, Tocqueville also noted some sore spots in American democratic values and practices. He commented on the plight of groups deprived of the right to vote; their lack of freedom stood out starkly in the otherwise **egalitarian** society of the

United States. He sympathized with the southeastern Indians uprooted from their homelands. He raised the possibility that conflicts between blacks and whites might eventually lead to bloodshed. He even contrasted the situation of young unmarried white women, who seemed so free-spirited, with that of wives, who appeared cautious and dull. He concluded, "In America a woman loses her independence forever in the bonds of matrimony." In other words, Tocqueville saw America for what it was: a blend of freedom and slavery, of independence and dependence.

Wards, Workers, and Warriors: Native Americans

2 Population growth in the United States—and on the borderlands between the United States and Mexican territory—put pressure on Indian societies. Yet different cultural groups responded in different ways to this pressure. Some, like the Cherokee, conformed to European American ways and became sedentary farmers. Others were forced to work for whites. Still others either waged war on white settlements and military forces or retreated farther and farther from European American settlements in the hope of avoiding clashes with the intruders.

3 Nevertheless, prominent whites continued to **denigrate** the humanity of all Indians. In the 1820s Henry Clay claimed that Indians were "essentially inferior to the Anglo-Saxon race . . . and their disappearance from the human family will be no great loss to the world." In 1828 the House of Representatives Committee on Indian Affairs surveyed the Indians of the South and concluded that "an Indian cannot work" and that all Indians were lazy and notable for their "thirst for spirituous liquours." According to the committee, when European American settlers depleted reserves of wildlife, Indians as a group would cease to exist.

Members of the Cherokee Nation bitterly denounced these assertions. "The Cherokees do not live upon the chase [for game]," they pointed out. Neither did the Creek, Choctaw, Chickasaw, and Seminole—the other members of the Five Civilized Tribes, so called for their varying degrees of conformity to white people's ways. 4

Charting a middle course between the Indian and European American worlds was Sequoyah, the son of a white Virginia trader-soldier and a Cherokee woman. A veteran of Andrew Jackson's campaign against the Creek in 1813–1814, Sequoyah moved to Arkansas in 1818, part of an early Cherokee migration west. In 1821 he finished a Cherokee syllabary (a written language consisting of syllables and letters, in contrast to pictures, or pictographs). The product of a dozen years' work, the syllabary consisted of 86 characters. In 1828 the Cherokee Phoenix, a newspaper based on the new writing system, began publication in New Echota, Georgia. 5

Sequoyah's written language enabled the increasingly dispersed Cherokee to remain in touch with each other on their own terms. At the same time, numerous Indian cultural groups lost their struggle to retain even modest control over their destinies. In some areas of the continent, smallpox continued to ravage native populations. In other regions, Indians became wards of, or dependent on, whites, living with and working for white families. Other groups, living close to whites, adopted their trading practices. In 6

Spanish California, the Muquelmne Miwok in the San Joaquin delta made a living by stealing and then selling the horses of Mexican settlers.

7 In other parts of California, Spanish missionaries conquered Indian groups, converted them to Christianity, and then forced them to work in the missions. In missions up and down the California coast, Indians worked as weavers, tanners, shoemakers, bricklayers, carpenters, blacksmiths, and other artisans. Some herded cattle and raised horses. Indian women cooked for the mission, cleaned, and spun wool. They wove cloth and sewed garments.

8 Nevertheless, even Indians living in or near missions resisted the cultural change imposed by the intruders. Catholic missionaries complained that Indian women such as those of the Chumash refused to learn Spanish. The refusal among some Indians to **assimilate** completely signaled persistent, deep-seated conflicts between native groups and incoming settlers. In 1824 a revolt among hundreds of newly converted Indians at the mission La Purisima Concepción north of Santa Barbara revealed a rising militancy among native peoples.

9 After the War of 1812, the U.S. government had rewarded some military veterans with land grants in the Old Northwest. Federal agents tried to clear the way for these new settlers by ousting Indians from the area. Overwhelmed by the number of whites, some Indian groups such as the Peoria and Kaskaskia gave up their lands to the **interlopers.** Others took a stand against the white intrusion. In 1826 and 1827 the Winnebago attacked white families and boat pilots living near Prairie du Chien, Wisconsin. Two years later, the Sauk chief Black Hawk (known to Indians as Maka-tai-me-she-kia-kiak) assembled a coalition of Fox, Winnebago, Kickapoo, and Potawatomi. Emboldened by the prospect of aid from British Canada, they clashed with federal troops and raided farmers' homesteads and miners' camps.

10 In August 1832 a force of 1,300 U.S. soldiers and volunteers struck back, killing 300 Indian men, women, and children encamped on the Bad Axe River in western Wisconsin. The massacre, the decisive point of what came to be called the Black Hawk War, marked the end of armed Indian resistance north of the Ohio River and east of the Mississippi.

Slaves and Free People of Color

11 In the 1820s the small proportion of free blacks within the southern population declined further. Southern whites perceived free blacks as an unwelcome and dangerous presence, especially given the possibility that they would conspire with slaves to spark a rebellion. For these reasons some states began to outlaw private manumissions (the practice of individual owners freeing their slaves) and to force free blacks to leave the state altogether.

12 One free black who inspired such fears was Denmark Vesey. Born on the Danish-controlled island of Saint Thomas in 1767, Vesey was a literate carpenter as well as a religious leader. In 1799 he won $1,500 in a Charleston, South Carolina, lottery and used some of the money to buy his freedom. In the summer of 1822, a Charleston court claimed to have unearthed evidence of a "diabolical plot" hatched by Vesey together with plantation slaves from the surrounding area.

13 Yet the historical record strongly suggests that no plot ever existed. Black "witnesses" who feared for their own lives provided inconsistent and contradictory testimony to a panel of judges. Authorities never located any material evidence of a plan, such as stockpiles of weapons. Under fire from other Charleston elites for rushing to judgment, the judges redoubled their efforts to **embellish** vague rumors of black discontent into a tale of a well-orchestrated uprising and to implicate growing numbers of black people. As a result of the testimony of several slaves, 35 black men were hanged and another 18 exiled outside the United States. Of those executed, Vesey and 23 other men said nothing to support even the vaguest charges of the court.

14 In the North, some blacks were granted the right to vote after emancipation in the late eighteenth century; however, many of those voting rights were lost in the early nineteenth century. New Jersey (in 1807), Connecticut (1818), New York (1821), and Pennsylvania (1838) all revoked the legislation that had let black men cast ballots. Free northern blacks continued to suffer under a number of legal restrictions. Most were not citizens and therefore perceived themselves as oppressed like the slaves in the South.

15 A new group of black leaders in the urban North began to link their fate to that of their enslaved brothers and sisters in the South. In Boston, North Carolina-born David Walker published his fiery *Walker's Appeal* to the Coloured Citizens of the World in 1829. Walker called for all blacks to integrate fully into American society, shunning racial segregation whether initiated by whites or by blacks themselves. Reminding his listeners of the horrors of the slave trade, he declared that black people were ready to die for freedom: "I give it as a fact, let twelve black men get well armed for battle, and they will kill and put to flight fifty whites."

Northern black leaders disagreed 16
among themselves on the issues of integration and black separatism—for example, whether blacks should create their own schools or press for inclusion in the public educational system. A few leaders favored leaving the country altogether, believing that black people would never find peace and freedom in the United States. Founded by whites in New Jersey in 1817, the American Colonization Society (ACS) paid for black Americans to settle Monrovia (later named Liberia) on the west coast of Africa. The ACS drew support from a variety of groups: whites in the upper South who wanted to free their slaves but believed that black and white people could not live in the same country, and some slaves and free people of color convinced that colonization would give them a fresh start. A small number of American-born blacks settled in Liberia. However, most black activists rejected colonization. They had been born on American soil, and their forebears had been buried there. Maria Stewart, an African American religious leader in Boston, declared, "But before I go [to Africa] the bayonet shall pierce me through."

Northern whites sought to control 17
black people and their movements. Outspoken black men and women such as Walker and Stewart alarmed northern whites who feared that if blacks could claim decent jobs,

white people would lose their own jobs. African Americans who worked outdoors as wagon drivers, peddlers, and street sweepers were taunted and in some cases attacked by whites who demanded **deference** from blacks in public. In October 1824 a white mob invaded a black neighborhood in Providence, Rhode Island. They terrorized its residents, destroyed buildings, and left the place "almost entirely in ruins." The catalyst for the riot had come the previous day, when a group of blacks had refused to yield the inside of the sidewalk—a cleaner place to walk—to white passersby.

18 The South's silence was broken in 1831 when white Southerners took steps to reinforce the institution of slavery, using both violent and legal means. That year Nat Turner, an enslaved preacher and mystic, led a slave revolt in Southampton, Virginia. In the 1820s the young Turner had looked skyward and had seen visions of "white spirits and black spirits engaged in battle . . . and blood flowed in streams." Turner believed that he had received divine instructions to lead other slaves to freedom, to "arise and prepare myself, and slay my enemies with their own weapons." In August he and a group of followers that eventually numbered 80 moved through the countryside, killing whites wherever they could find them. Ultimately, nearly 60 whites died at the hands of Turner's rebels. Turner himself managed to evade capture for more than two months. After he was captured, he was tried, convicted, and sentenced to death. A white man named Thomas Gray interviewed Turner in his jail cell and recorded his "confessions" before he was hanged.

Published in 1832 by Gray, *The Con-* 19
fessions of Nat Turner reached a large, horrified audience in the white South. According to Gray, Turner said that he had exhibited "uncommon intelligence" when he was a child. As a young man, he had received inspiration from the Bible, especially the passage "Seek ye the kingdom of Heaven and all things shall be added unto you." Perhaps most disturbing of all, Turner reported that, since 1830, he had been a slave of "Mr. Joseph Travis, who was to me a kind master, and placed the greatest confidence in me; in fact, I had no cause to complain of his treatment to me." Turner's "confessions" suggested the **subversive** potential of slaves who were literate and Christian and those who were treated kindly by their masters and mistresses.

After the Turner revolt, a wave of white 20
hysteria swept the South. In Virginia near where the killings had occurred, whites assaulted blacks with unbridled fury. The Virginia legislature seized the occasion to defeat various antislavery proposals. Thereafter, all the slave states moved to strengthen the institution of slavery. For all practical purposes, public debate over slavery ceased throughout the American South.

Legal and Economic Dependence: The Status of Women

In the political and economic realms, 21
the egalitarian impulse rarely affected the status of women in a legal or practical sense. Regardless of where they lived, enslaved women and Indian women had almost no rights under either U.S. or Spanish law. Still, legal systems in the United States and the Spanish borderlands differed

in their treatment of women. In the United States, most of the constraints that white married women had experienced in the colonial period still applied in the 1820s. A husband controlled the property that his wife brought to the marriage, and he had legal authority over their children. Indeed, the wife was considered her husband's possession. She had no right to make a contract, keep money she earned, vote, run for office, or serve on a jury. In contrast, in the Spanish Southwest, married women could own land and conduct business on their own. At the same time, however, husbands, fathers, and local priests continued to exert much influence over the lives of these women.

22 European American women's economic subordination served as a rationale for their political inferiority. The "common man" concept rested on the assumption that only men could ensure American economic growth and well-being. According to this view, men had the largest stake in society because only they owned property. That stake made them responsible citizens.

23 Yet women contributed to the economy in myriad ways. Although few women earned cash wages in the 1820s, almost all adult women worked. In the colonial period, society had highly valued women's labor in the fields, the garden, and the kitchen. However, in the early nineteenth century, work was becoming increasingly identified as labor that earned cash wages. This attitude proved particularly common in the Northeast, where increasing numbers of workers labored under the supervision of a boss. As this belief took root, men began valuing women's contributions to the household economy less and less. If women did not earn money, many men asked, did they really work at all?

In these years, well-off women in the 24
northeastern and mid-Atlantic states began to think of themselves as consumers and not producers of goods. They relied more and more on store-bought cloth and household supplies. Some could also afford to hire servants to perform housework for them. Privileged women gradually stopped thinking of their responsibilities as making goods or processing and preparing food. Rather, their main tasks were to manage servants and create a comfortable home for their husbands and children. They saw their labor as necessary to the well-being of their families, even if their compensation came in the form of emotional satisfaction rather than cash.

In contrast, women in other parts of 25
the country continued to engage in the same forms of household industry that had characterized the colonial period. In Spanish settlements, women played a central role in household production. They made all of their family's clothes by carding, spinning, and weaving the wool from sheep. They tanned cowhides and ground blue corn to make tortillas, or **atole**. They produced their own candles and soap, and they plastered the walls of the home.

Like women's work in general, the labor 26
of wives and mothers in Spanish-speaking regions had great cultural significance. In the Mexican territory of California, women engaged in backbreaking efforts so that members of their families could wear snow-white linen clothing. One community member recalled that "certainly to do so was one of

the chief anxieties" of well-to-do households: "There was sometimes a great deal of linen to be washed for it was the pride of every Spanish family to own much linen, and the mothers and daughters almost always wore white." Women used homemade soap to scrub the clothes on the rocks of a nearby spring. Then they spread out the wet garments to dry on the tops of bushes that grew on the mountainside.

27 In the Spanish mission of San Gabriel, California, the widow Eulalia Perez cooked, sewed, ministered to the ill, and instructed children in reading and writing. As housekeeper, Perez kept the keys to the mission storehouse. She also distributed supplies to the Indians and the vaqueros (cowboys) who lived in the mission. She supervised Indian servants as well as soap makers, wine pressers, and olive oil producers.

28 At Mission San Diego, Apolonaria Lorenzana worked as a healer and cared for the church sacristy and priestly vestments. From the time she arrived in Monterey at age 7 (in 1800) until her death in the late nineteenth century, Lorenzana devoted her life to such labors. Although the priests tried to restrict her to administering the mission hospital, she took pride in her nursing abilities, "even though Father Sanchez had told me not to do it myself, but to have it done, and only to be present so that the servant girls would do it well."

29 Indian women also engaged in a variety of essential tasks. Sioux and Mandan women, though of a social rank inferior to men, performed a great deal of manual labor in their own villages. They dressed buffalo skins that the men later sold to traders. They collected water and wood, cooked, dried meat and fruit, and cultivated maize (corn), pumpkins, and squash with hoes made from the shoulder blades of elk. These women worked collectively within a network of households rather than individually within nuclear families.

Many women, regardless of ethnicity, 30
were paid for their work with food and shelter but not money. Nevertheless, some women did work for cash wages during this era. New England women and children, for example, were the **vanguard** of factory wage-earners in the early manufacturing system. In Massachusetts in 1820, women and children constituted almost a third of all manufacturing workers. In the largest textile factories, they made up fully 80 percent of the workforce.

The business of textile manufacturing 31
took the tasks of spinning thread and weaving cloth out of the home, where such tasks often were performed by unpaid, unmarried daughters, and relocated those tasks in factories, where the same workers received wages. The famous "Lowell mill girls" are an apt example. Young, unmarried white women from New Hampshire, Vermont, and Massachusetts, these workers moved to the new company town of Lowell, Massachusetts, to take jobs as textile machine operatives. In New England, thousands of young men had migrated west, tipping the sex ratio in favor of women and creating a reserve of female laborers. But to attract young women to factory work, mill owners had to reassure them (and their parents) that they would be safe and well cared for away from home. To that end, they established boarding houses where employees could live together under the supervision of a matron—an older

woman who served as their mother-away-from-home.

32 Company towns set rules shaping employees' living conditions as well as their working conditions. In the early 1830s, a posted list of "Rules and Regulations" covered many aspects of the lives of the young women living at the Poignaud and Plant boardinghouse at Lancaster, Massachusetts. The list told the women how to enter the building (quietly, and then hang up "their bonnet, shawl, coat, etc. etc. in the entry") and where to sit at the dinner table (the two workers with greatest seniority were to take their places at the head of the table). Despite these rules, many young women valued the friendships they made with their coworkers and the money they made in the mills. Some of these women sent their wages back home so that their fathers could pay off the mortgage or their brothers could attend school.

33 But not all women wage-earners labored in large mills. In New York City, single women, wives, and widows toiled as needle workers in their homes. Impoverished, sewing in tiny attics by the dim light of candles, these women were at the mercy of jobbers—merchants who parceled out cuffs, collars, and shirt fronts that the women finished. Other urban women worked as street vendors, selling produce, or as cooks, nursemaids, or laundresses.

34 The new **delineation** between men's and women's work and workplaces intensified the drive for women's education begun after the Revolution. If well-to-do women were to assume domestic responsibilities while their husbands worked outside the home, then women must receive their own unique form of schooling, or so the reasoning went. Most ordinary women received little in the way of formal education. Yet elite young women had expanded educational opportunities, beginning in the early nineteenth century. Emma Willard founded a female academy in Troy, New York, in 1821, and Catharine Beecher established the Hartford Female Seminary two years later in Connecticut. For the most part, these schools catered to the daughters of wealthy families, young women who would never have to work in a factory to survive. Hailed as a means to prepare young women to serve as wives and mothers, the schools taught geography, foreign languages, mathematics, science, and philosophy, as well as the "female" pursuits of embroidery and music.

Out of this curriculum designed especially for women emerged women's rights activists, women who keenly felt both the potential of their own intelligence and the degrading nature of their social situation. Elizabeth Cady, an 1832 graduate of the Troy Female Seminary, later went on to marry Henry B. Stanton and bear seven children, but by the 1840s she strode onto the national stage as a tireless **advocate** of women's political and economic rights. 35

—From *Created Equal: A Social and Political History of the United States, Volume II* (From 1865), 2nd ed., pp. 373–381 by Jacqueline Jones, Peter H. Wood, Thomas Borstelmann, Elaine Tyler May, Vicki L. Ruiz.

VISUAL VOCABULARY

_________ is hard won in many places across the world.

a. suffrage
b. assimilation
c. subversion

Choose the best meaning of each word in **bold**. Use context clues to make your choice.

Vocabulary in Context ______ **1.** "Under fire from other Charleston elites for rushing to judgment, the judges redoubled their efforts to **embellish** vague rumors of black discontent into a tale of a well-orchestrated uprising and to implicate growing numbers of black people." (paragraph 13)

a. decorate
b. exaggerate
c. adorn
d. simplify

Vocabulary in Context ______ **2.** "The new **delineation** between men's and women's work and workplaces intensified the drive for women's education begun after the Revolution." (paragraph 34)

a. description
b. unity
c. discrimination
d. phase

Central Idea ______ **3.** Which of the following sentences from the first paragraph states the central idea of the passage?

a. "In the early 1830s, a wealthy Frenchman named Alexis de Tocqueville visited the United States and wrote about the contradictions he saw."
b. "With universal white manhood suffrage, white men could vote and run for office regardless of their class or religion."
c. "However, Tocqueville also noted some sore spots in American "democratic values and practices."
d. "In other words, Tocqueville saw America for what it was: a blend of freedom and slavery, of independence and dependence."

Central Idea ______ **4.** Which of the following sentences states the central idea of paragraphs 2 through 10?

a. "Population growth in the United States—and on the borderlands between the United States and Mexican territory—put pressure on Indian societies." (paragraph 2)

b. "Charting a middle course between the Indian and European American worlds was Sequoyah, the son of a white Virginia trader-soldier and a Cherokee woman." (paragraph 5)

c. "Overwhelmed by the number of whites, some Indian groups such as the Peoria and Kaskaskia gave up their lands to the interlopers." (paragraph 9)

d. "The massacre, the decisive point of what came to be called the Black Hawk War, marked the end of armed Indian resistance north of the Ohio River and east of the Mississippi." (paragraph 10)

Implied Central Idea ______ **5.** Which of the following sentences states the implied central idea of paragraphs 11 through 20?

a. Southern states enacted laws to support slavery and force free blacks to leave the region.

b. Northern states granted blacks the right to vote in the late eighteenth century only to revoke voting rights for blacks in the nineteenth century.

c. Black leaders disagreed about the issues of integration and black separation.

d. The United States enacted laws in the eighteenth and nineteenth century that enslaved blacks and limited the rights of free blacks.

Central Idea ______ **6.** Which of the following sentences states the central idea of paragraphs 21 through 35?

a. "In the political and economic realms, the egalitarian impulse rarely affected the status of women in a legal or practical sense." (paragraph 21)

b. "Yet women contributed to the economy in myriad ways." (paragraph 23)

c. "In these years, well-off women in the northeastern and mid-Atlantic states began to think of themselves as consumers and not producers of goods." (paragraph 24)

d. "The new delineation between men's and women's work and workplaces intensified the drive for women's education begun after the Revolution." (paragraph 34)

Supporting Details _______ **7.** According to the passage, Nat Turner led a slave revolt because he
a. was beaten as a slave.
b. was a literate Christian.
c. believed he had divine instructions to act.
d. had been sentenced to death.

Transitions _______ **8.** "Most were not citizens and therefore perceived themselves as oppressed like the slaves in the South." (paragraph 14)

The relationship of ideas within this sentence is
a. classification.
b. cause and effect.
c. comparison and contrast.
d. time order.

Transitions _______ **9.** "With universal white manhood suffrage, white men could vote and run for office regardless of their class or religion. However, Tocqueville also noted some sore spots in American democratic values and practices." (paragraph 1)

The relationship of ideas between these sentences is
a. cause and effect.
b. classification.
c. comparison and contrast.
d. generalization and example.

Thought Patterns _______ **10.** The overall thought pattern for the passage is
a. cause and effect.
b. classification.
c. comparison and contrast.
d. generalization and example

Thought Patterns _______ **11.** The thought pattern for paragraph 11 is
a. cause and effect.
b. classification.
c. comparison and contrast.
d. definition and example.

Fact and Opinion _______ **12.** "In the early 1830s, a wealthy Frenchman named Alexis de Tocqueville visited the United States and wrote about the contradictions he saw." (paragraph 1)

This sentence is a statement of
a. fact.
b. opinion.
c. fact and opinion.

Fact and Opinion _______ **13.** "Charting a middle course between the Indian and European American worlds was Sequoyah, the son of a white Virginia trader-soldier and a Cherokee woman." (paragraph 5)

This sentence is a statement of
a. fact.
b. opinion.
c. fact and opinion.

Fact and Opinion ______ **14.** "In Massachusetts in 1820, women and children constituted almost a third of all manufacturing workers." (paragraph 30)

This sentence is a statement of

a. fact.
b. opinion.
c. fact and opinion.

Tone ______ **15.** The overall tone of the passage is

a. angry.
b. neutral.
c. sympathetic.
d. cold.

Purpose ______ **16.** The primary purpose of the passage is

a. to inform.
b. to entertain.
c. to persuade.

Inferences ______ **17.** Based on the details in the passage, we can infer that

a. discrimination against minorities occurred mostly in the South during the Age of the Common Man.
b. the Age of the Common Man favored white males.
c. women had more rights than other minority groups during the Age of the Common Man.
d. all the various groups of people attained freedom during the Age of the Common Man.

Argument ______ **18.** Read the claim and supports taken from the passage. Then identify the detail that does not support the claim.

Claim: "In America a woman loses her independence forever in the bonds of matrimony." (paragraph 1)

a. "A husband controlled the property that his wife brought to the marriage, and he had legal authority over their children." (paragraph 21)
b. "Indeed, the wife was considered her husband's property." (paragraph 21)
c. "She had no right to make a contract, keep money she earned, vote, run for office, or serve on a jury." (paragraph 21)
d. "Elizabeth Cady, an 1832 graduate of the Troy Female Seminary, later went on to marry Henry B. Stanton and bear seven children, but by the 1840s she strode on the national stage as a tireless advocate of women's political and economic rights." (paragraph 35)

Argument **19.** Identify the persuasive technique used in the following sentence:

"In 1828 the House of Representatives Committee on Indian Affairs surveyed the Indians of the South and concluded that 'an Indian cannot work' and that all Indians were lazy and notable for their 'thirst for spirituous liquours.'" (paragraph 3)

a. false analogy
b. personal attack
c. begging the question
d. transfer

Argument ______ **20.** Identify the persuasive technique used in the following sentence:

"European American women's economic subordination served as a rationale for their political inferiority." (paragraph 22)

a. false analogy
b. personal attack
c. begging the question
d. transfer

Mapping

Complete the following concept map with information from the passage.

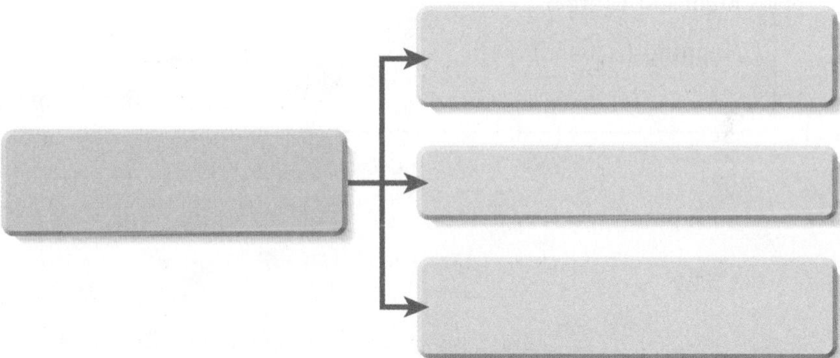

WHAT DO YOU THINK?

Now that you have read the passage, how would you define the Common Man in America during the first half of the 1800s? Do you think the idea of a Common Man still exists in America today? If so, how would you define the term today? Assume you are taking a college history course and you have been assigned an essay that connects what you are learning about history to current life. Choose one of the groups of people discussed in the passage and compare the status of that group during the Age of the Common Man to their current status today.

MASTER READER Scorecard

"Real People in the 'Age of the Common Man'"

Skill	Number Correct		Points		Total
Vocabulary					
Vocabulary in Context (2 items)	________	×	4	=	________
Comprehension					
Central Idea and Implied Central Idea (4 items)	________	×	4	=	________
Supporting Details (1 item)	________	×	4	=	________
Transitions (2 items)	________	×	4	=	________
Thought Patterns (2 items)	________	×	4	=	________
Fact and Opinion (3 items)	________	×	4	=	________
Tone and Purpose (2 items)	________	×	4	=	________
Inferences (1 item)	________	×	4	=	________
Argument (3 items)	________	×	4	=	________
Mapping (4 items)	________	×	5	=	________
			Comprehension Score		________

PART THREE

Combined-Skills Tests

Part Three contains 10 tests. The purpose of these tests is twofold: to track your growth as a reader and to prepare you for the formal tests you will face as you take college courses. Each test presents a reading passage and questions that cover some or all of the following skills: vocabulary in context, central ideas, supporting details, thought patterns, fact and opinion, tone and purpose, inferences, and argument.

TEST 1

Read the following passage, and then, answer the questions.

An Account of Alfred C. Cooley's Plight in the Face of Hurricane Katrina

by Sandra Offiah-Hawkins

[1]I am **aghast** by the sudden series of events. [2]There is no time to figure out the next day, for there is no time to visualize where we are going to all sleep this night—new, used, and old cars line up along Interstate 10 and Highway 1 as we all travel 10 miles an hour and wait in five-mile long lines in hopes of getting gas before the stations run out. [3]There are thirty-five of us traveling in a caravan: a Dodge Caravan, a Ford Taurus, a 2000 Grand Prix, and a Toyota CRV just to name a few. [4]All the relatives we can contact in New Orleans on this day August 27, 2005, are with us; we have been asked to leave our homes immediately, for the strong winds, the deluging rain, the terrible wrath of Hurricane Katrina is headed straight to New Orleans—our home.

[5]All of us leave our worldly possessions that day, bringing with us essentials for two days: I never think for a minute this is the end of what I have called home in the Ninth Ward for more than 50 years; all of us feel this is going to be a "James Brown turnaround." [6]I never think the wife and children I have been blessed with over the past 30 years would never again know life as they had known it in New Orleans.

***

[7]Throughout my life, I have known suffering: I suffer from Sickle Cell Anemia; my parents were told that I would not live past age 16, but here I am! [8]Within the last eight years, I have had two hip replacements as well as operations on my shoulder and on my left eye. [9]I have been placed in a situation of sometimes raising my four daughters as a single parent, and for many years, I have cared for my aging parents. [10]Now this—the unanticipated fury of Katrina.

***

[11]Once the storm hits, the power, as expected, is off for days in New Orleans, so it is Tuesday evening or Wednesday before we find out the Levee has broken. [12]Since we left home on Saturday, we have been traveling—first to Alexandria, Louisiana, where we sleep wherever there is a space on the floor at Dora Ann's house, my sister's former daughter-in-law's

home. [13]The next day, we plan our next move: my parents, sisters, and one brother stay there, but the remaining sixteen of us head East on I-10 towards Memphis, Tennessee, to the home of Stanley Trotter, my ex-wife's cousin, where there were now 20 people sharing a house.

[14]As the sixteen of us from New Orleans sit and watch television, we see the water begin to cover the roof of the familiar Circle Food Store, on North Clayborne Avenue and St. Bernard Avenue near the home we just abandoned; we realize now, we will not be returning home.

[15]It is extremely difficult for anyone watching the news, but it has to be most difficult for those of us witnessing everything we have worked for being destroyed right before our eyes; we have no idea what we should do or where we should go from here; we begin to mourn for friends and neighbors gone from our lives forever. [16]We continue watching as the Super Dome and Convention Center prove to be inadequate for the more than three thousand people seeking shelter and assistance; we listen to the reports of water leaks and criminals raping people in the Dome; we watch the people standing along the bridge in the smoldering heat with no sanitation, no food, no water; we see dead people floating in the water; we see another person dead in a wheelchair—not even covered. [17]"Why is help so slow to arrive?" we cry out at the television.

***

[18]My heart is truly heavy, and I can barely watch or believe what I am witnessing on television. [19]During this time, I think of my brother and his adult son, Robert Jr., both of whom remained in New Orleans when we left: I pray to God they survive—not knowing until days later that they have located a car and driven to Alexandria to be with my parents and other siblings; I pray to God they survive—not knowing that they have returned to help the people and the clean-up efforts in the areas hit hardest by Katrina and the tornadoes that followed.

***

[20]My mind drifts in time back to my uncles' and aunts' homes in Picayune, Mississippi, less than an hour's drive from New Orleans, and I think about all of my relatives living there: my cousins, my grandmothers, and paternal grandfather—I contemplate memories of Pilgrim Bound Baptist Church and other churches as the neighborhood readies for the 5th Sunday Singing. [21]Every fifth Sunday, all the churches in the area gather as members from the various congregations hope for an opportunity to be placed on the program to sing two selections each. [22]Reverend Woods, the announcer, is always filled with the Holy Spirit as he listens to the lyrics of songs and hymns both old and new: "What a Friend We Have

in Jesus," "Amazing Grace," "Do Lord Remember Me," and "Jesus Paid It All." [23]My father, brothers, and I would often travel to participate in the program as an all-male guest quartet from New Orleans, LA.

[24]Oh, I will never forget those days—they remain an inspiring part of my life because I had an opportunity to meet many people; some I had forgotten until now: Mrs. Jewel had a candy store next door to where my grandmother, aunt, and uncle lived; only a few blocks away, Bossie Boys, a place where teenagers would meet and dance on the weekends, was located just a few feet from my Uncle Bishop's barber shop.

***

[25]In the wake of Katrina, we move to Atoka, Tennessee, and it does not take long for people to hear about the sixteen people who fled New Orleans. [26]I shall never forget—Mrs. Snead of St. Mark Baptist Church and Pastor McGee alert church members, the fire department, and other social agencies about our **plight**; church members help us get food and clothing and raise money so that we can begin reestablishing ourselves; through the Red Cross, we are able to obtain proper documentation for identification. [27]No, I shall never forget—all the blessings that flowed from throughout the community to my family during these darkest hours.

[28]Finally people are allowed to return briefly to New Orleans to check out their property and help others during the aftermath of Katrina—I am anxious. [29]My son-in-law Javelle and I rent a van to go back and salvage what we can. [30]On the way, a police cruiser clocks my speed on the highway at 89 mph in a 70 mph zone; I tell the officer who I am, where I am going and why I am going there; he makes me prove that I have lived in New Orleans all of my life; then, he voids my ticket with a warning to slow down. [31]How ironic—I have to prove that I am a "real" life-long New Orleans "native."

[32]Getting off of Highway 610 at Franklin Avenue, we can see the marks of high water levels, and so much trash—garbage and debris strewn everywhere—we see the damage, the loss, and oh so much more—we see a holocaust: we see wind damage; we see cars flipped on their sides; we see homes with nothing more than a frame of what once was; we see X's marked on still-standing homes, with a number 1, 2, 3, or more to indicate the dead inside.

[33]When you have a near death experience, it is said, your life flashes before you. [34]That is exactly my experience during Katrina. [35]Burdens are heavy, I cannot sleep, and I worry about my family; days pass, and I know the end of my life is near: I now realize that it is the end of my life as I have known it—as I walk in what was once my parent's home on Clouet

Street. [36]I see just how high the water had come, and there, I raise my hands to lift several pictures off the wall.

Vocabulary ______ **1.** The best meaning of the word **aghast** as used in sentence 1 is
a. excited.
b. puzzled.
c. horrified.
d. unaffected.

Vocabulary ______ **2.** The best meaning of the word **plight** as used in sentence 26 is
a. promise.
b. troubles.
c. situation.
d. hope.

Implied Central Idea ______ **3.** Which sentence best states the author's implied central idea?
a. Hurricane Katrina destroyed Alfred C. Cooley's home in New Orleans.
b. Hurricane Katrina dramatically and unexpectedly changed the lives of Alfred C. Cooley, his family, and other victims of the storm.
c. Alfred C. Cooley and his family suffered more than most who lived through Hurricane Katrina.
d. The people of New Orleans did not receive adequate help during Hurricane Katrina.

Main Idea, Supporting Details ______ **4.** Sentence 3 states a
a. main idea.
b. major supporting detail.
c. minor supporting detail.

Thought Patterns ______ **5.** What is the overall thought pattern of the passage?
a. cause and effect
b. classification
c. generalization and example
d. time order

Transitions ______ **6.** What is the primary relationship of ideas within sentence 7?
a. cause and effect
b. classification
c. generalization and example
d. time order

Purpose ______ **7.** The overall purpose of the passage is
a. to entertain the reader with a personal story about living through Hurricane Katrina.
b. to inform the reader about the suffering caused by Hurricane Katrina.
c. to persuade the reader to give aid to the survivors of Hurricane Katrina.

Tone ______ **8.** The tone of sentence 31 is
a. sarcastic.
b. amused.
c. angry.
d. neutral.

Fact and Opinion ______ **9.** Overall, the passage uses
a. fact.
b. opinion.
c. fact and opinion.

Argument ______ **10.** Sentence 34 states a
a. claim.
b. support for a claim.

TEST 2

A Look at Man through The Vapid Eyes of His Captives

by Pierre Tristam

[1]Disturbingly recent exceptions aside, civilized nations now agree that burning fellow human beings at the stake, torturing them or enslaving them is inhuman. [2]The day will come when civilized nations will agree that imprisoning wild animals in zoos, whipping them about in circus acts from city to city or forcing them to do tricks for our amusement in such places as SeaWorld, Marineland and Epcot is as cruel to the animals as it is lewd of the people watching them.

[3]That day is far off, no doubt.[4] Pulling profits and emoting power over weaker creatures, vicariously enjoyed by those audiences that delight in the safe splashing of a killer whale or the harmlessness of a caged animal, are strong impulses. [5]Too strong to be outdone by notions of rights for beasts that don't speak English or pay taxes.

[6]Until then, handlers of animals forced into unnatural situations will continue to die, as SeaWorld's Dawn Brancheau did in February when a killer whale dragged her underwater after turning the tables and making her its plaything. [7]I keep reading references to Brancheau's death as "tragic." [8]What lazy news writers mean is that her death was sad, unfortunate, avoidable and, from the spectators' (but not the whale's) perspective, **lurid**, as it was for SeaWorld's PR.

[9]Brancheau's death was foretold. [10]Besides practicing drills by coaching their prisoners to follow a script, trainers like her practice not getting killed for a reason. [11]They presume at every moment to outwit a predator's instincts. [12]They can't outwit the law of averages. [13]For a brief moment, the whale that killed Brancheau went off script. [14]It acted in character. [15]Mauling her might have been the most natural thing the whale had done in years. [16]If it's an education spectators wanted, they finally got an authentic one.

[17]I don't mind the work with animals of people like the late Steve Irwin, the Australian of "Crocodile Hunter" fame killed by a stingray in 2006. [18]Irwin had his moments of cruelty when he wanted to prove that he could best a beast bigger, bitier or faster than him. [19]But mostly he worked on the animals' turf, on their terms. [20]He did not rearrange their nature for our amusement. [21]He risked his life to show us how wild these animals are, and how freely noble and untamable they should remain.

[22]This isn't to argue against domestication or even the slaughtering of animals. [23]We are animals and predators. [24]But domesticating an animal for help or companionship and certainly killing an animal for sustenance will always be more morally defensible than taming one for entertainment or "education." ([25]The less defensible gobs of cruelty in the chicken farms and the feedlots of the West, where cattle are turned into walking mummies of drugs and fat, have more to do with a nation's gluttony than sustenance. [26]But that's another story.)

[27]Places like SeaWorld love to claim that their shows give people a close-up of something unique that fosters an appreciation for nature and conservation. [28]Florida residents give the lie to that invention. [29]They've been converging on SeaWorld from subdivisions that have plowed under entire ecosystems and obliterated the habitats of 111 plants and animals (at last count). [30]That's not about to change.

[31]"The sensational Shamu show" itself plows under the killer whales' natural instincts to make them fit human **conceits**. [32]It doesn't honor the killer whales or their place in nature, since that place is nowhere at SeaWorld, so much as it pumps up man's capacity to synchronize dives, sentimental music and greeting-card philosophy with 5-ton creatures.

[33]The show, called "Believe," is motivational splashing. [34]And in an age when entire cable channels and other media by the ant pile are devoted to natural science, there's no justification anymore for amusement-park animal exploitation. [35]A minute's worth of a nature show like "Planet Earth," even on a television screen—which takes the viewer to untold places with an intimacy and humility that really does put humans'

insignificance in perspective—does more to inspire reverence for nature as it really is than any Shamu schmoozing that mostly warms up your gift shop's cockles.

[36]As for zoos, I've visited some of the great ones in the country. [37]I've also walked the wards of several jails and prisons, including death row in Nashville, Tenn. [38]In both places I saw the identical shuffling lethargy and vapid look of captivity. [39]Criminals are presumably imprisoned as a result of their own misdeeds. [40]Animals aren't. [41]The crime of their captivity, reflected in their eyes, is ours every time we visit a zoo or clap like dopes at a mud-wrestling show whenever Shamu flips. [42]Forget Willie. [43]Free them all.

—"A Look at Man Through The Vapid Eyes of His Captives" by Pierre Tristam from *The Daytona Beach News-Journal,* March 7, 2010, p. 10A. Reprinted by permission of Copyright Clearance Center.

VISUAL VOCABULARY

The natural ecosystem has been obliterated in preparation for development.

The best meaning of **obliterated** is ______.

a. overlooked.
b. restored.
c. destroyed.

Vocabulary ______ **1.** The best meaning of the word **lurid** as used in sentence 8 is

a. shocking.
b. colorful.
c. exciting.
d. expected.

Vocabulary ______ **2.** The best meaning of the word **conceits** as used in sentence 31 is

a. prideful ideas.
b. humility.
c. instincts.
d. sense of adventure.

Central Idea ______ **3.** Which sentence best states the author's central idea?

a. sentence 1
b. sentence 2
c. sentence 5
d. sentence 33

Supporting Details ______ **4.** Sentence 18 states a

a. main idea.
b. major supporting detail.
c. minor supporting detail.

Thought Patterns ______ **5.** What is the overall thought pattern of the passage?

a. time order
b. classification
c. cause and effect
d. generalization and example

Purpose ______ **6.** The overall purpose of the passage is

a. to entertain the reader with details about human treatment of captive animals.
b. to inform the reader about the problems associated with keeping wild animals captive for entertainment or educational purposes.
c. to argue against keeping wild animals captive for human entertainment or educational purposes.

Tone ______ **7.** The overall tone of the passage is

a. neutral.
b. biting.
c. sad.
d. reasonable.

Fact and Opinion ______ **8.** Sentences 3 and 4 (paragraph 2) are statements of

a. fact.
b. opinion.
c. fact and opinion.

Argument ______ **9.** In sentence 8, the author uses the propaganda technique of

a. false cause.
b. straw man.
c. false analogy.
d. personal attack.

Argument ______ **10.** In paragraph 10 (sentences 36–43), the author uses the propaganda technique of

a. false cause.
b. straw man.
c. false analogy.
d. personal attack.

TEST 3

Read the following passage, and then, answer the questions.

What Risks Are Involved in Tattooing?

[1]Despite the obvious popularity of body art, several complications can result from tattooing.

[2]Tattooing can cause infections. [3]Unsterile tattooing equipment and needles can transmit infectious diseases, such as hepatitis; thus the American Association of Blood Banks requires a one-year wait between getting a tattoo and donating blood. [4]Even if the needles are sterilized or never have been used, the equipment that holds the needles may not be sterilized reliably due to its design. [5]In addition, a tattoo must be cared for properly during the first week or so after the **pigments** are injected.

[6]Tattooing involves removal problems. [7]Despite advances in laser technology, removing a tattoo is a painstaking process, usually involving several treatments and considerable expense. [8]Complete removal without scarring may be impossible.

[9]Although allergic reactions to tattoo pigments are rare, when they happen they may be particularly troublesome because the pigments can be hard to remove. [10]Occasionally, people may develop an allergic reaction to tattoos they have had for years.

[11]Tattoos may also result in granulomas and keloids. [12]Granulomas are nodules that may form around material that the body perceives as foreign, such as particles of tattoo pigment. [13]If you are prone to developing keloids—scars that grow beyond normal boundaries—you are at risk of keloid formation from a tattoo. [14]Keloids may form any time you injure or traumatize your skin. [15]According to experts, tattooing or micropigmentation is a form of trauma, and keloids occur more frequently as a consequence of tattoo removal.

—Adapted from "Tattoos and Permanent Makeup." U.S. Food and Drug Administration Center for Food Safety and Applied Nutrition Office of Cosmetics and Colors Fact Sheet. 29 November 2000; Updated 14 July 2006. Online 8 July 2004. http://www.cfsan.fda.gov/~dms/cos-204.html

Vocabulary ______ **1.** The best meaning of the word **pigments** as used in sentence 5 is
a. infections.
b. protections.
c. dyes.
d. skin.

Central Idea ______ **2.** The sentence that best states the central idea of the passage is
a. sentence 1.
b. sentence 2.
c. sentence 5.
d. sentence 12.

Transitions ______ **3.** The relationship between sentences 11 and 12 is one of
a. definition.
b. cause and effect.
c. time order.
d. addition.

Transitions ______ **4.** The relationship of the ideas within sentence 3 is
a. definition.
b. cause and effect.
c. time order.
d. comparison and contrast.

Thought Patterns ______ **5.** What is the overall thought pattern of the passage?
a. comparison and contrast
b. cause and effect
c. time order
d. definition

Supporting Details ______ **6.** Sentence 6 is a
a. main idea.
b. major supporting detail.
c. minor supporting detail.

Purpose ______ **7.** The author's main purpose in the passage is
a. to inform.
b. to entertain.
c. to persuade.

Tone ______ **8.** The tone of the passage is
a. graphic.
b. judgmental.
c. objective.
d. pessimistic.

9–10. Complete the outline below with information from the passage.

Central idea: ____

I. Tattooing can cause infections.

II. Tattooing involves removal problems.

III. ____

IV. Tattooing may result in granulomas and keloids.

TEST 4

Read the passage from a history textbook, and then, answer the questions.

Textbook Skills

The Stewardship of Natural Resources History: Mapping Puget Sound and Western Washington

[1]Nestled between two dramatic mountain ranges, the Olympics and the Cascades, the lush and mild Puget Sound region provided abundant natural resources for fishing, timbering, and farming for Native Americans for thousands of years before the arrival of British Captain George Vancouver and Lieutenant Peter Puget in 1792. [2]Seattle grew in the twentieth century as

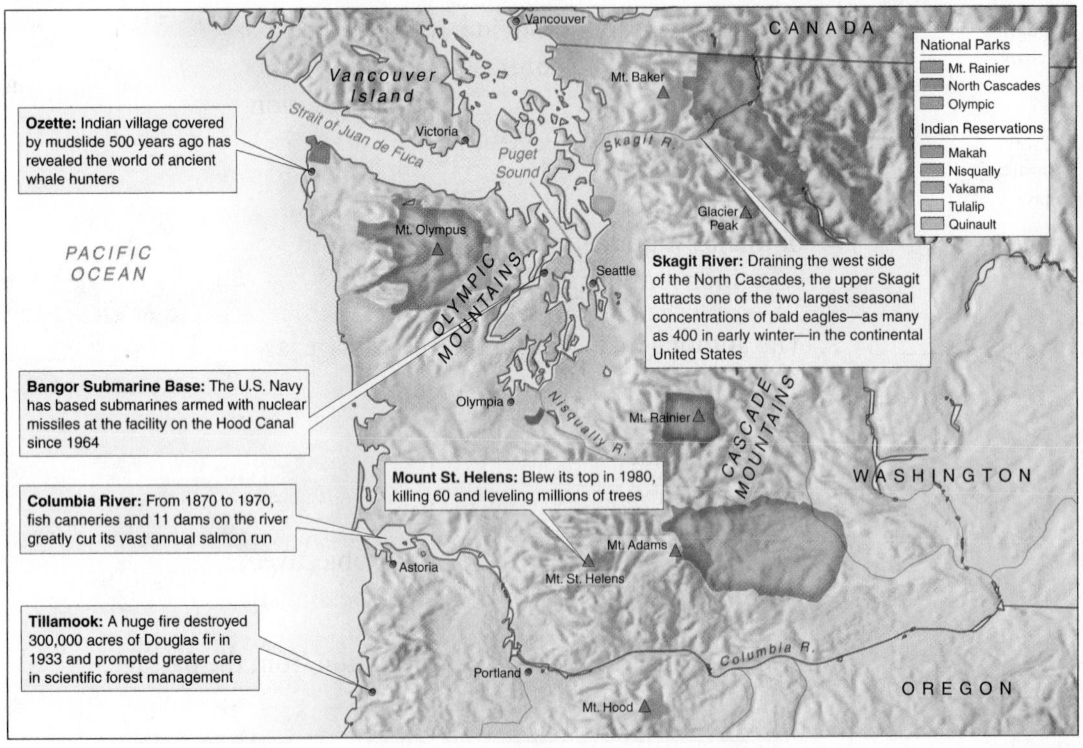

Vancouver
CANADA
National Parks
Mt. Rainier
North Cascades
Olympic
Indian Reservations
Makah
Nisqually
Yakama
Tulalip
Quinault
Vancouver Island
Mt. Baker
Strait of Juan de Fuca
Victoria
Puget Sound
Skagit R.
Ozette: Indian village covered by mudslide 500 years ago has revealed the world of ancient whale hunters
Glacier Peak
Mt. Olympus
PACIFIC OCEAN
OLYMPIC MOUNTAINS
Seattle
Skagit River: Draining the west side of the North Cascades, the upper Skagit attracts one of the two largest seasonal concentrations of bald eagles—as many as 400 in early winter—in the continental United States
Bangor Submarine Base: The U.S. Navy has based submarines armed with nuclear missiles at the facility on the Hood Canal since 1964
Olympia
Nisqually R.
Mt. Rainier
CASCADE MOUNTAINS
Mount St. Helens: Blew its top in 1980, killing 60 and leveling millions of trees
WASHINGTON
Columbia River: From 1870 to 1970, fish canneries and 11 dams on the river greatly cut its vast annual salmon run
Mt. Adams
Astoria
Mt. St. Helens
Tillamook: A huge fire destroyed 300,000 acres of Douglas fir in 1933 and prompted greater care in scientific forest management
Columbia R.
Portland
OREGON
Mt. Hood

a trading port, particularly with east Asian markets, and as the original aircraft manufacturing site of Boeing Corporation. [3]With the permanently snow-capped volcanic peaks of the Cascade range visible on clear days, Seattle continued to attract new residents and tourists alike. [4]The population of the former frontier state of Washington nearly doubled from 1970 to 2003, topping 6 million.

[5]Seattle was also the departure point for most Americans traveling north to Alaska. [6]Passenger ferries crisscrossed Puget Sound, providing an alternative method of commuting to work in Seattle for residents of other area communities. [7]In the 1980s the city became associated with the extraordinary business success of computer software giant Microsoft and coffee retailer Starbucks. [8]Recreation Equipment Incorporated (REI) began in Seattle as a consumer **cooperative** for buying products for outdoor activities such as rock-climbing and backpacking. [9]By 2000 REI had built stores across the United States.

—From *Created Equal: A Social and Political History of the United States, Volume II (from 1865)*, 2nd ed., p. 1019 by Jacqueline Jones, Peter H. Wood, Thomas Borstelmann, Elaine Tyler May, Vicki L. Ruiz.

Vocabulary ______ **1.** The best meaning of the word **stewardship** in the title is

a. duty.
b. need.
c. management.
d. benefit.

Vocabulary ______ **2.** The best meaning of the word **cooperative** as used in sentence 8 is

a. helper.
b. companion.
c. company.
d. product.

Implied Main Idea ______ **3.** Based on the title and the details, which sentence best states the implied main idea of the passage?

a. The Puget Sound region is a beautiful area that needs no stewardship.
b. Stewardship is needed because commerce and development are harming the Puget Sound region.
c. Under the stewardship of Native Americans, the Puget Sound region provided abundant natural resources for fishing, timbering, and farming.
d. The natural environment of the Seattle area has greatly influenced the economic development of the area.

Thought Patterns ______ **4.** The main thought pattern of the passage is

a. time order.
b. cause and effect.
c. classification.
d. listing.

Tone ______ **5.** The tone of the passage is

a. admiring.
b. neutral.
c. critical.
d. persuasive.

Answer the following questions based on information in the map.

Supporting Details ______ **6.** Based on the map, where does the vast salmon run occur?

a. Skagit River
b. Puget Sound
c. Tillamook
d. Columbia River

Supporting Details ______ **7.** Based on the map, how many Indian reservations are in the region?

a. two
b. three
c. five
d. six

Purpose ______ **8.** The purpose of the map is

a. to inform the reader about various historical resources, facts, and traits of the region.

b. to persuade the reader that, historically, human activity has harmed the natural resources of the region.
c. to entertain the reader with little known historical facts about the region.

Inference ______ **9.** Based on the information in the map, we can infer that
a. the salmon run is more vast than it once was.
b. Mount St. Helens is an active volcano.
c. the fire at Tillamook was started by human activity.
d. the Bangor Submarine Base has a nuclear power plant.

Argument ______ **10.** ***Claim***: The Puget Sound region is rich in natural resources.

Which of the following statements is not relevant to the claim?
a. An Indian village covered by mudslide 500 years ago has revealed the world of ancient whale hunters.
b. The U.S. Navy has based submarines with nuclear missiles at the facility on the Hood Canal since 1964.
c. The upper Skagit attracts one of the two largest seasonal concentrations of bald eagles—as many as 400 in early winter—in the continental United States.
d. Native Americans live on reservations that are smaller in size than the three national parks in the region.

TEST 5

Read the following passage, and then, answer the questions.

Medgar Evers

[1]Medgar Evers was one of the first martyrs of the civil-rights movement. [2]He was born in 1925 in Decatur, Mississippi, to James and Jessie Evers. [3]After a short stint in the army, he enrolled in Alcorn A&M College, graduating in 1952. [4]His first job out of college was traveling around rural Mississippi selling insurance. [5]He soon grew enraged at the **despicable** conditions of poor black families in his state, and joined the NAACP. [6]In 1954, he was appointed Mississippi's first field secretary.

[7]Evers was outspoken, and his demands were radical for his rigidly segregated state. [8]He fought for the enforcement of the 1954 court decision of *Brown* v. *Board of Education of Topeka,* which outlawed school segregation;

he fought for the right to vote, and he advocated boycotting merchants who discriminated. [9]He worked unceasingly despite the threats of violence that his speeches **engendered**. [10]He gave much of himself to this struggle, and in 1963, he gave his life. [11]On June 13, 1963, he drove home from a meeting, stepped out of his car, and was shot in the back.

[12]Immediately after Evers's death, the shotgun that was used to kill him was found in bushes nearby, with the owner's fingerprints still fresh. [13]Byron de la Beckwith, a vocal member of a local white-supremacist group, was arrested. [14]Despite the evidence against him, which included an earlier statement that he wanted to kill Evers, two trials with all-white juries ended in deadlock decisions, and Beckwith walked free. [15]Twenty years later, in 1989, information surfaced that suggested the jury in both trials had been tampered with. [16]The assistant District Attorney, with the help of Evers's widow, began putting together a new case. [17]On February 5, 1994, a multiracial jury re-tried Beckwith and found him guilty of the crime.

[18]The loss of Evers changed the tenor of the civil-rights struggle. [19]Anger replaced fear in the South, as hundreds of demonstrators marched in protest. [20]His death prompted President John Kennedy to ask Congress for a comprehensive civil-rights bill, which President Lyndon Johnson signed into law the following year. [21]Evers's death, as his life had, contributed much to the struggle for equality.

—"Medgar Evers" from *African American Almanac,* 7th ed. by Carney Smith and Chuck Stone. Belmont: Gale, a part of Cengage Learning, Inc., 1996.

Vocabulary ______ **1.** The synonym of the word **despicable** in sentence 5 is

a. contemptible. c. mean-spirited.
b. obvious. d. civilized.

Vocabulary ______ **2.** The best meaning of the word **engendered** as used in sentence 9 is

a. performed. c. provoked.
b. gratified. d. halted.

Main Idea, Details ______ **3.** Sentence 11 is

a. a main idea. c. a minor supporting detail.
b. a major supporting detail.

Central Idea ______ **4.** The central idea of the passage is best stated in sentence

a. 1. c. 5.
b. 2. d. 20.

Supporting Details ______ **5.** When did Medgar Evers become a martyr of the civil rights movement?

a. 1952
b. 1954
c. 1963
d. 1994

Transitions ______ **6.** What is the relationship of ideas between sentences 18 and 19?

a. cause and effect
b. comparison and contrast
c. classification
d. time order

Transitions ______ **7.** What is the relationship of ideas within sentence 14?

a. cause
b. contrast
c. definition
d. classification

Purpose ______ **8.** The author's main purpose is

a. to inform the reader about the contributions Medgar Evers made to the civil rights movement.
b. to inspire the reader with the story of a civil rights leader.
c. to persuade the reader that Medgar Evers is a martyr.

Tone ______ **9.** The tone of the passage is

a. detached.
b. admiring.
c. cynical.
d. unbiased.

Fact and Opinion ______ **10.** Sentence 7 states

a. a fact.
b. an opinion.
c. fact and opinion.

TEST 6

Read the following passage from the college textbook *Psychology and Life*, and then, answer the questions.

Textbook Skills

Mnemonics

[1]One memory-enhancing option is to draw on special mental strategies called *mnemonics* (from the Greek word meaning "to remember"). [2]**Mnemonics** are devices that **encode** a long series of facts by associating them with familiar and previously encoded information. [3]Many mnemonics work by giving you ready-made retrieval cues that help organize otherwise **arbitrary** information.

[4]Consider the *method of loci,* first practiced by ancient Greek orators. [5]The singular of *loci* is *locus,* and it means "place." [6]The method of loci is a means of remembering the order of a list of names or objects—or, for the orators, the individual sections of a long speech—by associating them with some sequence of places with which you are familiar. [7]To remember a grocery list, you might mentally put each item sequentially along the route you take to get from home to school and find the item associated with each spot.

[8]The *peg-word method* is similar to the method of loci, except that you associate the items on a list with a series of cues rather than with familiar locations. [9]Typically, the cues for the peg-word method are a series of rhymes that associate numbers with words. [10]For example, you might memorize "one is a *bun,*" "two is a *shoe,*" "three is a *tree,*" and so on. [11]Then you would associate each item on your list interacting with the appropriate cue. [12]For your grocery list, you might have the bread nestled among several buns, a shoe filled up with orange juice, a tree with ice cream cones rather than leaves, and so on. [13]You can see that the key to learning arbitrary information is to encode the information in such a fashion that you provide yourself with efficient retrieval cues.

Vocabulary ______ **1.** The best meaning of the term **encode** as used in sentence 2 is
a. approve.
b. understand.
c. unlock.
d. program.

Vocabulary ______ **2.** The best meaning of the word **arbitrary** as used in sentence 3 is
a. logical.
b. long-term.
c. random.
d. frivolous.

Main Idea, Details ______ **3.** Sentence 9 is
a. a main idea.
b. a major supporting detail.
c. a minor supporting detail.

Central Idea ______ **4.** Which sentence best states the central idea of the passage?
a. sentence 1
b. sentence 2
c. sentence 3
d. sentence 4

Transitions ______ **5.** What is the relationship of ideas within sentence 8?
a. comparison and contrast
b. addition
c. cause and effect
d. classification

Transitions ______ **6.** What is the relationship of ideas between sentences 6 and 7?
a. definition and example c. cause and effect
b. comparison and contrast d. time order

Purpose ______ **7.** The author's main purpose is
a. to inform the reader about two mnemonic techniques.
b. to entertain the reader with ways to remember information.
c. to persuade the reader to improve memory skills.

Tone ______ **8.** The tone of the passage is
a. bored. c. subjective.
b. challenging. d. objective.

Fact and Opinion ______ **9.** Overall, the details of this passage are
a. fact. c. fact and opinion.
b. opinion.

10. Complete the summary with information from the passage.

Mnemonics are mental strategies that encode facts by associating them with familiar and previously encoded information, such as the

____

TEST 7

Read the following excerpt from the speech President Bush delivered to honor novelist Harper Lee with the nation's highest award. Answer the questions.

Harper Lee and the Presidential Medal of Freedom

[1]"Harper Lee has made an outstanding contribution to America's literary tradition. [2]At a critical moment in our history, her beautiful book, *To Kill a Mockingbird,* helped focus the Nation on the turbulent struggle for equality."

—The text of Harper Lee's citation for the Presidential Medal of Freedom presented by President Bush on November 5, 2007.

President Bush: [3]"The story of an old order, and the glimmers of humanity that would one day overtake it, was unforgettably told in a book by Miss Harper Lee. [4]Soon after its publication a reviewer said this: 'A hundred pounds of sermons on tolerance, or an equal measure of **invective** deploring the lack of it, will weigh far less in the scale of enlightenment than a mere 18 ounces of a new fiction bearing the title *To Kill a Mockingbird*.'

[5]"Given her legendary **stature** as a novelist, you may be surprised to learn that Harper Lee, early in her career, was an airline reservation clerk. [6]Fortunately for all of us, she didn't stick to writing itineraries. [7]Her beautiful book, with its grateful prose and memorable characters, became one of the biggest-selling novels of the 20th century.

[8]"Forty-six years after winning the Pulitzer Prize, *To Kill a Mockingbird* still touches and inspires every reader. [9]We're moved by the story of a man falsely accused—with old prejudice massed against him, and an old sense of honor that rises to his defense. [10]We learn that courage can be a solitary business. [11]As the lawyer Atticus Finch tells his daughter, 'before I can live with other folks I've got to live with myself. [12]The one thing that doesn't abide by majority rule is a person's conscience.'

[13]"One reason *To Kill a Mockingbird* succeeded is the wise and kind heart of the author, which comes through on every page. [14]This daughter of Monroeville, Alabama had something to say about honor, and tolerance, and, most of all, love—and it still resonates. [15]Last year Harper Lee received an honorary doctorate at Notre Dame. [16]As the degree was presented, the graduating class rose as one, held up copies of her book, and cheered for the author they love.

[17]"*To Kill a Mockingbird* has influenced the character of our country for the better. [18]It's been a gift to the entire world. [19]As a model of good writing and humane sensibility, this book will be read and studied forever. [20]And so all of us are filled with admiration for a great American and a lovely lady named Harper Lee."

"President Bush Honors Medal of Freedom Recipients."
Office of the Press Secretary. 5 Nov. 2007.
http://www.whitehouse.gov/news/releases/2007/11/20071105-1.html.

Vocabulary ______ **1.** The best meaning of the word **invective** in sentence 4 is
a. criticism.
b. support.
c. wisdom.
d. emotion.

Vocabulary ______ **2.** The best meaning of the word **stature** as used in sentence 5 is
a. height.
b. status.
c. weight.
d. acceptance.

Central Idea ______ **3.** Which sentence best states the central idea of the passage?
a. sentence 1
b. sentence 3
c. sentence 8
d. sentence 20

Thought Patterns ______ **4.** The overall thought pattern used in the passage is
a. time.
b. listing.
c. cause and effect.
d. generalization and example.

Transitions ______ **5.** The relationship of ideas between sentence 10 and sentence 11 is
a. time.
b. listing.
c. cause and effect.
d. generalization and example.

Fact and Opinion ______ **6.** Sentence 2 is a statement of
a. fact.
b. opinion.
c. fact and opinion.

Purpose ______ **7.** The main purpose of the passage is
a. to inspire the reader to learn more about Harper Lee and read her novel *To Kill a Mockingbird.*
b. to entertain the audience by honoring the popular novelist, Harper Lee.
c. to inform the audience about and praise Harper Lee for her literary contribution: *To Kill a Mockingbird.*

8–10. Complete the summary based on information from the passage.

On November 5, 2007, President Bush awarded novelist Harper Lee the Presidential Medal of Freedom for her (8) "________________ book" *To Kill a Mockingbird.* The Pulitzer Prize-winning novel is "about a man falsely accused—with old (9) ________________ massed against him, and an old sense of (10) ________________ that rises to his defense."

TEST 8

Read the following passage from the college textbook *Educational Psychology,* and then, answer the questions.

Textbook Skills

Sources of Misunderstandings

[1]Some children are simply better than others at reading the classroom situation because the participation structures of the school match the structures they have learned at home. [2]The communication rules for

most school situations are similar to those in middle-class homes, so children from these homes often appear to be more **competent** communicators. [3]They know the unwritten rules. [4]Students from different cultural backgrounds may have learned participation structures that conflict with the behaviors expected in school. [5]For example, one study found that the home conversation style of Hawaiian children is to chime in with contributions to a story. [6]In school, however, this overlapping style is seen as "interrupting." [7]When the teachers in one school learned about these differences and made their reading groups more like their students' home conversation groups, the young Hawaiian children in their classes improved in reading.

[8]The source of misunderstanding can be a subtle sociolinguistic difference, such as how long the teacher waits to react to a student's response. [9]Researchers White and Tharp (1988) found that when Navajo students in one class paused in giving a response, their Anglo teacher seemed to think that they were finished speaking. [10]As a result, the teacher often unintentionally interrupted students. [11]In another study, researchers found that Pueblo Indian students participated twice as much in classes where teachers waited longer to react. [12]Waiting longer also helps girls to participate more freely in math and science classes.

—From *Educational Psychology,* 8th ed., p. 188 by Anita E. Woolfolk.

Vocabulary ______ **1.** The best meaning of the word **competent** in sentence 2 is
a. inept.
b. proficient.
c. thoughtful.
d. confident.

Thought Patterns ______ **2.** The thought pattern suggested by the organization of sentences 8 through 12 is
a. time.
b. cause and effect.
c. comparison and contrast.
d. definition.

Supporting Details ______ **3.** Sentence 12 is
a. a major supporting detail.
b. a minor supporting detail.

Central Idea ______ **4.** The central idea of the passage is best stated in sentence
a. Sentence 1.
b. Sentence 2.
c. Sentence 4.
d. Sentence 8.

Transitions ______ **5.** The relationship of ideas within sentence 2 is one of
a. cause and effect.
b. comparison and contrast.
c. time order.

Purpose ______ **6.** The author's main purpose is
a. to inform.
b. to persuade.
c. to entertain.

Fact and Opinion ______ **7.** Sentence 1 is a statement of
a. fact.
b. opinion.
c. fact and opinion.

8–10. Complete the summary with ideas from the passage.

Communication ________________ learned at home impact the student's ability to effectively ________________ in the classroom. Rules for participating in schools are similar to the communication rules in middle-class homes; thus students from different ________________ backgrounds may be misunderstood. For example, rules about contributing to the discussion and the length of pauses in response differ among cultures and can be misunderstood in a classroom situation.

TEST 9

Read the following passage from the college textbook *Life on Earth,* and then, answer the questions.

Textbook Skills

How Is Life on Land Distributed?

[1]On land, the crucial limiting factors are temperature and liquid water. [2]A **biome** comprises regions that have similar climates and vegetation owing to the interaction of temperature and rainfall or the availability of water. [3]Tropical forest biomes, located near the equator, vary in the amount of rainfall they receive. [4]Tropical deciduous forests occur in drier climates and contain deciduous trees that lose their leaves seasonally, typically in the driest months. [5]Tropical rain forests occur where rainfall is

plentiful and are dominated by huge broadleaf evergreen trees. [6]In rain forests, most nutrients are tied up in vegetation, and most animal life is arboreal. [7]Rain forests, home to at least 50% of all species, are rapidly being cut for agriculture, although the soil is extremely poor. [8]The African savanna, a grassland studded with scattered trees, has pronounced wet and dry seasons. [9]It is home to the world's most diverse and extensive herds of large mammals.

[10]Most deserts are located between 20° and 30° north and south latitude or in the rain shadows of mountain ranges. [11]In deserts, plants are often widely spaced and have adaptations to conserve water. [12]Most animals are small and nocturnal. [13]Chaparral exists in desert-like conditions that are moderated by proximity to a coastline, allowing small trees and bushes to thrive. [14]Grasslands, concentrated in the centers of continents, have a continuous grass cover without trees. [15]They have the world's richest soils and have largely been converted to agriculture. [16]Temperate deciduous forests, whose broadleaf trees drop their leaves in winter to conserve moisture, dominate the eastern half of the United States and are also found in Western Europe and East Asia. [17]Precipitation is higher than in grasslands. [18]The wet temperate rain forests, dominated by evergreens, are on the northern Pacific Coast of the United States. [19]The taiga, or northern coniferous forest, covers much of the northern United States, southern Canada, and northern Eurasia. [20]It is dominated by conifers whose small, waxy needles are adapted for water conservation and year-round photosynthesis. [21]The tundra is a frozen desert where permafrost prevents the growth of trees, and bushes remain stunted. [22]Nonetheless, diverse arrays of animal life and perennial plants flourish in this fragile biome, which is found on mountain peaks and in the Arctic.

—From *Life on Earth,* 5th ed., pp. 621–622 by Teresa Audesirk, Gerald Audesirk, and Bruce E. Byers.

Vocabulary ______ **1.** The best meaning of the word **biome** in the passage is
a. living organism. c. temperature.
b. environment. d. vegetation.

Transitions ______ **2.** The relationship of ideas within sentence 10 is one of
a. time order. c. cause and effect.
b. comparison and contrast. d. space order.

Supporting Details ______ **3.** Sentence 4 is
a. a major supporting detail. b. a minor supporting detail.

Supporting Details ______ **4.** Sentence 8 is
a. a major supporting detail. b. a minor supporting detail.

Central Idea ______ **5.** The best statement of the central idea is
a. sentence 1. c. sentence 3.
b. sentence 2. d. sentence 4.

6–10. Complete the outline with ideas from the passage.

Types of Biomes on Land

A. ______________

B. African ______________

C. Deserts

D. ______________

E. Grasslands

F. ______________

G. Taiga, Northern Coniferous Forest

H. ______________

TEST 10

Read the following passage from the college textbook *Politics in America,* and then, answer the questions.

Textbook Skills

[1]The United States is a nation of immigrants, from the first "boat people" (Pilgrims) to the latest Haitian refugees and Cuban *balseros* ("rafters"). [2]Historically, most of the people who came to settle in this country did so because they believed their lives would be better here, and American political culture today has been greatly affected by the beliefs and values they brought with them. [3]Americans are proud of their immigrant heritage and the freedom and opportunity the nation has extended to generations of "huddled masses yearning to be free"—words **emblazoned** on the Statue of Liberty in New York's harbor. [4]Today about 8 percent of the U.S. population is foreign-born.

[5]Immigration policy is a responsibility of the national government. [6]It was not until 1882 that Congress passed the first legislation restricting entry into the United States of persons alleged to be "undesirable" and virtually all Asians. [7]After World War I, Congress passed the comprehensive Immigration Act of 1921, which established maximum numbers of new immigrants each year and set a **quota** for immigrants for each foreign country at 3 percent of the number of that nation's foreign-born who were living in the United States in 1910, later reduced to 2 percent of the number living here in 1890. [8]These restrictions reflected anti-immigration feelings that were generally directed at the large wave of Southern and Eastern European Catholic and Jewish immigrants (from Poland, Russia, Hungary, Italy, and Greece) entering the United States prior to World War I. [9]It was not until the Immigration and Naturalization Act of 1965 that national origin quotas were abolished, replaced by preference categories for close relatives of U.S. citizens, professionals, and skilled workers.

—Dye, *Politics in America*, 5th ed., pp. 40–42.

Vocabulary ______ **1.** The best meaning of the word **emblazoned** in sentence 3 is

a. shouted.
b. inscribed.
c. required.
d. maintained.

Vocabulary ______ **2.** The best meaning of the word **quota** in sentence 7 is

a. statement.
b. allowance.
c. price.
d. expectation.

Thought Patterns ______ **3.** The overall thought pattern for the passage is

a. the effects of immigration on the United States.
b. the history of immigration in the United States.
c. the differences between United States natural-born citizens and immigrants.
d. the traits of an immigrant to the United States.

Supporting Details ______ **4.** Sentence 7 is a supporting detail that gives information about

a. why people immigrate into the United States.
b. the reasons for limiting immigration into the United States.
c. an effect of immigration on the United States.
d. the origin of an immigration law.

Central Idea ______ **5.** The best statement of the central idea of this passage is
a. sentence 1.
b. sentence 3.
c. sentence 5
d. sentence 8.

Transitions ______ **6.** The relationship of ideas between sentences 7 and 8 is one of
a. listing.
b. comparison and contrast.
c. cause and effect.
d. time order.

Purpose ______ **7.** The author's main purpose is
a. to inform the reader about the nature and history of immigration into the United States.
b. to amuse the reader with little known details about immigration into the United States.
c. to defend the immigration policies of the United States.

Tone ______ **8.** The tone of the first paragraph is
a. quarrelsome.
b. ironic.
c. assertive.
d. positive.

Fact and Opinion ______ **9.** Sentence 3 is a statement of
a. fact.
b. opinion.
c. fact and opinion.

Inference ______ **10.** The best title for the passage is
a. Restricting Immigration.
b. A Nation of Immigrants.
c. Quotas and Restrictions.
d. The Lure of Freedom and Democracy.

Text Credits

Aaron, Jane E. *The Little, Brown Compact Handbook,* 4th ed. Copyright © 2001 by Pearson Education, Inc. Reprinted by permission of Pearson Education, Inc., Glenview, IL.

Agee, Warren K., Philip H. Ault, and Edwin Emery. "Types of Media" and "Trash TV" from *Introduction to Mass Communications,* 12th ed. Boston: Allyn and Bacon, 1997.

Angelou, Maya. "Fannie Lou Hamer" from *Letter to My Daughter.* Copyright © 2008 by Maya Angelou. Used by permission of Random House, Inc.

Annan, Kofi. "The Truman Library Speech," December 11, 2006, http://www.un.org/News/ossg/sg/stories/statments_full.asp?statID=40. Reprinted by permission of the United Nations.

Audesirk, Teresa, Gerald Audesirk, and Bruce E. Byers. *Biology: Life on Earth,* 8th ed. Copyright © 2008 by Pearson Education, Inc. Printed and Electronically reproduced by permission of Pearson Education, Inc., Upper Saddle River, NJ.

Audesirk, Teresa, Gerald Audesirk, and Bruce E. Byers. *Life on Earth,* 5th ed. Copyright © 2009 by Pearson Education, Inc. Printed and Electronically reproduced by permission of Pearson Education, Inc., Upper Saddle River, NJ.

Barker, Larry L., and Deborah Gaut. "Initiating" from *Communication,* 8th ed. Boston: Allyn & Bacon, 2002.

Baron, Robert A., and Donn Byrne. *Social Psychology with Research Navigator,* 10th ed. Copyright © 2004 by Allyn & Bacon, Inc. Reproduced by permission of Pearson Education, Inc., Boston, MA.

Benokraitis, Nijole V. *Marriages and Families: Changes, Choices, and Constraints,* 4th ed. Copyright © 2002 by Pearson Education, Inc. Reproduced by permission of Pearson Education, Inc., Boston, MA.

Bergman, Edward, and William H. Renwick. *Introduction to Geography: People, Places, and Environment,* 4th ed. Copyright © 2008 by Pearson Education, Inc. Printed and Electronically reproduced by permission of Pearson Education, Inc., Upper Saddle River, NJ.

Bittinger, Marvin L., and Judith A. Beecher. *Introductory and Intermediate Algebra: A Combined Approach,* 2nd ed. Copyright © 2003 by Pearson Education, Inc. Reproduced by permission of Pearson Education, Inc., Boston, MA.

Blake, Joan Salge. *Nutrition and You,* 1st ed. Copyright © 2008 by Pearson Education, Inc. Printed and Electronically reproduced by permission of Pearson Education, Inc., Upper Saddle River, NJ.

Bradley, John P., Leo F. Daniels, and Thomas C. Jones. *The International Dictionary of Thoughts: An Encyclopedia of Quotations from Every Age for Every Occasion.* Chicago: J. G. Ferguson Pub. Co., 1969.

Brownell, Judi. *Listening: Attitudes, Principles, and Skills,* 2nd ed. Copyright © 2002 by Pearson Education, Inc. Reproduced by permission of Pearson Education, Inc., Boston, MA.

Carl, John D. *Think Social Problems.* Copyright © 2011 by Pearson Education, Inc. Reproduced by permission of Pearson Education, Inc., Boston, MA.

Carlson, Neil R., and William Buskist. *Psychology: The Science of Behavior,* 5th ed. Boston: Allyn & Bacon, 1997.

Carson, Clayborne, Emma J. Lapsansky-Werner, and Gary B. Nash. "The Jazz Age" from *The Struggle for Freedom: A History of African-Americans, Volume II.* White Plains: Longman, 2007.

Churchill, Winston S. "The Price of Greatness." Copyright © Winston S. Churchill. Reproduced with permission of Curtis Brown Ltd, London on behalf of the Estate of Sir Winston Churchill.

"Coacoochee" from *The Ormond Beach Historical Trust.* Ormond Beach: The Ormond Beach Historical Trust, 1999.

Cook, Roy A., Laura J. Yale and Joseph J. Marqua. *Tourism: The Business of Travel,* 4th ed. Copyright © 2010 by Pearson Education, Inc. Printed and Electronically reproduced by permission of Pearson Education, Inc., Upper Saddle River, NJ.

Danny Yee's Book Reviews, August 1992, http://dannyreviews.com/h/Animal_Farm.html. Reprinted by permission of Danny Yee.

Danny Yee's Book Reviews, June 1998, http://dannyreviews.com/h/Beyond_Boundary.html. Reprinted by permission of Danny Yee.

DeVito, Joseph A. *Essentials of Human Communication,* 4th ed. Copyright © 2002 by Pearson Education, Inc. Reproduced by permission of Pearson Education, Inc., Boston, MA.

DeVito, Joseph A. *The Interpersonal Communication Book,* 10th ed. Copyright © 2004 by Pearson Education, Inc. Reproduced by permission of Pearson Education, Inc., Boston, MA.

DeVito, Joseph A. *The Interpersonal Communication Book,* 11th ed. Copyright © 2007 by Pearson Education, Inc. Reproduced by permission of Pearson Education, Inc., Boston, MA.

DeVito, Joseph A. *Messages: Building Interpersonal Communication Skills.* Copyright © 1999 by Pearson Education, Inc. Reproduced by permission of Pearson Education, Inc., Boston, MA.

DeVito, Joseph A. *Messages: Building Interpersonal Communication Skills,* 6th ed. Boston: Allyn & Bacon, 2005.

Divine, Robert A., George M. Fredrickson, R. Hal Williams, and T. H. Breen. *The American Story,*1st ed. Copyright © 2002 by Pearson Education, Inc. Reprinted by permission of Pearson Education, Inc., Upper Saddle River, NJ.

Donatelle, Rebecca J., and Lorraine G. Davis. *Access to Health,* 7th ed. Copyright © 2002 by Pearson Education, Inc. Printed and Electronically reproduced by permission of Pearson Education, Inc., Upper Saddle River, NJ.

Donatelle, Rebecca. "Do Restaurants and Food Marketers Encourage Overeating?" from *Access to Health,* 11th ed., Green Edition. Copyright © 2010 by Pearson Education, Inc. Printed and Electronically reproduced by permission of Pearson Education, Inc., Upper Saddle River, NJ.

Donatelle, Rebecca J. "Eating Disorders" from *Health: The Basics,* 7th ed. Copyright © 2007 by Pearson Education, Inc. Printed and Electronically reproduced by permission of Pearson Education, Inc., Upper Saddle River, NJ.

Donatelle, Rebecca J. "Reinforcement" and "Caffeine Addiction" from *Health: The Basics,* 5th ed. Copyright © 2003 by Pearson Education, Inc. Printed and Electronically reproduced by permission of Pearson Education, Inc. Upper Saddle River, NJ.

Dye, Thomas R. *Politics in America, National Version,* 5th ed. Copyright © 2003 by Pearson Education, Inc. Reprinted by permission of Pearson Education, Inc., Glenview, IL.

Dye, Thomas R. *Politics in America,* 7th ed. Copyright © 2007 by Pearson Education, Inc. Reprinted by permission of Pearson Education, Inc., Glenview, IL.

Edgar, Robert R., Neil J. Hackett, George F. Jewsbury, Barbara S. Molony, and Matthew Gordon. "Cuba" and "The Iroquois of the Northeast Woodlands" from *Civilizations Past & Present, Combined Volume,* 12th ed. Copyright © 2008 by Pearson Education, Inc. Reprinted by permission of Pearson Education, Inc., Upper Saddle River, NJ.

Edwards, George C. III, Martin P. Wattenberg and Robert L. Lineberry. *Government in America: People, Politics, and Policy,* 5th ed., Brief Version. Copyright © 2000 by Pearson Education, Inc. Reprinted by permission of Pearson Education, Inc., Glenview, IL.

Eldin, Peter. *Jokes & Quotes for Speeches.* London: Ward Lock, Ltd., 1989.

Erdely, Sabrina Rubin. "Binge Drinking, A Campus Killer," *Reader's Digest,* November 1998. Reprinted by permission of author.

Faigley, Lester. *The Penguin Handbook.* White Plains: Longman, 2003.

Fernandez-Armesto, Felipe. *The World: A History, Combined Volume,* 2nd ed. Copyright © 2010 by Pearson Education, Inc. Reprinted by permission of Pearson Education, Inc., Upper Saddle River, NJ.

Folger, Joseph P., Marshall Scott Poole, and Richard K. Stutman. *Working through Conflict: Strategies for Relationships, Groups, and Organizations,* 4th ed. Boston: Allyn & Bacon, 2001.

Folkerts, Jean, and Stephen Lacy. *The Media in Your Life: An Introduction to Mass Communication,* 2nd ed. Boston: Allyn & Bacon, 2001.

Fowler, H. Ramsey, and Jane E. Aaron. *The Little, Brown Handbook,* 9th ed. Copyright © 2004 by Pearson Education, Inc. Reprinted by permission of Pearson Education, Inc., Glenview, IL.

Garraty, John A., and Mark A. Carnes. *The American Nation: A History of the United States to 1877, Volume I,* 10th ed. Copyright © 2000 by Pearson Education, Inc. Printed and Electronically reproduced by permission of Pearson Education, Inc., Upper Saddle River, NJ.

Gerrig, Richard J., and Phillip G. Zimbardo. *Psychology and Life,* 16th ed. Copyright © 2002 by Pearson Education, Inc. Reproduced by permission of Pearson Education, Inc., Boston, MA.

Gerrig, Richard J., and Philip G. Zimbardo. *Psychology and Life,* 19th ed. Copyright © 2010 by Pearson Education, Inc. Reproduced by permission of Pearson Education, Inc., Boston, MA.

Girdano, Daniel A., George S. Everly, Jr. and Dorothy E. Dusek. "Physical Effects of Modern Life" from *Controlling Stress and Tension,* 6th ed. Boston: Allyn & Bacon, 2001.

Griffin, Ricky W., and Ronald J. Ebert. *Business,* 8th ed. Copyright © 2006 by Pearson Education, Inc. Printed and Electronically reproduced by permission of Pearson Education, Inc., Upper Saddle River, NJ.

Harrell, Eben. "Think You're Operating on Free Will? Think Again" from *Time Magazine,* July 2, 2010. Copyright TIME INC. Reprinted by permission. TIME is a registered trademark of Time Inc. All rights reserved.

Harris, Marvin, and Oma Johnson. *Cultural Anthropology,* 7th ed. Copyright © 2007 by Pearson Education, Inc. Printed and Electronically reproduced by permission of Pearson Education, Inc., Upper Saddle River, NJ.

Henslin, James M. *Essentials of Sociology: A Down-to-Earth Approach,* 7th ed. Copyright © 2007 by Pearson Education, Inc. Reproduced by permission of Pearson Education, Inc., Boston, MA.

Henslin, James M. *Sociology: A Down-to-Earth Approach,* 9th ed. Copyright © 2008 by James Henslin. Reproduced by permission of Pearson Education, Inc., Boston, MA.

"Hidden Caffeine" by Ying Lou and Gia Kereselidze from USA TODAY, March 3, 2004. Reprinted by permission of USA Today.

Hughes, Langston. "I, Too" from *The Collected Poems of Langston Hughes* by Langston Hughes, edited by Arnold Rampersad with David Roessel, Associate Editor. Copyright © 1994 by the Estate of Langston Hughes. Used by permission of Alfred A. Knopf, a division of Random House, Inc.

"Introduction" from *The Cubist Paintings of Diego Maria Rivera.* http://www.nga.gov/exhibitions/2004/rivera/intro.shtm. Reproduced by permission of National Gallery of Art, Washington, D.C.

Jaffe, Michael L. *Understanding Parenting,* 2nd ed. Boston: Allyn & Bacon, 1997.

Jones, Jacqueline, Peter H. Wood, Thomas Borstelmann, Elaine Tyler May, and Vicki L. Ruiz. *Created Equal: A Social and Political History of the United States, Combined Volume,* 2nd ed. Copyright © 2006 by Pearson Education, Inc. Printed and Electronically reproduced by permission of Pearson Education, Inc., Upper Saddle River, NJ.

Jones, Jacqueline, C Peter H. Wood, Thomas Borstelmann, Elaine Tyler May, and Vicki L. Ruiz. *Created Equal: A Social and Political History of the United States, Volume II (From 1865),* 2nd ed. Copyright © 2006 by Pearson Education, Inc. Printed and Electronically reproduced by permission of Pearson Education, Inc., Upper Saddle River, NJ.

Joyce, James. "On the Beach at Fontana" from *Oxford Book of Modern Verse* by W. B. Yeats. New York: Oxford University Press, 1936.

Karren, Keith J., *Mind/Body Health: The Effects of Attitudes, Emotions, and Relationships*, 2nd ed. Copyright © 2002 by Pearson Education, Inc. Reprinted by permission of Pearson Education, Inc., Upper Saddle River, NJ.

Kennedy, X. J., and Dana Gioia. *Literature: An Introduction to Fiction, Poetry, and Drama,* 8th ed. Copyright © 2002 by Pearson Education, Inc. Reprinted by permission of Pearson Education, Inc., Glenview, IL.

Kennedy, X. J., and Dana Gioia. "Blues" from *Literature: An Introduction to Fiction, Poetry, and Drama,* 3rd Compact ed.. Copyright © 2003 by Pearson Education, Inc. Reprinted by permission of Pearson Education, Inc., Glenview, IL.

Kennedy, X. J., and Dana Gioia. *Literature: An Introduction to Fiction, Poetry, and Drama,* 4th ed. Copyright © 2005 by Pearson Education, Inc. Reprinted by permission of Pearson Education, Inc., Glenview, IL.

Kishlansky, Mark, Patrick Geary, and Patricia O'Brien. "God Kings" from *Civilization in the West,* 4th ed. White Plains: Longman, 2001.

Kosslyn, Stephen M., and Robin S. Rosenberg. "Semantic versus Episodic Memory" from *Fundamentals of Psychology: The Brain, the Person, the World.* Boston: Allyn & Bacon, 2002.

Kosslyn, Stephen M., and Robin S. Rosenberg. *Psychology: The Brain, the Person, the World.* Boston: Allyn & Bacon, 2002.

Kunz, Jennifer. "Characteristics of Successful Stepfamilies" from *Think Marriages & Families.* Copyright © 2011 by Pearson Education, Inc. Reproduced by permission of Pearson Education, Inc., Boston, MA.

Lee, Sharon, and Barry Edmonston. "New Marriages, New Families: U.S. Racial and Hispanic Intermarriage" from *Population Bulletin,* June 2005, Volume 60, Number 2. Washington, DC: Population Reference Bureau, 2005.

Lefton, Lester A., and Linda Brannon. *Psychology,* 8th ed. Boston: Allyn & Bacon, 2003.

Levack, Brian, Edward Muir, Meredith Veldman and Michael Maas. *The West: Encounters & Transformations, Atlas Edition, Combined Edition*, 2nd ed. Copyright © 2008 by Pearson Education, Inc. Printed and Electronically reproduced by permission of Pearson Education, Inc., Upper Saddle River, NJ.

Levinthal, Charles F. "Drugs in Our Lives" and "Differences in the Effects of Drugs" from *Drugs, Behavior, and Modern Society*, 3rd ed. Boston: Allyn and Bacon, 2002.

Lithwick, Dahlia. "Teens, Nude Photos and the Law: Ask Yourself: Should the Police Be Involved When Tipsy Teen Girls E-mail Their Boyfriends Naughty Valentine's Day Pictures?" from *Newsweek*, February 14 Copyright © 2009 Newsweek, Inc. All rights reserved. Used by permission and protected by the Copyright Laws of the United States. The printing, copying, redistribution, or retransmission of the Material without express written permission is prohibited.

Lutgens, Frederick K., Edward J. Tarbuck, and Dennis Tasa. *Foundations of Earth Science,* 5th ed. Copyright © 2008 by Pearson Education, Inc. Printed and Electronically reproduced by permission of Pearson Education, Inc., Upper Saddle River, NJ.

MacDougall, Alice Foote. *The Autobiography of a Business Woman*. Boston: Little, Brown and Company, 1928.

Madura, Jeff. *Personal Finance,* 2nd ed. Copyright © 2004 by Pearson Education, Inc. Reproduced by permission of Pearson Education, Inc., Boston, MA.

Madura, Jeff. *Personal Finance,* 3rd ed. Copyright © 2007 by Pearson Education, Inc. Reproduced by permission of Pearson Education, Inc., Boston, MA.

Maier, Richard. *Comparative Animal Behavior: An Evolutionary and Ecological Approach.* Boston: Allyn and Bacon, 1998.

Marieb, Elaine N. *Essentials of Human Anatomy and Physiology,* 9th ed. Copyright © 2009 by Pearson Education, Inc. Reprinted by permission of Pearson Education, Inc., Glenview, IL.

Martin, James Kirby, Randy J. Roberts, Steven Mintz, Linda O. McMurry and James H. Jones. *America and Its Peoples, Volume I: A Mosaic in the Making,* 3rd ed. Copyright © 1997 by Pearson Education, Inc. Printed and Electronically reproduced by permission of Pearson Education, Inc., Upper Saddle River, NJ.

Martin, James Kirby, Randy J. Roberts, Steven Mintz, Linda O. McMurry and James H. Jones. *America and Its Peoples: Volume II: A Mosaic in the Making,* 3rd ed. Copyright © 1997 by Pearson Education, Inc. Printed and Electronically reproduced by permission of Pearson Education, Inc., Upper Saddle River, NJ.

Maugham, W. Somerset. "An Appointment in Samarra" from *Sheppey* by W. Somerset Maugham. Copyright © 1933 by W. Somerset Maugham. Used by permission of Doubleday, a division of Random House, Inc. and A.P. Watt Ltd.

Mayer, Richard E. "Types of Learners Table" and "Three Facets of the Visualizer-Verbalizer Dimension" from *Journal of Educational Psychology*, 2003. Copyright © 2003 by the American Psychological Association. Adapted and Reprinted by permission of author and American Psychological Association, Washington, D.C.

McGuigan, F. J. *Encyclopedia of Stress,* 1st ed. Copyright © 1999 by Pearson Education, Inc. Printed and Electronically reproduced by permission of Pearson Education, Inc., Upper Saddle River, NJ.

Miller, Barbara. *Cultural Anthropology,* 4th ed. Copyright © 2007 by Pearson Education, Inc. Printed and Electronically reproduced by permission of Pearson Education, Inc., Upper Saddle River, NJ.

O'Connor, Karen J., and Larry J. Sabato. "Native Americans" from *American Government*, 2008 edition, 9th ed. Copyright © 2008 by Pearson Education, Inc. Reprinted by permission of Pearson Education, Inc., Glenview, IL.

Offiah-Hawkins, Sandra. "An Account of Alfred C. Cooley's Plight in the Face of Hurricane Katrina." Reprinted by permission of author.

"Opposing View: Don't Blame Us" by Pamela G. Bailey from USA TODAY, March 31, 2010. Reprinted by permission of USA Today.

"Our View on Obesity: Hooked on Junk Food" from USA TODAY, March 31, 2010. Reprinted by permission of USA Today.

"People," PBS: Sahara, PBS.org, December 29, 2003. Reproduced by permission of Telenova Productions.

Perry, John A., and Erna K. Perry. "The Impact of the Consumer Culture on the Growth of Suburbia" from *Contemporary Society: An Introduction to Social Science,* 10th ed. Boston: Allyn & Bacon, 2003.

Phillips, John C. *Sociology of Sport.* Copyright © 1993 by Allyn & Bacon. Reproduced by permission of Pearson Education, Inc., Boston, MA.

Pimentel, David. "Corn Can't Save Us: Debunking the Biofuel Myth" from *Kennebec Journal,* March 28, 2008. Reprinted by permission of Kennebec Journal.

Powers, Scott K., and Stephen L. Dodd. "Life Style Assessment" from *Behavior Change Log Book,* 3rd ed. Upper Saddle River: Pearson Education, Inc., 2003.

Powers, Scott K., and Stephen L. Dodd. *Total Fitness and Wellness Student Textbook Component,* 3rd ed. Copyright © 2003 by Pearson Education, Inc. Printed and Electronically reproduced by permission of Pearson Education, Inc., Upper Saddle River, NJ.

Powers, Scott K., Stephen L. Dodd and Virginia J. Noland. *Total Fitness and Wellness,* 4th ed. Copyright © 2006 by Pearson Education, Inc. Printed and Electronically reproduced by permission of Pearson Education, Inc., Upper Saddle River, NJ.

Pruitt, B. E., and Jane J. Stein. *Healthstyles: Decisions for Living Well,* 2nd ed. Copyright © 1999 by Pearson Education, Inc. Printed and Electronically reproduced by permission of Pearson Education, Inc., Upper Saddle River, NJ.

Queen Latifah and Karen Hunter. *Ladies First: Revelations of a Strong Woman.* New York: HarperCollins Publishers, 1999.

Ruggiero, Vincent R. *The Art of Thinking: A Guide to Critical and Creative Thought,* 7th ed. Copyright © 2004 by Pearson Education, Inc. Reprinted by permission of Pearson Education, Inc., Glenview, IL.

Schmalleger, Frank J. *Criminal Justice Today: An Introductory Text for the 21st Century,* 10th ed. Copyright © 2009 by Pearson Education, Inc. Printed and Electronically reproduced by permission of Pearson Education, Inc., Upper Saddle River, NJ.

"Seasonal Affective Disorder," Mental Health America, Copyright 2010 Mental Health America, http://www.mentalhealthamerica.net/go/sad.

Shiraev, Eric B., and David A. Levy. "Adolescence" from *Cross-Cultural Psychology: Critical Thinking and Contemporary Applications,* 3rd ed. Boston: Allyn and Bacon, 2007.

"Should We Be Criminalizing Bullies?" from *The Washington Post,* Copyright © April 7, 2010 The Washington Post. All rights reserved. Used by permission and protected by the Copyright Laws of the United States. The printing, copying, redistribution, or retransmission of the Material without express written permission is prohibited.

Simon, Todd F. "Levels of Protected Communication." Reprinted by permission of the author.

Smith, Carney, and Chuck Stone. "Medgar Evers" from *African American Almanac,* 7th ed. Belmont: Gale, a part of Cengage Learning, Inc., 1996.

Smith, Patrick D. *Forever Island.* Copyright © 1973 by W. W. Norton & Company, Inc. Used by permission of W. W. Norton & Company, Inc.

Smith, Thomas M., and Robert Leo Smith. *Elements of Ecology,* 4th ed. Copyright © 2000 by Pearson Education, Inc. Reprinted by permission of Pearson Education, Inc., Glenview, IL.

Smith, Thomas M., and Robert Leo Smith. *Elements of Ecology,* 6th ed. Copyright © 2006 by Pearson Education, Inc. Reprinted by permission of Pearson Education, Inc., Glenview, IL.

Sporre, Dennis J. *The Creative Impulse; An Introduction to the Arts,* 8th ed. Copyright © 2009 by Pearson Education, Inc. Reprinted by permission of Pearson Education, Inc., Upper Saddle River, NJ.

Stearns, Peter N., Michael B. Adas, Stuart B. Schwartz, and Marc Jason Gilbert. *World Civilizations: The Global Experience, Combined Volume, Atlas Edition,* 5th ed. Copyright © 2008 by Pearson Education, Inc. Printed and Electronically reproduced by permission of Pearson Education, Inc., Upper Saddle River, NJ.

Steinbeck, John. *East of Eden.* New York: Penguin Group (USA) Inc., 1952, 1980.

Sting. "The Mystery and Religion of Music." Boston: Berklee College of Music, 1994.

Tannahill, Neal R. "Taking Sides: Constitutional Principles" from *Think American Government,* 2nd ed. Copyright © 2011 by Pearson Education, Inc. Reprinted by permission of Pearson Education, Inc., Glenview, IL.

Tannen, Deborah. "Sex, Lies, Conversation" from *The Washington Post,* June 24, 1990, p. C3. Copyright © by Deborah Tannen. Reprinted by permission of author. This article is adapted from *You Just Don't Understand: Women and Men in Conversation.* New York: Ballantine, 1990. New paperback edition: New York: Quill, 2001.

"Tattoos: The Ancient and Mysterious History" by Cate Lineberry from Smithsonian.com, January 1, 2007. Copyright © 2010 by Smithsonian Institution. Reprinted by permission of Smithsonian Magazine. All rights reserved. Reproduction in any medium is strictly prohibited without permission from Smithsonian Institution. Such permission may be requested from Smithsonian Magazine.

Tristam, Pierre. "A Look at Man Through The Vapid Eyes of His Captives" from *The Daytona Beach News-Journal,* March 7, 2010, pp 10A. Reprinted by permission of Copyright Clearance Center.

Walvin, James. *Atlas of Slavery.* Copyright © 2006 by Pearson Education Limited. Reprinted by permission of Pearson Education Limited.

Wilen, William, Margaret Ishler, Janice Hutchison, and Richard Kindsvatter. *Dynamics of Effective Teaching,* 4th ed. White Plains: Longman, 2000.

Withgott, Jay H., and Scott R. Brennan. *Essential Environment: The Science Behind the Stories,* 3rd ed. Copyright © 2009 by Pearson Education, Inc. Reprinted by permission of Pearson Education, Inc., Upper Saddle River, NJ.

Wood, Samuel E., Ellen Green Wood, and Denise Boyd. *Mastering the World of Psychology*, 3rd ed. Copyright © 2008 by Pearson Education, Inc. Reproduced by permission of Pearson Education, Inc., Boston, MA.

Woolfolk, Anita E. *Educational Psychology,* 8th ed. Copyright © 2001 by Pearson Education, Inc. Reproduced by permission of Pearson Education, Inc., Boston, MA.

Woolfolk, Anita E. *Educational Psychology,* 10th ed. Copyright © 2007 by Pearson Education, Inc. Reproduced by permission of Pearson Education, Inc., Boston, MA.

Young, Billie Jean. "Fear Not the Fall" from *Fear Not the Fall: Poems and a Two-Act Drama.* Reprinted by permission of NewSouth Books, Montgomery, AL.

Photo Credits

p. 21, © Christopher Covey/Terra/Corbis; p. 24, © Troy Wayrynen/NewSport/Corbis; p. 30, © Arctic Images/ Corbis; p. 44, © Steve Hix/Somos Images/Corbis; p. 46, © Upton/Design Pics/Corbis; p. 50, © Steffen Schmidt /Keystone/Corbis; p. 90, vii, © Reunion des Musees Nationaux/Art Resource, Inc.; p. 94, © Erich Lessing /Alison Slagowitz/Art Resource, Inc.; p. 99, © Simon Potter/Taxi/Getty Images; p. 105, © Dennis Kunkel Microscopy, Inc./Corbis; p. 136, vii, © Stuart Westmorland/Corbis; p. 140, © Buddy Mays/Corbis; p. 154, © Kevin Lamarque/Reuters/Corbis; p. 198, © William Radcliffe/Science Faction/Corbis; p. 230, viii top, © Andrew Goetz/Corbis; p. 237, © Mika/Corbis; p. 245, © Corbis; p. 252, © Ron Watts/Corbis; p. 264, © Ludovic Maisant/Corbis; p. 266, © Greg Smith/Corbis News; p. 282 © Matt Herring/Ikon Images/Corbis; p. 286, © MedicalRF.com/Corbis <http://MedicalRF.com/Corbis>; p. 292, viii bottom, © DLILLC/Corbis; p. 299, © Philip Gendreau/Bettmann/Corbis; p. 311, © Martin B. Withers /Frank Lance Picture Agency/Corbis; p. 337, both, © Hugh Sitton/Zefa/Corbis; p. 341, © Images.com <http:// Images.com>; p. 346, © Will Hart/PhotoEdit; p. 369, © Ira Wyman/Sygma; p. 375, ix, © Tyrone Turner/National Geographic/Corbis; p. 379, © PBN Productions/Blend Images/Corbis; p. 403, © Blend Images/Corbis; p. 414, © whitehouse.gov <http://whitehouse.gov>; p. 419, left, © Louis DeLuca/Dallas Morning News/Corbis; p. 419, center, © Image Source/Corbis; p. 419, right, © Brooks Kraft/Corbis; p. 434, © Image Source/Corbis; p. 455, © Christine Schneider/Corbis; p. 460, © Etsa/Corbis; p. 467, The Museum of Modern Art. Licensed by SCALA, Art Resource.; p. 487, Courtesy of Edios Paramount Kobal Corporation; p. 498, © DLILLC/Flame/Corbis; p. 506, © Tom and Dee Ann McCarthy/Corbis; p. 524, © Burke/ Triolo Productions-Brand X-Royalty Free/Getty Creative; p. 527, x, © Corbis (RF)/Corbis; p. 540, © Matthias Ernert /DPA/Corbis; p. 571, xi, © Tim Pannell/Corbis; p. 604, © Medio Images/Corbis; p. 661, © Karen Kasmauski/Corbis; p. 683, © David Turnley/Corbis; p. 696, © Gabe Palmer/ Corbis; Chapter Review Card: © Don Mason/Blend Images/ Corbis.

Index